Charles Hebert Stitt

Applied Probability
Control
Economics
Information and Communication
Modeling and Identification
Numerical Techniques
Optimization

Applications of Mathematics

17

Edited by A. V. Balakrishnan

Applications of Mathematics

Lamberto Cesari

Optimization—Theory and Applications

Problems with Ordinary Differential Equations

With 82 Illustrations

Springer-Verlag
New York Heidelberg Berlin

Lamberto Cesari
Department of Mathematics
University of Michigan
Ann Arbor, MI 48104
U.S.A.

Editor

A. V. Balakrishnan
Systems Science Department
University of California
Los Angeles, CA 90024
U.S.A.

AMS Subject Classifications: 49-02, 93-02

Library of Congress Cataloging in Publication Data

Cesari, Lamberto.
Optimization-theory and applications.
(Applications of mathematics; 17)
Bibliography: p.
Includes index.
1. Calculus of variations. 2. Mathematical optimization. 3. Differential equations. I. Title. II. Series.
QA316.C47 515′.64 82-5776
AACR2

Typeset by Syntax International, Singapore.
Printed and bound by R. R. Donnelley & Sons, Harrisonburg, VA.
Printed in the United States of America.

9 8 7 6 5 4 3 2 1

ISBN 0-387-90676-2 Springer-Verlag New York Heidelberg Berlin
ISBN 3-540-90676-2 Springer-Verlag Berlin Heidelberg New York

Preface

This book has grown out of lectures and courses in calculus of variations and optimization taught for many years at the University of Michigan to graduate students at various stages of their careers, and always to a mixed audience of students in mathematics and engineering. It attempts to present a balanced view of the subject, giving some emphasis to its connections with the classical theory and to a number of those problems of economics and engineering which have motivated so many of the present developments, as well as presenting aspects of the current theory, particularly value theory and existence theorems. However, the presentation of the theory is connected to and accompanied by many concrete problems of optimization, classical and modern, some more technical and some less so, some discussed in detail and some only sketched or proposed as exercises.

No single part of the subject (such as the existence theorems, or the more traditional approach based on necessary conditions and on sufficient conditions, or the more recent one based on value function theory) can give a sufficient representation of the whole subject. This holds particularly for the existence theorems, some of which have been conceived to apply to certain large classes of problems of optimization.

For all these reasons it is essential to present many examples (Chapters 3 and 6) before the existence theorems (Chapters 9 and 11–16), and to investigate these examples by means of the usual necessary conditions, sufficient conditions, and value function theory.

This book only considers nonparametric problems of the calculus of variations in one independent variable and problems of optimal control monitored by ordinary differential equations. Multidimensional problems monitored by partial differential equations, parametric problems with simple and multiple integrals, parametric problems of optimal control, and related questions of nonlinear integration will be presented elsewhere.

Chapter 1 is introductory. The many types of problems of optimization are reviewed and their intricate relationships illustrated.

Chapter 2 presents the necessary conditions, the sufficient conditions, and the value function theory for classical problems of the calculus of variations. In particular, the Weierstrass necessary condition is being studied as a necessary condition for lower semicontinuity on a given trajectory.

Chapter 3 consists mainly of examples. In particular, it includes points of Ramsey's theory of economic growth, and points of theoretical mechanics.

Chapters 4 and 5 deal with problems of optimal control. They contain a statement of the necessary condition, a detailed discussion of the transversality relation in its generality, a discussion of Bellman's value function theory, and a statement of Boltyanskii's sufficient condition in terms of regular synthesis.

Chapter 6 consists mainly of examples. In particular, points of the neoclassical theory of economic growth are also studied.

Chapter 7 presents two proofs of the necessary condition for problems of optimal control.

Chapter 8 contains preparatory material for existence theorems, in particular, Kuratowski's and Ryll-Nardzewski's selection theorems, McShane's and Warfield's implicit function theorem, and some simple forms of the lower closure theorem for uniform convergence.

Chapter 9 deals with existence theorems for problems of optimal control with continuous data and compact control space. These are essentially Filippov's existence theorems. The proofs in this chapter are designed to be elementary in the sense that mere uniform convergence is involved, whereas in Chapters 10 and 11 use is made of weak convergence in L_1.

Chapter 10 presents the Banach–Saks–Mazur theorem, the Dunford-Pettis theorem, and closure, lower closure, and lower semicontinuity theorems for weak convergence in L_1.

Chapter 11 deals with existence theorems based on weak convergence. Existence theorems are proved for Lagrange problems with an integrand which is an extended function, and then existence theorems are derived for problems of optimal control. Moreover, existence theorems are proved for problems with comparison functionals, for isoperimetric problems, and specifically for problems which are linear in the derivatives, or in the controls. In particular, this chapter contains a present day version of the theorem established by Tonelli in 1914 for problems with a uniform growth property.

In Chapter 12 existence theorems are presented where a growth assumption fails at the points of a "slender" set. In Chapter 13 existence theorems under numerous analytical conditions are studied. Chapter 14 deals with existence theorems for problems without growth assumptions. Chapter 15 presents theorems based on mere pointwise convergence. Chapter 16 deals with Neustadt-type existence theorems for problems with no convexity assumptions.

Chapter 17 covers a few points of convex analysis including duality, and the equivalence of a certain concept of upper semicontinuity for sets with the concept of seminormality of Tonelli and McShane for functions, and suitable properties in terms of convex analysis.

Chapter 18 covers questions of approximation of usual and generalized trajectories.

Each chapter contains examples and exercises. Bibliographical notes at the end of each chapter provide some historical background and direct the reader to the literature in the field.

A number of parts in this book are in smaller print so as to facilitate, at a first reading, a faster perusal. The small-print passages include most of the examples and remarks, several of the complementary considerations, and a number of the more technical proofs.

I wish to thank the many associates and graduate students who, with their remarks and suggestions upon reading these notes, have contributed so much to make this presentation a reality.

Finally, I wish to express my appreciation to Springer-Verlag for their accomplished handling of the manuscript, their understanding and patience.

Contents

Chapter 8
The Implicit Function Theorem and the Elementary Closure Theorem 271

Chapter 9
Existence Theorems: The Bounded, or Elementary, Case 309

Chapter 10
Closure and Lower Closure Theorems under Weak Convergence 325

Chapter 11
Existence Theorems: Weak Convergence and Growth Conditions 367

To Isotta, always

CHAPTER 1
Problems of Optimization—A General View

1.1 Classical Lagrange Problems of the Calculus of Variations

Here we are concerned with minima and maxima of functionals of the form

(1.1.1) $$I[x] = \int_{t_1}^{t_2} f_0(t, x(t), x'(t))\,dt, \qquad (') = d/dt,$$

where we think of $I[x]$ as dependent on an n-vector continuous function $x(t) = (x^1, \dots, x^n)$, $t_1 \leq t \leq t_2$, or continuous curve of the form $C: x = x(t)$, $t_1 \leq t \leq t_2$, in R^{n+1}, in a suitable class. Actually the subject of our inquiry will go much farther than the mere analysis of minima and maxima of functionals.

Here t is the real or independent variable, $t \in R^1 = R$, usually called "time", and $x = (x^1, \dots, x^n) \in R^n, n \geq 1$, is a real vector variable, usually called the *space* or *phase* variable. Thus, we deal with continuous functions $x(t) = (x^1, \dots, x^n)$, $t_1 \leq t \leq t_2$, which we may call trajectories, or curves. Here $f_0(t, x, x')$ is a given real valued function defined on R^{1+2n}, or in whatever part of R^{1+2n} it is relevant and it will be called a *Lagrangian* function, or briefly a Lagrangian.

We may allow the variable (t, x) to vary only in a given set A of the tx-space R^{1+n}, possibly of the form $A = [t_0, T] \times A_0$, $A_0 \subset R^n$, and we do not exclude that A is the whole tx-space. Thus we may require that

(1.1.2) $$(t, x(t)) \in A, \qquad t_1 \leq t \leq t_2.$$

We may require the functions $x(t)$ to satisfy some boundary conditions. A typical one is "both end points fixed," or $x(t_1) = x_1$, $x(t_2) = x_2$ $(t_1, t_2, x_1, x_2$ fixed), $t_1 < t_2$, $x_1 = (x_1^1, \dots, x_1^n) \in R^n$, $x_2 = (x_2^1, \dots, x_2^n) \in R^n$. We may then say that we consider curves C "joining fixed points $1 = (t_1, x_1)$ and $2 = (t_2, x_2)$ in R^{1+n}".

A great variety of boundary conditions are of interest, e.g., C joins a fixed point $1 = (t_1, x_1)$ to a given curve $\Gamma: x = g(t)$, $t' \le t \le t''$, that is, $x(t_1) = x_1$, $x(t_2) = g(t_2)$, $t_1 < t_2$, $t' \le t_2 \le t''$. Alternatively, we may require that C join two given sets B_1 and B_2 in R^{n+1}. Thus, the boundary conditions concern the $2n + 2$ real numbers t_1, $x(t_1) = (x_1^1, \ldots, x_1^n)$, t_2, $x(t_2) = (x_2^1, \ldots, x_2^n)$, or the *ends* $e[x] = (t_1, x(t_1), t_2, x(t_2))$ of the trajectory x. Note that t_1 and t_2, in particular, need not be fixed. Often, these boundary conditions are expressed in terms of a set of equalities or inequalities concerning the $2n + 2$ numbers above. A general and compact way to express boundary conditions is to define a subset B of the $t_1x_1t_2x_2$-space R^{2n+2} and to require that

$$e[x] \in B, \quad \text{or} \quad (t_1, x(t_1), t_2, x(t_2)) \in B. \tag{1.1.3}$$

Thus, the case of both end points fixed, or t_1, x_1, t_2, x_2 fixed, corresponds to B being the single point (t_1, x_1, t_2, x_2) in R^{2n+2}; the case of fixed first end point (t_1, x_1) and second end point (t_2, x_2) on a given curve Γ corresponds to $B = (t_1, x_1) \times \Gamma$, a subset of R^{2n+2}.

Problems of minima and maxima for functionals (1.1.1) with only constrains as (1.1.2) and (1.1.3) are often referred to as Lagrange problems of the calculus of variations, and sometimes as free problems.

Besides (1.1.2), (1.1.3), another type of constraint is often required, namely

$$\int_{t_1}^{t_2} |x'(t)|^p \, dt \le C \tag{1.1.4}$$

for some constants $p \ge 1$, $C > 0$. More generally, we may require that for some "comparison functional" we have

$$\int_{t_1}^{t_2} H(t, x(t), x'(t)) \, dt \le C.$$

Alternatively, we may require that any number N of given analogous functionals have given values, say

$$J_j[x] = \int_{t_1}^{t_2} f_j(t, x(t), x'(t)) \, dt = C_j \; [\text{or} \le C_j], \qquad j = 1, \ldots, N.$$

These problems with equality signs are sometimes called *isoperimetric problems.* (See Section 3.6 for some examples). The same problems with $\le$ signs are sometimes called problems with comparison functionals.

And now a few words on the class of n-vector functions $x(t)$, $t_1 \le t \le t_2$, we shall take into consideration. One could expect to find the optimal solution in the class C^1 of all continuous functions $x(t) = (x^1, \ldots, x^n)$, $t_1 \le t \le t_2$, with continuous derivative $x'(t) = (x'^1, \ldots, x'^n)$. Very simple examples (see e.g. Section 2.6, Remark 2) show that it would be more realistic to search for optimal solutions in the class, say C_s, of all continuous functions $x(t) = (x^1, \ldots, x^n)$, $t_1 \le t \le t_2$, with sectionally continuous derivative. In such a situation, if we assume that $f_0(t, x, u)$ is defined and continuous in

$A \times R^n$, then $f_0(t, x(t), x'(t))$ would be sectionally continuous in $[t_1, t_2]$ and (1.1.1) would be a Riemann integral.

However, in view of other examples (see e.g. Section 2.6, Remark 1) in which the optimal solution is not in such a class C_s, and particularly because of exigencies related to the existence theorems (Chapters 9–16), it has been found more suitable to search for optimal solutions in the larger class of all *absolutely continuous* (AC) n-vector functions $x(t) = (x^1, \dots, x^n)$. (See Section 2.1 for definitions, and the Bibliographical notes at the end of this Chapter for historical views).

We only mention here that the class of AC functions is the largest class of continuous functions $x(t) = (x^1, \dots, x^n)$, $t_1 \le t \le t_2$, possessing derivative $x'(t) = (x'^1, \dots, x'^n)$ a.e. in $[t_1, t_2]$ and for which the fundamental theorem of calculus holds, i.e., $x(\beta) - x(\alpha) = \int_\alpha^\beta x'(t)\,dt$, the integral being a Lebesgue integral on each component (see Section 2.1 for the definition of AC functions). Conversely, if $g(t)$ is L-integrable, then $G(t) = \int_{t_1}^t g(\tau)\,d\tau$ is AC.

Again, if we assume that $f_0(t, x, u)$ is continuous in $A \times R^n$ and $x(t)$ is AC, then $f_0(\cdot, x(\cdot), x'(\cdot))$ is certainly measurable. In such a situation we shall explicitly require that $f_0(\cdot, x(\cdot), x'(\cdot))$ is L-integrable, and then (1.1.1) is an L-integral. We only mention here that a set E on the real line is said to be of measure zero if it can be covered by a countable collection of open intervals (α_i, β_i), $i = 1, 2, \dots$, possibly overlapping, whose total length $\sum_i (\beta_i - \alpha_i)$ is as small as we want. A property P then is said to hold almost everywhere (a.e.) if it holds everywhere but at the points of a set E of measure zero.

1.2 Classical Lagrange Problems with Constraints on the Derivatives

A very important recent extension of the concept above is to consider the same integral (1.1.1), with the same possible constraint (1.1.2) and boundary conditions (1.1.3), but now with restrictions concerning the possible values of x'. This can be understood by saying that, for every $(t, x) \in A$, a subset $Q(t, x)$ of R^n is assigned, and that we consider only n-vector AC functions $x(t) = (x^1, \dots, x^n)$, $t_1 \le t \le t_2$, whose derivative $x'(t) = (x'^1, \dots, x'^n)$ must belong to the corresponding set $Q(t, x(t))$. In other words, we may require that the n-vector AC function $x(t)$ satisfy

$$x'(t) \in Q(t, x(t)), \qquad t \in [t_1, t_2] \text{ (a.e.)} \tag{1.2.1}$$

This is called an orientor field, or an orientor field relation.

For instance, for $n = 1$ and $Q = Q(t, x) = [z \mid a \le z \le b]$, we would restrict ourselves to only those AC scalar functions $x(t)$ whose slope $x'(t)$ is between two fixed numbers a and b. For instance, for any $n \ge 1$ and $Q(t, x) = [z \in R^n \| z| \le a]$, we would restrict ourselves to only those AC n-vector functions $x(t) = (x^1, \ldots, x^n)$ whose tangent vector $x'(t) = (x'^1, \ldots, x'^n)$ belongs to a cone whose axis is parallel to the t-axis and of fixed opening a, thus $|x'(t)| \leq a$, $t \in [t_1, t_2]$ (a.e.).

As will become apparent in the next few pages this modification to the classical concept above has most striking interpretations and applications. In the present situation let M_0 denote the set $M_0 = [(t, x, z) | (t, x) \in A, z \in Q(t, x)] \subset R^{1+2n}$. Then it is enough to know that $f_0(t, x, z)$ is defined in M_0, and we shall take into consideration AC n-vector functions $x(t)$, $t_1 \leq t \leq t_2$, satisfying (1.1.2), (1.1.3), (1.2.1) and such that $f_0(\cdot, x(\cdot), x'(\cdot))$ is L-integrable in $[t_1, t_2]$.

Actually, we could even define $f_0(t, x, z)$ to be equal to $+\infty$ in $R^{1+2n} - M_0$, and then f_0 is said to be an extended function. If we do so, the sole requirement that $f_0(\cdot, x(\cdot), x'(\cdot))$ be L-integrable in $[t_1, t_2]$ automatically implies that $(t, x(t)) \in A$ and $x'(t) \in Q(t, x(t))$ for almost all $t \in [t_1, t_2]$.

Again, we consider the integral (1.1.1) with constraints (1.1.2) and boundary conditions (1.1.3), but now in addition we require that the AC trajectories $x(t) = (x^1, \ldots, x^n)$, $t_1 \leq t \leq t_2$, satisfy a system of differential equations

$$(1.2.2) \qquad G_j(t, x(t), x'(t)) = 0, \qquad t \in [t_1, t_2] \text{ (a.e.)}, \quad j = 1, \ldots, N < n,$$

where $G_j(t, x, u)$ are given real valued functions defined (say) in $A \times R^n$.

As an interpretation, let us consider, for any point $(t, x) \in A$, the set $Q(t, x) = [z \in R^n | G_j(t, x, z) = 0, j = 1, \ldots, N] \subset R^n$. Of course, the set $Q(t, x)$ may be empty. However, the constraint (1.2.2) for our AC n-vector functions $x(t)$ can be equivalently written in the form (1.2.1) above, or $x'(t) \in Q(t, x(t))$, where the sets Q are defined as stated. (See in Sections 1.7–13 further extensions and interpretations).

1.3 Classical Bolza Problems of the Calculus of Variations

We may consider here again given sets $A \subset R^{n+1}$, $B \subset R^{2n+2}$, and now we assume that real valued functions are given, $f_0(t, x, z)$ on $A \times R^n$ and $g(t_1, x_1, t_2, x_2)$ on B. Then instead of (1.1.1) we may consider problems of maximum and minimum for the functional

$$(1.3.1) \qquad I[x] = g(t_1, x(t_1), t_2, x(t_2)) + \int_{t_1}^{t_2} f_0(t, x(t), x'(t))\, dt,$$

with constraints (1.1.2), (1.1.3) as above, or $(t, x(t)) \in A$, $t \in [t_1, t_2]$, $(t_1, x(t_1), t_2, x(t_2)) \in B$.

Again, as above, we may assign a system of $N < n$ differential equations (1.3.1) to be satisfied by the AC n-vector function $x(t) = (x^1, \ldots, x^n)$, $t_1 \leq t \leq t_2$. Again, as above, we may assign, for every $(t, x) \in A$, a subset $Q(t, x)$ of R^n and require simply that $x'(t) \in Q(t, x(t))$, $t \in [t_1, t_2]$ (a.e.). (See Sections 1.7–12 for further extensions.)

1.4 Classical Problems Depending on Derivatives of Higher Order

Instead of functionals (1.1.1) we may have to deal with integrals depending on derivatives of higher order, namely,

$$I[x] = \int_{t_1}^{t_2} f_0(t, x(t), x'(t), \ldots, x^{(h)}(t))\, dt.$$

We shall see in Section 1.9 a reduction of these problems to other ones depending on derivatives of order one only, for which the usual conventions apply. Constraints as (1.2.2) in terms of higher order derivatives also can be reduced to orientor field equations (cf. Section 1.9).

1.5 Examples of Classical Problems of the Calculus of Variations

The concept of minimum and maximum of functionals will be made precise in Section 2.1 (absolute, strong local, weak local, in a given class, minima and maxima). Simple examples will be briefly mentioned in the nos. 1, 2, 3, 4 below of this Section. Let it be also mentioned here that not necessarily a functional takes on a minimum or a maximum (see no. 4), and, even if an optimal element exists, not necessarily the optimal element is unique (see no. 3).

1. *Classical brachistochrone problem.* Consider the problem of finding the curve $C: x = x(\xi)$, $\xi_1 \leq \xi \leq \xi_2$, in the vertical plane ξx, joining given points $1 = (\xi_1, x_1)$, $2 = (\xi_2, x_2)$, $\xi_1 < \xi_2$, $x_1 < x_2$, and such that a material point sliding along C without friction from 1 to 2, under gravity and with an initial speed $v_1 \geq 0$, reaches 2 in a minimum time. This problem was first solved by John Bernoulli in 1696. The problem

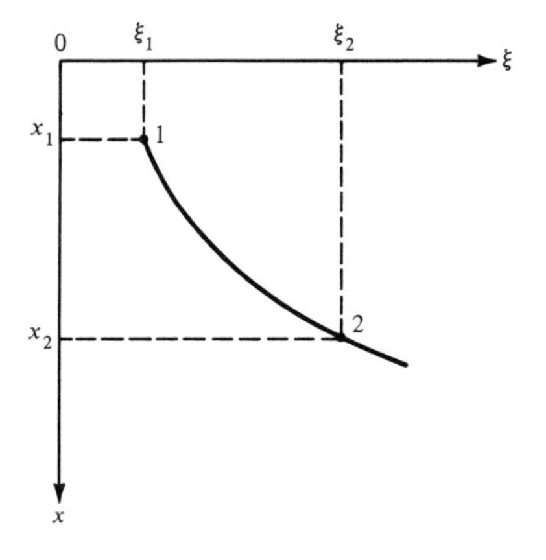

corresponds to the minimum of the functional

$$I[x] = \int_{\xi_1}^{\xi_2} (x - \alpha)^{-1/2}(1 + x'^2)^{1/2}\, d\xi, \qquad (') = d/d\xi,$$

in the class of all AC trajectories $x(\xi)$, $\xi_1 \le \xi < \xi_2$, $x(\xi_1) = x_1$, $x(\xi_2) = x_2$. Here α is the constant $\alpha = x_1 - v_1^2/2g$, and g is the gravity acceleration. The optimal solution exists and is unique, namely, an arc of a cycloid (see Section 3.12 for details). Here A is the set $[(\xi, x) | \xi_1 \le \xi \le \xi_2, x \ge x_1]$, and B is the singleton (ξ_1, x_1, ξ_2, x_2). If we consider the analogous problem of finding the curve C such that a material point sliding along C from $1 = (\xi_1, x_1)$ reaches the vertical line $\xi = \xi_2$ in a minimum time, then ξ_2 is fixed, x_2 is undetermined, A is the same as above, and $B = (\xi_1, x_1, \xi_2) \times R$, a subset of R^4. Concerning the integral $I[x]$ above, a derivation is given in Section 3.12.

2. *Problem of the surface of revolution of minimum area.* The problem of finding the curve $C: x = x(t)$, $t_1 \le t \le t_2$, $x(t) \ge 0$, in the tx-plane R^2, joining two given points $1 = (t_1, x_1)$, $2 = (t_2, x_2)$, $t_1 < t_2$, $x_1 > 0$, $x_2 > 0$, generating a surface S of revolution around the t-axis of minimum area, corresponds to the problem of finding the minimum of the functional

$$I[x] = \int_{t_1}^{t_2} x(t)(1 + x'^2(t))^{1/2}\, dt$$

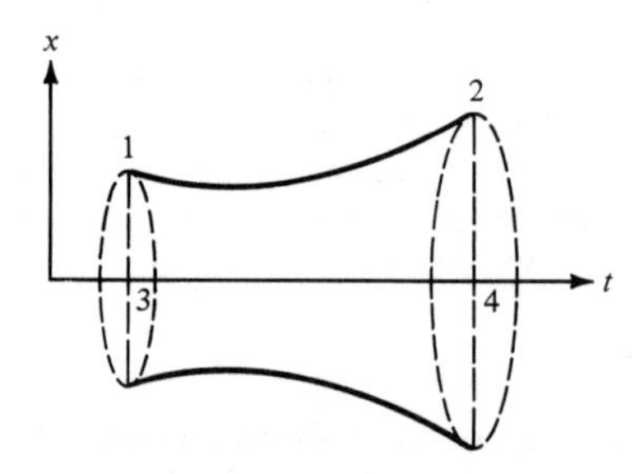

in the class of all AC trajectories x, lying in $A = [t_1, t_2] \times [0, +\infty)$, with both end points fixed. An optimal solution of the form $C: x = x(t)$, x AC, does not always exist. If it exists, then it is an arc of catenary (see Section 3.13 for details). There may be no minimum in the class of AC curves $x = x(t)$, $t_1 \le t \le t_2$, $x(t_1) = x_1$, $x(t_2) = x_2$. But in this case it is relevant to compare the areas of the surfaces of revolution described by these curves with the area of the surface of revolution described by the curve 1342 (two disks and the segment 34). The latter may be smaller and actually the optimal solution, but the arc 1342 is not of the form $x = x(t)$, $t_1 \le t \le t_2$ (see Section 3.13 for details).

3. *Paths of minimum length.* The problem of the nonparametric curve of the tx-plane R^2 of minimum length between two given points $1 = (t_1, x_1)$, $2 = (t_2, x_2)$, $t_1 < t_2$, corresponds to the problem of minimizing the functional

$$I[x] = \int_{t_1}^{t_2} (1 + x'^2(t))^{1/2}\, dt$$

with boundary conditions $x(t_1) = x_1$, $x(t_2) = x_2$, since the Jordan length of curves $C: x = x(t)$, $t_1 \le t \le t_2$, for x AC, is equal to the classical integral. The minimum of $I[x]$ is given by the segment $s = 12$, or $x(t) = x_1 + m(t - t_1)$, $t_1 \le t \le t_2$, with $m = (x_2 - x_1)/(t_2 - t_1)$. Of course, the same functional has no maximum, since there

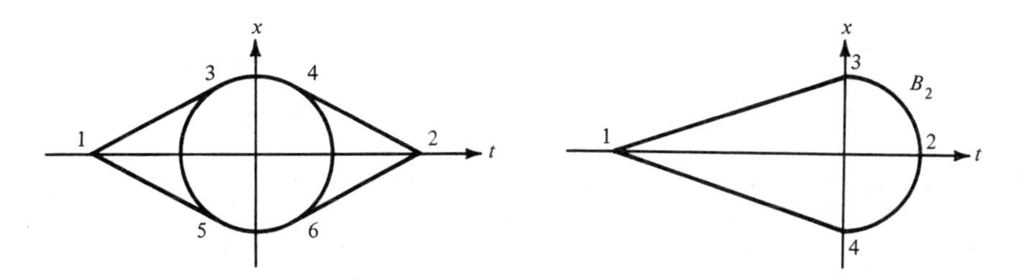

are polygonal lines $C: x = x(t)$, $t_1 \le t \le t_2$, joining 1 and 2 and of length as large as we want.

Note that, if we restrict the same functional $I[x]$ to the class of all functions $x(t)$ which are Lipschitzian of constant L, and hence AC with $|x'(t)| \le L$ a.e., then for $L > |m|$ the same functional $I[x]$ has still the same minimum given by the segment $s = 12$, and infinitely many maxima—certainly all polygonal lines S joining 1 and 2, whose segments have slopes $\pm L$ (as well as all AC functions $x(t)$ with $x'(t) = \pm L$ a.e.).

Note also that the same functional $I[x]$ in the class of all AC curves $C: x = x(t)$, $-2 \le t \le 2$, joining $1 = (-2, 0)$ to $2 = (2, 0)$ in the region $A = [(t, x) \in R^2 | t^2 + x^2 \ge 1]$ (that is, with no point in $t^2 + x^2 < 1$) has exactly two optimal solutions, namely, the curves 1342 and 1562 depicted in the left figure above. Finally, note that the same functional $I[x]$ in the class of all AC curves $C: x = x(t)$, $a \le t \le t_2$, joining $1 = (a, 0)$ to the arc $B_2 = [(t, x) | t^2 + x^2 = 1, t \ge 0]$, has exactly two minima given by $s = 13$ and $s = 14$ if $a < 0$, has infinitely many solutions if $a = 0$ (all radii from $(0, 0)$ to points $(s, y) \in B_2$, $s > 0$, and has exactly one optimal solution if $a > 0$ ($s = 12$).

4. *An example of a problem with no minimum.* The functional $I[x] = \int_0^1 tx'^2\, dt$ has no minimum in the class of all AC curves $C: x = x(t)$, $0 \le t \le 1$, joining $1 = (0, 1)$ to $2 = (1, 0)$.

To see this, we note that $f_0(t, x, x') = tx'^2$ is nonnegative for $0 \le t \le 1$; hence $i = \inf I[x] \ge 0$, where inf is taken in the class of all AC functions $x(t)$, $0 \le t \le 1$, with $x(0) = 1$, $x(1) = 0$. On the other hand, if we consider the AC functions $x_k(t)$, $0 \le t \le 1$, $k = 2, 3, \ldots$, defined by $x_k(t) = 1$ for $0 \le t \le k^{-1}$, $x_k(t) = -(\ln t)/(\ln k)$ for $k^{-1} \le t \le 1$, $k = 2, 3, \ldots$, then $x_k'(t) = 0$, or $= -(t \ln k)^{-1}$, respectively, and $I[x_k] = (\ln k)^{-1}$. Thus, $I[x_k] \to 0$ as $k \to \infty$, and hence $i = 0$.

The functional $I[x]$ cannot attain the value $i = 0$, since $I = 0$ would imply $tx'^2(t) = 0$ a.e. in $[0, 1]$, hence $x'(t) = 0$ a.e. in $[0, 1]$, or $x(t)$ constant, while we require here $x(0) = 1$, $x(1) = 0$. This proves that $I[x]$ has no absolute minimum in the class of all AC curves $C: x = x(t)$, $0 \le t \le 1$, joining $1 = (0, 1)$ to $2 = (1, 0)$. (Cf. Section 2.1C.)

5. *Another simple problem with no minimum.* Consider the integral $I[x] = \int_0^1 (x^2 + x'^2)^{1/2}\, dt$ in the class of all AC curves $x = x(t)$, $0 \le t \le 1$, joining the points $1 = (0, 0)$ to $2 = (1, 1)$. Here we have

$$I[x] = \int_0^1 (x^2 + x'^2)^{1/2}\, dt > \int_0^1 |x'|\, dt \ge \int_0^1 x'\, dt = x(1) - x(0) = 1.$$

Thus, for $i = \inf I[x]$ we have $i \ge 1$. If we consider now the sequence of AC functions $x_k(t)$, $0 \le t \le 1$, $k = 1, 2, \ldots$, defined by taking $x_k(t) = 0$ for $0 \le t \le 1 - k^{-1}$, $x_k(t) = 1 + k(t - 1)$ for $1 - k^{-1} \le t \le 1$, then $I[x_k] = \int_{1-k^{-1}}^1 (x_k^2 + k^2)^{1/2} \le (1 + k^2)^{1/2} k^{-1} \le (1 + k)k^{-1}$, that is, $I[x_k] \to 1$ as $k \to \infty$. Thus, $i = 1$; and we have seen above that I cannot take the value 1 in the class of all AC curves $x = x(t)$, $0 \le t \le 1$, joining $(0, 0)$ to $(1, 1)$.

1.6 Remarks

First of all, if we conceive our problem as the minimization of some functional in a class of curves C in the $tx^1 \cdots x^n$-space R^{n+1}, it is clear from the first lines of (1.1) that we have already restricted ourselves there to only those curves C in R^{n+1} which can be represented as AC n-vector functions $x = x(t)$, $t_1 \le t \le t_2$. Usually, t is time and $x = (x^1, \ldots, x^n)$ are space coordinates, and then the choice is most natural, since $x(t)$ then represents the law of motion of the point x along the curve C. In mechanics $x = (x^1, \ldots, x^n)$ may be the (Lagrangian) coordinates of a mechanical system, and again $x(t)$ represents the evolution of the system in time. However, if we think of both t and $x = (x^1, \ldots, x^n)$ as geometrical coordinates in R^{n+1}, the situation may be different. For instance, for $n = 1$, if we are looking for the shortest path from the point $(0, 0)$ to the straight line $B_2 = [x = 1, -\infty \le t \le +\infty]$, then the minimum is the segment s between $(0, 0)$ and $(0, 1)$, which has no representation $x = x(t)$. We should have considered curves $t = t(x)$, $0 \le x \le 1$, with $t(0) = 0$, $t(1) = 0$, and then s would have the representation $t = 0$, $0 \le x \le 1$. Obviously, the choice of coordinates may count.

Analogously, if for instance a given problem in the tx-plane R^2 has a solution which conceivably is the unit circle $\Gamma: t^2 + x^2 = 1$, obviously we should better think of Γ as the locus $\Gamma: (x^1)^2 + (x^2)^2 = 1$ in an x^1x^2-plane, and represent it in terms of some arbitrary parameter τ, say, as $\Gamma: x^1 = \cos \tau$, $x^2 = \sin \tau$, $0 \le \tau \le 2\pi$. By writing the functional in terms of the coordinate system $\tau x^1 x^2$ and considering it in a class of curves $x^1 = x^1(\tau)$, $x^2 = x^2(\tau)$, $a \le \tau \le b$, we may have a chance to state that Γ is the optimal solution in such a class.

It may happen that the functional we are dealing with does not depend on the particular parametrization, or representation, that we choose. In other words, it may occur that the functional depends only on the "path curve" C, or "parametric curve" C, and possibly its orientation. For instance, in mechanics a functional may depend only on the "oriented path curve" C, and not on the law of motion by which C is traveled (once and in a given sense). In these situation, we are dealing with a "parametric integral", depending only on the oriented parametric curve C, or oriented path curve C, in the x-space R^n, $x = (x^1, \ldots, x^n)$, in any of its possible representations $x = x(\tau)$, $a \le \tau \le b$. If (1.1.1) is the functional, then f_0 is called a "parametric" integrand. We shall mention in Chapter 14 that for this to occur f_0 must not depend on the independent variable and must be positive homogeneous of degree one in x', or $f_0(x, kx') = kf_0(x, x')$ for all $k \ge 0$. For parametric integrals the boundary conditions must be independent of the representation—namely, of the form $(x_1, x_2) \in B \subset R^{2n}$, or more particularly, $x_1 \in B_1 \subset R^n$, $x_2 \in B_2 \subset R^n$. Analogously, all constraints need to be independent of the representation. Though parametric integrals are particular cases of those we are considering in this book, most theorems can be given pertinent forms, usually stronger. We shall not present this aspect of optimization theory in this book. We shall discuss parametric problems in a forthcoming book [III].

All in all, there is a great flexibility in the choice of the form in which a given problem can be set, and the choice may make a difference. We shall encounter many problems which have no optimal solution as they are written, or in a given class, but they may have an optimal solution as soon as they are framed in a suitable form. For instance, we shall introduce in Section 1.14 the concept of "generalized solutions," or generalized curves, and we shall present very simple problems whose optimal solutions lie in such a class, and not in the class of usual solutions.

On the other hand, once a suitable frame has been found, we shall see in Chapter 2 that the optimal solutions and their characteristic properties have remarkable invariance properties with respect to general changes of coordinates $t = t(s, y)$, $x = x(s, y)$ under sole regularity conditions.

1.7 The Mayer Problems of Optimal Control

We shall deal here with a real variable $t \in R$, a real vector variable $x = (x^1, \ldots, x^n) \in R^n$, $n \geq 1$, and a real vector variable $u = (u^1, \ldots, u^m) \in R^m$, $m \geq 1$. Here t is the independent variable (time), x is the *space* variable (or state, or phase variable), and u is the *control* variable. Thus, we shall deal with functions $x(t) = (x^1, \ldots, x^n)$, $u(t) = (u^1, \ldots, u^m)$, $t_1 \leq t \leq t_2$, of which x will be called a trajectory, and u a *control*, or *steering function*, or *strategy*; and they will be required to satisfy a system of differential equations $dx/dt = f(t, x(t), u(t))$, $f(t, x, u) = (f_1, \ldots, f_n)$, or written componentwise $dx^i/dt = f_i(t, x(t), u(t))$, $t_1 \leq t \leq t_2$, $i = 1, \ldots, n$. In other words, for a chosen $u(t)$, we think of $x(t)$ as a solution of the differential system $dx/dt = f(t, x, u(t))$ in "normal form". As before, we allow the variable (t, x) to vary in a given subset A of R^{n+1}. We may allow the variable u to vary in a given set U of the u-space; possibly $U = R^m$, or $U = U(t, x)$ may depend in a well-prescribed manner on t, or x, or both. Here U is sometimes called the control space, or the restraint set.

As before, we may well require that x satisfy given boundary conditions, as in Sections 1.1–5, which we write in the short form $(t_1, x(t_1), t_2, x(t_2)) \in B$, where B is a given subset of R^{2+2n}. Finally, let $g(t_1, x_1, t_2, x_2)$ be a given real valued function defined on B, and we may seek the maxima and minima of the functional $I[x, u] = g(t_1, x(t_1), t_2, x(t_2))$. Such a functional is sometimes called the cost functional, or the cost, or the index of performance.

For instance, if (t_1, x_1) is fixed, and x_2 is fixed, but $g = t_2$, then we are simply required to determine the strategy $u(t)$ and corresponding trajectory $x(t)$ so that a moving point $P = (x_1, \ldots, x_n)$, monitored by the differential system (1.7.2) below, moves from x_1 to x_2 in the shortest possible time t_2. This is sometimes called a problem of minimum time, or a brachistochrone problem.

Summarizing, we have here the problem of finding minima or maxima of the functional

$$I[x, u] = g(t_1, x(t_1), t_2, x(t_2)) \tag{1.7.1}$$

with differential system

$$dx/dt = f(t, x(t), u(t)), \qquad t_1 \leq t \leq t_2, \tag{1.7.2}$$

boundary conditions

$$(1.7.3) \qquad e[x] = (t_1, x(t_1), t_2, x(t_2)) \in B,$$

and constraints

$$(1.7.4) \qquad (t, x(t)) \in A, \qquad t_1 \leq t \leq t_2,$$

$$(1.7.5) \qquad u(t) \in U(t, x(t)), \qquad t_1 \leq t \leq t_2.$$

This is called sometimes a Mayer problem in optimal control theory, or a canonical Mayer problem.

Constraints of the form

$$(1.7.6) \qquad \int_{t_1}^{t_2} |x'(t)|^p \, dt \leq C, \qquad p \geq 1,$$

or of the form

$$\int_{t_1}^{t_2} |u(t)|^q \, dt \leq D, \qquad q \geq 1,$$

or analogous ones, are also often required.

Again, as in Section 1.1, one expects to find the optimal solution x, u in the class of all pairs $x(t) = (x^1, \ldots, x^n)$, $u(t) = (u^1, \ldots, u^m)$, $t_1 \leq t \leq t_2$, with x continuous with sectionally continuous derivative and u sectionally continuous in $[t_1, t_2]$. However, because of many examples in which the optimal solution is not in such a class, and because of exigencies related to existence theorems, it has been found more suitable to search for optimal solutions in the larger class of all pairs x, u with x AC and u measurable in $[t_1, t_2]$. (See the Bibliographical Notes at the end of this Chapter).

With these conventions, we shall say that a pair $x(t)$, $u(t)$, $t_1 \leq t \leq t_2$, is *admissible* provided x is AC, u is measurable, and x, u satisfies (1.7.2), (1.7.4), (1.7.5). Finally, as we shall see in Section 1.11, the problems under consideration can be written in terms of orientor fields with no explicit mentioning of controls or strategies.

There are situations (f linear in u, U compact) where the maxima and minima of $I[x, u]$ are attained by control functions $u(t)$ taking their values on the boundary of U. For instance, suppose $m = 1$, $U = [-1 \leq u \leq 1]$, and that $u(t)$ assume only the values ± 1. These are called *bang-bang* solutions.

Lagrange and Bolza problems of the calculus of variations and isoperimetric problems can always be thought of as particular cases of problems of optimal control, as we shall easily see in Section 1.9. Mayer type problems of optimal control involving differential equations of higher order can always be reduced to Mayer problems with differential equations of order one. For instance the equation $x^{(h)}(t) = F(t, x, x', \ldots, x^{(h-1)}, u)$, by taking $x = y_1, x' = y_2, \ldots, x^{(h-1)} = y_h$, is reduced to the system $y_1' = y_2, \ldots, y_{h-1}' = y_h$, $y_h' = F(t, y_1, \ldots, y_h, u)$ of the form (1.7.2).

1.8 Lagrange and Bolza Problems of Optimal Control

Often the cost functional $I[x,u]$, or index of performance, is given in the form

$$I[x,u] = \int_{t_1}^{t_2} f_0(t, x(t), u(t))\, dt, \tag{1.8.1}$$

where $f_0(t,x,u)$ is a given real valued function defined on the set $M = [(t,x,u)|(t,x) \in A,\, u \in U(t,x)] \subset R^{1+n+m}$, and then we shall only consider pairs $x(t)$, $u(t)$, $t_1 \le t \le t_2$, x AC, u measurable, such that $f_0(\cdot, x(\cdot), u(\cdot))$ is measurable and Lebesgue integrable on $[t_1, t_2]$. Together with the cost functional (1.8.1) we shall also consider the system of differential equations (1.7.2), $dx/dt = f(t, x(t), u(t))$, $t_1 \le t \le t_2$; the boundary conditions (1.7.3), $(t_1, x(t_1), t_2, x(t_2)) \in B$, where B is a given subset of R^{2+2n}; and the constraints (1.7.4), (1.7.5), $(t, x(t)) \in A$, $u(t) \in U(t, x(t))$, where $A \subset R^{n+1}$, $U(t,x) \subset R^m$. This is called sometimes a Lagrange problem of control.

Alternatively, we may consider the cost functional

$$I[x,u] = g(t_1, x(t_1), t_2, x(t_2)) + \int_{t_1}^{t_2} f_0(t, x(t), u(t))\, dt, \tag{1.8.2}$$

where f_0 is as before and $g(t_1, x_1, t_2, x_2)$ is a given real valued function defined on B, together with differential equations (1.7.2), boundary conditions (1.7.3), and constraints (1.7.4), (1.7.5). This is called a Bolza problem of optimal control.

1.9 Theoretical Equivalence of Mayer, Lagrange, and Bolza Problems of Optimal Control. Problems of the Calculus of Variations as Problems of Optimal Control

Lagrange problems are readily reduced to Mayer problems by introducing an additional state variable x^0, the new state vector $\tilde{x} = (x^0, x) = (x^0, x^1, x^2, \ldots, x^n)$, an additional differential equation

$$dx^0/dt = f_0(t, x(t), u(t)),$$

and an additional initial condition $x^0(t_1) = 0$. Then the functional (1.8.1) becomes $I[x,u] = x^0(t_2)$, and we have a Mayer problem with $g = x^0(t_2)$, the $(n+1)$-vector $\tilde{x}$ replacing the n-vector x.

Analogously, Bolza problems are reduced to Mayer problems by introducing the state variable x^0, the additional differential equation $dx^0/dt = f_0$, and the additional initial condition $x^0(t_1) = 0$. Then the functional (1.8.2) becomes $I[x, u] = g(t_1, x(t_1), t_2, x(t_2)) + x^0(t_2)$, and again we have a Mayer problem.

It is evident that Bolza problems contain Mayer and Lagrange problems, since (1.8.2) reduces to (1.7.1) if $f_0 = 0$, and to (1.8.1) if $g = 0$.

It remains to show that Mayer and Bolza problems can be reduced to Lagrange problems. Indeed, by introducing an additional state variable x^0 with the additional differential equation and initial value

$$dx^0/dt = 0, \qquad x^0(t_1) = (t_2 - t_1)^{-1} g(t_1, x(t_1), t_2, x(t_2)),$$

the functional (1.7.1) becomes

$$I[x, u] = \int_{t_1}^{t_2} x^0(t)\, dt,$$

and functional (1.8.2) becomes

$$I[x, u] = \int_{t_1}^{t_2} [f_0(t, x(t), u(t)) + x^0(t)]\, dt.$$

Thus, Mayer, Lagrange, and Bolza problems can be said to be theoretically equivalent. However, the different necessary conditions (Chapters 4, 5) and the existence theorems may make it preferable to use one particular form or the other.

It is interesting to note that any Lagrange problem with $f_0 > 0$ can be reduced to a problem of minimum time. Indeed, if τ is a new time variable related to t by the relation $d\tau/dt = f_0(t, x(t), u(t))$, then the relation $t = t(\tau)$ can be inverted into a relation $\tau = \tau(t)$; the initial and terminal times t_1, t_2 become new times τ_1, τ_2, $\tau_1 = \tau(t_1)$, $\tau_2 = \tau(t_2)$; the differential system $dx/dt = f(t, x(t), u(t))$ becomes $dx/d\tau < f f_0^{-1}$; and the functional (1.8.1) becomes

$$I[x, u] = \int_{t_1}^{t_2} f_0\, dt = \int_{\tau_1}^{\tau_2} d\tau = \tau_2 - \tau_1.$$

Mayer, Lagrange, and Bolza problems are said to be autonomous provided the functions f, f_0 are independent of time.

Formally, any of the above problems can be reduced to an autonomous one by introducing an additional variable x^{n+1}, the additional differential equation $dx^{n+1}/dt = 1$, and the additional boundary condition $x^{n+1}(t_1) = t_1$, so that $x^{n+1} = t$.

The same problems can be even reduced to autonomous problems in a fixed interval, say $[0, 1]$. This can be done by introducing two new additional variables x^{n+1}, x^{n+2}, differential equations $dx^{n+1}/d\tau = x^{n+2}$, $dx^{n+2}/d\tau = 0$ (that is, $d^2x^{n+1}/d\tau^2 = 0$), and boundary conditions $x^{n+1}(0) = t_1$, $x^{n+1}(1) = t_2$. Then $t = x^{n+1} = t_1 + (t_2 - t_1)\tau$, $x^{n+2} = t_2 - t_1$, $0 \le \tau \le 1$, and t again becomes the state variable x^{n+1}.

The Lagrange problems of optimal control contain as particular cases the classical Lagrange problems of the calculus of variations. Indeed, if $m = n$, $f(t, x, u) = u$—that is, if $f_i(t, x, u) = u^i$, $i = 1, \ldots, n$, and $U = R^n$—then $x'(t) = u(t)$ and the problem (1.8.1) reduces to $I[x] = \int_{t_1}^{t_2} f_0(t, x(t), x'(t))\, dt$, with no differential equations, and only the constraints (1.7.3), (1.7.4), $(t_1, x(t_1), t_2, x(t_2)) \in B$, $(t, x(t)) \in A$. In other words, Lagrange problems of the calculus of variations as stated in Section 1.1 can always be thought of as Lagrange problems of optimal control (with $m = n$, $f = u$, $U = R^n$), and hence also as Mayer problems of control.

Analogously, classical Bolza problems of the calculus of variations as stated in Section 1.3 with sole constraints (1.1.2), (1.1.3) can always be thought of as Bolza problems of optimal control (again with $m = n$, $f = u$, $U = R^n$), and hence also as (say) Mayer problems, by the same devices we have used for Lagrange problems. Examples are mentioned below.

A few comments are needed here on the constraints (1.2.2) in the form of $N < n$ ordinary differential equations not necessarily in normal form,

$$G_j(t, x(t), x'(t)) = 0, \qquad j = 1, \dots, N < n.$$

We have seen in Section 1.2 that these constraints can be easily interpreted in terms of orientor fields. It is relevant we interpret them in terms of control parameters. This can be done by the following classical remark. Let us assume that at least locally the system of $N < n$ functions G_j, $j = 1, \dots, N$, can be completed into a system of n functions $G_1, \dots, G_N, G_{N+1}, \dots, G_n$ all of class C^1, with nonzero functional determinant $\partial G/\partial x'$, that is, the $n \times n$ matrix $[G_{jx's}, j, s = 1, \dots, n]$ has determinant different from zero. Then the system of n equations

$$G_1 = 0, \dots, \qquad G_N = 0, \qquad G_{N+1} = u^1, \dots, \qquad G_n = u^{n-N},$$

by the implicit function theorem of calculus, is locally solvable in x', that is, locally, there are functions $f_i(t, x, u)$, $i = 1, \dots, n$, with $u = (u^1, \dots, u^{n-N})$, such that

$$x'^1 = f_1(t, x, u), \dots, \qquad x'^n = f_n(t, x, u),$$

and here the variables $u^1, \dots, u^{n-N}$ have the role of control parameters.

Problems of the calculus of variations depending on derivatives of order higher than one, say

$$I[x] = \int_{t_1}^{t_2} f_0(t, x(t), \dots, x^{(h)}(t))\, dt,$$

can always be written in the form of a Lagrange problem of optimal control as stated in Section 1.4. Indeed, by taking $y_1 = x$, $y_2 = x'$, $\dots$, $y_h = x^{(h-1)}$, then $I[x]$ takes the form

$$I[y, u] = \int_{t_1}^{t_2} f_0(t, y(t), u(t))\, dt, \qquad y(t) = (y_1, \dots, y_h),$$

with control variable $u, m = 1$, and differential system

$$y_1' = y_2, \dots, \qquad y_{h-1}' = y_h, \qquad y_h' = u, \quad u \in U = R.$$

Finally, classical isoperimetric problems, as well as problems with comparison functionals, as stated in Section 1.1, are immediately written as

Lagrange problems of optimal control by introducing N auxiliary variables $y(t) = (y^1, \ldots, y^N)$, the $N + n$ differential equations $dx^i/dt = u^i$, $i = 1, \ldots, n$, $dy^j/dt = f_j(t, x, u)$, $j = 1, \ldots, N$, and the $2N$ extra boundary conditions $y^j(t_1) = 0$, $y^j(t_2) = C_j$, $j = 1, \ldots, N$.

On the other hand, as we shall see in Section 1.12, all problems of control can be reduced to classical problems with constraints on the derivatives as described in Sections 1.1 and 1.2, that is, in terms of orientor fields.

1.10 Examples of Problems of Optimal Control

Here are few examples of Lagrange and Bolza problems of optimal control.

1. *Minimum of* $I[x, u] = \int_0^1 u^2\,dt$ *with* $x' = x + u$, $x(0) = x_1 = 1$, $x(1) = x_2 = 0$. This is a Lagrange problem with $n = m = 1$, $U = R$. It is immediately reduced to the Mayer problem $I[x, y, u] = y(1)$ with $x' = x + u$, $y' = u^2$, $n = 2$, $m = 1$, $x(0) = 1$, $x(1) = 0$, $y(0) = y_1 = 0$, and in this Mayer problem $B = (0, 1, 0; 1, 0, y_2)$ is a straight line in R^6, since $y(1) = y_2 \in R$ is undetermined.

2. *Minimum of* $I[x, u] = \int_0^1 u^2\,dt + (x(1))^2$ *with* $x' = x + u$, $x(0) = x_1 = 1$. This is a Bolza problem with $n = m = 1$, $g = x_2^2 = (x_2(1))^2$, $U = R$. It is immediately reduced to the Mayer problem $I[x, y, u] = y(1) + (x(1))^2$ with $x' = x + u$, $y' = u^2$, $n = 2$, $m = 1$, $g = y_2 + x_2^2 = y(1) + (x(1))^2$, $x(0) = x_1 = 1$, $y(0) = y_1 = 0$.

3. A point P moves along the x-axis monitored by the equation $x'' = u$ with $|u| \leq 1$. Take P from any given state $(x = a, x' = b)$ to rest at the origin $(x = 0, x' = 0)$ in the shortest possible time (a brachistochrone problem). By using phase coordinates $x = x$, $y = x'$, we have a Mayer problem of minimum time with $n = 2$, $m = 1$, differential system $x' = y$, $y' = u$, control space $U = [u| -1 \leq u \leq 1] \subset R$, $A = R^3$, initial data $t_1 = 0$, $x(t_1) = a$, $y(t_1) = b$, terminal data $x(t_2) = 0$, $y(t_2) = 0$, $t_2 \geq 0$ undetermined, and functional $I[x, y, u] = g = t_2$. As we shall see (Sections 4.2B and 6.1), the optimal solution is unique, with strategy $u(t)$ taking only the values $u(t) = \pm 1$, and trajectory made up of two arcs of parabolas (stabilization of a point moving on a straight line under a limited external force).

4. A point P moves along the x-axis monitored by the equation $x'' + x = u$ with $|u| \leq 1$. Take P from any given state $(x = a, x' = b)$ to rest at the origin $(x = 0, x' = 0)$ in the shortest possible time. As in example 3, the problem is reduced to the Mayer problem with $n = 2$, $m = 1$, differential system $x' = y$, $y' = -x + u$, control space $U = [u| -1 \leq u \leq 1] \subset R$, $A = R^3$, initial data $t_1 = 0$, $x(t_1) = a$, $y(t_1) = b$, terminal data $x(t_2) = 0$, $y(t_2) = 0$, $t_2 \geq 0$, and functional $I[x, y, u] = g = t_2$. As we shall see in Section 6.2, the optimal solution is unique, with strategy $u(t)$ taking only the values $u(t) = \pm 1$, and trajectory made up of a finite number of arcs of circles in the xy-plane (stabilization of a point moving on a straight line under an elastic force and a limited external force).

1.11 Exercises

1. Consider the problem of the minimum of the functional $I[x, u] = \int_0^1 (1 + (x''')^2)\,dt$ with $x(0) = x_1$, $x(1) = x_2$, x_1 and x_2 fixed. Reduce this problem to (a) a Lagrange problem of optimal control; (b) a Mayer problem of optimal control.

2. Consider the problem of optimal control concerning the minimum of the functional $I[x,u]=g=x(1)-x'(1)$ with differential equation $x''=u-2$, where $u \in U=[-1,1]$, $x(0)=0$, $x'(0)=0$. Reduce this problem to (a) a Mayer problem of optimal control (b) a Lagrange problem of optimal control.
3. Consider the classical isoperimetric problem concerning the minimum of the integral $I[x]=\int_0^1 (1+(x'')^2)^{1/2}\,dt$ with constraints $x(0)=0$, $x(1)=0$, and $J[x]=\int_0^1 (1+(x')^2)^{1/2}\,dt=c$. Reduce this problem to (a) a Lagrange problem of optimal control; (b) a Mayer problem of optimal control.
4. Consider the problem of the path of minimum length between two points $1=(t_1,x_1,y_1)$, $2=(t_2,x_2,y_2)$ on the cylinder $x^2+y^2=r^2$.

   Write this problem as (a) a classical Lagrange problem; (b) a Lagrange problem of optimal control; (c) a Mayer problem of optimal control.
5. Consider the problem of optimal control concerning the minimum of the functional $I[x] < \int_0^{t_2}(1+t^2x^2+u^2)\,dt$ with constraints $x'(t)=u$, $|u|\leq 1$, $x(0)=x_1$, $x(t_2)=x_2$. Write this problem as (a) an autonomous problem; (b) an autonomous problem with fixed end times.

1.12 The Mayer Problems in Terms of Orientor Fields

Let us consider the Mayer problem of optimization of (1.7) with functional and constraints

$$
\begin{aligned}
&I[x]=g(t_1,x(t_1),t_2,x(t_2)),\\
&x'(t)=f(t,x(t),u(t)), \qquad u(t)\in U(t,x(t)), \quad t\in[t_1,t_2] \text{ (a.e.)},\\
&(t,x(t))\in A, \qquad (t_1,x(t_1),t_2,x(t_2))\in B,\\
&x=(x^1,\ldots,x^n), \qquad u=(u^1,\ldots,u^m).
\end{aligned}
\tag{1.12.1}
$$

For every $(t,x)\in A$ let us denote by $Q(t,x)$ the set of all $z=(z^1,\ldots,z^n)\in R^n$ such that $z=f(t,x,u)$ for some $u\in U(t,x)$, or $z^i=f_i(t,x,u)$, $i=1,\ldots,n$, $u\in U(t,x)$; in symbols

$$Q(t,x)=[z\in R^n \mid z=f(t,x,u),\, u\in U(t,x)]=f(t,x,U(t,x)).$$

The last notation is most common since it states that $Q(t,x)$ is the image in R^n of the set $U(t,x)$ in R^m, the image being obtained by means of the vector function $f(t,x,u)=(f_1,\ldots,f_n)$.

For every admissible pair $x(t)$, $u(t)$, $t_1\leq t\leq t_2$, that is, for every x AC, u measurable, x, u satisfying (1.11.1), we obviously have $x'(t)\in Q(t,x(t))$, $t\in[t_1,t_2]$ (a.e.). This remark suggests that instead of the problem (1.12.1), we may simply consider the problem

$$
\begin{aligned}
&I[x]=g(t_1,x(t_1),t_2,x(t_2)),\\
&x'(t)\in Q(t,x(t)), \qquad t\in[t_1,t_2] \text{ (a.e.)},\\
&(t,x(t))\in A, \qquad (t_1,x(t_1),t_2,x(t_2))\in B.
\end{aligned}
\tag{1.12.2}
$$

The second relation (1.12.2) means that at every point $(t, x) \in A$ we can choose any direction dx/dt of the set $Q(t, x)$. In other words, at every point $(t, x) \in A$ we have a possible set of directions $dx/dt \in Q(t, x)$, instead of only one direction (say $dx/dt = F(t, x)$) as for usual differential equations. We say that the second relation (1.12.2) defines an *orientor field* in A. This term is the analogue of "direction field" as defined by usual differential systems in normal form.

We have shown above that, for every admissible pair x, u, the trajectory $x(t)$, $t_1 \leq t \leq t_2$, can be interpreted as an (AC) solution of the orientor field $x' \in Q(t, x)$. We shall prove in Section 8.2 that, conversely, every (AC) solution $x(t)$, $t_1 \leq t \leq t_2$, of the orientor field $x'(t) \in Q(t, x(t))$ can be thought of as an (AC) solution of the differential system $dx/dt = f(t, x, u(t))$ for a suitable measurable control function $u = u(t)$, $t_1 \leq t \leq t_2$, or strategy, satisfying $u(t) \in U(t, x(t))$, $t \in [t_1, t_2]$ (a.e.) (under the sole hypotheses that A and M are closed (Section 8.2), or analogous hypotheses).

1.13 The Lagrange Problems of Control as Problems of the Calculus of Variations with Constraints on the Derivatives

Analogously, for the Lagrange problem of the minimum in optimal control with functional and constraints

$$I[x, u] = \int_{t_1}^{t_2} f_0(t, x(t), u(t))\, dt,$$

$$x'(t) = f(t, x(t), u(t)), \qquad u(t) \in U(t, x(t)), \quad t \in [t_1, t_2] \text{ (a.e.)},$$
$$(t, x(t)) \in A, \qquad (t_1, x(t_1), t_2, x(t_2)) \in B, \tag{1.13.1}$$
$$x = (x^1, \ldots, x^n), \qquad u = (u^1, \ldots, u^m),$$

we define, for every $(t, x) \in A$, the sets

$$Q(t, x) = [z \mid z = f(t, x, u),\ u \in U(t, x)] \subset R^n,$$
$$\tilde{Q}(t, x) = [(z^0, z) \mid z^0 \geq f_0(t, x, u),\ z = f(t, x, u),\ u \in U(t, x)] \subset R^{n+1}.$$

Here $Q(t, x)$ is the projection on the z-space R^n of the set $\tilde{Q}(t, x)$ in R^{n+1}. For every $(t, x) \in A$, we denote by $T(t, x, z)$ the (extended) real valued function

$$\begin{aligned} T(t, x, z) &= \inf[z^0 \mid (z^0, z) \in \tilde{Q}(t, x)] \\ &= \inf[z^0 \mid z^0 \geq f_0(t, x, u),\ z = f(t, x, u),\ u \in U(t, x)], \qquad z \in R^n, \end{aligned}$$

or $-\infty \leq T(t, x, z) \leq +\infty$. Here $T(t, x, u) = +\infty$ for every $z \notin Q(t, x)$, since the inf above is taken on an empty class of real numbers. On the other hand, $-\infty \leq T(t, x, z) < +\infty$ for $z \in Q(t, x)$ and $(t, x) \in A$.

The extended function $T(t, x, z)$ is often called the Lagrangian function associated to the problem of optimal control (1.13.1).

Then for any admissible pair $x(t)$, $u(t)$, $t_1 \leq t \leq t_2$, for the problem (1.13.1) we have

$$\begin{aligned} I_0[x] = \int_{t_1}^{t_2} T(t, x(t), x'(t))\, dt \leq I[x, u], \\ x'(t) \in Q(t, x(t)), \qquad t \in [t_1, t_2] \text{ (a.e.)}, \\ (t, x(t)) \in A, \qquad (t_1, x(t_1), t_2, x(t_2)) \in B, \end{aligned} \tag{1.13.2}$$

(provided, of course, that $T(t, x(t), x'(t))$ is measurable and L-integrable). In other words, we have here a problem of the calculus of variations with the constraint $x'(t) \in Q(t, x(t))$ on the derivative. Analogous considerations hold for Bolza problems of optimal control.

It may happens that the sets $\tilde{Q}(t, x)$ are closed and that $T(t, x, z)$ is finite, so that inf can be replaced by min in the definition of T, and then $(T(t, x, z), z) \in \tilde{Q}(t, x)$, (at least for $z \in Q(t, x)$ and all $(t, z) \in A$ but those whose abscissa t lies in a set of measure zero on the t-axis). Then, under mild assumptions, we shall prove in Section 8.2 that for any AC trajectory x for problem (1.13.2) there is a measurable function u such that x, u is admissible for problem (1.13.1) and $I_0 = I$. Thus, under the mentioned assumptions problems (1.13.1) and (1.13.2) are equivalent.

Here is an example of reduction of a problem (1.13.1) to a problem (1.13.2). For instance, for the problem with $n = 1$, $m = 2$,

$$I[x, u, v] = \int_0^1 (2u^2 + 2v^2)\, dt, \qquad dx/dt = u + v,$$

with $x(0) = x_1$, $x(1) = x_2$, and $(u, v) \in U = [u \geq 0, v \geq 0] \subset R^2$, we have

$$\begin{aligned} z^0 &= f_0(t, x, u, v) = 2u^2 + 2v^2 = (u + v)^2 + (u - v)^2, \\ z &= f(t, x, u, v) = u + v, \end{aligned}$$

and therefore $z^0 \geq T(t, x, z) = z^2$, while z ranges over $Q(t, x) = [z \mid z \geq 0]$. The given problem of optimal control has been reduced to the problem

$$I_0[x] = \int_0^1 (x'(t))^2\, dt, \qquad x(0) = x_1, \quad x(1) = x_2,$$

with the constraint

$$x'(t) \in Q(t, x(t)) = [z \mid z \geq 0] \subset R, \quad \text{that is, } x' \geq 0.$$

Given f_0 and f, not always the function T can be written in explicit form. However, theoretically, this function is available, and can be actually used. For instance, we shall use it in the proof of the existence theorems of Section 11.4.

1.14 Generalized Solutions

A. Generalized Solutions for Mayer Problems

Often a given problem has no optimal solution, but the mathematical problem and the corresponding set of solutions can be modified in such a way that an optimal solution exists, and yet neither the system of trajectories nor the corresponding values of the cost functional are essentially modified. The modified (or generalized) problem and its solutions are of interest in themselves, and often have relevant physical interpretations. Moreover, they have a theoretical relevance. Indeed, as we shall see in Chapter 16, a very simple proof of the existence of usual optimal solutions "without convexity hypotheses" (Neustadt type theorems) is to prove first the existence of generalized solutions and then to derive from them the corresponding usual and even bang-bang solutions.

Here we introduce generalized solutions as usual problems involving a finite number of ordinary strategies, which are thought of as being used at the same time according to some probability distribution (Gamkrelidze's chattering states and, from a different viewpoint, Young's generalized curves).

Briefly, instead of considering the usual cost functional, differential system, boundary conditions, and constraints

$$\begin{gathered} I[x,u] = g(t_1, x(t_1), t_2, x(t_2)), \\ dx/dt = f(t, x(t), u(t)), \qquad f = (f_1, \dots, f_n), \\ (t_1, x(t_1), t_2, x(t_2)) \in B, \qquad (t, x(t)) \in A, \qquad u(t) \in U(t, x(t)), \end{gathered} \tag{1.14.1}$$

we shall consider new cost functional, differential system, boundary conditions, and constraints

$$\begin{gathered} J[x,p,v] = g(t_1, x(t_1), t_2, x(t_2)), \\ dx/dt = h(t, x(t), p(t), v(t)), \qquad h = (h_1, \dots, h_n), \\ (t_1, x(t_1), t_2, x(t_2)) \in B, \quad (t, x(t)) \in A, \quad v(t) \in V(t, x(t)), \quad p(t) \in \Gamma, \end{gathered} \tag{1.14.2}$$

with $x = (x^1, \dots, x^n)$, $p = (p_1, \dots, p_\gamma)$, $v = (u^{(1)}, \dots, u^{(\gamma)})$;

$$h(t, x, p, v) = \sum_{j=1}^{\gamma} p_j f(t, x, u^{(j)}), \tag{1.14.3}$$

or in component form

$$h_i(t, x, p, v) = \sum_{j=1}^{\gamma} p_j f_i(t, x, u^{(j)}), \qquad i = 1, \dots, n;$$

and

$$v \in V(t,x) = [U(t,x)]^\gamma = U \times \cdots \times U \subset R^{m\gamma}, \tag{1.14.4}$$

$$p = (p_1, \dots, p_\gamma) \in \Gamma = [p \mid p_j \geq 0, j = 1, \dots, \gamma, p_1 + \cdots + p_\gamma = 1]. \tag{1.14.5}$$

Precisely, $v(t) = (u^{(1)}, \dots, u^{(\gamma)})$ represents a system of some γ ordinary strategies $u^{(1)}(t), \dots, u^{(\gamma)}(t)$, each $u^{(j)}(t)$ having its values in $U(t, x(t)) \subset R^m$. Thus, we think of $v = (u^{(1)}, \dots, u^{(\gamma)})$ as a vector variable whose components $u^{(1)}, \dots, u^{(\gamma)}$ are themselves vectors u with values in $U(t,x)$ as in (1.14.1) with $u^{(j)} \in U(t,x)$, $j = 1, \dots, \gamma$, and thus

$v \in V(t,x)$, where now $V = U \times \cdots \times U$ is a subset of $R^{m\gamma}$. In (1.14.2), $p = (p_1, \ldots, p_\gamma)$ represents a probability distribution. Hence, p is a point of the simplex Γ of the Euclidean space R^γ defined in (1.14.5). Thus, in (1.14.2) the new control variable is (p, v) with values $(p,v) \in \Gamma \times V(t,x) \subset R^{m\gamma+\gamma}$, that is, an $(m\gamma + \gamma)$-vector. Note that h in (1.14.3) can be thought of as a convex combination of $f(t,x,u^{(j)})$, $j = 1, \ldots, \gamma$, with coefficients p_j, and the $p_j(t)$ can well be thought of as probability distributions. We shall denote by M^* the set of all (t,x,p,v) with $(t,x) \in A$, $p \in \Gamma$, $v \in V(t,x)$.

A *generalized solution* $x(t)$, $p(t)$, $v(t)$, $t_1 \le t \le t_2$, is an admissible system for the new problem, that is, (a) $x(t)$ is AC in $[t_1,t_2]$; (b) $p(t)$, $v(t)$ are measurable in $[t_1,t_2]$; (c) $(t,x(t)) \in A$ for every $t \in [t_1,t_2]$; (d) $p(t) \in \Gamma$, $u^{(j)}(t) \in U(t,x(t))$ for $t \in [t_1,t_2]$ (a.e.), $j = 1, \ldots, \gamma$; (e) $dx/dt = h(t,x(t),p(t),v(t))$ for $t \in [t_1,t_2]$ (a.e.); (f) $(t_1,x(t_1),t_2,x(t_2)) \in B$. Thus, $x(t)$ is said to be a generalized trajectory generated by the γ strategies $u^{(j)}(t)$ with probability distribution $p(t) = (p_1, \ldots, p_\gamma)$, $t_1 \le t \le t_2$.

Every usual solution can be interpreted as a generalized solution. Indeed, if we take $p_1 = 1$, $p_2 = \cdots = p_\gamma = 0$, $u = u^{(1)}$, then obviously all relations (1.14.2–5) reduce to relations (1.14.1).

If we denote by Ω the class of all usual admissible pairs $x(t)$, $u(t)$, $t_1 \le t \le t_2$, and by Ω^* the class of all admissible generalized systems $x(t)$, $p(t)$, $v(t)$, $t_1 \le t \le t_2$, then we have just proved that $\Omega \subset \Omega^*$. If j denotes the infimum of $J[x,p,v] = g(e[x])$ in Ω^* (and i the infimum of $I[x,u] = g(e[x])$ in Ω), then $j \le i$.

As we shall see in Chapter 18, we have $j = i$ under very weak assumptions (usually satisfied in most applications). Under the same hypotheses we shall prove also that generalized trajectories can be uniformly approximated by usual trajectories.

We shall denote by $R(t,x)$ the set of all $z = (z^1, \ldots, z^n) \in R^n$ with $z = h(t,x,p,v)$ for $p \in \Gamma$, $v \in V(t,x)$. In view of (1.14.3) $R(t,x)$ is made up of convex combinations of γ points of $Q(t,x)$.

We shall choose for γ the minimum integer such that $R(t,x)$ is the "convex hull" of $Q(t,x)$, the smallest convex set containing $Q(t,x)$, briefly $R(t,x) = \operatorname{co} Q(t,x) \subset R^n$ for all $(t,x) \in A$ (cf. Carathéodory's theorem in Section 8.4) The number γ is always $\le n + 1$. Any higher value of γ will give rise to the same set $R(t,x)$, as we shall see in Section 8.4, and will not produce any new trajectory.

In terms of orientor fields relations (1.14.2) become

$$\begin{aligned} &J_0[x] = g(t_1,x(t_1),t_2,x(t_2)) = J[x,p,v], \\ &dx/dt \in \operatorname{co} Q(t,x(t)), \qquad t \in [t_1,t_2] \text{ (a.e.)}, \\ &(t,x(t)) \in A, \qquad t \in [t_1,t_2], \qquad e[x] \in B. \end{aligned} \tag{1.14.6}$$

We have shown that for any generalized system $x(t)$, $p(t)$, $v(t)$, $t_1 \le t \le t_2$, the generalized trajectory $x(t)$, $t_1 \le t \le t_2$, can be interpreted as an AC solution of the orientor field (1.14.6). Conversely, as we shall prove in Section 8.2 under mild assumptions (e.g., A and M^* closed), for any AC solution x of (1.14.6) there are some measurable functions p, v such that x, p, v is an admissible system for (1.14.2) and of course $J = J_0$.

Example

We give here an example of a problem which has no optimal usual solution but one well-determined optimal generalized solution. Let $m = 1$, $n = 2$, $A = R^2$, and U be made up of the two points $u = +1$ and $u = -1$. Consider the differential equations $dx/dt = y^2$, $dy/dt = u$, boundary conditions $x(0) = 0$, $y(0) = y(1) = 0$, and functional

$I[x, y, u] = g = x(1)$. Since $x(0) = 0$, $dx/dt = y^2 \geq 0$, we have $x(1) \geq 0$. If i denotes the infimum of $I[x, y, u]$ in the class Ω of all admissible pairs $x(t)$, $y(t)$, $u(t)$, $0 \leq t \leq 1$, then $i \geq 0$. On the other hand, let $x_k(t)$, $y_k(t)$, $u_k(t)$, $0 \leq t \leq 1$, $k = 1, 2, \ldots$, be the sequence of admissible pairs defined by $u_k(t) = +1$ for $j/k \leq t < j/k + 1/2k$, $u_k(t) = -1$ for $j/k + 1/2k \leq t < (j+1)/k$, $j = 0, 1, \ldots, k-1$, and hence $y_k(t) = t - j/k$ and $y_k(t) = (j+1)/k - t$ for t in the corresponding intervals; then $0 \leq y_k(t) \leq 1/2k$, $0 \leq y_k^2(t) \leq 1/4k^2$ for all $0 \leq t \leq 1$, and $x_k(1) \leq 1/4k^2$. Thus $0 \leq I[x_k, y_k, u_k] \leq 1/4k^2$ and $I[x_k, y_k, u_k] \to 0$ as $k \to +\infty$. This proves that $i = 0$. For no admissible pair $x(t)$, $y(t)$, $u(t)$, $0 \leq t \leq 1$, we can have $I[x, y, u] = 0$, since this would imply $x(1) = 0$, $x(t) = 0$ for all $0 \leq t \leq 1$, $y(t) = 0$ for all $0 \leq t \leq 1$, and finally $u(t) = 0$ a.e. in $[t_1, t_2]$, a contradiction, since $u = \pm 1$. We have proved that the problem above has no optimal usual solution. The corresponding generalized problem with $\gamma = 2$ now has differential equations $dx/dt = y^2$, $dy/dt = p_1 u^{(1)} + p_2 u^{(2)}$, boundary conditions $x(0) = 0$, $y(0) = y(1) = 0$, and functional $J = g = x(1)$. Moreover, $u^{(1)}, u^{(2)} = \pm 1$, $p_1 \geq 0$, $p_2 \geq 0$, $p_1 + p_2 = 1$. Since $w = p_1 u^{(1)} + p_2 u^{(2)}$ takes on all possible values between -1 and $+1$, we can replace the second equation by $dy/dt = w$ with $w \in W = [-1 \leq w \leq 1]$. The obvious optimal solution is here $x(t) = 0$, $y(t) = 0$, $w(t) = 0$ for all $0 \leq t \leq 1$, or $u^{(1)} = -1$, $u^{(2)} = +1$, $p_1 = \frac{1}{2}$, $p_2 = \frac{1}{2}$ for all $0 \leq t \leq 1$. Here the infimum of J is still zero, and $J_{\min} = j = i = 0$.

B. Generalized Solutions for Lagrange Problems

We have given above the definitions concerning generalized solutions for the Mayer problem. For problems written in the Lagrange form

$$I[x, u] = \int_{t_1}^{t_2} f_0(t, x, u)\, dt,$$

$$dx/dt = f(t, x, u), \quad (t, x) \in A, \quad u \in U(t, x), \quad e[x] \in B, \tag{1.14.7}$$

$$x = (x^1, \ldots, x^n), \qquad f = (f_1, \ldots, f_n), \qquad u = (u^1, \ldots, u^m),$$

the corresponding generalized problem is

$$J[x, p, v] = \int_{t_1}^{t_2} h_0(t, x, p, v)\, dt, \tag{1.14.8}$$

$$dx/dt = h(t, x, p, v), \quad (t, x) \in A, \quad (p, v) \in \Gamma \times V(t, x), \quad e[x] \in B,$$

where

$$x = (x^1, \ldots, x^n), \qquad h = (h_1, \ldots, h_n),$$

$$v = (u^{(1)}, \ldots, u^{(\gamma)}), \qquad u^{(j)} = (u^{j1}, \ldots, u^{jm}) \in U(t, x), \quad j = 1, \ldots, \gamma,$$

$$p = (p_1, \ldots, p_\gamma) \in \Gamma \equiv [p \mid p_j \geq 0, j = 1, \ldots, \gamma, p_1 + \cdots + p_\gamma = 1], \qquad V = U^m,$$

$$h_0(t, x, u) = \sum_{j=1}^{\gamma} p_j f_0(t, x, u^{(j)}),$$

$$h(t, x, u) = \sum_{j=1}^{\gamma} p_j f(t, x, u^{(j)}).$$

By reducing this problem to the Mayer form as mentioned in Section 1.9, we obtain a generalized Mayer problem as in Section 1.14A.

Let us consider the sets

$$
\begin{aligned}
Q(t,x) &= [z \mid z = f(t,x,u),\ u \in U(t,x)] \subset R^n \\
\tilde{Q}(t,x) &= [(z^0,z) \mid z^0 \geq f_0(t,x,u),\ z = f(t,x,u),\ u \in U(t,x)] \subset R^{n+1}, \\
R(t,x) &= [z \mid z = h(t,x,p,v),\ (p,v) \in \Gamma \times V(t,x)] \subset R^n, \\
\tilde{R}(t,x) &= [(z^0,z) \mid z^0 \geq h_0(t,x,p,v),\ z = h(t,x,p,v),\ (p,v) \in \Gamma \times V(t,x)] \subset R^{n+1}.
\end{aligned}
$$

We shall take γ in such a way that $\tilde{R}(t,x) = \text{co } \tilde{Q}(t,x)$, $R(t,x) = \text{co } Q(t,x)$, and by Carathéodory's theorem (Section 8.4) this is achieved for some $\gamma \leq n + 2$.

As in Section 1.13 we can associate a Lagrangian function to the problem of optimal control (1.14.8), namely

$$
\begin{aligned}
F_0(t,x,z) &= \inf[z^0 \mid (z^0,z) \in \text{co } \tilde{Q}(t,x)], \qquad (t,x) \in A, \quad z \in \text{co } Q(t,x). \\
&= \inf\left[z^0 \mid z^0 \geq \sum_j p_j f_0(t,x,u^{(j)}),\ z = \sum_j p_j f(t,x,u^{(j)}),\ p \in \Gamma,\ u^{(j)} \in U(t,x(t)) \right].
\end{aligned}
$$

Then we can associate to (1.14.8) the new problem in terms of orientor fields and the Lagrangian F_0,

$$
\begin{gathered}
J_0[x] = \int_{t_1}^{t_2} F_0(t,x(t),x'(t))\,dt, \\
(t,x(t)) \in A, \qquad x'(t) \in \text{co } Q(t,x(t)), \qquad t \in [t_1,t_2] \in \text{(a.e.)}, \qquad e[x] \in B.
\end{gathered} \tag{1.14.9}
$$

If it happens that the sets $\tilde{R}(t,x) = \text{co } \tilde{Q}(t,x)$ are closed and that $F_0(t,x,z)$ is finite, so that inf can be replaced by min in the definition of F_0, and then $(F_0(t,x,z),z) \in \tilde{R}(t,x)$ (at least for $z \in R(t,x)$ and all $(t,x) \in A$ but those whose abscissa t lies in a set of measure zero on the t-axis), then for any admissible system x, p, v for (1.14.8), the trajectory x is a solution for (1.14.9) and $J_0[x] \leq J[x,p,v]$. Conversely, we shall prove in Section 8.2 under mild hypotheses, that for any AC trajectory for (1.14.9) there are measurable functions p, v such that x, p, v is an admissible system for (1.14.8) and $J[x,p,v] < J_0[x]$. Thus under the mentioned assumptions problems (1.14.8) and (1.14.9) are equivalent.

As a particular case let us consider the classical Lagrange problems of the calculus of variations:

$$
I[x] = \int_{t_1}^{t_2} f_0(t,x(t),x'(t))\,dt, \qquad (t,x(t)) \in A, \qquad e[x] \in B. \tag{1.14.10}
$$

These problems can be written as Lagrange problems of optimal control:

$$
\begin{gathered}
I[x,u] = \int_{t_1}^{t_2} f_0(t,x(t),u(t))\,dt, \\
dx/dt = u, \quad u(t) \in U = R^n, \qquad (t,x(t)) \in A, \qquad e[x] \in B,
\end{gathered}
$$

with $m = n$, $U = R^n$, $f = u$. Now the corresponding generalized Lagrange problem of optimal control can be written in the form

$$
\begin{gathered}
J[x,p,v] = \int_{t_1}^{t_2} h_0(t,x(t),p(t),v(t))\,dt, \\
x'(t) = h(t,x(t),p(t),v(t)) \in R^n, \qquad (t,x(t)) \in A, \qquad e[x] \in B,
\end{gathered} \tag{1.14.11}
$$

where

$$x = (x^1, \ldots, x^n), \qquad (p, v) \in \Gamma \times R^{n\gamma},$$
$$p = (p_1, \ldots, p_\gamma) \in \Gamma = [p \mid p_j \geq 0, j = 1, \ldots, \gamma, p_1 + \cdots + p_\gamma = 1] \subset R^\gamma$$
$$v = (u^{(1)}, \ldots, u^{(\gamma)}), \qquad u^{(j)} \in U = R^n, \quad j = 1, \ldots, \gamma,$$
$$h_0(t, x, p, v) = \sum_j p_j f_0(t, x, u^{(j)}), \qquad h(t, x, p, v) = \sum_j p_j u^{(j)}.$$

Finally, the Lagrangian $F_0(t, x, z)$ associated to problem (1.14.11) is defined by

$$F_0(t, x, z) = \inf[z^0 \mid (z^0, z) \in \operatorname{co} \tilde{Q}(t, x)]$$
$$= \inf\left[z^0 \mid z^0 \geq \sum_j p_j f_0(t, x, u^{(j)}),\ z = \sum_j p_j u^{(j)},\ p \in \Gamma,\ u^{(j)} \in R^n\right],$$

and we have a new Lagrange problem

$$J_0[x] = \int_{t_1}^{t_2} F_0(t, x(t), x'(t))\, dt, \qquad (t, x(t)) \in A, \qquad e[x] \in B.$$

Example

We take here the Lagrange problem of the calculus of variations concerning the minimum of

$$I = \int_0^1 (x^2 + |x'^2 - 1|)\, dt$$

with $n = 1$, $x(0) = 0$, $x(1) = 0$. Here $x^2 + |x'^2 - 1| \geq 0$; hence $I \geq 0$, and, if i denotes the infimum of I in the class Ω of all AC functions $x(t)$, $0 \leq t \leq 1$, with $x(0) = 0$, $x(1) = 0$, then certainly $i \geq 0$. Let us consider now the sequence $[x_k]$ of trajectories defined by $x_k(t) = t - j/k$ for $j/k \leq t \leq j/k + 1/2k$, $x_k(t) = (j + 1)/k - t$ for $j/k + 1/2k \leq t \leq (j + 1)/k$, $j = 0, 1, \ldots, k - 1$. Then $0 \leq x_k(t) \leq 1/2k$, $x'(t) = \pm 1$ a.e. in $[0, 1]$, and $0 < I[x_k] \leq 1/4k^2$. Thus, $I[x_k] \to 0$ as $k \to \infty$, and hence $i = 0$. Obviously there is no AC function $x(t)$, $0 \leq t \leq 1$, with $x(0) = x(1) = 0$ and $I[x] = 0$, since this would imply $x(t) = 0$ for $0 \leq t \leq 1$, $x'(t) = \pm 1$ a.e. in $[0, 1]$, a contradiction. This proves that the problem above has no absolute minimum in Ω.

The corresponding generalized problem concerns the minimum of

$$J = \int_0^1 (x^2 + p_1|(u^{(1)})^2 - 1| + p_2|(u^{(2)})^2 - 1|)\, dt,$$

with $x(0) = 0$, $x(1) = 0$, $p_1 \geq 0$, $p_2 \geq 0$, $p_1 + p_2 = 1$, and differential equation

$$dx/dt = p_1 u^{(1)} + p_2 u^{(2)}, \qquad u^{(1)}, u^{(2)} \in R.$$

Obviously $w = p_1 u^{(1)} + p_2 u^{(2)}$ can take any possible value in R, as $u = x'$ in the original free problem. If j denotes the infimum of the new functional J, then obviously $j \geq 0$, since the integrand is still nonnegative. On the other hand, $j \leq i = 0$, since the class $\tilde{\Omega}$ of generalized solutions contains the class Ω of usual solutions. Thus, $j = i = 0$. An optimal solution is obviously given by $u^{(1)}(t) = 1$, $u^{(2)}(t) = -1$, $p_1(t) = \frac{1}{2}$, $p_2(t) = \frac{1}{2}$, $0 \leq t \leq 1$; hence $dx/dt = 0$, $x(t) = 0$ in $[0, 1]$, and $J_{\min} = j = i = 0$. The corresponding Lagrange problem of the calculus of variations is here

$$J[x] = \int_0^1 F_0(t, x, x')\, dt, \qquad x(0) = x(1) = 0,$$

with $F_0 = x^2$ for $-1 \leq x' \leq 1$, $F_0 = x^2 + (x'^2 - 1)$ for $|x'| \geq 1$.

Bibliographical Notes

Systematic use of Lebesgue integral and absolutely continuous trajectories in the calculus of variations was first made by L. Tonelli about 1918 (see the fundamental treatise [I], 1921–23), and then continued by N. Nagumo [1], E. J. McShane [5–10, 18], S. Cinquini [1–6], B. Mania [1–7], L. M. Graves [1], and many others. In 1955–60 L. S. Pontryagin and his school also made systematic use of Lebesgue integrals and AC trajectories in proposing and developing optimal control theory (see the volume L. S. Pontryagin, V. G. Boltyanskii, R. V. Gamkrelidze, E. F. Mishchenko [I], 1962).

For texts on classical calculus of variations and optimal control we mention here N. I. Akhiezer [I], M. Athans and P. L. Falb [I], A. V. Balakrishnan [I, II], G. A. Bliss [I, II], O. Bolza [I, II], C. Carathéodory [I], R. Conti [I], L. E. Elsgolc [I], W. M. Fleming and R. W. Rishel [I], A. R. Forsyth [I], C. Fox [I], I. M. Gelfand and S. V. Fomin [I], I. Gumowski and C. Mira [I], H. Hermes and J. P. LaSalle [I], M. R. Hestenes [I, II], E. B. Lee and L. Markus [I], G. Leitman [I, II], J. L. Lions [I], L. W. Neustadt [I], R. Pallu de la Barrière [I], L. S. Pontryagin, V. G. Boltyanskii, R. V. Gamkrelidze, E. F. Mishchenko [I], A. Strauss [I], L. Tonelli [I], J. Warga [I], R. Weinstock [I], L. C. Young [I].

The concept of generalized solutions was introduced by L. C. Young [I, 1–9] in 1937 from the viewpoint of functional analysis, and further developed by him for curves and surfaces. The theory of generalized solutions was continued later by E. J. McShane [12, 14, 16, 18, 20] and further developed by L. C. Young, W. H. Fleming, and others. The same concept was then presented by R. V. Gamkrelidze ([1], 1962) from a different point of view (sliding regimes, chattering states).

Our presentation in (1.14) relates to Gamkrelidze's viewpoint. The essential equivalence of this concept with the one of L. C. Young will be discussed elsewhere. When constraints other than differential equations are involved, L. C. Young's functional analysis approach is more suitable.

CHAPTER 2

The Classical Problems of the Calculus of Variations: Necessary Conditions and Sufficient Conditions; Convexity and Lower Semicontinuity

2.1 Minima and Maxima for Lagrange Problems of the Calculus of Variations

A. Absolute Minima and Maxima

We are concerned here with classical Lagrange problems of the calculus of variations (see Section 1.1). Precisely, we are concerned with minima and maxima of the functional

$$I[x] = \int_{t_1}^{t_2} f_0(t, x(t), x'(t))\, dt, \qquad x(t) = (x^1, \ldots, x^n), \tag{2.1.1}$$

with constraints and boundary conditions

$$(t, x(t)) \in A, \quad t \in [t_1, t_2], \qquad (t_1, x(t_1), t_2, x(t_2)) \in B. \tag{2.1.2}$$

Here we shall assume that A is a closed subset of the tx-space R^{n+1}, that B is a closed subset of the $t_1x_1t_2x_2$-spaceR^{2+2n}, and that $f_0(t, x, u)$ is a real valued continuous function in $A \times R^n$. We shall assume that A is the closure of its interior points.

As mentioned in (1.1), we shall consider n-vector functions $x(t) = (x^1, \ldots, x^n)$, $t_1 \leq t \leq t_2$, which are absolutely continuous (AC). It should also be stated here that any n-vector is thought of as a column vector. This will become relevant whenever matrix notations are used.

We shall consider here the class Ω of all n-vector AC functions $x(t) = (x^1, \ldots, x^n)$, $t_1 \leq t \leq t_2$, satisfying $(t, x(t)) \in A$ for $t \in [t_1, t_2]$, with

$f_0(\cdot, x(\cdot), x'(\cdot))$ L-integrable in $[t_1, t_2]$, and also satisfying the given boundary conditions $(t_1, x(t_1), t_2, x(t_2)) \in B$. We may say that this is the class of all admissible trajectories, or briefly trajectories.

Given the generality of the constraints under consideration, we must explicitely assume that they are compatible, in other words, that there is at least one admissible trajectory, or equivalently, that the class of the admissible trajectories is not empty.

Now a few definitions. A subset E of real numbers, $E \subset R$, is said to be of measure zero, or meas $E = 0$, provided for any $\varepsilon > 0$ there is a countable system of open intervals (α_i, β_i), $i = 1, 2, \ldots$, possibly overlapping, such that $E \subset \bigcup_i (\alpha_i, \beta_i)$, $\sum_i (\beta_i - \alpha_i) < \varepsilon$. A property P is said to hold almost every-here in R (briefly, a.e.), or for almost all $t \in R$ (briefly, a.e. $t \in R$), provided P holds everywhere but at the points of a set E of measure zero.

An open subset G of R is the countable union of nonoverlapping intervals, $G = \bigcup_s (A_s, B_s)$, and then meas $G = \sum_s (B_s - A_s)$ is the total length of its components (A_s, B_s), $s = 1, 2, \ldots$. Thus, a subset E of R has measure zero if and only if for every $\varepsilon > 0$ there is an open set G such that $E \subset G$, meas $G < \varepsilon$.

We just mention that $x(t)$, $t_1 \leq t \leq t_2$, is said to be AC in $[t_1, t_2]$ provided that, given $\varepsilon > 0$, there is some $\delta > 0$ such that for any finite system of nonoverlapping intervals $[\alpha_i, \beta_i]$, $i = 1, \ldots, N$, in $[a, b]$ with $\sum_i (\beta_i - \alpha_i) \leq \delta$, we have $\sum_i |x(\beta_i) - x(\alpha_i)| \leq \varepsilon$. If $x(t) = (x^1, \ldots, x^n)$ is AC in $[t_1, t_2]$, then its derivative $x'(t) = (x'^1, \ldots, x'^n)$ exists a.e. in $[t_1, t_2]$, and is Lebesgue integrable in $[t_1, t_2]$; that is, each component $x'^i(t)$ is L-integrable, and $x(\tau_2) - x(\tau_1) = \int_{\tau_1}^{\tau_2} x'(\tau)\, d\tau$ for all τ_1, τ_2 in $[t_1, t_2]$, where the integral is a Lebesgue integral. Conversely, any $G(t) = \int_{t_1}^{t} g(\tau)\, d\tau$ which is the Lebesgue integral function of an L-integrable function is AC.

For any trajectory x we denote by Γ the graph of x, that is, the set $\Gamma = [(t, x(t)), t_1 \leq t \leq t_2]$, $\Gamma \subset A \subset R^{1+n}$. For any $\delta > 0$ we denote by Γ_δ the δ-neighborhood of Γ, that is, the set of all $(t, x) \in A$ at a distance $\leq \delta$ from Γ; thus $\Gamma \subset \Gamma_\delta \subset R^{1+n}$.

A trajectory x is said to give (to be, for short) an *absolute* minimum for $I[x]$ in Ω, if $I[x] \leq I[y]$ for all trajectories y. It will be said to be a *proper* absolute minimum if $I[x] < I[y]$ for all y distinct from x in the class. Analogous definitions hold for maxima.

Actually, we could limit ourselves to a smaller class Ω of trajectories, say, by imposing further restrictions, e.g., passage through given points, or through a given set, or by requiring that certain components are monotone, or more generally by imposing constraints on the derivatives as mentioned in Sections 1.2, 1.12 and 1.13. Then we would have the concepts of absolute minimum and absolute maximum in any such given class Ω. This idea will be particularly used in connection with existence theorems in Chapters 9–16 where it is rather natural. In the present Chapter we may limit ourselves to the class Ω of all admissible trajectories.

B. Strong and Weak Local Minima and Maxima: The Case t_1, t_2 Fixed

Again, for the sake of simplicity, let us consider the class Ω of all (admissible) trajectories x for a given integral $I[x]$.

We limit ourselves to problems with t_1, t_2 fixed; thus $B = \{t_1\} \times \{t_2\} \times B_0$, where B_0 is a fixed subset of R^{2n}. For instance, for the problem with end-points fixed, i.e. $1 = (t_1, x_1)$, $2 = (t_2, x_2)$, then $B = \{t_1\} \times \{x_1\} \times \{t_2\} \times \{x_2\}$ is a singleton in R^{2n+2}. A trajectory $x \in \Omega$ is said to be a *strong local* minimum for $I[x]$ (in Ω) if there is some $\delta > 0$ such that $I[x] \leq I[y]$ for all trajectories $y \in \Omega$ with $|y(t) - x(t)| \leq \delta$, $t_1 \leq t \leq t_2$. A trajectory $x \in \Omega$ is said to be a *weak local* minimum for $I[x]$ if there are $\delta > 0$, $\sigma > 0$ such that $I[x] \leq I[y]$ for all trajectories in Ω with $|y(t) - x(t)| \leq \delta$ for $t \in [t_1, t_2]$, and $|y'(t) - x'(t)| \leq \sigma$ for $t \in [t_1, t_2]$ (a.e.). Analogous definitions of course hold also for maxima.

Thus, for local strong minima (and analogously for absolute minima) we completely disregard the values of the derivatives of the comparison elements y. We shall encounter striking examples where these distinctions are essential. For instance, the segment $x(t) = 0$, $0 \leq t \leq 1$, is the absolute minimum (as well as a strong local minimum) for the integral $I[x] = \int_0^1 x'^2\, dt$ in the class of all AC functions $x(t)$ with $x(0) = x(1) = 0$. The same segment $x(t) = 0$, $0 \leq t \leq 1$, is only a weak local minimum for the integral $I[x] = \int_0^1 (x'^2 - x'^4)\, dt$ in the same class. Indeed, for $|x'| \leq 1$ we certainly have $x'^2 - x'^4 \geq 0$ and thus $I[y] \geq I[x] = 0$ for all AC y with $y(0) = y(1) = 0$, $|y'(t)| \leq 1$, while $I[y]$ may take negative values if we allow y' to take values $|y'| > 1$ as large as we want. See Chapter 3 for other examples.

Remark. Note that for the local minima and maxima, we are only comparing $I[x]$ with the values taken by $I[y]$ for y in the same class, and y "sufficient close" x. In other words, we have introduced a topology, and in fact a metric or distance function $d(y, x)$. For strong extrema, indeed, we have taken $d(y, x) = \sup[|y(t) - x(t)|, t_1 \leq t \leq t_2]$; in other words, we have thought of Ω as a subset of $C([t_1, t_2], R^n)$, the space of all continuous functions from $[t_1, t_2]$ to R^n with the distance function just defined.

For weak extrema, the definition above corresponds to the choice in Ω of the distance function $d_1(y, x) = \sup|y(t) - x(t)| + \operatorname{ess\,sup}|y'(t) - x'(t)|$, both "sup" and "ess sup" being taken in $[t_1, t_2]$. In other words, we have thought of Ω as a subset of the space usually called $H^{1,\infty}([t_1, t_2], R^n)$. For instance, for $x(t) = 0$, $y_\varepsilon(t) = \varepsilon \sin(t/\varepsilon)$, $0 \leq t \leq 1$, then $d(y_\varepsilon, x) = \varepsilon$ and $d(y_\varepsilon, x) \to 0$ as $\varepsilon \to 0$, while $y_\varepsilon'(t) = \cos(t/\varepsilon)$ and $d_1(y_\varepsilon, x)$ does not tend to zero as $\varepsilon \to 0$. Note that, for $x(t) = 0$, $y_k(t) = t^k$, $0 \leq t \leq 1$, neither $d(y_k, x)$ nor $d_1(y_k, x)$ approach zero as $k \to \infty$. However, for the same functions in any fixed internal $0 \leq t \leq 1 - \sigma$, $0 < \sigma < 1$, both $d(y_k, x)$ and $d_1(y_k, x)$ approach zero as $k \to \infty$. Obviously, we may think of local maxima and minima in terms of other classes Ω and in terms of other topologies; for example, we could think of Ω as a subset of $H^{1,1}$, or $H^{1,2}$, or $H^{1,p}$, $p > 1$, among many others.

C. Minimum as Attained Infimum

The concepts above can be worded in a different, but equivalent way. Given the class Ω of elements x, let i denote $i = \inf_\Omega I[x]$. Thus, $-\infty \le i \le +\infty$, and $i = -\infty$ if $I[x]$ is not bounded below in Ω, and $i = +\infty$ if Ω is empty. Thus, $I[x]$ has an absolute minimum in Ω if and only if (a) i is finite, and (b) i is attained in Ω, that is, there is some element x in Ω such that $I[x] = i$. For local minima, the same is true provided we suitably restrict Ω. For maxima we have only to consider $j = \sup I[x]$, and the same definitions hold with all inequalities reversed.

D. Principle of Optimality

Note that for an element x to give a minimum or maximum to $I[x]$ appears to be a "global" property, or a property of the *whole* curve $C{:}x = x(t)$, $t_1 \le t \le t_2$. However, it is also a local property, as the following statement shows.

2.1.i (Local Property of Optimal Elements for Lagrange Problems, or Principle of Optimality). *Let Ω be the nonempty set of all admissible trajectories satisfying* (2.1.2), *that is, with the given boundary conditions and with their graphs lying in A). Let $x_0(t)$, $t_1 \le t \le t_2$, be optimal for the functional* (2.1.1) *in Ω. Let y_0 denote any subarc of the curve represented by x_0, i.e. $x_0(t)$, $\alpha \le t \le \beta$. Let Ω_0 denote the class of all admissible trajectories $y(t)$, $\alpha \le t \le \beta$, whose end values are the same as those of y_0 (or $e[y] = e[y_0]$) and whose graph is in A. Then y_0 is optimal for I in Ω_0.*

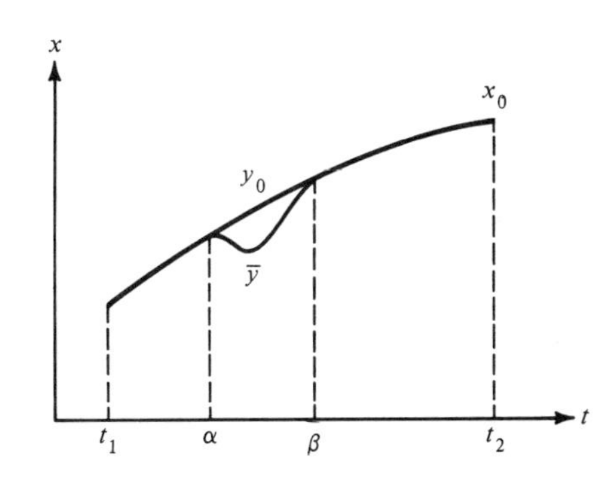

Proof. Let us consider a problem of minimum, for instance. First, the class Ω_0 is not empty, since $y_0 \in \Omega_0$. Now we assume, if possible, that y_0 is not optimal for I in Ω_0, and thus there is some element $\bar{y}$, or $\bar{y}(t)$, $\alpha \le t \le \beta$, with $\bar{y}(\alpha) = y_0(\alpha)$, $\bar{y}(\beta) = y_0(\beta)$, and with $I[\bar{y}] < I[y_0]$. Let $y(t)$, $t_1 \le t \le t_2$, denote the element defined by $y(t) = \bar{y}(t)$ for $t \in [\alpha, \beta]$, and $y(t) = x_0(t)$ for $t \in [t_1, t_2] - (\alpha, \beta)$. Then, $y \in \Omega$, and

$$\begin{aligned} I[y] &= \int_{t_1}^{\alpha} + \int_{\alpha}^{\beta} + \int_{\beta}^{t_2} f_0(t, y(t), y'(t))\, dt \\ &= I[x_0] + (I[\bar{y}] - I[y_0]) < I[x_0], \end{aligned}$$

a contradiction. In particular, for $\alpha = t_1$, $\beta = t_2$, x is optimal also in the (restricted) class of all elements y of Ω with $y(t_1) = x(t_1)$, $y(t_2) = x(t_2)$. □

E. Strong and Weak Local Minima and Maxima: The General Case

In problems of the calculus of variations we may be led in a natural way to consider classes Ω of AC functions $x(t)$, $t_1 \le t \le t_2$, defined in different intervals. For instance, if the problem is to join a fixed point $1 = (t_1, x_1)$ to a fixed set B of the tx-space, i.e. $x_1 = x(t_1)$ and $(t_2, x(t_2)) \in B$, then t_2 is any of the abscissas of the points (t, x) of B. If, more particularly, B is a curve $x = g(t)$, $t' \le t \le t''$, of the tx-space, then again $x(t_1) = t_1$, $t' \le t_2 \le t''$, $x(t_2) = g(t_2)$.

To define the concepts of local minima and maxima in such classes, we need to know, as in Section 2.1B above, what we mean by saying that a continuous curve $C: y = y(t)$, $c \le t \le d$, is "close" to a given continuous curve $C_0: x = x(t)$, $a \le t \le b$. This can be done by means of the following distance function:

$$\rho(y, x) = |c - a| + |d - b| + \sup|y(t) - x(t)|,$$

where now $x(t)$, $y(t)$ are thought of as extended by continuity to the whole interval $(-\infty, +\infty)$ by taking $x(t) = x(a)$ for $t \le a$, $x(t) = x(b)$ for $t \ge b$, $y(t) = y(c)$ for $t \le c$, $y(t) = y(d)$ for $t \ge d$, and then the sup is taken in $(-\infty, +\infty)$ (or equivalently in any interval $[a_0, b_0]$ containing both $[a, b]$ and $[c, d]$).

Thus, an element $x \in \Omega$ is said to be a strong local minimum for $I[x]$ in Ω if there is some $\delta > 0$ such that $I[x] \le I[y]$ for all $y \in \Omega$ with $\rho(y, x) \le \delta$. Analogously, we can define weak local minima, as well as strong and weak local minima, together with some of the generalizations mentioned in Section 2.1B.

Concerning the distance function $\rho(y, x)$ defined above, note that if $x(t) = 0$, $0 \le t \le 1$, and $y(t) = k^{-1}$, $0 \le t \le 1 + k^{-1}$, $k = 1, 2, \ldots$, then $\rho(y_k, x) \to 0$ as $k \to \infty$. The same is true for $y_k(t) = k^{-1} \sin kt$, $0 \le t \le 1 + k^{-1}$, $k = 1, 2, \ldots$. However, for $x(t) = 0$,

$0 \le t \le 1$, and $y_k(t) = t^k$, $0 \le t \le 1 - k^{-1}$, $k = 1, 2, \ldots$, we have $\rho(y_k, x) = k^{-1} + (1 - k^{-1})^k \to e^{-1}$ as $k \to \infty$.

For more details on the distance function ρ see Section 2.14.

F. Exercises

1. Show that the definition of weak local minimum given in Section 2.1B, and the same definition in terms of the distance function $d_1(y, x)$ in the Remark at the end of that subsection, are equivalent.
2. For functions $x(t)$, $y(t)$, $0 \le t \le 1$, in $H^{1,1}$, we may take $d_{1,1}(y, x) = \sup|y(t) - x(t)| + \int_0^1 |y'(t) - x'(t)|\,dt$. Give an example of functions $x(t)$ and $y_\varepsilon(t)$, $0 \le t \le 1$, y_ε depending on $\varepsilon > 0$, such that $d(y_\varepsilon, x) \to 0$ as $\varepsilon \to 0$, while $d_{1,1}(y_\varepsilon, x)$ does not.
3. The same as Exercise 2, but for functions $x, y \in H^{1,2}$ or $x, y \in H^{1,p}$, where we define $d_{1,2}$ and $d_{1,p}$ appropriately, $p \ge 1$.
4. For $x(t) = 0$, $y_\varepsilon(t) = \varepsilon^\alpha \sin(t/\varepsilon^\beta)$, $0 \le t \le 1$, $\alpha, \beta \ge 0$ real, and $p > 1$ real, state the ranges of α, β, p for which (a) $d(y_\varepsilon, x) \to 0$, or (b) $d_1(y_\varepsilon, x) \to 0$, or (c) $d_{1,1}(y_\varepsilon, x) \to 0$, or (d) $d_{1,p}(y_\varepsilon, x) \to 0$ as $\varepsilon \to 0$.
5. For $x(t) = 0$, $0 \le t \le 1$, $y_k(t) = t^k$, $0 \le t \le e^{\alpha_k}$, $\alpha_k < 0$, show that $\rho(y_k, x) \to 0$ or $k \to \infty$ provided $\alpha_k \to 0$, $k\alpha_k \to -\infty$ as $k \to \infty$.

G. Further Preliminaries to the Necessary Conditions

Note that the maxima [minima] of $I = \int_{t_1}^{t_2} f_0\,dt$ correspond to the minima [maxima] of $I = \int_{t_1}^{t_2} (-f_0)\,dt$. For the sake of simplicity, we shall refer below mostly to minima, the results for maxima being obtained by obvious changes in the inequalities. To express the well-known necessary conditions and the sufficient conditions we shall need to know that $f_0(t, x, x')$ is of class C^1 on $A \times R^n$, and we shall denote by $f_{0t} = \partial f_0/\partial t$, $f_{0x^i} = \partial f_0/\partial x^i$, $f_{0x'^i} = \partial f_0/\partial x'^i$ the first order partial derivatives of f_0. We shall also denote by f_{0x} and $f_{0x'}$ the n-vectors, or $n \times 1$ matrices, of the partial derivatives f_{0x^i} and $f_{0x'^i}$. Analogously, whenever we require f_0 to be of class C^2, we shall denote by $f_{0x^ix^j}$ and analogous symbols the second order partial derivatives, and by f_{0xx} and analogous symbols the $n \times n$ matrices of such partial derivatives.

Note that, for problems with both ends fixed, B is the single point (t_1, x_1, t_2, x_2) in R^{2+2n} and no further requirement is needed. In any other case, B is a proper set and some smoothness assumptions are also needed on B, at least in a neighborhood B_0 of the point $e[x] = (t_1, x(t_1), t_2, x(t_2)) = (t_1, x_1, t_2, x_2) \in B$.

In such cases all we shall need is that B_0, in a neighborhood of $e[x]$, is a manifold of class C^1 of some dimension k, $0 \le k \le 2n + 1$, possessing a tangent hyperplane B' at $e[x]$, whose vectors we shall denote by $h = (\tau_1, \xi_1, \tau_2, \xi_2)$,

$$\xi_1 = (\xi_1^1, \ldots, \xi_1^n) = (dx_1^1, \ldots, dx_1^n),\ \xi_2 = (\xi_2^1, \ldots, \xi_2^n) = (dx_2^1, \ldots, dx_2^n).$$

In the case, which is most common, that $B = B_1 \times B_2, B_1, B_2 \subset R^{n+1}$, all we need is that B_1 in a neighborhood of (t_1, x_1) is a manifold possessing a tangent hyperplane at (t_1, x_1), and the analogous requirement on B_2. No requirement is needed if B_1 or B_2 are single points.

For the formulation of the necessary conditions below we shall need the Hamiltonian function

$$\text{(2.1.3)} \quad H(t, x, x', \lambda) = f_0(t, x, x') + \sum_{i=1}^{n} \lambda_i x'^i, \qquad (t, x, x', \lambda) \in A \times R^{2n},$$

of $3n + 1$ arguments t, $x = (x^1, \ldots, x^n)$, $x' = (x'^i, \ldots, x'^n)$, $\lambda = (\lambda_1, \ldots, \lambda_n)$. We also need the Weierstrass function

$$\text{(2.1.4)} \quad E(t, x, x', X') = f_0(t, x, X') - f_0(t, x, x') - \sum_{i=1}^{n} (X'^i - x'^i) f_{0x'^i}(t, x, x'),$$

$$(t, x, x', X') \in A \times R^{2n},$$

also of $3n + 1$ arguments, t, x, x' as above, and $X' = (X'^1, \ldots, X'^n)$. In other words, E is the difference between $f_0(t, x, X')$, thought of as a function of X', and the first two terms of its Taylor expansion at x'.

2.2 Statement of Necessary Conditions

We state here a few necessary conditions for a maximum or minimum. Proofs and further necessary conditions will be given in Sections 2.3–10.

2.2.i (Theorem). *Under the assumptions of Section* 2.1 *let* $f_0(t, x, x')$ *be of class* C^1 *in* $A \times R^n$, *and let us consider* $I[x]$ *in the class of all trajectories* x *(that is, in the class* Ω *of all* AC *n-vector functions* $x(t) = (x^1, \ldots, x^n)$, $t_1 \leq t \leq t_2$, *satisfying* $(t, x(t)) \in A$, $(t_1, x(t_1), t_2, x(t_2)) \in B$, *and* $f_0(\cdot, x(\cdot), x'(\cdot)) \in L_1$. *Let* $x(t)$, $t_1 \leq t \leq t_2$, *be a given trajectory with derivative* $x'(t)$ *essentially bounded, and lying in the interior of* A *(i.e.,* $\Gamma_\delta \subset A$ *where* Γ_δ *is a* δ*-neighborhood of the graph* Γ *of* x*). Let us assume that* x *gives a strong local minimum for* $I[x]$ *(that is,* $I[x] \leq I[y]$ *for all trajectories* y *with graph contained in* Γ_δ *for some* δ*). Then, the following statements* (a)–(i) *hold ("necessary conditions"):*

(a) *The* n *functions* $f_{0x'^i}(t, x(t), x'(t))$, $t_1 \leq t \leq t_2$, *coincide* a.e. *in* $[t_1, t_2]$ *with* AC *functions, say* $-\lambda_i(t)$, $t_1 \leq t \leq t_2$, *and* $-d\lambda_i/dt = f_{0x^i}(t, x(t), x'(t))$ a.e. *in* $[t_1, t_2]$. *Thus, by identifying* $f_{0x'^i}(t, x(t), x'(t))$ *with* $-\lambda_i(t)$, *we write briefly*

$$\text{(2.2.1)} \qquad (d/dt) f_{0x'^i}(t, x(t), x'(t)) = f_{0x^i}(t, x(t), x'(t)),$$
$$t \in [t_1, t_2] \text{ (a.e.)}, \quad i = 1, \ldots, n.$$

In vectorial form, $\lambda(t) = -f_{0x'}(t, x(t), x'(t))$ *is* AC *and* $(d/dt) f_{0x'}(t, x(t), x'(t)) = f_{0x}(t, x(t), x'(t))$, $t \in [t_1, t_2]$ (a.e.).

This will be referred to as the *Euler necessary condition*, or as the Euler equations (E_i), $i = 1, \ldots, n$. See Section 2.4 for a proof.

For instance, for $n = 1$, $I[x] = \int_0^1 (x'^2 + x^2)\,dt$, then $f_0 = x'^2 + x^2$, $f_{0x'} = 2x'$, the function $2x'$ must coincide a.e. in $[0,1]$ with an AC function, and with this identification the Euler equation (E_1) is $(d/dt)(2x') = 2x$, or $x'' = x$. The optimal solutions, if any, are arcs of $x(t) = c_1 \cosh t + c_2 \sinh t$. See Remark 3 below for details on Euler equations, in particular their explicit form as second order differential equations, and Chapter 3, for instance Sections 3.1 and 3.2, for more examples. We may not repeat every time the need for the identification mentioned above (Cf. also (2.6.i, ii, iii)).

(b) *The function* $f_0(t, x(t), x'(t)) - \sum_{i=1}^n x'^i(t) f_{0x'^i}(t, x(t), x'(t))$, $t_1 \leq t \leq t_2$, *coincides* a.e. *in* $[t_1, t_2]$ *with an* AC *function* $M(t)$, *and* $dM/dt = f_{0t}(t, x(t), x'(t))$ a.e. *in* $[t_1, t_2]$. *Thus, by identifying the function above with* $M(t)$, *we write briefly*

$$(2.2.2)\qquad (d/dt)\left[f_0(t, x(t), x'(t)) - \sum_{i=1}^{n} x'^i(t) f_{0x'^i}(t, x(t), x'(t))\right]$$
$$= f_{0t}(t, x(t), x'(t)), \qquad t \in [t_1, t_2] \text{ (a.e.)}.$$

In other words, by using the Hamiltonian function (2.1.3) *and the n-vector function* $\lambda(t) = -f_{0x'}(t, x(t), x'(t))$, *we can say that the scalar function* $M(t) = H(t, x(t), x'(t), \lambda(t))$ *is* AC *and* $(d/dt)M(t) = f_{0t}(t, x(t), x'(t))$, $t \in [t_1, t_2]$ (a.e.).

This will be referred to as the *DuBois-Reymond necessary condition*, or the equation (E_0). See Section 2.4 for a proof. For instance, for $n = 1$, $I[x] = \int_0^1 (x'^2 + x^2)\,dt$, $f_0 = x'^2 + x^2$, and (E_0) reduces to $(d/dt)(x^2 - x'^2) = 0$, or $x^2 - x'^2 = c$, a costant. Here $M(t) = x^2(t) - x'^2(t)$. See Chapter 3 for more examples, particularly Section 3.4 and Exercises 8 and 9 in Section 3.7A.

(c) *The* AC $2n$-*vector function* $(x(t), \lambda(t))$, $t \in [t_1, t_2]$, $x(t) = (x^1, \ldots, x^n)$, $\lambda(t) = (\lambda_1, \ldots, \lambda_n)$, *satisfies the system of* $2n$ *equations*

$$(2.2.3)\qquad \frac{dx^i}{dt} = \frac{\partial H}{\partial \lambda_i}, \qquad \frac{d\lambda_i}{dt} = \frac{-\partial H}{\partial x^i}, \qquad t \in [t_1, t_2] \text{ (a.e.)},$$

for $i = 1, \ldots, n$ *(canonical equations), where* H *is the Hamiltonian function* (2.1.3).

In this form these relations are an immediate consequence of the Euler equations (see Section 2.4, where also a deeper version of the same relations is given). For instance, for $n = 1$, $I[x] = \int_0^1 (x'^2 + x^2)\,dt$, then $f_0 = x'^2 + x^2$, $H = f_0 + \lambda x' = x'^2 + x^2 + \lambda x'$, and the equations $x' = H_\lambda$, $\lambda' = -H_x$ reduce to the identity $x' = x'$ and $\lambda' = -2x$. Since $\lambda = -f_{0x'} = -2x'$, we again have the equation $x'' = x$ as in (a).

(d) *For* $t \in [t_1, t_2]$ (a.e.) *and all* $X' \in R^n$, *we have*

$$E(t, x(t), x'(t), X') \geq 0,$$

where E is the Weierstrass function (2.1.4), *or explicitly*

(2.2.4) $$E(t,x(t),x'(t),X') = f_0(t,x(t),X') - f_0(t,x(t),x'(t)) - \sum_i (X'^i - x'^i(t)) f_{0x'^i}(t,x(t),x'(t)) \geq 0.$$

This is the *Weierstrass necessary condition* for a local strong minimum. [For a local strong maximum, we have $E \leq 0$.] In other words, for $t \in [t_1, t_2]$ (a.e.) the function $f_0(t,x(t),u)$, $u \in R^n$, of u alone, is convex [concave] in u at the point $u = x'(t)$. See Section 2.10 for an elementary proof, and (2.19.ii) for another proof and a deeper understanding of the same condition. For instance, for $n=1$, $I=\int_{t_1}^{t_2}(x'^2+x^2)\,dt$, then $f_0=x'^2+x^2$, and $E=(X'^2+x^2)-(x'^2+x^2)-(X'-x')(2x')=(X'-x')^2 \geq 0$. The extrema of I can only be minima.

(e) *If f_0 is of class C^2, then for $t \in [t_1, t_2]$* (a.e.) *and all $\xi = (\xi_1, \ldots, \xi_n) \in R^n$ we have*

(2.2.5) $$Q = \sum_{i,j=1}^{n} f_{0x'^ix'^j}(t,x(t),x'(t))\xi_i\xi_j \geq 0,$$

that is, the quadratic form Q is positive semidefinite (Legendre necessary condition for a local minimum). (For a local maximum we have $Q \leq 0$.)

See Section 2.4 for an elementary proof. If $n = 1$, then $Q = f_{0x'x'}\xi^2$, and all is stated is that $f_{0x'x'}(t,x(t),x'(t)) \geq 0$ for $t \in [t_1, t_2]$ (a.e.). If $n = 1$ and $I[x] = \int_{t_1}^{t_2}(x'^2 + x^2)\,dt$, then $f_0 = x'^2 + x^2$, $f_{0x'x'} = 2 > 0$. Thus, the quadratic form $2\xi^2$ is positive definite. The extrema of I can only be minima.

(f) *For $t \in [t_1, t_2]$* (a.e.) *the function $H(t,x(t),u,\lambda(t))$, $u \in R^n$, of u alone, has an absolute minimum at $u = x'(t)$, where $\lambda(t) = (\lambda_1, \ldots, \lambda_n)$ are the* AC *functions defined under* (a). *The minimum of $H(t,x(t),u,\lambda(t))$ in R^n is then*

(2.2.6) $$M(t) = f_0(t,x(t),x'(t)) - \sum_{i=1}^{n} x'^i(t) f_{0x'^i}(t,x(t),x'(t)).$$

In other words,

$$f_0(t,x(t),x'(t)) + \sum_i \lambda_i(t)x'^i(t) = \min_{u \in R^n}\left[f_0(t,x(t),u) + \sum_i \lambda_i(t)u^i\right]$$

where $\lambda_i(t) = -f_{0x'^i}(t,x(t),x'(t))$, $i = 1, \ldots, n$.

See Section 2.4 for a derivation of (f) from (d). For instance, if $n = 1$ and $I[x] = \int_0^1 (x'^2 + x^2)\,dt$, then $H = f_0 + \lambda x' = x'^2 + x^2 + \lambda x'$. For every fixed x, H has a minimum (for $x' \in R$) at $x' = -\lambda/2$, or $\lambda = -2x'$, and $M = H_{\min} = x^2 - x'^2$. These are the same relations for λ and M we have seen in (a) and (b) above for the same example.

(g) *If x is continuous with sectionally continuous derivative in $[t_1, t_2]$, then at every (possible) first kind discontinuity point t_0 for x' (jump discontinuity of the derivative, or corner point $(t_0, x(t_0))$ of the trajectory x), we have (Erdman corner conditions)*

(2.2.7) $$f_{0x'^i}(t_0, x(t_0), x'(t_0 - 0)) = f_{0x'^i}(t_0, x(t_0), x'(t_0 + 0)), \qquad i = 1, \ldots, n,$$

(2.2.8) $$f_0(t_0, x(t_0), x'(t_0 - 0)) - \sum_{i=1}^{n} x'^i(t_0 - 0) f_{0x'^i}(t_0, x(t_0), x'(t_0 - 0))$$
$$= f_0(t_0, x(t_0), x'(t_0 + 0)) - \sum_{i=1}^{n} x'^i(t_0 + 0) f_{0x'^i}(t_0, x(t_0), x'(t_0 + 0)).$$

See Section 2.4 for an elementary proof. For instance, for $n = 1$, $I[x] = \int_0^1 (x'^2 + x^2)\,dt$, then $f_{0x'} = 2x$, $f_{0x'x'} = 2 > 0$, and $f_{0x'}$ cannot have the same value twice: the optimal solution, if any, cannot have corner points. (See Remark 3 in Section 2.2.) See the examples in Sections 3.10, 3.11, whose optimal solutions may exhibit corner points.

(h) *If x is continuous with sectionally continuous derivative in $[t_1, t_2]$, then statement* (d) *holds at every point $t_0 \in [t_1, t_2]$; in particular* (d) *holds at the points t_0 of jump discontinuity of x' with $x'(t_0)$ replaced there by $x'(t_0 - 0)$ as well as by $x'(t_0 + 0)$.*

(i) *For variable end points problems, let B' denote the tangential hyperplane to B at the point $e[x] = (t_1, x(t_1), t_2, x(t_2))$. Let $h = (\tau_1, \xi_1, \tau_2, \xi_2)$ denote any element of B', or in more usual notation, $h = (dt_1, dx_1, dt_2, dx_2)$. Then*

(2.2.9) $$\Delta = \left[\left(f_0 - \sum_{i=1}^{n} x'^i f_{0x'^i} \right) dt + \sum_{i=1}^{n} f_{0x'^i}\, dx^i \right]_1^2 = 0$$

for all tangent vectors $h = (dt_1, dx_1, dt_2, dx_2) \in B'$, the tangent hyperspace to B at $(t_1, x(t_1), t_2, x(t_2))$. Here the coefficients of dt, dx^i are the AC *functions $M(t)$ and $-\lambda_i(t)$ computed at the the points $1 = (t_1, x(t_1))$ and $2 = (t_2, x(t_2))$. If B possesses at (t_1, x_1, t_2, x_2) only a convex tangent cone B', then $\Delta \geq 0$ for all $h \in B'$.*

See Section 2.4 for a proof. The relation (2.2.9) is called the *transversality* relation, and this relation is obviously trivial for fixed end points problems (since $dt_1 = 0$, $dx_1 = 0$, $dt_2 = 0$, $dx_2 = 0$).

For the case of the first end point $1 = (t_1, x_1)$ fixed, or $B = (t_1, x_1) \times B_2$, B_2 a smooth manifold, relation (2.2.9) reduces to

$$\left(f_0 - \sum_{i=1}^{n} x'^i f_{0x'^i} \right) dt_2 + \sum_{i=1}^{n} f_{0x'^i}\, dx_2^i = 0,$$

or $M(t_2)\,dt_2 - \sum_i \lambda_i(t_2)\,dx_2^i = 0$, where dt_2, dx_2^i are computed on B_2 at (t_2, x_2). We also say that the trajectory x is *transversal* to B_2 at the point $2 = (t_2, x_2)$. Often this relation reduces to an orthogonality relation. For instance, for

$n = 1, f_0 = (1 + x'^2)^{1/2}$, then $f_{0x'} = x'(1 + x'^2)^{-1/2}, f_0 - x'f_{0x'} = (1 + x'^2)^{-1/2}$ and the relation above reduces to $dt_2 + x'\,dx_2 = 0$, or $x'(dx_2/dt_2) = -1$. Statement (i) will be expanded by numerous examples (see Section 3.1B and Section 3.7C, problems 1–8). Since the transversality relation is preserved unchanged in problems of optimal control, a further discussion of this relation is presented in Sections 4.4 and 5.3B.

The statements above concerning the Hamiltonian, i.e., (c), (f), (i) are particularly relevant because they are preserved unchanged in the more general problems of optimal control (Chapters 4, 5). Other necessary conditions for problems of the calculus of variations will be stated and proved in Sections 2.3–5.

Remark 1. For any given classical Lagrange problem concerning the functional $I[x] = \int_{t_1}^{t_2} f_0\,dt$, any arc of solution $x(t)$, $\alpha \le t \le \beta$, of class C^1 of the system of Euler equations (2.2.1) is called an *extremal arc*, or an *extremal* of the given problem. Optimal solutions are often expected to be made up of such extremal arcs. The integral $I[x]$ is said to be *stationary* at any solution x of the Euler equations, even if x may not be optimal.

Remark 2. For classical isoperimetric problems as stated in Section 1.5 concerning the minima and maxima of $I[x] = \int_{t_1}^{t_2} f_0\,dt$ with N integral constraints of the form $J_j[x] = \int_{t_1}^{t_2} f_j\,dt = C_j$, C_j given constants, $j = 1, \dots, N$, we shall prove in Section 4.8 that the optimal solutions satisfy the necessary conditions for the auxiliary Lagrange problem $\int_{t_1}^{t_2} F\,dt$ with $F = f_0 + \lambda_1 f_1 + \cdots + \lambda_N f_N$, $\lambda_1, \dots, \lambda_N$ undetermined constants.

For problems concerning minima and maxima of $I[x]$ with constraints given by $N < n$ differential equations of the form (1.2.2), or $G_s(t, x(t), x'(t)) = 0$, $s = 1, \dots, N$, the optimal solutions satisfy the necessary conditions for the auxiliary Lagrange problem $\int_{t_1}^{t_2} F\,dt$ with $F = f_0 + p_1(t)G_1 + \cdots + p_N(t)G_N$ and where $p_1, \dots, p_N$ are functions (see Section 4.8).

Remark 3. If $f_0(t, x')$ does not depend on x, Euler's equations reduce to

$$f_{0x'^i}(t, x'(t)) = C_i, \qquad i = 1, \dots, n, \tag{2.2.10}$$

where the C_i are arbitrary constants. In other words, the second order Euler equations posess the "first integrals" (2.2.10), a system of n first order differential equations with n arbitrary constants.

If $f_0(x, x')$ does not depend on t, DuBois-Reymond's equation (2.2.2) reduces to

$$f_0(x(t), x'(t)) - \sum_{i=1}^{n} x'^i(t) f_{0x'^i}(x(t), x'(t)) = C_0, \tag{2.2.11}$$

where C_0 is an arbitrary constant, or $H(x(t), x'(t), \lambda(t)) = C_0$ where $\lambda(t) = f_{0x'}(t, x(t), x'(t))$. In other words, if $f_0(x, x')$ does not depend on t, then the Hamiltonian is constant on every optimal trajectory.

For $n = 1$, Euler's equation (2.2.1) is

$$(d/dt) f_{0x'}(t, x(t), x'(t)) = f_{0x}(t, x(t), x'(t)). \tag{2.2.12}$$

For f of class C^2 and x of class C^2 we have the explicit form

$$f_{0tx'} + x' f_{0xx'} + x'' f_{0x'x'} - f_{0x} = 0. \tag{2.2.13}$$

For $n = 1$ and $f_0(x, x')$ not depending on t, the DuBois-Reymond relation (2.2.2) reduces to

$$f_0(x(t), x'(t)) - x'(t)f_{0x'}(x(t), x'(t)) = C_0. \tag{2.2.14}$$

Note that for $n = 1$, $f_0(x, x')$ not depending on t, f_0 of class C^2, and x continuous with x' continuous and x'' sectionally continuous, the relation (2.2.14) can be derived from Euler's equation (2.2.12) by the simple remark that

$$\begin{aligned}(d/dt)[f_0 - x'f_{0x'}] &= x'f_{0x} + x''f_{0x'} - x''f_{0x'} - x'^2 f_{0x'x} - x'x''f_{0x'x'} \\ &= x'[f_{0x} - x'f_{0x'x} - x''f_{0x'x'}] = 0.\end{aligned}$$

For $n \geq 1$, $f_0(t, x, u)$ of class C^2, and x also of class C^2, then the explicit form of Euler's equations (2.2.1) is

$$f_{0x^i} = f_{0x'^i t} + \sum_j f_{0x'^i x^j} x'^j + \sum_j f_{0x'^i x'^j} x''^j, \qquad i = 1, \ldots, n. \tag{2.2.15}$$

For $n = 1$ the Erdman corner conditions become

$$f_{0x'}(t_0, x(t_0), x'(t_0 - 0)) = f_{0x'}(t_0, x(t_0), x'(t_0 + 0)),$$

$$\begin{aligned}&f_0(t_0, x(t_0), x'(t_0 - 0)) - x'(t_0 - 0)f_{0x'}(t_0, x(t_0), x'(t_0 - 0)) \\ &\quad = f_0(t_0, x(t_0), x'(t_0 + 0)) - x'(t_0 + 0)f_{0x'}(t_0, x(t_0), x'(t_0 + 0)).\end{aligned}$$

Thus, if we know that $f_{0x'}$ is a strictly monotone function of x', then $f_{0x'}$ cannot take twice the same value, and we conclude that the optimal trajectory x cannot have corner points. The same conclusion can be derived if $f_0 - x'f_{0x'}$ is strictly monotone in x'.

For arbitrary $n \geq 1$, we can say, analogously, that if for each t, $t_1 \leq t \leq t_2$, the n-vector function of u given by $[f_{0x'^i}(t, x(t), u), i = 1, \ldots, n]$ never takes the same value more than once as u describes R^n, then there cannot be corner points. In other words, we require that $t_1 \leq t \leq t_2$, $u, v \in R^n$, $u \neq v$, imply

$$[f_{0x'^i}(t, x(t), u), i = 1, \ldots, n] \neq [f_{0x'^i}(t, x(t), v), i = 1, \ldots, n].$$

Conditions for this to occur will be mentioned in Section 2.6.

Remark 4. If f_0 is of class C^m, $m > 2$, if $R(t)$ denotes the $n \times n$ matrix $f_{0x'x'}(t, x(t), x'(t))$ (i.e. $R(t) = [f_{0x'^i x'^j}]$), and if $\det R(t_0) \neq 0$ at some t_0, then by the implicit function theorem of calculus, the system (2.2.15) can be written in normal form $x'' = F(t, x, x')$, $F = (F_1, \ldots, F_n)$, with F of class C^{m-2} in a neighborhood of $(t_0, x(t_0), x'(t_0))$. Then $\det R(t_0) \neq 0$ guarantees the local uniqueness of the extremal of given initial data $x(t_0)$, $x'(t_0)$ at t_0.

Remark 5. For $n = 1$ and $f_0(t, x)$ not depending on x', the Euler equation reduces to $f_{0x}(t, x) = 0$ which is not even a differential equation. The only possible extremals (if any) are made up of zeros of $f_{0x}(t, x)$. For instance, for $I[x] = \int_{t_1}^{t_2} (t - x)^2\,dt$, $f_{0x} = -2(t - x)$, and the only extremal is the straight line $x = t$.

For $n = 1$, and $f_0(t, x, x') = A(t, x) + B(t, x)x'$ linear in x', then the Euler equation reduces to $A_t(t, x) - B_x(t, x) = 0$, again not a differential equation. Note that here $I[x]$ is the line integral $\int_{t_1}^{t_2} (A\,dt + B\,dx)$. If $A_t \equiv B_x$, i.e., I is an exact differential, then the Euler equation is an identity, and all trajectories of class C^1 are extremals, in harmony with the fact that $I[x]$ depends only on the end points of the trajectories (within any simply connected region V containing the given trajectory $C: x = x(t)$, $t_1 \leq t \leq t_2$).

Remark 6. For $n = 1$, and f_0 of class C^m, $m > 1$, as in Remark 4, $R(t)$ is a scalar, and for $R(t_0) \neq 0$ the Euler equation can be locally written in the form $x'' = F(t, x, x')$. Whenever the Euler equation can be written in the normal form $x'' = F(t, x, x')$ for all $t_1 \leq t \leq t_2$, $x, x' \in R$, the question whether there is a solution $x(t)$, $t_1 \leq t \leq t_2$, satisfying a two point boundary condition $x(t_1) = x_1$, $x(t_2) = x_2$ may be answered, for instance, by the following statement: If $F, F_x, F_{x'}$ are continuous, if there are a constant $k > 0$ and continuous functions $a(x, y) \geq 0$, $b(x, y) \geq 0$ such that $F_x(t, x, x') \geq k$, $|F(t, x, x')| \leq ax'^2 + b$ for all $t_1 \leq t \leq t_2$, $x, x' \in R$, then for any given x_1, x_2 real there is one and only one solution $x(t)$, $t_1 \leq t \leq t_2$, of $x'' = F(t, x, x')$ with $x(t_1) = x_1$, $x(t_2) = x_2$. (S. N. Bernstein [1]). For instance, for $f_0(t, x, x') = e^{-2x^2}(x'^2 - 1)$, then the Euler equation is $x' = 2xx'^2 + 4x = F$, $F_x = 2x'^2 + 4 \geq 4$, and Bernstein's theorem holds with $k = 4$, $a = 2x$, $b = 4x$.

Remark 7. If $\phi(t, x) = \phi(t, x^1, \ldots, x^n)$ is any functions of class C^2, say in R^{n+1}, let

$$F_0(t, x, x') = \phi_t + \sum_i \phi_{x^i} x'^i.$$

Then the two integrals

$$I[x] = \int_{t_1}^{t_2} f_0(t, x, x')\, dt, \qquad H[x] = \int_{t_1}^{t_2} [f_0(t, x, x') + F_0(t, x, x')]\, dt$$

have the same Euler equations. Hint: Show by direct computation that $F_{0x^i} - (d/dt)F^0{}_{x'^i} \equiv 0$. Also, $H[x] = I[x] + \phi(t_2, x_2) - \phi(t_1, x_1)$.

Remark 8. We have already mentioned in Section 1.6 that

$$I[x] = \int_a^b f_0(x(t), x'(t))\, dt, \qquad (t, x(t)) \in A,\ x = (x^1, \ldots, x^n),$$

is a parametric integral, that is, does not depend on the particular AC representation of the path curve $C: x = x(t)$, $a \leq t \leq b$, if and only if f_0 is independent of t and positively homogeneous of degree one in x', i.e., $f_0(x, kx') = kf_0(x, x')$ for all $k \geq 0$, $x \in A$, $x' \in R^n$. Then, the first derivatives $f_{0x'^i}$, $i = 1, \ldots, n$, are positively homogeneous of degree zero in x', and by Euler's theorem on homogeneous functions, we have $f_0(x, x') = \sum_{i=1}^n x'^i f_{0x'^i}(x, x')$. As a consequence the Weierstrass function E takes the form

$$\begin{aligned} E(x, x', X') &= f_0(x, X') - f_0(x, x') - \sum_i (X'^i - x'^i) f_{0x'^i}(x, x') \\ &= f_0(x, X') - \sum_i X'^i f_{0x'^i}(x, x') \end{aligned}$$

and consequently $E(x, x', X')$ is positively homogeneous of degree one in X' and of degree zero in x'. However, the requirement that f_0 is of class C^1 is unrealistic here since already $f_0 = |x'|$, the integrand function of the length integral, has discontinuous first partial derivatives at the origin 0. We shall only assume that f_0 is of class C^1 in $A \times (R^n - 0)$. Only rectifiable path curves $C: x = x(t)$, $a \leq t \leq b$, should be considered, namely, continuous curves of finite Jordan length L. These curves always have an AC representation, namely the one in terms of their arc length parameter s, or $C: x = X(s)$, $0 \leq s \leq L$, with $X(s)$ Lipschitzian of constant one and $|X'(s)| = 1$ a.e..

For $n = 2$, it is convenient to write (x, y) for x, and (x', y') for x', so that the integral is now

$$J[x, y] = \int_a^b f_0(x(t), y(t), x'(t), y'(t))\, dt,$$

with $(x(t), y(t)) \in A$ and $f_0(x, y, x', y')$ positively homogeneous of degree one in (x', y'). By Euler's theorem we have

$$f_0 = x'f_{0x'} + y'f_{0y'}, \qquad f_{0x} = x'f_{0xx'} + y'f_{0xy'}, \qquad f_{0y} = x'f_{0yx'} + y'f_{0yy'}.$$

If f_0 is of class C^2 in $A \times (R^2 - 0)$, then by differentiation of the first relation above with respect to x' and y' we also have

$$x'f_{0x'x'} + y'f_{0x'y'} = 0, \qquad x'f_{0x'y'} + y'f_{0y'y'} = 0.$$

For $(x', y') \neq (0, 0)$ then

$$f^*(x, y, x', y') = \frac{f_{0x'x'}}{y'^2} = -\frac{f_{0x'y'}}{x'y'} = \frac{f_{0y'y'}}{x'^2}.$$

This new function is well defined for $(x', y') \neq (0, 0)$ since at least one of the denominators is always different from zero. Moreover, F is positively homogeneous of order -3 in (x', y'). For instance, for $f_0 = (x'^2 + y'^2)^{1/2}$, then $f^* = (x'^2 + y'^2)^{-3/2}$.

The Euler equations of the integral $J[x, y]$ are

$$f_{0x} - (d/dt)f_{0x'} = 0, \qquad f_{0y} - (d/dt)f_{0y'} = 0,$$

and if f_0 and x'' are both of class C^2, also

$$f_{0x} - x'f_{0x'x} - y'f_{0x'y} - x''f_{0x'x'} - y''f_{0x'y'} = 0,$$
$$f_{0y} - x'f_{0y'x} - y'f_{0y'y} - x''f_{0y'x'} - y''f_{0y'y'} = 0.$$

By using the expressions for f_{0x}, f_{0y} we have given above and the definition of the function f^*, we obtain, after simplification,

$$f_{0x} - (d/dt)f_{0x'} = y'[f_{0y'x} - f_{0x'y} + (x'y'' - x''y')f^*] = 0,$$
$$f_{0y} - (d/dt)f_{0y'} = -x'[f_{0y'x} - f_{0x'y} + (x'y'' - x''y')f^*] = 0.$$

Since $(x', y') \neq (0, 0)$, this is possible only if the bracket is zero. On the other hand, we know that $(x'^2 + y'^2)^{-3/2}(y''x' - x''y') = 1/r$, where r is the radius of curvature of C. With respect to the length parameter s for the representation of C we have then $x'^2 + y'^2 = 1$ and hence,

$$\frac{1}{r} = \frac{(f_{0x'y} - f_{0y'x})}{f^*}.$$

This is the Weierstrass form of Euler's equations for parametric integrals (for $n = 2$, f_0 of class C^2, and C with length parameter representation of class C^2).

2.3 Necessary Conditions in Terms of Gateau Derivatives

We shall need here a lemma concerning differentiation under the integral sign, which we take from analysis.

2.3.i (Lemma). *Let $F(t, a)$ be defined for all t in a set E and all a in an interval (α, β). Let $F(t, a)$ be L-integrable over E for each fixed $a \in (\alpha, \beta)$, and take*

$\zeta(a) = \int_E F(t,a)\,dt$. For all $a \in (\alpha, \beta)$ and all $t \in E - E_0$, where meas *$E_0 = 0$, let $F_a(t,a)$ be defined and smaller in absolute value than an L-integrable function $g(t)$. Then for all $a \in (\alpha, \beta)$ the derivative $\zeta'(a)$ exists and $\zeta'(a) = \int_E F_a(t,a)\,dt$.*

For this statement we refer to McShane [I, p. 217, 39.2]. The statement extends to a closed interval $[\alpha, \beta]$ with the derivatives with respect to a at α and β being right and left derivatives respectively. Lemma (2.3.i) contains the usual theorem of calculus for differentiation under the integral sign.

Given a function $F(x)$, $x = (x^1, \ldots, x^m) \in R^m$, beside the notation F_{x^i}, $F_{x^i x^j}$ for partial derivatives, we shall use the notation $F_\alpha = D^\alpha F$, $\alpha = (\alpha_1, \ldots, \alpha_m)$, $\alpha_i \geq 0$ integers, $k = \alpha_1 + \cdots + \alpha_m = |\alpha|$, for partial derivatives of order k. Below we shall require a function F to be of some class C^r in a closed subset D of R^m. By this we shall mean that F is defined in an open subset U of R^m containing D, and that F has continuous partial derivatives F_α of orders $|\alpha|$, $0 \leq |\alpha| \leq r$, in U.

This convention, which we use here only to simplify notation, has its motivation in the result due to Whitney [1–4] that if, for a real valued function $F(x)$ defined on a closed set K of R^m, there are functions F_α, $0 \leq |\alpha| \leq r$, also continuous on K, such that the usual Taylor polynomials centered at any point of K approximate F (as usual), then F can be extended to a function of class C^r on an open neighborhood U of K so that $D^\alpha F = F_\alpha$ everywhere on K.

We now return to the usual notation of the previous section. Thus we shall denote by u^* the transpose of any matrix u. The inner product $\sum_i x^i y^i$ of two elements $x = \text{col}(x^1, \ldots, x^n)$, $y = \text{col}(y^1, \ldots, y^n)$ in R^n will be denoted by $x \cdot y$, or $x^* y$, or sometimes simply xy. Also, for $f_0(t, x, x')$ of class C^1 or C^2 in some subset of R^{1+2n}, we shall denote by f_{0x}, $f_{0x'}$ the n-vectors $f_{0x} = \text{col}(f_{0x^1}, \ldots, f_{0x^n})$, $f_{0x'} = \text{col}(f_{0x'^1}, \ldots, f_{0x'^n})$, by f_{0xx} the $n \times n$ matrix $f_{0xx} = (f_{0x^i x^j}, i, j = 1, \ldots, n)$, and analogously for $f_{0x'x'}$, $f_{0xx'}$. With this notation Euler's equations for any $n \geq 1$ become $(d/dt) f_{0x'} = f_{0x}$.

Let $x(t) = (x^1, \ldots, x^n)$, $t_1 \leq t \leq t_2$, be a given AC n-vector function with x' essentially bounded and whose graph lies in the interior of the closed set A. Then there is some $\delta > 0$ such that the whole set $\Gamma_\delta = [(t, y) | t_1 \leq t \leq t_2, |y - x(t)| \leq \delta]$ is contained in A. We assume as usual that $f_0(t, x, x')$ is of class C^1 in $A \times R^n$. Let Ω_0 denote the set of all AC n-vector functions $y(t) = (y^1, \ldots, y^n)$, $t_1 \leq t \leq t_2$, with $y(t_1) = x(t_1)$, $y(t_2) = x(t_2)$, with y' essentially bounded and whose graph lies in Γ_δ. Let v^0 denote the collection of all AC function $\eta(t) = (\eta^1, \ldots, \eta^n)$, $t_1 \leq t \leq t_2$, with essentially bounded derivative and $\eta(t_1) = \eta(t_2) = 0$. Then, for every $\eta \in v^0$ there is some $a_0 > 0$ such that, for $-a_0 \leq a \leq a_0$, all points $(t, x(t) + a\eta(t))$, $t_1 \leq t \leq t_2$, lie in Γ_δ.

2.3.ii (Theorem: A Necessary Condition). *If f_0 is of class C^1 in $A \times R^n$, if $x(t)$, $t_1 \leq t \leq t_2$, is* AC *with essentially bounded derivative, if the graph of x lies in the interior of A, and if x is a weak local minimum for I in Ω_0, then* *for every* $\eta \in v^0$ *the function*

$$\psi(a) = \int_{t_1}^{t_2} f_0(t, x(t) + a\eta(t), x'(t) + a\eta'(t))\, dt \tag{2.3.1}$$

has a derivative for every $-a_0 \le a \le a_0$, *and*

$$\begin{aligned} \psi'(0) = J_1[\eta] = J_1[\eta; x] &= \int_{t_1}^{t_2} [\eta^* f_{0x}^0(t) + \eta'^* f_{0x'}^0(t)]\, dt \\ &= \int_{t_1}^{t_2} \sum_{i=1}^{n} [f_{0x^i}(t, x(t), x'(t))\eta^i(t) + f_{0x'^i}(t, x(t), x'(t))\eta'^i(t)]\, dt = 0, \end{aligned} \tag{2.3.2}$$

where $f_{0x}^0(t) = f_{0x}(t, x(t), x'(t))$, $f_{0x'}^0(t) = f_{0x'}(t, x(t), x'(t))$.

Proof. Let $F(t, a)$ be the integrand in (2.3.1). If E_0 denotes the set of measure zero of all $t \in E = [t_1, t_2]$ where $x'(t)$ or $\eta'(t)$ may be undefined or infinite, then the partial derivative

$$\begin{aligned} F_a(t, a) = \eta(t)^* f_{0x}(t, x(t) + \eta(t), x'(t) + a\eta'(t)) \\ + \eta'(t)^* f_{0x'}(t, x(t) + a\eta(t), x'(t) + a\eta'(t)) \end{aligned}$$

also exists at every $t \in [t_1, t_2] - E_0$. Since f_0 is of class C^1, since x, η are bounded in $[t_1, t_2]$, and since x', η' are essentially bounded, we conclude that $F_a(t, a)$ has a bound $|F_a(t, a)| \le M$ for $(t, a) \in (E - E_0) \times [-a_0, a_0]$. By lemma (2.3.i), $\psi'(a) = \int_{t_1}^{t_2} F_a(t, a)\, dt$. Since $\psi(a)$, $-a_0 \le a \le a_0$, has a minimum at $a = 0$ we conclude that $\psi'(0) = 0$, or $J_1[\eta; x] = 0$.

In (2.3.2) $\eta^* f_{0x}$ and $\eta'^* f_{0x'}$ denote inner products in R^n; we may denote them also by $\eta \cdot f_{0x}$ and $\eta' \cdot f_{0x'}$, or simply ηf_{0x} and $\eta' f_{0x'}$.

Note that $J_1[\eta; x] = \lim_{a \to 0} a^{-1}(I[x + a\eta] - I[x])$ considered above, a linear form in η, is sometimes called the Gateau derivative of $I[x]$ with respect to η at x, and also the *first variation* of $I[x]$ at x.

In (2.3.i) the requirement that f_0 is of class C^1 is excessive. Much less is needed. For instance, assume that $A = [t_0, T] \times R^n$. Then it is enough to assume first that f_0 is a Carathéodory function, namely, for a.a. $t \in [t_0, T]$, f_0 is of class C^1 in (x, x') in $R^n \times R^n$, and that, for all (x, x'), f_0 is measurable in t in $[t_0, T]$. (Cf. McShane [I]). Moreover, we need to know that for all a of some interval $(-a_0, a_0)$ and every $\eta \in v^0$, $F(t, a) = f_0(t, x(t) + a\eta(t), x'(t) + a\eta'(t))$ is L-integrable in $[t_1, t_2]$, and that $F_a(t, a)$ is in absolute value below some L-integrable function $g(t)$ in $[t_1, t_2]$. Then, lemma (2.3.i) applies as usual and (2.3.ii) holds.

2.3.iii (Theorem: A Necessary Condition). *If* f_0 *is of class* C^2 *in* $A \times R^n$, *if* $x(t)$, $t_1 \le t \le t_2$, *is* AC *with essentially bounded derivative, and if* x *is a weak local minimum for* I *in* M, *then the function* $\psi(a)$ *in* (2.3.ii) *has derivatives*

$\psi'(a)$, $\psi''(a)$ for every a in a neighborhood of $a = 0$, $\psi'(0) = J_1[\eta] = 0$, and

$$\psi''(0) = J_2[\eta] = J_2[\eta; x] = \int_{t_1}^{t_2} 2\omega(t, \eta, \eta')\, dt \geq 0, \tag{2.3.3}$$

$$\begin{aligned} 2\omega(t, \eta, \eta') &= \eta'^*[R(t)\eta' + Q(t)\eta] + \eta^*[Q^*(t)\eta' + P(t)\eta] \\ &= \eta'^*[f^0_{0x'x'}(t)\eta' + f^0_{0x'x}(t)\eta] + \eta^*[f^0_{0xx'}(t)\eta' + f^0_{0xx}(t)\eta] \\ &= \sum_{i,j=1}^{n} [f^0_{0x'^ix'^j}\eta^{i'}\eta^{j'} + f^0_{0x'^ix^j}\eta^{i'}\eta^j + f^0_{0x^ix'^j}\eta^i\eta^{j'} + f^0_{0x^ix^j}\eta^i\eta^j], \end{aligned} \tag{2.3.4}$$

where $R(t) = f^0_{0x'x'}(t) = f_{0x'x'}(t, x(t), x'(t))$, $Q(t) = f^0_{0x'x}(t)$, $P(t) = f^0_{0xx}(t)$ are $n \times n$ matrices with measurable essentially bounded entries, and $R^* = R$, $P^* = P$.

Proof. The proof that $\psi''(0)$ exists and is equal to J_2 is the same as for $\psi'(0)$ in (2.3.ii). Since $\psi(a)$ has a minimum at $a = 0$, we must have $\psi''(a) \geq 0$. □

In the formula (2.3.4) $\eta'^*R(t)\eta'$ denote the matrix product of $1 \times n$, $n \times n$, $n \times 1$ matrices, as is clear from the explicit definition.

The integral $J_2[\eta; x]$, a quadratic form in η, is often called the *accessory integral* of $I[x]$, and also the *second variation* of $I[x]$ at x.

Remark 1. Theorem (2.3.ii) holds even in a slightly more general situation. Indeed, let $z(a, t) = (z^1, \ldots, z^n)$, $|a| \leq a_0$, $t_1 \leq t \leq t_2$, be a function continuous in $[-a_0, a_0] \times [t_1, t_2]$ together with the partial derivatives z_a, z_t, z_{at} and with $z(0, t) = 0$, $z_t(0, t) = 0$, $\eta(t) = z_a(0, t)$, $\eta'(t) = z_{at}(0, t)$. Then, $\psi'(0) = \lim_{a \to 0} a^{-1}(I[x + z] - I[x])$ exists and is given by (2.3.2). The proof is the same with $F(t, a) = f_0(t, x(t) + z(a, t),\ x'(t) + z_t(a, t))$. Actually, it is enough to know that there are a set $E_0 \subset [t_1, t_2]$ of measure zero, an AC function $\eta(t)$, $t_1 \leq t \leq t_2$, with essentially bounded derivative $\eta'(t)$, and a constant N such that z_a, z_t, z_{at} exist for all $t \in [t_1, t_2] - E_0$ and all $|a| \leq a_0$, with $z(0, t) = 0$, $z_t(0, t) = 0$, $|z(a, t)| \leq |a|N$, $|z_t(a, t)| \leq N$, $|z_a(a, t)| \leq N$, $|z_{at}(a, t)| \leq N$, and $a^{-1}z(a, t) \to z_a(0, t) = \eta(t)$, $a^{-1}z_t(a, t) \to z_{at}(0, t) = \eta'(t)$ as $a \to 0$. This situation includes the case $z(a, t) = a\eta(t)$ which is the one contemplated in (2.3.ii).

Remark 2. The requirement in (2.3.ii) that x' be essentially bounded can be omitted under suitable assumptions. For instance, let $x(t)$, $t_1 \leq t \leq t_2$, be a given AC n-vector function, whose graph Γ lies in A and let $\Gamma_\delta = [(x, y) | t_1 \leq t \leq t_2, |y - x(t)| \leq \delta]$. We shall assume that $x' \in L_p[t_1, t_2]$, for some $p \geq 1$. We may note here that if we know that $f_0(t, x, x') \geq \mu|x'|^p + m_0$ for some constants $\mu > 0$, m_0 real and all (t, x, x') on Γ or in some region containing Γ, then the mere existence and finiteness of $I[x]$ implies that $x' \in L_p$. Indeed $\int_{t_1}^{t_2} |x'|^p\, dt \leq \mu^{-1}(I[x] - m_0(t_2 - t_1))$. We shall further assume that $\Gamma_\delta \subset A$ for some $\delta > 0$, and that for suitable constants $M, m \geq 0$ we have $|f_0|, |f_{0x}|, |f_{0x'}| \leq M|x'|^p + m$ for all $(t, x, x') \in \Gamma_\delta \times R^n$. Let Ω_p denote the family of all AC n-vector functions $y(t)$, $t_1 \leq t \leq t_2$, with $y(t_1) = x(t_1)$, $y(t_2) = x(t_2)$, whose graph lies in Γ_δ and with $y' \in L_p[t_1, t_2]$. Let v^0 as above be the family of all AC n-vector functions $\eta(t)$, $t_1 \leq t \leq t_2$, with η' essentially bounded and $\eta(t_1) = \eta(t_2) = 0$. Then for every $\eta \in v^0$ there is some $a_0 > 0$ such that $x + a\eta$ has graph in Γ_δ for all $|a| \leq a_0$.

Let $L_0 = \max|\eta|$, $L = \text{ess sup}|\eta'|$. First $I[x]$ exists since $|f_0(t, x(t), x'(t))| \leq M|x'(t)|^p + m$, hence f_0 is L_1-integrable in $[t_1, t_2]$. For any $|a| \leq a_0$ then $x(t) + a\eta(t)$ has graph in

Γ_δ, $|x' + a\eta'| \leq |x'| + |a|L$, $|x' + a\eta'|^p \leq 2^p|x'|^p + 2^p|a|^p L^p$, and $x + a\eta \in \Omega_p$. Moreover

$$|f_0(t, x(t) + a\eta(t), x'(t) + a\eta'(t))| \leq 2^p M|x'(t)|^p + 2^p a_0^p M L^p + m,$$

hence $I[x + a\eta]$ exists for all $|a| \leq a_0$. The same relation holds for f_{0x} and for $f_{0x'}$. As in the proof of (2.3.ii) the integrand function $F(t, a)$ in (2.3.1) has derivative $F_a(t, a)$ for all $t \in [t_1, t_2] - E_0$ where E_0 is a set of measure zero as in the proof of (2.3.ii). From the expression of $F_a(t, a)$ given there we derive now

$$|F_a(t, a)| \leq n(L_0 + L)(2^p M|x'(t)|^p + 2^p a_0^p M L^p + m$$

for all $t \in [t_1, t_2] - E_0$ and $|a| \leq a_0$. Thus, F_a is smaller in absolute value than an L_1-integrable function $g(t)$. By (2.3.i) the derivative $\psi'(a)$ exists for all $|a| \leqq a_0$, and $\psi'(0)$ is given by (2.3.2). If x is a weak local minimum (or maximum) for $I[x]$ in Ω_p then $\psi'(0) = J_1[x;\eta] = 0$ as in (2.3.ii).

Remark 3. Statement (2.3.ii) has a more general formulation in which we need not assume that the graph of x lies in the interior of the closed set A. For this formulation some preparation is needed.

Let D denote a given set of the txx'-space R^{1+2n}; let $x(t) = (x^1, \ldots, x^n)$, $t_1 \leq t \leq t_2$, be a given AC n-vector function with x' essentially bounded; and assume that $(t, x(t), x'(t)) \in D$, and that for every $t \in [t_1, t_2]$ the section $D(t) = [(x, x')|(t, x, x') \in D]$ is convex. Moreover, we assume that $f_0(t, x, x')$ is of class C^1 on $\bar{D}$; that is, f_0 is defined on an open subset U of R^{1+2n} containing the closure $\bar{D}$ of D, and f_0 has continuous first order partial derivatives on U.

Let M denote the class of all functions $y(t) = (y^1, \ldots, y^n)$, $t_1 \leq t \leq t_2$, AC in $[t_1, t_2]$ with y' essentially bounded, and with $(t, y(t), y'(t)) \in D$, $y(t_1) = x(t_1)$, $y(t_2) = x(t_2)$. Also, let $v^0 = v^0[x]$ denote the class of all $\eta(t) = (\eta^1, \ldots, \eta^n)$, $t_1 \leq t \leq t_2$, also AC in $[t_1, t_2]$ with η' essentially bounded and $\eta(t_1) = \eta(t_2) = 0$, for each of which there is some $a_0 > 0$ such that $(t, x(t) + a_0\eta(t), x'(t) + a_0\eta'(t)) \in D$. Then, by the convexity assumption on the sections $D(t)$, we derive that $(t, x(t) + a\eta(t), x'(t) + a\eta'(t)) \in D$ for all $0 \leq a \leq a_0$ and $t \in [t_1 t_2]$ (a.e.).

Note that it may well occur that both η and $-\eta$ belong to $v^0 = v^0[x]$. Moreover, if x lies in the interior of D, there is some $\delta > 0$ such that the whole set $\Gamma_\delta[x] = [(t, y)|t_1 \leq t \leq t_2, |y - x(t)| \leq \delta]$ is contained in D, and then v^0 is the entire collection of AC functions $\eta(t) = (\eta^1, \ldots, \eta^n)$, $t_1 \leq t \leq t_2$, with $\eta(t_1) = \eta(t_2) = 0$ and essentially bounded derivative η'.

In the general situation, statement (2.3.ii) holds in the weaker form that $\psi'^+(0)$, the right derivative of $\psi(a)$ at $a = 0$, exists, is given by the integral expressions (2.3.2), and $\psi'^+(0) \geq 0$. The proof is the same as for (2.3.ii) and still based on lemma (2.3.i).

If both η and $-\eta$ belong to $v^0 = v^0[x]$, then $\psi'(0)$ exists and $\psi'(0) = J_1[\eta; x] = 0$. In particular, if x lies in the interior of A, then $v^0 = v^0[x]$ is the class of all AC $\eta(t)$, $t_1 \leq t \leq t_2$, with $\eta(t_1) = \eta(t_2) = 0$ and η' essentially bounded, and then certainly $\psi'(0) = J_1[\eta; x] = 0$.

Remark 4. In Remark 3 the hypothesis that $f_0(t, x, x')$ is of class C^1 in D can be replaced by a much weaker one. Namely, it is enough to know that $f_0(t, x, x')$ is measurable in t for every (x, x'), that for almost all t, $f_0(t, x, x')$ is continuous with continuous derivatives f_{0x^i}, $f_{0x'^i}$, $i = 1, \ldots, n$, and that f_0, f_{0x}, $f_{0x'}$ are bounded in every bounded part of V. The proof is the same and still based on lemma (2.3.i).

2.4 Proofs of the Necessary Conditions and of Their Invariant Character

A. Proofs of the Necessary Conditions

2.4.i (LEMMA). *Let $h(t)$, $a \le t \le b$, $h \in L_1[a,b]$, be a given real valued function such that*

$$\int_a^b h(t)\eta'(t)\,dt = 0 \tag{2.4.1}$$

for all AC *real valued functions $\eta(t)$, $a \le t \le b$, with $\eta(a) = 0$, $\eta(b) = 0$ and $\eta'(t)$ essentially bounded. Then $h(t) = C$, a constant, $t \in [a,b]$* (a.e.).

Proof.

(a) First assume that $h(t)$ be essentially bounded. For every real constant C we have $C\int_a^b \eta'(t)\,dt = C[\eta(b) - \eta(a)] = 0$ and thus

$$\int_a^b [h(t) - C]\eta'(t)\,dt = 0 \tag{2.4.2}$$

for all η as in the lemma and any constant C. If we take

$$\eta(t) = \int_a^t [h(\alpha) - C]\,d\alpha, \qquad C = (b-a)^{-1}\int_a^b h(t)\,dt, \tag{2.4.3}$$

then η is AC in $[a,b]$ with $\eta(a) = \eta(b) = 0$, with $\eta'(t) = h(t) - C$ essentially bounded, and by substitution in (2.4.2) we have $\int_a^b [h(t) - C]^2\,dt = 0$. Thus, $h(t) = C$ in $[a,b]$ (a.e.).

(b) If we only know that h is L_1 integrable, the above proof does not apply. The following proof can be traced back to work of DuBois-Reymond [1]. First we know that for almost all $t \in [a,b]$, $h(t)$ is finite and the derivative of its indefinite integral $\int_{t_1}^t h(s)\,ds$. If E_0 denote the complementary set, then meas $E_0 = 0$. Let τ_1, τ_2 be any two points in $[t_1, t_2] - E_0$, $\tau_1 < \tau_2$, and let δ be any number $0 < \delta < 2^{-1}(\tau_2 - \tau_1)$. Let us define $\eta(t)$, $a \le t \le b$, by taking $\eta(t) = 0$ for $t \in [a, \tau_1]$ and for $t \in [\tau_2, b]$; $\eta(t) = \delta$ for $t \in [\tau_1 + \delta, \tau_2 - \delta]$; $\eta(t) = t - \tau_1$ for $t \in [\tau_1, \tau_1 + \delta]$; $\eta(t) = \tau_2 - t$ for $t \in [\tau_2 - \delta, \tau_2]$. Then (2.4.1) yields

$$\int_{\tau_1}^{\tau_1+\delta} h(t)\,dt = \int_{\tau-\delta}^{\tau_2} h(t)\,dt,$$

and by division by δ and passage to the limit as $\delta \to 0+$, we derive $h(\tau_1) = h(\tau_2)$. Since τ_1 and τ_2 are any two points of $[a,b]$, we conclude that $h(t) = C$ for $t \in [a,b]$ (a.e.). □

Note that it is enough to require that (2.4.1) is true for all η continuous with sectionally continuous derivative (and $\eta(t_1) = \eta(t_2) = 0$).

2.4.ii (LEMMA). *Let $h(t)$, $a \le t \le b$, $h \in L_1[a,b]$, be a given real valued L-integrable function such that*

$$\int_a^b h(t)\eta(t)\,dt = 0 \tag{2.4.4}$$

for all real valued functions $\eta(t)$, $a \le t \le b$, of class C^∞ in $[a,b]$ with $\eta(a) = \eta(b) = 0$. Then $h(t) = 0$ in $[a,b]$ (a.e.).

Proof.

(a) First, let us assume that h is continuous in $[a,b]$ and that (2.4.4) holds for all functions $\eta(t)$, $a \le t \le b$, continuous in $[a,b]$ with $\eta(a) = \eta(b) = 0$ [alternatively, for all η of some class C^m, $m \ge 1$]. Then, $h(t)$ must be identically zero in $[a,b]$. Indeed, in the contrary case, there would be some $t_0 \in (a,b)$ with $h(t_0) \ne 0$, say $h(t_0) > 0$, and thus an interval $[t_0 - c, t_0 + c] \subset [a,b]$ where $h(t) > 0$. Then, we take $\eta(t) = 0$ for $|t - t_0| \ge c$, and $\eta(t) = c - |t - t_0|$ for $|t - t_0| \le c$ [alternatively, we take, for $|t - t_0| \le c$, either $\eta(t) = (c^2 - (t - t_0)^2)^m$, or $\eta(t) = \exp[-(c^2 - (t - t_0)^2)^{-1}]$. In any case, we have

$$0 = \int_a^b h(t)\eta(t)\,dt = \int_{t_0 - c}^{t_0 + c} h(t)\eta(t)\,dt > 0,$$

a contradiction. The last inequality follows from the fact that both $h(t)$ and $\eta(t)$ are positive in $(t_0 - c, t_0 + c)$.

(b) Now let h be only of class $L_1[a,b]$. Then, for almost all $t_0 \in (a,b)$, $h(t_0)$ is the derivative of $H_0(t) = \int_a^t h(\tau)\,d\tau$ at t_0. Let t_0 be any such point, and assume, if possible, that $h(t_0) = m \ne 0$, say $m > 0$. Then, we can take $c > 0$ sufficiently small so that $[t_0 - c, t_0 + c] \subset [a,b]$ and $H(t) = \int_{t_0}^t h(\tau)\,d\tau$ lies between $2^{-1}m(t - t_0)$ and $2^{-1}(3m)(t - t_0)$ for $|t - t_0| \le c$. We take now η of class C^∞ with $\eta(t) = 0$ for $|t - t_0| \ge c$, $\eta(t_0) = 1$, and $-\eta'$ of the same sign of $t - t_0$ for $0 < |t - t_0| < c$. Then, by integration by parts, we have

$$0 = \int_a^b h(t)\eta(t)\,dt = -\int_{t_0 - c}^{t_0 + c} H(t)\eta'(t)\,dt > 0,$$

a contradiction. Thus, $h(t_0) = 0$, and we have proved that h is zero a.e. in $[a,b]$. □

Proof of Euler's equations for a weak local minimum and x' essentially bounded. We prove here Euler's equations (E_i) not only for a strong local minimum as stated in (2.2.i), but also for any weak local minimum. Thus, we assume here that the vector function $x(t) = (x^1, \ldots, x^n)$, $t_1 \le t \le t_2$, is AC with x' essentially bounded, and that x is a weak local minimum for the functional. We do not repeat the assumptions of (2.2.i). Let $\eta(t) = (\eta^1, \ldots, \eta^n)$, $t_1 \le t \le t_2$, be any given AC vector function with $\eta(t_1) = \eta(t_2) = 0$ and η' essentially bounded. Then there is some $a_0 > 0$ such that $y(t) = x(t) + a\eta(t)$, $t_1 \le t \le t_2$, has graph in A for all $|a| \le a_0$, and there is an a_1, $0 < a_1 \le a_0$, such that $I[y] \ge I[x]$ for all $|a| \le a_1$. Then $\psi(a) \ge \psi(0)$ for $|a| \le a_1$, where

ψ is the function defined in (2.3.1) and finally $\psi'(0) = 0$, or $J_1[\eta] = 0$ as stated in (2.3.ii).

If, for fixed i, we take $\eta(t) = (\eta^1, \ldots, \eta^n)$ with $\eta^j = 0$ for all $j \neq i$, then the relation $J_1[\eta] = 0$ reduces to

$$\int_{t_1}^{t_2} [f_{0x^i}(t, x(t), x'(t))\eta^i(t) + f_{0x'^i}(t, x(t), x'(t))\eta'^i(t)]\, dt = 0$$

for all AC real valued functions $\eta^i(t)$ with $\eta^i(t_1) = \eta^i(t_2) = 0$ and η'^i essentially bounded. By integration by parts (McShane [I, 36.1, p. 209]) we derive

$$\int_{t_1}^{t_2} \left[f_{0x'^i}(t, x(t), x'(t)) - \int_{t_1}^{t} f_{0x^i}(\tau, x(\tau), x'(\tau))\, d\tau \right] \eta'^i(t)\, dt = 0,$$

and by (2.4.i), therefore,

$$f_{0x'^i}(t, x(t), x'(t)) - \int_{t_1}^{t} f_{0x^i}(\tau, x(\tau), x'(\tau))\, d\tau = C_i, \qquad t \in [t_1, t_2] \text{ (a.e.)}, \tag{2.4.5}$$

C_i a constant; and this relation holds for $t \in [t_1, t_2]$, and thus $f_{0x'^i}(t, x(t), x'(t))$ must coincide a.e. in $[t_1, t_2]$ with an AC function, say $-\lambda_i(t)$. Moreover, by identifying $f_{0x'^i}(t, x(t), x'(t))$ with $-\lambda_i(t)$, and differentiation, (2.4.5) yields $(d/dt)f_{0x'^i}(t, x(t), x'(t)) = f_{0x^i}(t, x(t), x'(t))$ a.e. in $[t_1, t_2]$. We have proved Euler's equations (E_i), $i = 1, \ldots, n$. □

Remark 1. Note that the hypothesis that x' be essentially bounded can be removed under the assumptions used in Section 2.3, Remark 3, namely that $x' \in L_p[t_1, t_2]$ for some $p \geq 1$, and that in a neighborhood $\Gamma_\delta \subset A$ of the graph Γ of x we have $|f_0|$, $|f_{0x}|$, $|f_{0x'}| \leq M|x'|^p + m$ for some constants M, $m \geq 0$ and all $(t, x, x') \in \Gamma_\delta \times R^n$. Indeed, as we have seen in Section 2.3, Remark 2, the functions $f_{0x}(t, x(t), x'(t))$, $f_{0x'}(t, x(t), x'(t))$ are L_1-integrable in $[t_1, t_2]$, and we still have $J_1[\eta; x] = 0$ for all AC n-vector functions $\eta(t)$, $t_1 \leq t \leq t_2$, with $\eta(t_1) = \eta(t_2) = 0$ and η' essentially bounded. The proof of the Euler equation above is still valid. Indeed, we still can integrate by parts, and we can still use the lemma (2.4.i) because the first member of (2.4.5) is certainly L_1-integrable in $[t_1, t_2]$.

Two examples are needed here. More examples will be discussed in Chapter 3.

Example 1. $I[x] = \int_{t_1}^{t_2} [a(t)x'^2 + 2b(t)xx' + c(t)x^2]\, dt$, x scalar, with $A = [t_1, t_2] \times R$, $n = 1$, and a, b, c constants or given bounded measurable functions in $[t_1, t_2]$. If $x(t)$, $t_1 \leq t \leq t_2$, is AC with essentially bounded derivative, and if $I[y] \geq I[x]$, say for all $y(t)$, $t_1 \leq t \leq t_2$, AC with essentially bounded derivative and $y(t_1) = x(t_1)$, $y(t_2) = x(t_2)$, then we know from the above that $2^{-1}f_{0x'} = ax' + bx$ is AC (or coincides with such a function a.e. in $[t_1, t_2]$), and the Euler equation $(d/dt)f_{0x'} = f_{0x}$ yields $(ax' + bx)' = bx' + cx$ a.e. in $[t_1, t_2]$. Thus, for the integral $I[x] = \int_{t_1}^{t_2} x'^2\, dt$, the Euler equation is $x'' = 0$. Let us take constants $\delta > 0$, m, $m_0 \geq 0$, such that $\Gamma_\delta \subset A$, $m_0 = \delta + \max x(t)$, and such that a, b, c are in absolute value less than m. If $a(t) \geq \mu > 0$ for some constant $\mu > 0$, then $f_0(t, x, x') = ax'^2 + 2bxx' + cx^2 \geq x'^2 - 2mm_0x' - mm_0^2$, and for some constant C we also have $f_0 \geq (\mu/2)x'^2 - C$ for $(t, x) \in \Gamma_\delta$ and all $x' \in R$. Thus, the mere existence and finiteness of $I[x]$ implies that $x'(t)$ is L_2-integrable. For some constants

$M, m_1 \geq 0$ we certainly have $|f_0|, |f_{0x}|, |f_{0x'}| \leq Mx'^2 + m_1$ for $(t, x) \in \Gamma_\delta$ and all $x' \in R$. By Remark 1, the restrictions above concerning the boundedness of x' and y' can be disregarded.

EXAMPLE 2. Let A^* denote the transpose of any matrix A. Let $R(t) = [r_{ij}]$, $Q(t) = [Q_{ij}]$, $P(t) = [P_{ij}]$ be $n \times n$ matrices with entries measurable and essentially bounded in $[t_1, t_2]$, $R^* = R$, $P^* = P$, and let $I[x]$ denote the quandratic integral

$$\begin{aligned} I[x] &= \int_{t_1}^{t_2} f_0(t, x, x')\, dt \\ &= \int_{t_1}^{t_2} [x'^*(Rx' + Qx) + x^*(Q^*x' + Px)]\, dt \\ &= \int_{t_1}^{t_2} \sum_{i,j=1}^{n} [r_{ij}x'^i x'^j + q_{ij}x'^i x^j + q_{ji}x^i x'^j + p_{ij}x^i x^j]\, dt, \end{aligned}$$

with $r_{ij} = r_{ji}$, $p_{ij} = p_{ji}$. If $x(t) = (x^1, \ldots, x^n)$, $t_1 \leq t \leq t_2$, is AC with derivative x' essentially bounded, and $I[y] \geq I[x]$ for all $y(t) = (y^1, \ldots, y^n)$, $t_1 \leq t \leq t_2$, also AC with essentially bounded derivatives, and $y(t_1) = x(t_1)$, $y(t_2) = x(t_2)$, then we know from the above that $f_{0x'^i}(t, x(t), x'(t))$ is AC in $[t_1, t_2]$ (or coincide with an AC function a.e. in $[t_1, t_2]$), and the Euler equations hold $(d/dt) f_{0x'^i} = f_{0x^i}$, $i = 1, \ldots, n$, that is, since $R = R^*$, $P = P^*$, $T(t) = R(t)x' + Q(t)x$ is AC and

$$\begin{aligned} (d/dt) &\sum_{j=1}^{n} [r_{ij}x'^j + r_{ji}x'^j + q_{ij}x^j + q_{ji}x^j] \\ &= \sum_{j=1}^{n} [q_{ji}x'^j + q_{ji}x'^j + p_{ij}x^j + p_{ji}x^j], \qquad i = 1, \ldots, n, \end{aligned}$$

or in vector form

$$\text{(E)} \qquad (d/dt)(R(t)x' + Q(t)x) - (Q^*(t)x' + P(t)x) = 0, \qquad t \in [t_1, t_2] \text{ (a.e.)}.$$

The n rows of the $n \times 1$ matrix $R(t)x' + Q(t)x$ are AC functions in $[t_1, t_2]$ (or coincide with such functions a.e. in $[t_1, t_2]$).

Whenever $R(t)$, $Q(t)$, $P(t)$ are known to have continuous entries in $[t_1, t_2]$, the following remark is relevant and indeed we shall need it in Section 2.5. If $t_0 \in (t_1, t_2)$ is a point of jump discontinuity for $x'(t)$, then by the Erdman corner condition we derive that

$$R(t_0)x'(t_0 - 0) + Q(t_0)x(t_0) = R(t_0)x'(t_0 + 0) + Q(t_0)x(t_0),$$

and hence $R(t_0)[x'(t_0 + 0) - x'(t_0 - 0)] = 0$. Thus, if $\det R(t_0) \neq 0$, then $x'(t_0 + 0) = x'(t_0 - 0)$. In other words, for R, Q, P continuous and $R(t)$ nonsingular, x' cannot have points of jump discontinuity.

Now let us take constants $\delta > 0$, $m, m_0 > 0$, such that $\Gamma_\delta \subset A$, $m_0 = \delta + \max|x(t)|$, and such that every entry of the matrices R, Q, P is in absolute value less than m. If $R(t)$ is not only nonsingular, but for every t also positive definite, and there is a constant $\mu > 0$ such that $\xi^*R(t)\xi \geq \mu|\xi|^2$ for all $\xi \in R^n$ and all $t \in [t_1, t_2]$, then we also have

$$f_0 = x'^*(R(t)x' + Q(t)x) + x^*(Q^*(t)x' + P(t)x) \geq \mu|x'|^2 - 2n^2mm_0|x'| - n^2mm_0^2,$$

and hence f_0 is also larger than $(\mu/2)|x'|^2 - C$ for some constant C, for $(t, x) \in \Gamma_\delta$, and all $x' \in R^n$. Analogously, for some constants $M, m_1 \geq 0$ we certainly have $|f_0(t, x, x')| \leq M|x'|^2 + m_1$ for the same t, x, x', and an analogous relation holds for f_{0x} and $f_{0x'}$. By Remark 1, the restrictions above concerning the boundedness of x' and y' can be disregarded.

Proof of DuBois-Reymond equation. We have already proved this equation in (2.2) in a particular situation. Here is a general proof.

Let us assume first that $x(t)$, $t_1 \le t \le t_2$, is AC with derivative x' essentially bounded, and gives a strong local minimum for (2.1.1). We want to prove that $f_0 - x'^* f_{0x'}$ coincide a.e. in $[t_1, t_2]$ with an AC function with derivative f_{0t}.

We need consider an arbitrary change of the independent variable t, or $t = t(\tau)$, $\tau_1 \le \tau \le \tau_2$, $t(\tau_1) = t_1$, $t(\tau_2) = t_2$, $t(\tau)$ AC with $|t'(\tau) - 1| \le \frac{1}{2}$. Then the functional $I[x]$ is changed into

$$
\begin{aligned}
I[x] &= \int_{t_1}^{t_2} f_0(t, x(t), x'(t))\,dt \\
&= \int_{\tau_1}^{\tau_2} f_0[t(\tau), x(t(\tau)), (x(t(\tau)))'/t'(\tau)]t'(\tau)\,d\tau.
\end{aligned}
$$

where $(') = d/dt$ in the first integral, and $(') = d/d\tau$ in the second integral. If $\tilde{y}(\tau)$ denotes the $(n+1)$-vector function $\tilde{y}(\tau) = [t(\tau), y(\tau)]$, $\tau_1 \le \tau \le \tau_2$, $t(\tau)$, $y(\tau)$ AC in $[t_1, t_2]$ with derivatives t', y' essentially bounded, then we may well consider the auxiliary integral

$$J[\tilde{y}] = \int_{\tau_1}^{\tau_2} F_0(t, y, t', y')\,d\tau$$

with $F_0(t, y, t', y') = f_0(t, y, y'/t')t'$. For $t(\tau) = \tau$, then J reduces to I. Note that $|t'(\tau) - 1| \le \frac{1}{2}$ guarantees that $t(\tau)$ has an inverse function $\tau = \tau(t)$, $t_1 \le t \le t_2$, $\tau(t)$ AC with $\frac{2}{3} \le \tau'(t) \le 2$, $t \in [t_1, t_2]$, (a.e.), $\tau(t_1) = t_1$, $\tau(t_2) = t_2$, and by the change $\tau = \tau(t)$, $\tilde{y}$ is carried over to $(t, y(t))$. Note that, if we consider a one parameter family of vector functions $\tilde{y}_a(\tau)$ of the form

$$\tilde{y}_a(\tau) = [\tau + a\sigma_0(\tau), y(\tau) + a\sigma(\tau)]$$

where σ, σ_0 are AC in $[t_1, t_2]$, with essentially bounded derivatives, and $\sigma_0(t_1) = \sigma_0(t_2) = 0$, $\sigma(t_1) = \sigma(t_2) = 0$, then we certainly have $|t'(\tau) - 1| = |a\sigma_0'(\tau)| \le \frac{1}{2}$ for all $|a|$ sufficiently small, and then $t = \tau + a\sigma_0(\tau)$ has an inverse $\tau = T(t, a)$, $T(t, a)$ AC in t, $T(t, 0) = t$, $T'(t, a)$ essentially bounded, and $T(t, a) \to t$, $T'(t, a) \to 1$ uniformly as $a \to 0$. Then $\tilde{y}_0(t)$ is transformed into $(t, x(t))$ by $\tau = T(t, 0)$, and $\tilde{y}_a(\tau)$, or $(\tau + a\sigma_0(\tau), x(\tau) + a\sigma(\tau))$, is transformed by $\tau = T(t, a)$ into $(t, X(t, a))$ with

$$X(t, a) = x(T(t, a))) + a\sigma(T(t, a)), \qquad t_1 \le t \le t_2. \tag{2.4.6}$$

Here $X(t, a) \to x(t)$ uniformly in $[t_1, t_2]$ as $a \to 0$ and $X'(t, a)$ is uniformly bounded. Since $x(t)$ is a strong local minimum for $I[x]$, we have $I[X(t, a)] \ge I[x]$ for $|a|$ sufficiently small, and then $J[y_a] \ge J[x]$. By (2.2.i), part (a), the Euler equation holds for J. Since

$$F_{0t'} = \partial F_0/\partial t' = f_0 - (t')^{-1} \sum_i y'^i f_{0x'^i}, \qquad F_{0t} = \partial F_0/\partial t = f_{0t} t',$$

we conclude that $F_{0t'}$, computed along $\tilde{y}_0$, is AC, and that $(d/d\tau)F_{0t'} = F_{0t}$ along $\tilde{y}_0$. Since $d/d\tau = (dt/d\tau)(d/dt) = t'(d/dt)$, we also have $t'(d/dt)[f_0 - (t')^{-1} \sum_i y'^i f_{0x'^i}] = f_{0t} t'$, and finally $(d/dt)[f_0 - \sum_i x'^i f_{0x'^i}] = f_{0t}$. This is the DuBois-Reymond equation (E_0). We have proved (2.2.2) under the assumption made in Section 2.2 that x_0 is a strong local minimum, x AC with x' essentially bounded. □

Remark 2. If we assume that x is continuous with continuous derivative, then $X'(t, a)$ approaches $x'(t)$ as $a \to 0$ uniformly, and the argument above then proves (E_0) also for a weak local minimum x which is continuous with continuous derivative.

Proof of Erdman's corner condition. As mentioned in proving Euler's equations, for every $i = 1, \ldots, n$, the function $f_{0x'^i}(t, x(t), x'(t))$ coincides with an AC function $\phi(t)$ in $[t_1, t_2]$, and by usual convention we shall write $f_{0x'^i}(t, x(t), x'(t)) = \phi(t)$. Thus ϕ is continuous at t_0, or $\phi(t_0 - 0) = \phi(t_0 + 0)$. But f is of class C^1, $f_{0x'^i}$ is continuous in its $2n + 1$ arguments, $x(t)$ is continuous at t_0, and x' has limits $x'(t_0 \mp 0)$ at the point t_0 both finite. Thus

$$f_{0x'^i}(t_0, x(t_0), x'(t_0 - 0)) = \phi(t_0 - 0) = \phi(t_0 + 0) = f_{0x'^i}(t_0, x(t_0), x'(t_0 + 0)).$$

This is the relation (2.2.7). The same argument can be used for (2.2.8). □

Proof of the Legendre condition. Let us assume that $x(t)$, $t_1 \le t \le t_2$, AC with essentially bounded derivative, gives a weak local minimum. Let us write $2\omega(t, \eta, \eta')$ in the form

$$(2.4.7) \qquad 2\omega(t, \eta, \eta') = \eta'^*[R(t)\eta' + Q(t)\eta] + \eta^*[Q^*(t)\eta' + P(t)\eta],$$

where R, Q, P are the $n \times n$ matrices indicated in (2.3.iii). Their entries are all measurable essentially bounded functions in $[t_1, t_2]$. Thus, for almost all $t_0 \in (t_1, t_2)$ these matrices (that is, all their entries) are the derivatives of their indefinite Lebesgue integrals. Let t_0 be any such point. Let $\zeta(t)$, $-\infty < t < +\infty$, denote the scalar function defined by $\zeta(t) = 1 - |t|$ for $|t| \le 1$, $\zeta(t) \equiv 0$ for $|t| \ge 1$; let $\xi \in R^n$ be any fixed n-vector; let $0 < \varepsilon \le \min[t_0 - t_1, t_2 - t_0]$, and take $\eta_\varepsilon(t) = \varepsilon\xi\zeta(\varepsilon^{-1}(t - t_0))$, $t_1 \le t \le t_2$. Then, for $|\varepsilon|$ sufficiently small $\eta_\varepsilon \in v^0$, by (2.3.iii) $J_2[\eta_\varepsilon] \ge 0$, and hence

$$(2.4.8) \qquad \lim_{\varepsilon \to 0+} (2\varepsilon)^{-1} J_2[\eta_\varepsilon] = \lim_{\varepsilon \to 0+} (2\varepsilon)^{-1} \int_{t_1}^{t_2} 2\omega(t, \eta_\varepsilon, \eta'_\varepsilon)\, dt \ge 0$$

if this limit exists, where 2ω is given by (2.4.7). Let us show that this limit is $\xi^* R(t_0)\xi$. Let $P = [p_{ij}]$, $Q = [q_{ij}]$, $R = [r_{ij}]$, and let M denote any essential bound for the entries p_{ij}, q_{ij}, r_{ij} in $[t_1, t_2]$. In (2.4.8), $2\omega = 0$ for $|t - t_0| > \varepsilon$. With $c = |\xi|$, then $|\eta_\varepsilon(t)| \le \varepsilon|\xi| = \varepsilon c$ for $t_0 - \varepsilon \le t \le t_0 + \varepsilon$. For $t_0 < t < t_0 + \varepsilon$ we also have $\eta'_\varepsilon(t) = -\xi$; for $t_0 - \varepsilon < t < t_0$ we have $\eta'(t) = \xi$. Hence, for $\xi = (\xi_1, \dots, \xi_n)$ we have $\eta'^*_\varepsilon R \eta'_\varepsilon = \sum_{ij} \xi_i r_{ij}(t)\xi_j$ for both $t_0 - \varepsilon < t < t_0$ and for $t_0 < t < t_0 + \varepsilon$. Since $(2\varepsilon)^{-1} \int_{t_0-\varepsilon}^{t_0+\varepsilon} r_{ij}(t)\, dt \to r_{ij}(t_0)$ as $\varepsilon \to 0+$, we see that the contribution of $R(t)$ in the limit (2.4.8) is $\sum_{ij} \xi_i r_{ij}(t_0)\xi_j = \xi^* R(t_0)\xi$. On the other hand, $|\eta^*_\varepsilon P \eta_\varepsilon| \le n^2\varepsilon^2 c^2 M$, and an analogous estimate holds for $\eta'_\varepsilon Q \eta_\varepsilon$, $\eta^*_\varepsilon Q^* \eta'_\varepsilon$, so that the corresponding terms in the limit (2.4.8) approach zero as $\varepsilon \to 0+$. Thus, the limit (2.4.8) is $\xi^* R(t_0)\xi$. We have proved $\xi^* R(t_0)\xi \ge 0$ for all n-vectors ξ, where $R(t_0)$ is the $n \times n$ matrix $f_{0x'x'}(t_0, x(t_0), x'(t_0))$. This proves Legendre condition (e) of (2.2.i).

Proof of the canonical equations (2.2.3). The first n equations (2.2.3) are identities, since $\partial H/\partial \lambda_i = x'^i = dx^i/dt$; the next n equations are a consequence of the Euler equations and the properties of the AC functions $\lambda_i(t)$ defined and discussed under (a) in Theorem (2.2.i), namely

$$-\partial H/\partial x^i = -f_{0x^i} = -(d/dt) f_{0x'^i} = d\lambda_i/dt, \qquad i = 1, \dots, n. \qquad \square$$

Derivation of (f) *from* (d) *in Theorem* (2.2.i). Indeed, if x is a local strong minimum, then from the definitions (2.2.1), (2.2.2) of the functions H, E, we have

$$\begin{aligned}
&H(t, x(t), u, \lambda(t)) - H(t, x(t), x'(t), \lambda(t)) \\
&\quad = f_0(t, x(t), u) + \sum_{i=1}^{n} \lambda_i(t) u^i - f_0(t, x(t), x'(t)) - \sum_{i=1}^{n} \lambda_i(t) x'^i(t) \\
&\quad = f_0(t, x(t), u) - f_0(t, x(t), x'(t)) - \sum_{i=1}^{n} (u^i - x'^i(t)) f_{0x'^i}(t, x(t), x'(t)) \\
&\quad = E(t, x(t), x'(t), u) \ge 0
\end{aligned}$$

for all $u \in R^n$ [≤ 0 for a local strong maximum]. □

Remark 3. We have proved above statements (a), (b), (c), (e), (g) of Theorem (2.2.i) as stated for strong local minima and maxima. We have proved (a), (e), (g) even for weak local minima and maxima. Statement (h) is an obvious consequence of (g). We have proved these statements under the assumption that $x'(t)$ is essentially bounded. We have also proved (a) for x' unbounded but in L_p under suitable assumptions on f_0. We shall prove (a) and (b) again for x' unbounded under different assumptions on f_0 in Section 2.9.

Statements (d) and (f) hold for local strong minima and maxima as stated in (2.2.i). Actually, (f) is a consequence of (d), and we shall give an elementary proof of (d) in Section 2.11. However, statement (d), or the Weierstrass necessary condition, is a corollary of a statement of functional analysis we shall prove in Section 2.19: the convexity of $f_0(t, x(t), u)$ with respect to u at $u = x'(t)$ for almost all $t \in [t_1, t_2]$ is a necessary condition for the lower semicontinuity of $I[x]$ at any given trajectory x with respect to the uniform convergence. ($I[x]$ is certainly lower semicontinuous at any local strong minimum.)

Statement (i) will be proved in Section 2.8.

B. Stationary Solutions and Invariant Character of the Euler Equations

Let $x(t)$, $t_1 \le t \le t_2$, be as usual an AC trajectory with graph Γ in the interior of A and x' essentially bounded. From the proof of the Euler equations above it is clear that they are equivalent to the statement that $\psi'(a) = 0$, or equivalently that the first variation $J_1[x;\eta]$ is zero for all AC functions $\eta(t)$, $t_1 \le t \le t_2$, with η' essentially bounded and $\eta(t_1) = \eta(t_2) = 0$. In the terms of Remark 1 of Section 2.2 we may also say that x is stationary if and only if $J_1[x;\eta] = 0$ for all η just stated.

Euler equations (E_i) have an invariant character which we present here briefly. Let us consider the usual integral

$$I[x] = \int_{t_1}^{t_2} f_0(t, x(t), x'(t))\, dt, \qquad (\,') = d/dt,$$

and let us assume that we perform a change of variables

$$t = t(s, y), \qquad x = x(s, y),$$
$$t, s \in R, \qquad x = (x^1, \dots, x^n) \in R^n, \qquad y = (y^1, \dots, y^n) \in R^n,$$

so that trajectories $x(t)$, $t_1 \le t \le t_2$, are mapped into trajectories $y(s)$, $s_1 \le s \le s_2$, and the integral $I[x]$ is transformed into the integral

$$H[y] = \int_{s_1}^{s_2} F_0(s, y, y')\, ds, \qquad (\,') = d/ds,$$

where now

$$F_0(s, y, y') = f_0(t(s, y), x(s, y), (x_t + x_y y')/(t_s + t_y y'))(t_s + t_y y').$$

It is relevant that the Euler equations for I,

$$(d/dt) f_{0x'^i} - f_{0x^i} = 0, \qquad i = 1, \dots, n,$$

are transformed into the Euler equations for H,

$$(d/ds)F_{0y'^i} - F_{0y^i} = 0, \qquad i = 1, \dots, n.$$

Let us explain this in terms of first variations. We shall understand here that the change of variables $(t, x) \to (s, y)$ is one-one with inverse map $s = s(t, x)$, $y = y(t, x)$, and that both direct and inverse are of class C^1 at least in a neighborhood Γ_δ of Γ. Moreover we assume that $s = s(t, x(t))$ is strictly increasing in $[t_1, t_2]$ with $s'(t) = s_t + s_x x' \geq \mu > 0$ for some constant $\mu > 0$. Thus, if $\eta(t)$, $t_1 \leq t \leq t_2$, is also AC with essentially bounded derivative and $\eta(t_1) = \eta(t_2) = 0$, then $x(t) + a\eta(t)$ has graph in Γ_δ for all $|a|$ sufficiently small, and for $S(t) = s(t, x(t) + a\eta(t))$ we also have $S'(t) = s_t + s_x(x' + a\eta') \geq \mu/2$. Then $S(t)$ is strictly increasing along $x + a\eta$, and as t describes $[t_1, t_2]$, then both $s(t)$ and $S(t)$ describe $[s_1, s_2]$ with inverses $t = t(s)$ on x, and $t = T(a, s)$ on $x + a\eta$. We can now say that the trajectory x is transformed into the trajectory $y(s) = [y(t, x(t))]_{t=t(s)}$, and that the trajectory $x + a\eta$ is transformed into the trajectory

$$y(a, s) = [y(t, x(t) + a\eta(t))]_{y=T(a,s)} = y(s) + Y(a, s), \qquad s_1 \leq s \leq s_2.$$

By limiting ourselves to trajectories x and functions η all of class C^1, for the sake of simplicity, we see that $Y(0, s) = 0$, $Y_s(0, s) = 0$, and $a^{-1}Y(a, s) \to \sigma(s)$, $a^{-1}Y_s(a, s) \to \sigma'(s)$ as $a \to 0$, where $\sigma(s)$, $s_1 \leq s \leq s_2$, depends on η. Here we have $I[x] = H[y]$, $I[x(t) + a\eta(t)] = H[y(s) + Y(a, s)]$, and by Remark 1 of Section 2.3 we also have

$$\psi'(0) = \lim_{a\to 0} a^{-1}(I[x + a\eta] - I[x]) = \lim_{a \to 0} a^{-1}(H[y + Y] - H[y]) = 0,$$

and by our initial remark also

$$\psi'(0) = J_1[x; \eta] = K_1[y; \sigma] = 0,$$

where J_1 is the first variation of I, and K_1 the first variation of H. Analogously, if we start from $H[y + a\sigma]$ for some arbitrary σ, then we obtain a function $X(a, t)$ and corresponding η, for which the relation above holds for J_1 and K_1. The Euler equations for I and H then hold correspondingly.

To see an application of the above remarks, let us consider the case where $n = 1$, (r, φ) are polar coordinates in R^2, $s = r \cos \varphi$, $y = r \sin \varphi$, and

$$I[r] = \int_{\varphi_1}^{\varphi_2} (r^2 + r'^2)^{1/2}\, d\varphi, \qquad (') = d/d\varphi,$$

with $r = r(\varphi)$. Then I is transformed into

$$H[y] = \int_{s_1}^{s_2} (1 + y'^2)^{1/2}\, ds, \qquad (') = d/ds,$$

with $y = y(s)$. The Euler equation for $I[r]$ is

$$r(r^2 + r'^2)^{-1/2} - (d/d\varphi)(r'(r^2 + r'^2)^{-1/2}) = 0,$$

while the Euler equation for $H[y]$ is $y'' = 0$. The extremals $y = as + b$ for H correspond to the extremals $r \sin \varphi = ar \cos \varphi + b$ for I.

C. The Legendre Transformation and a New Version of the Canonical Equations

Let us assume that $f_0(t, x, x')$ is of class C^2, and that det $f_{0x'x'} \neq 0$ in a region of the txx' space. Then, locally, the relations $\lambda_i = -f_{0x'^i}(t, x, x')$, $i = 1, \ldots, n$, (or briefly $\lambda = -f_{0x'}$) can be inverted yielding $x'^i = p_i(t, x, \lambda)$, $i = 1, \ldots, n$, (or briefly $x' = p(t, x, \lambda)$). Then the Hamiltonian

$$H(t, x, x', \lambda) = f_0(t, x, x') + \sum_{i=1}^{n} \lambda_i x'^i$$

can be written as a function of t, x, λ, say,

$$H_0(t, x, \lambda) = f_0(t, x, p(t, x, \lambda)) + \sum_{i=1}^{n} \lambda_i p_i(t, x, \lambda).$$

The transformation $x' = p(t, x, \lambda)$ is called the Legendre transformation. Note that because of $\lambda_i = -f_{0x'^i}$ we have

$$\begin{aligned} dH_0 &= f_{0t}\, dt + \sum_i f_{0x^i}\, dx^i + \sum_i f_{0x'^i}\, dx'^i + \sum_i x'^i\, d\lambda_i + \sum_i \lambda_i\, dx'^i \\ &= f_{0t}\, dt + \sum_i f_{0x^i}\, dx^i + \sum_i x'^i\, d\lambda_i, \end{aligned}$$

and hence

$$H_{0t} = f_{0t}, \qquad H_{0x^i} = f_{0x^i} \qquad H_{0\lambda_i} = x'^i, \qquad i = 1, \ldots, n.$$

From the Euler equations (E_i), or $d\lambda_i/dt = -(d/dt)f_{0x'^i} = f_{0x^i}$, we derive now

$$\frac{dx^i}{dt} = \frac{\partial H_0}{\partial \lambda_i}, \qquad \frac{d\lambda_i}{dt} = -\frac{\partial H_0}{\partial x^i}, \qquad i = 1, \ldots, n,$$

These are again the canonical equations (2.2.3), where now the Hamiltonian $H_0(t, x, \lambda)$ is a function of t, x, λ.

For instance, for $n = 1$, $f_0(t, x, x') = (1 + x'^2)^{1/2}$, we have

$$\lambda = -x'(1 + x'^2)^{-1/2},$$

hence

$$x' = -\lambda(1 - \lambda^2)^{-1/2},$$

and from $H(t, x, x', \lambda) = (1 + x'^2)^{1/2} + \lambda x'$, we derive

$$H_0(t, x, \lambda) = (1 - \lambda^2)^{1/2}.$$

The canonical equations are now

$$dx/dt = \partial H_0/\partial \lambda = -\lambda(1 - \lambda^2)^{-1/2}, \qquad d\lambda/dt = \partial H_0/\partial x = 0,$$

from which we derive as expected

$$\lambda = c, \qquad dx/dt = m, \qquad m = -c(1 - c^2)^{-1/2} \text{ constants}, \qquad |c| < 1.$$

It is easy to see that the Legendre transformation is an involution. Indeed, first from $\lambda_i = -f_{0x'^i}(t, x, x')$ we see that $D = \det[f_{0x'^i x'^j}]$ is the functional determinant $D =$

$\partial(\lambda_1, \ldots, \lambda_n)/\partial(x'_1, \ldots, x'_n)$. If $D \neq 0$, then from $H_{\lambda_i} = x'^i = p_i(t, x, \lambda)$ we see that $D^{-1} = \partial(x'_1, \ldots, x'_n)/\partial(\lambda_1, \ldots, \lambda_n) \neq 0$, and moreover

$$H_0(t, x, \lambda) - \sum_i \lambda_i H_{0\lambda_i}(t, x, \lambda) = f_0(t, x, x') + \sum_i \lambda_i x'^i - \sum_i \lambda_i H_{0\lambda_i} = f_0(t, x, x'),$$

which proves the statement.

We note here that, if $f_0(x, x')$ is independent of t, then by relations (2.4.9) and (2.4.10), and along any extremal $x(t)$, we have

$$\begin{aligned} dH_0/dt = dH_0(x(t), \lambda(t)) &= \sum_i f_{0x^i}(dx^i/dt) + \sum_i x'^i(d\lambda_i/dt) \\ &= \sum_i (\partial H/\partial x^i)(\partial H/\partial \lambda_i) - (\partial H/\partial \lambda_i)(\partial H/\partial x^i) = 0, \end{aligned}$$

and then $H_0(x(t), \lambda(t)) = C_0$, which is a different form of relation (2.2.11). In other words, if $f_0(x, x')$ does not depend on t, then $H_0(x, \lambda) = C_0$ is a first integral of the canonical equations.

More generally, let us consider any function $F(x, \lambda)$ of class C^1 and let us compute F along any extremal $x(t)$ with $\lambda(t) = -f_{0x'}(t, x(t), x'(t))$. Then

$$\begin{aligned} (d/dt)F(x(t), \lambda(t)) &= \sum_i (F_{x^i} x'^i + F_{\lambda_i} \lambda'_i) \\ &= \sum_i (\partial F/\partial x^i)(\partial H_0/\partial \lambda_i) - (\partial F/\partial \lambda_i)(\partial H_0/\partial x^i) = [F, H_0], \end{aligned}$$

where the last expression, usually denoted $[F, H_0]$, is often called the Poisson bracket. Thus, F is a first integral of the canonical equations if and only if $[F, H_0] = 0$ along any extremal. Since for every point (x, λ) there certainly passes one and only one extremal $x(t)$ with corresponding $\lambda(t) = -f_{0x'}$, we conclude that $F(x, \lambda)$ is a first integral of the canonical equations if and only if $[F, H_0] = 0$.

Now, together with the integral

$$I[x] = \int_{t_1}^{t_2} f_0(t, x, x')\, dt$$

we consider the integral

$$J[x, \lambda] = \int_{t_1}^{t_2} \left[H_0(t, x, \lambda) - \sum_i \lambda_i x_i'^i \right] dt,$$

where x and λ are thought of as independent variables, and where the integrand function does not depend on the derivatives λ'_i. The Euler equations of J are now

$$\frac{d\lambda_i}{dt} = \frac{\partial H_0}{\partial x^i}, \qquad \frac{dx^i}{dt} = \frac{\partial H_0}{\partial \lambda_i}, \qquad i = 1, \ldots, n,$$

and these are again the canonical equations, which appear now as the Euler equations of the integral J with x and λ as independent variables and $H_0 = H_0(t, x, \lambda)$. We have already shown that these equations are equivalent to the Euler equations for I.

Example. For $n = 1$ and $I[x] = \int_{t_1}^{t_2} (P(x)x'^2 + Q(x)x^2)\, dt$, we have $f_0 = Px'^2 + Qx^2$, and the Euler equation is $2Px'' = -Px'^2 + Q'x^2 + 2Qx$. Also $\lambda = -2Px'$, and $H = Px'^2 + Qx^2 + \lambda x'$. For $P \neq 0$, then $x' = -2^{-1}P^{-1}\lambda$, and

$$H_0(x, \lambda) = Qx^2 - 4^{-1}P^{-1}\lambda^2.$$

The canonical equations in terms of $H(x, x', \lambda)$ are $dx/dt = x'$, $d\lambda/dt = -P'x'^2 - Q'x^2 - 2Qx$. In terms of $H_0(x, \lambda)$ they are instead

$$dx/dt = -2^{-1}P^{-1}\lambda, \qquad d\lambda/dt = -Q'x^2 - 2Qx - 4^{-1}P^{-2}P'\lambda^2.$$

Of course, for $\lambda = -2Px'$ we have again the previous relations.

D. The Canonical Transformations

We consider now transformations of variables of the general form

(2.4.11) $$X^i = X^i(t, x, \lambda), \qquad P_i = P_i(t, x, \lambda), \qquad i = 1, \ldots, n.$$

Transformation of this form for which the canonical equations

(2.4.12) $$\frac{dx^i}{dt} = \frac{\partial H_0}{\partial \lambda_i}, \qquad \frac{d\lambda_i}{dt} = -\frac{\partial H_0}{\partial x^i}, \qquad i = 1, \ldots, n,$$

are invariant, are said to be *canonical transformations*. Thus, if $H_0^*(t, X, P)$ is the new Hamiltonian in the variables t, X, P, then we want (2.14.12) to be transformed into

(2.4.13) $$\frac{dX^i}{dt} = \frac{\partial H_0^*}{\partial P_i}, \qquad \frac{dP_i}{dt} = -\frac{\partial H_0^*}{\partial X^i}, \qquad i = 1, \ldots, n.$$

If we think of the canonical equations (2.4.12) as the Euler equations of the integral $J[x, \lambda]$, then the equations (2.4.13) must be the Euler equations of the integral

$$J^*[X, P] = \int_{t_1}^{t_2} \left(H^*(t, X, P) + \sum_i P_i X'^i \right) dt.$$

In Remark 7 at the end of Section 2.2 we noticed that two integrals certainly have the same Euler equations if they differ by an exact differential. Thus if the transformation (2.4.11) is such that

(2.4.14) $$H_0\, dt + \sum_i \lambda_i\, dx^i = H_0^*\, dt + \sum_i P_i\, dX^i + d\Phi(t, x, \lambda),$$

for some function $\Phi(t, x, \lambda)$ of class C^1, then the transformation is certainly canonical. In this situation Φ is said to be the generating function of the canonical transformation. Actually, if we give any function Φ as stated, then a corresponding canonical transformation can be defined of which Φ is the generating function. Indeed,

$$d\Phi = (H_0 - H_0^*)\, dt + \sum_i \lambda_i\, dx^i - \sum_i P_i\, dX^i,$$

and then

(2.4.15) $$\partial\Phi/\partial t = H - H^*, \qquad \partial\Phi/\partial x^i = \lambda_i, \qquad \partial\Phi/\partial X^i = -P_i, \qquad i = 1, \ldots, n.$$

These are canonical transformations. Indeed, these $2n + 1$ relations establish the connection between the old variables x^i, λ_i, $i = 1, \ldots, n$, and the new variables X^i, P_i, $i = 1, \ldots, n$, and they further give the expression of the new Hamiltonian $H^* = H - \partial\Phi/\partial t$. Here Φ is thought of as a function of t, x, X.

If we want to express Φ in terms of t, x, P, then we rewrite (2.4.14) in the form

$$\begin{aligned} d\left(\Phi + \sum_i P_i X^i\right) &= d\Phi + \sum_i X^i \, dP_i + \sum_i P_i \, dX^i \\ &= (H_0 - H_0^*)\, dt + \sum_i \lambda_i \, dx^i - \sum_i P_i \, dX^i + \sum_i X^i \, dP_i + \sum_i P_i \, dX^i \\ &= (H_0 - H_0^*)\, dt + \sum_i \lambda_i \, dx^i + \sum_i X^i \, dP_i. \end{aligned}$$

Now $\Psi = \Phi + \sum_i P_i X^i$ is the new generating function, and we have

$$\partial \Psi / \partial t = H_0 - H_0^*, \qquad \partial \Psi / \partial x^i = \lambda_i, \qquad \partial \Psi / \partial P_i = X^i, \qquad i = 1, \ldots, n.$$

2.5 Jacobi's Necessary Condition

Let us assume that f_0 is of class C^2 and that $x(t)$, $t_1 \le t \le t_2$, is AC with essentially bounded derivative, and gives a weak local minimum to the functional (2.1.1). Then we know from (2.3.iii) that

$$(2.5.1) \quad J_2[\eta] = \int_{t_1}^{t_2} [\eta'^*(R(t)\eta' + Q(t)\eta) + \eta^*(Q^*(t)\eta' + P(t)\eta)]\, dt \ge 0$$

where $R(t) = f_{0x'x'}(t) = f_{0x'x'}(t, x(t), x'(t))$ and analogously $Q(t) = f_{0x'x}$, $P(t) = f_{0xx}$. Here R, Q, P are $n \times n$ matrices with measurable essentially bounded entries. Then $J_2[0] = 0$, that is, $\eta(t) \equiv 0$ is an absolute minimum for J_2. Consequently (from (2.2.i)(a)) $\eta(t) \equiv 0$ is a solution of the Euler system for the integral J_2, or

$$(2.5.2) \qquad (d/dt)\omega_{\eta'^i} = \omega_{\eta^i}, \qquad i = 1, \ldots, n,$$

or by explicit computations

$$(2.5.3) \quad (d/dt)[R(t)u' + Q(t)u] = Q^*(t)u' + P(t)u, \qquad t \in [t_1, t_2] \text{ (a.e.)}.$$

This of course is trivial, since this system is linear. However, system (2.5.2) is important, as we shall see, and is called the *Jacobi* (or *accessory*) *differential system* for $I[x]$.

By a solution $u(t)$, $t_1 \le t \le t_2$, of the Jacobi system (2.5.2), we understand any AC vector function $u(t) = (u_1, \ldots, u_n)$, $t_1 \le t \le t_2$, such that for some other vector function $v(t) = (v_1, \ldots, v_n)$ also AC in $[t_1, t_2]$, we have

$$(2.5.4) \qquad \begin{aligned} v(t) &= R(t)u' + Q(t)u, \\ v'(t) &= Q^*(t)u' + P(t)u, \qquad t \in [t_1, t_2] \text{ (a.e.)} \end{aligned}$$

Note that 2ω is a quadratic form in η, η'; hence, by Euler's theorem on homogeneous functions, $2\omega = \sum_i (\eta^i \omega_{\eta^i} + \eta'^i \omega_{\eta'^i})$. This implies that for any solution $u(t)$, $t_1 \le t \le t_2$, of the accessory equations we have from (2.5.2)

$$(2.5.5) \qquad J_2[u] = \int_{t_1}^{t_2} 2\omega\, dt = \int_{t_1}^{t_2} \sum_i (u^i \omega_{\eta^i} + u'^i \omega_{\eta'^i})\, dt$$

$$= \int_{t_1}^{t_2} \sum_i [u^i(d\omega_{\eta'^i}/dt) + u'^i \omega_{\eta'^i}]\, dt = \left[\sum_i u^i \omega_{\eta'^i}\right]_{t_1}^{t_2},$$

where we have used the Euler equations to pass from the third to the fourth member. Note that whenever $\det R(t) \neq 0$, $t \in [t_1, t_2]$ (a.e.), then system (2.5.4) yields

$$u'(t) = -R^{-1}Qu + R^{-1}v,$$
$$v' = Q^*[-R^{-1}Qu + R^{-1}v] + Pu,$$

and finally, since $(Q^*R^{-1})^* = R^{-1}Q$, we have

$$(2.5.6) \qquad \begin{aligned} u'(t) &= A(t)u + B(t)v, \\ v'(t) &= C(t)u - A^*(t)v, \end{aligned}$$

where

$$A(t) = -R^{-1}Q, \qquad B(t) = R^{-1}, \qquad C(t) = P - Q^*R^{-1}Q,$$
$$u(t) = (u_1, \ldots, u_n), \qquad v(t) = (v_1, \ldots, v_n).$$

Here u and v are said to be the *canonical* variables, and (2.5.6) is the Jacobi system written in terms of such variables when $\det R(t) \neq 0$ in $[t_1, t_2]$ (a.e.).

As in Remark 1 of Section 2.2, we say that any C^1 solution x of the Euler equations is an extremal. We say that the extremal arc $E_{12}: x = x(t)$, $t_1 \le t \le t_2$, is *nonsingular* if $\det R(t) \neq 0$ for all $t_1 \le t \le t_2$, $R(t) = f_{0x'x'}(t, x(t), x'(t))$. Let $1 = (t_1, x(t_1))$, $2 = (t_2, x(t_2))$ denote the end points of E_{12}. We say that $\bar{t}$, $t_1 \le \bar{t} \le t_2$, is *conjugate* to t_1, or that the point $3 = (\bar{t}, x(\bar{t}))$ is conjugate to $1 = (t_1, x(t_1))$ on E_{12}, if there is a solution $u(t) = (u^1, \ldots, u^n)$, $t_1 \le t \le \bar{t}$, of the accessory system (2.5.2) or (2.5.3) with $u(t_1) = u(\bar{t}) = 0$ and not identically zero in $[t_1, \bar{t}]$,

2.5.i (Jacobi's Necessary Condition). *If f_0 is of class C^2, and if the extremal arc $E_{12}: x = x(t)$, $t_1 \le t \le t_2$, is nonsingular and a weak minimum for $I[x]$, then there is no point $\bar{t} \in (t_1, t_2)$ conjugate to t_1, that is, there is no point $3 = (\bar{t}, x(\bar{t}))$, $t_1 < \bar{t} < t_2$, conjugate to 1 on E_{12} between 1 and 2.*

Proof. Since f is of class C^2 and $x(t)$, $x'(t)$ are continuous by hypothesis, the $n \times n$ matrix $R(t) = f_{0x'x'}(t, x(t), x'(t))$ has continuous entries, and then $\det R(t)$ is continuous in $[t_1, t_2]$. Thus, either $\det R(t) \ge \mu > 0$, or $\det R(t) \le -\mu < 0$ for some constant $\mu > 0$ and $t_1 \le t \le t_2$. Hence, $B = R^{-1}(t)$ also has continuous entries, and the same holds for the matrices $Q = f_{0x'x}$, $P = f_{0xx}$, $A = -R^{-1}Q$, $C = P - Q^*R^{-1}Q$. Thus, system (2.5.6) satisfies the conditions of the existence and uniqueness theorems. Assume, if possible, that (2.5.6) has a non-identically-zero solution $u(t)$, $v(t)$, $t_1 \le t \le t_2$, with $u(t_1) = u(\bar{t}) = 0$ for some $t_1 < \bar{t} < t_2$. Then we must have $u'(\bar{t}) \neq 0$, since $u'(\bar{t}) = 0$ would imply $v(\bar{t}) = 0, v'(\bar{t}) = 0$, and then $u(t), v(t)$ would be identically zero in $[t_1, \bar{t}]$. Now, let us consider the function $\bar{u}(t) = u(t)$ for $t_1 \leq t \leq \bar{t}$, $\bar{u}(t) = 0$ for $\bar{t} \leq t \leq t_2$, which is still a continuous solution of (2.5.6), and therefore of (2.5.2) and (2.5.3) with sectionally continuous derivative, and thus a corner point at $\bar{t}$. From (2.5.5) we derive that $J_2[\bar{u}] = 0$, and thus $\bar{u}$ is an absolute minimum for $J_2[\eta]$. Since $\omega_{\eta'} = R(t)u' + Q(t)u$ and $u(\bar{t}) = 0$, by the Erdman corner condition we have $R(\bar{t})u'(\bar{t} + 0) = R(\bar{t})u'(\bar{t} - 0)$ with $u'(\bar{t} - 0) \neq 0 = u'(\bar{t} - 0)$. Here $\det R(\bar{t}) \neq 0$; thus $R(\bar{t})u$ cannot take the same value for distinct n-vectors u. We have reached a contradiction. This proves (2.5.i). □

For $n = 1$ the accessory system reduces to the equation

$$(d/dt)[f_{0x'x'}\eta' + f_{0xx'}\eta] = f_{0xx'}\eta' + f_{0xx}\eta,$$

where $f_{0x'x'} = f_{0x'x'}(t, x(t), x'(t))$ and analogously for $f_{0xx'}$ and f_{0xx}.

For instance, for $n = 1$, $I[x] = \int_{t_1}^{t_2} (x'^2 - x^2)\, dt$, $f_0 = x'^2 - x^2$, $f_{0x'x'} = 2$, $f_{0x'x} = 0, f_{0xx} = -2$, and the accessory equation is $\eta'' + \eta = 0$, with solutions $C \sin(t - \gamma)$, C, γ arbitrary. Thus, for any t_1, the point $t_1 + \pi$ is conjugate to t_1 (see Section 3.8 for more details on this example).

The following further remarks are relevant. First, let us assume that the extremal arc $E_{12}: x = x(t) = (x^1, \ldots, x^n)$, $t_1 \leq t \leq t_2$, is imbedded in a one parameter family $x(t, \alpha)$, $\alpha' < \alpha < \alpha''$, that is, $x(t) = x(t, \alpha_0)$ for some $\alpha_0 \in (\alpha', \alpha'')$.

2.5.ii. *If the extremal arc $E_{12}: x = x(t)$, $t_1 \leq t \leq t_2$, is imbedded in a one parameter family of extremals $x(t, \alpha)$, with x, $x' = \partial x/\partial t$, $x_\alpha = \partial x/\partial \alpha$, x'_α of class C^1, then $u(t) = x_\alpha(t, \alpha_0) = (u^1, \ldots, u^n)$ is a solution of the accessory system relative to E_{12}.*

Proof. In vector notation,

$$(d/dt)f_{0x'}(t, x(t, \alpha), x'(t, \alpha)) = f_{0x}(t, x(t, \alpha), x'(t, \alpha)),$$

and by differentiation with respect to α also,

$$(d/dt)[f_{0x'x}x_\alpha + f_{0x'x'}x'_\alpha] = f_{0xx}x_\alpha + f_{0xx'}x'_\alpha.$$

Taking $\alpha = \alpha_0$ and using the same notation used above, we have

$$(d/dt)[R(t)u' + Q(t)u] = Q^*(t)u' + P(t)u.$$

If $\det R(t) \neq 0$ in $[t_1, t_2]$, and $v(t) = R(t)u' + Q(t)u$, then relations (2.5.6) also hold. □

Now, let us assume that the nonsingular extremal arc $E_{12}: x = x(t)$, $t_1 \leq t \leq t_2$, is imbedded in a $2n$ parameter family $x(t, c)$, $c = (c_1, \ldots, c_{2n})$, of extremals $x(t) = x(t, c_0)$, $c_0 = (c_{01}, \ldots, c_{0,2n})$, and that x, x', x_{c_s}, x'_{c_s}, $s = 1, \ldots, 2n$, are all of class C^1. Then $u_s(t) = x_{c_s}(t, c_0)$, $s = 1, \ldots, 2n$, are $2n$ solutions of the accessory equation (2.5.3) relative to E_{12}.

In (2.11) and (2.12) we will discuss the question of imbedding an extremal arc E_{12} in a family of extremals.

Here we simply consider suitable systems of solutions of the accessory equation (2.5.3).

Now let $E_{12}: x = x(t)$, $t_1 \le t \le t_2$, be a nonsingular extremal. Let $u_s(t)$, $t_1 \le t \le t_2$, be $2n$ solutions of the accessory equations (2.5.3), and let us consider the $2n \times 2n$ determinant

$$d(t) = \det[(u_s(t), u_s'(t)),\ s = 1, \ldots, 2n].$$

Precisely as for our general convention, we think as usual of the $2n$-vector $(u_s(t), u_s'(t))$ as a column $2n$-vector made up first of the n components of u_s and then of the n components of u_s'. Thus, the sth column of $d(t)$ is made up first of the n components of u_s and then of the n components of u_s'. Let us prove that $d(t)$ is either identically zero in $[t_1, t_2]$, or always different from zero. Indeed, if $d(t_0) = 0$ at some $t_0 \in [t_1, t_2]$, then there is a system of $2n$ constants $c = (c_1, \ldots, c_{2n})$ not all zero, such that for

$$u(t) = \sum_{s=1}^{2n} c_s u_s(t), \qquad u'(t) = \sum_{s=1}^{2n} c_s u_s'(t),$$

we have $u(t_0) = 0$, $u'(t_0) = 0$. But $u(t)$ is also a solution of the linear accessory equation (2.5.3), and by the uniqueness theorem for ordinary differential equations we have $u(t) = 0$, $u'(t) = 0$ for all $t \in [t_1, t_2]$. Then the $2n$ columns of $d(t)$ are linearly dependent in $[t_1, t_2]$, and $d(t) = 0$ for all $t_1 \le t \le t_2$. If $d(t)$ is not zero, then the $2n$ solutions $u_s(t)$, $s = 1, \ldots, 2n$, are linearly independent in $[t_1, t_2]$.

Let us now consider the following $2n \times 2n$ determinant:

$$D(t, t_1) = \det[(u_s(t), u_s(t_1)),\ s = 1, \ldots, 2n].$$

Precisely as above, we think of the $2n$-vector $(u_s(t),\ u_s(t_1))$ as a column $2n$-vector made up first of the n components of $u_s(t)$ and then of the n components of $u_s(t_1)$. Thus, the sth column of $D(t, t_1)$ is made up first of the n components of $u_s(t)$, and then of the n components of $u_s(t_1)$.

2.5.iii. *If $E_{12}: x = x(t)$, $t_1 \le t \le t_2$, is a nonsingular extremal arc, and $D(t, t_1)$ is not identically zero, then $\bar{t} \in (t_1, t_2)$ is conjugate to t_1 if and only if $D(\bar{t}, t_1) = 0$.*

Proof. Let $\bar{t}$ be conjugate to t_1. The solutions u_s are certainly linearly independent in $[t_1, t_2]$ and form a fundamental system, since otherwise D would be identically zero. Now a particular solution u characterizing the conjugate point $\bar{t}$, that is, $u(t_1) = u(\bar{t}) = 0$, u not identically zero in $[t_1, t]$, must be a linear combination $u = \sum_s c_s u_s$ of the solutions above with coefficients c_s not all zero. Since $u(t_1) = u(\bar{t}) = 0$, then $D(\bar{t}, t_1) = 0$.

Conversely, suppose $D(\bar{t}, t_1) = 0$ without D being identically zero. Then the constants c_s can be determined so that $u(t_1) = u(\bar{t}) = 0$ and u is not identically zero in $(t_1, \bar{t})$; otherwise the same u_s would be linearly dependent. Thus, $\bar{t}$ is conjugate to t_1. □

2.5.iv. *Let $E_{12}: x = x(t)$, $t_1 \le t \le t_2$, be a nonsingular extremal, and let $[u_s(t), t = 1, \ldots, n]$ be an $n \times n$ matrix whose n columns are solutions $u_s(t)$ of the accessory system relative to E_{12} with $u_s(t_1) = 0$, $s = 1, \ldots, n$,* $\det[u_s(t)]$ *not identically zero. Then any $\bar{t} \in (t_1, t_2)$ is conjugate to t_1 if and only if* $\det[u_s(\bar{t})] = 1$.

The proof is similar to the previous one.

2.6 Smoothness Properties of Optimal Solutions

A. Existence and Continuity of the First Derivative

Let $f_0(t, x, x')$ be of class C^1 on $A \times R^n$, A closed, and let $x(t)$, $a \le t \le b$, be an AC n-vector function with graph in A. We shall need the simple hypothesis:

2.6.1. For each t, $a \le t \le b$, the n-vector function of u

$$[f_{0x'^i}(t, x(t), u),\ i = 1, \ldots, n]$$

never takes twice the same value as u describes R^n. In other words, $a \le t \le b$, $u, v \in R^n$, $u \neq v$ implies

$$[f_{0x'^i}(t, x(t), u),\ i = 1, \ldots, n] \neq [f_{0x'^i}(t, x(t), v),\ i = 1, \ldots, n].$$

This hypothesis (2.6.1) is certainly satisfied if

$$E(t, x(t), u, v) > 0 \tag{2.6.2}$$

for all $a \le t \le b$ and all $u, v \in R^n$, $u \neq v$. Indeed, assume, if possible, that for given $u, v \in R^n$, $u \neq v$, we have

$$l_v = f_{0x'^i}(t, x(i), u) = f_{0x'^i}(t, x(t), v), \qquad i = 1, \ldots, n.$$

Then

$$f_0(t, x(t), v) - f_0(t, x(t), u) - \sum_i l_i(v^i - u^i) = E(t, x(t), u, v) > 0,$$
$$f_0(t, x(t), u) - f_0(t, x(t), v) - \sum_i l_i(u^i - v^i) = E(t, x(t), v, u) > 0,$$

and by addition we obtain $0 > 0$, a contradiction. We have proved that (2.6.2) implies (2.6.1).

In turn (2.6.2) is certainly satisfied if f_0 has continuous second order partial derivatives $f_{0x'^ix'^j}$ and

$$Q(t, x(t), u, \xi) = \sum_{i,j=1}^{n} f_{0x'^ix'^j}(t, x(t), u)\xi_i\xi_j > 0 \tag{2.6.3}$$

for all t, u, ξ with $a \le t \le b$, $u, \xi \in R^n$, $\xi \neq 0$. To prove that (2.6.3) implies (2.6.2) we note that, by definition (2.1.2) and Taylor's formula, we have

$$E(t, x(t), u, v) = \int_0^1 \sum_i (v^i - u^i)[f_{0x'^i}(t, x(t), u + \alpha(v - u)) - f_{0x'^i}(t, x(t), u)]\, d\alpha$$
$$= \int_0^1 \int_0^1 \alpha \sum_{i,j} (v^i - u^i)(v^j - u^j) f_{0x'^ix'^j}(t, x(t), u + \alpha\beta(v - u))\, d\alpha\, d\beta.$$

For the statement and proof of the theorems below it is convenient to denote by $f_{0x'}$, f_{0x} the following n-vector functions on $A \times R^n$:

$$f_{0x'}(t, x, u) = [f_{0x'^i}(t, x, u), i = 1, \ldots, n],$$
$$f_{0x}(t, x, u) = [f_{0x^i}(t, x, u), i = 1, \ldots, n].$$

We may also need the further hypothesis:

2.6.4. For $u \in R^n$, $|u| \to +\infty$, we have $|f_{0x'}(t, x(t), u)| \to +\infty$ uniformly in $[a, b]$. In other words, we assume that, given $N > 0$, there is another constant $R \geq 0$ such that $t \in [a, b]$, $u \in R^n$, $|u| \geq R$ implies $|f_{0x'}(t, x(t), u)| \geq N$.

Note that (2.6.4) is not a consequence of (2.6.1). For instance, for $n = 1$, $f_0 = (1 + x'^2)^{1/2}$, we have $f_{0x'} = x'(1 + x'^2)^{-1/2}$. This is a strictly increasing bounded function of x' in $(-\infty, +\infty)$, and thus f_0 satisfies (2.6.1) but not (2.6.4).

On the other hand, $f_0 = -x'^2 + x'^4$ satisfies (2.6.4), but does not satisfy (2.6.1).

We shall now assume that the AC trajectory $x(t)$, $a \leq t \leq b$, satisfies the Euler equations (E_i), $i = 1, \ldots, n$, of (2.2.i). Namely, we need to express this requirement as precisely as in Section 2.2:

2.6.5. There is an AC n-vector function $\phi(t) = (\phi_1, \ldots, \phi_n)$, $a \leq t \leq b$, such that almost everywhere in $[a, b]$ we have

$$f_{0x'}(t, x(t), x'(t)) = \phi(t), \qquad (d/dt)\phi(t) = f_{0x}(t, x(t), x'(t)).$$

We begin with the following statement:

2.6.i (Boundedness of the First Derivatives). *If f_0 is of class C^1 in $A \times R^n$ and satisfies* (2.6.4), *and if $x(t)$, $a \leq t \leq b$, is any* AC *n-vector function satisfying* (2.6.5), *then x' is essentially bounded and x is Lipschitzian in $[a, b]$.*

Proof. The n-vector function ϕ in (2.6.5) is AC, and hence continuous and bounded in $[a, b]$, say $|\phi(t)| \leq N$. By (2.6.4) there is $R \geq 0$ such that $t \in [a, b]$, $u \in R^n$, $|u| \geq R$ implies $|f_{0x'}(t, x(t), u)| \geq N + 1$. Since $|f_{0x'}(t, x(t), x'(t))| = |\phi(t)| \leq N$ a.e. in $[a, b]$, we conclude that $|x'(t)| \leq R$ also a.e. in $[a, b]$. Since x is AC, we derive that x is Lipschitzian of constant R in $[a, b]$. □

We are now in a position to state and prove the first main theorem concerning the smoothness of trajectories:

2.6.ii (Theorem (Tonelli): Continuity of the First Derivative). *If $x(t)$, $a \leq t \leq b$, is* AC *with graph in A and essentially bounded derivative x' in $[a, b]$, if f_0 is of class C^1 in $A \times R^n$, and* (2.6.1), (2.6.5) *hold, then x' exists everywhere in $[a, b]$ and is continuous in $[a, b]$, that is, x is of class C^1.*

If it is not known that x' is essentially bounded, then the conclusion of (2.6.ii) is still valid under the additional hypothesis (2.6.4) concerning f_0.

Proof. If $x(t)$ is known to be continuous in $[a, b]$ with sectionally continuous derivative $x'(t)$, we have only to prove that x has no corner point. This is simply a consequence of the Erdmann corner condition (f) of (2.2.i). As mentioned in Remark 1 in Section 2.2 under conditions (2.6.1) there cannot be corner points.

If x is AC in $[a, b]$ with x' essentially bounded, then $f_0(t, x(t), x'(t))$ is also essentially bounded, measurable, and L-integrable in $[a, b]$. In this situation, to prove (2.6.ii) we have to prove that (α) $x'(t)$ exists everywhere in $[a, b]$, and (β) $x'(t)$ is continuous in $[a, b]$.

Since x is AC in $[a, b]$, the derivative x' exists almost everywhere in $[a, b]$. First we assume that x' is essentially bounded, say $|x'(t)| \leq m$ for almost all $t \in [a, b]$.

Let S denote the set of all $t \in [a, b]$ where $x'(t)$, $\phi'(t)$ are defined, where $|x'(t)| \leq m$, where each of the n functions $f_{0x'^i}(t, x(t), x'(t))$ coincides with $\phi_i(t)$, $i = 1, \ldots, n$, and where the relations $(d/dt)\phi_i(t) = f_{0x^i}(t, x(t), x'(t))$ hold. Then $S \subset [a, b]$, meas $S = b - a$, and hence S is everywhere dense in $[a, b]$. If t_0 is any point of $[a, b]$, then t_0 is a point of accumulation of points $t \in S$ with $t \neq t_0$. Let us prove first that $x'(t)$ has a limit as $t \to t_0$ with $t \in S$. Suppose this is not true. Then there are sequences $[t_k]$, $[t'_k]$ of points of S with $t_k \to t_0$, $t'_k \to t_0$, and such that $[x'(t_k)]$, $[x'(t'_k)]$ have distinct limits. Since x' is bounded in S, then we can assume that both limits are finite, say $x'(t_k) \to u$, $x'(t'_k) \to v$, $u \neq v$, u, v finite, $u, v \in R^n$. Since the relations $f_{0x'^i}(t, x(t), x'(t)) = \phi_i(t)$, $i = 1, \ldots, n$, hold at every point $t \in S$, in particular at $t = t_k$ and $t = t'_k$, then as $k \to \infty$ we obtain two relations which by comparison yield

$$f_{0x'^i}(t_0, x_0, u) = f_{0x'^i}(t_0, x_0, v), \qquad i = 1, \ldots, n,$$

where $x_0 = x(t_0)$. The hypothesis (2.6.1) implies $u = v$, a contradiction. This proves that $u(t_0) = \lim x'(t)$ exists and is finite as $t \to t_0$ along points of S, and this holds at every $t_0 \in [a, b]$. The same argument shows that u is a continuous function on $[a, b]$.

Let us prove that $u(t) = x'(t)$ a.e. in $[t_1, t_2]$. Indeed, x' is measurable, and hence continuous on certain closed subsets K_s of $[a, b]$ with meas $K_s > b - a - s^{-1}$, and we know that almost every point of K_s is a point of density one for K_s. Hence, for every fixed s and $t_0 \in K_s$, there is a sequence $[t_k]$ of points $t_k \in S \cap K_s$ with $t_k \to t_0$, $x'(t_0) = \lim x'(t_k) = u(t_0)$ as $k \to \infty$. Thus, $x'(t) = u(t)$ a.e. on each K_s, and also $x'(t) = u(t)$ a.e. in $[a, b]$. Finally, for every $t \in [a, b]$ we have $x(t) - x(a) = \int_a^t x'(\tau)\, d\tau = \int_a^t u(\tau)\, d\tau$; hence x is continuously differentiable in $[a, b]$, $x' = u$ everywhere, and x is of class C^1 as stated. □

If x' is not known to be essentially bounded, but (2.6.4) holds, then from (2.6.i) we derive that x' is essentially bounded, and the argument above applies.

Remark 1. (A counterexample for theorem (2.6.i)). As we shall see in Section 3.10, the absolute minimum of the functional $I[x] = \int_{t_1}^{t_2} xx'^2\,dt$ with $x \ge 0$, $x(t_1) = 0$, $x(t_2) = x_2 > 0$, $t_1 < t_2$, is of the form $x(t) = k(t - t_1)^{2/3}$, $t_1 \le t \le t_2$, and x' is unbounded. Here $f_0 = xx'^2$ does not satisfy (2.6.1) along $x(t)$.

Remark 2. (A counterexample for theorem (2.6.ii)). The absolute minimum of the functional $I[x] = \int_{-1}^{1} x^2(1 - x')^2\,dt$ with $x(-1) = 0$, $x(1) = 1$, is certainly given by the trajectory x defined by $x(t) = 0$ for $-1 \le t \le 0$, $x(t) = t$ for $0 \le t \le 1$, and x is AC and x' is discontinuous at $t = 0$. Here $f_0(t, x, u) = x^2(1 - u)^2$ does not satisfy (2.6.1) along $x(t)$.

B. Existence and Continuity of the Second and Higher Derivatives

We shall prove here that under mild hypotheses, any arc of class C^1 satisfying the Euler equation is actually of class C^2 or higher. Precisely, we shall prove the statement

2.6.iii (THEOREM (WEIERSTRASS): CONTINUITY OF THE SECOND DERIVATIVE). *If $x(t)$, $a \le t \le b$, is of class C^1 with graph in A, if f_0 is of class C^2 $[C^m, m \ge 2]$, if (2.6.5) holds and if $\det f_{0x'x'}$ is never zero along x, that is,*

$$\det(f_{0x'^ix'^j}(t, x(t), x'(t)), i, j = 1, \ldots, n) \ne 0 \quad \text{for all } a \le t \le b, \tag{2.6.6}$$

then x is of class C^2 $[C^m]$ in $[a, b]$.

Condition (2.6.6) is certainly satisfied if

$$Q = \sum_{ij} f_{0x'^ix'^j}(t, x(t), x'(t))\xi_i\xi_j > 0 \tag{2.6.7}$$

for all $\xi = (\xi_1, \ldots, \xi_n) \ne 0$, $\xi \in R^n$, and all $t \in [a, b]$.

Proof of (2.6.iii). Let t be any point of $[a, b]$. If t is replaced by some $t + \Delta t$ also in $[a, b]$, $\Delta t \ne 0$, then the vectors $x = x(t)$, $x' = x'(t)$ are replaced by certain vectors $x + \Delta x = x(t + \Delta t)$, $x' + \Delta x' = x'(t + \Delta t)$, where $\Delta x, \Delta x' \to 0$ as $\Delta t \to 0$, since $x(t)$, $x'(t)$ are continuous at t by hypothesis. Also, $\Delta x/\Delta t \to x' = x'(t)$ as $\Delta t \to 0$. Finally, $f_{0x'}(t, x(t), x'(t))$, which we shall denote simply by $f_{0x'}(t, x, x')$, is replaced by $f_{0x'}(t + \Delta t, x + \Delta x, x' + \Delta x')$. By Taylor's formula we have

$$\begin{aligned} \Delta f_{0x'^i} &= f_{0x'^i}(t + \Delta t, x + \Delta x, x' + \Delta x') - f_{0x'^i}(t, x, x') \\ &= f_{0x'^it}\Delta t + \sum_j f_{0x'^ix^j}\Delta x^j + \sum_j f_{0x'^ix'^j}\Delta x'^j, \qquad i = 1, \ldots, n, \end{aligned} \tag{2.6.8}$$

where the arguments of all $f_{0x'^it}, f_{0x'^ix^j}, f_{0x'^ix'^j}$ are $t + \theta\Delta t$, $x + \theta\Delta x$, $x' + \theta\Delta x'$ for some θ, $0 < \theta < 1$, which depends on i, t, and Δt. Dividing the equation (2.6.8) by Δt we obtain

$$\frac{\Delta f_{0x'^i}}{\Delta t} = f_{0x'^it} + \sum_j f_{0x'^ix^j}\frac{\Delta x^j}{\Delta t} + \sum_j f_{0x'^ix'^j}\frac{\Delta x'^j}{\Delta t}, \qquad i = 1, \ldots, n, \tag{2.6.9}$$

and we interpret these equations as a linear algebraic system in the n unknowns $\Delta x'^j/\Delta t, j = 1, \ldots, n$. The determinant D of such a system has limit $D_0 \neq 0$ as $\Delta t \to 0$ with $D_0 = \det f_{0x'^ix'^j}(t, x(t), x'(t))$. Thus, for Δt sufficiently small, $D \neq 0$, and we can solve (2.6.9) with respect to the n quotients $\Delta x'^j/\Delta t$, $j = 1, \ldots, n$. As we know from Cramer's rule, each $\Delta x'^j/\Delta t$ is then the quotient of two determinants, the one in the denominator being D. We do not need their explicit expressions. We need only know that the n quotients above are of the form

$$\frac{\Delta x'^i}{\Delta t} = D^{-1}R_i\left(\frac{\Delta f_{0x'^j}}{\Delta t}, f_{0x'^jt}, f_{0x'^jx^s}, f_{0x'^jx'^s}, \frac{\Delta x^j}{\Delta t}\right), \tag{2.6.10}$$

where R_i is a polynomial with constant coefficients in the arguments listed in parentheses. In (2.6.10) we have $D \to D_0 \neq 0$ as $\Delta t \to 0$, and also $\Delta f_{0x'^j}/\Delta t \to (d/dt)f_{0x'^j}$, and these last derivatives exist and equal f_{0x^j} by virtue of (2.6.5). Also, $\Delta x^j/\Delta t \to x'^j(t)$ as $\Delta t \to 0$, and $f_{0x'^jt}, f_{0x'^jx^s}, f_{0x'^jx'^s}$ converge as $\Delta t \to 0$ to the same expressions with arguments $t, x(t), x'(t)$. This proves that $\Delta x'^j/\Delta t$ has a finite limit as $\Delta t \to 0$, that is, $x''^i(t)$ exists and is finite, and

$$x''^i(t) = D_0^{-1}R_i(f_{0x^j}, f_{0x'^jt}, f_{0x'^jx^s}, f_{0x'^jx^s}, x), \qquad i = 1, \ldots, n.$$

Since $D_0 \neq 0$ in $[a, b]$, we conclude that x''^i too is a continuous function of t in $[a, b]$. Note that we can perform now the same limit as $\Delta t \to 0$ in (2.6.9), and we obtain

$$(d/dt)f_{0x'^i} = f_{0x'^it} + \sum_j f_{0x'^ix^j}x'^j + \sum_j f_{0x'^ix'^j}x''^j, \qquad i = 1, \ldots, n.$$

Using (2.6.5) we obtain the Euler equations in their explicit form (2.2.15). □

Remark 3 (A Counterexample for Theorem (2.6.iii)). The absolute minimum of the functional $I[x] = \int_{-1}^{1} x^2(2t - x')^2\, dt$ with $x(-1) = 0$, $x(1) = 1$, is certainly given by the trajectory x defined by $x(t) = 0$ for $-1 \leq t \leq 0$, $x(t) = t^2$ for $0 \leq t \leq 1$, and x, x' are AC, but x'' is discontinuous at $t = 0$. Here $f_0(t, x, u) = x^2(2t - u)^2$ does not satisfy (2.6.6) along $x(t)$.

2.7 Proof of the Euler and DuBois-Reymond Conditions in the Unbounded Case

A. The Condition (S)

We shall use the same notation and general hypotheses as in Section 2.2, but now we allow the optimal AC solution $x(t)$, $t_1 \leq t \leq t_2$, to have unbounded derivative $x'(t)$. We shall need further requirements on the function $f_0(t, x, u)$. Namely, we shall assume here that:

(S_i) There is a continuous function $S(t,x') \geq 0$, $(t,x') \in R^{n+1}$, and some $\delta > 0$, such that $S(t,x'(t))$ is L-integrable on $[t_1,t_2]$ and $|f_{0x^i}(t,y,x')| \leq S(t,x')$ for all $t \in [t_1,t_2]$, $x' \in R^n$, $y = (y^1,\ldots,y^n) \in R^n$, $y^j = x^j(t)$ for $j \neq i$, $|y^i - x^i(t)| \leq \delta$. Here $i = 1,\ldots,n$.

(S_0) There is a continuous function $S(t,x') \geq 0$, $(t,x') \in R^{n+1}$, and some $\delta > 0$, such that $S(t,x'(t))$ is L-integrable in $[t_1,t_2]$, and $|f_t(\tau,x,x')| \leq S(t,x')$ for all $t \in [t_1,t_2]$, $|\tau - t| \leq \delta$, $x = x(t)$.

Note that a good candidate for S is often of the form

$$S(t,x') = M_1|f_0(t,x(t),x')| + M_2|x'| + M_3$$

for suitable constants M_1, M_2, M_3. Note that (S_i) is certainly satisfied if $f_0(t,x,x')$ does not depend on x^i, and (S_0) is satisfied if $f_0(t,x,x')$ does not depend on t.

Note that conditions (S_0), (S_1), ..., (S_n) are certainly satisfied if x' is essentially bounded in $[t_1,t_2]$, say $|x'(t)| \leq M$, $t \in [t_1,t_2]$ (a.e.). Then, all f_{0x^i}, f_{0t}, $i = 1,\ldots,n$, are continuous in the compact set $\Gamma_\delta \times V_M$, V_M the solid ball of center of the origin and radius M in R^n. If $|f_{0t}|, |f_{0x^i}| \leq N$ in $\Gamma_\delta \times V_M$, then we can take $S(t,x') = N$.

Note that conditions (S_0), (S_1), ..., (S_n) could be worded as only one condition, say, (S), but we shall use them separately to reach separate conclusions.

As in Section 2.4 for the proof of the DuBois-Reymond equation, we proceed to a change of independent variable. Let s be the arc length parameter on the curve $C_0: x = x(t)$, $t_1 \leq t \leq t_2$, so that $s(t) = \int_{t_1}^t (1 + (x'(\tau))^2)^{1/2}\, d\tau$ with $s(t_1) = 0$, $s(t_2) = L$, the (Jordan) length of C_0, and $s(t)$ is AC with $s'(t) \geq 1$ a.e. Thus $s(t)$ is AC and has an AC inverse $t(s)$, $0 \leq s \leq L$, with $t'(s) > 0$ (a.e.) in $[0,L]$. If $X(s) = x(t(s))$, $0 \leq s \leq L$, then $t(s)$, $X(s)$ are Lipschitzian of constant one in $[0,L]$. By the usual change of variable (E. J. McShane [I, p. 211]), we have

$$I[x] = \int_{t_1}^{t_2} f_0(t,x(t),x'(t))\,dt = \int_0^L f_0(t(s),X(s),X'(s)/t'(s))t'(s)\,ds.$$

By taking $F_0(t,x,t',x') = f_0(t,x,x'/t')t'$, we have

$$I[x] = \mathfrak{I}[C] = \mathfrak{I}[X] = \int_0^L F_0(t(s),X(s),t'(s),X'(s))\,ds. \tag{2.7.1}$$

Let us prove that hypothesis (S_i) implies

$$\phi_i(s) = F_{0x'^i} - \int_0^s F_{0x^i}\,d\sigma = c_i, \qquad 0 \leq s \leq L, \quad i = 1,\ldots,n, \tag{2.7.2}$$

for some constant c_i, where the arguments of $F_{0x'^i}$ are $t(s)$, $X(s)$, $t'(s)$, $X'(s)$, and the same for F_{0x^i} with σ replacing s. In what follows we disregard the set of measure zero of all s where $(t'(s), X'(s))$ may not exist. The proof of (2.7.2) is by contradiction. Assume (2.7:2) is not true. Then there are constants $d_1 < d_2$ and disjoint sets E_1^*, E_2^* of positive measure such that $\phi_i(s) \leq d_1$ for $s \in E_1^*$ and $\phi_i(s) \geq d_2$ for $s \in E_2^*$, while $t'(s) > 0$ a.e. in $[0,L]$. Here we shall denote by $|E|$ the Lebesgue measure of a measurable set E in R. Then there is some constant $k > 0$ and two subsets E_1, E_2 of E_1^*, E_2^*, also of positive measure such that

$$\begin{aligned} t'(s) &\geq k > 0, &\quad \phi_i(s) &\leq d_1 &\quad &\text{for } s \in E_1, |E_1| > 0,\\ t'(s) &\geq k > 0, &\quad \phi_i(s) &\geq d_2 &\quad &\text{for } s \in E_2, |E_2| > 0. \end{aligned} \tag{2.7.3}$$

Denoting by χ_i the function defined by $\chi_i(s) = 1$ for $s \in E_i$, $\chi_i(s) = 0$ for $s \in [0,L] - E_i$, $i = 1, 2$, let us take

$$\psi(s) = \int_0^s [|E_2|\chi_1(\sigma) - |E_1|\chi_2(\sigma)]\,d\sigma, \qquad 0 \leq s \leq L, \tag{2.7.4}$$

so that ψ is AC in $[0, L]$ with $\psi(0) = \psi(L) = 0$. Moreover

$$\psi'(s) = -|E_1| \quad \text{a.e. in } E_2, \qquad \psi'(s) = |E_2| \quad \text{a.e. in } E_1,$$
$$\psi'(s) = 0 \qquad \text{a.e. in } [0, L] - (E_1 \cup E_2).$$

For $i = 1, \ldots, n$, and $-1 \leq \alpha \leq 1$, we consider now the curve $C_\alpha: t = t_\alpha(s)$, $x = X_\alpha(s)$, $0 \leq s \leq L$, with $t_\alpha(s) = t(s)$, $X^i_\alpha(s) = X^i(s) + \alpha\psi(s)$, $X^j_\alpha(s) = X^j(s)$, $j \neq i$. We have $|\psi'(s)| \leq L$ (a.e.), and if $N = \max |\psi(s)|$, then for $|\alpha| \leq \alpha_0 = \min[1, \delta N^{-1}, L^{-1}]$, we have $|X^i_\alpha(s) - X^i(s)| \leq \delta$, $|X'^i_\alpha(s) - X'^i(s)| \leq 1$ (a.e.), and C_α lies in Γ_δ.

Since $t'_\alpha(s) = t'(s) > 0$ a.e. in $[0, L]$, the curve C_α has an AC representation $x = x_\alpha(t)$, $t_1 \leq t \leq t_2$, C_α lies in Γ_δ, and C_α has the same end points as C, since $\psi(0) = 0$, $\psi(L) = 0$. Thus, $\mathfrak{I}[C_\alpha] \geq \mathfrak{I}[C]$ for all $|\alpha| \leq \alpha_0$. Moreover $\Phi(s, \alpha) = F_0(t_\alpha(s), X_\alpha(s), t'_\alpha(s), X'_\alpha(s))$ has partial derivative $\phi_\alpha(s, \alpha) = F_{0x^i}\psi(s) + F_{0x'^i}\psi'(s)$ for $s \in [0, L]$ (a.e.) and $|\alpha| \leq \alpha_0$.

For $s \in E_1 \cup E_2$ we have $t'_\alpha(s) = t'(s) \geq k > 0$, $|X'_\alpha(s)| \leq 2$, $(t_\alpha(s), X_\alpha(s)) \in \Gamma_\delta$, $|\psi'(s)| \leq L$, $|\psi(s)| \leq N$, and thus both terms $F_{0x^i}\psi$, $F_{0x'^i}\psi$ are bounded in $E_1 \cup E_2$. For $s \in [0, L] - (E_1 \cup E_2)$ we have $\psi'(s) = 0$ and $F_{0x'^i}\psi' = 0$, while, by hypothesis (S_i),

$$|F_{0x^i}| \leq S(t, X'(s)/t'(s))t'(s),$$

which is L-integrable in $[0, L]$, while $|\psi(s)| \leq N$. Thus, for $|\alpha| \leq \alpha_0$, $\phi_\alpha(s, \alpha)$ is in absolute value less than a fixed L-integrable function on $[0, L]$. By usual rule of differentiation under the sign of integral we have

$$(d\mathfrak{I}(C_\alpha)/d\alpha)_{\alpha=0} = \int_0^L (F_{0x^i}\psi(s) + F_{0x'^i}\psi'(s))\, ds,$$

where the left hand side is zero, so that, by integration by parts (E. J. McShane [I, 36.1, p. 209]) we also have, by comparison with (2.7.3),

$$0 = \int_0^L \phi_i(s)\psi'(s)\, ds = \left(\int_{E_1} + \int_{E_2}\right)\phi_i(s)\psi(s)\, ds$$
$$\leq |E_1||E_2| d_1 - |E_1||E_2| d_2 < 0,$$

a contradiction. We have proved relation (2.7.2).

Let us prove that hypothesis (S_0) implies

$$\phi_0(s) = F_{0t'} - \int_0^s F_{0t}\, d\sigma = c_0, \qquad 0 \leq s \leq L, \tag{2.7.5}$$

for some constant c_0. The proof by contradiction is analogous to the one above. We define ψ as before, and C_α: $t = t_\alpha(s)$, $x = X_\alpha(s)$, $0 \leq s \leq L$, by taking $t_\alpha(s) = t(s) + \alpha\psi(s)$, $X_\alpha(s) = X(s)$, $0 \leq s \leq L$, $|\alpha| \leq 1$.

Let $\rho > 0$ be so chosen that $t, \tau \in [t_1, t_2]$, $|t - \tau| \leq \rho$ implies $|x(t) - x(\tau)| \leq \delta$ where δ is the constant in condition (S_0). Now we take $|\alpha| \leq \alpha_0 = \min[1, 2^{-1}kN^{-1}, \rho N^{-1}, 2^{-1}kL^{-1}]$. For $|\alpha| \leq \alpha_0$ then we have $t'_\alpha(s) = t'(s) + \alpha\psi'(s) \geq k - k/2 = k/2 > 0$ for $s \in E_1 \cup E_2$, while $\psi'(s) = 0$, $t'_\alpha(s) = t'(s) > 0$ a.e. in $[0, L]$ and C_α has an AC representation $x = x_\alpha(t)$, $t_1 \leq t \leq t_2$. We also have $|t_\alpha(s) - t(s)| \leq |\alpha| N \leq \rho$, and hence $|x_\alpha(t) - x(t)| = |x(t_\alpha(s)) - x(t(s))| < \delta$, so that C_α lies in Γ_δ, and finally $\mathfrak{I}[C_\alpha] \geq \mathfrak{I}[C]$.

The argument continues as before, where now, for $|\alpha| \leq \alpha_0$, $\phi_\alpha(s, \alpha) = F_{0t}\psi + F_{0t'}\psi'$, and ϕ_α is bounded in $E_1 \cup E_2$. As before, $F_{0t'}\psi' = 0$ in $[0, L] - (E_1 \cup E_2)$, and by hypothesis S_0, $F_{0t}\psi$ is in absolute value less than a fixed L-integrable function on $[0, L] - (E_1 \cup E_2)$. The same argument as before proves relation (2.7.5).

Relations (2.7.2), (2.7.5) by the transformation $s = s(t)$ yield relations (E_i), $i = 1, \ldots, n$, and (E_0), respectively.

Remark. The functions η above have been defined in such a way to be zero at the end points t_1 and t_2. If it is required instead that $\eta(t_1) = \xi_1, \eta(t_2) = \xi_2$ for particular n-vectors ξ_1, ξ_2, then we have only to take two n-vector constants $c_1 = (c_1^1, \ldots, c_1^n)$, $c_2 = (c_2^1, \ldots, c_2^n)$, with $c_1|E_1| + c_2|E_2| = \xi_2 - \xi_1$, define $\chi(t; c_1, c_2) = c_1$ for $t \in E_1$, $\chi(t, c_1, c_2) = c_2$ for $t \in E_2$, $\chi(t; c_1, c_2) = 0$ otherwise, and then take $\eta(t; c_1, c_2) = \xi_1 + \int_{t_1}^{t} \chi(\tau; c_1, c_2)\, d\tau$ instead of (2.7.4). Then relation (2.7.2) holds as before. This remark will be useful in the next section.

2.8 Proof of the Transversality Relations

First we assume t_1 and t_2 fixed. Again, $x(t)$, $t_1 \le t \le t_2$, is an AC optimal solution in the class of all such AC trajectories in the interior of A and with $(t_1, x(t_1), t_2, x(t_2)) \in B$. Certainly, x is optimal in the smaller class of all X with $X(t_1) = x(t_1)$, $X(t_2) = x(t_2)$; hence, $f_{0x'^i}(t, x(t), x'(t))$ coincides a.e. in $[t_1, t_2]$ with an AC function, say $-\lambda_i(t)$, $t_1 \le t \le t_2$, $i = 1, \ldots, n$, and Euler's equations (2.2.1) hold. We take $\delta > 0$ so small that the set $\Gamma_\delta = [(t, y) | t_1 \le t \le t_2, |y - x(t)| \le \delta]$ is contained in A. We take in B an arbitrary curve $\Xi(\xi) = (t_1, X_1(\xi), t_2, X_2(\xi))$, $-1 \le \xi \le 1$, of class C^1, passing through $e[x]$ in B, $e[x] = (t_1, x(t_1), t_2, x(t_2))$, namely, $\Xi(\xi) \in B$ for all $-1 \le \xi \le 1$, $X_1(0) = x_1 = x(t_1)$, $X_2(0) = x_2 = x(t_2)$. We take $\bar{\xi}$, $0 < \bar{\xi} \le 1$, so small that $|X_1(\xi) - x_1| \le \delta$, $|X_2(\xi) - x_2| \le \delta$ for $|\xi| \le \bar{\xi}$. For any constant $c > 0$, and any $\bar{a}$, $0 < \bar{a} \le 1$, such that $\bar{a}c \le \bar{\xi}$, we take

$$\mathscr{X}(a, t) = (t_2 - t_1)^{-1}[(X_1(ca) - x_1)(t_2 - t) + (X_2(ca) - x_2)(t - t_1)]$$

so that $\mathscr{X}, \mathscr{X}_a, \mathscr{X}_t, \mathscr{X}_{at}$ are continuous in $[-\bar{a}, \bar{a}] \times [t_1, t_2]$, and $\mathscr{X}(0, t) = \mathscr{X}_t(0, t) = 0$, $\mathscr{X}(a, t_1) = X_1(ac) - x_1$, $\mathscr{X}(a, t_2) = X_2(ac) - x_2$. As we have proved in (2.3.ii) and subsequent Remark 1, the quotient $a^{-1}(I[x + \mathscr{X}] - I[x])$ has limit $J_1[\eta]$ as $a \to 0$, namely,

$$J_1[\eta] = \int_{t_1}^{t_2} \sum_{i=1}^{n} [f_{0x^i}(t, x(t), x'(t))\eta^i(t) + f_{0x'^i}(t, x(t), x'(t))\eta'^i(t)]\, dt = 0,$$

where

$$\eta(t) = (\eta^1, \ldots, \eta^n) = \mathscr{X}_a(0, t) = (t_2 - t_1)^{-1}[cX_1'(0)(t_2 - t) + cX_2'(0)(t - t_1)],$$
$$\eta'(t) = \mathscr{X}_{at}(0, t) = (t_2 - t_1)^{-1}[cX_1'(0) + cX_2'(0)],$$

and hence $\eta(t_1) = cX_1'(0)$, $\eta(t_2) = cX_2'(0)$. Since $f_{0x'^i}(t, x(t), x'(t))$ coincides with an AC function, say $-\lambda_i(t)$, in $[t_1, t_2]$ (a.e.), by integration by parts, we have

$$\int_{t_1}^{t_2} \sum_{i=1}^{n} \left[f_{0x^i}(t, x(t), x'(t)) - (d/dt) f_{0x'^i}(t, x(t), x'(t)) \right] \eta^i(t)\, dt$$
$$- \sum_{i=1}^{n} [\lambda_i(t_2)\eta^i(t_2) - \lambda_i(t_1)\eta^i(t_1)] = 0,$$

The bracket in the integral is identically zero, and therefore we derive

$$- \sum_{i=1}^{n} [\lambda_i(t_2)(cX_2'^i(0)) - \lambda_i(t_1)(cX_1'^i(0))] = 0.$$

By writing $cX'_2(0) = dx_2$, $cX'_1(0) = dx_1$, we obtain (2.2.9) for $dt_1 = dt_2 = 0$ and (dx_1, dx_2) any element of B' namely,

$$\left[\sum_{i=1}^{n} f_{0x'} dx^i\right]_1^2 = 0.$$

If t_1, t_2 are not fixed, then we proceed as in the proof of the DuBois-Reymond condition in Section 2.4, by changing $I[x]$ into the autonomous integral $J[\tilde{y}]$ in a fixed interval $\tau_1 \le \tau \le \tau_2$, say the interval $[t_1, t_2]$ of the optimal solution x. Then applying our result to J and returning to the variable t we derive (2.2.9). Indeed,

$$J[\tilde{y}] = \int_{\tau_1}^{\tau_2} F_0(t, y, t', y')\, d\tau$$

with $F_0(t, y, t', y') = f_0(t, y, y'/t')t'$ and $t = t(\tau)$, $\tau_1 \le \tau \le \tau_2$ as in Section 2.4. Here we take an arbitrary curve $\tilde{\Xi}(\xi) = (T_1(\xi), X_1(\xi), T_2(\xi), X_2(\xi))$, $-1 \le \xi \le 1$, of class C^1, passing through $e[x] = (t_1, x(t_1), t_2, x(t_2))$, namely, $\tilde{\Xi}(\xi) \in B$ for all $-1 \le \xi \le 1$, $T_1(0) = t_1$, $T_2(0) = t_2$, $X_1(0) = x_1 = x(t_1)$, $X_2(0) = x_2 = x(t_2)$, and we proceed as above. We conclude that

$$-\lambda_0(\tau_2)(cT'_2(0)) - \sum_{i=1}^{n} \lambda_i(\tau_2)(cX'^i_2(0)) + \lambda_0(\tau_1)(cT'_1(0)) + \sum_{i=1}^{n} \lambda_i(\tau_1)(cX'^i_1(0)) = 0,$$

where

$$\lambda_0(\tau) = -F_{0t'} = -f_0(t, y, y'/t') + \sum_i f_{0x'^i}(t, y, y'/t')(y'^i/t'),$$

$$\lambda_i(\tau) = -F_{0y'^i} = -f_{0x'^i}(t, y, y'/t'), \quad i = 1, \ldots, n,$$

and where $t = t(\tau)$. By returning to the original variables, we obtain

$$\left[(f_0 - \sum_i x'^i f_{0x'^i})(cT'(0)) + \sum_i f_{0x'^i}(cX'^i(0))\right]_1^2 = 0,$$

or

$$\left[(f_0 - \sum_i x'^i f_{0x'^i})\, dt + \sum_i f_{0x'^i}\, dx^i\right]_1^2 = 0.$$

If x' is unbounded, then we use the special variations defined in Section 2.7, Remark, under the same specific assumptions of that Section.

2.9 The String Property and a Form of Jacobi's Necessary Condition

A. A Lemma on Extremal Arcs

We shall consider two parametric curves

$$C: t = t_3(a), \quad x = x_3(a), \qquad a' \le a \le a'',$$
$$D: t = t_4(a), \quad x = x_4(a), \qquad a' \le a \le a'',$$

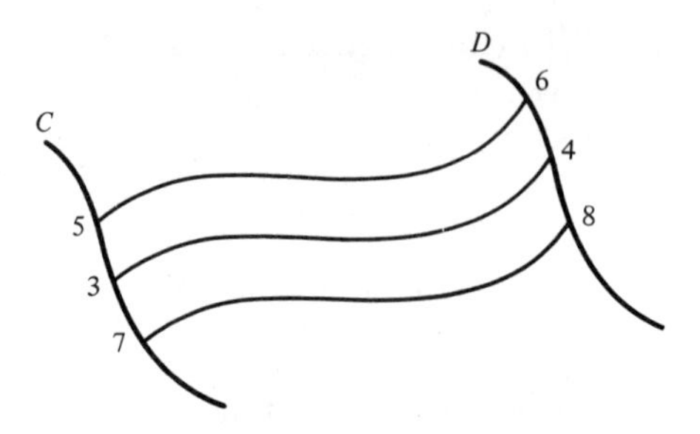

in the tx-space R^{n+1} (thus $x_3 = (x_3^1, \ldots, x_3^n)$, $x_4 = (x_4^1, \ldots, x_4^n)$) and a family of curves

$$E_a: x = x(t, a), \quad t_3(a) \le t \le t_4(a),$$

for $a' \le a \le a''$, with $x = (x^1, \ldots, x^n)$, each E_a joining the points $3 = [t_3(a), x_3(a)]$ and $4 = [t_4(a), x_4(a)]$. Thus

$$x(t_3(a), a) = x_3(a), \qquad x(t_4(a), a) = x_4(a) \tag{2.9.1}$$

for every $a' \le a \le a''$. We shall assume $t_3(a)$, $x_3(a)$, $t_4(a)$, $x_4(a)$ of class C^1, and $x(t, a)$ continuous with $x' = \partial x/\partial t$, $x_a = \partial x/\partial a$, $x'_a = \partial^2 x/\partial t\, \partial a$ for $t_3(a) \le t \le t_4(a)$, $a' \le a \le a''$. Note that we shall denote by x' and x_a partial derivatives with respect to t and a respectively. We shall denote by I and I^* the line integrals

$$\begin{aligned} I &= \int f_0\, dt = \int f_0(t, x(t), x'(t))\, dt, \\ I^* &= \int \left[f_0\, dt + \sum_i (dx^i - x'^i\, dt) f_{0x'^i} \right], \end{aligned} \tag{2.9.2}$$

where I is the usual integral under consideration, and I^* is often called the Hilbert integral associated with I.

In this Section we shall compute $I[x]$ on the curves E_a, so that I becomes a function of a, namely,

$$I(a) = I[E_a] = \int_{t_3(a)}^{t_4(a)} f_0(t, x(t, a), x'(t, a))\, dt, \qquad a' \le a \le a''.$$

Then $I(a)$ is of class C^1 in $[a', a'']$ and its derivative has the usual expression

$$I'(a) = dI/da = [f_0\, dt/da]_3^4 + \int_{t_3(a)}^{t_4(a)} \sum_i (f_{0x^i} x_a^i + f_{0x'^i} x_a'^i)\, dt, \tag{2.9.3}$$

where the arguments of f_{0x^i}, $f_{0x'^i}$ are t, $x(t, a)$, $x'(t, a)$, where in the bracket dt/da denote $dt_3(a)/da$ and $dt_4(a)/da$ respectively, and where the arguments of f_0 are as above t, $x(t, a)$, $x'(t, a)$ at $t = t_3(a)$ and $t = t_4(a)$ respectively.

We shall discuss I^* in more detail in Section 2.11. In this Section we need only $I^*[C]$ and $I^*[D]$. Namely, under $I^*[C]$ we mean the line integral I^* where f_0, $f_{0x'^i}$ are computed at t, $x(t, a)$, $x'(t, a)$ for $t = t_3(a)$, (thus, $x'^i = x'^i(t_3(a), a)$), dx^i, dt are computed along C, i.e., $dx^i = (dx_3^i/da)\, da$, $dt = (dt_3/da)\, da$. Analogous definitions hold for $I^*[D]$.

Note that, in the present context, it may well occur that the curve C, or D, are reduced to a single point x_0, say, $x_3(a) = x(t_3(a), a) = x_0$ for all a, while $x'(t_3(a), a)$ may depend on a. We shall need this particular case.

2.9.i (LEMMA). *If the functions* $t_3(a), x_3(a) = (x_3^1, \ldots, x_3^n), t_4(a), x_4(a) = (x_4^1, \ldots, x_4^n)$, $a' \leq a \leq a''$, *are of class* C^1, *if* $x(t,a) = (x^1, \ldots, x^n), t_3(a) \leq t \leq t_4(a), a' \leq a \leq a''$, *is continuous together with* x', x_a, x'_a, *and if* (2.9.1) *holds, then for every particular value* $a = a_0$ *for which* $E_{34} = E_a: x = x(t,a), t_3(a) \leq t \leq t_4(a)$, *satisfies the Euler equations* $(d/dt)f_{0x'^i} = f_{0x^i}, i = 1, \ldots, n$, *we have*

$$dI = \left[f_0\, dt + \sum_i (dx^i - x'^i\, dt) f_{0x'^i} \right]_3^4, \tag{2.9.4}$$

where $x' = (x'^1, \ldots, x'^n)$ *denotes the slope of* E_{34} *at 3 and at 4, where* $dt, dx = (dx^1, \ldots, dx^n)$ *denote the differential of* $t_3(a), x_3(a), t_4(a), x_4(a)$ *at* $a = a_0$, *and the arguments* t, x, x' *in* $f_0, f_{0x'^i}$ *correspond to* E_{34}, *that is, are* $t, x(t,a), x'(t,a)$ *at* $t = t_3(a), t = t_4(a)$, *and* $a = a_0$.

Proof. For $a = a_0$ the expression (2.9.3), by virtue of Euler's equations above and manipulation, becomes

$$\begin{aligned} dI/da &= [f_0\, dt/da]_3^4 + \int_{t_3(a)}^{t_4(a)} \sum_i (x_a'^i f_{0x^i} + x_a^i (d/dt) f_{0x'^i})\, dt \\ &= [f_0\, dt/da]_3^4 + \int_{t_3(a)}^{t_4(a)} (d/dt)\left(\sum_i x_a^i f_{0x'^i} \right) dt \\ &= \left[f_0\, dt/da + \sum_i x_a^i f_{0x'^i} \right]_3^4. \end{aligned}$$

On the other hand, from the equations (2.9.1) written in component form and differentiation, we have

$$dx_3^i/da = x_3'^i\, dt_3/da + x_{3a}^i, \qquad dx_4^i/da = x_4'^i\, dt_4/da + x_{4a}^i,$$

and hence

$$dI/da = \left[f_0\, dt/da + \sum_i (dx^i/da - x'^i\, dt/da) f_{0x'^i} \right]_3^4,$$

from which (2.9.4) follows. □

2.9.ii. *Under the conditions of* (2.9.i), *if along every arc* $E_a: x = x(t,a), t_3(a) \leq t \leq t_4(a)$, *the Euler equations* $(d/dt)f_{0x'^i} = f_{0x^i}, i = 1, \ldots, n$, *are satisfied, and (say)* $E_{56} = E_{a'}$, $E_{78} = E_{a''}$, *then*

$$I[E_{78}] - I[E_{56}] = I^*[D_{68}] - I^*[C_{57}]. \tag{2.9.5}$$

Proof. The relation (2.9.4) holds for every $a' \leq a \leq a''$, and by integration on $[a', a'']$ and use of line integrals (2.9.2) we obtain (2.9.5). □

Remark. If the curve C is reduced to a single point (that is, $x_3(a)$ is constant), then $3 = 5 = 7$, $I^*[C_{57}] = 0$, and (2.9.5) reduces to

$$I[E_{38}] - I[E_{36}] = I^*[D_{68}].$$

Finally, if in addition the curves E_a are all tangent to D at $t = t_4(a)$ (that is, $dx^i/dt = x'^i$, $i = 1, \ldots, n$, at $t = t_4(a)$), then $I^* = I$ on D, and the latter relation reduces further to

$$I[E_{38}] = I[E_{36}] + I[D_{68}].$$

This relation is often called the *string property* of I (see the following figure).

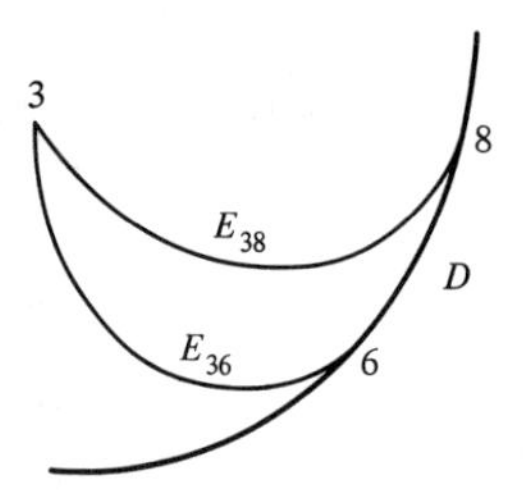

B. A Variant of Jacobi's Necessary Condition

Let us assume that the extremal $E_{12}:x = x(t)$, $t_1 \leq t \leq t_2$, is imbedded in a one parameter family of extremal arcs all passing through $1 = (t_1, x(t_1))$, and each having a point of contact $(\bar{t}, x(\bar{t}))$ with some curve D in the tx-space as in the figure, that is, D is an envelope for the family. A point $6 = (\bar{t}, x(\bar{t}))$ of contact of E_{12} with D is sometime called a "geometric conjugate point to 1 on E_{12}." Here we assume that D is of class C^1 and D has an arc, say 56, projecting from 6 backward toward 1.

2.9.iii (Jacobi's Necessary Condition—Geometric Form). *If f_0 is of class C^3, if the extremal arc $E_{12}:x = x(t)$, $t_1 \leq t \leq t_2$, is nonsingular, if E_{12} is a weak minimum for $I[x]$, and if E_{12} is imbedded in a one parameter family of extremals possessing an envelope D, then there cannot be any point of contact $(\bar{t}, x(\bar{t}))$ of E_{12} on D with $t_1 < \bar{t} < t_2$.*

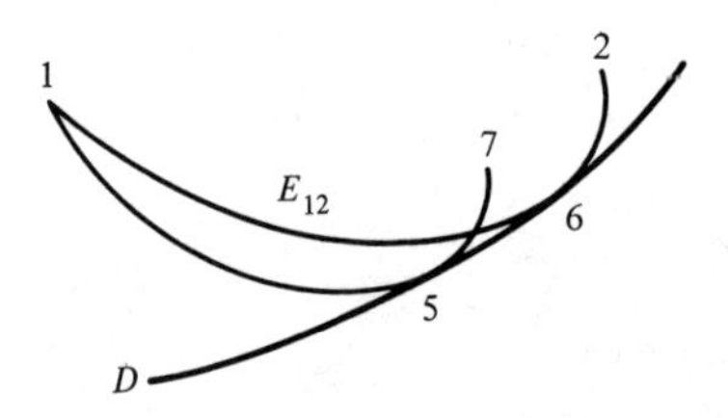

Proof. From the string property we have $I[E_{15}] + I[D_{56}] = I[E_{16}]$; hence, all curves $E_{15} + D_{56} + E_{62}$ give to the functional the same value, consequently they are all optimal. Thus, D_{56} is optimal, and being of class C^1, satisfies the Euler equations and is. Now at 5 there are two extremals E_{57} and D_{56}. This is impossible, since they both satisfy the Euler equation, they both have the same initial values $x(t_5)$, $x'(t_5)$, and $\det R(t_5) \neq 0$, this last guaranteeing uniqueness by usual uniqueness theorems for ordinary differential equations (cf. Remark 4 at the end of Section 2.2). □

Remark. Note that the argument above shows that even the point $2 = (t_2, x(t_2))$ cannot be a point of contact of E_{12} with the envelope D provided D has a branch projecting backward toward 1. It does not forbid 2 to be such a point of contact with D if D has a cusp at 2 with both branches projecting away from 1. The argument does not apply either if D is a single point.

2.10 An Elementary Proof of Weierstrass's Necessary Condition

We assume here that all requirements of Section 2.1 and of Theorem (2.2.i) hold and in addition that x' is sectionally continuous. Concerning f_0 we actually need only here that f_0 is continuous in $A \times R^n$ with its partial derivatives $f_{0x^i}, f_{0x'^i}$, $i = 1, \ldots, n$. We shall make use of the Weierstrass E-function (2.1.4):

$$E(t, x, x', X') = f_0(t, x, X') - f_0(t, x, x') - \sum_{i=1}^{n} (X'^i - x'^i) f_{0x'^i}(t, x, x').$$

2.10.i. *If $I[x]$ has a strong local minimum at some trajectory $x(t)$, $t_1 \le t \le t_2$, continuous with sectionally continuous derivative x' and graph Γ in the interior of A, then at every point $(t, x(t))$ we have*

$$E(t, x(t), x'(t), X') \ge 0 \tag{2.10.1}$$

for all n-vectors $X' = (X'^1, \ldots, X'^n) \in R^n$. If t is a point of jump discontinuity for x', then (2.10.1) *holds with $x'(t)$ replaced by $x'(t - 0)$ as well as with $x'(t)$ replaced by $x'(t + 0)$.*

Proof. Let E denote the curve 12, or $x = x(t)$, $t_1 \le t \le t_2$. First let $t = t_3$ be a point interior to $[t_1, t_2]$ and of continuity for $x'(t)$. Let t_3', t_4 be two points close to t_3, say $t_1 < t_3 < t_3' < t_4 < t_2$, such that x' is continuous in $[t_3, t_4]$.

Given any $X' \in R^n$, let

$$X(t) = x(t_3) + (t - t_3)X', \qquad t_3 \le t \le t_3'.$$

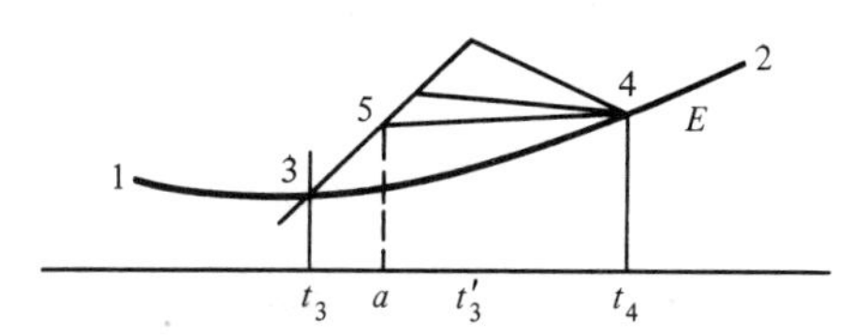

Let $x(t, a) = (x^1, \ldots, x^n)$ be a vector valued function of (t, a), continuous, with continuous x_t, x_a, x_{at}, for $t_3 \le a \le t_3'$, $a \le t \le t_4$, such that

$$X(a) = x(a, a), \qquad x(t, t_3) = x(t), \qquad x(t_4, a) = x(t_4).$$

In other words $x = x(t, a)$, $a \le t \le t_4$, describes a one parameter family of arcs joining $5 = (a, x(a, a)) = (a, X(a))$ to the fixed point $4 = (t_4, x(t_4))$ (see picture). Such a family is easy to construct, since we can take

$$x = x(t, a) = x(t) + (t_4 - a)^{-1}(t_4 - t)(X(a) - x(a)).$$

We are now in the situation described in (2.9.i) with the curve C reduced to the segment $x = X(a)$, $t_3 \le a \le t_3'$, and the curve D reduced to the single point 4. For $a = t_3$, the arc $x = x(t, t_3)$, $a = t_3 \le t \le t_4$, reduces to the arc 34, or $x(t)$, $t_3 \le t \le t_4$, which certainly satisfies Euler's equations. On the other hand, if C_a denotes the arc 13542, then

$I[C_a] \geq I[E]$ for $a \geq t_3$ sufficiently close to t_3, $C_{t_3} = E$, and

$$0 \leq I[C_a] - I[E] = \int_{t_3}^{a} f_0(t, X(t), X')\, dt + \int_{a}^{t_4} f_0(t, x(t, a), x'(t, a))\, dt$$
$$- \int_{t_3}^{t_4} f_0(t, x(t), x'(t))\, dt,$$

where only the first two terms depend on a. The (right) derivative of this expression at $a = t_3$ is therefore nonnegative. Here the derivative with respect to a of the first term in the last member at $a = t_3$ is $f_0(t_3, x(t_3), X')$; the derivative of the second term in the last member at $a = t_3$ is given by (2.9.i). Actually, the terms in (2.9.4) at 4 are identically zero, and $da = dt$, $dx^i/dt = X'^i$ at 3. Thus

$$0 \leq f_0(t_3, x(t_3), X') - \left[f_0(t_3, x(t_3), x'(t_3)) + \sum_{i=1}^{n} (X'^i - x'^i(t_3)) f_{0x'^i}(t_3, x(t_3), x'(t_3)) \right].$$

By using (2.1.4) and writing t for t_3, we have $E(t, x(t), x'(t), X') \geq 0$, and (2.10.1) is proved if $t_1 < t < t_2$ and t is a point of continuity for x'. If $t = \bar{t}$ is a point of jump discontinuity for x', then all points $t \neq \bar{t}$ of a neighborhood of t are points of continuity for x', and (2.10.1) holds at these points. By a passage to the limit as $t \to \bar{t} - 0$ or $t \to \bar{t} + 0$, we prove (2.10.1) for $t = \bar{t}$ and $x'(t \pm 0)$ replacing $x'(t)$. The same argument holds at $t = t_1$ and $t = t_2$. (Cf. Section 2.19 for a proof and a deeper interpretation of the Weierstrass condition in the general case).

2.11 Classical Fields and Weierstrass's Sufficient Conditions

A. Fields

Let R be a *simply connected* region of the tx-space R^{n+1}, $x = (x^1, \ldots, x^n)$, say R is open, and assume that the integrand function $f_0(t, x, x')$ of the given integral $I[C] = \int_C f_0\, dt$ is of class C^1 in $R \times R^n$.

By R being simply connected we mean that any two continuous curves in R having the same end points can be deformed one into another by continuity in R. Thus, any closed curve in R can be deformed by continuity in R to a single point.

We shall think of R as simply covered by a family E_a: $x = x(t, a)$, $t'(a) \leq t \leq t''(a)$, of extremals of the integral I, that is, such that there passes one and only one extremal E_a of the family through every point (t, x) of R. Thus, $a = a(t, x)$ is a single valued function on R, and the parameter, or index, $a = (a^1, \ldots, a^n)$ may describe a connected region R_0 of the index space. Moreover the slope $p = p(t, x) = x'(t, a(t, x))$ of the extremal E_a through (t, x) is also a single valued function on R, and the identity holds $x(t, a(t, x)) = x$. For the present Subsections A and B, all we need to know is that a function

$p(t,x) = (p_1, \ldots, p_n)$ is given in R with the following properties:

(α) the functions p_i are single valued and of class C^1 in R;
(β) the line integral

$$(2.11.1)\quad I^*[C] = \int_C \left[f_0(t,x,p(t,x))\,dt + \sum_i (dx^i - p_i(t,x)\,dt) f_{0x'^i}(t,x,p(t,x)) \right]$$

depends only upon the end points of C for C in R.

Then we say that R is a field (or a Weierstrass field) with slope function $p(t,x)$ (or slope functions $p_i(t,x)$, $i = 1, \ldots, n$) and the differential equations $dx/dt = p(t,x)$, componentwise

$$(2.11.2)\qquad dx^i/dt = p_i(t,x), \qquad i = 1, \ldots, n,$$

are said to be the differential equations of the field.

Note that, since $p(t,x)$ is of class C^1, then (2.11.2) satisfies the hypotheses of the existence and uniqueness theorem for ordinary differential systems. Thus, through each point (t_0, x_0) in the interior of R there passes one and only one solution $E: x = x(t)$, $t' \le t \le t''$, of (2.11.2), whose end points, if finite, are on the boundary of R. These trajectories x are said to be the *extremals* of the field, and certainly they simply cover R. We shall see in subsections C and E below that they are actually extremals of the integral $I[C]$. Note that from $x'(t) = p(t, x(t))$ we derive $x'' = p_t + p_x x' = p_t + p_x p$, that is, x'' exists and is continuous along each extremal E of the field.

The line integral $I^*[C]$ above is said to be the *Hilbert integral* relative to f_0 and the field R with slope function $p(t,x)$. Note that $I^*[C]$ can be written in the equivalent and more familiar form

$$(2.11.3)\qquad I^*[C] = \int_C [A_0\,dt + A_1\,dx^1 + \cdots + A_n\,dx^n]$$

$$= \int_{t_1}^{t_2} \left[A_0(t,x(t)) + \sum_{i=1}^{n} A_i(t,x(t)) x'^i(t) \right] dt,$$

where

$$A_0 = A_0(t,x) = f_0(t,x,p(t,x)) - \sum_i f_{0x'^i}(t,x,p(t,x)) p_i(t,x),$$

$$A_i = A_i(t,x) = f_{0x'^i}(t,x,p(t,x)), \qquad i = 1, \ldots n.$$

2.11.i. *On any extremal $E_{12}: x = x(t)$, $t_1 \le t \le t_2$, of the field R we have $I[E_{12}] = I^*[E_{12}]$.*

Indeed, by definition, we have $dx^i/dt = p_i(t,x)$ everywhere on E_{12}, and the Hilbert integral (2.11.1) then coincides with $I[C]$.

2.11.ii. *The Hilbert integral $I^*[C]$ depends only on the end points of the curve C in R.*

Proof. By hypothesis $f_0(t, x, x')$ is of class C^1 for $(t, x, x') \in R \times R^n$. Thus the partial derivatives f_{0t}, f_{0x^i} are continuous, and the functions $A_0(t, x)$, $A_i(t, x)$ are continuous in R. If $C: x = x(t)$, $t_1 \le t \le t_2$, is any curve with x AC in $[t_1, t_2]$, and graph in R, then $x'(t)$ is L_1-integrable, the functions $A_0(t, x(t))$, $A_i(t, x(t))$, $i = 1, \ldots, n$, are continuous in $[t_1, t_2]$, the linear combination $A_0 + \sum_i A_i x'^i$ is also L_1-integrable, and $I^*[C]$ as given by the second expression in (2.11.3) is finite. Since R is simply connected, condition (β) for a field is equivalent to the requirement that $A_0(t, x) + \sum_i A_i(t, x)\, dx^i$ is the exact differential of a function $F(t, x)$, that is, $F_t = A_0$, $F_{x^i} = A_i$, $i = 1, \ldots, n$, and then F is of class C^1. As a consequence $F(t, x(t))$ is AC with

$$(d/dt)F(t, x(t)) = A_0(t, x(t)) + \sum_{i=1}^{n} A_i(t, x(t))x'^i(t), \qquad t \in [t_1, t_2] \text{ (a.e.)}.$$

Hence, $I^*[C] = \int_{t_1}^{t_2} (d/dt)F(t, x(t)) = F(t_2, x(t_2)) - F(t_1, x(t_1))$. This proves that $I^*[C]$ depends only on the end points of the curve C lying in R. □

2.11.iii. *If $E_{12}: x = x(t)$, $t_1 \le t \le t_2$, $x = (x^1, \ldots, x^n)$, is an arc of an extremal of a field R with end points $1 = (t_1, x_1)$, $2 = (t_2, x_2)$, $x_1 = (x_1^1, \ldots, x_1^n)$, $x^2 = (x_2^1, \ldots, x_2^n)$, and if $C_{12}: x = X(t)$, $t_1 \le t \le t_2$, $X = (X^1, \ldots, X^n)$, is any other curve lying in R and joining 1 and 2, then*

$$\begin{aligned} I[C_{12}] - I[E_{12}] &= \int_{t_1}^{t_2} E(t, X, p, X')\, dt \\ &= \int_{t_1}^{t_2} E(t, X(t), p(t, X(t)), X'(t))\, dt, \end{aligned} \tag{2.11.4}$$

where E is the Weierstrass function defined in (2.1.4).

Proof. By (2.11.i) we have $I[E_{12}] = I^*[E_{12}]$, since E_{12} is an extremal of the field, and by (2.11.ii) we have $I^*[E_{12}] = I^*[C_{12}]$, since E_{12} and C_{12} are curves in R with the same end points 1 and 2. Finally, by the use of the definition of the function E we have

$$\begin{aligned} I[C_{12}] - I[E_{12}] &= I[C_{12}] - I^*[E_{12}] = I[C_{12}] - I^*[C_{12}] \\ &= \int_{t_1}^{t_2} \left[f_0(t, X, X') - f_0(t, X, p) - \sum_i (X'^i - p_i) f_{0x'^i}(t, X, p) \right] dt \\ &= \int_{t_1}^{t_2} E(t, X(t), p(t, X(t)), X'(t))\, dt. \end{aligned}$$

□

B. Weierstrass's Sufficient Condition for an Extremum

We shall assume below that a field R is given with its family of extremals simply covering R, say $E_a: x(t, a)$, $t'(a) \le t \le t''(a)$, and relative function $a(t, x)$ and slope function $p(t, x)$, so that, as stated, $x(t, a(t, x)) = x$, $x'(t, a(t, x)) = p(t, x)$. We shall further assume that a given extremal $E_{12}: x = x(t)$, $t_1 \le t \le t_2$, of the integral $I[x]$ is an arc of an extremal of the field, that is, there is some

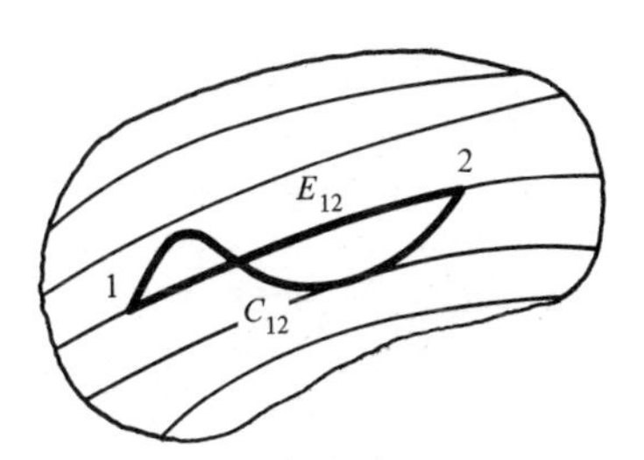

a_0 such that $x(t) = x(t, a_0)$, and $x'(t) = p(t, x(t)) = x'(t, a_0)$ for $t'(a_0) \le t_1 \le t \le t_2 \le t''(a_0)$. For the sake of brevity we shall simply say that the extremal E_{12} is *imbedded* in the field R. We further assume that for some $\delta > 0$ we have $\Gamma_\delta \subset R$ where $\Gamma_\delta = [(t,x) | t_1 \le t \le t_2, |x - x(t)| \le \delta]$. Here we understand as usual that $f_0(t, x, x')$ is of class C^1, and by a curve $C: x = x(t)$, $t_1 \le t \le t_2$, in R we understand as usual that x is AC with graph Γ in R, and with $f_0(t, x(t), x'(t))$ L_1 integrable in $[t_1, t_2]$.

2.11.iv (Sufficient Condition for a Strong Local Minimum). *If the extremal E_{12} is imbedded in a field R with slope function $p(t,x)$, and if $E(t, x, p(t,x), X') \ge 0$ for all vectors X' and all points $(t,x) \in R$, then for every curve C_{12} lying in R and having the same end points* 1 *and* 2 *of E_{12} we have $I[C_{12}] \ge I[E_{12}]$. If $E(t, x, p(t,x), X') > 0$ for all $(t,x) \in R$ and $X' \ne p(t,x)$, then $I[C_{12}] > I[E_{12}]$ for all curves C_{12} not identical to E_{12}.*

Proof. By (2.11.iii) we have

$$I[C_{12}] - I[E_{12}] = \int_{t_1}^{t_2} E(t, X(t), p(t, X(t)), X'(t))\, dt \ge 0. \tag{2.11.5}$$

On the other hand, $I[C_{12}] - I[E_{12}] = 0$ implies $E = 0$ for all $t \in [t_1, t_2]$; hence $X'(t) = p(t, X(t))$ for $t \in [t_1, t_2]$. This shows that C_{12} is a solution of the differential system $dx/dt = p(t,x)$ with $X(t_1) = x(t_1)$. By the uniqueness theorem for solutions of differential systems, we have $X(t) = x(t)$ for all $t \in [t_1, t_2]$, or $C_{12} = E_{12}$. Thus, $I[C_{12}] > I[E_{12}]$ for every curve C_{12} not identical to E_{12}. □

2.11.v (Sufficient Condition for a Weak Local Minimum). *If f_0 is of class C^2, if the extremal E_{12} is imbedded in a field R with slope function $p(t,x)$ and if the quadratic form Q of* (2.2.5) *taken along E_{12} is positive definite, that is, $\sum_{ij} f_{0x'^i x'^j}(t, x(t), x'(t))\xi_i\xi_j > 0$ for all $\xi = (\xi_1, \ldots, \xi_n) \ne 0$ and $t_1 \le t \le t_2$, then there is an $\varepsilon > 0$ such that for any trajectory $C_{12}: x = X(t)$, $t_1 \le t \le t_2$, having the same end points* 1 *and* 2 *as E_{12} with $|X(t) - x(t)| \le \varepsilon$, $|X'(t) - x'(t)| \le \varepsilon$, $t_1 \le t \le t_2$, we have $I[C_{12}] \ge I[E_{12}]$, and equality holds only if C_{12} is identical with E_{12}. For* n = 1 *assumption* (a) *is replaced by the assumption $f_{0x'x'} > 0$ along E_{12}, or $f_{0x'x'}(t, x(t), x'(t)) > 0$ for $t_1 \le t \le t_2$.*

Proof. For $n = 1$ we know that $x(t)$, $x'(t)$ are continuous functions in $[t_1, t_2]$ and that $f_{0x'x'}(t, x(t), x'(t)) > 0$ for all $t \in [t_1, t_2]$; hence by continuity $f_{0x'x'}(t, x(t), x'(t))$ has a positive minimum in $[t_1, t_2]$, and again by continuity there is some δ', $0 < \delta' \leq \delta$, such that $f_{0x'x'}(t, x, x') > 0$ for all t, x, x' with $t_1 \leq t \leq t_2$, $|x - x(t)| \leq \delta'$, $|x' - x'(t)| \leq \delta'$. For $n > 1$ we know that $Q(t; \xi) = \sum_{ij} f_{0x'^i x'^j}(t, x(t), x'(t))\xi_i\xi_j$ is positive definite in ξ for every $t \in (t_1, t_2]$. By an analogous continuity argument we conclude that there is a δ', $0 < \delta' \leq \delta$, such that the quadratic forms $Q = \sum_{ij} f_{0x'^i x'^j}(t, x, x')\xi_i\xi_j$ are positive definite in ξ for all (t, x, x') with $t_1 \leq t \leq t_2$, $|x - x(t)| \leq \delta'$, $|x' - x'(t)| \leq \delta'$. Let us take ε, $0 < \varepsilon \leq \delta' \leq \delta$, so small that $|p(t, x) - p(t, x(t)| \leq \delta'$ for all $t_1 \leq t \leq t_2$, $|x - x(t)| \leq \varepsilon$. Finally, let $C_{12}: x = X(t)$, $t_1 \leq t \leq t_2$, be a curve in R with $|X(t) - x(t)| \leq \varepsilon$, $|X'(t) - x'(t)| \leq \varepsilon$. Then $|X(t) - x(t)| \leq \varepsilon \leq \delta' \leq \delta$, and

$$|p(t, X(t)) - x'(t)| = |p(t, X(t)) - p(t, x(t))| \leq \delta' \leq \delta,$$
$$|X'(t) - p(t, x(t))| = |X'(t) - x'(t)| \leq \varepsilon \leq \delta' \leq \delta.$$

By the second remainder in Taylor's formula we have now

$$\begin{aligned} &E(t, X(t), p(t, X(t)), X'(t)) \\ &\quad = f_0(t, X(t), X'(t)) - f_0(t, X(t), p(t, X(t))) \\ &\qquad - \sum_i (X^{i'}(t) - p_i(t, X(t)))f_{0x'^i}(t, X(t), p(t, X(t))) \\ &\quad = 2^{-1} \sum_{ij} (X'^i - p_i(t, X(t))(X'^j - p_j(t, X(t))f_{0x'^i x'^j}(t, X(t), \theta), \end{aligned}$$

where θ is some point between $p(t, X(t))$ and $X'(t)$. Since both these points are within the sphere in R^n of center $x'(t)$ and radius δ', we also have $|\theta - x'(t)| \leq \delta'$, and hence the quadratic expression above is positive definite and $E \geq 0$ for every $t_1 \leq t \leq t_2$. From (2.11.5) we conclude that

$$I[C_{12}] - I[E_{12}] = \int_{t_1}^{t_2} E(t, X(t), p(t, X(t)), X'(t))\, dt \geq 0.$$

Again, as in the proof of (2.11.iv), $I[C_{12}] - I[E_{12}] = 0$ implies $E = 0$ for all $t \in [t_1, t_2]$; hence $X'^i(t) = p_i(t, X(t))$, $i = 1, \ldots, n$, for every $t \in [t_1, t_2]$, that is, $X(t)$ is the solution of the system $dx/dt = p(t, x)$ with $X(t_1) = x(t_1)$. By the uniqueness theorem we deduce $X(t) = x(t)$ for all $t_1 \leq t \leq t_2$, so that $C_{12} = E_{12}$. □

2.11.vi (Sufficient Condition for a Strong Local Minimum). *If the extremal $E_{12}: x = x(t)$, $t_1 \leq t \leq t_2$, is imbedded in a field R of slope function $p(t, x)$, and if $E(t, x, x', X') \geq 0$ for all $t_1 \leq t \leq t_2$, $|x - x(t)| \leq \delta'$, $|x' - x'(t)| \leq \delta'$, all vectors $X' \in R^n$ and some $\delta' > 0$, then there is an $\varepsilon > 0$, such that for any trajectory $C_{12}: x = X(t)$, $t_1 \leq t \leq t_2$, having the same end points 1 and 2 as E_{12} with $|X(t) - x(t)| \leq \varepsilon$ for $t_1 \leq t \leq t_2$, we have $I[C_{12}] \geq I[E_{12}]$. If $E > 0$ for all t, x, x', X' as above with $X' \neq x'$, then $I[C_{12}] > I[E_{12}]$ for every curve C_{12} as above distinct from E_{12}.*

Proof. We may assume $\delta' \leq \delta$. By continuity, there is a number ε, $0 < \varepsilon \leq \delta$, so small that

$$|p(t, x) - x'(t)| = |p(t, x) - p(t, x(t))| \leq \delta' \leq \delta$$

for all $t_1 \le t \le t_2$, $|x - x(t)| \le \varepsilon$. Now, if $C_{12}: x = X(t)$, $t_1 \le t \le t_2$, is any curve joining 1 and 2 as E_{12} with $|X(t) - x(t)| \le \varepsilon$, then $|X(t) - x(t)| \le \delta'$, and hence C_{12} lies in R, $|p(t, X(t)) - x'(t)| \le \delta'$, and

$$E(t, X(t), p(t, X(t)), X'(t)) \ge 0$$

for all $t_1 \le t \le t_2$. From (2.11.5) we conclude that

$$I[C_{12}] - I[E_{12}] = \int_{t_1}^{t_2} E(t, X(t), p(t, X(t)), X'(t))\, dt \ge 0.$$

If $E > 0$ for all (t, x, x', X') as above with $X' \ne x'$, then $I[C_{12}] = I[E_{12}]$ implies $C_{12} = E_{12}$ as in the proof of (2.11.iv). □

C. Characterization of the Extremals for $n = 1$

From here to the end of this Section we assume that f_0 is of class C^2 and we consider only extremals of class C^2. In this Subsection x is a scalar, $p(t, x)$ a scalar function in R, and the integrals under consideration are

$$I[C] = \int_C f_0(t, x, x')\, dt,$$

$$\text{(2.11.6)} \quad I^*[C] = \int_C [f_0(t, x, p(t, x))\, dt + f_{0x'}(t, x, p(t, x))(dx - p(t, x)\, dt)]$$

$$= \int_C [(f_0 - p f_{0x'})\, dt + f_{0x'}\, dx] = \int_C A_0\, dt + A_1\, dx.$$

2.11.vii. *For $n = 1$, if R is a simply connected region in R^2 and $p(t, x)$ a single valued function of class C^1 in R, then the solutions E of the equation $dx/dt = p(t, x)$ are extremals of $I[C]$ if and only if I^* is the integral of an exact differential.*

Proof. The slope of the field extremal $E: x = x(t)$ passing through (t, x) is $p(t, x)$. Hence,

$$\text{(2.11.7)} \quad x' = p, \qquad x'' = dx'/dt = (d/dt)p(t, x(t)) = p_t + p_x x' = p_t + p_x p.$$

Since R is simply connected, then condition (β) for a field at the beginning of Subsection A is satisfied if and only if I^* is the integral of an exact differential, that is, if $\partial A_0/\partial x = \partial A_1/\partial t$ everywhere in the interior of R, where

$$\begin{aligned} \partial A_0/\partial x &= (\partial/\partial x)[f_0(t, x, p(t, x)) - p(t, x) f_{0x'}(t, x, p(t, x))] \\ &= f_{0x} + f_{0x'} p_x - p_x f_{0x'} - p f_{0x'x} - p f_{0x'x'} p_x \\ &= f_{0x} - p f_{0x'x} - p p_x f_{0x'x'}, \\ \partial A_1/\partial t &= (\partial/\partial t) f_{0x'}(t, x, p(t, x)) = f_{0x't} + f_{0x'x'} p_t. \end{aligned}$$

Thus, the equality $\partial A/\partial x = \partial A_1/\partial t$ is equivalent to

$$f_{0x't} + f_{0x'x'} p_t - f_{0x} + p f_{0x'x} + p p_x f_{0x'x'} = 0,$$

and by using (2.11.7) also equivalent to

$$f_{0x't} + f_{0xx'}x' + f_{0x'x'}x'' - f_{0x} = 0, \tag{2.11.8}$$

which is the Euler equation for f_0. In other words, the extremals of a field, as solutions of the differential equation $dx/dt = p(t, x)$, are extremal, in the usual sense, that is, are smooth solutions of the Euler equation. Statement (2.11.vii) is thereby proved. □

D. Construction of a Field for $n = 1$

Let us consider any one parameter family $E_a: x = x(t, a)$ of arcs of extremals of $I[C]$ filling once a simply connected region R of the tx-plane R^2, that is, such that there is one and only one extremal $E_a: x = x(t, a)$, $t'(a) \le t \le t''(a)$ through every point $(t, x) \in R$. Let $a = a(t, x)$ denote the value of the parameter a corresponding to the extremal E_a through (t, x). Then the slope $p(t, x)$ of E_a at (t, x) is given by $p(t, x) = x'(t, a(t, x))$, and both $a(t, x)$, and $p(t, x)$ are single valued functions of (t, x) in R. If $x(t, a)$ and $a(t, x)$ are continuous in R together with x', x'', x'_a, a_t, a_x, then $p(t, x)$ as well as $p_t = x'' + x'_a a_t$ and $p_x = x'_a a_x$ are continuous in R. Thus condition (α) for a field holds, and so condition (β) by virtue of (2.11.vii). We have proved that R is a field.

Note that $a = a(t, x)$ is the inverse function of $x = x(t, a)$; hence, if $x(t, a)$ is continuous in R together with x', x'', x_a, and if x_a has a constant sign in R, then the inverse function $a(t, x)$ is known to exist and be continuous together with a_t and a_x. Note that $x = x(t, a(t, x))$ is an identity, hence $1 = x_a a_x$, $0 = x' + x_a a_t$, and hence $a_x = (x_a)^{-1}$, $a_t = -x'(x_a)^{-1}$.

For instance, if $n = 1$ and $f_0 = f_0(x')$ depends on x' only and $f_{0x'x'} \ne 0$, then the Euler equation (2.11.8) is $x'' = 0$, and hence all straight lines $x'' = 0$, or $x = ht + k$, (h, k constants) are extremals. If we fill $R = R^2$ with a family E_a of parallel lines, say $E_a: x = x(t, a) = ht + a$ (h a fixed constant), then $a = a(t, x) = x - ht$, $p(t, x) = x' = h$ are certainly single valued and continuous together with $p_t = 0$, $p_x = 0$ in $R = R^2$. Analogously, if R is not the whole plane R^2 and (t_0, x_0) is a point outside R such that R is completely at the right of some straight line $t = t_0 + \varepsilon$, $\varepsilon > 0$, then we can fill R with a family of straight lines through (t_0, x_0) or $E_a: x = x(t, a) = x_0 + a(t - t_0)$, and then $a = a(t, x) = (x - x_0)(t - t_0)^{-1}$, $p(t, x) = a(t, x)$ are single valued functions in R together with p_t and p_x.

Examples

1. Consider the integral $I[C] = \int_{t_1}^{t_2} x'^2(1 + x')^2\, dt$. We have here $n = 1$, $f_0 = x'^2(1 + x')^2$, and the Euler equation again is $x'' = 0$. If $1 = (0, 1)$, $2 = (2, 0)$, then E_{12}: $x = x(t) = 1 - 2^{-1}t$, $0 \le t \le 2$, is an extremal through 1 and 2, and a field of extremals covering the whole plane $R = R^2$ containing E_{12} is $x = x(t, a) = a - 2^{-1}t$, so that

$a = a(t,x) = x + 2^{-1}t$, $p = p(t,x) = 2^{-1}$. Here $f_0 = x'^2 + 2x'^3 + x'^4$, $f_{0x'x'} = 2 + 12x' + 12x'^2$, and $f_{0x'x'} = -1 < 0$ for $x' = -\frac{1}{2}$, that is, along E_{12}. The Legendre condition is certainly satisfied on E_{12}. By (2.12.v) we conclude that there is an $\varepsilon > 0$ such that, for any curve $C_{12}: x = X(t)$, $t_1 \le t \le t_2$, distinct from E_{12} joining 1 and 2 in R, with $|X(t) - x(t)| \le \varepsilon$, $|X'(t) - x'(t)| \le \varepsilon$, we have $I[C_{12}] < I[E_{12}]$. The restrictions on X and X' can be written now in the form

$$1 - 2^{-1}t - \varepsilon \le X(t) \le 1 - 2^{-1}t + \varepsilon, \quad -2^{-1} - \varepsilon \le X'(t) \le -2^{-1} + \varepsilon, \quad 0 \le t \le 2.$$

Here E_{12} is a weak local maximum for $I[C]$ with $I[E_{12}] = \frac{1}{8}$. Note that the problem under consideration has no absolute maximum, since I can take values as large as we want. The present problem has an absolute minimum, since if we take $E_{12}: x = X(t)$, $0 \le t \le 2$, with X defined by $X(t) = 1$ for $0 \le t \le 1$, $X(t) = 2 - t$ for $1 \le t \le 2$, then $I(E_{12}) = 0$. For more details on the integral above see Chapter 3.

2. Consider the integral $I[C] = \int_{t_1}^{t_2} xx'^2\,dt$. We have here $n = 1$, $f_0 = xx'^2$, $f_{0x} = x'^2$, $f_{0x'} = 2xx'$, $f_{0x'x'} = 2x$; and the DuBois-Reymond equation (2.2.14), or $x'f_{0x'} - f_0 = c$, is $xx'^2 = c$, c constant. If $1 = (0,1)$, $2 = (1,1)$, then $E_{12}: x = x(t) = 1$, $0 \le t \le 1$, is an extremal joining 1 and 2, and E_{12} is contained in the field of extremals $x = a$, a constant, covering the whole tx-plane $R = R^2$, with $a = a(t,x) = x$, $p = p(t,x) = 0$. We have $f_{0x'x'} = 2 > 0$ for $x = 1$, $x' = 0$, that is, along E_{12}. Thus the Legendre condition is certainly satisfied on E_{12}. We have now

$$E(t,x,x',X') = xX'^2 - xx'^2 - (X' - x')2xx' = x(X' - x')^2,$$

and thus $E \ge 0$ for all $x \ge 0$, and $E > 0$ for $x > 0$, $X' \ne x'$. By (2.11.vi), we have $I[C_{12}] > I[E_{12}]$ for any curve $C_{12}: x = X(t)$, $0 \le t \le 1$, distinct from E_{12} joining 1 and 2 as E_{12}, with $X(t) \ge 0$. Thus E_{12} is a strong local minimum for $I[C]$, and $I[E_{12}] = 0$. Here E_{12} is certainly not an absolute minimum for $I[C]$ (in the whole tx-plane). Indeed, if we take $C_{12}: x = X(t)$, $0 \le t \le 1$, with $X(t) = 1 - mt$ for $0 \le t \le 2^{-1}$, $X(t) = 1 - m(1 - t)$ for $2^{-1} \le t \le 1$, $m > 4$ a constant, then $X'(t) = \pm m$, and by calculation we have

$$I[C_{12}] = 2m^2 \int_0^{2^{-1}} (1 - mt)\,dt = m^2(1 - 4^{-1}m) < 0.$$

This example shows also that $I[C]$ has no absolute minimum (in the whole tx-plane), since the last integral can take negative values as large in absolute value as we want. See Chapter 3 for more details on this integral, where we shall consider also the same problem in the half plane $x \ge 0$.

Exercises

1. Consider the problem of minimizing the integral $I[x] = \int_{t_1}^{t_2} x'^{2m}\,dt$, $n = 1$, $m > 1$ integer, with fixed end points $1 = (t_1, x_1)$, $2 = (t_2, x_2)$, $t_1 < t_2$. The line segment $s = 12$ gives the absolute minimum of I. Prove this directly, or by constructing a field of extremals.
2. The same as in Exercise 1, for the integral $I[x] = \int_{t_1}^{t_2} (1 + x'^2)^{1/2}\,dt$, $n = 1$. A direct analysis will be exhibited in Section 3.1.

See Chapter 3 for more exercises.

E. Characterization of the Extremals for $n > 1$

Since K is simply connected, requirement (β) for a field is equivalent to the requirement that I^* as defined by (2.11.1) is the integral of an exact differential, or that the $n(n+1)/2$ equations

$$\partial A_0/\partial x^i = \partial A_i/\partial t, \quad \partial A_i/\partial x^j = \partial A_j/\partial x^i, \qquad i \neq j, \quad i, j = 1, \ldots, n, \tag{2.11.9}$$

hold everywhere in R. We shall need these equations in a more explicit form. To this purpose we note the identities

$$\begin{aligned}\partial A_i/\partial x^j - \partial A_j/\partial x^i &= \partial f_{0x'^i}/\partial x^j - \partial f_{0x'^j}/\partial x^i \\ &= f_{0x'^ix^j} - f_{0x'^jx^i} + \sum_h (f_{0x'^ix'^h}\partial p_h/\partial x^j - f_{0x'^jx'^h}\partial p_h/\partial x^i),\end{aligned} \tag{2.11.10}$$

where $i \neq j$, $i, j = 1, \ldots, n$. Also we have

$$\begin{aligned}\partial A_0/\partial x^i - \partial A_i/\partial t &= (\partial/\partial x^i)(f_0 - \sum_j f_{0x'^j}p_j) - (\partial/\partial t)f_{0x'^i} \\ &= f_{0x^i} + \sum_j f_{0x'^j}\partial p_j/\partial x^i - f_{0x'^it} - \sum_j f_{0x'^ix'^j}\partial p_j/\partial t \\ &\quad - \sum_j f_{0x'^jx^i}p_j - \sum_j\sum_h f_{0x'^jx'^h}p_j\partial p_h/\partial x^i - \sum_j f_{0x'^j}\partial p_j/\partial x^i.\end{aligned}$$

Here the second and the seventh term cancel. By adding and subtract terms, in particular noting that

$$\sum_j\sum_h f_{0x'^ix'^j}p_h(\partial p_j/\partial x^h) = \sum_j\sum_h f_{0x'^ix'^h}p_j(\partial p_h/\partial x^j),$$

we also have

$$\begin{aligned}&\partial A_0/\partial x^i - \partial A_i/\partial t \\ &\quad = f_{0x^i} - \left[f_{0x'^it} + \sum_j f_{0x'^ix^j}p_j + \sum_j f_{0x'^ix'^j}\left(\partial p_j/\partial t + \sum_h p_h\partial p_j/\partial x^h\right)\right] \\ &\qquad + \sum_j\left[f_{0x'^ix^j} - f_{0x'^jx^i} + \sum_h(f_{0x'^ix'^h}\partial p_h/\partial x^j - f_{0x'^jx'^h}\partial p_h/\partial x^i)\right]p_j, \\ &\quad = f_{0x^i} - \left[f_{0x'^it} + \sum_j f_{0x'^ix^j}p_j + \sum_j f_{0x'^ix'^j}\left(\partial p_j/\partial t + \sum_h p_h\partial p_j/\partial x^h\right)\right] \\ &\qquad + \sum_j p_j(\partial A_i/\partial x^j - \partial A_j/\partial x^i).\end{aligned} \tag{2.11.11}$$

2.11.viii. *For $n > 1$, if R is simply connected and $p(t, x) = (p_1, \ldots, p_n)$ is of class C^1 in R, then condition (β) for a field holds if and only if the solutions $E: x = x(t)$ to the differential system $dx^i/dt = p_i(t, x)$, $i = 1, \ldots, n$, in R are extremals of $I[C]$ and the $n(n-1)/2$ equations are satisfied everywhere in R:*

$$f_{0x'^ix^j} - f_{0x'^jx^i} + \sum_h [f_{0x'^ix'^h}\partial p_h/\partial x^j - f_{0x'^jx'^h}\partial p_h/\partial x^i] = 0, \qquad i \neq j, \quad i, j = 1, \ldots, n. \tag{2.11.12}$$

Proof. Along the solutions E of the system above we have

$$\begin{aligned}dx^i/dt &= p_i(t, x), \qquad i = 1, \ldots, n, \\ d^2x^i/dt^2 &= (d/dt)p_i(t, x(t)) = \partial p_i/\partial t + \sum_h p_h\partial p_i/\partial x^h.\end{aligned} \tag{2.11.13}$$

Now, if (β) holds, then the equations (2.11.9) hold, and relations (2.11.12) hold because of the identities (2.11.10). Moreover, from (2.11.13) and the identity (2.11.11), we obtain

$$0 = f_{0x^i} - \left[f_{0x'^i t} + \sum_j f_{0x'^i x^j} p_j + \sum_j f_{0x'^i x'^j} x''^j \right], \qquad i = 1, \ldots, n,$$

and these are the n Euler equations (E_i) in explicit form.

Conversely, if the solutions E are extremals of $I[C]$ and the equations (2.11.12) hold, then the second relations (2.11.9) hold, and because of (2.11.12), (2.11.13), also the first relations (2.11.9) hold. Condition (β) for a field is satisfied. □

F. Construction of a Field for $n > 1$

The equations (2.11.12) are not easy to verify, and therefore geometrical processes have been devised for the construction of a field.

Let E_a be an n-parameter family of extremals of $I[C]$, say

$$E_a : x = x(t, a), \qquad a = (a_1, \ldots, a_n),$$

simply covering the region R. Thus, through each point (t, x) of R there passes one and only one extremal of the family, say E_a, with $a = a(t, x)$, (a single valued function in R). In other words, there is some region $\bar{R}$ of the auxiliary $ta_1 \cdots a_n$-space which is mapped one-one and onto R by the relation $t = t, x = x(t, a)$ (and thus there is an inverse map $t = t$, $a = a(t, x)$).

Let us assume that $x(t, a)$ is continuous with continuous partial derivatives x', x'', x_{a_i}, x'_{a_i}, $i = 1, \ldots, n$, and that $a(t, x)$ is continuous with continuous partial derivatives a', a_{x^i}, $i = 1, \ldots, n$. Then the slope function $p = p(t, x) = (p_1, \ldots, p_n)$, $(t, x) \in R$ is given by

$$p_i(t, x) = x'^i(t, a(t, x)), \qquad i = 1, \ldots, n,$$

and by virtue of our assumption $p(t, x)$ is single valued and of class C^1 in R.

However we do not exclude above that the extremals of the family extend well beyond R, and in this case we shall allow the curves E_a to intersect if needed. In this situation the functions $a(t, x)$, $p(t, x)$ may not be defined outside R, though we still assume that the functions $x(t, a)$, $x'(t, a)$ are of class C^1 whenever we need them.

The following geometrical considerations often allow us to guarantee that condition (β) also is satisfied.

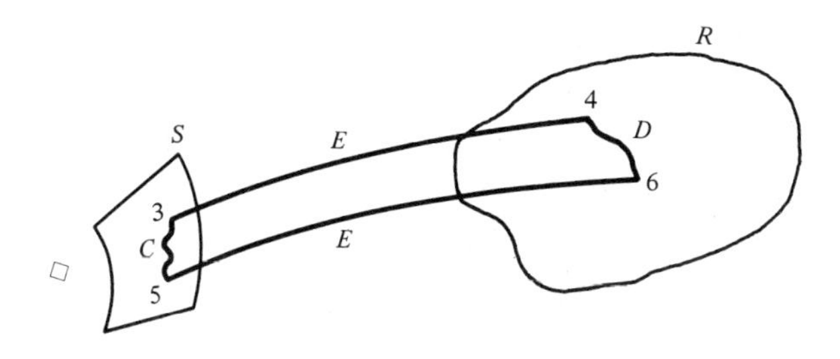

Let us assume that every extremal E_a of the given family cuts, possibly outside R, a given surface S at some point (t_0, x_0) depending on a, $t_0 = t_0(a)$, $x_0 = x(t_0(a), a)$, and assume that $t_0(a)$ is a function of class C^1 in a. Then, also $x_0 = x_0(t_0(a), a)$ is a function of class C^1 in a. Through every point $(t, x) \in R$ there passes one and only one extremal E_a with $a = a(t, x)$. The corresponding point (t_0, x_0) that E_a has on S then depends on (t, x). Precisely, $t_0 = t_0(a(t, x))$, $x_0 = x_0[t_0(a(t, x)), a(t, x)]$. These functions too are of class C^1 in R. Now, if (t, x) describes a parametric curve in R, say $D_{46}: t = t(\alpha)$, $x = x(\alpha)$, $\alpha' \le \alpha \le \alpha''$, of class C^1 (that is, $t(\alpha)$, $x(\alpha)$ are of class C^1), then the corresponding point (t_0, x_0) describes a curve C_{35} also of class C^1 and lying on S. The integral I^* on both C_{35} and D_{46} is defined by (2.9.2). Namely the point (t, x) on D_{46} is given by $t = t(\alpha)$, $x = x(\alpha)$, the corresponding extremal through (t, x) is E_a with $a = a(t, x)$, or $a(\alpha) = a(t(\alpha), x(\alpha))$, and E_a has a slope at (t, x) given by $x'(t, a) = p(t, x) = p(t(\alpha), x(\alpha))$, as in (2.11). The point (t_0, x_0) on C_{35} is given by $t_0 = t_0(a)$, $x_0 = x(t_0(a), a)$ with $a = a(\alpha)$, and the extremal E_a through (t_0, x_0) has slope at (t_0, x_0) given by $x'(t_0(a), a)$ with $a = a(\alpha)$. With these conventions the following statement holds:

2.11.ix. *If the extremals E_a cut a given surface S, and for every curve D_{46} in R, the corresponding integral $I^*[C_{35}]$ depends only upon the end points of the curve C_{35}, then the field condition (β) is satisfied in R.*

Proof. We have only to use the relation (2.9.5) in the form

$$I^*[D_{46}] = I^*[C_{35}] + I[E_{56}] - I[E_{34}].$$

For given points 4 and 6 in R, both $I[E_{56}]$ and $I[E_{34}]$ are fixed, and by hypothesis, $I^*[C_{35}]$ depends only upon 3 and 5, that is, 4 and 6. Thus, $I^*[D_{46}]$ has values which depend only upon the end points 4 and 6 of D_{46} and not on the actual curve D_{46} in R. □

Statement (2.11.ix) applies particularly well to the case where all extremals E_a pass through the same point $P_0 = (t_0, x_0)$ outside R. Then $I^*[C_{35}] = 0$ for every curve D_{46} in R, the hypersurface S being here the single point P_0.

G. The Hamilton-Jacobi Partial Differential Equation

If R is a Weierstrass field with slope function $p(t, x)$, then we know that

$$(f_0(t, x, p(t, x)) - \sum_i p_i(t, x) f_{0x'^i}(t, x, p(t, x))\, dt + \sum_i f_{0x'^i}(t, x, p(t, x))\, dx^i$$

is the exact differential of some function, say $-W(t, x)$, that is,

$$-\partial W/\partial t = f_0(t, x, p(t, x)) - \sum_i p_i(t, x) f_{0x'^i}(t, x, p(t, x)),$$

$$-\partial W/\partial x^i = f_{0x'^i}(t, x, p(t, x)), \qquad i = 1, \dots, n, \quad (t, x) \in R.$$

This is the same as saying that the surfaces $W(t, x) = \text{constant}$ in R are transversal to the marked trajectories of the field (see Section 2.2). Also we see that W satisfies the relation

$$\partial W/\partial t + f_0(t, x, p(t, x)) + \sum_i p_i(t, x)\, \partial W/\partial x^i = 0, \qquad (t, x) \in R.$$

By the definition of the Hamiltonian function $H(t,x,x',\lambda) = f_0(t,x,x') + \sum_i \lambda_i x'^i$, we also have

$$\partial W/\partial t + H(t,x,p(t,x),\partial W/\partial x) = 0,$$

where $\partial W/\partial x = (W_{x^1}, \ldots, W_{x^n})$.

If we assume that $\det f_{0x'x'} \neq 0$ in $R \times R^n$, then the relations $\lambda_i = -f_{0x'^i}(t,x,x')$, $i = 1, \ldots, n$, can be inverted, at least locally, yielding $x' = p(t,x,\lambda)$, that is, $x'^i = p_i(t,x,\lambda)$, $i = 1, \ldots, n$ (Legendre transformation, see Section 2.4C). If we take as in Section 2.4C

$$H_0(t,x,\lambda) = H(t,x,p(t,x,\lambda),\lambda) = f_0(t,x,p(t,x)) + \sum_i \lambda_i p_i(t,x,\lambda),$$

then we have the Hamilton-Jacobi partial differential equation

$$\partial W/\partial t + H_0(t,x,\partial W/\partial x) = 0, \qquad (t,x) \in R.$$

Example

If $I[x] = \int (1 + x'^2)^{1/2}\,dt$, $n = 1$, and $R = [(t,x) \mid t \leq a_0]$, let (a,b) be any point with $a_0 < a$. Then all straight lines through (a,b) form a field in R with marked trajectories $x = b - ma + mt$, $t \leq a_0$, the constant m ranging in R. These trajectories fill once R with slope function $p(t,x) = (a-t)^{-1}(b-x)$. Here $f_{0x'} = x'(1 + x'^2)^{-1/2}$, $f_0 - x'f_{0x'} = (1 + x'^2)^{-1/2}$, and for $x' = p(t,x)$, also $f_{0x'} = (b-x)((a-t)^2 + (b-x)^2)^{-1/2}$, $f_0 - x'f_{0x'} = (a-t)((a-t)^2 + (b-x)^2)^{-1/2}$. Then,

$$((a-t)^2 + (b-x)^2)^{-1/2}((a-t)\,dt + (b-x)\,dx)$$

is the exact differential of $-((a-t)^2 + (b-x)^2)^{1/2}$. Then, $W(t,x) = ((a-t)^2 + (b-x)^2)^{1/2}$ and we can easily verify that

$$W_t + H(t,x,p(t,x),W_x) = W_t + (1 + p^2(t,x))^{1/2} + p(t,x)W_x = 0.$$

Here the relation $\lambda = -x'(1 + x'^2)^{-1/2}$ is invertible yielding $x' = p(t,x,\lambda) = -\lambda(1 - \lambda^2)^{-1/2}$, and then $H_0 = H(t,x,p(t,x,\lambda),\lambda) = f_0 + p\lambda = (1 - \lambda^2)^{1/2}$, and the Hamilton-Jacobi partial differential equation is $W_t + (1 - W_x^2)^{1/2} = 0$, or

$$W_t^2 + W_x^2 = 1.$$

It is immediately verified that $W(t,x)$ satisfies this relation. Note that $f_0 - x'f_{x'} = (1 + x'^2)^{-1/2}$ is positive, while W is decreasing as t increases, so that the factor -1 in $-W_t = f_0 - x'f_{0x'}$ is essential.

In Section 2.13. we shall see a theory of fields from a quite different angle. In particular we shall see there that the "value function" is always a solution of the Hamilton-Jacobi first order partial differential equation

$$\partial S/\partial t + H_0(t,x,\partial S/\partial x) = 0. \tag{2.11.14}$$

Let us prove two relevant properties of this equation. To this effect we assume that $H_0(t, x, \lambda)$, $\lambda = (\lambda_1, \ldots, \lambda_n)$, is of class C^1.

2.11.x. *If $S(t, x, \alpha)$ is any solution of class C^2 of (2.11.14) depending on k parameters $\alpha = (\alpha_1, \ldots, \alpha_k)$, then the k partial derivatives $\partial S/\partial\alpha_j$, $j = 1, \ldots, k$, are constant along any AC solution $x(t)$, $\lambda(t)$ of the canonical equations.*

Proof. For S replaced by $S(t, x, \alpha)$ in (2.11.14) and by differentiation with respect to α_j we have

$$\partial^2 S/\partial t\,\partial\alpha_j = -\sum_i (\partial H_0/\partial\lambda_i)(\partial^2 S/\partial x_i\,\partial\alpha_j).$$

Now by simple computation we obtain

$$\begin{aligned}(d/dt)(\partial S/\partial\alpha_j) &= \partial^2 S/\partial t\,\partial\alpha_j + \sum_{i=1}^{n} (\partial^2 S/\partial x_i\,\partial\alpha_j)(dx^i/dt) \\ &= \sum_{i=1}^{n} (\partial^2 S/\partial x^i\,\partial\alpha_j)(dx^i/dt - \partial H_0/\partial\lambda_i) = 0,\end{aligned}$$

and then $\partial S/\partial\alpha_j = \beta$, a constant, $j = 1, \ldots, k$. □

2.11.xi. *If $S(t, x, \alpha)$ is any solution of class C^2 of (2.11.14) depending on n parameters $\alpha = (\alpha_1, \ldots, \alpha_n)$ such that $\det[\partial^2 S/\partial x^i\,\partial\alpha_j] \neq 0$, then the n relations*

$$(\partial/\partial\alpha_j)S(t, x, \alpha) = \beta_j, \qquad j = 1, \ldots, n, \tag{2.11.15}$$

define locally an n-vector function $x = x(t, \alpha, \beta) = (x^1, \ldots, x^n)$ with $\alpha = (\alpha_1, \ldots, \alpha_n)$, $\beta = (\beta_1, \ldots, \beta_n)$, which together with

$$\lambda(t) = (\lambda_1, \ldots, \lambda_n), \qquad \lambda_j = (\partial/\partial x^j)S(t, x, \alpha), \qquad j = 1, \ldots, n, \tag{2.11.16}$$

constitute a general solution of the canonical equations

$$dx^i/dt = \partial H_0/\partial\lambda_i, \qquad d\lambda_i/dt = -\partial H_0/\partial x^i, \qquad i = 1, \ldots, n. \tag{2.11.17}$$

Proof. We know already that the first n equations (2.11.15) hold when x is replaced by any solution $x(t)$ of (2.11.17). Since $\det[\partial^2 S/\partial x^i\,\partial\alpha_j] \neq 0$, we can locally determine functions $x_j(t)$, $j = 1, \ldots, n$, satisfying relations (2.11.15). We claim that these functions, together with the corresponding $\lambda_j(t)$ defined by relations (2.11.16), solve equations (2.11.17). To show this, let us differentiate (2.11.15) with respect to t and where the x^i are thought of as functions of t. We have as before

$$\begin{aligned}0 = (d/dt)(\partial S/\partial\alpha_j) &= \partial^2 S/\partial t\,\partial\alpha_j + \sum_{i=1}^{n} (\partial^2 S/\partial x^i\,\partial\alpha_j)(dx^i/dt) \\ &= \sum_{i=1}^{n} (\partial^2 S/\partial x^i\,\partial\alpha_j)(dx^i/dt - \partial H_0/\partial\lambda_i), \qquad j = 1, \ldots, n.\end{aligned}$$

Since $\det[\partial^2 S/\partial x^i\,\partial\alpha_j] \neq 0$, we conclude that $dx^i/dt - \partial H_0/\partial\lambda_i) = 0$, $i = 1, \ldots, n$, which is the first set of equations (2.11.17). Analogously, by differentiating

(2.11.16) with respect to t we obtain

$$d\lambda_i/dt = (d/dt)(\partial S/\partial x^i) = \partial^2 S/\partial t\,\partial x^i + \sum_{j=1}^{n} (\partial^2 S/\partial x^i\,\partial x^j)(dx^j/dt)$$

$$= \partial^2 S/\partial t\,\partial x^i + \sum_{j=1}^{n} (\partial^2 S/\partial x^i\,\partial x^j)(\partial H_0/\partial \lambda_j).$$

By differentiating (2.11.14) with respect to x^i we also have

$$\partial^2 S/\partial t\,\partial x^i = -\partial H_0/\partial x^i - \sum_{j=1}^{n} (\partial H_0/\partial \lambda_j)(\partial^2 S/\partial x^j\,\partial x^i),$$

and by comparison we have $d\lambda_i/dt = -\partial H_0/\partial x^i$, $i = 1, \dots, n$. □

2.12 More Sufficient Conditions

Let A be a given subset of the tx-space R^{n+1}, and let $f_0(t, x, x')$ be of class C^m, $m \geq 2$, in $A \times R^n$.

2.12.i (AN IMBEDDING THEOREM). *Let $E_{12}: x = x(t)$, $t_1 \leq t \leq t_2$, be a given nonsingular extremal, lying in the interior of A, and assume that there are no points $\bar{t}$, $t_1 < \bar{t} \leq t_2$, conjugate to t_1 in $(t_1, t_2]$. Then E_{12} can be imbedded in an n-parameter family $x(t, c)$, $c = (c_1, \dots, c_n)$, whose trajectories fill once some neighborhood R of E_{12}, defining a field in R.*

Proof. From Section 2.6B we know that x'' is continuous in $[t_1, t_2]$. Hence, the Euler differential system can be written in the form (2.2.15). Since $\det R(t) \neq 0$, $t_1 \leq t \leq t_2$, that is, $\det f_{0x'x'}(t, x(t), x'(t)) \neq 0$, by the implicit function theorem of calculus, system (2.2.15) can be thought of as written in normal form $x_i'' = F_i(t, x, x')$, $i = 1, \dots, n$, with all F_i of class C^{m-2} in all their arguments, at least for all $(t, x, x') \in R^{2n+1}$ of a neighborhood N_δ of the set of points $[t, x(t), x'(t), t_1 \leq t \leq t_2]$. Moreover, by the existence theorems for ordinary differential equations (e.g., McShane [I], Ch. 9) we can think of the extremal E_{12} as extended to an interval $[t_0, t_1]$ with $t_0 < t_1 < t_2$, with t_0 sufficiently close to t_1, and with preservation of the condition $\det f_{0x'x'}(t, x(t), x'(t)) \neq 0$ for all $t_0 \leq t \leq t_2$.

Let us prove that, for t_0 sufficiently close to t_1, there cannot be points $\bar{t}$, $t_0 < \bar{t} \leq t_2$, conjugate to t_0 in $(t_0, t_2]$. To see this, let us consider the accessory linear differential equation (2.5.3) relative to E_{12}, an equation which again can be written in the form

$$R(t)u'' + [R'(t) + Q(t) - Q^*(t)]u' + [Q'(t) - P(t)]u = 0, \qquad t \in [t_1, t_2],$$

where $\det R(t) \neq 0$ for all $t_1 \leq t \leq t_2$, and hence can be also written in normal form $u'' = M(t)u' + N(t)u$. Thus, there is a system $u_s(t)$, $s = 1, \dots, 2n$, of linearly independent solutions in $[t_1, t_2]$. With these solutions we form the $2n \times 2n$ determinants $d(t)$ and $D(t, t_0)$ we have encountered in (2.5):

$$d(t) = \det[(u_s(t), u_s'(t)), s = 1, \dots, 2n],$$
$$D(t, t_1) = \det[(u_s(t), u_s(t_1)), s = 1, \dots, 2n].$$

If in $D(t, t_0)$ we subtract the last n rows from the first ones and make use of the Taylor expansions $u^i(t) = u^i(t_0) + (t - t_0)\int_0^1 u'^i(t_0 + \theta(t - t_0))\,d\theta$, $i = 1, \ldots, n$, then $D(t, t_0) = (t - t_0)^n A(t, t_0)$, where now $A(t, t_0)$ is the determinant of the $2n \times 2n$ matrix whose sth column is made up of the n integrals $\int_0^1 u_s'^i(t_0 + \theta(t - t_0))\,d\theta$, $i = 1, \ldots, n$, and of the n-vector $u_s^i(t_1)$, $i = 1, \ldots, n$. Thus, $A(t_0, t_0) = (-1)^n d(t_0)$. For t_0 replaced by t_1 we then have $A(t_1, t_1) = (-1)^n d(t_1) \neq 0$. On the other hand, $A(t, t_1) \neq 0$ also for $t_1 < t \leq t_2$ because there are no points conjugate to t_1 in $(t_1, t_2]$. By continuity, we can take $t_0 < t_1$ so close to t_1 so that $A(t, t_0) \neq 0$, and hence $D(t, t_0) \neq 0$ for all $t_0 < t \leq t_2$.

Again by differential equation theory we know that if (ξ, η) is any $2n$-vector in a sufficiently small neighborhood of (ξ_0, η_0), $\xi_0 = x(t_0)$, $\eta_0 = x'(t_0)$, then there is one and only one solution $x(t)$, or $x(t; t_0, \xi, \eta)$, with $x(t_0) = \xi$, $x'(t_0) = \eta$. For $c = (\xi, \eta) = (\xi_1, \ldots, \xi_n, \eta_1, \ldots, \eta_n)$ we have a $2n$-parameter family of extremals. Now the $2n \times 2n$ matrix whose columns are (x_{c_s}, x'_{c_s}), $s = 1, \ldots, 2n$, has determinant $S(t)$ which is $\neq 0$ for all $t_0 \leq t \leq t_2$. To see this, first we note that the initial data are represented by the equations $\xi = x(t_0; t_0, \xi, \eta)$, $\eta = x'(t_0; t_0, \xi, \eta)$, which are identities, and where x and x' are functions of class C^{m-2} in their arguments. By differentiation we have then for $t = t_0$, $0 = x^i_{\xi_j} = x'^i_{\eta_j}$ for $i \neq j$, and analogous relations, briefly, $x^i_{\xi_j} = \delta_{ij}$, $x'^i_{\eta_j} = \delta_{ij}$, $i, j = 1, \ldots, n$; hence $S(t_0) = 1 \neq 0$. By differential equation theory we know that $S(t)$ is either identically zero or always different from zero; hence, $S(t) \neq 0$ for all $t_0 \leq t \leq t_2$. If we take $\xi = \xi_0 = x(t_0)$, then we have an n-parameter family of extremals, say $x(t, \eta)$, for every $\eta \in R^n$ with $|\eta - x'(t_0)| < \varepsilon$ for some $\varepsilon > 0$. The rank of the $2n \times n$ matrix $[x(t, \eta), x'(t, \eta)]$ must be n, since this matrix is only made up of the last n columns of the $2n \times 2n$ matrix we had before.

Finally, the $n \times n$ determinant $\det[x(t, \eta)]$ cannot be identically zero, since otherwise, its n columns would be linearly dependent, and then a linear combination of its columns, say $x(t) = c_1 x_1 + \cdots + c_n x_n = 0$, $c = (c_1, \ldots, c_n) \neq 0$, $t_0 \leq t \leq t_2$, would be identically zero, and hence $x'(t) = c_1 x'_1 + \cdots + c_n x'_n = 0$ would be also identically zero, and this is impossible, since $x'(t_0)$ is the same linear combination of the n columns of the unit matrix. From (2.5.ii), the n columns of $[x_\eta]$ are n solutions of the accessory system.

From (2.5.iii) we know now that the zeros of $\det[x_\eta(t, \eta)]$ are the conjugate points to t_0. Thus, $\det[x_\eta(t, \eta)] \neq 0$ for all $t_1 < t \leq t_2$.

We can now prove that for all $\bar{t}$, $t_0 < \bar{t} \leq t_2$, $x(\bar{t}, \eta)$ describes a neighborhood of $x(\bar{t})$ as η describes a sufficiently small neighborhood of $\eta_0 = x'(t_0)$. Indeed, $x(\bar{t}, \eta_0) = x(\bar{t})$ and $\det[x_\eta(\bar{t}, \eta_0)] \neq 0$, so the contention is only a corollary of the implicit function theorem of calculus. Thus, the extremals $x(t, \eta)$ cover a simply connected neighborhood R of E_{12}. By the uniqueness theorem for differential equations no two such extremals can pass through the same point $(\bar{t}, x) \in R$.

Furthermore, since $x(t, \eta)$, $x'(t, \eta)$ are functions of class C^1 in t and x, then the equation $x = x(t, \eta)$, $x \in R$, defines a function $\eta = \eta(t, x)$, $(t, x) \in R$, also of class C^1 in R, by the implicit function theorem of calculus. Finally, $p = x'(t, \eta)$ yields $p = p(t, x) = x'(t, \eta(t, x))$ as a function of class C^1 in R. In other words, there is defined in R a function $p = p(t, x)$ of class C^1, which gives the slope of the only trajectory $x(t, \eta)$ passing through (t, x). Condition (α) of Subsection 2.11A for a field in R is satisfied. Moreover, by the last remark in the proof of (2.11.ix) we know that this family of extremals $x = x(t, \eta)$ satisfies also condition (β) of Subsection 2.11A. Thus we have a field of extremals in R. □

2.12.ii (A SUFFICIENT CONDITION FOR A WEAK LOCAL MINIMUM). *If $E_{12}: x = x(t)$, $t_1 \leq t \leq t_2$, is an extremal, if $R(t)$ is positive definite (that is, $Q = u^* R(t) u > 0$ for all $u \in R^n$, $u \neq 0$, and $t_1 \leq t \leq t_2$), and if there is no point $\bar{t}$, $t_1 < \bar{t} \leq t_2$, conjugate to t_1 in $(t_1, t_2]$, then E_{12} is a weak local minimum for $I[x]$.*

Proof. By (2.12.i) and the last remark in Section 2.11 we know that E_{12} can be imbedded in a field R. Statement (2.12.ii) is now a corollary of (2.11.v). Let us note that here $R(t)$ positive definite for all $t_1 \le t \le t_2$ implies that $\det R(t) > 0$ for the same t, and E_{12} is nonsingular. □

2.12.iii (A Sufficient Condition for a Strong Local Minimum). *If $E_{12}: x = x(t)$, $t_1 \le t \le t_2$, is a nonsingular extremal, if for some $\varepsilon > 0$ we have $E(t, x, x', X') \ge 0$ for all $t_1 \le t \le t_2$, $|x - x(t)| \le \varepsilon$, $|x' - x'(t)| \le \varepsilon$, and $X' \in R^n$, and if there is no $\bar{t}$, $t_1 < \bar{t} \le t_2$, conjugate to t_1 in $(t_1, t_2]$, then E_{12} is a strong local minimum for $I[x]$, and moreover $I[C_{12}] > I[E_{12}]$ for C_{12} sufficiently close to E_{12} and distinct from E_{12}.*

Proof. As in the previous proof, E_{12} can be imbedded in a field R, and we need only apply (2.11.vi). We have proved that E_{12} is a strong local minimum. To prove the last part we need only to show that $E(t, x, x', X') > 0$ for all t, x, x' as stated and $X' \ne x'$. Indeed, let us suppose that $E(t, x, x', X') = 0$ for some $X' \ne x'$. Then E has a minimum at x', and the partial derivatives of E with respect to x' must be zero at that point. By computation we obtain $\sum_j f_{0x'^i x'^j}(X'^j - x'^j) = 0$ for all $i = 1, \dots, n$, and with $X' \ne x'$. This is impossible, since $\det R(t) = \det[f_{0x'^i x'^j}] \ne 0$. □

2.12.iv (A Sufficient Condition for a Strong Local Minimum). *If $E_{12}: x = x(t)$, $t_1 \le t \le t_2$, is an extremal, if for some $\varepsilon > 0$ we have $E(t, x, x', X') > 0$ for all $t_1 \le t \le t_2$, $|x - x(t)| \le \varepsilon$, $|x' - x'(t)| \le \varepsilon$, and all $X' \in R^n$, $X' \ne x'$, if $R(t)$ is positive definite, (that is, $Q = u^* R(t) u > 0$ for all $u \in R^n$, $u \ne 0$, and $t_1 \le t \le t_2$), and if there is no point $\bar{t}$, $t_1 < \bar{t} \le t_2$, conjugate to t_1, then E_{12} is a strong local minimum for $I[x]$.*

Proof. Since $R(t)$ is positive definite, we have $\det R(t) > 0$ for all $t_1 \le t \le t_2$. The statement is now a corollary of (2.12.iii). □

2.12.v (A Sufficient Condition for a Strong Local Minimum). *If $E_{12}: x = x(t)$, $t_1 \le t \le t_2$, is an extremal, if there is $\varepsilon > 0$ such that $Q(t, x, x', \xi) = \xi^* f_{0x'x'}(t, x, x')\xi > 0$ for all $t_1 \le t \le t_2$, $|x - x(t)| < \varepsilon$, and all ξ, $x' \in R^n$, $\xi \ne 0$, and if there is no point $\bar{t}$, $t_1 \le \bar{t} \le t_2$, conjugate to t_1, then E_{12} is a strong local minimum.*

Proof. First, $\det R(t) > 0$ for all $t_1 \le t \le t_2$, and E_{12}, therefore, is nonsingular. Furthermore, $E(t, x, x', X') = 2^{-1}(X' - x')^* f_{0x'x'}(t, x, \theta)(X' - x')$, where $f_{0x'x'}$ is computed at t, x, θ, and θ is some point on the segment between x' and X'. Thus, $E > 0$ for all $X' \ne x'$. The statement is now a corollary of (2.12.iii). □

Let A be a closed subset of the tx-space R^{n+1}, and let $f_0(t, x, x')$ denote a real valued function of class C^2 in $A \times R^n$. Moreover we assume here that for a given interval $[t_1, t_2]$ and all $t \in [t_1, t_2]$ the set $A(t) = [x \in R^n | (t, x) \in A]$ is convex. In other words, for every $\bar{t} \in [t_1, t_2]$ the intersection $A(\bar{t})$ of A with the hyperspace $t = \bar{t}$ is convex.

Let T denote the class of all AC n-vector functions $x(t)$, $t_1 \le t \le t_2$, with derivative $x'(t)$ essentially bounded, and with $(t, x(t)) \in A$, $t_1 \le t \le t_2$, $x(t_1) = x_1$, $x(t_2) = x_2$ for fixed x_1, x_2.

For every element $x \in T$ let V_x denote the class of all AC n-vector functions $\eta(t)$, $t_1 \le t \le t_2$, with η' essentially bounded, with $\eta(t_1) = 0$, $\eta(t_2) = 0$, and such that $(t, x(t) + a\eta(t)) \in A$ for all $t_1 \le t \le t_2$ and all real a of some interval $[0, a_0]$, $a_0 > 0$. Let $J_1[\eta; x]$ and $J_2[\eta; x]$ denote the linear and quadratic forms defined by (2.3.2) and (2.3.3).

2.12.vi (A SUFFICIENT CONDITION FOR AN ABSOLUTE MINIMUM). *Let $E_{12}: x = x_0(t)$, $t_1 \le t \le t_2$, be an extremal, and assume that $J_1[\eta; x_0] \ge 0$ for all $\eta \in V_{x_0}$ and $J_2[\eta; x] \ge 0$ for all $x \in T$ and $\eta \in V_x$. Then x_0 is an absolute minimum for $I[x]$ in T. If $J_2[\eta; x] > 0$ for all $x \in T$ and $\eta \in V_x$, $\eta \not\equiv 0$, then x_0 is a proper absolute minimum for $I[x]$ in T.*

Proof. Let $x(t)$, $t_1 \le t \le t_2$, be any element of T. Then, by the convexity hypothesis on A, we see that $\eta(t) = x(t) - x_0(t)$, $t_1 \le t \le t_2$, belongs to V_{x_0}, and that for any $0 \le \theta \le 1$, the element $x_\theta(t) = (1-\theta)x_0(t) + \theta x(t) = x_0(t) + \theta[x(t) - x_0(t)]$, $t_1 \le t \le t_2$, also belongs to T. Moreover, for $0 \le \theta \le 1$, $\eta_\theta = x - x_\theta$ also belongs to V_{x_0}. Now, by Taylor's formula with integral form of the remainder we have

$$I[x] - I[x_0] = J_1[\eta; x_0] + \tilde{J}_2[\eta; x_0]$$

with

$$\tilde{J}_2[\eta; x_0] = \int_{t_1}^{t_2} 2\tilde{\omega}(t, \eta(t), \eta'(t))\, dt,$$

$$\tilde{\omega}(t, \eta(t), \eta'(t)) = \eta^{*\prime}[\tilde{f}_{0x'x'}\eta' + \tilde{f}_{0x'x}\eta] + \eta^*[\tilde{f}_{0xx'}\eta' + \tilde{f}_{0xx}\eta],$$

$$\tilde{f}_{0x'x'} = \int_0^1 (1-\theta) f_{0x'x'}(t, x_\theta(t), x_\theta'(t))\, d\theta$$

and analogous expressions hold for $\tilde{f}_{0xx'}$, $\tilde{f}_{0x'x}$, $\tilde{f}_{0xx}$. Thus,

$$\tilde{J}_2[\eta; x_0] = \int_0^1 (1-\theta) J_2[\eta; x_\theta]\, d\theta.$$

Since $J_2[\eta; x_\theta] \ge 0$ for all $0 \le \theta \le 1$, we conclude that $I[x] - I[x_0] \ge 0$ for all $x \in T$. If $x \not\equiv x_0$, then $\eta \not\equiv 0$, $J_2[\eta; x_\theta] > 0$ for $0 \le \theta \le 1$, and $I[x] - I[x_0] > 0$. □

Remark 1. The geometrical configuration described above is only a particularization of the one we have considered in Remark 2 of Section 2.3, namely, with $D = A \times R^n$, $M = T$, $v^0[x] = V_x$, where D, M, v^0 are defined there. Thus, in the notation of (2.12.v), if both η and $-\eta$ belong to V_{x_0}, then $J_1[x_0, \eta] = 0$ is a necessary condition for x_0 to be a minimum of I. In particular, if x_0 lies in the interior of A, then V_{x_0} is the class of all AC $\eta(t)$, $t_1 \le t \le t_2$, with η' essentially bounded, and $\eta(t_1) = \eta(t_2) = 0$, and then $J_1[x_0, \eta] = 0$ for all such η is a necessary condition for a minimum.

Remark 2 (A CONDITION IN TERMS OF A RICCATI EQUATION). For $n = 1$, the following transformation of the second variation is relevant. First note that for any scalar functions $u(t)$, $\eta(t)$, $t_1 \le t \le t_2$, with $\eta(t_1) = \eta(t_2) = 0$, and of class C^1, we have

$$\int_{t_1}^{t_2} (d/dt)(u\eta^2)\, dt = [u\eta^2]_{t_1}^{t_2} = 0.$$

Now let $V_x \subset R^2$ be as in (2.12.vi), let f_0 be of class C^2, let $E_{12}: x = x(t)$, $t_1 \le t \le t_2$, be a given extremal, and assume that for the scalar $r(t) = f_{0x'x'}(t, x(t), x'(t))$ we have $r(t) > 0$ for all $t_1 \le t \le t_2$. Let q and p be the scalars $q(t) = f_{0x'x}$, $p(t) = f_{0xx}$. Then for the second variation $J_2[\eta; x]$ we have, from (2.3.3) and by direct computation,

$$\begin{aligned} J_2[\eta; x] &= \int_{t_1}^{t_2} [r\eta'^2 + 2q\eta\eta' + p\eta^2]\, dt \\ &= \int_{t_1}^{t_2} [r\eta'^2 + 2q\eta\eta' + p\eta^2 - (d/dt)(u\eta^2)]\, dt \\ &= \int_{t_1}^{t_2} [r(\eta' + r^{-1}(q-u)\eta)^2 + (-r^{-1}(q-u)^2 + p - u')\eta^2]\, dt. \end{aligned}$$

Thus, if $u(t)$, $t_1 \leq t \leq t_2$, is any solution, continuous in $[t_1, t_2]$, of the Riccati equation

$$u' = p - r^{-1}(q - u)^2,$$

then

$$J_2[\eta; x] = \int_{t_1}^{t_2} r(t)(\eta' + r^{-1}(q - u)\eta)^2 \, dt \geq 0.$$

For $n > 1$, let $W(t)$ be any $n \times n$ matrix whose entries are of class C^1, and let $\eta(t)$ be an n-vector also of class C^1, with $\eta(t_1) = \eta(t_2) = 0$. Then

$$0 = [\eta^* W\eta]_{t_1}^{t_2} = \int_{t_1}^{t_2} (d/dt)(\eta^* W\eta) \, dt = \int_{t_1}^{t_2} (\eta'^* W\eta + \eta^* W\eta' + \eta^* W'\eta) \, dt.$$

Let $V_x \subset R^{n+1}$ be as in (2.12.vi), let $f_0(t, x, x')$ be of class C^2, let $E_{12}: x = x(t)$, $t_1 \leq t \leq t_2$, be a given extremal, and suppose that $R(t) = f_{0x'x'}(t, x(t), x'(t))$ is positive definite for all $t_1 \leq t \leq t_2$. First, by direct computation we see that for any $n \times n$ symmetric matrix $W(t)$ of class C^1 we have

$$\begin{aligned}
&[\eta' + R^{-1}(Q - W)\eta]^* R[\eta' + R^{-1}(Q - W)\eta] \\
&\quad = (\eta'^* R\eta' + \eta'^* Q\eta + \eta^* Q^*\eta' + \eta^* P\eta) - (\eta'^* W\eta + \eta^* W\eta') \\
&\quad\quad + \eta^*[Q^* R^{-1} Q - Q^* R^{-1} W - W^* R^{-1} Q + W^* R^{-1} W - P]\eta.
\end{aligned}$$

By the remark above and by (2.3.3), then

$$\begin{aligned}
J_2[\eta; x] &= \int_{t_1}^{t_2} (\eta'^* R\eta' + \eta'^* Q\eta + \eta^* Q^*\eta' + \eta^* P\eta) \, dt \\
&= \int_{t_1}^{t_2} [\eta' + R^{-1}(Q - W)\eta]^* R[\eta' + R^{-1}(Q - W)\eta] \, dt + S, \\
S &= \int_{t_1}^{t_2} \eta^*[-Q^* R^{-1} Q + Q^* R^{-1} W + W^* R^{-1} Q - W^* R^{-1} W + P - W']\eta \, dt.
\end{aligned}$$

Thus, if W is a symmetric matrix solution of class C^1 of the matrix Riccati equation

$$W' = P - Q^* R^{-1} Q + Q^* R^{-1} W + W^* R^{-1} Q - W^* R^{-1} W,$$

we also have

$$J_2[\eta; x] = \int_{t_1}^{t_2} [\eta' + R^{-1}(Q - W)\eta]^* R[\eta' + R^{-1}(Q - W)\eta] \, dt \geq 0.$$

For a study of these transformations leading to sufficient conditions for a strong, or a weak, minimum, and for their connections with the Jacobi accessory system, we refer to T. W. Reid [1].

Remark 3. It may well occur that the functional $I[x]$ for which we seek extrema depend also on a real parameter $w \in R$ [or on a number q of parameters $w = (w^1, \ldots, w^q)$ with $w \in W$, a region in R^q), say

$$I[x, w] = g(t_1, x_1, t_2, x_2, w) + \int_{t_1}^{t_2} f_0(t, x(t), x'(t), w) \, dt,$$

under the usual constraints $(t, x(t)) \in A \subset R^{n+1}$, $(t_1, x(t_1), t_2, x(t_2)) \in B \subset R^{2n+2}$, and where $x(t) = (x^1, \ldots, x^n)$ is AC in $[t_1, t_2]$. We assume as in Section 2.2 that f_0 and g are of class C^1. If a pair $x(t)$, $t_1 \leq t \leq t_2$, $w \in W$, gives a maximum or a minimum to $I[x, w]$, x an AC function as in (2.2.i) and w in the interior of W, then the same necessary conditions as in (2.2.i) hold, and in addition

$$g_w(t_1, x(t_1), t_2, x(t_2), w) + \int_{t_1}^{t_2} f_{0w}(t, x(t), x'(t), w) \, dt = 0.$$

If $q > 1$ then the same holds for each of the q first order partial derivatives of g and f_0 with respect to $w^1, \ldots, w^q$.

Remark 4. In Section 2.2, Remark 5, we have already noticed that the case $n = 1$, f_0 linear in x', is exceptional, that is, the case where $f_0 = F(t, x) + G(t, x)x'$. Assume F, G to be of class C^2. First the Euler equation $(d/dt)f_{0x'} = f_{0x}$ reduces after simplifications to $G_t - F_x = 0$, which is not a differential equation. This is often called a singular case. Moreover $f_{0x'x'} \equiv 0$, $E \equiv 0$. The following remarks show that an optimal solution can still be determined.

First we note that

$$I[C] = \int_{t_1}^{t_2} (F(t, x) + G(t, x)x')\,dt = \int_C F(t, x)\,dt + G(t, x)\,dx$$

is a usual linear integral along the curve C, and thus I is defined also for parametric curves C in A. We shall assume below that A is actually the region within and on a simple closed curve C_0 of the tx-plane R^2, and that the extrema of I are sought in the class of the curves C in A joining two given points $1 = (t_1, x_1)$ and $2 = (t_2, x_2)$ on the boundary C_0 of A. For any two curves $C: x = x(t)$ and $C': x = y(t)$ joining 1 and 2 in A and not having other points in common, then C and $-C'$ form the boundary of a simple region $R \subset A$, and by Green's theorem we have

$$I[C] - I[C'] = \left(\int_C + \int_{-C'}\right)(F\,dt + G\,dx) = \iint_R (G_t - F_x)\,dt\,dx = \iint_R \omega(t, x)\,dt\,dx,$$

where we have assumed that the curve C' is above the curve C, and thus the boundary of R is traveled counterclockwise. If, say, C' is below C, then a sign minus must preceed the double integrals above. Thus, if $\omega(t, x) < 0$ in A, that is, $G_t - F_x < 0$ in A, then $I[C] < I[C']$ in the situation depicted in the illustration. A slightly more detailed

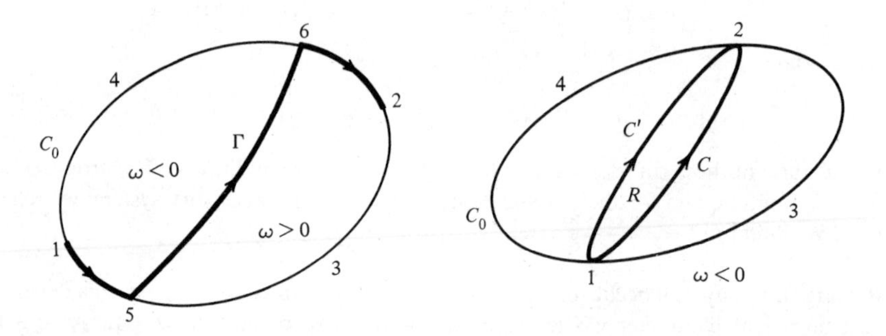

analysis shows that the arc 132 of C_0 gives the minimum for $I[C]$ in A, while the arc 142 of C_0 gives the maximum for $I[C]$ in A. The opposite occurs if $\omega > 0$ in A. Let us now assume that ω changes sign in A, namely, that the locus of $\omega(t, x) = 0$ is a simple arc Γ joining the points 5 and 6 of C_0 and dividing A into two regions, with $\omega < 0$ above Γ, and $\omega > 0$ below Γ (see illustration). A simple analysis shows that the curve 1562 gives a minimum for $I[C]$ in A. In other words, the optimum for I in A may well be made of two arcs 15 and 62 of the boundary of A and of the singular arc 56 on which $G_t - F_x = 0$. These considerations can be extended to many configurations concerning the zeros of $G_t - F_x$ in A, and have had the most remarkable applications to problems of technology (See A. Miele, [I, 1–3]). See Section 3.7A, nos. 12–15 for examples. For extensions of the considerations above to the case $n > 1$ we refer to A. Miele (loc. cit.) and G. W. Haynes [1].

2.13 Value Function and Further Sufficient Conditions

We consider the case where $B = B_2$ is a given subset of $V \subset A \subset R^{1+n}$ and we are interested in the minimum of the integral in the class $\Omega_{t_1x_1}$ of all admissible (AC) trajectories $x(t)$, $t_1 \le t \le t_2$, transferring any given point $(t_1, x_1) \in V$ to B in V, and thus $x(t_1) = x_1$, $(t_2, x(t_2)) \in B$, $(t, x(t)) \in V$. For every $(t_1, x_1) \in V$ we take now

$$\omega(t_1, x_1) = \inf_{\Omega t_1 x_1} I[x] = \inf_{\Omega t_1 x_1} \int_{t_1}^{t_2} f_0(t, x(t), x'(t))\, dt. \tag{2.13.1}$$

The function $\omega(t_1, x_1)$ is thus defined in V, and is called the value function. Then $\omega(t_1, x_1)$ may have the values $-\infty$ and $+\infty$, the latter whenever the class $\Omega_{t_1x_1}$ is empty. We shall assume, however, that every point (t_1, x_1) of V can be transferred to B by admissible (AC) trajectories, and thus $\Omega_{t_1x_1} \neq \emptyset$ for all $(t_1, x_1) \in V$, and $\omega(t_1, x_1)$ may still have the value $-\infty$, but cannot have the value $+\infty$. We also agree that $(t, x(t)) \in V - B$ for every $t_1 \le t < t_2$, so that $\omega(t, x) = 0$ for $(t, x) \in B$. To simplify notation we shall assume that V is a closed subset of R^{n+1}, and that V is the closure of its interior points.

2.13.i. *If $x(t), t_1 \le t \le t_2$, is optimal for $(t_1, x(t_1))$, then for every $\bar{t}, t_1 \le \bar{t} \le t_2$, the trajectory $x(t), \bar{t} \le t \le t_2$, is optimal for $(\bar{t}, x(\bar{t}))$.*

Proof. If for some $\bar{t}$, $t_1 < \bar{t} < t_2$, there were an admissible trajectory $\bar{x}(t)$, $\bar{t} \le t \le t_2^*, \bar{x}(\bar{t}) = x(\bar{t})$, transferring $(\bar{t}, x(\bar{t}))$ to B in V, with

$$\int_{\bar{t}}^{t_2^*} f_0(t, \bar{x}(t), \bar{x}'(t))\, dt < \int_{\bar{t}}^{t_2} f_0(t, x(t), x'(t))\, dt,$$

then for the extended trajectory $\bar{x}(t)$, $t_1 \le t \le t_2^*$, defined in $[t_1, \bar{t}]$ by taking $\bar{x}(t) = x(t)$, $t_1 \le t < \bar{t}$, we would have $I[\bar{x}] < I[x]$, a contradiction. □

2.13.ii. *If $x(t)$, $t_1 \le t \le t_2$, is any admissible trajectory in V, $x(t_1) = x_1$, $(t_2, x(t_2)) \in B$, then the function*

$$S(t) = \omega(t, x(t)) - \int_t^{t_2} f_0(s, x(s), x'(s))\, ds, \qquad t_1 \le t \le t_2, \tag{2.13.2}$$

is not decreasing in $[t_1, t_2]$. If x is an optimal trajectory for (t_1, x_1), then $S(t)$ is constant ($\equiv 0$) in $[t_1, t_2]$ (and $x(t)$, $\bar{t} \le t \le t_2$, is optimal for $(\bar{t}, x(\bar{t}))$ for every $\bar{t}$, $t_1 \le \bar{t} \le t_2$).

Proof. For any two points $t_1 \le \tau_1 < \tau_2 \le t_2$, $\xi_1 = x(\tau_1)$, $\xi_2 = x(t_2)$, and any admissible trajectory $\bar{x}(t)$, $\tau_2 \le t \le t_2^*$, transferring (τ_2, ξ_2) to B in V, $\bar{x} \in \Omega_{\tau_2\xi_2}$, we consider the extension $\bar{x}(t)$, $\tau_1 \le t \le t_2^*$, defined by taking $\bar{x}(t) = x(t)$ for $\tau_1 \le t \le \tau_2$. Then $\bar{x} \in \Omega_{\tau_1\xi_1}$ and by (2.13.1) we have

$$\omega(\tau_1, \xi_1) \le \int_{\tau_1}^{\tau_2} f_0(t, x(t), x'(t))\, dt + \omega(\tau_2, \xi_2).$$

By subtracting the analogous integral from τ_2 to t_2 from both members and transferring the integral above to the left hand side, we obtain

$$\omega(\tau_1, \xi_1) - \int_{\tau_1}^{t_2} f_0(t, x(t), x'(t)\, dt \le \omega(\tau_2, \xi_2) - \int_{\tau_2}^{t_2} f_0(t, x(t), x'(t))\, dt,$$

and the first part of (2.13.ii) is proved. If x is optimal, then $S(t)$ is constant (and equal to zero) along x, and equality holds in this relation. □

2.13.iii (A SUFFICIENT CONDITION FOR OPTIMALITY FOR A SINGLE TRAJECTORY). *If* (t_1, x_1) *is a point of* V, *and there is a function* $w(t,x)$ *in* V, *with* $w(t,x) = 0$ *for* $(t,x) \in B$, *such that* (a) *for every (admissible) trajectory* $x(t)$, $t_1 \le t \le t_2$, $x_1 = x(t_1)$, $(t_2, x(t_2)) \in B$, *then* $S(t) = w(t, x(t)) - \int_t^{t_2} f_0(s, x(s), x'(s))\, ds$ *is not decreasing in* $[t_1, t_2]$; *and* (b) *for some admissible trajectory* $x^*(t)$, $t_1 \le t \le t_2^*$, $x_1 = x^*(t_1)$, $(t_2^*, x^*(t_2^*)) \in B$, *the analogous expression* $S(t)$ *is constant (and* $\equiv 0$) *in* $[t_1, t_2^*]$, *then* x^* *is optimal for* (t_1, x_1), *and* $w(t_1, x_1) = \omega(t_1, x_1)$.

Proof. From the above we have

$$w(t_1, x_1) - \int_{t_1}^{t_2} f_0(t, x(t), x'(t))\, dt \le w(t_2, x(t_2)) = 0,$$

$$w(t_1, x_1) - \int_{t_1}^{t_2^*} f_0(t, x^*(t), x'^*(t))\, dt = w(t_2^*, x^*(t_2^*)) = 0,$$

and hence

$$\int_{t_1}^{t_2^*} f_0(t, x^*(t), x'^*(t))\, dt \le \int_{t_1}^{t_2} f_0(t, x(t), x'(t))\, dt.$$

Thus, x^* is optimal. Moreover $S(t_1) = 0$ implies

$$w(t_1, x_1) = \int_{t_1}^{t_2} f_0(t, x^*(t), x'^*(t))\, dt$$

where x^* is optimal, and hence $w(t_1, x_1) = \omega(t_1, x_1)$. □

2.13.iv (A SUFFICIENT CONDITION FOR OPTIMALITY FOR A SINGLE TRAJECTORY). *If* (t_1, x_1) *is a point of* V, *and there is a function* $w(t,x)$ *in* V, *continuous in* V, *continuously differentiable in* $V - B$, *with* $w(t,x) = 0$ *on* B *and such that* (a) *we have*

$$w_t(t,x) + \sum_{i=1}^{n} w_{x^i}(t,x) u^i + f_0(t,x,u) \ge 0 \tag{2.13.3}$$

for all $(t,x) \in V - B$ *and all* $u = (u^1, \ldots, u^n) \in R^n$, *and* (b) *for some admissible trajectory* $x^*(t)$, $t_1 \leq t \leq t_2^*$, *the expression*

$$S(t) = w(t, x^*(t)) - \int_t^{t_2^*} f_0(s, x^*(s), x^{*\prime}(s))\,ds$$

is constant on $[t_1, t_2^*]$, *that is, for* $t \in [t_1, t_2^*]$(a.e.),

$$w_t(t, x^*(t)) + \sum_{i=1}^n w_{x^i}(t, x^*(t)) x^{*\prime i}(t) + f_0(t, x^*(t), x^{*\prime}(t)) = 0,$$

then x^* *is optimal for* (t_1, x_1) *in* V, *and* $w(t_1, x_1) = \omega(t_1, x_1)$.

This is a corollary of (2.13.iii).

Let V be any set in R^{1+n} as stated at the beginning of this Section, let B be the target, a subset of V, and assume that B, if not a single point, has the same smoothness conditions we have required in Sections 2.1–2, in particular, for the transversality relations (2.2.9).

Let us assume that each point (t,x) of V can be taken to the target B by means of trajectories $x(\tau)$, $t \leq \tau \leq t_2$, which are continuous with sectionally continuous derivatives (briefly, of class C_s). Then the following holds:

2.13.v. *If* $\omega(t,x)$ *is of class* C^1 *in* V, *then at every interior point* (t,x) *of* V *we have*

$$\omega_t(t,x) + \sum_i \omega_{x^i}(t,x)u^i + f_0(t,x,u) \geq 0 \quad \text{for all } u \in R^n.$$

If (t,x) *can be taken to* B *by an optimal trajectory* $x^*(\tau)$, $t \leq \tau \leq t_2$, *of class* C_s, *and* $u_0 = x'^*(t+0)$, *then*

$$\omega_t(t,x) + \sum_i \omega_{x^i}(t,x)u_0^i + f_0(t,x,u_0) = 0.$$

Proof. If (s,y) is any point in the interior of V, let $\bar{x}(t)$, $s \leq t \leq t_2$, denote the trajectory we obtain by first taking the segment $\sigma: x = \bar{x}(t) = y + u(t-s)$, $s \leq t \leq s + \varepsilon$, with $u \in R^n$ and ε sufficiently small so that σ is in the interior of V, and then taking any trajectory $\bar{x}(t)$, $s + \varepsilon \leq t \leq t_2$, which transfers $(s+\varepsilon, y + u\varepsilon)$ to B in V. By (2.13.ii) the function $S(t)$, defined by (2.13.2) and computed along $\bar{x}$, is nondecreasing; hence $S'(s+0) \geq 0$, or

$$\omega_t(s,y) + \sum_i \omega_{x^i}(s,y)u^i + f_0(s,y,u) \geq 0,$$

and this is our first statement. If there is an optimal trajectory $x^*(t)$, $s \leq t \leq t_2$, of class C_s, then the same function $S(t)$ computed along x^* is constant ($\equiv 0$), and $S'(s+0) = 0$, or

$$\omega_t(s,y) + \sum_i \omega_{x^i}(s,y)u_0^i + f_0(s,y,u_0) = 0, \qquad u_0 = x'^*(s+),$$

and this is our second statement. □

Statement (2.13.v) can be reworded by saying that if ω is of class C^1, a necessary condition in order that any point $(t, x) \in V$ can be taken to B by an optimal trajectory x^* of class C_s is that

$$\min_{u \in R^n} \left[\omega_t(t, x) + \sum_i \omega_{x^i}(t, x)u^i + f_0(t, x, u) \right] = 0. \tag{2.13.4}$$

Equation (2.13.4) is called the "partial differential equation of dynamic programming".

Statement (2.13.v) can be inverted:

2.13.vi. *If the function $w(t, x)$ is continuous in V, is of class C^1 in $V - B$, with $w(t, x) = 0$ on B, and satisfies (2.13.4) in $V - B$, if (t, x) is any point interior to V, and $x(\tau)$, $t \leq \tau \leq t_2$, is any trajectory of class C_s transferring (t, x) to B in V, then*

$$w_t(\tau, x(\tau)) + \sum_i w_{x^i}(\tau, x(\tau))x'^i(\tau) + f_0(\tau, x(\tau), x'(\tau)) \geq 0,$$

and the function $S(\tau) = w(\tau, x(\tau)) - \int_\tau^{t_2} f_0(s, x(s), x'(s))\, ds$ is nondecreasing on $[t, t_2]$. If $x^*(\tau)$, $t \leq \tau \leq t_2^*$, is a trajectory of class C_s transferring (t, x) to B for which

$$w_t(\tau, x^*(\tau)) + \sum_i w_{x^i}(\tau, x^*(\tau))x'^{*i}(\tau) + f_0(\tau, x^*(\tau), x'^*(\tau)) = 0,$$

then x^* is optimal and $w(t, x) = \omega(t, x)$.

Proof. Since w is of class C^1 and x is of class C_s, then $w(\tau, x(\tau))$ is also of class C_s, and

$$S'(\tau) = w_t(\tau, x(\tau)) + \sum_i w_{x^i}(\tau, x(\tau))x'^i(\tau) + f_0(\tau, x(\tau), x'(\tau)) \geq 0.$$

Thus, $S(\tau)$ is nondecreasing in $[t, t_2]$. On the trajectory x^* the same expression is zero; hence $S(\tau)$ is constant on x^*, $S(\tau) \equiv 0$. The statement is now a consequence of (2.13.iii). □

A stronger form of (2.13.v) is as follow:

2.13.vii. *If $\omega(t, x)$ is of class C^2 and every point $(s, y) \in V$ can be taken to B by an optimal trajectory $x(t)$, $s \leq t \leq t_2$, of class C^1 and interior to V, then along any of these optimal trajectories we have*

$$(d/dt)\omega_{x^i}(t, x(t)) = -f_{0x^i}(t, x(t), x'(t)) = -(d/dt)f_{0x'^i}(t, x(t), x'(t)),$$

$$f_0(t, x(t), x'(t)) + \sum_i \omega_{x^i}(t, x(t))x'^i(t) = \min_{u \in R^n} \left[f_0(t, x(t), u) + \sum_i \omega_{x^i}(t, x(t))u^i \right].$$

Proof. The partial derivatives $\omega_{x^j}(t, x)$ are of class C^1, and so is $x(t)$, so that

$$(d/dt)\omega_{x^j}(t, x(t)) = \omega_{x^j t}(t, x(t)) + \sum_i \omega_{x^i x^j}(t, x(t))x'^i(t). \tag{2.13.5}$$

Now from (2.13.v) we have, for all $(t, x) \in V$ and $u \in R^n$,

$$\omega_t(t, x) + \sum_i \omega_{x^i}(t, x)u^i + f_0(t, x, u) \geq 0,$$

(2.13.6)

$$\omega_t(t, x(t)) + \sum_i \omega_{x^i}(t, x(t))x'^i(t) + f_0(t, x(t), x'(t)) = 0.$$

This means that $x(t)$ minimizes the expression

$$\omega_t(t, x) + \sum_i \omega_{x^i}(t, x)x'^i(t) + f_0(t, x, x'(t)) \quad \text{for } x \in V.$$

Because of the differentiability assumptions, the first order partial derivatives of the same expression with respect to all x^j must be zero at $x = x(t)$

$$\omega_{tx^j}(t, x(t)) + \sum_i \omega_{x^i x^j}(t, x(t))x'^i(t) + f_{0x^j}(t, x(t), x'(t)) = 0, \qquad j = 1, \ldots, n,$$

and by comparison with (2.13.4) also

$$(d/dt)\omega_{x^j}(t, x(t)) = -f_{0x^j}(t, x(t), x'(t)), \qquad j = 1, \ldots, n,$$

and this is the first statement in (2.13.vii). Now the relations (2.13.6) state that $u = x'(t)$ minimizes the expression

$$\omega(t, x(t)) + \sum_i \omega_{x^i}(t, x(t))u^i + f_0(t, x(t), u) \quad \text{for } u \in R^n,$$

or, equivalently, the second statement in (2.13.vii) holds. □

Statement (2.13.vii) shows that the n-vector $[\omega_{x^i}(t, x(t)), i = 1, \ldots, n]$ shares with the n-vector $\lambda(t) = -[f_{0x'^i}(t, x(t), x'(t)), i = 1, \ldots, n]$ the property of satisfying the canonical equations and the property of minimizing the Hamiltonian. Under more stringent requirements we shall prove in (2.13.viii) that actually $\omega_x(t, x(t)) = \lambda(t)$, and hence $f_0 + \sum_i \omega_{x^i}x'^i = f_0 + \sum_i \lambda_i x'^i = H$.

Now let us assume that a family of trajectories $x(t)$ in V is given, covering V and with terminal points $(t_2, x(t_2))$ on B. We say that these trajectories are the *marked trajectories* of the field V. Let $p(t, x)$ denote the slope $x'(t)$ of the trajectory $x(t)$ through (t, x).

Thus, each point $(s, y) \in V - B$ is taken to B by a unique trajectory $x = x(t; s, y)$, $s \leq t \leq t_2(s, y)$, with $x(s; s, y) = y$, $(t, x(t; s, y)) \in V$, $(t_2, x(t_2; s, y)) \in B$ for $t_2 = t_2(s, y)$, and $p(t, x(t; s, y)) = x'(t; s, y)$. Also, we denote by $w(s, y)$ the function

$$w(s, y) = \int_s^{t_2(s,y)} f_0(t, x(t; s, y), x'(t; s, y))\, dt, \qquad (s, y) \in V.$$

We assume that B has the smoothness properties required for (2.2.i), and we shall also assume that each marked trajectory hits the target B at a positive angle and nonzero speed. We further assume that f_0 is of class C^2; that $w(s, y)$ is continuous in V, zero on B, and of class C^2 in $V - B$; that $p(s, y)$ is of class C^1 in $V - B$, and that $x'(t; s, y)$ is of class C^1 for $(t, s) \in V - B$ and $s \leq t \leq t_2(s, y)$. Then we say that a field has been effected in V, and the function $p(t, x) = (p_1, \ldots, p_n)$ is often called the *slope function* and sometimes the *feedback function* of the field.

2.13.viii. *Let a field be effected in V with functions $w(s, y)$, $p(s, y)$, $x(t; s, y)$, $x'(t; s, y)$ satisfying the smoothness hypotheses above, and such that the marked trajectories satisfy the necessary conditions in (2.2.i). Then all marked trajectories are optimal, the partial differential equation (2.13.4) is satisfied in V, $w(s, y) = \omega(s, y)$, and on each of the marked trajectories we have*

$$\omega_{x^i}(t, x(t)) = \lambda_i(t) = -f_{0x'^i}(t, x(t), x_t(t)), \qquad i = 1, \ldots, n, \tag{2.13.7}$$

$$\omega_t(t, x(t)) = -M(t) = -\left[f_0(t, x(t), x'(t)) - \sum_i x'^i(t) f_{0x'^i}(t, x(t), x'(t))\right] \tag{2.13.8}$$

Proof. For every point $(s, y) \in V$ there is a marked trajectory $x(t; s, y)$, $s \le t \le t_2(s, y)$, taking (s, y) to the target B and whose slope is p, that is,

$$x(s; s, y) = y, \qquad (t_2(s, y), x_2(s, y)) \in B, \qquad x_2(s, y) = x(t_2(s, y); s, y), \tag{2.13.9}$$

$$x'(t; s, y) = p(t, x(t; s, y)), \qquad s \le t \le t_2(s, y). \tag{2.13.10}$$

Let $x^0(t; s, y) = -\int_t^{t_2(s,y)} f_0(\alpha, x(\alpha; s, y), p(\alpha, x(\alpha; s, y))\, d\alpha$, so that

$$x'^0(t; s, y) = f_0(t, x(t; s, y), p(t, x(t; s, y))), \qquad s \le t \le t_2(s, y). \tag{2.13.11}$$

We assume that the target B is a manifold of class C^1 and that each marked trajectory hits the target at a positive angle and nonzero speed. If $T(t, x) = 0$ is the equation of B, then $T(t_2(s, y), x(t_2(s, y); s, y)) = 0$. To prove that $t_2(s, y)$ is of class C^1 we have only to apply the implicit function theorem of calculus to the equation $T(t, x(t; s, y)) = 0$ in the unknown t. Indeed, $(\partial/\partial t)T(t, x(t; s, y)) = T_t + \sum_i T_{x_i}\, dx^i/dt = T_t + \sum_i T_{x_i} x'^i(t; s, y)$, and this derivative is $\neq 0$, since we have assumed that the trajectories hit B at an angle different from zero. Thus, $t_2(s, y)$ is of class C^1, and so is $x(t_2(s, y); s, y)$.

Below, by F_t and $F_y = (F_{y^1}, \ldots, F_{y^n})$ we denote as usual the first order partial derivatives of any function F with respect to t and $y = (y^1, \ldots, y^n)$. By (2.13.10–11) we have

$$\begin{aligned} \Lambda &= (d/dt)\left(x_y^0 + \sum_i \lambda_i(t) x_y^i\right) \\ &= \sum_i f_{0x^i} x_y^i + \sum_i f_{0x'^i}(p_i(t, x(t; s, y)))_y + \sum_i \lambda_i'(t) x_y^i + \sum_i \lambda_i(t) x_y'^i, \end{aligned}$$

where, because of the canonical equations (2.2.3), $d\lambda_i/dt = -H_{x^i}$, $dx^i/dt = p_i$, and then also $x_y'^i = (p_i)_y$, $i = 1, \ldots, n$. Here for the sake of brevity we write p_i for $p_i(t, x(t; s, y)) = x'^i(t; s, y)$. Now, from the minimum property of H, the expression $H = f_0(t, x, u) + \sum_i \lambda_i u^i$ has its minimum value at $u = p(t, x)$, and hence $H_{u^i} = f_{0x'^i} + \lambda_i = 0$, $i = 1, \ldots, n$. Now Λ becomes

$$\begin{aligned} \Lambda &= \sum_i f_{0x^i} x_y^i + \sum_i f_{0x'^i}(p_i)_y - \sum_i f_{0x^i} x_y^i + \sum_i \lambda_i (p_i)_y \\ &= \sum_i (f_{0x'^i} + \lambda_i)(p_i)_y = 0. \end{aligned}$$

Thus, the expression $x_y^0 + \sum_i \lambda_i x_y^i$ is constant on the interval $[s, t_2(s, y)]$, or

$$(2.13.12)\quad x_y^0(s) + \sum_i \lambda_i(s)x_y^i(s) = x_y^0(t_2) + \sum_i \lambda_i(t_2)x_y^i(t_2), \qquad t_2 = t_2(s, y).$$

On the other hand, we have the relations

$$0 = x_2^0(s, y) = x^0(t_2(s, y); s, y), \qquad x_2(s, y) = x(t_2(s, y); s, y).$$

and by differentiation we obtain the vector and matrix relations

$$0 = x_t^0 t_{2y} + x_y^0(t_2) = f_0 t_{2y} + x_y^0(t_2), \qquad x_{2y}(s, y) = pt_{2y} + x_y(t_2),$$

or

$$x_y^0(t_2) = -f_0 t_{2y}, \qquad x_y(t_2) = -pt_{2y} + x_{2y}.$$

where the second relation can also be written in the form $x_y^i(t_2) = -p_i t_{2y} + x_{2y}^i$, $i = 1, \ldots, n$,. By adding to the first equation the second one multiplied by $\lambda^*(t_2) = \text{row}(\lambda_1, \ldots, \lambda_n)$, we obtain the relation

$$(2.13.13)\quad \begin{aligned} x_y^0(t_2) + \lambda^*(t_2)x_y(t_2) &= -f_0 t_{2y} - \lambda^*(t_2)pt_{2y} + \lambda^*(t_2)x_{2y} \\ &= -(f_0 + \lambda^* p)t_{2y} + \lambda^* x_{2y}. \end{aligned}$$

Each marked trajectory must satisfy the transversality relation (2.2.9) with (s, y) fixed and $(t_2(s, y), x_2(s, y))$ on the manifold B. Hence

$$\left(f_0 - \sum_i x'^i f_{0x'^i}\right) dt_2(s, y) + \sum_i f_{0x'^i}\, dx_2^i = 0,$$

where f_0 and all other expressions are computed at $(t_2(s, y), x_2(s, y))$. This relation can also be written in the form

$$\left(f_0 + \sum_i \lambda_i(t_2)p_i(t_2)\right) dt_2 - \sum_i \lambda_i(t_2)\, dx_2^i = 0.$$

If we keep s fixed and we think of y as variable, then $dt_2 = t_{2y}\, dy$, $dx_2 = x_{2y}\, dy$, where dy is arbitrary, and the last relation yields

$$(f_0 + \lambda^* p)t_{2y} - \lambda^* x_{2y} = 0.$$

By comparing with (2.13.13) we see that the second member of (2.13.12) is zero. The first member of (2.13.12) is easy to compute. First, we have $x^0(s) = -w(s, y)$, and hence $x_y^0(s) = -w_y(s, y)$. From (2.13.9), i.e. $x(s; s, y) = y$, we derive $x_y(s; s, y) = I$. Thus, the first member of (2.13.12) is $-w_y + \lambda(s)$. We have proved that $-w_y(s, y) + \lambda(s) = 0$, and this is the relation (2.13.7) for w. Now on each marked trajectory $x(t)$, $t_1 \le t \le t_2$, we have $w(t, x(t)) = \int_t^{t_2} f_0(s, x(s), x'(s))\, ds$, and by differentiation

$$w_t(t, x(t)) + \sum_i w_{x^i}(t, x(t))x'^i(t) + f_0(t, x(t), x'(t)) = 0.$$

Since $w_{x^i} = \lambda_i$, this relation becomes $w_t + H(t, x(t), x'(t), \lambda(t)) = 0$, or $w_t + M(t) = 0$, and this is relation (2.13.8) for w.

By the minimum property of the Hamiltonian H(see (2.2.i)), we have for all $u \in R^n$

$$
\begin{aligned}
w_t(t, x(t)) &+ \sum_{i=1}^{n} w_{x^i}(t, x(t))p_i(t, x(t)) + f_0(t, x(t), p(t, x(t))) \\
&= -M(t) + f_0(t, x(t), p(t, x(t)) + \sum_{i=1}^{n} \lambda_i(t)p_i(t, x(t)) \\
&= -M(t) + H(t, x(t), p(t, x(t)), \lambda(t)) = 0 \\
&\leq -M(t) + H(t, x(t), u, \lambda(t)) \\
&= w_t(t, x(t)) + \sum_{i=1}^{n} w_{x^i}(t, x(t))u^i + f_0(t, x(t), u)
\end{aligned}
$$

This proves (2.13.4). The optimality of all marked trajectories is now a corollary of (2.13.iii), and $w(t, x) = \omega(t, x)$. This completes the proof of (2.13.viii). □

Remark. The smoothness assumptions for theorem (2.13.viii) can be reduced as we shall see in Chapter 4 in a more general context for problems of optimal control, where we will allow for lines of discontinuities for the function p. The proofs are essentially the same, and we have presented them here in a particularly simple situation.

Note that we have assumed p of class C^2 in $V - B$, not in B. Indeed p may not be even defined on B. For instance, for $f_0 = (1 + x'^2)^{1/2}$, $n = 1$, $B = (1, 0)$, $A = [(t, x)|t \leq 1]$, then the optimal trajectories are the segments joining (s, y) to $(1, 0)$, $s < 1$. Thus, $x(t; s, y) = y - (1 - s)^{-1}y(t - s)$, $x'(t; s, y) = -(1 - s)^{-1}y$, $s < 1$, $p(t, x) = -(1 - t)^{-1}x$ $t < 1$, and $p(t, x)$ is not defined for $t = 1$.

2.13.ix. *If a field has been effected in V with functions $\omega(t, x)$ continuous in V and continuously differentiable on $V - B$, and $p(t, x)$ continuous on $V - B$, then the value function $\omega(t, x)$ satisfies the first order partial differential equation in $V - B$.*

$$\omega_t(t, x) + H(t, x, p(t, x), \omega_x(t, x)) = 0. \tag{2.13.14}$$

In other words, the value function satisfies the Hamilton-Jacobi equation we encountered in Section 2.11G.

Proof. From (2.13.7) we have $\omega_{x^i}(t, x(t)) = \lambda_i(t)$, $i = 1, \ldots, n$. Hence, from (2.13.8) we also have

$$
\begin{aligned}
0 &= \omega_t(t, x(t)) + f_0(t, x(t), p(t, x(t))) + \sum_{i=1}^{n} \lambda_i(t)p_i(t, x(t)) \\
&= \omega_t(t, x(t)) + H(t, x(t), p(t, x(t)), \omega_x(t, x(t))), \qquad t_1 \leq t \leq t_2,
\end{aligned}
$$

or

$$\omega_t(t,x) + H(t,x,p(t,x),\omega_x(t,x)) = 0, \qquad (t,x) \in V.$$

In brief, we have proved the above relation (2.13.14) together with the relations

$$\omega_x(t,x(t)) = \lambda(t) = -f_{0x'}(t,x(t),x'(t)), \qquad \omega_t(t,x(t)) = -M(t),$$
$$M(t) = f_0(t,x(t),x'(t)) - (x'(t))^* f_{0x'}(t,x(t),x'(t)), \qquad d\lambda/dt = -f_{0x}(t,x(t),x'(t)),$$

where, as usual,

$$\omega_x = (\omega_{x^1}, \ldots, \omega_{x^n}), \qquad x'(t) = (x'^1, \ldots, x'^n), \qquad \lambda(t) = (\lambda_1, \ldots, \lambda_n),$$
$$f_{0x'} = (f_{0x'^1}, \ldots, f_{0x'^n}), \qquad f_{0x} = (f_{0x^1}, \ldots, f_{0x^n}). \qquad \square$$

For the integral $I[x] = \int (1 + x'^2)^{1/2}\, dt$, $n = 1$, with fixed end point $t = a$, $x = b$, we have $f_0(t,x,x') = (1 + x'^2)^{1/2}$, $f_{0x'} = x'(1 + x'^2)^{-1/2}$, and a field V is made up of the marked trajectories $x = b - ma + mt$, $t \leq a$, with the constant m ranging in R. These trajectories fill once the open set $[(t,x) | t < a]$, with slope function $p(t,x) = m = (a - t)^{-1}(b - x)$. The value function is $\omega(t,x) = ((a - t)^2 + (b - x)^2)^{1/2}$, and $S(t)$ is identically zero along each marked trajectory. Also, $\omega_t = -(a - t)((a - t)^2 + (b - x)^2)^{-1/2}$, $\omega_x = -(b - x)((a - t)^2 + (b - x)^2)^{-1/2}$. The Hamiltonian is here $H(t,x,x',\lambda) = (1 + x'^2)^{1/2} + \lambda x'$, and we can verify that

$$\omega_t + H(t,x,p(t,x),\omega_x(t,x)) = \omega_t + f_0(t,x,p(t,x)) + p(t,x)\omega_x = 0,$$
$$\lambda(t) = -f_{0x'}(t,x,p(t,x)) = -(b - x)((a - t)^2 + (b - x)^2)^{-1/2} = \omega_x.$$

Under the assumptions already made on V, we have in V the slope function $u = p = p(t,x)$, $(t,x) \in V$, representing at any point of V the slope of the optimal trajectory through (t,x). Under one further assumption, the partial differential equation can be given another form.

Let us assume that, for any $(t,x) \in V$, the relation

$$\lambda = -f_{0x'}(t,x,p), \qquad \lambda = (\lambda_1, \ldots, \lambda_n), \qquad p = (p_1, \ldots, p_n),$$

or $\lambda_i = -f_{0x'^i}(t,x,p)$, $i = 1, \ldots, n$, represents a continuous 1-1 transformation which, therefore, we can invert, obtaining $p = p(t,x,\lambda)$, or $p_i = p_i(t,x,\lambda)$, $i = 1, \ldots, n$ (Legendre transformation, see Section 2.4C). Then, equation (2.13.14) takes the form

$$\text{(2.13.15)} \qquad \omega_t(t,x) + f_0(t,x,p(t,x,\lambda)) + \sum_{i=1}^{n} \omega_{x^i}(t,x) p_i(t,x,\lambda) = 0.$$

If H_0 denotes $H_0(t,x,\lambda) = f_0(t,x,p(t,x,\lambda)) + \sum_{i=1}^{n} \lambda_i p_i(t,x,\lambda)$, then (2.13.15) becomes

$$\text{(2.13.16)} \qquad \omega_t(t,x) + H_0(t,x,\omega_x(t,x)) = 0, \qquad (t,x) \in V - B.$$

This is the classical Weierstrass–Jacobi partial differential equation.

Example

For the case of the length integral $I = \int (1 + x'^2)^{1/2}\,dt$ in the tx-plane, we have $n = 1$, $f_0 = (1 + x'^2)^{1/2}$, $f_{0x'} = x'(1 + x'^2)^{-1/2}$, and

$$H(t, x, p, \lambda) = (1 + p^2)^{1/2} + \lambda p.$$

In a field V with slope function $p = p(t, x)$, the partial differential equation $\omega_t + H(t, x, p, \omega_x) = 0$ is now

(2.13.17) $$\omega_t + (1 + p^2)^{1/2} + p\omega_x = 0$$

with $\omega_x = \lambda$. Here the relation $\lambda = -f_{0x'}(p)$, or $\lambda = -p(1 + p^2)^{-1/2}$, is indeed a continuous one-one transformation from $p \in R$ onto $\lambda \in (-1, 1)$, with inverse function $p = -\lambda(1 - \lambda^2)^{-1/2}$. Then

$$H_0(t, x, \lambda) = H(t, x, p(t, x, \lambda)) = (1 - \lambda^2)^{1/2}$$

and (2.13.17) becomes $\omega_t + (1 - \omega_x^2)^{1/2} = 0$, or

(2.13.18) $$\omega_t^2 + \omega_x^2 = 1.$$

For instance, for $V = R^2$, $B = \{(a, b)\}$ a single point, then $\omega(t, x) = ((t - a)^2 + (x - b)^2)^{1/2}$, and one can verify that this function ω satisfies (2.13.18) in $R^2 - \{(a, b)\}$.

2.14 Uniform Convergence and Other Modes of Convergence

As we have stated all along in this chapter, we consider the integrals $I[x] = \int_{t_1}^{t_2} f_0(t, x(t), x'(t))\,dt$ as functionals in some class Ω of the family, say T, of of all of AC n-vector functions $x(t) = (x^1, \ldots, x^n)$, $t_1 \leq t \leq t_2$, or curves $C: x = x(t)$, $t_1 \leq t \leq t_2$, for which $f_0(\cdot, x(\cdot), x'(\cdot))$ is L-integrable in $[t_1, t_2]$.

As we have seen in Section 2.1A and E, for elements $x(t)$, $y(t)$, $t_1 \leq t \leq t_2$, defined in the same interval, the usual distance function $d(x, y) = \sup|x(t) - y(x)|$ is a natural way to measure their apartness. If $x(t)$, $a \leq t \leq b$, and $y(t)$, $c \leq t \leq d$, are defined in intervals which may be distinct, then the following distance function ρ has been proposed:

$$\rho(x, y) = |a - c| + |b - d| + \sup_t |x(t) - y(t)|,$$

where now x and y are thought of as having been extended to $(-\infty, +\infty)$ by continuity and constancy outside their original intervals of definition (cf. Section 2.1E). By doing this, we think of T as imbedded in the metric space of all n-vector continuous functions $x(t)$, $a \leq t \leq b$, defined in arbitrary intervals. Thus, a sequence of elements of T, say $x_k(t)$, $t_{1k} \leq t \leq t_{2k}$, $k = 1, 2, \ldots$, is said to be *convergent in the ρ-metric*, or briefly *in the uniform topology*, or *uniformly*, toward an element of T, say $x(t)$, $t_1 \leq t \leq t_2$, provided $\rho(x_k, x) \to 0$ as $k \to \infty$, or equivalently if $t_{1k} \to t_1$, $t_{2k} \to t_2$ as $k \to \infty$, and $x_k \to x$ uniformly (in some interval $[a_0, b_0]$ containing all $[t_{1k}, t_{2k}]$, and after the extension already mentioned). We may even refer to this convergence as convergence in the metric space C with distance function ρ. Again, for all elements x_k and x of T, whenever the derivatives need to taken into consideration, we may require as before that $x_k \to x$ in the ρ-metric *and* that $x'_k \to x'$ in L_1 (and then we may say briefly that $x_k \to x$ in $H^{1,1}$). Alternatively we may require as before that $x_k \to x$ in the ρ-metric *and* that $x'_k \to x'$ in L_p, that is, $\|x'_k - x'\|_{L_p} = (\int_{a_0}^{b_0} |x'_k - x'|^p \, dt)^{1/p} \to 0$ as $k \to \infty$ (and then we say briefly that $x_k \to x$ in $H^{1,p}$). (Here again x_k and x are thought of as having been extended to some interval $[a_0, b_0]$ as before).

Finally, we may require for $x_k(t)$, $t_{1k} \le t \le t_{2k}$, $k = 1, 2, \dots$, and $x(t)$, $a \le t \le b$, all elements of T, that $x_k \to x$ in the ρ-metric and that $x'_k \to x'$ *weakly* in L_1, and then we say briefly that $x_k \to x$ weakly in $H^{1,1}$. The definition of weak convergence in L_1 and its main properties will be mentioned in Section 10.3.

For the sake of simplicity we shall concentrate on two modes of convergence:

(a) $x_k \to x$ in the ρ-metric; briefly, $x_k \to x$ "in the uniform convergence of the trajectories", or "uniformly".
(b) $x_k \to x$ in the ρ-metric and $x'_k \to x'$ weakly in L_1; briefly, $x_k \to x$ "in the weak convergence of the derivatives", or "weakly in $H^{1,1}$".

As an example, we may consider the sequence $x_k(t) = k^{-s} \sin k^2 t$, $k^{-1} \le t \le 1 + k^{-1}$, $k = 2, 3, \dots$, and the element $x_0(t) = 0$, $0 \le t \le 1$. For any $s > 0$, then x_k converges to x_0 as $k \to \infty$ in the mode (a), or "uniform convergence of the trajectories". For $s > 2$, then x_k converges to x_0 as $k \to \infty$ also in the mode (b), or "weak convergence of the derivatives". (Note that, for $s < 2$, the norms $\|x'_k\|$ in L_1 diverge, and no subsequence can be weakly convergent in L_1. For $s > 2$, $x'_k \to 0$ uniformly. For $s = 2$ the derivatives $x'_k = \cos k^2 t$, $k^{-1} \le t \le 1 + k^{-1}$, are equibounded, and hence equiabsolutely integrable, and by the Dunford–Pettis lemma and other statements (see Section 10.3) $x'_k \to 0$ weakly in L_1 as $k \to \infty$.

Exercises

1. Let B be a closed subset of the $t_1 x_1 t_2 x_2$-space R^{2n+2}, and $g(t_1, x_1, t_2, x_2)$ any given continuous real valued function defined on B. If $x_k(t)$, $t_{1k} \le t \le t_{2k}$, $k = 1, 2, \dots$, is a sequence of n-vector continuous functions converging in the ρ-metric toward a (continuous) n-vector function $x(t)$, $t_1 \leqq t \leqq t_2$, and $(t_{1k}, x_k(t_{1k}), t_{2k}, x_k(t_{2k})) \in B$ for all k, then $(t_1, x(t_1), t_2, x(t_2)) \in B$ and $g(t_{1k}, x_k(t_{1k}), t_{2k}, x_k(t_{2k})) \to g(t_1, x(t_1), t_2, x(t_2))$ as $k \to \infty$.
2. Let $x(t)$, $t_1 \le t \le t_2$, $x_k(t)$, $t_{1k} \le t \le t_{2k}$, $k = 1, 2, \dots$, be AC functions with $t_{1k} \to t_1$, $t_{2k} \to t_2$, $x'_k \to x'$ weakly in L_1, and suppose there is some sequence $[\bar{t}_k]$ of points $t_{1k} \le \bar{t}_k \le t_{2k}$, with $\bar{t}_k \to \bar{t}$, $t_1 \le \bar{t} \le t_2$, $x_k(\bar{t}_k) \to x(\bar{t})$. Then $x_k \to x$ in the ρ-metric as well as in the weak convergence of the derivatives (modes (a) and (b)). (Use the Dunford–Pettis theorem in Section 10.3.

2.15 Semicontinuity of Functionals

Let us recall here that given any space, or set S, of elements x, any concept of convergence of a sequence x_k of elements of S to an element x of S must satisfy the following basic axioms; (a) If x_k converges to x then any subsequence x_{k_s} must also converge to x; (b) any sequence of repetitions $x, x, \ldots, x, \ldots$ must converge to x, where x is any element of S.

Let σ denote any given concept of convergence in a set S, and let $F[x]$, or $F: S \to R$, be any functional on S, or real valued function defined on S. Then, $F[x]$ is said to be lower semicontinuous at an element $\bar{x}$ of S with respect to the concept σ of convergence, provided for every sequence x_k of elements of S converging to $\bar{x}$ we have $F[\bar{x}] \leq \liminf_{k \to \infty} F[x_k]$. For upper semicontinuity we require that $F[\bar{x}] \geq \limsup_{k \to \infty} F[x_k]$. Though functionals F need not have either of these properties, we shall see, both in Sections 2.16–19 and in Chapters 8 to 16, how relevant is lower semicontinuity to the existence of minima (and upper semicontinuity to maxima). We shall see, in Section 2.16 and later in Section 8.1 remarkable results concerning lower and upper semicontinuous functionals in rather abstract settings. Let it be clear that if it happens that a functional F is both upper and lower semicontinuous at some $\bar{x}$, then F is continuous at $\bar{x}$, that is, $F(\bar{x}) = \lim_{k \to \infty} F[x_k]$ for sequences $x_k \to x$ in the given topology σ. It is also clear that if F has an absolute minimum [maximum] in S at the point $\bar{x}$, then certainly F is lower [upper] semicontinuous at $\bar{x}$. In other words, lower [upper] semicontinuity at $\bar{x}$ in S is a necessary condition for F to have an absolute minimum [maximum] at $\bar{x}$, and this holds no matter what concept of convergence σ we take into consideration.

For the moment, we are concerned with functionals

$$I[x] = \int_{t_1}^{t_2} f_0(t, x(t), x'(t))\, dt \tag{2.15.1}$$

defined for elements x of some class Ω of AC vector functions $x(t)$, $t_1 \leq t \leq t_2$, with $(t, x(t)) \in A$ and $f_0(\cdot, x(\cdot), x'(\cdot))$ L-integrable in $[t_1, t_2]$. (Here A is a subset of R^{1+n}, and f_0 is a given real valued function defined, say, on $A \times R^n$).

Limiting ourselves to the two main modes of convergence (a) and (b) mentioned in (2.14), we then have corresponding concepts of say, lower semicontinuity. Thus, we say that $I[x]$ is lower semicontinuous at a given element $x_0(t)$, $t_1 \leq t \leq t_2$, of Ω with respect to "uniform convergence of the trajectories" provided

$$I[x_0] \leq \liminf_{k \to \infty} I[x_k] \tag{2.15.2}$$

for every sequence $x_k(t)$, $t_{1k} \leq t \leq t_{2k}$, $k = 1, 2, \ldots$, of elements x_k of Ω with $x_k \to x_0$ in the ρ-metric (mode (a), or, briefly, uniformly).

Analogously, we say that $I[x]$ is lower semicontinuous at an element x_0 of Ω with respect to "weak convergence of the derivatives" provided (2.15.2) holds for all sequences x_k above with $x_k \to x_0$ in the ρ-metric (briefly, uniformly) and $x'_k \to x'_0$ weakly in L_1 (mode (b)). The same definitions hold for upper semicontinuity. Note that if $x_0(t)$, $t_1 \leq t \leq t_2$, is an absolute minimum for $I[x]$ in Ω, or at least x_0 is a strong local minimum for $I[x]$ in Ω (cf. Section 2.1), then certainly $I[x]$ is lower semicontinuous with respect to the uniform convergence of the trajectories (mode (a)). An analogous statement holds for a strong local maximum and upper semicontinuity with respect to mode (a) of convergence.

We shall see in Section 12.17 important statements concerning lower and upper semicontinuity of the functionals (2.15.1).

2.16 Remarks on Convex Sets and Convex Real Valued Functions

Let us recall here that a subset M of points in R^n is said to be convex if $u_1, u_2 \in M, 0 \leq \alpha \leq 1$, implies that $\alpha u_1 + (1 - \alpha)u_2 \in M$, that is, for any points u_1, u_2 of M, the entire segment $s = u_1u_2$ lies in M. We shall prove in Section 8.4 that if M is convex, if $u_1, \ldots, u_m$ are points of M, and if $\alpha_i \geq 0$, $i = 1, \ldots, m$, are numbers with $\alpha_1 + \cdots + \alpha_m = 1$, then $u_0 = \sum \alpha_i u_i$ is also a point of M. Any sum of $u_0 = \sum \alpha_i u_i$ with such coefficients is said to be a convex combination of points u_i.

A real valued function $f_0(u)$, $u \in M$, is said to be convex in M if M is convex and u_1, $u_2 \in M$, $0 \leq \alpha \leq 1$, imply that

$$f_0(\alpha u_1 + (1 - \alpha)u_2) \leq \alpha f_0(u_1) + (1 - \alpha)f_0(u_2),$$

that is, the value of f_0 at any point of $s = u_1u_2$ is not greater than the corresponding value of the linear function between $f_0(u_1)$ and $f_0(u_2)$. Of course, the functions f_0 for which analogous relations holds with $\geq$ sign are said to be concave functions (thus, $-f_0$ is convex).

We shall prove in Section 2.17 that if f_0 is convex on the convex subset M of R^n, if $u_1, \ldots, u_m$ is any finite set of points of M, and if $\alpha_i \geq 0$, $i = 1, \ldots, m$, are real numbers with $\sum_i \alpha_i = 1$, then $u_0 = \sum_i \alpha_i u_i \in M$, and

$$f_0\left(\sum_i \alpha_i u_i\right) \leq \sum_i \alpha_i f_0(u_i).$$

We say that f_0 is convex at the point $u_0 \in M$ if for all finite systems $u_1, \ldots, u_m$ of points of M and numbers $\alpha_i \geq 0$, $i = 1, \ldots, m$, with $\sum \alpha_i = 1$, $u_0 = \sum \alpha_i u_i$, we also have

$$f_0(u_0) \leq \sum \alpha_i f_0 u_i). \tag{2.16.1}$$

We shall prove in Section 2.17 that (a) f_0 is convex on the convex subset M of R^n if and only if f_0 is convex at every point u_0 of M, and (b) for local convexity at any given point u_0 it is enough to verify that (2.16.1) holds for all convex combinations $u_0 = \sum \alpha_i u_i$ of at most $m \leq n + 1$ points u_i of M.

As we shall see in Section 8.4 convexity of $f_0(u)$ at u_0 simply means geometrically that f_0 is never below any of its supporting hyperplanes at u_0. In particular, if f_0 is of class C^1 and convex at u_0, then for $u = (u^1, \dots, u^n)$, $u_0 = (u_0^1, \dots, u_0^n)$, we have

$$f_0(u) \geq f_0(u_0) + \sum_{i=1}^{n} (u^i - u_0^i) f_{0u^i}(u_0) \quad \text{for all } u \in R^n. \tag{2.16.2}$$

We shall discuss at length, in Sections 8.1, 8.4 and 17.1–8, properties of convex sets and of convex (or concave) functions. Here we need to consider the integrand function $f_0(t, x, u)$ which appears in (2.15.1), which we assume here to be defined in $A \times R^n$ for the sake of simplicity. Thus, we say that for any $(t, x) \in A$, $f_0(t, x, u)$ is convex in u provided $u_1, u_2 \in R^n, 0 \leq \alpha \leq 1$ implies

$$f_0(t, x, \alpha u_1 + (1 - \alpha) u_2) \leq \alpha f_0(t, x, u_1) + (1 - \alpha) f_0(t, x, u_2).$$

We say that $f_0(t, x, u)$ is convex in u at some u_0 provided

$$f_0(t, x, u_0) \leq \sum \alpha_i f_0(t, x, u_i)$$

for all finite system of points $u_i \in R^n$ and numbers $\alpha_i \geq 0$, $i = 1, \dots, m$, with $u_0 = \sum \alpha_i u_i$, $\sum \alpha_i = 1$.

From (2.16.2) we derive that if $f_0(t, x, u)$ is of class C^1 and convex in u at some point u_0, then

$$f_0(t, x, u) \geq f_0(t, x, u_0) + \sum_{i=1}^{n} (u^i - u_0^i) f_{0u^i}(t, x, u_0).$$

Thus, $E(t, x, u_0, u) \geq 0$ for all $u \in R^n$, and the converse is also true. We shall see in the following Sections important relations between convexity of the integrand functions $f_0(t, x, u)$ in u and properties of the functional $I[x]$.

2.17 A Lemma Concerning Convex Integrands

A property of the integrand f_0 in (2.15.1) of which we shall see the relevance below is that $f_0(t, x, x')$ is convex in x' for every $(t, x) \in A$. As a preliminary we prove the following statement:

2.17.i. *If A is a closed subset of the tx-space R^{n+1} and $f_0(t, x, x')$ is a continuous function in $A \times R^n$ which is convex in x' for every $(t, x) \in A$, and if $x(t) =$*

$(x^1, \dots, x^n)$, $t_1 \le t \le t_2$, *is any AC n-vector function with graph in A, then* (2.15.1), *as a Lebesgue integral, exists and is finite or* $+\infty$.

Proof. Let $\delta_{ij} = 0$ for $i \neq j$, $= 1$ for $i = j$, where $i, j = 1, \dots, n$, and for every i let v_i denote the constant n-vector $v_i = (\delta_{ij}, j = 1, \dots, n)$, $i = 1, \dots, n$. Let us consider now the $2n + 1$ real valued continuous functions

$$\alpha_0(t) = f_0(t, x(t), 0), \qquad \alpha_i(t) = f_0(t, x(t), v_i),$$
$$\beta_i(t) = f_0(t, x(t), -v_i), \qquad t_1 \le t \le t_2, \quad i = 1, \dots, n.$$

For every fixed $(t, x(t))$, we consider $f_0(t, x(t), u)$ as a convex function of u only, and we note that it must have a supporting hyperplane (at least) at the point $u = 0$ (in R^n):

$$\pi : z^0 = \mathscr{L}_t(u) = \alpha_0(t) + \sum_{i=1}^{n} b_i(t) u^i, \qquad u = (u^1, \dots, u^n) \in R^n,$$

where the nonzero n-vector $b(t) = (b_1, \dots, b_n)$ in general depends on t, that is, a linear expression $\mathscr{L}_t(u)$ as above such that $f_0(t, x(t), u) \ge \mathscr{L}_t(u)$ for all $u \in R^n$. (See Sections 8.1 and 8.4 for details on convex functions and the existence of supporting hyperplanes.) In particular for $u = v_i$ and $u = -v_i$, $i = 1, \dots, n$, we have

$$\alpha_i(t) \ge \alpha_0(t) + b_i(t), \qquad \beta_i(t) \ge \alpha_0(t) - b_i(t),$$

so that

$$\alpha_0(t) - \beta_i(t) \le b_i(t) \le \alpha_i(t) - \alpha_0(t), \qquad t_1 \le t \le t_2.$$

Thus, the functions $b_i(t)$ are certainly bounded in $[t_1, t_2]$. If $L \ge |b(t)|$ for all $t \in [t_1, t_2]$, then, by the Schwartz inequality in R^n, we have

$$\left| \sum_i b_i(t) x'^i(t) \right| \le |b(t)| \, |x'(t)| \le L |x'(t)|$$

and

$$f_0(t, x(t), x'(t)) \ge \alpha_0(t) + \sum_i b_i(t) x'^i(t) \ge \alpha_0(t) - L|x'(t)|,$$

where α_0 is continuous and $|x'|$ certainly L-integrable. This yields statement (2.17.i). $\square$

2.18 Convexity and Lower Semicontinuity: A Necessary and Sufficient Condition

We are concerned here with the functional

$$I[x] = \int_{t_1}^{t_2} f_0(t, x(t), x'(t))\, dt$$

in the class T of all AC n-vector functions $x(t) = (x^1, \dots, x^n)$, $t_1 \le t \le t_2$,

t_1, t_2 arbitrary, with graph $\Gamma = [(t, x(t)), t_1 \le t \le t_2]$ contained in a given subset A of the tx-space R^{n+1}, and for which $f_0(\cdot, x(\cdot), x(\cdot))$ is L-integrable in $[t_1, t_2]$. We shall assume A to be closed, and moreover the closure of its interior: $A = \mathrm{cl}(\mathrm{int}\, A)$.

We begin by stating the following simple theorem which shows, rather forcefully and in a rather general situation, the strict relation between the global property of $f_0(t, x, u)$ being convex in u, and the property of $I[x]$ being lower semicontinuous.

2.18.i (Theorem). *If $f_0(t, x, u)$ is continuous in $A \times R^n$, then $I[x]$ is lower semicontinuous in the class T with respect to weak convergence of the derivatives (mode* (b) *of Section* 2.14) *if and only if $f_0(t, x, u)$ is convex in u for every $(t, x) \in A$.*

Both the necessity part and the sufficiency part are consequences of more general statements which we shall prove in subsequent sections. Actually, the necessity of the convexity of $f_0(t, x, x')$ will follow from statement (2.19.ii). The sufficiency will follow from statement (10.8.ii) and subsequent Remark 6.

2.19 Convexity as a Necessary Condition for Lower Semicontinuity

2.19.i (Theorem: A Necessary Condition for Lower Semicontinuity on Every Trajectory). *Let A be closed in R^{n+1} and $f_0(t, x, u)$ continuous on $A \times R^n$. If $I[x]$ is lower semicontinuous at all elements $x_0 \in T$ with graph in the interior of A and with respect to uniform convergence of the trajectories, then $f_0(t, x, u)$ is convex in u for every (t, x) in the closure of the interior of A* (Tonelli [I]). *The same occurs for weak convergence of the derivatives (mode* (b) *of Section* 2.14).

Remark. Before proceeding, we must point out that, by changing all the signs, (2.19.i) also states that if $I[x]$ is upper semicontinuous at every element $x_0 \in T$ with graph in the interior of A, then $f_0(t, x, u)$ is concave in u at least at every $(t, x) \in \mathrm{cl}(\mathrm{int}\, A)$. Thus, if $I[x]$ is continuous at the same elements x_0, then $f_0(t, x, u)$ is both convex and concave in u at every (t, x) as before, that is, f_0 is *linear* in u, that is, of the form

$$f_0(t, x, u) = F_0(t, x) + \sum_{i=1}^{n} F_i(t, x)u^i.$$

In other words, the only functionals $I[x]$ which are continuous in the uniform may be continuous are necessarily linear (continuous with respect to either mode (a) or (b) of Section 2.14). These linear functionals are actually continuous with respect to weak convergence of the derivatives (mode (b) of Section 2.14). Cf. Remark 3 of Section 10.8.

Proof of (2.19.i). Suppose that there is a point (t_0, x_0) in the interior of A at which $f_0(t_0, x_0, u)$ is not convex in u. This means that there is a constant $\sigma > 0$ and a convex combination $u_0 = \sum_{i=1}^m \lambda_i u_i$, $\lambda_i \geq 0$, $u_i \in R^n$, $i = 1, \dots, m$, $\lambda_1 + \cdots + \lambda_m = 1$, such that

$$f_0(t_0, x_0, u_0) \geq \sum_{i=1}^{m} \lambda_i f_0(t_0, x_0, u_i) + \sigma.$$

Since $f_0(t, x, u)$ is continuous, there is a $\delta > 0$ such that all (t, x) with $t_0 \leq t \leq t_0 + \delta$, $|x - x_0| < \delta$ are in A, and

$$|f_0(t, x, u_i) - f_0(t_0, x_0, u_i)| < \sigma/3, \qquad i = 0, 1, \dots, m,$$

for all these points (t, x). Let $\gamma = \delta$ if $|u_0| \leq \frac{1}{2}$, $\gamma = \delta/(2|u_0|)$ if $|u_0| \geq \frac{1}{2}$.

Let C_0 be the segment $C_0: x = x_0(t) = x_0 + (t - t_0)u_0$, $t_0 \leq t \leq t_0 + \gamma$. Then for all $t \in [t_0, t_0 + \gamma]$ we have $|x_0 + (t - t_0)u_0 - x_0| \leq \gamma|u_0| \leq \delta/2$. Let us divide $[t_0, t_0 + \gamma]$ into k equal parts by means of points $t_0 + j\gamma/k$, $j = 0, 1, \dots, k$. Let us divide each of these intervals into m parts, whose lengths are proportional to $\lambda_1, \dots, \lambda_m$, by using the points $t_0 + (j + \sum_{i=1}^r \lambda_i)\gamma/k$, $r = 0, 1, \dots, m$. Let us define a curve $C_k: x = x_k(t)$, $t_0 \leq t \leq t_0 + \gamma$, by requiring that $x_k(t_0) = x_0(t_0) = x_0$ and that $dx_k/dt = u_r$ when

$$t_0 + \left(j + \sum_{i=1}^{r-1} \lambda_i\right)\gamma/k < t < t_0 + \left(j + \sum_{i=1}^{r} \lambda_i\right)\gamma/k, \qquad r = 1, \dots, m.$$

Then

$$\begin{aligned} x_k(t_0 + j\gamma/k) &= x_0 + \int_{t_0}^{t_0 + j\gamma/k} x_k'(t)\,dt = x_0 + j\left(\sum_{i=1}^{m} \lambda_i u_i\right)\gamma/k \\ &= x_0 + j\gamma u_0/k = x_0(t_0 + j\gamma/k). \end{aligned}$$

Thus, C_k is a polygonal line whose points of abscissa $t_0 + j\gamma/k$, $j = 0, 1, \dots, k$, are on C_0 (see illustration, where we have assumed $n = 1$, $k = 3$, $m = 3$).

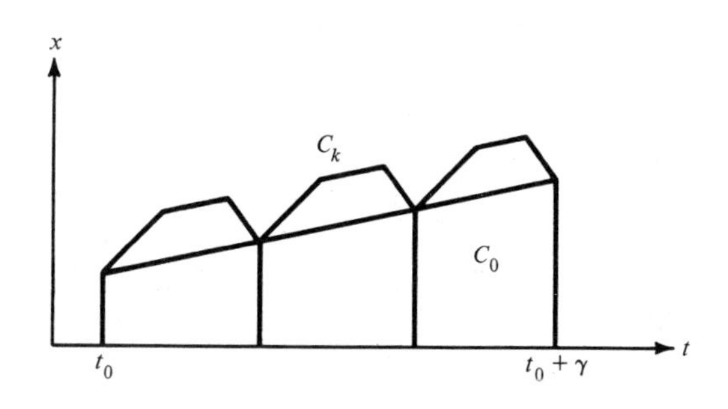

Moreover, every $t \in [t_0, t_0 + \gamma]$ has distance $\leq \gamma/2k$ from some point $t_0 + j\gamma/k$; hence,

$$|x_k(t) - x_0(t)| \leq |x_k(t) - x_k(t_0 + j\gamma/k)| + |x_k(t_0 + j\gamma/k) - x_0(t_0 + j\gamma/k)| + |x_0(t_0 + j\gamma/k) - x_0(t)| \leq |(\gamma/2k) \max|u_i| + 0 + (\gamma/2k)|u_0|.$$

Thus, $x_k(t)$ converges uniformly to $x_0(t)$ as $k \to \infty$. Let t_{jr} denote the points of subdivision $t_{jr} = t_0 + (j + \sum_{i=1}^{r} \lambda_i)\gamma/k$, $r = 0, 1, \ldots, m$, $j = 0, 1, \ldots, k$. Then for all k sufficiently large the curve C_k is contained in the set $[(t, x) | t_0 \leq t \leq t_0 + \gamma, |x - x_0| \leq \delta]$, and

$$\begin{aligned} I[x_k] &= \sum_{j=1}^{k} \sum_{r=1}^{m} \int_{t_{j-1,r-1}}^{t_{j-1,r-1}+\gamma\lambda_r/k} f_0(t, x_k(t), u_r)\, dt \\ &\leq \sum_{j=1}^{k} \sum_{r=1}^{m} (f_0(t_0, x_0, u_r) + \sigma/3)(\gamma\lambda_r/k) \\ &\leq \sum_{j=1}^{k} (f_0(t_0, x_0, u_0) - \sigma)(\gamma/k) + \sigma\gamma/3 \\ &= \sum_{j=1}^{k} (\gamma/k) f_0(t_0, x_0, u_0) - 2\sigma\gamma/3 \\ &\leq \sum_{j=0}^{k} \int_{t_0+(j-1)\gamma/k}^{t_0+j\gamma/k} (f_0(t, x_0(t), u_0) + \sigma/3)\, dt - 2\sigma\gamma/3 \\ &= I[x_0] - \sigma\gamma/3. \end{aligned}$$

Note that $x_k'(t)$ equals one of the numbers u_j, $j = 1, \ldots, m$, for all $t \in [t_0, t_0 + \gamma]$ (but finitely many); hence, the functions $x_k'(t)$, $t_0 \leq t \leq t_0 + \gamma$, $k = 1, 2, \ldots$, are equibounded. There is a subsequence, say $[k_s]$, therefore, such that x_{k_s}' converges weakly in L_∞, and hence in L_1, toward a function $y(t)$, $t_0 \leq t \leq t_0 + \gamma$, $y \in L_1$. Finally, as $s \to \infty$ and for any $t \in [t_0, t_0 + \sigma]$, we have

$$x_{k_s}(t) - x_{k_s}(t_0) = \int_a^t x_{k_s}'(\tau)\, d\tau \to \int_a^t y(\tau)\, d\tau,$$

$$x_{k_s}(t) - x_{k_s}(t_0) \to x_0(t) - x_0(t_0) = \int_a^t x_0'(\tau)\, d\tau;$$

hence $\int_{t_0}^t y\, d\tau = \int_{t_0}^t x_0'\, d\tau$ for all $t \in [t_0, t_0 + \gamma]$, and $y(t) = x_0'(t)$ a.e. in $[t_0, t_0 + \gamma]$. We have proved that $x_{k_s}' \to x_0'$ weakly in L_1 as $s \to \infty$.

By assuming that for some $(t_0, x_0) \in \operatorname{int} A$, $f_0(t, x_0, u)$ is not convex in u, we have constructed a sequence $x_{k_s}(t)$, $t_0 \leq t \leq t_0 + \gamma$, $s = 1, 2, \ldots$, which converges to $x_0(t)$, $t_0 \leq t \leq t_0 + \gamma$, in either of the modes (a), (b) of Section 2.14, and for which $I[x_0] > \liminf_{s\to\infty} I[x_{k_s}]$, a contradiction. Thus, $f_0(t_0, x_0, u)$ is convex in u for every $(t_0, x_0) \in \operatorname{int} A$, and by the continuity of f_0, the same holds for every (t_0, x_0) of the closure of the interior of A. Statement (2.19.i) is thereby proved. □

Statement (2.19.i) has an analogue concerning the lower semicontinuity property on a given trajectory $x_0(t)$, $t_1 \le t \le t_2$, $x_0 \in T$. We still assume here that A is a closed subset of the tx-space R^{n+1} with $A = \text{cl}(\text{int } A)$, and that f_0 is a continuous real function in $A \times R^n$.

2.19.ii (Theorem: A Necessary Condition for Lower Semicontinuity on a Given Trajectory). *Let $x_0(t)$, $a \le t \le b$, be a trajectory of class T, whose graph therefore is in A. If $I[x]$ is lower semicontinuous at x_0 with respect to uniform convergence of the trajectories* (*or mode* (a) *of Section* 2.14), *then at almost all $\bar{t} \in [a, b]$ such that $(\bar{t}, x_0(\bar{t}))$ is an interior point of A and $x_0'(\bar{t})$ exists and is finite, the scalar function $f_0(\bar{t}, x_0(\bar{t}), u)$, $u \in R^n$, is convex in u at the point $u = x_0'(\bar{t})$* (Tonelli [I]). *The same holds for weak convergence of the derivatives, or mode* (b) *of Section* 2.14.

Proof.

(a) The proof is by contradiction. Let E denote the subset of all $t \in [a, b]$, if any, such that $(t, x_0(t))$ is an interior point of A, $x_0'(t)$ exists and is finite, and $f_0(t, x_0(t), u)$ is not convex in u at $u = x_0'(t)$. Because of the continuity of f_0 the set E is measurable. Let us assume, if possible, that E has positive measure. We can even select a subinterval $[a', b']$ of $[a, b]$ whose points are all interior to $[a, b]$, containing a part of E still of positive measure. Let this part be denoted by E. Thus $E \subset [a', b']$, $|E| = 2\mu > 0$, and we take a number $\delta^* > 0$ such that $(t, x(t) + y) \in A$ for all $a' \le t \le b'$, $y \in R^n$, $|y| \le 2\delta^*$. By definition x_0' and $f_0(t, x_0(t), x_0'(t))$ are L-integrable in $[a', b']$. Also, at almost every point $t \in [a', b']$, $x_0'(t)$ exists and is finite and

$$\varepsilon^{-1} \int_t^{t+\varepsilon} f_0(\tau, x_0(\tau), x_0'(\tau))\, d\tau \to f_0(t, x_0(t), x_0'(t)) \quad \text{as } \varepsilon \to 0.$$

Let F_0 be the set of all points $t \in [a', b']$ (if any) where this does not occur, and include a' and b' in F_0. Here again as in Section 2.7 we denote by $|E|$ the measure of a measurable set E. Then $|F_0| = 0$. Also, $|E - F_0| = 2\mu > 0$, and there is a compact subset $E_1 \subset E - F_0$ of measure $|E_1| = \mu > 0$. By Lusin's Theorem (cf., e.g., E. J. McShane [I, p. 236]), we may choose E_1 in such a way that $x_0'(t)$ is continuous in E_1. Note that, for every $t_0 \in E_1$, there are a number $\sigma_0 > 0$ and a convex combination $u_0 = \sum_i \lambda_i u_i$ of points $u_j \in R^n$, $\lambda_j \ge 0$, $j = 1, \dots, \nu \le m$, $\sum_j \lambda_j = 1$, such that $u_0 = x_0'(t_0)$ and

$$f_0(t_0, x_0, u_0) \ge \sum_j \lambda_j f_0(t_0, x_0, u_j) + 2\sigma_0, \tag{2.19.1}$$

where $x_0 = x_0(t_0)$, and where, by Section (17.1), we can take $\nu \le m = n + 1$. By allowing some of the λ_j to be zero and corresponding u_j arbitrary, we can assume $\nu = m$. Note that there is a number δ_0, $0 < \delta_0 \le \delta^*$, such that $|t - t_0| \le \delta_0$, $|x - x_0| \le \delta_0$, $|u - u_0| \le \delta_0$ implies

$$|f_0(t, x, u) - f_0(t_0, x_0, u_0)| \le \sigma_0/4,$$
$$|f_0(t, x, u_j) - f_0(t_0, x_0, u_j)| \le \sigma_0/4, \qquad j = 1, \dots, m.$$

Finally, there is a number c_0, $0 < c_0 \le \delta_0$, such that $|t - t_0| \le c_0$ implies $|x(t) - x(t_0)| \le \delta_0$, and $|t - t_0| \le c_0$, $t \in E_1$, implies $|x'(t) - x'(t_0)| \le \delta_0$. To each point $t_0 \in E_1$ we associate the interval $(t_0 - c_0, t_0 + c_0)$. We have now an open cover of the compact set

E_1, and by the Borel covering theorem, finitely many of these intervals cover E_1. Let $(t_i - c_i, t_i + c_i)$, $i = 1, \ldots, N$, be this cover of E_1, and let $x_i = x(t_i)$, $u_i = x'(t_i)$, u_{ij}, λ_{ij}, $j = 1, \ldots, m$, δ_i, c_i be the corresponding elements, so that, for instance,

$$\sum_{j=1}^{m} \lambda_{ij} u_{ij} = u_i, \qquad f_0(t_i, x_i, u_i) \geq \sum_{j=1}^{m} \lambda_{ij} f_0(t_i, x_i, u_{ij}) + 2\sigma_i, \tag{2.19.2}$$

$$i = 1, \ldots, N.$$

(b) Each point $\bar{t} \in E_1$ certainly belongs to one, say $(t_i - c_i, t_i + c_i)$, of these intervals, and then $\bar{t} \in E_1$, $t_i \in E_1$, $|\bar{t} - t_i| \leq c_i \leq \delta_i$. If $\bar{x} = x(\bar{t})$, $\bar{u} = x'(\bar{t})$, then

$$\begin{aligned}
&|x(\bar{t}) - x(\bar{t}_i)| = |\bar{x} - x_i| \leq \delta_i \leq \delta^*,\\
&|\bar{u} - u_i| = |x'(\bar{t}) - x'(t_i)| \leq \delta_i,\\
&|f_0(\bar{t}, \bar{x}, \bar{u}) - f_0(t_i, x_i, u_i)| \leq \sigma_i/4,\\
&|f_0(\bar{t}, \bar{x}, u_{ij}) - f_0(t_i, x_i, u_{ij})| \leq \sigma_i/4, \qquad j = 1, \ldots, m.
\end{aligned}$$

From (2.19.2) we then derive

$$f_0(\bar{t}, \bar{x}, \bar{u}) \geq \sum_{j=1}^{m} \lambda_{ij} f_0(\bar{t}, \bar{x}, u_{ij}) + 3\sigma_i/2. \tag{2.19.3}$$

Let $\Delta = 1 + \max_{ij}|u_{ij}|$, and $\sigma = \min \sigma_i$. For every $j = 1, \ldots, m$, let $\{u_j\}$ denote the collection of points $u_{ij} \in R^n$, $i = 1, \ldots, N$. We have now the following proposition: For each point $\bar{t} \in E_1$ there is a system of m points $\bar{u}_j \in \{u_j\}$, $j = 1, \ldots, m$, and numbers $\lambda_j \geq 0$, $j = 1, \ldots, m$, $\sum_j \lambda_j = 1$, such that

$$f_0(\bar{t}, x(\bar{t}), x'(\bar{t})) \geq \sum_{j=1}^{m} \lambda_j f_0(\bar{t}, x(\bar{t}), \bar{u}_j) + 3\sigma/2. \tag{2.19.4}$$

(c) Let ν be any integer. For every $\bar{t} \in E_1$ and corresponding points $\bar{u}_j \in R^n$ and numbers $\lambda_j \geq 0$, $j = 1, \ldots, m$, $\sum_j \lambda_j = 1$, let $\bar{\delta}$ be the number δ_i stated above, and let δ' be a number $0 < \delta' \leq \bar{\delta} \leq \delta^*$, $\delta' \leq 1$, such that $|t - \bar{t}| \leq \delta'$, $|x - \bar{x}| \leq \delta'$, $|u - \bar{u}| \leq \delta'$ implies $|f_0(t, x, u) - f_0(\bar{t}, \bar{x}, \bar{u})| \leq \sigma/4$, and $|t - \bar{t}| \leq \delta'$, $|x - \bar{x}| \leq \delta'$, $|u - u_j| \leq \delta'$ implies $|f_0(t, x, u) - f_0(t, \bar{x}, u_j)| \leq \sigma/4$, $j = 1, \ldots, m$. Also, let $d > 0$ be a number such that $d\Delta \leq \min[1/\nu, \delta']$, and such that

$$|(t - \bar{t})^{-1}(x(t) - x(\bar{t})) - x'(\bar{t})| \leq \min[1/\nu, \delta'] \quad \text{for } |t - \bar{t}| \leq d. \tag{2.19.5}$$

Since $\bar{t} \in E_1 \subset E - F_0$, then $\bar{t} \notin F_0$, and we can choose $d > 0$ so that we have also

$$\left| (t - \bar{t})^{-1} \int_{\bar{t}}^{t} f_0(\tau, x_0(\tau)), x_0'(\tau)) \, d\tau - f_0(\bar{t}, x_0(\bar{t}), x_0'(\bar{t})) \right| < \sigma/4. \tag{2.19.6}$$

To each point $\bar{t} \in E_1$ we associate now the interval $(\bar{t} - d, \bar{t} + d)$. We have an open cover of the compact set E_1, and finitely many of these intervals cover E_1. Let $(\bar{t}_h - d_h, \bar{t}_h + d_h)$, $h = 1, \ldots, M$, be this cover of E_1, and let $\bar{x}_h = x(\bar{t}_h)$, $\bar{u}_h = x_0'(\bar{t}_h)$, u_{hj}, λ_{hj}, $j = 1, \ldots, M$, δ_h', d_h be the corresponding elements, $\delta_h' \leq 1$.

(d) We shall now select and relabel some of these intervals. Let $(\bar{t}_1 - d_1, \bar{t} + d_1)$ denote the one of these finitely many intervals which cover the first point of E_1 on $[a, b]$, and has maximum $\bar{t}_1 + d_1$. If $\bar{t}_1 + d_1 \in E_1$, let $(\bar{t}_2 - d_2, \bar{t}_2 + d_2)$ be the interval of the finite collection which covers $\bar{t}_1 + d_1$ and has maximum $t_2 + d_2$. If $\bar{t}_1 + d_1$ does not belong to E_1, let $(\bar{t}_2 - d_2, \bar{t}_2 + d_2)$ denote the interval of the finite collection which covers the first point of E_1 after $\bar{t} + d_1$ and has maximum $\bar{t}_2 + d_2$. By repeating this process finitely many times we obtain a finite collection of intervals $(\bar{t}_h - d_h, \bar{t}_h + d_h)$, $h = 1, \dots, M'$, $M' \leq M$, which covers E_1, and for which $\bar{t}_h < \bar{t}_{h+1}$, $h = 1, \dots, M' - 1$. Indeed, otherwise $(\bar{t}_h - d_h, \bar{t}_h + d_h)$ would not have the maximal property for which it has been chosen.

Now we need to reduce the same intervals to a new finite collection, say $I_i = [\alpha_i, \beta_i]$, $k = 1, \dots, m'$, of nonoverlapping closed intervals. To this effect, we proceed as follows. If $\bar{t}_h + d_h \leq \bar{t}_{h+1} - d_{h+1}$, we consider the intervals $[\bar{t}_h, \bar{t}_h + d_h]$, $[\bar{t}_{h+1} - d_{h+1}, \bar{t}_{h+1}]$. If $\bar{t}_h < \bar{t}_{h+1} < \bar{t}_h + d_h$, we consider the interval $[\bar{t}_h, \bar{t}_{h+1}]$. By doing this for $h = 1, \dots, M'$, and finally taking the parts that $[\bar{t}_1 - d_1, \bar{t}_1]$ and $[\bar{t}_{M'}, \bar{t}_{M'} + d_{M'}]$ have in $[a, b]$, we obtain the finite collection $J_i = [\alpha_i, \beta_i]$, $i = 1, \dots, m'$, which we shall index in such a way that $a \leq \alpha_1 < \beta_1 \leq \alpha_2 < \beta_2 \leq \cdots \leq \alpha_N < \beta_N \leq b$. Note that for every J_i there is a point t_i (either $t_i = \alpha_i$, or $t_i = \beta_i$) such that $t_i = \bar{t}_h$ where $\bar{t}_h$ is one of the points above, and (α_i, β_i) is entirely contained in $(\bar{t}_h - d_h, \bar{t}_h + d_h)$. Now let $x_i = x_0(t_i) = x_0(\bar{t}_h)$, $u_i = x'(t_i) = x'(t_h)$, $\gamma_i = \beta_i - \alpha_i$; let u_{ij}, $j = 1, \dots, m$, be the collection of points $u_j \in R^n$, $j = 1, \dots, m$, relative to $t_i = t_h$, and $\lambda_{ij} \geq 0$ the relative constants. Then, $\gamma_i \leq d_h \leq d_h \Delta \leq \min[1/v, \delta'_h]$, and from (2.19.4) also

$$f_0(t_i, x_i, u_i) \geq \sum_j \lambda_{ij} f_0(t_i, x_i, u_{ij}) + \sigma. \tag{2.19.7}$$

On the other hand, the intervals $[\alpha_i, \beta_i]$, $i = 1, \dots, m'$, cover E_1, so that

$$\sum_j (\beta_j - \alpha_j) = \sum_j \gamma_j \geq |E_1| = \mu > 0. \tag{2.19.8}$$

Since $\gamma_i \leq d_h$, we certainly have, by (2.19.6),

$$\left| \gamma_i^{-1} \int_{\alpha_i}^{\beta_i} f_0(t, x_0(t), x'_0(t))\, dt - f_0(t_i, x_0(t_i), x_0(t_i)) \right| < \sigma/4, \tag{2.19.9}$$

and hence

$$\int_{\alpha_i}^{\beta_i} f_0(t, x_0(t), x'_0(t))\, dt \geq \gamma_i f_0(t_i, x_i, u_i) - \gamma_i \sigma/4.$$

If we take

$$u_i^* = \gamma_i^{-1}(x_0(\beta_i) - x_0(\alpha_i)), \quad u_{ij}^* = u_{ij} + (u_i^* - u_i), \qquad j = 1, \dots, m,$$

then by (2.19.5),

$$\sum_{j=1}^{m} \lambda_{ij} u_{ij}^* = \left(\sum_j \lambda_{ij} u_{ij} - u_i \right) + u_i^* = u_i^*,$$

$$|u_i^* - u_i| = |\gamma_i^{-1}(x_0(\beta_i) - x_0(\alpha_i)) - x'_0(t_i)| \leq \delta'_h,$$

$$|u_{ij}^* - u_{ij}| \leq \delta'_h \leq 1, \qquad j = 1, \dots, m.$$

(e) Let us divide each interval $J_i = [\alpha_i, \beta_i]$ into m parts by means of the points $t_{ir} = \alpha_i + (\lambda_{i1} + \cdots + \lambda_{ir})\gamma_i$, $r = 0, 1, \dots, m$; hence $t_{i0} = \alpha_i$, $t_{im} = \beta_i$. Let us define a new continuous vector function $x_v(t)$, $a \leq t \leq b$, $x_v \in T$, by taking $x_v(t) = x_0(t)$ for all $t \in [a, b] - J_i$, and $dx/dt = u_{ij}^*$ for all $t_{i,j-1} < t < t_{ij}$, $j = 1, \dots, m$, $i = 1, \dots, M'$. Note that

$$t_{ij} - t_{i,j-1} = \lambda_{ij}\gamma_i = \lambda_{ij}(\beta_i - \alpha_i), \qquad j = 1, \dots, m. \tag{2.19.10}$$

Obviously,

$$x_v(\beta_i) = x_v(\alpha_i) + \gamma_i(\lambda_{i1} u_{i1}^* + \cdots + \lambda_{im} u_{im}^*) = x_v(\alpha_i) + \gamma_i u_i^* = x_0(\beta_i).$$

Hence, x_ν is continuous as stated. Also, for $t \in [\alpha_i, \beta_i]$ we have

$$\begin{aligned}|x_\nu(t) - x_i| = |x_\nu(t) - x_\nu(t_i)| &\le \gamma_i(\max|u^*_{ij}|)\\ &\le \gamma_i(1 + \max|u_{ij}|) = \gamma_i \Delta \le d_h \Delta \le \min[1/\nu, \delta'_h],\end{aligned}$$

so that, for $t \in [\alpha_i, \beta_i]$ and hence $|t - t_i| \le \gamma_i \le d_h \le \delta'_h$, we also have

(2.19.11) $$|f_0(t, x_\nu(t), u^*_{ij}) - f_0(t_i, x_i, u_{ij})| < \sigma/4.$$

We have now, first by (2.19.9) and (2.19.11), and then by (2.19.10),

$$\begin{aligned}I[x_\nu] - I[x_0] &= \sum_i \left[\sum_j \int_{t_{i,j-1}}^{t_{ij}} f_0(t, x_\nu(t), u^*_{ij})\,dt - \int_{\alpha_i}^{\beta_i} f_0(t, x_0(t), x'_0(t))\,dt\right]\\ &\le \sum_i \left[\sum_j \int_{t_{i,j-1}}^{t_{ij}} f_0(t_i, x_i, u_{ij})\,dt - \gamma_i f_0(t_i, x_i, u_i) + \sigma\gamma_i/2\right]\\ &= \sum_i \gamma_i \left[\sum_j \lambda_{ij} f_0(t_i, x_i, u_{ij}) - f_0(t_i, x_i, u_i)\right] + \sum_i \gamma_i \sigma/2.\end{aligned}$$

Finally, by (2.19.7) and (2.19.8) we have

(2.19.12) $$I[x] - I[x_0] \le \sum_i (-\gamma_i \sigma + \gamma_i \sigma/2) = -(\sigma/2) \sum_i \gamma_i \le -\sigma\mu/2,$$

and this holds for every $\nu = 1, 2, \dots$, with constant $-\sigma\mu/2$ independent of ν.

For every $t \in [\alpha_i, \beta_i]$ we have also

$$\begin{aligned}|x_\nu(t) - x_0(t)| &= |(x_\nu(t) - x_\nu(t_i)) + (x_\nu(t_i) - x_0(t_i)) + (x_0(t) - x_0(t_i))|\\ &\le \delta'_h + 0 + \delta'_h \le 2\delta^*,\end{aligned}$$

and thus $(t, x_\nu(t)) \in A$ for all $a \le t \le b$ and $\nu = 1, 2, \dots$. On the other hand we have also $|x_\nu(t) - x_0(t)| \le 1/\nu + 0 + 1/\nu = 2/\nu$, and thus $x \to x_0$ uniformly in $[a, b]$ as $\nu \to \infty$.

Also note that for $x'_\nu(t)$ we have either $x'_\nu(t) = x'_0(t)$, or $|x'_\nu(t)| = |u^*_{ij}| \le \Delta$. Thus, the functions $x'_\nu(t)$, $a \le t \le b$, $\nu = 1, 2, \dots$, are equiabsolutely integrable in $[a, b]$. Hence, by (10.3.i), there is a subsequence, say still $[\nu]$, such that x'_ν converges weakly in L_1 to some vector function $y(t)$, $a \le t \le b$, $y \in (L_1[a, b])^n$. For every $t \in [a, b]$ and as $\nu \to \infty$, we have

$$x_\nu(t) - x_\nu(a) = \int_a^t x'_\nu(\tau)\,d\tau \to \int_a^t y(\tau)\,d\tau,$$

$$x_\nu(t) - x_\nu(a) \to x_0(t) - x_0(a) = \int_a^t x'_0(\tau)\,d\tau.$$

Thus, $\int_a^t y\,d\tau = \int_a^t x'_0\,d\tau$, $x'_0(t) = y(t)$ a.e. in $[a, b]$, and $x'_\nu \to x'_0$ weakly in $(L_1[a, b])^n$. We have proved that $x_\nu \to x_0$ in the "weak convergence of the derivatives". Thus, $x_\nu \to x_0$ in both modes (a) and (b) of Section 2.14 but $I[x]$ is not lower semicontinuous in either mode of convergence. We have reached a contradiction. Theorem (2.19.ii) is thereby proved. □

Remark. As mentioned in Section 2.2, Theorem (2.19.ii) implies Weierstrass's necessary condition for a strong local minimum (or maximum). Indeed, if $x_0(t)$, $t_1 \le t \le t_2$, is a strong local minimum, then certainly $I[x]$ is lower semicontinuous at x_0 with respect to the "uniform convergence of the trajectories" (mode (a) of Section 2.14). Hence, for almost all $t \in [t_1, t_2]$, $f_0(t, x_0(t), u)$ is a convex function of u at $u_0 = x_0'(t)$. If $f_0(t, x_0(t), u)$ is of class C^1 in u, then $z = f_0(t, x_0(t), u)$, $u \in R^n$, is never below its tangent hyperplane at $u_0 = x_0'(t)$, so that for $u = (u^1, \ldots, u^n) \in R$, $u_0 = (u_0^1, \ldots, u_0^n)$, then

$$f_0(t, x_0(t), u) \geq f_0(t, x_0(t), u_0) + \sum_{i=1}^{n} (u^i - u_0^i) f_{0x'^i}(t, x_0(t), x_0'(t)),$$

or with the notation of Section 2.2

$$f_0(t, x_0(t), X') \geq f_0(t, x_0(t), x_0'(t)) + \sum_{i=1}^{n} (X'^i - x_0'(t)) f_{0x'^i}(t, x_0(t), x_0'(t)),$$

and hence

$$E(t, x_0(t), x_0'(t), X') \geq 0 \quad \text{for all } X' \in R^n.$$

It is now seen that the Weierstrass condition is not only a necessary condition for a strong local minimum, as stated in (2.2.i)(d), but also a necessary condition for the functional $I[x]$ to be lower semicontinuous (say, with respect to uniform convergence of the trajectories).

2.20 Statement of an Existence Theorem for Lagrange Problems of the Calculus of Variations

A. The Statement

We state briefly here existence theorems for Lagrange problems of the calculus of variations considered in this Chapter 2. These theorems will be contained as particular cases in the much more general existence theorems of Chapter 11.

Thus, we are concerned with integrals

$$I[x] = \int_{t_1}^{t_2} f_0(t, x(t), x'(t))\, dt,$$

and we shall state theorems guaranteeing the existence of an absolute minimum of $I[x]$ in the class Ω of all AC vector functions $x(t)$, $t_1 \leq t \leq t_2$, for which $f_0(\cdot, x(\cdot), x'(\cdot))$ is L-integrable, and which satisfy the constraints

$$(t, x(t)) \in A \subset R^{n+1}, \qquad (t_1, x(t_1), t_2, x(t_2)) \in B \subset R^{2n+2}.$$

We assume that the class Ω is nonempty, that is, the requirements are compatible. The elements x of the class Ω will be called trajectories.

2.20.i (TONELLI'S EXISTENCE THEOREM). *If A is compact, B is closed, if $f_0(t,x,x')$ is continuous in $A \times R^n$ and convex in x' for every $(t,x) \in A$, and if* ($\gamma 1$) *there is some real function $\Phi(\xi)$, $0 \le \xi \le +\infty$, bounded below and such that $\Phi(\xi)/\xi \to +\infty$ as $\xi \to +\infty$, and $f_0(t,x,x') \ge \Phi(|x'|)$ for all $(t,x,x') \in A \times R^n$, then $I[x]$ has an absolute minimum in Ω.*

The condition that A is compact (that is, closed and bounded), can be easily reduced. Indeed, if A is closed but contained in some slab $[t_0, T] \times R^n$, then (2.20.i) still holds if, for instance, we know that

(C_1) every trajectory $x \in \Omega$ contains at least one point $(t^*, x(t^*))$ in some compact subset P of R^{n+1}.

This point $(t^*, x(t^*))$ may well depend on the trajectory x. This condition (a) is certainly satisfied if, say, the first end point is fixed or the second end point is fixed.

If A is closed but not contained in any slab as before, then (2.20.i) still holds if (a) holds and in addition

(C_2) There are constants $\mu > 0$, $C > 0$ such that $f_0(t,x,x') \ge \mu$ for all $(t,x,x') \in A \times R^n$ with $|t| \ge C$.

2.20.ii (AN EXISTENCE THEOREM). *Statement* (2.20.i) *holds even if the growth condition* ($\gamma 1$) *is replaced by the weaker requirement* ($\gamma 2$): *For every $\varepsilon > 0$ there is a locally L-integrable function $\psi_\varepsilon(t) \ge 0$, $t \in R$, such that $|x'| \le \psi_\varepsilon(t) + \varepsilon f_0(t,x,x')$ for all $(t,x,x') \in A \times R^n$.*

Condition ($\gamma 2$) is equivalent to the analogous condition ($\gamma 3$): For every $p \in R^n$ there is a locally L-integrable function $\phi_p(t) \ge 0$, $t \in R^n$, such that $f_0(t,x,x') \ge p \cdot x' - \phi_p(t)$ for all $(t,x,x') \in A \times R^n$.

Further existence theorems with or without growth conditions will be proved in Chapter 11. In Chapter 12 we shall prove existence theorems in which the growth conditions are relaxed on certain "slender" sets of A, for instance, countably many straight lines not orthogonal to the t-axis, or countably many AC curves $x = \varphi(t)$ in A. In Chapter 14 we shall prove existence theorems in which the growth conditions (γ) are replace by Lipschitz-type or other growth-type conditions. In Chapter 15 we shall prove existence theorems for problems without growth conditions at all. One of these theorems will apply to the brachistochrone problem. Other existence theorems with no growth condition will be proved in Chapter 15.

We have already noticed in Chapter I that the convexity condition can be always incorporated in a given Lagrange problem if we allow for generalized solutions, and all the existence theorems now mentioned hold of course for problems written in terms of generalized solutions. However, in Chapter 16 we shall prove existence theorems for usual solutions of Lagrange problems with $f_0(t,x,x') = A(t)x + C(t,x')$, linear in x and not necessarily convex in x'.

B. Examples

The Lagrange problems below of the calculus of variations have an absolute minimum by theorems (2.20.i–ii) and variants.

1. $I = \int_0^1 x'^2\,dt$, $x(0) = x_1$, $x(1) = x_2$, $n = 1$.

2. $I = \int_0^1 (1 + t + x^2)x'^2\,dt$, $x(0) = 1$, $x(1)$ undetermined, $n = 1$. Here A is the slab $[0 \le t \le 1, x \in R]$, $P = \{(0,1)\}$, and F_0 satisfies the growth condition $(\gamma 1)$. Conditions (h_1) and (C_1) hold.

3. $I = \int_0^{t_2} (1 + |t| + |x|)|x'|^p\,dt$, $n = 1$, $p > 1$, with $x(0) = 1$, and (t_2, x_2) on the locus $\Gamma = [t = (1 + y^2)^{-1}, -\infty < y < +\infty]$. Here we can take for A the slab $[0 \le t \le 1, x \in R]$, $P = \{(0,1)\}$, and F_0 satisfies $(\gamma 1)$. Again (h_1), (C_1) hold.

4. $I = \int_0^1 t^q x'^2\,dt$, $x(0) = 1$, $x(1) = 0$, $0 < q < 1$, $n = 1$. Here $F_0 = t^q x'^2$ satisfies condition $(\gamma 2)$. Indeed, given $\varepsilon > 0$, take $\psi_\varepsilon(t) = \varepsilon^{-1}t^{-q}$, an L-integrable function in $[0,1]$ since $0 < q < 1$. Then for $|z| \le \varepsilon^{-1}t^{-q}$ we have $|z| \le \psi_\varepsilon(t) \le \psi_\varepsilon(t) + \varepsilon F_0$; if $|z| \ge \varepsilon^{-1}t^{-q}$, then $\varepsilon t^q|z| \ge 1$, and $|z| \le \varepsilon t^q z^2 = \varepsilon F_0 \le \psi_\varepsilon(t) + \varepsilon F_0$. Here $A = [0 \le t \le 1, z \in R]$, $P = \{(0,1)\}$, and again conditions (h_1), (C_1) are satisfied.

5. $I = \int_0^{t_2} (1 + t^q x'^2)\,dt$, $n = 1$, $x(0) = 0$, (t_2, x_2) on the locus $[tx = 1, t > 0, x > 0]$, $0 < q < 1$. Here we can take $A = [(t,x) | t \ge 0, x \ge 0]$, $P = \{(0,0)\}$. Conditions $(\gamma 2)$, (h_1), (C_1), (C_2) are all satisfied.

6. $I = \int_0^{t_2} (1 + x'^2)^q\,dt$, $n = 1$, $q > \frac{1}{2}$, $x(0) = 0$, (t_2, x_2) on the half straight line $\Gamma = [0 \le t < +\infty, x = 1]$. Here $F_0 \ge 1$, $B = (0,0) \times \Gamma$ is closed, and we can take $A = [(t,x) | t \ge 0, 0 \le x \le 1]$. Conditions $(\gamma 1)$, (h_1), (C_1), (C_2) are all satisfied.

7. $I = \int_0^{t_2} (1 + x'^2 + y'^2)^q\,dt$, $n = 2$, $q > \frac{1}{2}$, $x(0) = y(0) = 0$, (t_2, x_2, y_2) on the locus $\Gamma = [y \ge 1, (y - 1)^2 = t^2 + x^2]$, a nappe of a cone. Here $F_0 \ge 1$ satisfies condition $(\gamma 1)$ with $\phi(\zeta) = (1 + \zeta^2)^q$, and $B = (0,0,0) \times \Gamma$ is closed, and $A = [y \ge 0, (t,x) \in R^2]$ is closed. Conditions (h_1), (C_1), (C_2) hold.

C. Exercises

The reader can easily verify that the following problems satisfy the conditions of theorems (2.20.i–ii).

1. $I = \int_{t_1}^{t_2} (1 + x'^2)\,dt$, $I = \int_{t_1}^{t_2} x'^4\,dt$, with fixed and points $1 = (t_1, x_1)$, $2 = (t_2, x_2)$, $t_1 < t_2$.
2. $I = \int_{t_1}^{t_2} (atx' + x'^2)\,dt$ with fixed end points $1 = (t_1, x_1)$, $2 = (t_2, x_2)$, $t_1 < t_2$, a constant. Also, the same integral with fixed first end point $1 = (t_1, x_1)$, and second end point $2 = (t_2, x_2)$ on the vertical straight line $t = t_2$, that is, t_2 fixed, $t_2 > t_1$, x_2 arbitrary.
3. $I = \int_{t_1}^{t_2} (1 + (1 + t^2)x'^2)\,dt$ with fixed end points $1 = (t_1, x_1)$, $2 = (t_2, x_2)$, $t_1 < t_2$. Also, the same integral with $1 = (t_1, x_1)$ fixed, and $2 = (t_2, x_2)$ on the straight line $B_2 = [(t, x_2), t \ge t]$, x_2 fixed, $x_2 \ne x_1$.
4. $I = \int_{t_1}^{t_2} t^{-1}x'^2\,dt$, with fixed first end point $1 = (t_1, x_1)$ and second end point. $2 = (t_2, x_2)$ $0 < t_1 < t_2$, on a given closed subset B_2 of the strip $t_1 \le t \le N$, $x \in R$ (B_2 not entirely lying on the straight line $t = t_1$).
5. $I = \int_{t_1}^{t_2} tx'^2\,dt$, with fixed end points $1 = (t_1, x_1)$, $2 = (t_2, x_2)$, $0 < t_1 < t_2$. (For $t_1 = 0$ we have seen in Section 1.6, Example 4, that an absolute minimum need not exist).
6. $I = \int_{t_1}^{t_2} (1 + x'^2)\,dt$ with fixed first end point $1 = (t_1, x_1)$ and the second end point $2 = (t_2, x_2)$ on the locus $B_2 = [(t, x_2), t \ge t_1]$, $x_2 \ne x_1$, x_2 fixed. Same problem with $1 = (0,1)$ and $B_2 = [(t,0) | 3 \le t \le 4]$.

7. $I = \int_{t_1}^{t_2}(1 + x'^2)\,dt$, fixed first end point $1 = (0,0)$, second end point on the locus $B_2 = [(t,x)|tx = 1, t > 0, x > 0]$.
8. $I[x] = \int_{t_1}^{t_2}(x^2 + x'^2)\,dt$, $n = 1$, $x(t_1) = x_1$, $x(t_2) = x_2$, $t_1 < t_2$, x_1, x_2 fixed. Here A is the slab $[t_1, t_2] \times R$, and B is the point $(t_1, x_1, t_2, x_2) \in R^4$.
9. $I[x] = \int_0^{t_2}(1 - x' + x'^2)\,dt$, $n = 1$, $x(0) = 1$, $(t_2, x_2) \in B_2 = [t \geq 0, x \geq 0, tx = 1]$. Here $A = [0, +\infty)$, $B = \{0\} \times \{1\} \times B_2$.
10. $I[x] = \int_{t_1}^{t_2}(1 + t + t^2x + e^{|x'|})\,dt$, $n = 1$, $x(t_1) = x_1$, t_1, x_1, t_2 fixed, $t_1 < t_2$, $x_2 = x(t_2)$ undetermined. Here $A = [t_1, t_2] \times R$, $B = \{t_1\} \times \{x_1\} \times \{t_2\} \times R$.
11. $I[x] = \int_{t_1}^{t_2}(1 + t^2 + x^2 + |x'|^{3/2})\,dt$, $n = 1$, $x(t_1) = x_1$, $x(t_2) = 0$, t_1, x_1 fixed, t_2 undetermined, $t_2 \geq t_1$. Here $A = [t_1, +\infty) \times R$, $B = \{t_1\} \times \{x_1\} \times [t_1, \infty) \times \{0\}$.
12. $I[x, y] = \int_{t_1}^{t_2}(1 + t^2 + x^2 + y^2 + x'^2 + y'^2)\,dt$, $n = 2$, $x(t_1) = x_1$, $y(t_1) = y_1$, $x(t_2) = x_2$, $y(t_2) = y_2$, t_1, x_1, v_1, x_2, y_2 fixed, $(x_1, y_1) \neq (x_2, y_2)$, $t_2 \geq t_1$ undetermined. Here $A = [t_1, +\infty) \times R^2$, $B = (t_1, x_1, y_1, x_2, y_2) \times [t_1, +\infty)$.
13. $I[x, y] = \int_0^1(1 - t + t^2 + (1 + t^2)x'^2 + (2 + t^2)y'^2 + tx'y')\,dt$, $n = 2$, $t_1 = 0$, $t_2 = 1$, $x_1^2 + y_1^2 = 1$, $x_2^2 + y_2^2 = 2$. Here $A = [0,1] \times R^2$, $B = \{0\} \times B_1 \times \{1\} \times B_2$, $B_1 = [x^2 + y^2 = 1]$, $B_2 = [x^2 + y^2 = 2]$.

Bibliographical Notes

The main necessary conditions have been proved in Sections 2.2–5, first for AC trajectories with essentially bounded derivatives (Lipschitzian trajectories), and then for AC trajectories with possible unbounded derivatives (Section 2.9). In the latter case the proof is modeled after L. Tonelli [I]. The proof of statement (2.6.i) concerning the existence and continuity of the first derivative of AC optimal solutions is also modeled after L. Tonelli [I], and so is the straightforward proof in (2.9) of the transversality relation. The elementary proof of Weierstrass's necessary condition and related sufficient conditions (Sections 2.11–12) are modeled after G. A. Bliss [I, II]. For further discussion of the Riccati matrix equation, beyond the few points given in Section 2.12 we refer to T. W. Reid [1, 2].

We found it suitable to present both the concept of a field in the sense of Weierstrass Section 2.11, and the parallel concept of field from the view point of the value function Section 2.13. The considerations concerning the value function in (2.13) belong to Bellman's theory of dynamic programming for classical problems of the calculus of variations. The same theory will be seen in a more general setting, namely for problems of optimal control, in Chapter 4, with the additions due to Boltyanskii. Most of the classical sufficient conditions are proved in terms of Weierstrass fields (2.11–12). The rather subtle proof that the marked trajectories of a field are optimal is presented in (2.13) for a smooth slope function and in Section (4.5) for sectionally continuous slope functions (Boltyanskii regular synthesis). For dynamic programming see Bellman [II, III].

The concept of functional appeared for the first time in the work of V. Volterra [1, 2] (1887). Volterra conceived them as "functions of lines". His examples, mostly taken from applications, were all nonlinear, and the functionals of the calculus of variations were of course among them. He conceived, already in 1887, a formal "calculus" for them (see, e.g., V. Volterra [I, III], and V. Volterra and J. Peres [I]).

Soon after Baire had introduced in 1908 the concept of semicontinuity for real valued functions, Tonelli, in 1914, recognized semicontinuity as one of the relevant properties of the functionals of the calculus of variations. In particular Tonelli recognized that the Weierstrass condition, namely the convexity of $f_0(t, x, u)$ as a function of u, is not only a necessary condition for minima, but also a necessary and in a sense sufficient condition for lower semicontinuity, in the frame of functional analysis (L. Tonelli [I], [III]).

In the present exposition—mainly for problems with one independent variable—we limit ourselves to the consideration of two types of convergence (cf. Section 2.14), or topologies: (a) uniform convergence, or convergence in C, and (b) the weak convergence of the derivatives with uniform convergence of the trajectories, or convergence in $H^{1,1}$. For the general concepts of σ-space and σ-convergence, briefly mentioned in Section 2.15 we refer to V. Volterra and J. Peres [I]. Some results based on mere σ-convergence will be stated and proved in Section 8.1.

In Section 2.19 we have given Tonelli's proof of convexity as a necessary condition for lower semicontinuity, both globally (Theorem (2.19.i)) and on any single trajectory (Theorem (2.19.ii)), the latter including therefore also a proof of Weierstrass's condition in the calculus of variations.

For the functionals of the calculus of variations with continuous integrands, and with respect to weak convergence of the derivatives (convergence in $H^{1,1}$), mere convexity of $f_0(t, x, u)$ as a function of u is both necessary and sufficient for lower semicontinuity, a very simple result which remained unnoticed until recently (L. Cesari [23]), and which is stated in Section 2.18 and will be proved later in Section 10.7. See the end of Chapter 10 for further bibliographical information.

For problems of the calculus of variations for multiple integrals, necessary conditions are known extending some of those of this chapter. However, the optimal solutions may not be as smooth as we would expect—in other words, the theorems of Section 2.6 do not extend to multiple integrals. In Section 8.1 we give abstract versions of necessary conditions, and refer the reader to the extensive expositions of C. B. Morrey [I] and of J. L. Lions [I].

CHAPTER 3

Examples and Exercises on Classical Problems

3.1 An Introductory Example

Let us briefly consider, in terms of the integral $I = \int_{t_1}^{t_2} (1 + x'^2)^{1/2}\, dt$, the question of the path of minimum length between two fixed points, or between a fixed point and a given curve in the tx-plane. Here $f_0 = (1 + x'^2)^{1/2}$ depends on x' only, and satisfies condition (S) of Section 2.8. Any extremal must satisfy Euler's equation in the reduced form (2.2.10), $f_{0x'} = c$, or $x'(1 + x'^2)^{-1/2} = c$. Here $f_{0x'} = x'(1 + x'^2)^{-1/2}$ is a strictly increasing function of x' with range $(-1, 1)$. Thus $-1 < c < 1$, and there is one and only one value $x' = m$, depending on c, such that $x'(1 + x'^2)^{-1/2} = c$. From (2.6.iii) we know that any optimal solution is of class C^2 and therefore an extremal. Thus an optimal solution must be a segment $x' = m$, $x(t) = mt + b$, m, b constants.

(a) *Fixed end points.* For both end points fixed, $1 = (t_1, x_1)$, $2 = (t_2, x_2)$, $t_1 < t_2$, then $x(t) = x_1 + m(t - t_1)$, $t_1 \leq t \leq t_2$, $m = (x_2 - x_1)(t_2 - t_1)^{-1}$, corresponding to the segment $s = 12$ of length $L = ((t_2 - t_1)^2 + (x_2 - x_1)^2)^{1/2}$. That this is optimal can be seen as follows. Since $f_{0x'x'} = (1 + x')^{-3/2} > 0$, we have, by Taylor's formula,

$$f_0(x') - f_0(m) - (x' - m) f_{0x'}(m) = 2^{-1}(x' - m)^2 f_{0x'x'}(\tilde{m}) > 0$$

for any x' and where $\tilde{m}$ is some number between m and x'. Thus, for any AC arc $C: x = x(t)$, $t_1 \leq t \leq t_2$, $x(t_1) = x_1$, $x(t_2) = x_2$, we have

$$I[C] - I[s] \geq \int_{t_1}^{t_2} (x'(t) - m) f_{0x'}(m)\, dt = f_{0x'}(m)[x(t_2) - x(t_1) - m(t_2 - t_1)] = 0,$$

and equality holds if and only if $x'(t) = m$ in $[t_1, t_2]$ (a.e.), that is, if C is the segment $s = 12$.

Another way to show that $s = 12$ is the optimal path, is to imbed $s = 12$ in the field of extremals made up of all straight lines parallel to s and covering the whole tx-plane. Here $f_{0x'x'} > 0$; hence $E(x', X') > 0$ for all $X' \neq x'$, and $s = 12$ is the path of minimum length between 1 and 2 by Weierstrass's statement (2.11.iv), so $I[C_{12}] \geq I[s]$

for all $C_{12}: x = x(t)$, $t_1 \le t \le t_2$, x AC, joining 1 and 2, with equality holding only if $C_{12} = s$.

(b) *Variable end point*. In the case of the first end point $1 = (t_1, x_1)$ being fixed, and the second end point $2 = (t_2, x_2)$ being on any arc $\Gamma: x = g(t)$, $t' \le t \le t''$, of class C^1, then either $t' < t_2 < t''$, or $t_2 = t'$, or $t_2 = t''$. If $t' < t_2 < t''$, then B' is the linear manifold $(dt_2, dx_2 = g'(t_2)\,dt_2)$, dt_2 arbitrary (the tangent line to Γ at $(t_2, g(t_2))$). Here $f_{0x'} = x'(1 + x'^2)^{-1/2}$, $f_0 - x'f_{0x'} = (1 + x'^2)^{-1/2}$, and (2.2.9) yields $\Delta = dt_2[1 + x'(t_2)g'(t_2)] = 0$, dt_2 arbitrary; thus $x'(t_2)g'(t_2) = -1$. The segment $s = 12$ is therefore orthogonal to Γ at 2.

If $t_2 = t'$, then B' is the cone $dx_2 = g'(t_2)\,dt_2$ with $dt_2 \ge 0$, and (2.2.i) yields $\Delta = dt_2[1 + x'(t_2)g'(t_2)] \ge 0$ with $dt_2 \ge 0$, or $x'(t_2)g'(t_2) \ge -1$. Thus the angle α between the oriented tangents τ to $s = 12$ and τ' to Γ is $0 \le \alpha \le \pi/2$.

If $t_2 = t''$, then B' is the cone $dx_2 = g'(t_2)\,dt_2$ with $dt_2 \le 0$, and (2.2.i) yields $\Delta = dt_2[1 + x'(t_2)g'(t_2)] \ge 0$ with $dt_2 \le 0$, or $x'(t_2)g'(t_2) \le -1$. The angle α is $\pi/2 \le \alpha \le \pi$.

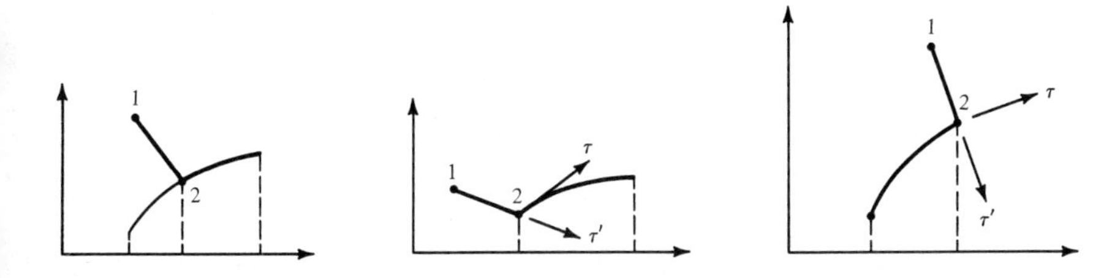

Remark. Critical examples concerning the same length integral I above have been mentioned in Section 1.6, Example 3.

3.2 Geodesics

A. The Equation of the Geodesics

Let us briefly consider the problem of joining two given points 1 and 2 on a surface S by an arc C lying on the surface and having the shortest possible length. Any such arc is called a *geodesic* on S.

Let us assume that S is given in parametric form

$$x = x(u, v), \quad y = y(u, v), \quad z = z(u, v), \qquad (u, v) \in D,$$

as a map of class C^1 from a fixed connected domain in D in the uv-plane. Then on any curve C of class C^1 on S the arc length parameter s has differential ds given by

$$\begin{aligned}(ds)^2 &= (dx)^2 + (dy)^2 + (dz)^2\\ &= P(u, v)(du)^2 + 2Q(u, v)\,du\,dv + R(u, v)(dv)^2,\end{aligned}$$

$$P = x_u^2 + y_u^2 + z_u^2, \qquad R = x_v^2 + y_v^2 + z_v^2, \qquad Q = x_u x_v + y_u y_v + z_u z_v.$$

If we think of the given points 1 and 2 on S as the images of certain points on D, say briefly $1 = (u_1, v_1)$, $2 = (u_2, v_2)$, $u_1 < u_2$, and C as the image of some path C_0 in D, say of the form $v = v(u)$, $u_1 \le u \le u_2$, $v(u_1) = v_1$, $v(u_2) = v_2$, $(u, v(u)) \in D$, then we have the problem of the minimum of the functional

$$I = \int_{u_1}^{u_2} (P(u,v) + 2Q(u,v)v' + R(u,v)v'^2)^{1/2}\, du,$$

with $v' = dv/du$. Here $f_0(u, v, v') = (P + 2Qv' + Rv'^2)^{1/2}$, and the Euler equation (2.2.12), $f_{0v} = (d/du) f_{0v'}$, is

$$\begin{aligned} 2^{-1}(P_v + 2Q_v v' + R_v v'^2)(P + 2Qv' + Rv'^2)^{-1/2} \\ = (d/du)(Q + Rv')(P + 2Qv' + Rv'^2)^{-1/2}. \end{aligned} \tag{3.2.1}$$

If P, Q, R depend on u only, then the reduced Euler equation (2.2.10), or $f_{0v'} = C_1$, becomes

$$Q + Rv' = C_1(P + 2Qv' + Rv'^2)^{1/2}.$$

If, in addition, $Q = 0$, that is, the lines $u =$ constant and $v =$ constant on S are orthogonal, then we have the equation

$$Rv' = C_1(P + Rv'^2)^{1/2}, \tag{3.2.2}$$

from which we derive $(R^2 - C_1^2 R)v'^2 = C_1^2 P$, and then

$$v = C_1 \int P^{1/2}(R^2 - C_1^2 R)^{-1/2}\, du, \tag{3.2.3}$$

where P and R are functions of u only.

Remark. The existence of at least one parametric path curve of minimum Jordan length joining two given points of S follows easily from general considerations (Cf. Section 14.1A. For the purpose of framing the problem in the present nonparametric discussion we have assumed that a curve of minimum length on S has a nonparametric representation $v = v_0(u)$, $u_1 \le u \le u_2$, and we compare its length only with the length of nonparametric curves on S having the same end points. Existence theorem for nonparametric integrals including the integral I are discussed in Section 14. For a discussion of the problem only in terms of parametric integrals see L. Tonelli [I].

B. Geodesics on a Sphere

By using polar coordinates, the sphere S of radius $a > 0$ has equations

$$x = a \sin u \cos v, \qquad y = a \sin u \sin v, \qquad z = a \cos u, \tag{3.2.4}$$

where $0 \le u \le \pi$, $0 \le v \le 2\pi$ represent colatitude and longitude respectively (and $u = 0$ and $u = \pi$ represent the poles). Here

$$\begin{aligned} P &= x_u^2 + y_u^2 + z_u^2 = a^2, \\ Q &= x_u x_v + y_u y_v + z_u z_v = 0, \\ R &= x_v^2 + y_v^2 + z_v^2 = a^2 \sin^2 u \end{aligned}$$

depend only on u. Thus, for curves $C:v = v(u)$, $u_1 \le u \le u_2$, on S we have

$$I = a \int_{u_1}^{u_2} (1 + v'^2 \sin^2 u)^{1/2}\, du,$$

$f_0(u, v')$ depends only on u and v', and the Euler equation $f_{0v'} = C_1$, i.e. (3.2.2), becomes $v' \sin^2 u = C_1(1 + v'^2 \sin^2 u)^{1/2}$, or

$$v'^2 \sin^2 u(\sin^2 u - C_1^2) = C_1^2. \tag{3.2.5}$$

Equation (3.2.3) now is

$$v = C_1 \int (\sin^4 u - C_1^2 \sin^2 u)^{-1/2}\, du;$$

hence, $v = -\arcsin(\gamma^{-1} \cot u) + C_2$, or $\sin(C_2 - v) = \gamma^{-1} \cot u$, and finally

$$(\sin C_2)(a \sin u \cos v) - (\cos C_2)(a \sin u \sin v) - \gamma^{-1}(a \cos u) = 0, \tag{3.2.6}$$

where γ is a constant with $\gamma^2 = C_1^{-2} - 1$. In other words, any optimal path is on a plane $Ax + By + Cz = 0$ through $(0, 0, 0)$. Thus, any geodesic on a sphere S is an arc of a great circle.

A few details on the integration above should be mentioned. First we have taken points $1 = (u_1, v_1)$, $2 = (u_2, v_2)$ not at antipodes on S, and we have chosen the polar representation in such a way that $0 < u_1 < u_2 < \pi$, $-\pi/2 < v_1$, $v_2 < \pi/2$ (or even $v_1 = v_2 = 0$). Thus, we seek optimal paths of the form $C:v = v(u), u_1 \le u \le u_2, v(u_1) = v_1$, $v(u_2) = v_2$. Now we note that $C_1 = 0$ in equation (3.2.5) corresponds to the solutions $v' = 0$, or $v =$ constant, the arcs of meridians on S. For $C_1 \ne 0$, then we must have $|C_1| < 1$, and actually, if $\sigma = \min[\sin u_1, \sin u_2]$, then $\sin u \ge \sigma > 0$ as u describes $[u_1, u_2]$. From equation (3.2.5) we derive now $|C_1| \le \sigma < 1$, and $C_1^{-2} \ge \sigma^{-2} > 1$, $\gamma^2 \ge \sigma^{-2} - 1 > 0$. Moreover, for $u_1 < u < u_2$, we have $\sin u > \sigma > 0, 1 + \cot^2 u = \sin^{-2} u < \sigma^{-2}$, $\cot^2 u < \sigma^{-2} - 1 \le \gamma^2$, and $\gamma^{-1}|\cot u| < 1$. Thus, $\arcsin(\gamma^{-1} \cot u)$ is defined in (u_1, u_2) with values in $(0, \pi)$, and

$$\begin{aligned}(d/du) \arcsin(\gamma^{-1} \cot u) &= -(1 - \gamma^{-2} \cot^2 u)^{-1/2} \gamma^{-1} \csc^2 u \\ &= -(\gamma^2 + 1 - \sin^{-2} u)^{-1/2} \sin^{-2} u \\ &= -C_1(\sin^4 u - C_1^2 \sin^2 u)^{-1/2},\end{aligned}$$

from which (3.2.6) follows. We have proved that any geodesic between two points 1 and 2 on S not at antipodes is the smaller of the two arcs of the great circle which joins them. This statement can be extended to points 1 and 2 at antipodes on S by the remark that any subarc also must be a geodesic, but now there are infinitely-many such arcs of great circles joining 1 and 2.

Since we know from existence theorems (cf. Section 14.1A) that paths of minimum length between any two points on S exist, we have also proved that the arcs of great circles mentioned above are the geodesics on S.

Note that, without the use of existence theorems, one can obtain the same results from sufficient conditions and the theory of fields. For instance, for points 1 and 2 not at antipodes, it is not restrictive to assume $1 = (u_1, 0)$, $2 = (u_2, 0)$, $0 < u_1 < u_2 < \pi$. We take $C_{12}:v = 0$, $u_1 \le u \le u_2$, and then C_{12} is imbedded in the field of extremals $v =$ constant (the family of meridians). Finally, C_{12} is optimal, as follows from Section 2.11, since $f_{0v'v'} = (1 + v'^2 \sin^2 u)^{-3/2} \sin^2 u > 0$ and $E(u, v', V') > 0$ for all $V' \ne v'$.

3.3 Exercises

1. Study the geodesics on a surface of revolution $y^2 + z^2 = [g(x)]^2$, or in parametric form

$$x = u, \qquad y = g(u) \cos v, \qquad z = g(u) \sin v. \tag{3.3.1}$$

2. Show that the family of geodesics of the paraboloid of revolution

$$x = u, \qquad y = u^{1/2} \cos v, \qquad z = u^{1/2} \sin v$$

has the form

$$u - C^2 = u(1 + 4C^2) \cos^{-2} \{v + 2C \log[k(2(u - C^2)^{1/2} + (4u + 1)^{1/2})]\}.$$

where C and k are arbitrary constants.

3. Prove that any geodesic on one nappe of the right circular cone

$$x^2 = b^2(y^2 + z^2) \tag{3.3.2}$$

has the following property: If the nappe is cut from the vertex along a generator and the surface of the cone is made to lie flat on a plane surface, then the geodesic becomes a straight line. *Hint*: use the following representation of the cone:

$$x = (1 + b^2)^{-1} br, \quad y = (1 + b^2)^{-1} r \cos(\theta(1 + b^2)^{1/2}), \quad z = (1 + b^2)^{-1} r \sin(\theta(1 + b^2)^{1/2})$$

in terms of the parameters r and θ. Then r and θ become usual polar coordinates on the flattened surface of the cone.

4. Prove the property analogous to the one in Exercise 3 for geodesics on a right circular cylinder.
5. Prove the same for an arbitrary cylindrical surface.
6. Show that for integrals $I = \int_{t_1}^{t_2} f_0(t, x')\,dt$, with f_0 of class C^2 in x' and $f_{0x'x'} \geq 0$, the function $I(\varepsilon) = \int_{t_1}^{t_2} f_0(t, X')\,dt$ is of class C^2 in ε if $X = x(t) + \varepsilon\eta(t)$, where $\eta(t_1) = \eta(t_2) = 0$, and x, η are continuous with sectionally continuous derivative. Moreover, $I''_{(0)} = \int_{t_1}^{t_2} f_{0x'x'}(t, x')\eta'^2\,dt$. Finally, show that if x satisfies the Euler equation for I, then x is an absolute minimum for I.
7. Show that the function f_0 for the geodesics on a surface of revolution satisfies the property of Exercise 6.

3.4 Fermat's Principle

Fermat expressed the principle that the time elapsed in the passage of light from a source A to an observer B should be the minimum possible. Thus, in an optical homogeneous material where the velocity is constant, the light path is the segment AB.

(a) If A and B are in two different media, each optically homogeneous with light velocities v_1 and v_2 respectively, and they are separated by a plane π, then the Fermat principle implies the usual law of refraction, or Snell's law. Indeed, we can choose xyz coordinates so that π is the plane $y = 0$, and A and B are on the plane $z = 0$, say $A = (x_1, y_1)$, $B = (x_2, y_2)$, with $z_1 = z_2 = 0$, $x_1 < x_2$, $y_1 < 0$, $y_2 > 0$; then the light path

ACB must be in the same plane and made up of two segments AC and CB, $C = (x,0)$ again with $z = 0$. Then for the time T from A to B we have

$$T = v_1^{-1}[(x - x_1)^2 + y_1^2]^{1/2} + v_2^{-1}[(x_2 - x)^2 + y_2^2]^{1/2},$$

and the minimum of T is attained for x satisfying

$$dT/dx = v_1^{-1}[(x - x_1)^2 + y_1^2]^{-1/2}(x - x_1) - v_2^{-1}[(x_2 - x)^2 + y_2^2]^{-1/2}(x_2 - x) = 0$$

or

$$\frac{\sin \phi_1}{v_1} = \frac{\sin \phi_2}{v_2}, \tag{3.4.1}$$

where ϕ_1 is the angle between the normal to the interface $y = 0$ and the path AC, and ϕ_2 is the correspondent angle for CB. This is the law of refraction.

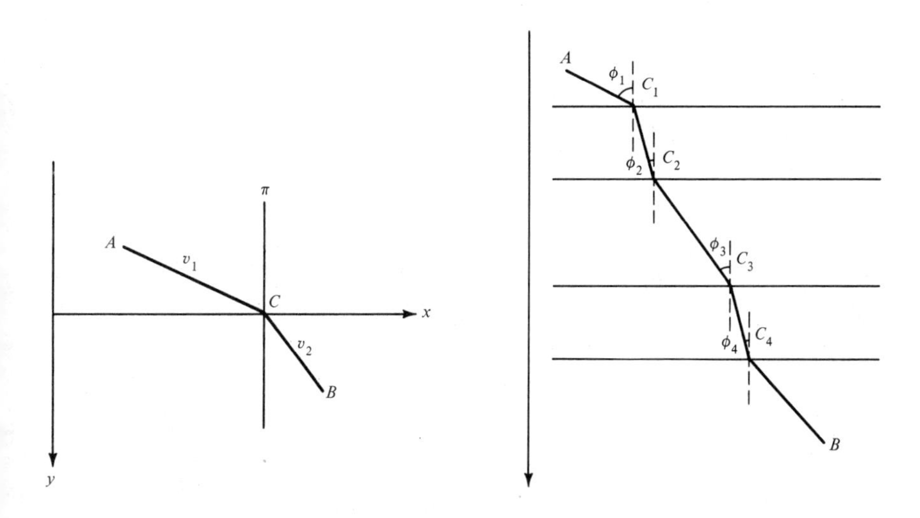

(b) If we have now a set of N contiguous parallel faced homogeneous media in which the light velocities are $v_1, v_2, \ldots, v_N$, and if $C_1, C_2, \ldots, C_{N-1}$ are the points at which the light path crosses the separation planes $y = h_1, y = h_2, \ldots, y = h_{N-1}, h_1 < h_2 < \ldots < h_{N-1}$, then the same principle will require that the light path $AC_1C_2 \cdots C_{N-1}B$ be a path of minimum time, and thus a polygonal line with vertices $C_1, C_2, \ldots, C_{N-1}$. Consequently, each subarc $AC_1C_2, C_1C_2C_3, \ldots, C_{N-2}C_{N-1}B$ must be an optimal path between A and C_2, C_1 and $C_3, \ldots, C_{N-2}$ and B respectively, and from the above we derive that the path $AC_1C_2 \cdots C_{N-1}B$ is contained in a plane orthogonal to the interfaces and satisfies

$$v_1^{-1} \sin \phi_1 = v_2^{-1} \sin \phi_2 = \cdots = v_N^{-1} \sin \phi_N, \tag{3.4.2}$$

where $\phi_1, \phi_2, \ldots, \phi_N$ are the angles of the consecutive segments with the normal to the interfaces. In other words, $v_i^{-1} \sin \phi_i = K$, $i = 1, \ldots, N$, K a constant.

(c) Let us assume now that the medium is such that the light velocity $v = v(y)$ is a continuous positive function. Then the Fermat principle stating that the light path from A to B is a path of minimum time, is reducible here to the minimization of the integral

$$I[y] = \int_A^B v^{-1}\,ds = \int_{x_1}^{x_2} v^{-1}(y)(1 + y'^2)^{1/2}\,dx,$$

where $y' = dy/dx$ and we think of I as depending on the admissible trajectory $y = y(x)$, $x_1 \le x \le x_2$, with $y(x_1) = y_1$, $y(x_2) = y_2$, $A = (x_1, y_1)$, $B = (x_2, y_2)$, $x_1 < x_2$. Then $f_0 = v^{-1}(y)(1 + y'^2)^{1/2}$, and the DuBois-Reymond equation (2.2.14), or $f_0 - y'f_{0y'} = C$, reduces here, after simplification to

(3.4.3) $$v^{-1}(y)(1 + y'^2)^{-1/2} = C.$$

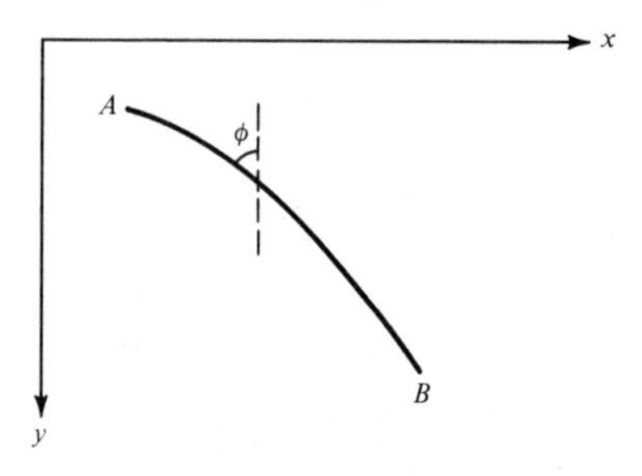

If ϕ denotes the angle of the tangent t to the path at $P = (x, y(x))$ with the normal to the planes y = constant, we have $\sin\phi = (1 + y'^2)^{-1/2}$, and hence

(3.4.4) $$v^{-1}\sin\phi(x) = C.$$

This extends the law of refraction, as stated by (3.4.1) and (3.4.2) to the continuous case. For instance, for $v = a$, then (3.4.3) yields $y' = m$ and $y(x) = mx + c$, a, m, c constants. For instance, for $v = ay^{-1/2}$, then (3.4.3) reduces to $y' = (Ca)^{-1}(y - C^2a^2)^{1/2}$, and then $x = 2Ca(y - C^2a^2)^{1/2} + D$, $y > Ca$, a, C, D constants, $C, a > 0$.

Note that the problem of the minimum time of descent (the classical brachistochrone problem of Section 1.6, Example 1) is a particular case of the problem above. Indeed, we have only to take $v = (y - \alpha)^{1/2}$, $y \ge \alpha$, whence $y_2 > y_1 \ge \alpha$, and then the functional above reduces to the one we have mentioned in Section 1.6, Example 1 (see Section 3.12 for more details). Actually, John Bernoulli in 1696 solved the classical brachistochrone problem as a problem of optics. The existence of the minimum for integrals of the type above is proved in general in Section 14.3, and in detail for the brachistochrone problem in Section 14.4, Example 3.

Exercises

1. Write the integral which must be extremized, according to Fermat's principle, if the light paths are not restricted to plane curves, and with $v = v(x, y, z)$. Let x be the independent variable.
2. Describe the plane paths of light in the two dimensional media in which the light velocities are given respectively by (a) $v = ay$; (b) $v = a/y$; (c) $v = ay^{1/2}$, where $a > 0$, $y > 0$.

3.5 The Ramsay Model of Economic Growth

We consider here an economy in which a single homogeneous good is produced with the aid of a capital $K(t)$ which may depend on t, and in which the total output $Y(t)$ is either consumed or invested. Thus, if $C(t)$ is the total consumption, we have $Y(t) = C(t) + K'(t)$. It is assumed that there is no deterioration or depreciation of capital, and that the production $Y(t)$ is a known function $Y = \Psi(K)$ of the capital. We shall require $C \geq 0$, $\Psi(K) \geq 0$, while K' may be positive or negative: since there is no deterioration or depreciation, the capital can be consumed.

Since the economic objective of any planning concerns the standard of living, we assume that a utility function $U(C)$ is known which measures the instantaneous well-being of the economy, and we assume that in any planning we should try to maximize the global utility

$$W = \int_{t_1}^{t_2} U(C(t))\,dt$$

in a finite interval of time $[t_1, t_2]$ (though we need not exclude $t_2 = +\infty$, infinite horizon). Here $C(t) = Y(t) - K'(t)$, and $Y = \Psi(K)$. Thus, we have the problem of the calculus of variations concerning the maximum of

$$W = \int_{t_1}^{t_2} U(\Psi(K(t)) - K'(t))\,dt.$$

Here we assume that $U(C)$ is a smooth, positive, nondecreasing function of C for $C \geq 0$, and two obvious constraints are here that $K(t) \geq 0$ and $K' \leq \Psi(K(t))$. However, neither of the two extreme cases $K = 0$ and $K' = \Psi(K)$ should be taken into consideration. Let us see whether an optimal solution can be visualized in the interior of the domain, that is, with $K(t) > 0$, $K(t_1) = K_1 > 0$ initial capital, and $K'(t) < \Psi(K(t))$. In this sense, we have a free problem of the calculus of variations with

$$f_0(K, K') = U(\Psi(K)) - K'), \qquad f_{0K'} = -U'(\Psi(K) - K'),$$

and the DuBois-Reymond equation $f_0 - K'f_{0K'} = c$ becomes

$$U(\Psi(K) - K') + K'U'(\Psi(K) - K') = c, \tag{3.5.1}$$

where c is a constant.

Here we consider only one particular choice for the function $U(C)$, namely, $U(C) = U^* - \alpha(C - C^*)^2$ for $0 \leq C \leq C^*$, $\alpha > 0$, and $U(C) = U^*$ for $C \geq C^*$. (Here it can be said that the utility function U "saturates" at $C = C^*$.) For the case t_1, t_2 finite, the value of U^* is not relevant, and thus we may take $U^* = 0$. Also, we assume that the economy is well below the point of saturation: $0 < C(t) < C^*$ for $t_1 \leq t \leq t_2$. We shall choose for the production function $\Psi(K)$ a linear one, $Y = \Psi(K) = \beta K$, β a positive constant. With these conventions and choices, equation (3.5.1) with $C = \Psi(K) - K' = \beta K - K'$ and with $\gamma = c/\alpha$ becomes

$$-\alpha(\beta K - K' - C^*)^2 - 2\alpha K'(\beta K - K' - C^*) = c,$$

or

$$K'^2 - (\beta K - C^*)^2 = \gamma. \tag{3.5.2}$$

In the phase plane (K, K'), equation (3.5.2) represents a family of hyperbolas with center $K' = 0$, $K = K^* = C^*/\beta$, for $\gamma \neq 0$, and corresponding asymptotes $K' = \pm(\beta K - C^*)$.

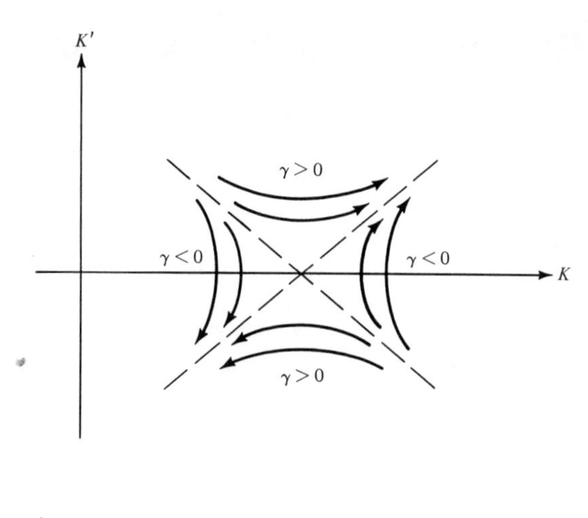

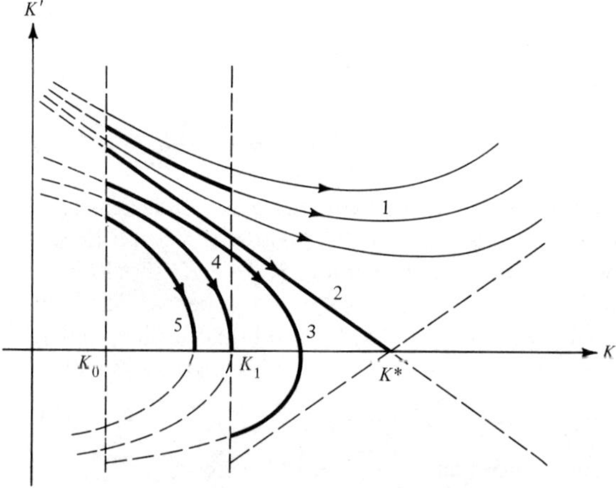

All these curves are traveled in the sense indicated in the picture. The hyperbolas above and below the asymptotes correspond to positive constants $\gamma > 0$; those on the right and left, to $\gamma < 0$.

Now let us consider specific values $K_0 < K_1$, say $K_0 < K_1 \leq K^*$. There are infinitely many such curves along which the economy could move from K_0 to K_1 (e.g., the bold arc on trajectory 1 in the second illustration). These differ one from the other, however, in the time T required to go from K_0 to K_1. The higher curves have larger values of K' all the way and hence reach K_1 in a shorter time than the lower curves. We can label these curves with the value of T, the time required to go from K_0 to K_1. Thus, the specification of T determines the value of γ. A small T corresponds to a large positive value of γ. As T increases, γ decreases and finally becomes negative. For T large, $\gamma < 0$, the curves take the value K twice, at two different times T_1 and T_2 (e.g., the bold arc on trajectory 3 in the second illustration): K first rises above K_1 and then decreases, returning to K_1. Note that, in general, K is positive at $t = T_1$, that is, at $t = T_1$ the economy is not consuming all it produces. If $K_1 = K^*$, then $K' \to 0$ as $K \to K^*$, and this occurs for $\gamma = 0$ (bold arc on the trajectory 2). In this situation, $K' = \beta(K^* - K)$ and by integration $K(t) = K^* - D\exp(-\beta t)$, D a constant, and this shows that $K \to K^*$ as $t \to +\infty$. For $\gamma > 0$, the point of equilibrium $K' = 0$ is reached by an optimal path. Note that at this point $dK'/dK = (dK'/dt)(dt/dK) = K''/K'$, or $K'' = 0$, an inflection point on the trajectories $K = K(t)$ depicted in the third illustration. Five arcs in the phase plane (K, K') have been labeled 1 to 5, and the corresponding arcs of trajectories $K(t)$ are depicted in the third illustration.

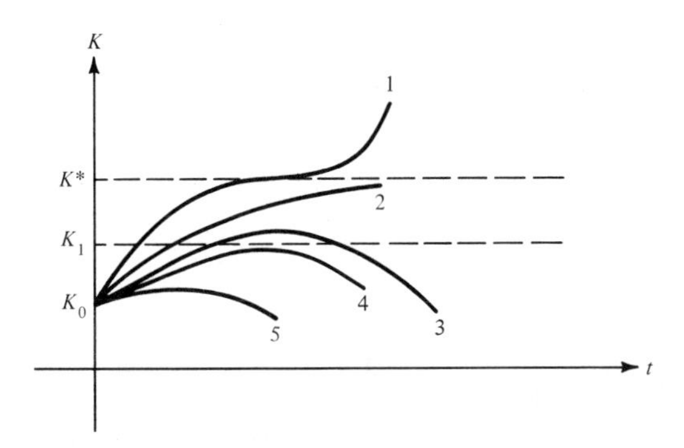

The results above have been derived directly by an analysis of the differential equation (3.5.2) in the phase plane (K, K'). However, the same equation can be integrated formally. Thus, for $y = \beta K - C^*$ and $\gamma = 0$, we have the equation $y' = \pm\beta y$ and the arc of trajectory 2 above is given by

$$K(t) = K^* - (K^* - K_1)\exp(-\beta(t - t_1)), \qquad t_1 \le t < +\infty,$$

with $K(t_1) = K_1$ and $K(+\infty) = K^*$. For $\gamma > 0$, we take $y = K - C^*/\beta = K - K^*$, $\gamma = \beta^2 L^2$ and (3.5.2) becomes $y'^2 = \beta^2(L^2 + y^2)$. Hence, $y = -L \sinh \beta(t^* - t)$ and the arc of trajectory 1 above with $K(t_1) = K_1$, $K(t^*) = K^*$ is given by

$$K(t) = K^* - L \sinh \beta(t^* - t), \qquad t_1 \le t \le t^*,$$

where $L = \gamma^{1/2}\beta^{-1}$ and t^* is determined by $L^{-1}(K^* - K_1) = \sinh \beta(t^* - t_1)$.

Exercise

1. Discuss Ramsay's model with $\Psi(K) = \beta K$ and $U(C) = -U_0 e^{-\alpha C}$, α constant.

3.6 Two Isoperimetric Problems

1. *The classical isoperimetric problem.* We are to find the curves C in R^2 of given distint end points A, B, of given length L, and such that the area between C and the chord AB is maximum. We may well assume $A = (t_1, 0)$, $B = (t_2, 0)$, $t_2 - t_1 = l > 0$,

and we want to maximize

$$I[y] = \int_{t_1}^{t_2} y\,dt$$

with

$$J[y] = \int_{t_1}^{t_2} (1 + y'^2)\,dt = L.$$

We have the classical isoperimetric problem with $f_0 = y, f_1 = (1 + y'^2)^{1/2}$. By Section 4.8 the optimal solutions (if any) are among those satisfying the Euler equation and the other necessary conditions corresponding to $F = f_0 + \lambda f_1 = y + \lambda(1 + y'^2)^{1/2}$, where λ is an undetermined constant. Since F does not depend on t, the DuBois-Reymond equation $F - y'F_{y'} = C$ yields

$$(y - C)(1 + y'^2)^{1/2} + \lambda = 0.$$

Hence,

$$\pm(y - C)(\lambda^2 - (y - C)^2)^{-1/2}\,dy = dt,$$

$$\mp(\lambda^2 - (y - C)^2)^{1/2} = t - D,$$

and finally

$$(y - C)^2 + (t - D)^2 = \lambda^2.$$

These are all the circles in R^2 (for $\lambda \neq 0$) and all the straight lines $y = C$ (for $\lambda = 0$). The passage trough A, B implies that $D = 2^{-1}(t_1 + t_2)$. If $l \leq L \leq \pi l/2$ then there is one and only one arc $y = y(t)$, $t_1 \leq t \leq t_2$, from such loci of length L. For $L < l$ and $L > \pi l/2$ the problem has no solutions (in nonparametric form). This problem will be resumed in Section 6.8.

2. *The shape of a hanging rope.* We are to find the slope of a heavy rope or chain, of length l extended between two fixed points A and B. Since in the rest position the center of gravity takes the lowest possible position, the problem is reduced to that of finding the minimum of the static moment I with respect to the t-axis, which is directed horizontally. Thus, if $1 = (t_1, y_1)$, $2 = (t_2, y_2)$, $t_1 < t_2$, and if $C: y = y(t)$, $t_1 \leq t \leq t_2$, $y(t_1) = y_1$, $y(t_2) = y_2$, denotes any AC real function, or curve joining 1 to 2, we have the problem of minimizing

$$I = \int_{t_1}^{t_2} y(1 + y'^2)^{1/2}\,dt$$

with

$$J = \int_{t_1}^{t_2} (1 + y'^2)^{1/2}\,dt = l.$$

We have here a classical isoperimetric problem with $f_0 = y(1 + y'^2)^{1/2}, f_1 = (1 + y'^2)^{1/2}$. By Section 4.8 optimal solutions are among those satisfying the Euler equation corresponding to $F = f_0 + \lambda f_1 = (y + \lambda)(1 + y'^2)^{1/2}$, where λ is an undetermined constant. Since F does not depend on t, the DuBois-Reymond equation $F - y'F_{y'} = C_1$ yields

$$y + \lambda = C_1(1 + y'^2)^{1/2}.$$

By taking $y' = \sinh v$, we have $y + \lambda = C_1 \cosh v$, $dy/dv = C_1 \sinh v$, and $dt = (\sinh v)^{-1}\,dy = C_1\,dv$. Hence

$$t = C_1 v + C_2, \qquad y + \lambda = C_1 \cosh v,$$

and by eliminating v we obtain

$$y + \lambda = C_1 \cosh((t - C_2)/C_1),$$

a family of catenaries. The undetermined constants λ, C_1, C_2 should be determined in terms of t_1, y_1, t_2, y_2, l. A necessary condition for compatibility is of course that $l \geq ((t_2 - t_1)^2 + (y_2 - y_1)^2)^{1/2}$. This problem will be resumed in Section 6.8.

3. *The classical isoperimetric problem in parametric form.* We are to find the closed curves C in R^2 of given length L which enclose the maximum area. In other words, we want to maximize

$$I[x, y] = 2^{-1} \int_0^a (y'x - x'y)\, dt$$

with

$$J[x, y] = \int_0^a (x'^2 + y'^2)^{1/2}\, dt = L,$$

in the class of all path curves $C: x = x(t)$, $y = y(t)$, $0 \leq t \leq a$, with $x(0) = x(a)$, $y(0) = y(a)$. (Here x, y are AC with $x'^2 + y'^2 \neq 0$. If we take for t the arc length parameter s, then $x'^2 + y'^2 = 1$ a.e., and $a = L$). Both I and J are parametric integrals as mentioned in Remark 8 of Section 2.2 since $f_0 = 2^{-1}(y'x - x'y)$ and $f_1 = (x'^2 + y'^2)^{1/2}$ do not depend on the independent variable and are positively homogeneous of degree one in x', y'. By Section 4.8 the optimal solutions (if any) are among those satisfying the Euler equations and the other necessary conditions for $F = f_0 + \lambda f_1 = 2^{-1}(y'x - x'y) + \lambda(x'^2 + y'^2)^{1/2}$ for some constant λ. Here we have

$$F_{x'} = -2^{-1}y + \lambda x'(x'^2 + y'^2)^{-1/2}, \qquad F_{y'} = 2^{-1}x + \lambda y'(x'^2 + y'^2)^{-1/2},$$

$F_{x'y} = -2^{-1}, F_{y'x} = 2^{-1}$, and the function F^* relative to F defined in Remark 8 of Section 2.2 is

$$F^* = (y')^{-2}F_{x'x'} = -(x'y')^{-1}F_{x'y'} = (x')^{-2}F_{y'y'} = \lambda(x'^2 + y'^2)^{-3/2}.$$

If we take s, the arc length parameter, as independent variable, we have $x'^2 + y'^2 = 1$, and the Weierstrass form of the Euler equations is now $1/r = -\lambda^{-1}$, a constant. Thus any optimal solution has constant curvature, and is therefore a circle in R^2.

3.7 More Examples of Classical Problems

A. Fixed End Point Problems

We briefly consider here a few problems for which we seek the minimum in the class of all trajectories with fixed end points.

1. $\int_{t_1}^{t_2} x'^2\, dt$, $\int_{t_1}^{t_2} (1 + x'^2)\, dt$. Here f_0 depends on x' only, and also satisfies condition (S). Any optimal solution must satisfy Euler's equation in the reduced form $f_{0x'} = c$, or $x' = C$, c, C constants. Since $f_{0x'x'} = 2 > 0$, by Section 2.6 we know that any optimal solution is of class C^2 and an extremal. (From Section 2.2 we know that no corner is possible). Thus any optimal arc is a segment. For both end points fixed, $1 = (t_1, x_1)$, $2 = (t_2, x_2)$, the absolute minimum is known to exist (Section 2.20, Example 1 and Exercise 1). Thus it must be given by the segment $s = 12$.

2. $\int_{t_1}^{t_2} (tx' + x'^2)\, dt$. Here f_0 does not depend on x and satisfies condition (S) of Section 2.7. Hence any optimal solution satisfies the Euler equation $f_{0x'} = C$, or $t + 2x = C$. Since $f_{0x'x'} = 2 > 0$, by Section 2.6 any optimal solution is of class C^2 and an extremal. By integration we have $x(t) = -4^{-1}(t - a)^2 + b$, a, b constants. For both end points fixed, $1 = (t_1, x_1)$, $2 = (t_2, x_2)$, $t_1 < t_2$, the absolute minimum is

known to exist (Section 2.20C, Exercise 2). Thus, it must be given by an arc of one of the above parabolas joining 1 and 2. For instance for $1 = (0, 1)$, $2 = (1, 0)$, the absolute minimum is given by $x(t) = 16^{-1}(25 - (2t + 3)^2)$, $0 \le t \le 1$.

3. $\int_{t_1}^{t_2} (1 + (1 + t^2)x'^2)\,dt$. Here f_0 does not depend on x and satisfies condition (S) of Section 2.7. Hence, any optimal solution satisfies the Euler equation $f_{0x'} = C$, or $2(1 + t^2)x' = C$. Since $f_{0x'x'} = 2(1 + t^2) > 0$, by Section 2.6 any optimal solution is of class C^2 and an extremal. By integration $x(t) = a$ arctant $t + b$, a, b constants. For both end points fixed, $1 = (t_1, x_1)$, $2 = (t_2, x_2)$, $t_1 < t_2$, the absolute minimum is known to exist (Section 2.20C, Exercise 3). Thus, it must be given by such an arc joining 1 to 2.

4. $\int_{t_1}^{t_2} t^{-1}x'^2\,dt$, $0 < t_1 < t_2$, $x(t_1) = x_1$, $x(t_2) = x_2$. Here f_0 does not depend on x and satisfies condition (S) of Section 2.7. Hence, any optimal solution satisfies the reduced Euler equation $f_{0x'} = C$, or $2t^{-1}x' = C$. Since $f_{0x'x'} = 2t^{-1} > 0$ for $t_1 < t < t_2$, any optimal solution is of class C^∞ and an extremal. Then $x(t) = at^2 + b$, a, b constants. For both end points fixed, $1 = (t_1, x_1)$, $2 = (t_2, x_2)$, $0 < t_1 < t_2$, the absolute minimum is known to exist (Section 2.20C, Exercise 4). Thus, it must be given by such an arc joining 1 to 2.

5. $\int_{t_1}^{t_2} t^{1/2}x'^2\,dt$, $0 = t_1 \le t_2$, $x(t_1) = x_1$, $x(t_2) = x_2$. Here f_0 does not depend on x and satisfies (S) of Section 2.7. Hence, any optimal solution satisfies the reduced Euler equation $f_{0x'} = C$, or $2t^{1/2}x' = C$. Since $f_{0x'x'} = 2t^{1/2} > 0$ for $t > 0$, by Section 2.6 any optimal solution $x(t)$, $0 \le t \le t_2$, must be of class C^2 (actually C^∞) for $0 < t \le t_2$ and actually an extremal in such an half open interval. Then, $x(t)$ is such an extremal in the closed interval $[0, t_2]$. Thus, $x(t) = at^{1/2} + b$, a, b $0 \le t_1 < t_2$, the absolute minimum is known to exist (Section 2.20B, Example 4). Thus, it must be given by such an arc joining 1 to 2.

6. $\int_{t_1}^{t_2} tx'^2\,dt$, $0 < t_1 < t_2$, $x(t_1) = x_1$, $x(t_2) = x_2$. The extremals are segments $x = C$ and arcs of $x = a + b \log t$, a, b constants. From Section 1.5, Example 4, we know that this problem with fixed end points $1 = (0, 1)$, $2 = (1, 0)$ has no absolute minimum. Actually, no such arc can join 1 and 2. On the other hand, for the same problem with fixed end points $1 = (t_1, x_1)$, $2 = (t_2, x_2)$, $0 < t_1 < t_2$, any optimal arc is of class C^∞ and an extremal. From Section 2.20C, Exercise 5, we know that an optimal solution exists, and therefore it is given by such an arc joining 1 and 2.

7. $\int_{t_1}^{t_2} x'^2(1 + x')^2\,dt$. The extremals are segments. There may be corner points, since $f_{0x'} = 2x' + 6x'^2 + 4x'^3$ is not monotone. See Section 3.11 for more details.

8. $\int_{t_1}^{t_2} xx'^2\,dt$, $x \ge 0$. The extremals are solutions of the DuBois-Reymond equation (2.2.14), $f_0 - x'f_{0x'} = C$, and are segments $x = C$ and arcs of $x = a(t - b)^{2/3}$. See Section 3.10 for more details.

9. $\int_{t_1}^{t_2} (x^2 + x'^2)^{1/2}\,dt$. The extremals are solutions of the DuBois-Reymond equation $f_0 - x'f_{0x'} = C$, or $C^2x'^2 = x^2(x^2 - C^2)$, and are arcs of $x \sin(t - a) = b$, a, b constants. For fixed end points $1 = (0, 1)$, $2 = (\pi/2, 1)$ we can take $a = -\pi/4$, $b = 2^{-1/2}$, and the extremal $E_{12}: x(t) = 2^{-1/2} \csc(t + \pi/4)$, $0 \le t \le \pi/2$, joins 1 and 2. The family of extremals $x(t) = b \csc(t + \pi/4)$, $0 \le t \le \pi/2$, covers simply the strip $R = [0, \pi/2] \times R$ as b ranges over R. We have a Weierstrass field in R. Since $f_{0x'x'} = 2 > 0$, by (2.11.iv) the extremal E_{12} gives an absolute minimum. Cf. Section 1.5, Example 5, where the same problem with fixed end points $1 = (0, 0)$, $2 = (1, 1)$ was shown to have no absolute minimum. Indeed no arc $x = b \csc(t - a)$ can join 1 and 2.

10. $\int_{t_1}^{t_2} (x - \alpha)^{-1/2}(1 + x'^2)^{1/2}\,dt$ (brachistochrone: minimum time of descent). The extremals are arcs of cycloids. See Section 3.12 for details.

11. $\int_{t_1}^{t_2} x(1 + x'^2)^{1/2}\,dt$, $x \ge 0$ (surface of revolution with minimum area). The extremals are arcs of a catenary. See Section 3.13 for details.

12. $\int_0^3 (x^2 + (t^2 - 2t)x')\,dt$, both end points fixed $1 = (0,0)$, $2 = (3,1)$, in the rectangle $A = [0 \le t \le 3, 0 \le x \le 1]$. Here $n = 1$, $f_0 = x^2 + (t^2 - 2t)x'$ is linear in x', and in the terms of Remark 4 of Section 2.12, we have $F = x^2$, $G = t^2 - 2t$, $\omega(t,x) = G_t - F_x = 2t - 2 - 2x$. Thus, $\omega = 0$ on the line $\Gamma: x = t - 1$, and $\omega < 0$ above Γ, and $\omega > 0$ below Γ. The minimum of the integral in A is given by the polygonal line 1342 made up of the segments 13 and 42 on the boundary of A and of the singular arcs 3, and $I_{\min} = \frac{2}{3}$. The polygonal lines 152 and 162 are parametric curves in A on which $I = 3$, and represent the maximum of I in the class of the parametric curves in A. The same integral has no maximum in the class of the AC curves $x = x(t)$, $x(0) = 0$, $x(3) = 1$, since for these curves I can only take values lower than 3 and as close to 3 as we want, but this value 3 cannot be attained.

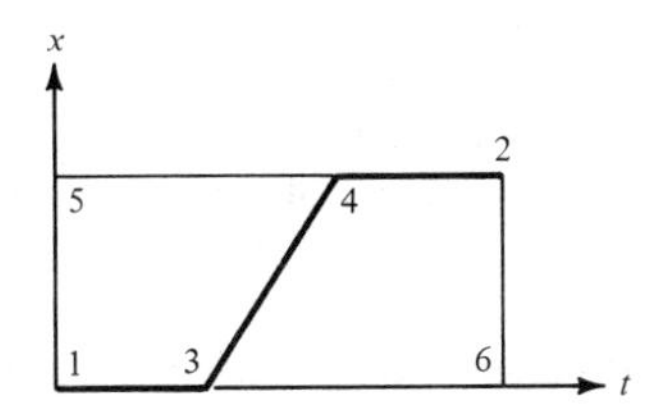

13. Show that the integral $\int_0^3 (x^2 + 2tx')\,dt$ with constraints $x(0) = 0$, $x(3) = 1$, $(t, x(t)) \in A = [0 \le t \le 3, 0 \le x \le 1]$, has no minimum and no maximum in the class of the AC curves $x = x(t)$, $0 \le t \le 3$, under the constraints.

14. $\int_{-1}^{1} [(2^{-1}x^2 + 3^{-1}x^3) + (tx^2 + 2t^2)x']\,dt$, both end points fixed $1 = (-1,-1)$, $2 = (1,1)$, in the class of the AC curves $x = x(t)$, $-1 \le t \le 1$, in the region $A = [(t,x) | \, |t| + |x| \le 2]$. Show that the integral has a minimum in this class but no maximum. Compute $I_{\min}$.

15. Determine the maximum and the minimum of $I = \int_0^2 (-x - x^2 - t^2x')\,dt$ in the class of all AC curves $x = x(t)$, $0 \le t \le 1$, joining $B_1 = [t = 0, 0 \le x \le 1]$ to $2 = (2,1)$, within the rectangle $A = [0 \le t \le 2, 0 \le x \le 1]$. Ans.: $I_{\max} = -35/12$, $I_{\min} = -4$.

B. Computation of the Functions E and H

1. $\int_{t_1}^{t_2} x'^2\,dt$. Here $E(x', X') = X'^2 - x'^2 - 2x'(X' - x') = (X' - x')^2 \ge 0$, $H(x', \lambda) = x'^2 + \lambda x'$, and H as a function of x' has a minimum at $x' = -\lambda/2$.

2. $\int_{t_1}^{t_2} x'^4\,dt$. Here $E(x', X^1) = (X' - x')^2(X'^2 + 2X'x' + 3x'^2) \ge 0$, $H(x', \lambda) = x'^4 + \lambda x'$, and H as a function of x' has a minimum at $x' = (-\lambda/4)^{1/3}$.

3. $\int_{t_1}^{t_2} (x^2 + x'^2)^{1/2}$. Here $E(x, x', X') = (x^2 + x'^2)^{-1/2}[(x^2 + X'^2)^{1/2}(x^2 + x'^2)^{1/2} - x^2 - x'X')] \ge 0$, $H(x, x', \lambda) = (x^2 + x'^2)^{1/2} + \lambda x'$, and H as a function of x' has a minimum at $x' = -\lambda|x|(1 - \lambda^2)^{-1/2}$ for $|\lambda| < 1$.

C. Variable End Point Problems

1. $\int_{t_1}^{t_2} (1 + x'^2)\,dt$, first end point $1 = (t_1, x_1)$ fixed, second end point $2 = (t_2, x_2)$ on the straight line $B_2 = [(t, x_2) | t \ge t_1, x_2 \text{ fixed}, x_2 \ne x_1]$. Then from Subsection A, Exercise 1, any optimal arc is a segment. Now $f_{0x'} = 2x'$, $f_0 - x' f_{0x'} = 1 - x'^2$, and the

transversality relation in (2.2.i) yields $(1 - x'^2)\,dt_2 = 0$, dt_2 arbitrary. Hence, $x'(t_2) = \pm 1$. We see that the segment $s = 12$ from 1 to B_2 at an angle $\pi/4$ with B_2 is the only element satisfying the necessary conditions. From Section 2.20C, Exercise 6, we know that the minimum exists. Thus, the minimum is given by the indicated segment $s = 12$.

2. $\int_{t_1}^{t_2} (1 + x'^2)\,dt$, first end point $1 = (0, 1)$, second end point on the segment $B_2 = [(t, 0) \mid 3 \le t \le 4]$. Obviously, no segment $s = 12$ can hit B_2 in an interior point at an angle $\pi/4$, so, from the above, the second end point in the optimal path is not in the interior of B. Thus $s = 12$ must be the segment from $1 = (0, 1)$ to either $(3, 0)$, or $(4, 0)$. For $2 = (3, 0)$ we have $x' = -\frac{1}{3}$; hence $\Delta = (1 - x'^2)\,dt_2 = (8/9)\,dt_2 \ge 0$ for all $dt_2 \ge 0$, as required by the transversality relation in (2.2.i). The point $(4, 0)$ does not satisfy the analogous relation. Since from Section 2.20C, Exercise 6, we know that the minimum exists, it must be given by $s = 12$ with $2 = (3, 0)$.

3. $\int_{t_1}^{t_2} (1 + x'^2)\,dt$, fixed first end point $1 = (0, 0)$, second end point $2 = (t_2, x_2)$ on the locus $B_2 = [(t, x) \mid tx = 1, t > 0, x > 0]$. As in no. 1 the minimum is given by a segment $s = 12$, or $x = mt$, and by the transversality relation $(1 - x'^2)\,dt_2 + 2x'\,dx_2 = 0$, where dt_2, dx_2 are computed on B_2, while x' is the slope of the trajectory at 2. The equations $x\,dt + t\,dx = 0$ and $(1 - x'^2)\,dt + 2x'\,dx = 0$ yield $2xx' - t(1 - x'^2) = 0$ where $x' = x/t$. Thus, $3x^2/t = t$ with $tx = 1$, and finally $t^4 = 3$. Since from Section 2.20C, Exercise 7, we know that the minimum exists, it must be given by the segment $s = 12$, $1 = (0, 0)$, $2 = (3^{1/4}, 3^{-1/4})$, or $x = 3^{-1/2}t$, $0 \le t \le 3^{1/4}$.

4. $\int_{t_1}^{t_2} (tx' + x'^2)\,dt$, first end point $1 = (0, x_1)$ fixed, second end point on the straight line $B_2 = [(t_2, x) \mid t_2 > 0$ fixed, x arbitrary$]$. Here B' is the linear manifold $B' = [dt_2 = 0$, dx_2 arbitrary$]$. Then (2.2.i) yields $(t + 2x')\,dx_2 = 0$, and $x' = -t/2$. In other words, the angle α between the locus 12 and B_2 at 2 depends on t_2. From Subsection A, Example 2 above, $x(t) = -4^{-1}(t - a)^2 + b$. Then a and b are determined by the system of equations $-4^{-1}a^2 + b = x_1$, $-2^{-1}(t_2 - a) = -2^{-1}t_2$. Thus, $a = 0$, $b = x_1$. Since we know from Section 2.20C, Exercise 2, that the absolute minimum exists, it must be given by $x(t) = -4^{-1}t^2 + x_1$, $0 \le t \le t_2$.

5. $\int_{t_1}^{t_2} (1 + (1 + t^2)x'^2)\,dt$, first end point $1 = (0, x_1)$, $x_1 > 0$, second end point on the half straight line $B_2 = [(t, 0) \mid t \ge 0]$. Then (2.2.i) yields $[1 - (1 + t^2)x'^2]\,dt_2 = 0$, dt_2 arbitrary, or $x' = \pm(1 + t^2)^{-1/2}$ at the point $2 = (t_2, 0)$. From Subsection A, Example 3 above, $x(t) = a \arctan t + b$. Thus, $b = x_1$, and t_2, a are determined by the system of equations $a \arctan t_2 + x_1 = 0$, $a(1 + t_2^2)^{-1} = \pm(1 + t_2^2)^{-1/2}$. We know that the absolute minimum exists. Thus, it must be given by the corresponding arc. For instance, for $x_1 = 2\pi/3$, we have $t_2 = 3^{1/2}$, $a = -2$, and $x(t) = x(t) = -2 \arctan t + 2\pi/3$, $0 \le t \le 3^{1/2}$, is optimal.

6. $\int_{t_1}^{t_2} t^{-1}x'^2\,dt$, first end point $(1, 0)$, second end point on the locus $B_2 = [(t, x) \mid x = (13/3) + (t - 2)^2, t \ge 1]$. From Subsection A above, Example 4, we obtained the parabolas $x = at^2 + b$. Thus $b = -a$ for the passage through $(1, 0)$, and $x(t) = a(t^2 - 1)$, $1 \le t \le t_2$. The transversality relation in (2.2.i) yields $(-t^{-1}x'^2)\,dt_2 + (2t^{-1}x')\,dx_2 = 0$ with $dx_2 = 2(t - 2)\,dt_2$; hence, $x' = 4(t - 2)$, where $x' = 2at$. Now a and t_2 are determined by the system of equations $a(t_2^2 - 1) = (13/3) + (t_2 - 2)^2$, $2at_2 = 4(t_2 - 2)$. Thus $t_2 = -4(a - 2)^{-1}$, and by computation we obtain a third degree equation in a, namely $3a^3 + 13a^2 - 88a + 52 = 0$, with three real roots $a = \frac{2}{3}$, $a = 3.17891$, $a = -8.17891$. Each of the corresponding parabolas $x(t) = a(t^2 - 1)$ has two real intersections with the parabola B_2. For instance, for $a = \frac{2}{3}$, $x(t) = \frac{2}{3}(t^2 - 1)$ has two intersections $(3, 16/3)$ and $(9, 160/3)$ with B_2, and at these points $x'(3) = 4$, $x'(9) = 12$. Only for $t_2 = 3$ is the equation $x'(t_2) = 4(t_2 - 2)$ satisfied with the common value 4. For analogous reasons we must discard all the other cases. The absolute minimum is known to exist (Section 2.20C, Exercise 4

and remark below). It must be given by $x(t) = \frac{2}{3}(t^2 - 1)$, $0 \le t \le 3$. Thus, the only candidate for optimality is $E_{12}: x = \frac{2}{3}(t^2 - 1)$, $0 \le t \le 3$, with $I[E_{12}] = 8$, $t_2 = 3$.

As noticed in Section 2.20, Exercise 4, the problem above has certainly an absolute minimum if we limit $2 = (t_2, x_2)$ to be on any arc $x = (13/3) + (t - 2)^2$, $1 \le t \le N$, N finite. Now t_2 can be bounded by the following argument. Any minimum better than E_{12}, if any, must be of the form $E: x = x_0(t) = a(t^2 - 1)$, $1 \le t \le t_2$, with $I[x_0] \le 8$, $a(t_2^2 - 1) \in B_2$, hence, $2a^2 t_2^2 \le 8$, and $a(t - 1) = (13/3) + (t_2 - 2)^2$. As $t_2 \to \infty$ we derive $a \to 0$ from the first relation and $a \to 1$ from the second one. Thus, we must have $t_2 \le N$ for some constant N, and an optimal solution exists by (2.20.i). This is given by E_{12}.

7. $\int_{t_1}^{t_2} (x^2 + x'^2)^{1/2}\, dt$ first end point fixed $1 = (t_1, x_1)$, second end point $2 = (t_2, x_2)$ on the straight line $B_2: t = t_2$, t_2 fixed, $t_2 > t_1$. We have already computed the extremals in the Subsection A above, Example 9. Now $f_{0x'} = x'(x^2 + x'^2)^{-1/2}$, B' is the linear manifold $dt_2 = 0$, dx_2 arbitrary, and the transversality relation in (2.2.i) yields $x_2'\, dx_2 = 0$, and $x'(t_2) = 0$. The optimal curve $C = 12$ must be orthogonal to B_2 at 2.

8. $\int_{\xi_1}^{\xi_2} (x - \alpha)^{-1/2}(1 + x'^2)^{1/2}\, d\xi$, first end point $1 = (\xi_1, x_1)$, fixed, second end point $2 = (\xi_2, x_2)$ on the parabola $\Gamma: \xi = \xi_0 - x^2$, ξ_0 fixed, $\xi_1 < \xi_0 - x_1^2$, $x_1 > 0$ (brachistochrone: minimum time of descent from 1 to the parabola Γ). (Cf. Section 1.5, Example 1, and Subsection A above, Example 10). The extremals are arcs of cycloids. Here B' is the linear manifold $B' = [d\xi_2, dx_2 = (2x_2)\, d\xi_2,\ d\xi_2$ arbitrary$]$, and the transversality relation in (2.2.i) yields $d\xi_2 + x'\, dx_2 = 0$, or $(dx_2/d\xi_2)x'(\xi_2) = -1$. We see that the cycloid C_{12} must be orthogonal to Γ at 2.

3.8 Miscellaneous Exercises

1. (a) Regarding the left-hand member of the obvious inequality

$$\int_{t_1}^{t_2} [g(t) + \sigma h(t)]^2\, dt \ge 0$$

as a quadratic function of the parameter σ, prove the Schwartz inequality

$$\int_{t_1}^{t_2} h^2\, dt \int_{t_1}^{t_2} g^2\, dt \ge \left[\int_{t_1}^{t_2} gh\, dt\right]^2, \tag{3.8.1}$$

where equality holds if and only if $g(t) = Ah(t)$ for some constant A.

(b) Given $x(t_1) = x_1$, $x(t_2) = x_2$, and $p(t)$ a known function, $p(t) > 0$, use (3.8.1) to prove that the absolute minimum of $J = \int_{t_1}^{t_2} p^2 x'^2\, dt$ is $m = (x_2 - x_1)^2 [\int_{t_1}^{t_2} p^{-2}\, dt]^{-1}$, and that this minimum is attained if and only if $x' = Ap^{-2}$ and then $A = (x_2 - x_1)[\int_{t_1}^{t_2} p^{-2}\, dt]^{-1}$.

(c) Show that $x' = Ap^{-2}$ is a first integral of the Euler equation associated with the integral J of part (b). Verify that m is the actual value of J when $x' = Ap^{-2}$.

2. (a) Show that if $x(t)$ satisfies the Euler equation associated with $J = \int_{t_1}^{t_2} (p^2 x'^2 + q^2 x^2)\, dt$, where p and q are known functions of t, then J has the value $[p^2 x x']_{t_1}^{t_2}$.

(b) Show that if x satisfies the Euler equation associated with J, if $\eta(t)$ is an arbitrary continuous function with sectionally continuous derivative, and if $\eta(t_1) = \eta(t_2) = 0$, then

$$\int_{t_1}^{t_2} (p^2 x' \eta' + q^2 x \eta)\, dt = 0.$$

(c) Show that by replacing x in J with $x + \eta$ (thus x and $x + \eta$ have the same end values), the value of J is increased by the nonnegative amount

$$\int_{t_1}^{t_2} (p^2\eta'^2 + q^2\eta^2)\, dt.$$

3. Investigate the extrema of the following functionals:
   (a) $I[x] = \int_1^2 x'(1 + t^2x')\, dt$, $x(1) = -1$, $x(2) = 1$.
   (b) $I[x] = \int_0^{\pi/4} (4x^2 - x'^2 + 8x)\, dt$, $x(0) = -1$, $x(\pi/4) = 0$.
   (c) $I[x] = \int_1^2 (t^2x'^2 + 12x^2)\, dt$, $x(1) = 1$, $x(2) = 8$.
   (d) $I[x] = \int_0^1 (x'^2 + x^2 + 2xe^{2t})\, dt$, $x(0) = \frac{1}{3}$, $x(1) = (\frac{1}{3})e^2$.
4. Prove that $y = bx/a$ is a weak minimum but not a strong minimum of the functional

$$I[x] = \int_0^a x'^3\, dt,$$

   where $x(0) = 0$, $x(a) = b$, $a > 0$, $b > 0$.
5. Prove that the extrema of the functional

$$\int_a^b \eta(t, x)(1 + x'^2)^{1/2}\, dt$$

   are always strong minima if $\eta(t, x) \geq 0$ for all t and x.
6. Show that for the same integral as in Exercise 5, if the first end point is fixed and the second end point is on a curve Γ, then the transversality relation implies the orthogonality of the optimal path with Γ.
7. $\int_{t_1}^{t_2} t^{1/3}x'^2\, dt.$
8. $\int_{t_1}^{t_2} t^2x'^2\, dt.$
9. $\int_{t_1}^{t_2} x'(1 + t^2x')\, dt.$
10. $\int_{t_1}^{t_2} t^3x'^2\, dt.$
11. $\int_{t_1}^{t_2} t^{-2}x'^2\, dt.$
12. $\int_{t_1}^{t_2} (x'^2 + x^2)\, dt.$
13. $\int_{t_1}^{t_2} x^{-1}(1 + x'^2)^{1/2}\, dt.$
14. $\int_{t_1}^{t_2} [(1 + x'^2)^{1/2} - tx]\, dt.$
15. $\int_{t_1}^{t_2} (x'^2 - 1)(x'^2 - 4)\, dt.$

3.9 The Integral $I = \int (x'^2 - x^2)\, dt$

Here $f_0 = x'^2 - x^2$ is independent of t, $f_{0x} = -2x$, $f_{0x'} = 2x'$, $f_{0x'x'} = 2 > 0$, and any extremum for I is a minimum. The Euler equation is $x = -(d/dt)x'$, hence $x'' + x = 0$, and the extremals are of the form $x = c_1 \sin t + c_2 \cos t$, or $x = c \cos(t - \alpha)$, c_1, c_2, c, α

arbitrary constants. Note that by Section 2.7 any minimum for I satisfies the DuBois-Reymond equation $f_0 - x' f_{0x'} = C$; hence, $x'^2 + x^2 = c^2$, and again $x = c\cos(t - \alpha)$. If the end point conditions are $x(t_1) = x_1$, $x(t_2) = x_2$, $t_1 < t_2$, t_1, x_1, t_2, x_2 fixed, then we must determine c_1, c_2 in such a way that $c_1 \sin t_1 + c_2 \cos t_1 = x_1$, $c_1 \sin t_2 + c_2 \cos t_2 = x_2$, and the determinant of this linear system is $D = \sin(t_1 - t_2)$. If $D \neq 0$, then

$$E_{12}\colon x(t) = (\sin(t_1 - t_2))^{-1}((x_1 \cos t_2 - x_2 \cos t_1)\sin t + (x_2 \sin t_1 - x_1 \sin t_2)\cos t).$$

(a) *Case* $0 < t_2 - t_1 < \pi$. Here $D \neq 0$, and then the extremal $E_{12}\colon x = x_0(t)$, $t_1 \le t \le t_2$, through $1 = (t_1, x_1)$ and $2 = (t_2, x_2)$ is uniquely determined. A field of extremals containing E_{12} is then defined by $E_a\colon x = x(t,a) = x_0(t) + a\sin(t - t_1 + m)$ for some fixed $m > 0$ chosen so that $t_2 - t_1 + m < \pi$. Then $x_a = \sin(t - t_1 + m)$, and as t describes $t_1 \le t \le t_2$, the argument $t - t_1 + m$ varies between m and $t_2 - t_1 + m$, with $0 < m < t_2 - t_1 + m < \pi$. Thus x_a has a positive minimum $x_a = \sin(t - t_1 + m) \ge \mu > 0$ in $[t_1, t_2]$. Thus, for every t, $x(t,a)$ is a strictly increasing function of a, and the extremals $E_a\colon x = x(t,a)$, $t_1 \le t \le t_2$, $-\infty < a < +\infty$, cover simply the strip $R = [t_1 \le t \le t_2, -\infty < x < +\infty]$. In addition $a(t,x) = [x - x_0(t)](\sin(t - t_1 + m))^{-1}$ and $p(t,x) = x'(t,a) = x_0'(t) + a(t,x)\cos(t - t_1 + m)$ are single valued continuous functions in R (possessing continuous derivatives of all orders). Since $f_{x'x'} = 2 > 0$, we have $E_{12}(t,x,p,x') > 0$ for all $(t,x) \in R$, p, x' reals, and by (2.11.iv) $I[C_{12}] > I[E_{12}]$ for every curve $C_{12}\colon x = x(t)$, $t_1 \le t \le t_2$, $x(t)$ AC in $[t_1, t_2]$, $x'(t)$ essentially bounded, having the same end points as E_{12}. Thus, for $0 < t_2 - t_1 < \pi$, the problem has one and only one optimal solution E_{12}. For instance, for $1 = (0,0)$, $2 = (\pi/3, \sqrt{3}/2)$, the optimal solution is E_{12}: $x = x_0(t) = \sin t$, $0 \le t \le \pi/3$, and $I[E_{12}] = \int_0^{\pi/3}(\cos^2 t - \sin^2 t)\,dt = \sqrt{3}/4 = 0.434$. For instance, for $1 = (-\delta, x_0 - m\delta)$, $2 = (\delta, x_0 + m\delta)$, $m \neq 0$, $0 < \delta < \pi/2$, the optimal solution is

$$E_{12}\colon x = m(2\delta/\sin 2\delta)\cos\delta \sin t + x_0(2\sin\delta/\sin 2\delta)\cos t, \qquad -\delta \le t \le \delta,$$

whose derivative is

$$x'(t) = m(2\delta/\sin 2\delta)\cos\delta\cos t - x_0(2\sin\delta/\sin 2\delta)\sin t, \qquad -\delta \le t \le \delta,$$

and we see that for $m \neq 0$ and x_0 in absolute value below a given constant, then $|x'(t) - m| \le K\delta$ for some constant K; in other words, $x'(t)$ can be made as close to m as we want by taking δ sufficiently small.

(b) *Case* $t_2 - t_1 > \pi$. The system of equations determining c_1 and c_2 may or may not have solutions, but even if it has a solution and an extremal $E_{12}\colon x = x_0(t)$, $t_1 \le t \le t_2$, joining 1 and 2 exists, $I[x]$ has no minimum, since the Jacobi necessary condition (Section 2.5) is not satisfied. Indeed, $f_{xx} = -2$, $f_{x'x'} = 2$, $f_{xx'} = 0$, and the accessory equation (relative to E_{12}) is $\eta'' + \eta = 0$. The solution $\eta = \sin(t - t_1)$, $t_1 \le t \le t_2$, is zero at $t = t_1$ and at $t = t_1 + \pi < t_2$. Thus, the point 3 on E_{12} of abscissa $t_3 = t_1 + \pi$ is conjugate to 1 on E_2 and is between 1 and 2. For $x_2 - x_1 > \pi$, the integral $I[x]$ has no minimum.

For instance, for $t_1 = 0$, $x_1 = 0$, $t_2 = a$, $x_2 = 0$, $\pi < a < 2\pi$, the only solution of the Euler equation with these boundary data is $x_0(t) = 0$, $0 \le t \le a$, with $I[x_0] = 0$. This is no minimum, since for $x(t) = \sin(\pi a^{-1} t)$, $0 \le t \le a$, we have $I[x] = (2a)^{-1}(\pi^2 - a^2) < 0$.

(c) *Case* $t_2 - t_1 = \pi$, $x_2 \neq -x_1$. The integral I has no minimum since the system determining c_1 and c_2 has no solution.

(d) *Case* $t_2 - t_1 = \pi$, $x_2 = -x_1$. This case is more difficult to discuss. The system determining c_1 and c_2 reduces to $c_1 \sin t_1 + c_2 \cos t_1 = x_1$ and has infinitely many solutions. By a translation we may always assume $t_1 = 0$ and $t_2 = \pi$, and then the system reduces to $c_2 = x_1$. Thus, for $t_1 = 0$, $t_2 = \pi$, $x_1 = -x_2 = k$, we have infinitely many extremals joining $(0, k)$ and $(\pi, -k)$:

$$E: x = k \cos t + c \sin t, \qquad 0 \le t \le \pi,$$

where c is an arbitrary constant. By direct computation we have

$$I[E] = \int_0^\pi [(c \cos t - k \sin t)^2 - (c \sin t + k \cos t)^2]\, dt = 0$$

for every value of the constant c. We shall prove that each one of the extremals E gives an absolute minimum of $I[x]$, that is, we shall prove that $I[C_{12}] \ge I[E] = 0$ for every $C: x = x(t)$ joining $1 = (0, k)$ and $2 = (\pi, -k)$.

Let $R > |k|$ be any given constant and let use denote by A_0 the region $A_0 = [0, \pi] \times [-R, R]$. The integral $I = \int_0^\pi (x'^2 - x^2)\, dt$ with the constraints $x(0) = k$, $x(\pi) = -k$, $(t, x(t)) \in A_0$ has an absolute minimum $C_{12}: x = x(t), 0 \le t \le \pi$, by the existence theorem (2.20.i) (since now $f_0 = x'^2 - x^2 \ge x'^2 - R^2$ and $(\gamma 1)$ holds). If $|x(t)| < R$ for all $t \in [0, \pi]$, then by (2.6.iii) x is of class C^∞ and hence an extremal, namely one of the extremals E above. If C_{12} has two points in common with the straight line $x = R$, say $P' = (t', R)$, $P'' = (t'', R)$, then the entire segment $P'P''$ belongs to C_{12}, since $x' \ge 0$, $|x| \le R$ on C_{12} and $x' = 0$, $x = R$ on the segment, and C_{12} is optimal. Let us assume that $P'P''$ is the maximal segment (or single point) that C_{12} has in common with $x = R$, $0 < t' \le t'' < \pi$. Let us prove that the two arcs of C_{12} ending at P' and P'' are tangent to $x = R$ at these points. The argument is by contradiction. Indeed, if this is not true for P' and $t' < t''$, then for $\delta > 0$ sufficiently small we could take two points $P_1 = (t' - \delta, x(t' - \delta))$ and $P_2 = (t' + \delta, R)$ in such a way that the arc P_1P' of C_{12} has only the second end point on the boundary of A_0 and hence by (2.6.i) is of class C^1 with $m = x'(t' - 0) > 0$, and the segment P_1P_2 has slope m_0 as close as we want to $m/2$ if δ is sufficiently small. But then the minimum of the integral $\int (x'^2 - x^2)\, dt$ between P_1 and P' is not the arc $P_1P'P_2$ of C_{12} but the arc indicated at the end of part (a) above whose slope can be made as close to m_0 as we want by taking δ sufficiently small, and therefore completely contained in A_0, a contradiction. Thus the arc P_1P' is tangent to $x = R$ at P' and $m = 0$, and an analogous argument holds for P''. Also an analogous argument holds for the case $t' = t''$, as well as for the contacts of C_{12} with $x = -R$. We see that the minimizing curve C_{12} is made up of arcs $c \cos(t - \alpha)$ tangent to the straight lines $x = R$ and $x = -R$ and of segments of these lines. The first of such arcs, starting from the point $(0, k)$, must be of the form $k \cos t + c \sin t$. But there is only one such curve say E in this family tangent to $x = R$, and such curve is in A_0. By part (a) C_{12} must coincide with E. Since R is arbitrary, and the value of the integral on all curves E is the same, we see that the minimum of the integral in A is given by the infinitely many curves E.

(e) (A remark concerning cases (b) and (c)). Let us prove that $\inf I[x] = -\infty$ for $t_2 - t_1 > \pi$, as well as for $t_2 - t_1 = \pi$, $x_2 \ne -x_1$. Let $0 = t_1 < \pi < t_2 < 2\pi$, $x_1 = x_2 = 0$. Let $\delta > 0$ be any number, $0 < \delta \le (2\pi - t_2)/2$, and let $C_{12}: x = x(t)$, $0 \le t \le t_2$, be defined by taking $x(t) = \sin t$ for $0 \le t \le t_3$, and $x(t) = C \sin(t_2 - t)$ for $t_3 \le t \le t_2$, where $t_3 = t_2 - \pi + \delta$, $C = (\sin t_3)/(\sin \delta)$. Obviously C joins $1 = (0,0)$ to $2 = (t_2, 0)$, and is made up of two arcs of extremals with a corner point at $3 = (t_3, x(t_3))$. On the other hand, $C_{12} = C_{13} + C_{32}$, and $I[C_{13}]$ has a finite limit as $\delta \to 0$. By direct computations we have

$$I[C_{32}] = 2^{-1} C^2 \sin 2(t_2 - t_3) = -\cos \delta \sin^2(t_2 + \delta)/\sin \delta,$$

and $I[C_{32}] \to -\infty$ as $\delta \to 0 + 0$. Finally $I[C_{12}] \to -\infty$ as $\delta \to 0 + 0$, and $\inf I[C]$ in the class of curves joining 1 and 2 is $-\infty$. An analogous proof holds in the other cases.

3.10 The Integral $I = \int xx'^2\,dt$

(a) First, let us consider the class Ω of all curves $C: x = x(t), t_1 \le t \le t_2$, with $x(t) \ge 0$, $x(t)$ AC in $[t_1, t_2]$, joining two given points $1 = (t_1, x_1)$, $2 = (t_2, x_2)$, $t_1 < t_2$, $x_1 \ge 0$, $x_2 \ge 0$. Then A is the strip $[t_1 \le t \le t_2,\ 0 \le x < +\infty]$, and then $f_0 = xx'^2 \ge 0$, and f_0 is continuous with its partial derivatives, $f_{0x} = x'^2$, $f_{0x'} = 2xx'$, $f_{0x'x'} = 2x \ge 0$, everywhere in A. Let C_0 denote any optimal solution, if any. Since $f_{0x'x'} > 0$ for $x > 0$, by (2.6.iii) we conclude that any arc of C_0 lying above the t-axis, is of class C^∞ and an extremal arc (corners may occur only on the t-axis). The Euler equation yields $2xx'' = -x'^2$, and since $x'' = (dx'/dx)(dx/dt)$, also $2xx'(dx'/dx) = -x'^2$. Thus, either $x' = 0$ and $x(t) = b$, or $x' \ne 0$ and $2x(dx'/dx) = -x'$, from which $xx'^2 = C$, C constant, $C \ge 0$. Then, either $C = 0$ and $x(t) = 0$, or $C > 0$ and $(2/3)x^{3/2} = \pm at + b$ and $x(t) = k|t - t_0|^{2/3}$, k, t_0 constants, $k \ge 0$. These are all possible extremals.

If $1 = (t_1, c)$, $2 = (t_2, c)$, $c \ge 0$, then $E_{12}: x = x_0(t) = c$, $t_1 \le t \le t_2$, is an extremal arc without corners, and obviously optimal, since $I[x] \ge I[x_0] = 0$.

If $1 = (t_1, 0)$, $2 = (t_2, x_2)$, $x_2 > 0$, then $E_{12}: x = x_0(t) = k(t - t_1)^{2/3}$ with $k = x_2(t_2 - t_1)^{-2/3}$ is the only possible optimal solution, and E_{12} is of class C^∞ with the exception of the point t_1 where E_{12} has a vertical tangent. Analogously, if $1 = (t_1, x_1)$, $x_1 > 0$, $2 = (t_2, 0)$, then $E_{12}: x = x(t) = k(t_2 - t)^{2/3}$ with $k = x_1(t_2 - t_1)^{-2/3}$ is the only possible optimal solution, and E_{12} possesses a vertical tangent at t_2.

If $1 = (t_1, x_1)$, $2 = (t_2, x_2)$, $x_1 > 0$, $x_2 > 0$, $x_1 \ne x_2$, we consider only the case $x_1 < x_2$, the other one being analogous. Here there is an extremal of class C^∞ joining 1 and 2, namely $E_{12}: x = x_0(t) = k(t - t_0)^{2/3}$, with k and t_0 determined by $k(t_1 - t_0)^{2/3} = x_1$, $k(t_2 - t_0)^{2/3} = x_2$, or $(t_2 - t_0)/(t_1 - t_0) = (x_2/x_1)^{3/2}$; hence $t_0 < t_1 < t_2$. By computation, we find $I[E_{12}] = (\frac{4}{9})k^3(t_2 - t_1)$ and finally $I[E_{12}] = (\frac{4}{9})(x_2^{3/2} - x_1^{3/2})^2(t_2 - t_1)^{-1}$. Thus, $I[E_{12}]$ is a continuous function of the end points 1 and 2 of E_{12}.

Actually, in the last considered case $0 < x_1 < x_2$ there are other solutions C_{12}: $x = x(t)$, made up of two arcs of extremals with a cusp at some point $3 = (0, t_3)$, $t_1 < t_3 < t_2$, with vertical tangent there. Indeed for any such t_3 we can take $x(t) = k_1(t_3 - t)^{2/3}$ for $t_1 \le t \le t_3$, and $x(t) = k_2(t - t_3)^{2/3}$ for $t_3 \le t \le t_2$, provided $k_1 = x_1(t_3 - t_1)^{-2/3}$, $k_2 = x_2(t_1 - t_3)^{-2/3}$.

Let us prove that in all cases E_{12} is the only optimal solution.

We begin with the latter case: $x_1 > 0$, $x_2 > 0$, $x_1 < x_2$, where we have $t_0 < t_1 < t_2$.

The family of extremal arcs $E_a: x = x(t, a) = a|t - t_0|^{2/3}$, $t_1 \le t \le t_2$ (the same t_0 as for E_{12}) describes the region $R = A$ as a describes $[0, +\infty)$. The corresponding functions are $a = a(t, x) = x(t - t_0)^{-2/3}$, $p = p(t, x) = (2/3)a(t - t_0)^{-1/3} = (2x/3)(t - t_0)^{-1}$, and since t_0 is outside $[t_1, t_2]$, both functions $a(t, x)$, $p(t, x)$ are continuous with all their partial derivatives in A. Here $E(t, x, p, x') = 2x(x' - p)^2$ and therefore we have $E \ge 0$ for all $(t, x) \in R$, $E > 0$ for $x > 0$, $x' \ne p$. By the usual formula (2.11.5) we derive $I[C] - I[E_{12}] > 0$ for any curve C_{12} joining 1 and 2 in R. In particular this is true for the solutions $\bar{E}_{12}$ with a cusp at a point $(t_3, 0)$.

The same reasoning when x_1, x_2 are not both positive runs into trouble, but can be modified. Assume, for instance $x_1 = 0$, $x_2 > 0$. Then $E_{12}: x = x_0(t) = k(t - t_1)^{2/3}$, $t_1 \le t \le t_2$, and the family $E_a: x = x(t, a) = a(t - t_1)^{2/3}$ defines a field in every region $R_\delta \equiv [t_1 + \delta \le t \le t_2, 0 \le x < +\infty]$, but not in A.

Let $C_{12}: x = x(t)$, $t_1 \le t \le t_2$, be any curve in R joining $1 = (t_1, 0)$ and $2 = (t_2, x_2)$, $x_2 > 0$, with x AC and xx'^2 L_1-integrable in $[t_1, t_2]$. Let us consider a point $3 = (t_1 + \delta, x(t_1 + \delta))$ on C_{12}, with $\delta > 0$ as small as we want. Let E_{32} denote the optimal solution for the two points 3 and 2 as fixed end points. If C_{12} is distinct from E_{12}, so are C_{32} and E_{32} for δ sufficiently small, and $I[C_{32}] > I[E_{32}]$. On the other hand, $I[C_{32}] \to I[C_{12}]$ as $\delta \to 0$ by the L_1-integrability of xx'^2 in $[t_1, t_2]$; and also $I[E_{32}] \to I[E_{12}]$ as $\delta \to 0$ by the remark made above concerning the continuity of $I[E_{12}]$ as a function of the end points 1 and 2. Actually, given $\varepsilon > 0$ we can take δ so small that both differences $I[C_{32}] - I[C_{12}]$ and $I[E_{32}] - I[E_{12}]$ are in absolute value less than ε, while the difference $I[C_{32}] - I[E_{32}]$ is given by formula (2.11.5) and has a positive limit as $\delta \to 0$. Thus, $I[C_{12}] > I[E_{12}]$ for any curve C_{12} distinct from E_{12}. The case $1 = (t_1, c)$, $2 = (t_2, c)$, $c \geq 0$ does not present difficulties since E_{12} is the segment $x = c$ between 1 and 2, and a field is easily constructed by taking all the segments $x = a$, $0 \leq a < +\infty$. Thus $I[C_{12}] > I[E_{12}]$ for any curve C_{12} in R between 1 and 2.

(b) The same problem above but in the whole strip $t_1 \leq t \leq t_2$, $-\infty < x < +\infty$, has no minimum, since $\inf I[C] = -\infty$, as already shown in Section 2.11C, Example 2.

3.11 The Integral $I = \int x'^2(1 + x')^2\, dt$

We shall consider the extrema of this integral in the class Ω of all curves $C: x = x(t)$, $t_1 \leq t \leq t_2$ ($x(t)$ AC in $[t_1, t_2]$, x' L^4-integrable) joining two given fixed points $1 = (t_1, x_1)$, $2 = (t_2, x_2)$, $t_1 < t_2$. Let m denote the number $m = (x_2 - x_1)/(t_2 - t_1)$.

(a) Here $f_0 = x'^2(1 + x')^2 = x'^4 + 2x'^3 + x'^2$ depends on x' only, $f_0 \geq 0$,

$$f_{0x'} = 4x'^3 + 6x'^2 + 2x' = 4x'(x' + 1)(x' + \tfrac{1}{2}),$$
$$f_{0x'x'} = 2(6x'^2 + 6x' + 1) = 2(x' - m_1)(x' - m_2),$$

with $m_1, m_2 = 2^{-1}(-1 \pm 3^{-1/2})$, $m_1 = -0.7887$, $m_2 = -0.2113$. Thus $f_{0x'x'} > 0$ for $x' < m_1$ and $x' > m_2$, $f_{0x'x'} < 0$ for $m_1 < x' < m_2$. Since f_0 does not depend on x, condition (S) of Section 2.7 holds, and hence any optimal solution satisfies the Euler equation $f_{0x'}(x') = C$. Thus, any extremal arc is a segment $x' = a$, $x(t) = at + b$, a, b constants. The Weierstrass function $E(t, x, x', X') = f_0(X') - f_0(x') - (X' - x')f_{0x'}(x')$ is here $E = (X' - x')^2[(X' + x' + 1)^2 + 2x'(x' + 1)]$.

(b) Assume $-1 \leq m \leq 0$, $m = (x_2 - x_1)/(t_2 - t_1)$. Since $f_0 = 0$ for $x' = 0$ and $x' = -1$, and $f_0 \geq 0$ otherwise, we see that for any polygonal line E^* whose sides have slopes 0 and -1, as well as for any curve $E^*: x = x(t)$, $t_1 \leq t \leq t_2$, with $x(t)$ AC in $[t_1, t_2]$, and $x'(t) = 0$ or -1 a.e. in $[t_1, t_2]$, we have $I[E^*] = 0$, while $I[C] > 0$ for any other curve. Since $-1 \leq m \leq 0$, there is always such a curve, or polygonal line, E_{12} joining 1 and 2. Indeed, for $m = -1$ or $m = 0$, E_{12}^* is necessarily the segment 12. For $-1 < m < 0$, there are infinitely many polygonal lines and curves E_{12}^* joining 1 and 2. Indeed, it is enough to take $\phi(t) = -1$ on an arbitrary measurable set $H \subset [t_1, t_2]$ with meas $H = (-m)(t_2 - t_1)$,

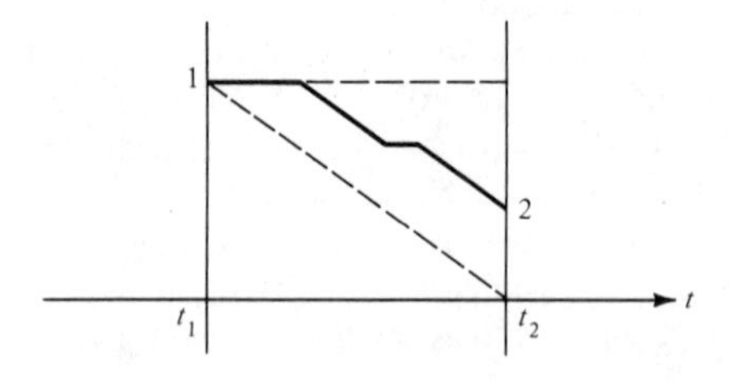

$\phi(t) = 0$ on the complementary set, and $x(t) = x_1 + \int_{t_1}^{t} \phi(\tau)\,d\tau$. Then $x'(t) = -1$ or 0 a.e. in $[t_1, t_2]$, and $x(t_1) = x_1$, $x(t_2) = x_1 + (-1)(\text{meas } H) = x_1 + m(t_2 - t_1) = x_2$. All these curves E_{12}^* give the *absolute minimum* to $I[x]$ in Ω, since $I[E_{12}^*] = 0$, and $I[C_{12}] > 0$ on every other curve C_{12} of Ω.

Note that along E_{12}^* we have $x' = -1$ or $x' = 0$, $f_0(-1) = f_0(0) = 0$, $f_{0x'}(-1) = f_{0x'}(0) = 0$, $f_{0x'x'}(-1) = 2 > 0$, $f_{0x'x'}(0) = 2 > 0$, $E = X'^2(X' + 1)^2 \geq 0$, for $X' \neq x'$, and Euler, DuBois-Reymond, Legendre, and Weierstrass conditions are all satisfied.

(c) Assume that $m_1 < m < m_2$ where $m = (x_2 - x_1)/(t_2 - t_1)$. Then $E_{12}: x = x_1 + m(t - t_1)$, $t_1 \leq t \leq t_2$, is an extremal arc joining 1 and 2, along which $f_{0x'x'}$ has a constant negative value. The accessory equation reduces here to $\eta'' = 0$ and any of its solutions which is zero at $t = t_1$ and which is not identically zero has no other zeros. Thus, there are no conjugate points on E_{12}. By (2.11.ii), E_{12} is a weak local maximum for $I[x]$ in Ω.

If we assume $m < m_1$, or $m > m_2$, then the same argument proves that $E_{12}: x = x_1 + m(t - t_1)$, $t_1 \leq t \leq t_2$, gives a weak local minimum for $I[x]$ in Ω.

This result can be improved. Again assume first $m_1 < m < m_2$, and consider the field of extremals $x = a + m(t - t_1)$, $t_1 \leq t \leq t_2$, a real, covering the strip $A = [t_1 \leq t \leq t_2, -\infty < x < +\infty]$, all of constant slope m. By the usual formula (2.11.5), the difference $I[C_{12}] - I[E_{12}]$ is expressed as the integral along C_{12} of $E(t, x, m, x') = 2^{-1}(x' - m)^2 \times f_{0x'x'}(\theta)$, where θ is some number between m and x'. Since m is in (m_1, m_2), if we assume $m_1 \leq x' \leq m_2$, then θ is necessarily in (m_1, m_2), $f_{0x'x'}(\theta) < 0$, and $E < 0$ for $x' \neq m$. We conclude that $I[C_{12}] < I[E_{12}]$ for every curve C_{12} in Ω with slope $x'(t)$ satisfying $m_1 \leq x'(t) \leq m_2$ a.e. in $[t_1, t_2]$. The same argument shows that if $m < m_1$ (or $m > m_2$), then $I[C_{12}] > I[E_{12}]$ for every curve C_{12} of Ω satisfying $x'(t) \leq m_1$ (or $x'(t) \geq m_2$) a.e. in $[t_1, t_2]$.

This result can be further improved. Again assume $m_1 < m < m_2$. By using the Weierstrass function $E = (x' - m)^2[(x' + m + 1)^2 + 2m(m + 1)]$, where the expression in brackets reduces to $6m^2 + 6m + 1 = 2^{-1}f_{0x'x'}$ for $x' = m$, and hence is negative for $x' = m$ with $m_1 < m < m_2$. On the other hand, $E = (x' - m)^2(6\theta^2 + 6\theta + 1)$ for some θ between m and x'. We see that $E < 0$ for all $x' \neq m$ of an interval (p_1, p_2) containing (m_1, m_2). Thus, $I[C_{12}] < I[E_{12}]$ for every curve C_{12} in Ω, distinct from E_{12}, joining 1 and 2 and with slope $p_1 \leq x'(t) \leq p_2$. For instance, for $m = -\frac{1}{2}$, we have $E = (x' + \frac{1}{2})^2((x' + \frac{1}{2})^2 - \frac{1}{2})$, $p_1 = 2^{-1}(-1 - 2^{1/2}) = -1.2071$, $p_2 = 2^{-1}(-1 + 2^{1/2}) = 0.2071$, and (p_1, p_2) is an interval larger than (m_1, m_2). Note that for $m_1 < m < m_2$, the extremal E_{12} is a weak local maximum, but not an absolute maximum, nor a strong local maximum.

Assume now $-1 < m < m_1$. Then $E > 0$ for all $-\infty < x' < m_1$, $x' \neq m$, but the expression for E given above shows that actually $E > 0$ for all $-\infty < x' < p_1$, $x' \neq m$, where $p_1 = -m - 1 - (-2m(m + 1))^{1/2}$, and this value of p_1 must be $> m_1$. Then $I[C_{12}] > I[E_{12}]$ for all curves C_{12} of Ω distinct from E_{12} with slopes $x'(t) \leq p_1$. Analogously, we prove that for $m_2 < m < 0$ we have $I[C_{12}] > I[E_{12}]$ for all curves C_{12} of Ω distinct from E_{12} and slope $x'(t) \geq p_2$, where $p_2 = -m - 1 + (-2m(m + 1))^{1/2}$. In either case E_{12} is a weak local minimum, but not an absolute minimum, since $I[E_{12}] > 0$, while $I[E_{12}^*] = 0$.

Assume now $m < -1$ or $m > 0$. Then $E > 0$ has a constant sign (for $x' \neq m$); hence $I[C_{12}] > I[E_{12}]$ for all curves C_{12} in Ω distinct from E_{12}, and hence E_{12} is an absolute minimum, and this holds also for $m = -1$ and $m = 0$.

Summarizing, the extremal arc E_{12} is an absolute minimum if $m \leq -1$ or $m \geq 0$; E_{12} is a local weak minimum if $-1 < m < m_1$ or $m_2 < m < 0$; E_{12} is a local weak maximum if $m_1 < m < m_2$.

(d) We shall now discuss possible solutions with corner points. At any corner point (t, x) let s_1, s_2 be the two slopes, left and right. Then s_1 and s_2 must satisfy Erdman corner conditions or $f_{0x'}(s_1) = f_{0x'}(s_2)$, and $f_0(s_1) - s_1 f_{0x'}(s_1) = f_0(s_2) - s_2 f_{0x'}(s_2)$. By substitution of the expressions above we obtain $4s_1^3 + 6s_1^2 + 2s_1 = 4s_2^3 + 6s_2^2 + 2s_2$, and $3s_1^4 + 4s_1^3 + s_1^2 = 3s_2^4 + 4s_2^3 + s_2^2$. By algebraic manipulations we find $s_1 + s_2 = -1$, $s_1 s_2 = 0$, and hence either $s_1 = -1, s_2 = 0$, or $s_1 = 0, s_2 = -1$. We obtain again the curves E_{12} we have discussed in (b), where we proved that each one of these curves gives an absolute minimum for $I[x]$ in A.

There are other solutions. For instance, assume $m_1 < m < m_2$, and take any two numbers s_1, s_2 such that $f_{0x'}(s_1) = f_{0x'}(s_2)$, and $s_1 < -1 < m_1 < m < m_2 < s_2 < 0$. This is possible since $f_{0x'}(s) < 0$ for $s < -1$ and for $-\frac{1}{2} < s < 0$, and then $c = f_{0x'}(s_1) = f_{0x'}(s_2) < 0$, $f_{0x'x'}(s_1) > 0$, $f_{0x'x'}(s_2) > 0$. Take any two complementary subsets E_1, E_2 in $[t_1, t_2]$ with meas $E_1 = (t_2 - t_1)(s_2 - m)(s_2 - s_1)^{-1}$ and then meas $E_2 = (t_2 - t_1)(m - s_1)(s_2 - s_1)^{-1}$. Take $\phi(t) = s_1$ for $t \in E_1$, $\phi(t) = s_2$ for $t \in E_2$, and let E'_{12} be the trajectory $E'_{12}: x = x(t) = x_1 + \int_{t_1}^{t} \phi(\tau)\, d\tau$. Then $x(t_1) = x_1$, $x(t_2) = x_2$, x is AC with $x'(t) = s_1$ for $t \in E_1$ (a.e.), $x'(t) = s_2$ for $t \in E_2$ (a.e.), and $f_{0x'}(t) = c$ for $t \in [t_1, t_2]$ (a.e.), i.e., the Euler equation is satisfied. If $C_{12}: x = X(t)$, $t_1 \le t \le t_2$, is any other curve with $X(t_1) = x_1$, $X(t_2) = x_2$, $|X'(t) - x'(t)| < \delta$ and $\delta = \min[m_1 - s_1, s_2 - m_2]$, then by Taylor's formula

$$I[C_{12}] - I[E'_{12}] = \int_{t_1}^{t_2} (f_0(X') - f_0(x'))\, dt$$

$$= c \int_{t_1}^{t_2} (X'(t) - x'(t))\, dt + 2^{-1} \int_{t_1}^{t_2} (X'(t) - x'(t))^2 f_{0x'x'}(\theta)\, dt,$$

where θ is between $X'(t)$ and $x'(t)$. Hence, either $\theta < m_1$ or $\theta > m_2$. Thus, the first integral is zero and the second one is positive. This proves that any of the curves E'_{12} is a weak local minimum for $I[x]$. Since $f_0(s_1) > 0$, $f_0(s_2) > 0$, then $I[E'_{12}] > 0 = I[E^*_{12}]$. As a particular case we can take $E_1 = [t_1, t]$, $E_2 = [t, t_2]$, two consecutive intervals of the indicated measures, and then E'_{12} is a polygonal line made up of two segments of slopes s_1 and s_2. In the limiting case with $s_1 = -1$, $s_2 = 0$, then we have the curves E^*_{12} of part (b) which are all absolute minima for $I[x]$.

Again for $m_1 < m < m_2$, if we take s_1, s_2 in such a way that $-1 < s_1 < m_1 < m < m_2 < 0 < s_2$ and $f_{0x'}(s_1) = f_{0x'}(s_2)$, the same construction yields curves E'_{12} which are weak local minima for $I[x]$.

For $-1 < m < m_1$ we take $-1 < s_1 < m < m_1 < m_2 < 0 < s_2$; for $m_2 < m < 0$ we take $-1 < s_1 < m_1 < m_2 < m < 0 < s_2$, and each of the corresponding curves E'_{12} is a weak local minimum for I.

Note that these weak local minima do not satisfy the DuBois-Reymond condition in the usual form. This is not contradictory. Indeed we have proved this conditions to be necessary in Section 2.2 Remark 3, as a consequence of the Euler equation for $n = 1$, f_0 and x of class C^2; hence even for weak extrema since the Euler equation holds. We have also proved the necessity of the DuBois-Reymond condition in Section 2.4 and in Section 2.7 for strong extrema for $n \ge 1$, f_0 of class C^1 and x AC (respectively for x' essentially bounded, and for x' unbounded but under condition (S)). In Section 2.4 we have proved the DuBois-Reymond condition even for weak extrema for $n \ge 1$, f_0 of class C^1, and x' continuous. The weak local minima we have encountered above do not fall in any of these situations.

(e) We shall now discuss the generalized solutions for the problem of the absolute minimum of the same integral $I = \int_{t_1}^{t_2} f_0(x(t))\, dt$. As we have seen in Section 1.14, we are concerned with the minimum of the functional

$$J[x,p,v] = \int_{t_1}^{t_2} [p_1 f_0(u_1) + p_2 f_0(u_2)]\, dt, \qquad p = (p_1, p_2), \qquad v = (u_1, u_2),$$

in the class Ω^* of all AC scalar functions $x(t)$, $t_1 \le t \le t_2$, with $x(t_1) = x_1$, $x(t_2) = x_2$, all measurable functions $p_1(t) \ge 0$, $p_2(t) \ge 0$, $p_1(t) + p_2(t) = 1$, and all measurable functions $u_1(t)$, $u_2(t)$. As mentioned in Section 1.14, the problem reduces to the minimum of the integral

$$I^*[x] = \int_{t_1}^{t_2} F_0(x'(t))\, dt,$$

where F_0 is defined by taking $F_0(x') = f_0(x')$ for $x' \le -1$ and $x' \ge 0$, and by taking $F_0(x') = 0$ for $-1 \le x' \le 0$. Here F_0 is of class C^1 and the Euler equation $F_{0x'} = C$ is the same as before if $C > 0$, while for $C = 0$ it leaves x' arbitrary, $-1 \le x' \le 0$. If $-1 < m < 0$, any curve $E_{12}^*: x = x(t)$, $t_1 \le t \le t_2$, x AC with $x(t_1) = x_1$, $x(t_2) = x_2$, $-1 \le x'(t) \le 0$, gives the absolute minimum for I^* with common value zero. If $m \le -1$ and $m \ge 0$ the absolute minimum of I^* is given by the segment E_{12} of slope m joining 1 and 2 as before.

The curves E_{12} of part (b) are obviously curves E_{12}^*. One can also say that any of generalized solutions E_{12}^*, with x' taking arbitrary values between -1 and 0, is equivalent to any of the curves E_{12} of part (b) with x' taking only the values -1 and 0.

3.12 Brachistochrone, or Path of Quickest Descent

(a) Given two points 1 and 2 in a vertical plane π, 2 at a level lower than 1 and not on the same vertical line as 1, we are to determine the curve C joining 1 and 2 such that a material point P, starting at 1 with velocity v_1, will glide from 1 to 2 along C, under the force of gravity only and without friction, in a minimum time T. Let us take in π a Cartesian system of reference so that $1 = (x_1, y_1)$, $2 = (x_2, y_2)$, $x_1 < x_2$, $y_1 < y_2$. Let Ω be the collection of all (nonparametric) curves $C: y = y(x)$, $x_1 \le x \le x_2$, with $y(x_1) = y_1$, $y(x_2) = y_2$, and $y(x)$ AC in $[x_1, x_2]$. Let s denote the arc length on C from 1 to 2, $0 \le s \le L$; let t be time, $0 \le t \le T$; and let v be the instantaneous velocity of P. Then $v = ds/dt$, and $T = \int_0^L ds/v = \int_{x_1}^{x_2} (1 + y'^2)^{1/2}\, dx/v$, where $(') = d/dx$. On the other hand, the increase in kinetic energy in the interval $[0, t]$ must be equated to the decrease in potential energy; hence $(mv^2/2) - (mv_1^2/2) = mg(y - y_1)$, where g is the constant of acceleration. Hence, $v^2 = 2gy - 2gy_1 + v_1^2$, so $v^2 = 2g(y - \alpha)$, with $\alpha = y_1 - v_1^2/2g$. Finally

$$\sqrt{2g}\,T = I[y] = \int_{x_1}^{x_2} (y - \alpha)^{-1/2} (1 + y'^2)^{1/2}\, dx, \qquad (') = d/dx,$$

and we have to minimize the integral $I[y]$ in the class Ω.

(b) Let us assume $v_1 > 0$. Then $y - \alpha = y - y_1 + v_1^2/2g \ge v_1^2/2g > 0$. We may take for A the region $A = [x_1 \le x \le x_2, y \ge y_1]$, so that A is a closed set and $y \ge y_1$ implies $(y - \alpha)^{-1/2} \le \sqrt{2g}/v_1$. Thus $f_0 = (y - \alpha)^{-1/2}(1 + y'^2)^{1/2}$ does not depend on the independent variable x, but only on y and $y' = dy/dx$, and f_0 is continuous in A together with all its partial derivatives.

By the existence theorem (14.3.ii) (see Example 3 at the end of Section 14.4), $I[y]$ has an absolute minimum in Ω, any minimizing curve is of class C^2 and satisfies Euler's equation.

(c) Here we have $f_{0y'} = y'(y - \alpha)^{-1/2}(1 + y'^2)^{-1/2}$, and the DuBois-Reymond equation (2.2.14), or $f_0 - y'f_{0y'} = c$, yields

$$(y - \alpha)^{-1/2}(1 + y'^2)^{1/2} - y'^2(y - \alpha)^{-1/2}(1 + y'^2)^{-1/2} = c,$$

and after simplification and taking $c = 1/\sqrt{2b}$,

$$(y - \alpha)(1 + y'^2) = 2b. \tag{3.12.1}$$

The introduction of a parameter τ by means of the equation $y' = -\tan(\tau/2)$ simplifies computations. We have $y' = -\sin\tau(1 + \cos\tau)^{-1}$, and $1 + y'^2 = \cos^{-2}(\tau/2)$. Then equation (3.12.1) yields

$$y - \alpha = 2b(1 + y'^2)^{-1} = 2b\cos^2(\tau/2) = b(1 + \cos\tau).$$

Hence, $dy/d\tau = -b\sin\tau$ and

$$dx/d\tau = (dx/dy)(dy/d\tau) = (1/y')(dy/d\tau) = b(1 + \cos\tau) = 2b\cos^2(\tau/2).$$

By immediate integration we have now

$$x = a + b(\tau + \sin\tau), \qquad y = \alpha + b(1 + \cos\tau), \tag{3.12.2}$$

where a and b are constants of integration ($b > 0$). This shows that any minimizing curve is an arc 12 of a *cycloid*—precisely, the locus of a point fixed on the circumference of a circle of radius b as the circle rolls on the lower side of the line $y = \alpha = y_1 - v_1^2/2g$. The value of the integral $I[y]$ on any arc E_{12} of these cycloids passing through 1 and 2 can be easily obtained by the use of the variable τ, and $I[E_{12}] = (2b)^{1/2}(\tau_2 - \tau_1)$. We prove below that there is one and only one of these cycloids passing through 1 and 2. Analytically this boundary value problem reduces to the determination of four real numbers a, b, τ_1, τ_2 satisfying the four equations

$$x_1 - a = b(\tau_1 + \sin\tau_1), \qquad x_2 - a = b(\tau_2 + \sin\tau_2)$$
$$y_1 - \alpha = b(1 + \cos\tau_1), \qquad y_2 - \alpha = b(1 + \cos\tau_2).$$

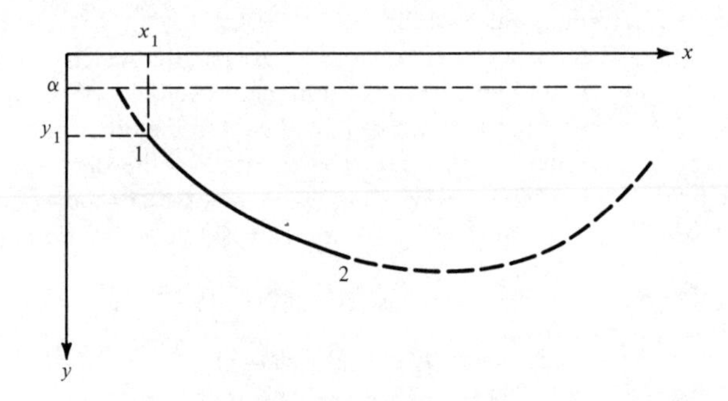

(d) Let us prove that there is one and only one cycloid (3.12.2) passing through 1 and 2. Let us draw an arbitrary one of the cycloids (3.12.2) and intersect it by a line 1′2′ parallel to the straight line 12 as shown in the illustration. If we move the line 1′2′, keeping it always parallel to 12, from the position L' to the position L'', then the ratio of the length of the segment 0′1′ to that of the segment 1′2′ increases from 0 to ∞ and

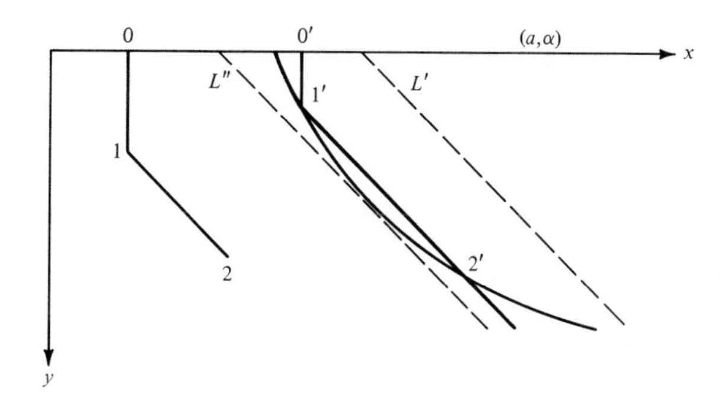

passes once only through the value of the corresponding ratio 01 to 12. At the position which gives the equality of these ratios, the lengths of 0′1′ and 1′2′ are not necessarily equal to 01 and 12 respectively. By changing the value of b, however, the cycloid can be expanded or contracted into another similar to itself, having the same center (a, α), and the new segments, say 0″1″ and 1″2″, corresponding to 0′1′ and 1′2′, will have the same ratio as before. By properly choosing the value of b, the segments 0″1″ and 1″2″ can be made exactly equal to 01 and 12 respectively. Finally, changing the value of a, we can slide the cycloid along the fixed line $y = y_1$, and 1″ and 2″ can be made to coincide with 1 and 2 respectively. This shows that there is one and only one cycloid (3.12.2) through 1 and 2.

(e) Concluding the argument in subsections (b–d), we can finally state that the integral $I[y]$ has an absolute minimum in Ω, and that this absolute minimum is given by the only cycloid (3.12.2) passing through the two points 1 and 2. This cycloid is thus the curve along which the point P slides without friction from 1 to 2 under the sole action of gravity in a minimum time.

(f) We could reach the same statement by means of sufficient conditions and without the use of existence theorems.

If E_{12} is the particular arc of the only cycloid (3.12.2) passing through 1 and 2, say for $a = a_0$ and $b = b_0$, let us keep a_0 fixed, and let b vary from 0 to $+\infty$. Then the family of cycloids

$$x = a_0 + b(\tau + \sin \tau), \qquad y = \alpha + b(1 + \cos \tau), \tag{3.12.3}$$

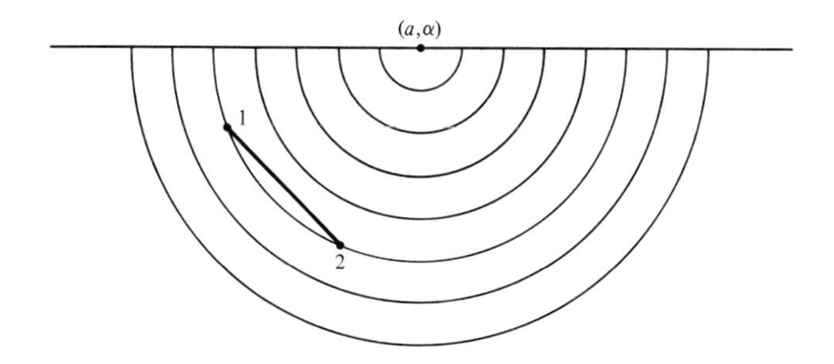

where $-\pi < \tau < \pi, 0 < b < +\infty$, fills once the whole half plane $R = [-\infty < x < +\infty, y > \alpha]$. Indeed, for the same τ and different values of b, say b' and b'', we have points (x', y'), (x'', y'') with $(x' - a_0)/(x'' - a_0) = b'/b'' = (y' - \alpha)/(y'' - \alpha)$. This shows that (x', y'), (x'', y'') are on the same half straight line issuing from (a_0, α). The cycloids of the family are similar figures, and cover simply R. For each point $(x, y) \in R$, there is a well-determined value $b = b(x, y)$ such that the corresponding cycloid passes through (x, y) for a value $\tau = \tau(x, y)$ of the parameter τ, and with slope $p(x, y)$ given by $p(x, y) = -\tan(\tau(x, y)/2)$. We have to prove that $\tau(x, y)$, $p(x, y)$, $b(x, y)$ are continuous functions of (x, y) in R. Now τ is given by the equation

$$\frac{x - a_0}{y - \alpha} = \frac{\tau + \sin \tau}{1 + \cos \tau},$$

which has the form $g(x, y) = \phi(\tau)$, and

$$\phi'(\tau) = \cos^{-2}(\tau/2)[1 + (\tau/2)\tan(\tau/2)].$$

For $-\pi < \tau < \pi$, $\phi'(\tau) > 0$; hence $\phi(\tau)$ is always increasing, and $\phi(-\pi + 0) = -\infty$, $\phi(\pi - 0) = +\infty$. Therefore $s = \phi(\tau)$ has an inverse function $\tau = \psi(s)$, $\psi(-\infty) = -\pi$, $\psi(+\infty) = \pi$, and $\tau = \psi[(x - a_0)/(y - \alpha)]$. This shows that τ is a continuous function of (x, y) in R. On the other hand, $\phi'(\tau) > 0$ is continuous in $[-\pi, \pi]$; hence $\psi'(s)$ exists for all s, $-\infty < s < +\infty$, $\psi'(s) = 1/\phi'(\tau)$, thus ψ is continuous with its first derivative, and thus $\tau(x, y) = \psi[(x - a_0)/(y - \alpha)]$ has continuous first partial derivatives in R. Finally, $b = b(x, y) = (y - \alpha)/(1 + \cos \tau)$, and thus $b(x, y)$ is also continuous in R with its first partial derivatives.

We have shown that R is a field of extremals. Since $f_{y'y'} > 0$ in R, by (2.11.iv) we have $I[C_{12}] \geq I[E_{12}]$ for every curve C_{12} in Ω joining 1 and 2, and equality holds if and only if $C_{12} \equiv E_{12}$.

(g) We shall now discuss the case $v_1 = 0$. Then $\alpha = y_1$, and $f_0 = (y - y_1)^{-1/2}(1 + \dot{y}^2)^{1/2}$ has the singular line $y = y_1$. The conditions of the existence theorem (14.3.ii) (see Example 3 at the end of Section 14.4) are all satisfied, and therefore, $I[y]$ has an absolute minimum. On the other hand, the considerations of subsection (d) above can be repeated with obvious changes, and we conclude as we did there that there is one and only one cycloid E_{12} through 1 and 2, but now E_{12} has a vertical tangent at 1. We conclude as in subsection (e) that $I[y]$ has an absolute minimum in Ω given by a cycloid E_{12}.

To reach the same result via sufficient conditions in the present case $v_1 = 0$ requires more work.

(h) Again, as in Subsection (d), the equations (3.12.1), where now $-\pi \leq \tau \leq \pi$, $0 \leq b < +\infty$, give a family of cycloid which fills the half plane $R_0 = [-\infty < x < \infty, y \geq y_1]$. Nevertheless, the corresponding functions $p(x, y)$, $u(x, y)$, $b(x, y)$ are continuous only for $y > y_1$, and hence in each closed region $R_1 = [-\infty < x < \infty, y \geq y_1 + \delta]$, $\delta > 0$. The arc E_{12} does not belong to C^1 since its slope is $+\infty$ at the point 1; nevertheless, $I[E_{12}]$ has still the finite value $I[E_{12}] = \sqrt{2b}(\tau_2 - \tau_1)$. Let K_0 be the collection of all continuous curves $C_{12}: y = y(x)$, $x_1 \leq x \leq x_2$, passing through 1 and 2, with $y(x) > \alpha$ in $(x_1, x_2]$, of class C_s when restricted to $(x_1, x_2]$, and such that $I[y]$ has a finite value (as a generalized Riemann integral of a nonnegative function having at most one point of infinity at $x = x_1$). Then $E_{12} \in K_0$. If C_{12} is any curve in K_0 distinct from E_{12}, take any point, say 3, on C_{12}. Through 3 there passes a unique cycloid E_3 from 3 to the vertical line through (a_0, α). Let 7 denote the analogous point of the cycloid E_{12}, also continued up to the vertical line through (a_0, α). Then the sum $I[C_{13}] + I[E_3]$

varies continuously as the point 3 moves from 1 to 2 along C_{12}, beginning with the value $I[E_{12}] + I[E_{27}]$ and ending with the value $I[C_{12}] + I[E_{27}]$. Hence, if we show that this sum does not decrease, we conclude that $I[C_{12}] \geq I[E_{12}]$. Take a second point 4 on C_{12} near 3, and let 6 be the corresponding point on the vertical line $x = a_0$. By (2.9.5) we have

$$I[E_{46}] - I[E_{35}] = I^*[D_{56}] - I^*[C_{34}],$$

and hence

$$(I[C_{14}] + I[E_{46}]) - (I[C_{13}] + I[E_{35}]) = I[C_{34}] + I^*[D_{56}] - I^*[C_{34}].$$

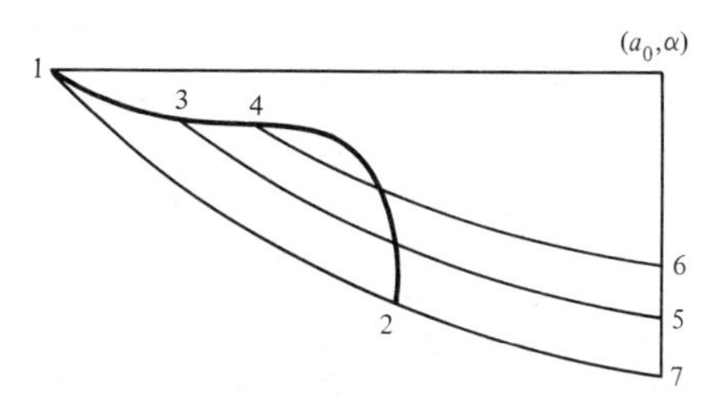

The vertical line D_{56} is orthogonal to the cycloids of the field; hence $dx + p\,dy = 0$ along D_{56}. On the other hand

$$\begin{aligned} I^*[D_{56}] &= \int [f\,dx + (dy - p\,dx)f_{y'}] \\ &= \int [(y-\alpha)^{-1/2}(1+p^2)^{1/2}\,dx + (dy - p\,dx)p(y-\alpha)^{-1/2}(1+p^2)^{1/2}] \\ &= \int (y-\alpha)^{-1/2}(1+p^2)^{-1/2}(dx + p\,dy) = 0. \end{aligned}$$

(In other words, D is transversal to the extremals of the family.) Finally, by Section 2.11,

$$I[C_{34}] - I^*[C_{34}] = \int_{x_3}^{x_5} E(x, Y, p, Y')\,dx \geq 0,$$

and equality holds if and only if the arc C_{35} is an arc of extremal. If every partial arc C_{34} of C_{12} is an arc of one of the cycloids of the field, then C_{12} is such an arc and $C_{12} = E_{12}$. Otherwise, $I[C_{12}] > I[E_{12}]$. Thus $E_{12} \in K_0$ and is the absolute minimum for $I[y]$ in K_0. For the previous analysis, see G. A. Bliss [I].

3.13 Surface of Revolution of Minimum Area

(a) We are to determine the curve of the xy-plane joining two points $1 = (x_1, y_1)$, $2 = (x_2, y_2)$, $x_1 < x_2$, $y_1 \geq 0$, $y_2 \geq 0$, lying entirely in the half plane $y \geq 0$ and such that the surface of revolution generated by rotation about the x-axis has minimum area. Let Ω be the collection of all curves $C: y = y(x), x_1 \leq x \leq x_2$, $y(x_1) = y_1$, $y(x_2) = y_2$, $y(x)$ AC in $[x_1, x_2]$, $y(x) \geq 0$. We may as well study the integral

$$I[y] = \int_{x_1}^{x_2} y(1 + y'^2)^{1/2}\,dx, \qquad y \in \Omega, \quad (') = d/dx,$$

since the area above is 2π times the value of this integral. Here A is the strip $A = [x_1 \le x \le x_2, 0 \le y \le +\infty]$.

(b) Note that if $y_1 = y_2 = 0$, then $I[y]$ has an absolute minimum given by the trivial solution $y \equiv 0$. Let y_1, or y_2, or both be positive. Here we have $f_0 = f_0(y, y') = y(1 + y'^2)^{1/2}$, $f_{0y'} = yy'(1 + y'^2)^{-1/2}$, $f_{0y'y'} = y(1 + y'^2)^{-3/2} \ge 0$. Here f_0 does not depend on t; hence, by Section 2.7, if $C_0: x = x(t)$, $t_1 \le t \le t_2$, is any optimal solution, then any arc of C_0 with $y(x) > 0$ satisfies the DuBois-Reymond equation. By Section 2.6 any such arc is also of class C^2, and is an extremal.

The DuBois-Reymond equation (2.2.14), $f_0 - y'f_{0y'} = C$, yields, after simplification, $y(1 + y'^2)^{-1/2} = b$, and by integration

$$y = b\cosh(b^{-1}(x - a)), \tag{3.13.1}$$

where a and b are constants of integration, and as usual $\cosh z = 2^{-1}(e^z + e^{-z})$, $\sinh z = 2^{-1}(e^z - e^{-z})$ for any z. Since there are no corner points with $y > 0$, we conclude that there can be a minimum if and only if either $y_1 = y_2 = 0$, and the minimum is given by a segment of the x-axis, or both y_1, y_2 are positive, and the minimum (if any) is given by an arc of a *catenary*. Assume y_1, $y_2 > 0$, and let us discuss whether there is a catenary (3.13.1) through 1 and 2. The passage through 1 gives the equation $y_1 = b\cosh(b^{-1}(x_1 - a))$; hence, if we take $y_1 = b\cosh\alpha$, we obtain for a the explicit value $a = x_1 - y_1\alpha/\cosh\alpha$. Hence, $x_1 - a = y_1\alpha/\cosh\alpha$, $b^{-1}(x_1 - a) = \alpha$, $y_1 = b\cosh\alpha = b\cosh(b^{-1}(x_1 - a))$, and we obtain the family of all catenaries (3.13.1) through 1 in the form

$$y = y_1(\cosh\alpha)^{-1}\cosh(\alpha + (x - x_1)y_1^{-1}\cosh\alpha) = y(x, \alpha). \tag{3.13.2}$$

We shall denote by ($'$) the operation of differentiation with respect to x, and by the subscript α the operation of differentiation with respect to α. Also, the subscript 1 denotes the value taken by the corresponding variable when $x = x_1$. By computation, we have

$$y' = \sinh(\alpha + (x - x_1)y_1^{-1}\cosh\alpha), \qquad y_1' = \sinh\alpha,$$

$$y_\alpha = y'y_1'(\cosh\alpha)^{-1}\left(x - \frac{y}{y'} - x_1 + \frac{y_1}{y_1'}\right). \tag{3.13.3}$$

The tangent to any of the catenaries (3.13.2) at the point 1 has the equation $Y - y_1 = y_1'(X - x_1)$, and the tangent at any point $P = (x, y)$ to the same catenary has the equation $Y - y = y'(X - x)$. Their point of intersection (X, Y) has ordinate Y given by

$$Y = y'y_1'(y' - y_1')^{-1}\left(x - \frac{y}{y'} - x_1 + \frac{y_1}{y_1'}\right). \tag{3.13.4}$$

Thus, we obtain from (3.13.3) that $y_\alpha = Y(y' - y_1)/\cosh\alpha$. We shall see that on each catenary E of the family (3.13.2) there is a point $P = (x, y)$, or $P = (x, y(x, \alpha))$, at which $y_\alpha = 0$, and this point P will generate the envelope Γ of the family (3.13.2), and will indeed be the point of contact of E with Γ, and thus the conjugate of 1 on E. From the latter formula we see that y_α and Y vanish together.

(c) This remark justifies the following simple construction of the conjugate point P to 1 on a given extremal E through 1. If M is the intersection with the x-axis of the tangent to E at 1, then P is the point of contact of the tangent to E from M.

If a point (x, y) moves from 1 along a catenary $E: y(x, \alpha)$, then y_α is at first positive and then changes sign when (x, y) passes the point P conjugate to 1 on E. We shall need now the second derivatives of $y(x,\alpha)$ at the conjugate points P. Thus we shall differentiate y' and y_α and use consistently the fact that $y_\alpha = 0$ at these points, and hence, by (3.13.3) with $y_\alpha = 0$, $x - x_1$ can be replaced by $(y/y') - (y_1/y_1')$. Also, we know that $y_1' < 0$, and that $y' > 0$ at P. We obtain

$$y_\alpha' = (y'y_1^2)^{-1}(y^2 y_1') \cosh \alpha < 0, \qquad y'' = y_1^{-2} y \cosh^2 \alpha < 0,$$

$$y_{\alpha\alpha} = y_1^{-2} y' y_1'^2 \left(\left(\frac{y}{y'}\right)^3 - \left(\frac{y_1}{y_1'}\right)^3 \right) > 0.$$

The last evaluation, $y_{\alpha\alpha} > 0$, is in agreement with the fact that at the points of contact P, we have $y_\alpha' = 0$ and a minimum for y as a function of α.

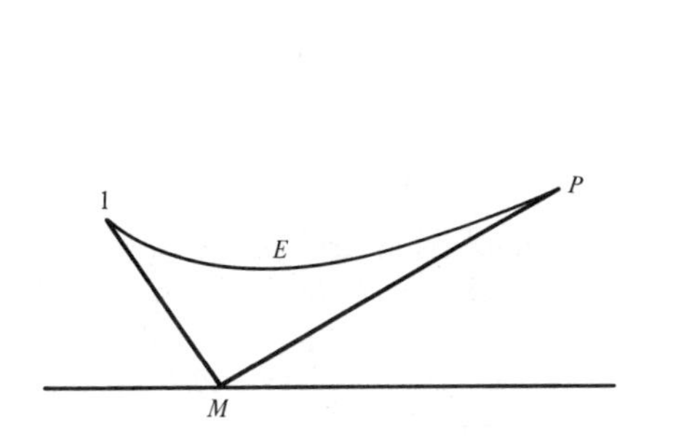

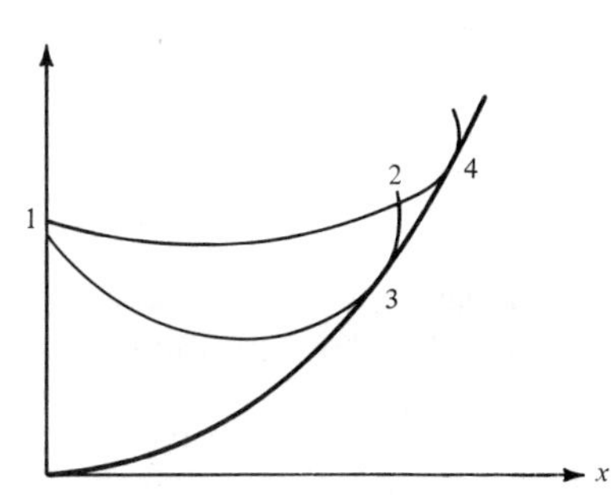

We are now in a position to study further the family of catenaries (3.13.2) and the locus of the conjugate point P to 1 on the catenaries of the family. First of all, we have

$$y(x,\alpha) = \frac{y_1}{\cosh \alpha} \cosh\left(\alpha + \frac{x - x_1}{y_1} \cosh \alpha\right)$$

$$= \left(\frac{\alpha}{\cosh \alpha} + \frac{x - x_1}{y_1}\right) \frac{y_1 \cosh\left[\left(\dfrac{\alpha}{\cosh \alpha} + \dfrac{x - x_1}{y_1}\right) \cosh \alpha\right]}{\left(\dfrac{\alpha}{\cosh \alpha} + \dfrac{x - x_1}{y_1}\right) \cosh \alpha},$$

where $\alpha/\cosh \alpha \to 0$, $[(\alpha/\cosh \alpha) + (x - x_1)/y_1] \cosh \alpha \to +\infty$ as $\alpha \to \infty$, and finally the last fraction in the formula above $\to +\infty$; hence, $y(x,\alpha) \to +\infty$ as $\alpha \to +\infty$ for every fixed $x > x_1$, and analogously $y(x,\alpha) \to +\infty$ as $\alpha \to -\infty$. We should note that y_α changes from negative to positive whenever it vanishes, since $y_{\alpha\alpha} > 0$; hence y_α can vanish only once. Thus, for every $x > x_1$ fixed, $y(x,\alpha)$ diminishes from $-\infty$ to a minimum and then increases to $+\infty$ again, since y_α varies from negative values to zero once and then to positive values. Let us denote by $g(x)$ the minimum of $y(x,\alpha)$ for any $x > x_1$. The curve $G: y = g(x)$, $x > x_1$, is the locus of the conjugate points P. Through every point 2 above G there pass exactly two catenaries of the family, say 132, 124; on one of these (132) there is a conjugate point 3 to 1, and on the other (124) there is none. Every point 3 on the curve G is joined to 1 by one and only one catenary of the family for which 3 is the conjugate point. Every point below G is joined to 1 by no catenary of the family (3.13.2). Thus, if the point 2 is below the curve G, the integral $I[y]$ has no minimum in K, since there is no catenary joining 1 and 2. If the point 2 is on the curve G, the integral $I[y]$ has no minimum in K either, since 2 is the conjugate to 1 on E_{12} and the reasoning of Section 2.9 applies.

(c) For any given $x > x_1$ the value of α for which $y_\alpha = 0$ will be denoted by $\alpha(x)$; hence $y_\alpha(x, \alpha(x)) = 0$, and $g(x) = y(x, \alpha(x))$. Since $y_{\alpha\alpha} > 0$ at $(x, \alpha(x))$, by the implicit function theorem we conclude that $\alpha(x)$ is differentiable and hence $y'_\alpha + y_{\alpha\alpha}\alpha' = 0$, and $g'(x) = y' + y_\alpha\alpha' = y' > 0$. On the other hand, by using the expressions of y'' and y'_α above we have

$$g''(x) = y'' + y'_\alpha\alpha' = yy_1y'^3(\cosh\alpha)^2(y_1^3y'^3 - y^3y_1'^3)^{-1} > 0.$$

The two relations $g'(x) > 0, g''(x) > 0$ show that G is concave upward, and that $g(x) \to +\infty$ as $x \to +\infty$. To show that also $g'(x) \to +\infty$ as $x \to +\infty$, we should note that the slope $g'(x)$ of G at P is the same as the slope $y'(x, \alpha(x))$ to the catenary (3.13.2) through P, and this slope is given, because of (3.13.3), by

$$\begin{aligned} y'(x, \alpha(x)) &= \sinh\left[\alpha + y_1^{-1}(\cosh\alpha)\left(\frac{y_1}{y_1'} - \frac{y}{y'}\right)\right] \\ &= \sinh[\alpha - (\sinh\alpha)^{-1}\cosh\alpha + y_1^{-1}(y/y')\cosh\alpha]. \end{aligned}$$

This number approaches $+\infty$ as $\alpha \to +\infty$. We shall now show that $g(x) \to 0$, $g'(x) \to 0$ as $x \to x_1 + 0$. The vertex (a, b) of each catenary (3.13.2) is above G, and $a = x_1 - (y_1\alpha/\cosh\alpha)$, $b = y_1/\cosh\alpha$. Hence, as $\alpha \to -\infty$, we see that $a \to x_1$, $b \to 0$, and finally $(a, b) \to (x_1, 0)$. This proves that $g(x) \to 0$ as $x \to x_1 + 0$. Also, the slope of the segment joining $(x_1, 0)$ to (a, b) is $-1/\alpha$, and $-1/\alpha \to 0$ as $\alpha \to -\infty$. This proves that also $g'(x) \to 0$ as $x \to x_1 + 0$.

Let 2 be a point above the curve G, and let E_{12} be the arc of the unique catenary $C: y = b_0\cosh(b_0^{-1}(x - a_0))$ joining 1 to 2 and containing no point conjugate to 1. The region above G covered by the curves (3.13.2) is no field (since they all pass through 1). To define a field containing E_{12} we may take a point 0 on the extension C of the catenary E_{12} at the left of 1 and so close to 1 that the conjugate point 3 to 0 on C, according to the construction above, is still at the right of 2. The tangents to C at 0 and 3 meet at a point $4 = (x_4, 0)$ of the real axis. The similarity transformation of center 4, say $x - x_4 = (b_0/b)(X - x_4)$, $y = (b_0/b)Y$, transforms E_{12} into another catenary

$$E_b: Y = y(x, b) = b\cosh b^{-1}(X - x_4 + (b/b_0)(x_4 - a_0))$$

that is, another catenary of the form $y = b\cosh(b^{-1}(x - a))$ with parameters $a = x_4 - (b/b_0)(x_4 - a_0)$ and b. If b is thought of as a parameter, we see that the catenaries E_b fill the V-shaped region V between the radii 40 and 43. These catenaries are all tangent to the two radii 40 and 43. Through each point $(x, y) \in V$ there passes one and only one catenary E for $b = b(x, y)$ with slope $p(x, y)$. It is left to the reader to prove

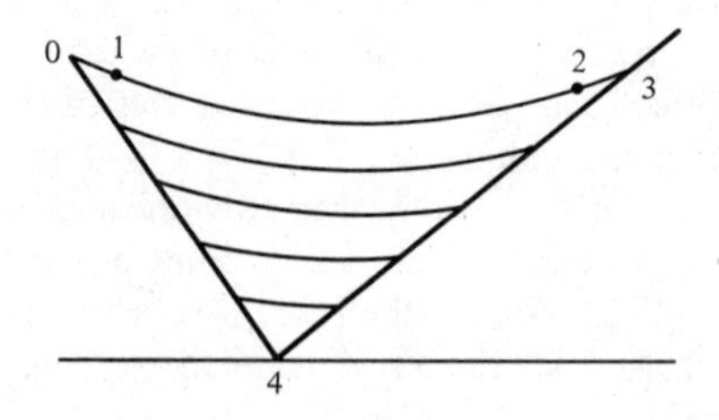

that, in V, $b(x, y)$ and $p(x, y)$ are continuous with their first partial derivatives in V. Since $f_{yy} > 0$ in V, by (2.11.iv) we conclude that $I[C_{12}] \geq I[E_{12}]$ for every curve C_{12} of the class Ω in V, and that equality holds if and only if $C_{12} = E_{12}$. Thus E_{12} is a strong local minimum for $I[y]$ in V.

(d) The reader may ask whether the relation $I[C_{12}] > I[E_{12}]$, which is valid for all curves C_{12} joining 1 and 2 in the V-shaped region V, is still valid in the larger region $R = [x_1 \leq x \leq x_2, 0 \leq y < +\infty]$. The answer, as we shall see below, can be negative.

We shall first consider rectifiable continuous path curves $C: x = x(t)$, $y = y(t)$, $t_1 \leq t \leq t_2$, in the xy-plane, that is, $x(t)$, $y(t)$ AC in $[t_1, t_2]$. Then the integral corresponding to $I[C]$ in this more general class of curves is

$$I[C] = \int_C y(x'^2 + y'^2)^{1/2}\,dt = \int_C y\,ds,$$

where s is the arc length parameter along C (see Chapter 2 for more details). If we denote by C_{12} now a segment parallel to the y-axis, say $[x = x_1, y_2 \leq y \leq y_1]$, and by C_{13} any given regular parametric curve $(X(t), Y(t))$ having one end point at 1 and the same length $L = y_1 - y_2$ as s, then

$$I[C_{13}] - I[C_{12}] = \int_0^L Y\,ds - \int_0^L y\,ds = \int_0^L (Y - y)\,ds \geq 0$$

since obviously $Y(s) \geq y(s)$ for every s, and hence equality holds above if and only if $Y(s) \equiv y(s)$, $C_{13} \equiv C_{12}$, $3 = 2$. Now let us consider the usual two points $1 = (x_1, y_1)$, $2 = (x_2, y_2)$, $x_1 < x_2$, $y_1 > 0$, $y_2 > 0$, and denote by P_{1342} the polygonal made up of two segments 13, 24 normal to the x-axis from 1 and 2, and of the segment 34 of the x-axis. Then $I[P_{34}] = 0$, and $I[P_{1342}] = (y_1^2 + y_2^2)/2$. We shall denote by r the region of the xy-plane made up of the three rectangles $[x_i - \delta \leq x \leq x_i + \delta, 0 \leq y \leq y_i + \delta]$, $i = 1, 2$, and $[x_1 \leq x \leq x_2, 0 \leq y \leq \delta]$. The remark above shows that if $\delta > 0$ is sufficiently small, then for any curve C_{12} joining 1 and 2 in r and distinct from P_{1342} we have $I[C_{12}] > I[P_{1342}]$. In other words, P_{1342} gives a strong relative minimum for $I[C]$, or more precisely, an absolute minimum in the class of all parametric (regular) curves joining 1 to 2 in r.

We can prove that $I[C_{12}] > I[P_{1342}]$ for all parametric regular curves C_{12} joining 1 and 2 in the upper half plane $y \geq 0$ provided $2 = (x_2, y_2)$ is below or on the curve G. Precisely, we prove the relation $I[C_{12}] > I[P_{1342}]$ for every curve C_{12} having at least one point, say 5, on G. We may well assume that 5 is the first point of C_{12} on G. If $5'$ is any point on C_{12} between 1 and 5, then $5'$ is above G, and we have $I[C_{15'}] \geq I[E_{15'}]$.

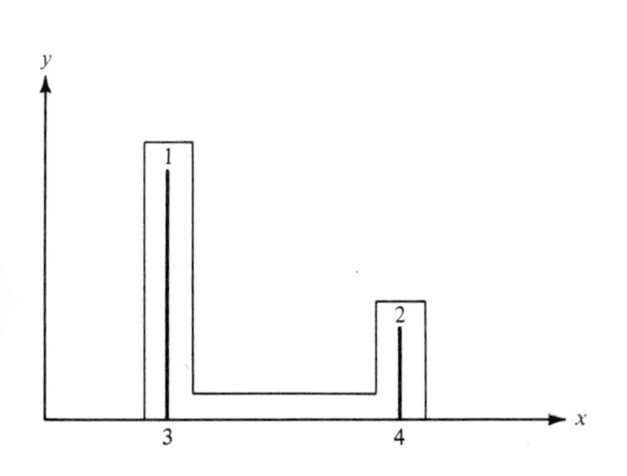

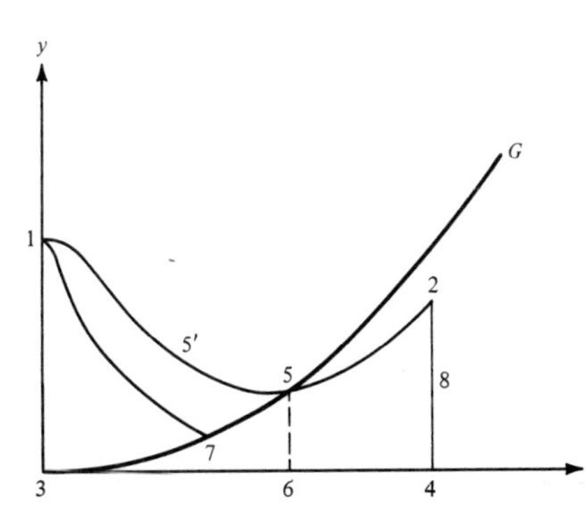

As 5′ approaches 5 on C, we obtain $I[C_{15}] \geq I[E_{15}]$. Let 6 be the foot of the perpendicular from 5 on the x-axis, and 7 any point on G close to 3. By the string property (Section 2.9), we have

$$I[C_{15}] \geq I[E_{15}] = I[E_{17} + G_{75}].$$

If we take the point 7 sufficiently close to 3, then the length of the path $E_{17} + G_{75}$ is larger than $y_1 + y_2$, and hence $I[E_{17} + G_{75}] > I[P_{1365}]$ and finally $I[C_{15}] > I[P_{1365}]$. On the other hand, if 8 is the point on P_{24} having the same ordinate as 5, then the length of C_{52} is certainly larger than the length of P_{28}, and hence $I[C_{52}] > I[P_{82}]$, $I[P_{65}] = I[P_{648}]$, and finally

$$I[C_{12}] = I[C_{15}] + I[C_{52}] > I[P_{136}] + I[P_{648}] + I[P_{82}] = I[P_{1342}].$$

We conclude that for points $2 = (x_2, y_2)$ on or below the curve G, the polygonal line P_{1342} gives the absolute minimum for the integral I in the class of all parametric regular curves joining 1 to 2.

When 2 is above the curve G, then we have already denoted by E_{12} that one catenary which gives a strong relative minimum for I in a V-shaped region V, and we have now the polygonal line P_{1342} which gives a strong relative minimum for I in a region r. We should compare $I[E_{12}]$ and $I[P_{1432}]$. Their difference Δ is given by

$$I[E_{12}] - I[P_{1432}] = \int_0^{s_2} y\,ds - 2^{-1}(y_1^2 + y_2^2),$$

where s_2 is the length of E_{12}, that is, the value of the arc length parameter s along E_{12} at 2. If we denote by E the catenary containing the arc E_{12}, the expression above can be thought of as a function of s (that is, 2 moves along E), and we have $d\Delta/ds_2 = y_2(1 - dy_2/ds_2) > 0$. Thus Δ is an increasing function as 2 moves along any catenary E from 1 to 2. Obviously $\Delta < 0$ at 1, since $I[E_{12}]$ is zero there. On the other hand, $\Delta > 0$ at the point of contact 5 of E with G. Thus, there is on each E_{15} a well-determined point 3 where $\Delta = 0$. It can be proved that the locus of 3 is a curve $H: y = h(x)$, $x \geq x_1$, with $h > 0$, $h' > 0$, $h > g$, $h(+\infty) = h'(+\infty) = +\infty$, $h(x_1 + 0) = h'(x_1 + 0) = 0$.

We conclude that for $2 = (x_2, y_2)$ above or on H, the catenary E_{12} gives the absolute minimum for I. For $2 = (x_2, y_2)$ between G and H we have $I[E_{12}] > I[P_{1432}]$, and thus E_{12} is only a strong local minimum. For 2 below G the absolute minimum (in the class of all parametric regular curves joining 1 to 2) exists and is given by the polygonal line P_{1432}, while I has no absolute minimum in the class of the nonparametric curves joining 1 and 2.

Various parametric representations of the two curves G and H are known; see W. S. Kimball [I], who also gives numerical tables of $g(x)$ and $h(x)$. (For the previous analysis, see G. A. Bliss [I]). For a discussion of the same problem of the surface of revolution of minimum area, only in terms of parametric problems, see L. Tonelli [I].

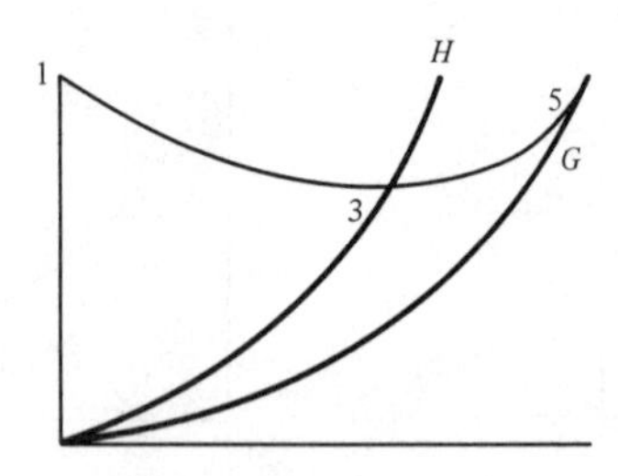

3.14 The Principles of Mechanics

A. D'Alembert's Principle and Lagrange's Equations of Motion

As usual in classical mechanics let us consider a finite system of N material points P_j of mass m_j, variously interconnected, say $P_j = (x_j, y_j, z_j)$, $j = 1, \dots, N$, with respect to an inertial system $Oxyz$. The principle of virtual work in statics concerns with real displacements dP_j and virtual displacements δP_j (see below), away from boundaries (reversible displacements), and in contact with boundaries (nonreversible displacements). The principle states that the system is in equilibrium if and only if the virtual work δL of the active forces F_j applied to the points P_j is zero for all possible reversible virtual displacements, and nonpositive for all possible nonreversible virtual displacements. In symbols:

$$\delta L = \sum_{j=1}^{N} F_j \cdot \delta P_j = 0 \qquad [\leq 0].$$

In dynamics the inertial forces $\bar{I}_j = -m_j a_j = (-m_j x_j'', -m_j y_j'', -m_j z_j'')$ must be added to the other forces, and the principle then states that the actual motions and changes in the configuration of the system are monitored by the sole requirement that

$$\delta L = \sum_{j=1}^{N} (F_j - m_j a_j) \cdot \delta P_j = 0 \qquad [\leq 0], \tag{3.14.1}$$

where a_j is the vector acceleration of P_j (D'Alembert's principle).

Let us assume that the configurations of the system are representable in terms of n parameters $q_1, \dots, q_n$ (Lagrangean coordinates) and time t,

$$P_j = P_j(q_1, \dots, q_n, t), \qquad i = 1, \dots, N,$$

in such a way that each P_j, actually each coordinate x_j, y_j, z_j, is a function of class C^2 of $q_1, \dots, q_n, t$ in some domain $D \subset R^{n+1}$, and that the representation is 1-1 at least locally. Actually, we shall require that the $3N \times n$ matrix $M = [\partial x_j/\partial q_s, \partial y_j/\partial q_s, \partial z_j/\partial q_s]$, $j = 1, \dots, N$, $s = 1, \dots, n$, has maximum rank at every point. Then the velocity v_j of P_j is

$$v_j = (dx_j/dt, dy_j/dt, dz_j/dt) = dP_j/dt = \sum_{s=1}^{n} (\partial P_j/\partial q_s) q_s' + \partial P_j/\partial t, \tag{3.14.2}$$

and the real displacements dP_j and the virtual displacements δP_j are respectively

$$\begin{aligned} dP_j &= \sum_{s=1}^{n} (\partial P_j/\partial q_s)\, dq_s + (\partial P_j/\partial t)\, dt, \\ \delta P_j &= \sum_{s=1}^{n} (\partial P_j/\partial q_s)\, dq_s, \qquad j = 1, \dots, N, \qquad s = 1, \dots, n. \end{aligned} \tag{3.14.3}$$

In other words, the virtual displacements δP_j are computed disregarding the direct dependence of P_j on t, or equivalently, disregarding instant by instant the possible dependence on time of the constraints. For constraints independent of time, then $\partial P_j/\partial t = 0$, and real and virtual displacements coincide. Moreover, for $q = (q_1, \dots, q_n)$, in the interior of D and under the assumptions, the displacements are certainly reversible.

We shall now derive the Lagrange equations of motion for systems with reversible displacements (away from boundaries). For the sake of simplicity we shall limit ourselves to systems with constraints independent of time. Then (3.14.3) becomes

$$v_j = \sum_s (\partial P_j/\partial q_s)q'_s, \qquad j = 1, \ldots, n, \tag{3.14.4}$$

and, if we note that the P_j depend only on the q_s and that we can interchange the differentiations with respect to t and the q_s, we have

$$\begin{aligned} &\partial v_j/\partial q'_s = \partial P_j/\partial q_s, \\ &(d/dt)(\partial P_j/\partial q_s) = (\partial/\partial q_s)(dP_j/dt) = \partial v_j/\partial q_s, \qquad j = 1, \ldots, N, \qquad s = 1, \ldots, n. \end{aligned} \tag{3.14.5}$$

First we need a few remarks on the kinetic energy $T = 2^{-1}\sum_j m_j|v_j|^2$. From (3.14.4) we derive

$$\begin{aligned} |v_j|^2 = v_j \cdot v_j &= \left(\sum_s (\partial P_j/\partial q_s)q'_s\right)\left(\sum_k (\partial P_j/\partial q_k)q'_k\right) \\ &= \sum_{s,k} (\partial P_j/\partial q_s)(\partial P_j/\partial q_k)q'_s q'_k, \end{aligned}$$

$$T = 2^{-1}\sum_j m_j|v_j|^2 = 2^{-1}\sum_{s,k} T_{sk}q'_s q'_k, \tag{3.14.6}$$

with

$$T_{sk} = \sum_j m_j(\partial P_j/\partial q_s)(\partial P_j/\partial q_k), \qquad T_{sk} = T_{ks}.$$

Thus, T is a quadratic form in the q'_s with coefficients T_{sk} depending only on the q_s. We shall also need the derivatives

$$p_s = \partial T/\partial q'_s = \sum_k T_{sk}q'_k, \qquad s = 1, \ldots, n,$$

and we denote by p the n-vector $p = (p_1, \ldots, p_n)$. By Euler theorem on homogeneous functions we have

$$2T = \sum_{s=1}^{n} (\partial T/\partial q'_s)q'_s = \sum_{s=1}^{n} p_s q'_s. \tag{3.14.7}$$

Here $T = 0$ if and only if $q' = (q'_1, \ldots, q'_n) = 0$. The condition is obviously sufficient. To prove the necessity, note that in the opposite case there would be a system of q'_s not all zero for which $T = 0$, that is, $v_j = 0$ for all $j = 1, \ldots, N$, hence dx_j/dt, dy_j/dt, dz_j/dt, $j = 1, \ldots, n$, all zero, and from relations (3.14.4) we would derive that the matrix M above is not of maximum rank.

Let us prove that $\det[T_{sk}] \neq 0$. Indeed, in the opposite case there would be a system of q'_s not all zero with all p_s equal to zero, hence from (3.14.7), $T = 0$, a contradiction. Thus, $T > 0$ for all $q' \neq 0$, in other words, T is a positive definite quadratic form.

Now we have only to use relation (3.14.1) (with equality sign) after some transformations. First we write the second relation (3.14.3) in the form $\delta P_j = \sum_s (\partial P_j/\partial q_s)\,\delta q_s$, with δq_s instead of dq_s since they represent arbitrary displacements, and then we have

$$\sum_j F_j \cdot \delta P_j = \sum_j F_j \cdot \sum_s (\partial P_j/\partial q_s)\,\delta q_s = \sum_s Q_s\,\delta q_s,$$

where

$$Q_s = \sum_j F_j \cdot (\partial P_j/\partial q_s).$$

These Q_s are often called the generalized forces. Also

$$\sum_j m_j a_j \cdot \delta P_j = \sum_j m_j a_j \cdot \sum_s (\partial P_j/\partial q_s)\,\delta q_s = \sum R_s \delta q_s,$$

where

$$R_s = \sum_j m_j a_j \cdot (\partial P_j/\partial q_s),$$

and now (3.14.1) becomes

$$\sum_s (Q_s - R_s)\,\delta q_s = 0$$

for all δq_s. This implies that $Q_s = R_s$, $s = 1, \ldots, n$. By manipulation and relations (3.14.5) we have now

$$\begin{aligned}
R_s &= \sum_j m_j a_j \cdot (\partial P_j/\partial q_s) = \sum_j m_j (dv_j/dt) \cdot (\partial P_j/\partial q_s) \\
&= \sum_j m_j[(d/dt)(v_j \cdot \partial P_j/\partial q_s) - (v_j \cdot (d/dt)(\partial P_j/\partial q_s)] \\
&= \sum_j m_j[(d/dt)(v_j \cdot \partial v_j/\partial q_s') - (v_j \cdot \partial v_j/\partial q_s)] \\
&= 2^{-1} \sum_j m_j (d/dt)(\partial |v_j|^2/\partial q_s') - 2^{-1} \sum_j m_j (\partial |v_j|^2/\partial q_s) \\
&= (d/dt)(\partial/\partial q_s') 2^{-1} \sum_j m_j |v_j|^2 - (\partial/\partial q_s) 2^{-1} \sum_j m_j |v_j|^2 .
\end{aligned}$$

Thus, relations $Q_s = R_s$ become

$$\frac{d}{dt}\frac{\partial T}{\partial q_s'} - \frac{\partial T}{\partial q_s} = Q_s, \qquad s = 1, \ldots, n. \tag{3.14.8}$$

These are the Lagrange equations of motion. Whenever the generalized forces Q_s derive from a potential $V = -U$, i.e., $Q_s = \partial U/\partial q_s$, $s = 1, \ldots, n$, depending on the q_s only, then the Lagrange equations can be written in the form

$$\frac{d}{dt}\frac{\partial(T + U)}{\partial q_s'} - \frac{\partial(T + U)}{\partial q_s} = 0, \qquad s = 1, \ldots, n.$$

If $L = T + U$, then these relations are the Euler equations of the Lagrange, or "action" integral

$$J = \int_{t_1}^{t_2} L\,dt.$$

Thus, the Lagrange equations of motion can be reworded by saying that, for the case of forces F_i depending on a potential, then in the actual motion of the system the action integral is stationary (cf. Section 2.2, Remark 1). If we note that

$$L = T + U = \sum_{k,s} T_{ks} q_k' q_s' + U,$$

where all T_{ks} and U depend only on the q_s and the quadratic form is definite positive, we see that along the actual motion of the system the Legendre necessary condition for a minimum is certainly satisfied (cf. Section 2.2, part (e) of statement (2.2.i)). This statement can be improved. Indeed, if $q(t) = (q_1, \ldots, q_n)$, $t_1 \le t \le t_2$, represents the actual motion, not only $T_{sk}(q(t))$ are the coefficients of a definite positive quadratic form, but the $T_{sk}(q)$ are the coefficients of a definite positive quadratic form for all q. Finally, if

t_1 is fixed and we take $t_2 > t_1$ sufficiently close to t_1 so that the interval $(t_1, t_2]$ is free of points conjugate to t_1, then by (2.12.v) the actual motion represents a strong local minimum for the action integral J. In other words, along the actual motion the action integral is stationary in the large and a strong local minimum in the small.

The action integral may not have any minimum in the large, as the following example shows. Let P be a material point of mass m free to move along the x-axis under the elastic force $-kx$. Then $T = 2^{-1}mx'^2$, $U = -2^{-1}kx^2$ the Euler equation is $mx'' + kx = 0$, or $x'' + \omega^2 x = 0$ with $\omega^2 = k/m$, and $J = 2^{-1}m \int_{t_1}^{t_2} (x'^2 - \omega^2 x^2)\,dt$. The accessory equation is $\eta'' + \omega^2\eta = 0$, so that $\bar{t} = t_1 + \pi/\omega$ is conjugate to t_1. We know from Section 3.9 that J has no minimum for $t_2 > \bar{t}$.

The result obtained above that the Lagrange equations of motion are the Euler equations of the action integral is important because so much of Chapter 2 holds in theoretical mechanics as well as in the calculus of variations. In particular the Hamilton-Jacobi partial differential equation holds

$$\partial S/\partial t + H_0(t, q, \partial S/\partial q) = 0, \qquad q = (q_1, \ldots, q_n),$$

and the same theorems (2.11.x), (2.11.xi) relate this partial differential equation to the Lagrange equations of motion.

B. The Theorems of the Quantities of Motion and of Kinetic Moments

Let us denote here by $\{\vec{F}_j^e\}$ and $\{\vec{F}_j^i\}$ the finite collections of all the exterior and all interior forces acting at each instant on the point P_j. Then the equations of motion are as we know $m_j d^2P_j/dt^2 = \sum \vec{F}_j^e + \sum \vec{F}_j^i$, where $\sum$ simply denotes the vectorial sum of all such forces acting on P_j. If $\vec{F}_j^e = (X_j^e, Y_j^e, Z_j^e)$, $\vec{F}_j^i = (X_j^i, Y_j^i, Z_j^i)$, then the same equations in component form are

$$m_j d^2x_j/dt^2 = \sum X_j^e + \sum X_j^i, \qquad m_j d^2y_j/dt^2 = \sum Y_j^e + \sum Y_j^i$$
(3.14.9)
$$m_j d^2z_j/dt^2 = \sum Z_j^e + \sum Z_j^i.$$

By addition we have

$$\sum_j m_j d^2P_j/dt^2 = \sum_j\sum F_j^e + \sum_j\sum F_j^i,$$

where the last sum is certainly zero because the interior forces are two by two equal and of opposite signs. Thus, $\sum_j m_j d^2P_j/dt^2 = \sum\sum_j F_j^e$, or

$$\sum_j m_j d^2x_j/dt^2 = \sum_j\sum X_j^e, \qquad \sum_j m_j d^2y_j/dt^2 = \sum_j\sum Y_j^e, \qquad \sum_j m_j d^2z_j/dt^2 = \sum_j\sum Z_j^e.$$

These equations can be written in the equivalent form $(d/dt)\sum_j m_j dP_j/dt = \sum\sum_j F_j^e$, or

$$(d/dt)\sum_j m_j dx_j/dt = \sum_j\sum X_j^e, \qquad (d/dt)\sum_j m_j dy_j/dt = \sum_j\sum Y_j^e,$$
(3.14.10)
$$(d/dt)\sum_j m_j dz_j/dt = \sum_j\sum Z_j^e.$$

Since $m_j dP_j/dt$ is called the quantity of motion of P_j, the last relations express the *theorems of quantities of motion*:

3.14.i (THEOREM). *The derivative with respect to time of the sum of the quantities of motion of the N points of the system is equal to the sum of all exterior forces acting on the system.*

3.14.ii (THEOREM). *The derivative with respect to time of the projection of the sum of the quantities of motion with respect to any fixed axis is equal to the projection of the resultant of all the exterior forces on that axis.*

Moreover, if $M = \sum_j m_j$ is the total mass of the system, and $G = (\xi, \eta, \zeta)$ denote the center of gravity of the system, then

$$M\xi = \sum_j m_j x_j, \qquad M\eta = \sum_j m_j y_j, \qquad M\zeta = \sum_j m_j z_j,$$

and the relations (3.14.10) yield

$$M\, d^2\xi/dt^2 = \sum_j\sum X_j^e, \qquad M\, d^2\eta/dt^2 = \sum_j\sum Y_j^e, \qquad M\, d^2\zeta/dt^2 = \sum_j\sum Z_j^e.$$

The last relation expresses the theorem:

3.14.iii (THEOREM). *The motion of the center of mass of a system is the same as if it were a material point of mass M to which are applied forces equal and parallel to all exterior forces applied to the single points P_j of the system.*

Finally, from (3.14.9), by multiplications and additions, we have, for instance,

$$\sum_j m_j(x_j d^2y_j/dt^2 - y_j d^2x_j/dt^2) = \sum_j\sum (x_j Y_j^e - y_j X_j^e) + \sum_j\sum (x_j Y_j^i - y_j X_j^i), \tag{3.14.11}$$

where the last parentheses represent moments of the interior forces with respect to the Oz axis. Note that, for each force, say (X_1^i, Y_1^i) from P_2 applied to P_1, or $(X_1^i, Y_1^i) = (\sigma(x_2 - x_1), \sigma(y_2 - y_1))$, there is another force $(X_2^i, Y_2^i) = (-\sigma(x_2 - x_1), -\sigma(y_2 - y_1))$ applied to P_2, and then $(x_1 Y_1^i - y_1 X_1^i) + (x_2 Y_2^i - y_2 X_2^i) = 0$. Thus, the last sum in (3.14.11) is zero, and (3.14.11) reduces to

$$(d/dt) \sum_j m_j(x_j dy_j/dt - y_j dx_j/dt) = \sum_j\sum (x_j Y_j^e - y_j X_j^e).$$

This last relation expresses the theorem of the kinetic moments:

3.14.iv (THEOREM). *The derivative with respect to time of the sum of the moments of the quantities of motion (kinetic moments) of the points of a system with respect to any fixed axis is equal to the sum of the moments of all external forces with respect to the same axis.*

C. Instantaneous Axis of Rotation of a Rigid Body

If $Oxyz$ is a system of coordinates attached to a rigid body B, or a mobile reference system, and $Ox_1y_1z_1$ is a fixed system of coordinates, then the usual transformation of coordinates holds at any instant t:

$$\begin{aligned} x_1 &= x_0 + \alpha x + \alpha_1 y + \alpha_2 z, \\ y_1 &= y_0 + \beta x + \beta_1 y + \beta_2 z, \\ z_1 &= z_0 + \gamma x + \gamma_1 y + \gamma_2 z, \end{aligned} \tag{3.14.12}$$

where (x_0, y_0, z_0) are the coordinates of O with respect to $Ox_1y_1z_1$, and $(\alpha, \alpha_1, \alpha_2; \beta, \beta_1, \beta_2; \gamma, \gamma_1, \gamma_2)$ are the cosines of the nine angles of the axes $Oxyz$ with the axes

$Ox_1y_1z_1$. If B is in motion, then $x_0, y_0, z_0, \alpha, \ldots, \gamma_2$ are functions of t, while x, y, z are fixed, since these represent the coordinates of a point P of the rigid body B with respect to $Oxyz$ attached to B. By differentiating, the velocity $V = (V_{x_1}, V_{y_1}, V_{z_1}) = (dx_1/dt, dy_1/dt, dz_1/dt)$ of P is given by

$$\text{(3.14.13)} \qquad \left.\begin{aligned} V_{x_1} &= x_0' + x\alpha' + y\alpha_1' + z\alpha_2', \\ V_{y_1} &= y_0' + x\beta' + y\beta_1' + z\beta_2', \\ V_{z_1} &= z_0' + x\gamma' + y\gamma_1' + z\gamma_2' \end{aligned}\right\} \qquad (') = d/dt).$$

On the other hand, if V_x, V_y, V_z denote the projections of V on the $Oxyz$ axes, we have

$$\text{(3.14.14)} \qquad \begin{aligned} V_x &= \alpha V_{x_1} + \beta V_{y_1} + \gamma V_{z_1}, \\ V_y &= \alpha_1 V_{x_1} + \beta_1 V_{y_1} + \gamma_1 V_{z_1}, \\ V_z &= \alpha_2 V_{x_1} + \beta_2 V_{y_1} + \gamma_2 V_{z_1}. \end{aligned}$$

In substituting (3.14.13) in (3.14.14) we must first take note of the identity $\alpha^2 + \beta^2 + \gamma^2 = 1$ and analogous ones, which by differentiation yield $\alpha\alpha' + \beta\beta' + \gamma\gamma' = 0$ etc. On the other hand, from the identities $\alpha_1\alpha_2 + \beta_1\beta_2 + \gamma_1\gamma_2 = 0$, $\alpha\alpha_2 + \beta\beta_2 + \gamma\gamma_2 = 0$, $\alpha\alpha_1 + \beta\beta_1 + \gamma\gamma_1 = 0$, we obtain by differentiation

$$\text{(3.14.15)} \qquad \begin{aligned} p &= \alpha_2\alpha_1' + \beta_2\beta_1' + \gamma_2\gamma_1' = -(\alpha_1\alpha_2' + \beta_1\beta_2' + \gamma_1\gamma_2'), \\ q &= \alpha\alpha_2' + \beta\beta_2' + \gamma\gamma_2' = -(\alpha_2\alpha' + \beta_2\beta' + \gamma_2\gamma'), \\ r &= \alpha_1\alpha' + \beta_1\beta' + \gamma_1\gamma' = -(\alpha\alpha_1' + \beta\beta_1' + \gamma\gamma_1'). \end{aligned}$$

With these definitions and identities the substitution of (3.14.13) in (3.14.14) yields

$$\text{(3.14.16)} \qquad \begin{aligned} V_x &= V_x^0 + qz - ry, \\ V_y &= V_y^0 + rx - pz, \\ V_z &= V_z^0 + py - qx, \end{aligned}$$

where $V^0 = (V_x^0, V_y^0, V_z^0)$ simply indicates the projections of the velocity V of O on the moving axes. The other terms in (3.14.16) indicate that the motion of B with respect to any of its own points O can be interpreted as due to an instantaneous rotation vector $O\omega = (p, q, r)$ of components p, q, r given by (3.14.15). The vector $(qz - ry, rx - pz, py - qx)$ is said to be the moment of the vector $O\omega$ with respect to the point $P = (x, y, z)$ of B.

Concerning the kinetic energy T of the system B, first assume that O is fixed, that is, the rigid body B moves around the fixed point O. Then $V_x^0 = V_y^0 = V_z^0 = 0$, and

$$\begin{aligned} T &= 2^{-1} \sum_j m_j(x_j'^2 + y_j'^2 + z_j'^2) \\ &= 2^{-1} \sum_j m_j[(qz_j - ry_j)^2 + (rx_j - pz_j)^2 + (py_j - qx_j)^2] \\ &= 2^{-1}[Ap^2 + Bq^2 + Cr^2 - 2Dqr - 2Erp - 2Fpq], \end{aligned}$$

where

$$A = \sum_j m_j(y_j^2 + z_j^2), \qquad B = \sum_j m_j(z_j^2 + x_j^2), \qquad C = \sum_j m_j(x_j^2 + y_j^2),$$

$$D = \sum_j m_j y_j z_j, \qquad E = \sum_j m_j z_j x_j, \qquad F = \sum_j m_j x_j y_j.$$

Now let us assume instead that also O is moving, but that O is the center of mass of P, so that $\sum_j m_j x_j = 0, \sum_j m_j y_j = 0, \sum_j m_j z_j = 0$. From the relations (3.14.16) we derive that

$$T = 2^{-1} \sum m_j(x_j'^2 + y_j'^2 + z_j'^2)$$

$$= 2^{-1}\left(\sum_j m_j\right)(V_x^{02} + V_y^{02} + V_z^{02})$$

$$+ 2^{-1}[Ap^2 + Bq^2 + Cr^2 - 2Dqr - 2Erp - 2Fpq].$$

Finally, if O is again the center of masses of B and $Oxyz$ are the principal axes, then $D = E = F = 0$, and if v denotes the velocity of O and $M = \sum_j m_j$, then

$$T = 2^{-1}Mv^2 + 2^{-1}[Ap^2 + Bq^2 + Cr^2].$$

D. Euler's Equations for the Motion of a Rigid Body around Its Center of Mass

Let us compute, at any instant t, the resultant moment $O\sigma$ of the quantities of motion of the points P_j of B relative to O. Let $\sigma_x, \sigma_y, \sigma_z$ denote the components of $O\sigma$ on the mobile axes $Oxyz$. Thus, for instance, σ_x is the sum of the moments relative to Ox. The quantities of motion of the points P_j have projections $m_j v_{jx}, m_j v_{jy}, m_j v_{jz}$. The sum σ_x of the moments of motion is, therefore, by (3.14.16),

$$\sigma_x = \sum_j m_j(y_j v_{jz} - z_j v_{jy})$$

$$= \sum_j m_j[y_j(py_j - qx_j) - z_j(rx_j - pz_j)]$$

$$= \sum_j m_j[p(y_j^2 + z_j^2) - qx_j y_j - rx_j z_j]$$

$$= Ap - Fq - Er.$$

This expression equals $\partial T/\partial p$. Hence, by symmetry we also have

$$\sigma_x = \partial T/\partial p, \qquad \sigma_y = \partial T/\partial q, \qquad \sigma_z = \partial T/\partial r.$$

Now let L, M, N be the sums of the moments of all exterior forces with respect to the axes $Oxyz$. We know that the resultant moment of all these forces with respect to O is a vector OS whose components are still L, M, N. We also know from (3.14.iv) that the absolute velocity $\vec{u}$ of the point σ is equal to $\overrightarrow{OS}$. The projections of this velocity $\vec{u}$ thus are equal to those of $\overrightarrow{OS}$, that is, are equal to L, M, N.

The point σ has coordinates $\sigma_x, \sigma_y, \sigma_z$ with respect to $Oxyz$, and thus, as t varies, the components of $\vec{u}$ with respect to $Oxyz$ are $d\sigma_x/dt, d\sigma_y/dt, d\sigma_z/dt$. By increasing these components by the quantities $q\sigma_z - r\sigma_y, r\sigma_x - p\sigma_z, p\sigma_y - q\sigma_x$ respectively, we obtain the components of $\vec{u}$ with respect to the fixed axes. We have then the equations

$$\text{(3.14.17)} \qquad \begin{gathered} d\sigma_x/dt + q\sigma_z - r\sigma_y = L, \qquad d\sigma_y/dt + r\sigma_x - p\sigma_z = M, \\ d\sigma_z/dt + p\sigma_y - qr_x = N. \end{gathered}$$

By taking for $Oxyz$ a system of axes with origin O the center of mass, and for axes the principal axes through O, then $T = 2^{-1}(Ap^2 + Bq^2 + Cr^2)$, $\sigma_x = \partial T/\partial p = Ap$, $\sigma_y = Bq$, $\sigma_z = Cr$, and the equations (3.14.17) become the Euler equations for the motion of a rigid body around its center of gravity O and principal axes $Oxyz$:

$$\begin{aligned} A\,dp/dt + (C - B)qr &= L, \\ B\,dq/dt + (A - C)rp &= M, \\ C\,dr/dt + (B - A)pq &= N. \end{aligned} \tag{3.14.18}$$

We shall see another derivation of these equations in subsection F. We shall use these relations in a number of examples and exercises in Sections 6.1–3.

E. The Euler Angles for the Motion of a Rigid Body about One of Its Points

The Euler angles θ, φ, ψ can be thought of as independent parameters, or Lagrange coordinates, for the description of the motion of a rigid body B about one of its points.

Let us consider a fixed orthogonal system $Ox_1y_1z_1$, and another system $Oxyz$, similarly oriented, attached to the rigid body and thus in motion with respect to $Ox_1y_1z_1$. Let the orientation be chosen in such a way that a rotation of 90° in the positive sense around the z-axis takes the x-axis into the y-axis, and the same occurs for $Oxyz$.

Let OI denote the intersection of the xy-plane with the x_1y_1-plane, and on this straight line define the positive direction to be from O to I. Let ψ be the angle, measured in the positive direction with respect to the z_1-axis, which takes the axis Ox_1 into the direction OI. Now the axis Oz is orthogonal to the plane IOx; let φ denote the angle, measured in the positive direction with respect to the z-axis, which takes OI into Ox. Finally, the axis OI is orthogonal to the plane Ozz_1; let θ denote the angle, measured in the positive direction with respect to the OI-axis, which takes Oz_1 to Oz. According to all these conventions, the angle IOy is $\varphi + \pi/2$. The angles θ, φ, ψ are independent. To each set of values of these three angles, there corresponds one and only one position of $Oxyz$ with respect to $Ox_1y_1z_1$.

We may think of θ, φ, ψ as functions of time, and their derivatives are then θ', φ', ψ'. Actually, each of these derivatives represents a rotation, and thus a rotation vector, for which we use the same symbol. The instantaneous rotation $\vec{\omega}$ of the system $Oxyz$ with respect to $Ox_1y_1z_1$ is the resultant of these rotations ψ', θ', φ' around Oz_1, OI, Oz. These three rotation components are represented in the diagram by vectors equal to ψ', θ', φ'

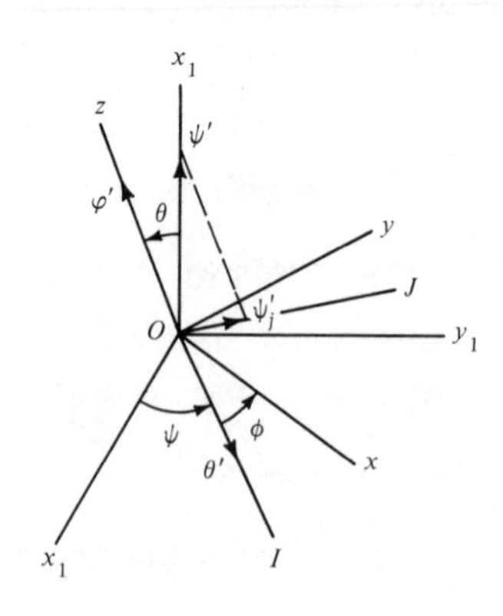

on the axes Oz_1, OI, Oz. The resultant rotational ω is the geometric sum of these vectors: its projection on any axis is equal to the sum of the projections of the components ψ', θ', φ' on the same axis.

First we determine the projection of $\vec{\omega}$ on the three orthogonal axes OI, OJ, Oz, where OJ is on the xOy plane and forms the angle $+\pi/2$ with OI. Let ω_i, ω_j, ω_z denote the three projections. Note that the vector $\vec{\psi}'$ is in the plane zOJ, and thus $\vec{\psi}'$ can be decomposed into its components ψ'_j and ψ'_z on OJ and z, namely, $\psi'_j = \psi' \sin\theta$, $\psi'_z = \psi' \cos\theta$. Thus, the three components ω_i, ω_j, ω_z on the axes OI, OJ, Oz are

$$\omega_i = \theta', \qquad \omega_j = \psi' \sin\theta, \qquad \omega_z = \psi' \cos\theta + \varphi'.$$

In order to obtain p and q, it suffices to take the sum of the projections of the components ω_i, ω_j on Ox and Oy. Since the orthogonal axes IOJ are on the plane xOy, and Ox makes an angle φ with OI, we have

$$p = \omega_x = \omega_i \cos\varphi + \omega_j \sin\varphi,$$
$$q = \omega_y = \omega_i \cos(\varphi + \pi/2) + \omega_j \cos\varphi.$$

Finally, we have the relations

$$\begin{aligned} p &= \psi' \sin\theta \sin\varphi + \theta' \cos\varphi, \\ q &= \psi' \sin\theta \cos\varphi - \theta' \sin\varphi, \\ r &= \psi' \cos\theta + \varphi'. \end{aligned} \tag{3.14.19}$$

F. Derivation of the Euler Equations as Lagrange Equations of Motion

Having determined a system of Lagrange coordinates ψ, θ, φ for the motion of a rigid body B with respect to its center of mass, we have only to apply the Lagrange equations (3.14.8). Choosing for the axes $Oxyz$ the principal axes through the center of mass O, we have $T = 2^{-1}(Ap^2 + Bq^2 + Cr^2)$, with p, q, r given by the relations (3.14.19). For ψ, θ, φ as Lagrange coordinates, let Ψ, Θ, Φ denote the corresponding functions Q_j of Subsection A.

The Lagrange equation relative to the variable φ is now

$$(d/dt)(\partial T/\partial\varphi') - \partial T/\partial\varphi = \Phi, \tag{3.14.20}$$

where now, from the expression for T above and the relations (3.14.19), we derive

$$\begin{aligned} \partial T/\partial\varphi' &= (\partial T/\partial r)(\partial r/\partial\varphi') = Cr, \\ \partial p/\partial\varphi &= q, \qquad \partial q/\partial\varphi = -p, \\ \partial T/\partial\varphi &= (\partial T/\partial p)(\partial p/\partial\varphi) + (\partial T/\partial q)(\partial q/\partial\varphi) \\ &= Ap(\partial p/\partial\varphi) + Bq(\partial q/\partial\varphi) = (A - B)pq. \end{aligned}$$

Thus, (3.14.20) becomes

$$C(dr/dt) + (B - A)pq = N.$$

It remains to show that $\Phi = N$, the moment of all given forces with respect to Oz. To this effect, note that $\Phi\,\delta\varphi$ is the sum of all virtual works of forces in an elementary displacement that leaves ψ, θ constant, i.e. a rotation $\delta\varphi$ around Oz. Thus

$$\Phi\,\delta\varphi = \sum_j (X_j\,\delta x_j + Y_j\,\delta y_j + Z_j\,\delta z_j) = \sum (x_j Y_j - y_j X_j)\,\delta\varphi,$$

or

$$\Phi = \sum (x_j Y_j - y_j X_j) = N,$$

and (3.14.20) becomes

$$C(dr/dt) + (B - A)pq = N.$$

The quantities ψ, θ, φ do not appear. Since p, q, r play exactly the same roles, by symmetry the analogous equations must also hold:

$$A(dp/dt) + (C - B)qr = L,$$
$$B(dq/dt) + (A - C)rp = M.$$

We have again obtained the Euler equations (3.14.18) for the motion of a rigid body.

Bibliographical Notes

For the classical examples in Sections 3.1–4, 3.6–7 the reader may compare R. Weinstock [I], C. Fox [I], I. M. Gelfand and S. V. Fomin [I], L. E. Elsgolc [I]. For the example in Section 3.5 and a great many applications of the calculus of variations to economics we refer to G. Hadley and M. C. Kemp [I], M. D. Intriligator [I], and P. Newman [I]. For the discussion of the example in Section 3.9 the reader may compare L. Tonelli [I]. For the examples in Sections 3.12–13 more details are given in G. A. Bliss [I]. In Section 3.14 we have presented a brief introduction to rational mechanics, both as an illustration of a subject whose development has been in the past inextricably related to the calculus of variations, and for the derivation of the equations of motion of rigid bodies, equations which will be used in problems of optimal control in Chapter 6 and which are, after all, the necessary (Euler) conditions for certain problems of the calculus of variations.

The few topics of theoretical mechanics in Section 3.14 are modeled on P. Appell [I, Vol. 2]. For further references on theoretical mechanics we also mention D. Graffi [I], S. Goldstein [I], and C. Lanczos [I].

CHAPTER 4

Statement of the Necessary Condition for Mayer Problems of Optimal Control

4.1 Some General Assumptions

We consider here Mayer problems of optimization. Precisely, we are concerned with the problem of the minimum of a functional

$$I[x, u] = g(t_1, x(t_1), t_2, x(t_2)) \tag{4.1.1}$$

with differential equations, constraints, and boundary conditions

$$dx/dt = f(t, x(t), u(t)), \qquad t \in [t_1, t_2] \text{ (a.e.)}, \tag{4.1.2}$$

$$(t, x(t)) \in A, \qquad t \in [t_1, t_2], \tag{4.1.3}$$

$$u(t) \in U(t), \qquad t \in [t_1, t_2] \text{ (a.e.)}, \tag{4.1.4}$$

$$e[x] = (t_1, x(t_1), t_2, x(t_2)) \in B, \tag{4.1.5}$$

in the class Ω of *all* admissible pairs $x(t) = (x^1, \ldots, x^n)$, $u(t) = (u^1, \ldots, u^m)$, $t_1 \leq t \leq t_2$. Again, $f(t, x, u) = (f_1, \ldots, f_n)$ is a given vector function, and the system (4.1.2) can be written equivalently in the form

$$dx^i/dt = f_i(t, x(t), u(t)), \qquad t \in [t_1, t_2] \text{ (a.e.)}, \ i = 1, \ldots, n.$$

Here U will be assumed to be either a fixed subset of the u-space R^m, or the whole space R^m, or depending on t *only*. For the sake of simplicity, we shall refer mostly to problems of minimum, since the same will hold for problems of maximum or, what is the same, problems of maximum become problems of minimum by changing g into $-g$.

Given the generality of the constraints (4.1.2–5) under consideration, we must explicitely assume that they are compatible, that is, that there is at least one pair $x(t)$, $u(t)$, $t_1 \leq t \leq t_2$, x AC, u measurable, satisfying (4.1.2–5).

We say that such pairs are admissible. Thus, we assume explicitly that the data are compatible, that is, that the class Ω of all admissible pairs is not empty. In any particular problem this may have to be verified.

Also, as mentioned, we assume that the minimum of the functional is being sought in the whole class Ω of all admissible pairs (x, u). We say that $I[x, u]$ has an absolute minimum (in Ω), and that this minimum is attained at an element $(x, u) \in \Omega$ (any admissible pair), provided $I[x, u] \leq I[\tilde{x}, \tilde{u}]$ for all elements $(\tilde{x}, \tilde{u}) \in \Omega$ (admissible pairs). In other words, if i denotes the infimum of $I[x, u]$ in Ω, then i is finite, and $I[x, u] = \mathrm{i}$ for some element $(x, u) \in \Omega$.

In no way do we expect that the minimum of the functional is attained at only one element of Ω, though this happens in many cases.

For the sake of simplicity, we formulate below the necessary condition under a simple set of assumptions, (a)–(e), and we shall then mention alternate assumptions. Different proofs of the necessary condition—of various degrees of sophistication—will be given in Chapter 7 (see also 4.2D).

To formulate the necessary condition we shall also need new variables $\lambda = (\lambda_1, \ldots, \lambda_n)$, called multipliers, and an auxiliary function $H(t, x, u, \lambda)$, the Hamiltonian, defined in $M \times R^n$ by taking

$$H(t, x, u, \lambda) = \lambda_1 f_1 + \cdots + \lambda_n f_n = \sum_{j=1}^{n} \lambda_j f_j(t, x, u). \tag{4.1.6}$$

Thus, H is linear in the multipliers $\lambda_1, \ldots, \lambda_n$. Finally, for every $(t, x, \lambda) \in A \times R^n$ we shall consider H as a function of u only, with u in U (or $U(t)$), and search for its minimum value in U (or $U(t)$). If this minimum is lacking we shall search for its infimum in U or $U(t)$, and say in any case

$$M(t, x, \lambda) = \inf_{u \in U} H(t, x, u, \lambda). \tag{4.1.7}$$

In this Chapter we shall think of A as a closed subset of the tx-space R^{1+n} and, if M denotes the set of all $(t, x, u) \in R^{1+n+m}$ with $(t, x) \in A$, $u \in U(t)$, we shall assume that M is closed and that $f(t, x, u) = (f_1, \ldots, f_n)$ is of class C^1 on M. We shall denote as usual by f_{it}, f_{ix^j} the partial derivatives of f_i with respect to t and x^j. Also, we shall denote by $H_{x^i} = \partial H/\partial x^i$, $H_t = \partial H/\partial t$, $H_{\lambda_i} = \partial H/\partial \lambda_i$, $i = 1, \ldots, n$, the partial derivatives of H with respect to x^i, t, λ_i. Obviously

$$H_{x^i} = \sum_j \lambda_j f_{jx^i}, \quad H_t = \sum_j \lambda_j f_{jt}, \quad H_{\lambda_i} = f_i, \qquad i = 1, \ldots, n, \tag{4.1.8}$$

where $\sum_j$ will always denote a sum ranging over all $j = 1, \ldots, n$.

We shall list now a few specific hypotheses for our first presentation below of the necessary condition. First we shall assume that

(a) A certain admissible pair $x(t) = (x^1, \ldots, x^n)$, $u(t) = (u^1, \ldots, u^m)$, $t_1 \leq t \leq t_2$, gives a minimum of $I[x, u] = g(e[x])$ in the class Ω of all admissible pairs x, u, that is, $I[x,u] \le I[\tilde{x},\tilde{u}]$, or $g(e[x]) \le g(e[\tilde{x}])$, for every admissible pair $\tilde{x}$, $\tilde{u}$.

We shall assume, more specifically, that

(b) the graph $[(t,x(t)) | t_1 \le t \le t_2]$ of the optimal trajectory x is made up of only interior points of A: briefly, x is made up of interior points of A.

Finally, we shall assume, for the moment, that

(c) U is a *fixed closed* subset of R^m, and the optimal control function u is bounded, that is, $|u(t)| \le N, t_1 \le t \le t_2$, for some constant N (though U may be unbounded, and we do not exclude the case $U = R^m$).

Condition (c) is certainly satisfied if U is a fixed compact subset of R^m, that is, U is fixed, bounded, and closed, since then $|u| \le N$ for all $u \in U$, and hence $|u(t)| \le N$, $|\tilde{u}(t)| \le N$, $t_1 \le t \le t_2$, for all strategies, optimal or not. We shall list below in Section 4.2C, Remark 5, other possible assumptions which may replace (c) above.

Some general assumptions are needed on the smoothness of B and g, since now B is not a "single point," and we must have some control over how g varies when (t_1, x_1, t_2, x_2) describes B. We shall assume

(d) that the end point $e[x] = (t_1, x(t_1), t_2, x(t_2))$ of the optimal trajectory x is a point of B, where B possesses a "tangent linear variety" B' (of some dimension k, $0 \le k \le 2n + 2$; see Section 4.4 below for examples and details), whose vectors will be denoted by $h = (\tau_1, \xi_1, \tau_2, \xi_2)$ with $\xi_1 = (\xi_1^1, \ldots, \xi_1^n)$, $\xi_2^n = (\xi_2^1, \ldots, \xi_2^n)$, or in differential form $h = (dt_1, dx_1, dt_2, dx_2)$ with $dx_1 = (dx_1^1, \ldots, dx_1^n)$, $dx_2 = (dx_2^1, \ldots, dx_2^n)$.

(e) g possesses a differential dg at $e[x]$, say

$$dg = g_{t_1}\tau_1 + \sum_{i=1}^{n} g_{x_1^i}\xi_1^i + g_{t_2}\tau_2 + \sum_{i=1}^{n} g_{x_2^i}\xi_2^i,$$

or

$$dg = g_{t_1}\,dt_1 + \sum_{i=1}^{n} g_{x_1^i}\,dx_1^i + g_{t_2}\,dt_2 + \sum_{i=1}^{n} g_{x_2^i}\,dx_2^i,$$

where $g_{t_1}, \ldots, g_{x_2^n}$ denote partial derivatives of g with respect to $t_1, \ldots, x_2^n$, all computed at $e[x]$.

In many cases most of these differentials $dt_1, \ldots, dx_2^n$ are zero except for a few which are arbitrary, or satisfy simple relations, as we shall see by a great many examples in Section 4.4, where we shall discuss these assumptions and their implications in the transversality relations.

We shall also discuss in Remark 10 of Section 4.2C the case where B possesses at $e[x]$, not a full tangent hyperspace B' of tangent vectors h, as assumed in (d), but only a convex cone of tangent vectors h, as at end points, edges, or vertices of B.

4.2 The Necessary Condition for Mayer Problems of Optimal Control

A. The Necessary Condition

4.2.i (THEOREM). *Under the hypotheses* (a)–(e) *listed in Section* 4.1 *let* $x(t) = (x^1, \ldots, x^n)$, $u(t) = (u^1, \ldots, u^m)$, $t_1 \leq t \leq t_2$, *be an optimal pair, that is, an admissible pair* x, u *such that* $I[x, u] \leq I[\tilde{x}, \tilde{u}]$ *for all pairs* $\tilde{x}, \tilde{u}$ *of the class* Ω *of all admissible pairs. Then the optimal pair* x, u *necessarily has the following properties*:

(P1) *There is an absolutely continuous vector function* $\lambda(t) = (\lambda_1, \ldots, \lambda_n)$, *(multipliers), such that*

$$d\lambda_i/dt = -H_{x^i}(t, x(t), u(t), \lambda(t)), \qquad i = 1, \ldots, n,$$

for t *in* $[t_1, t_2]$ (a.e.). *If* dg *is not identically zero at* $e[x]$, *then* $\lambda(t)$ *is never zero in* $[t_1, t_2]$.

(P2) *For almost any fixed* t *in* $[t_1, t_2]$ (a.e.), *the Hamiltonian* $H(t, x(t), u, \lambda(t))$—*thought of as a real valued function of* u *only with* u *in* U—*takes its minimum value in* U *at the optimal strategy* $u = u(t)$, *or*

$$M(t, x(t), \lambda(t)) = H(t, x(t), u(t), \lambda(t)),$$

and this relation holds for any t *in* $[t_1, t_2]$ (a.e.).

(P3) *The function* $M(t) = M(t, x(t), \lambda(t))$ *is absolutely continuous in* $[t_1, t_2]$ *(more specifically,* $M(t)$ *coincides* a.e. *in* $[t_1, t_2]$ *with an* AC *function), and (with this identification)*

$$dM/dt = (d/dt)M(t, x(t), \lambda(t)) = H_t(t, x(t), u(t), \lambda(t))$$

for t *in* $[t_1, t_2]$ (a.e.).

(P4) *Transversality relation. There is a constant* $\lambda_0 \geq 0$ *such that*

$$(\lambda_0 g_{t_1} - M(t_1))\, dt_1 + \sum_{i=1}^{n} (\lambda_0 g_{x_1^i} + \lambda_i(t_1))\, dx_1^i$$

$$+(\lambda_0 g_{t_2} + M(t_2))\, dt_2 + \sum_{i=1}^{n} (\lambda_0 g_{x_2^i} - \lambda_i(t_2))\, dx_2^i = 0$$

for every vector $h = (\tau_1, \xi_1, \tau_2, \xi_2) \in B'$, *or briefly* $h = (dt_1, dx_1, dt_2, dx_2) \in B'$, *that is,*

$$\lambda_0\, dg + \left[M(t)\, dt - \sum_{i=1}^{n} \lambda_i(t)\, dx^i \right]_1^2 = 0. \tag{4.2.1}$$

This form is classical, and in each particular situation yields precise information on boundary values of the multipliers λ_i and of the function $M(t)$, as we shall see in detail in Section 4.4 below in a number of typical and rather general situations.

Note that x, u above is an admissible pair, so that the differential equations

$$dx^i/dt = f_i(t, x(t), u(t)), \qquad i = 1, \dots, n, \tag{4.2.2}$$

are certainly satisfied for t in $[t_1, t_2]$ (a.e.). Note that these equations and the n equations (P1) can be written, in view of (4.1.2), in the symmetric form

$$\frac{dx^i}{dt} = \frac{\partial H}{\partial \lambda_i}, \quad \frac{d\lambda_i}{dt} = -\frac{\partial H}{\partial x^i}, \qquad i = 1, \dots, n. \tag{4.2.3}$$

These are the so-called *canonical equations.*

The equations (4.1.2), (P1), and (P3) (that is, the equations (4.2.3)) can be given the equivalent integral form

$$\begin{aligned} x^i(t) &= x^i(t_1) + \int_{t_1}^{t} f_i(\tau, x(\tau), u(\tau))\, d\tau, \qquad i = 1, \dots, n, \\ \lambda_i(t) &= \lambda_i(t_1) - \int_{t_1}^{t} H_{x^i}(\tau, x(\tau), u(\tau), \lambda(\tau))\, d\tau, \qquad i = 1, \dots, n, \\ M(t) &= M(t_1) + \int_{t_1}^{t} H_t(\tau, x(\tau), u(\tau), \lambda(\tau))\, d\tau, \end{aligned} \tag{4.2.4}$$

which hold for all t, $t_1 \leq t \leq t_2$. Using the expressions for H and the expressions (4.1.8) for the partial derivatives H_t, H_{x^i}, we can write the equations (P1), (P3) also in the explicit form

$$d\lambda_i/dt = -\sum_{j=1}^{n} \lambda_j f_{jx^i}(t, x(t), u(t)), \; i = 1, \dots, n, \tag{4.2.5}$$

$$dM/dt = \sum_{j=1}^{n} \lambda_j f_{jt}(t, x(t), u(t)). \tag{4.2.6}$$

Thus, we see from (4.2.5) that the multipliers $\lambda_i(t)$, $i = 1, \dots, n$, are the solutions in $[t_1, t_2]$ of a system of linear homogeneous differential equations. We can always multiply them, therefore, by an arbitrary nonzero constant—actually, an arbitrary positive constant—and still preserve both properties (P1) and (P2). Note that for autonomous problems (that is, when f is independent of t), all f_{jt} are zero, $dM/dt = 0$, and M is a constant.

The transversality relation (P4) is essentially an orthogonality relation. Indeed, it can be written in the form

$$A_{10} dt_1 + \sum_{i=1}^{n} A_{1i}\, dx_1^i + A_{20}\, dt_2 + \sum_{i=1}^{n} A_{2i}\, dx_2^i = 0,$$

where

$$\begin{aligned} A_{10} &= \lambda_0 g_{t_1} - M(t_1), \qquad & A_{1i} &= \lambda_0 g_{x_1^i} + \lambda_i(t_1), \quad i = 1, \dots, n, \\ A_{20} &= \lambda_0 g_{t_2} + M(t_2), \qquad & A_{2i} &= \lambda_0 g_{x_2^i} - \lambda_i(t_2), \quad i = 1, \dots, n. \end{aligned}$$

Thus, if A denotes the $(2n+2)$-vector

$$A = (A_{10}, A_{1i}, i = 1, \ldots, n, A_{20}, A_{2i}, i = 1, \ldots, n),$$

then (P4) states that A is orthogonal to B at the point $e[x] \in B$, that is, A is orthogonal to the hyperplane B' tangent to B at $e[x]$. We shall discuss (P4) in detail in Section 4.4 with many examples, for some of which (P4) has further striking geometric interpretations.

As mentioned above, if dg is not identically zero, then $\lambda(t) = (\lambda_1, \ldots, \lambda_n)$ itself is never zero in $[t_1, t_2]$. Finally, whenever $\lambda_0 > 0$, we can always multiply the $(n+1)$-vector $(\lambda_0, \lambda_1(t), \ldots, \lambda_n(t))$ by a positive constant and make $\lambda_0 = 1$. There are criteria which guarantee that $\lambda_0 > 0$, and then we can take $\lambda_0 = 1$. (See Remark in Section 7.4E.)

If we denote by $A(t) = [a_{ij}(t)]$ the $n \times n$ matrix whose entries are $a_{ij}(t) = f_{ix^j}(t, x(t), u(t))$, $i, j = 1, \ldots, n$, then the system (4.2.5) can be written in the compact form

$$d\lambda/dt = -A^*(t)\lambda, \tag{4.2.7}$$

where A^* is the transpose of the matrix A.

Note that conditions (P1)–(P4) above are necessary conditions for a minimum. The necessary conditions for a maximum are essentially the same, and can be obtained by replacing g with $-g$ in Mayer problems (f_0 with $-f_0$ in Lagrange problems).

It may well occur that the control variable, as determined by the necessary condition (4.2.i) has values on the boundary of U, say, $u(t) \in bdU$, as in the example below, and in most of the examples we shall discuss in the Sections 6.1–6. In these cases we say that the optimal control is *bang-bang* (as in the case in which $U = [-1 \le u \le 1]$ and $u(t)$ takes only the values -1 and $+1$).

Alternatively, it may occur that $u(t)$ takes values in the interior of U for an entire arc of the trajectory, say, $u(t) \in \operatorname{int} U$ for all $\alpha < t < \beta$. In this case it follows from (P2) that

$$H_{u^j}(t, x(t), u(t), \lambda(t)) = 0, \qquad j = 1, \ldots, m, \qquad \alpha < t < \beta. \tag{4.2.8}$$

In this situation, if in addition we know that $H(t, x(t), u, \lambda(t))$ has second order partial derivatives with respect to $u^1, \ldots, u^m$, at least in a neighborhood of $u(t)$, then we must also have

$$\sum_{j,h=1}^{m} H_{u^j u^h} \xi_j \xi_h \ge 0 \tag{4.2.9}$$

for all $\xi = (\xi_1, \ldots, \xi_m) \in R^m$, all $\alpha < t < \beta$, and where the derivatives $H_{u^j u^h}$ are computed at $(t, x(t), u(t), \lambda(t))$. Relation (4.2.9) is often called the *Legendre-Clebsch* necessary condition.

The situation we have just depicted may occur rather naturally, if for instance, U is an open set, or in particular if $U = R^m$ is the whole u-space.

Finally, we shall see in the Sections 4.7 and 4.8 that the classical necessary conditions for problems of the calculus of variations and for classical isoperimetric problems can be derived from the necessary condition (4.2.i) (and variants) for Mayer problems of optimal control. In either case we shall have $U = R^m$.

B. Example

We consider the problem of the stabilization of a point moving on a straight line under a limited external force. A point P moves along the x-axis, governed by the equation $x'' = u$ with $|u| \leq 1$. The problem is to take P from any given state ($x = a$, $x' = b$) to rest at the origin ($x = 0$, $x' = 0$) in the shortest time. As mentioned at the end of Section 1.10, Example 3, we have here a Mayer problem of minimum time with $n = 2$, $m = 1$, $I[x, y, u] = g = t_2$, with the system $dx/dt = y$, $dy/dt = u$, $u \in U = [-1 \leq u \leq 1] \subset R$, $t_1 = 0$, $x(t_1) = a$, $y(t_1) = b$, $x(t_2) = 0$, $y(t_2) = 0$, $t_2 \geq t_1$. Here we have

$$H = H(t, x, y, u, \lambda_1, \lambda_2) = \lambda_1 y + \lambda_2 u,$$

$$M = M(t, x, y, \lambda_1, \lambda_2) = \begin{cases} \lambda_1 y - \lambda_2 & \text{if } \lambda_2 > 0, \\ \lambda_1 y + \lambda_2 & \text{if } \lambda_2 < 0. \end{cases}$$

In particular, if $\lambda_2 > 0$, then H attains its minimum for $u = -1$; if $\lambda_2 < 0$, then H attains its minimum for $u = +1$. If we use the "signum function" $\alpha = \operatorname{sgn} \beta$ defined by $\alpha = 1$ for $\beta > 0$, and $\alpha = -1$ for $\beta < 0$; (α any value between -1 and 1 for $\beta = 0$), then the possible optimal strategy $u(t)$ is related to the multiplier $\lambda_2(t)$ by the relation

$$u(t) = -\operatorname{sgn} \lambda_2(t), \qquad t_1 \leq t \leq t_2.$$

The equations for the multipliers λ_1, λ_2 are

$$d\lambda_1/dt = -H_x = 0, \quad d\lambda_2/dt = -H_y = -\lambda_1, \qquad t_1 \leq t \leq t_2.$$

Thus, $\lambda_1 = c_1$, $\lambda_2 = -c_1 t + c_2$, $t_1 \leq t \leq t_2$, c_1, c_2 constants. Now c_1, c_2 cannot be both zero, since $\lambda(t) = (\lambda_1, \lambda_2)$ would be both zero in $[t_1, t_2]$, in contradiction with (P1). Thus, either $c_1 \neq 0$ and λ_2 is a nonzero linear function, or $c_1 = 0$ and $\lambda_2 = c_2$ is a nonzero constant. Thus, λ_2 changes sign at most once in $[t_1, t_2]$, and so $[t_1, t_2]$ can be divided at most into two subintervals, in one of which $u = 1$ and in the other $u = -1$.

If $u = 1$, then $y = t - \alpha$, $x = 2^{-1}(t - \alpha)^2 + \beta$, α, β constants, that is, we have the parabolas $x = 2^{-1}y^2 + C$, along which $P = (x, y)$ moves in such a way that y is increasing. If $u = -1$, then $y = -(t - \alpha)$, $x = 2^{-1}(t - \alpha)^2 + \beta$, α, β constants, that is, we have the parabolas $x = -2^{-1}y^2 + C$, along which $P = (x, y)$ moves in such a way that y is decreasing. Thus, the optimal solution, if any, must be made up of at most two arcs of such parabolas. Since $x(t_2) = y(t_2) = 0$, there are only two arcs of such parabolas reaching $(0, 0)$,

$$AO: y = -(t - t_2), \qquad x = -2^{-1}(t - t_2)^2, \quad u = -1,$$
$$BO: y = (t - t_2), \qquad x = 2^{-1}(t - t_2)^2, \quad u = 1,$$

and they are graphically represented with an arrow in the illustration denoting the sense along which they are traveled. Now any point (a, b) above the line BOA can be joined to BO by an arc of a parabola $x = -2^{-1}y^2 + C$ along which $u = -1$; any point (a, b) below the line BOA can be joined to AO by an arc of a parabola $x = 2^{-1}y^2 + C$ along which $u = 1$.

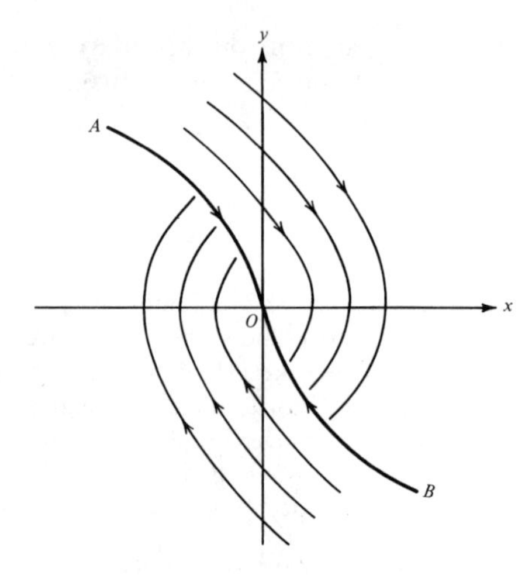

The illustration sketches the family of solutions. Every point $P = (a, b)$ can be taken to $(0, 0)$ by a unique path PAO, or PBO, satisfying the necessary conditions (P1)–(P3), as well as (P4) as we shall see in Section 6.1. Since we shall see from Sections 4.3 and 9.4 that an optimal solution exists, the uniquely determined path and corresponding strategy are optimal. In Section 4.5, by sufficient conditions (and "feedback" and "synthesis" considerations) we shall see another reason why the uniquely determined paths PAO, or POB, are optimal.

Exercise

Write the problem of the curve $C: x = x(t), t_1 \le t \le t_2$, of minimum length between two points $1 = (t_1, x_1), 2 = (t_2, x_2), t_1 < t_2$, as a Mayer problem of optimal control, and use the necessary conditions (4.2.i) to obtain the results in Section 3.1.

C. Remarks

Remark 1. For the use of the necessary condition (4.2.i), or (P1)–(P4), the reader should see the numerous examples and exercises in Chapter 6, particularly Sections 6.1–5. In all those examples we shall determine one, or a few, admissible pairs which satisfy the necessary condition, and are therefore good candidates for optimality. Alternatively, the opposite conclusion can be reached, that is, that no admissible pair satisfies the necessary condition, and then certainly the minimum of the functional is not attained (at least under the given set of assumptions). On the other hand, it may even occur that there are infinitely many optimal solutions and that the necessary condition (4.2.i) yields no information on them. Examples of such an occurrence are given in Remark 8 below. The relevance of the necessary condition (4.2.i) lies in the fact that it is valid even if the optimal strategy $u(t)$ lies for most t on the boundary of the control space U (though we must assume that the trajectory x lies in the interior of A).

Remark 2. Note that if $u(t)$ is continuous in $[t_1, t_2]$, then $x(t)$ is continuous in $[t_1, t_2]$ together with x', and so are $\lambda(t)$, $\lambda'(t)$, and all statements (P1)–(P4) hold for all t in $[t_1, t_2]$. If $u(t)$ is sectionally continuous in $[t_1, t_2]$, then $x(t)$ is continuous in $[t_1, t_2]$ with sectionally continuous derivative x', and so is $\lambda(t)$, and all statements (P1)–(P4) hold at all points t of continuity of $u(t)$, as well as at the points t of jump discontinuity of $u(t)$ if in each relation we replace $u(t)$, $x'(t)$, $\lambda'(t)$, $M'(t)$ by $u(t+0)$, $x'(t+0)$, $\lambda'(t+0)$, $M'(t+0)$ as well as by $u(t-0)$, $x'(t-0)$, $\lambda'(t-0)$, $M'(t-0)$. If $u(t)$ is merely measurable, then $x(t)$, $\lambda(t)$ are only absolutely continuous in $[t_1, t_2]$, x', λ' may not exist in a subset of measure zero in $[t_1, t_2]$, and actually all the relations (P1)–(P4) may not be satisfied in a set of measure zero. Note that we can always change the values of a strategy in a set of measure zero; the trajectory $x(t)$ and the multipliers $\lambda(t)$ are not modified.

Remark 3. Among the general assumptions for the necessary condition, we required the set A to be closed. Since the graph $[(t, x(t)) | t_1 \leq t \leq t_2]$ of the optimal trajectory x is certainly compact (and by hypothesis made up of points all interior to A), there is certainly some number $\delta > 0$ such that all points (t, y) at a distance $\leq \delta$ from the graph of $x(t)$ are also interior to A. If we denote this set of points by A_0, we see that the graph of x is made up of interior points of A_0, and A_0 is now compact. We see, therefore, that it is not restrictive to assume A compact instead of closed in the general assumptions for the necessary condition, or alternatively, that A is an open bounded subset of the tx-space R^{n+1}.

Remark 4. If $x(t)$, $u(t)$, $t_1 \leq t \leq t_2$ of Section 4.1, is any admissible pair, with $u(t)$ bounded, or essentially bounded, then $f(t, x(t), u(t)) = (f_1, \ldots, f_n)$ is bounded and therefore $x(t)$ is Lipschitzian. Analogously, the partial derivatives $f_{ix^j}(t, x(t), u(t))$ are bounded, and then the multipliers $\lambda(t) = (\lambda_1, \ldots, \lambda_n)$ are also Lipschitzian. Note that if $x(t)$, $u(t)$, $t_1 \leq t \leq t_2$, is any admissible pair, and $(t_0, x(t_0))$ is an interior point of A, if $u(t)$ is bounded and W is any bounded neighborhood of (t_0, x_0) in A, then $f(t, x, u(t)) = (f_1, \ldots, f_n)$ and the partial derivatives $f_{ix^j}(t, x, u(t))$ are bounded for $(t, x) \in W$, they are continuous in x for a.a. t, and measurable in t for all x, and finally $f(t, x, u(t))$ is uniformly Lipschitzian in x for $(t, x) \in W$. Thus, the differential system $dx/dt = f(t, x, u(t))$ satisfies usual conditions for the local existence and uniqueness theorem (see e.g., E. J. McShane [*I*, pp. 344–345]).

Hence, for any $(\bar{t}, \bar{x})$ in W there is one and only one solution $\bar{x}(t)$ in a neighborhood *of* $\bar{t}$ with $\bar{x}(\bar{t}) = \bar{x}$.

Remark 5 (Alternate Hypotheses for (c) of Section 4.1). There are situations where the optimal trajectory is not Lipschitzian (for instance, the trajectory $x(t) = (1 - t^2)^{1/2}$, $-1 \leq t \leq 1$, a semicircle, certainly is not Lipschitzian). In such situations condition (c) of Section 4.1 cannot hold. We shall denote by $x(t)$, $u(t)$, $t_1 \leq t \leq t_2$, a given optimal pair for which we assume, as in Section 4.1, that all points $(t, x(t))$, $t_1 \leq t \leq t_2$, are interior to A. Thus, there is some $\delta_0 > 0$ such that all points $(t', x') \in R^{n+1}$ with $|t' - t| \leq \delta_0$, $|x' - x(t)| \leq \delta_0$ for some $t \in [t_1, t_2]$ are also all interior to A. An assumption wider than (c) and under which the conclusions of (4.2.i) still hold is as follows:

(c′) U is a fixed closed subset of R^m. There is a number δ, $0 < \delta \leq \delta_0$, and a scalar function $S(t) \geq 0$, $t_1 \leq t \leq t_2$, such that (c_1') $S(t)$ is L-integrable in $[t_1, t_2]$; and (c_2')

for every $t \in [t_1, t_2]$ and point $(t', x') \in A$ with $|t' - t| \leq \delta$, $|x' - x(t)| \leq \delta$, we have

$$|f_{it}(t', x', u(t))|, |f_{ix_j}(t', x', u(t))| \leq S(t), \qquad i, j = 1, \dots, n. \tag{4.2.10}$$

This is often called a condition (S).

Another assumption replacing (c) or (c′) is as follows:

(c″) $U(t)$ is a closed subset of R^m for every t and: (c_1'') the optimal control function u is bounded, that is, $|u(t)| \leq N$, $t_1 \leq t \leq t_2$; and (c_2'') for almost all $\bar{t} \in [t_1, t_2]$ and every point $\bar{u} \in U(\bar{t})$ and number $\varepsilon > 0$, there is some $v \in R^m$ and numbers $\sigma > 0$, $\sigma < \delta_0$, such that $|v - \bar{u}| < \varepsilon$, $v \in U(t)$ for all $t \in [\bar{t} - \sigma, \bar{t} + \sigma]$.

Let us denote by I the projection of the set A on the t-axis, that is, $I = [t | (t, x) \in A$ for some $x \in R^n]$. Condition (c″) is certainly satisfied if $U(t)$ is compact for every t and contained in a fixed compact set U_0 of R^m, if the set V of all (t, u) with $t \in I$, $u \in U(t)$ is the closure of an open subset of R^{m+1}, and if every $(\bar{t}, \bar{u}) \in (\bar{t}, U(\bar{t}))$ is a point of accumulation of points $(\bar{t}, v)$ interior to V.

Another assumption replacing (c), (c′), or (c″) is as follows:

(c‴) $U(t)$ is an open subset of R^m for every t, such that: (c_1''') the set V of all (t, u) with $t \in I$, $u \in U(t)$ is open (relatively to $I \times R^m$); and (c_2''') the optimal control function u is bounded, that is, $|u(t)| \leq N$, $t_1 \leq t \leq t_2$.

Another assumption replacing (c), (c′), (c″), or (c‴), is as follows:

($c^{(iv)}$) $U(t)$ is a closed subset of R^m for every t, such that: ($c_1^{(iv)}$) for almost all $\bar{t} \in [t_1, t_2]$, every point $\bar{u} \in U(\bar{t})$, and $\varepsilon > 0$ there is some $v \in R^m$ and numbers $\sigma > 0$, $\sigma < \delta_0$, such that $|v - \bar{u}| < \varepsilon$, $v \in U(t)$ for all $t \in [\bar{t} - \sigma, \bar{t} + \sigma]$; ($c_2^{(iv)}$) there is a number δ, $0 < \delta \leq \delta_0$, and a scalar function $S(t) \geq 0$, $t_1 \leq t \leq t_2$, L-integrable in $[t_1, t_2]$, such that for every $t \in [t_1, t_2]$ and points $(t', x') \in A$ with $|t' - t| \leq \delta$, $|x' - x(t)| \leq \delta$, the relations (4.2.10) hold.

Finally, another assumption replacing (c), (c′), (c″), (c‴), or ($c^{(iv)}$) is as follows:

($c^{(v)}$) $U(t)$ is an open subset of R^m for every t, such that: ($c_1^{(v)}$) the set V of all (t, u) with $t \in I$, $u \in U(t)$ is open (relatively to $I \times R^m$); and ($c_2^{(v)}$) statement ($c_2^{(iv)}$) holds.

Under any one of these hypotheses, the conclusions in (4.2.i) still hold as stated. The situations described by properties (c) to ($c^{(v)}$) cover essentially all practical cases. Proofs are given in Chapter 7. Finally, for control sets $U(t)$ depending on t, a weaker form (P_2^*) of (P2) will be stated and proved in Section 7.1.

Remark 6. In (e) of Section 4.1 we have assumed that the scalar function g possesses a differential dg at a point of a given set B. We remind the reader here that given any subset K of the x-space E_n, $x = (x^1, \dots, x^n)$, we say that a scalar function $g(x)$ possesses a differential $dg = \sum_{i=1}^n a_i\, dx^i$ at a given point $\bar{x} = (\bar{x}^1, \dots, \bar{x}^n)$ of K provided there are numbers $a_1, \dots, a_n$ such that

$$g(x) - g(\bar{x}) = \sum_{i=1}^{n} a_i(x^i - \bar{x}^i) + |x - \bar{x}|\varepsilon(x) \tag{4.2.11}$$

for every point x in K, where $\varepsilon(x) \to 0$ as $|x - \bar{x}| \to 0$.

In Section 4.1 we have also assumed that the scalar functions f_i possess continuous partial derivatives on a given set M. We recall here that given any closed subset K of the x-space E_n, $x = (x^1, \dots, x^n)$, we say that a scalar function $g(x)$ possesses continuous first order partial derivatives $g_1(x), \dots, g_n(x)$ in K provided that: (a) the functions $g_i(x)$ are continuous in K, $i = 1, \dots, n$, and (b) for every point $\bar{x}$ of K a relation (4.2.11) holds with $a_i = g_i(\bar{x})$, $i = 1, \dots, n$. It is known (Whitney [4]) that there is then an extension $G(x)$ of g in the whole of R^n which is continuously differentiable in R^n and $G = g$, $\partial G/\partial x^i = g_i$ in K, $i = 1, \dots, n$. We refer to Whitney [4] for the analogous definitions and statements concerning partial derivatives of higher order.

Remark 7. We have assumed so far that the pair $x(t)$, $u(t)$, $t_1 \le t \le t_2$, gives an absolute minimum for the functional in Ω, that is, $I[x, u] \le I[\tilde{x}, \tilde{u}]$ for all pairs $\tilde{x}$, $\tilde{u}$ in Ω. Nevertheless, the entire statement (P1)–(P4) still holds even if we know only that $I[x, u] \le I[\tilde{x}, \tilde{u}]$ holds only for the pairs $\tilde{x}$, $\tilde{u}$ with the trajectory $\tilde{x}$ satisfying the same boundary conditions as x, and lying in any small neighborhood N_δ of the trajectory x. If this occurs we say that the pair x, u is a *strong local minimum*, and thus (P1)–(P4) constitute a necessary condition for strong local minima.

Remark 8. The optimal solution is not necessarily unique, and it may well occur that not enough information can be gathered from the necessary condition to characterize candidates for the optimal solutions. Here are a few examples.

1. Find the minimum time t_2 for a moving point $P = (x, y)$ governed by $x' = y' = u$, $|u| \le 1$, starting from $(0, 0)$ at time $t_1 = 0$, to hit a moving target Q moving on a trajectory Γ with law of motion $x = h(t)$, $y = k(t)$, $t' \le t \le t''$. The only possible trajectories for P lie on the straight line $r: x = y$. Assume that r crosses the locus Γ at only one point $R = (a, a)$ in the first quadrant. Then $x = y = t$, $u(t) = 1$, $0 \le t \le a$ is the trajectory by which P reaches the locus Γ in a minimum time a. We assume that Q reaches R at a time $t_2 > a$, and thus certainly $t' \le t_2 \le t''$. Then there are infinitely many laws of motion for P to reach R at the time t_2; $u(t)$ can take arbitrary values in $[-1, 1]$ (provided $\int_0^{t_2} u(t)\,dt = a$). All these trajectories are admissible and also optimal with $I[x, u] = t_2$. One of these trajectories is of course $x = y = t$, $u = 1$ for $0 \le t \le a$; $x = y = a$, $u = 0$ for $a \le t \le t_2$.

2. Here we want that the point P as in Example 1 to hit the locus Γ in such a way to minimize $I[x, u] = (t_2 - b)^2$, where now b is a fixed number, $b > a$. Again, there are

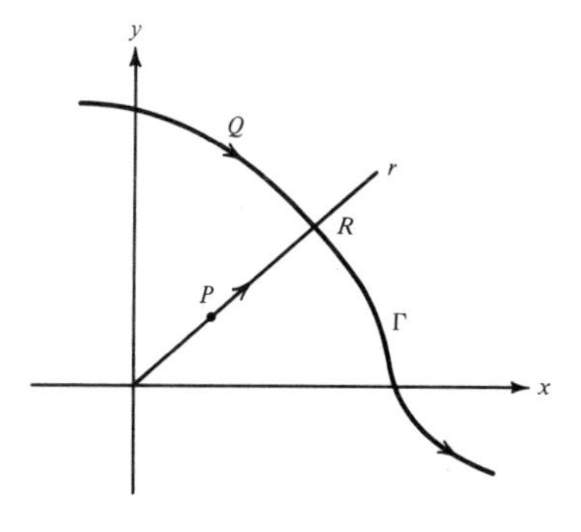

infinitely many laws of motion for P to reach R at the time $t_2 = b$; $u(t)$ can take arbitrary values in $[-1,1]$ (provided $\int_0^{t_2} u(t)\,dt = a$). All these trajectories are admissible and optimal with $I[x,u] = 0$. The same trajectory singled out in Example 1 is also of interest here.

In both examples, $H = \lambda_1 u + \lambda_2 u$, $d\lambda_1/dt = d\lambda_2/dt = 0$, $\lambda_1 = c_1$, $\lambda_2 = c_2$, c_1, c_2 constants, and for $\lambda_1 + \lambda_2 = c_1 + c_2 = 0$, H is identically zero, and the minimum property of H yields no information on the control u.

3. Problems as the one above, where the necessary condition (4.2.i) does not determine the optimal strategy, are often called *singular* problems. Often this situation occurs only for certain arcs of the optimal solution which then are called singular arcs. All this happens as a rule if the control variable $u = (u^1, \dots, u^m)$ enters linearly in the functions f_i, that is $f_i = F_i(t,x) + \sum_{j=1}^m G_{ij}(t,x)u^j$, and for those arcs of the optimal trajectory for which the control variable $u(t)$ is expected to take values in the interior of U (or U is open, or $U = R^m$), as mentioned at the end of Subsection 4.2A. Indeed then

$$H = \sum_{i=1}^n \lambda_i F_i + \sum_{j=1}^m \left(\sum_{i=1}^n \lambda_i G_{ij}\right) u^j,$$

and the equations (4.2.8), or $H_{u^j} = 0$, reduce to

$$\sum_{i=1}^n \lambda_i G_{ij}(t,x) = 0, \qquad j = 1, \dots, m,$$

which do not depend on u, and thus leave $u = (u^1, \dots, u^m)$ undetermined.

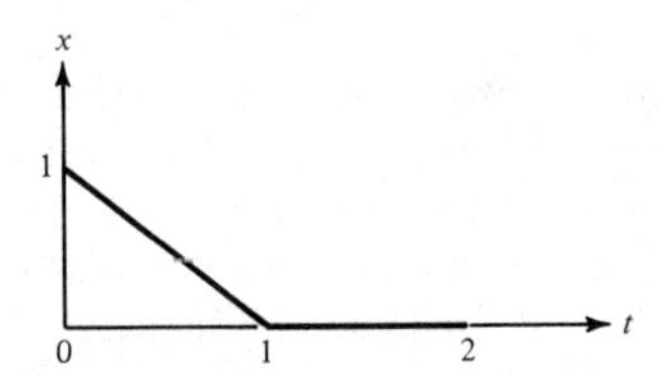

Here is a specific example. Let us consider the problem of the minimum of $I[x,u] = x_2^2 + \int_0^2 x^2\,dt$, under the constraints $dx/dt = u$, $|u| \le 1$, $x(0) = 1$, $n = m = 1$. By taking $dy/dt = x^2$, $y(0) = 0$, we have the Mayer problem $I[x,y,u] = x_2^2 + y_2$, under the constraints $dy/dt = x^2$, $dx/dt = u$, $|u| \le 1$, $x(0) = 1$, $y(0) = 0$. Then $H = \Lambda x^2 + \lambda u$, and we take $u = -1$ if $\lambda > 0$, $u = 1$ if $\lambda < 0$, while u remains undetermined if $\lambda = 0$. Moreover, $d\Lambda/dt = -H_y = 0$, $d\lambda/dt = -H_x = -2\Lambda x$, and $\Lambda = C$, a constant. By the transversality relation with $dg = 2x_2 dx_2 + dy_2$, $dt_1 = dt_2 = 0$, $dx_1 = dy_1 = 0$, dx_2, dy_2 arbitrary, we have $(1 - \Lambda)\,dy_2 + (2x_2 - \lambda(t_2))dx_2 = 0$, hence $\Lambda = 1$, $\lambda(t_2) = 2x_2$. If we take $u(t) = -1$, hence $x(t) = 1 - t$, $0 \le t \le 1$, we see that $x(1) = 0$, and for the minimum of I we can only take $x(t) = 0$ for $1 \le t \le 2$, $x_2 = x(2) = 0$, and $u(t) = 0$ for $1 \le t \le 2$. Then, $d\lambda/dt = -2(1-t)$ for $0 \le t \le 1$, $d\lambda/dt = 0$ for $1 \le t \le 2$, and we take $\lambda(t) = (1-t)^2$ for $0 \le t \le 1$, $\lambda(t) = 0$ for $1 \le t \le 2$. The optimal solution is now depicted in the illustration, and $I_{\min} = \frac{1}{3}$. The value $u(t) = 0$ for $1 \le t \le 2$ is not determined by (4.2.i) since $\lambda(t) = 0$ in this interval. The arc $x(t)$, $1 \le t \le 2$, of the optimal solution is said to be a singular arc.

Remark 9. For $m = 1$, $n \ge 1$, and all f_i linear in u, or $f_i = F_i(t,x) + G_i(t,x)u$, $i = 1, \dots, m$, then $H(t,x,u,\lambda) = \sum_i \lambda_i F_i + (\sum_i \lambda_i G_i)u$. The canonical equations become $x_i' = F_i + G_i u$, $\lambda_i' = -\sum_j \lambda_j F_{jx^i} + (\sum_j \lambda_j G_{jx^i})u$. On a singular arc, with $u(t) \in \operatorname{int} U$, $U \subset R$, then $H_u = 0$, and $H_u = \sum_j \lambda_j G_j(t,x) = 0$ does not contain u. Let us assume here that all F_i and

G_i have continuous partial derivatives of orders as large as we need. Let us take now successive total derivatives of H_u with respect to t, say $H^{(1)} = (d/dt)H_u$, where now we replace $\lambda'(t)$ and $x'(t)$ by their expressions above. Let us proceed computing $H^{(2)} = (d/dt)H^{(1)}$ with the same substitutions, or briefly $H^{(2)} = (d^2/dt^2)H_u$, etc. It may occur that for some minimal p an expression $H^{(2p)}$ is obtained which actually contains u, of course linearly. Under various assumptions it has been proved that

$$(-1)^p(\partial/\partial u)(d^{2p}/dt^{2p})H_u > 0$$

is a necessary condition for a minimum of I. This condition is often called the *generalized Legendre-Clebsch* necessary condition.

For $n \geq 1$, $m > 1$, the generalized Legendre-Clebsch condition becomes, in matrix notation and the same conventions as before,

$$(\partial/\partial u)(d^q/dt^q)H_u = 0 \quad \text{for } q \text{ odd}, \qquad (-1)^{2p}(\partial/\partial u)(d^{2p}/dt^{2p})H_u \geq 0$$

in $[t_1, t_2]$(H. J. Kelley [1,2] and H. J. Kelley, R. E. Kopp, and H. G. Moyer [1]). Two more necessary conditions for singular arcs will be stated and proved in Section 7.3K. For more details, further necessary conditions, and important technical applications we refer to D. J. Bell and H. Jacobson [I].

In the example of Remark 8, on the singular arc $x(t) = 0$, $1 \leq t \leq 2$, we have $H_u = \lambda$, $H^{(1)} = (d/dt)H_u = \lambda' = -2x$, $H^{(2)} = (d/dt)H^{(1)} = -2x' = -2u$, $p = 1$, and finally $-(\partial/\partial u)H^{(2)} = 2 > 0$. The generalized Legendre-Clebsch condition is satisfied.

Remark 10. THE CASE OF B' A CONVEX CONE. In Section 4.2 we have considered the set B' of all tangent vectors $h = (\tau_1, \xi_1, \tau_2, \xi_2)$ at $e[x]$ in B. The set B' certainly is a cone, since $h \in B'$ and $\mu \geq 0$ imply that $\mu h \in B'$. This follows from the definition of tangent vector, and B' is said to be a convex cone if B', as a subset of the linear space R^{2n+2}, is convex, that is, $h_1, h_2 \in B$, $\mu_1, \mu_2 \geq 0$, $\mu_1 + \mu_2 = 1$ implies that $\mu_1 h_1 + \mu_2 h_2 \in B'$. Finally, if B' is a linear space, then we say that B has a full tangent space B' at $\eta(x)$.

If B' is only a closed convex cone, the necessary condition (4.2.i) still holds, except for the statement (4.2.1) in P4, which is replaced by

$$\lambda_0\, dg + \left[M(t)\, dt - \sum_{i=1}^{n} \lambda_i(t)\, dx^i \right]_1^2 \geq 0. \tag{4.2.13}$$

We denote this modified statement by P4*. Note that whenever B' is a linear space, P4* reduces to P4. Indeed, (4.2.13) must hold for both $(dg, dt_1, dx_1, dt_2, dx_2)$ and $(-dg, -dt_1, -dx_1, -dt_2, -dx_2)$, and this implies equality in (4.2.13) that is, (4.2.1).

D. An Elementary Proof of (P2) of (4.2.i) in a Particular Case

While we shall give in Chapter 7 two proofs of the necessary condition (4.2.i) or (P1-4) in the general case, we anticipate here an elementary proof of (P2) for the important linear case with initial data. Namely, we assume that t_1, $x(t_1) = x_1$ and t_2 are fixed, that $x(t_2) = x_2$ is arbitrary, that g is linear in x_2, that f is linear in x and u, and that U is a fixed set. In other words, we consider the Mayer problem of the minimum of $I[x, u]$ with

$$I[x, u] = g(x_2) = Cx_2, \qquad dx/dt = A(t)x + B(t)u, \qquad u \in U,$$

where $C = \text{row}(\gamma_1, \ldots, \gamma_n)$, $A(t) = [a_{ij}(t)]$, $B(t) = [b_{is}(t)]$ are given $1 \times n$, $n \times n$, $n \times m$ matrices respectively, γ_i constants, $a_{ij}(t)$, $b_{is}(t)$ continuous functions on $[t_1, t_2]$. Thus,

$$I[x, u] = \sum_{i=1}^{n} \gamma_i x^i(t_2),$$

$$dx^i/dt = \sum_{j=1}^{n} a_{ij}(t)x^j(t) + \sum_{s=1}^{m} b_{is}(t)u^s(t), \tag{4.2.11}$$

$$H(t, x, u, \lambda) = \sum_{ij} \lambda_i a_{ij} x^j + \sum_{is} \lambda_i b_{is} u^s,$$

$$d\lambda_i/dt = -H_{x^i} = -\sum_j \lambda_j a_{ji}, \qquad i = 1, \ldots, n. \tag{4.2.12}$$

Here $dt_1 = dt_2 = 0$, $dx_1^i = 0$, $g_{x^i} = \gamma_i$, and the transversality relation (P4) reduces to $\sum_i (\lambda_0 \gamma_i - \lambda_i(t_2))\, dx_2^i = 0$ where all dx_2^i are arbitrary. Thus, $\lambda_0 \gamma_i - \lambda_i(t_2) = 0$, $i = 1, \ldots, n$. For any given admissible pair $x_0(t)$, $u_0(t)$, $t_1 \le t \le t_2$, we take for the multipliers $\lambda_i(t)$, $t_1 \le t \le t_2$, the unique AC solution $\lambda(t)$ of the linear system (4.2.12) with terminal data $\lambda_i(t_2) = \lambda_0 \gamma_i$ where $\lambda_0 > 0$ is an arbitrary constant.

For any other admissible pair $x(t)$, $u(t)$, $t_1 \le t \le t_2$, we have now $x^i(t_1) = x_0^i(t_1)$ and

$$\begin{aligned} \Delta = \lambda_0(I[x, u] - I[x_0, u_0]) &= \lambda_0 \sum_i \gamma_i (x^i(t_2) - x_0^i(t_2)) \\ &= \sum_i \lambda_i(t_2)(x^i(t_2) - x_0^i(t_2)) \\ &= \int_{t_1}^{t_2} (d/dt) \sum_i \lambda_i(t)(x^i(t) - x_0^i(t)). \end{aligned}$$

By (4.2.11) and (4.2.12) we derive

$$\begin{aligned} \Delta &= \int_{t_1}^{t_2} \left[\sum_i \lambda_i'(t)(x^i(t) - x_0^i(t)) + \sum_i \lambda_i(t)(x'^i(t) - x_0'^i(t) \right] dt \\ &= \int_{t_1}^{t_2} \left[-\sum_{ij} \lambda_j a_{ji}(x^i - x_0^i) + \sum_{ij} \lambda_i a_{ij}(x^j - x_0^j) + \sum_{is} \lambda_i b_{is}(u^s - u_0^s) \right] dt. \end{aligned}$$

By exchanging i with j we see that the second term in the last expression cancels the first one. By adding and subtracting terms and manipulation we have

$$\begin{aligned} \Delta &= \int_{t_1}^{t_2} \left[\left(\sum_{ij} \lambda_i a_{ij} x_0^j + \sum_{is} \lambda_i b_{is} u^s \right) - \left(\sum_{ij} \lambda_i a_{ij} x_0^j + \sum_{is} \lambda_i b_{is} u_0^s \right) \right] dt \\ &= \int_{t_1}^{t_2} [H(t, x_0(t), u(t), \lambda(t)) - H(t, x_0(t), u_0(t), \lambda(t))]\, dt. \end{aligned} \tag{4.2.13}$$

Let us assume that x_0, u_0 is optimal. First, x_0 and λ_0 are AC, and u_0 is measurable, and also bounded as in the main assumptions of (4.2.i). Thus, $H(t, x_0(t), u_0(t), \lambda(t))$ is a bounded measurable function in $[t_1, t_2]$, and hence L-integrable. Then, at almost all t, $H(t, x_0(t), u_0(t), \lambda(t))$ is the derivative of its indefinite integral. If $t_1 \leq \bar{t} \leq t_2$ is any such point, if $\varepsilon > 0$ is such that $t_1 \leq \bar{t} - \varepsilon < \bar{t} < t_2$, and if $\bar{u}$ is any point of U, let us take $u(t) = \bar{u}$ for $\bar{t} - \varepsilon \leq t \leq \bar{t}$, and $u(t) = u_0(t)$ otherwise. Then

$$\varepsilon^{-1}\Delta = \varepsilon^{-1} \int_{t-\varepsilon}^{t} [H(t, x_0(t), \bar{u}, \lambda(t)) - H(t, x_0(t), u_0(t), \lambda(t))]\, dt,$$

and as $\varepsilon \to 0+$ we also have

$$\lim_{\varepsilon \to 0+} \lambda_0 \varepsilon^{-1}(I[x, u] - I[x_0, u_0]) = H(\bar{t}, x_0(\bar{t}), \bar{u}, \lambda(\bar{t})) - H(\bar{t}, x_0(\bar{t}), u_0(\bar{t}), \lambda(\bar{t}))$$

$$\geq 0.$$

This difference is nonnegative since $I[x, u] \geq I[x_0, u_0]$. This holds for almost all $\bar{t} \in (t_1, t_2]$ and $\bar{u} \in U$; hence $H(t, x_0(t), u, \lambda(t)) \geq H(t, x_0(t), u_0(t), \lambda(t))$ for all $u \in U$ and almost all t. We have proved (P2) in the present situation.

An important further conclusion of the previous remarks is that, in the linear situation depicted above, condition (P2) is not only necessary but also sufficient for optimality. Indeed, if x, u and x_0, u_0 are admissible pairs, and $\lambda(t)$ is chosen as stated above, then (4.2.13) holds, (P2) implies that $H(t, x_0(t), u(t), \lambda(t)) \geq H(t, x_0(t), u_0(t), \lambda(t))$ for almost all t, and $\Delta \geq 0$, that is, $I[x, u] \geq I[x_0, u_0]$.

4.3 Statement of an Existence Theorem for Mayer's Problems of Optimal Control

We state here briefly a particular but useful existence theorem for the problems we are considering in this Chapter. This theorem will be contained in the statements we shall prove in Chapter 9 and in the more general statements of Chapters 11–16.

Thus, we are concerned with functionals of the form $I[x, u] = g(t_1, x(t_1), t_2, x(t_2))$ and we shall state a theorem guaranteeing the existence of an absolute minimum of $I[x, u]$ in the class Ω of all pairs $x(t)$, $u(t)$, $t_1 \leq t \leq t_2$, x AC, u measurable, with $dx/dt = f(t, x(t), u(t))$, $t \in [t_1, t_2]$ (a.e.), and with $(t, x(t)) \in A \subset R^{n+1}$, $u(t) \in U \subset R^m$, $(t_1, x(t_1), t_2, x(t_2)) \in B \subset R^{2n+2}$.

4.3.i (FILIPPOV'S EXISTENCE THEOREM). *If A and U are compact, B is closed, f is continuous on $A \times U$, g is continuous on B, Ω is not empty, and for every $(t, x) \in A$ the set $Q(t, x) = f(t, x, U) \subset R^n$ is convex, then $I[x, u]$ has an absolute minimum in Ω.*

If A is not compact, but closed and contained in a slab $t_0 \leq t \leq T$, $x \in R^n$, t_0, T finite, then the statement still holds if we know for instance that (b) there is a compact subset P of A such that every trajectory in Ω has at least one point $(t^*, x(t^*)) \in P$; and (c) there is a constant C such that $x^1 f_1 + \cdots + x^n f_n \leq C(|x|^2 + 1)$ for all $(t, x, u) \in A \times U$.

If A is contained in no slab as above but A is closed, then the statement still holds if (b) and (c) hold, and in addition (d) $g(t_1, x_1, t_2, x_2) \to +\infty$ as $t_2 - t_1 \to +\infty$, uniformly with respect to x_1 and x_2 and for $(t_1, x_1, t_2, x_2) \in B$.

We note that condition (b) is certainly satisfied if all trajectories x in Ω have either the first end point (t_1, x_1) fixed, or the second end point (t_2, x_2) fixed.

We note here that if f is linear in x, precisely, if $f(t, x, u) = A(t, u) + B(t, u)x$, where $A = [a_i]$ is an $n \times 1$ matrix and $B = [b_{ij}]$ is an $n \times n$ matrix with all entries continuous and bounded in $A \times U$, then certainly condition (c) is satisfied. Indeed if $|a_i(t, u)|$, $|b_{ij}(t, u)| \leq c$ then

$$\sum_i x^i f_i = \sum_i a_i x^i + \sum_i \sum_j b_{ij} x^i x^j \leq nc|x| + n^2 c|x|^2 \leq 2n^2 c(|x|^2 + 1).$$

Both statement (4.3.i) and conditions (b), (c), (d) will be restated in more general forms in Sections 9.2 and 9.4.

Example

Let us consider the same example of Section 4.2B, with $n = 2$, $m = 1$, $g = t_2$, $x' = y$, $y' = u$, $t_1 = 0$, $x_1 = a$, $y_1 = b$, $x_2 = 0$, $y_2 = 0$, $t_2 \geq 0$ undetermined, $U = [-1 \leq u \leq 1]$ is compact, $B = (0, a, b, 0, 0) \times [t \geq 0]$ is closed, and we have seen in Section 4.2B that Ω is not empty. Here all trajectories start at (a, b) and (b) holds with P the single point $(0, a, b)$. Moreover, $xf_1 + yf_2 = xy + yu < x^2 + y^2 + 1$, and condition (c) holds with $C = 1$. Finally $g \to +\infty$ as $t_2 \to +\infty$, and (d) also holds. Thus, the absolute minimum exists.

More examples are discussed in Chapter 6.

4.4 Examples of Transversality Relations for Mayer Problems

We shall now apply the transversality relation (P4) of the necessary condition (4.2.i) to a number of particular but rather typical cases. In each case, we shall write explicitly the vectors h of the linear space (tangent space) B'. In each case we shall deduce from (P4) a number of finite relations concerning the values of the multipliers $\lambda_1(t), \ldots, \lambda_n(t)$ and of the function $M(t)$ at the end points t_1 and t_2.

We mention in passing that, as proved in Section 4.2 the function $M(t)$ is actually a constant in $[t_1, t_2]$ whenever the problem is autonomous, that i', when $f_1, \ldots, f_n$ depend on x and u only, and not on t.

A. The Transversality Relation in General

We consider here the case where B is defined by a certain number of equations, say $k-1$, or

$$\phi_j(t_1, x_1, t_2, x_2) = 0, \qquad j = 2, \ldots, k.$$

In this situation, it will be convenient to denote $g(t_1, x_1, t_2, x_2)$ by $\phi_1(t_1, x_1, t_2, x_2)$, or $g = \phi_1$. We assume here all $\phi_1, \ldots, \phi_k$ are of class C^1. The transversality relation (P4),

$$\lambda_0\, dg + \left[M(t)\, dt - \sum_{i=1}^{n} \lambda_i(t)\, dx^i \right]_1^2 = 0,$$

or

$$\text{(4.4.1)} \qquad dT = [\lambda_0 g_{t_1} - M(t_1)]\, dt_1 + \sum_{i=1}^{n} [\lambda_0 g_{x_1^i} + \lambda_i(t_1)]\, dx_1^i$$
$$+ [\lambda_0 g_{t_2} + M(t_2)]\, dt_2 + \sum_{i=1}^{n} [\lambda_0 g_{x_2^i} - \lambda_i(t_2)]\, dx_2^i = 0$$

must be valid for all $(2n+2)$-vectors $h = (dt_1, dx_1, dt_2, dx_2)$ satisfying

$$\text{(4.4.2)} \qquad d\phi_j = \phi_{jt_1}\, dt_1 + \sum_{i=1}^{n} \phi_{jx_1^i}\, dx_1^i + \phi_{jt_2}\, dt_2 + \sum_{i=1}^{n} \phi_{jx_2^i}\, dx_2^i = 0, \qquad j = 2, \ldots, k.$$

In other words, equation (4.4.1) must be an algebraic consequence of the $k-1$ linear equations (4.4.2), that is, there is a numerical $(k-1)$-vector $\Lambda_2, \ldots, \Lambda_k$, such that

$$\text{(4.4.3)} \qquad \begin{aligned} \lambda_0 g_{t_1} - M(t_1) &= -\sum_{j=2}^{k} \Lambda_j \phi_{jt_1}, \\ \lambda_0 g_{x_1^i} + \lambda_i(t_1) &= -\sum_{j=2}^{k} \Lambda_j \phi_{jx_1^i}, \qquad i = 1, \ldots, n, \\ \lambda_0 g_{t_2} + M(t_2) &= -\sum_{j=2}^{k} \Lambda_j \phi_{jt_2}, \\ \lambda_0 g_{x_2^i} - \lambda_i(t_2) &= -\sum_{j=2}^{n} \Lambda_j \phi_{jx_2^i}, \qquad i = 1, \ldots, n. \end{aligned}$$

By writing Λ_1 for λ_0, and ϕ_1 for g, these relations become

$$\text{(4.4.4)} \qquad \begin{aligned} \lambda_i(t_1) &= -\sum_{j=1}^{k} \Lambda_j \phi_{jx_1^i}(\eta), \qquad & \lambda_i(t_2) &= \sum_{j=1}^{k} \Lambda_j \phi_{jx_2^i}(\eta), \\ M(t_1) &= \sum_{j=1}^{k} \Lambda_j \phi_{jt_1}(\eta), \qquad & M(t_2) &= -\sum_{j=1}^{k} \Lambda_j \phi_{jt_2}(\eta), \end{aligned}$$

where $\eta = e[x] = (t_1, x_1, t_2, x_2)$.

In the situation under consideration, the transversality relation (P4) yields the following statement: There is a k-vector $\Lambda = (\Lambda_1, \ldots, \Lambda_k)$, nonzero with $\Lambda_1 \geq 0$, such that the relations (4.4.4) hold at $t = t_1$ and $t = t_2$.

B. The Transversality Relation for Unilateral Constraints

We consider here the case where B is defined by a number of equalities and inequalities, say

$$\phi_j(t_1, x_1, t_2, x_2) = 0, \qquad j = 2, \ldots, k',$$
$$\phi_j(t_1, x_1, t_2, x_2) \geq 0, \qquad j = k' + 1, \ldots, k.$$

Let $x(t)$, $t_1 \leq t \leq t_2$, be the given optimal trajectory and let $e[x] = (t_1, x(t_1), t_2, x(t_2))$ as usual. Then $\phi_j(e[x]) = 0$ for $j = 1, \ldots, k'$, and $\phi_j(e[x]) \geq 0$ for $j = k' + 1, \ldots, k$. Some of the last relations may be satisfied with an equal sign. By a relabeling we may assume that

$$\phi_j(e[x]) = 0 \quad \text{for } j = 1, \ldots, k', k' + 1, \ldots, k'',$$
$$\phi_j(e[x]) > 0 \quad \text{for } j = k'' + 1, \ldots, k, \qquad 0 \leq k' \leq k'' \leq k.$$

It is often said that the constraints $\phi_j, j = k' + 1, \ldots, k''$, are active, or briefly that the indices $j = k' + 1, \ldots, k''$ are active for the trajectory x. As in Subsection A we denote by $d\phi_j$ the differential of ϕ_j at $e[x]$. Then B' is the set of all dt_1, dx_1, dt_2, dx_2 satisfying

$$d\phi_j = 0 \quad \text{for } j = 1, \ldots, k', \qquad d\phi_j \geq 0 \quad \text{for } j = k' + 1, \ldots, k'',$$

the remaining $\phi_j, j = k'' + 1, \ldots, k$, having no relevance on B'. Here B' is a convex cone and the transversality relation in the form mentioned in Section 4.2, Remark 10, yields $dT - X_0 = 0$ with $X_0 \geq 0$ for all $dt_1, dx_1^i, dt_2, dx_2^i$ satisfying $d\phi_j = 0$ for $j = 1, \ldots, k'$ and $d\phi_j - X_j = 0$ with $X_j \geq 0$ for $j = k' + 1, \ldots k''$. Then as in Subsection A, there are $k'' - 1$ numbers $\Lambda_2, \ldots, \Lambda_{k''}$ such that relations (4.4.3) hold and $-X_0 = -\sum_{j=k'+1}^{k''} \Lambda_j(-X_j)$, that is, $X_0 = -\sum_{j=k'+1}^{k''} \Lambda_j X_j$, and since $X_0 \geq 0$ and the X_j can take any nonnegative value, the multipliers $\Lambda_j, j = k' + 1, \ldots, k''$, must be nonpositive. By writing Λ_1 for λ_0, and ϕ_1 for g as in Section A, we obtain relations (4.4.4).

In the situation under consideration, the transversality relation of Section 4.2C, Remark 10, yields the following statement: there is a k''-vector $(\Lambda_1, \ldots, \Lambda_{k''})$, non zero and with $\Lambda_1 \geq 0$, $\Lambda_j \leq 0$ for $j = k' + 1, \ldots, k''$, such that the relations (4.4.4) hold at $t = t_1$ and $t = t_2$.

EXAMPLE 1. Take $n = 2$, t_1, $x_1 = (x_1^1, x_1^2)$, t_2 fixed, $g = g(x_2^2)$, and further constraints: $x_2 = (x_2^1, x_2^2)$ with $x_2^1 = 0$, $x_2^2 \geq a$ for some fixed a. Then B is the half straight line $x_2^2 \geq a$. If the second end point of the trajectory $x(t)$, $t_1 \leq t \leq t_2$, is $x(t_2) = (0, a)$, then $B' = (dx_2^1, dx_2^2)$ is the cone $dx_2^2 \geq 0$. The transversality relation is here $(\lambda_0 g_{x_2^2} - \lambda_2(t_2))\, dx_2^2 \geq 0$ for all $dx_2^2 \geq 0$, and we obtain directly $\lambda_0 g_{x_2^2} - \lambda_2(t_2) \geq 0$. The rule stated above yields the same result.

EXAMPLE 2. Take $n = 2$, t_1, $x_1 = (x_1^1, x_1^2)$, t_2 fixed $g = g(x_2^1, x_2^2)$, and further constraint: $x_2 = (x_2^1, x_2^2)$ with $x_2^2 \geq a$ for some fixed a. Then B is the half plane x_2^1 arbitrary, $x_2^2 \geq 0$. If the second end point of the trajectory $x(t)$, $t_1 \leq t \leq t_2$, is $x(t_2) = (x^1(t_2), a)$, then $B' = (dx_2^1, dx_2^2)$ is made up of all dx_2^1 arbitrary, and $dx_2^2 \geq 0$. The transversality relation is now

$$(\lambda_0 g_{x_2^1} - \lambda_1(t_2))\, dx_2^1 + (\lambda_0 g_{x_2^2} - \lambda_2(t_2))\, dx_2^2 \geq 0.$$

By taking first $dx_2^2 = 0$ and dx_2^1 arbitrary, and then $dx_2^1 = 0$ and $dx_1^2 \geq 0$ arbitrary, we obtain directly

$$\lambda_0 g_{x_2^1} - \lambda_1(t_2) = 0, \qquad \lambda_0 g_{x_2^2} - \lambda_2(t_2) \geq 0.$$

The rule stated above yields the same result.

C. The Transversality Relation in Specific Cases

(a) Let us consider the case of the first end point $1 = (t_1, x_1)$ being fixed, $x_1 = (x_1^1, \ldots, x_1^n)$, and the second end point x_2 also fixed in R^n, $x_2 = (x_2^1, \ldots, x_2^n)$, to be reached at some undetermined time $t_2 \geq t_1$ (as in the example of Section 4.2B). We certainly have $dt_1 = 0$, $dx_1 = 0$ (that is, $dx_1^i = 0$, $i = 1, \ldots, n$), $dx_2 = 0$ (that is, $dx_2^i = 0$, $i = 1, \ldots, n$). If we assume $t_2 > t_1$, then dt_2 is arbitrary. Here $B = (t_1, x_1) \times (t \geq t_1) \times (x_2)$ (a half straight line, dimension $k = 1$), and $B' = (0, 0, dt_2, 0)$ (that is, $B' = R$). (See illustration where we have taken $n = 2$.) For a problem of minimum time we have $I[x, u] = g = t_2$, and hence $dg = dt_2$, or

$$g_{t_1} = 0, \quad g_{t_2} = 1, \quad g_{x_1^i} = 0, \quad g_{x_2^i} = 0,$$

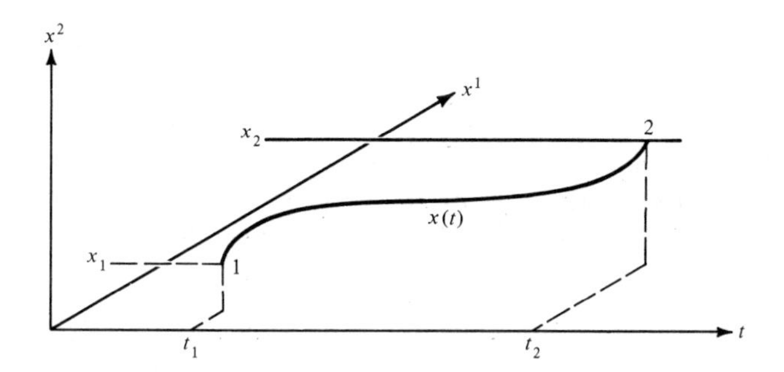

$i = 1, \ldots, n$. The transversality relation (P4) of Section 4.2, or formula (4.2.1), reduces now to $(\lambda_0 + M(t_2))\, dt_2 = 0$, where dt_2 is arbitrary. We have then the finite relation

$$\lambda_0 + M(t_2) = 0,$$

or $M(t_2) = -\lambda_0 \leq 0$, which must be satisfied at t_2 together with the equations $x^i(t_2) = x_2^i$, $i = 1, \ldots, n$.

(b) Let us consider the case of the first end point $1 = (t_1, x_1)$ being fixed, $x_1 = (x_1^1, \ldots, x_1^n)$, and the second end point $2 = (t_2, x(t_2))$, $x(t_2) = (x^1(t_2), \ldots, x^n(t_2))$, being on a given curve $\Gamma: x = b(t)$, $b(t) = (b_1(t), \ldots, b_n(t))$, $t' \leq t \leq t''$; hence $x(t_2) = b(t_2)$, or $x^i(t_2) = b_i(t_2)$, $i = 1, \ldots, n$, with $t' \leq t_2 \leq t''$, $t_2 \geq t_1$. (See illustration, where we have taken $n = 2$.) This is the problem of a material point P in R^n starting at $x = x_1$ at time $t = t_1$, required to hit a moving target point Q during a time interval $[t', t'']$.

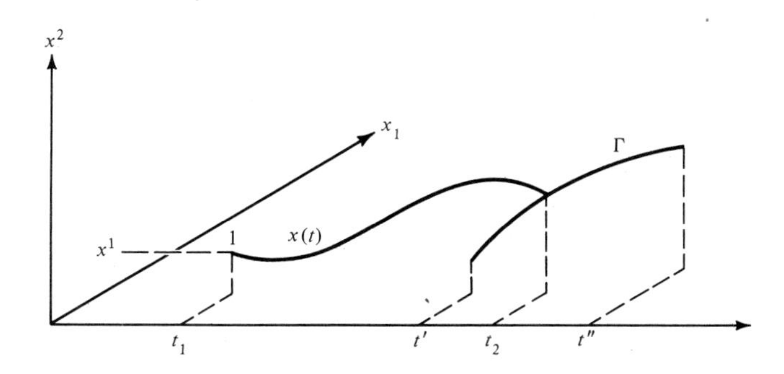

The next illustration is to be thought of as depicting the actual situation in R^{n+1}. Here $b(t)$ then represents the known law of motion of the point Q during the time interval $[t', t'']$, and $x(t)$ the law of motion of the point P from the time t_1 to the time t_2 when P hits Q.

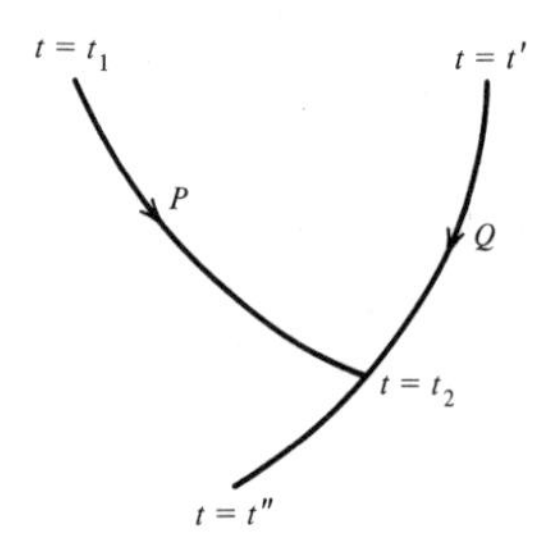

Let us assume that $t_2 > t_1$, that 2 is not an end point of Γ (that is, $t' < t_2 < t''$), and that b is differentiable at t_2..Here $dt_1 = 0$, and $dx_1 = 0$ (that is, $dx_1^i = 0$, $i = 1, \ldots, n$). Also, Γ has a tangent line l at $(t_2, b(t_2))$ of slope $b'(t_2) = (b_1'(t_2), \ldots, b_n'(t_2))$; hence, $dx_2 = b'(t_2)\,dt_2$, that is, $dx_2^i = b_i'(t_2)\,dt_2$, $i = 1, \ldots, n$. Here $B = (t_1, x_1) \times \Gamma$, $B' = (0, 0, dt_2, b'(t_2)\,dt_2)$, and dt_2 is arbitrary. Finally, $I = g = g(t_2, x_2) = g(t_2, x_2^1, \ldots, x_2^n)$, and

$$dg = g_{t_2}\,dt_2 + \sum_1^n g_{x_2^i}\,dx_2^i, \quad dx_2^i = b_i'(t_2)\,dt_2, \qquad i = 1, \ldots, n.$$

Then the transversality relation (P4) of Section 4.2, formula (4.2.1), reduces to

$$\left[\lambda_0\left(g_{t_2} + \sum_i g_{x_2^i} b_i'(t_2)\right) + M(t_2) - \sum_i \lambda_i(t_2) b_i'(t_2)\right] dt_2 = 0,$$

where dt_2 is arbitrary. Hence, we have the finite relation

$$\lambda_0\left(g_{t_2} + \sum_i g_{x_2^i} b_i'(t_2)\right) + M(t_2) - \sum_i \lambda_i(t_2) b'(t_2) = 0. \tag{4.4.5}$$

If the problem is a problem of minimum time, that is, $g = t_2$, $I[x, u] = t_2$, then $dg = dt_2$, or $g_{t_2} = 1$, $g_{x_2^i} = 0$, $i = 1, \ldots, n$, and the relation (4.4.5) reduces to

$$\lambda_0 + M(t_2) - \sum_i \lambda_i(t_2) b'(t_2) = 0. \tag{4.4.6}$$

(c) Let us consider the case of the first end point $1 = (t_1, x_1)$ being fixed, $x_1 = (x_1^1, \ldots, x_2^n)$, and the second end point x_2 on a given fixed set S in R^n (target), $x_2 = (x_2^1, \ldots, x_2^n)$, to be reached at some indeterminate time $t_1 \le t' \le t_2 \le t''$. Let us consider first the case, say (c1), where $S = R^n$ is the whole x-space, and assume $t' < t_2 < t''$, t_2 undetermined. (In the illustration we have taken $n = 2$.) Then, we have $dt_1 = 0$, $dx_1 = 0$ (that is, $dx_1^i = 0$, $i = 1, \ldots, n$), while dt_2, dx_2 are arbitrary, $dx_2 = (dx_2^1, \ldots, dx_2^n)$—that is, $dt_2, dx_2^1, \ldots, dx_2^n$ are all arbitrary. Here $B = (t_1, x_1) \times (t_2 \ge t_1) \times R^n$ (dimension $k = n + 1$), and $B' = (0, 0, dt_2, dx_2)$. Then the transversality relation (P4) of Section 4.2 reduces to

$$(\lambda_0 g_{t_2} + M(t_2))\,dt_2 + \sum_{j=1}^n (\lambda_0 g_{x_2^j} - \lambda_j(t_2))\,dx_2^j = 0. \tag{4.4.7}$$

By taking dt_2, dx_2^j, $j = 1, \ldots, n$, all equal to zero but one which is left arbitrary, we conclude that the following $n + 1$ (finite) equations must hold:

(4.4.8) $$\lambda_0 g_{t_2} + M(t_2) = 0, \qquad \lambda_0 g_{x_2^i} - \lambda_i(t_2) = 0, \qquad i = 1, \ldots, n.$$

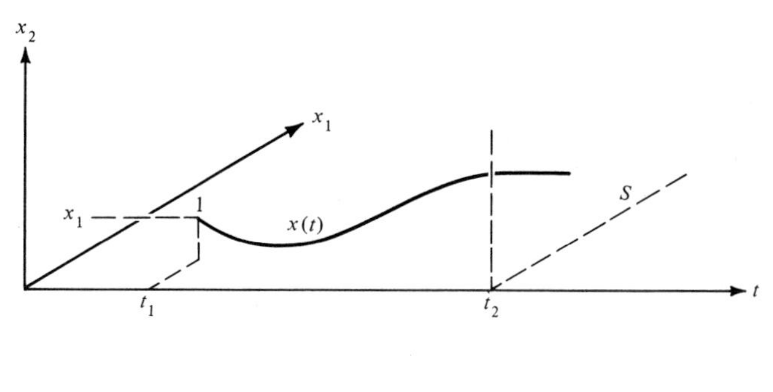

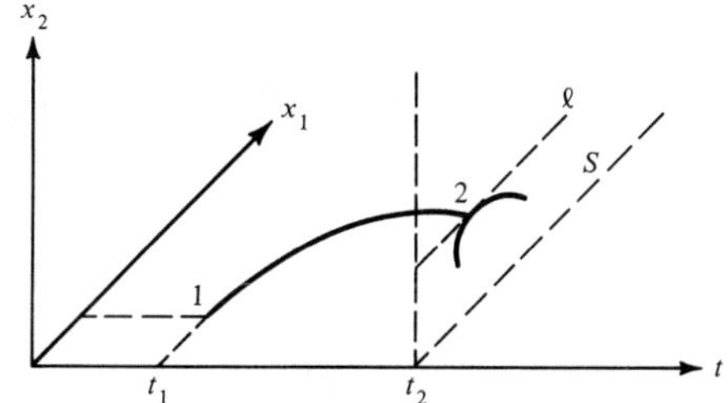

Let us consider now the case, say (c2), where S is a given curve in R^n, say $S: x = b(\tau)$, $\tau' \le \tau \le \tau''$, τ a parameter, $b(\tau) = (b_1, \ldots, b_n)$, and assume that x hits the target S at a time $t_2 > t_1$ (as above) and at a point x_2 which is not an end point of S: precisely, $x_2 = x(t_2) = b(\tau)$ for some $\tau = \tau_0$, $\tau' < \tau_0 < \tau''$. Then, as before, $dt_1 = 0$, $dx_1 = 0$ (or $dx_1^i = 0$, $i = 1, \ldots, n$), dt_2 arbitrary, and now $dx_2^i = b_i'(\tau)\,d\tau$, $d\tau$ arbitrary. Here $B = (t_1, x_1) \times (t_2 \ge t_1) \times S$ (dimension $k = 2$), and $B' = (0, 0, dt_2, b'(\tau)\,d\tau)$, dt_2, $d\tau$ arbitrary. The transversality relation (P4) of Section 4.2 reduces to

$$(\lambda_0 g_{t_2} + M(t_2))\,dt_2 + \left[\sum_{j=1}^{n} (\lambda_0 g_{x_2^j} - \lambda_j(t_2))b_j'(\tau)\right] d\tau = 0,$$

where dt_2, $d\tau$ are both arbitrary. By taking either dt_2 or $d\tau$ equal to zero and leaving the other one arbitrary, we conclude that the following two (finite) equations must hold:

(4.4.9) $$\lambda_0 g_{t_2} + M(t_2) = 0, \qquad \sum_{j=1}^{n} (\lambda_0 g_{x_2^j} - \lambda_j(t_2))b_j'(\tau) = 0.$$

The second relation states that the n-vector $\lambda_0 g_{x_2^i} - \lambda_i(t_2)$, $i = 1, \ldots, n$, is orthogonal at 2 to the tangent l to the curve S at the same point.

If the problem is a problem of minimum time, then $I = g = t_2$, $dg = dt_2$, or $g_{t_2} = 1$, $g_{x_2^i} = 0$, $i = 1, \ldots, n$, and (4.4.9) reduces to

(4.4.10) $$\lambda_0 + M(t_2) = 0, \qquad \sum_{j=1}^{n} \lambda_j(t_2)b_j'(\tau) = 0.$$

Let us consider now the case, say (c3), where S is a given k-dimensional manifold in R^n. If we assume that S is given parametrically as before, say, $S{:}x = b(\tau)$, $b(\tau) = (b_1, \ldots, b_n)$, $\tau = (\tau_1, \ldots, \tau_k)$, then $x_2 = x(t_2) = b(\tau)$ for some $\tau = \tau_0 = (\tau_{10}, \ldots, \tau_{k0})$. We shall assume that $b(\tau)$ is defined in a neighborhood V of τ_0, and that $b(\tau)$ is of class C^1 in V. Then, as before, $dt_1 = 0$, $dx_1 = 0$ (or $dx_1^i = 0$, $i = 1, \ldots, n$), dt_2 arbitrary, and

$$dx_2^i = \sum_{s=1}^{k} b_{i\tau_s}(\tau)\, d\tau_s, \qquad i = 1, \ldots, n, \tag{4.4.11}$$

where $d\tau_1, \ldots, d\tau_k$ are arbitrary. Here $B = (t_1, x_1) \times (t \geq t_1) \times S$ (dimension $k + 1$), and $B' = (0, 0, dt_2, dx_2)$, with dt_2 arbitrary, $dx_2 = (dx_2^1, \ldots, dx_2^n)$, dx_2^i given by (4.4.11), and $d\tau_1, \ldots, d\tau_k$ arbitrary. The transversality relation (P4) of Section 4.2 is now (4.4.7) with dx_2^i given by (4.4.11); hence

$$(\lambda_0 g_{t_2} + M(t_2))\, dt_2 + \sum_{s=1}^{k} \left[\sum_{j=1}^{n} (\lambda_0 g_{x_2^j} - \lambda_j(t_2)) b_{j\tau_s}(\tau) \right] d\tau_s = 0,$$

with dt_2, $d\tau_1, \ldots, d\tau_k$ arbitrary. This yields $k + 1$ (finite) equations

$$\lambda_0 g_{t_2} + M(t_2) = 0, \qquad \sum_{j=1}^{n} (\lambda_0 g_{x_2^j} - \lambda_j(t_2)) b_{j\tau_s}(\tau) = 0, \quad s = 1, \ldots, k. \tag{4.4.12}$$

Again the last k relations state that the n-vector $\lambda_0 g_i - \lambda_i(t_2)$, $i = 1, \ldots, n$, is orthogonal at 2 to the manifold S.

In the particular case in which $I = g = t_2$ (problem of minimum time), then $dg = dt_2$, or $g_{t_2} = 1$, $g_{x_2^i} = 0$, $i = 1, \ldots, n$, and (4.4.12) reduces to

$$\lambda_0 + M(t_2) = 0, \qquad \sum_{j=1}^{n} \lambda_j(t_2) b_{j\tau_s}(\tau) = 0, \quad s = 1, \ldots, k. \tag{4.4.13}$$

Let us consider now the case, say (c4), where S is given by $n - k$ equations $\beta_\sigma(x) = 0$, $\sigma = 1, \ldots, n - k$, and then $dx_2 = (dx_2^1, \ldots, dx_2^n)$ denotes any arbitrary vector satisfying

$$\sum_j \beta_{\sigma x^j}(x_2)\, dx_2^j = 0, \qquad \sigma = 1, \ldots, n - k. \tag{4.4.14}$$

Then (P4) again has the form (4.4.7), where dt_2 is arbitrary and $dx_2^1, \ldots, dx_2^n$ satisfy (4.4.14). As before, we conclude that $\lambda_0 g_{t_2} + M(t_2) = 0$, and that

$$\sum_{j=1}^{n} (\lambda_0 g_{x_2^j} - \lambda_j(t_2))\, dx_2^j = 0$$

for all $dx_2^1, \ldots, dx_2^n$ satisfying (4.4.14). The latter states again that the n-vector $\lambda_0 g_{x_2^j} - \lambda_j(t_2)$, $j = 1, \ldots, n$, is orthogonal at 2 to S.

In the particular case where $I = g = t_2$ (problem of minimum time), we have

$$\lambda_0 + M(t_2) = 0, \qquad \sum_{j=1}^{n} \lambda_j(t_2)\, dx_2^j = 0,$$

for all $dx_2^1, \ldots, dx_2^n$ satisfying (4.4.14).

(d) Let us consider the case of the first end point $1 = (t_1, x_1)$ being fixed, the second end point $2 = (t_2, x(t_2))$ being on a "moving target". We consider only the case where the "moving target" is represented by $n - k$ equations:

$$\beta_\sigma(t, x) = 0, \qquad \sigma = 1, \ldots, n - k. \tag{4.4.15}$$

In other words, we require that at time t_2 the $n-k$ equations

(4.4.16) $$\beta_\sigma(t_2, x(t_2)) = 0, \qquad \sigma = 1, \ldots, n-k,$$

are satisfied. If we denote by $S(t)$ the set of points $x \in R^n$ satisfying the $n-k$ equations (4.4.15), then the requirement can be written in the form $x(t_2) \in S(t_2)$, and this variable set $S(t)$ is the "moving target". We assume that the functions $\beta_\sigma(t, x)$ are of class C^1, that $t_2 > t_1$, and that the point $(t_2, x(t_2))$ is in the interior of the region $R \subset R^{n+1}$ where the functions $b_\sigma(t, x)$ are defined. Then the equations above yield

(4.4.17) $$\beta_{\sigma t_2}(t_2, x_2)\, dt_2 + \sum_{i=1}^{n} \beta_{\sigma x_2^i}(t_2, x_2)\, dx_2^i = 0, \qquad \sigma = 1, \ldots, n-k,$$

where $x_2 = x(t_2)$, and $\beta_{\sigma t}$, $\beta_{\sigma x_2^i}$ denote the partial derivatives of $\beta_\sigma(t_2, x_2)$ with respect to t_2 and x_2^i. The transversality relation (P4) of Section 4.2, formula (4.2.1), becomes

(4.4.18) $$(\lambda_0 g_{t_2} + M(t_2))\, dt_2 + \sum_{i=1}^{n} (\lambda_0 g_{x_2^i} - \lambda_i(t_2))\, dx_2^i = 0$$

for all systems of numbers $(dt_2, dx_2^i, i = 1, \ldots, n)$ satisfying (4.4.17). In particular, for $I = g = t_2$ (problem of minimum time), we have $dg = dt_2$, or $g_{t_2} = 1$, $g_{x_2^i} = 0$, $i = 1, \ldots, n$, and (4.4.18) reduces to

(4.4.19) $$(\lambda_0 + M(t_2))\, dt_2 - \sum_{i=1}^{n} \lambda_i(t_2)\, dx_2^i = 0.$$

D. Exercise

Derive the transversality relation in finite form for the Mayer problem, with first end point $1 = (t_1, x_1)$ fixed, terminal time t_2 fixed, $t_2 > t_1$, and second end point x_2 on a given set B in R^n (target). Consider the same cases (c1)–(c4).

Ans.: Same as under (c), but with no conclusion on $M(t_2)$.

4.5 The Value Function

We consider now the problem of transferring any given fixed point (t_1, x_1) in a set V to a target set $B \subset V$ by means of admissible trajectories $x(t)$, $t_1 \le t \le t_2$, in V, that is, $x(t_1) = x_1$, $(t_2, x(t_2)) \in B$, $(t, x(t)) \in V$. We want to minimize $g(t_2, x(t_2))$, where g is thought of as a continuous real valued function $g(t, x)$ in B.

For any $(t_1, x_1) \in V$ we denote by $\Omega_{t_1 x_1}$ the family of all admissible pairs $x(t)$, $u(t)$, $t_1 \le t \le t_2$, transferring (t_1, x_1) to B in V, and by $\Sigma_{t_1 x_1}$ the subset of all points $(t_2, x(t_2)) \in B$ which are terminal points of trajectories x in $\Omega_{t_1 x_1}$. For every $(t_1, x_1) \in V$ we take now

(4.5.1) $$\omega(t_1, x_1) = \inf_{(x,u) \in \Omega_{t_1 x_1}} g(t_2, x(t_2)) = \inf_{(t_2, x_2) \in \Sigma_{t_1 x_1}} g(t_2, x_2).$$

The function $\omega(t_1, x_1)$ is called the *value function* in V, and may have the value $-\infty$. Also, $\omega(t_1, x_1) = +\infty$ whenever $\Omega_{t_1 x_1}$ is empty. Moreover, $\omega(t, x) = g(t, x)$ for all $(t, x) \in B$.

4.5.i. *For every admissible trajectory $x(t)$, $t_1 \le t \le t_2$, transferring any point $(t_1, x(t_1))$ to B in V, the function $\omega(t, x(t))$ is not decreasing in $[t_1, t_2]$.*

Proof. For any two points $t_1 \le \tau_1 \le \tau_2 \le t_2$, and $\xi_1 = x(\tau_1)$, $\xi_2 = x(\tau_2)$, we consider any admissible trajectory $\bar{x}(t)$, $\tau_2 \le t \le t_2^*$, transferring (τ_2, ξ_2) to B in V. This can be extended to $[\tau_1, t_2^*]$ by taking $\bar{x}(t) = x(t)$ for $[\tau_1, \tau_2]$, and now $\bar{x}$ transfers (τ_1, ξ_1) to B in V. Let $\Omega'_{\tau_2 \xi_2}$ denote the class of the trajectories in $\Omega_{\tau_2 \xi_2}$, each trajectory increased, or augmented, by the fixed arc $\xi_1 \xi_2$. We have proved that $\Omega_{\tau_1 \xi_1} \supset \Omega_{\tau_2 \xi_2}$ and $\Sigma_{\tau_1 \xi_1} \supset \Sigma_{\tau_2 \xi_2}$. By the definition (4.5.1) we derive that $\omega(\tau_1, \xi_1) \le \omega(\tau_2, \xi_2)$. □

4.5.ii. *If $(t_1, x_1) \in V$, if $x(t)$, $t_1 \le t \le t_2$, is any admissible trajectory transferring (t_1, x_1) to B in V, $x(t_1) = x_1$, $(t_2, x(t_2)) \in B$, $(t, x(t)) \in V$, and if x is optimal for (t_1, x_1), then $\omega(t, x(t))$ is constant on $[t_1, t_2]$. Moreover, the trajectory x is optimal for each of its points $(\bar{t}, x(\bar{t}))$.*

Proof. Indeed, $\omega(t, x(t))$ is not decreasing on $[t_1, t_2]$, by (4.5.i). On the other hand, if x is optimal, then $\omega(t_1, x(t_1)) = \omega(t_2, x(t_2)) = \inf[g(t^*, x^*) | (t^*, x^*) \in \Sigma_{t_1 x_1}]$. Hence $\omega(t, x(t))$ is constant on $[t_1, t_2]$. Conversely, if $\omega(t, x(t))$ is constant on $[t_1, t_2]$, then $\omega(t_1, x(t_1)) = \omega(t_2, x(t_2))$, $(t_2, x(t_2)) \in B$, $\omega(t_1, x(t_1)) = \inf[g(t^*, x^*) | (t^*, x^*) \in \Sigma_{t_1 x_1}]$, that is, x is optimal. □

From now on in this section we assume that (a) any point $(t_1, x_1) \in V$ can be transferred to B in V by some admissible pair $x(t)$, $u(t)$, $t_1 \le t \le t_2$, $u(t)$ sectionally continuous, $x(t_1) = x_1$, $(t_2, x(t_2)) \in B$, $u(t) \in U$, $(t, x(t)) \in V$; and (b) U is a fixed subset of R^m.

4.5.iii. *Under hypotheses* (a), (b), *if (t, x) is any interior point of V where the value function $\omega(t, x)$ is differentiable, then*

$$\omega_t(t, x) + \sum_{i=1}^{n} \omega_{x^i}(t, x) f_i(t, x, v) \ge 0$$

for all $v \in U$. Moreover, if there is an optimal pair $x^*(s)$, $u^*(s)$, $t \le s \le t_2^*$, $x^*(t) = x$, $(t_2^*, x^*(t_2^*)) \in B$, $(s, x^*(s)) \in V$, $u^*(s)$ sectionally continuous, then

$$\min_{v \in U} \left[\omega_t(t, x) + \sum_{i=1}^{n} \omega_{x^i}(t, x) f_i(t, x, v) \right] = 0, \tag{4.5.2}$$

and the minimum is attained by $v = u^*(t + 0)$.

Proof. For any fixed $v \in U$ let us consider the trajectory issuing from x with constant strategy $u(s) = v$, defined by the differential system $x'(s) = f(s, x(s), v)$, $t \le s \le t + \varepsilon$, with $x(t) = x$, and $\varepsilon > 0$ sufficiently small so that the arc $x(s)$,

$t \le s \le t + \varepsilon$, is completely contained in a neighborhood of (t, x) interior to V. If $(t + \varepsilon, x + \sigma)$ is the terminal point of this arc, then we can transfer this point to B in V by some admissible pair $x(s)$, $u(s)$, $t + \varepsilon \le s \le t^*$. Now the pair $x(s)$, $u(s)$, $t \le s \le t^*$, transfers (t, x) to B in V, and by (4.5.i), $\omega(t, x)$ is monotone nondecreasing along $x(s)$. Hence

$$0 \le d\omega(s, x(s))/ds|_{s=t+0} = \omega_t(t, x) + \sum_{i=1}^{n} \omega_{x^i}(t, x) f_i(t, x, v) \ge 0.$$

If there is an optimal pair x^*, u^* as stated, then $\omega(t, x)$ is constant along x^* and the same expression is zero. This proves (4.5.iii). □

4.5.iv. *Under hypotheses* (a) *and* (b), *if all points $(t, x) \in V$ can be optimally transferred to B in V by an admissible optimal pair $x(s)$, $u(s)$, $t \le s \le t_2$, if there is a function $p(t, x)$ in $V - B$, continuous in $V - B$, such that $u(s) = p(s, x(s))$ for any such pair, and if the value function ω is continuous in V and of class C^1 in $V - B$ ($\omega = g$ on B), then $\omega(t, x)$ satisfies in $V - B$ the partial differential equation*

$$\omega_t(t, x) + \sum_{i=1}^{n} \omega_{x^i}(t, x) f_i(t, x, p(t, x)) = 0,$$

or

$$\omega_t(t, x) + H(t, x, p(t, x), \omega_x(t, x)) = 0. \tag{4.5.3}$$

This is a corollary of (4.5.iii).

4.5.v. *Under hypotheses* (a), (b), *let (t_1, x_1) be an interior point of V, let $x^*(t)$, $u^*(t)$, $t_1 \le t \le t_2$, be an optimal pair, transferring (t_1, x_1) to B in V, and such that the entire trajectory $(t, x^*(t))$ is made up of points interior to V, except perhaps the terminal point $(t_2, x^*(t_2)) \in B$. Suppose that the value function $\omega(t, x)$ is continuous on V and of class C^2 on $V - B$. Then the functions $\lambda_i(t) = \omega_{x^i}(t, x(t))$, $t_1 \le t \le t_2$, $i = 1, \ldots, n$, satisfy the relations $d\lambda_i/dt = -H_{x^i}(t, x^*(t), u^*(t), \lambda(t))$ and $H(t, x^*(t), u^*(t), \lambda(t)) \le H(t, x^*(t), u, \lambda(t))$ for all $u \in U$ and $t \in [t_1, t_2]$* (a.e.). *In other words, $\lambda(t) = (\lambda_1, \ldots, \lambda_n)$ are multipliers as in* (4.2.i).

Proof. Here (t_1, x_1) is an interior point of V, and $x^*(t)$, $u^*(t)$, $t_1 \le t \le t_2$, is an optimal pair transferring (t_1, x_1) to B in V, and with trajectory x^* entirely in the interior of V, except perhaps the terminal point $(t_2, x^*(t_2))$. Thus, $(d/dt)x^{i*}(t) = f_i(t, x^*(t), u^*(t))$, $t \in [t_1, t_2]$ (a.e.). Here the value function $\omega(t, x)$ is twice continuously differentiable in $V - B$. Thus, at each point t of continuity for u^*, then x^* has a continuous derivative, and by the chain rule of calculus we have

$$(d/dt)\omega_{x^j}(t, x^*(t)) = \omega_{tx^j}(t, x^*(t)) + \sum_{i=1}^{n} \omega_{x^i x^j}(t, x^*(t)) f_i(t, x^*(t), u^*(t)). \tag{4.5.4}$$

From (4.5.iii) we know that

$$\omega_t(t,x) + \sum_{i=1}^{n} \omega_{x^i}(t,x) f_i(t,x,u) \geq 0$$

for all $(t,x) \in V$, $u \in U$. In particular

$$\omega_t(t,x) + \sum_{i=1}^{n} \omega_{x^i}(t,x) f_i(t,x,u^*(t)) \geq 0, \tag{4.5.5}$$

$$\omega_t(t,x^*(t)) + \sum_{i=1}^{n} \omega_{x^i}(t,x^*(t)) f_i(t,x^*(t),u^*(t)) = 0, \tag{4.5.6}$$

$$\omega_t(t,x^*(t)) + \sum_{i=1}^{n} \omega_{x^i}(t,x^*(t)) f_i(t,x^*(t),u) \geq 0 \tag{4.5.7}$$

for all $u \in U$. In other words, for t fixed, $x = x^*(t)$ is a minimum for the expression (4.5.5) thought of as a function of x. Thus, the n first order partial derivatives of (4.5.5) with respect to x^j, $j = 1, \ldots, n$, are zero at $x = x^*(t)$, or

$$\omega_{tx^j}(t,x^*(t)) + \sum_{i=1}^{n} \omega_{x^i x^j}(t,x^*(t)) f_i(t,x^*(t),u^*(t))$$

$$+ \sum_{i=1}^{n} \omega_{x^i}(t,x^*(t)) f_{ix^j}(t,x^*(t),u^*(t)) = 0.$$

By comparison with (4.5.4) we derive

$$\begin{aligned}(d/dt)\omega_{x^j}(t,x^*(t)) &= -\sum_{i=1}^{n} \omega_{x^i}(t,x^*(t)) f_{ix^j}(t,x^*(t),u^*(t)) \\ &= -H_{x^j}(t,x^*(t),u^*(t),\omega_x(t,x^*(t))),\end{aligned}$$

where $\omega_x = (\omega_{x^1}, \ldots, \omega_{x^n})$. Moreover, by comparison of (4.5.6) and (4.5.7) we derive that

$$H(t,x^*(t),u,\omega_x(t,x^*(t)) \geq H(t,x^*(t),u^*(t),\omega_x(t,x^*(t)))$$

for all $u \in U$. This proves (4.5.v). □

4.6 Sufficient Conditions

A. Sufficient Conditions for a Single Trajectory

We begin by showing that statement (4.5.iii) has a converse.

4.6.i (A SUFFICIENT CONDITION FOR OPTIMALITY FOR A SINGLE TRAJECTORY). *Let $w(t,x)$ be a function on V such that $w(t,x) = g(t,x)$ for $(t,x) \in B \subset V$, and let (t_1,x_1) be a point of V such that for every trajectory $x(t), t_1 \leq t \leq t_2$, in*

$\Omega_{t_1 x_1}$, $w(t, x(t))$ is finite and nondecreasing, and for some pair $x^*(t)$, $u^*(t)$, $t_1 \leq t \leq t_2^*$, in $\Omega_{t_1 x_1}$, $w(t, x^*(t))$ is constant on $[t_1, t_2^*]$. Then the pair x^*, u^* is optimal for (t_1, x_1) in V.

Proof. If Σ is the set of points of B reachable from (t_1, x_1), then $g(t_2^*, x^*(t_2^*)) = w(t_2^*, x^*(t_2^*)) = w(t_1, x_1) \leq w(t_2, x_2) = g(t_2, x(t_2))$ for every pair $x(t)$, $u(t)$, $t_1 \leq t \leq t_2$ of $\Omega_{t_1 x_1}$. Thus, $g(t_2^*, x(t_2^*))$ is the minimum of g on $\sum_{t_1 x_1}$, that is, x^*, u^* is optimal. □

4.6.ii (A Sufficient Condition for a Single Trajectory). *Let $w(t, x)$ be a given function in V with $w(t, x) = g(t, x)$ for $(t, x) \in B$, $w(t, x)$ continuous in V, continuously differentiable in $V - B$, satisfying*

$$w_t(t, x) + \sum_{i=1}^{n} w_{x^i}(t, x) f_i(t, x, u) \geq 0$$

for all $(t, x) \in V - B$ and $u \in U$. If $x^*(t)$, $u^*(t)$, $t_1 \leq t \leq t_2^*$, is an admissible pair transferring (t_1, x_1) to B in V, and

$$w_t(t, x^*(t)) + \sum_{i=1}^{n} w_{x^i}(t, x^*(t)) f_i(t, x^*(t), u^*(t)) = 0, \qquad t \in [t_1, t_2^*] \text{ (a.e.)},$$

(in other words, $w(t, x^*(t))$, $t_1 \leq t \leq t_2^*$, is constant), then (x^*, u^*) is optimal.

This is a corollary of (4.6.i).

We consider now the Hamiltonian for the Mayer problem under consideration, $H(t, x, u, \lambda) = \sum_{i=1}^{n} \lambda_i f_i(t, x, u)$. Let $x(t)$, $u(t)$, $t_1 \leq t \leq t_2$, be an admissible pair transferring the point $(t_1, x_1) \in V$ to the target B in V; that is, $x(t_1) = x_1$, $(t_2, x(t_2)) \in B$, $(t, x(t)) \in V$.

4.6.iii. *If $w(t, x)$ is a continuous function in V, of class C^1 in $V - B$, with $w(t, x) = g(t, x)$ on B, if $x(t)$, $u(t)$, $t_1 \leq t \leq t_2$, is an admissible pair in V, $x(t_1) = x_1$, $(t_2, x(t_2)) \in B$, $(t, x(t)) \in V$, and if $w(t, x(t))$ is constant on $[t_1, t_2]$, then*

$$w_t(t, x(t)) + H(t, x(t), u(t), w_x(t, x(t))) = 0, \qquad t \in [t_1, t_2] \text{ (a.e.)}.$$

Proof. Since w is of class C^1 in $V - B$ and continuous in V, we have

$$\begin{aligned} 0 = dw(t, x(t))/dt &= w_t(t, x(t)) + \sum_{i=1}^{n} w_{x^i}(t, x(t)) x'^i(t), \\ &= w_t(t, x(t)) + \sum_{i=1}^{n} w_{x^i}(t, x(t)) f_i(t, x(t), u(t)) \\ &= w_t(t, x(t)) + H(t, x(t), u(t), w_x(t, x(t))), \\ &\qquad t_1 \leq t \leq t_2 \text{ (a.e.)}. \end{aligned}$$

□

The requirement in (4.6.ii) that w be continuously differentiable in $V - B$ is unrealistic. In most applications the function w is only continuous in V

and continuously differentiable in $V - S$, where S is a set of Lebesgue measure zero in R^{n+1}, $B \subset S \subset V \subset R^{n+1}$. The following extension of (4.6.ii) is relevant.

4.6.iv (BOLTYANSKII'S SUFFICIENT CONDITION FOR A SINGLE TRAJECTORY). *Let V, B, and S be as above, with* meas $S = 0$. *Let $w(t, x)$ be a given function in V with $w(t, x) = g(t, x)$ for $(t, x) \in B$, $w(t, x)$ continuous in V and continuously differentiable in $V - S$, such that*

$$w_t(t, x) + \sum_{i=1}^{n} w_{x^i}(t, x) f_i(t, x, u) \geq 0 \tag{4.6.1}$$

for all $(t, x) \in V - S$ and all $u \in U$, and such that $w(t, x^*(t))$, $t_1 \leq t \leq t_2^*$, is constant on a given admissible trajectory $x^*(t)$ transferring a point (t_1, x_1) to B in V. Then x^* is optimal for (t_1, x_1) in V.

The proof of this theorem will be given in Chapter 7.

B. Synthesis

We assume now that any point $(t, x) \in V$ can be transferred to B in V by a well-determined admissible pair $x(s)$, $u(s)$, $t \leq s \leq t_2$, in such a way that for some function $p(t, x) = (p_1, \ldots, p_m)$ in V we have $u(s) = p(s, x(s))$, $t \leq s \leq t_2$. We may say that each of the above trajectories is a marked trajectory. It is often said that a *synthesis* has been effected in V, and that $p(t, x)$ is a *feedback* control function in V. Note that we do not exclude that a point may be taken to B by more than one marked trajectory in V. The two statements below attempt to decide under what additional requirements the marked trajectories are optimal in V.

If $(t, x) \in V$ and $x(s)$, $u(s)$, $t \leq s \leq t_2$, are any marked trajectory and corresponding strategy transferring (t, x) to $(t_2, x(t_2)) \in B$, then we take

$$w(t, x) = g(t_2, x(t_2)), \qquad (t, x) \in V. \tag{4.6.2}$$

This function w is of course constant on each marked trajectory in V.

It is relevant to state and prove that simple requirements, easily verifiable, guarantee that all marked trajectories are optimal in V. As proved by Boltyanskii, the same conditions (P1)–(P4) of the necessary statement in (4.2), plus suitable regularity and smoothness hypotheses, suffice. Under these hypotheses we shall say with Boltyanskii that we have in V a *regular synthesis*. We shall indeed require that $p(t, x)$ be sectionally smooth in the precise sense below, that the function w defined by (4.6.1) be continuous in V, and that the marked trajectories hit B and certain critical lines and surfaces at angles all different from zero. Then the same function (4.6.2) will be proved to be sectionally smooth, and all marked trajectories are optimal.

C. Regular Syntheses

Let K be some bounded, s-dimensional, convex polyhedron, $0 \leq s \leq n$, in the ξ-space, $\xi = (\xi^1, \ldots, \xi^n)$. We shall think of K as closed, that is, $\partial K \subset K$. Let us assume that ϕ, or $x = \phi(\xi)$, or $x^i = \phi^i(\xi)$, $i = 1, \ldots, n$, is a given transformation of C^1 in K from K to $V \subset R^n$, which is 1-1, and whose $n \times s$ matrix of first order partial derivatives $[\partial\phi^i/\partial\xi^j]$ is of rank s at every point of K. Then the image $L = \phi(K)$ of K in R^n is said to be a *curvilinear s-dimension* $\leq n$ in V, which may be empty.

If a set $S \subset V$ is the union of finitely many or countably many curvilinear polyhedra arranged in such a way that only finitely many of these polyhedra intersect every closed bounded set lying in V, then S will be called a *piecewise smooth set* in V. (The polyhedra may "cluster" at the boundary of the set V if V is not closed.) If among the curvilinear polyhedra whose union is S there is some polyhedron of dimension k while all others have dimension $\leq k$, then we say that S is k-dimensional. We know that any set of dimension less than n in R^n which is piecewise smooth does not contain interior points.

Let V be open in R^{n+1}, let $B \subset V$ be a closed piecewise smooth set in V of dimension $\leq n$, let S be any given piecewise smooth set in V also of dimension $\leq n$, and $B \subset S \subset V$, and let (t_1, x_1) be any point of V. We assume from here on that U is a fixed subset of R^m. We do not exclude that B may be the union of parts as stated above. Also, let N be a given piecewise smooth set of dimension $\leq n$ in V, which may be empty.

We assume now that S is the union of parts $P^k \subset P^{k+1} \subset \cdots \subset P^n$ contained in V, each P^i of dimension i. If $P^{n+1} = V$, $P^{k-1} = \emptyset$, then we assume that for each $i = k, \ldots, n+1$, the set $P^i = (P^{i-1} \cup N)$ has only finitely many components, each of which is an i-dimensional smooth manifold in V; we shall call these components i-dimensional cells. We assume that the function $p(t, x)$ is continuous and continuously differentiable in each cell. More specifically, we assume that, if $u_\sigma(t, x)$ denotes the restriction of $u(t, x)$ in the open cell σ, then u_σ can be extended to a neighborhood U of $\bar{\sigma}$ so as to be continuously differentiable in U, say the δ-neighborhood of $\bar{\sigma}$. We also assume that $f(t, x, u)$ can be extended as a function of class C^1 to a neighborhood W of its domain of definition M. Then, for $\delta > 0$ sufficiently small, $f(t, x, u_\sigma(t, x))$ is defined and of class C^1 in U. We further need the following specific assumptions.

(A) All cells are grouped into cells of the first, second, and third kind. All $(n+1)$-dimensional cells are of the first kind; B is the union of cells of the third kind. If σ is any i-dimensional cell of the first kind, $k \leq i \leq n+1$, then through every point of the cell there passes a unique trajectory of the system $dx/dt = f(t, x, p(t, x))$. Furthermore, there exists an $(i-1)$-dimensional cell, say $\Pi\sigma$, such that every trajectory in σ leaves the cell σ in a finite time by striking against the cell $\Pi\sigma$ at a nonzero angle and nonzero velocity.

(B) If σ is any i-dimensional cell of the second kind, then there is an $(i + 1)$-dimensional cell $\Sigma\sigma$ of the first kind such that from any point P of the cell σ there issues a unique trajectory moving into the cell $\Sigma\sigma$ and having only the point P on σ. We assume that $p(t, x)$ is continuously differentiable in $\sigma \cup \Sigma\sigma$ if σ is of the second kind.

(C) The conditions above guarantee the possibility of continuing the trajectories from cell to cell: from the cell σ to the cell $\Pi\sigma$ if σ is of the first kind, and from σ to $\Pi\Sigma\sigma$ if σ is of the second kind. It is required that any trajectory go through only finitely many cells (ending in B). All the mentioned trajectories are called *marked trajectories*. Thus, from every point of the set $V - N$ there issues a unique trajectory that leads to B. We require that also from the points of N there issue some trajectory ending in B, which is not necessarily unique; these are also said to be marked. All marked trajectories satisfy the necessary conditions (P1)–(P4) of Section 4.2.

When all previous assumptions hold, we say that we have effected a *regular synthesis* in V. Note that the possible discontinuities of $p(t, x)$ in V are only at points P of V belonging to finitely many manifolds of dimension $\leq n$ (or countably many such manifolds, but only finitely many of them having a nonzero intersection with any compact subset of V). Note that along any marked trajectory $x(t)$, the feedback function $p(t, x(t))$ has at most finitely many points of jump discontinuity.

4.6.v (THEOREM: BOLTYANSKII [1]). *Assume that a regular synthesis has been effected in V with exceptional set S, $V \supset S \supset B$, with a possible set N, with cells σ_i, with feedback control function $p(t, x)$, so that each point (s, y) of V can be taken to B by some marked trajectory $x(t)$ and corresponding control $u(t)$, $s \leq t \leq t_2(s, y)$, $x(s) = y$, $u(t) = p(t, x(t))$, $(t_2, x(t_2)) \in B$, and each marked trajectory x and corresponding control u satisfy the necessary conditions* (P1)–(P4) *of Section* 4.2 *with multipliers $\lambda(t) = (\lambda_1, \ldots, \lambda_n)$, $\lambda(t) \neq 0$, constant $\lambda_0 > 0$, Hamiltonian H and minimized Hamiltonian $M(t)$. Then the function $w(t, x)$ defined by* (4.6.2) *is of class C^1 in each cell σ_i, the constant λ_0 can be taken to be the same positive constants for all marked trajectories, and in each cell σ_i we have*

$$\begin{aligned} &\lambda_j(t) = \lambda_0 w_{x^j}(t, x(t)), \quad j = 1, \ldots, n, \qquad M(t) = \lambda_0 w_t(t, x(t)), \\ &w_t(t, x) + H(t, x, p(t, x), w_x(t, x)) = 0, \qquad (t, x) \in \sigma_i. \end{aligned} \tag{4.6.3}$$

Moreover, each marked trajectory is optimal.

The proof of relations (4.6.3) will be given in Chapter 7. The last statement is a consequence of (4.6.iv).

D. Example

We consider here the same example we have used in Section 4.2 to illustrate the necessary condition. We shall discuss a few more details on the same example in Section 6.1 (Example 1). In Section 4.2 we have already effected a regular synthesis covering the whole xy-plane with target the single point $(0,0)$. Actually, the problem is an autonomous one, V is the whole xyt-space, and the target B is the t-axis, or $R = [x = 0, y = 0, -\infty < t < +\infty]$. Let us denote by A^+, A^- the parts of the xy-plane above and below the switch line $\Gamma = AOB$ of Section 4.2B, and by $\Gamma^+ = AO$, $\Gamma^- = BO$ the two parts of Γ. We have seen in Section 4.2B that any point (a,b) of the plane is taken to $(0,0)$ in a time $t_2(a,b)$ by the trajectories $x(t)$, $y(t)$ determined there, and we shall compute an expression for $t_2(a,b)$ in Section 6.1, Example 1. Thus, any point (t,x,y) is taken to the t-axis R, by the corresponding trajectory $(t, X(t), Y(t))$ in a time $t_2(x,y)$. Thus, $w(t,x,y) = t + t_2(x,y)$ is constant along the trajectory. Note that all these (marked) trajectories hit the boundaries of the successive cells at positive angles, and the necessary conditions (P1)–(P4) are satisfied (see Section 6.1, Example 1). By (4.6.v) we conclude that each marked trajectory transferring any of its points to $B = R$ is optimal in $V = R^3$. (As mentioned in Section 4.2 the optimality of the marked trajectories can also be obtained by the existence theorems of Sections 4.3 and 9.4.)

Further examples and many exercises on the material above will be mentioned in Sections 6.1–6.

4.7 Appendix: Derivation of Some of the Classical Necessary Conditions of Section 2.1 from the Necessary Condition for Mayer Problems of Optimal Control

The classical problem of the minimum of a functional

$$I[x] = \int_{t_1}^{t_2} f_0(t, x(t), x'(t))\, dt, \qquad x = (x^1, \ldots x^n),$$

with boundary conditions $(t_1, x(t_1), t_2, x(t_2)) \in B \subset R^{2n+2}$ is immediately reduced to the Mayer problem of control with AC trajectories $\tilde{x}(t) = (x^0, x^1, \ldots, x^n) = (x^0, x)$, $t_1 \le t \le t_2$, controls $u(t) = (u^1, \ldots, u^n)$, differential equations

$$dx^0/dt = f_0(t, x(t), u(t)) \qquad dx^i/dt = u(t), \qquad t \in [t_1, t_2] \text{ (a.e.)}, i = 1, \ldots, n,$$

functional $I[x,u] = g = x^0(t_2)$, and boundary conditions and constraints

$$x^0(t_1) = 0, \qquad (t_1, x(t_1), t_2, x(t_2)) \in B \subset R^{2n+2}.$$

For $\tilde{\lambda} = (\lambda_0, \lambda_1, \ldots, \lambda_n) = (\lambda_0, \lambda)$ we have

$$H(t, \tilde{x}, u, \tilde{\lambda}) = \lambda_0 f_0(t, x, u) + \lambda_1 u^1 + \cdots + \lambda_n u^n. \tag{4.7.1}$$

For t, x, λ fixed, H can have a minimum at some $u \in R^n$ only if $\partial H/\partial u^i = 0$, $i = 1, \ldots, n$, since we have assumed f_0 of class C^1. From (P2) we have $\lambda_0 f_{0x'^i} + \lambda_i = 0$, or

$$\lambda_i = -\lambda_0 f_{0x'^i}(t, x, u), \qquad i = 1, \ldots, n. \tag{4.7.2}$$

From (P1), $\tilde{\lambda}(t)$ must be an AC vector function satisfying $d\lambda_i/dt = -H_{x^i}$, $i = 0, 1, \ldots, n$, or $d\lambda_0/dt = 0$, $d\lambda_i/dt = -\lambda_0 f_{0x^i}$. Thus, λ_0 is a constant, and since $\lambda_i = -\lambda_0 f_{0x'^i}(t, x, u)$, we see that λ_0 must be a nonzero constant, since otherwise $\lambda(t)$ would be identically zero, contrary to (P_1) (here $dg = dx_2^0$ is not identically zero). Thus,

$$\lambda_i(t) = -\lambda_0 f_{0x'^i}(t, x(t), u(t)), \tag{4.7.3}$$

$$d\lambda_i/dt = -\lambda_0 f_{0x^i}(t, x(t), u(t)) \tag{4.7.4}$$

for $t_1 \le t \le t_2$, λ_i AC in $[t_1, t_2]$, λ_0 a nonzero constant. Thus, $f_{0x'^i}(t, x(t), u(t))$ is an AC function in $[t_1, t_2]$, and

$$(d/dt)(-\lambda_0 f_{0x'^i}(t, x(t), u(t)) = -\lambda_0 f_{0x^i}(t, x(t), u(t)),$$

and finally

$$(d/dt) f_{0x'^i}(t, x(t), x'(t)) = f_{0x^i}(t, x(t), x'(t)), \qquad t \in [t_1, t_2] \text{ (a.e.)}, \quad i = 1, \ldots, n.$$

These are the Euler equations. Now from (P_3),

$$\begin{aligned} M(t) &= \min_{u \in R^n} H(t, x(t), u, \lambda(t)) \\ &= \lambda_0 \left[f_0(t, x(t), u(t)) - \sum_{i=1}^n x'^i(t) f_{0x'^i}(t, x(t), u(t)) \right] \end{aligned}$$

is an AC function in $[t_1, t_2]$, with $dM/dt = H_t$, that is,

$$M_0(t) = \lambda_0 \left[f_0(t, x(t), x'(t)) - \sum_{i=1}^n x'^i(t) f_{0x'^i}(t, x(t), x'(t)) \right] \tag{4.7.5}$$

is AC with

$$(d/dt) \left[f_0(t, x(t), x'(t)) - \sum_{i=1}^n x'^i(t) f_{0x'^i}(t, x(t), x'(t)) \right] = f_{0t}(t, x(t), x'(t)).$$

This is the DuBois-Reymond condition. Now the transversality relation (P_4) states that there is a constant $\bar{\lambda}_0 \ge 0$ such that

$$\bar{\lambda}_0 \, dg + \left[M_0(t)\, dt - \sum_{i=0}^n \lambda_i(t)\, dx^i \right]_1^2 = 0,$$

with $dg = dx_2^0$, $\lambda_0(t) = \lambda_0$ a nonzero constant, $dx_1^0 = 0$, and thus

$$(\bar{\lambda}_0 - \lambda_0)\, dx_2^0 = 0, \qquad \left[M_0(t)\, dt - \sum_{i=1}^n \lambda_i(t)\, dx^i \right]_1^2 = 0.$$

Here the second part is independent of dx_2^0. Thus, $\bar{\lambda}_0 = \lambda_0$, $\bar{\lambda}_0 \geq 0$, $\lambda_0 \neq 0$, hence $\bar{\lambda}_0 = \lambda_0 > 0$, and by dividing the second relation by λ_0, we have

$$\left[\lambda_0^{-1} M_0(t)\, dt + \sum_{i=1}^{n} f_{0x'^i}(t, x(t), x'(t))\, dx^i\right]_1^2 = 0,$$

where $\lambda_0^{-1} M_0(t)$ is the expression in brackets in (4.7.5). This is the classical transversality relation. Now from the minimum property of H we have

$$H(t, x(t), u(t), \lambda(t)) \leq H(t, x(t), u, \lambda(t)), \qquad t \in [t_1, t_2] \text{ (a.e.)},$$

for all $u \in R^n$, or

$$\lambda_0\left[f_0(t, x(t), u(t)) - \sum_{i=1}^{n} u^i(t) f_{0x'^i}(t, x(t), u(t))\right]$$
$$\leq \lambda_0\left[f_0(t, x(t), u) - \sum_{i=1}^{n} u^i f_{0x'^i}(t, x(t), u(t))\right]$$

for all $u \in R^n$, where λ_0 is a positive constant. Thus

$$f_0(t, x(t), u) - f_0(t, x(t), u(t)) - \sum_{i=1}^{n} (u^i - u^i(t)) f_{0x'^i}(t, x(t), u(t)) \geq 0,$$

and by writing X' for u and $x'(t)$ for $u(t)$, we have

$$E(t, x(t), x'(t), X') = f_0(t, x(t), X') - f_0(t, x(t), x'(t))$$
$$- \sum_{i=1}^{n} (X'^i - x'^i(t)) f_{0x'^i}(t, x(t), x'(t)) \geq 0.$$

This is the Weierstrass condition.

4.8 Appendix: Derivation of the Classical Necessary Condition for Isoperimetric Problems from the Necessary Condition for Mayer Problems of Optimal Control

We consider here the problem of the maxima and minima of the integral

$$I[x] = \int_{t_1}^{t_2} f_0(t, x(t), x'(t))\, dt \qquad x(t) = (x^1, \ldots, x^n),$$

with usual boundary conditions and constraints

$$(t, x(t)) \in A \subset R^{n+1}, \qquad (t_1, x(t_1), t_2, x(t_2)) \in B \subset R^{2n+2},$$

and N side conditions of the form

$$J_s[x] = \int_{t_1}^{t_2} f_s(t, x(t), x'(t))\, dt = L_s, \qquad s = 1, \ldots, N,$$

where L_s are given numbers, and $J_1, \ldots, J_N$ are given functionals. The classical theory shows e.g. that the optimal solutions of class C^1 are to be found among the extremals of the auxiliary problem

$$P[x] = \int_{t_1}^{t_2} [\eta_0 f_0 + \cdots + \eta_N f_N]\, dt,$$

where $\eta_0, \ldots, \eta_N$ are $N + 1$ undetermined constants, $\eta_0 \geq 0$. Actually, the above problem can be shown to be equivalent to a Mayer problem of optimal control, and thus the entire necessary condition (4.2.i) applies, and the classical statements can be immediately derived. Indeed, as in Section 4.7, the above problem is transformed into the Mayer problem concerning the functional $I[x, y, u]$ below, with differential equations, boundary conditions, and constraints

$$\begin{gathered} dy^0/dt = f_0(t, x, u), \quad dy^s/dt = f_s(t, x, u), \quad s = 1, \ldots, N, \quad dx^i/dt = u^i, \quad i = 1, \ldots, n, \\ y^0(t_1) = 0, \qquad y^s(t_1) = 0, \qquad y^s(t_2) = L_s, \qquad s = 1, \ldots, N, \\ I[x, y, u] = y^0(t_2), \qquad (t_1, x(t_1), t_2, x(t_2)) \in B, \qquad u \in U = R^n, \\ x = (x^1, \ldots, x^n), \qquad y = (y^0, y^1, \ldots, y^N), \qquad u = (u^1, \ldots, u^n). \end{gathered}$$

Here the Hamiltonian is

$$\begin{gathered} H = H(t, x, y, u, \lambda, \eta) = \lambda_1 u^1 + \cdots + \lambda_n u^n + \eta_0 f_0 + \eta_1 f_1 + \cdots + \eta_N f_N, \\ \lambda = (\lambda_1, \ldots, \lambda_n), \qquad \eta = (\eta_0, \eta_1, \ldots, \eta_N), \end{gathered}$$

and the minimum of H as u describes $U = R^n$ must occur at points $u \in R^n$ where $H_{u^i} = 0$, $i = 1, \ldots, n$, or

$$\lambda_i + \eta_0 f_{0x'^i} + \eta_1 f_{1x'^i} + \cdots + \eta_N f_{Nx'^i} = 0, \qquad i = 1, \ldots, n.$$

On the other hand, $\lambda_1(t), \ldots, \lambda_n(t), \eta_0(t), \ldots, n_N(t)$ satisfy the differential equations

$$\begin{aligned} d\lambda_i/dt &= -H_{x^i} = -(\eta_0 f_{0x^i} + \eta_1 f_{1x^i} + \cdots + \eta_N f_{Nx^i}), \qquad i = 1, \ldots, n, \\ d\eta_s/dt &= -H_{y^s} = 0, \qquad s = 0, 1, \ldots, N. \end{aligned}$$

Thus, the multipliers $\eta_0, \eta_1, \ldots, \eta_N$ are constants. By taking $F(t, x, u, \eta) = \eta_0 f_0 + \cdots + \eta_N f_N$, $\eta = (\eta_0, \eta_1, \ldots, \eta_N)$, we see that the equations above for the multipliers λ_i become

$$\lambda_i = -F_{x'^i}, \quad d\lambda_i/dt = -F_{x^i}, \qquad i = 1, \ldots, n.$$

Thus, as in Section 4.7,

$$(d/dt)F_{x'^i} = F_{x^i}, \qquad i = 1, \ldots, n.$$

These are the Euler equations for $F = \eta_0 f_0 + \cdots + \eta_N f_N$, $\eta_0, \eta_1, \ldots, \eta_N$ undetermined constants.

Moreover, again for x optimal,

$$M(t) = \sum_{j=0}^{N} \eta_j f_j + \sum_{i=1}^{n} u^i \lambda_i = \sum_{j=0}^{N} \eta_j f_j - \sum_{i=1}^{n} x'^i \sum_{j=0}^{N} \eta_j f_{jx'^i}$$

is AC in $[t_1, t_2]$ with $(d/dt)M(t) = H_t = \sum_{j=0}^{N} \eta_j f_{jt}$. Hence,

$$(d/dt)F - \sum_{i=1}^{n} x'^i F_{x'^i} = (d/dt)M(t) = H_t = F_t,$$

and this is the DuBois-Reymond equation for F. By (P4) the transversality relation

$$\bar{\lambda}_0\, dg + \left[M(t)\, dt - \sum_{i=1}^{n} \lambda_i(t)\, dx^i - \sum_{j=0}^{N} \eta_j\, dy^j \right]_1^2 = 0$$

holds for some constant $\bar{\lambda}_0 \geq 0$, with $dy_0^1 = 0$, $dy_1^j = dy_2^j = 0$, $j = 1, \ldots, N$, $dg = dy_2^0$ arbitrary. Thus, $\bar{\lambda}_0 - \eta_0 = 0$, $\eta_0 = \bar{\lambda}_0 \geq 0$, and the classical transversality relation for F is obtained:

$$\left[\left(\sum_{j=0}^{N} \eta_j \left(f_j - \sum_{i=1}^{n} x'^i f_{jx'^i} \right) \right) dt + \sum_{i=1}^{n} \left(\sum_{j=0}^{N} \eta_j f_{jx'^i} \right) dx^i \right]_1^2 = 0.$$

4.9 Appendix: Derivation of the Classical Necessary Condition for Lagrange Problems of the Calculus of Variations with Differential Equations as Constraints

We already stated in Section 2.2 that the optimal solutions for functionals

$$I[x] = \int_{t_1}^{t_2} f_0(t, x(t), x'(t))\, dt, \qquad x(t) = (x^1, \ldots, x^n), \tag{4.9.1}$$

with constraints

$$(t_1, x(t_1), t_2, x(t_2)) \in B, \qquad G_j(t, x(t), x'(t)) = 0, \qquad j = 1, \ldots, N < n, \tag{4.9.2}$$

must satisfy the Euler equations for the auxiliary problem

$$J[x] = \int_{t_1}^{t_2} F_0(t, x(t), x'(t))\, dt,$$

$$F_0(t, x, x') = \lambda_0 f_0(t, x, x') + \sum_{j=1}^{N} p_j(t) G_j(t, x, x'),$$

for $\lambda_0 \geq 0$ a constant and $p_j(t)$, $j = 1, \ldots, N$, suitable functions.

We shall prove this for the case in which system (4.9.2) is given as explicitly solved with respect to N of the derivatives x'^i. After relabeling we can write the system in the form

$$dx^j/dt = f_j(t, x(t), dx^{N+1}/dt, \ldots, dx^n/dt), \qquad j = 1, \ldots, N,$$

or

$$G_j(t, x, x') = x'^j - f_j(t, x^1, \ldots, x^n, x'^{N+1}, \ldots, x'^n), \qquad j = 1, \ldots, N. \tag{4.9.3}$$

We can rewrite problem (4.9.1), (4.9.3) as a problem of control

$$J[x, v] = \int_{t_1}^{t_2} L_0(t, x(t), x'(t))\, dt,$$

$$dx^i/dt = f_i(t, x(t), v(t)), \qquad i = 1, \ldots, N,$$

$$dx^{N+j}/dt = v^j(t), \qquad j = 1, \ldots, n - N,$$

with state variables $x = (x^1, \ldots, x^n)$, control variables $v = (v^1, \ldots, v^{n-N})$, $v \in R^{n-N}$, and

$$L_0(t, x, v) = f_0(t, x^1, \ldots, x^n, f_1(t, x, v), \ldots, f_N(t, x, v), v^1, \ldots, v^{n-N}).$$

For this Lagrange problem of optimal control, and for the sake of simplicity, we take from Section 5.1 the form of the Hamiltonian

$$H(t, x, v, \lambda) = \lambda_0 L_0 + \lambda_1 f_1 + \cdots + \lambda_N f_N + \lambda_{N+1} v^1 + \cdots + \lambda_n v^{n-N},$$

with $\lambda_0 \geq 0$ a constant, $\lambda_1, \ldots, \lambda_n$ AC functions satisfying the differential equations $d\lambda_i/dt = -\partial H/\partial x^i$, $i = 1, \ldots, n$. We have here

$$d\lambda_i/dt = -\lambda_0\left(f_{0x^i} + \sum_{s=1}^{N} f_{0x'^s} f_{sx^i}\right) - \sum_{s=1}^{N} \lambda_s f_{sx^i},$$

or

$$\text{(4.9.4)} \quad \lambda_i(t) = \lambda_i(t_1) - \int_{t_1}^{t} \left[\lambda_0 f_{0x^i} + \sum_{s=1}^{N} f_{sx^i}(\lambda_0 f_{0x'^s} + \lambda_s)\right] d\tau, \qquad i = 1, \ldots, n.$$

For the minimum of H we must have $\partial H/\partial v^j = 0$, $j = 1, \ldots, n - N$, or

$$\lambda_0 f_{0v^j} + \lambda_0 \sum_{s=1}^{N} f_{0x'^s} f_{sv^j} + \sum_{s=1}^{N} \lambda_s f_{sv^j} + \lambda_{N+j} = 0, \qquad j = 1, \ldots, n - N,$$

or

$$\text{(4.9.5)} \qquad \lambda_{N+j} = -\lambda_0 f_{0v^j} - \sum_{s=1}^{N} f_{sv^j}(\lambda_0 f_{0x'^s} + \lambda_s), \qquad j = 1, \ldots, n - N.$$

By comparing (4.9.5) and the last $n - N$ relations (4.9.4), we have

$$\text{(4.9.6)} \qquad \lambda_0 f_{0v^{i-N}} + \sum_{s=1}^{N} f_{sv^{i-N}}(\lambda_0 f_{0x'^i} + \lambda_s)$$

$$= -\lambda_i(t_1) + \int_{t_1}^{t} \lambda_0 f_{0x^i} + \sum_{s=1}^{N} f_{sx^i}(\lambda_0 f_{0x'^i} + \lambda_s)\, d\tau.$$

If we take

$$p_i(t) = \lambda_i(t) + \lambda_0 f_{0x'^i}, \qquad i = 1, \ldots, N,$$

then relations (4.9.4) become

$$\lambda_i(t) = \lambda_i(t_1) - \int_{t_1}^{t} \left(\lambda_0 f_{0x^i} + \sum_{s=1}^{N} p_s f_{sx^i}\right) d\tau, \qquad i = 1, \ldots, n,$$

and if we also write $f_{0x'^i}$ for $f_{0v^{i-N}}$ then relations (4.9.6) become

$$\lambda_0 f_{0x'^i} + \sum_{s=1}^{N} p_s f_{sx'^i} = -\lambda_i(t_1) + \int_{t_1}^{t} \left(\lambda_0 f_{0x^i} + \sum_{s=1}^{N} p_s f_{sx^i}\right) d\tau, \qquad i = N + 1, \ldots, n.$$

Finally, if we take

$$F_0(t, x, x') = -\lambda_0 f_0(t, x, x') + \sum_{s=1}^{N} p_s(t) G_s(t, x, x'),$$

we have for $i = 1, \ldots, N$,

$$F_{0x'^i} = -\lambda_0 f_{0x'^i} + \sum_{s=1}^{N} p_s G_{sx'^i} = -\lambda_0 f_{0x'^i} + p_i$$

$$= \lambda_i = \lambda_i(t_1) + \int_{t_1}^{t} \left(-\lambda_0 f_{0x^i} - \sum_{s=1}^{N} p_s f_{sx^i} \right) d\tau$$

$$= \lambda_i(t_1) + \int_{t_1}^{t} \left(-\lambda_0 f_{0x^i} + \sum_{s=1}^{N} p_s G_{sx^i} \right) d\tau$$

$$= \lambda_i(t_1) + \int_{t_1}^{t} F_{0x^i} \, d\tau.$$

For $i = N + 1, \ldots, n$, we also have

$$F_{0x'^i} = -\lambda_0 f_{0x'^i} - \sum_{s=1}^{N} p_s f_{sx'^i}$$

$$= \lambda_i(t_1) - \int_{t_1}^{t} \left(\lambda_0 f_{0x^i} + \sum_{s=1}^{N} p_s f_{sx^i} \right) d\tau$$

$$= \lambda_i(t_1) + \int_{t_1}^{t} \left(-\lambda_0 f_{0x^i} + \sum_{s=1}^{N} p_s G_{sx^i} \right) d\tau$$

$$= \lambda_i(t_1) + \int_{t_1}^{t} F_{0x^i} \, d\tau.$$

Thus, the partial derivatives $F_{0x'^i}(t, x(t), x'(t))$ coincide almost everywhere with AC functions. By identifying these expressions with such AC functions we can write briefly that $(d/dt)F_{0x'^i} = F_{0x^i}$, $t \in [t_1, t_2]$ (a.e.), $i = 1, \ldots, n$.

Bibliographical Notes

The necessary condition (4.2.i) is essentially due to L. S. Pontryagin [1], who recognized that this formulation holds even if the control function $u(t)$ lies on the boundary of the control space U (while the state variable $x(t)$ must be in the interior of A). However, the transversality relation (P4) is classical, and so are the canonical equations. Basically, there is here a return to the consideration of the function $M(t, x, z)$, which is the classical "Hamiltonian". It is remarkable that a previous partial form of the necessary condition (4.2.i) had been stated by M. R. Hestenes [1] several years before Pontryagin, but passed unnoticed.

The examples in Remark 8 of Section 4.2C show that there are situations in which the necessary condition (4.2.i) gives no information concerning the possible optimal solution (if any). The first example is taken from H. Hermes and J. P. LaSalle [I]. For singular optimal control problems we refer to D. J. Bell and D. H. Jacobson [I].

For the alternate conditions to (4.2.i) we refer to E. J. McShane [18].

The value function has been given much emphasis by R. Bellman [I, II]. The general concept of synthesis is of course a present day form of the idea of a field. The precise formulation of regular synthesis and the related theorem (4.6.v) are due to V. G. Boltyanskii [1]. For further bibliographical information see the end of Chapter 7.

CHAPTER 5

Lagrange and Bolza Problems of Optimal Control and Other Problems

As we have shown in Section 1.9, Lagrange, Bolza, and Mayer problems can be essentially transformed one into the other by simple changes of space variables. Thus, the necessary condition and the other statements of Chapter 4 for Mayer problems have their counterpart for Lagrange and Bolza problems. We present here briefly the main statements for the Bolza problems for easy reference, those for Lagrange problems being the same (with $g = 0$).

5.1 The Necessary Condition for Bolza and Lagrange Problems of Optimal Control

We consider here a Bolza problem concerning the minimum of the functional

$$I[x,u] = g(t_1, x(t_1), t_2, x(t_2)) + \int_{t_1}^{t_2} f_0(t, x(t), u(t))\, dt, \qquad g, f_0 \text{ scalars}, \tag{5.1.1}$$

with $x(t) = (x^1, \ldots, x^n)$, $u(t) = (u^1, \ldots, u^m)$, satisfying the differential system

$$dx/dt = f(t, x(t), u(t)), \qquad f = (f_1, \ldots, f_n), \tag{5.1.2}$$

the boundary conditions $(t_1, x(t_1), t_2, x(t_2)) \in B \subset R^{2n+2}$, and the constraints $(t, x(t)) \in A$, $u(t) \in U$.

Here all $n + 1$ scalar functions $f_i(t, x, u)$, $i = 0, 1, \ldots, n$, are assumed to be defined and continuous in the set M of all (t, x, u) with $(t, x) \in A$, $u \in U(t)$ (or $M = A \times U$ if U is a fixed set), together with their partial derivatives f_{it}, f_{ix^j}, $i = 0, 1, \ldots, n$. Also, g is a real valued function of class C^1, and $g \equiv 0$ for Lagrange problems. The class Ω of the admissible pairs $x(t)$, $u(t)$, $t_1 \le t \le t_2$, is defined as in Section 4.1, though we must now require explicitly that $f_0(t, x(t), u(t))$ be integrable in $[t_1, t_2]$. As in Section 4.1, we assume that $x(t)$, $u(t)$, $t_1 \le t \le t_2$, is an optimal pair and we need not repeat the assumptions (a)–(d) of Section 4.1 for the situation under consideration. In particular, the graph of the optimal trajectory x is in the interior of A, U is a fixed set (not necessarily bounded), and the optimal strategy u is bounded. (Alternatively, assumptions for the case of u unbounded and for other variants can be made.)

We shall need now an $(n+1)$-vector $\tilde{\lambda} = (\lambda_0, \lambda_1, \ldots, \lambda_n)$, and the Hamiltonian function

$$H(t, x, u, \tilde{\lambda}) = \lambda_0 f_0(t, x, u) + \lambda_1 f_1(t, x, u) + \cdots + \lambda_n f_n(t, x, u). \tag{5.1.3}$$

Also, we shall need the related function

$$M(t, x, \tilde{\lambda}) = \inf_{u \in U} H(t, x, u, \tilde{\lambda}). \tag{5.1.4}$$

The necessary condition for Lagrange problems now takes the form below, which we shall deduce explicitly from (4.2.i) in Section 5.2. A direct proof is indicated in Section 7.3I.

5.1.i (Theorem). *Under the hypotheses listed above, let* $x(t) = (x^1, \ldots, x^n)$, $u(t) = (u^1, \ldots, u^m)$, $t_1 \le t \le t_2$, *be an optimal pair, that is, an admissible pair* x, u *such that* $I[x, u] \le I[\tilde{x}, \tilde{u}]$ *for all pairs* $\tilde{x}, \tilde{u}$ *of the class* Ω *of all admissible pairs. Then the optimal pair* x, u *has the following properties*:

(P1′) *There is an absolutely continuous vector function* $\tilde{\lambda}(t) = (\lambda_0, \lambda_1, \ldots, \lambda_n)$, $t_1 \le t \le t_2$ *(multipliers), which is never zero in* $[t_1, t_2]$, *with* λ_0 *a constant in* $[t_1, t_2]$, $\lambda_0 \ge 0$, *such that*

$$d\lambda_i/dt = -H_{x^i}(t, x(t), u(t), \tilde{\lambda}(t)), \qquad i = 1, \ldots, n, \quad t \in [t_1, t_2] \text{ (a.e.)}.$$

(P2′) *For every fixed* t *in* $[t_1, t_2]$ (a.e.), *the Hamiltonian* $H(t, x(t), u, \tilde{\lambda}(t))$ *as a function of* u *only (with* u *in* U*) takes its minimum value in* U *at* $u = u(t)$:

$$M(t, x(t), \lambda(t)) = H(t, x(t), u(t), \lambda(t)), \qquad t \in [t_1, t_2] \text{ (a.e.)}.$$

(P3′) *The function* $M(t) = M(t, x(t), \tilde{\lambda}(t))$ *is absolutely continuous in* $[t_1, t_2]$ *(more specifically,* $M(t)$ *coincides* a.e. *in* $[t_1, t_2]$ *with an* AC *function), and*

$$\begin{aligned} dM/dt &= (d/dt)M(t, x(t), \lambda(t), u(t)) \\ &= H_t(t, x(t), u(t), \tilde{\lambda}(t)), \qquad t \in [t_1, t_2] \text{ (a.e.)}. \end{aligned}$$

(P4′) *Transversality relation*:

$$\lambda_0\, dg - M(t_1)\, dt_1 + \sum_{j=1}^{n} \lambda_j(t_1)\, dx_1^j + M(t_2)\, dt_2 - \sum_{j=1}^{n} \lambda_j(t_2)\, dx_2^j = 0$$

for every $(2n+2)$*-vector* $h = (dt_1, dx_1, dt_2, dx_2) \in B'$, *that is,*

$$\lambda_0\, dg + \left[M(t)\, dt - \sum_{j=1}^{n} \lambda_j(t)\, dx^j \right]_1^2 = 0. \tag{5.1.5}$$

The transversality relation is identically satisfied if t_1, x_1, t_2, x_2 are fixed, that is, for the boundary conditions which correspond to the case that both end points and times are fixed ($dt_1 = dx_1^i = dt_2 = dx_2^i = 0$, $i = 1, \ldots, n$). For Lagrange problems of course $g \equiv 0$, $dg \equiv 0$.

Here x, u is an admissible pair itself, so that the differential equations

$$dx^i/dt = f_i(t, x(t), u(t)), \qquad i = 1, \ldots, n, \tag{5.1.6}$$

hold, and these equations, together with (P1), yield the canonical equations

$$\frac{dx^i}{dt} = \frac{\partial H}{\partial \lambda_i}, \quad \frac{d\lambda_i}{dt} = -\frac{\partial H}{\partial x^i}, \qquad i = 1, \ldots, n. \tag{5.1.7}$$

These same equations as well as (P3) can also be written in the explicit forms

$$dx^i/dt = f_i(t, x(t), u(t)), \qquad i = 1, \ldots, n,$$

$$d\lambda_i/dt = -\sum_{j=0}^{n} \lambda_j(t) f_{jx^i}(t, x(t), u(t)), \qquad i = 1, \ldots, n,$$

$$dM/dt = \sum_{j=0}^{n} \lambda_j(t) f_{jt}(t, x(t), u(t)), \qquad t_1 \le t \le t_2.$$

If we denote by x^0 an auxiliary variable satisfying the differential equation $dx^0/dt = f_0(t, x, u)$ and initial condition $x^0(t_1) = 0$, then

$$I[x, u] = \int_{t_1}^{t_2} f_0 \, dt = x^0(t_2) - x^0(t_1) = x^0(t_2),$$

and the two equations

$$\frac{dx^0}{dt} = \frac{\partial H}{\partial \lambda_0}, \qquad \frac{d\lambda_0}{dt} = -\frac{\partial H}{\partial x^0}$$

can be added to the canonical equations (5.1.7), since $\partial H/\partial \lambda_0 = f_0$ and $H_{x^0} = 0$, and hence λ_0 is a constant, as stated.

Finally, let us note that for *autonomous* problems, that is, for problems where $f_0, f_1, \ldots, f_n$ do not depend on t, but only on x and u, we have $f_{it} = \partial f_i/\partial t = 0$. Hence, (P3′) yields $dM/dt = 0$, and the function $M(t)$ is a *constant* in $[t_1, t_2]$. In particular, for autonomous problems between two fixed points $x_1 = (x_1^i, i = 1, \ldots, n)$ and $x_2 = (x_2^i, i - 1, \ldots, n)$, in an undetermined time, then we can take $t_1 = 0$, $dx_1^i = dx_2^i = 0$, $i = 1, \ldots, n$, dt_2 arbitrary. Hence, (P4′) yields $M(t_2)\,dt_2 = 0$, or $M(t_2) = 0$, and $M(t)$ is the constant zero in $[t_1, t_2]$. This occurs, for instance, in a problem of minimum time (between two fixed points), where $f_0 = 1$, and then $I = t_2$ (see Section 5.3A).

Remark 1. There are cases where we can guarantee that $\lambda_0 \neq 0$, and hence $\lambda_0 > 0$ and we can take $\lambda_0 = 1$. For instance, if $U = R^m$, $m \ge n$, and the $m \times n$ matrix $[f_{iu_s}(t, x(t), u(t)), i = 1, \ldots, n, s = 1, \ldots, m]$ has rank n, then $\lambda_0 \neq 0$. Indeed, for the property minimum of H we need that, along $x(t)$, $u(t)$, we have $\lambda_0 f_{0u_s} + \sum_{i=1}^{n} \lambda_i f_{iu_s} = 0$, $s = 1, \ldots, m$, and $\lambda_0 = 0$ would imply either that the n vectors f_{iu_s}, $s = 1, \ldots, m$, are linearly dependent, or that $\lambda_0 = 0$, $\lambda_1 = \cdots = \lambda_n = 0$, in both cases a contradiction. For instance, if $U = R^m$, $m \ge n$, and all f_i, $i = 1, \ldots, n$, are linear in u, then $f_i = A_i(t, x) + \sum_{s=1}^{m} B_{is}(t, x)u^s$, and we require that the $m \times n$ matrix $[B_{is}(t, x(t))]$ have rank n along $x(t)$. This is exactly what occurs for Lagrange problems of the calculus of variations treated as problems of optimal control, that is, $m = n$, $U = R^n$, $f = u$, i.e., $f_i = u^i$, $i = 1, \ldots, n$, $A_i = 0$, $B_{is} = \delta_{is}$, $\det[B_{is}] = 1$. On the same question of $\lambda_0 > 0$, cf. Remark in Section 7.4E for problems written in the equivalent Mayer form.

Remark 2. As for Mayer problems in Section 4.3, we briefly state here a particular existence theorem for Lagrange and Mayer problems of optimal control. This theorem will be contained in the more general statements we shall proved in Chapter 9 and in the even more general statements in Chapters 11–16. Thus, we are concerned here with a functional $I[x,u]$ of the form (5.1.1) and we state a theorem guaranteeing the existence of an absolute minimum of $I[x,u]$ in the class Ω, we assume not empty, of all pairs $x(t)$, $u(t)$, $t_1 \le t \le t_2$, x AC, u measurable, satisfying (5.1.2) and the constraints $(t, x(t)) \in A \subset R^{n+1}$, $u(t) \in U \subset R^m$, and $(t_1, x(t_1), t_2, x(t_2)) \in B \subset R^{2n+2}$. We shall need the sets $\tilde{Q}(t,x) = [(z^0, z) | z^0 \ge f_0(t,x,u),\ z = f(t,x,u),\ u \in U] \subset R^{n+1}$.

5.1.ii (Filippov's Existence Theorem). *If A and U are compact, B is closed, f_0, f are continuous on $M = A \times U$, g is continuous on B, Ω is not empty, and for every $(t,x) \in A$ the set $\tilde{Q}(t,x)$ is convex, then $I[x,u]$ has an absolute minimum in Ω.*

If A is not compact, but closed and contained in a slab $t_0 \le t \le T$, $x \in R^n$, t_0, T finite, then the statement still holds if the conditions (b), (c) *of Section* 4.3 *hold.*

If A is contained in no slab as above, but A is closed, then the statement still holds if (b), (c) *hold, and in addition if* (d′) *$g(t_1, x_1, t_2, x_2) \to +\infty$ as $t_2 - t_1 \to +\infty$ uniformly with respect to x_1 and x_2, and for $(t_1, x_1, t_2, x_2) \in B$; and* (d″) *there are numbers $\mu > 0$, $M_0 \ge 0$ such that $f_0(t,x,u) \ge \mu$ for all $|t| \ge M_0$ $(t,x) \in A$, $u \in U$. Actually, if $\bar{x}$, $\bar{u}$ is a known admissible pair for which I has a value $I[\bar{x},\bar{u}] = m$, then all we need to know instead of* (d) *is that there is some M such that for all admissible pairs x, u with $t_2 - t_1 > M$ we have $I[x,u] > m$.*

Example

Take $n = 1$, $m = 1$, $f_0 = 1 + x^4 + u^4$, differential equation $x' = f = tx + t^2 + tu + u$, $U = [-1 \le u \le 1]$, $t_1 = 0$, $x(0) = 1$, $x(t_2) = 0$, t_2 undetermined, $t_2 \ge 0$, $g = 0$. For every (t,x) the set $\tilde{Q}(t,x) = [(z^0, z) | z^0 \ge 1 + x^4 + u^4,\ z = tx + t^2 + tu + u,\ -1 \le u \le 1]$ is certainly convex since f is linear in u and f_0 is convex in u. We note that $u = -1$, $x = 1 - t$, $0 \le t \le 1$, is an admissible pair, with $I = \frac{11}{5}$, and thus Ω is certainly not empty. Moreover, since $f_0 \ge 1$, any admissible pair with $t_2 > \frac{11}{5}$ gives to I a value larger than $\frac{11}{5}$ and can be disregarded. Therefore, we can limit ourselves to admissible pairs with $0 \le t_2 \le \frac{11}{5}$. With this remark, if $A_0 = [0 \le t \le \frac{11}{5}]$, then $A = A_0 \times R$, $B = (0) \times (1) \times A_0 \times (0)$, and thus A and B are closed, while U is compact. Condition (b) is trivially satisfied since the initial point $(0, 1)$ is fixed. Condition (c) is satisfied since f is linear in x, and both t and $t^2 + tu + u$ are continuous and bounded in A. All requirements of Theorem (5.1.ii) are satisfied. The integral has an absolute minimum under the constraints.

5.2 Derivation of Properties (P1′)–(P4′) from (P1)–(P4)

We shall now deduce (P1′)–(P4′) from the analogous relations (P1)–(P4) of Section 4.2. To this purpose let us first transform the Bolza problem above into a Mayer problem. As mentioned in Section 1.9, we introduce an auxiliary variable x^{n+1}, the extra differential equation $dx^{n+1}/dt = f_0(t, x(t), u(t))$, and the extra boundary condition $x^{n+1}(t_1) = 0$.

Thus

$$I[x,u] = g + \int_{t_1}^{t_2} f_0\,dt = g(t_1, x(t_1), t_2, x(t_2)) + x^{n+1}(t_2).$$

We denote by $\tilde{x}$ the $(n+1)$-vector $\tilde{x} = (x^1, \ldots, x^n, x^{n+1})$, and by $\tilde{f}(t,x,u)$ the $(n+1)$-vector function $\tilde{f}(t,x,u) = (f_1, \ldots, f_n, f_0)$. We have now the Mayer problem of the minimum of the functional

$$J[\tilde{x},u] = g + x^{n+1}(t_2) = I[x,u]$$

with differential equations

$$dx^i/dt = f_i(t,x,u), \quad i = 1, \ldots, n, \qquad dx^{n+1}/dt = f_0(t,x,u),$$

boundary conditions $(t_1, x(t_1), t_2, x(t_2)) \in B$, $x^{n+1}(t_1) = 0$, and the same constraints as before: $(t, \tilde{x}(t)) \in \tilde{A}$, $u(t) \in U$, where $\tilde{A} = A \times R$. The boundary conditions can be expressed in the equivalent form $(t_1, \tilde{x}(t_1), t_2, \tilde{x}(t_2)) \in \tilde{B}$, where $\tilde{B}$ is the closed subset of the space R^{2n+4} defined by $\tilde{B} = B \times (x_1^{n+1} = 0) \times (x_2^{n+1} \in R)$, since we have assigned the fixed value $x_1^{n+1} = 0$ for $x^{n+1}(t_1)$, and we have left undetermined the value $x_2^{n+1} = x^{n+1}(t_2)$. Thus, in any case, we have $dx_1^{n+1} = 0$, and dx_2^{n+1} is arbitrary. In other words, the new set $\tilde{B}'$ is the set $\tilde{B}' = B' \times (0) \times R$, since dx_2^{n+1} is arbitrary. According to Section 4.2, we need now an $(n+1)$-vector $\lambda = (\lambda_1, \ldots, \lambda_n, \lambda_{n+1})$, the Hamiltonian function

$$\tilde{H}(t,x,u,\lambda) = \lambda_1 f_1 + \cdots + \lambda_n f_n + \lambda_{n+1} f_0,$$

and the related function

$$\tilde{M}(t,x,\tilde{\lambda}) = \inf_{u \in U} \tilde{H}(t,x,u,\tilde{\lambda}),$$

where we have written x instead of $\tilde{x}$, since these functions do not depend on x^{n+1}. Also, we shall need a nonnegative constant $\lambda_0 \geq 0$.

We are now in a position to write explicitly statements (P1)–(P4) of (4.2.i), for the so obtained Mayer problem, where $n+1$ replaces n of (4.2.i). The following remarks are relevant. First, from (P1) of (4.2.i), we obtain for the multiplier λ_{n+1} the equation $d\lambda_{n+1}/dt = -H_{x^{n+1}}$, where now $H_{x^{n+1}} = 0$; hence $d\lambda_{n+1}/dt = 0$, and $\lambda_{n+1}(t)$ is a constant in $[t_1, t_2]$. Secondly, let us write (P4) for the new problem, namely,

$$\lambda_0[dg + dx_2^{n+1}] + \left[\tilde{M}(t)\,dt - \sum_{i=1}^{n+1} \lambda_i(t)\,dx^i\right]_1^2 = 0,$$

or

$$\lambda_0\,dg + (\lambda_0 - \lambda_{n+1})\,dx_2^{n+1} + \lambda_{n+1}\,dx_1^{n+1} + \left[\tilde{M}(t)\,dt - \sum_{i=1}^{n} \lambda_i(t)\,dx^i\right]_1^2 = 0,$$

where $x_1^{n+1} = 0$, $dx_1^{n+1} = 0$, dx_2^{n+1} arbitrary. Hence $\lambda_{n+1} = \lambda_0 \geq 0$, a constant, so that $\tilde{H}$ as defined above reduces to H as defined by (5.1.3), and $\tilde{M}$ as defined above reduces to M as defined by (5.1.4). Then, relation $d\tilde{M}/dt = \tilde{H}_t$ reduces to $dM/dt = H_t$, that is, (P3′). The transversality relation becomes now

$$\lambda_0\,dg + \left[M(t)\,dt - \sum_{i=1}^{n} \lambda_i(t)\,dx^i\right]_1^2 = 0.$$

We have derived relations (P1′)–(P4′) from (P1)–(P4).

5.3 Examples of Applications of the Necessary Conditions for Lagrange Problems of Optimal Control

A. The Example of Section 4.2B

The example discussed in Section 4.2B can be written as the Lagrange problem of the minimum of the integral

$$I[x, y, u] = \int_{t_1}^{t_2} dt = t_2 - t_1$$

(where we can take $t_1 = 0$), with differential equations, constraints, and boundary conditions

$$dx/dt = y, \qquad dy/dt = u, \quad u \in U = [-1 \leq u \leq 1],$$
$$x(t_1) = a, \quad y(t_1) = b, \quad x(t_2) = 0, \quad y(t_2) = 0.$$

Then $f_0 = 1$, $f_1 = y$, $f_2 = u$, and the Hamiltonian is $H = \lambda_0 + \lambda_1 y + \lambda_2 u$, with λ_0 a constant. The analysis is now the same as for the example in Section 4.2B.

Note that here the Hamiltonian is $H = \lambda_0 + \lambda_1 y + \lambda_2 u$, with $\lambda_0 \geq 0$ constant, and λ_1, λ_2 satisfying $d\lambda_1/dt = 0$, $d\lambda_2/dt = -\lambda_1$; hence $\lambda_1 = c_1$, $\lambda_2 = -c_1 t + c_2$, c_1, c_2 constant, as for the example in Section 4.2B. Here again $H_t = 0$, $dM/dt = 0$, and M is a constant. On the other hand, $g = 0$, $dt_1 = 0$, $dx_1 = dx_2 = dy_1 = dy_2 = 0$, dt_2 arbitrary, and (P4′) yields $M(t_2)\,dt_2 = 0$, or $M(t_2) = 0$.

B. Exercises

1. Write the problem of the curve $C{:}x = x(t)$, $t_1 \leq t \leq t_2$, of minimum length between two points $1 = (t_1, x_1)$, $2 = (t_2, x_2)$, $t_1 < t_2$ (or between a point and a curve) as a Lagrange problem of optimal control, and use the necessary conditions of this chapter to obtain the results in Section 3.1.
2. In analogy with Section 4.4A, write the transversality relations in finite form for the Bolza problem when B is defined by $k - 1$ equations $\phi_j(t_1, x_1, t_2, x_2) = 0, j = 2, \ldots, k$, and $I[x, u] = \phi_1(t_1, x_1, t_2, x_2) + \int_{t_1}^{t_2} f_0\, dt$.
3. In analogy with 4.4C(c), derive the transversality relations in finite form for the Lagrange problem with first end point $1 = (t_1, x_1)$ fixed, second end point x_2 on a given set B in R^n (target), when B has to be reached at some undetermined time $t_2 > t_1$. Case (c1): $B = R^n$ is the whole x-space. Case (c2): B is a given curve in R^n, B: $x = b(\tau)$, $\tau' \leq \tau \leq \tau''$, τ a parameter, $b(\tau) = (b_1, \ldots, b_n)$, and x hits B at a time $t_2 > t_1$, $x_2 = x(t_2) = b(\tau_0)$, $\tau' < \tau_0 < \tau''$. Case (c3): B is a given k-dimensional manifold in R^n, $B{:}x = b(\tau)$, $b(\tau) = (b_1, \ldots, b_n)$, $\tau = (\tau_1, \ldots, \tau_k)$, $x = x(t_2) = b(\tau_0)$, $t_2 > t_1$, $\tau_0 = (\tau_{10}, \ldots, \tau_{k0})$ interior to the region of definition of $b(\tau)$.

   Ans.: (c1) $M(t_2) = 0$, $\lambda_i(t_2) = 0$, $i = 1, \ldots, n$; (c2) $M(t_2) = 0, \sum_{j=1}^n \lambda_j(t_2) b_j'(\tau_0) = 0$; (c3) $M(t_2) = 0$, $\sum_{j=1}^n \lambda_j(t_2) b_{j\tau_s}(\tau_0) = 0$, $s = 1, \ldots, k$.

4. The same as Exercise 3, case (c3), where B is given by $n-k$ equations $\beta_0(x)=0$, $\sigma=1,\ldots,n-k$. (Note that $M(t_2)=0$, and that $\sum_{j=1}^{n}\lambda_j(t_2)\,dx_2^j=0$ for all $dx_2=(dx_2^1,\ldots,dx_2^n)$ satisfying $\sum_{j=1}^{n}\beta_{\sigma x^j}(x_2)\,dx_2^j=0$, $\sigma=1,\ldots,n-k$.)
5. In analogy with Section 4.4D derive the transversality relations in finite form for the Lagrange problem, with first end point $1=(t_1,x_1)$ fixed, terminal time $t_2>t_1$ fixed, and second end point x_2 on a given set B in R^2 (target). Same cases as (c 1–3) above.

   Ans.: Same as in Exercises 3 and 4 with no conclusion on $M(t_2)$.

5.4 The Value Function

We consider here the Bolza problem in Section 5.1 for the case in which any given fixed point (t_1,x_1) in a set V has to be transferred to a target set $B\subset V$ by means of an admissible trajectory $x(t)$, $t_1\le t\le t_2$, in V, and corresponding control function $u(t)$, for which the functional

$$I[x,u]=g(t_2,x(t_2))+\int_{t_1}^{t_2}f_0(t,x(t),u(t))\,dt$$

has its minimum value under the same requirements as in Section 5.1. In other words, g is here a given function of (t,x) on B, and for the AC trajectory x and the measurable control function u we require as usual $u(t)\in U$, $x(t_1)=x_1$, $(t_2,x(t_2))\in B$, and $(t,x(t))\in V$.

For any $(t_1,x_1)\in V$ we denote by $\Omega_{t_1x_1}$ the family of all admissible pairs $x(t)$, $u(t)$, $t_1\le t\le t_2$, transferring (t_1,x_1) to B in V; we shall always assume below that $(t_2,x(t_2))$ is the only point on B of the trajectory $(t,x(t))$; and we take

$$\omega(t_1,x_1)=\inf_{\Omega_{t_1x_1}}I[x,u]=\inf_{\Omega_{t_1x_1}}\left[g(t_2,x(t_2))+\int_{t_1}^{t_2}f_0(t,x(t),u(t))\,dt\right].$$

The function $\omega(t_1,x_1)$ is thus defined in V and may have the value $-\infty$. Also $\omega(t_1,x_1)=+\infty$ whenever $\Omega_{t_1x_1}$ is empty.

First we shall consider all points (t_1,x_1) for which the problem above has a minimum.

Let $3=(\bar t,\bar x)$ be any point of an optimal pair $x_0(t)$, $u_0(t)$, $t_1\le t\le t_2$, say $\bar x=x_0(\bar t)$, $t_1\le\bar t\le t_2$, with $x(t_1)=x_1$. Clearly the point $3=(\bar t,\bar x)$ can be transferred to B in V, since the pair $x_0(t)$, $u_0(t)$, $\bar t\le t\le t_2$—that is, the restriction of x_0, u_0 to the subinterval $[\bar t,t_2]$—evidently performs the task. Thus, the class $\bar\Omega$ of all (admissible) pairs x, u transferring $3=(\bar t,\bar x)$ to B is not empty.

5.4.i (Property of Optimality). *Every point $(\bar t,\bar x)$ of the trajectory x_0, say $\bar x=x_0(\bar t)$, $t_1\le\bar t\le t_2$, is transferred to B in V optimally by the pair $x_0(t)$, $u_0(t)$, $\bar t\le t\le t_2$.*

Proof. Let $1=(t_1,x_1)$, $2=(t_2,x_2(t_2))$. If the statement were not true, then there would be another admissible pair $x_1(t)$, $u_1(t)$, $\bar t\le t\le t_2'$, transferring $3=(\bar t,\bar x)$ to some point, say $4=(t_2',x_1(t_2'))$, on B, with (briefly) $I_{34}<I_{32}$, or (briefly) $J_{34}+g(4)<J_{32}+g(2)$, where J_{34}, J_{32} are the values of the integrals on 34 and 32, and $g(2)$, $g(4)$ are the values of g at 2 and 4. Now let us consider the admissible pair $X(t)$, $v(t)$, $t_1\le t\le t_2$, defined by $X=x_0$, $v=u_0$ on $[t_1,\bar t]$, $X=x_1$, $v=u_1$ on $[\bar t,t_2']$. Then X, v transfers $1=(t_1,x_1)$

to 4 in V with $J_{13} + J_{34} + g(4) < J_{13} + J_{32} + g(2)$, that is, $I_{14} < I_{12}$, a contradiction, since x_0, u_0 transfers 1 to 2 in V optimally. This proves (5.4.i). □

We now state a number of theorems concerning the value function for the Bolza problem. Clearly, they can all be derived from those in Section 4.5 for Mayer problems. The proofs, therefore, are omitted, and left as an exercise for the reader.

5.4.ii. *For every admissible trajectory $x(t)$, $t_1 \leq t \leq t_2$, transferring any point $(t_1, x(t_1))$ to B in V, the function*

$$S(t) = \omega(t, x(t)) - \int_t^{t_2} f_0(s, x(s), u(s))\, ds, \qquad t_1 \leq t \leq t_2,$$

is not decreasing in $[t_1, t_2]$ and is equal to $g(t_2, x(t_2))$ for $t = t_2$. If x, u is an optimal pair for (t_1, x_1) in V, then $S(t)$ is constant in $[t_1, t_2]$.

From now on in this section we assume that (a) any point $(t_1, x_1) \in V$ can be transferred to B in V by some admissible pair $x(t)$, $u(t)$, $t_1 \leq t \leq t_2$, $u(t)$ sectionally continuous, $u(t) \in U$; and (b) U is a fixed subset of R^m.

5.4.iii. *Under hypotheses* (a), (b), *if (t, x) is any interior point of V where the value function $\omega(t, x)$ is differentiable, then*

$$\omega_t(t, x) + f_0(t, x, v) + \sum_{i=1}^{n} \omega_{x^i}(t, x) f_i(t, x, v) \geq 0$$

for all $v \in U$. Moreover, if there is an optimal pair $x^*(s)$, $u^*(s)$, $t \leq s \leq t_2^*$, transferring (t, x) to B in V, u^* also sectionally continuous, then

$$\min_{v \in U} \left[\omega_t(t, x) + f_0(t, x, v) + \sum_{i=1}^{n} \omega_{x^i}(t, x) f_i(t, x, v) \right] = 0$$

and the minimum is attained by $v = u^*(t + 0)$.

5.4.iv. *Under hypotheses* (a) *and* (b), *if all points $(t, x) \in V$ can be optimally transferred to B in V by an admissible optimal pair $x(s)$, $u(s)$, $t \leq s \leq t_2$, if there is a function $p(t, x)$ in $V - B$, continuous in $V - B$, such that $u(s) = p(s, x(s))$ for any such pair, and if the value function $\omega(t, x)$ is continuous in V and of class C^1 in $V - B$ ($\omega = g$ on B), then $\omega(t, x)$ satisfies in $V - B$ the partial differential equation*

$$\omega_t(t, x) + f_0(t, x, u) + \sum_{i=1}^{n} \omega_{x^i}(t, x) f_i(t, x, p(t, x)) = 0,$$

or

$$\omega_t(t, x) + H(t, x, p(t, x), \omega_x(t, x)) = 0.$$

Here H is the Hamiltonian defined by (5.1.3) (*with $\lambda_0 = 1$*; cf. *Section* 5.1).

5.4.v. *Under hypotheses* (a), (b), *let (t_1, x_1) be an interior point of V, let $x^*(t)$, $u^*(t)$, $t_1 \leq t \leq t_2$, be an optimal pair transferring (t_1, x_1) to B in V, and such that the entire trajectory $(t, x^*(t))$ is made up of points interior to V, except perhaps the terminal point $(t_2, x^*(t_2)) \in B$. Suppose that the value function $\omega(t, x)$ is continuous on V and of class C^2 on $V - B$. Then* *the functions* $\lambda_i(t) = \omega_{x^i}(t, x(t))$, $t_1 \le t \le t_2$, $i = 1, \ldots, n$, *satisfy the relations* $d\lambda_i/dt = -H_{x^i}(t, x^*(t), u^*(t), \lambda(t))$ *and* $H(t, x^*(t), u^*(t), \lambda(t)) \le H(t, x^*(t), u, \lambda(t))$ *for all* $u \in U$ *and* $t \in [t_1, t_2]$ (a.e.). *In other words,* $\lambda(t) = (\lambda_1, \ldots, \lambda_n)$ *are the multipliers.*

5.5 Sufficient Conditions for the Bolza Problem

As in Section 4.6, we begin with the converse of (5.4.iii).

5.5.i (A SUFFICIENT CONDITION FOR OPTIMALITY FOR A SINGLE TRAJECTORY). *Let* $w(t, x)$ *be a function on* V *such that* $w(t, x) = g(t, x)$ *for* $(t, x) \in B \subset V$, *and let* (t_1, x_1) *be a point of* V *such that, for every pair* $x(t), u(t)$, $t_1 \le t \le t_2$, *in* $\Omega_{t_1 x_1}$, $w(t, x(t))$ *is finite and* $S(t) = w(t, x(t)) - \int_t^{t_2} f_0(s, x(s), u(s))\, ds$ *is nondecreasing, and for some pair* $x^*(t), u^*(t)$, $t_1 \le t \le t_2$, *in* $\Omega_{t_1 x_1}$, $S(t) = w(t, x^*(t)) - \int_t^{t_2} f_0(s, x^*(s), u^*(s))\, ds$ *is constant on* $[t_1, t_2]$. *Then, the pair* x^*, u^* *is optimal for* (t_1, x_1) *in* V.

5.5.ii (A SUFFICIENT CONDITION FOR OPTIMALITY FOR A SINGLE TRAJECTORY). *Let* $w(t, x)$ *be a given function in* V *with* $w(t, x) = g(t, x)$ *on* B, $w(t, x)$ *continuous in* V *and continuously differentiable in* $V - B$, *satisfying*

$$w_t(t, x) + f_0(t, x, u) + \sum_{i=1}^{n} w_{x^i}(t, x) f_i(t, x, u) \ge 0$$

for all $(t, x) \in V - B$ *and* $u \in U$. *If* $x^*(t), u^*(t)$, $t_1 \le t \le t_2$, *is an admissible pair transferring* (t_1, x_1) *to* B *in* V, *and*

$$w_t(t, x^*(t)) + f_0(t, x^*(t), u^*(t)) + \sum_{i=1}^{n} w_{x^i}(t, x^*(t)) f_i(t, x^*(t), u^*(t)) = 0, \qquad t \in [t_1, t_2] \text{ (a.e.)},$$

(in other words, $w(t, x^*(t)) - \int_t^{t_2} f_0(s, x^*(s), u^*(s))\, ds$, $t_1 \le t \le t_2$, *is constant), then* (x^*, u^*) *is optimal.*

5.5.iii. *If* $w(t, x)$ *is a continuous function in* V, *of class* C^1 *in* $V - B$, *with* $w(t, x) = g(t, x)$ *on* B, *if* $x(t), u(t)$, $t_1 \le t \le t_2$, *is an admissible pair in* V *transferring* (t_1, x_1) *to* B *in* V, *and if* $S(t) = w(t, x(t)) - \int_t^{t_2} f_0(s, x(t), u(s))\, ds$ *is constant on* $[t_1, t_2]$, *then*

$$w_t(t, x(t)) + H(t, x(t), u(t), w_x(t, x(t))) = 0, \qquad t \in [t_1, t_2] \text{ (a.e.)}.$$

5.5.iv (BOLTYANSKII'S SUFFICIENT CONDITION FOR A SINGLE TRAJECTORY). *Let* V, B, *and* S *be subsets of* R^{1+n}, $B \subset S \subset V$, S *of measure zero. Let* $w(t, x)$ *be a given function in* V, *with* $w(t, x) = g(t, x)$ *on* B, $w(t, x)$ *continuous in* V *and continuously differentiable in* $V - S$, *such that*

$$w_t(t, x) + f_0(t, x, u) + \sum_{i=1}^{n} w_{x^i}(t, x) f_i(t, x, u) \ge 0$$

for all $(t, x) \in V - S$ *and all* $u \in U$, *and such that* $S(t) = w(t, x^*(t)) - \int_t^{t_2} f_0(s, x^*(s), u^*(s))\, ds$, $t_1 \le t \le t_2$, *is constant on a given admissible pair* $x^*(t), u^*(t)$, $t_1 \le t \le t_2$, *transferring* (t_1, x_1) *to* B *in* V. *Then* x^*, u^* *is optimal for* (t_1, x_1) *in* V.

Remark. The concepts of synthesis, feedback control, marked trajectories, and regular synthesis are completely analogous to those we have discussed in Section 4.6 for Mayer problems. We leave their formulation as an exercise for the reader, together with the analogous statement of Boltyanskii, (4.6.v).

As a further exercise the reader may derive the statements of Sections 5.4–5 from those of Sections 4.5–6. Alternatively, the reader may prove the statements of Sections 5.4–5 directly. The proofs are analogous to the ones for Sections 4.5–6.

Bibliographical Notes

The statements of this chapter are only technical variants of those of Chapter 4, to which we refer.

CHAPTER 6

Examples and Exercises on Optimal Control

6.1 Stabilization of a Material Point Moving on a Straight Line under a Limited External Force

A point P moves along the x-axis governed by the equation $x'' = u$ with $|u| \leq 1$. We are to take P from any given state $x = a$, $x' = b$ to rest at the origin $x = 0$, $x' = 0$ in the shortest time. By introducing phase coordinates $x = x$, $y = x'$, we have the Mayer problem of minimum time:

$$\begin{aligned}
&dx/dt = y, \quad dy/dt = u, \\
&0 \leq t \leq t_2, \quad u \in U = [-1 \leq u \leq 1], \\
&I[x, y, u] = g = t_2, \\
&t_1 = 0, \quad x(t_1) = a, \quad y(t_1) = b, \quad x(t_2) = 0, \quad y(t_2) = 0,
\end{aligned} \tag{6.1.1}$$

where we seek the minimum of the functional I under the constraints.

We have already initiated the discussion of this problem in Sections 4.2B and 4.6D where we showed that any state (a, b) can be taken to $(0, 0)$ by a unique pair $x(t)$, $u(t)$, $0 \leq t \leq t_2$, satisfying the requirements of the necessary conditions (P1), (P2) of (4.2.i). Already in Section 4.2B we concluded that this unique pair was the optimal solution because of the existence theorems. Independently, in Section 4.6D, we reached the same conclusion by regular synthesis and Boltyanskii's theorem.

It is interesting—and also good training for more difficult problems—to see what further information (P3), (P4) would entail. In Section 4.2B we computed H and the unique pair x, u satisfying (P1), (P2). Now $H_t = 0$, since the problem is autonomous; hence, from (P3), $dM/dt = 0$, M a constant along the optimal trajectory. Here $t_1 = 0$, $x(t_1) = a$, $y(t_1) = b$, $x(t_2) = 0$, $y(t_2) = 0$, $g = t_2$; hence $0 = dt_1 = dx_1 = dy_1 = dx_2 = dy_2$, dt_2 arbitrary, $dg = dt_2$, and (P4) yields $(\lambda_0 + M(t_2))\, dt_2 = 0$. Thus, $\lambda_0 = -M(t_2)$.

For $P = (a, b)$ above the switch line BOA (see figure in Section 4.2B), the control function is $u = -1$, the arc PB has the equation $x = -2^{-1}(t - \alpha)^2 + \beta$, $y = \alpha - t$, where α, β can be determined from the initial data $x(0) = a$, $y(0) = b$. We obtain PB: $x = -2^{-1}(b-t)^2 + a + 2^{-1}b^2$, $y = b - t$, $u = -1$, $0 \le t \le \bar{t}$. Here $\bar{t}$ is the time in which the arc PB intersects the arc $BO: x = 2^{-1}y^2$, so that

$$-2^{-1}(b-\bar{t})^2 + a + 2^{-1}b^2 = 2^{-1}(b-\bar{t})^2,$$
$$\bar{t} = b + (a + 2^{-1}b^2)^{1/2} \qquad \text{(nonnegative square root)},$$
$$B = (x(\bar{t}), y(\bar{t})) = (2^{-1}a + 4^{-1}b^2, -(a + 2^{-1}b^2)^{1/2}).$$

Finally, $BO: x = 2^{-1}(t-t_2)^2$, $y = t - t_2$, $u = 1$, and t_2 can be determined by requiring again $B = (x(\bar{t}), y(\bar{t}))$, or

$$2^{-1}a + 4^{-1}b^2 = 2^{-1}(b + (a + 2^{-1}b^2)^{1/2} - t_2)^2,$$
$$-(a + 2^{-1}b^2)^{1/2} = b + (a + 2^{-1}b^2)^{1/2} - t_2,$$

from which

$$t_2 = b + 2(a + 2^{-1}b^2)^{1/2}.$$

Since $\lambda_1(t) = c_1$, and $\lambda_2(t) = -c_1t + c_2$ is zero at $t = \bar{t}$ and positive for $0 < t < \bar{t}$, we have $c_2 = c_1\bar{t}$, $\lambda_2(t) = c_1(\bar{t} - t)$, $0 \le t \le t_2$, $c_1 > 0$. For $0 \le t \le \bar{t}$, we have $u = -1$, and

$$M(t) = \lambda_1 y + \lambda_2 u = c_1(b-t) - c_1(\bar{t}-t) = c_1(b-\bar{t})$$
$$= -c_1(a + 2^{-1}b^2)^{1/2};$$

for $\bar{t} \le t \le t_2$, we have $u = +1$, and

$$M(t) = \lambda_1 y + \lambda_2 u = c_1(t-t_2) + c_1(\bar{t}-t)$$
$$= c_1(\bar{t}-t_2) = -c_1(a + 2^{-1}b^2)^{1/2}.$$

Thus, $M(t)$ is a negative constant along PBO, and $\lambda_0 = -M(t_2) = c_1(a + 2^{-1}b^2)^{1/2} > 0$, as expected. We can take $\lambda_0 = 1$, $c_1 = (a + 2^{-1}b^2)^{-1/2}$, $c_2 = c_1\bar{t} = 1 + b(a + 2^{-1}b^2)^{-1/2}$, and $M(t) = -1$ along the entire trajectory PBO. Moreover, $\lambda_2(t) = c_1(\bar{t} - t)$, and $\lambda_2 > 0$ at $t = 0$, that is, at the point (a, b), we have $\lambda_2 = c_2 > 0$.

For $P = (a, b)$ below the switch line, the control function is $u = +1$, and for the arcs PA, AO and times $\bar{t}$ and t_2 we find, by analogous computations,

$$PA: x = 2^{-1}(t+b)^2 + a - 2^{-1}b^2,\ y = t + b,\ u = 1,\ 0 \le t \le \bar{t},$$
$$\bar{t} = -b + (2^{-1}b^2 - a)^{1/2} \qquad \text{(nonnegative square root)},$$
$$AO: x = -2^{-1}(t_2 - t)^2, \quad y = t_2 - t, \quad u = -1, \quad \bar{t} \le t \le t_2,$$
$$A = (x(\bar{t}), y(\bar{t})) = [-2^{-1}(2^{-1}b^2 - a), (2^{-1}b^2 - a)^{1/2}],$$
$$t_2 = -b + 2(2^{-1}b^2 - a)^{1/2}.$$

Since $\lambda_1(t) = c_1$, and $\lambda_2(t) = -c_1t + c_2$ is zero at $t = \bar{t}$ and negative for $0 \le t \le \bar{t}$, we have $c_2 = c_1\bar{t}$, $\lambda_2(t) = c_1(\bar{t} - t)$, $0 \le t \le t_2$, $c_1 < 0$. Also, $M(t) = \lambda_1 y + \lambda_2 u = c_1(b + \bar{t}) = c_1(2^{-1}b^2 - a)^{1/2}$, a negative constant, and $\lambda_0 = -M(t_2) = -c_1(2^{-1}b^2 - a)^{1/2} > 0$, as expected. We can take $\lambda_0 = 1$, $c_1 = -(2^{-1}b^2 - a)^{1/2}$, $c_2 = -1 + b(2^{-1}b^2 - a)^{-1/2}$, and again $M(t) = -1$ along the entire trajectory PAB.

We have shown above that we are dealing with a feedback situation, as discussed in Section 4.6 or of *synthesis*, as it is often called. Here V is the whole (t, x, y)-space R^3, and $B = B_2$ is the positive t-axis, or $B = [(t, 0, 0)|, t \ge 0]$. Now, if we write (x, y) for

(a, b), then t_2, c_1, c_2 become functions of x, y, namely,

$$\begin{aligned} t_2 &= y + 2(x + 2^{-1}y^2)^{1/2}, \\ c_1 &= (x + 2^{-1}y^2)^{-1/2} > 0, \\ c_2 &= 1 + y(x + 2^{-1}y^2)^{-1/2} > 0 \qquad \text{above } AOB, \\ t_2 &= -y + 2(2^{-1}y^2 - x)^{1/2}, \\ c_1 &= -(2^{-1}y^2 - x)^{1/2} < 0, \\ c_2 &= -1 + y(2^{-1}y^2 - x)^{-1/2} < 0 \qquad \text{below } AOB, \end{aligned}$$

and the first relations hold also on OA, and the second relations hold also on OB.

Thus, the function w defined by

$$\begin{aligned} w(t, x, y) &= t + t_2 = t + y + 2(x + 2^{-1}y^2)^{1/2} \qquad \text{above } AOB \text{ and on } OA, \\ w(t, x, y) &= t + t_2 = t - y + 2(2^{-1}y^2 - x)^{1/2} \qquad \text{below } AOB \text{ and on } OB, \end{aligned} \tag{6.1.2}$$

computed along any trajectory $x(t)$, $y(t)$, $0 \le t \le t_2$ (optimal or not), has derivative

$$D = (d/dt)w(t, x(t), y(t)) = 1 + w_x x' + w_y y' = 1 + w_x y + w_y u,$$

namely

$$\begin{aligned} D &= 1 + y(x + 2^{-1}y^2)^{-1/2} + [1 + y(x + 2^{-1}y^2)^{-1/2}]u \qquad \text{above } AOB, \\ D &= 1 - y(2^{-1}y^2 - x)^{-1/2} + [-1 + y(2^{-1}y^2 - x)^{-1/2}]u \qquad \text{below } AOB, \end{aligned}$$

where the bracket is λ_2 at (x, y), and $\lambda_2 > 0$ above AOB, $\lambda_2 < 0$ below AOB. Thus, for any fixed (x, y), D takes its minimum value zero for $u = -1$ above AOB, for $u = +1$ below AOB, and in any case

$$D = (d/dt)w(t, x(t), y(t)) \ge D_{\min} = 0.$$

In other words, the function $w(t, x, y)$ is nondecreasing on any trajectory $x(t)$, $y(t)$, $0 \le t \le t_2$, and for any fixed (a, b), $w(t, x, y)$ is constant ($D \equiv 0$) on the unique trajectory determined above transferring (a, b) to $(0, 0)$ and satisfying the necessary condition. By virtue of (4.6.v), we conclude that this is an optimal trajectory.

In view of the above, $w(t, x, y)$ as defined by (6.1.2) is the value function and we can denote it by $\omega(t, x, y)$: it is constant along each optimal trajectory in V and its value $\omega(0, x, y) = \omega(t_2, 0, 0) = t_2$ is actually the minimum value of the functional I thought of as a function of the initial point (x, y). Above, we have also verified that ω satisfies the partial differential equation

$$\begin{aligned} \omega_t + y\omega_x - \omega_y &= 0 \qquad \text{above } AOB, \\ \omega_t + y\omega_x + \omega_y &= 0 \qquad \text{below } AOB, \end{aligned}$$

that is, the Hamilton–Jacobi equation. The reader may verify that ω is continuous in R^2, but its partial derivatives have jumps across the line AOB.

Remark. The proposed equation $x'' = u$, $|u| \le 1$, is the nondimensional form for the actual problem $my'' = f$, (y the space variable, m the mass, f the force in the direction of the y-axis, $|f| \le F$, m, F positive constants). By taking $x = (m/F)y$, $u = f/F$, we obtain the equations $x'' = u$, $|u| \le 1$, as proposed. The magnitude constraint $|f| \le F$ can be thought of as due to physical limitations in the thrust available with given equipment. The same equation is also the nondimensional form for the analogous problem of taking a shaft to rest in a fixed position under limited torque action. The equation is then $Iy'' = f$ (I the moment of inertia, y the angular displacement, f the torque, $|f| \leq F$, I, F given positive constants).

The line AOB is called a *switch curve*. If we denote by γ^- the arc OA, by γ^+ the arc OB, by R^- the part of the xy-plane R^2 above AOB, and by R^+ the part of the xy-plane R^2 below AOB, then we see that the optimal value p of the control variable u is now proved to be a single valued function $p(x, y)$ in R^2, namely $p = -1$ on $\gamma^- \cup R^-$, and $p = +1$ on $\gamma^+ \cup R^+$. This situation with one or a few switch curves is common to a great many problems (see examples below).

There may be more switch curves, and in higher dimensions there may be switch surfaces or hypersurfaces.

The practical importance of having reduced the optimal control variable to a function $p(x, y)$ of the state variables should also be pointed out. If some device is created which automatically feeds into the system the value $u = p(x, y)$ of the control variable, then an automatic feedback optimal control system has been designed. In the present situation (and in many similar ones), all the device needs to do is read whether the point (x, y) is above or below the switch curve, and feed the correct value $p = -1$ or $p = +1$ into the system.

6.2 Stabilization of a Material Point under an Elastic Force and a Limited External Force

A point P moves along the x-axis governed by the equation $x'' + x = u$ with $|u| \leq 1$. We are to take P from any given state $x = a$, $x' = b$ to rest at the origin $x = 0$, $x' = 0$ in the shortest time. By introducing phase coordinates $x = x$, $y = x'$ we have a Mayer problem of minimum time:

$$\begin{gathered} dx/dt = y, \qquad dy/dt = -x + u, \quad u \in U = [-1 \leq u \leq 1], \\ t_1 = 0, \quad x(t_1) = a, \quad y(t_1) = b, \quad x(t_2) = 0, \quad y(t_2) = 0, \\ I[x, y, u] = g = t_2. \end{gathered} \tag{6.2.1}$$

Here we have

$$\begin{aligned} H &= H(t, x, y, u, \lambda_1, \lambda_2) = \lambda_1 y + \lambda_2(-x + u), \\ M &= M(t, x, y, \lambda_1, \lambda_2) = \min_{-1 \leq u \leq 1} H \\ &= \begin{cases} \lambda_1 y - \lambda_2 x - \lambda_2 & \text{if } \lambda_2 > 0, \\ \lambda_1 y - \lambda_2 x + \lambda_2 & \text{if } \lambda_2 < 0. \end{cases} \end{aligned}$$

Precisely, if $\lambda_2 > 0$, then the minimum of H is taken for $u = -1$; if $\lambda_2 < 0$, then the minimum of H is taken for $u = 1$. In other words, the optimal strategy $u(t)$ is related to the continuous multiplier $\lambda_2(t)$ by the relation $u(t) = -\operatorname{sgn} \lambda_2(t)$, $t_1 \leq t \leq t_2$. The equations for λ_1, λ_2 are

$$d\lambda_1/dt = \lambda_2, \qquad d\lambda_2/dt = -\lambda_1.$$

Thus, $\lambda_1 = \alpha \sin(t + \beta)$, $\lambda_2 = \alpha \cos(t + \beta)$, α, β constants. Since $dg = dt_2$ is not identically zero, λ_1 and λ_2 cannot be both zero. Thus $\alpha \neq 0$, and we can assume $\alpha > 0$. Here λ_2 changes sign together with $\cos(t + \beta)$, and in the corresponding intervals u has the values $+1$ and -1. Precisely, an optimal solution is given by a strategy $u(t)$ which is sectionally continuous with values alternately $+1$ and -1, and the time intervals of constancy are of length π except for the first and the last ones, which may be shorter. In the intervals where $u=1$ the equations (6.2.1) yield $x=1+\gamma\sin(t+\delta)$, $y=\gamma\cos(t+\delta)$, γ, δ constants, or $(x-1)^2+y^2=\gamma^2$, that is, all circles of center $O_1=(1,0)$ traversed in the clockwise sense. In the intervals where $u=-1$ the equations (6.2.1) yield $x=-1+\gamma\sin(t+\delta)$, $y=\gamma\cos(t+\delta)$, or $(x+1)^2+y^2=\gamma^2$, that is, all circles of center $O_2=(-1,0)$, also traversed in the clockwise sense. We give in the figure the two families of circles.

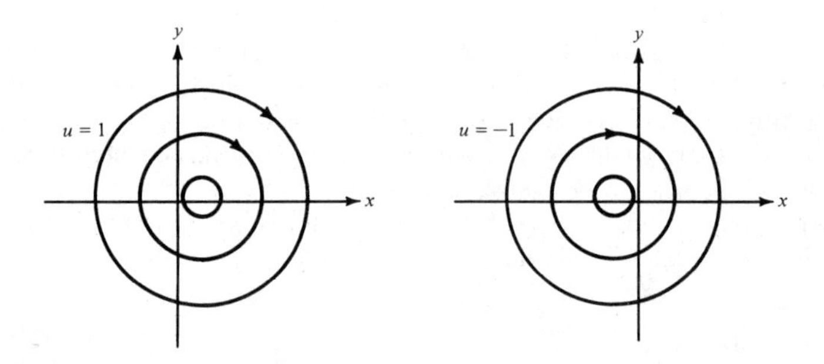

An optimal path must be made of a final arc reaching the origin at the time t_2, and thus be either of the two arcs A_1O or B_1O with centers O_1 and O_2, of radius 1, and reaching the origin. But such an arc, say A_1O, must be preceded by an arc A_2A_1 with center O_2 and of opening exactly π. Then the point A_2 is symmetric with A_1 with respect to O_2 and hence lies on the semicircle $N_1A_2N_2$ of radius 1, identical to OA_1M_1. Now the arc A_2A_1 must be preceded by an arc A_3A_2 with center O_1 and opening exactly π. Then the point A_3 is symmetric with A_2 with respect to O_1 and hence lies on the semicircle $M_1A_3M_2$ of radius 1 and identical to $N_1A_2N_2$. We can continue thus with arcs alternately with center O_1 and O_2. Analogously, we can continue any final arc as B_1O. Then we see that a unique solution satisfying the necessary condition is given by a trajectory $PA_nA_{n-1}\cdots A_2A_1O$ or $PB_nB_{n-1}\cdots B_2B_1O$ made up of arcs with centers O_1 and O_2 alternately. All the corner points are on the dashed curve $\cdots N_3N_2N_1OM_1M_2M_3\cdots$ of the figure.

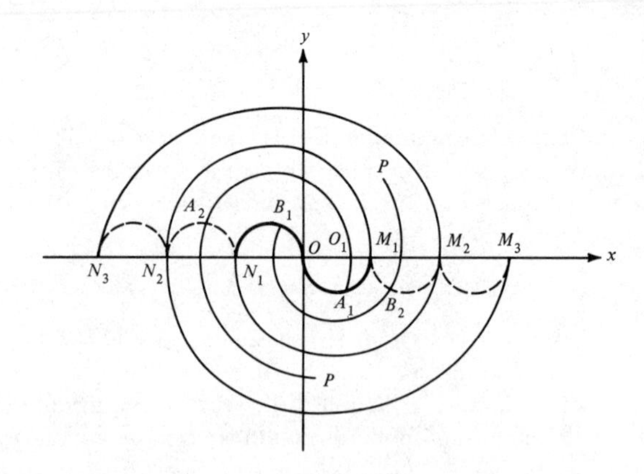

Every point (a, b) can be transferred to $(0, 0)$ by one and only one of these paths satisfying necessary condition. Here $A = R^3$, $U = [-1 \le u \le 1]$ is compact, $B = (0, a, b, 0, 0) \times [t \ge 0]$ is closed, $g = t_2$ is continuous, and Ω is not empty. Also, $P = \{(a, b)\}$ is compact, $xf_1 + yf_2 = xy + y(x + u) \le 2x^2 + 2y^2$, and $g \to +\infty$ as $t_2 \to +\infty$. By the existence theorem (4.3.i) an optimal solution exists. Thus, the trajectory from (a, b) to $(0, 0)$, we have uniquely determined above, is optimal. The dashed line $\cdots N_3N_2N_1OM_1M_2M_3 \cdots$ is the switch line separating the two possible values $u = \pm 1$ for the optimal strategy.

Remark 1. The proposed equation $x'' + x = u$, $|u| \le 1$, of the present problem is a nondimensional form for the problem of the mass–spring system monitored by equations $my'' + ay = f$ (m the mass, y the displacement from the rest position, f the external force, $|f| \le F$, a the spring constant, m, a, F positive constants). By taking $\omega^2 = a/m$, $u = f/F$, we have the equations $y'' + \omega^2 y = ku$, $|u| \le 1$, with $k = F/m$ (see Exercise 13 in Section 6.6A).

In these equations, we can always assume $k = 1$ by replacing y with the new variable $z = k^{-1}y$. Finally, in the equations $z'' + \omega^2 z = u$, $|u| \le 1$, we can always assume $\omega = 1$ by a change in both time and space variables, namely, $\tau = \omega t$, $z = \omega^2 x$. We are led to the same equations in the case of the small displacement of a pendulum and of the torsion pendulum.

Remark 2. Given a rigid body B in space, let a_1, a_2, a_3 denote the principal axes through the center of mass of B, let I_1, I_2, I_3 be the moments of inertia of B with respect to a_1, a_2, a_3, and let y_1, y_2, y_3 be angular velocities. We assume that suitably located jets impart torques τ_1, τ_2, τ_3 on B. Then the differential equations of motion are, as we know from Sections 3.14D,F, equations (3.14.18),

$$
\begin{aligned}
I_1\, dy_1/dt &= (I_2 - I_3)y_2y_3 + \tau_1,\\
I_2\, dy_2/dt &= (I_3 - I_1)y_3y_1 + \tau_2,\\
I_3\, dy_3/dt &= (I_1 - I_2)y_1y_2 + \tau_3.
\end{aligned}
$$

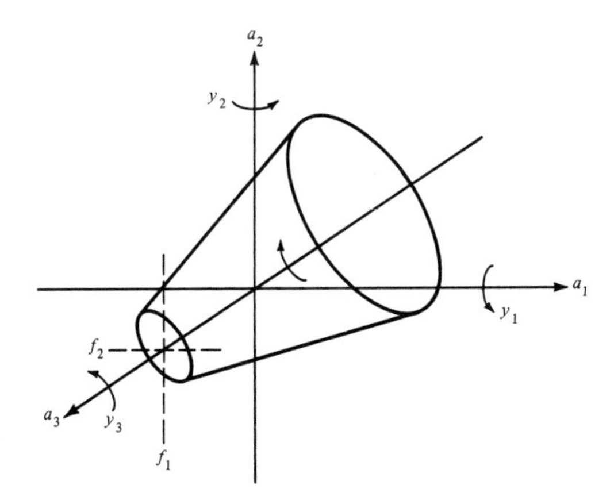

If we assume that the body has an axis of symmetry, say the axis a_3, then $I_1 = I_2$. If I denotes their common value, then the equations become

$$\begin{aligned} I\,dy_1/dt &= (I - I_3)y_2y_3 + \tau_1, \\ I\,dy_2/dt &= (I_3 - I)y_3y_1 + \tau_2, \\ I_3\,dy_3/dt &= \tau_3. \end{aligned}$$

We assume that on the axis a_3 of the body there are two gas jets at a fixed distance $c > 0$ from the center of mass, of thrust f_1, f_2, producing torques $\tau_1 = cf_1$, $\tau_2 = cf_2$. Also we assume that there is a third jet on the plane y_1y_2 at a distance $c_3 > 0$ from the axis a_3 and orthogonal to the radius from the center of mass. Thus $\tau_3 = c_3f_3$, and the equations above reduce to

$$\begin{aligned} dy_1/dt &= I^{-1}(I - I_3)y_3y_2 + I^{-1}cf_1, \\ dy_2/dt &= -I^{-1}(I - I_3)y_3y_1 + I^{-1}cf_2, \\ dy_3/dt &= I_3^{-1}c_3f_3. \end{aligned}$$

Note that if at $t = 0$ the three velocities have values $y_1(0) = \xi_1$, $y_2(0) = \xi_2$, $y_3(0) = \xi_3$, then the thrust $f_3(t)$ determines $y_3(t)$:

$$y_3(t) = \xi_3 + \int_0^t I^{-1}c_3f_3(t)\,dt.$$

In other words, the third equation can be solved independently of the other two, and once $y_3(t)$ is known, then we can solve the system of the first two equations. For physical reasons it may be convenient to keep y_3 constant (say for aerodynamical stability in the reentry of a capsule). Thus, let us assume that $y_3(t)$ has been brought to the desired constant value by acting on the third jet alone, so that from a given instant on, say for $t \geq 0$, we take $f_3(t) = 0$, $y_3(t) = \xi_3 = \text{constant}$. Then the first two equations above become linear equations in y_1, y_2:

$$dy_1/dt = I^{-1}(I - I_3)\xi_3y_2 + I^{-1}cf_1, \qquad dy_2/dt = -I^{-1}(I - I_3)\xi_3y_1 + I^{-1}cf_2,$$

or

$$dy_1/dt = \omega y_2 + u, \qquad dy_2/dt = -\omega y_1 + v,$$

with $\omega = I^{-1}(I - I_3)\xi_3$, $u = I^{-1}cf_1$, $v = I^{-1}cf_2$, $|u| \leq I^{-1}cF_1$, $|v| \leq I^{-1}cF_2$.

As a further specialization, let us assume now that $f_1 = 0$ and hence $u = 0$. Thus, only the jet j_2 is functioning, the one which is parallel to the axis a_1, and the system reduces to

$$y_1' = \omega y_2, \qquad y_2' = -\omega y_1 + v,$$

that is, $y_1'' + \omega^2 y_1 = \omega v$, with ω as above, $v = I^{-1}cf_2$, $|v| \leq I^{-1}cF_2$. By the use of suitable units it is possible to make $\omega = 1$. We have now the desired interpretation of the example of this Section 6.2: the problem of reducing the initial velocities $y_1(0) = \xi_1$, $y_2(0) = \xi_2$ to zero, $y_1(t_2) = 0$, $y_2(t_2) = 0$, in a minimum time t_2 (while $y_3 = \xi_3$ remains constant).

The example discussed in the following Section 6.3 correspond to the situation in which both jets j_1 and j_2 are functioning with $F_1 = F_2$. By the use of suitable units it is possible to make $I^{-1}cF_1 = I^{-1}cF_2 = 1$.

6.3 Minimum Time Stabilization of a Reentry Vehicle

Let us consider a system governed by the differential equations

$$dx/dt = \omega y + ku, \quad dy/dt = -\omega x + kv, \quad |u| \leq 1, \quad |v| \leq 1. \tag{6.3.1}$$

We are to take any state (a, b) to $(0, 0)$ in the minimum time. Here ω and k are positive constants. In other words, we have a Mayer problem with equation (6.3.1) and further data

$$\begin{aligned} U = [-1 \leq u, v \leq 1], \quad t_1 = 0, \quad x(t_1) = a, \quad y(t_1) = b, \\ x(t_2) = 0, \quad y(t_2) = 0, \qquad I[x, y, u, v] = g = t_2. \end{aligned} \tag{6.3.2}$$

Here we have

$$\begin{aligned} H &= H(t, x, y, u, v, \lambda_1, \lambda_2) = \omega(\lambda_1 y - \lambda_2 x) + k(\lambda_1 u + \lambda_2 v), \\ M &= M(t, x, y, \lambda_1, \lambda_2) = \min H, \end{aligned}$$

where min is taken for $(u, v) \in U$, and hence as before the optimal strategy $u(t)$, $v(t)$ is related to the continuous multipliers $\lambda_1(t)$, $\lambda_2(t)$ by

$$u(t) = -\operatorname{sgn} \lambda_1(t), \quad v(t) = -\operatorname{sgn} \lambda_2(t), \qquad t_1 \leq t \leq t_2, \tag{6.3.3}$$

whence

$$M(t, x, y, \lambda_1, \lambda_2) = \omega(\lambda_1 y - \lambda_2 x) - k(|\lambda_1| + |\lambda_2|).$$

The equations for λ_1, λ_2 are

$$d\lambda_1/dt = \omega\lambda_2, \qquad d\lambda_2/dt = -\omega\lambda_1;$$

hence

$$d^2\lambda_1/dt^2 = -\omega^2\lambda_1,$$

and

$$\lambda_1(t) = C \sin(\omega t + \alpha), \qquad \lambda_2(t) = C \cos(\omega t + \alpha), \tag{6.3.4}$$

C, α constants. These constants C, α determine $\lambda_1(t)$, $\lambda_2(t)$ as well as $u(t)$, $v(t)$ by means of the relations (6.3.3).

We see from (6.3.4) that each $\lambda_1(t)$ and $\lambda_2(t)$, $0 \leq t \leq t_2$, have constant signs in intervals of length π/ω (except for the two terminal intervals, which may be shorter), and that in half of each such interval, that is, in each time interval $\pi/2\omega$, either $\lambda_1(t)$ or $\lambda_2(t)$ will change sign alternately. If we set $u(t) = \Delta_1$, $v(t) = \Delta_2$, $\Delta_1, \Delta_2 = \pm 1$, we see from (6.3.3), (6.3.4) that each $u(t)$ and $v(t)$, $0 \leq t \leq t_2$, is constant in intervals of length π/ω (except for the two terminal intervals), and that in each time interval $\pi/2\omega$ either $u(t)$ or $v(t)$ switches between the values ± 1 alternately. In each interval of length $\pi/2\omega$ in which both u and v are constants, the equations (6.3.1) become

$$dx/dt = \omega y + k\Delta_1, \qquad dy/dt = -\omega x + k\Delta_2;$$

hence

$$x(t) = D \sin(\omega t + \beta) + k\omega^{-1}\Delta_2, \qquad y(t) = D \cos(\omega t + \beta) - k\omega^{-1}\Delta_1,$$

D, β constants, and

$$(x - k\omega^{-1}\Delta_2)^2 + (y + k\omega^{-1}\Delta_1)^2 = D^2.$$

Precisely, in each such interval of length $\pi/2\omega$, the point (x, y) describes exactly one fourth of the circumference of such a circle in the clockwise sense. We have now enough information to describe graphically the optimal trajectories and strategies. If we take $c = k\omega^{-1}$, note that the values $(-1, +1)$, $(-1, -1)$, $(+1, -1)$, $(+1, +1)$ for u and v correspond to arcs of circles with centers $(-c, -c)$, $(-c, c)$, (c, c), $(c, -c)$.

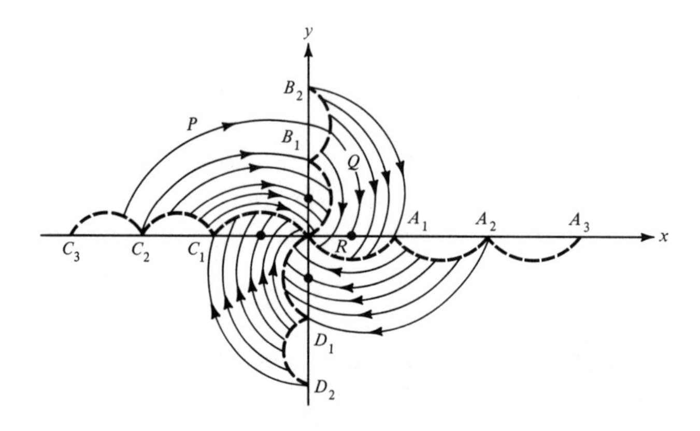

If the origin $O = (0, 0)$ is reached, then it can be reached only by means of the oriented arcs A_1O, B_1O, C_1O, D_1O from the circles with centers $T_1 = (c, c)$, $T_2 = (-c, c)$, $T_3 = (-c, -c)$, $T_4 = (c, -c)$ respectively (see illustration), and with the indicated orientation. Each of these arcs is one fourth of the corresponding circumference. These arcs can be reached in turn by means of analogous arcs, also one fourth of the circles with centers T_2, T_3, T_4, T_1 respectively, as in the illustration, and whose initial points are on arcs B_2B_1, C_2C_1, D_2D_1, A_2A_1 similar to the previous ones (because of symmetry). In turn these four arcs can be reached by means of analogous arcs of circles with centers T_3, T_4, T_1, T_2 respectively, whose initial points are on arcs C_3C_2, D_3D_2, A_3A_2, B_3B_2 similar to the previous ones; and so on. In the illustration the path $PQRO$ depicts the only possible optimal trajectory from the state $P = (a, b)$ to the equilibrium state $(0, 0)$.

The entire xy-plane is now divided by four lines Γ_1, Γ_2, Γ_3, Γ_4 into four regions R_1, R_2, R_3, R_4, where $\Gamma_1 = OA_1A_2A_3, \ldots$, etc. In each region R_i the trajectories are arcs with of center T_i, oriented clockwise, $i = 1, 2, 3, 4$, which are generated by constant values of (u, v), say $(1, -1)$, $(-1, -1)$, $(-1, 1)$, $(1, 1)$ respectively. The Γ_i are the switch lines.

Here $A = R^3$, $U = [-1 \leq u,\ v \leq +1]$ is compact, $B = (0, a, b, 0, 0) \times [t \geq 0]$ is closed, $g = t_2$ is continuous, and Ω is not empty. Also, $P = \{(a, b)\}$ is compact, $xf_1 + yf_2 = x(\omega y + ku) + y(-\omega x + kv) \leq k(x^2 + y^2 + 1)$, and $g \to \infty$ as $t_2 \to +\infty$. By the existence theorem (4.3.i) an optimal solution exists. Thus, the uniquely determined path and corresponding strategy are optimal.

6.4 Soft Landing on the Moon

If h denotes the height, v the vertical velocity, and m the variable mass of a spacecraft attempting to land on the moon, g the acceleration of gravity on the moon, u the variable thrust, and μ the maximal thrust attainable by the spacecraft's jet, then the equations of motion are

$$h' = v, \quad v' = -g + m^{-1}u, \quad m' = -ku, \qquad (') = d/dt,$$

where k is a constant. The third equation simply states that the loss of mass per second, that is, the fuel used by the jet per second, is proportional to the thrust of the jet. We shall assume that at time $t = 0$ the variables h, v, m have initial values $h_0 > 0$, $v_0 \leq 0$, and $M + F$, where M is the mass of the spacecraft without fuel, and F the total amount of fuel. We have to arrange that at the time $t_2 > 0$ of landing we have $h = 0$, $v = 0$, and that the amount of fuel used, or $M + F - m(t_2)$, is a minimum, that is, $-m(t_2)$ is a minimum.

By changing h, v, m into x_1, x_2, x_3 we have the Mayer problem of the minimum of the functional $I[x, u] = -x_3(t_2)$ under the constraints

$$\begin{aligned} x_1' &= x_2, \qquad x_2' = -g + x_3^{-1}u, \qquad x_3' = -ku, \\ x_1(0) &= h_0, \qquad x_2(0) = v_0, \qquad x_3(0) = M + F, \\ x_1(t_2) &= x_2(t_2) = 0, \qquad 0 \leq u \leq \mu, \qquad x_3(t_2) \geq M. \end{aligned}$$

The Hamiltonian and the equations for the multipliers are here

$$\begin{gathered} H = \lambda_1 x_2 + \lambda_2(-g + x_3^{-1}u) - \lambda_3 ku = (\lambda_1 x_2 - \lambda_2 g) + (\lambda_2 x_3^{-1} - \lambda_3 k)u, \\ \lambda_1' = 0, \qquad \lambda_2' = -\lambda_1, \qquad \lambda_3' = \lambda_2 x_3^{-2}u, \end{gathered}$$

so that $\lambda_1 = c_1$, $\lambda_2 = -c_1 t + c_2$, $0 \leq t \leq t_2$, where c_1, c_2 are suitable constants. Now, for $\lambda_2 x_3^{-1} - \lambda_3 k > 0$ the minimum of H is attained for $u = 0$, and then $H = \lambda_1 x_2 - \lambda_2 g$. For $\lambda_2 x_3^{-1} - \lambda_3 k < 0$ the minimum of H is attained for $u = \mu$, and then $H = (\lambda_1 x_2 - \lambda_2 g) + \mu(\lambda_2 x_3^{-1} - \lambda_3 k)$. Note that if in an interval $[\bar{t}, T]$ we have $u = 0$, and hence $x_1' = x_2$, $x_2' = -g$, $x_3' = 0$, $\lambda_3' = 0$ (corresponding to free fall), then for $t \in [\bar{t}, T]$, we have $x_3(t) = x_3(\bar{t})$ and $x_2(t) = x_2(\bar{t}) - g(t - \bar{t})$, $x_1(t) = -2^{-1}g(t - \bar{t})^2 + x_2(\bar{t})(t - \bar{t}) + x_1(\bar{t})$, and (x, x_2) describes an arc of a parabola $x_1 = -2^{-1}g^{-1}x_2^2 + cx_2 + d$, with x_2 negative and decreasing. If in an interval $[\bar{t}, T]$ we have $u = \mu$, and hence $x_1' = x_2$, $x_2' = -g + x_3^{-1}\mu$, $x_3' = -k\mu$ (corresponding to maximal thrust), then by integration we have

$$\begin{aligned} & x_3(t) = \bar{x}_3 - k\mu(t - \bar{t}), \\ & x_2(t) = \bar{x}_2 - g(t - \bar{t}) - k^{-1} \log[\bar{x}_3^{-1}(\bar{x}_3 - k\mu(t - \bar{t}))], \\ & x_1(t) = \bar{x}_1 + \bar{x}_2(t - \bar{t}) - 2^{-1}g(t - \bar{t})^2 \\ & \qquad + (k^2\mu)^{-1}[\bar{x}_3 - k\mu(t - \bar{t})] \log[\bar{x}_3^{-1}(\bar{x}_3 - k\mu(t - \bar{t}))] + k^{-1}(t - \bar{t}), \end{aligned} \tag{6.4.1}$$

where $\bar{x}_1, \bar{x}_2, \bar{x}_3$ are the values of x_1, x_2, x_3 at $\bar{t}$.

We obtain now the locus OA of the points (x_1, x_2), or (h, v), which can be driven to $(0, 0)$ by an arc of a trajectory of maximal thrust. We have only to take, in (6.4.1), $t - \bar{t} = \tau$, write m for x_3, and then derive that $v = 0$, $h = 0$ imply

$$\begin{aligned} \bar{x}_2 &= g\tau + k^{-1} \log[(m - k\mu\tau)/m], \\ \bar{x}_1 &= 2^{-1}g\tau^2 - k^{-2}\mu^{-1}(m - k\mu\tau) \log[(m - k\mu\tau)/m] - (v + k^{-1})\tau. \end{aligned}$$

This is the arc OA in the illustration. Any initial state $(k_0, v_0) = P$ above the arc OA can then be taken to $O = (0, 0)$ by a unique arc of a parabola corresponding to free fall, and by a unique arc OA corresponding to maximal thrust. We see that the state

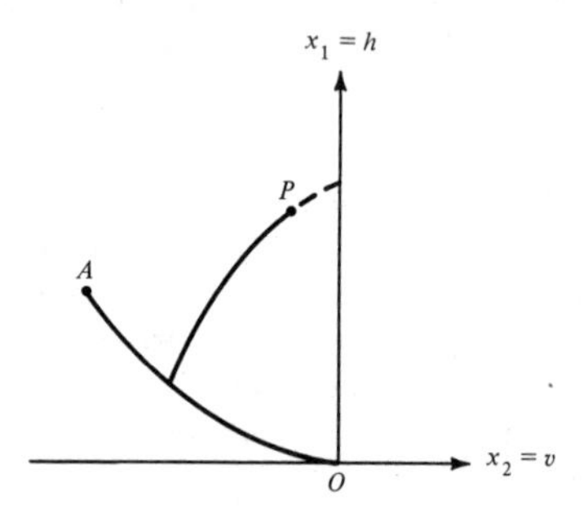

$P = (h_0, v_0)$ above or on the curve OA can be taken to $O = (0,0)$ by a unique trajectory satisfying the necessary condition.

Let us show that the constraints impose an upper bound for t_2. To this purpose let us take a number ε, $0 < \varepsilon < g(M + F)^{-1}$, and let E_1 and E_2 denote the subsets of all $t \in [0, t_2]$ at which $\varepsilon \le u \le \mu$, and $0 \le u < \varepsilon$, respectively. Then

$$F = \int_0^{t_2} ku\,dt = \left(\int_{E_1} + \int_{E_2}\right) ku\,dt \ge k\varepsilon \text{ meas } E_1,$$

and hence meas $E_1 \le k^{-1}\varepsilon^{-1}F$. On the other hand,

$$x_2(t) = v_0 + \int_0^t (-g + x_3^{-1}u)\,d\beta,$$

$$0 = x_1(t_2) = h_0 + \int_0^{t_2} x_2(\alpha)\,d\alpha = h_0 + \int_0^{t_2} d\alpha \left[v_0 + \int_0^{\alpha} (-g + x_3^{-1}u)\,d\beta \right]$$

$$= h_0 + v_0 t_2 + \left(\int_{E_1} + \int_{E_2}\right)(t_2 - \beta)(-g + x_3^{-1}u)\,d\beta$$

$$\le h_0 + 0 + t_2(g + M^{-1}\mu) \text{ meas } E_1 + \int_0^{t_2 - k^{-1}\varepsilon^{-1}F} (t_2 - \beta)(-g + (M+F)^{-1}\varepsilon)\,d\beta.$$

Finally,

$$0 \le h_0 + t_2(k^{-1}\varepsilon^{-1}F)(g + M^{-1}\mu) + (-g + (M + F)^{-1}\varepsilon)(-2^{-1}(k^{-1}\varepsilon^{-1}F)^2 + 2^{-1}t_2^2),$$

where $-g + (M + F)^{-1}\varepsilon < 0$. This is impossible for t_2 sufficiently large. Thus, $0 < t_2 \le T$ for some constant T which depends only on h_0, k, F, M, g.

We are in a position to apply now the existence theorem (4.3.i). Here A is the set product of intervals $[0, T] \times [0, h_0] \times [-\infty, 0] \times [M, M + F]$, therefore closed and contained in the slab $[0, T] \times R^3$. Also, $U = [0, \mu]$ is compact, $B = (0, h_0, v_0, M + F, 0, 0) \times [m \ge M]$ is closed, and Ω is not empty. Moreover, every trajectory starts from the fixed point $(0, h_0, v_0, M + F)$ and condition (b) of (4.3.i) is satisfied. Finally,

$$\begin{aligned} x_1 f_1 + x_2 f_2 + x_3 f_3 &= x_1 x_2 + x_2(-g + x_3^{-1}u) + x_3(-ku) \\ &\le x_1^2 + x_2^2 + (g + M^{-1}\mu)(x_2^2 + 1) + k\mu(x_3^2 + 1). \end{aligned}$$

and condition (c) also is satisfied. By (4.3.i) we conclude that an absolute minimum exists, and thus the uniquely determined trajectory and strategy are optimal.

6.5 Three More Problems on the Stabilization of a Point Moving on a Straight Line

1. A point P moves along the x-axis governed by the equation $x'' = u$ with $|u| \le 1$. We are to take P from any given state ($x = a$, $x' = b$) to rest ($x' = 0$) in the shortest time. If x and $y = x'$ are the phase coordinates, we have the Mayer's problem of minimum time: problem):

$$
\begin{aligned}
&dx/dt = y, \qquad dy/dt = u, \quad u \in U = [-1 \le u \le 1], \\
&t_1 = 0, \quad x(t_1) = a, \quad y(t_1) = b, \quad y(t_2) = 0, \quad t_2 \ge t_1, \\
&I[x, y, u] = g = t_2 = \min.
\end{aligned}
\tag{6.5.1}
$$

The problem is similar to Section 6.1, but here the goal is to land (x, y) on the x-axis in the shortest time. Here B is the set $t_1 = 0, x_1 = a, y_1 = b, y_2 = 0, t_2, x_2$ undetermined, $t_2 \ge t_1$. Thus, if $t_2 > t_1 = 0$, both dt_2, dx_2 can have arbitrary values, that is, $B' = R^2$. Also $dg = dt_2$. The transversality relation (P4) of (4.2.i) reduces to $(\lambda_0 + M(t_2))\,dt_2 - \lambda_1(t_2)\,dx_2 = 0$, where $\lambda_0 \ge 0$. Since dt_2, dx_2 are arbitrary, we have $0 \le \lambda_0 = -M(t_2)$, $\lambda_1(t_2) = 0$. From Section 6.1, we have $\lambda_1 = c_1 = 0$, $\lambda_2 = -c_1 t + c_2$; hence $\lambda_2 = c_2 \ne 0$. We conclude that $\lambda_2(t)$ has a constant sign in $[0, t_2]$. As in Section 6.1, either $u = +1$ in $(0, t_2)$, in which case the trajectory is an arc of the parabola $x = 2^{-1}y^2 + C$, $y \le 0$, along which $P(x, y)$ moves in such a way that y increases (arcs PA of the figure); or $u = -1$ in $(0, t_2)$, in which case the trajectory is an arc $x = -2^{-1}y^2 + C$, $y \ge 0$, along which $P(x, y)$ moves in such a way that y decreases (arcs QB of the figure).

2. We consider the same Problem 1 above, with the goal of taking P to the origin in the shortest time. The problem is the same as in (6.5.1) except that

$$t_1 = 0, \quad x(t_1) = a, \quad y(t_1) = b, \quad x(t_2) = 0, \quad t_2 \ge t_1.$$

Here B is the set $[t_1 = 0, x_1 = a, y_1 = b, x_2 = 0, t_2, y_2$ undetermined, $t_2 \ge t_1]$. If $t_2 > t_1$, then dt_2, dy_2 are arbitrary, that is, $B' = R^2$. Also $dg = dt_2$. The transversality relation (P4) of (4.2.i) reduces to $(\lambda_0 + M(t_2))\,dt_2 - \lambda_2(t_2)\,dx_2 = 0$, where $\lambda_0 \ge 0$. Since dt_2, dx_2 are arbitrary, we have $0 \le \lambda_0 = -M(t_2)$, $\lambda_2(t_2) = 0$. From Section 6.1 we have $\lambda_1 = c_1$, $\lambda_2 = -c_1 t + c_2$; hence $-c_1 t_2 + c_2 = 0$. Here $c_1 = 0$ would imply $c_2 = 0$; hence $\lambda_1 = \lambda_2 = 0$, in contradiction with (P1). Thus $c_1 \ne 0$, and λ_2 has a constant sign in $[t_1, t_2)$. As in Section 6.1, either $u = +1$ in $(0, t_2)$ and the trajectory is an arc of parabola $x = 2^{-1}y^2 + C$, $x \le 0$, along which $P(x, y)$ moves in such a way that y increases (arcs PA of the figure), or $u = -1$ in $(0, t_2)$ and the trajectory is an arc $x = -2^{-1}y^2 + C$, $x \ge 0$, along which $P(x, y)$ moves in such a way that y decreases (arcs QB of the figure).

3. We consider the same Problem 1 above with the goal now of taking P to rest in a segment $-d \le x \le d$ of the x-axis. The problem is as in (6.5.1) except that

$$t_1 = 0, \qquad x(t_1) = a, \qquad y(t_1) = b, \qquad -d \le x(t_2) \le d, \qquad y(t_2) = 0, \qquad t_2 \ge t_1.$$

In other words, we seek to take $P(x, y)$ to the segment $[y = 0, -d \le x \le d]$ (LL' in the illustration) in the shortest time. The optimal trajectory is clearly the only one in the figure at left passing through a given point (a, b). Thus, we have the types of trajectories exemplified by PA, QB, KL, ML, $K'L'$, $M'L'$, $P'RL$, $Q'R'L'$.

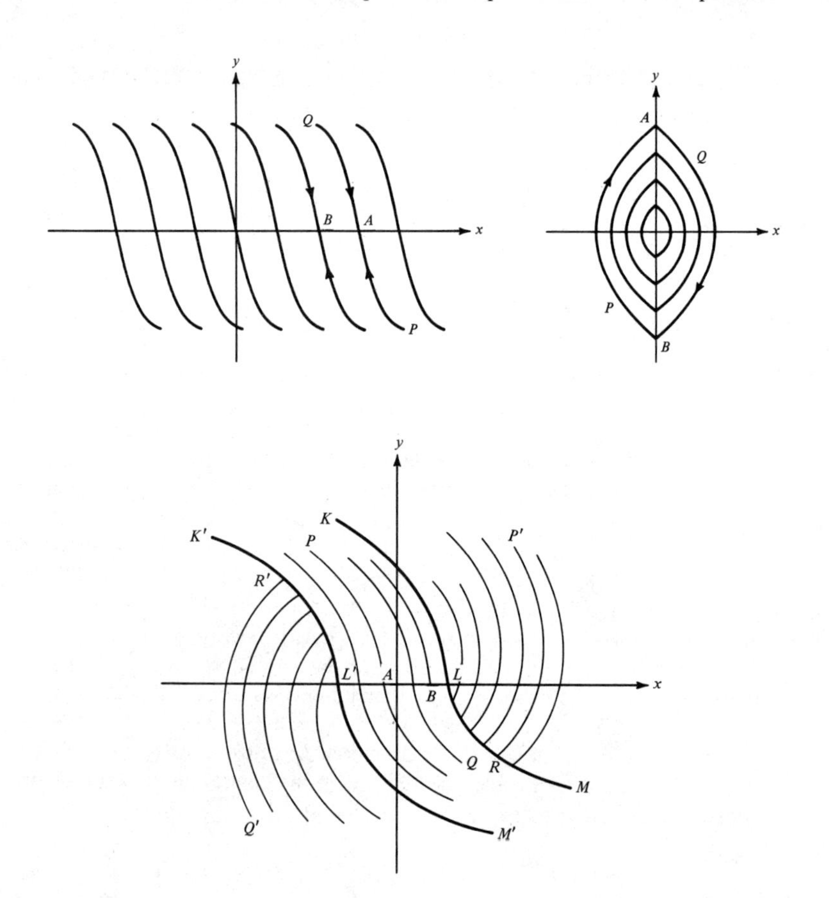

6.6 Exercises

1. Consider the problem, analogous to Section 6.1 above, of the minimum time to transfer P from the state (a, b) to the state (c, d), a, b, c, d arbitrary. Show that for suitable relative positions of (a, b) and (c, d) in R^2 there may be two strategies satisfying the necessary condition, but only one is optimal.
2. The same as Exercise 1, with $c = 1$, $d = 1$, or $c = 1$, $d = -1$. Compute the optimal time $t_2(a, b)$, and show that this time is sectionally continuous, but not continuous in R^2. Why?
3. Consider the problem, analogous to Section 6.1, where now the target is the set made up of two points, say $S = \{(-c, 0), (c, 0)\}$ for some $c > 0$. Show that there may be two optimal strategies.
4. Consider the problem, analogous to Section 6.1, where the target is the circle $s = [x^2 + y^2 = 1]$.
5. A system is governed by the differential equations $x' = \alpha x + \alpha u$, $y' = \beta y + \beta u$, $|u| \leq 1$, where $0 > \alpha > \beta$ are given constants. Find the optimal strategy in order to steer any state (a, b) to $(0, 0)$ in the minimum time.

Remark. The differential system in Exercise 5 is a canonical form for the differential equation $y'' + ay' + by = Ku$, $|u| \le 1$, where a, b, K are given constants, $K > 0$, when we assume that the characteristic equation $\rho^2 + a\rho + b = 0$ has distinct real roots $\alpha, \beta, \alpha \ne \beta$. To see this, first write the equation as a system $y_1' = y_2, y_2' = -by_1 - ay_2 + Ku$, where $y_1 = y$, $y_2 = y'$. If we take $y_1 = z_1 + z_2$, $y_2 = \alpha z_1 + \beta z_2$, then $z_1 = (\beta - \alpha)^{-1}(\beta y_1 - y_2)$, $z_2 = (\beta - \alpha)^{-1}(y_2 - \alpha y_1)$, and after simple computations using $a = -\alpha - \beta$, $b = \alpha\beta$, we have the system

$$z_1' = \alpha z_1 - (\beta - \alpha)^{-1}Ku, \qquad z_2' = \beta z_2 + (\beta - \alpha)^{-1}Ku.$$

Finally, if we take $x_1 = -\alpha(\beta - \alpha)K^{-1}z_1$, $x_2 = \beta(\beta - \alpha)K^{-1}z_2$, we have the proposed system

$$x_1' = \alpha x_1 + \alpha u, \quad x_2' = \beta x_2 + \beta u, \qquad |u| \le 1.$$

In Exercise 5 above we have considered only the (stable) case where both roots are negative.

6. A point P moves along the x-axis monitored by the equations $z'' + \alpha z' = ku$, $|u| \le 1$. Take P from any state $z = a$, $z' = b$ to rest at the origin $z = 0$, $z' = 0$ in a minimum time. Here a and b are positive constants. It is convenient to reduce the equation to the system $x' = u$, $y' = -\alpha y + u$, $|u| \le 1$, by taking $x = (z' + \alpha z)/k$, $y = z'/k$.
7. The same as Exercise 6 with the differential equation $z''' + mz'' + nz' = u$ and the hypothesis that the roots α, β of the equations $\rho^2 + m\rho + n = 0$ are real, $0 > \alpha > \beta$.
8. The same as Exercise 6 with the differential equation $z^{(4)} = u$, $|u| \le 1$.
9. In Section 6.2 above, determine the optimal time $t_2(a, b)$.
10. The same as Section 6.2, except that P has to be transferred from any state (a, b) to any state (c, d).
11. The same as Exercise 10, with $c = 1$, $d = 1$. Determine the optimal time $t_2(a, b)$. Is this function continuous? Is the optimal strategy unique?
12. The same as Section 6.2, or Exercise 10 above, where the target S is the set made up of the two points $(c = 1, d = 0)$ and $(c = -1, d = 0)$.
13. The same as Section 6.2, but with differential equation $x'' + \omega^2 x = ku$, $|u| \le 1$, ω, k positive constants. (A convenient form for the corresponding system is $x_1' = \omega x_2$, $x_2' = -\omega x_1 + ku$, with $x_1 = \omega x$, $x_2 = x'$.)
14. A system is governed by the differential equations

$$x' = -\alpha x + \omega y, \qquad y' = -\omega x + \alpha y + u, \quad |u| \le 1.$$

Take any state (a, b) to $(0, 0)$ in the minimum time. (The discussion of this problem is similar to Exercise 13, and hence to Section 6.2. Here α, ω are positive constants.)

Remark. The equations of this exercise are a canonical form for the problem concerning the second order differential equation $y'' + ay' + by = Ku$, $|u| \le 1$, a, b, K given constants, $K < 0$, when the characteristic equation $\rho^2 + a\rho + b = 0$ has complex roots $-\alpha + i\omega$, $-\alpha - i\omega$, $\alpha > 0$. Indeed, then $a = 2\alpha$, $b = \alpha^2 + \omega^2$, and the equation $y'' + ay' + by = Ku$ is immediately reduced to the system

$$y_1' = y_2, \qquad y_2' = -(\alpha^2 + \omega^2)y_1 - 2\alpha y_2 + Ku, \quad |u| \le 1.$$

If we take $x = (x_1, x_2)$, $y = (y_1, y_2)$, $x = Py$, where P is the 2×2 matrix $P = K^{-1}(\omega, 0; \alpha, 1)$, then immediate computations yield $x_1' = -\alpha x_1 + \omega x_2$, $x_2' = -\omega x_1 - \alpha x_2 + u$, which is the system proposed in Exercise 14. The hypothesis $\alpha > 0$ corresponds to a stability assumption for the point of equilibrium $(0,0)$.

15. The same as Exercise 14 with α replaced by $-\alpha$. (Not all states (a,b) can be taken to the origin. The origin $(0,0)$ is now an unstable point of equilibrium.)
16. The same as Section 6.3 with the differential system

$$x' = \omega y + u, \quad y' = \omega x + v, \qquad |u| \le 1, \quad |v| \le 2.$$

17. The same as Exercise 16 with the differential system

$$x' = \omega y + u, \quad y' = -\omega x + v, \qquad |u| \le m, \quad |v| \le n, \quad m, n > 0.$$

18. The same as Exercise 16 with the differential system

$$x' = -x + y + u, \quad y' = -x - y + v, \qquad |u| \le 1, \quad |v| \le 1.$$

19. The same as Exercise 16 with the differential system

$$\begin{pmatrix} x' \\ y' \end{pmatrix} = A\begin{pmatrix} x \\ y \end{pmatrix} + B\begin{pmatrix} u \\ v \end{pmatrix}, \qquad |u| \le 1, \quad |v| \le 1,$$

in the following cases:

(a) $A = (-1,2;0,-2)$, $B = (1,2;1,1)$.
(b) $A = (-1,0;0,-2)$, $B = (1,0;0,1)$.
(c) $A = (0,0;0,1)$, $B = (1,1;1,-1)$.
(d) $A = (0,1;0,-1)$, $B = (0,1;1,0)$.

Show the switch curves in the various cases.

20. A system is governed by the equations $x' = u$, $|u| \le 1$. Show that the minimum time required to transfer the state $x = \xi$ to $x = 0$ is $|\xi|$.
21. A system is governed by the equations $x' = -ax + u$, $|u| \le 1$ (a positive constant). Find the minimum time required to transfer the state $x = \xi$ to $x = 0$.
22. The same as Exercise 21, with equations $x' = ax + Ku$, $|u| \le 1$, where a, K are positive constants. (Not all states $a = \xi$ can be transferred to the origin. The origin is here an unstable point of equilibrium.)
23. The same as Exercise 21, with equations $x' = -ax|x| + u$, $|u| \le 1$, where a is a positive constant.
24. A system is governed by the nonlinear second order equation $x'' + x'|x'| = u$ with $|u| \le 1$. Take the state $x = a$, $x' = b$ to the state $x = 0$, $x' = 0$ in a minimum time. (*Hint*: Find and sketch first the two arcs of trajectories leading to the origin and forming the switch curve.) The qualitative picture is similar to the ones in Section 6.1 and most of the exercises above. The cases $x > 0$ and $x < 0$ need different integration patterns, but trajectories may cross the line $x = 0$ and are made up of arcs obtained by different devices. The family of trajectories can be sketched.
25. The same as Exercise 24, with the nonlinear equation $x'' + (x')^3 = u$, and $|u| \le 1$.

Remark. These nonlinear problems present difficulties of integration, but give rise to the same qualitative picture as in the linear ones. In actual problems, the switch curve is important, and is often obtained by numerical methods or mechanical devices. Thus the design of an automatic feedback optimal control system need not be essentially different from the one mentioned in the Remark in Section 6.1.

26. $I = \int_{t_1}^{t_2} |u|\, dt$ with $x'' = u$, $|u| \leq 1$, $x(t_1) = x_1$, $x(t_2) = x_2$.
27. $I = \int_0^1 u^2\, dt$ with $x' = x + u$, $x(0) = 1$, $x(1) = 0$.
28. High speed trains are being planned between Boston and Washington, a flat distance of 400 miles. (a) What is the shortest possible duration of the trip if the only constraint is that the maximum acceptable acceleration and deceleration are $2g$ ($g = 32$ feet/sec^2, acceleration of gravity)? (b) What is the shortest possible duration of the trip if, in addition to the acceleration constraint, there is also the constraint that velocity cannot exceed 360 miles/hour ($=528$ feet/sec)?
29. Find the advertising policy which maximizes sales over a period of time where the rate of change of sales decreases at a rate proportional to sales but increases at a rate proportional to the rate of advertising as applied to the share of the market not already purchasing the product. The problem is

$$I = \int_{t_1}^{t_2} S(t)\, dt, \qquad S' = -aS + bA[1 - S/M],$$

$$S(t_0) = S_0, \qquad 0 \leq A \leq \bar{A},$$

where S is sales, A is advertising, M is the extent of the market, and t_1, t_2, S_0, A are given positive numbers, $t_1 < t_2$.
30. $I = \int_0^2 (2x - 3u)\, dt$ with $x' = x + u$, $x(0) = 5$, $0 \leq u \leq 2$, x scalar.
31. $I = \int_0^2 (2x - 3u - u^2)\, dt$ with $x' = x + u$, $x(0) = 5$, $0 \leq u \leq 2$, x scalar.
32. $I = \int_0^1 (x^2 - u^2)\, dt$ with $x' = y$, $y' = -2x - 3y - u$, $0 \leq u \leq 1$, $x(0) = y(0) = 1$, $x(1) = y(1) = 0$, x, y scalar.
33. $I = t_2$ with $x' = ux$, $y' = v$, $|u| \leq 1$, $|v| \leq 1$, $t_1 = 0$, $x(0) = a$, $y(0) = b$, x, y scalar.
34. $I = t_2$ with $x' = f(t, x) + u$, $|u| \leq 1$, $x = (x^1, \ldots, x^n)$, $u = (u^1, \ldots, u^n)$.

6.7 Optimal Economic Growth

We consider here an economy in which a single homogeneous good is produced, in which the total output $Y(t)$ is either consumed or invested, and thus, if $C(t)$ and $I(t)$ are the total consumption and the total investment, we have $Y(t) = C(t) + I(t)$. It is assumed that investment is used both to augment the stock of capital $K(t)$, and to replace depreciated capital. If capital depreciates at a constant rate μ, then $\mu K(t)$ is the depreciated capital, and $I(t) = K'(t) + \mu K(t)$.

Let $L(t)$ denote the labor force at time t. We assume that L grows at known exponential rate n, that is, $L' = nL$.

Now $y = Y/L$, $c = C/L$, $i = I/L$, $k = K/L$ are the output, consumption, investment, and capital per worker respectively. Then

$$y(t) = c + i, \qquad i(t) = K'/L + \mu k,$$

with

$$dk/dt = (d/dt)(K/L) = K'/L - (K/L)(L'/L),$$

or $K'/L = k' + nk$, and

$$i(t) = k' + (\mu + n)k$$

or $i(t) = k' + mk$, with $m = \mu + n$.

The output per worker, y, is assumed to be a known function $f(k)$ of the capital per worker, with f smooth, $f > 0$, $f' > 0$ for all $k > 0$ not too large, $f'' < 0$ for all $k > 0$, $f(0+0) = 0$, $f'(0+0) = +\infty$, $f(+\infty) = -\infty$. From $y = f(k)$, $y = c + i$, and $i = k' + mk$, we derive the basic differential equation of economic growth:

$$f(k(t)) = c(t) + mk(t) + k'(t), \qquad m = \mu + n. \tag{6.7.1}$$

Thus, the output per worker, $f(k(t))$, is allocated among three uses: consumption per worker, $c(t)$; maintenance of the level of capital per worker, $mk(t)$, due to depreciation and dilution of capital in an increasing labor force; and net increase $k'(t)$ of capital per worker. In the first figure the graphs of $f(k)$ and mk are given; in the second figure the graph of $f(k) - mk$ is given. Here $\hat{k}$ and $\tilde{k}$ denote the values of k corresponding to the maximum of $f(k) - mk$, and to the zero of $f(k) - mk$. Thus, $f'(\hat{k}) = m$, $f(\tilde{k}) = m\tilde{k}$, and $f(k) - mk = c + k'$ for all k.

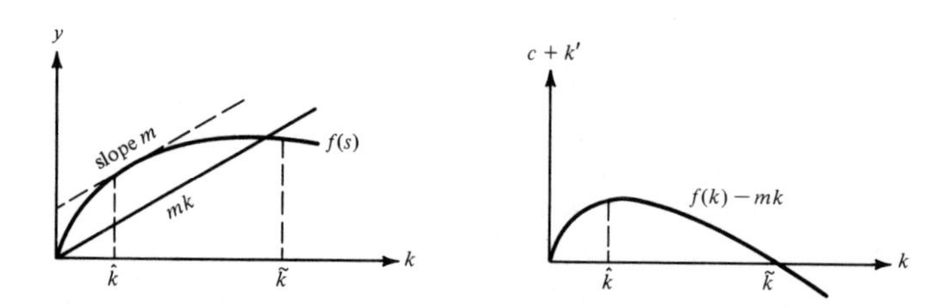

In the limiting situation $c = 0$ (case (a)), then the same graph is a phase diagram for k, and $\tilde{k}$ is a point of stable equilibrium, since if $k > \tilde{k}$ then $k' < 0$, and if $k < \tilde{k}$ then $k' > 0$ (see arrows).

In case (b), the consumption per worker is at its maximum level $c = \hat{c}$, with $f'(k) = m = \mu + n$. This is the equilibrium point which maximizes the sustainable level of consumption per worker, with $\hat{c} = f(\hat{k}) - m\hat{k}$, $k' = 0$. It is called the golden rule level of consumption per worker. It is a point of instability, since though deviations to the right may be compensated, deviations to the left are not (see arrows).

In case (c), with a fixed consumption per worker, $\bar{c}$, $0 < \bar{c} < \hat{c}$, there are two points of equilibrium k_L and k_U, $0 < k_L < k_U$, with k_L unstable, k_U stable (see arrows).

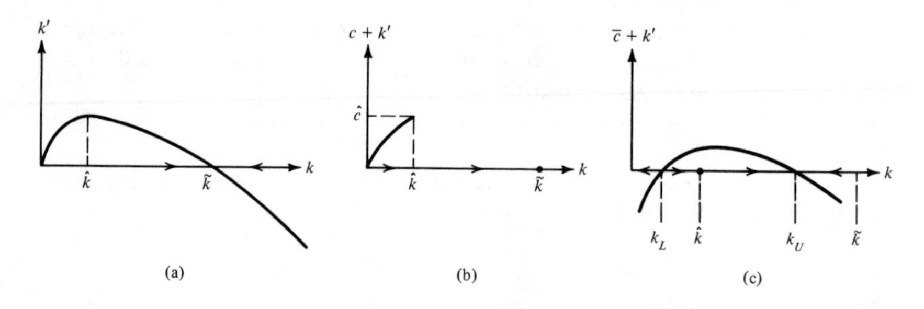

The economic objective of any planning must concern the standard of living, and thus we shall need a utility function $U(c)$ which measures the instantaneous economic well-being, and we shall assume that in any planning we should try to maximize the global utility

$$W = \int_{t_1}^{t_2} e^{-\delta(t-t_1)} U(c(t))\, dt$$

in an interval of time $[t_1, t_2]$. Here $\delta > 0$ represents the rate of interest, or discount, and a larger δ means that we favor the near over the distant future.

The time t_2 is often called the horizon, and we need not exclude $t_2 = +\infty$ (infinite horizon). It is assumed that U is a known function of class C^2, with $U(c) > 0$, $U'(c) > 0$, $U''(c) < 0$, $U'(0+0) = +\infty$, $U(+\infty) = +\infty$.

Let $k_1 = k(t_1)$ be the initial value of $k(t)$. If t_2 is finite, then we should also require $k(t_2) \geq k_2$ to be not less than some positive value k_2. We shall treat c as the control variable, and k as the state variable. Thus, we have the Lagrange problem

$$W[k, c] = \int_{t_1}^{t_2} e^{-\delta(t-t_1)} U(c(t))\, dt,$$

$$k' = f(k) - mk - c, \qquad k(t_1) = k_1,$$
$$k \geq 0, \qquad 0 \leq c \leq f(k),$$

and obviously, only solutions with $0 < c < f(k)$ should be taken into consideration. Here the Hamiltonian is

$$H(t, k, c, \lambda) = e^{-\delta(t-t_1)} U(c) + \lambda[f(k) - mk - c],$$

and if H has to have a minimum at some $c \in (0, f(k))$, then we must have $H_c = 0$, or

$$e^{-\delta(t-t_1)} U'(c) - \lambda = 0.$$

The equation for the multiplier λ is $d\lambda/dt = -H_k$, or

$$d\lambda/dt = -\lambda[f'(k) - m].$$

By the change of variable $\lambda = e^{-\delta(t-t_1)} q$, we have for q the equations $q = U'(c)$ and $-\delta q + q' = -qf' + mq$, and in final form

$$f'(k) + q'/q = \mu + n + \delta. \tag{6.7.2}$$

The expression $\sigma(c) = -cU''(c)/U'(c)$ is interpreted as the "elasticity of marginal utility", and from $q = U'(c)$, or $q' = U''(c)c'$, we derive

$$q'/q = (U''/U')c' = -\sigma(c)(c'/c). \tag{6.7.3}$$

Here q'/q is interpreted as a capital gain in terms of the utility function U, and equation (6.7.2) can be interpreted by saying that the marginal product $f'(k)$ plus the capital gain q'/q is equal to the loss $\mu + n + \delta$ due to depreciation μ, the dilution of equity via population growth n, and the interest δ. Finally, equations (6.7.1), (6.7.2) and (6.7.3) yield the differential equations for an optimal path:

$$\begin{aligned} c' &= (\sigma(c))^{-1}(f'(k) - \mu - n - \delta)c, \\ k' &= f(k) - (\mu + n)k - c. \end{aligned} \tag{6.7.4}$$

One possible solution of (6.7.4) is $c' = 0$, $k' = 0$ (constant consumption per worker and constant capital per worker); then $c = c^*$, $k = k^*$, with $f'(k^*) = \mu + n + \delta$, $c^* = f(k^*) - (\mu + n)k^*$. This state of equilibrium is called the balanced growth path, with

$$L(t) = L_1 e^{n(t-t_1)}, \qquad C(t) = c^* L = c^* L_1 e^{n(t-t_1)},$$
$$K(t) = k^* L = k^* L_1 e^{n(t-t_1)},$$

$Y = Lf(k^*) = L_1 f(k^*) e^{n(t-t_1)}$, and $0 < k^* < \hat{k}$, $0 < c^* < \hat{c}$. This state of equilibrium is also called the modified golden rule growth path, since it modifies the golden rule to allow for a nonzero discount rate δ.

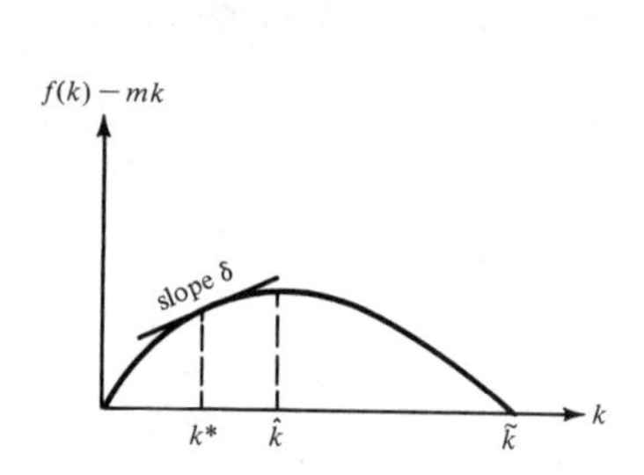

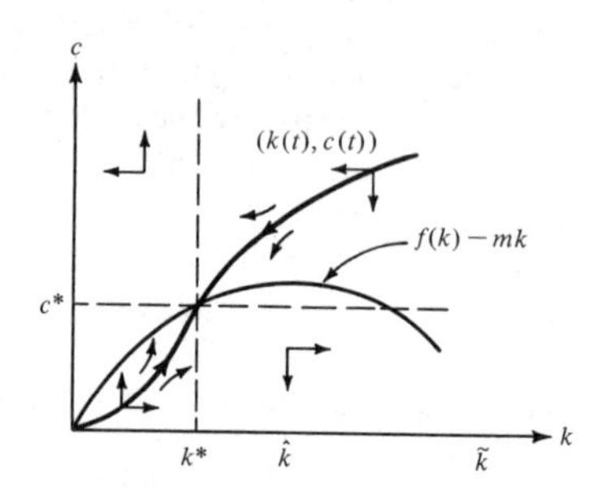

Note that $c' \gtreqless 0$ according as $k \lesseqgtr k^*$, and that $k' \gtreqless 0$ according as $c \lesseqgtr f(k) - mk$. Thus the loci $k = k^*$ and $z = f(k) - mk$ divide the quadrant $k > 0, c > 0$ into four regions characterized by different signs for c' and k' (see arrows). The point (k^*, c^*) is a point of equilibrium for (6.7.4). To determine its nature, let us consider the linearized system at (k^*, c^*) obtained by taking the linear terms in the Taylor expansions of the second members of (6.7.4) at (k^*, c^*):

$$c' = (\sigma(c^*))^{-1}c^*f''(k^*)(k - k^*)$$
$$k' = \delta(k - k^*) - (c - c^*).$$

Since $f'' < 0$, the characteristic equation $\rho^2 - \delta\rho + (\sigma(c^*))^{-1}c^*f''(k^*) = 0$ has two real roots of opposite signs. By a theorem of Poincaré (cf., e.g., Cesari [II, p. 162]), the point (k^*, c^*) is a saddle point for (6.7.4), and thus an unstable point of equilibrium. There are only two paths leading to (k^*, c^*) which are indicated in the figure; one of them is traveled from $(0, 0)$ to (k^*, c^*). This is the optimal path leading the economy from capital per worker $k_1 < k^*$ to as close as we want to k^*. The arrows indicate its instability. (Some readers may well conclude that the way to utopia is highly unstable, and that an economy could be steered to it only by a government whose wisdom is truly exceptional. It is revealing that the word wisdom has disappeared from our everyday language.)

6.8 Two More Classical Problems

1. *A problem of length and area.* We consider again the problem in Section 3.6 concerning the maximum of the integral

$$I[x] = \int_{t_1}^{t_2} x\,dt$$

with

$$J[x] = \int_{t_1}^{t_2} (1 + x'^2)^{1/2}\,dt = L.$$

The problem can be written as the Mayer problem of the minimum of $I[x, y, z, u] = g = -y(t_2)$ with differential equations, boundary conditions, and constraints

$$dx/dt = u, \quad dy/dt = x, \quad dz/dt = (1 + u^2)^{1/2}, \qquad u \in U = R,$$
$$x(t_1) = 0, \quad x(t_2) = 0, \quad y(t_1) = 0, \qquad z(t_1) = 0, \quad z(t_2) = L,$$

In this example U is unbounded, and therefore we do not know a priori that the optimal strategy u is bounded—in other words, the assumption (c) of Section 4.1 cannot be verified. Nevertheless, here $f_1 = u$, $f_2 = x$, $f_3 = (1 + u^2)^{1/2}$ have first order partial derivatives with respect to t, x, y, z, all of which are zero except $f_{2x} = 1$. Here condition (c') of Remark 5 of Section 4.2C is satisfied with $S(t, u) = 1$ for all $(t, u) \in [t_0, t_1] \times R$. Thus, the necessary condition holds in the usual form.

The Hamiltonian is

$$H(x, y, z, u, \lambda_1, \lambda_2, \lambda_3) = \lambda_1 u + \lambda_2 x + \lambda_3(1 + u^2)^{1/2},$$

and H has a minimum as a function of u for $-\infty < u < +\infty$ only if $\lambda_3 > |\lambda_1|$, and the minimum is given by the value of u satisfying $\lambda_1 + \lambda_3 u(1 + u^2)^{-1/2} = 0$. Here H may have a minimum also in the trivial case $\lambda_3 = \lambda_1 = 0$, with $H_{\min} = \lambda_2 x$ and u undetermined. The equations for the multipliers are

$$d\lambda_1/dt = -\lambda_2, \qquad d\lambda_2/dt = 0, \qquad d\lambda_3/dt = 0,$$

so that

$$\lambda_2 = c_2, \quad \lambda_3 = c_3, \quad \lambda_1 = c_1 - c_2 t, \qquad c_1, c_2, c_3 \text{ constants.}$$

The case of λ_1 and λ_3 both identically zero in $[t_1, t_2]$ would require $c_1 = c_2 = c_3$ and $\lambda_1 = \lambda_2 = \lambda_3 = 0$, in contradiction with (P1). Thus, we must have $\lambda_3 > |\lambda_1|$; hence $\lambda_3 = c_3 > 0$, and

$$u(1 + u^2)^{-1/2} = -\lambda_1/\lambda_3 = -c_3^{-1}(c_1 - c_2 t) = c^{-1}(t - \alpha)$$

for some constants c, α. By taking squares and noting that $u = x' = dx/dt$, we have

$$\begin{aligned} x'^2(1 + x'^2)^{-1} &= c^{-2}(t - \alpha)^2, \\ dx/dt &= -(t - \alpha)[c^2 - (t - \alpha)^2]^{-1/2}, \\ x - \beta &= [c^2 - (t - \alpha)^2]^{1/2}, \\ (t - \alpha)^2 + (x - \beta)^2 &= c^2, \end{aligned}$$

the same result we obtained before.

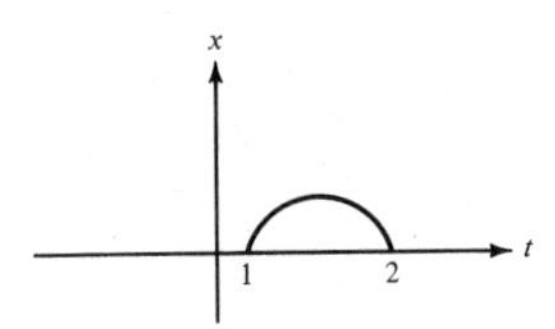

2. *The shape of a hanging rope.* As we have seen in Section 3.6 this problem concerns the minimum of the integral

$$I[y] = \int_{t_1}^{t_2} y(1 + y'^2)\, dt$$

with

$$J[y] = \int_{t_1}^{t_2} (1 + y'^2)\, dt = l, \qquad y(t_1) = y_1, \qquad y(t_2) = y_2.$$

This problem can be written as the Mayer problem concerning the minimum of $I[x, y, z, u] = x(t_2)$ with differential equations, boundary conditions, and constraints

$$dx/dt = y(1 + u^2)^{1/2}, \qquad dy/dt = u, \qquad dz/dt = (1 + u^2)^{1/2},$$
$$x(t_1) = 0, \quad y(t_1) = y_1, \quad z(t_1) = 0, \quad y(t_2) = y_2, \quad z(t_2) = l,$$

with $y \geq 0$ and $u \in U = R$. Here the Hamiltonian is

$$H = H(x, y, z, u, \lambda_1, \lambda_2, \lambda_3) = \lambda_1 y(1 + u^2)^{1/2} + \lambda_2 u + \lambda_3(1 + u^2)^{1/2},$$

and H may have a minimum when u describes R only if $\lambda_1 y + \lambda_3 > |\lambda_2|$ and $H_u = 0$, or

$$(\lambda_1 y + \lambda_3)u(1 + u^2)^{-1/2} + \lambda_2 = 0. \tag{6.8.1}$$

The equations for the multipliers are

$$d\lambda_1/dt = -H_x = 0, \qquad d\lambda_2/dt = -H_y = -\lambda_1(1 + u^2)^{1/2},$$
$$d\lambda_3/dt = -H_z = 0,$$

and thus $\lambda_1 = c_1$, $\lambda_3 = c_3$ (constants). By comparing the remaining equation with (6.8.1) we have

$$(d/dt)[(c_1 y + c_3)u(1 + u^2)^{-1/2}] = c_1(1 + u^2)^{1/2}.$$

Since $u = y'$, this is the Euler equation for $c_1 f_0 + c_3 f_1$, where f_0 and f_1 are the integrands of I and J.

Actually, (6.8.1) yields

$$y' = u = -\lambda_2((c_1 y + c_3)^2 - \lambda_2^2)^{-1/2}, \tag{6.8.2}$$

hence, by computation,

$$M(t) = H_{\min} = (c_1 y + c_3)(1 + u^2)^{1/2} + \lambda_2 u = ((c_1 y + c_3)^2 - \lambda_2^2)^{1/2}.$$

By (4.2.i)(P3) then $dM/dt = H_t = 0$, or $M(t) = C$, hence $((c_1 y + c_3)^2 - \lambda_2^2)^{1/2} = C$ and by (6.8.2), $u = -\lambda_2 C^{-1}$, $\lambda_2 = -Cu$. Now again by (6.8.1) we have

$$(c_1 y + c_3)y'(1 + y'^2)^{-1/2} = Cy'.$$

Thus, either $y' = 0$, $y = D$, $u = 0$, or $y' \neq 0$ and

$$(c_1 y + c_3)(1 + y'^2)^{-1/2} = C$$

from which, by integration,

$$y + D = C\cosh((x - a)/C),$$

a family of catenaries.

Exercises

1. It is required to extremize

$$I = \int_{t_1}^{t_2} f(t, x, x')\,dt + F(w)$$

with respect to functions $x(t)$ and real w for which

$$J = \int_{t_1}^{t_2} g(t, x, x')\,dt + G(w)$$

has a prescribed value. Here we think of x as having prescribed values at t_1 and t_2.

You may use optimal control, or the classical theory. You will see that for an optimal solution it is required that

$$(d/dt)f^*_{x'} = f^*_x, \qquad F^*_w = 0,$$

where $f^* = f + \lambda g$, $F^* = F + \lambda G$, and λ is a constant parameter.

2. Apply Exercise 1 to the following problem: A flexible uniform rope of length L hangs in unstable equilibrium with one end fixed at (t_1, x_1), so that it passes over a frictionless pin at (t_2, x_2). We know that the position of the rope extended between the two given points hangs in the form of a catenary. What is the position of the free end of the rope?

6.9 The Navigation Problem

A. The Elementary Navigation Problem (a Problem of Calculus)

A boat P moves with constant velocity V, keeping a constant angle θ with a fixed direction. There is a stream of constant velocity, and the boat is supposed to go from a point A (which we may think of as fixed) to a point B which is either fixed or in uniform motion with respect to A. Determine the angle θ in such a way the boat reaches B.

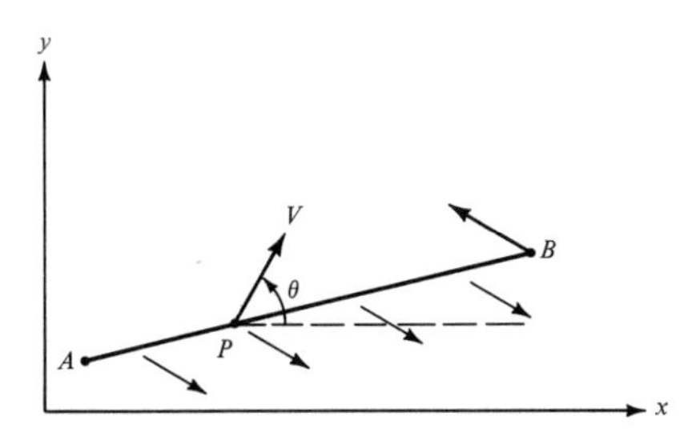

Let us denote by p, q the components of the velocity of the stream in the direction of the x- and y-axes. If B is in motion, let us denote by $x = at + b$, $y = ct + d$, the x and y coordinates of B at time t. Thus, (b, d) is the position of B at $t = 0$. The components of the velocity of P with respect to the x and y axes are $x' = p + V\cos\theta$, $y' = q + V\sin\theta$. If we assume, to simplify notation, that A is the origin of the coordinates and that the point P leaves A at the time $t = 0$, the coordinates of P at time t are

$$x = (p + V\cos\theta)t, \qquad y = (q + V\sin\theta)t$$

and P will meet B at the time t determined by the equations

$$(p + V\cos\theta)t = at + b, \qquad (q + V\sin\theta)t = ct + d.$$

To eliminate t, we have the relations

$$\begin{aligned} &(p - a + V\cos\theta)t = b, \qquad (q - c + V\sin\theta)t = d,\\ &d(p - a + V\cos\theta) = b(q - c + V\sin\theta),\\ &d\cos\theta - b\sin\theta = V^{-1}[b(q - c) - d(p - a)]. \end{aligned} \tag{6.9.1}$$

If we introduce the constant α in such a way that

$$\sin \alpha = (b^2 + d^2)^{-1/2} d, \qquad \cos \alpha = (b^2 + d^2)^{-1/2} b,$$

and we assume $V > 0$, then

$$\sin(\theta - \alpha) = V^{-1}(b^2 + d^2)^{-1/2}[d(p - a) - b(q - c)]. \tag{6.9.2}$$

Thus, we have first to require that

$$(b^2 + d^2)^{-1/2}|d(p - a) - b(q - c)| \leq V, \tag{6.9.3}$$

and then equation (6.9.2) gives θ, either of the equations (6.9.1) gives t, and we have to require that this value of t be nonnegative.

For instance, for $c = d = 0$, $b > 0$, $q = 0$, the inequality (6.9.3) is satisfied, we have $\alpha = 0$, $\sin \theta = 0$; hence $\theta = 0$, or $\theta = \pi$, the second equation (6.9.1) is trivial, and the first equation (6.9.1) is reduced to $(p - a \pm V)t = b$. Thus, for $V > a - p$, we take $\theta = 0$ and $t = t' = b(V + p - a)^{-1}$. Note that for $V > a - p$ and $V < p - a$, $\theta = \pi$ also gives a time $t = t'' = b(-V + p - a)^{-1} > 0$ but larger than t'. For $V > a - p$, $V > p - a$, we have $t' > 0$, $t'' < 0$; for $V < a - p$, $V > p - a$, both times t', t'' are negative, and the latter case is not to be considered.

B. The Navigation Problem with Constant Boat Speed and Variable Steering Function (a Mayer Problem)

Again we assume that the boat P moves with constant speed V (with respect to the water) and an angle $\theta = \theta(t)$ with respect to the positive x-axis. The stream velocity now has components

$$p = p(t, x, y), \qquad q = q(t, x, y) \tag{6.9.4}$$

with respect to the positive x- and y-axes. The boat leaves the point $1 = (x_1, y_1)$ at a time $t = t_1$, and is required to reach the point $2 = (x_2, y_2)$, which we assume now as fixed, at a time $t = t_2$. We wish to determine $\theta(t)$, $t_1 \leq t \leq t_2$, in such a way that t_2 is a minimum. We have here three unknown functions $x(t)$, $y(t)$, $\theta(t)$, $t_1 \leq t \leq t_2$, satisfying the equations and the end conditions

$$\begin{gathered} x' = p + V \cos \theta, \qquad y' = q + V \sin \theta, \\ x(t_1) = x_1, \quad y(t_1) = y_1, \quad x(t_2) = x_2, \quad y(t_2) = y_2, \end{gathered} \tag{6.9.5}$$

and we have to determine $x(t)$, $y(t)$, $\theta(t)$ in such a way to make t_2 a minimum. We shall consider x, y as the state variables and θ as the control variable, $n = 2$, $m = 1$, and we take the Hamiltonian

$$H = H(t, x, y, \theta, \lambda_1, \lambda_2) = \lambda_1(p + V \cos \theta) + \lambda_2(q + V \sin \theta). \tag{6.9.6}$$

Then for $H_{\min}$ we must have $\partial H/\partial \theta = V(-\lambda_1 \sin \theta + \lambda_2 \cos \theta) = 0$, and the equations for λ_1 and λ_2 are $\lambda_1' = -\partial H/\partial x$, $\lambda_2' = -\partial H/\partial y$. Thus, we have five differential equations

$$\begin{gathered} \lambda_1' + \lambda_1 p_x + \lambda_2 q_x = 0, \qquad \lambda_2' + \lambda_1 p_y + \lambda_2 q_y = 0, \qquad \lambda_1 \sin \theta = \lambda_2 \cos \theta, \\ x' = p + V \cos \theta, \qquad y' = q + V \sin \theta, \end{gathered} \tag{6.9.7}$$

in the five unknowns x, y, θ, λ_1, λ_2. Note that if p, q do not depend explicitly upon time, that is, $p = p(x, y)$, $q = q(x, y)$ (steady stream), then the problem is autonomous and we may as well take $t_1 = 0$. For instance, if p, q are constants, then also λ_1, λ_2 are constants, θ is constant, and x, y are linear functions of t as in the elementary case (subsection A).

The case $p = mx + n$, $q = 0$, $t_1 = 0$, $x_1 = y_1 = 0$, $x_2 > 0$, $m > 0$ is more interesting. The equations above yield $\lambda_1' + m\lambda_1 = 0$, $\lambda_2' = 0$; hence $\lambda_1 = C_1 e^{-mt}$, $\lambda_2 = C_2$, C_1, C_2 constants. It is convenient to assume $C_2 = kh$, $C_1 = k$; hence $\lambda_1 = ke^{-mt}$, $\lambda_2 = kh$, and the third equation (6.9.7) yields $\tan\theta = he^{mt}$,

$$\sin\theta = he^{mt}(1 + h^2e^{2mt})^{-1/2}, \qquad \cos\theta = (1 + h^2e^{2mt})^{-1/2}.$$

The last two equations (6.9.7) are then linear equations

$$x' = mx + n + V(1 + h^2e^{2mt})^{-1/2}, \qquad y' = Vhe^{mt}(1 + h^2e^{2mt})^{-1/2}.$$

By integration we have

$$\begin{aligned} x &= C_3e^{mt} + m^{-1}n(e^{mt} - 1) - m^{-1}V(1 + h^2e^{2mt})^{1/2}, \\ y &= C_4 + m^{-1}V\log[he^{mt} + (1 + h^2e^{2mt})^{1/2}], \end{aligned}$$

where C_3, C_4 are constants. The conditions $x(0) = 0$, $y(0) = 0$ determine C_3, C_4, and we have

$$\begin{aligned} x &= m^{-1}n(e^{mt} - 1) + m^{-1}V[(1 + h^2)^{1/2}e^{mt} - (1 + h^2e^{2mt})^{1/2}], \\ y &= m^{-1}V\log\{(h + (1 + h^2)^{1/2})^{-1}[he^{mt} + (1 + h^2e^{2mt})^{1/2}]\}, \end{aligned} \tag{6.9.8}$$

where there appears only the constant of integration h. The conditions $x(t_2) = x_2$, $y(t_2) = y_2$ yield the relations

$$\begin{aligned} (1 + h^2)^{1/2}e^{mt_2} - (1 + h^2e^{2mt_2})^{1/2} + V^{-1}n(e^{mt_2} - 1) &= mV^{-1}x_2, \\ he^{mt_2} + (1 + h^2e^{2mt_2})^{1/2} &= (h + (1 + h^2)^{1/2})e^{mV^{-1}y_2}, \end{aligned}$$

which may be used for the determination of t_2 and h. Then the trajectory is given by the formula (6.9.8), and the steering function by

$$\theta = \arctan(he^{mt}).$$

C. Variants of the Navigation Problem with Constant Speed

It is clear that the previous analysis leading to the five differential equations (6.9.7) does not depend upon the boundary conditions, nor on what we actually minimize.

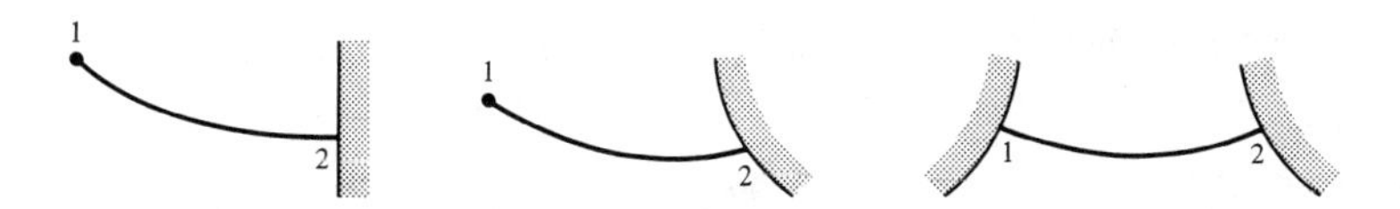

I. We may require that 2 be on a straight line parallel to the y-axis (a shore) and that the boat reach the shore in a minimum time t_2.

II. We may require instead that 2 is on a curve $Q(x, y) = 0$ (shore), and that the boat reach the shore in a minimum time.

III. We may suppose explicitly that the stream does depend on time, and that t_1 is undetermined as well as t_2, and we may require that the boat reach the shore $Q(x, y) = 0$ as in case II in a minimum time $t_2 - t_1$.
IV. We may require that the boat leave a shore $P(x, y) = 0$ at a fixed time t_1, and reach the shore $Q(x, y) = 0$ in a minimum time t_2.
V. Alternatively, both t_1, t_2 may be undetermined, and we may require that the boat leave the shore $P = 0$ and reach the shore $Q = 0$ in a minimum time $t_2 - t_1$.
VI. We may require that the boat leave a fixed point 1 at a fixed time t_1 and meet a moving point B in a minimum time t_2, where B is supposed to move on a given path with a fixed itinerary $x = X(t)$, $y = Y(t)$.
VII. Finally, we could require that the boat leave the fixed point 1 at a fixed time t_1, and reach in a fixed time t_2 a point $2 = (x_2, y_2)$ of maximum abscissa x_2, or maximum ordinate y_2, or maximum distance $x_2^2 + y_2^2$ from the initial point $x_1 = y_1 = 0$.

In all these and analogous situations the five differential equations (6.9.7) are the same. The given boundary conditions and the transversality relations will be used to determine the solutions. The transversality relations (P4) are here

$$\begin{aligned} -M(t_1)\,dt_1 + \lambda_1(t_1)\,dx_1 + \lambda_2(t_1)\,dy_1 \\ + M(t_2)\,dt_2 - \lambda_1(t_2)\,dx_2 - \lambda_2(t_2)\,dy_2 + \lambda_0\,dg = 0, \end{aligned}$$

where $M(t) = H(t, x(t), y(t), \theta(t), \lambda_1(t), \lambda_2(t))$.

In case I we have t_1 fixed, x_1, y_1, x_2 fixed, $g = t_2$; hence $dg = dt_2$, $dt_1 = dx_1 = dy_1 = dx_2 = 0$, and the transversality relation reduces to

$$M(t_2)\,dt_2 - \lambda_2(t_2)\,dy_2 + \lambda_0\,dt_2 = 0$$

for all dt_2, dy_2. Hence $M(t_2) + \lambda_0 = 0$, $\lambda_2(t_2) = 0$. Since λ_0 is undetermined, the only boundary conditions are here

$$t_1 \text{ fixed}, \quad x(t_1) = x_1, \quad y(t_1) = y_1, \quad x(t_2) = x_2, \quad \lambda_2(t_2) = 0.$$

In case II we have t_1 fixed, x_1, y_1 fixed, $Q(x_2, y_2) = 0$. Hence $dx_1 = dy_1 = 0$, $dt_1 = 0$, $Q_{x_2}\,dx_2 + Q_{y_2}\,dy_2 = 0$, and the transversality relation reduces to

$$M(t_2)\,dt_2 - \lambda_1(t_2)\,dx_2 - \lambda_2(t_2)\,dy_2 + \lambda_0\,dt_2 = 0$$

for all dt_2, dx_2, dy_2 as above. Thus, $M(t_2) + \lambda_0 = 0$, and $\lambda_1(t_2)\,dx_2 + \lambda_2(t_2)\,dy_2 = 0$ for all dx_2, dy_2 satisfying $Q_{x_2}\,dx_2 + Q_{y_2}\,dy_2 = 0$. Since λ_0 remains undetermined, the only boundary conditions are

$$\begin{aligned} &t_1 \text{ fixed}, \qquad x(t_1) = x_1, \qquad y(t_1) = y_1, \\ &Q(x(t_2), y(t_2)) = 0, \qquad Q_{y_2}\lambda_1(t_2) - Q_{x_2}\lambda_2(t_2) = 0. \end{aligned}$$

In case III we have x_1, y_1 fixed, $g = t_2 - t_1$, $Q(x_2, y_2) = 0$. Hence $dx_1 = dy_1 = 0$, $dg = dt_2 - dt_1$, $Q_{x_2}\,dx_2 + Q_{y_2}\,dy_2 = 0$, and the transversality relation reduces to

$$-M(t_1)\,dt_1 + M(t_2)\,dt_2 - \lambda_1(t_2)\,dx_2 - \lambda_2(t_2)\,dy_2 + \lambda_0(dt_2 - dt_1) = 0$$

for all dt_1, dt_2, dx_2, dy_2 as above. Thus, $M(t_1) + \lambda_0 = 0$, $M(t_2) + \lambda_0 = 0$, and $\lambda_1(t_2)\,dx_2 + \lambda_2(t_2)\,dy_2 = 0$ for all dx_2, dy_2 satisfying $Q_{x_2}\,dx_2 + Q_{y_2}\,dy_2 = 0$. Eliminating λ_0, the only boundary conditions are

$$\begin{aligned} &x(t_1) = x_1, \quad y(t_1) = y_1, \qquad Q(x(t_2), y(t_2)) = 0, \\ &M(t_1) = M(t_2), \qquad Q_{y_2}\lambda_1(t_2) - Q_{x_2}\lambda_2(t_2) = 0. \end{aligned}$$

In case IV we have t_1 fixed, $P(x_1, y_1) = 0$, $Q(x_2, y_2) = 0$, $g = t_2$. Hence, $dt_1 = 0$, $dg = dt_2$, $P_{x_1} dx_1 + P_{y_1} dy_1 = 0$, $Q_{x_2} dx_2 + Q_{y_2} dy_2 = 0$, and the transversality relation reduces to

$$\lambda_1(t_1)\,dx_1 + \lambda_2(t_1)\,dy_1 + M(t_2)\,dt_2 - \lambda_1(t_2)\,dx_2 - \lambda_2(t_2)\,dy_2 + \lambda_0\,dt_2 = 0$$

for all dx_1, dy_1, dt_2, dx_2, dy_2 as above, with λ_0 undetermined. Thus, the only boundary conditions are

$$\begin{aligned} &t_1 \text{ fixed}, \quad P(x(t_1), y(t_1)) = 0, \quad Q(x(t_2), y(t_2)) = 0, \\ &P_{y_1}\lambda_1(t_1) - P_{x_1}\lambda_2(t_1) = 0, \quad Q_{y_2}\lambda_1(t_2) - Q_{x_2}\lambda_2(t_2) = 0. \end{aligned}$$

Briefly, in case V the boundary conditions are

$$\begin{aligned} &P(x(t_1, y(t_1)) = 0, \quad Q(x(t_2), y(t_2)) = 0, \quad M(t_1) = M(t_2), \\ &P_{y_1}\lambda_1(t_1) - P_{x_1}\lambda_2(t_1) = 0, \quad Q_{y_2}\lambda_1(t_2) - Q_{x_2}\lambda_2(t_2) = 0. \end{aligned}$$

In case VI the boundary conditions are

$$t_1 \text{ fixed}, \quad x(t_1) = x_1, \quad y(t_1) = y_1, \quad x(t_2) = X(t_2), \quad y(t_2) = Y(t_2).$$

In case VII with, say, $x_2 = \max$, $g = x_2$, $dg = dx_2$, the boundary conditions are

$$t_1 \text{ fixed}, \quad x(t_1) = x_1, \quad y(t_1) = y_1, \quad \lambda_2(t_2) = 0.$$

D. The Navigation Problem with Variable Speed, Mass of Boat, and Steering Function

We have now a new function, namely z, the mass of the boat at time t, and we denote by $z(t_1) = z_1$ its initial value. Again, $p(t, x, y)$, $q(t, x, y)$ are the components of the stream velocity. Now the force applied by the propeller depends upon the fuel consumption, represented by the rate of diminution of the mass, namely, $w = -z'$. Thus, the velocity V is now a function $V = V(z, w)$ of z, w for $0 \le z \le z_1$, $0 \le w \le w_0$, z_1, w_0 fixed. If we require the boat to leave a fixed point $1 = (x_1, y_1)$ at a fixed time t_1 and to reach a fixed point $2 = (x_2, y_2)$ in a minimum time t_2, we have the equations

$$\begin{aligned} &x' = p + V\cos\theta, \quad y' = q + V\sin\theta, \quad z' = -w, \\ &t_1 \text{ fixed}, \quad x(t_1) = x_1, \quad y(t_1) = y_1, \quad z(t_1) = z_1, \quad x(t_2) = x_2, \quad y(t_2) = y_2. \end{aligned} \tag{6.9.9}$$

We consider x, y, z as the state variables, and θ, w as the control variables, $n = 3$, $m = 2$. We have a Mayer problem with $g = t_2$, and we take the Hamiltonian

$$H(t, x, y, z, \theta, w, \lambda_1, \lambda_2, \lambda_3) = \lambda_1(p + V\cos\theta) + \lambda_2(q + V\sin\theta) - \lambda_3 w.$$

For $H_{\min}$ we should have $\partial H/\partial\theta = V(-\lambda_1\sin\theta + \lambda_2\cos\theta) = 0$, and of course $w = w_0$ for $\lambda_3 > 0$, $w = 0$ for $\lambda_3 < 0$. The equations for the multipliers are

$$\begin{aligned} &\lambda_1' + \lambda_1 p_x + \lambda_2 q_x = 0, \quad \lambda_2' + \lambda_1 p_y + \lambda_2 q_y = 0, \\ &\lambda_3' + \lambda_1 V_z \cos\theta + \lambda_2 V_z \sin\theta. \end{aligned} \tag{6.9.10}$$

Again as, in subsection C we may change the end conditions and the quantity to minimize or maximize without changing the equations (6.9.9)–(6.9.10). For instance, we could try to minimize the consumption of fuel, which would correspond to the requirement $z(t_2) = \max$. We do not discuss this general problem in more detail.

Bibliographical Notes

For the problems in Sections 6.1–6 and many more analogous problems the reader may consult M. Athans and P. L. Falb [I] and W. H. Fleming and R. W. Rishel [I]. The problems in Sections 6.1–6 are often used as simple models of problems of space mechanics (cf. W. C. Nelson and E. E. Loft [I]). For the problem in Section 6.7 see M. D. Intriligator [I] pp. 405–416. We refer to this book and to the book by G. Hadley and M. C. Kemp [I] for many other problems of optimization in economics. The problem in Section 6.9, which is only sketched here, has a wide literature. We mention here only E. J. McShane [8] and B. Manià [5].

CHAPTER 7

Proofs of the Necessary Condition for Control Problems and Related Topics

7.1 Description of the Problem of Optimization

Let A denote the constraint set, a closed subset of the tx-space, with t in R, and the space variable $x = (x^1, \ldots, x^n)$ in R^n. Let $U(t)$, the control set, be a subset of the u-space R^m, $u = (u^1, \ldots, u^m)$ the control variable. Let $M = [(t, x, u) | (t, x) \in A,\ u \in U(t)]$ be a closed subset of R^{1+n+m}, and let $f = (f_1, \ldots, f_n)$ be a continuous vector function from M into R^n. Let the boundary set B be a closed set of points (t_1, x_1, t_2, x_2) in R^{2n+2}, $x_1 = (x_1^1, \ldots, x_1^n)$, $x_2 = (x_2^1, \ldots, x_2^n)$. Let g be a continuous function from B into R.

We shall consider the class Ω of all pairs $x(t)$, $u(t)$, $t_1 \le t \le t_2$, called admissible pairs, satisfying the following conditions:

(a) $x(t)$ is absolutely continuous in $[t_1, t_2]$;
(b) $u(t)$ is measurable in $[t_1, t_2]$;
(c) $(t, x(t)) \in A$, $t_1 \le t \le t_2$;
(d) $(t_1, x(t_1), t_2, x(t_2)) \in B$:
(e) $u(t) \in U(t)$ a.e. in $[t_1, t_2]$;
(f) the state equation $dx(t)/dt = f(t, x(t), u(t))$ is satisfied a.e. in $[t_1, t_2]$.

Let $e[x] = (t_1, x(t_1), t_2, x(t_2))$. The functional $I[x, u] = g(e[x]) = g(t_1, x(t_1), t_2, x(t_2))$ is called the cost functional.

We seek the absolute minimum of $I[x, u]$ in the class Ω. If (x_0, u_0) has the property that $I[x_0, u_0] \le I[x, u]$ for all $(x, u) \in \Omega$, then we say that x_0, u_0 is an optimal pair, and we may say that u_0 is an optimal control, and x_0 is an optimal trajectory. Though the optimal pair x_0, u_0 may not be unique in Ω, the value of the cost functional $I[x_0, u_0]$ is the same for all optimal pairs.

We now state necessary conditions for a pair $(x_0, u_0) \in \Omega$ to be an optimal pair.

7.1.i (THEOREM: A NECESSARY CONDITION). *Given a control system as described above, assume that $f(t,x,u)$ possesses continuous partial derivatives $f_t = (f_{it} = \partial f_i/\partial t, i = 1, \ldots, n)$, $f_x = (f_{ixj} = \partial f_i/\partial x^j, i,j = 1, \ldots, n)$ in M, and that the set $Q(t,x) = f(t,x,U(t)) = [z \in R^n | z = f(t,x,u)$ for some u in $U(t)]$ is convex in R^n for each (t,x) in A (see Remark 1(b) below).*

Let $x_0(t), u_0(t), t_1 \le t \le t_2$, denote an optimal pair for which:

(α) *The graph of x_0, $[(t, x_0(t)), t_1 \le t \le t_2]$, is interior to A.*
(β) *$u_0(t)$ is bounded in $[t_1, t_2]$; that is, $|u_0(t)| \le d$, $t_1 \le t \le t_2$, for some constant d (see Remark 1(c)).*
(γ) *The end point $e[x_0] = (t_1, x_0(t_1), t_2, x_0(t_2))$ of the optimal trajectory x_0 is a point of B, at which B possesses a tangent hyperplane B' of some dimension k, $0 \le k \le 2n+2$, whose vectors will be denoted by $h = (\tau_1, \xi_1, \tau_2, \xi_2)$, with $\xi_1 = (\xi_1^1, \ldots, \xi_1^n)$, $\xi_2 = (\xi_2^1, \ldots, \xi_2^n)$, or in differential form, $h = (dt_1, dx_1, dt_2, dx_2)$, with $dx_1 = (dx_1^1, \ldots, dx_1^n)$, $dx_2 = (dx_2^1, \ldots, dx_2^n)$.*
(δ) *g possesses a differential dg at $e[x_0]$, say*

$$dg = g_{t_1}\tau_1 + \sum_{i=1}^{n} g_{x_1^i}\xi_1^i + g_{t_2}\tau_2 + \sum_{i=1}^{n} g_{x_1^i}\xi_2^i,$$

or

$$dg = g_{t_1}\,dt_1 + \sum_{i=1}^{n} g_{x_1^i}\,dx_1^i + g_{t_2}\,dt_2 + \sum_{i=1}^{n} g_{x_2^i}\,dx_2^i,$$

where $g_{t_1}, \ldots, g_{x_2^n}$ denote partial derivatives of g with respect to $t_1, \ldots, x_2^n$, all computed at $e[x_0]$.

Let the Hamiltonian H be defined by

$$H(t,x,u,\lambda) = \lambda f(t,x,u) = \lambda_1 f_1 + \cdots + \lambda_n f_n.$$

Then there exists a family of vector functions

$$\lambda(t) = (\lambda_1(t), \ldots, \lambda_n(t)), \qquad t_1 \le t \le t_2,$$

which we shall call multipliers, with the following properties:

(P_1) *$\lambda(t)$ is absolutely continuous in $[t_1, t_2]$, and satisfies*

$$d\lambda_i(t)/dt = -H_{x^i}(t, x_0(t), u_0(t), \lambda(t)), \qquad i = 1, 2, \ldots, n,$$

for almost all t in $[t_1, t_2]$. If dg is not identically zero at $e[x_0]$, then $\lambda(t)$ is never zero in $[t_1, t_2]$.
(P_2^*) *Weak minimum principle: Given any bounded, measurable function $u(t)$, $u(t) \in U(t)$ a.e. in $[t_1, t_2]$, then for a.a. t in $[t_1, t_2]$, $H(t, x_0(t), u_0(t), \lambda(t)) \le H(t, x_0(t), u(t), \lambda(t))$.*
(P_2) *Usual minimum principle: Let $U(t) = U$, $t_1 \le t \le t_2$, be a fixed closed subset of R^m. Then, $M(t) = M(t, x_0(t), \lambda(t)) = H(t, x_0(t), u_0(t), \lambda(t))$ for a.a.*

t in $[t_1, t_2]$, where $M(t, x, \lambda)$ is defined by $M(t, x, \lambda) = \inf_{u \in U(t)} H(t, x, u, \lambda)$, $(t, x, \lambda) \in A \times R^n$.

(P_3) *The function $M(t) = M(t, x_0(t), \lambda(t))$ coincides with an AC function a.e. in $[t_1, t_2]$, and with this identification then $dM(t)/dt = (d/dt)M(t, x_0(t), \lambda(t)) = H_t(t, x_0(t), u_0(t), \lambda(t))$, $t \in [t_1, t_2]$ (a.e.).*

(P_4) *Transversality relation: There is a constant $\lambda_0 \geq 0$ such that*

$$(\lambda_0 g_{t_1} - M(t_1))\, dt_1 + \sum_{i=1}^{n} (\lambda_0 g_{x_1^i} + \lambda_i(t_1))\, dx_1^i$$

$$+ (\lambda_0 g_{t_2} + M(t_2))\, dt_2 + \sum_{i=1}^{n} (\lambda_0 g_{x_2^i} - \lambda_i(t_2))\, dx_2^i = 0$$

for every vector $h = (dt_1, dx_1, dt_2, dx_2)$ in B', and where $dg = g_{t_1}\, dt_1 + \sum_i g_{x_1^i}\, dx_1^i + g_{t_2}\, dt_2 + \sum_i g_{x_2^i}\, dx_2^i$ is assumed to be not identically zero.

Remark 1.

(a) We shall prove (P1)–(P4) first under the simplifying assumption that t_1, t_2 are fixed. For the extension of this proof with this restriction removed, see Section 7.3G below.

(b) We shall remove the restriction that $Q(t, x)$ is convex by requiring $U(t) = U$, a fixed closed subset of R^m (see our second proof in Section 7.4A–D below), or by requiring other properties of $U(t)$ (see Section 7.4E).

(c) We shall remove the restriction that $u(t)$ is bounded under the additional assumptions (S) of Section 4.2C, Remark 5(c′) (see Section 7.3H below).

(d) If $u_0(t)$ is bounded, then instead of the closed sets $U(t)$ or U we can always restrict ourselves in the proofs to the compact sets which are the intersections of $U(t)$ or U with a fixed closed ball in R^m of center the origin and radius R sufficiently large.

(e) We have already anticipated in Section 4.2D an elementary partial proof of the necessary condition for the linear case.

7.2 Sketch of the Proofs

We shall give below two proofs (Sections 7.3 and Section 7.4) of statements (P1)–(P4). These proofs, like most proofs of the necessary condition, have something in common. The basic idea indeed is very simple, and we wish to present it in a form which is easy to grasp, before we embark in all the technicalities. To do this let us assume here that t_1, x_1, and t_2 are fixed. Let $x_0(t)$, $t_1 \leq t \leq t_2$, be a trajectory that we know is optimal, and let us compare it with all the other trajectories, say $x(t)$, $t_1 \leq t \leq t_2$, starting at the same initial point $x(t_1) = x_0(t_1) = x_1$. Thus, B is reduced to a set B_2 of the x-space, and $g = g(x_2)$ is a real continuous function of x_2 which we may well think of as defined in all of the x-space. At the end time $t = t_2$, we consider the vector $\tilde{Y} = (Y^0, Y)$ with $Y^0 \geq g(x(t_2))$, $Y = x(t_2)$. In the $(n+1)$-dimensional space R^{n+1} (see diagram with $n = 1$), these points $\tilde{Y} = (Y^0, Y)$ form a set W (see diagram with B_2 reduced to the single point x_{20} on the x-axis).

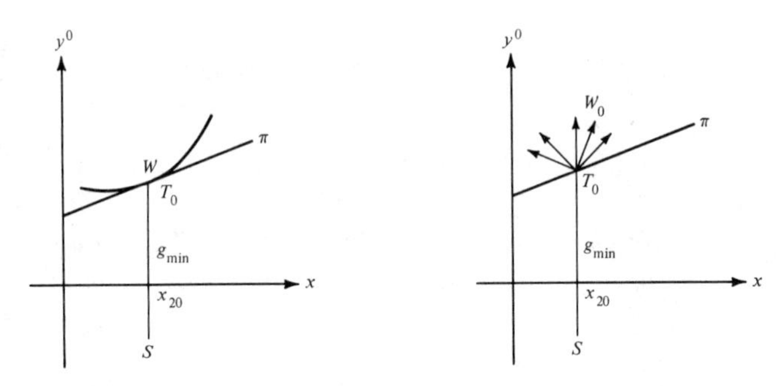

If $T_0 = (x_0(t_2), g_{\min})$, $y^0 = g_{\min} = g(x_0(t_2))$, $x_{20} = x_0(t_2)$, then W has no point in common with the half straight line $S = [(y^0, x),\ y^0 < g_{\min},\ x = x_{20}]$. If it happens that the sets $U(t)$ are all convex, if the system $x' = f(t, x, u)$, is linear in x and u, or $x' = A(t)x + B(t)u$ (A an $n \times n$ matrix, B an $n \times m$ matrix with continuous entries), and if B_2 is convex, then W is also convex, and there is a hyperplane $\pi: \chi_0(y^0 - g_{\min}) + \sum_i \chi_i(y^i - x_{20}^i) = 0$ separating W from S. As we shall see (Section 7.3C–F), this is essentially what is needed to prove (P1)–(P4). Unfortunately, in general, W is not a convex set. To overcome this difficulty, the problem under consideration is linearized, replacing, that is, B_2 with its tangent plane at the point $x_{20} = x_0(t_2)$, and the system $x' = f(t, x, u)$ by a suitable linearization in what we shall call the *variations*. Then, W is replaced by a convex cone W_0 of vertex T_0, and we take a hyperplane π separating W_0 from S. [By this linearization process, the point T_0 will be also transferred to the origin (Section 7.3B), W_0 then will be a cone whose vertex is the origin, and π_0 will have the equation $\chi_0 y^0 + \chi_1 y^1 + \cdots + \chi_n y^n = 0$.]

7.3 The First Proof

A. The Class of Variations

As in most proofs of the necessary condition, we introduce a class of variations such that the corresponding trajectories satisfy exactly the differential system, the constraints, and the initial conditions, but not necessarily the terminal boundary conditions. As stated, we assume here that t_1, t_2 are fixed.

(a) Let h be a $(2n+2)$-vector $h = (\tau_1, \xi_1, \tau_2, \xi_2) \in B'$, that is, a tangent vector to B at the point $e[x_0]$. Here $e[x_0] = (t_1, x_0(t_1), t_2, x_0(t_2))$, where t_1, t_2 are fixed; hence $\tau_1 = \tau_2 = 0$. We can think of $h = (0, \xi_1, 0, \xi_2)$ as the vector tangent to a curve C of class C^1 lying in B and issuing from $e[x_0]$, say in this case $C:(t_1, X_1(z), t_2, X_2(z))$, $0 \le z \le 1$, where $(t_1, X_1(z), t_2, X_2(z)) \in B$ for $0 \le z \le 1$, and $x_0(t_1) = X_1(0)$, $x_0(t_2) = X_2(0)$, $\xi_1 = X_1'(0)$, $\xi_2 = X_2'(0)$ where $X_1(z)$, $X_2(z)$ are continuously differentiable in $[0,1]$. We may even extend $X_1(z)$, $X_2(z)$ to $[-1,1]$ as continuously differentiable functions, if we no longer require that the new part of the curve $C:(t_1, X_1(z), t_2, X_2(z))$ lie in B. With the same conventions we can say that ξ_1 is the tangent to the curve $C':x = X_1(z)$ at $z = 0$, and ξ_2 is the tangent to the curve $C'':x = X_2(z)$ at $z = 0$. Let $x(t,z)$ denote the solution to the differential equation $dx/dt = f(t, x, u_0(t))$ with initial value $x(t_1, z) = X_1(z)$. Thus, $x(t_1, 0) = X_1(0) = x_0(t_1)$. Now $f(t, x, u_0(t))$ is bounded together with the partial derivatives $f_{ix^j}(t, x, u_0(t))$ for (t,x) in a δ-neighborhood Γ_δ of the graph Γ of x_0, $\Gamma_\delta \subset A$. Thus, $f(t, x, u_0(t))$ is uniformly Lipschitzian in x only for a.a. t, and of course measurable in t for all x. Now, by uniqueness theorem for differential systems (see, e.g., McShane [I, p. 345]), we have $x(t,0) = x_0(t)$, $t_1 \le t \le t_2$. Moreover, by [loc. cit. p. 352], we know that for every $z > 0$ sufficiently small, $x(t,z)$ exists in the whole interval $[t_1, t_2]$ with graph in $\Gamma_\delta \subset A$, and with $x(t,z) \to x_0(t)$ uniformly in $[t_1, t_2]$ as $z \to 0$. Finally, by [loc. cit. pp. 356–357], $x(t,z)$ is continuously differentiable in z, and since $dx(t,z)/dt = f(t, x(t,z), u_0(t))$, $x(t_1, z) = X_1(z)$, then $(d/dt)(\partial x(t,z)/\partial z) = f_x(t, x(t,z), u_0(t))(\partial x(t,z)/\partial z)$, $t_1 \le t \le t_2$, $\partial x(t_1, z)/\partial z = X_1'(z)$. If we take $y(t) = (\partial x(t,z)/\partial z)_{z=0}$, then we also have

$$(7.3.1) \qquad dy/dt = f_x(t, x_0(t), u_0(t))y(t), \qquad t_1 \le t \le t_2, \quad y(t_1) = \xi_1.$$

In other words, $y(t) = (y^1, \ldots, y^n)$ satisfies a system of linear differential equations. This system (7.3.1) is often called the "variational system" relative to the solution $x_0(t)$, $u_0(t)$ of the basic system $dx/dt = f(t, x, u)$, and the "variation" represented by the displacement of the initial point from $(t_1, x_0(t_1)) = (t_1, X_1(0))$ to $(t_1, X_1(z))$ for z close to zero. Note that if $A(t)$ denotes the $n \times n$ matrix $A(t) = f_x(t, x_0(t), u_0(t))$, or $A(t) = [f_{ix^j}(t, x_0(t), u_0(t))]$, then $dy/dt = A(t)y$, $y(t_1) = \xi_1$. If $\Phi(t)$ is any fundamental $n \times n$ matrix of the linear homogeneous differential system $dy/dt = A(t)y$, then $y(t) = \Phi(t)\Phi^{-1}(t_1)\xi_1$, $t_1 \le t \le t_2$. The latter is often said to represent the law by which the perturbation ξ_1 at t_1 is *transferred* along the trajectory x_0.

(b) By a *variation* we shall denote a triple $U = (c, u, h)$ made up of a nonnegative number c, a bounded measurable m-vector function $u(t) = (u^1, \ldots, u^m)$, $t_1 \le t \le t_2$, with $u(t) \in U(t)$ for almost all $t \in [t_1, t_2]$, and a $(2n+2)$-vector $h = (\tau_1, \xi_1, \tau_2, \xi_2) \in B'$, that is, a tangent vector to B at the point $e[x_0]$. Note that $\tau_1 = \tau_2 = 0$, since we have assumed t_1, t_2 fixed. As before, we think of $h = (0, \xi_1, 0, \xi_2)$ as the tangent vector at $P_1 = (t_1, x_0(t_1), t_2, x_0(t_2))$ to a continuously differentiable curve $C = [t_1, X_1(z), t_2, X_2(z)]$, $0 \le z \le 1$, issuing from P_1 and lying on B; thus $X_1(0) = x_0(t_1)$, $X_2(0) = x_0(t_2)$, $X_1'(0) = \xi_1$, $X_2'(0) = \xi_2$. As before, we may extend $X_1(z)$, $X_2(z)$ to

$[-1, 1]$ as continuously differentiable functions if we do not require that the new part of C lie in B.

Now, for any z, $-1 \leq z \leq 1$, we consider the following linear combination of $f(t, x, u_0(t))$ and $f(t, x, u(t))$, namely

$$q_z(t, x) = (1 - cz)f(t, x, u_0(t)) + czf(t, x, u(t)), \qquad t_1 \leq t \leq t_2.$$

Let $x(t, z)$ denote the solution of the differential system $dx/dt = q_z(t, x)$, $x = (x^1, \ldots, x^n)$, with initial values $x(t_1, z) = X_1(z)$.

Here the graph Γ of x_0 lies in the interior of A, so that there is some $\delta > 0$ such that the set Γ_δ of all (t, x) with $t_1 \leq t \leq t_2$, $|x - x_0(t)| \leq \delta$ lies in A. Since $f(t, x, u) = (f_1, \ldots, f_n)$ is continuous in M together with the partial derivatives f_{ix^j}, and $u_0(t)$, $u(t)$ are bounded, say $|u_0|, |u| \leq N$, then all f_i, f_{ix^j} are bounded for $(t, x) \in \Gamma_\delta$, $|u| \leq N$, $u \in U(t)$, and so are $f_i(t, x, u_0(t))$, $f_i(t, x, u(t))$ in Γ_δ together with the corresponding partial derivatives. Finally, $q_z(t, x) = q(t, x; z) = (q_1, \ldots, q_n)$ is bounded for $(t, x) \in \Gamma_\delta$, $-1 \leq z \leq 1$, and so are the partial derivatives q_{ix^j}, q_{iz}. Since $q_z(t, x) = f(t, x, u_0(t))$ for $z = 0$, then $x(t, 0) = x_0(t)$, $t_1 \leq t \leq t_2$, by the uniqueness theorem. Moreover as before, $x(t, z)$ exists in all of the interval $[t_1, t_2]$ for $z > 0$ sufficiently small, $x(t, z) \to x_0(t)$ uniformly on $[t_1, t_2]$ as $z \to 0+$, and finally $x(t, z)$ is continuously differentiable with respect to z. Thus

$$dx(t, z)/dt = (1 - cz)f(t, x(t, z), u_0(t)) + czf(t, x(t, z), u(t)),$$
$$t_1 \leq t \leq t_2, \qquad x(t_1, z) = X_1(z),$$

and by differentiation also

$$(d/dt)(\partial x(t, z)/\partial z) = -cf(t, x(t, z), u_0(t)) + cf(t, x(t, z), u(t))$$
$$+ [(1 - cz)f_x(t, x(t, z), u_0(t)) + czf_x(t, x(t, z), u(t)](\partial x(t, z)/\partial z),$$
$$t_1 \leq t \leq t_2, \qquad dx(t_1, z)/dz = X_1'(z).$$

Thus, if we take $y(t) = \partial x(t, z)/\partial z)_{z=0}$, then

$$\text{(7.3.2)} \qquad \begin{aligned} dy/dt = -cf(t, x_0(t), u_0(t)) + cf(t, x_0(t), u(t)) + f_x(t, x_0(t), u_0(t))y(t),\\ t_1 \leq t \leq t_2, \qquad y(t_1) = \xi_1, \end{aligned}$$

or $dy/dt = A(t)y + b(t)$, $y(t_1) = \xi_1$, with $A(t)$ as before, and $b(t) = -cf(t, x_0(t), u_0(t)) + cf(t, x_0(t), u(t))$. Again, as before, $y(t) = (y^1, \ldots, y^n)$ satisfies a system of linear differential equations. The system (7.3.2) can be called the "variational system" relative to the solution $x_0(t), u_0(t)$ by the displacement of the initial point from $(t_1, x(t_1)) = (t_1, X_1(0))$ to $(t_1, X_1(z))$ and the modification represented by the change from $f(t, x, u_0(t))$ to $q_z(t, x)$ for z close to zero. With the same notation as before, we have now

$$y(t) = \Phi(t)\left\{\Phi^{-1}(t_1)\xi_1 + \int_{t_1}^{t} \Phi^{-1}(\tau)[-cf(\tau, x_0(\tau), u_0(\tau)) + cf(\tau, x_0(\tau), u(\tau))]\, d\tau\right\}.$$

As we shall see below, for every z sufficiently small, $x(t, z)$ defined as a solution of $dx/dt = q_z(t, x)$, is actually a trajectory of the problem under consideration, in the sense that there is a measurable function of t, or $u(t, z)$ with $u(t, z) \in U(t)$ such that $dx/dt = q_z(t, x) = f(t, x(t, z), u(t, z))$, $t \in [t_1, t_2]$ (a.e.).

(c) The following extension is also needed. Let us consider an arbitrary system of $s \geq 1$ variations as before $v_\sigma = (c_\sigma, u_\sigma, h_\sigma)$, $\sigma = 1, \ldots, s$. Then $h_\sigma = (0, \xi_{1\sigma}, 0, \xi_{2\sigma}) \in B'$, $\sigma = 1, \ldots, s$, and each h_σ is a tangent vector to B at $e[x_0]$. As before we can think of each h_σ as the tangent vector to a curve C_σ of class C^1 lying in B and issuing from $e[x_0]$, $\sigma = 1, \ldots, s$. It is convenient to think of those s curves C_σ as belonging to a suitable manifold of dimension $\leq s$ in B. In other words, we introduce the vector variable $z = (z_1, \ldots, z_s)$ varying in the hypercube $I = [z | 0 \leq z_\sigma \leq 1, \sigma = 1, \ldots, s]$, and vector functions $X_1(z)$, $X_2(z)$ defined in I such that $(t_1, X_1(z), t_2, X_2(z)) \in B$ for $z \in I$, $X_1(0) = x_0(t_1)$, $X_2(0) = x_0(t_2)$, $X_{1z_\sigma}(0) = \xi_{1\sigma}$, $X_{2z_\sigma}(0) = \xi_{2\sigma}$, where $X_1(z)$, $X_2(z)$ are continuously differentiable in I and $X_{1z_\sigma}(0)$, $X_{2z_\sigma}(0)$ denote the partial derivatives of X_1, X_2 with respect to z_σ at $0 = (0, \ldots, 0)$. Note that X_1, X_2 represent the curve C_σ above as z_σ describes the interval $0 \leq z_\sigma \leq 1$ of the z_σ-axis, $\sigma = 1, \ldots, s$. As before, we extend the functions $X_1(z)$, $X_2(z)$ to the whole hypercube $V = [z | -1 \leq z_\sigma \leq 1, \sigma = 1, \ldots, s]$ so that $X_1(z)$, $X_2(z)$ are still continuous differentiable in the whole of V (we do not require that the new parts of the manifold so added lie in B). For every $z \in V$ we consider now the differential system with initial conditions,

$$dx/dt = q_z(t, x), \qquad t_1 \leq t \leq t_2, \quad x(t_1, z) = X_1(z), \tag{7.3.3}$$

where $q_z(t, x) = q(t, x; z) = (q_1, \ldots, q_n)$ is the vector function of t and x, depending on the parameter z in V, defined by

$$q_z(t, x) = \left(1 - \sum_\sigma c_\sigma z_\sigma\right) f(t, x, u_0(t)) + \sum_\sigma c_\sigma z_\sigma f(t, x, u_\sigma(t)), \qquad t_1 \leq t \leq t_2. \tag{7.3.4}$$

Let $x(t, z)$ denote the solution of system (7.3.4) with the indicated initial values.

The graph $[(t, x_0(t)), t_1 \leq t \leq t_2]$ of x_0 has been assumed to lie in the interior of A. Thus, there is some $\delta > 0$ such that the set Γ_δ of all (t, x) with $t_1 \leq t \leq t_2$, $|x - x_0(t)| \leq \delta$, lies in A. We have assumed that $u_0(t)$, $u_1(t), \ldots, u_s(t)$ are all bounded, say in absolute value $\leq M'$, and that f is continuous, together with the partial derivatives f_{ix^j}. Thus, for $(t, x) \in \Gamma_\delta$, $|u| \leq M'$, $u \in U(t)$, then all $f_i(t, x, u), f_{ix^j}(t, x, u)$ are bounded, say $|f_i|$, $|f_{ix^j}| \leq M''$, and hence all $q_i(t, x, z)$, q_{ix^j}, q_{iz} are bounded for $(t, x) \in \Gamma_\delta$, $z \in V$, say $|q_i|, |q_{ix^j}|, |q_{iz}| \leq M'''$ for the same t, x and $z \in V$. Finally, $X_1(z)$ is a continuous function of z, and $q_z(t, x)$ is a continuous function of (x, z) for every t and is a measurable function of t for every (x, z).

Note that for $z = 0 = (0, \ldots, 0)$ we have $q_0(t, x) = f(t, x, u_0(t))$, $X_1(0) = x(t_1, 0) = x_0(t_1)$. Thus, for $z = 0$, $x(t, 0)$ satisfies the same differential system and initial conditions as $x_0(t)$, and by the usual uniqueness theorem we have

$x(t,0) = x_0(t)$, $t_1 \le t \le t_2$, (see, e.g. McShane [I, loc. cit.]). Moreover, there is some number γ, $0 < \gamma \le 1$, such that for $|z_\sigma| \le \gamma$, $\sigma = 1, \ldots, s$, the solution $x(t,z)$ of the differential system and initial data (7.3.3) exists in the whole interval $[t_1, t_2]$ and $|x(t,z) - x(t,0)| = |x(t,z) - x_0(t)| \le \delta$, $t_1 \le t \le t_2$. Thus, if we denote by V_γ the hypercube $V_\gamma = [z | -\gamma \le z_\sigma \le \gamma, \sigma = 1, \ldots, s]$, we conclude that for every $z \in V_\gamma$, certainly $x(t,z)$ exists in the whole interval $[t_1, t_2]$ and its graph lies in A. Also, $x(t,z) \to x_0(t)$ uniformly in $[t_1, t_2]$ as $z \to 0$, $x(t,z)$ is continuously differentiable with respect to z, and for $t_1 \le t \le t_2$,

$$(7.3.5) \quad dx(t,z)/dt = \left(1 - \sum_\sigma c_\sigma z_\sigma\right) f(t, x(t,z), u_0(t) + \sum_\sigma c_\sigma z_\sigma f(t, x(t,z), u_\sigma(t)).$$

Then, the partial derivatives $\partial x(t,z)/\partial z_\sigma$, $\sigma = 1, \ldots, s$, satisfy the equations

$$(7.3.6) \quad \begin{aligned} (d/dt)(\partial x(t,z)/\partial z_\sigma) = &- c_\sigma f(t, x(t,z), u_0(t)) + c_\sigma f(t, x(t,z), u_\sigma(t)) \\ &+ \left(1 - \sum_\sigma c_\sigma z_\sigma\right) f_x(t, x(t,z), u_0(t))(\partial x(t,z)/\partial z_\sigma) \\ &+ \sum_\sigma c_\sigma z_\sigma f_x(t, x(t,z), u_\sigma(t))(\partial x(t,z)/\partial z_\sigma), \end{aligned}$$

$$t_1 \le t \le t_2, \qquad \partial x(t_1, z)/\partial z_\sigma = X_{1z_\sigma}(z), \quad \sigma = 1, \ldots, s.$$

If we take $y_\sigma(t) = (\partial x(t,z)/\partial z_\sigma)_{z=0}$, $t_1 \le t \le t_2$, $\sigma = 1, \ldots, s$, then $y_\sigma(t) = (y^1, \ldots, y^n)$ satisfies the relations

$$(7.3.7) \quad \begin{aligned} dy_\sigma/dt = &- c_\sigma f(t, x_0(t), u_0(t)) + c_\sigma f(t, x_0(t), u_\sigma(t)) \\ &+ f_x(t, x_0(t), u_0(t)) y_\sigma(t), \\ t_1 \le t \le t_2, \quad & y_\sigma(t_1) = X_{1z_\sigma}(0) = \xi_{1\sigma}, \quad \sigma = 1, \ldots, s. \end{aligned}$$

Again, $y_\sigma(t) = (y^1, \ldots, y^n)$ satisfies a system of linear differential equations. The system (7.3.7) can be called the variational system relative to the solution $x_0(t)$, $u_0(t)$ of the basic system $dx/dt = f(t,x,u)$, and describes the variation represented by the displacement of the initial point from $(t_1, x(t_1)) = (t_1, X_1(0))$ to $(t_1, X_1(z))$ and the modification represented by the change from $f(t, x, u_0(t))$ to $q_z(t,x)$ as defined by (7.3.4) for $z = z_\sigma$ close to zero.

From the second part of (7.3.6) we see that

$$h_{1z} = [(d/dz)X_1(c_1 z, \ldots, c_s z)]_{z=0} = \sum_{\sigma=1}^{s} c_\sigma X_{1z_\sigma}(0) = \sum_{\sigma=1}^{s} c_\sigma \xi_{1\sigma}.$$

Let h_{2z} denote the analogous expression in terms of $X_2(z)$.

It remains to show that for every $z \in V_\gamma \cap I$, $x(t,z)$ defined as a solution of $dx/dt = q_z(t,x)$ is actually a trajectory of the problem under consideration in the sense that there is a measurable function of t, or $u(t,z)$, with $u(t,z) \in U(t)$, such that

$$(7.3.8) \qquad dx/dt = q_z(t,x) = f(t, x(t,z), u(t,z)), \qquad t \in [t_1, t_2] \text{ (a.e.)}$$

Indeed, for $c = (c_1, \ldots, c_s)$ fixed, $\gamma > 0$ sufficiently small, and $z = (z_1, \ldots, z_s)$ with $0 \le z_\sigma \le \gamma$, then $c_\sigma z_\sigma \ge 0$, $\sigma = 1, \ldots, s$, $1 - \sum_\sigma c_\sigma z_\sigma \ge 0$. Hence, $q_z(t, x)$ is a convex combination of the $j + 1$ points $f(t, x, u_0(t))$, $f(t, x, u_\sigma(t))$, $\sigma = 1, \ldots, s$, all in the convex set $Q(t, x)$ of R^n. In other words, $q_z(t, x) \in Q(t, x)$, or $dx(t, z)/dt \in Q(t, x(t, z))$, $t \in [t_1, t_2]$ (a.e.), for every $z \in V_\gamma \cap I$. By the implicit function theorem for orientor fields (see (8.2.iii)) we see that for every $z \in V_\gamma \cap I$, there is a measurable function of t, namely $u(t, z)$, $t_1 \le t \le t_2$, with $u(t, z) \in U(t)$, for which (7.3.8) holds. In other words, for every $z \in V_\gamma \cap I$, $x(t, z)$ is the trajectory relative to the variation $V_z = (1, u(t, z), h_z)$ with $h_z = (0, h_{1z}, 0, h_{2z})$.

B. The Cone K

We shall now consider the cone K in R^{n+1} made up of the terminal points of the linearized trajectories in R^n corresponding to all possible variations defined above, and associated values of the linearized cost function. We shall prove that the point $(-1, 0, \ldots, 0)$ in R^{n+1} is not an interior point of K. The argument is by contradiction, showing that in the opposite case, there would exist an admissible trajectory giving a lower cost than the optimal cost. The proof of the necessary condition then follows by taking a supporting hyperplane to K.

For every variation $v = (c, u, h)$ with $h = (0, \xi_1, 0, \xi_2)$, let us consider the $(n + 1)$-vector $\tilde{Y}(v) = (Y^0, Y) = (Y^0, Y^1, \ldots, Y^n)$ defined by $Y^0(v) = g_{x_1}\xi_1 + g_{x_2}\xi_2$, $Y(v) = y(t_2; v) - \xi_2$, where g_{x_1}, g_{x_2} are the $1 \times n$ matrices of the partial derivatives of $g(t_1, x_1, t_2, x_2)$ with respect to the arguments $x_1^1, \ldots, x_1^n$, or $x_2^1, \ldots, x_2^n$ respectively, these partial derivatives being evaluated at the point $e[x_0] = (t_1, x_0(t_1), t_2, x_0(t_2))$, and where $y(t; v)$ simply denotes the solution of the differential system and initial data (7.3.7) with $s = 1$, $v_\sigma = v = (c, u, h)$, $h = (0, \xi_1, 0, \xi_2)$. We shall denote by $K \subset R^{n+1}$ the set of all such vectors $\tilde{Y}(v)$ in the $y^0 y^1 \cdots y^n$-space R^{n+1}.

7.3.i (Lemma). *The set K is a convex cone with vertex at $(0, \ldots, 0)$; that is, if $\tilde{Y}(v_1)$, $\tilde{Y}(v_2) \in K$, and a_1, $a_2 \ge 0$, then there is a variation $v = (c, y, h)$, $h = (0, \xi_1, 0, \xi_2)$ such that $\tilde{Y}(v) = a_1 \tilde{Y}(v_1) + a_2 \tilde{Y}(v_2)$.*

Proof. Let $v_\sigma = (c_\sigma, u_\sigma, h_\sigma)$, $h_\sigma = (0, \xi_{1\sigma}, 0, \xi_{2\sigma})$, $\sigma = 1, 2$, be two given variations and $Y(v_1)$, $Y(v_2)$ the corresponding vectors in R^{n+1}. Assume first $a_1 c_1 + a_2 c_2 \ne 0$, and hence (since $a_1, a_2, c_1, c_2 \ge 0$) $a_1 c_1 + a_2 c_2 > 0$. Take $h = a_1 h_1 + a_2 h_2$, $c = a_1 c_1 + a_2 c_2$; hence, if $h = (0, \xi_1, 0, \xi_2)$, then $\xi_\sigma = a_1 \xi_{1\sigma} + a_2 \xi_{2\sigma}$, $\sigma = 1, 2$. Let us consider the convex combination of $f(t, x_0(t), u_\sigma(t))$, $\sigma = 1, 2$,

$$q(t) = (a_1 c_1 + a_2 c_2)^{-1}[a_1 c_1 f(t, x_0(t), u_1(t)) + a_2 c_2 f(t, x_0(t), u_2(t))]. \tag{7.3.9}$$

Since $f(t, x_0(t), u_\sigma(t)) \in Q(t, x_0(t))$, $\sigma = 1, 2$, and $Q(t, x_0(t))$ is convex, we see that $q(t) \in Q(t, x_0(t))$ for almost all $t \in [t_1, t_2]$. By the implicit function theorem for orientor fields (see Section 8.2) there is a measurable control function $u(t)$, $t_1 \le t \le t_2$, $u(t) \in U(t)$, such that

$$q(t) = f(t, x_0(t), u(t)), \quad t \in [t_1, t_2] \text{ (a.e.)}. \tag{7.3.10}$$

We denote by v the new variation $v = (c, u, h)$ with $h = (0, \xi_1, 0, \xi_2)$. If $y(t; v_\sigma)$, $\sigma = 1, 2$, denote the solutions of the variational equation and initial data (7.3.7) relative to the variation v_σ, and if y denotes the linear combination $y(t) = a_1 y(t; v_1) + a_2 y(t; v_2)$, then by linear combination of the relevant equations (7.3.10) with coefficients a_1, a_2 and the use of (7.3.7), (7.3.9) and the definitions of c, h, we obtain for y the equation and initial data

$$\begin{aligned} dy/dt = -cf(t, x_0(t), u_0(t)) + cf(t, x_0(t), u(t)) + f_x(t, x_0(t), u_0(t))y(t), \\ t_1 \le t \le t_2, \qquad y(t_1) = \xi_1; \end{aligned} \tag{7.3.11}$$

that is, $y(t) = y(t; v)$ is the unique solution of (7.3.7) relative to the new variation v.

From (7.3.11) we obtain

$$\begin{aligned} a_1 Y^0(v_1) + a_2 Y^0(v_2) &= a_1(g_{x_1}\xi_{11} + g_{x_2}\xi_{21}) + a_2(g_{x_1}\xi_{12} + g_{x_2}\xi_{22}) \\ &= g_{x_1}\xi_1 + g_{x_2}\xi_2 = Y^0(v), \\ a_1 Y(v_1) + a_2 Y(v_2) &= a_1(y(t_2; v_1) - \xi_{21}) + a_2(y(t_2; v_2) - \xi_{22}) \\ &= y(t_2; v) - \xi_2 = Y(v), \end{aligned}$$

or $a_1 \tilde{Y}(v_1) + a_2 \tilde{Y}(v_2) = \tilde{Y}(v)$. If $a_1 c_1 + a_2 c_2 = 0$, then $a_1 c_1 = a_2 c_2 = 0$, and (7.3.11) become $dy/dt = f_x(t, x_0(t), u_0(t))y(t)$, $y(t_1) = \xi_1$, $t_1 \le t \le t_2$, and hence the above argument holds for the variation $v = (0, u_0, h)$. We have thus proved that K is a convex cone. □

7.3.ii (LEMMA). *The point $(-1, 0, \ldots, 0)$ is not interior to K.*

Proof. Assume, if possible, that $(-1, 0, \ldots, 0)$ is interior to K. Then for some $\delta > 0$ sufficiently small the $n + 1$ points in R^{n+1}

$$\begin{aligned} (-1, -\delta, 0, \ldots, 0), (-1, 0, -\delta, 0, \ldots, 0), \ldots, \\ (-1, 0, \ldots, 0, -\delta), (-1, \delta, \delta, \ldots, \delta) \end{aligned} \tag{7.3.12}$$

certainly belong to K, and hence there are variations $v_1, v_2, \ldots, v_{n+1}$ such that the corresponding vectors $\tilde{Y}(v_1), \ldots, \tilde{Y}(v_{n+1})$ are exactly the corresponding vectors (7.3.12) with $v_\sigma = (c_\sigma, u_\sigma, h_\sigma)$, $h_\sigma = (0, \xi_{1\sigma}, 0, \xi_{2\sigma})$, $\sigma = 1, \ldots, n + 1$. We shall now take $s = n + 1$ in Subsection 7.3A(c) and denote by $X_1(z)$ $X_2(z)$, $z = (z_1, \ldots, z_{n+1}) \in V$, the corresponding functions, and by $x(t, z) = (x^1, \ldots, x^n)$ the corresponding solution of (7.3.3) with initial values $x(t_1, z) = X_1(z)$. We now have to compare the end values of $x(t, z)$, or $x(t_2, z)$, with $X_2(z)$, and the value of the functional $g(e[x(t, z)])$ with $g(e[x_0]) = g_{\min}$. Here $x(t_1, z) = X_1(z)$ for $z \in V_\gamma \cap I$, while in general $x(t_2, z)$ does not coincide with

$X_2(z)$. In other words, we may be interested in determining $z \in V_\gamma \cap I$ in such a way that $x(t_2, z) - X_2(z) = 0$. Concerning the difference $z_0 = g_{\min} - g(t_1, X_1(z), t_2, X_2(z))$ all we can say is that $z_0 \le 0$ whenever there is some trajectory x joining $X_1(z)$ to $X_2(z)$. Thus, we have the $n + 1$ equations

$$(7.3.13) \qquad \begin{aligned} &g(t_1, X_1(z), t_2, X_2(z)) + z_0 - g_{\min} = 0, \\ &x^i(t_2, z) - X_2^i(z) = 0, \quad i = 1, \dots, n. \end{aligned}$$

These $n + 1$ equations (in the $n + 2$ unknowns $z_1, \dots, z_{n+1}, z_0$) are obviously satisfied by $z_1 = \cdots = z_{n+1} = z_0 = 0$, since then $x(t_1, 0) = x_0(t_1) = X_1(0)$, $x(t_2, 0) = x_0(t_2) = X_2(0)$, and $g(e[x_0]) = g_{\min}$. At the point $(0, \dots, 0, 0)$ the partial derivatives of the first members with respect to, say, z_σ are respectively

$$\sum_{j=1}^{n} (g_{x_1^j} X_{1z_\sigma}^j + g_{x_2^j} X_{2z_\sigma}^j)_{z=0} = \sum_{j=1}^{n} (g_{x_1^j} \xi_1^j + g_{x_2^j} \xi_2^j) = Y^0(v_\sigma),$$

$$(\partial x^i(t_2, z)/\partial z_\sigma)_{z=0} - (X_{2z_\sigma}^i(z))_{z=0} = y^i(t_2; v_\sigma) - \xi_2^i = Y^i(v_\sigma),$$
$$i = 1, \dots, n, \quad \sigma = 1, \dots, n + 1.$$

In other words, the $(n + 1) \times (n + 1)$ functional determinant of the $n + 1$ equations (7.3.13) with respect to the $n + 1$ variables $z_1, z_2, \dots, z_{n+1}$ is the determinant of the $n + 1$ vectors (7.3.12), and this determinant is $(-1)^{n+1}(n + 1)\delta^n \ne 0$. By the implicit function theorem of calculus we conclude that for every $z_0 \ne 0$ and sufficiently small, the $n + 1$ equations (7.3.13) can be solved with respect to $z_1, \dots, z_{n+1}$, and that again for z_0 sufficiently small, the solutions

$$z_\sigma = Z_\sigma(z_0), \quad \sigma = 1, 2, \dots, n + 1, \quad \text{or} \quad Z(z_0) = (Z_1, \dots, Z_{n+1}),$$

are continuously differentiable functions of z_0. In other words, there is a neighborhood $(-\lambda, \lambda)$ of $z_0 = 0$ such that for $-\lambda \le z_0 \le \lambda$, the functions $Z_\sigma(z_0)$ satisfy the equations

$$(7.3.14) \qquad \begin{aligned} &g(t_1, X_1(Z(z_0)), t_2, X_2(Z(z_0))) + z_0 - g_{\min} = 0, \\ &x^i(t_2, Z(z_0)) - X_2^i(Z(z_0)) = 0. \qquad i = 1, \dots, n. \end{aligned}$$

Note that we have also

$$x^i(t_1, Z(z_0)) - X_1^i(Z(z_0)) = 0, \qquad i = 1, \dots, n,$$

since this relation holds for all z as stated in (7.3.3). Thus, for every $z = Z(z_0)$, $-\lambda_0 \le z_0 \le \lambda_0$, we have

$$(7.3.15) \qquad (t_1, x(t_1, z), t_2, x(t_2, z)) = (t_1, X_1(z), t_2, X_2(z)) \in B.$$

Again, by the implicit function theorem, the derivatives of the functions $Z_\sigma(z_0)$, $\sigma = 1, \dots, n + 1$, at $z_0 = 0$ can be obtained by differentiating the relations (7.3.14) with respect to z_0 and taking $z_0 = 0$. We obtain

$$(7.3.16) \quad \sum_{\sigma=1}^{n+1} Y^0(v_\sigma) Z'_\sigma(0) + 1 = 0, \qquad \sum_{\sigma=1}^{n+1} Y^i(v_\sigma) Z'_\sigma(0) = 0, \quad i = 1, \dots, n,$$

where the coefficients of this system are given by (7.3.12), that is,

$$-Z_1'(0) - Z_2'(0) - \cdots - Z_{n+1}'(0) + 1 = 0,$$
$$-\delta Z_1'(0) + \delta Z_{n+1}'(0) = 0, \ldots, \qquad -\delta Z_n'(0) + \delta Z_{n+1}'(0) = 0;$$

and hence,

$$Z_1'(0) = Z_2'(0) = \cdots = Z_{n+1}'(0) = (n+1)^{-1} > 0. \tag{7.3.17}$$

Since the matrix of the coefficients of (7.3.16) is nonsingular, this is the only solution. We conclude that for z_0 positive and sufficiently small, say again $0 < z_0 < \lambda$, the numbers $z_\sigma = Z_\sigma(z_0)$, $\sigma = 1, \ldots, n+1$, are all positive and as close to zero as we want, since $Z_\sigma(0) = 0$. Thus $z = Z(z_0) = (Z_1, \ldots, Z_{n+1}) \in V_\gamma \cap I$ for $0 < z_0 \leq \lambda$. From (7.3.8) and (7.3.14) we conclude that for $0 < z_0 \leq \lambda$ and $z = Z(z_0)$, the pair $x(t, z)$, $u(t, z)$, $t_1 \leq t \leq t_2$, is admissible. From (7.3.14) we now have

$$I[x(t,z), u(t,z)] = e[x(t,z)] = g_{\min} - z_0 < g_{\min} \quad \text{for } 0 < z_0 \leq \lambda,$$

and this contradicts the definition of $g_{\min}$. We have proved that $(-1, 0, \ldots, 0)$ is not an interior point of K. This completes the proof of Lemma (7.3.ii). □

7.3.iii (Lemma). *There are numbers $\chi_0, \chi_1, \ldots, \chi_n$ not all zero, $\chi_0 \geq 0$, such that $\sum_{i=0}^n \chi_i Y^i(v) \geq 0$ for all variations v.*

Proof. If K has no interior points, then the lemma follows immediately. If K has interior points, then by (7.3.i), (7.3.ii), and Section 8.4C, K possesses a supporting hyperplane through $(0, \ldots, 0)$, say $\sum_{i=0}^n \chi_i z^i = 0$, with K contained in $\sum_{i=0}^n \chi_i z^i \geq 0$, and $(-1, 0, \ldots, 0)$ contained in $\sum_{i=0}^n \chi_i z^i \leq 0$; in particular $\chi_0 \geq 0$, with $(\chi_0, \ldots, x_n)$ not identically zero. Lemma (7.3.iii) is thereby proved. □

Remark. Let us prove that, if dg is not identically zero, then the n-vector $\chi = (\chi_1, \ldots, \chi_n)$ must be nonzero. Indeed, whenever $\chi_1 = \cdots = \chi_n = 0$, then $\chi_0 > 0$, the hyperplane becomes $z^0 = 0$, and K is contained in the half space $z^0 \geq 0$, or $Y^0(v) \geq 0$ for all variations v. If we assume that g possesses a differential dg at $e[x_0]$, then Y^0 is this differential $dg = g_{x_1}\xi_1 + g_{x_2}\xi_2$, a linear function, which changes sign by changing the sign of (ξ_1, ξ_2). Thus, $dg \geq 0$ implies $dg = 0$ (identically). Thus, for dg not identically zero at $e[x_0]$, the n-vector $(\chi_1, \ldots, \chi_n)$ must be nonzero.

C. Proof of (P_1)

Given any variation $v = (c, u, h)$ with $h = (0, \xi_1, 0, \xi_2)$, the corresponding variational equation and initial data are

$$dy/dt = -cf(t, x_0(t), u_0(t)) + cf(t, x_0(t), u(t)) + f_x(t, x_0(t), u_0(t))y(t),$$
$$t_1 \leq t \leq t_2, \qquad y(t_1) = \xi_1.$$

If $A(t)$ denotes the $n \times n$ matrix $A(t) = f_x(t, x_0(t), u_0(t))$, i.e., the matrix

$[f_{ix^j}(t, x_0(t), u_0(t)), i, j = 1, \ldots, n]$, then y is the unique solution of the differential system with initial data

$$y' - A(t)y = -cf(t, x_0(t), u_0(t)) + cf(t, x_0(t), u(t)), \qquad t_1 \le t \le t_2, \quad y(t_1) = \xi_1.$$

Let $A^*(t)$ denote the transpose of the matrix $A(t)$. For $a, b \in R^n$ let $a \cdot b = \sum_i^n a_i b_i$ denote the inner product in R^n. For $z_1(t)$, $z_2(t)$, $t_1 \le t \le t_2$, two n-vector functions with $z_1(t) \cdot z_2(t)$ L-integrable in $[t_1, t_2]$, let (z_1, z_2) denote the usual integral $\int_{t_1}^{t_2} z_1(t) \cdot z_2(t)\,dt$. For $y(t)$, $\lambda(t)$, $t_1 \le t \le t_2$, AC n-vector functions, let L and L^* denote the operators defined by $Ly = y' - A(t)y$, $L^*\lambda = -\lambda' - A^*(t)\lambda$. By integration by parts we have now

$$\begin{aligned}(Ly, \lambda) &= \int_{t_1}^{t_2} y'(t) \cdot \lambda(t)\,dt - \int_{t_1}^{t_2} A(t)y(t) \cdot \lambda(t)\,dt \\ &= y(t_2) \cdot \lambda(t_2) - y(t_1) \cdot \lambda(t_1) - \int_{t_1}^{t_2} y(t) \cdot \lambda'(t)\,dt - \int_{t_1}^{t_2} y(t) \cdot A^*(t)\lambda(t)\,dt \\ &= y(t_2) \cdot \lambda(t_2) - y(t_1) \cdot \lambda(t_1) + (y, L^*\lambda).\end{aligned} \tag{7.3.18}$$

Because of this relation, L^* is called the dual operator of L.

Let us take for λ the AC n-vector function which is the unique solution of the differential system $L^*\lambda = 0$ with $\lambda(t_2) = \chi = (\chi_1, \ldots, \chi_n)$. Then

$$\begin{aligned}y(t_2) \cdot \lambda(t_2) - y(t_1) \cdot \lambda(t_1) &= (Ly, \lambda) \\ &= \int_{t_1}^{t_2} \lambda(t) \cdot [-cf(t, x_0(t), u_0(t)) + cf(t, x_0(t), u(t))]\,dt,\end{aligned}$$

where $\lambda' = -A^*(t)\lambda, \lambda(t_2) = \chi$; hence,

$$d\lambda_i/dt = -\sum_{j=1} \lambda_j f_{jx^i}(t, x_0(t), u_0(t)), \qquad t_1 \le t \le t_2.$$

For $\chi \ne 0$ then $\lambda(t) \ne 0$ at any $t \in [t_1, t_2]$, because, by uniqueness theorem, $\lambda(t) = 0$ for some t would imply that λ is identically zero, hence $\chi = 0$, a contradiction. We have proved (P_1).

D. Proof of (P_4)

Let $\lambda_0 = \chi_0 \ge 0$ and note that $\lambda(t_2) = \chi$. Let v denote any variation $v = (c, u, h)$, $h = (0, \xi_1, 0, \xi_2)$. Let us replace $y(t_2; v)$ in the inequality $\sum_i \chi_i Y^i(v) \ge 0$ in (7.3.iii). We obtain

$$\begin{aligned}0 \le \sum_{i=0}^{n} \chi_i Y^i(v) &= \chi_0 Y^0(v) + \chi \cdot Y(v) \\ &= \lambda_0(g_{x_1} \cdot \xi_1 + g_{x_2} \cdot \xi_2) + \lambda(t_2) \cdot [y(t_2) - \xi_2] \\ &= \lambda_0(g_{x_1} \cdot \xi_1 + g_{x_2} \cdot \xi_2) - \lambda(t_2) \cdot \xi_2 \\ &\quad + [\lambda(t_2) \cdot y(t_2) - \lambda(t_1) \cdot y(t_1)] + \lambda(t_1) \cdot y(t_1) \\ &= \lambda_0(g_{x_1} \cdot \xi_1 + g_{x_2} \cdot \xi_2) - \lambda(t_2) \cdot \xi_2 + \lambda(t_1) \cdot \xi_1 \\ &\quad + \int_{t_1}^{t_2} c\lambda(t) \cdot [-f(t, x_0(t), u_0(t)) + f(t, x_0(t), u(t))]\,dt.\end{aligned}$$

By using the definition of the Hamiltonian we obtain

$$\begin{aligned}(7.3.19)\quad &\sum_{i=1}^{n}[\lambda_0 g_{x_1^i} + \lambda_i(t_1)]\xi_1^i + \sum_{i=1}^{n}[\lambda_0 g_{x_2^i} - \lambda_i(t_2)]\xi_2^i \\ &+ c\int_{t_1}^{t_2}[H(t, x_0(t), u(t), \lambda(t)) - H(t, x_0(t), u_0(t), \lambda(t))]\,dt \geq 0.\end{aligned}$$

For $c = 0$ we obtain

$$(7.3.20)\qquad \sum_{i=1}^{n}[\lambda_0 g_{x_1^i} + \lambda_i(t_1)]\xi_1^i + \sum_{i=1}^{n}[\lambda_0 g_{x_1^i} - \lambda_i(t_2)]\xi_2^i \geq 0.$$

Since B possesses a tangent plane B' at $e[x_0]$, (7.3.20) holds with equality for any $(0, \xi_1, 0, \xi_2) \in B'$, and this is relation (P_4) of the necessary condition when $\tau_1 = \tau_2 = 0$.

E. Proofs of (P_2^*) and (P_2)

From (7.3.19) for $h = 0$, that is, $\xi_1 = 0$, $\xi_2 = 0$, and $c = 1$, we have

$$(7.3.21)\qquad \int_{t_1}^{t_2}[H(t, x_0(t), u(t), \lambda(t)) - H(t, x_0(t), u_0(t), \lambda(t))]\,dt \geq 0.$$

Let $u(t)$ be any bounded measurable function with $u(t) \in U(t)$ a.e. in $[t_1, t_2]$. Let

$$\Delta_u(t) = H(t, x_0(t), u(t), \lambda(t)) - H(t, x_0(t), u_0(t), \lambda(t)).$$

Then $\Delta_u(t)$ is measurable and L-integrable in $[t_1, t_2]$, and for almost all t in $[t_1, t_2]$,

$$(7.3.22)\qquad (d/dt)\int_{t_1}^{t}\Delta_u(\tau)\,d\tau = \Delta_u(t).$$

Let $\bar{t}$ be such a point in (t_1, t_2). We wish to show that $\Delta_u(\bar{t}) \geq 0$. To this end, let us choose an arbitrarily small positive h, $t_1 \leq \bar{t} - h < \bar{t}$, and consider the "mixed control" $u_h(t)$,

$$u_h(t) = \begin{cases} u_0(t) & \text{if } t \in [t_1, t_2] - [\bar{t} - h, \bar{t}], \\ u(t) & \text{if } t \in [\bar{t} - h, \bar{t}]. \end{cases}$$

Then $v = (1, u_h, 0)$ is a variation, and hence (7.3.21) and (7.3.22) yield

$$0 \leq \int_{t_1}^{t_2}\Delta_{uh}(t)\,dt = \int_{\bar{t}-h}^{\bar{t}}\Delta_u(t)\,dt = h(\Delta_u(\bar{t})) + o(h).$$

Dividing by $h > 0$, we obtain

$$0 \leq \Delta_u(\bar{t}) + o(h)/h,$$

and hence, by taking $h \to 0+$, this yields $\Delta_u(\bar{t}) \geq 0$. Statement (P_2^*) is thereby proved. Property (P_2^*) is of some relevance, since no requirement was needed for its proof on the variable closed set $U(t) \subset R^m$, but the general requirements in Section 7.1, namely, that A is closed in R^{1+n} and $M = [(t, x, u) | (t, x) \in A, u \in U(t)]$ is closed in R^{1+n+m}.

Let us now prove (P2). Since $U(t) = U$ is a closed subset of R^m, there is a countable subset U_c of U such that the closure of U_c, cl U_c, is U. Let $U_c = \{u_1, u_2, \ldots, u_k, \ldots\}$. Consider the constant controls $u_i(t) = u_i$, $t_1 \le t \le t_2$, $i = 1, 2, \ldots$. Then for each i, $u_i(t)$ is a measurable bounded function in $[t_1, t_2]$, with $u_i \in U$. Hence (P_2^*) applies to each of these controls. In particular, for each i, there exists a set $K_i \subset [t_1, t_2]$, meas $K_i = 0$, possibly empty, such that

$$H(t, x_0(t), u_0(t), \lambda(t)) \le H(t, x_0(t), u(t), \lambda(t)) \tag{7.3.23}$$

holds for $u(t) = u_i(t)$ in $[t_1, t_2] - K_i$. Let $K = \bigcup_i K_i$. Then meas $K = 0$. Let $G = [t_1, t_2] - K$. We shall now show that (7.3.23) holds for $t \in G$. Choose any $t_0 \in G$. Since cl $U_c = U$, there exists a (minimizing) subsequence $[u_{k_j}]$ of $[u_k]$ such that

$$\begin{aligned} H(t_0, x_0(t_0), u_{k_j}, \lambda(t_0)) &\to \inf_{u \in U} H(t_0, x_0(t_0), u, \lambda(t_0)) \\ &= M(t_0, x_0(t_0), \lambda(t_0)) \end{aligned} \tag{7.3.24}$$

as $j \to \infty$. Moreover, from (7.3.23),

$$\begin{aligned} H(t_0, x_0(t_0), u_{k_j}, \lambda(t_0)) &= H(t_0, x_0(t_0), u_{k_j}(t_0), \lambda(t_0)) \\ &\ge H(t_0, x_0(t_0), u_0(t_0), \lambda(t_0)), \end{aligned} \tag{7.3.25}$$

$j = 1, 2, \ldots$. Hence, (7.3.24) and (7.3.25) yield

$$H(t_0, x_0(t_0), u_0(t_0), \lambda(t_0)) \le \inf_{u \in U} H(t_0, x_0(t_0), u, \lambda(t_0)). \tag{7.3.26}$$

Since $u(t_0) \in U$, (7.3.26) holds with equality. Since t_0 was chosen arbitrarily in G and meas $G = t_2 - t_1$, (P_2) is thereby proved.

F. Proof of (P_3)

(a) This property is a consequence of the following lemma which concerns autonomous problems.

7.3.iv (Lemma). *Assume that the control space U is a fixed compact subset of the u-space R^m, and that $f(x, u) = (f_1, \ldots, f_n)$ is a continuous vector function of x, u, with continuous first order partial derivatives $f_x = (f_{ix^j}, i, j = 1, \ldots, n)$ on $A_0 \times U$, where A_0 is a compact subset of the x-space R^n. Let $x(t)$, $\lambda(t)$ be* AC *vector functions, and $u(t)$ measurable in $[t_1, t_2]$, $u(t) \in U$, and assume that $dx^i/dt = \partial H/\partial \lambda_i$, $d\lambda_i/dt = -\partial H/\partial x^i$, $i = 1, \ldots, n$, and that $H(x(t), u(t), \lambda(t)) = M(x(t), \lambda(t))$* a.e. *in $[t_1, t_2]$, where $M(x, \lambda) =$* min $H(x, u, \lambda)$ *and* min *is taken for $u \in U$. Then $M(x(t), \lambda(t))$ is constant in $[t_1, t_2]$.*

Proof. Let Λ be the set $\Lambda = [\lambda \,|\, |\lambda| \le \bar{\lambda}] \in R^n$, where $\bar{\lambda}$ is the maximum of $|\lambda(t)|$ in $[t_1, t_2]$. Then $H(x, u, \lambda) = \sum_i \lambda_i f_i(x, u)$ is continuous in the compact set $A_0 \times U \times \Lambda$ together with its first order partial derivatives H_λ and H_x. Thus, there is a constant $K \ge 0$ such that $|H(x, u, \lambda) - H(x', u, \lambda')| \le K\delta$ for all pairs of points (x, u, λ), $(x', u, \lambda') \in A_0 \times U \times \Lambda$ at a distance $\leq \delta$. If (x, λ), (x', λ') are any two points of $A_0 \times \Lambda$ at a distance $\leq \delta$, and u, u' are points of U such that $M(x, \lambda) = H(x, u, \lambda)$, $M(x', \lambda') = H(x', u', \lambda')$, then we have

$$\begin{aligned} M(x, \lambda) &= H(x, u, \lambda) \geq H(x', u, \lambda') - k\delta \\ &\geq H(x', u', \lambda') - k\delta = M(x', \lambda') - k\delta, \end{aligned}$$

and analogously $M(x', \lambda') \geq M(x, \lambda) - k\delta$. Thus, $|M(x, \lambda) - M(x', \lambda')| \leq k\delta$, and this proves that $M(x, \lambda)$ is Lipschitzian in $A_0 \times \Lambda$. As a consequence, $M(x(t), \lambda(t))$ is AC in $[t_1, t_2]$, and hence possesses a derivative a.e. in $[t_1, t_2]$. Let us prove that this derivative is zero a.e.

Let $\bar{t} \in (t_1, t_2)$ be any point where such a derivative exists, where the canonical equations hold as assumed in the statement, and where $M(x(\bar{t}), \lambda(\bar{t})) = H(x(\bar{t}), u(\bar{t}), \lambda(\bar{t}))$. Let $\bar{u} = u(\bar{t})$. Then for every t', $\bar{t} < t' < t_2$, we also have $M(x(t'), \lambda(t')) \leq H(x(t'), \bar{u}, \lambda(t'))$. Hence $M(x(t'), \lambda(t')) - M(x(\bar{t}), \lambda(\bar{t})) \leq H(x(t'), \bar{u}, \lambda(t')) - H(x(\bar{t}), \bar{u}, \lambda(\bar{t}))$, and by division by $t' - \bar{t} > 0$ and passage to the limit as $t' \to \bar{t} + 0$, we derive

$$\begin{aligned} (d/dt)M(x(\bar{t}), \lambda(\bar{t})) &\leq (d/dt)H(x(\bar{t}), \bar{u}, \lambda(\bar{t})) \\ &= \sum_i [(\partial H/\partial x^i)(dx^i/dt) + (\partial H/\partial \lambda_i)(d\lambda_i/dt)] = 0, \end{aligned}$$

where $\sum_i$ ranges over all $i = 1, \ldots, n$, and we have used the canonical equations. Thus, $(d/dt)M(x(t), \lambda(t)) \leq 0$ whenever this derivative exists in (t_1, t_2). By repeating this argument (using points t', $t_1 < t' < \bar{t}$, with division by $t' - \bar{t} < 0$, and passage to the limit as $t' \to \bar{t} - 0$), we prove that the same derivative is ≥ 0 whenever it exists. Thus, $M(x(t), \lambda(t))$ is an AC function in $[t_1, t_2]$ with zero derivative a.e., and hence a constant in $[t_1, t_2]$. □

(b) If $f(t, x, u)$ does not depend on t, then by previous lemma we conclude that $M(x(t), \lambda(t)) = c$ in $[t_1, t_2]$, and (P_3) is proved for f independent of t. Let us assume that f depends on t, x, u, but f is continuous in $A \times U$ together with its partial derivatives f_t, f_x. We can reduce the given problem to one which is autonomous by introducing the auxiliary variable x^{n+1} with the additional differential equation and boundary conditions $dx^{n+1}/dt = 1$, $x^{n+1}(t_1) = t_1$. Then $x^{n+1}(t_2) = t_2$ and the new problem is an autonomous one; we have one more component, say λ_{n+1}, for the vector λ, and the new Hamiltonian, say H_1, is

$$H_1 = H_1(x, x^{n+1}, u, \lambda, \lambda_{n+1}) = \lambda_1 f_1 + \cdots + \lambda_n f_n + \lambda_{n+1}.$$

We have also two more canonical equations, say

$$\begin{aligned} dx^{n+1}/dt &= \partial H_1/\partial \lambda_{n+1} = 1, \\ d\lambda_{n+1}/dt &= -\partial H_1/\partial x^{n+1} = -\partial H/\partial t = -(\lambda_1 f_{1t} + \cdots + \lambda_n f_{nt}). \end{aligned}$$

Thus,

$$\lambda_{n+1}(t) = c' - \int_{t_1}^{t} (\lambda_1 f_{1t} + \cdots + \lambda_n f_{nt})\, dt$$

for some constant c'. The minimum M_1 of H_1 in U is the same as the minimum of H in U augmented by the term λ_{n+1}, which does not depend on u. Thus, $M_1 = M + \lambda_{n+1}$. On the other hand, M_1 is now constant along the present solution, so that $M + \lambda_{n+1} = c$, a constant, or

$$M = c - \lambda_{n+1} = c - c' + \int_{t_1}^{t} (\lambda_1 f_{1t} + \cdots + \lambda_n f_{nt})\, dt, \qquad t_1 \leq t \leq t_2.$$

This proves relation (P_3).

G. Removal of the Restriction that t_1 and t_2 Are Fixed

Let us prove the necessary condition when t_1 and t_2 are not fixed. To do this, we transform the given problem into another one with fixed initial and final times, and treat t as another space variable. Let $x_0(t)$, $u_0(t)$, $t_{10} \leq t \leq t_{20}$, be the given optimal pair, whose existence we assume in (7.1.i), and we assume therefore that $I[x, u] \geq I[x_0, u_0]$ for all admissible pairs $x(t)$, $u(t)$, $t_1 \leq t \leq t_2$, certainly at least for all such pairs with t_1, t_2 arbitrarily close to t_{10}, t_{20} respectively. Thus, we shall assume that the quotient $\mu = (t_2 - t_1)/(t_{20} - t_{10})$ is close to one, say between $\frac{3}{4}$ and $\frac{5}{4}$. We may then consider the transformation $t = t_1 + \mu(\tau - t_{10})$ with τ varying in the fixed interval $t_{10} \leq \tau \leq t_{20}$. Actually, it is advantageous to consider more general transformation laws, namely, all those represented by an equation of the form

$$dt/d\tau = \tfrac{1}{2}\alpha + \tfrac{3}{2}\beta, \qquad t(t_{10}) = t_1, \quad t(t_{20}) = t_2, \tag{7.3.27}$$

where $(\alpha, \beta) \in \Gamma = [\alpha \geq 0, \beta \geq 0, \alpha + \beta = 1]$. Thus, t is actually treated as a new space variable, satisfying differential equation and boundary conditions (7.3.27) and where α, β are two new control variables. In other words, in (7.3.27) we may take for α and β arbitrary measurable functions of τ in $[t_{10}, t_{20}]$ provided $\alpha \geq 0$, $\beta \geq 0$, $\alpha + \beta = 1$. Note that (7.3.27) yields $\frac{1}{2} \leq dt/d\tau \leq \frac{3}{2}$, so that $t(\tau)$ is a Lipschitz strictly increasing function, and so is its inverse $\tau(t)$. We shall denote by w the expression $w = \frac{1}{2}\alpha + \frac{3}{2}\beta$. We can write now the new Mayer problem relative to the $n + 1$ space variables $\tilde{x} = (x^1, \ldots, x^n, t)$, the $m + 2$ control variables $\tilde{u} = (u^1, \ldots, u^m, \alpha, \beta)$, control space $\tilde{U} = U \times \Gamma$, and the differential system

$$d\tilde{x}/d\tau = \tilde{f}(\tilde{x}, \tilde{u}), \qquad \tilde{f} = (\tilde{f}_1, \ldots, \tilde{f}_n, \tilde{f}_{n+1}).$$

Since $dx^i/d\tau = (dx^i/dt)(dt/d\tau)$, we can write this system in more explicit form:

$$dx^i/d\tau = wf_i, \quad i = 1, \ldots, n, \qquad dt/d\tau = w,$$

and write $\tilde{x}^i(\tau) = x^i(t(\tau))$, $i = 1, \ldots, n$, $t = t(\tau)$, $t_{10} \leq \tau \leq t_{20}$. The new set $\tilde{B}$ is now the set of all points $(t_{10}, t_1, x_1, t_{20}, t_2, x_2) \in R^{2n+4}$ with t_{10}, t_{20} fixed and $(t_1, x_1, t_2, x_2) \in B$, or $\tilde{B} = \{t_{10}\} \times \{t_{20}\} \times B$. The function g and the functional are the same as before, though the functional is now written in the form

$$\tilde{I}[\tilde{x}, \tilde{u}] = g(\tilde{x}_1(t_{10}), \tilde{x}_2(t_{20})) = g(t_1, x(t_1), t_2, x(t_2)).$$

Note that now we have two new arbitrary measurable functions $\alpha(\tau)$, $\beta(\tau)$, $t_{10} \leq \tau \leq t_{20}$, satisfying $\alpha \geq 0$, $\beta \geq 0$, $\alpha + \beta = 1$. The transformations $t = t_1 + \mu(\tau - t_{10})$ are only a particular case of the transformations above, and the latter can be realized, for instance, by taking $\alpha(\tau) = \alpha_0$, $\beta(\tau) = \beta_0$, α_0, β_0 constants, with $\alpha_0 \geq 0$, $\beta_0 \geq 0$, $\alpha_0 + \beta_0 = 1$, $\frac{1}{2}\alpha_0 + \frac{3}{2}\beta_0 = \mu$, and these equations define α_0, β_0 univocally, with $0 < \alpha_0, \beta_0 < 1$, since

$\frac{3}{4} \le \mu \le \frac{5}{4}$. In particular, for $\mu = 1$, we have $\alpha_0 = \beta_0 = \frac{1}{2}$. The new problem is autonomous, since τ does not appear explicitely in $\tilde{f}_1, \ldots, \tilde{f}_{n+1}$, $\tilde{U}$ is a fixed set, and t_{10}, t_{20} are fixed numbers.

The new problem certainly has an optimal solution, namely

$$\tilde{x}(\tau) = [x_0(t(\tau)), t(\tau)], \qquad \tilde{u}(\tau) = [u(t(\tau)), \alpha_0, \beta_0],$$

where now $t(\tau) = t_{10} + \mu(\tau - t_{10})$ reduces to $t(\tau) = \tau$, and $1 = \mu = \frac{1}{2}\alpha_0 + \frac{3}{2}\beta_0, \alpha_0 = \beta_0 = \frac{1}{2}$. We can now apply the necessary condition for autonomous problems to the pair $\tilde{x}, \tilde{u}$. For this we need multipliers $\tilde{\lambda}(t) = (\tilde{\lambda}_1, \ldots, \tilde{\lambda}_n, \tilde{\lambda}_{n+1})$, a new Hamiltonian $\tilde{H}$, and a new function $\tilde{M}$,

$$\tilde{H}(\tilde{x}, u, \alpha, \beta, \tilde{\lambda}) = w(\tilde{\lambda}_1 f_1 + \cdots + \tilde{\lambda}_n f_n + \tilde{\lambda}_{n+1}),$$
$$\tilde{M}(\tilde{x}, \tilde{\lambda}) = \inf_{u,\alpha,\beta,} \tilde{H}(\tilde{x}, u, \alpha, \beta, \tilde{\lambda}),$$

where the infimum is taken for $(u, \alpha, \beta) \in U \times \Gamma$. The multipliers satisfy the equations

$$\text{(7.3.28)} \qquad \begin{aligned} d\tilde{\lambda}_i/d\tau &= -w \sum_{j=1}^{n} f_{jx^i}\tilde{\lambda}_j, \qquad i = 1, \ldots, n, \\ d\tilde{\lambda}_{n+1}/d\tau &= -w \sum_{j=1}^{n} f_{jt}\tilde{\lambda}_j, \end{aligned}$$

and property (P_2), already proved for fixed t_{10}, t_{20}, yields

$$\text{(7.3.29)} \qquad \begin{aligned} \tilde{M}(\tau) &= \tilde{M}(\tilde{x}(\tau), \tilde{\lambda}(\tau)) \\ &= \min_{u,\alpha,\beta} \tilde{H}(\tilde{x}(\tau), u, \alpha, \beta, \tilde{\lambda}(\tau)) \\ &= \min_{\alpha,\beta} (\tfrac{1}{2}\alpha + \tfrac{3}{2}\beta) \min_{u} \left[\sum_{i=1}^{n} \tilde{\lambda}_i(\tau)\tilde{f}_i(\tilde{x}(\tau), u) + \tilde{\lambda}_{n+1}(\tau) \right] \end{aligned}$$

and

$$\text{(7.3.30)} \qquad \begin{aligned} \tilde{M}(\tau) &= \tilde{M}(\tilde{x}(\tau), \tilde{\lambda}(\tau)) \\ &= \tilde{H}(\tilde{x}(\tau), u_0(\tau), \alpha_0, \beta_0, \tilde{\lambda}(\tau)) \\ &= (\tfrac{1}{2}\alpha_0 + \tfrac{3}{2}\beta_0) \left[\sum_{i=1}^{n} \tilde{\lambda}_i(\tau)\tilde{f}_i(\tilde{x}(\tau), \tilde{u}_0(\tau)) + \tilde{\lambda}_{n+1}(\tau) \right] \end{aligned}$$

with $\alpha_0 = \beta_0 = \frac{1}{2}$. Comparing (7.3.30) and (7.3.29) we see that this is possible only if the expression in brackets is zero, or

$$-\tilde{\lambda}_{n+1}(\tau) = \sum_{i=1}^{n} \tilde{\lambda}_i(\tau)\tilde{f}_i(\tilde{x}(\tau), \tilde{u}_0(\tau)),$$

or taking $\tau = \tau(t)$,

$$\text{(7.3.31)} \qquad -\lambda_{n+1}(t) = \sum_{i=1}^{n} \lambda_i(t) f_i(t, x(t), u_0(t)).$$

Since $\frac{1}{2}\alpha_0 + \frac{3}{2}\beta_0 > 0$, equations (7.3.30) and (7.3.31) show that the minimum of $\sum_i \lambda_i f_i$ is attained by $u_0(\tau) = u_0(t(\tau))$, that is, the strategy $u(t)$ is the same as for the original problem. Hence, the equations (7.3.28) yield

$$\text{(7.3.32)} \qquad d\lambda_i/dt = -\sum_j f_{jx^i}\lambda_j, \quad i = 1, \ldots, n, \qquad d\lambda_{n+1}/dt = -\sum_j f_{jt}\lambda_j.$$

Thus, the multipliers $\lambda_1, \ldots, \lambda_n$ satisfy the same differential equations as the original problem. Finally, since the bracket in (7.3.29) is zero, we have $\tilde{M}(\tau) = 0$, and the original function $M(t)$ coincides with $-\lambda_{n+1}(t)$ by force of (7.3.31). The transversality relation for end times fixed yields now

$$[\lambda_0 f_{t_1} + \lambda_{n+1}(t_1)]\tau_1 + \sum_{j=1}^{n} [\lambda_0 g_{x_1^j} + \lambda_j(t_1)]\xi_1^j$$
$$+ [\lambda_0 g_{t_0} - \lambda_{n+1}(t_2)]\tau_2 + \sum_{j=1}^{n} [\lambda_0 g_{x_2^j} - \lambda_j(t_2)]\xi_2^j \geq 0.$$

By using $-\lambda_{n+1} = M$, we see that this relation reduces to (P_4). We have proved the transversality relation for t_1, t_2 variable.

Remark. Note that this section is completely independent of Subsection F, where we proved (P_3). Actually, we have here a new proof of that statement. Indeed, from the relation $-\lambda_{n+1}(t) = M(t)$ we conclude that $M(t)$ is AC in $[t_1, t_2]$, and (7.3.32) yields

$$dM/dt = -d\lambda_{n+1}/dt = H_t(t, x_0(t), u_0(t), \lambda(t)).$$

H. Removal of the Restriction that $u_0(t)$ Is Bounded

Here the optimal strategy $u_0(t)$, $t_1 \leq t \leq t_2$, is assumed to be only measurable, but not bounded, nor essentially bounded. The main proof above of the necessary condition remains essentially the same with a few modifications, using the alternate assumption (c′) of Remark 5 of Subsection 4.2C, or condition (S). The main modification occurs in Subsection A, parts (a), (b), (c), where none of the functions $f_i(t, x_0(t), u_0(t))$, $f_{ix^j}(t, x_0(t), u_0(t))$ is now necessarily bounded in $[t_1, t_2]$. However, by using hypothesis (S), and by taking a smaller δ so that property (S) holds, we can arrange that for every $t \in [t_1, t_2]$ and $x \in R^n$ with $|x - x_0(t)| \leq 4\delta$ we have $|f_{it}(t, x, u_0(t))|$, $|f_{ix^j}(t, x, u_0(t))| \leq S(t)$, where $S(t)$ is a given L-integrable function in $[t_1, t_2]$. Then

$$|f_i(t, x, u_0(t))| = \left| f_i(t, x_0(t), u_0(t)) + \sum_{j=1}^{n} f_{ix^j}(t, \tilde{x}, u_0(t))(x^j - x_0^j(t)) \right|$$
$$\leq |f_i(t, x_0(t), u_0(t))| + 4n\delta S(t) = N_i(t), \qquad t_1 \leq t \leq t_2.$$

where $\tilde{x}$ denotes a point on the segment between $x_0(t)$ and x, all $|f_i(t, x_0(t), u_0(t))|$ are L-integrable functions in $[t_1, t_2]$, and then the $N_i(t)$ are also fixed L-integrable functions in $[t_1, t_2]$. Now in the argument in Subsection A we noticed that, given any system of bounded measurable functions $u_\sigma(t)$, $t_1 \leq t \leq t_2$, with $u_\sigma(t) \in U(t)$, then there is some constant M'' such that $|f_i(t, x, u(t))|$, $|f_{ix^j}(t, x, u(t))| \leq M''$, $t_1 \leq t \leq t_2$, for $|z_\sigma| \leq \gamma$, $\sigma = 1, \ldots, s$, for all x with $|x - x_0(t)| \leq \delta$, and all i, j, σ. In the present situation we have instead $|f_i| \leq N_i(t)$, and then, for $c = \max|c_\sigma|$, we also have

$$|q_i(t, x; z)| \leq \left| \left(1 - \sum_\sigma c_\sigma z_\sigma\right) f_i(t, x, u_0(t)) + \sum_\sigma c_\sigma z_\sigma f_i(t, x, u_\sigma(t)) \right| \leq N_i(t),$$

a fixed L-integrable function of t in $[t_1, t_2]$, and analogous relations hold for the partial derivatives $|q_{ix^j}| \leq S(t)$, $|q_{iz_\sigma}| \leq cN_i(t)$. The second members of system (7.3.5) are then in absolute value below a fixed L-integrable function, independent of $z = (z_1, \ldots, z_\sigma) \in V$, with $|z|$ sufficiently small. The theorems from differential equation theory we have used in Subsection A still hold in the present circumstances (see, e.g., McShane [I, pp. 345, 352, 356]). For other alternative conditions see Section 7.4E. This completes the first proof of the necessary condition.

Remark. The assumption at the beginning of Section 7.1 that the set M be closed in R^{1+n+m} was needed in Section 7.3A in order to make use of statement (8.2.iii). In view of Exercise 5 of Section 8.2, much less is needed. Namely, it is enough to know that for some neighborhood Γ_δ of the graph Γ of $x_0(t)$, $t_1 \leq t \leq t_2$, and for any $\varepsilon > 0$ there is a compact subset K_ε of $[t_1, t_2]$, meas $K_\varepsilon > t_2 - t_1 - \varepsilon$, such that the set $M_\varepsilon = [(t, x, u) | (t, x) \in \Gamma_\delta, u \in U(t), t \in K_\varepsilon]$ is closed (and f is continuous on M_ε).

I. Extension of Proof to Bolza Problems

In Section 5.1 we stated the necessary condition (5.1.i) for Bolza problems. In Section 5.2 we have proved (5.1.i) by deriving it from the analogous statement (4.2.i) for Mayer problems. Here is a direct proof of (5.1.i), which is a modification of the one given above in Subsections A–H.

The entire argument in Subsection A is the same, and now we define K as the set of all points $z = \tilde{Y}(v) = (Y^0(v), Y(v))$ in R^{n+1} with

$$\begin{aligned} Y^0(v) &= g_{x_1} \cdot \xi_1 + g_{x_2} \cdot \xi_2 \\ &\quad + \int_{t_1}^{t_2} \{c[-f_0(t, x_0(t), u_0(t)) + f_0(t, x_0(t), u(t))] \\ &\quad + f_{0x}(t, x_0(t), u_0(t)) \cdot y(t, v)\}\, dt, \\ Y(v) &= y(t_2, v) - \xi_2, \end{aligned}$$

for all variations $v = (c, u, h)$, $h = (0, \xi_1, 0, \xi_2)$, and t_1, t_2 fixed.

Again, K is a convex cone of vertex the origin, and $(-1, 0, \ldots, 0)$ is not an interior point of K. Thus, as in Subsection B, the cone K has a supporting hyperplane through the origin $\sum_{i=0}^{n} \chi_i z^i = 0$, and K is contained in the half space $\sum_{i=0}^{n} \chi_i z^i \geq 0$ with $\chi_0 \geq 0$. We take now $\lambda_0 = \chi_0$, and for any variation $v = (c, u, h)$ the corresponding variational system and initial data are

$$\text{(7.3.33)} \quad Ly \equiv dy/dt - A(t)y = -cf(t, x_0(t), u_0(t)) + cf(t, x_0(t), u(t)), \qquad y(t_1) = \xi_1,$$

with $A(t) = [f_{ix^j}(t, x_0(t), u_0(t))]$. As in Subsection C, the dual operator is $L^*\lambda = -d\lambda/dt - A^*(t)\lambda$, and we define the multipliers $\lambda(t) = (\lambda_1, \ldots, \lambda_n)$ as the unique AC solution of the differential system and data at $t = t_2$,

$$\text{(7.3.34)} \quad L^*\lambda \equiv -d\lambda/dt - A^*(t)\lambda = \lambda_0 f_{0x}(t, x_0(t), u_0(t)), \qquad \lambda(t_2) = \chi = (\chi_1, \ldots, \chi_n).$$

As in Subsection D, we replace $y(t_2, v)$ in the inequality $\sum_i \chi_i Y^i(v) \geq 0$ to obtain

$$\begin{aligned} 0 \leq \sum_{i=0}^{n} \chi_i Y^i(v) &= \chi_0 Y^0(v) + \chi \cdot Y(v) \\ &= \lambda_0[g_{x_1} \cdot \xi_1 + g_{x_2} \cdot \xi_2] + \lambda(t_2) \cdot [y(t_2) - \xi_2] \\ &\quad + \lambda_0 \int_{t_1}^{t_2} c[-f_0(t, x_0(t), u_0(t) + f_0(t, x_0(t), u(t))]\, dt \\ &\quad + \lambda_0 \int_{t_1}^{t_2} f_{0x}(t, x_0(t), u_0(t)) \cdot y(t)\, dt. \end{aligned}$$

By (7.3.18) we introduce the operators L and L^* as in Subsection D to obtain

$$\begin{aligned}0 \le \lambda_0[g_{x_1} \cdot \xi_1 + g_{x_2} \cdot \xi_2] - \lambda(t_2) \cdot \xi_2 + \lambda(t_1) \cdot y(t_1) \\ + \int_{t_1}^{t_2} (\lambda \cdot Ly - y \cdot L^*\lambda)\, dt \\ + \lambda_0 \int_{t_1}^{t_2} c[-f_0(t, x_0(t), u_0(t)) + f_0(t, x_0(t), u(t))]\, dt \\ + \lambda_0 \int_{t_1}^{t_2} f_{0x}(t, x_0(t), u_0(t)) \cdot y(t)\, dt.\end{aligned}$$

By (7.3.33) and (7.3.34), and noting that $\xi_1 = y(t_1)$, we finally have after simplification,

$$\begin{aligned}(\lambda_0 g_{x_1} + \lambda(t_1)) \cdot \xi_1 + (\lambda_0 g_{x_2} - \lambda(t_2)) \cdot \xi_2 \\ + c \int_{t_1}^{t_2} (\lambda_0[-f_0(t, x_0(t), u_0(t)) + f_0(t, x_0(t), u(t))] \\ + \lambda \cdot [-f(t, x_0(t), u_0(t)) + f(t, x_0(t), u(t))])\, dt \ge 0.\end{aligned}$$

This is relation (7.3.19) with the definition of the Hamiltonian (5.1.3). The proof of (5.2.i) is now the same as in Subsections A–H.

J. Two Necessary Conditions for Optimal Singular Arcs

(a) For the problem

$$\begin{gathered}I[x, u] = g(x(t_2)), \quad dx/dt = f(t, x, u), \\ t_1, x_1, t_2 \text{ fixed}, \quad x_2 = x(t_2) \in R^n, \quad u(t) \in U(t), \\ x = (x^1, \ldots, x^n), \quad f = (f_1, \ldots, f_n), \quad u = (u^1, \ldots, u^m),\end{gathered}$$

we have

$$H = \sum_{j=1}^{n} \lambda_j f_j, \qquad d\lambda_i/dt = -H_{x^i} = -\sum_{j=1}^{n} \lambda_j f_{jx^i}.$$

Here $B = (t_1, x_1, t_2) \times R^n$, and the transversality relation yields $\lambda_i(t_2) = \lambda_0 g_{x_2^i}$, $i = 1, \ldots, n$, $\lambda_0 \ge 0$. We take here $\lambda_0 = 1$. For any variation $v = (1, u, h)$, $x(t, \varepsilon)$ satisfies

$$dx/dt = (1 - \varepsilon) f(t, x(t, \varepsilon), u_0(t)) + \varepsilon f(t, x(t, \varepsilon), u(t)), \qquad x(t_1, \varepsilon) = 0,$$

and hence

$$\begin{aligned}(d/dt)(\partial x(t, \varepsilon)/\partial\varepsilon) = -f(t, x(t, \varepsilon), u_0(t)) + f(t, x(t, \varepsilon), u(t)) \\ + [(1 - \varepsilon) f_x(t, x(t, \varepsilon), u_0(t)) + \varepsilon f_x(t, x(t, \varepsilon), u(t))](\partial x(t, \varepsilon)/\partial\varepsilon),\end{aligned}$$

with $\partial x(t_1, \varepsilon)/\partial\varepsilon = 0$ as in Subsection A. Hence, for $y(t) = (\partial x(t, \varepsilon)/\partial\varepsilon)_{\varepsilon=0}$, we have the variational system

$$\begin{gathered}dy/dt - A(t)y = \Delta_u f, \qquad y(t_1) = 0, \\ A(t) = [f_{ix^j}(t, x_0(t), u(t))], \quad \Delta_u f = f(t, x_0(t), u(t)) - f(t, x_0(t), u_0(t)).\end{gathered}$$

Moreover, $I_\varepsilon(u) = g(x(t_2, \varepsilon))$.

$$0 \le I_\varepsilon(u) - I(u_0) = g_x(x_0(t_2)) \cdot y(t_2) + o(\varepsilon).$$

and for the first variation J_1 we have now

$$\begin{aligned} J_1 &= g_x(x_0(t_2)) \cdot y(t_2) = \lambda(t_2) \cdot y(t_2) \\ &= \lambda(t_1) \cdot y(t_1) + \int_{t_1}^{t_2} (\lambda' \cdot y + \lambda \cdot y')\, dt \\ &= \int_{t_1}^{t_2} (-H_x \cdot y + \lambda \cdot Ay + \lambda \cdot \Delta_u f)\, dt = \int_{t_1}^{t_2} \lambda \cdot \Delta_u f\, dt \\ &= \int_{t_1}^{t_2} \Delta_u H\, dt \geq 0. \end{aligned} \tag{7.3.35}$$

Thus, by the assumption that $x_0(t)$, $t_1 \leq t \leq t_2$, is a singular arc, we have $\Delta_u H = 0$ and $J_1 = 0$.

Note that if f and g are of class C^2, then $x(t, \varepsilon)$ is also of class C^2, and

$$\begin{aligned} (d/dt)(\partial^2 x(t,\varepsilon)/\partial\varepsilon^2) = &-2f_x(t, x(t,\varepsilon), u_0(t))(\partial x(t,\varepsilon)/\partial\varepsilon) \\ &+ 2f_x(t, x(t,\varepsilon), u(t))(\partial x(t,\varepsilon)/\partial\varepsilon) \\ &+ (1-\varepsilon)(\partial x(t,\varepsilon)/\partial\varepsilon)^* f_{xx}(t, x(t,\varepsilon), u_0(t))(\partial x(t,\varepsilon)/\partial\varepsilon) \\ &+ (1-\varepsilon) f_x(t, x(t,\varepsilon), u_0(t))(\partial^2 x(t,\varepsilon)/\partial\varepsilon^2) \\ &+ \varepsilon(\partial x(t,\varepsilon)/\partial\varepsilon)^* f_{xx}(t, x(t,\varepsilon), u(t))(\partial x(t,\varepsilon)/\partial\varepsilon) \\ &+ \varepsilon f_x(t, x(t,\varepsilon), u(t))(\partial^2 x(t,\varepsilon)/\partial\varepsilon^2). \end{aligned}$$

Hence, for $z(t) = (\partial^2 x(t,\varepsilon)/\partial\varepsilon)_{\varepsilon=0}$ we have the variational system

$$dz/dt - A(t)z = 2\Delta_u f_x y + y^* f_{xx} y, \qquad z(t_1) = 0,$$

where the terms of the form $y^* f_{xx} y$ here and above denote n-vectors whose ith component is $y^* f_{ixx} y$. Moreover,

$$\begin{aligned} 0 \leq I_\varepsilon(u) - I(u_0) &= g_x(x_0(t_2)) \cdot y(t_2) \\ &+ 2^{-1}\varepsilon^2[y^*(t_2) g_{xx}(x_0(t_2)) y(t_2) + g_x(x_0(t_2)) \cdot z(t_2)] + o(\varepsilon^2). \end{aligned}$$

Since

$$\begin{aligned} g_x(x_0(t_2)) \cdot z(t_2) &= \lambda(t_2) \cdot z(t_2) \\ &= \lambda(t_1) \cdot z(t_1) + \int_{t_1}^{t_2} (\lambda' \cdot z + \lambda \cdot z')\, dt \\ &= \int_{t_1}^{t_2} (-H_x \cdot z + \lambda \cdot f_x z + 2\lambda \cdot \Delta_u f_x y + \lambda \cdot y^* f_{xx} y)\, dt \\ &= \int_{t_1}^{t_2} (2\Delta_u H_x \cdot y + y^* H_{xx} y)\, dt. \end{aligned}$$

Thus, for the second variation J_2 we have now

$$\begin{aligned} J_2 &= 2^{-1}(y^* g_{xx} y + g_x(x_0(t_2)) \cdot z(t_2)) \\ &= 2^{-1} y^*(t_2) g_{xx}(x_0(t_2)) y(t_2) + \int_{t_1}^{t_2} (\Delta_u H_x \cdot y + 2^{-1} y^* H_{xx} y)\, dt \geq 0. \end{aligned} \tag{7.3.36}$$

Now let $\mu(t)$ denote the AC vector function defined by

$$d\mu/dt = -f_x^* \mu - H_{xx} y, \quad \mu(t_2) = g_{xx}(x_0(t_2)) y(t_2) \tag{7.3.37}$$

Then,

$$y^*(t_2)g_{xx}(x_0(t_2))y(t_2) = \mu(t_2) \cdot y(t_2) = \mu(t_1) \cdot y(t_1) + \int_{t_1}^{t_2} (\mu' \cdot y + \mu \cdot y')\, dt$$
$$= \int_{t_1}^{t_2} (-f_x^*\mu \cdot y - y^*H_{xx}y + \mu \cdot f_x y + \mu \cdot \Delta_u f)\, dt$$
$$= \int_{t_1}^{t_2} (-y^*H_{xx}y + \mu \cdot \Delta_u f)\, dt,$$

and finally the second variation J_2, as given by (7.3.36), becomes

$$\int_{t_1}^{t_2} [\Delta_u H_x(t) \cdot y(t) + 2^{-1}\mu(t) \cdot \Delta_u f(t)]\, dt \geq 0.$$

By the same argument as in Subsection E we also have, for $t \in [t_1, t_2]$ (a.e.),

$$2\Delta_u H_x(t, x_0(t), u_0(t), \lambda(t)) \cdot y(t) + \mu(t) \cdot \Delta_u f(t, x_0(t), u_0(t)) \geq 0. \tag{7.3.38}$$

Finally, for U a fixed subset of R^m, we also have (cf. Subsection E)

$$\begin{aligned} &2H_x(t, x_0(t), u_0(t), \lambda(t)) \cdot y(t) + \mu(t) \cdot f(t, x_0(t), u_0(t)) \\ &\quad = \min_{u \in U}[2H_x(t, x_0(t), u, \lambda(t)) + \mu(t) \cdot f(t, x_0(t), u)]. \end{aligned} \tag{7.3.39}$$

Condition (7.3.37) is the necessary condition for singular controls recently proved by Kazemi [1] by the present argument, and independently by Gilbert and Bernstein [1]. For the alternative forms (7.3.38) and (7.3.39) see Kazemi [1].

(b) We shall assume here that $U(t)$ is convex. Thus, $u_0(t) \in U(t)$, and for any other $u(t) \in U(t)$ the entire segment $(1 - \varepsilon)u_0(t) + \varepsilon u(t)$, $0 \leq \varepsilon \leq 1$, belongs to $U(t)$. Now we define $x(t, \varepsilon)$ by taking

$$\begin{aligned} dx(t,\varepsilon)/dt &= f\,(t, x(t,\varepsilon), u_0(t) + \varepsilon v(t)) \\ &= f(t, x(t,\varepsilon), (1-\varepsilon)u_0(t) + \varepsilon u(t)), \qquad x(t_1, \varepsilon) = x_1. \end{aligned}$$

Then, under the same assumptions as under (a), we have

$$\begin{aligned} (d/dt)(\partial x(t,\varepsilon)/\partial\varepsilon) &= f_x(t, x(t,\varepsilon), u_0(t) + \varepsilon v(t))(\partial x(t,\varepsilon)/\partial\varepsilon) \\ &\quad + f_u(t, x(t,\varepsilon), u_0(t) + \varepsilon v(t))v(t), \end{aligned}$$

so that, for $y(t) = (\partial x(t,\varepsilon)/\partial\varepsilon)_{\varepsilon=0}$, we have

$$dy/dt = f_x y + f_u v, \qquad y(t_1) = 0.$$

For the first variation J_1 we have now as in (7.3.35)

$$J_1 = \lambda(t_1) \cdot y(t_1) + \int_{t_1}^{t_2} (\lambda' \cdot y + \lambda \cdot y')\, dt$$
$$= \int_{t_1}^{t_2} (-H_x \cdot y + \lambda \cdot Ay + \lambda \cdot f_u v)\, dt$$
$$= \int_{t_1}^{t_2} \lambda \cdot f_u v\, dt = \int_{t_1}^{t_2} H_u \cdot v\, dt.$$

Thus, if x_0 is singular, then certainly $H_u \cdot v = 0$ and $J_1 = 0$.

As before for f and g of class C^2, and denoting $\partial x(t,\varepsilon)/\partial\varepsilon$ by x_ε, we have

$$(d/dt)(\partial^2 x(t,\varepsilon)/\partial\varepsilon^2) = x_\varepsilon^* f_{xx} x_\varepsilon + 2v^* f_{xu} x_\varepsilon + v^* f_{uu} v + f_x(\partial^2 x(t,\varepsilon)/\partial\varepsilon^2),$$

and for $z(t) = (\partial^2 x(t,\varepsilon)/\partial\varepsilon^2)_{\varepsilon=0}$ we also have

$$dz/dt - A(t)z = y^* f_{xx} y + 2v^* f_{xu} y + v^* f_{uu} v, \qquad z(t_1) = 0.$$

Again

$$g_x(x_0(t_2)) \cdot z(t_2) = \lambda(t_2) \cdot z(t_2) = \lambda(t_1) \cdot z(t_1) + \int_{t_1}^{t_2} (\lambda' \cdot z + \lambda \cdot z')\, dt$$

$$= \int_{t_1}^{t_2} (-H_x \cdot z + \lambda \cdot f_x z + \lambda \cdot y^* f_{xx} y + 2\lambda \cdot v^* f_{xu} y + \lambda \cdot v^* f_{uu} v)\, dt$$

$$= \int_{t_1}^{t_2} (y^* H_{xx} y + 2v^* H_{xu} y + v^* H_{uu} v)\, dt$$

so that the second variation J_2 has now the expression

$$J_2 = 2^{-1} y^*(t_2) g_{xx}(x_0(t_2)) y(t_2) + \int_{t_1}^{t_2} (2^{-1} y^* H_{xx} y + v^* H_{xu} y + 2^{-1} v^* H_{uu} v)\, dt \geq 0.$$

For $Q = H_{xx}$, $C = H_{xu}$, $S = g_{xx}$, $A = f_x$, $B = f_u$, and assuming that x_0 is singular, then $H_u = 0$, $H_{uu} = 0$, then

$$J_2 = 2^{-1} y^*(t) S y(t_2) + \int_{t_1}^{t_2} (2^{-1} y^* Q y + v^* C y)\, dt,$$

where $y' = Ax + Bu$. Hence, for any $n \times n$ symmetric matrix $W(t)$ we have

$$J_2 = 2^{-1} y^*(t_2) S y(t_2) + \int_{t_1}^{t_2} (2^{-1} y^* Q y + v^* C y + 2^{-1} y^* W(Ay + Bv - y'))\, dt$$

and by integration by parts of the term $y^* W y'$ we derive

$$J_2 = 2^{-1} y^*(t_2)[S - W(t_2)] y(t_2)$$
$$+ \int_{t_1}^{t_2} (2^{-1} y^*(W' + Q + A^* W + WA) y + v^*(C + B^* W) y)\, dt.$$

If W satisfies the Riccati equation $-W' = Q + A^* W + WA$, with $W(t_2) = S$, then

$$J_2 = \int_{t_1}^{t_2} v^*(C + B^* W) y\, dt \geq 0.$$

For U independent of t, and by taking $v = k$, a constant vector in any interval $[t, t + \delta]$, and $v = 0$ otherwise, we have $J_2 = k^*(C + B^* W) B k\, \delta^2 \geq 0$. In other words, the $m \times m$ matrix $(C + B^* W) B$ is positive definite, that is,

$$(H_{xu} + f_u^* W) f_u \geq 0,$$

where W denotes the solution of the Riccati equation $-W' = H_{xx} + f_x^* W + W f_x$ with $W(t_2) = g_{xx}(x_0(t_2))$. This is Jacobsen's necessary condition for singular arcs (cf. Bell and Anderson [I]).

7.4 Second Proof of the Necessary Condition

In this second proof (subsections A–D below) we do not need the assumption in (7.1.i) that the sets $Q(t, x)$ are convex in R^n, nor that the set M is closed, but we do need throughout that U is a fixed subset of R^m. For alternative assumptions see subsection 7.4E.

A. Needle-like Perturbations

In this second proof no assumption is made concerning the convexity of the sets $Q(t, x) = f(t, x, U)$, but we assume that U is a fixed subset of R^m. Again, as in the first proof, we assume first that t_1 and t_2 are fixed; that the graph of the optimal trajectory $x_0(t)$, $t_1 \le t \le t_2$, is interior to A; that the optimal strategy $u(t)$ is bounded; that the end point $e[x_0] = (t_1, x_0(t_1), t_2, x_0(t_2))$ of the optimal trajectory $x_0(t)$ is a point of B where B possesses a tangent hyperplane B' of some dimension k, $0 \le k \le 2n + 2$, (or a convex cone), and that g possesses a differential at $e[x_0]$.

Let $\bar{t}$ be any point of $(t_1, t_2]$, or $t_1 < \bar{t} \le t_2$. Let $\bar{u}$ be any point of U, c any positive constant, and $\varepsilon > 0$ a parameter. Then, for $\varepsilon > 0$ sufficiently small we have $t_1 < \bar{t} - c\varepsilon < \bar{t} \le t_2$, and we define the perturbed strategy $u_\varepsilon(t)$, $t_1 \le t \le t_2$, by taking $u_\varepsilon(t) = \bar{u}$ for $\bar{t} - c\varepsilon < t \le \bar{t}$, and $u_\varepsilon(t) = u_0(t)$ otherwise. Then, $u_\varepsilon(t) \in U$ for all $t \in [t_1, t_2]$. If $x(t, \varepsilon)$, $t_1 \le t \le t_2$, is the trajectory corresponding to $u_\varepsilon(t)$ obtained by integrating $dx/dt = f(t, x, u_\varepsilon(t))$ from t_1 and initial value $x(t_1)$, then $x(t, \varepsilon) = x_0(t)$ for all $t_1 \le t \le \bar{t} - c\varepsilon$. We say that $x(t, \varepsilon)$ is the perturbed trajectory.

7.4.i (Lemma). *For $\varepsilon \ge 0$ sufficiently small, $x(t, \varepsilon)$ exists for all $t \in [t_1, t_2]$ and $x(t, \varepsilon) \to x_0(t)$ as $\varepsilon \to 0+$ uniformly in $[t_1, t_2]$.*

Proof. First, $x(t, \varepsilon) = x_0(t)$ for $t_0 \le t \le \bar{t} - c\varepsilon$. Since x_0 is continuous in $[t_1, t_2]$, the graph of x_0, or $[(t, x_0(t)), t_1 \le t \le t_2]$, is a compact set made up of points interior to A. Thus, there is some $\delta > 0$ such that all points (t', x') at a distance $\le 4\delta$ from the graph of x_0 are interior to A, and form a set $\bar{A}$ which is also compact. Then the continuous function $f(t, x, u)$ is bounded in the compact set $\bar{A} \times U$, say $|f(t, x, u)| \le M$ in $\bar{A} \times U$. We shall take $\varepsilon > 0$ sufficiently small so that $c\varepsilon < \delta$, $Mc\varepsilon < \delta$. For $\bar{t} - c\varepsilon \le t \le \bar{t}$ then

$$x(t, \varepsilon) = x_0(t - c\varepsilon) + \int_{t - c\varepsilon}^{t} f(\tau, x(\tau, \varepsilon), \bar{u})\, d\tau;$$

hence $|t - (\bar{t} - c\varepsilon)| \le c\varepsilon < \delta$, and $|x(t, \varepsilon) - x_0(\bar{t} - c\varepsilon)| \le Mc\varepsilon < \delta$. Since analogous relations hold for $x_0(t)$, we see that $|x(t, \varepsilon) - x_0(t)| < 2\delta$ for $\bar{t} - c\varepsilon \le t \le \bar{t}$. In $[\bar{t}, t_2]$, $x(t, \varepsilon)$ and $x_0(t)$ are solutions of the same equation $dx/dt = f(t, x, u_0(t))$ with initial values $x(\bar{t}, \varepsilon)$ and $x_0(\bar{t})$. By differential equation theory we know that for $\varepsilon > 0$ sufficiently small, $x(t, \varepsilon)$ is defined in all of $[\bar{t}, t_2]$ and converges to $x_0(t)$ uniformly as $\varepsilon \to 0$. This holds in the whole interval $[t_1, t_2]$, and (7.4.i) is proved. □

For $\varepsilon > 0$ sufficiently small, the graph of $x(t, \varepsilon)$ is contained in A. We shall now assume that $\bar{t}$ is a point of $(t_1, t_2]$ where

$$(d/dt) \int_{t_1}^{t} f(\tau, x_0(\tau), u_0(\tau))\, d\tau = f(t, x_0(t), u_0(t)),$$

that is where $f(t, x_0(t), u_0(t))$ is the derivative of its integral function. It is known that almost all points t of (t_1, t_2) have this property. If $\bar{t}$ is such a point, then

$$\lim_{h \to 0+} h^{-1} \int_{\bar{t} - h}^{\bar{t}} f(t, x_0(t), u_0(t))\, dt = f(\bar{t}, x_0(\bar{t}), u_0(\bar{t}));$$

hence

$$\int_{\bar{t} - h}^{\bar{t}} f(t, x_0(t), u_0(t))\, dt = hf(\bar{t}, x_0(\bar{t}), u_0(\bar{t})) + o(h),$$

where the symbol $o(h)$ simply means that $o(h)/h \to 0$ as $h \to 0+$. We refer to these points $\bar{t}$ of (t_1, t_2) as Lebesgue points. Also, let $A(t)$ denote the $n \times n$ matrix $A(t) = [a_{ij}(t)] = [f_x(t, x_0(t), u_0(t)]$, that is, $a_{ij} = f_{ix^j}$, $i, j = 1, \ldots, n$, and let us consider the homogeneous linear system $y' = A(t)y$, whose coefficients $a_{ij}(t)$ are measurable and bounded in $[t_1, t_2]$. Let $\Phi(t)$ be any $n \times n$ matrix whose n columns are linearly independent solutions of $y' = A(t)y$ in $[t_1, t_2]$.

7.4.ii (LEMMA). *If $\bar{t}$ is a Lebesgue point of $(t_1, t_2]$, then*

$$\xi = (\partial x(\bar{t}, \varepsilon)/\partial \varepsilon)_{\varepsilon=0} = c[f(\bar{t}, x_0(\bar{t}), \bar{u}) - f(\bar{t}, x_0(\bar{t}), u_0(\bar{t}))], \tag{7.4.1}$$

$$\tilde{\xi} = (\partial x(t, \varepsilon)/\partial \varepsilon)_{\varepsilon=0} = \Phi(t)\Phi^{-1}(\bar{t})\xi, \qquad \bar{t} \leq t \leq t_2. \tag{7.4.2}$$

Proof. We have here

$$x(\bar{t}, \varepsilon) = x_0(\bar{t}) + \int_{\bar{t}-c\varepsilon}^{\bar{t}} f(t, x(t, \varepsilon), \bar{u})\, dt - \int_{\bar{t}-c\varepsilon}^{\bar{t}} f(t, x_0(t), u_0(t))\, dt. \tag{7.4.3}$$

Let $\sigma(\varepsilon) = \max|x(t, \varepsilon) - x_0(\bar{t})|$, $\delta(\varepsilon) = \max|f(t, x(t, \varepsilon), \bar{u}) - f(\bar{t}, x_0(\bar{t}), \bar{u})|$, where max is taken as t describes $[\bar{t} - c\varepsilon, \bar{t}]$. Then $\sigma(\varepsilon) \to 0$ as $\varepsilon \to 0+$, and, because of the uniform continuity of $f(t, x, \bar{u})$ as a function of (t, x) in the compact set $\bar{A}$, we also have $\delta(\varepsilon) \to 0$. Then

$$\left| \int_{\bar{t}-c\varepsilon}^{\bar{t}} f(t, x(t, \varepsilon), \bar{u})\, dt - c\varepsilon f(\bar{t}, x_0(\bar{t}), \bar{u}) \right| \leq c\varepsilon\delta(\varepsilon),$$

that is, the difference in $|\ |$ is $o(\varepsilon)$. On the other hand, since $\bar{t}$ is a Lebesgue point, we also have

$$\int_{\bar{t}-c\varepsilon}^{t} f(t, x_0(t), u_0(t))\, dt = c\varepsilon f(\bar{t}, x_0(\bar{t}), u_0(\bar{t})) + o(\varepsilon).$$

Thus (7.4.3) yields

$$x(\bar{t}, \varepsilon) - x_0(\bar{t}) = c\varepsilon[f(\bar{t}, x_0(\bar{t}), \bar{u}) - f(\bar{t}, x_0(\bar{t}), u_0(\bar{t}))] + o(\varepsilon)$$

where $o(\varepsilon)/\varepsilon \to 0$ as $\varepsilon \to 0+$. By dividing this relation by ε and taking the limit as $\varepsilon \to 0$ we prove (7.4.1). The relation (7.4.2) is only the law by which the perturbation ξ is transferred along x_0 in the interval $[\bar{t}, t_2]$ (see Section 7.3A). □

The considerations above can be generalized as follows. Let $z = (z_1, \ldots, z_s)$, and let R be the hypercube $[z | 0 \leq z_\sigma \leq 1, \sigma = 1, \ldots, s]$ in the z-space R^s. Let $\bar{u}_1, \ldots, \bar{u}_s$ be s points of U, let $c_1, \ldots, c_s$ be s nonnegative constants, and let $\bar{t}$ be a point of (t_1, t_2). Then, for all $z \in R$ with $|z|$ sufficiently small, all points $\bar{t} - c_1 z_1$, $\bar{t} - c_1 z_1 - c_2 z_2, \ldots, \bar{t} - \sum_\sigma c_\sigma z_\sigma$ are between t_1 and $\bar{t}$, and we can define the strategy $u(t, z)$, $t_1 \leq t \leq t_2$, by taking $u(t, z) = \bar{u}_1$ if $t - c_1 z_1 < t \leq \bar{t}$, $u(t, z) = \bar{u}_2$ if $\bar{t} - c_1 z_1 - c_2 z_2 < t \leq \bar{t} - c_1 z_1, \ldots, u(t, z) = \bar{u}_s$ if $\bar{t} - \sum_\sigma c_\sigma z_\sigma < t \leq \bar{t} - \sum_\sigma c_\sigma z_\sigma + c_s z_s$, and by taking $u(t, z) = u_0(t)$ otherwise. Then $u(t, z) \in U$ for every $t \in [t_1, t_2]$. Note that the interval where we have taken, say, $u(t, z) = \bar{u}_\sigma$ is zero if $z_\sigma = 0$, or if $c_\sigma = 0$, $\sigma = 1, \ldots, s$. Finally, let $x(t, z)$, $t_1 \leq t \leq t_2$, be the trajectory we obtain by integrating $dx/dt = f(t, x, u(t, z))$ from t_1 with initial value $x(t_1)$. Then $x(t, z) = x_0(t)$ for $t_1 \leq t \leq \bar{t} - \sum_\sigma c_\sigma z_\sigma$. Statements (7.4.i–ii) can now be reformulated as follow.

7.4.iii (LEMMA). *For all $z \in R$ with $|z|$ sufficiently small and $\bar{t}$ a Lebesgue point of $(t_1, t_2]$, $x(t, z)$ exists in all of $[t_1, t_2]$, and $x(t, z) \to x_0(t)$ uniformly in $[t_1, t_2]$ as $z \to 0$ (that is, $(z_1, \ldots, z_s) \to (0, \ldots, 0)$, or $|z| \to 0$).*

7.4.iv (Lemma). *For $z \in R$ with z sufficiently small, and $\bar{t}$ a Lebesgue point of $(t_1, t_2]$, the function $x(t, z)$ is differentiable at $z = 0$ and*

$$\xi = (\partial x(\bar{t}, z)/\partial z_\sigma)_{z=0} = c_\sigma[f(\bar{t}, x_0(\bar{t}), \bar{u}_\sigma) - f(\bar{t}, x_0(\bar{t}), u_0(\bar{t})],$$
$$\tilde{\xi} = (\partial x(t, z)/\partial z_\sigma)_{z=0} = \Phi(t)\Phi^{-1}(\bar{t})\xi, \qquad \bar{t} \leq t \leq t_2, \quad \sigma = 1, \ldots, s.$$

The proofs are only modifications of those for (7.4.i) and (7.4.ii), and are left as exercises for the reader.

B. McShane's Variations

By a variation v we shall now mean any finite system

$$v = (\bar{t}_r, c_r, \bar{u}_r, r = 1, \ldots, k; h) \tag{7.4.4}$$

made up of k Lebesgue points $\bar{t}_r$ of $(t_1, t_2]$, $t_1 < \bar{t}_1 \leq \bar{t}_2 \leq \cdots \leq \bar{t}_k \leq t_2$, of k nonnegative constants c_r, of k points $\bar{u}_r$ of U, and of one $(2n+2)$-vector $h = (\tau_1, \xi_1, \tau_2, \xi_2) \in B'$, that is, a vector tangent to B at the point $e[x_0]$. Here $e[x_0] = (t_1, x_0(t_1), t_2, x_0(t_2))$, where t_1, t_2 are fixed; hence, $\tau_1 = \tau_2 = 0$. As in Section 7.3A, we think of $h = (0, \xi_1, 0, \xi_2)$ as the vector tangent to a curve C of class C^1 lying in B and issuing from $e[x_0]$ (say in this case $C = (t_1, X_1(z), t_2, X_2(z))$, $0 \leq z \leq 1$), and hence

$$(t_1, X_1(z), t_2, X_2(z)) \in B, \qquad 0 \leq z \leq 1,$$
$$x_0(t_1) = X_1(0), \quad x_0(t_2) = X_2(0), \qquad \xi_1 = X_1'(0), \quad \xi_2 = X_2'(0),$$

where $X_1(z), X_2(z)$ are continuously differentiable in $[0, 1]$. As in Section 7.3A we extend $X_1(z), X_2(z)$ to the whole interval $[-1, 1]$ so that they are still continuously differentiable in $[-1, 1]$, and by the same conventions. Again we can say that ξ_1 is the tangent to the curve $C': x = X_1(z)$ at $z = 0$, and ξ_2 is the tangent to the curve $C'': x = X_2(z)$ at $z = 0$.

If a point $\bar{t}_r$ is not repeated, that is, $\bar{t}_{r-1} < \bar{t}_r < t_{r+1}$, then let I_r denote the interval $I_r = [\bar{t}_r - zc_r, \bar{t}_r]$ for $0 < z \leq 1$, and $z > 0$ so small that $\bar{t}_{r-1} < \bar{t}_r - zc_r$, If a point t is repeated, say s times, $\bar{t}_{r-1} < \bar{t}_r = \bar{t}_{r+1} = \cdots = \bar{t}_{r+s-1} < \bar{t}_{r+s}$, then we consider the s consecutive intervals $I_r = [\bar{t}_r - zc_r, \bar{t}_r]$, $I_{r+1} = [\bar{t}_r - zc_r - zc_{r+1}, \bar{t}_r - zc_r], \ldots, I_{r+s-1} = [\bar{t}_r - zc_r - \cdots - zc_{r+s-1}, \bar{t}_r - zc_r - \cdots - zc_{r+s-2}]$. Again, we can take $z > 0$ so small that $\bar{t}_{r-1} < t_r - zc_r - \cdots - zc_{r+s-1}$. Thus, we have, for $z > 0$ sufficiently small, k intervals, all contained in $(t_1, t_2]$, and of lengths $zc_1, zc_2, \ldots, zc_k$ respectively. Moreover, these k intervals are nonoverlapping, that is, they intersect at most at their end points. We define now the strategy $u(t, z)$ by taking $u(t, z) = \bar{u}_r$ in the corresponding interval I_r, $r = 1, \ldots, k$, and by taking $u(t, z) = u_0(t)$ otherwise. We shall then denote by $x(t, z, v)$, $t_1 \leq t \leq t_2$, the trajectory that we obtain by integrating $dx/dt = f(t, x, u(t, z))$ from t_1 with initial value $X_1(z)$. From the discussion in subsection A we know that, for $z > 0$ sufficiently small, $x(t, z, v)$ exists in all of $[t_1, t_2]$, and that $x(t, z, v) \to x_0(t)$ as $z \to 0+$ uniformly in $[t_1, t_2]$ (for a given v). In addition $x(t, z, v)$ has derivative with respect to z at $z = 0$ for almost all $t \in [t_1, t_2]$, and if $y(t, v) = (\partial x(t, z, v)/\partial z)_{z=0}$, $t_1 \leq t \leq t_2$, then

$$\tilde{\xi} = y(t_2, v)$$
$$= \Phi(t_2)\left\{\Phi^{-1}(t_1)\xi_1 + \sum_{r=1}^{k} c_r\Phi^{-1}(\bar{t}_r)[f(\bar{t}_r, x_0(\bar{t}_r), \bar{u}_r) - f(\bar{t}_r, x_0(\bar{t}_r), u_0(\bar{t}_r))]\right\}. \tag{7.4.5}$$

Just as in subsection A, we may take $z \in R^s$ and consider the following generalization of the above considerations. We take systems

$$v = (\bar{t}_r, c_{r\sigma}, \bar{u}_{r\sigma}, r = 1, \dots, k, h_\sigma, \sigma = 1, \dots, s) \tag{7.4.6}$$

made up of k Lebesgue points $\bar{t}_r$ of $(t_1, t_2]$, $t_1 < \bar{t}_1 \le \bar{t}_2 \le \cdots \le \bar{t}_k \le t_2$, of ks nonnegative constants $c_{r\sigma} \ge 0$, of ks points $\bar{u}_{r\sigma}$ of U, and of s vectors $h_\sigma = (0, \xi_{1\sigma}, 0, \xi_{2\sigma}) \in B'$, $\sigma = 1, \dots, s$, all tangent to B at $e[x_0]$. Let $z = (z_1, \dots, z_s) \in R$, where R is the hypercube $[0 \le z_\sigma \le 1, \sigma = 1, \dots, s]$ in R^s.

Let us suppose, for a moment, that the points $\bar{t}_1, \dots, \bar{t}_k$ are distinct. Then, for $|z|$ sufficiently small the ks intervals $[\bar{t}_r - c_{r1}z_1, \bar{t}_r]$, $[\bar{t}_r - c_{r1}z_1 - c_{r2}z_2, \bar{t}_r - c_{r1}z_1], \dots,$ $[\bar{t}_r - \sum_\sigma c_{r\sigma}z_\sigma, \bar{t}_r - \sum_\sigma c_{r\sigma}z_\sigma + c_{rs}z_s]$, $r = 1, \dots, k$, $\sigma = 1, \dots, s$, are certainly nonoverlapping and contained in $[t_1, t_2]$. We now define $u(t, z)$ by taking $u(t, z)$ equal to $\bar{u}_{r1}$, $\bar{u}_{r2}, \dots, \bar{u}_{rs}$, $r = 1, \dots, k$, in the ks intervals above respectively, and equal to $u_0(t)$ otherwise. If the points $\bar{t}_1, \dots, \bar{t}_r$ are not all distinct, then we take adjacent systems of intervals as indicated above, and we define $u(t, z)$ accordingly.

Each vector $h_\sigma = (0, \xi_{1\sigma}, 0, \xi_{2\sigma})$ is tangent to a curve $C_\sigma = (t_1, X_{1\sigma}(z_\sigma), t_2, X_{2\sigma}(z_\sigma))$ issuing from $e[x_0]$, lying in B, with $X_1'(0) = \xi_{1\sigma}$, $X_2'(0) = \xi_{2\sigma}$, $\sigma = 1, \dots, s$. As we know from Section 7.3A(c), there are functions $X_1(z)$, $X_2(z)$, $z \in R$, of class C^1 in R, such that $X_\alpha(0, \dots, 0, z_\sigma, 0, \dots, 0) = X_{\alpha\sigma}(z_\sigma)$ for $z_\sigma \ge 0$ sufficiently small, such that $(t_1, X_1(z), t_2, X_2(z)) \in B$ for all $z = (z_1, \dots, z_s) \in R$, and

$$(\partial X_\alpha / \partial z_\sigma)_{z=0} = X'_{\alpha\sigma}(0), \qquad \alpha = 1, 2, \quad \sigma = 1, \dots, s.$$

Finally, we denote by $x(t, v, z)$ the trajectory obtained by integrating the differential equation $dx/dt = f(t, x, u(t, z))$ from t_1 with initial value $X(z)$, $z = (z_1, \dots, z_s)$. Then, for $z \in R$, $|z|$ sufficiently small, $x(t, v, z)$ exists in all of $[t_1, t_2]$ and is of class C^1 in z for every t. If we take

$$y_\sigma(t, v) = (\partial x(t, z)/\partial z_\sigma)_{z=0}, \qquad t_1 \le t \le t_2,$$

then we have for all $\sigma = 1, \dots, s$

$$\tilde{\xi}_\sigma = y_\sigma(t_2, v) \tag{7.4.7}$$

$$= \Phi(t_2)\left\{\Phi^{-1}(t_1)\xi_{1\sigma} + \sum_{r=1}^{k} \Phi^{-1}(\bar{t}_r)[f(\bar{t}_r, x_0(\bar{t}_r), \bar{u}_{r\sigma}) - f(\bar{t}_r, x_0(\bar{t}_r), u_0(\bar{t}_r))]\right\}.$$

C. The New Cone K

We shall now consider the cone K in R^{n+1} made up of the terminal points of the linearized trajectories in R^n corresponding to all possible variations defined above, and associated values of the linearized cost functional. We shall prove that K is a convex cone (Lemma (7.4.v)), and that the point $(-1, 0, \dots, 0)$ in R^{n+1} is not an interior point of K (Lemma (7.4.vi)). The latter is proved by contradiction, showing that in the opposite case there would exist an admissible trajectory giving a lower cost than the optimal one. The proof of the necessary condition then follows by taking a supporting hyperplane to K.

For every variation $v = (\bar{t}_1, \dots, \bar{t}_k, c_1, \dots, c_k, \bar{u}_1, \dots, \bar{u}_k, h)$ with $h = (0, \xi_1, 0, \xi_2)$, let us consider the $(n+1)$-vector $\tilde{Y}(v) = (Y^0, Y) = (Y^0, Y^1, \dots, Y^n)$ defined by $Y^0(v) = g_{x_1}\xi_1 + g_{x_2}\xi_2$, $Y(v) = y(t_2, v) - \xi_2$, where g_{x_1}, g_{x_2} are the $1 \times n$ matrices of the partial derivatives of $g(t_1, x_1, t_2, x_2)$ with respect to the arguments $x_1^1, \dots, x_1^n$, or $x_2^1, \dots, x_2^n$

respectively, these partial derivatives being evaluated at the point

$$e[x_0] = (t_1, x_0(t_1), t_2, x_0(t_2)).$$

We shall denote by $K \subset R^{n+1}$ the set of all such vectors $\tilde{Y}(v)$ in the $y^0 y^1 \cdots y^n$-space R^{n+1}.

7.4.v (LEMMA). *The set K is a convex cone with vertex at* $(0, \ldots, 0)$, *that is, if* $\tilde{Y}(v_1)$, $\tilde{Y}(v_2) \in K$ *and* $a_1, a_2 \geq 0$, *then there is a variation* v *such that* $\tilde{Y}(v) = a_1 \tilde{Y}(v_1) + a_2 \tilde{Y}(v_2)$.

Proof. It is not restrictive to assume that v_1 and v_2 are variations as defined by (7.4.4) and relative to the same Lebesgue points of $[t_1, t_2]$; thus

$$v_\sigma = (\bar{t}_1, \ldots, \bar{t}_k, c_{1\sigma}, \ldots, c_{k\sigma}, \bar{u}_{1\sigma}, \ldots, \bar{u}_{k\sigma}, h_\sigma), \qquad \sigma = 1, 2.$$

Indeed, if v_1 and v_2 were defined by means of different sets of points $\bar{t}$, we would have to take for $\bar{t}, \ldots, \bar{t}_k$ above the union of the two sets of points $\bar{t}$, and take in each v_1 and v_2 constants $c = 0$ in correspondence to all new points. Let v be the variation defined in subsection B above by means of the same points $\bar{t}_1, \ldots, \bar{t}_k$, constants $c_{r\sigma}$, points $\bar{u}_{r\sigma} \in U$, $r = 1, \ldots, k$, $\sigma = 1, 2$, and vectors h_σ, $\sigma = 1, 2$. We now apply the process discussed in subsection B with $s = 2$. For $z = (z_1, z_2) \in R$, $R = [0 \leq z_1, z_2 \leq 1]$, vector h defined by

$$h = a_1 h_1 + a_2 h_2 = [0, a_1 \xi_{11} + a_2 \xi_{12}, 0, a_1 \xi_{21} + a_2 \xi_{22}] = (0, \xi_1, 0, \xi_2),$$

and $z_1 = a_1 \varepsilon$, $z_2 = a_2 \varepsilon$ (i.e. $z = (a_1 \varepsilon, a_2 \varepsilon)$), $\varepsilon \geq 0$, sufficiently small, we have a unique variation v with points $\bar{t}_1, \ldots, \bar{t}_k$, constants $a_\sigma c_{r\sigma}$, $r = 1, \ldots, k$, $\sigma = 1, 2$, points $\bar{u}_{r\sigma} \in U$, $r = 1, \ldots, k$, $\sigma = 1, 2$, for which

$$\xi = y(t_2 v) = a_1 y(t_2, v_1) + a_2 y(t_2, v_2).$$

Then we have

$$\begin{aligned}(7.4.8) \qquad Y(v) &= y(t_2, v) - \xi_2 = a_1(y(t_2, v_1) - \xi_{21}) + a_2(y(t_2, v_2) - \xi_{22}) \\ &= a_1 Y(v_1) + a_2 Y(v_2),\end{aligned}$$

$$\begin{aligned}(7.4.9) \qquad Y^0(v) &= g_{x_1} \xi_1 + g_{x_2} \xi_2 = g_{x_1}(a_1 \xi_{11} + a_2 \xi_{12}) + g_{x_2}(a_1 \xi_{21} + a_2 \xi_{22}) \\ &= a_1(g_{x_1} \xi_{11} + g_{x_2} \xi_{21}) + a_2(g_{x_1} \xi_{12} + g_{x_2} \xi_{22}) \\ &= a_1 Y^0(v_1) + a_2 Y^0(v_2).\end{aligned}$$

This proves that K is a convex cone, with vertex the origin. All variations v with zero constants c give $Y^0(v) = 0$. □

7.4.vi (LEMMA). *The point* $(-1, 0, \ldots, 0)$ *is not interior to K.*

The proof is identical to the one for (7.3.ii) where we use $n + 1$ variations v_σ as defined for the present proof. As before, we can assume that these $n + 1$ variations v_σ correspond to the same system of Lebesgue points in $(t_1, t_2]$, or

$$v_\sigma = [\bar{t}_1, \ldots, \bar{t}_k, c_{1\sigma}, \ldots, c_{k\sigma}, \bar{u}_{1\sigma}, \ldots, \bar{u}_{k\sigma}, h_\sigma), \qquad \sigma = 1, 2, \ldots, n + 1.$$

7.4.vii (LEMMA). *There are numbers* $\chi_0, \chi_1, \ldots, \chi_n$ *not all zero,* $\chi_0 \geq 0$, *such that* $\sum_{i=0}^{n} \chi_i Y^i(v) \geq 0$ *for all variations* v.

The proof is the same as for (7.3.iii). The same remark holds as at the end of Section 7.3B namely, if dg is not identically zero, then the n-vector $(\chi_1, \ldots, \chi_n)$ is not zero.

D. Completion of Proof

Proof of (P_1). For the variation v made up of only a vector $h = (0, \xi_1, 0, \xi_2) \in B'$ (that is, $c = 0$), then $y(t; v) = \Phi(t)\Phi^{-1}(t_1)\xi_1$. As in Section 7.3C we take for $\lambda(t)$, $t_1 \le t \le t_2$, the AC n-vector solution of $\lambda' = -A^*(t)\lambda$ with $\lambda(t_2) = \chi$, that is,

$$\lambda_i'(t) = -\sum_{j=1}^{n} \lambda_j f_{jx^i}(t, x_0(t), u_0(t)), \qquad t_1 \le t \le t_2, \quad \lambda(t_2) = \chi,$$

and this proves (P1). Let us take $\lambda_0(t) = \lambda_0 = \chi_0$. Now let us prove that

$$\lambda(t) = (\chi^*\Phi(t_2)\Phi^{-1}(t))^*, \qquad t_1 \le t \le t_2.$$

First for the function λ defined by this relation we have $\lambda(t_2) = \chi$ as required. Secondly, from $\Phi' = A(t)\Phi$, and the well known relation $(d/dt)\Phi^{-1} = -\Phi^{-1}\Phi'\Phi^{-1}$, we obtain

$$\begin{aligned}\lambda'(t) &= (\chi^*\Phi(t_2)(\Phi^{-1}(t))')^* = -(\chi^*\Phi(t_2)\Phi^{-1}(t)\Phi'(t)\Phi^{-1}(t))^* \\ &= -(\Phi^{-1}(t))^*(\Phi'(t))^*(\chi^*\Phi(t_2)\Phi^{-1}(t))^* \\ &= -(\Phi^{-1}(t))^*(A(t)\Phi(t))^*\lambda(t) = -(\Phi^{-1}(t))^*\Phi^*(t)A^*(t)\lambda(t) \\ &= -A^*(t)\lambda(t).\end{aligned}$$

By the uniqueness theorem for differential equations the proof is complete. □

Proof of (P_4). Let v denote any variation $v = (\bar{\tau}, c, \bar{u}, h)$, $h = (0, \xi_1, 0, \xi_2)$, with only one Lebesgue point $\bar{\tau} \in (t_1, t_2]$, constant $c \ge 0$, point $\bar{u} \in U$, and vector $h \in B'$. We shall now replace in (7.4.vii), or $\sum_{i=0}^{n} \chi_i Y^i(v) \ge 0$, each $Y^i(v)$ by its expression (7.4.8–9) in terms of g and $y(t_2, v)$, and replace $y(t_2, v)$ by its expression (7.4.5). We have

$$\begin{aligned}&\lambda_0(g_{x_1}\xi_1 + g_{x_2}\xi_2) + \chi \cdot \Phi(t_2)\{\Phi^{-1}(t_1)\xi_1 \\ &\quad + c\Phi^{-1}(\bar{\tau})[f(\bar{\tau}, x_0(\bar{\tau}), \bar{u}) - f(\bar{\tau}, x_0(\bar{\tau}), u_0(\bar{\tau}))] - \Phi^{-1}(t_2)\xi_2\} \ge 0\end{aligned} \tag{7.4.10}$$

where

$$\begin{aligned}\chi \cdot \Phi(t_2)\Phi^{-1}(t_1)\xi_1 &= \chi^*(\Phi(t_2)\Phi^{-1}(t_1)\xi_1) \\ &= (\chi^*\Phi(t_2)\Phi^{-1}(t_1))\xi_1 \\ &= (\lambda(t_1))^*\xi_1 = \lambda(t_1) \cdot \xi_1,\end{aligned}$$

and $a \cdot b$ is the usual inner product of two vectors, $a \cdot b = a^*b$. Analogously, we have $\chi \cdot \Phi(t_2)\Phi^{-1}(\bar{\tau})f = \lambda(\bar{\tau}) \cdot f$. Hence, the relation (7.4.10) yields

$$\lambda_0(g_{x_1}\xi_1 + g_{x_2}\xi_2) - \lambda(t_2) \cdot \xi_2 + \lambda(t_1) \cdot \xi_1 + c\lambda(\bar{\tau}) \cdot [f(\bar{\tau}, x_0(\bar{\tau}), \bar{u}) - f(\bar{\tau}, x_0(\bar{\tau}), u_0(\bar{\tau}))] \ge 0.$$

Using the definition of the Hamiltonian, we obtain

$$\begin{aligned}&\sum_{i=1}^{n} [\lambda_0 g_{x_1^i} + \lambda_i(t_1)]\xi_1^i + \sum_{i=1}^{n} [\lambda_0 g_{x_2^i} - \lambda_i(t_2)]\xi_2^i \\ &\quad + c[H(\bar{\tau}, x_0(\bar{\tau}), \bar{u}, \lambda(\bar{\tau})) - H(\bar{\tau}, x_0(\bar{\tau}), u_0(\bar{\tau}), \lambda(\bar{\tau}))] \ge 0.\end{aligned} \tag{7.4.11}$$

For $c = 0$ this relation yields

$$\sum_{i=1}^{n} [\lambda_0 g_{x_1^i} + \lambda_i(t_1)]\xi_1^i + \sum_{i=1}^{n} [\lambda_0 g_{x_2^i} - \lambda_i(t_2)]\xi_2^i \ge 0, \tag{7.4.12}$$

which is (4.2.13) of Section 4.2C, Remark 10, when $\tau_1 = \tau_2 = 0$. As mentioned there this form yields (P4) of Section 4.2A when B' is a linear space, and $\tau_1 = \tau_2 = 0$. □

Proof of (P_2). By taking $c = 1$ and $\xi_1 = \xi_2 = 0$ in (7.4.11) we obtain

$$H(\bar{t}, x_0(\bar{t}), \bar{u}, \lambda(\bar{t})) - H(\bar{t}, x_0(\bar{t}), u_0(\bar{t}), \lambda(\bar{t})) \geq 0,$$

and this relation is valid for all $\bar{u} \in U$. Since $u_0(t) \in U$ also, we conclude that

$$H(\bar{t}, x_0(\bar{t}), u_0(\bar{t}), \lambda(\bar{t})) = \min_{u \in U} H(\bar{t}, x_0(\bar{t}), u, \lambda(\bar{t})) = M(\bar{t}, x_0(\bar{t}), \lambda(\bar{t})).$$

This relation is valid for all Lebesgue points $\bar{t} \in (t_1, t_2]$, that is, for almost all points $\bar{t} \in [t_1, t_2]$. This proves (P_2). □

The proofs of (P_3) is the same as in Section 7.3F. The restriction that t_1 and t_2 are fixed can be removed exactly as in Section 7.3G. The case of unbounded $u_0(t)$ can be handled as in Section 7.3H.

E. Alternate Assumptions and Other Remarks

First let us assume that condition (c″) of Section 4.2C, Remark 5, holds. Now $U(t)$ may depend on t. We know already that we can disregard sets of points $\bar{t}$ of measure zero. For any $\bar{t}$ for which (c″) holds, and which is a Lebesgue point, we take now intervals $[\bar{t} - \sigma, \bar{t}]$ and points $u \in U(t)$ which belong to $U(t)$ for all $\bar{t} - \sigma \leq t \leq \bar{t}$. All variations used in the proof above (Subsections A–D) must be taken now with values u as described. Then we prove properties (P1–4) only for such points $u \in U(\bar{t})$. Since f and H are continuous functions of their arguments, the same properties hold for all $\bar{u} \in U(\bar{t})$, since these are points of accumulation of points $u \in U(t)$ as above. The same arguments hold also under the assumptions (c‴), (c^{iv}), (c^{v}).

Remark. We take up here the question mentioned in Sections 4.2A and 5.1 as to whether $\lambda_0 \neq 0$, and hence $\lambda_0 > 0$ and we can take $\lambda_0 = 1$ in the necessary condition. We restrict ourselves to Theorem (4.2.i) for Mayer problems.

First, note that the case $\lambda_0 = 0$ may actually occur. Indeed, consider the Mayer problem with $n = 2$, $m = 1$, $dx/dt = u$, $dy/dt = u^2$, $u \in U = R$, $t_1 = 0$, $t_2 = 1$, $x(0) = x_1 = 0$, $y(0) = y_1 = 0$, $y(1) = y_2 = 0$, $g = x_2 = x(1)$. Then $H = \lambda_1 u + \lambda_2 u^2$, and H has a minimum only for $\lambda_2 > 0$, $u = -\lambda_1/2\lambda_2$, with $H_{\min} = -\lambda_1^2/4\lambda_2$. Also, $d\lambda_1/dt = 0$, $d\lambda_2/dt = 0$, $\lambda_1 = c_1$, $\lambda_2 = c_2 > 0$, c_1, c_2 constants, $M(t) = -c_1^2/4c_2$, $dt_1 = dt_2 = 0$, $dx_1 = dy_1 = dy_2 = 0$, and (P4) yields $\lambda_0\, dx_2 - c_1\, dx_2 = 0$; hence $\lambda_0 = c_1$. Now, the only solution is $u(t) = 0$, $0 \leq t \leq 1$; hence $c_1 = 0$, $\lambda_0 = 0$.

To state a criterion under which $\lambda_0 \neq 0$, let us first note that, in the notation of Section 7.3B, $\lambda_0 = 0$ corresponds to a vertical supporting hyperspace for the convex cone K. There are situations where this can be excluded. For instance, if we can produce $n + 1$ variations $v_s = (c_s, u_s, h_s)$, $s = 1, \ldots, n + 1$, such that for $t = t_2$, the end points $P_s = y_s(t_2)$ of the solutions y_s of the corresponding linear variational equations form an n-dimensional (not degenerated) simplex in R^n containing $P_0 = 0$ in its interior, then certainly $\lambda_0 \neq 0$. Indeed, the $n+1$ points in R^{n+1} given by $(Y^0(v_s), P_s)$, $s = 1, \ldots, n+1$, belong to the convex cone K of vertex $(0, P_0)$, and this cone cannot have a vertical supporting space at P_0.

For problems with t_1 fixed and t_2 undetermined, the variations v_s can be chosen with controls $u_s(t)$, $t_1 \leq t \leq t_2'$, defined in intervals not necessarily the same as for $u_0(t)$, $t_1 \leq t \leq t_2$.

For instance, for the simple problem considered in Section 4.2 and also in Section 6.1, with $n = 2$, $m = 1$, $U = [-1 \le u \le 1]$, $g = t_2$, $t_1 = 0$, $x_1 = a$, $y_1 = b$, $x_2 = 0$, $y_2 = 0$, $f_1 = y$, $f_2 = u$, let us consider any optimal solution as depicted in Section 4.2, made up of two arcs $[t_1, \bar{t}]$ and $[\bar{t}, t_2]$ and with controls $u_0 = -1$, $u_0 = 1$ respectively. The variational equation for a variation $v = (0, u, c)$ reduces to $y_1' = y_2$, $y_2' = c(u - u_0)$ with $y_1(0) = 0$, $y_2(0) = 0$. For a variation v_1 with $c = 1$, and $u(t)$ defined by $u = 1$ in an interval $(\bar{t} - \varepsilon, \bar{t})$ and $u = u_0$ otherwise, then $u = 1$, $u_0 = -1$, $u - u_0 = 2$ in $(\bar{t} - \varepsilon, \bar{t})$, and $y_2(t_2) = 2\varepsilon$, $y_1(t_2) = \varepsilon^2 + 2\varepsilon(t_2 - \bar{t})$. For a variation v_2 defined by taking $u(t)$, $0 \le t \le t_2$, with $u(t) = -1$ in an interval $(\bar{t}, \bar{t} + \varepsilon)$ and $u = u_0$ otherwise, then $u = -1$, $u_0 = 1$, $u - u_0 = -2$ in $(\bar{t}, \bar{t} + \varepsilon)$, and $y_2(t_2) = -4\varepsilon$, $y_1(t_2) = -\varepsilon^2 - 2\varepsilon(t_2 - \bar{t} - \varepsilon)$. For a variation v_3 defined by taking $u(t)$, $0 \le t \le t_2 + 4\varepsilon$, with $u(t) = -1$ for $\bar{t} \le t \le \bar{t} + \varepsilon$, then $y_2(t_2 + 4\varepsilon) = 2\varepsilon$, $y_1(t_2 + 4\varepsilon) = -\varepsilon^2 - 2\varepsilon(t_2 + 3\varepsilon - \bar{t})$. If P_1, P_2, P_3 denote the terminals of the y-trajectories just now determined, we see that P_1P_2 passes below P_0, and P_1P_3 passes above P_0. In other words, $P_1P_2P_3$ is a simplex containing P_0 in its interior. Hence $\lambda_0 \ne 0$.

7.5 Proof of Boltyanskii's Statements (4.6.iv–v)

(a) Here we assume that a regular synthesis has been effected in V with feedback control function $p(t, x)$, target B, and exceptional set S, $B \subset S \subset V$, as stated in Section 4.6.

Let σ be a cell of $R^n - S$, and consider a trajectory $x(t; s, y)$ as a function of its initial point (s, y) in V, that is, $x(s; s, y) = y$ for $(s, y) \in \sigma$. Then, if the trajectory x crosses the cell σ_j, we shall denote by $\tau_{j-1}(s, y)$ and $\tau_j(s, y)$ the times at which x enters and leaves σ_j. We shall prove first the following statements:

7.5.i. *The time $\tau_j(s, y)$ and corresponding positions $x_j(s, y) = x(\tau_j(s, y), s, y)$ at which x leaves the cell σ_j are of class C^1 on σ.*

7.5.ii. *The function $x(t; x\ y)$ is of class C^1 in $(t; s, y)$ in the set $[(t; s, y) | (s, y) \in \sigma, \tau_{j-1}(s, y) \le t \le \tau_j(s, y)]$. Moreover, the $n \times n$ matrix $x_y = [\partial x^i/\partial y^j, i, j = 1, \dots, n]$ satisfies the matrix relation $(d/dt)x_y = (f_x + f_u u_x)x_y$; precisely, with initial values $x_y(s; s, y) = I$, the identity matrix, and $x_s(s; s, y) = -f(s, x(s; s, y), u(s, y))$, we have*

$$(d/dt)x_y(t; s, y) = [f_x(t, x(t; s, y), u(t, x(t; s, y)) + f_u(t, x(t; s, y), u(t, x(t; s, y)))u_x(t, x(t; s, y))]x_y(t; s, y),$$

and an analogous relation holds for x_s.

7.5.iii. *At time $\tau_{j-1} = \tau_{j-1}(s, y)$ and $\tau_j = \tau_j(s, y)$ the $n \times n$ matrix $x_y(t; s, y)$ has right and left limits respectively given by*

$$x_y(\tau_{j-1} + 0) = -f(\tau_{j-1}, x_{j-1}, u_j(\tau_{j-1}, x_{j-1}))\tau_{j-1,y} + x_{j-1,y},$$
$$x_y(\tau_j - 0) = -f(\tau_j, x_j, u_j(\tau_j, x_j))\tau_{j,y} + x_{j,y},$$

where $u(t, x) = u_j(t, x)$ is a continuous function in $\text{cl}\,\sigma_j$, where $x_{j-1} = x(\tau_{j-1}(t, s); t, s)$, $x_j = x(\tau_j(t, s); t, s)$; where $\tau_{j-1,y} = (\tau_{j-1}(s, y))_y$, $\tau_{j,y} = (\tau_j(s, y))_y$; and where $x_{j,y} = (x(\tau_j(s, y), s, y))_y$. Analogously, at $t = t_2(s, y) = \tau_q(s, y)$, we have $x_y(\tau_q - 0) = -f(\tau_q, x_q,$

$u(\tau_q, x_q))\tau_{q,y} + x_{q,y}$, where $\tau_{q-1}(s, y) < t < \tau_q(s, y)$ is the arc of $x(t; s, y)$ in the last cell σ_q transversed by the trajectory, where $u = u_q(t, x)$ is continuous in cl σ_q, where $x_y(\tau_q - 0)$ is the limiting value of $x_y(t; s, y)$ as $t \to \tau_q(s, y)$, and where $x_{q,y} = (x(\tau_q(s, y); s, y))_y$. Analogous relations hold for the derivative x_s.

Remark. Before we prove these statements, note the following simple example. Let $n = 1$, let $\sigma = \sigma_1$ be the cell $[(t, x)|t > 0, x > t]$, let $\sigma_2 = [(t, x)|t > 0, 0 < x < t]$, $\sigma_3 = [(t, x)|t > 0, x < 0]$, and take $f = -1$ in σ_1, $f = -2$ in σ_2, $f = -3$ in σ_3. For any $(s, y) \in \sigma$, or $s > 0$, $y > s$, we have $x(t; s, y) = -t + (y + s)$ in σ_1; hence, $\tau_1(s, y) = 2^{-1}(y + s)$, $x_1(t, s) = 2^{-1}(y + s)$. Then $x(t; s, y) = -2t + \frac{3}{2}(y + s)$ in σ_2; hence, $\tau_2(s, y) = \frac{3}{4}(y + s)$, $x_2(t, s) = 0$. Finally, $x(t; s, y) = -3t + \frac{9}{4}(y + s)$ in σ_3. Thus, $x_y = 1$ in σ_1, $x_y = \frac{3}{2}$ in σ_2, and $x_y = \frac{9}{4}$ in σ_3. In particular, $x_y(\tau_1-) = 1$, $x_y(\tau_1+) = \frac{3}{2}$, $x_y(\tau_2-) = \frac{3}{2}$, $x_y(\tau_2+) = \frac{9}{4}$. On the other hand, $x(\tau_1(s, y); s, y) = 2^{-1}(y + s)$, $x_{1,y} = 2^{-1}$, and $x(\tau_2(s, y); s, y) = 0$, $x_{2,y} = 0$.

Proof of (7.5.i–iii). Let $\sigma_1 = \sigma, \sigma_2, \ldots, \sigma_q$ denote the finite sequence of cells of the first kind which the trajectory $x(t; s, y)$ beginning at $(s, y) \in \sigma$ will go through. As a convention, we take $\tau_0(s, y) = s$, $x_0(s, y) = x(s; s, y) = y$. The proof is by induction on the integer j. Thus, we suppose that (7.5.i–iii) hold for $j - 1$, and we shall prove them for j. Note that the arc $x(t; s, y)$ for $\tau_{j-1}(s, y) < t < \tau_j(s, y)$ is in σ_j with initial data in σ_j given by $w = \tau_{j-1}(s, y)$, $z = x_{j-1} = x(\tau_{j-1}(s, y); s, y)$. Moreover, within σ_j, we have assumed $u(t, x)$ Lipschitz continuous and f of class C^1. Thus, the differential system in σ_j, $dx/dt = f(t, x, u(t, x))$ with initial data $x(w) = z$, has a unique solution $x(t; w, z)$ which exists at least for t in a neighborhood $(w - \varepsilon, w + \varepsilon)$ of w, and $x(t; w, z)$ is of class C^1 in (t, w, z) in a neighborhood of $(\tau_{j-1}, \tau_{j-1}, x_{j-1})$. Since w, z are, by the induction hypothesis, of class C^1 in (s, y), we conclude that the composite function $(s, y) \to (w, z) \to x(t; w(s, y), z(s, y))$ is also of class C^1 in (s, y).

Now we have to see what happens at $t = \tau_j(s, y)$. If $T(t, x) = 0$ denotes the equation of the cell covered by $(\tau_j(s, y), x(\tau_j(s, y); s, y))$, then $T(\tau_j(s, y), x(\tau_j(s, y); s, y)) = 0$. We have only to apply the implicit function theorem to see whether the solution $t = \tau_j(s, y)$ of the equation $T(t, x(t; s, y)) = 0$ in t is of class C^1. This is true because

$$(\partial/\partial t)T(t, x(t; s, y)) = T_t + \sum_{i=1}^{n} T_{x^i}\, dx^i/dt = T_t + \sum_{i=1}^{n} T_{x^i} f_i \neq 0,$$

and this derivative is $\neq 0$, since we have assumed that the trajectories hit the walls of the cells at an angle $\neq 0$. This proves that $\tau_j(s, y)$ is of class C^1, and then $x(\tau_j(s, y); s, y)$ also is of class C^1. By the induction argument, (7.5.i) is proved. □

Now we have

$$x_t(t; s, y) = f(t, x(t; s, y), u(t, x(t; s, y))), \tag{7.5.1}$$

and by differentiation with respect to y and s we have the matrix relations

$$x_{ty} = (f_x + f_u u_x)x_y, \qquad x_{ts} = (f_x + f_u u_x)x_s, \tag{7.5.2}$$

where the arguments of f_x, f_u, u_x, x_y, x_s are the same as in (7.5.1) and t ranges in $[s, t_2]$.

Since $x(s; s, y) = y$, we have $x_y(s; s, y) = I$, the identity matrix. Also, $x_t(s; s, y) + x_s(s; s, y) = 0$, or $x_s(s; s, y) = -x_t = -f(s, x(s, y), u(s, y))$. This proves (7.5.ii). □

Let us denote by $u_j(t, x)$ the function, continuous on the closure of σ_j, which coincides with $u(t, x)$ in σ_j. Let $P_j = [\tau_j(s, y), x(\tau_j(s, y); s, y)]$. Then the arc $P_{j-1}P_j$ of $x(t; s, y)$ for

$\tau_{j-1}(s, y) \le t \le \tau_j(s, y)$ is of class C^1 with velocity given by

$$x_t(t; s, y) = f(t, x(t; s, y), u(t, x(t; s, y))), \tag{7.5.3}$$

where $u = u_j$. Thus, at each point P_j we must expect a corner point for the trajectory $x(t; s, y)$ with derivatives left and right given by (7.5.3), where we write for u the limiting values at P_j of u_j and u_{j+1} respectively.

Thus, for $x_j(s, y) = x_j(\tau_j(s, y); s, y)$ we must expect analogous jump discontinuities in the derivatives $(x_j(s, y))_y$ and $(x_j(s, y))_s$. The limiting values left and right for these derivatives are easy to compute.

To this effect we recall that we have assumed that the function $u_j(t, x)$, $(t, x) \in \sigma_j$, can be extended as a continuously differentiable function, say still u_j, in the neighborhood $U_{j\delta}$ of σ_j of radius $\delta > 0$. Also, we have assumed that $f(t, x, u)$ can be extended as a function of class C^1 in a neighborhood W of its domain $M \subset R^{1+n+m}$. Now, for $\delta > 0$ sufficiently small, $f(t, x, u(t, x))$ is defined for $(t, x) \in U_{j\delta}$ and is a function of class C^1, and the solution $x(t; s, y)$ of the differential problem

$$x_t(t; s, y) = f(t, x, u_j(t, x)), \qquad x(s; s, y) = y,$$

can be extended to a function $X_j(t; s, y)$, briefly $x(t; s, y)$, of class C^1 in $(t; s, y)$. Now, from $x_j(s, y) = x(\tau_j(s, y); s, y)$ we obtain by differentiation

$$(x_j(s, y))_y = x_t(\tau_j(s, y); s, y)(\tau_j(s, y))_y + x_y(\tau_j(s, y); s, y),$$
$$(x_j(s, y))_s = x_t(\tau_j(s, y); s, y)(\tau_j(s, y))_s + x_s(\tau_j(s, y); s, y),$$

where $x_t = f$, and

$$x_y(\tau_j -) = -f(\tau_j, x_j, u(\tau_j, x_j))\tau_{jy} + x_{jy},$$
$$x_s(\tau_j -) = -f(\tau_j, x_j, u(\tau_j, x_j))\tau_{js} + x_{js}.$$

Analogously, $x_{j-1}(s, y) = x_t(\tau_{j-1}(s, y); s, y)$, and hence,

$$x_y(\tau_{j-1} +) = -f(\tau_{j-1}, x_{j-1}, u(\tau_{j-1}, x_{j-1}))\tau_{j-1,y} + x_{j-1,y},$$
$$x_s(\tau_{j-1} +) = -f(\tau_{j-1}, x_{j-1}, u(\tau_{j-1}, x_{j-1}))\tau_{j-1,s} + x_{j-1,s},$$

where $u = u_j(t, s)$ is continuously differentiable on cl σ_j. This proves (7.5.iii). □

(b) If $\lambda(t) = (\lambda_1, \dots, \lambda_n)$ is an AC vector, or $n \times 1$ matrix, such that $\min_{u \in U} \lambda^*(t) f(t, x(t; s, y), u) = \lambda^*(t) f(t, x(t; s, y), u(t, x(t; s, y)))$, then we also have

$$\lambda^*(t) f_u(t, x(t; s, y), u(t, x(t; s, y)) u_x(t, x(t; s, y))) = 0. \tag{7.5.4}$$

Indeed, let $(t, x(t; s, y))$ be in the cell σ_j for $\tau_{j-1}(s, y) < t < \tau_j(s, y)$. For $(w, z) \in \sigma_j$ then $u = u_j(w, z)$ is continuously differentiable in cl σ_j, and the expression

$$S(w, z) = \lambda^*(t) f(t, x(t; s, y), u(w, z))$$

is a continuously differentiable function of (w, z) in σ_j. For $w = t$, then $S(t, z) = H(t, x(t; s, y), u(t, z))$, and by (P_2) this function (of z only) takes its minimum value at $z = x(t; s, y)$. Thus, its partial derivatives with respect to $z = (z_1, \dots, z_m)$ must be zero at $z = x(t; s, y)$, or

$$\lambda^*(t) f_u(t, x(t; s, y), u(t, x(t; s, y))) u_z(t, x(t; s, y)) = 0$$

and this is the relation (7.5.4) but for notational differences.

(c) As we have proved before, $t_2(s, y)$, $x(t_2(s, y))$ are of class C^1 in each cell; hence

(7.5.5) $$w(s, y) = g(t_2(s, y), x(t_2(s, y); s, y))$$

is also of class C^1, and

(7.5.6) $$w_y(s, y) = g_t t_{2y}(s, y) + g_x(x(t_2(s, y); s, y))_y,$$

where the arguments of g_t and g_x are $t_2(s, y), x(t_2(s, y); s, y)$. By multiplication of the relation stated in (7.5.2) by $\lambda^*(t)$ we have

(7.5.7) $$\lambda^*(t)x_{ty} = \lambda^*(t)[f_x + f_u u_x]x_y.$$

Since $\lambda_t^* = -\lambda^* f_x$, we also have, by multiplication by x_y,

(7.5.8) $$\lambda_t^* x_y = -\lambda^* f_x x_y,$$

and now by (7.5.2), (7.5.4), (7.5.8) we have

$$\begin{aligned}(d/dt)(\lambda^*(t)x_y(t)) &= \lambda_t^*(t)x_y + \lambda^*(t)x_{ty}\\ &= -\lambda^* f_x x_y + \lambda^*[f_x + f_u u_x]x_y = (\lambda^* f_u u_x)x_y = 0.\end{aligned}$$

By integration in $[s, \tau_1]$, $[\tau_{j-1}, \tau_j]$, $[\tau_{q-1}, t_2]$, we have

$$\begin{aligned}\lambda^*(\tau_1)x_y(\tau_1-) - \lambda^*(s)x_y(s) &= 0,\\ \lambda^*(\tau_j)x_y(\tau_j-) - \lambda^*(\tau_{j-1})x_y(\tau_{j-1}+) &= 0,\\ \lambda^*(t_2)x_y(t_2-) - \lambda^*(\tau_{q-1})x_y(\tau_{q-1}+) &= 0,\end{aligned}$$

and by addition,

(7.5.9) $$\lambda^*(s)x_y(s) = \sum_j \lambda^*(\tau_j)[x_y(\tau_j-) - x_y(\tau_j+)] + \lambda^*(t_2)x_y(t_2-),$$

with $t_2 = t_2(s, y) = \tau_q(s, y)$, and where $x_y(t_2-)$ is the limiting value of $x_y(t; s, y)$ as $t \to t_2(s, y) - 0$ in the last cell σ_q of the first kind transversed by the trajectory before hitting B. By (7.5.iii) we have $x_y(\tau_q-) = -f(\tau_q, x_q, u(\tau_q, x_q))\tau_{q,y} + x_{q,y}$, or in the present notation $x_y(t_2-) = -f(t_2, x_2, u(t_2, x_2))t_{2y} + x_{2y}$; hence

(7.5.10) $$\lambda^*(t_2)x_y(t_2-) = \lambda^*(t_2)x_{2y} - \lambda^*(t_2)f t_{2y},$$

with $t_2 = t_2(s, y)$, $x_2 = x(t_2(s, y); s, y)$, $u = u_q(t_2, x_2)$, $t_{2y} = (t_2(s, y))_y$, $x_{2y} = (x(t_2(s, y), s, y))_y$. Now, transversality relation (P_4) (first end point (s, y) fixed, second end point on a manifold) yields

$$\lambda_0[g_t\, dt_2 + g_x\, dx_2] + M(t_2)\, dt_2 - \lambda^*(t_2)\, dx_2 = 0,$$

where $M(t_2) = H(t_2, x_2, u(t_2, x_2), \lambda(t_2)) = \lambda^*(t_2)f(t_2, x_2, u(t_2, x_2))$; thus briefly

$$\lambda_0[g_t\, dt_2 + g_x\, dx_2] = \lambda^*(t_2)\, dx_2 - \lambda^*(t_2)f\, dt_2.$$

Noting that, as y describes a neighborhood of a given point, say $\bar{y}$, in R^n, then the point $(t_2(s, y), x(t_2(s, y); s, y))$ certainly moves in a neighborhood of the corresponding point in B, we have $dt_2 = t_{2y}(s, y)\, dy$, $dx_2 = x_{2y}\, dy$, dy arbitrary, and the relation above yields

(7.5.11) $$\lambda_0[g_t t_{2y} + g_x x_{2y}] = \lambda^*(t_2)x_{2y} - \lambda^*(t_2)f t_{2y},$$

where $t_{2y} = (t_2(s, y))_y$, $x_2 = x(t_2(s, y); s, y)$, $x_{2y} = (x(t_2(s, y); s, y))_y$, $f = f(t_2, x_2, u(t_2, x_2))$ with $u(t, x) = u_q$, and g_t, g_x are computed at (t_2, x_2).

From (7.5.6), (7.5.10), (7.5.11) we derive now

$$\begin{aligned}\lambda^*(t_2)x_{2y} &= \lambda_0[g_t t_{2y} + g_x x_{2y}] + \lambda^*(t_2)ft_{2y}\\ &= \lambda_0[g_t t_{2y} + g_x x_{2y}] + \lambda^*(t_2)x_{2y} - \lambda^*(t_2)x_y(t_2-),\end{aligned}$$

or

(7.5.12)
$$\lambda^*(t_2)x_y(t_2-) = \lambda_0 w_y(s, y).$$

Now, again from the relations stated in (7.5.iii) and the continuity of $M(t)$, we have

(7.5.13)
$$\begin{aligned}&\sum_j \lambda^*(\tau_j)[x_y(\tau_j-) - x_y(\tau_j+)]\\ &\quad = \sum_j \lambda^*(\tau_j)[f(\tau_j, x_j, u_{j+1}(\tau_j, x_j)) - f(\tau_j, x_j, u_j(\tau_j, x_j))]\tau_{jy}\\ &\quad = \sum_j [M(\tau_j+) - M(\tau_j-)]\tau_{jy} = 0.\end{aligned}$$

On the other hand, by (7.5.9), (7.5.12), (7.5.13) and noting that $x_y(s; s, y) = I$, the identity matrix, we have

$$\begin{aligned}\lambda^*(s) = \lambda^*(s)x_y(s) &= \sum_j \lambda^*(\tau_j)[x_y(\tau_j-) - x_y(\tau_j+)] + \lambda^*(t_2)x_y(t_2-)\\ &= \lambda^*(t_2)x_y(t_2-) = \lambda_0 w_y(s, y).\end{aligned}$$

Here $\lambda_0 \geq 0$ is a constant. We certainly have $\lambda_0 > 0$: otherwise we would have $\lambda(s) = 0$, a contradiction, since dg is assumed to be not identically zero. With the notation of (4.6.v), we have proved that $\lambda^*(t) = \lambda_0 w_y(t, x(t))$, or $\lambda_j(t) = \lambda_0 w_{x^j}(t, x(t))$, $j = 1, \ldots, n$.

(d) We proceed now as in (c) but differentiating with respect to s. First from (7.5.5) we derive

(7.5.14)
$$w_s(s, y) = g_t t_{2s}(s, y) + g_x(x(t_2(s, y); s, y))_s.$$

From (7.5.2) by multiplication by λ^*, and from $\lambda_t^* = -\lambda^* f_x$ by multiplication by x_s, we derive

$$\lambda^*(t)x_{ts} = \lambda^*(t)[f_x + f_u u_x]x_s, \qquad \lambda_t^*(t)x_s = -\lambda^*(t)f_x x_s,$$

And then, in each cell σ_j we have

$$\begin{aligned}(d/dt)(\lambda^*(t)x_s(t; s, y)) &= \lambda_t^*(t)x_s + \lambda^*(t)x_{ts}\\ &= -\lambda^*(t)f_x x_s + \lambda^*(t)[f_x + f_u u_x]x_s\\ &= (\lambda^*(t)f_u u_x)x_s = 0.\end{aligned}$$

By integration in each cell and addition as in (c), we obtain now, instead of (7.5.9),

(7.5.15)
$$\lambda^*(s)x_s(s) = \sum_j \lambda^*(\tau_j)[x_s(\tau_j-) - x_s(\tau_j+)] + \lambda^*(t_2)x_s(t_2-).$$

From (7.5.iii) we have $x_s(t_2-) = -f(t_2, x_2, u(t_2, x_2))t_{2s} + x_{2s}$, and hence

(7.5.16)
$$\lambda^*(t_2)x_s(t_2-) = \lambda^*(t_2)x_{2s} - \lambda^*(t_2)ft_{2s}.$$

The same transversality relation used in (c) with $dt_2 = t_{2s}(s, y)\,ds$, $dx_2 = x_{2s}ds$, ds arbitrary, yields now

(7.5.17)
$$\lambda_0[g_t t_{2s} + g_x x_{2s}] = \lambda^*(t_2)x_{2s} - \lambda^*(t_2)ft_{2s}.$$

From (7.5.14), (7.5.16), (7.5.17) we derive as before

$$\lambda^*(t_2)x_{2s} = \lambda_0[g_t t_{2s} + g_x x_{2s}] + \lambda^*(t_2) f t_{2s}$$
$$= \lambda_0[g_t t_{2s} + g_x x_{2s}] + \lambda^*(t_2)x_{2s} - \lambda^*(t_2)x_s(t_2-),$$

or

$$\lambda^*(t_2)x_s(t_2-) = \lambda_0 w_s(s, y). \tag{7.5.18}$$

Again, from (7.5.iii) and the continuity of $M(t)$ we have

$$\begin{aligned}\sum_j \lambda^*(\tau_j)[x_s(\tau_j-) - x_s(\tau_j+)]\\ &= \sum_j \lambda^*(\tau_j)[f(\tau_j, x_j, u_{j+1}(\tau_j, x_j)) - f(\tau_j, x_j, u_j(\tau_j, x_j))]\tau_{js}\\ &= \sum_j [M(\tau_j+) - M(\tau_j-)]\tau_{js} = 0.\end{aligned} \tag{7.5.19}$$

On the other hand, by (7.5.15), (7.16.18), (7.5.19), and noting that $x_s(s; s, y) = -f(s, y, u(s, y))$, we have

$$\begin{aligned}-\lambda^*(s)f(s, y, u(s, y)) &= \lambda^*(s)x_s(s; s, y)\\ &= \sum_j \lambda^*(\tau_j)[x_s(\tau_j-) - x_s(\tau_j+)] + \lambda^*(t_2)x_s(t_2-)\\ &= \lambda^*(t_2)x_s(t_2-) = \lambda_0 w_s(s, y).\end{aligned}$$

We have proved that $M(s) = -\lambda_0 w_s(s, y)$.

(e) In each cell σ_j, by using (P_2) and the relations $\lambda_i = \lambda_0 w_{y^i}$, $M = -\lambda_0 w_s$ with $\lambda_0 > 0$, we have

$$\begin{aligned}\min_{u\in U}\left[w_s(s, y) + \sum_i w_{y^i}(s, y) f_i(s, y, u)\right]\\ &= \lambda_0^{-1} \min_{u\in U}\left[-M(s) + \sum_i \lambda_i(s) f_i(s, y, u)\right]\\ &= \lambda_0^{-1}\left[-M(s) + \sum_i \lambda_i(s) f_i(s, y, u(s, y))\right] = 0.\end{aligned}$$

In other words, $w(t, x(t))$ is monotone nondecreasing (or $(d/dt)w(t, x(t)) \geq 0$) on each trajectory in σ_j, and $w(t, x(t))$ is constant (or $(d/dt)w(t, x(t)) = 0$) on each marked trajectory. Since w is continuous in V, w has the same properties in the whole of V. The optimality of all marked trajectories is now a consequence of (4.6.i). Statement (4.6.v) is thereby proved.

Bibliographical Notes

The first general proofs of the necessary condition were obtained by a group of mathematicians in Moscow under the leadership of L. S. Pontryagin: cf. V. G. Boltyanskii [2], L. S. Pontryagin [1], and L. S. Pontryagin, V. G. Boltyanskii, R. V. Gamkrelidze, and E. F. Mishchenko [I]. Some tools for these original proofs can be traced in earlier work of McShane [11] on problems of the calculus of variations. In Chapter IV we already mentioned that H. R. Hestenes had anticipated parts of the necessary condition. Many other proofs followed in a short time. Let us mention here those of H. Halkin [2, 4],

H. Halkin and L. W. Neustadt [1], M. R. Hestenes and E. J. McShane [1], E. O. Roxin [2], A. Strauss [I], H. Hermes and J. P. LaSalle [I]. Let us mention also the proofs of E. J. McShane [18] for generalized solutions in terms of linear functionals, and of R. V. Gamkrelidze [2] in terms of generalized solutions as chattering states (Section 1.14). Let us mention here also the formulations and proofs of A. I. Egorov [1] and L. I. Rozonoer [1, 2]. For abstract formulations of the necessary condition we refer to L. W. Neustadt [I, 2–7] also in connection to the problem of expressing the necessary condition in such a way to include the case in which the trajectory may have parts on the boundary of the domain. In this connection see also R. V. Gamkrelidze [2], J. Warga [I, 4, 6–8], and H. Halkin [5].

The proof in Section 7.3A–H may be compared with L. Cesari and R. F. Baum [1] and with the previous proof of E. J. McShane [18]. In the paper by L. Cesari and R. F. Baum [1] the necessary condition is also proved by the same process for control problems whose state variable has its values in C. The second proof in Section 7.4A–E may be compared with the original proof by L. S. Pontryagin, V. G. Boltyanskii, R. V. Gamkrelidze, and E. F. Mishchenko [I]. We have already presented Boltyanskii's sufficiency theory [2] in Section 4.6. Some of Boltyanskii's proofs have been given in detail in Section 7.5 above. Some abstract formulations of the necessary condition will be presented in Section 8.1. As we mentioned in Section 4.2, it may well occur that the necessary condition (4.2.i) gives no information on the optimal solution, or on certain arcs of it. These arcs are often called singular arcs. For linear problems it is possible to characterize when the optimal solution is unique, when it is determined by the necessary condition, and when the necessary condition is also sufficient for an optimal solution. We refer for this analysis to H. Hermes and J. P. LaSalle [I]. A great deal of work has been done to obtain further necessary conditions which may be of help in determining the singular arcs. Here we can only refer to work done by W. F. Powers and his school (W. F. Powers and J. P. McDanell [1–3]; W. F. Powers and E. R. Edge [1]; W. F. Powers, Bang Der Cheng, and E. R. Edge [1]). Also the reader is referred to the exposition D. J. Bell and D. H. Jacobson [I] and to the recent work of E. G. Gilbert and D. S. Bernstein [1] and of M. Kazemi [1]. In particular, the latter (cf. Section 7.3K(a)), starting with the process in L. Cesari and R. F. Baum [1], has proved necessary conditions for optimal solutions, singular or not, for ordinary and partial differential equations.

Problems where the cost functional has its values in R^n, or in a Hilbert or Banach space, have been discussed by many authors, as Pareto problems in economics or under other interpretations. Necessary conditions for an optimal solution of such problems have been proved by J. P. Aubin [1], P. L. Yu [1], and L. A. Zadeh [1], while C. Olech [5, 7, 8], N. O. Dacunha and E. Polak [1], and L. Cesari and M. B. Suryanarayana [4–7] have focused on the question of the existence of an optimal solution for such problems.

For multidimensional problems of optimal control, possibly monitored by partial differential equations or other functional relations we refer the reader to the extensive expositions of C. B. Morrey [I] and of J. L. Lions [I], with references to the vast literature on this subject.

For the case of partial differential equations in the Dieudonné–Rashevsky form, L. Cesari [34] has also discussed forms of necessary conditions for multidimensional problems. Furthermore, M. B. Suryanarayana has established necessary conditions for problems monitored by hyperbolic equations in [2], and by total partial differential equations in [5]. Finally, R. F. Baum [5] has determined necessary conditions for multidimensional problems with lower dimensional controls.

CHAPTER 8

The Implicit Function Theorem and the Elementary Closure Theorem

8.1 Remarks on Semicontinuous Functionals

As in Section 2.15 let us consider briefly an abstract space S of elements x, and let us assume that a concept σ of convergence of sequences x_k of elements of S has been defined, satisfying the two main axioms: (a) If $[x_k]$ converges to x in S, then any subsequence $[x_{k_s}]$ also converges to x; (b) Any sequence of repetitions $[x, x, \ldots, x, \ldots]$ must converge to x, where x is any element of S. Any such space is called a σ-limit space. In Section 2.15 we introduced the concepts of σ-lower and σ-upper semicontinuity of a functional $F: S \to$ reals. A functional which is both upper and lower semicontinuous is said to be continuous. Let us show here that, already at this level of generality, quite relevant theorems can be proved. To this effect, let us carry over the usual concepts. Thus, we say that a subset A of S is σ-closed if all elements of accumulation of A in S belong to A; that is, if $x_0 \in S$ is the σ-limit of elements x_k of A, then $x_0 \in A$. We say that a subset A of S is relatively sequentially σ-compact if every sequence $[x_k]$ of elements of A possesses a subsequence $[x_{k_s}]$ which is σ-convergent to an element x of S.

8.1.i. *If S is a σ-limit space, if $F: S \to R$ is lower semicontinuous on S, then for every real number a, the set $M_a = [x \in S \mid F(x) \leq a]$ is closed. If F is upper semicontinuous, then the sets $M'_a = [x \in S \mid F(x) \geq a]$ are closed.*

Indeed, if x_0 is a point of accumulation of M_a, then there is a sequence x_k of elements $x_k \in M_a$ with $F(x_k) \leq a$, $x_k \to x_0$, and then $F(x_0) \leq \liminf_k F(x_k) \leq a$. The same proof works for upper semicontinuity. In the usual terminology this theorem can be reworded by saying that lower and upper semicontinuous functionals are B-measurable. The same statement (8.1.i) holds for functionals F defined on a σ-closed subset A of S where the M_a are the corresponding subsets of A.

8.1.ii. *Let S be a σ-limit space, let A be a nonempty σ-closed and relatively sequentially σ-compact subset of S, and assume that F is lower semicontinuous at every $x_0 \in A$ with respect to σ-convergence. Then F is bounded below in A and has an absolute minimum in A.*

Analogously, if F is upper semicontinuous, then F is bounded above and has an absolute maximum in A.

Proof. Let $m = \inf_A F(x)$, $-\infty \leq m < +\infty$, and take any sequence $[x_k]$ of elements of A such that $F(x_k) \to m$ as $k \to \infty$. We may well assume that $F(x_k) \leq m + 1/k$ if m is finite, and $F(x_k) \leq -k$ if $m = -\infty$. First, A is σ-compact; hence there is a subsequence $[x_{k_s}]$ which is σ-convergent to an element x_0 of S. Since A is σ-closed, then x_0 belongs to A, so $F(x_0)$ is defined and is a real number. Then, by lower semicontinuity we have $-\infty < F(x_0) \leq \liminf_{k\to\infty} F(x_k) = m < +\infty$. Thus, $F(x_0)$ is finite, and so is m. Since $x_0 \in A$, also $F(x_0) \geq m$. By comparison, we have $F(x_0) = m$, and the existence of the minimum for F on A is proved. An analogous proof holds for upper semicontinuous functionals and maxima. □

As a consequence of (8.1.ii) we derive that any continuous functional on a σ-closed and relatively sequentially σ-compact set A has both an absolute minimum and an absolute maximum.

Statement (8.1.ii) holds even under weaker hypotheses. Indeed, we could assume that F may take on A the value $+\infty$, with $F(x)$ not identically $+\infty$ on A, and $F(x) > -\infty$ for all $x \in A$. Moreover, we could assume that A is nonempty and σ-closed, and only that the sets $M_a = [x \in A \,|\, F(x) \leq a]$, if not empty, are relatively sequentially σ-compact. Some authors call such sets A "inf-compact", but we shall not need this terminology. Analogous remarks hold for upper semicontinuous functionals.

The following particularization of the above concepts and statements is important.

8.1.iii. *Let S be a real reflexive Banach space of elements x with norm $\|x\|$, and take in S for σ-convergence the weak convergence in S. Let A be a nonempty closed convex subset of S. Let $F: A \to R$ be a functional which is lower semicontinuous in A with respect to weak convergence, and such that $F(x) \to +\infty$ as $\|x\| \to +\infty$, $x \in A$. Then, F is bounded below in A and has an absolute minimum in A.*

Proof. First we note that the convex set A is closed in the weak as well as in the strong topology in S. Indeed, by the Banach–Saks–Mazur theorem (cf. Section 10.1), weak and strong closures of a convex set in any Banach space coincide. Let $i = \inf[F(x) \,|\, x \in A]$, $-\infty \leq i < +\infty$, and let N be any real number $N > i$. Then, the set $A_N = [x \in A \,|\, F(x) \leq N]$ is nonempty and bounded, since $F(x) \to +\infty$ as $\|x\| \to +\infty$ in A. From functional analysis we know that A_N, as a bounded subset of a reflexive Banach space, is sequentially compact with respect to weak convergence. Thus, for any minimizing sequence x_k, that is, $F(x_k) \to i$ as $k \to +\infty$, $x_k \in A$, we certainly have $x_k \in A_N \subset A$ for all k sufficiently large, and we can choose a subsequence, say still $[k]$, such that $x_k \to x_0 \in A$ in the weak convergence of S. By the lower semicontinuity in A we have $-\infty < F(x_0) \leq \liminf_k F(x_k) = i < N$. Hence, $x_0 \in A_N$. As in the proof of (8.1.ii), we also have $F(x_0) \geq i$, and finally $F(x_0) = i$.

A functional $F: A \to R$ on a convex set A of a linear space S is said to be convex in A provided $x_1, x_2 \in A$, $0 \leq \alpha \leq 1$, implies $F((1-\alpha)x_1 + \alpha x_2) \leq (1-\alpha)F(x_1) + \alpha F(x_2)$. The same functional is said to be strictly convex in A provided F is convex in A and strict inequality holds above for all $0 < \alpha < 1$.

8.1.iv. *Under the conditions of* (8.1.iii), *if* $F: A \to$ *reals is strictly convex on* A, *then the element* $x \in A$ *at which* $F(x) = i$ *(equivalently,* $F(x) \leq F(y)$ *for all* $y \in A$*) is unique.*

Proof. If there were two such elements x_1, x_2 with $F(x_1) = F(x_2) = i$, $x_1, x_2 \in A$, then $F(2^{-1}(x_1 + x_2)) < 2^{-1}F(x_1) + 2^{-1}F(x_2) = i$, a contradiction. □

Again, let A be a convex subset of the real Banach space S. A functional $F: A \to R$ is said to have a Gateau derivative $F'_x h$ at a point $x \in A$ provided the limit

$$\lim_{\alpha \to 0+} \alpha^{-1}[F(x + \alpha h) - F(x)] = F'_x h$$

exists for every h such that $x + h \in A$. The same functional F is said to be differentiable in A if $F'_x h$ exists for all $x, x + h \in A$ (and $F'_x(h) = -F'(-h)$ at every x in the interior of A).

A stronger concept is often used. Let S^* denote the dual space of S, that is, the space of all linear continuous operators $z: S \to R$ on S. Then the same functional $F: A \to R$ above is said to have a Fréchet derivative at a point $x \in A$ provided there is an element F'_x of S^*, and for every $\varepsilon > 0$ a number $\delta = \delta(\varepsilon, x) > 0$ such that $|F(x+h) - F(x) - F'_x h| \leq \varepsilon \|h\|$ for all $\|h\| \leq \delta$, $x + h \in A$.

8.1.v. *If* $F: A \to R$ *is a convex differentiable functional on a convex subset* A *of a Banach space* S, *and if* $x \in A$ *is any point of* A *where* $F(x) \leq F(y)$ *for all* $y \in A$, *then we have also*

$$F'_x(y - x) \geq 0 \quad \text{for all } y \in A. \tag{8.1.1}$$

Conversely, if this relation holds at some $x \in A$, *then* $F(x) \leq F(y)$ *for all* $y \in A$.

Proof. If $x \in A$ is such that $F(x) \leq F(y)$ for all $y \in A$, then for $y = x + h \in A$, the entire segment $x + \alpha h$, $0 \leq \alpha \leq 1$, lies in the convex set A, and $F(x) \leq F(x + \alpha h)$. For $0 < \alpha \leq 1$, we also have

$$\alpha^{-1}(F(x + \alpha h) - F(x)) \geq 0,$$

and by taking the limit as $\alpha \to 0+$ we derive $F'_x h \geq 0$, or $F'_x(y - x) \geq 0$ for all $y \in A$. Conversely, if $x \in A$ and (8.1.1) holds for all $y \in A$, then by convexity

$$F((1 - \alpha)x + \alpha y) \leq (1 - \alpha)F(x) + \alpha F(y), \qquad 0 \leq \alpha \leq 1,$$

or

$$F(y) - F(x) \geq \alpha^{-1}(F(x + \alpha(y - x)) - F(x)), \qquad 0 < \alpha < 1.$$

Since the limit $F'_x(y - x)$ exists by hypothesis and is nonnegative, we derive, as $\alpha \to 0+$, that

$$F(y) - F(x) \geq F'_x(y - x) \geq 0,$$

and this holds for all $y \in A$. □

The relation (8.1.1) is called a "variational inequality", and, as we have proved under the assumed hypotheses, (8.1.1) is a characteristic property of the elements $x \in A$ at which F has its minimum value. If $A = S$, then we can take in (8.1.1) $y = x \pm h$, $h \in S$, and (8.1.1) yields $F'_x h = 0$ for all $h \in S$. This is the abstract form of Theorem (2.3.ii). Some authors refer to the relation $F'_x h = 0$ for all h as the "Euler equation" for F.

As a further particularization of the above considerations, let us assume that $\pi(x, y)$ is a given symmetric continuous bilinear form on S, that is, $\pi: S \times S \to R$, $\pi(x, y) = \pi(y, x)$, and $\pi(x, y)$ is linear and continuous in x for every y, and in y for every x. Let $L: S \to R$ be a given linear continuous functional on S.

If $\pi(x, y)$ is coercive, that is, $\pi(x, x) \geq c\|x\|^2$ for all $x \in S$ and a constant $c > 0$, then the functional

$$F(x) = \pi(x, x) - 2L(x), \qquad x \in S,$$

has all the properties we have requested in (8.1.iii) and (8.1.iv). Indeed, $|L(x)| \leq M\|x\|$ for some constant M, and $F(x) \geq c\|x\|^2 - 2M\|x\|$. Then, $F(x) \to +\infty$ as $\|x\| \to +\infty$. Moreover, F is convex and strictly convex in S. Indeed, for $0 \leq \alpha \leq 1$ and $x_1 \neq x_2$, we have

$$\begin{aligned} F((1-\alpha)x_1 + \alpha x_2) &= F(x_1 + \alpha(x_2 - x_1)) \\ &= \pi(x_1 + \alpha(x_2 - x_1), x_1 + \alpha(x_2 - x_1)) - 2L(x_1 + \alpha(x_2 - x_1)) \\ &= \pi(x_1, x_1) + 2\alpha\pi(x_1, x_2 - x_1) + \alpha^2\pi(x_2 - x_1, x_2 - x_1) \\ &\quad - 2L(x_1) - 2\alpha L(x_2 - x_1) = m + n\alpha + p\alpha^2 = P(\alpha). \end{aligned}$$

Here $P(0) = F(x_1)$, $P(1) = F(x_2)$, and since $p = \pi(x_2 - x_1, x_2 - x_1) > 0$, the polynomial P is strictly convex in α for $0 \leq \alpha \leq 1$, so that

$$F((1 - \alpha)x_1 + \alpha x_2) = P(\alpha) < (1 - \alpha)P(0) + \alpha P(1) = (1 - \alpha)F(x_1) + \alpha F(x_2)$$

for all $0 < \alpha < 1$. We have proved that F is convex and strictly convex in S. It is easily seen that the Gateau derivative of F is $F'_x h = 2[\pi(x, h) - L(h)]$. As a corollary of (8.1.iii–v), we have

8.1.vi. *For π bilinear, symmetric, continuous, and coercive, and L linear and continuous, then $F(x) = \pi(x, x) - 2L(x)$ has an absolute minimum in every convex closed subset A of S. The unique point $x \in A$ for which $F(x) \leq F(y)$ for all $y \in A$ is characterized by the inequality $\pi(x, y - x) \geq L(y - x)$ for all $y \in A$.*

If $A = S$, then the equality $F'_x h = 0$ reduces to $\pi(x, y - x) = L(y - x)$ for all $y \in S$.

8.2 The Implicit Function Theorem

A. An Abstract Form of the Implicit Function Theorem

Given any two metric spaces X, Y and any single valued function $f: X \to Y$, we denote by Y_0 the image of f, or $Y_0 = [y \in Y \,|\, y = f(x), x \in X]$, and for every subset F of Y we denote by $f^{-1}F$ the set $f^{-1}F = [x \in X \,|\, f(x) \in F]$. If f is continuous, then for every closed set F in Y, $f^{-1}F$ is also closed; if G is open in Y, then $f^{-1}G$ is also open. Thus, if f is continuous, then any point y of Y_0 has a counterimage $f^{-1}y$ which is a closed subset of X.

Thus, f^{-1} is a set valued function, and we shall see in Section 8.5 that f^{-1} is upper semicontinuous if f is continuous.

A single valued function $\varphi: Y_0 \to X$ such that $\varphi(y) \in f^{-1}y$ for every $y \in Y_0$ is called a *partial inverse* of f, and for any such φ we have $f[\varphi(y)] = y$ for all $y \in Y_0$, that is, $f\varphi$ is the identity on Y_0. (In the terminology of Section 8.3 φ is a "selection" of the set valued function f^{-1}). Here we discuss the question whether, for any continuous single valued $f: X \to Y$, there is a B-measurable single valued partial inverse $\varphi: Y_0 \to X$. Here we show that the answer is affirmative under some assumptions on X.

8.2.i (A Partial Inverse Theorem). *Given any two metric spaces X, Y, where X is the countable union of compact subspaces of X, let $f: X \to Y$ be any continuous mapping, and let $Y_0 = f(X)$ be the image of X in Y. Then there is a B-measurable map $\varphi: Y_0 \to X$ such that $f\varphi$ is the identity map on Y_0, that is, $f[\varphi(y)] = y$ for every $y \in Y_0$.*

Proof.

(a) Let us suppose that X is replaced by a closed set $L \subset [0, +\infty)$. In this situation, let $T(y) = \inf f^{-1}(y)$ for $y \in Y_0$. Here $f^{-1}(y)$ is a nonempty set of real nonnegative numbers, and the operator inf applies. Actually, f is a

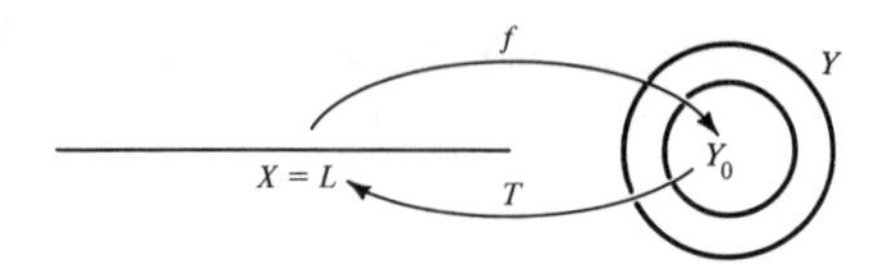

continuous map, hence $f^{-1}(y)$ is a closed nonempty subset of $[0, +\infty)$, and hence $f^{-1}(y)$ has a minimum $T(y) = \min f^{-1}(y)$, whence $T(y) \in f^{-1}(y)$, and $f[T(y)] = y$ for every $y \in Y_0$. Let us prove that $T: Y_0 \to L$ is a lower semicontinuous (real single valued) function. Suppose this is not the case. Then, there is a point $y_0 \in Y_0$, a number $\varepsilon > 0$, and a sequence $[y_k]$ such that $y_k \in Y_0$, $y_k \to y_0$, $T(y_k) \leq T(y_0) - \varepsilon$, or $0 \leq x_k \leq x_0 - \varepsilon$, with $x_k = T(y_k)$, $f(x_k) = y_k$, $x_0 = T(y_0)$, $f(x_0) = y_0$. Here $[x_k]$ is a bounded sequence of real numbers; hence there is a subsequence, say $[x_{k_s}]$, with $x_{k_s} \to \bar{x}$ and $0 \leq \bar{x} \leq x_0 - \varepsilon$. Here $x_{k_s} = T(y_{k_s}) \in L$, and thus $\bar{x} \in L$, since L is closed. Also, $f(x_{k_s}) = f(T(y_{k_s})) = y_{k_s} \to y_0$, $x_{k_s} \to \bar{x}$, hence $f(\bar{x}) = y_0$, since f is continuous on L. Thus, $\bar{x} \in f^{-1}(y_0) \leq x_0 - \varepsilon$, a contradiction, since $x_0 = \min f^{-1}(y_0)$. We have proved that $T: Y_0 \to L$ is lower semicontinuous in Y_0, and hence B-measurable because of (8.1.i), and $f(T(y)) = y$ for every $y \in Y_0$. Theorem (8.2.i) is proved for X replaced by any closed subset L of $[0, +\infty)$.

(b) Let us consider now the general case. By general topology (Kelley [1, Theorem 3.28]) we know that any compact metric space X is the continuous image $\Lambda: K \to X$ of some closed subset K of $[0, 1]$, $\Lambda K = X$. (Actually, under some restriction on the compact set X—namely connectedness and local connectedness—we even know that X is the continuous image of an interval, say $[0, 1]$, or any interval, and if X is, say, a square or a cube, we say that Λ is the Peano curve filling a square or a cube. The restrictions on the compact set X are only that X must be connected and locally connected. But we shall not need these particularizations.) In our case $X = \bigcup X_\alpha$, where X_α, $\alpha = 1, 2, \ldots$, is a sequence of compact subsets of X, and we can think of each X_α as being the continuous image $l_\alpha: L_\alpha \to X_\alpha$, $\alpha = 1, 2, \ldots$, of some closed subset of $[0, 1]$, say $L_\alpha \subset [\varepsilon, 1 - \varepsilon]$ for some $\varepsilon > 0$. Let us denote by L the set which coincides with $L'_\alpha = \alpha + L_\alpha$ in $[\alpha, \alpha + 1]$, that is, the displacement L'_α of L_α in $[\alpha, \alpha + 1]$. Then L is a closed subset of $[0, +\infty)$, and we shall denote by $l: L \to X$ the map which coincides with l_α on L'_α. Then l is a continuous map of L onto X. We have the situation shown in the picture, and, by (a), there is a B-measurable map $T: Y_0 \to L$ such that $(fl)(T(y)) = y$ for every $y \in Y_0$, or $(flT)(y) = y$, since fl is a continuous map, and $Y_0 = f(X) = f(l(L)) = (fl)(L)$. If we take $\varphi = lT$, $\varphi: Y_0 \to X$, we have $f(\varphi(y)) = y$

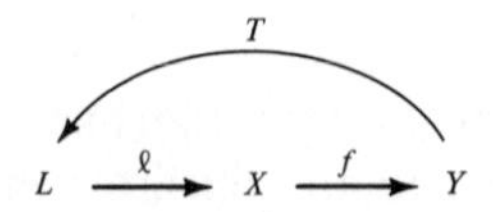

for every $y \in Y_0$. Obviously, φ is a B-measurable map, as it is the superposition of a continuous map l on the B-measurable map T. Theorem (8.2.i) is now completely proved. Note that T is a lower semicontinuous function. □

We just mention here the general concept of measure space $(X, \mathscr{A}, m)$, that is, a space X, a σ-ring $\mathscr{A}$ of subsets A of X, and a real valued function $m: \mathscr{A} \to R$ with the following properties: (a) $\bigcup A = X$, and m is a measure, that is, (b) $m(\emptyset) = 0$, where $\emptyset$ is the empty set; (c) $m(A) \geq 0$ for all $A \in \mathscr{A}$; (d) $E_i \in \mathscr{A}$, $i = 1, 2, \ldots$, $E_i \cap E_j = \emptyset$ for all $i \neq j$, implies $m(\bigcup_{i=1}^{\infty} E_i) = \sum_{i=1}^{\infty} m(E_i)$. Then, a real valued function $f(x)$, $x \in X$, is said to be measurable with respect to the measure space $(X, \mathscr{A}, m)$ if for every real a the set $[x \in X \mid f(x) < a]$ is in the σ-ring $\mathscr{A}$. For a vector valued function $f(x) = (f_1, \ldots, f_n)$ we say that f is measurable with respect to $(X, \mathscr{A}, m)$ if each component f_i of f is measurable. The most common example of a measure space is that $X = R$, m is the Lebesgue measure, and $\mathscr{A}$ is the collection of all Lebesgue measurable subsets of R.

8.2.ii (AN ABSTRACT FORM OF THE IMPLICIT FUNCTION THEOREM). *Let S be a measure space, let X and Y be metric spaces where X is the countable union of compact subspaces, let $f: X \to Y$ be a continuous mapping, let $Y_0 = f(X)$, and let $\sigma: S \to Y$ be a measurable map such that $\sigma(S) \subset f(X) = Y_0$. Then there is a measurable map $\psi: S \to X$ such that $f(\psi(t)) = \sigma(t)$ for all $t \in S$.*

Proof. Indeed, by (8.2.i) there is a B-measurable map $\varphi: Y_0 \to X$ such that $f[\varphi(y)] = y$ for every $y \in Y_0$. Then, the map $\psi = \varphi\sigma: S \to X$ is measurable, $f\psi: S \to Y$, and $f\psi = f(\varphi\sigma) = (f\varphi)\sigma = \sigma$ on S. □

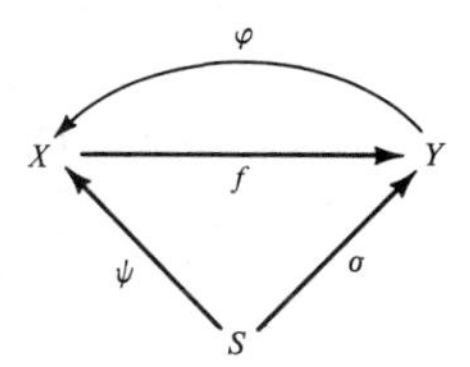

Remark. The same theorem (8.2.ii) holds also for any topological space X which is the countable union of compact metrizable subsets, and any Hausdorff space Y. The theorem was proved by E. J. McShane and R. B. Warfield [1].

B. Orientor Fields and the Implicit Function Theorem

We assume that a set A is given in the tx-space R^{1+n}, $x = (x^1, \ldots, x^n)$, and that for every $(t, x) \in A$ a nonempty set $Q(t, x)$ of points $z = (z^1, \ldots, z^n)$ of the z-space R^n is assigned, or $Q(t,x) \subset R^n$, and this set may depend on (t,x). As mentioned in Section 1.2, we refer to the relation

$$dx/dt \in Q(t,x) \tag{8.2.1}$$

as an *orientor field* (or a differential equation with multivalued second member, or a contingent equation, or a differential inclusion). A solution $x(t)$. $t_1 \leq t \leq t_2$, of (8.2.1) is any vector valued function $x(t) = (x^1, \ldots, x^n)$ such that (a) $x(t)$ absolutely continuous (AC) in $[t_1, t_2]$; (b) $(t, x(t)) \in A$ for all $t \in [t_1, t_2]$; (c) $dx/dt \in Q(t, x(t))$ a.e. in $[t_1, t_2]$. Thus, for almost all $t \in [t_1, t_2]$ the direction $dx/dt = (x'^1, \ldots, x'^n)$ of the curve $x = x(t)$ at $(t, x(t))$ is one of the "allowable directions" $z \in Q(t, x(t))$ assigned at $(t, x(t))$.

An orientor field will be said to be *autonomous* if $Q(t,x)$ depends on x only and not on t. Nevertheless, every orientor field can be written as an autonomous one by a change of coordinates. Indeed, if we add the vector variable x^0 satisfying the differential equation $dx^0/dt = 1$ and initial condition $x^0(t_1) = t_1$, and if we then use the $(n+1)$-vector $\tilde{x} = (x^0, x^1, \ldots, x^n)$ and direction set $\tilde{Q}(\tilde{x}) = [\tilde{z} = (z^0, z^1, \ldots, z^n) = (z^0, z),\ z \in Q(x^0, x),\ z^0 = 1]$, then the system (8.2.1) becomes $d\tilde{x}/dt \in \tilde{Q}(\tilde{x})$. We may use this remark in proofs in order to simplify notations.

We return now to the notation of Section 1.12, where we have seen that, if an AC vector function $x(t) = (x^1, \ldots, x^n)$, $t_1 \leq t \leq t_2$, is a solution of the differential system $x'(t) = f(t, x(t), u(t))$, $t_1 \leq t \leq t_2$, for some $u(t)$ measurable, $u(t) \in U(t, x(t))$, then it can always be written in the form of an AC solution of the orientor field $x'(t) \in Q(t, x(t))$ where $Q(t,x) = f(t, x, U(t,x))$. We are in a position to show that the converse is also true.

8.2.iii (AN IMPLICIT FUNCTION THEOREM FOR ORIENTOR FIELDS). *If A is a closed subset of the tx-space R^{1+n}, if $U(t,x)$ is a subset of R^m for every $(t,x) \in A$; if the set M of all $(t,x,u) \in R^{1+n+m}$ with $(t,x) \in A$, $u \in U(t,x)$, is closed; if $f(t,x,u) = (f_1, \ldots, f_n)$ is continuous on M and $Q(t,x)$ denotes the set $Q(t,x) = f(t,x,U(t,x))$ in R^n; and if $x(t)$, $t_1 \leq t \leq t_2$, is an* AC *vector function such that $(t, x(t)) \in A$ for all $t \in [t_1, t_2]$ and $x'(t) \in Q(t, x(t))$ for almost all $t \in [t_1, t_2]$, then there is a measurable function $u(t)$, $t_1 \leq t \leq t_2$, such that $u(t) \in U(t, x(t))$ and $x'(t) = f(t, x(t), u(t))$ for almost all $t \in [t_1, t_2]$.*

Note that A is the projection of M on the tx-space, and that for every $(\bar{t}, \bar{x}) \in A$ the set $U(\bar{t}, \bar{x})$ is the projection on the u-space of the intersection of M with the subspace $[t = \bar{t},\ x = \bar{x}]$ in R^{1+n+m}. Thus, the assumption that M is closed certainly implies that $U(\bar{t}, \bar{x})$ is a closed subset of R^m. Thus, for every $(t,x) \in A$, the set $U(t,x)$ is necessarily closed. Moreover, if A_0 is the projection of the sets A and M on the t-axis, then $A_0 \supset [t_1, t_2]$.

Proof of (8.2.iii). As usual we denote by M the set of all (t,x,u) with $(t,x) \in A$ and $u \in U(t,x)$, hence $M \subset R^{1+n+m}$. Also, we denote by N the set of all (t,x,z) with $(t,x) \in A$, $z = f(t,x,u)$, $u \in U(t,x)$; hence $N \subset R^{1+2n}$. Let $F: M \to N$ denote the continuous map defined by $(t,x,u) \to (t,x,z)$ with $z = f(t,x,u)$. Here

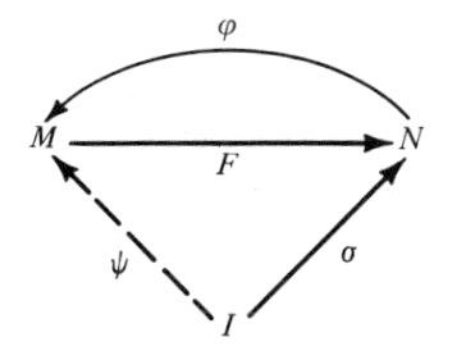

M and N are metric spaces since $M \subset R^{1+n+m}$, $N \subset R^{1+2n}$; also, M is closed by hypothesis, and M is the countable union of compact subsets, say $M_\alpha = [(t,x,u) \in M \,|\, |t| + |x| + |u| \leq \alpha]$, $\alpha = 1, 2, \ldots$. Finally, $N = F(M)$. By (8.1.ii) there is a B-measurable map $\varphi: N \to M$ such that $F\varphi$ is the identity map, that is, $F\varphi(t,x,z) = (t,x,z)$ for every $(t,x,z) \in N$. Now let us consider the map $\sigma: I \to N$ on $I = [t_1, t_2]$, defined by $t \to (t, x(t), x'(t))$, where $x(t)$, $t_1 \leq t \leq t_2$, is an AC solution of the orientor field $x' \in Q(t,x) = f(t,x,U(t,x))$. Then $\psi = \varphi\sigma: I \to M$, and ψ maps t into some $(t, x(t), u(t)) \in M$, with $F(t, x(t), u(t)) = (t, x(t), x'(t))$, or $x'(t) = f(t, x(t), u(t))$. Actually, σ is defined not on all of I, but in the subset I_0 of I where $x'(t)$ exists, and meas I_0 = meas $I = t_2 - t_1$, and thus the concluding relation holds in I_0, that is, $x'(t) = f(t, x(t), u(t))$ a.e. in $[t_1, t_2]$. On the other hand, $x'(t)$ is measurable, that is, σ is measurable, φ is B-measurable, and hence $u(t)$ is measurable. The implicit function theorem (8.2.iii) is thereby proved. □

C. Exercises

The following statements are often used. In this book we shall apply them to the case $\nu = 1$, $G = (a,b) \subset R$, x AC in $[a,b]$, and $\xi(t) = x'(t)$, $t \in [a,b]$ (a.e.), $r = n$.

1. Prove the following more general form of (8.2.iii): Let A be a closed subset of the tx-space $R^{\nu+n}$; for every $(t,x) \in A$ let $U(t,x)$ be any subset of R^m, assume that the set $M = [(t,x,u) \,|\, (t,x) \in A,\, u \in U(t,x)]$ be closed in $R^{\nu+n+m}$, let $f(t,x,u) = (f_1, \ldots, f_r)$ be a continuous function on M, and for every $(t,x) \in A$ let $Q(t,x) = f(t,x,U(t,x)) = [z \,|\, z = f(t,x,u),\, u \in U(t,x)] \subset R^r$. Let G be a measurable subset of R^ν, and $x(t) = (x^1, \ldots, x^n)$, $\xi(t) = (\xi^1, \ldots, \xi^r)$, $t \in G$, be measurable functions on G such that $(t, x(t)) \in A$, $\xi(t) \in Q(t, x(t))$, $t \in G$ (a.e.). Then there is a measurable function $u(t) = (u^1, \ldots, u^m)$, $t \in G$, such that $u(t) \in U(t, x(t))$, $\xi(t) = f(t, x(t), u(t))$, $t \in G$ (a.e.).
2. Let A be a closed subset of the tx-space $R^{\nu+n}$. For every $(t,x) \in A$ let $U(t,x)$ be any subset of R^m. Assume that the set $M = [(t,x,u) \,|\, (t,x) \in A,\, u \in U(t,x)]$ is closed in $R^{\nu+n+m}$. Let $f_0(t,x,u)$ and $f(t,x,u) = (f_1, \ldots, f_r)$ be continuous functions on M, and for every $(t,x) \in A$, let $\tilde{Q}(t,x) = [(z^0, z) \,|\, z^0 \geq f_0(t,x,u),\, z = f(t,x,u),\, u \in U(t,x)] \subset R^{r+1}$. Let G be a measurable subset of the t-space R^ν, and let $\eta(t)$, $\xi(t) = (\xi_1, \ldots, \xi_r)$, $x(t) = (x^1, \ldots, x^n)$, $t \in G$, be measurable functions such that $(t, x(t)) \in A$, $(\eta(t), \xi(t)) \in \tilde{Q}(t, x(t))$, $t \in G$ (a.e.). Then there is a measurable function $u(t) = (u^1, \ldots, u^m)$, $t \in G$, such that $u(t) \in U(t, x(t))$, $\eta(t) \geq f_0(t, x(t), u(t))$, $\xi(t) = f(t, x(t), u(t))$, $t \in G$ (a.e.).
3. Let M be a compact subset of the xu-space R^{n+m}, let $f_0(x,u)$, $f(x,u) = (f_1, \ldots, f_r)$ be continuous functions on M, and for every $(x,z) \in R^{n+r}$ let

$$T(x,z) = \inf[z^0 \,|\, z^0 \geq f_0(x,u),\, z = f(x,u),\, (x,u) \in M], \qquad -\infty \leq T(x,z) \leq +\infty.$$

Let $R(x,z) = [u \mid z = f(x,u), (x,u) \in M]$. Then (a) $T(x,z) = +\infty$ if $R(x,z)$ is empty; (b) $T(x,z) = \min[z^0 \mid z^0 \geq f_0(x,u), (x,u) \in R(x,z)]$ if $R(x,z)$ is not empty; (c) $T(x,z)$ is lower semicontinuous in R^{n+r}. The same statements hold even if f is continuous and f_0 is only lower semicontinuous on M.

A further extension is as follows: Let M be a compact subset of the xu-space R^{n+m}, let $f_{01}(x,u), \ldots, f_{0\alpha}(x,u)$ be lower semicontinuous functions on M, let $f_1(x,u), \ldots, f_r(x,u)$ be continuous functions on M, and for every $(x,z) \in R^{n+r}$ and $i = 1, \ldots, \alpha$, let $T_i(x,z) = \inf[\mathfrak{z}^i \mid (\mathfrak{z}^1, \ldots, \mathfrak{z}^\alpha, z^1, \ldots, z^r),\ \mathfrak{z}^j \geq f_{0j}(x,u),\ j = 1, \ldots, \alpha,$ $z^s = f_s(x,u), s = 1, \ldots, r, (x,u) \in M]$. Then the extended functions $T_i(x,z), i = 1, \ldots, \alpha$, are lower semicontinuous in R^{n+r}.

4. Let M be a closed subset of the xu-space R^{n+m}, and let $f_0(x,u)$, $f(x,u) = (f_1, \ldots, f_r)$ and $T(x,z)$ be as in Exercise 3. Then $T(x,z)$ is B-measurable in R^{n+r}.
5. The same as Exercise 1, except that now G is a measurable subset of R^ν with finite measure, and on A, M, f we make the following assumption: For every $\varepsilon > 0$ there is a compact subset K of G such that $\text{meas}(G - K) < \varepsilon$, the set $A_K = [(t,x) \in A \mid t \in K]$ is closed, the set $M_K = [(t,x,u) \in M \mid t \in K]$ is closed, and the function f is continuous on M_K. The conclusion is the same. Hint: For every $\varepsilon = k^{-1}$, $k = 1, 2, \ldots$, there is a compact set K_k as above, and by Exercise 1 there is a measurable function $u_k(t)$, $t \in K_k$, with $u_k(t) \in U(t, x(t))$, $\xi(t) = f(t, x(t), u_k(t))$ for $t \in K_k$ (a.e.). Now take $u(t) = u_k(t)$ for $t \in K_k - (K_1 \cup \cdots \cup K_{k-1})$, $k = 1, 2, \ldots$.
6. The same as Exercise 2, except that now G is a measurable subset of R^ν with finite measure, and on A, M, f_0, f we make only the following assumption: For every $\varepsilon > 0$ there is a compact subset K of G such that $\text{meas}(G - K) < \varepsilon$, the set $A_K = [(t,x) \in A \mid t \in K]$ is closed, the set $M_K = [(t,x,u) \in M \mid t \in K]$ is closed, and the functions f_0, f are continuous on M_K. The conclusion is the same.
7. The same as Exercises 2 and 6, except that now G is a measurable subset of R^ν with finite measure, and we have defined $T(t,x,z)$, $(t,x,z) \in R^{\nu+n+m}$, by taking

$$T(t,x,z) = \inf[z^0 \mid z^0 \geq f_0(t,x,u),\ z = f(t,x,u),\ u \in U(t,x)]$$
$$= \inf[z^0 \mid (z^0, z) \in \tilde{Q}(t,x)].$$

Let $\xi(t) = (\xi^1, \ldots, \xi^r)$, $x(t) = (x^1, \ldots, x^n)$, $t \in G$, be measurable functions such that $(t, x(t)) \in A$, $T(t, x(t), \xi(t)) \in \tilde{Q}(t, x(t))$, $t \in G$ (a.e.). Prove that (a) $T(t, x(t), \xi(t))$ is measurable in G; and (b) there is a measurable function $u(t) = (u^1, \ldots, u^m)$, $t \in G$, such that $u(t) \in U(t, x(t))$, $\xi(t) = f(t, x(t), u(t))$, $T(t, x(t), \xi(t)) = f_0(t, x(t), u(t))$, $t \in G$ (a.e.).

8.3 Selection Theorems

A. A General Selection Theorem

The question we treat here is most relevant and, as we shall see, will allow a different interpretation and a generalization of what we have discussed in Section 8.2.

Let X, Y be two arbitrary sets or spaces. Let us consider a set valued map F, or $x \to F(x)$, $x \in X$, $F(x) \subset Y$, mapping any element x of X into a subset $F(x)$ of Y. Sometimes F is also called a multifunction from X to Y. Alternatively, we may think of $F(x)$ as a variable subset of Y depending on the point, or parameter, x in X.

We say that a single valued map $f: X \to Y$, or $x \to f(x)$, is a *selector* of F provided $f(x) \in F(x)$ for every $x \in X$. Under some assumptions on F, X and Y, we will be able to prove the existence of selectors f having relevant properties.

Now let X be any given set, Y be a metric space, and S be a countably additive family of subsets of X, that is, such that, if $A_n \in S$ for $n = 1, 2, \ldots$, then $\bigcup_{n=1}^{\infty} A_n \in S$. Let $\rho(p, q)$ denote the distance function in Y.

8.3.i (Lemma). *If $f_n: X \to Y$, $n = 1, 2, \ldots$, is a sequence of maps converging uniformly to a map $f: X \to Y$, and such that $f_n^{-1}(G) \in S$, $n = 1, 2, \ldots$, for every open subset G of Y, then $f^{-1}(G) \in S$ for every open subset G of Y.*

Proof. For any open set G of Y let us consider the closed set $Y - G$, and the closed set $(Y - G)_n$ of all $y \in Y$ at a distance $\leq 1/n$ from $Y - G$. For every n let m_n be an integer such that $m_1 < m_2 < \cdots$, $m_n \to \infty$, and $\rho(f_{m_n}(x), f(x)) \leq 1/n$ for all $x \in X$. Let us prove the formula

$$f^{-1}(G) = \bigcup_{n=1}^{\infty} f_{m_n}^{-1}(Y - (Y - G)_n). \tag{8.3.1}$$

Indeed, if $\bar{x} \in f^{-1}(G)$, then $\bar{y} = f(\bar{x}) \in G$ and there is $\delta > 0$ such that $N_\delta(\bar{y}) \subset G$, that is, the entire open neighborhood $N_\delta(\bar{y})$ of $\bar{y}$ is contained in G, since G is open. Now for any n with $1/n < \delta/2$, or $n > 2/\delta$, we have

$$\rho(f_{m_n}(\bar{x}), \bar{y}) \leq \rho(f_{m_n}(\bar{x}), f(\bar{x})) + \rho(f(\bar{x}), \bar{y}) \leq n^{-1} + 0 = n^{-1} < \delta/2,$$

while all points of $(Y - G)_n$ are at a distance from $\bar{y}$ which is $\geq \delta - 1/n > 2/n - 1/n = 1/n$. Thus, $\bar{x} \in f_{m_n}^{-1}(Y - (Y - G)_n)$ for all $n > 2/\delta$, and we have proved that in (8.3.1) the relation $\subset$ certainly holds. Conversely, let $\bar{x}$ be a point in the second member of (8.3.1). Then $\bar{x} \in f_{m_n}^{-1}(Y - (Y - G)_n)$ for some n; hence $f_{m_n}(\bar{x}) \in Y - (Y - G)_n$, that is, $f_{m_n}(\bar{x})$ is at a distance $> 1/n$ from $Y - G$; and therefore $f(\bar{x})$ is at a distance > 0 from $Y - G$, that is, $f(\bar{x}) \in G$, or $\bar{x} \in f^{-1}(G)$. We have proved that in (8.3.1) also the relation $\supset$ holds. Thus, (8.3.1) is proved. Now $Y - G$ is closed, $(Y - G)_n$ is also closed, $Y - (Y - G)_n$ is open. Hence, $f_{m_n}^{-1}(Y - (Y - G)_n)$ is in the class S, and so is $f^{-1}(G)$, since S is countably additive. This proves (8.3.i). □

Let L be a field of subsets of X, that is, L is a collection of subsets of X with the property that $A, B \in L$ implies that $A \cup B$, $A \cap B$, $X - A$ also belong to L. Let S denote the countably additive family induced by L, that is, briefly, the family of all countable unions of elements of L.

8.3.ii (Theorem: Kuratowski and Ryll-Nardzewski [1]). *Let Y be a complete metric space which is countably separable, that is, there is a countable subset of Y which is everywhere dense in Y. Let X be any set, or space, with a field L of subsets inducing a countably additive family S of subsets of X. Let $x \to F(x)$, $x \in X$, be a set valued function such that* (a) *$F(x)$ is a closed subset of Y for every $x \in X$;* (b) *for every open subset G of Y the set $[x \in X \mid F(x) \cap G \neq \emptyset] \in S$. Then there exists a selector $f: X \to Y$ such that $f^{-1}(G) \in S$ for every open subset G of Y.*

Proof. Let $R = [r_i, i = 1, 2, \ldots]$ be a countable subset of distinct points of Y, everywhere dense in Y. Thus we understand that $r_i \neq r_s$ for $i \neq s$. By modifying the metric space if needed, we may assume that Y has diameter < 1. We shall obtain f as the limit of maps $f_n: X \to Y$ with the following properties:

(8.3.2) $$f_n^{-1}(G) \in S \quad \text{for every open subset } G \text{ of } Y;$$

(8.3.3) $$\rho(f_n(x), F(x)) < 1/2^n \quad \text{for every } x \in X;$$

(8.3.4) $$\rho(f_n(x), f_{n-1}(x)) < 1/2^{n-1} \quad \text{for all } x \in X \text{ and } n = 1, 2, \ldots.$$

Let us proceed by induction. First, take $f_0 = r_1$ for all $x \in X$, so that (8.3.2–4) are trivially satisfied. Assume that f_{n-1} satisfying (8.3.2–4) has been found, and let us determine f_n. To do this we take

(8.3.5) $$C_i^n = [x \mid \rho(r_i, F(x)) < 1/2^n],$$

(8.3.6) $$D_i^n = [x \mid \rho(r_i, f_{n-1}(x)) \le 1/2^{n-1}],$$

(8.3.7) $$A_i^n = C_i^n \cap D_i^n.$$

Let us prove that

$$X = \bigcup_{i=1}^{\infty} A_i^n \quad \text{for every } n.$$

Indeed, for every point $x \in X$ we have $\rho(f_{n-1}(x), F(x)) < 1/2^{n-1}$; hence there is some $y \in F(x)$ with $\rho(y, f_{n-1}(x)) < 1/2^{n-1}$. Since $Y = \text{cl } R$, there is some i such that

$$\rho(r_i, y) \le \min[1/2^n, 1/2^{n-1} - \rho(y, f_{n-1}(x))].$$

Then,

$$\rho(r_i, F(x)) \le \rho(r_i, y) \le 1/2^n,$$
$$\rho(r_i, f_{n-1}(x)) \le \rho(r_i, y) + \rho(y, f_{n-1}(x)) \le 1/2^{n-1};$$

hence $x \in A_n^i$. Thus, $X = \bigcup_i A_i^n$.

Denote by B_i^n the open ball $[y \in Y \mid \rho(y, r_i) < 1/2^n]$. Then (8.3.5) and (8.3.6) become

(8.3.8) $$C_i^n = [x \mid F(x) \cap B_i^n \neq \emptyset],$$

(8.3.9) $$D_i^n = f_{n-1}^{-1}(B_i^{n-1}).$$

From (b) we have $C_i^n \in S$, and from (8.3.2) and induction at $n-1$ we have $D_i^n \in S$; hence, $A_i^n \in S$. Consequently, each set $A_i^n = \bigcup_j E_{ij}^n$ is the countable union of subsets E_{ij}^n of the field L, and $X = \bigcup_i \bigcup_j E_{ij}^n$. We may well arrange the double sequence E_{ij}^n, $i, j = 1, 2, \ldots$, into a simple sequence $E_{k_s m_s}^n$, $s = 1, 2, \ldots$, so that if $E_s^n = E_{k_s m_s}^n$, then

$$X = E_1^n \cup E_2^n \cup \cdots \cup E_s^n \cup \cdots.$$

We shall now define $f_n: X \to R$ as follows: $f_n(x) = r_{k_s}$ if $x \in E_s^n - (E_1^n \cup \cdots \cup E_{s-1}^n)$. It remains to show that f_n satisfies the relations (8.3.2–4). By definition $f_n^{-1}(r_{k_s}) = E_s^n - (E_1^n \cup \cdots \cup E_{s-1}^n)$. Since L is a field, it follows that $f_n^{-1}(r_{k_s}) \in L$, and since $f_n^{-1}(r_i) = \bigcup_{k_s = i} f_n^{-1}(r_{k_s})$, we also have $f_n^{-1}(r_i) \in S$ for each i, and finally $f_n^{-1}(Z) \in S$ for every subset Z of R, since R is countable, and S is countably additive. We have proved (8.3.2).

Now, for a given $x \in X$, let $x \in E_s^n - (E_1^n \cup \cdots \cup E_{s-1}^n)$, let $k_s = i$, and note that $x \in E_s^n$ implies $x \in E_s^n \subset A_i^n = C_i^n \cap D_i^n$, and (8.3.3), (8.3.4) follow from (8.3.5) and (8.3.6), since $f_n(x) = r_i$. We have proved that the sequence f_n, $n = 1, 2, \ldots$, is completely defined and satisfies (8.3.2–4). By (8.3.4) the sequence f_n is Cauchy, and since Y is complete, f_n converges uniformly to a function $f: X \to Y$. By Lemma (8.3.i), f satisfies the condition (8.3.2). Finally, $f(x) \in F(x)$ by (8.3.3). Theorem (8.3.ii) is thereby proved. □

8.3.iii (Corollary). *Theorem* (8.3.ii) *still holds even if* (b) *is replaced by* (c): *for every closed subset K of Y the set* $\{x \in X \mid F(x) \cap K \neq \emptyset\} \in L$.

Proof. Since Y is a metric space, every open set $G \subset Y$ is an F_s-set, that is, $G = K_1 \cup K_2 \cup \cdots$ is the countable union of closed sets K_s. Then

$$\{x \mid F(x) \cap G \neq \emptyset\} = \bigcup_{s=1}^{\infty} \{x \mid F(x) \cap K_s \neq \emptyset\}. \qquad \square$$

Remark 1. If $S = L$ the proof of (8.3.ii) can be slightly simplified: namely, the decomposition of A_i^n into sets E_{ij}^n is not needed, and we can simply define f_n by taking $f_n(x) = r_i$ for $x \in A_i^n - (A_1^n \cup \cdots \cup A_{i-1}^n)$.

B. L-measurable Set Valued Functions and L-measurable Selectors

We shall now consider the case where X is an interval $[a, b]$ and L is the field of all Lebesgue measurable subsets of $[a, b]$.

For $Y = R$, we know that a real valued map $f: X \to Y$ is said to be L-measurable if for every real α the set $[x \in X \mid f(x) < \alpha]$ is L-measurable, that is, belongs to L. Then, by taking complements, intersections, countable unions, and countable intersections, we immediately show that all sets $[x \mid f(x) \geq \alpha]$, $[x \mid f(x) \leq \alpha]$, $[x \mid f(x) > \alpha]$, $[x \mid \alpha < f(x) < \beta]$, and finally $[x \mid f(x) \in G]$ where G is any open subset of R, are measurable. Note that here $S = L$.

For Y any topological space it is now natural to say that a single valued map $f: X \to Y$ is L-measurable if for every open subset G of Y the set $[x \in X \mid f(x) \in G]$ belongs to L. Finally, again for Y any topological space, we shall say that a set valued map $x \to F(x)$, $x \in X$, $F(x) \subset Y$, is L-measurable if for every open subset G of Y the set $[x \in X \mid F(x) \cap G \neq \emptyset]$ belongs to L.

8.3.iv Theorem (Kuratowski and Ryll-Nardzewski [1]). *Let Y be a separable complete metric space, and let $x \to F(x) \subset Y$, be a set valued map whose values $F(x)$ are closed subsets of Y. If F is L-measurable, then there is an L-measurable selector $f: X \to Y$ with $f(x) \in F(x)$ for all $x \in X$.*

This is an immediate consequence of (8.3.ii).

We know that any measurable single valued function $f: X \to Y$, $X = [a, b]$, has the Lusin property, that is, given $\varepsilon > 0$, there is a compact subset K of $X = [a, b]$ such that $\text{meas}(X - K) < \varepsilon$, and f restricted to K is continuous. It can be proved (see, e.g., Castaing and Valadier [I]) that any L-measurable set valued map $x \to F(x)$, $x \in X = [a, b]$, whose values $F(x)$ are closed subsets of Y, has an analogous Lusin type property, namely, given $\varepsilon > 0$, there is a compact subset K of X such that $\text{meas}(X - K) < \varepsilon$, and for every open subset G of Y the set $[x \in K \mid F(x) \cap G \neq \emptyset]$ is open relative to K.

We shall not need in the sequel the full strength of the theory of measurable set valued functions.

C. Carathéodory's Functions

Let $f(x, y)$ be a given real valued function defined in a product space $G \times R^n$, where G is a given L-measurable subset of some space R^s, $s \geq 1, n \geq 1$. We say that f is a Carathéodory function if f is continuous in y for almost all $x \in G$, and is measurable in x for every $y \in R^n$.

8.3.v (Scorza-Dragoni [1]). *If G is any measurable subset of R^s and if $f(x, y)$ is defined in $G \times R^n$, is continuous in y for almost all $x \in G$, and is measurable in x for all $y \in R^n$, then for any $\eta > 0$ there is a closed subset K of G such that* meas$(G - K) < \eta$ *and f is continuous on $K \times R^n$.*

Proof. For the sake of simplicity we take $s = 1, n = 1, G = [0, 1], B = [0, 1]$, f defined on $G \times B$; and if E_0 is some subset of measure zero on G, we assume f continuous in y for every $x \in G - E_0$, and measurable in x for every $y \in B$.

For any given $\varepsilon > 0$ and integer m let $E_{\varepsilon m}$ be the set of all $x \in [0, 1]$ such that y_1, $y_2 \in [0, 1]$, $|y_1 - y_2| \leq m^{-1}$ implies $|f(x, y_1) - f(x, y_2)| \leq 3^{-1}\varepsilon$. Let us prove that $E_{\varepsilon m}$ is measurable. It is enough to prove that $[0, 1] - E_{\varepsilon m}$ is measurable, and since meas $E_0 = 0$, it is enough to prove that $D = [0, 1] - E_{\varepsilon m} - E_0$ is measurable. Now if $x \in D$, then there is a pair of real numbers y_1, y_2 with $|y_1 - y_2| \leq m^{-1}$ and $|f(x, y_1) - f(x, y_2)| > 3^{-1}\varepsilon$. Since $x \notin E_0$, $f(x, y)$ is continuous in y, and therefore we may well assume that y_1 and y_2 are rational. In other words, D is the union of sets of the form $[x| |f(x, y_1) - f(x, y_2)| > 3^{-1}\varepsilon]$ where y_1, y_2 are rational numbers with $|y_1 - y_2| \leq m^{-1}$. This is a countable family of measurable sets, and thus, D is measurable, and $E_{\varepsilon m}$ is measurable.

Here $E_{\varepsilon m} \subset E_{\varepsilon, m+1}$, and if E_ε is the union of all the sets $E_{\varepsilon m}$, $m = 1, 2, \ldots$, then E_ε is measurable, and we take $F = [0, 1] - E_0$. Note that for every $x \in F$, $f(x, y)$ is continuous in y on the compact set B, and hence uniformly continuous in y. In other words, every $x \in F$ belongs to some $E_{\varepsilon m}$, or $[0, 1] - E_0 = E_\varepsilon = \bigcup_m E_{\varepsilon m}$, meas $E_\varepsilon = 1$. We conclude that there is some m_0 such that meas $E_{\varepsilon m_0} > 1 - 2^{-1}\varepsilon\eta$, and for all $x \in E_{\varepsilon m_0}$ and all $y_1, y_2 \in [0, 1]$, $|y_1 - y_2| \leq m_0^{-1}$, we also have $|f(x, y_1) - f(x, y_2)| \leq 3^{-1}\varepsilon$.

Let $0 = u_0 < u_1 < \cdots < u_q = 1$ be $q + 1$ equidistant points in $[0, 1]$ with $q \geq m_0$, so that $u_j - u_{j-1} = q^{-1} \leq m_0^{-1}$. For every $j = 1, \ldots, q$, the function $f(x, u_j)$ is measurable in x. Hence, by Lusin's theorem, there is some closed set $F_j \subset [0, 1]$ with meas $F_j > 1 - (2q)^{-1}\varepsilon\eta$ and $f(x, u_j)$ is continuous in x on F_j—in fact uniformly continuous. If $V = \bigcap_{j=1}^q F_j$, then meas $V > 1 - 2^{-1}\varepsilon\eta$ and all q functions of x, $f(x, u_j)$, $j = 1, \ldots, q$, are continuous on V—in fact uniformly continuous. Thus, there is some $\delta_1 = \delta_1(\varepsilon)$ such that $x_1, x_2 \in V$, $|x_1 - x_2| \leq \delta_1(\varepsilon)$ implies $|f(x_1, u_j) - f(x_2, u_j)| \leq 3^{-1}\varepsilon$, $j = 1, \ldots, q$. Finally, for $E'_\varepsilon = V \cap E_{\varepsilon m_0}$ and $\delta(\varepsilon) = \min[m_0^{-1}, \delta_1(\varepsilon)]$, we see that for any two points $(x_1, y_1), (x_2, y_2)$ with $x_1, x_2 \in E'_\varepsilon$, $|x_1 - x_2| \leq \delta(\varepsilon)$, $y_1, y_2 \in [0, 1]$, $|y_1 - y_2| \leq \delta(\varepsilon)$, there is some j with $|y_1 - u_j|, |y_2 - u_j| \leq m_0^{-1}$ and

$$|f(x_1, y_1) - f(x_2, y_2)| \leq |f(x_1, y_1) - f(x_1, u_j)| + |f(x_1, u_j) - f(x_2, u_j)| + |f(x_2, u_j) - f(x_2, y_2)| \leq \varepsilon/3 + \varepsilon/3 + \varepsilon/3 = \varepsilon.$$

The set E'_ε depends also on η, but we shall keep η fixed. Now we take $\varepsilon = \varepsilon_i = 2^{-i-1}$, $E'_i = E'_{\varepsilon_i}$, $i = 1, 2, \ldots$, and $E^* = \bigcap_{i=1}^\infty E'_i$. Then meas $E^* > 1 - \eta \sum_{i=1}^\infty 2^{-i-1} > 1 - \eta$. Note that f is continuous on $E^* \times B$. Indeed, given $\gamma > 0$, take $i = i(\gamma)$, so that $\varepsilon_i < \gamma$. If (x_1, y_1), (x_2, y_2) are in $E^* \times B$, and their distance is $< \delta(\varepsilon_i) = \delta(\gamma)$. then certainly $x_1, x_2 \in E'_i$, and $|f(x_1, y_1) - f(x_2, y_2)| \leq \varepsilon_i < \gamma$. Now E^* is measurable, meas $E^* > 1 - \eta$. Thus, there is some closed subset K of E^* with meas $K > 1 - \eta$, and f is continuous on $K \times B$. Statement (8.3.v) is thereby proved under the mentioned restrictions. We leave to the exercises to prove it under the stated general hypotheses. □

Remark 2. The statement (8.3.5) will have a role in what follows. Here we mention only an immediate application. If $f(x, y)$ satisfies the conditions of (8.3.v), and $y(x)$, $x \in G$, is measurable with values in R^n, then $f(x, y(x))$ is measurable in G. Indeed, for every $\eta > 0$, there is K compact, $K \subset G$, with $\operatorname{meas}(G - K) < \eta$, such that $f(x, y)$ is continuous in $K \times R^n$, and there is K' compact, $K' \subset G$, with $\operatorname{meas}(G - K') < \eta$, such that $y(x)$ is continuous on K'. Then $f(x, y(x))$ is continuous in $K \cap K'$, and since $\operatorname{meas}(G - K \cap K') < 2\eta$ with η arbitrary, we conclude that $f(x, y(x))$ is measurable in G.

D. Another Form of the Implicit Function Theorem

As in Section 8.2B, let $A = [a, b] \times R^n$ (a closed subset of the tx-space R^{n+1}), let U be a fixed closed subset of the u-space R^m, and let $f(t, x, u) = (f_1, \ldots, f_n)$ be a Carathéodory function defined on $[a, b] \times R^n \times U$, that is, f is measurable in t for every $(x, u) \in R^n \times U$, and is continuous in (x, u) for almost all $t \in [a, b]$. As usual, for every $(t, x) \in [a, b] \times R^n$, let $Q(t, x)$ denote the subset of all $z = (z^1, \ldots, z^n) \in R^n$ such that $z = f(t, x, u)$, $u \in U$, that is, $Q(t, x) = f(t, x, U)$. Let $x(t) = (x^1, \ldots, x^n)$, $a \le t \le b$, be any AC function, such that

$$x'(t) \in Q(t, x(t)), \qquad t \in [a, b] \text{ (a.e.)}.$$

In other words, $x(t)$ is an AC solution for the orientor field $dx/dt \in Q(t, x)$.

8.3.vi (AN IMPLICIT FUNCTION THEOREM). *Under the above assumptions there is an L-measurable $u(t)$, $a \le t \le b$, with $u(t) \in U$, $x'(t) = f(t, x(t), u(t))$, $t \in [a, b]$* (a.e.).

Proof. Two proofs are given here of this statement.

I. By (8.3.v), for $\eta = k^{-1}$, $k = 1, 2, \ldots$, there is a closed subset F_k of $[a, b]$ such that $f(t, x, u)$ is continuous in $F_k \times R^n \times U$ and $\operatorname{meas} F_k > b - a - k^{-1}$. If $F = \bigcup_{k=1}^{\infty} F_k$, then $F \subset [a, b]$, $\operatorname{meas} F = b - a$. From the implicit function theorem (8.2.iii) there is a measurable function $u_k(t)$, $t \in F_k$, such that $u_k(t) \in U$ for $t \in F_k$, $x'(t) = f(t, x(t), u_k(t))$, $t \in F_k$ (a.e.). Then, if we take $u(t) = u_k(t)$ for $t \in F_k - (F_1 \cup \cdots \cup F_{k-1})$, then $u(t)$ is defined a.e. in $[a, b]$, it is measurable in $[a, b]$, and $u(t) \in U$, $x'(t) = f(t, x(t), u(t))$, $t \in [a, b]$ (a.e.).

II. Let F_0 denote the set of all $t \in [a, b]$ for which either $x'(t)$ does not exist, or it is not finite, or $x'(t) \in Q(t, x(t))$ does not hold, or $f(t, x, u)$ is not a continuous function of (x, u). Then $\operatorname{meas} F_0 = 0$. Again, by (8.3.v), for $\eta = k^{-1}$, $k = 1, 2, \ldots$, there is a closed subset F_k of $[a, b]$ such that $f(t, x, u)$ is continuous in $F_k \times R^n \times U$, and $\operatorname{meas} F_k > b - a - k^{-1}$. We may well assume that $F_0 \cap F_k = \emptyset$, that is, $F_0 \subset [a, b] - F_k$ for all k. Let $F = \bigcup_{k=1}^{\infty} F_k$, so that $\operatorname{meas} F = b - a$, $F_0 \subset [a, b] - F$. For $t \in F$ let $\tilde{U}(t)$ denote the subset of all $u \in U$ such that $f(t, x(t), u) = x'(t)$. We know that $\tilde{U}(t)$ is not empty, since $x'(t) \in Q(t, x(t)) = f(t, x(t), U)$ for $t \in F$. Let us prove that the set-valued map $t \to \tilde{U}(t)$, $t \in F$, $\tilde{U}(t) \subset U \subset R^m$, is L-measurable. Let B_s, $s = 1, 2, \ldots$, denote the closed ball with center the origin and radius s in R^m. Let G be any given open subset of R^m. Since $f(t, x(t), u)$ is a continuous function on the compact set $F_k \times U \cap B_s$, and $x'(t)$ is continuous on F_k, the set $F_{ks} = [t \in F_k | f(t, x(t), u) = x'(t)$ for some $u \in U \cap B_s]$ is closed, and the set $F'_{ks} = [t \in F_k | f(t, x(t), u) = x'(t)$ for some $u \in G \cap U \cap B_s]$ is open in F_{ks}, that is, open relative to F_{ks}. Hence, the set $F'_k = \bigcup_s F'_{ks} = [t \in F_k | f(t, x(t), u) = x'(t)$

for some $u \in G \cap U]$ is B-measurable, and the set $F'' = \bigcup_k F'_k$ is also B-measurable. Finally, F'' differs from the set $V = [t \in [a,b] \mid f(t, x(t), u) = x'(t)$ for some $u \in G \cap U]$ by at most a set of measure zero ($\subset [a,b] - F$). Thus, V is measurable, and $t \to \tilde{U}(t)$ is a set valued L-measurable map. By (8.3.iv) there is an L-measurable $u(t)$, $t \in [a,b]$, such that $u(t) \in \tilde{U}(t)$, that is, $u(t) \in U$, $x'(t) = f(t, x(t), u(t))$, $t \in [a,b]$ (a.e.). □

Remark 3. Statement (8.3.vi) is a particular case of Section 8.2.C, Exercise 5. Indeed, under the conditions of (8.3.vi), by (8.3.v), for every $\varepsilon > 0$ there is a compact subset K of $[a,b]$ with $\text{meas}([a,b] - K) < \varepsilon$ such that $f(t,x,u)$ is continuous on the closed set $M_K = K \times R^n \times U$, and $A_K = K \times R^n$ is also closed. The statement in Section 8.2C, Exercise 5 is more general, since there we allow (t,x) to cover an arbitrary set A, U is an arbitrary set $U(t,x)$ which may depend on t and x, and f is an arbitrary function on $M = [(t,x,u) \mid (t,x) \in A, u \in U(t,x)]$, with the sole restriction that for $\varepsilon > 0$ there is a compact subset K of $[a,b]$ such that $\text{meas}([a,b] - K) < \varepsilon$, $A_K = [(t,x) \in A \mid t \in K]$ is closed, $M_K = [(t,x,u) \in M \mid t \in K]$ is closed, and $f(t,x,u)$ is continuous on M_K.

8.4 Convexity, Carathéodory's Theorem, Extreme Points

A. Convexity

Let X be any linear space over the reals (cf. Dunford and Schwartz, [I, p. 49]). A subset K of X is said to be *convex* provided $x_1, x_2 \in K$, $0 \le \alpha \le 1$, implies that $x = (1-\alpha)x_1 + \alpha x_2$ is also a point of K; that is, the entire segment $s = x_1x_2$ between two points of K is contained in K.

Given any m points, $x_1, \ldots, x_m$ in X, any point $x = \lambda_1 x_1 + \cdots + \lambda_m x_m$ with $\lambda_s \ge 0$, $s = 1, \ldots, m$, $\lambda_1 + \cdots + \lambda_m = 1$, is said to be a *convex* combination of $x_1, \ldots, x_m$.

Given any subset A of X, we shall denote by co A the smallest convex set in X containing A. This set co A is said to be the *convex hull* of A. We also denote by $\tilde{A}$ the set of all convex combinations of arbitrarily many points of A. Finally, for every fixed integer $m \ge 2$, we denote by A_m the set of all combinations of at most m points of A. Thus, $A_m \subset A_{m+1} \subset \tilde{A}$, $\tilde{A} = \bigcup A_m$, where $\bigcup$ ranges over all $m = 2, 3, \ldots$.

8.4.i. *A subset A of a linear space X is convex if and only if every convex combination $x = \sum_{i=1}^m \lambda_i x_i$ of points of A belongs to A.*

The sufficiency is trivial, since the requirement for $m = 2$ reduces to the definition of convex set. The necessity is a consequence of the statement:

8.4.ii. *For any subset A of a linear space X we have $\bigcup_m A_m = \tilde{A} = \text{co } A$.*

Proof.

(a) Let us prove that $\tilde{A} \subset \text{co } A$. It is enough to prove that $A_m \subset \text{co } A$ for every m. That $A_2 \subset \text{co } A$ is a consequence of the definition of convex set. Let us assume that we have proved that $A_{m-1} \subset \text{co } A$, and let us prove that $A_m \subset \text{co } A$. Indeed, if $x = \lambda_1 x_1 + \cdots + \lambda_m x_m$, $0 \le \lambda_s \le 1$, $s = 1, \ldots, m$, $\lambda_1 + \cdots + \lambda_s = 1$, either all λ_s are zero but one, say $\lambda_m = 1$, and then $x = x_m \in A \subset \text{co } A$; or at least two of the λ_s are positive, say λ_{m-1} and λ_m, and then $\alpha = \lambda_1 + \cdots + \lambda_{m-1} > 0$, $1 - \alpha = \lambda_m > 0$, and $x = \alpha(\sum_1^{n-1}(\lambda_s/\alpha)x_s) + (1 - \alpha)x_m$. Here, $x_m \in A$, and the expression in parenthesis is a point of A_m, and hence of co A by the induction hypothesis; and therefore $x \in \text{co } A$. We have proved that $A_m \subset \text{co } A$ for every m, and hence $\tilde{A} \subset \text{co } A$.

(b) Let us prove that $\tilde{A}$ is a convex set, and that $\tilde{A} = \text{co } A$. Indeed, if $x, y \in \tilde{A}$ and $z = \alpha x + (1 - \alpha)y$, $0 \leq \alpha \leq 1$, then $x = p_1 v_1 + \cdots + p_m v_m$, $y = q_1 w_1 + \cdots + q_M w_M$ with $v_s \in A$, $p_s \geq 0$, $s = 1, \ldots, m$, $w_s \in A$, $q_s \geq 0$, $s = 1, \ldots, M$, and $p_1 + \cdots + p_m = 1$, $q_1 + \cdots + q_M = 1$. As a consequence, we have also

$$z = \alpha x + (1 - \alpha)y = \alpha p_1 v_1 + \cdots + \alpha p_m v_m + (1 - \alpha)q_1 w_1 + \cdots + (1 - \alpha)q_M w_M.$$

Thus, z is a point of $\tilde{A}$, and $\tilde{A}$ is convex, $A \subset \tilde{A} \subset \text{co } A$. Since co A is the smallest convex set containing A, we conclude that $\tilde{A} = \text{co } A$.

8.4.iii (THEOREM: CARATHÉODORY [I]). *For any subset A of R^n, every point of the convex hull of A is the convex combination of at most $n + 1$ suitable points of A. In symbols,* co $A = \tilde{A} = A_{n+1}$.

Proof. It is enough we prove that any convex combination of $m \geq n + 2$ points of R^n is also a convex combination of at most $n + 1$ of the same points. For this it is enough to prove that any convex combination of $m \geq n + 2$ points of R^n is also the convex combination of at most $m - 1$ of the same points. Let $v_1, \ldots, v_m$ be the m points of R^n, and x any convex combination of them. It is not restrictive to assume $x = 0$; hence

$$0 = \lambda_1 v_1 + \cdots + \lambda_m v_m, \qquad \lambda_s \geq 0, \quad s = 1, \ldots, m, \quad \lambda_1 + \cdots + \lambda_m = 1. \tag{8.4.1}$$

If some of the numbers λ_s are zero, then $x = 0$ is the convex combination of fewer than m points v_s. Thus, we assume $\lambda_s > 0$, $s = 1, \ldots, m$. Analogously, we can assume all $v_s \neq 0$ and distinct. For $m \geq n + 2$, we have $m - 1 \geq n + 1$, and there must be a linear combination

$$c_1 v_1 + \cdots + c_{m-1} v_{m-1} = 0 \tag{8.4.2}$$

of the $m - 1$ nonzero vectors $v_1, \ldots, v_m$ with coefficients $c_1, \ldots, c_{m-1}$ real and not all zero. Indeed, if $v_i = (v_1^i, \ldots, v_n^i)$, then the system of n linear homogeneous algebraic equations in the $m - 1 \geq n + 1$ unknowns $c_1, \ldots, c_{m-1}$,

$$c_1 v_1^s + \cdots + c_{m-1} v_{m-1}^s = 0, \qquad s = 1, \ldots, n,$$

must have a solution $c_1, \ldots, c_{m-1}$ with numbers c_i not all zero. Thus, for every real α we also have from (8.4.1) and (8.4.2)

$$0 = (\lambda_1 + \alpha c_1)v_1 + \cdots + (\lambda_{m-1} + \alpha c_{m-1})v_{m-1} + \lambda_m v_m,$$

and for all α of a maximal interval $\bar{\alpha} < \alpha < \bar{\beta}$, $\bar{\alpha} < 0$, $\bar{\beta} > 0$, we still have

$$\lambda_1 + \alpha c_1 > 0, \ldots, \lambda_{m-1} + \alpha c_{m-1} > 0, \qquad \lambda_m > 0.$$

Thus, for $\alpha = \bar{\alpha}$ and $\alpha = \bar{\beta}$, at least one of the numbers $\lambda_1 + \alpha c_1, \ldots, \lambda_{m-1} + \alpha c_{m-1}$ is zero, say for $\alpha = \bar{\alpha}$, and $\lambda_1 + \bar{\alpha} c_1 = 0$, $\lambda_s + \bar{\alpha} c_s \geq 0$, $s = 2, \ldots, m - 1$. Then,

$$0 = (\lambda_2 + \bar{\alpha} c_2)v_2 + \cdots + (\lambda_{m-1} + \bar{\alpha} c_{m-1})v_{m-1} + \lambda_m v_m,$$
$$C = (\lambda_2 + \bar{\alpha} c_2) + \cdots + (\lambda_{m-1} + \bar{\alpha} c_{m-1}) + \lambda_m > 0.$$

For $q_s = (\lambda_s + \bar{\alpha}c_s)/C$, $s = 2, \ldots, m-1$, $q_m = \lambda_m/C$, all numbers q_s are nonnegative with $q_2 + \cdots + q_m = 1$, and $0 = q_2v_2 + \cdots + q_mv_m$, that is, $x = 0$ is a convex combination of $m-1$ of the original m vectors $v_1, \ldots, v_m$. Theorem (8.4.iii) is thereby proved. □

B. The Closed Convex Hull of a Set

If we denote by cl A and co A, respectively, the closure and the convex hull of a set A in X, then we may well consider also the sets cl co A, co cl A.

8.4.iv. *For any set A in R^n, the sets A,* cl A, co A, co cl A *are all contained in* cl co A.

Proof. It is enough to prove that co cl $A \subset$ cl co A. Indeed, by (8.4.iii), any point $x \in$ co cl A can be written in the form $x = \sum_1^\nu \lambda_i x_i$ where $0 \leq \lambda_i \leq 1$, $i = 1, \ldots, \nu = n+1$, $\sum_1^\nu \lambda_i = 1$, and $x_i \in$ cl A, $i = 1, \ldots, \nu$. Hence, there are ν sequences $[x_{ik}, k = 1, 2, \ldots]$ of points $x_{ik} \in A$, $k = 1, 2, \ldots$, with $x_{ik} \to x_i$ as $k \to \infty$, $i = 1, \ldots, \nu$. If we take $x_k = \sum_1^\nu \lambda_i x_{ik}$, then

$$|x_k - x| = \left|\sum_1^\nu \lambda_i x_{ik} - x\right| = \left|\sum_1^\nu \lambda_i(x_{ik} - x_i)\right| \leq \sum_1^\nu \lambda_i |x_{ik} - x_i|.$$

Hence, $|x_k - x| \to 0$ as $k \to \infty$, or $x_k \to x$, where now $x_k \in$ co A. Thus, $x \in$ cl co A. We have proved that every point $x \in$ co cl A is also a point of cl co A, or co cl $A \subset$ cl co A. □

The set cl co A is often called the closed convex hull of A.

Note that co cl A may well be actually smaller than cl co A. For instance, if $A = [(x, y) | 0 < x < +\infty, y = \pm 1/x]$, then co cl $A = [(x, y) | 0 < x < +\infty, -\infty < y < +\infty]$, while cl co $A = [(x, y) | 0 \leq x < +\infty, -\infty < y < +\infty]$. However, if A is a bounded subset of R^n, then co cl A = cl co A. The proof of this last statement is left as an exercise for the reader. Also note that for any compact subset A of R^n we have co A = co(∂A), where ∂A denotes the boundary of A.

Note that if $f(t) = (f_1, \ldots, f_n)$, $t \in G$, is any L-integrable function on a measurable set $G \subset R$, $0 <$ meas $G < \infty$ (that is, each component f_i is L-integrable), and if the values $f(t)$ of f belong to some subset A of R^n, then the mean value of f on G belongs to cl co A. In other words, $f \in (L(G))^n$, $f(t) \in A \subset R^n$, implies

$$(\text{meas } G)^{-1} \int_G f(t)\, dt \in \text{cl co } A.$$

Concerning the last statement, let us consider first any step function on G (cf. McShane [I, p. 54]), that is, a function $f: G \to A \subset R^n$ with values in A such that f is constant on each set G_i, $i = 1, \ldots, N$, of a finite decomposition of G into disjoint measurable subsets G_i, each of positive measure. Then, if f_i denotes the constant value of f on G_i, then by the definition of L-integral of a step function we have

$$\int_G f(t)\, dt = \sum_{i=1}^N f_i(\text{meas } G_i).$$

If $\gamma_i =$ meas G_i/meas G, then $0 < \gamma_i \leq 1$, $i = 1, \ldots, N$, $\sum_{i=1}^N \gamma_i = 1$, and the mean value $m(f)$ of f, or $m(f) = (\text{meas } G)^{-1} \int_G f(t)\, dt = \sum_{i=1}^N \gamma_i f_i \in$ co A, is a convex combination of the values $f_i \in A$, and $m(f) \in$ co A. For any measurable L-integrable function

$f: G \to A \subset R^n$, the L-integral of f is the limit of the integral defined on suitable step functions f_k, and then $m(f) = \lim_{k\to\infty} m(f_k) \in \text{cl co } A$. Thus, $m(f) \in A$ if A is already a closed convex set.

C. Supporting Hyperspaces

If $l(x)$ denotes any real valued nonzero linear function on a real linear space X, then the set $S = [x \mid l(x) = c]$ where l has a constant value c is said to be a hyperspace of X. Then a hyperspace S divides X into two half spaces, say $S^+ = [x \mid l(x) \geq c]$ and $S^- = [x \mid l(x) \leq c]$. A hyperspace S is said to be a supporting hyperspace for a convex set K if K is contained in one of the two half spaces S^+ or S^-. For the sake of simplicity we shall assume from now on that X is a linear topological space.

8.4.v (EXISTENCE OF SUPPORTING HYPERSPACES) (cf. Dunford and Schwartz [I, pp. 412, 418]). *For any convex subset K of a linear topological space X, possessing interior points, and any point $x_0 \in X - K$, there is some supporting hyperspace for K through x_0, or S: $l(x) = l(x_0)$, where l is a continuous linear real valued nonzero functional on X.*

A supporting hyperspace through a point need not be unique, as we can see by considering a convex polygon K in R^2 and taking for x_0 one of its vertices or any exterior point. Note that if a convex set K has interior points, then the set $K_0 = \text{int } K$ of all its interior points and the closure cl K of K are also convex.

If $K \subset X = R^n$ and K has dimension $\leq n - 1$, then K is contained in a hyperspace S of $X = R^n$, which also is a supporting hyperspace for K through each of its points. If $K \subset X = R^n$ has interior points, then K_0 is convex, and each supporting space for K_0 through any point $x_0 \in \text{bd } K$ is also a supporting space for K through x_0.

8.4.vi. *If K is any convex subset of X, then* cl $K = \bigcap S^+$, *where the intersection is taken over all half spaces S^+ containing K.*

Proof. Since cl $K \subset S^+$ for every S^+ as above, we certainly have cl $K \subset \bigcap S^+$. Let us prove the opposite relation, cl $K \supset \bigcap S^+$. Let x_0 denote any point $x_0 \in \bigcap S^+$. Let $\bar{x}$ be any fixed point of K, and let s be the closed segment $s = [x_0, \bar{x}]$. Then $s \cap \text{cl } K$ is a closed subset of s, and we denote by x_1 the point of $s \cap \text{cl } K$ closest to x_0. If $x_1 = x_0$ then $x_0 = x_1 \in \text{cl } X$.

Now suppose, if possible, that $x_0 \neq x_1$. Then there must be a circular neighborhood U of x_0 not containing points of K, since otherwise $x_0 \in \text{cl } K$, and $x_1 = x_0$, which is not the case. Then the entire half-closed, half-open segment $[x_0 x_1)$ does not contain points of K; in particular $\bar{\bar{x}} = 2^{-1}(x_0 + x_1) \notin K$. Then there is a hyperplane l through $\bar{\bar{x}}$ with $l(\bar{\bar{x}}) = c$, $l(\bar{x}) > c$, and $l(x_0) < c$. In other words, $x_0 \notin S^+$; hence $x_0 \notin \bigcap S^+$, a contradiction. Thus, $x_0 = x_1$, and $x_0 \in \text{cl } K$. We have proved that $\bigcap S^+ \subset \text{cl } K$, and (8.4.vi) is proved. □

Remark. Statement (8.4.vi) has also a more general version: For any two disjoint convex subsets K and L of the linear topological space X of which one at least has interior points, there is some separating hyperspace $S: l(x) = c$, where l is a continuous linear real valued nonzero functional on X, and $l(x) \geq c$ for $x \in K$, $l(x) \leq c$ for $x \in L$ [Dunford and Schwartz, loc. cit.]

D. Extreme Points

If K is a convex compact subset of any linear space X, a point $\bar{x} \in K$ is said to be an extreme point of K provided: if $\bar{x} = \alpha x_1 + (1 - \alpha)x_2$ for some two points x_1, x_2 of K and $0 < \alpha < 1$, then $x_1 = x_2$, and hence $\bar{x} = x_1 = x_2$. The set of all extreme points of K is denoted extr K.

For instance, if K is a convex polygon in R^2, then the extreme points are the vertices of K. If K is a ball in R^n, then all its boundary points are extreme points.

The existence and main properties of extreme points are specified in a statement of Krein-Milman which holds in any locally convex linear topological Hausdorff space X:

8.4.vii (KREIN-MILMAN) (cf. Dunford and Schwartz I, vol. 1. p. 440). *Any convex compact nonempty subset K of a locally convex linear topological Hausdorff space X possesses extreme points, and* $K =$ co extr K.

Another interesting property of the extreme points is as follows:

8.4.viii (cf. Dunford and Schwartz, loc. cit.). *If K is any compact subset of X, then* extr co $K \subset K$.

Note that the set extr K of a convex compact set need not be a closed set, as the following example in R^3 shows. Let us consider the disk $S = [(x, y)|(x - 1)^2 + y^2 \leq 1, z = 0]$ and let K be the solid body we obtain by projecting S from the points $(0, 0, 1)$ and $(0, 0, -1)$. Then extr K is made up of the points $(0, 0, 1)$ and $(0, 0, -1)$ and all (x, y, z) with $(x - 1)^2 + y^2 = 1$, $z = 0$, $x > 0$. This set is not closed. (It can be proved that extr K of a compact set K is a G_δ-set.)

8.5 Upper Semicontinuity Properties of Set Valued Functions

A. Upper Semicontinuity by Set Inclusion, and Properties (K) and (Q)

We have already encountered sets $U(t, x)$, $Q(t, x)$ depending on the pair (t, x), that is, set valued functions, or multifunctions, $(t, x) \to U(t, x)$, $(t, x) \to Q(t, x)$, where the independent variable (t, x) ranges over a given set A of the tx-space R^{n+1}. To simplify notation, we should now denote the independent variable simply by x, and consider set valued functions, or multifunctions, $x \to Q(x)$, $x \in A \subset R^\nu$, $Q(x) \subset R^n$. This is even more to the point in that we shall have occasion later to consider set valued maps $t \to Z(t)$, or $x \to Z(x)$. On the other hand, our considerations in the present section hold for any set valued map $x \to Q(x)$, where x ranges in any metric space X, and the sets $Q(x)$ are subsets of any topological space, or linear topological space, Y. We shall use therefore this terminology in the present section. In Section 8.6 we shall need Y to be a finite dimensional space, and we shall return there to the notation $x \to Q(x)$, $x \in A \subset R^\nu$, $Q(x) \subset R^n$.

First, let us assume that both X and Y are given metric spaces.

Given a metric space X, a point $x_0 \in X$, and a number $\delta > 0$, we denote by $N_\delta(x_0)$, called the δ-neighborhood of x_0 in X, the set of all $x \in X$ at a distance $\leq \delta$ from x_0; thus $N_\delta(x_0) \subset X$. Let $x \to Q(x)$ denote a set valued map, or multifunction, from a metric space X to a metric space Y, that is, $Q(x) \subset Y$ for every $x \in X$. Given $x_0 \in X$ and $\delta > 0$, we denote by $Q(x_0; \delta)$ the union of all sets $Q(x)$ with $x \in N_\delta(x_0)$, or $Q(x_0; \delta) = \bigcup[Q(x), x \in N_\delta(x_0)]$. Also, given $\varepsilon > 0$, we denote by $[Q(x_0)]_\varepsilon$ the ε-neighborhood of $Q(x_0)$, i.e. the set of all points of Y at a distance $\leq \varepsilon$ from $Q(x_0)$, that is, at a distance $\leq \varepsilon$ from points of $Q(x_0)$.

We say that a set valued map $x \to Q(x)$, $x \in X$, $Q(x) \subset Y$, is *upper semicontinuous by set inclusion* at x_0 provided, given $\varepsilon > 0$ there is some $\delta > 0$, $\delta = \delta(x_0, \varepsilon)$, such that $Q(x) \subset [Q(x_0)]_\varepsilon$ for all $x \in N_\delta(x_0)$, that is, $Q(x_0; \delta) \subset [Q(x_0)]_\varepsilon$. We say that $x \to Q(x)$ is upper semicontinuous by set inclusion in X if it has this property at every point $x_0 \in X$. For the sake of brevity, we may simply say that the sets $Q(x)$ have such a property. In simple words, we could say that, for upper semicontinuity at x_0, the nearby sets $Q(x)$ cannot be much "larger" than $Q(x_0)$, though some or even all of them could be much "smaller".

8.5.i. *If X and Y are metric spaces, X is compact, and $f: X \to Y$ is a single valued continuous map, then the set valued function $f^{-1}(y)$, $y \in Y_0 = f(X)$, is upper semicontinuous by set inclusion.*

Proof. If this were not the case, then there would be $x_0 \in X$, $y_0 = f(x_0) \in Y_0$, $\varepsilon > 0$, and sequences $[x_k]$ in X and $[y_k]$ in Y_0 with $y_k = f(x_k)$, $y_k \to y_0$ as $k \to \infty$, and $\operatorname{dist}\{x_k, f^{-1}(y_0)\} \geq \varepsilon$. By the compactness of X there is some $\bar{x} \in X$ and a subsequence, say still $[x_k]$, with $x_k \to \bar{x}$ as $k \to \infty$. By the continuity of f, $f(\bar{x}) = y_0$, or $\bar{x} \in f^{-1}(y_0)$ with $x_k \to \bar{x}$, $\operatorname{dist}\{\bar{x}, f^{-1}(y_0)\} \geq \varepsilon$, a contradiction. □

Note that (8.5.i) does not necessarily hold if X is not compact, as the following example shows. Take $X = Y = R$, $f(x) = xe^{-x}$, so that $f(0) = 0$, $f^{-1}(0) = \{0\}$, and for $x_k \to +\infty$, $y_k = f(x_k) \to 0$. The points x_k escape any given neighborhood of 0 in X, and the sets $f^{-1}(y)$ are not upper semicontinuous by set inclusion at $y = 0$.

The property of upper semicontinuity by set inclusion is well suited for compact sets and spaces. For the "unbounded" case, other properties, essentially more general, have been proposed.

Given any set Z in a linear topological space Y, we shall denote by cl Z, bd Z, co Z the closure of Z, the boundary of Z, and the convex hull of Z respectively. Thus, cl co Z denotes the closure of the convex hull of Z, or briefly, the closed convex hull of Z (cf. Section 8.4B).

Let $x \to Q(x)$, $x \in X$, $Q(x) \subset Y$, be a set valued map from a metric space X to a linear topological space Y. Let x_0 be a point of X. Kuratowski's concept of upper semicontinuity, or property (K), is relevant. We say that the map $x \to Q(x)$ has property (K) at x_0 provided $Q(x_0) = \bigcap_\delta \operatorname{cl} Q(x_0; \delta)$, that is,

$$Q(x_0) = \bigcap_{\delta > 0} \operatorname{cl} \bigcup_{x \in N_\delta(x_0)} Q(x). \tag{8.5.1}$$

Here $Q(x_0)$, as the intersection of closed sets, is certainly a closed set.

We shall need also the following variant. We say that the map $x \to Q(x)$ has property (Q) at x_0 provided $Q(x_0) = \bigcap_\delta \operatorname{cl} \operatorname{co} Q(x_0; \delta)$, that is,

$$Q(x_0) = \bigcap_{\delta > 0} \operatorname{cl} \operatorname{co} \bigcup_{x \in N_\delta(x_0)} Q(x). \tag{8.5.2}$$

Here $Q(x_0)$, as the intersection of closed convex sets, is certainly closed and convex.

Again, we shall say that the map $x \to Q(x)$ has property (K) [or property (Q)] in X if it has property (K) [or (Q)] at every point of X. For brevity, we may also say that the sets $Q(x)$ have property (K) [or (Q)] at x_0, or in X. The indication "with respect to x" may be needed if the sets depend also on other parameters.

Remark 1. Note that both in (8.5.1) and (8.5.2) the inclusion $\subset$ is trivial, since the second member always contains the entire set $Q(x_0)$. Thus, what is actually required in (8.5.1) and (8.5.2) is that the inclusion $\supset$ hold.

Note that both in (8.5.1) and (8.5.2) we do not exclude the case of sets Q empty. Then (8.5.1) becomes $Q(x_0) = \emptyset = \bigcap_\delta \operatorname{cl} Q(x_0; \delta)$, and (8.5.2) becomes $Q(x_0) = \emptyset = \bigcap_\delta \operatorname{cl} \operatorname{co} Q(x_0; \delta)$.

Also, note that properties (K) and (Q) are often expressed in terms of "a given sequence $x_k, z_k, k = 1, 2, \ldots,$" with $x_k \in X, x_k \to x_0$ in X, $z_k \in Q(x_k) \subset Y$, and in that case all is required is that, say, for property (Q),

$$Q(x_0) \supset \bigcap_{h=1}^{\infty} \operatorname{cl} \operatorname{co} \left\{ \bigcup_{s=h}^{\infty} z_s \right\}.$$

For instance, suppose that G is a given measurable subset of points $t \in R^\nu$, and for every $t \in G$ that $x \to Q(t, x)$ is a given set valued function from R^n to R^r and that $x_0(t)$, $x_k(t)$, $z_k(t)$, $t \in G$, $k = 1, 2, \ldots,$ are given measurable functions with values in R^n and R^r respectively, with $x_k(t) \to x_0(t)$ pointwise a.e. in G, and $z_k(t) \in Q(t, x_k(t))$ for all k and $t \in G$. Then all that need be required for a global property (Q) in G with respect to the sequence $[x_k, z_k]$ is that for almost all $t \in G$ we have

$$Q(t, x_0(t)) \supset \bigcap_{h=1}^{\infty} \operatorname{cl} \operatorname{co} \left\{ \bigcup_{s=h}^{\infty} z_s(t) \right\}. \tag{8.5.3}$$

If $x_k(t) \to x_0(t)$ in measure, we need only require that (8.5.3) hold for a suitable subsequence $[z_{k_s}]$. Below we shall refer to the definitions (8.5.1–2).

8.5.ii. *Property* (Q) *implies property* (K).

Indeed,

$$Q(x_0) \subset \bigcap_{\delta > 0} \text{cl } Q(x_0;\delta) \subset \bigcap_{\delta > 0} \text{cl co } Q(x_0;\delta) = Q(x_0),$$

and thus equality holds throughout in this relation.

8.5.iii. *Let $x \to Q(x)$, $x \in X$, $Q(x) \subset Y$, be a set valued map, and let $M = [(x, y) | x \in X, y \in Q(x)] \subset X \times Y$, that is, M is the graph of the set valued map. Then $x \to Q(x)$ has property* (K) *in X if and only if M is closed in the product space $X \times Y$.*

Proof. Suppose that the sets $Q(x)$ have property (K) in X, and let us prove that M is closed. Let $(\bar{x}, \bar{y})$ be a point of accumulation of M. Then there is $[(x_k, y_k)]$ with $(x_k, y_k) \in M$, $(x_k, y_k) \to (\bar{x}, \bar{y})$, $y_k \in Q(x_k)$. Thus, $x_k \to \bar{x}$ in X, and $y_k \to \bar{y}$ in Y with $y_k \in Q(x_k)$. Thus, $\bar{y} \in \bigcap_{\delta > 0} \text{cl } Q(\bar{x};\delta) = Q(\bar{x})$, or $(\bar{x}, \bar{y}) \in M$, and M is closed. Conversely, assume that M is closed in the product space, and let us prove that the sets $Q(x)$ have property (K) at every point $\bar{x} \in X$. Let $\bar{x}$ be any point of X, and take any $\bar{y} \in \bigcap_{\delta > 0} \text{cl } Q(\bar{x};\delta)$. Then there is $[(x_k, y_k)]$ with $x_k \to \bar{x}$ in X, $y_k \to \bar{y}$ in Y, with $y_k \in Q(x_k;\delta_k)$, $\delta_k \to 0$, and the distance of x_k from $\bar{x}$ is $\leq \delta_k$. Then $(x_k, y_k) \in M$, $(x_k, y_k) \to (\bar{x}, \bar{y})$ in $X \times Y$, where M is closed. We conclude that $(\bar{x}, \bar{y}) \in M$, or $\bar{y} \in Q(\bar{x})$, and property (K) holds at $\bar{x}$. Here we have assumed that $\bigcap_{\delta > 0} \text{cl } Q(\bar{x};\delta)$ is not empty. If this set is empty, then certainly $Q(\bar{x})$ is empty, and property (K) holds at $\bar{x}$.

Remark 2. We have noticed in Remark 1 that (8.5.iii) (direct and inverse) holds even if some of the sets $Q(x)$ are empty (but property (K) is verified at every $x \in X$). If A denotes the set of all $x \in X$ where $Q(x)$ is not empty, then (8.5.iii) can be reworded by saying that M is closed in $X \times Y$ if and only the map $x \to Q(x)$ has property (K) in cl A (in A if A is closed). The requirement involving the closure of A is needed, as the following example shows: $x \in A = [0 < x < 1] \subset R$, $Q(x) = \{0\} \subset R$. The sets Q have property (K) in A, but the graph $M = [(x, 0) | 0 < x < 1]$ is not closed in R^2. Statement (8.5.iii) with this remark and the parallel statement (8.5.v) below will be summarized in Remark 3. As a parenthetical remark, note that if we take $Q(x) = [y | y \geq x^{-1}(1 - x)^{-1}]$, $0 < x < 1$, $Q(x)$ the empty set otherwise, then A is open in R but M is closed in R^2, and the map $x \to Q(x)$ has property (K) at all $x \in$ cl A(and at all $x \in R$).

Concerning the relations between upper semicontinuity by set inclusion and properties (K) and (Q), we shall limit ourselves to set valued maps $x \to Q(x)$, $x \in A \subset R^\nu$, $Q(x) \subset R^n$.

8.5.iv. *Let $x \to Q(x)$, $x \in A \subset R^\nu$, $Q(x) \subset R^n$ be a set valued map which is upper semicontinuous by set inclusion at $x_0 \in A$. If the set $Q(x_0)$ is closed, then the* *sets $Q(x)$ have property* (K) *at* x_0; *if the set* $Q(x_0)$ *is closed and convex, the sets* $Q(x)$ *have property* (Q) *at* x_0.

Indeed, given $\varepsilon > 0$, there is $\delta > 0$ such that

$$Q(x_0) \subset Q(x_0;\delta) \subset [Q(x_0)]_\varepsilon.$$

Hence

$$Q(x_0) \subset \bigcap_{\delta>0} \operatorname{cl} Q(x_0;\delta) \subset \operatorname{cl}[Q(x_0)]_\varepsilon,$$

and since $Q(x_0)$ is closed and ε arbitrary, we also have

$$Q(x_0) \subset \bigcap_{\delta>0} \operatorname{cl} Q(x_0;\delta) \subset Q(x_0),$$

and thus equality holds throughout in this relation. Analogous proof holds for property (Q) if the set $Q(x_0)$ is closed and convex.

B. The Function F and Related Sets $\tilde{Q}(x)$ and M

First let U be a given subset of R^n, and $F(u)$, $u \in U$, be a real valued function, finite everywhere in U. The epigraph of F, or epi F, is defined as the set

$$\tilde{Q} = \text{epi } F = [(z^0, u) \,|\, z^0 \geq F(u),\ u \in U] \subset R^{n+1}.$$

We shall think of F as defined everywhere in R^n by taking $F = +\infty$ for $u \in R^n - U$ and then we say that F is an extended function. This does not change the set $\tilde{Q} = \text{epi } F$.

8.5.v. *If F is an extended function, then* epi F *is closed in* R^{n+1} *if and only if F is lower semicontinuous in* R^n. *This is certainly the case if U is closed in* R^n *and F is lower semicontinuous on U.*

Proof. Let us assume that F is lower semicontinuous in R^n and let us prove that epi F is closed. Let $(\bar{z}, \bar{u}) \in \operatorname{cl}(\text{epi } F)$. Then there is a sequence (z_k, u_k), $k = 1, 2, \ldots$, of points $(z_k, u_k) \in \text{epi } F$ with $z_k \to \bar{z}$, $u_k \to \bar{u}$ as $k \to \infty$, with $\bar{u} \in R^n$ and $z_k \geq F(u_k)$. By lower semicontinuity $F(\bar{u}) \leq \liminf F(u_k) \leq \lim z_k = \bar{z}$, and $(\bar{z}, \bar{u}) \in \text{epi } F$, that is, epi F is closed. Assume that epi F is closed, and let us prove that F is lower semicontinuous. Negate. Then there are points $\bar{u}, u_1, u_2, \ldots$, such that $u_k \to \bar{u}$, $F(u_k) \to F(\bar{u}) - \varepsilon$ for some $\varepsilon > 0$. Then $(F(u_k), u_k) \in \text{epi } F$, $(F(u_k), u_k) \to (F(\bar{u}) - \varepsilon, \bar{u})$, and thus $(F\bar{u} - \varepsilon, \bar{u}) \notin \text{epi } F$, that is, epi F is not closed, a contradiction. Here we have assumed $F(u)$ finite. Analogous argument holds if $F(u) = +\infty$ or $F(u) = -\infty$. □

Remark 3. Note that if U is not closed, the mere lower semicontinuity of F in U does not imply the lower semicontinuity of F in R^n. For instance, for $U = [u \,|\, -\infty < u < 0] \subset R$, $F(u) = 0$ if $u < 0$, $F(u) = +\infty$ if $u \geq 0$, then F is lower semicontinuous in U but not in R^n, and epi F is not closed. Note the following slight extension of (8.5.v): If $F_1(u), \ldots, F_\alpha(u)$ are α extended functions on R^n, and $\tilde{Q}$ denotes the set $\tilde{Q} = [(z,u) = (z^1, \ldots, z^\alpha, u^1, \ldots, u^n) | z^i \geq F_i(u), i = 1, \ldots, \alpha, u \in R^n]$, then $\tilde{Q}$ is closed in $R^{n+\alpha}$ if and only if the α functions $F_1, \ldots, F_\alpha$ are lower semicontinuous in R^n. We leave the proof of this statement as an exercise for the reader.

Remark 4. Let A be any subset of R^ν, let $x \to U(x)$, $x \in A \subset R^\nu$, $U(x) \subset R^n$, and $M = [(x,u) | x \in A, u \in U(x)] \subset R^{\nu+n}$. Let $F(x,u)$ be a real valued function defined on M, finite everywhere on M, and let

$$\tilde{M} = [(x, z^0, u) | x \in A, u \in U(x), z^0 \geq F(x,u)].$$

For any $x \in A$, let $\tilde{Q}(x)$ denote the set $\tilde{Q}(x) = [(z^0, u) | z^0 \geq F(x,u), u \in U(x)] \subset R^{n+1}$. Again, let us think of F as defined everywhere in $R^{\nu+n}$ by taking $F = +\infty$ in $R^{\nu+n} - M$ so that $\tilde{Q}(x)$ is the empty set for $x \in R^\nu - A$. Combining (8.5.iii) and (8.5.v), we can now state that $\tilde{M}$ is closed if and only if the extended function $F(x,u)$ is lower semi-continuous in $R^{\nu+n}$, and if and only if the sets $\tilde{Q}(x)$, $x \in R^\nu$, have property (K) in R^ν (that is, the set valued map $x \to \tilde{Q}(x)$ has property (K) in R^ν, or equivalently in cl A).

C. The Functions f_0, f and the Related Function T and Sets $\tilde{Q}(x)$ and $\tilde{M}$.

Let A be a closed subset of R^ν, let $x \to U(x)$, $x \in A \subset R^\nu$, $U(x) \subset R^m$, and $M = [(x,u) | x \in A, u \in U(x)] \subset R^{\nu+m}$. Let $f_0(x,u)$ and $f(x,u) = (f_1, \ldots, f_n)$ be functions defined on M, and for every $x \in A$, let $Q(x) = f(x, U(x)) = [z | z = f(x,u), u \in U(x)] \subset R^n$, and, as before, let $M_0 = [(x,z) | x \in A, z \in Q(x)] = [(x,z) | x \in A, z = f(x,u), u \in U(x)] \subset R^{\nu+n}$. Again, for every $x \in A$, let $\tilde{Q}(x)$ denote the set $\tilde{Q}(x) = [(z^0, z) | z^0 \geq f_0(x,u), z = f(x,u), u \in U(x)] \subset R^{n+1}$, and let $\tilde{M}_0 = [(x, z^0, z) | x \in A, (z^0, z) \in \tilde{Q}(x)]$. In other words, M is the graph of the sets $U(x)$, M_0 the graph of the sets $Q(x)$, and $\tilde{M}_0$ the graph of the sets $\tilde{Q}(x)$. Finally, for every $x \in A$ and $z \in Q(x)$, let $T(x,z)$ be the extended real function

$$\begin{aligned} T(x,z) &= \inf[z^0 | (z^0, z) \in \tilde{Q}(x)] = \inf[z^0 | (x, z^0, z) \in \tilde{M}_0] \\ &= \inf[z^0 | z^0 \geq f_0(x,u), z = f(x,u), u \in U(x)]. \end{aligned}$$

Whenever inf above can be replaced by min, then $(T(x,z), z) \in \tilde{Q}(x)$. This is certainly the case if the sets $\tilde{Q}(x)$ are closed (in particular if they have property (K) in the closed set A), and then $\tilde{M}_0 = \text{epi } T(x,z)$ and $\tilde{Q}(x) = \text{epi}_z\, T(x,z)$ for every $x \in A$. For $\tilde{M}_0 = \text{epi } T(x,z)$, and by combining (8.5.iii) and (8.5.v), we can say that $\tilde{M}_0$ is closed in $R^{\nu+n+1}$ if and only if the extended function $T(x,z)$ is lower semicontinuous in $R^{\nu+n}$, and if and only if the sets $\tilde{Q}(x)$ have property (K) in A (that is, the set valued map $x \to \tilde{Q}(x)$ has property (K) in A).

Remark 5. The following example shows that M closed (but not compact) and f_0, f continuous on M does not imply that $\tilde{M}_0$ is closed, that T is lower semicontinuous, or that the sets $\tilde{Q}(x)$ have property (K). Take $\nu = 1$, $m = 2$, $f(x, u, v) = |u|$, $f_0(x, u, v) = -|x|\,|v| + 4^{-1}x^2v^2$, $x \in A = R$, $(u, v) \in U = R^2$, so that $M = R^3$ is certainly closed. Note that $\psi(\zeta) = -a\zeta + 4^{-1}a^2\zeta^2$ has minimum -1 at $\zeta = 2/a$ if $a > 0$, and $\psi \equiv 0$ if $a = 0$. Thus, $\tilde{Q}(x) = [(z^0, z)|z^0 \geq -1, z \geq 0]$ if $x \neq 0$, $\tilde{Q}(0) = [(z^0, z)|z^0 \geq 0, z \geq 0]$ if $x = 0$. The set $\tilde{M}_0$ is not closed, the function T is not lower semicontinuous, and the sets $\tilde{Q}(x)$ do not have property (K) at $x = 0$.

For M compact, the following simple statements hold, which we will have occasion to use.

8.5.vi. (a) *Let $x \to U(x)$, $x \in A \subset R^\nu$, $U(x) \subset R^m$, and $M = [(x, u)|x \in A, u \in U(x)] \subset R^{\nu+m}$. Let $f(x, u) = (f_1, \ldots, f_n)$ be a function defined in M, and for every $x \in A$ let $Q(x) = f(x, U(x)) = [z|z = f(x, u), u \in U(x)] \subset R^n$. Also, let $M_0 = [(x, z)|x \in A, z \in Q(x)] = [(x, z)|x \in A, z = f(x, u), u \in U(x)] \subset R^{\nu+n}$. If M is compact and f continuous on M, then the sets $Q(x)$ are all compact and contained in a fixed ball in R^n, are upper semicontinuous by set inclusion, and hence have property* (K) *in A, and if convex, also property* (Q) *in A. Moreover, M_0 is compact.*

(b) *Let A, $U(x)$, M, f, $Q(x)$, and M_0 be as above. Let $f_0(x, u)$ be a real valued function on M, and for every $x \in A$ let $\tilde{Q}(x) = [(z^0, z)|z^0 \geq f_0(x, u)$, $z = f(x, u)$, $u \in U(x)]$, and let $\tilde{M}_0 = [(x, z^0, z)|x \in A, (z^0, z) \in \tilde{Q}(x)] = [(x, z^0, z)|$ $x \in A$, $z^0 \geq f_0(x, u)$, $z = f(x, u)$, $u \in U(x)] \subset R^{\nu+1+n}$. If M is compact and f_0, f are continuous on M, then all sets $\tilde{Q}(x)$ are closed with compact projection $Q(x)$ on the z-space R^n; they are upper semicontinuous by set inclusion, and hence have property* (K) *in A, and if convex, also property* (Q) *in A. Moreover, $\tilde{M}_0$ is closed with compact projection M_0 on the xz-space $R^{\nu+n}$.*

(c) *For $(x, z) \in M_0$ take*

$$T(x, z) = \inf[z^0|(z^0, z) \in \tilde{Q}(x)] = \inf[z^0|(x, z^0, z) \in \tilde{M}_0].$$

If M is compact and f_0, f continuous on M, then inf *can be replaced by* min, *and $\tilde{M}_0 = [(x, z^0, z)|z^0 \geq T(x, z), (x, z) \in M_0]$. Moreover, T is lower semicontinuous on M_0. If T is defined in all of $R^{\nu+n}$ by taking $T = +\infty$ in $R^{\nu+n} - M_0$, then T is certainly lower semicontinuous in $R^{\nu+n}$.*

Proof. The compactness of M implies the compactness of A and of every set $U(\bar{x})$. Indeed, $U(\bar{x})$ is the intersection of M with the hyperspace $x = \bar{x}$ in $R^{\nu+m}$ and A is the projection of M on the hyperspace $u = 0$ of $R^{\nu+m}$. For any $x_0 \in A$ let us consider the set $Q(x_0) = f(x_0, U(x_0))$. Then $Q(x_0)$, as the continuous image of a compact set $U(x_0)$, is closed and compact.

Let us prove that the sets $Q(x)$ are upper semicontinuous by set inclusion at x_0. Suppose this is not true. Then there are an $\varepsilon > 0$, points $x_k \in A$ with $x_k \to x_0$, and points $z_k \in Q(x_k)$, all z_k at a distance $\geq \varepsilon$ from $Q(x_0)$. But $z_k = f(x_k, u_k)$ for some $u_k \in U(x_k)$, so $(x_k, u_k) \in M$, a compact set. Thus there is some $u_0 \in R^m$, and a subsequence, say still $[k]$ for brevity, with $(x_k, u_k) \to (x_0, u_0)$ in $R^{\nu+n}$. But $(x_0, u_0) \in M$ since M is closed, so $u_0 \in U(x_0)$, and $z_0 = f(x_0, u_0) \in Q(x_0)$, and by the continuity of f, also $z_k = f(x_k, u_k) \to f(x_0, u_0) = z_0$, with $z_0 \in Q(x_0)$, a contradiction. Thus, the sets $Q(x)$ are upper semicontinuous by set inclusion, and hence have properties (K) and (Q) by (8.5.iv). This proves part (a) of (8.5.vi). We leave the proof of parts (b) and (c) as an exercise for the reader. □

Remark 6. Under the assumption that f is continuous on M, then M compact as in (8.5.vi) implies that all sets $Q(x)$ are compact and all contained in some ball in R^n, but the converse of course is not true. It is left as an exercise for the reader to see which parts of (8.5.vi) are still valid under the sole hypothesis that the sets $Q(x)$ are all compact and contained in a fixed ball in R^n.

Remark 7. By using here the same notation as in (8.5.vi), the first conclusion in (c) can be summed up by saying that $\tilde{M}_0$ is the epigraph of T, or $M_0 = \text{epi } T$. The second conclusion in (c) cannot be improved, that is, T may not be continuous on A, but only lower semicontinuous as stated. This is shown by the following example. Let $\nu = n = m = 1$, and take $A = [x \mid 0 \le x \le 1]$, $U(x) = \{0\}$ if $0 < x \le 1$, $U(0) = \{0\} \cup \{1\}$, so that $M = [(x, u) \mid u = 0 \text{ if } 0 < x \le 1, u = 0 \text{ and } u = 1 \text{ if } x = 0]$. Let $f(x, u) = x$, $f_0(x, u) = -u$, so that $Q(x) = [z = x, 0 \le x \le 1]$, $M_0 = [(x, z) \mid 0 \le x \le 1, z = x]$. Finally, $\tilde{Q}(x) = [z^0 \ge 0, z = x]$ if $0 < x \le 1$, $\tilde{Q}(0) = [z^0 \ge -1, z = 0]$, $\tilde{M}_0 = [(x, z^0, z) \mid z = x, z^0 \ge 0 \text{ if } 0 < x \le 1, z^0 \ge -1 \text{ if } x = 0]$, and $T(x, z) = T(x, x) = 0$ if $0 < x \le 1$ and $T(0, 0) = -1$. Here T is lower semicontinuous but not continuous, M is compact, and f_0, f are continuous on M.

8.5.vii. *If A is closed, M is closed, $f_0(x, u)$, $f(x, u)$ are continuous on M, and either $f_0 \to +\infty$, or $|f| \to +\infty$ uniformly on A as $|u| \to +\infty$, then the set $\tilde{M}_0$ is closed, that is, the set valued function $x \to \tilde{Q}(x)$, $x \in A \subset R^\nu$, $\tilde{Q}(x) \subset R^{n+1}$, has property* (K) *in A.*

The proof is left as an exercise for the reader.

Here is an example of unbounded closed and convex sets possessing properties (K) and (Q) but not upper semicontinuous by set inclusion:

$$Q(t) = [(x, y) \mid x \ge 0, 0 \le y \le tx], \qquad 0 \le t \le 1.$$

Here $t \to Q(t)$, $t \in [0, 1] \subset R$, $Q(t) \subset R^2$, each $Q(t)$ is a cone in R^2 (an angle), and obviously, for $t > t_0 \ge 0$, $Q(t)$ is not contained in any $[Q(t_0)]_\varepsilon$, no matter how close t is to t_0.

Remark 8. Here is a situation under which property (Q) holds in the weak form as in Remark 1 of this section. Let A be a given subset of the tx-space $R^{\nu+n}$, let A_0 be the projection of A on the t-space R^ν, and for $t \in A_0$ let $A(t) = [x \mid (t, x) \in A]$. Let $G \subset A_0 \subset R^\nu$ be a given measurable subset of finite measure of A_0, and let $x(t)$, $\xi_k(t)$, $\bar{\xi}_k(t)$, $t \in G$, $k = 1, 2, \ldots$, be given measurable functions.

8.5.viii. *If* (a) *for almost all $t \in G$, we have $x(t) \in A(t)$, $Q(t, x(t))$ is closed and convex, and $\bar{\xi}_k(t) \in Q(t, x(t))$, $k = 1, 2, \ldots$; and* (b) *the differences $\delta_k(t) = \xi_k(t) - \bar{\xi}_k(t)$, $t \in G$, $k = 1, 2, \ldots$, approach zero pointwise a.e. in G as $k \to \infty$, then*

$$Q(t, x(t)) \supset \bigcap_{h=1}^{\infty} \operatorname{cl} \operatorname{co} \left\{ \bigcup_{s=h}^{\infty} \xi_s(t) \right\}, \quad t \in G \text{ (a.e.)}.$$

For instance, we may assume that $\xi_k(t) \in Q(t, x_k(t))$, that $x_k(t) \to x(t)$ pointwise a.e. in G as $k \to \infty$, and we shall see in Section 13.4 specific hypotheses under which $\delta_k \to 0$ as needed.

Proof. Given $\eta > 0$, there are a compact subset K of G and an integer k_0 such that $\operatorname{meas}(G - K) \leq \eta$ and $|\xi_k(t) - \bar{\xi}_k(t)| \leq \eta$ for all $t \in K$, $k \geq k_0$. Since $\bar{\xi}_k(t) \in Q(t, x(t))$ and $Q(t, x(t))$ is closed and convex, we also have

$$\operatorname{cl} \operatorname{co} \left\{ \bigcup_{s=h}^{\infty} \xi_s(t) \right\} \subset Q(t, x(t)),$$

and for $t \in K$ and $h \geq k_0$ also

$$\operatorname{cl} \operatorname{co} \left\{ \bigcup_{s=h}^{\infty} \xi_s(t) \right\} \subset [Q(t, x(t))]_\eta.$$

Hence,

$$\bigcap_{h=1}^{\infty} \operatorname{cl} \operatorname{co} \left\{ \bigcup_{s=h}^{\infty} \xi_s(t) \right\} \subset [Q(t, x(t))]_\eta,$$

and this is true for all $t \in K$ with $\operatorname{meas}(G - K) \leq \eta$. Since η is arbitrary, we easily derive from this that for almost all $t \in G$ we have

$$\bigcap_{h=1}^{\infty} \operatorname{cl} \operatorname{co} \left\{ \bigcup_{s=h}^{\infty} \xi_s(t) \right\} \subset Q(t, x(t)). \qquad \square$$

8.6 The Elementary Closure Theorem

We consider here an orientor field equation as defined in Section 1.12

$$(8.6.1) \qquad x'(t) \in Q(t, x(t)), \quad x(t) = (x^1, \ldots, x^n), \qquad (t, x(t)) \in A,$$

where A is a given subset of the tx-space R^{n+1}, and we assume that to every $(t, x) \in A$ a subset $Q(t, x)$ of R^n is assigned. Then a solution of (8.6.1) is an AC n-vector function $x(t) = (x^1, \ldots, x^n)$, $t_1 \leq t \leq t_2$, satisfying (8.6.1) a.e. in $[t_1, t_2]$.

We are here interested in the following question: Given a sequence $x_k(t)$, $t_1 \leq t \leq t_2$, $k = 1, 2, \ldots$, of AC solutions of (8.6.1) convergent in some mode of convergence toward an AC function $x(t)$, $t_1 \leq t \leq t_2$, can we conclude that x is a solution of (8.6.1)? In this section we give sufficient conditions for the question to have an affirmative answer in connection with uniform convergence, namely the mode (a) of convergence in Section 2.14. In Chapter 10 we will see that the question has an affirmative answer also in connection with the mode (b) of convergence in Section 2.14 under weaker assumptions. An example at the end of this section will show that the question may not always have a positive answer.

8.6.i (A CLOSURE THEOREM). *Let A be a closed subset of the tx-space R^{n+1}, for every $(t,x) \in A$ let $Q(t,x)$ be a given subset of points $z = (z^1, \ldots, z^n) \in R^n$, and let $x_k(t) = (x_k^1, \ldots, x_k^n)$, $t_{1k} \le t \le t_{2k}$, $k = 1, 2, \ldots$, be a sequence of* AC *solutions of the orientor field* (8.6.1) *convergent in the ρ-metric to an* AC *function $x(t) = (x^1, \ldots, x^n)$, $t_1 \le t \le t_2$. Let us assume that for almost all $\bar{t} \in [t_1, t_2]$ the sets $Q(t,x)$ have property* (Q) *with respect to (t,x) at $(\bar{t}, x(\bar{t}))$. Then $x(t)$, $t_1 \le t \le t_2$, is also a solution of the orientor field* (8.6.1).

In other words, we know that each $x_k(t)$, $t_{1k} \le t \le t_{2k}$, $k = 1, 2, \ldots$, is AC, that $(t, x_k(t)) \in A$ for every $t \in [t_{1k}, t_{2k}]$, and that $dx_k/dt \in Q(t, x_k(t))$ a.e. in $[t_{1k}, t_{2k}]$; we know that $\rho(x_k, x) \to 0$, hence $t_{1k} \to t_1$, $t_{2k} \to t_2$, as $k \to \infty$, and that $x(t)$ is AC in $[t_1, t_2]$, and we want to prove that $(t, x(t)) \in A$ for all $t \in [t_1, t_2]$, and that $dx/dt \in Q(t, x(t))$ a.e. in $[t_1, t_2]$.

Proof of (8.6.i). The vector functions $x'(t)$, $t_1 \le t \le t_2$, $x_k'(t)$, $t_{1k} \le t \le t_{2k}$, are defined a.e. in $[t_1, t_2]$ and $[t_{1k}, t_{2k}]$ respectively, $k = 1, 2, \ldots$, and are L-integrable in the respective intervals (that is, each component is L-integrable).

Now $\rho(x_k, x) \to 0$, hence $t_{1k} \to t_1$, $t_{2k} \to t_2$ and

$$\max[|x_k(t) - x(t)|, \ -\infty < t < +\infty] \to 0$$

as $k \to \infty$ (after extension of x_k and x to $(-\infty, +\infty)$ by continuity and constancy of these functions outside their intervals of definition). Thus if $t \in (t_1, t_2)$, or $t_1 < t < t_2$, then $t_{1k} < t < t_{2k}$ for all k sufficiently large, $(t, x_k(t)) \in A$ for the same k, and $x_k(t) \to x(t)$ as $k \to \infty$. Therefore, we have $(t, x(t)) \in A$ for all $t_1 < t < t_2$ since A is closed. Because $x(t)$ is continuous in $[t_1, t_2]$ and hence continuous at t_1 and t_2, and again A is closed, we conclude that $(t, x(t)) \in A$ for every $t_1 \le t \le t_2$.

For almost all $t \in [t_1, t_2]$ the derivative $x'(t)$ exists and is finite. Let t_0 be such a point with $t_1 < t_0 < t_2$. Then, there is a $\sigma > 0$ with $t_1 < t_0 - \sigma < t_0 + \sigma < t_2$, and for some k_0 and all $k \ge k_0$, also $t_{1k} < t_0 - \sigma < t_0 + \sigma < t_{2k}$. Let $x_0 = x(t_0)$. We have $x_k(t) \to x(t)$ uniformly in $[t_0 - \sigma, t_0 + \sigma]$, and all functions $x(t)$, $x_k(t)$ are continuous in the same interval. Thus, they are equicontinuous in $[t_0 - \sigma, t_0 + \sigma]$. Given $\varepsilon > 0$, there is $\delta > 0$ such that $t, t' \in [t_0 - \sigma, t_0 + \sigma]$, $|t - t'| \le \delta$, $k \ge k_0$ implies

$$|x(t) - x(t')| \le \varepsilon/2, \qquad |x_k(t) - x_k(t')| \le \varepsilon/2. \tag{8.6.2}$$

We can assume $0 < \delta < \sigma$, $\delta \le \varepsilon$. For any h, $0 < h < \delta$, let us consider the averages

$$m_h = h^{-1} \int_0^h x'(t_0 + s)\, ds = h^{-1}[x(t_0 + h) - x(t_0)], \tag{8.6.3}$$

$$m_{hk} = h^{-1}\int_0^h x_k'(t_0+s)\,ds = h^{-1}[x_k(t_0+h) - x_k(t_0)]. \tag{8.6.4}$$

Given $\tau > 0$, we can take h so small that

$$|m_h - x'(t_0)| \le \tau. \tag{8.6.5}$$

Having so fixed h, let us take $k_1 \ge k_0$ so large that

$$|m_{hk} - m_h| \le \tau, \qquad |x_k(t_0) - x(t_0)| \le \varepsilon/2 \tag{8.6.6}$$

for all $k \ge k_1$. This is possible because $x_k(t) \to x(t)$ as $k \to \infty$ both at $t = t_0$ and $t = t_0 + h$. Finally, for $0 \le s \le h$,

$$\begin{aligned} |x_k(t_0+s) - x(t_0)| &\le |x_k(t_0+s) - x_k(t_0)| + |x_k(t_0) - x(t_0)| \\ &\le \varepsilon/2 + \varepsilon/2 = \varepsilon, \end{aligned} \tag{8.6.7}$$

$$\begin{aligned} &|(t_0+s) - t_0| = s \le h \le \delta \le \varepsilon, \\ &x_k'(t_0+s) \in Q(t_0+s,\, x_k(t_0+s)) \quad \text{a.e.} \end{aligned} \tag{8.6.8}$$

Hence, for almost all s, $0 \le s \le h$, $x_k'(t_0+s) \in Q(t_0, x_0; 2\varepsilon)$ and consequently

$$x_k'(t_0+s) \in \text{cl co } Q(t_0, x_0; 2\varepsilon) \quad \text{a.e. in } [0,h].$$

The average m_{hk} as defined by (8.6.4) is then also a point of the same closed and convex set, or

$$m_{hk} \in \text{cl co } Q(t_0, x_0; 2\varepsilon)$$

for the chosen h and every $k \ge k_1$. By the relations (8.6.5) and (8.6.6) we derive

$$|x'(t_0) - m_{hk}| \le |x'(t_0) - m_h| + |m_h - m_{hk}| \le 2\tau,$$

and hence

$$x'(t_0) \in [\text{cl co } Q(t_0, x_0; 2\varepsilon)]_{2\tau}.$$

Here τ is an arbitrary number, and the set in brackets is closed; hence

$$x'(t_0) \in \bigcap_\tau [\text{cl co } Q(t_0, x_0; 2\varepsilon)]_{2\tau} = \text{cl co } Q(t_0, x_0; 2\varepsilon),$$

for every $\varepsilon > 0$. Thus, by property (Q),

$$x'(t_0) \in \bigcap_\varepsilon \text{cl co } Q(t_0, x_0; 2\varepsilon) = Q(t_0, x_0).$$

We have proved that for almost all $t \in [t_1, t_2]$, we have $dx/dt \in Q(t, x(t))$. The closure theorem (8.6.i) is thereby proved. □

The following example illustrates the closure theorem (8.6.i). Let $n = 1$, $A = R^2$, $Q = Q(t,x) = [z \mid -1 \le z \le 1]$, and $x_k(t)$, $0 \le t \le 1$, $k = 1, 2, \dots$, be defined by $x_k(t) = t - ik^{-1}$ if $ik^{-1} \le t \le ik^{-1} + (2k)^{-1}$, $x_k(t) = (i+1)k^{-1} - t$ if $ik^{-1} + (2k)^{-1} \le t \le (i+1)k^{-1}$, $i = 1, 2, \dots, k-1$. Then $x_k(t) \to x_0(t) = 0$ uniformly in $[0,1]$. On the other hand, $x_k'(t) = \pm 1$ according as t is an interior point of one or the other of the two sets of intervals above, $x_0'(t) = 0$, and $x_k'(t)$, $x_0'(t) \in Q$ for almost all t. Here Q is a closed convex set. If we had taken $Q = Q(t,x) = [z \mid z = -1 \text{ and } z = +1]$, then obviously $x_k'(t) \in Q$ while $x_0'(t) \notin Q$. Here Q is closed but not convex.

8.7 Some Fatou-Like Lemmas

8.7.i (Fatou's Lemma). *If $\eta_k(t) \geq 0$, $a \leq t \leq b$, $k = 1, 2, \ldots$, is a sequence of nonnegative L-integrable functions, and*

$$\liminf_{k\to\infty} \int_a^b \eta_k(t)\,dt = i < +\infty,$$

then $\eta(t) = \liminf_{k\to\infty} \eta_k(t)$, $a \leq t \leq b$, is L-integrable, and $\int_a^b \eta(t)\,dt \leq i$. Under the same hypotheses, if the functions $\eta_k(t)$ converge in measure toward a function $\eta_0(t)$ [or they converge pointwise a.e. *to $\eta_0(t)$], then $\eta_0(t) \geq 0$ is L-integrable, and again $\int_a^b \eta_0(t)\,dt \leq i$.*

We refer for this lemma to McShane [I, p. 167]. The same statement holds if $\eta_k(t) \geq -\psi(t)$ for all t and k, where $\psi \geq 0$ is a fixed L-integrable function. It is enough to apply the statement above to the functions $\eta_k(t) + \psi(t) \geq 0$. The same statement (8.7.i) holds even if $\psi(t)$, $\psi_k(t)$, $k = 1, 2, \ldots$, are L-integrable functions with $\eta_k(t) \geq -\psi_k(t)$, $\psi_k(t) \to \psi(t)$ as $k \to \infty$ a.e. in $[a, b]$, and $\int_a^b \psi_k(t)\,dt \to \int_a^b \psi(t)\,dt$ as $k \to \infty$. Again, it is enough to apply (8.7.i) to the functions $\eta_k(t) + \psi_k(t) \geq 0$.

Under the conditions of Fatou's lemma, let us consider for each $h > 0$ the same function η above and the following functions $\tilde{\eta}_k$ and $\tilde{\eta}$:

$$\eta(t) = \liminf_{k\to\infty} \eta_k(t), \qquad \tilde{\eta}_h(t) = \liminf_{k\to\infty} h^{-1} \int_t^{t+h} \eta_k(s)\,ds, \tag{8.7.1}$$
$$\tilde{\eta}(t) = \liminf_{h\to 0^+} \tilde{\eta}_h(t), \qquad a \leq t \leq b,$$

where in the second relation we understand that $\eta_k(s) = 0$ for s outside $[a, b]$.

8.7.ii (A Variant of Fatou's Lemma). *Under the conditions of* (8.7.i), *for almost all $t \in [a, b]$ we have $0 \leq \eta(t) \leq \tilde{\eta}(t)$, all functions $\tilde{\eta}_h$ and $\tilde{\eta}$ are L-integrable, and $\int_a^b \tilde{\eta}_h(t)\,dt \leq i$, $\int_a^b \tilde{\eta}(t)\,dt \leq i$.*

Proof.

(a) Since $\eta_k(t) \geq 0$, we certainly have $\eta(t) \geq 0$. Let us define η_k and $\tilde{\eta}_h$ to be equal to zero for $t \geq b$. Then by Fatou's lemma in the interval $[t, t + h]$ we have

$$\int_t^{t+h} \eta(s)\,ds \leq \liminf_{k\to\infty} \int_t^{t+h} \eta_k(s)\,ds,$$

and by multiplication by h^{-1}

$$h^{-1} \int_t^{t+h} \eta(s)\,ds \leq \tilde{\eta}_h(t)$$

for all $t \in [a, b]$ and any $h > 0$. Since η is L-integrable, for almost all t, $\eta(t)$ is the derivative of its indefinite integral. In other words, for almost all t, there is $h_0 = h_0(t, \varepsilon) > 0$ such that for $0 < h \leq h_0(t, \varepsilon)$ we have

$$\left| h^{-1} \int_t^{t+h} \eta(s)\, ds - \eta(t) \right| \leq \varepsilon,$$

and hence

$$\eta(t) - \varepsilon \leq \tilde{\eta}_h(t), \qquad 0 < h \leq h_0(t, \varepsilon).$$

As $h \to 0$, by keeping t fixed, we have $\eta(t) - \varepsilon \leq \tilde{\eta}(t)$, and this holds for every $\varepsilon > 0$. Thus, $\eta(t) \leq \tilde{\eta}(t)$ for almost all t.

(b) For every k and for every h, $0 < h < b - a$, we have now

$$\int_a^b dt \left(h^{-1} \int_t^{t+h} \eta_k(s)\, ds \right) = h^{-1} \left[\int_a^{a+h} (s - a)\eta_k(s)\, ds + \int_{a+h}^b h\eta_k(s)\, ds \right]$$

$$\leq \int_a^b \eta_k(s)\, ds;$$

thus, each function $h^{-1} \int_t^{t+h} \eta_k(s)\, ds$ is L-integrable in $[a, b]$, and by Fatou's lemma we have

$$\int_a^b \left(\liminf_{k \to \infty} h^{-1} \int_t^{t+h} \eta_k(s)\, ds \right) dt \leq \liminf_{k \to \infty} \int_a^b dt \left(h^{-1} \int_t^{t+h} \eta_k(s)\, ds \right),$$

and

$$\int_a^b \tilde{\eta}_h(t)\, dt \leq \liminf_{k \to \infty} \int_a^b \eta_k(t)\, dt \leq i.$$

Finally, again by Fatou's lemma,

$$\int_a^b \tilde{\eta}(t)\, dt \leq \liminf_{h \to 0+} \int_a^b \tilde{\eta}_h(t)\, dt \leq i,$$

and (8.7.ii) is thereby proved. □

8.8 Lower Closure Theorems with Respect to Uniform Convergence

Problems of control of the Lagrange and Bolza types are usually reduced to orientor fields of the form

$$(8.8.1) \qquad (\eta(t), x'(t)) \in \tilde{Q}(t, x(t)), \quad x(t) = (x^1, \ldots, x^n), \qquad t_1 \leq t \leq t_2,$$

where $\int_{t_1}^{t_2} \eta(t)\, dt$ is the value of the functional, and the subsets $\tilde{Q}(t, x)$ of R^{n+1} have the property (a) in (8.8.i) below. The problem of closure in Section 8.6 is replaced here by the following question which is a combination of closure and lower semicontinuity: Given a sequence $\eta_k(t)$, $x_k(t)$, $t_1 \leq t \leq t_2$, $k = 1, 2, \ldots$, of functions, $\eta_k(t)$ L-integrable, $x_k(t)$ AC, satisfying (8.8.1) (a.e.), with $x_k(t)$ converging in some mode of convergence toward an AC function $x(t)$, $t_1 \leq t \leq t_2$, is there an L-integrable function $\eta(t)$, $t_1 \leq t \leq t_2$, such that the pair η, x satisfies (8.8.1) a.e. in $[t_1, t_2]$ and such that $\int_{t_1}^{t_2} \eta(t)\, dt \leq \liminf_{k \to \infty} \int_{t_1}^{t_2} \eta_k(t)\, dt$? This problem is often called a problem of "lower closure". In this section we discuss it in relation to uniform convergence of the trajectories.

8.8.i (A LOWER CLOSURE THEOREM). *Let A be a closed subset of the tx-space R^{n+1} and for every $(t,x) \in A$ let $\tilde{Q}(t,x)$ be a given subset of points $z = (z^0, z) = (z^0, z^1, \ldots, z^n) \in R^{n+1}$, with the following properties*: (a) *if $(z^0, z) \in \tilde{Q}(t,x)$ and $z^0 \le z^{0\prime}$, then $(z^{0\prime}, z) \in \tilde{Q}(t,x)$*; (b) *there is a real valued function $\psi(t) \ge 0$, $t \in R$, locally integrable, such that if $(z^0, z) \in \tilde{Q}(t,x)$ then $z^0 \ge -\psi(t)$. Let $\eta_k(t)$, $x_k(t)$, $t_{1k} \le t \le t_{2k}$, $k = 1, 2, \ldots$, be a sequence of functions, $\eta_k(t)$ scalar, L-integrable, $x_k(t) = (x_k^1, \ldots, x_k^n)$* AC *in $[t_{1k}, t_{2k}]$, such that*

$$(t, x_k(t)) \in A, \quad (\eta_k(t), x_k'(t)) \in \tilde{Q}(t, x_k(t)), \qquad t_{1k} \le t \le t_{2k} \text{ (a.e.)}, \quad k = 1, 2, \ldots,$$

$$-\infty < \liminf_{k\to\infty} \int_{t_{1k}}^{t_{2k}} \eta_k(t)\,dt = i < +\infty,$$

and such that the functions x_k converge in the ρ-metric to an AC *function $x(t) = (x^1, \ldots, x^n)$, $t_1 \le t \le t_2$. Let us assume that for almost all $\bar{t} \in [t_1, t_2]$ the sets $\tilde{Q}(t,x)$ have property (Q) with respect to (t,x) at $(\bar{t}, x(\bar{t}))$. Then there is a real valued L-integrable function $\eta(t)$, $t_1 \le t \le t_2$, such that $(t, x(t)) \in A$, $(\eta(t), x'(t)) \in \tilde{Q}(t, x(t))$, $t_1 \le t \le t_2$* (a.e.), *and*

$$-\infty < \int_{t_1}^{t_2} \eta(t)\,dt \le i < +\infty.$$

Proof. First, we extend the functions $\eta_k(t)$ by taking them equal to zero for $t \ge t_{2k}$ and $t \le t_{1k}$. Then, these functions are all defined in $[t_1, t_2]$, and we construct the functions $\tilde{\eta}_h(t)$ and $\tilde{\eta}(t)$, $t_1 \le t \le t_2$, as in Lemma (8.7.ii). Here $\eta_k(t) \ge -\psi(t)$ for all t and k; hence, if $\eta_0(t) = \liminf \eta_k(t)$, we have $\tilde{\eta}(t) \ge \eta_0(t) \ge -\psi(t)$ for all t, where $\tilde{\eta}$ is the function defined in (8.7.1). For almost all $t \in (t_1, t_2)$ the derivative $x'(t)$ exists and is finite, and $\eta_0(t)$ and $\tilde{\eta}(t)$ are finite. Let t_0 be such a point, $t_1 < t_0 < t_2$. Then there is a $\sigma > 0$ with $t_1 < t_0 - \sigma < t_0 + \sigma < t_2$, and for some k_0 and for all $k > k_0$, also $t_{1k} < t_0 - \sigma < t_0 + \sigma < t_{2k}$. Let $x_0 = x(t_0)$, $x_0' = x'(t_0)$. We have $x_k(t) \to x(t)$ uniformly in $[t_0 - \sigma, t_0 + \sigma]$. Given $\varepsilon > 0$, there is $\delta > 0$ such that $t, t' \in [t_0 - \sigma, t_0 + \sigma]$, $|t - t'| \le \delta$, $k \ge k_0$ implies

$$|x(t) - x(t')| \le \varepsilon/2, \qquad |x_k(t) - x_k(t')| \le \varepsilon/2. \tag{8.8.2}$$

We can assume $0 < \delta < \sigma$, $\delta \le \varepsilon$. For any h, $0 < h \le \delta$, we consider the averages

$$m_h = h^{-1} \int_0^h x'(t_0 + s)\,ds = h^{-1}[x(t_0 + h) - x(t_0)], \tag{8.8.3}$$

$$m_{hk} = h^{-1} \int_0^h x_k'(t_0 + s)\,ds = h^{-1}[x_k(t_0 + h) - x_k(t_0)]. \tag{8.8.4}$$

Given $\tau > 0$, we know that for $h > 0$ sufficiently small we have $|m_h - x_0'| < \tau$. On the other hand, $\tilde{\eta}(t_0)$ is finite, and $\tilde{\eta}(t_0) = \liminf \tilde{\eta}_h(t_0)$ as $h \to 0$. Thus, we can choose h in such a way that

$$|m_h - x_0'| \le \tau, \qquad |\tilde{\eta}(t_0) - \tilde{\eta}_h(t_0)| \le \tau. \tag{8.8.5}$$

Having so fixed h, let us take $k_1 \ge k_0$ so large that

$$|m_{hk} - m_h| \le \tau, \qquad |x_k(t_0) - x(t_0)| \le \varepsilon/2 \tag{8.8.6}$$

for all $k \ge k_1$. This is possible because $x_k(t) \to x(t)$ as $k \to \infty$ both at $t = t_0$ and $t = t_0 + h$. Finally, for $0 \le s \le h$,

$$|x_k(t_0 + s) - x(t_0)| \le |x_k(t_0 + s) - x_k(t_0)| + |x_k(t_0) - x_0| \le \varepsilon/2 + \varepsilon/2 = \varepsilon,$$
$$|(t_0 + s) - t_0| = s \le h \le \delta \le \varepsilon,$$
$$(\eta_k(t_0 + s), x_k'(t_0 + s)) \in \tilde{Q}(t_0 + s, x_k(t_0 + s)) \quad \text{a.e.},$$
$$(\eta_k(t_0 + s), x_k'(t_0 + s)) \in \tilde{Q}(t_0, x_0; 2\varepsilon),$$

where $\tilde{Q}(t_0, x_0; 2\varepsilon)$ is the union of all sets $Q(t, x)$ with $(t, x) \in A$, $|t - t_0| \leq \varepsilon$, $|x - x_0| \leq \varepsilon$. Finally, by the remark at the end of Section 8.4B concerning the mean value of vector valued functions, we have

$$(8.8.7) \qquad \left(h^{-1} \int_0^h \eta_k(t_0 + s)\, ds,\ h^{-1} \int_0^h x_k'(t_0 + s)\, ds\right) \in \operatorname{cl} \operatorname{co} \tilde{Q}(t_0, x_0; 2\varepsilon).$$

Concerning the first term in the parentheses in this relation, we know that $\liminf h^{-1} \int_0^h \eta_k(t_0 + s)\, ds = \tilde{\eta}_h(t_0)$ as $k \to \infty$. Thus, there are infinitely many k such that

$$\left|h^{-1} \int_0^h \eta_k(t_0 + s)\, ds - \tilde{\eta}_h(t_0)\right| \leq \tau$$

and by comparison with (8.8.5) also

$$\left|h^{-1} \int_0^h \eta_k(t_0 + s)\, ds - \tilde{\eta}(t_0)\right| < 2\tau \quad \text{for infinitely many } k.$$

The second term in parenthesis in (8.8.7) is the average m_{hk}, and by (8.8.5) and (8.8.6) we derive that

$$|m_{kh} - x_0'| \leq 2\tau \quad \text{for all } k \text{ sufficiently large.}$$

Thus, (8.8.7) yields

$$(\tilde{\eta}(t_0), x_0') \in [\operatorname{cl} \operatorname{co} \tilde{Q}(t_0, x_0; 2\varepsilon)]_{4\tau}.$$

Here $\tau > 0$ is arbitrary, and thus

$$(\tilde{\eta}(t_0), x_0') \in \operatorname{cl} \operatorname{co} \tilde{Q}(t_0, x_0; 2\varepsilon).$$

Here $\varepsilon > 0$ is also arbitrary, and by property (Q) we derive that

$$(\tilde{\eta}(t_0, x'(t_0)) \in \tilde{Q}(t_0, x(t_0)).$$

Here t_0 is any point of (t_1, t_2) not in a set of measure zero. We have proved that

$$(\tilde{\eta}(t), x'(t)) \in \tilde{Q}(t, x(t)), \qquad t \in [t_1, t_2] \text{ (a.e.)},$$

and from Lemma (8.7.ii) we know that

$$\int_{t_1}^{t_2} \tilde{\eta}(t)\, dt \leq i.$$

The lower closure theorem (8.8.i) is thereby proved. □

We may remark here that the various scalar functions we have been dealing with are in the relation

$$-\psi(t) \leq T(t, x(t), x'(t)) \leq \eta(t) \leq \tilde{\eta}(t), \qquad t_1 \leq t \leq t_2,$$

where $T(t, x, z) = \inf[z^0 | (z^0, z) \in \tilde{Q}(t, x)]$ is the scalar function defined in Section 8.5C.

Remark 1. Note that in the closure theorem (8.6.i) and in the lower closure theorem (8.8.i) we assume that $x_k \to x$ in the ρ-metric, that is, uniformly (mode (a) of Section 2.14), and no requirement is made concerning the derivatives x_k'. In this situation, the requirement in (8.6.i) that the sets $Q(t, x)$ have property (Q) with respect to (t, x) (and the analogous requirement on the sets $\tilde{Q}(t, x)$ in statement (8.8.i)) cannot be reduced. This is shown by the following example. In Sections 10.6 and 10.8 (statements (10.6.i) and

(10.8.i)) we shall assume weak convergence of the derivatives and we will be able to dispense with explicitly requiring property (Q). However, in the proofs we shall still make use of a "reduced property (Q) with respect to x only" of certain auxiliary sets, and such property (Q) will be a consequence of the other assumptions.

Let $n = 1$ and $A = [0, 1] \times R$, let C be a closed Cantor subset of $[0, 1]$ whose measure, meas C, is positive, and let $C' = [0, 1] - C$. Then C' is the countable union of disjoint subintervals of $[0, 1]$. Let $\sigma(t)$ be a continuous function on C' which is positive and integrable on C' and which tends to $+\infty$ whenever t tends to an end of any interval component of C'. Let $m = 1$, and define

$$\begin{aligned} Q(t, x) = U(t, x) = U(t) &= \{-1\} && \text{if } t \in C, \\ &= \{u \in R \mid u \geq \sigma(t)\} && \text{if } t \in C', \end{aligned}$$

and take $\tilde{Q}(t, x) = [(z^0, z) \mid z^0 \geq 0,\ z \in Q(t, x)]$, $(t, x) \in A$. Let us extend the function σ by taking $\sigma(t) = 0$ for $t \in C$, and consider the decomposition of $[0, 1]$ into k intervals of equal length: $J_k = [t_{k,s-1}, t_{ks}]$, $s = 1, \ldots, k$, $t_{ks} = s/k$. Define $\xi_k(t)$ by taking $\xi_k(t) = \sigma(t) + v_k(t)$, where $v_k(t) = -1$ if $t \in C$, and $v_k(t) = \text{meas}(C \cap J_k)/\text{meas}(C' \cap J_k)$ if $t \in C' \cap J_{ks}$. Then $\xi_k(t)$ is integrable in $[0, 1]$, and $\xi_k(t) \in U(t)$ for every $t \in [0, 1]$ and k. Let $x_k(t) = \int_0^t \xi_k(\tau)\, d\tau$, $0 \leq t \leq 1$, or

$$x_k(t) = x(t) + y_k(t) = \int_0^t \sigma(\tau)\, d\tau + \int_0^t v_k(\tau)\, d\tau.$$

Here $y_k(t_{ks}) = 0$ for all s and k, and $|y_k(t)| \leq 1/k$. Hence, $x_k \to x$ uniformly on $[0, 1]$ as $k \to \infty$, and x_k and x are AC, with $x'(t) = \sigma(t)$, $t \in [0, 1]$ (a.e.). We also take $\eta_k(t) = 0$, $\eta(t) = 0$, $t \in [0, 1]$. Now $x'(t) = 0$ a.e. in C, while $U(t) = \{-1\}$ for $t \in C$. Thus $x'(t) \notin Q(t)$, $(\eta(t), x'(t)) \notin \tilde{Q}(t)$ on a subset C of positive measure in $[0, 1]$.

In this example $Q(t)$, $\tilde{Q}(t)$ have property (Q) (with respect to t) on the set C as well as on the set C', but not in $[0, 1]$.

8.8.ii (A LOWER CLOSURE THEOREM). *Let A be a closed subset of the tx-space R^{n+1}, and for every $(t, x) \in A$ let $\tilde{Q}(t, x)$ be a given nonempty subset of points $z = (z^0, z) = (z^0, z^1, \ldots, z^n) \in R^{n+1}$, with the following property*: (a) *if $(z^0, z) \in \tilde{Q}(t, x)$ and $z^0 \leq z^{0\prime}$, then $(z^{0\prime}, z) \in \tilde{Q}(t, x)$. Let $\eta_k(t)$, $x_k(t)$, $t_{1k} \leq t \leq t_{2k}$, $k = 1, 2, \ldots$, be a sequence of functions, $\eta_k(t)$ real valued and L-integrable, $x_k(t) = (x_k^1, \ldots, x_k^n)$* AC *in $[t_{1k}, t_{2k}]$, such that*

$$(t, x_k(t)) \in A,\ (\eta_k(t), x_k'(t)) \in \tilde{Q}(t, x_k(t)),\quad t_{1k} \leq t \leq t_{2k}, \qquad k = 1, 2, \ldots,$$

$$-\infty < \liminf_{k \to \infty} \int_{t_{1k}}^{t_{2k}} \eta_k(t)\, dt \leq i < +\infty,$$

and the functions x_k converge in the ρ-metric to an AC *function $x(t) = (x^1, \ldots, x^n)$, $t_1 \leq t \leq t_2$. Let us assume that* (b) *for every $\bar{t} \in [t_1, t_2]$, the sets $\tilde{Q}(t, x)$ have property* (Q) *with respect to (t, x) at $(\bar{t}, x(\bar{t}))$. Then there is a real valued L-integrable function $\eta(t)$, $t_1 \leq t \leq t_2$, such that $(t, x(t)) \in A$, $(\eta(t), x'(t)) \in \tilde{Q}(t, x(t))$, $t_1 \leq t \leq t_2$* (a.e.), *$-\infty < \int_{t_1}^{t_2} \eta(t)\, dt \leq i$.*

Proof. For $(t, x) \in A$ let $Q(t, x)$ denote the projection of $\tilde{Q}(t, x)$ on the z-space R^n, and note that, for

$$T(t, x, z) = \inf[\eta \mid (\eta, z) \in \tilde{Q}(t, x)]$$

we have $-\infty \leq T(t, x, z) < +\infty$ for $z \in Q(t, x)$, and $T(t, x, z) = +\infty$ for $z \in R^n - Q(t, x)$. For $\bar{t} \in [t_1, t_2]$, $\bar{x} = x(\bar{t})$, we have $(\bar{t}, \bar{x}) \in A$, and the sets $\tilde{Q}(\bar{t}, \bar{x})$, $Q(\bar{t}, \bar{x})$ are not empty.

If $x'(\bar{t})$ exists and is finite, then we take $\bar{z} = x'(\bar{t})$; if $x'(\bar{t})$ does not exist, or is infinite, then we take for z any point $\bar{z} \in Q(\bar{t}, \bar{x})$. By (17.5.i), $T(t, x, z)$ is seminormal at $(\bar{t}, \bar{x})$, and thus certainly property (X) of Section 17.3 must hold at $(\bar{t}, \bar{x}, \bar{z})$. For $\varepsilon = 1$, then there are numbers $\bar{\delta} > 0$, $\bar{r}$ real, $\bar{b} = (\bar{b}_1, \ldots, \bar{b}_n) \in R^n$, such that, for $\bar{h}(z) = \bar{r} + \bar{b} \cdot z$ we have

$$T(t, x, z) \geq \bar{h}(z) \quad \text{for all } (t, x) \in A, |t - \bar{t}| \leq \bar{\delta}, |x - \bar{x}| \leq \bar{\delta}, z \in R^n.$$

Let $\bar{\rho}$, $0 < \bar{\rho} \leq \bar{\delta}/2$, be a number such that $|t - \bar{t}| \leq \bar{\rho}$, $t \in [t_1, t_2]$, implies $|x(t) - x(\bar{t})| \leq \bar{\delta}/2$. Now we consider the open intervals $(\bar{t} - \bar{\rho}, \bar{t} + \bar{\rho})$ as an open cover of $[t_1, t_2]$. By the Borel covering theorem there is a finite system $\bar{t}_i, \rho_i, \delta_i, r_i, b_i, h_i(z) = r_i + b_i \cdot z$, $i = 1, \ldots, N$, such that the intervals $(\bar{t}_i - \rho_i, \bar{t}_i + \rho_i)$, $i = 1, \ldots, N$, cover $[t_1, t_2]$. Let $\rho = \min \rho_i$, $\delta = \min \delta_i$. Let Γ denote the graph of x, or $\Gamma = [(t, x) | t_1 \leq t \leq t_2, x = x(t)]$, and let Γ_ρ be the ρ-neighborhood of Γ in R^{n+1}. Let a denote the maximum of $t_1 - \rho$ and of those $\bar{t}_i - \rho_i$ which are $< t_1$; let b denote the minimum of $t_2 + \rho$ and of those $\bar{t}_i + \rho_i$ which are $> t_2$. Then $a < t_1 < t_2 < b$. Now we use the end points of the intervals $(\bar{t}_i - \rho_i, \bar{t}_i + \rho_i)$, $i = 1, \ldots, N$, to define a finite partition, say $a = \tau_0 < \tau_1 < \cdots < \tau_{M+1} = b$, of $[a, b]$. Let us prove that if $(t, x) \in \Gamma_\rho \cap A$, $a \leq t \leq b$, then $t \in [\tau_s, \tau_{s+1}]$ for some s, $[\tau_s, \tau_{s+1}]$ is contained in some $[\bar{t}_i - \rho_i, \bar{t}_i + \rho_i]$, and $|t - \bar{t}_i| \leq \delta_i$, $|x - x(\bar{t}_i)| \leq \delta$. Indeed, either $t_1 \leq t \leq t_2$, and then $|t - \bar{t}_i| \leq \rho_i \leq \delta_i/2 < \delta_i$, $|x - x(t)| \leq \rho_i \leq \delta_i/2$, $|x(t) - x(\bar{t}_i)| \leq \delta_i/2$, and $|x - x(\bar{t}_i)| \leq \delta_i/2 + \delta_i/2 = \delta_i$; or $a \leq t \leq t_1$, and then again $|t - \bar{t}_i| < \delta_i$, $|x - x(t_1)| \leq \rho \leq \delta/2$, $|x(t_1) - x(\bar{t}_i)| \leq \delta_i/2$, and $|x - x(\bar{t}_i)| \leq \delta/2 + \delta_i/2 \leq \delta_i$; or finally $t_2 < t \leq b$, and the analogous argument holds. If we denote r_i and b_i by r_s and b_s, we conclude that

$$(8.8.8) \qquad (t, x) \in \Gamma_\rho \cap A, \ a \leq t \leq b, \quad \text{implies} \quad \tau_s \leq t \leq \tau_{s+1} \text{ for some } s,$$

and $T(t, x, z) \geq r_s + b_s \cdot z$ for all $z \in R^n$. For any s, and $\tau_s \leq t \leq \tau_{s+1}$, $(t, x) \in \Gamma_\rho \cap A$, let $\tilde{Q}^{(s)}(t, x)$ denote the set

$$\tilde{Q}^{(s)}(t, x) = [(\bar{z}^0, z) | \bar{z}^0 = z^0 - h_s(z), (z^0, z) \in \tilde{Q}(t, x)], \qquad h_s = r_s + b_s \cdot z.$$

Then $(\bar{z}^0, z) \in \tilde{Q}^{(s)}$ implies $\bar{z}^0 \geq 0$. Moreover, for $\tau_s \leq t \leq \tau_{s+1}$, $(t, x) \in \Gamma_\rho \cap A$, the sets $\tilde{Q}^{(s)}(t, x)$ satisfy property (Q) with respect to (t, x) at every $(\bar{t}, x(\bar{t}))$.

Let $x_k^{(s)}$, $x^{(s)}$ denote the restrictions of x_k, x on the interval $[\tau_s, \tau_{s+1}]$, and let I_{ks}, I_s be the respective intervals of definition. Since $x_k \to x$ in the ρ-metric, then also $x_k^{(s)} \to x^{(s)}$ in the ρ-metric as $k \to \infty$, $s = 0, 1, \ldots, M$, and for k sufficiently large we have $(t, x_k(t)) \in \Gamma_\rho$, $[t_{1k}, t_{2k}] \subset [a, b]$. Finally, for $\eta_k^{(s)}(t) = \eta_k(t) - h_s[x_k'(t)]$, we have

$$(t, x_k(t)) \in \Gamma_\rho \cap A, \quad (\eta_k^{(s)}(t), x_k'(t)) \in \tilde{Q}^{(s)}(t, x_k(t)), \qquad t \in I_{ks}.$$

If $i_s = \liminf_k \int_{I_{ks}} \eta_k^{(s)}(t)\, dt$, $s = 0, 1, \ldots, M$, we have $\eta_k^{(s)}(t) \geq 0$, $i_s \geq 0$, and by (8.8.i) there is an L-integrable function $\eta^{(s)}(t)$, $t \in I_s$, with

$$(t, x(t)) \in \Gamma_\rho \cap A, \quad (\eta^{(s)}(t), x'(t)) \in \tilde{Q}^{(s)}(t, x(t)), \qquad t \in I_s,$$

$$\int_{I_s} \eta^{(s)}(t)\, dt \leq i_s, \qquad s = 0, 1, \ldots, M.$$

Let $\eta(t)$, $t_1 \leq t \leq t_2$, be the function and Δ_{ks} the numbers defined by

$$\eta(t) = \eta^{(s)}(t) + h_s[x'(t)], \quad t \in I_s, \quad \Delta_{ks} = \int_{I_{ks}} h_s[x_k'(t)]\, dt - \int_{I_s} h_s[x'(t)]\, dt.$$

Then $(\eta(t), x'(t)) \in \tilde{Q}(t, x(t))$, $t \in [t_1, t_2]$ (a.e.), and we also have

$$(8.8.9)\quad \int_{t_1}^{t_2} \eta(t)\,dt = \sum_s \int_{I_s} (\eta^{(s)}(t) + h_s[x'(t)])\,dt$$

$$\leq \sum_s i_s + \sum_s \int_{I_s} h_s[x'(t)]\,dt$$

$$\leq \sum_s \liminf_k \left[\int_{I_{ks}} \eta_k(t)\,dt - \int_{I_{ks}} h_s[x_k'(t)]\,dt\right] + \sum_s \int_{I_s} h_s[x'(t)]\,dt$$

$$\leq \liminf_k \left[\sum_s \int_{I_{ks}} \eta_k(t)\,dt - \sum_s \Delta_{ks}\right] \leq i + \liminf_k \left|\sum_s \Delta_{ks}\right|.$$

Now, $h_s(z) = r_s + b_s \cdot z$, and we take $R = \max|r_s|$, $B = \max_s|b_s|$. Then, for $s = 1, 2, \ldots, M-1$, we have $I_{ks} = I_s = [\tau_s, \tau_{s+1}]$, and

$$|\Delta_{ks}| = \left|b_s \cdot \left(\int_{I_{ks}} x_k'\,dt - \int_{I_s} x'\,dt\right)\right|$$

$$\leq |b_s|\,|(x_k(\tau_{s+1}) - x(\tau_{s+1})) - (x_k(\tau_s) - x(\tau_s))|$$

$$\leq 2B\rho(x_k, x).$$

For $s = 0$ we have $I_{k1} = [t_{1k}, \tau_1]$, $I_1 = [t_1, \tau_1]$, and

$$|\Delta_{k1}| \leq |b_1|\,|(x_k(\tau_1) - x(\tau_1)) - (x_k(t_{1k}) - x(t_1))| \leq 2B\rho(x_k, x),$$

and analogously $|\Delta_{kM}| \leq 2B\rho(x_k, x)$. Thus $|\sum_s \Delta_{ks}| \leq 2(M+1)B\rho(x_k, x)$ approaches zero as $k \to \infty$, and from (8.8.9) we derive $\int_{t_1}^{t_2} \eta(t)\,dt \leq i$. This proves Theorem (8.8.ii). □

Remark 2. Theorems (8.8.i) and (8.8.ii) are independent. In (8.8.i) we assume $z^0 \geq -\psi(t)$, $\psi \in L$, for every $(z^0, z) \in \tilde{Q}(t, x)$, and property (Q) is required for all $(t, x(t))$ but a set of points t of measure zero. In (8.8.ii) no lower bound ψ is known, but property (Q) is required at *all* $t \in [t_1, t_2]$. Note that in (8.8.i) we could have specified that property (Q) is required for almost all $\bar{t}$, but at any other point $\bar{t}$ we require that either (a) there are a number $\delta > 0$ and a linear function $h(z) = r + b \cdot z$, $r \in R$, $b = (b_1, \ldots, b_n) \in R^n$, such that $(t, x) \in A$, $|t - \bar{t}| \leq \delta$, $|x - x(\bar{t})| \leq \delta$ imply $T(t, x, z) \geq h(z)$ for all $z \in R^n$ or equivalently $(t, x) \in A$, $|t - \bar{t}| \leq \delta$, $|x - x(\bar{t})| \leq \delta$, $(z^0, z) \in \tilde{Q}(t, x)$ implies $z^0 \geq h(z)$; or (b) there are a number $\delta > 0$ and an L-integrable function $\psi(t) \geq 0$, $\bar{t} - \delta \leq t \leq \bar{t} + \delta$, such that $(t, x) \in A$, $|t - \bar{t}| \leq \delta$, $|x - x(\bar{t})| \leq \delta$ implies $T(t, x, z) \geq -\psi(t)$.

Bibliographical Notes

As stated in Section 8.1, for the concepts of σ-space and σ-convergence we refer to V. Volterra and J. Peres [I]. The very general theorems (8.1.i) and (8.1.ii) are based on mere σ-convergence, and therefore certainly apply to the two modes of convergence of interest here: uniform convergence, and weak convergence in $H^{1,1}$. As soon as we deal with a Banach space, namely a reflexive Banach space X, and σ-convergence is the weak convergence in X, then the much stronger theorems (8.1.iii–vi) hold, which we also have easily proved in Section 8.1. (For further developments along this line we refer to J. L. Lions [I]).

The implicit function theorem, in the forms (8.2.ii) and (8.2.iii), is due to E. J. McShane and R. B. Warfield [1]. The selection theorems as presented in Section 8.3 are due to K. Kuratowski and C. Ryll-Nardzewski [1]. On implicit function theorems and measurable selection theorems we mention here the work of M. Q. Jacobs [1, 2], A. P. Robertson [1], N. U. Ahmed and K. L. Teo [1], J. K. Cole [1], C. J. Himmelberg, M. Q. Jacobs, and F. S. Van Vleck [1], F. V. Chong [1], A. Plis [1–3], T. Wazewski [1–4], A. F. Filippov [2], S. K. Zaremba [1], and the recent monographs of C. Castaing and M. Valadier [I] and of C. Berge [I]. A bibliography on this subject has been collected by D. H. Wagner [1], and a supplement by A. D. Ioffe [2].

In Section 8.3 we have presented forms of the theorem of G. Scorza-Dragoni [1] in connection with the implicit function theorem in optimal control theory. For further extensions of this theorem we refer to G. S. Goodman [2], and on the same general topic we mention here the work of E. Baiada [1].

The concept of upper semicontinuity by set inclusion can be traced in F. Hausdorff [I]. K. Kuratowski introduced his concept of upper semicontinuity for closed set valued functions, property (K), in [1] in 1932. The variant called property (Q) for closed convex set valued functions was proposed by Cesari [6] in 1966. The preliminary properties in Section 8.5, and the more specific properties which will be proved in Section 10.5, were proved by Cesari in [6, 8, 13].

Property (Q) was used by J. D. Schuur and S. N. Chow [1] to prove the existence and main properties of solutions of the Cauchy problem $x'(t) \in Q(t, x(t))$, $x(t_0) = x_0$, in Banach spaces.

Property (Q) was used by Cesari [12–23], L. Cesari and D. E. Cowles [1], M. B. Suryanarayana [5–7], and L. Cesari and M. B. Suryanarayana [1–8] to prove theorems of lower semicontinuity and existence in problems of optimization with ordinary and partial differential equations, with single valued as well as multivalued functionals (Pareto problems). Some of the results will be presented in Chapter 10 of this book, where also more bibliographic references will be given.

Property (Q) was used by R. F. Baum [1–4] in problems of optimization with ordinary differential equations in infinite intervals (infinite horizon in economics), and in problems of optimization with partial differential equations where the controls are functions in R^k and the state variables are functions in R^n with $k < n$.

Property (Q) has been used by T. S. Angell [1–4] in problems of optimization with functional differential equations and in problems with lags.

Property (Q) can be thought of as a generalization of Minty's and Brezis's maximal monotonicity property as proved by M. B. Suryanarayana [8, 10], and we shall present a proof of this result in Section 17.8. As such, property (Q) has been recently used by S. H. Hou [3] for the proof of existence theorems for boundary value problems for ordinary and partial differential equations (controllability), and by T. S. Angell [8, 9] for nonlinear Volterra equations and hereditary systems.

The elementary closure theorem (8.6.i) was proved by L. Cesari [6]. The Fatou-like lemma (8.7.ii) and subsequent elementary proof of the lower closure theorem (8.8.i) appear here for the first time.

CHAPTER 9

Existence Theorems: The Bounded, or Elementary, Case

9.1 Ascoli's Theorem

We shall need Ascoli's compactness theorem, a well-known elementary form of which is as follows:

9.1.i. (ASCOLI'S COMPACTNESS THEOREM) (cf., e.g., McShane [I, p. 336]). *Given any sequence of equibounded and equicontinuous functions $x_k(t) = (x_k^1, \dots, x_k^n)$, $a \le t \le b$, $k = 1, 2, \dots$, there is a subsequence $[k_s]$ and a continuous function $x(t) = (x^1, \dots, x^n)$, $a \le t \le b$, such that $x_{k_s} \to x$ uniformly in $[a, b]$ as $s \to \infty$.*

By equiboundedness we mean that there is an $N > 0$ such that $|x_k(t)| \le N$ for all t and k. By equicontinuity we mean that given $\varepsilon > 0$ there is $\delta = \delta(\varepsilon) > 0$ such that for all $t', t'' \in [a, b]$ with $|t'' - t'| \le \delta$ and all k we also have $|x_k(t'') - x_k(t')| \le \varepsilon$.

For functions $x_k(t)$, $t_{1k} \le t \le t_{2k}$, $k = 1, 2, \dots$, which may be defined in different intervals, then (9.1.i) can be reworded as follows in terms of the ρ-distance function of Section 2.14:

9.1.ii. *Given any sequence of equibounded and equicontinuous functions $x_k(t)$, $t_{1k} \le t \le t_{2k}$, $k = 1, 2, \dots$, defined in intervals $[t_{1k}, t_{2k}]$ all contained in some fixed interval $[a, b]$, then there is a subsequence $[k_s]$ and a continuous function $x(t)$, $t_1 \le t \le t_2$, such that $t_{1k_s} \to t_1$, $t_{2k_s} \to t_2$, and $x_{k_s} \to x$ in the ρ-metric as $s \to \infty$.*

Indeed, first we extract the subsequence $[k_s]$ so that $t_{1k_s} \to t_1$, $t_{2k_s} \to t_2$, and then $a \le t_1 \le t_2 \le b$; further we extend all x_k to $[a, b]$ as usual by continuity and constancy outside their original intervals of definition. This preserves equibounded and equicontinuity. It remains only to apply (9.1.i) to the so obtained sequence $x_{k_s}(t), a \leq t \leq b, s = 1, 2, \ldots$, so as to obtain a further subsequence which has the required properties. Theorem (9.1.i) is the usual forms of Ascoli's theorem in $C([a,b], R^n)$. Statement (9.1.ii) is the corresponding form in the metric space C.

In either form it is clear that AC functions $x_k(t)$ with equibounded derivatives, say $|x_k'(t)| \leq N$ for all t and k, are equi-Lipschitzian, $|x_k(t'') - x_k(t')| = |\int_{t'}^{t''} x_k'(t)\,dt| \leq N|t'' - t'|$, and hence equicontinuous.

9.2 Filippov's Existence Theorem for Mayer Problems of Optimal Control

Let A be a subset of the tx-space R^{1+n}, let A_0 denote the projection of A on the t-axis, and for $t \in A_0$ let $A(t) = [x \in R^n | (t,x) \in A]$. For every $(t,x) \in A$ let $U(t,x)$ be a given subset of the u-space R^m, and let $M \subset R^{1+n+m}$ be the set of all (t,x,u) with $(t,x) \in A$, $u \in U(t,x)$. Let $f(t,x,u) = (f_1, \ldots, f_n)$ be a given function on M. For every $(t,x) \in A$ let $Q(t,x) = f(t,x,U(t,x)) \subset R^n$ be the set of all $z = (z^1, \ldots, z^n)$ with $z = f(t,x,u)$ for some $u \in U(t,x)$. Let B be a given subset of the $t_1x_1t_2x_2$-space R^{2+2n}, and $g(t_1,x_1,t_2,x_2)$ a real valued function defined on B. We consider the problem of the minimization of the functional

$$I[x,u] = g(t_1, x(t_1), t_2, x(t_2)) \tag{9.2.1}$$

for pairs of functions $x(t) = (x^1, \ldots, x^n)$, $u(t) = (u^1, \ldots, u^m)$, $t_1 \leq t \leq t_2$, x AC, u measurable, satisfying

$$\begin{aligned} &x'(t) = f(t, x(t), u(t)), \qquad t \in [t_1, t_2] \text{ (a.e.)},\\ &(t, x(t)) \in A, \;\; u(t) \in U(t, x(t)), \quad t \in [t_1, t_2] \text{ (a.e.)};\\ &(t_1, x(t_1), t_2, x(t_2)) \in B. \end{aligned} \tag{9.2.2}$$

A pair $x(t)$, $u(t)$, $t_1 \leq t \leq t_2$, x AC, u measurable, satisfying all requirements (9.2.2), is said to be admissible for the problem (9.2.1–2), and then x is said to be an admissible trajectory, and u an admissible strategy, or control function. If Ω is a class of admissible pairs x, u, let $\Omega_x = \{x\}$ denote the family of the corresponding trajectories, that is, $\Omega_x = [x | (x,u) \in \Omega$ for some $u]$.

We may consider the problem of the minimum of $I[x,u]$ in the class of all admissible pairs. Alternatively, we may consider the problem of the minimum of $I[x,u]$ in a smaller class, say the class of all admissible pairs x, u, whose trajectories x pass through a given point (t_0, x_0) of A, or through finitely many of such points, or through a given closed subset of A, or so that some components x^i of the trajectories are monotone, or analogous classes. In this case, we need only require Ω to satisfy a mild closedness condition. Namely, we shall say that Ω is Γ_u-closed if Ω has the following property: if $x_k(t)$, $u_k(t)$, $t_{1k} \le t \le t_{2k}$, $k = 1, 2, \ldots$, are admissible pairs all in Ω, if $x(t)$, $u(t)$, $t_1 \le t \le t_2$, is an admissible pair, if $x_k \to x$ in the ρ-metric, or uniformly, then (x, u) belongs to Ω. Thus, the class of all admissible pairs, with A and B closed sets, is certainly Γ_u-closed. Again, with A and B closed sets, the classes of all admissible pairs whose trajectories pass through a given point $(t_0, x_0) \in A$, or through finitely many of such points of A, or through a given closed subset P of A, or for which one or more components x^i are monotone, and other analogous classes, are all Γ_u-closed according to the definition above. Note that the closedness property Γ_u is a property of the family Ω_x, or $\{x\}$, of admissible trajectories.

In any case we assume below that the class Ω of all admissible pairs x, u is nonempty, an assumption which can be expressed equivalently by saying that the requirements (9.2.2) are *compatible*, or that the system is *controllable* for what concerns the problem under consideration.

9.2.i (THE FILIPPOV EXISTENCE THEOREM FOR MAYER PROBLEMS). *Let A be compact, B closed, M compact, g lower semicontinuous on B, $f(t, x, u) = (f_1, \ldots, f_n)$ continuous on M, and let us assume that for almost all t the sets $Q(t, x)$, $x \in A(t)$, are convex. Then the functional $I[x, u]$ given by* (9.2.1) *has an absolute minimum in the nonempty class Ω of all admissible pairs* (*as well as in any nonempty Γ_u-closed class Ω of admissible pairs*).

Proof. Since A is compact and M is compact, we know from Section 1.11 and the implicit function theorem Section 8.2 that the problem (9.2.1–2) is equivalent to the orientor field problem

$$
\begin{gathered}
I[x] = g(t_1, x(t_1), t_2, x(t_2)), \\
x'(t) \in Q(t, x(t)), \qquad t \in [t_1, t_2] \text{ (a.e.)}, \qquad x \text{ AC in } [t_1, t_2], \\
(t, x(t)) \in A, \qquad (t_1, x(t_1), t_2, x(t_2)) \in B,
\end{gathered} \tag{9.2.3}
$$

where $Q(t, x) = f(t, x, U(t, x)) \subset R^n$, that is, $Q(t, x) = [z \in R^n \mid z = f(t, x, u), u \in U(t, x)]$.

Here the set A is compact, M is compact, and $f(t, x, u)$ is continuous on M. Then, by (8.5.vi)(a), the sets $Q(t, x)$ are all compact, are all contained in a fixed ball in R^n, are upper semicontinuous by set inclusion (with respect to $(t, x) \in A$), and have property (K) on A, and since for almost all t they are convex, they also have property (Q) in (t, x) at all $(t, x) \in A$, except perhaps on a set of points whose abscissas t form a set of measure zero on the t-axis.

Now we note that for any admissible x, the point $(t_1, x(t_1), t_2, x(t_2))$ must be in the compact set $B \cap (A \times A)$, and thus the lower semicontinuous function g has a minimum in $B \cap (A \times A)$, and hence is certainly bounded below there. Since, moreover, the class Ω is nonempty, the infimum i of $I[x] = g(t_1, x(t_1), t_2, x(t_2))$ in Ω is finite, $-\infty < i < +\infty$. Let $x_k(t)$, $u_k(t)$, $t_{1k} \le t \le t_{2k}$, $k = 1, 2, \ldots$, be any sequence of elements of Ω with $g(t_{1k}, x_k(t_{1k}), t_{2k}, x_k(t_{2k})) \to i$ as $k \to \infty$. Since $x_k'(t) \in Q(t, x_k(t))$ a.e. in $[t_{1k}, t_{2k}]$, and since all these sets $Q(t, x_k(t))$ are contained in a fixed ball in R^n, we have $|x_k'(t)| \leq N$ for some N, for all k and almost all $t \in [t_{1k}, t_{2k}]$. Hence, the trajectories x_k are equi-Lipschitzian, and hence equicontinuous. Since A is compact and $(t, x_k(t)) \in A$, we can take N in such a way that we also have $-N \leq t_{1k} < t_{2k} \leq N$, $|x_k(t)| \leq N$ for all k and $t \in [t_{1k}, t_{2k}]$. By Ascoli's theorem there is a subsequence $x_{k_s}(t)$, $t_{1k_s} \leq t \leq t_{2k_s}$, such that $t_{1k_s} \to t_1$, $t_{2k_s} \to t_2$, and x_{k_s} converges in the ρ-metric toward a continuous vector function $x(t)$, $t_1 \leq t \leq t_2$. Since the trajectories x_{k_s} are equi-Lipschitzian, the limit element x is Lipschitzian in $[t_1, t_2]$, and hence AC.

Since $x'_{k_s}(t) \in Q(t, x_{k_s}(t))$ a.e. in $[t_{1k_s}, t_{2k_s}]$ for all s, by the closure theorem (8.6.i) we derive that $x'(t) \in Q(t, x(t))$ a.e. in $[t_1, t_2]$.

Since $(t, x_{k_s}(t)) \in A$ with A closed, then $(t, x(t)) \in A$ for all $t \in [t_1, t_2]$. Since $(t_{1k}, x_k(t_{1k}), t_{2k}, x_k(t_{2k})) \in B$ with B closed, then $(t_1, x(t_1), t_2, x(t_2)) \in B$. We have proved that $x(t)$, $t_1 \leq t \leq t_2$, is a solution of the orientor field problem (9.2.3). That there exists a measurable $u(t)$, $t_1 \leq t \leq t_2$, such that (9.2.2) holds is a consequence of the implicit function theorem (8.2.iii). Thus, $(x, u) \in \Omega$. Since g is lower semicontinuous on B, we have

$$\begin{aligned} I[x, u] &= g(t_1, x(t_1), t_2, x(t_2)) \\ &\leq \liminf_{s \to \infty} g(t_{1k_s}, x_{k_s}(t_{1k_s}), t_{2k_s}, x_{k_s}(t_{2k_s})) = i. \end{aligned}$$

Since $(x, u) \in \Omega$, we have $I[x, u] \geq i$, and by comparison $I[x, u] = i$. If we were dealing with the problem of minimum of I in a Γ_u-closed class Ω, then we would take for i the infimum of $I[x]$ in Ω_x, and the argument would be the same. Since all x_k are admissible and $x_k \to x$ in the ρ-metric, then as before we know that x is admissible, that there is a measurable u for which (9.2.2) hold, and that $I[x, u] \leq i$. On the other hand, all x_k are in Ω_x, Ω_x is Γ_u-closed, and hence $(x, u) \in \Omega$, and $I[x, u] \geq i$. Thus $I[x, u] = i$ as before, and (9.2.i) is thereby proved. □

Remark 1. The hypotheses concerning the compactness of A and M will be removed in Section 9.4. It is clear that all we have to do is to guarantee that, under hypotheses, a minimizing sequence $x_k(t)$, $t_{1k} \leq t \leq t_{2k}$, $k = 1, 2, \ldots$, such as the one above remains in a compact part A_1 of A, and that the set $M_1 = [(t, x, u) | (t, x) \in A_1, u \in U(t, x)]$, or $M_1 = [(t, x, u) \in M | (t, x) \in A_1]$, is also compact.

Remark 2. If the sets $Q(t, x)$ are not convex, then (9.2.i) still guarantees the existence of an optimal generalized solution. By a generalized solution (cf. Section 1.14) we mean a solution of a new problem in which relations (9.2.1–2) are replaced by

$$J[x, p, v] = g(t_1, x(t_1), t_2, x(t_2)),$$

$$x'(t) = \sum_{s=1}^{\gamma} p_s(t) f(t, x(t), u^{(s)}(t)), \qquad u^{(s)}(t) \in U(t, x(t)),$$

$$(t, x(t)) \in A, \qquad (t_1, x(t_1), t_2, x(t_2)) \in B, \qquad \sum_{s=1}^{\gamma} p_s(t) = 1, \qquad p_s(t) \geq 0,$$

$$t \in [t_1, t_2] \text{ (a.e.)}, \qquad x(t) \text{ AC}, \qquad p_s(t), u^{(s)}(t) \text{ measurable}.$$

Let $p = (p_1, \dots, p_\gamma)$, $v = (u^{(1)}, \dots, u^{(\gamma)})$, $\Gamma = [p_s \geq 0, \sum_s p_s = 1] \subset R^\gamma$, $V = U^\gamma$, $\gamma = n + 1$, and $h(t, x, p, v) = \sum_s p_s f(t, x, u^{(s)})$. For the new problems the sets $Q(t, x)$ are replaced by the sets $R(t, x) = \text{co } Q(t, x) = h(t, x, \Gamma \times V)$, where h is a continuous function. By (8.v.vi)(a) the sets $R(t, x)$ are all compact, convex, all contained in a fixed ball, in R^n, are upper semicontinuous by set inclusion, have property (K) and since they are convex also property (Q) (at all (t, x) except perhaps on a set of points whose abscissas t form a set of measure zero). The proof is now the same as before. Note that the sets Q are in R^n and hence $R(t, x) = \text{co } Q(t, x)$ for $\gamma = n + 1$ by Carathéodory's theorem (8.4.iii).

9.3 Filippov's Existence Theorem for Lagrange and Bolza Problems of Optimal Control

For A, B, $U(t, x)$, M, g as before, let $f(t, x, u) = (f_1, \dots, f_n)$, and $f_0(t, x, u)$ be functions defined on M. For every $(t, x) \in A$ let $Q(t, x) = f(t, x, U(t, x)) \subset R^n$ be the same set we had before, and let $\tilde{Q}(t, x) \subset R^{n+1}$ be the set of all (z^0, z) with $z^0 \geq f_0(t, x, u)$, $z = f(t, x, u)$ for some $u \in U(t, x)$. We consider the problem of the minimization of the functional

$$I[x, u] = g(t_1, x(t_1), t_2, x(t_2)) + \int_{t_1}^{t_2} f_0(t, x(t), u(t))\, dt \tag{9.3.1}$$

for pairs of functions $x(t) = (x^1, \dots, x^n)$, $u(t) = (u^1, \dots, u^m)$, $t_1 \leq t \leq t_2$, x AC, u measurable in $[t_1, t_2]$, satisfying

$$\begin{aligned} x'(t) &= f(t, x(t), u(t)), \quad t \in [t_1, t_2] \text{ (a.e.)}, \\ (t, x(t)) &\in A, \ u(t) \in U(t, x(t)), \quad t \in [t_1, t_2] \text{ (a.e.)}, \\ (t_1\, x(t_1), t_2, x(t_2)) &\in B, \ f_0(\cdot, x(\cdot), u(\cdot)) \ L\text{-integrable in } [t_1, t_2]. \end{aligned} \tag{9.3.2}$$

A pair $x(t)$, $u(t)$, $t_1 \leq t \leq t_2$, x AC, u measurable, satisfying all requirements (9.3.2) is said to be admissible for the problem (9.3.1–2), and x is then an admissible trajectory and u an admissible strategy. If Ω is a class of admissible pairs (x, u), let $\Omega_x = \{x\}$ as before be the class of the corresponding trajectories.

As before, we may try to minimize $I[x, u]$ in the class Ω of all admissible pairs, and we explicitly assume that Ω is nonempty, that is, the requirements (9.3.2) are compatible, or equivalently, the system is controllable. Alternatively, we may want to minimize $I[x, u]$ in a smaller class Ω of admissible pairs, and in this case we need only to know that Ω possesses a closedness property similar to the one we stated in Section 9.2. Here we shall say that Ω is Γ_u-closed for the problem (9.3.1–2) if Ω has the following property: if $x_k(t)$, $u_k(t)$, $t_{1k} \leq t \leq t_{2k}$, $k = 1, 2, \dots$, are admissible pairs all in Ω, if $x(t)$, $u(t)$, $t_1 \leq t \leq t_2$, is also an admissible pair, and if $x_k \to x$ in the ρ-metric, then (x, u) belongs to Ω.

9.3.i (The Filippov Existence Theorem for Bolza and Lagrange Problems). *Let A be compact, B closed, M compact, g lower semicontinuous on B, and $f_0(t,x,u)$, $f(t,x,u) = (f_1, \ldots, f_n)$ continuous on M. Let us assume that for almost all t the sets $\tilde{Q}(t,x)$, $x \in A(t)$, are convex. Then the functional $I[x,u]$ given by* (9.3.1) *has an absolute minimum in the nonempty class Ω of all admissible pairs (as well as in any nonempty Γ_u-closed class Ω of admissible pairs).*

We shall give two proofs (I and II) of this theorem, both elementary. The first proof simply shows that (9.3.i) is a corollary of (9.2.i). The second one is based again on uniform convergence and the elementary lower closure theorem (8.8.ii). As for (9.2.i), this theorem too is discussed as a problem for orientor fields.

Proofs of (9.3.i).

I. Since M is compact and f_0, f are continuous on M, there is a constant N such that $|f_0(t,s,u)|, |f(t,x,u)| \le N$ for all $(t,x,u) \in M$.

Let $\tilde{u} = (u^0, u) = (u^0, u^1, \ldots, u^m)$ denote a new control variable, let $\tilde{x} = (x^0, x) = (x^0, x^1, \ldots, x^n)$ denote a new space variable, and let $J[\tilde{x}, \tilde{u}] = x^0(t_2)$ be a new functional. We consider now the Mayer problem of optimal control

$$(9.3.3)\qquad \begin{gathered} J[\tilde{x},\tilde{u}] = x^0(t_2) + g(t_1, x(t_1), t_2, x(t_2)), \\ dx^0/dt = u^0, \qquad dx/dt = f(t,x,u), \\ x^0(t_1) = 0, \qquad (t_1, x(t_1), t_2, x(t_2)) \in B, \\ \tilde{u} \subset \tilde{U}(t,\tilde{x}) = [N \ge u^0 \ge f_0(t,x,u),\ u \in U(t,x)]. \end{gathered}$$

For this problem we have

$$\begin{aligned} J[\tilde{x},\tilde{u}] &= x^0(t_2) + g(e[x]) = \int_{t_1}^{t_2} u^0(t)\,dt + g(e[x]) \\ &\ge \int_{t_1}^{t_2} f_0(t, x(t), u(t))\,dt + g(e[x]) = I[x,u], \end{aligned}$$

and equality holds whenever $u^0(t) = f_0(t, x(t), u(t))$ a.e. in $[t_1, t_2]$. Let $Q^*(t,x)$ denote the usual sets $Q(t,x)$ for this Mayer problem. We then have

$$Q^*(t,\tilde{x}) = [(z^0, z) \,|\, N \ge z^0 \ge f_0(t,x,u),\ z = f(t,x,u),\ u \in U(t,x)],$$

and $\tilde{Q}(t,x)$ convex certainly implies that $Q^*(t,\tilde{x})$ is convex. Since $|u^0| \le N$, we have $|x^0(t)| \le DN$, where D is the diameter of A_0, the projection of A on the t-axis. For this Mayer problem we take $\tilde{A} = A \times [-DN, DN]$, so that $\tilde{A}$ is compact, and $(t, x^0(t), x(t)) \in \tilde{A}$ represents no other restriction than $(t, x(t)) \in A$ in the original problem. Finally, the set $\tilde{M}$ for this Mayer problem is the set we have allowed for (t, x^0, x, u^0, u), namely, $(t,x,u) \in M$, $f_0(t,x,u) \le u^0 \le N$, $-DN \le x^0 \le DN$, certainly a compact set, since M is compact and f_0 is continuous on M.

We can now apply (9.2.i) to the new Mayer problem, which thereby has an absolute minimum, and for this minimum we must have $u^0(t) = f_0(t, x(t), u(t))$ a.e. in $[t_1, t_2]$; hence $J[\tilde{x}, \tilde{u}] = I[x, u] = i$.

II. For every admissible pair $(x, u) \in \Omega$, the point $(t_1, x(t_1), t_2, x(t_2))$ is in the compact set $B \cap (A \times A)$. The lower semicontinuous function g has a minimum m in the compact set $B \cap (A \times A)$. Since M is compact and f_0, f are continuous on M, there is a constant N such that $|f_0(t, x, u)|, |f(t, x, u)| \leq N$ on M, and we can take N so that we also have $|t|, |x|, |u| \leq N$ for $(t, x, u) \in M$. Thus, the set A_0, the projection of the compact set A on the t-axis, is contained in the interval $[-N, N]$. Let D denote the diameter of A_0.

Note that, if $(t, x) \in A$ and $z \in Q(t, x) = f(t, x, U(t, x)) \subset R^n$, then $(z^0, z) \in \tilde{Q}(t, x)$ implies $z^0 \geq -N$, and certainly $(N, z) \in \tilde{Q}(t, x)$.

From (8.5.vi)(b) with x replaced by (t, x), we derive that the sets $\tilde{Q}(t, x)$ are closed in R^{n+1} with compact projection $Q(t, x)$ in the z-space R^n, and that the same sets $\tilde{Q}(t, x)$ are upper semicontinuous by set inclusion and hence have property (K) in A (with respect to (t, x)). Moreover, for almost all $\bar{t}$ and any $\bar{x} \in A(\bar{t})$, the set $\tilde{Q}(\bar{t}, \bar{x})$ is convex, and the same sets $\tilde{Q}(t, x)$ have property (Q) at $(\bar{t}, \bar{x})$ (with respect to (t, x)).

Let us consider now the following minimum problem, expressed in terms of orientor fields: Determine a pair $x(t), \eta(t), t_1 \leq t \leq t_2$, x AC, η L-integrable, for which the functional

$$J[x, \eta] = g(t_1, x(t_1), t_2, x(t_2)) + \int_{t_1}^{t_2} \eta(t)\, dt \tag{9.3.4}$$

has its minimum value under the constraints

$$\begin{aligned} &(t, x(t)) \in A, \quad (\eta(t), x'(t)) \in \tilde{Q}(t, x(t)), \qquad t \in [t_1, t_2] \text{ (a.e.)}, \\ &\qquad\qquad (t_1, x(t_1), t_2, x(t_2)) \in B. \end{aligned} \tag{9.3.5}$$

Let Ω' denote the class of all pairs $x(t), \eta(t), t_1 \leq t \leq t_2$, for which all the requirements above are satisfied. From the above, we know that for $(x, \eta) \in \Omega'$ we have $\eta(t) \geq -N$ for all t, and moreover $(x, \eta_0) \in \Omega'$ for $\eta_0(t) \equiv N$.

Finally, since $\eta(t) \geq f_0(t, x(t), u(t))$, $t_2 - t_1 \leq D$, $g \geq m$ in $B \cap (A \times A)$, and $f_0 \geq -N$ on M, we have, from (9.3.2) and (9.3.4), $J[x, \eta] \geq I[x, u] \geq m - DN$, and also $J = I$ whenever $\eta(t) = f_0(t, x(t), u(t))$, $t \in [t_1, t_2]$ (a.e.). If $j = \inf_{\Omega'} J[x, \eta]$, and $i = \inf_{\Omega'} I[x, u]$, then $j = i$ and both are finite.

Let $x_k(t), \eta_k(t), t_{1k} \leq t \leq t_{2k}, k = 1, 2, \ldots$, be a minimizing sequence for J, that is, $(x_k, \eta_k) \in \Omega'$ and $J[x_k, \eta_k] \to j$ as $k \to \infty$. Thus, $(t, x_k(t)) \in A$, $(\eta_k(t), x_k'(t)) \in \tilde{Q}(t, x_k(t))$, $x_k'(t) \in Q(t, x_k(t))$, $t \in [t_{1k}, t_{2k}]$ (a.e.), $(t_{1k}, x_k(t_{1k}), t_{2k}, x_k(t_{2k})) \in B$ for all k, and

$$g(t_{1k}, x_k(t_{1k}), t_{2k}, x_k(t_{2k})) + \int_{t_{1k}}^{t_{2k}} \eta_k(t)\, dt \to j \quad \text{as } k \to \infty.$$

Since both parts in this expression are certainly bounded below, and j is finite, both parts are bounded, and we can take a subsequence, say still $[k]$ for the sake of simplicity, such that $g(t_{1k}, x_k(t_{1k}), t_{2k}, x_k(t_{2k})) \to j'$, $\int_{t_{1k}}^{t_{2k}} \eta_k(t)\, dt \to j''$, both j' and j'' finite with $j' + j'' = j$. Since $Q(t, x_k(t)) = f(t, x_k(t), U(t, x_k(t)))$, we have $|x_k'(t)| \leq N$ for all t and k; hence, the functions $x_k(t)$ are equi-Lipschitzian, and therefore equicontinuous. Since $(t, x_k(t)) \in A$, we have $-N \leq t_{1k} < t_{2k} \leq N$, $|x_k(t)| \leq N$ for all t and k.

By Ascoli's theorem, there is a subsequence $[k_s]$ such that $t_{1k_s} \to t_1$, $t_{2k_s} \to t_2$ as $s \to \infty$, and there is a continuous function $x(t), t_1 \leq t \leq t_2$, such that $x_{k_s}(t) \to x(t)$ in the ρ-metric. Since A and B are closed, $(t, x(t)) \in A$ and $(t_1, x(t_1), t_2, x(t_2)) \in B$. Since g

is lower semicontinuous, we also have $g(t_1, x(t_1), t_2, x(t_2)) \leq j'$. Since the functions x_k are equi-Lipschitzian, x too is Lipschitzian and hence AC. By the lower closure theorem (8.8.i) with $\psi(t) = -N$, there is an L-integrable function $\eta(t)$, $t_1 \leq t \leq t_2$, such that $(\eta(t), x'(t)) \in \tilde{Q}(t, x(t))$, $t \in [t_1, t_2]$ (a.e.), and $\int_{t_1}^{t_2} \eta(t)\,dt \leq j''$. Hence

$$J[x, \eta] = g(t_1, x(t_1), t_2, x(t_2)) + \int_{t_1}^{t_2} \eta(t)\,dt \leq j' + j'' = j.$$

But $(x, \eta) \in \Omega'$, so that $J[x, \eta] \geq j$, and this shows that $J[x, \eta] = j$.

Note that $\tilde{Q}(t, x)$ is the continuous image of $U(t, x) \times [0, +\infty)$ under the map $(u, v) \to [f(t, x, u, f_0(t, x, u) + v]$, and $(\eta(t), x'(t)) \in \tilde{Q}(t, x(t))$. Also the sets $\tilde{Q}(t, x)$ have property (K), and moreover, for almost all $\bar{t} \in [t_1, t_2]$, the sets $\tilde{Q}(t, x)$ have property (Q) with respect to (t, x) at $(\bar{t}, x(\bar{t}))$. By the form of the implicit function theorem expressed in Section 8.2, Exercise 2 (with $\eta(t) = f(t, x(t), u(t)) + v(t)$, $v(t) \geq 0$), we derive that there are measurable functions $u(t)$, $v(t)$, $t_1 \leq t \leq t_2$, $u(t) \in U(t, x(t)$, $v(t) \geq 0$, such that $x'(t) = f(t, x(t), u(t))$, $\eta(t) = f_0(t, x(t), u(t)) + v(t)$. By the minimum property of j we must have $v(t) = 0$ a.e. in $[t_1, t_2]$. Hence,

$$\begin{aligned} J[x, \eta] &= I[x, u] \\ &= g(t_1, x(t_1), t_2, x(t_2)) + \int_{t_1}^{t_2} f_0(t, x(t), u(t))\,dt = j = i. \end{aligned} \tag{9.3.6}$$

This completes the second proof of (9.3.i) in the case where Ω is the class of all admissible pairs x, u.

If Ω is any Γ_u-closed class of admissible pairs x, u, let Ω_x denote the class of the corresponding trajectories, or $\Omega_x = \{x\} = [x | (x, u) \in \Omega$ for some $u]$. The argument above is the same, where we take for Ω' the class of all pairs x, η as stated but with $x \in \Omega_x$. At the end, we must only add that $x_k \to x$ in the ρ-metric, and that Ω is a Γ_u-closed class; hence, $(x, u) \in \Omega$. Theorem (9.3.i) is thereby proved. □

Remark. If the sets $\tilde{Q}(t, x)$ are not convex, then (9.3.i) still guarantees the existence of an optimal generalized solution. By a generalized solution of the present Lagrange and Bolza problems (cf. Section 1.14) we mean a solution of a new problem in which relations (9.3.1–2) are replaced by

$$J[x, p, v] = g(t_1, x(t_1), t_2, x(t_2)) + \int_{t_1}^{t_2} \sum_{s=1}^{\gamma} p_s(t) f_0(t, x(t), u^{(s)}(t))\,dt,$$

$$x'(t) = \sum_{s=1}^{\gamma} p_s(t) f(t, x(t), u^{(s)}(t)), \qquad u^{(s)}(t) \in U(t, x(t)),$$

$$(t, x(t)) \in A,\ (t_1, x(t_1), t_2, x(t_2)) \in B, \qquad \sum_{s=1}^{\gamma} p_s(t) = 1,\ p_s(t) \geq 0,$$

$$t \in [t_1, t_2] \text{ (a.e.)}, \qquad x(t) \text{ AC}, \qquad p_s(t), u^{(s)}(t) \text{ measurable}, \qquad s = 1, \ldots, \gamma.$$

Let $p = (p_1, \ldots, p)$, $v = (u^{(1)}, \ldots, u^{(\gamma)})$, $\Gamma = [p_s \geq 0, \sum p_s = 1] \subset R^\gamma$, $V = U^\gamma$, $\gamma = n + 2$ and $h(t, x, p, v) = \sum_s p_s f(t, x, u^{(s)})$, $h_0(t, x, p, v) = \sum_s p_s f_0(t, x, u^{(s)})$. For the new problem the sets $\tilde{Q}(t, x)$ are replaced by the sets $\tilde{R}(t, x) = \text{co}\, \tilde{Q}(t, x)$, namely the sets of all $(z^0, z) \in R^{n+1}$ with $z^0 \geq h_0(t, x, p, v)$, $z = h(t, x, p, v)$, $(p, v) \in \Gamma \times V$. By (8.5.vi)(b) the sets $R(t, x)$ are all closed, convex, are upper semicontinuous by set inclusion, have property (K), and since they are convex, also property (Q) (at all (t, x) except perhaps on a set of points whose abscissas form a set of measure zero.) The proof is now the same as before. Note that the sets Q are in R^{n+1} and hence $R(t, x) = \text{co}\, Q(t, x)$ for $\gamma = n + 2$ by Carathéodory's theorem (8.4.iii).

9.4 Elimination of the Hypothesis that A Is Compact in Filippov's Theorem for Mayer Problems

If A is not compact but closed, then M in Theorem (9.2.i) cannot be compact but at best only closed. Nevertheless, all that we need to know in the proof of Theorem (9.2.i) is that a minimizing sequence $[x_k]$ lies in a bounded closed subset A_1 of A, and then that the part of M relative to A_1, say $M_1 = [(t, x, u) \in M \,|\, (t, x) \in A_1]$, is compact.

Very often, the fact that a minimizing sequence $[x_k]$ remains in a bounded (closed) part A_1 of A is a trivial consequence of the data, and then all that is needed is to verify that the part M_1 of M relative to A_1 is compact. For instance, if we want to minimize the length of certain trajectories having a fixed end point (t_0, x_0) in A, then we may well limit ourselves to those trajectories whose length is less than some fixed constant l, and then the same trajectories lie in a fixed bounded closed part A_1 of A, namely $A_1 = A \cap S_l$, where S_l is the closed ball of center (t_0, x_0) and radius l.

Whenever the needed information ($[x_k]$ lies in a bounded closed part A_1 of A and M_1 is compact) is not trivially evident, there are a variety of conditions which guarantee it. In other words, if A is not compact but closed, then Theorem (9.2.i) still holds under a variety of simple additional conditions and simple changes, which we list and prove below.

If A is not compact, but A is closed and contained in a slab $[t_0 \leq t \leq T,\, x \in R^n]$, t_0, T finite, then Theorem (9.2.i) still holds provided: (a) the hypothesis M compact is replaced by the weaker hypothesis that for every $N \geq 0$ the set M_N of all $(t, x, u) \in M$ with $|x| \leq N$ is compact; (b) there is a compact subset P of A such that every trajectory x of Ω possesses at least one point $(t^*, x(t^*)) \in P$; and (c) there is a constant $C \geq 0$ such that $x^1 f_1 + \cdots + x^n f_n \leq C(|x|^2 + 1)$ for all $(t, x, u) \in M$.

Condition (b) is certainly satisfied if for instance the initial point $(t_1, x(t_1))$ is fixed, or the end point $(t_2, x(t_2))$ is fixed, or, more generally, if $B = B_1 \times B_2$, where B_1, B_2 are closed subsets of the $t_1 x_1$- and $t_2 x_2$-spaces respectively, at least one of which is compact.

Condition (c) is certainly satisfied if (c_0) $|f| \leq c(|x| + 1)$ for $(t, x, u) \in M$ and some constant $c \geq 0$. Indeed (c_0) implies

$$\text{(9.4.1)} \qquad \begin{aligned} x^1 f_1 + \cdots + x^n f_n &\leq nc|x|(|x| + 1) \leq nc(|x|^2 + 2^{-1}|x|^2 + 2^{-1}) \\ &\leq (3nc/2)(|x|^2 + 1), \end{aligned}$$

and (c) holds with $C = 3nc/2$.

On the other hand, condition (c) can be replaced by the more general condition: (c') there is a scalar function $V(t, x)$ of class C^1 in A and a positive constant C such that

$$\text{(9.4.2)} \qquad \operatorname{grad}_x V(t, x) \cdot f(t, x, u) + \partial V/\partial t \leq CV(t, x)$$

for all $(t, x, u) \in M$, and for every a, b, α real the set $[x \,|\, V(t, x) \leq \alpha$ for $(t, x) \in A$, $a \leq t \leq b]$ is compact.

Note that, if we take $V(t, x) = |x|^2 + 1$, then condition (c') reduces to condition (c). Condition (c') belongs to Lyapunov's theory of ordinary differential equations, and $V(t, x)$ is often called a Lyapunov function.

Condition (c) can be replaced by the more general hypothesis: (c_g) there is a locally integrable scalar function $g(t) \geq 0$ such that

$$x^1 f_1 + \cdots + x^n f_n \leq g(t)(|x|^2 + 1)$$

for all $(t, x, u) \in M$. Analogously, condition (c') can be replaced by a similar condition (c'_g) where the constant C in (9.4.2) is replaced by a locally integrable scalar function $g(t) \geq 0$. Here $g(t)$ locally integrable means that g is L-integrable in any finite interval.

Finally, if A is not compact, nor contained in any slab as above, but A is closed, then Theorem (9.2.i) still holds provided conditions (a), (b), (c) (or alternates) hold, and also (d) $g(t_1, x_1, t_2, x_2) \to +\infty$ as $t_2 - t_1 \to \infty$ uniformly for all (t_1, x_1, t_2, x_2) such that the slab $[t_1 \leq t \leq t_2, x \in R^n]$ contains P.

Condition (d) can be replaced by the following condition which is much weaker: (d') there is some pair $\bar{x}(t), \bar{u}(t), \bar{t}_1 \leq t \leq \bar{t}_2$, in Ω, and some fixed interval $[a, b]$ containing the projection P_0 of P on the t-axis, such that if $l = g(e[\bar{x}])$, then for every pair $x(t), u(t), t_1 \leq t \leq t_2$, in Ω with $g(e[x]) \leq l$ we have $a \leq t_1 \leq t_2 \leq b$.

Let us prove that under hypotheses (b), (c) every trajectory $x(t), t_1 \leq t \leq t_2$, with $t_0 \leq t_1 < t_2 \leq T$, lies in some set $S = [(t, x) | t_0 \leq t \leq T, |x| \leq D]$. To prove this, let $Z = |x|^2 + 1 = (x^1)^2 + \cdots + (x^n)^2 + 1$. Then $Z(t) \geq 1$ and $dZ/dt = 2(x^1 f_1 + \cdots + x^n f_n) \leq 2CZ$; hence, by integration from t^* to t, where we recall that $(t^*, x(t^*)) \in P$, also $1 \leq Z(t) \leq Z(t^*) \exp(2C|t - t^*|)$. If N is such that $|x| \leq N$ for every $(t, x) \in P$, then

$$Z(t) \leq (N^2 + 1) \exp(2C(T - t_0)),$$

and finally

$$|x(t)| \leq Z^{1/2}(t) \leq (N^2 + 1)^{1/2} \exp(C(T - t_0)) = D.$$

Let us prove that the same holds under conditions (b), (c'). Indeed, if $Z(t) = V(t, x(t))$, then $dZ/dt = \operatorname{grad}_x V(t, x(t)) \cdot x'(t) + (\partial V/\partial t)_{x=x(t)} = \operatorname{grad}_x V(t, x(t)) \cdot f(t, x(t), u(t)) + (\partial V/\partial t)_{x=x(t)} \leq CV(t, x(t)) = CZ(t)$, and, by integration from t^* to t, also $V(t, x(t)) \leq V(t^*, x(t^*)) \exp(C(T - t_0)) \leq \alpha$.

Now the set of all x such that $V(t, x) \leq \alpha$ for some $(t, x) \in A$ and $t_0 \leq t \leq T$, is compact and hence bounded, say $|x| \leq D$. Thus, for every trajectory $x(t), t_1 \leq t \leq t_2$, in Ω, we have $t_0 \leq t_1 \leq t_2 \leq T, |x(t)| \leq D$.

We leave as an exercise for the reader to prove that the same conclusion holds under hypotheses (b), (c_g) or (b), (c'_g).

Let us suppose that A is closed but not contained in any slab $[t_0 \leq t \leq T, x \in R^n]$. Let $\bar{x}(t), \bar{u}(t), \bar{t}_1 \leq t \leq \bar{t}_2$, be some admissible pair in Ω, and let $l = g(e[x])$. Then for every other admissible pair $x(t), u(t), t_1 \leq t \leq t_2$, in Ω with at least one end point t_1 or t_2 not in $[a, b]$, we have $g(e[x]) \geq l$ by condition (d'). If we take any minimizing sequence $[x_k, u_k]$ and we replace any element x_k, u_k with $g(e[x_k]) \geq l$ by the fixed element $\bar{x}, \bar{u}$, we still have a minimizing sequence. All elements of the new sequence have now end points t_1, t_2 satisfying $a \leq t_1 \leq t_2 \leq b$, and we are in the same situation as above with t_0, T replaced by a, b.

9.5 Elimination of the Hypothesis that A Is Compact in Filippov's Theorem for Lagrange and Bolza Problems

If A is not compact, but closed, we shall assume, as in Section 9.4, that M is closed, but for every compact part A_1 of A the set M_1 of all (t, x, u) with $(t, x) \in A_1, u \in U(t, x)$ is compact.

If A is closed and contained in a slab $[t_0 \le t \le T, x \in R^n]$, t_0, T finite, then the existence theorem (9.3.i) still holds under the additional requirements (a), (b), (c) or alternates, as in Section 9.4.

Indeed, under conditions (a), (b), (c) (or alternates), the trajectories $x(t)$, $t_1 \le t \le t_2$ (with $t_0 \le t_1 \le t_2 \le T$), $x \in R^n$, are contained in a suitable compact set $A_1 = A \cap [t_0 \le t \le T, |x| \le N]$. Then the corresponding set $M_1 = [(t, x, u) | (t, x) \in A, t_0 \le t \le T, |x| \le N, u \in U(t, x)]$ is also compact by hypothesis, and thus $|f_0| \le K$ for all $(t, x, u) \in M_1$ and suitable constant K. Also, g has a minimum m in the compact set $B \cap (A_1 \times A_1)$. Finally, the functional is $\ge m - K(T - t_0)$, and hence bounded below as before.

Under conditions (a), (b), (c) we denote by P_0 the compact subset which is the projection of P on the t-axis.

If A is not compact, nor contained in any slab as above, but *closed*, then the existence theorem (9.3.i) still holds if we add to the requirements (a), (b), (c) of Section 9.4 the additional requirement: (d″) there are constants μ' real, $\mu > 0, D > 0$, such that $f_0(t, x, u) \ge \mu$ for all $(t, x) \in A$, $u \in U(t, x)$ with $|t| \ge D$, and such that $g(t_1, x(t_1), t_2, x(t_2)) \ge \mu'$ for all $t_1, t_2, t_2 - t_1 \ge D$, $[t_1, t_2] \cap P_0 \ne \emptyset$. For Lagrange problems, $g \equiv 0$, $\mu' = 0$.

Indeed, we can take D sufficiently large so that the projection P_0 of the set P on the t-axis is contained in $[-D, D]$. Then (a), (b) and (c) guarantee, as above, that the parts of the trajectories x lying in $A \cap [-D \le t \le D, x \in R^n]$ are actually in the compact set $A_1 = A \cap [-D \le t \le D, |x| \le N]$, and then the corresponding set $M_1 = [(t, x, u) | (t, x) \in A, |t| \le D, |x| \le N, u \in U(t, x)]$ is compact, and there is a constant K such that $|f_0(t, x, u)| \le K$ on M_1. Now, for every interval $[t_1, t_2]$ with $[t_1, t_2] \cap P_0 \ne \emptyset, t_2 - t_1 \ge 3D$, the interval $[t_1, t_2]$ contains parts of total length $\ge t_2 - t_1 - 2D$, which are outside $[-D, D]$, and where $f_0 \ge \mu > 0$, while $f_0 > -K$ in $[-D, D]$. Thus

$$I = g + \int_{t_1}^{t_2} f_0 \, dt \ge \mu' - 2DK + (t_2 - t_1 - 2D)\mu,$$

and $I \to +\infty$ as $t_2 - t_1 \to +\infty$. We have proved that the present condition (d″), (together with (a), (b), (c)) implies condition (d) of Section 9.4 for the corresponding Mayer problem, and our contention is proved.

Finally, condition (d′) can be replaced by the following more general condition: (d‴) There are constants μ' real, $D > 0$, and a locally L-integrable function $\mu(t)$, $-\infty < t < +\infty$, with $\int_{-\infty}^{0} \mu(t)\, dt = +\infty$, $\int_0^{+\infty} \mu(t)\, dt = +\infty$, and such that $f_0(t, x, u) \ge \mu(t)$ for all $(t, x) \in A$, $u \in U(t, x)$, and such that $g(t_1, x_1, t_2, x_2) \ge \mu'$ for all t_1, t_2 with $t_2 - t_1 \ge D$, $[t_1, t_2] \cap P_0 \ne \emptyset$.

For Lagrange problems we take $g \equiv 0$, $\mu' = 0$. The proof that Theorem (9.3.i) still holds under the additional conditions (b), (c), (d‴) is left as an exercise for the reader.

9.6 Examples

A. Examples of Application of Filippov's Theorem to Mayer Problems

In Section 4.3 and in Sections 6.1–5 we have already seen several examples of problems where Filippov's existence theorem is applicable. Here are a few others.

1. A point P moves along the x-axis monitored by the second order differential equation $x'' - x = u$ with $|u| \le 1$. We are to take P from any given state $x = a$, $x' = b$

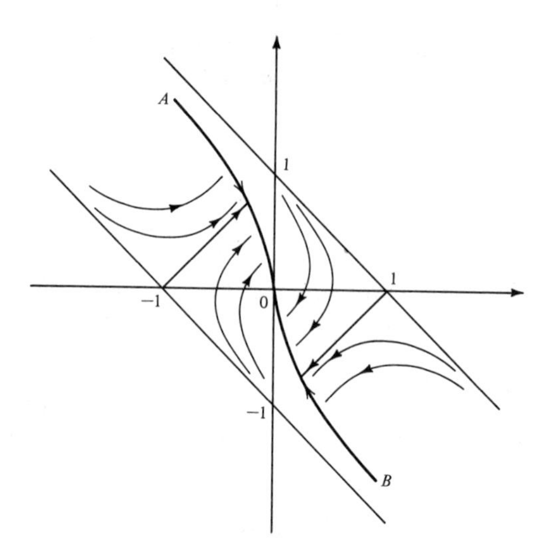

to rest at the origin $x = 0, x' = 0$ (if possible) in the shortest time. By using state variables $x = x$, $y = x'$, we have a Mayer problem of minimum time with differential system, constraint, and boundary conditions

$$dx/dt = y, \qquad dy/dt = x + u, \quad u \in U = [-1 \le u \le 1],$$
$$t_1 = 0, \qquad x(t_1) = a, \quad y(t_1) = b, \quad x(t_2) = 0, \quad y(t_2) = 0, \qquad t_2 \ge t_1, g = t_2,$$

and functional $I[x, y, u] = g = t_2$ which we want to minimize. Here $n = 2$, $m = 1$, $f_1 = y$, $f_2 = x + u$; hence $|xf_1 + yf_2| = |2xy + yu| \le 2(x^2 + y^2 + 1)$; condition (a) is satisfied, and so is condition (b), since $x(t_2) = y(t_2) = 0$. We shall prove below that only the points (a, b) with $-1 < a + b < 1$ can be steered to the origin $(0, 0)$ in some finite time t_2. Thus, for any such point (a, b) we can take $0 = t_1 \le t_2 \le T$ for some T (which may depend on (a, b)). Now $A = [0, T] \times R^2$, $M = A \times [-1, 1]$, and for any (t, x, y) the set Q is a segment, a compact convex set. By the existence theorem (9.2.i) (and conditions (a), (b), (c)) the problem under consideration has an optimal solution (for $-1 < a + b < 1$).

Let us prove that points (a, b) with $a + b \ge 1$, or $a + b \le -1$, cannot be taken to the origin. Indeed, $d(x + y)/dt = (x + y) + u$; hence, by integration with initial value $a + b \ge 1$, we have $d(x + y)/dt \ge a + b - 1 \ge 0$, and $x + y$ remains constant or increases, so that (a, b) cannot be steered to the origin. Analogously, by integration with initial value $a + b \le -1$, we have $d(x + y)/dt \le a + b + 1 \le 0$, and $x + y$ remains constant or decreases. Let us prove that the points (a, b) with $-1 < a + b < 1$ can be steered to the origin. First note that for $u = 1$ the point (x, y) travels along arcs F^+ of hyperbolas $(x + 1)^2 - y^2 = \pm C^2$, or segments of the straight lines $|x + 1| = |y|$. For $u = -1$ the point (x, y) travels along arcs F^- of hyperbolas $(x - 1)^2 - y^2 = \pm C^2$, or segments of the straight lines $|x - 1| = |y|$. Now, any point (a, b) on the half hyperbola $OB: (x + 1)^2 - y^2 = 1$, $y \le 0$, $x \ge 0$ traveled from B to $O = (0, 0)$ can be taken to the origin in a finite time by the strategy $u = +1$; the points (a, b) of the half hyperbola $OA: (x - 1)^2 - y^2 = 1$, $y \ge 0$, $x \le 0$ traveled from A to O can be taken to the origin in a finite time by the strategy $u = -1$. In turn any point (a, b) with $a + b < 1$ and lying above the line AOB

can be taken to OB by means of an arc F^- and strategy $u = -1$. Analogously, any point (a, b) with $a + b > -1$ and lying below the line AOB can be taken to OA by means of an arc F^+ and strategy $u = 1$. Thus, all points (a, b) with $-1 < a + b < 1$ can be steered to the origin.

2. Take $n = 2$, $m = 1$, state variables x, y, control variable u, system $x' = f_1 = xu + y^2 + u$, $y' = f_2 = -xyu + y^3 - x$, $u \in U = [-1 \le u \le 1]$, $g = t_2$, $x(0) = 0$, $y(0) = 0$, end point $(x(t_2), y(t_2))$ on the locus $\Gamma: x^2 + y^2 = 1$. This is a problem of minimum time. Note that for $u = 1$, $y = 0$, the first equation reduces to $x' = x + 1$ whose solution $x = -1 + e^t, 0 \le t < +\infty$, is positive, increasing, and reaches $x = 1$ in a time $T = \log 2$. Actually, $y^2 \ge 0$, hence $x' \ge x + 1$, and the point (x, y) certainly reaches Γ in a time $t_2 \le T$. Thus, at least one trajectory reaches the target in a finite time, and Ω is not empty. We can take here $A = [0 \le t \le 2] \times [x^2 + y^2 \le 1]$, a compact set in R^3. Since u enters linearly in f_1 and f_2, then for every (x, y) the set $Q(t, x)$ is a segment, certainly a convex set. Here $B = (0, 0) \times \Gamma$ is certainly a closed set. All conditions in Theorem (9.2.i) are satisfied. The problem has an optimal solution.

3. Take $n = 2$, $m = 1$, state variables x, y, system $x' = f_1 = xu + y$, $y' = f_2 = 1 + u$, $U = [u | -1 \le u \le 1]$, $x(0) = y(0) = 1$, $t_2 = 1$, both $x(t_2) = x_2$ and $y(t_2) = y_2$ undetermined, $g = (x_2^2 + y_2^2 + 1)^{-1}$. Here $A = [0, 1] \times R^2$, $B = (0) \times (1) \times (1) \times (1) \times R^2$, the initial point is fixed, and $|xf_1 + yf_2| \le x^2 + |xy| + |2y| \le \frac{3}{2}(x^2 + y^2 + 1)$. All sets $Q(x, y)$ are segments, A and B are closed, and all conditions in Theorem (9.2.i) as well as (a), (b), (c) of Section 9.4 are satisfied. An optimal solution exists.

4. Take $n = 2$, $m = 2$, state variables x, y, control variables u, v, system $x' = f_1 = xu + yv$, $y' = f_2 = u + v$, $U = [(u, v) | -1 \le u, v \le 1]$, $t_1 = 0$, $x(0) = 0$, $y(0) = 0$, $x_2 = x(t_2)$, $y_2 = y(t_2)$ on the locus $\Gamma: t_2 = (1 + x_2^2 + y_2^2)^{-1}$, with $g = (x_2^4 + y_2^4 + 1)^{-1}$. Then $0 \le t_2 \le 1$, $A = [0, 1] \times R^2$, $B = (0) \times (0) \times (0) \times \Gamma$, $|xf_1 + yf_2| \le \frac{3}{2}(x^2 + y^2 + 1)$, all $Q(x, y)$ are parallelograms, and the initial point is fixed. All conditions of Theorem (9.2.i) and (a), (b), and (c) of Section 9.4 are satisfied. Certainly, we can steer the initial state $(0, 0)$ at time $t_1 = 0$ to the locus Γ of the $t_2 x_2 y_2$-space. Indeed, by taking $u = 0$, $v = 1$, $x = t^2/2$, $y = t$, then $t(1 + x^2 + y^2)$ grows from 0 to $+\infty$ as t increases, and then equals 1 for some $t_2 > 0$. An optimal solution certainly exists.

5. Take $n = 2$, $m = 2$, state variables x, y, control variables u, v, system $x' = f_1 = xu + yv + y^3$, $y' = f_2 = xv - yu - x^3$, (x_2, y_2) on Γ, initial point $t_1 = 0$, $x_1 = y_1 = 0$, and U, g and Γ as in Example 4. Again, A and B are closed, all sets $Q(x, y)$ are parallelograms as before, and conditions (a) and (b) of Section 9.4 are satisfied. Take $V = x^4 + y^4 + 2$. Then $V_t = 0$, $|V_x f_1 + V_y f_2| = |4x^4 u + 4x^3 yv + 4y^3 xv - 4y^4 u| \le 4(x^4 + y^4 + |xy|(x^2 + y^2)) \le 8(x^4 + y^4) \le 8V$, and condition (c′) of Section 9.4 is satisfied. Certainly, we can steer $(0, 0)$ at $t_1 = 0$ to Γ. Indeed, by the argument concerning V in Section 9.4 we conclude that the solution $(x(t), y(t))$ starting at $(0, 0)$ at $t = 0$ exists in all of $0 \le t \le 1$. Thus, for some $t = t_2$, $0 < t_2 \le 1$, the trajectory $(x(t), y(t))$, $0 \le t \le 1$, must cross the locus $\Gamma: t_2 = (1 + x_2^2 + y_2^2)^{-1}$, $0 < t_2 \le 1$, $(x_2, y_2) \in R^2$. An optimal solution certainly exists.

B. Counterexamples

6. Let us show that an optimal solution may fail to exist under the conditions of Theorem (9.2.i) if A is unbounded and (a) and (c) hold, but not (b). Take $n = 1$, $m = 1$, system $x' = f = u$, $U = [-1 \le u \le 1]$, $t_1 = 0$, $t_2 = 1$, $x_1 = x(0)$, $x_2 = x(1)$ undetermined, $g = (x_1^2 + x_2^2 + 1)^{-1}$. Here $A = [(t, x) | 0 \le t \le 1, -\infty < x < +\infty]$, $B = (0) \times$

$R \times (1) \times R$, $|f| \leq 1$, the sets Q are all segments, A and B are closed, and the conditions of Theorem (9.2.i) as well as (a) and (c) of Section 9.4 are satisfied. If we take the sequence of admissible pairs $x_k(t) = k$, $u_k(t) = 0$, $0 \leq t \leq 1$, $k = 1, 2, \ldots$, then $I[x_k, u_k] \to 0$ as $k \to \infty$; hence $i = 0$. The infimum $i = 0$ is not attained. In this problem no point of the trajectories is required to be on some compact set.

7. Let us show that an optimal solution may fail to exist under the conditions of Theorem (9.2.i) if A is unbounded and (a) and (b) hold, but not (c) of Section 9.4. Take $n = 1$, $m = 1$, $f = ux^2$, $U = [-1 \leq u \leq 1]$, $t_1 = 0$, $t_2 = 1$, $x(0) = 1$, $x_2 = x(1)$ undetermined, $g = (1 + x_2^2)^{-1}$. Here $A = [(t, x) | 0 \leq t \leq 1, -\infty < x < +\infty]$, $B = (0) \times (1) \times (1) \times R$, the initial point is fixed, A and B are closed, the sets Q are all segments, and (a) and (b) are satisfied. If we take the sequence of admissible pairs $x_k(t) = (1 - \alpha_k t)^{-1}$, $u_k(t) = \alpha_k = 1 - k^{-1}$, $k = 1, 2, \ldots$, then $x_k(1) = k \to \infty$, $I[x_k, u_k] = (1 + k^2)^{-1} \to 0$ as $k \to \infty$; hence $i = 0$. The infimum $i = 0$ is not attained.

8. Let us show that an optimal solution may fail to exist, for A unbounded and not contained in a slab as above, under the conditions of Theorem (9.2.i) and (a), (b), and (c), but not (d). Take $n = 2$, $m = 1$, state variables x, y, $U = [-1 \leq u \leq 1]$, system $x' = f_1 = u$, $y' = (1 + t)^{-1}|x|$, $t_1 = 0$, $t_2 \geq 0$, $x(0) = y(0) = 0$, $x(t_2) = 1$, $g = y_2 = y(t_2)$. We have here the problem of transferring (x, y) from $(0, 0)$ at $t_1 = 0$ to the locus $\Gamma = [x = 1, y \in R]$ at a point of minimum coordinate y in an undetermined time $t_2 \geq 0$. Thus, if R^+ denotes $R^+ = [0 \leq t < +\infty]$, we have $A = R^+ \times R^2$, $M = A \times U$, $B = (0) \times (0) \times (0) \times R^+ \times (1) \times R$, all closed. The sets Q are all segments. The initial point is fixed. All conditions of Theorem (9.2.i) and (a), (b), and (c) of Section 9.4 are satisfied. If we take the admissible pairs $x_k(t)$, $y_k(t)$, $u_k(t)$, $0 \leq t \leq k + 1$, defined by $u_k(t) = 0$, $x_k(t) = 0$, $y_k(t) = 0$ for $0 \leq t \leq k$, and $u_k(t) = 1$, $x_k(t) = t - k$, $y_k(t) = \int_k^t (1 + \tau)^{-1}(\tau - k)\, d\tau$ for $k \leq t \leq k + 1 = t_2$, $k = 1, 2, \ldots$, then $|y_k(t_2)| \leq (1 + k)^{-1}$, $I[x_k, y_k, u_k] \to 0$ as $k \to \infty$; hence $i = 0$. The infimum $i = 0$ is not attained.

C. Examples of Application of Filippov's Theorem to Lagrange and Bolza Problems

1. Take $n = 1, m = 1, t_1 = 0, 0 \leq t_2 \leq 1, x(0) = 0, x(t_2) = 1, g = 0, f_0 = t^2 + x^2 + u^2$, system $x' = f = x^2 u + u$, $U = [-1 \leq u \leq 1]$. The boundary conditions can be satisfied, since, by taking $u = 1$, we have $x' \geq 1$ and the point $x = 1$ is reached in a time $t_2 \leq 1$. Here we can well limit ourselves to trajectories x which are monotone nondecreasing in $[0, t_2]$, since if $x(\alpha) = x(\beta) = \bar{x}$ for some $\alpha < \beta$ and $x(t) > \bar{x}$ in (α, β), then by taking $u = 0$ in (α, β) we keep $x(t)$ constant in (α, β) and we reduce the value of the functional. Thus we can take $A = [0 \leq t \leq 1, 0 \leq x \leq 1]$, and then A is compact, $M = A \times U$ is compact, and $B = (0) \times (0) \times R \times (1)$ is closed. The sets $\tilde{Q}(t, x)$ are all convex. The functional $\int_0^{t_2} (t^2 + x^2 + u^2)\, dt$ has an absolute minimum under the constraints.

2. Take $n = 2$, $m = 1$, state variables x, y, system $x' = f_1 = xu + y$, $y' = f_2 = 1 + u$, take $g = 0$, $f_0 = x^2 + y^2 + u^2$, $U = [-1 \leq u \leq 1]$, $x(0) = y(0) = 1$, $x(1) = y(1) = 1$. Here $A = [0, 1] \times R^2$, $B = (0) \times (1) \times (1) \times (1) \times (1) \times (1)$, the initial point is fixed, and $|xf_1 + yf_2| \leq x^2 + |xy| + 2|y| \leq \frac{3}{2}(x^2 + y^2 + 1)$. All sets $\tilde{Q}(x, y)$ are convex; A is closed and contained in a slab. B is a single point. The boundary and other data can be satisfied, as the following admissible pair shows: $x(t) = y(t) = 1, u(t) = -1, 0 \leq t \leq 1$. All conditions of theorem (9.3.i) and (a), (b), (c) are satisfied. The functional $I = \int_0^1 (t^2 + x^2 + u^2)\, dt$ has an absolute minimum under the constraints.

3. Take $n = 2$, $m = 2$, state variables x, y, control variables u, v, $f_0 = x^2 + y^2 + u^2 + v^2$, system $x' = f_1 = xu + yv$, $y' = f_2 = u + v$, $U = [(u,v)|-1 \le u, v \le 1]$, $t_1 = 0$, $x(0) = y(0) = 1$, $t_2 \ge 0$, $(x_2, y_2) = (x(t_2), y(t_2))$ on the locus $\Gamma: t_2 = (1 + x_2^2 + y_2^2)^{-1}$. Here $A = [0,1] \times R^2$, $B = (0) \times (1) \times (1) \times \Gamma$, $|xf_1 + yf_2| \le \frac{3}{2}(x^2 + y^2 + 1)$, and all sets $\tilde{Q}(t,x,y)$ are convex. All conditions of Theorem (9.3.i) and (a), (b), (c) are satisfied. The functional $I = \int_0^{t_2} (x^2 + y^2 + u^2 + v^2)\,dt$ has an absolute minimum under the constraints.

4. Take $n = 2$, $m = 2$, state variables x, y, control variables u, v, $f_0 = x^2 + y^2 + u^2 + v^2$, system $x' = f_1 = xu + yv + y^3$, $y' = f_2 = xv - yu - x^3$, initial point $t_1 = 0$, $x_1 = y_1 = 1$, (x_2, y_2) on Γ as in Example 3 above, U as in Example 3. All sets $\tilde{Q}(t,x,y)$ are convex. Conditions (a) and (b) are satisfied. If we take $V = x^4 + y^4 + 1$, condition (c′) is also satisfied. (Compare with Example 5 in subsection A.) The functional $I = \int_0^{t_2} (x^2 + y^2 + u^2 + v^2)\,dt$ has an absolute minimum under the constraints.

D. Counterexamples

5. Let us show that an optimal solution may fail to exist if the condition of convexity of the sets $\tilde{Q}$ in Theorem (9.3.i) is not satisfied. Take $n = 1$, $m = 1$, $x(0) = 0$, $x(1) = 0$, $f_0 = x^2$, system $x' = f = u$, $U = [u = -1 \text{ and } u = +1]$. For every $k = 1, 2, \dots$, let $x_k(t) = t - i/k$ if $i/k \le t \le i/k + 1/2k$, and $x_k(t) = (i+1)/k - t$ if $i/k + 1/2k \le t \le (i+1)/k$, $i = 0, 1, \dots, k-1$, so that $u_k(t) = x_k'(t) = \pm 1$. Then $I[x_k, u_k] = 3^{-1}2^{-2}k^{-2}$ and $I[x_k, u_k] \to 0$ as $k \to \infty$. Since $I[x,u] \ge 0$, we have $i = \inf I[x,u] = 0$. The value $i = 0$ cannot be attained, since $I = 0$ would require $x(t) \equiv 0$, $u(t) = x'(t) = 0$, and this is not an admissible pair. The functional $\int_0^1 x^2\,dt$ has no absolute minimum under the constraints. The sets $\tilde{Q}(x) = [(z^0, z)|z^0 \ge x^2, z = 1 \text{ and } z = -1]$ are not convex.

6. An example analogous to Example 6 is as follows. Take $n = 2$, $m = 2$, state variables x, y, controls u, v, $f_0 = x^2 + y^2$, system $x' = f_1 = u$, $y' = f_2 = v$, $U = [(u,v)|u^2 + v^2 = 1]$, $t_1 = 0$, $t_2 = 1$, $x(0) = y(0) = 0$, $x(1) = y(1) = 0$. If we take $x_k(t) = (2k\pi)^{-1} \sin 2k\pi t$, $y_k(t) = (2k\pi)^{-1}(1 - \cos 2k\pi t)$, $0 \le t \le 1$, $k = 1, 2, \dots$, then $u_k = x_k' = \cos 2k\pi t$, $v_k = y_k' = \sin 2k\pi t$, $u_k^2 + v_k^2 = 1$, $I[x_k, u_k] = (2k^2\pi^2)^{-1}$, $I[x_k, u_k] \to 0$ as $k \to 0$. Thus, $i = \inf I[x,u] = 0$. The value $i = 0$ cannot be attained, since $I = 0$ would imply $x(t) \equiv 0$, $y(t) \equiv 0$, $u \equiv v \equiv 0$, and this is not an admissible system. The integral $\int_0^1 (x^2 + y^2)\,dt$ has no absolute minimum under the constraints. The sets $\tilde{Q}(x,y) = [(z^0, z_1, z_2)|z^0 \ge x^2 + y^2, z_1^2 + z_2^2 = 1]$ are not convex.

7. Let us show that an optimal solution may fail to exist under the conditions of (9.3.i), if A is unbounded and contained in a slab as above, and (a) holds, but not (b). Take $n = 1$, $m = 1$, system $x' = u$, $U = [-1 \le u \le 1]$, $t_1 = 0$, $t_2 = 1$; $x_1 = x(0)$, $x_2 = x(1)$ both undetermined; $f_0 = (1 + x^2)^{-1}$. The integral $I = \int_0^1 (1 + x^2)^{-1}\,dt$ has no absolute minimum under the constraints. Indeed, for the admissible pairs x_k, u_k defined by $x_k(t) = k$, $u_k(t) = 0$, $0 \le t \le 1$, $k = 1, 2, \dots$, we have $I[x_k, u_k] \to 0$ as $k \to \infty$ and hence $i = 0$, but $i = 0$ is not attainable.

8. Let us show that an optimal solution may fail to exist under the conditions of (9.3.i), if A is unbounded and contained in a slab, and if (b) holds but not (c). Take $n = 1$, $m = 1$, $f_0 = (1 - (1 - t)x)^2$, system $x' = ux^2$, $U = [0 \le u \le 1]$, $t_1 = 0$, $t_2 = 1$; $x(0) = 1$, $x_2 = x(1)$ undetermined. By taking x_k, u_k, $k = 2, 3, \dots$, as in Example 7 of subsection B, we have $I[x_k, u_k] = (k-1)^{-3}(k^2 - 1 - 2k \log k) \to 0$ as $k \to +\infty$, and $i = 0$. The infimum $i = 0$ is not attained, since $1 - (1 - t)x = 0$ for $x = (1 - t)^{-1}$, and

this trajectory, though generated by $u(t) = 1$, is not AC in $[0, 1]$, actually not even continuous.

9. Let us show that an optimal solution may fail to exist for A unbounded, not contained in a slab, under the conditions of (9.3.i) and (a), (b), (c), but not (d″). Take $n = 1$, $m = 1$, $U = [-1 \le u \le 1]$, $f_0 = (1 + t)^{-1}|x|$, system $x' = f = u$, $t_1 = 0$, $t_2 \ge 0$, $x(0) = 0$, $x(t_2) = 1$. (Compare with the Example 8 in subsection B.) The functional $I = \int_0^{t_2} (1 + t)^{-1}|x|\, dt$ has no absolute minimum under the constraints.

Bibliographical Notes

The existence theorems (9.2.i) and (9.3.i) for problems of optimal control with bounded strategies and compact control space were proved by A. F. Filippov [1] in 1962. They could well be considered as particular cases of the existence theorems of Chapter 11, but we have felt it necessary to give of them simple, elementary, straightforward proofs. Indeed, we have proved them by mere uniform topology, and not even the concept of lower semicontinuity needed to be mentioned. They are suited for problems of applications as many of those listed in Chapter 6. The present proof of (9.2.i) for the Mayer problem is similar to the one of Filippov, though his closure argument is replaced here by the analogous closure theorem (8.6.i), and his selection argument is replaced by the much more straightforward McShane–Warfield implicit function theorem (8.2.iii). For Lagrange and Bolza problems, Theorem (9.3.i), proof I is simply a standard reduction to the Mayer problem as in Filippov; proof II makes use of the lower closure theorem (8.8.i), which in turn is based on the Fatou-like statement (8.7.ii). In any case we have tried to preserve the elementary character of Filippov's proofs.

CHAPTER 10

Closure and Lower Closure Theorems under Weak Convergence

10.1 The Banach–Saks–Mazur Theorem

If X is a normed linear space over the reals with norm $\|x\|$, let X^* be the dual of X, that is, the space of all linear bounded operators x^* on X, the linear operation being denoted by (x^*, x), or $X^* \times X \to R$. A sequence $[x_k]$ of elements of X then is said to be convergent in X to x provided $\|x_k - x\| \to 0$ as $k \to \infty$. A sequence $[x_k]$ of elements of X is said to be weakly convergent in X to x provided $(x^*, x_k) \to (x^*, x)$ as $k \to \infty$ for all $x^* \in X^*$.

Both convergences are examples of σ-limits in the sense of Sections 2.15 and 8.1.

Let us mention here the following theorem:

10.1.i (Banach, Saks, Mazur). *If X is any normed space over the reals with dual X^* and norm $\|x\|$ in X, if x_k, $k = 1, 2, \ldots$, is any sequence of elements $x_k \in X$ convergent weakly in X to an element $x \in X$, then there is a system of real numbers $c_{ki} \geq 0$, $i = 1, \ldots, k$, $k = 1, 2, \ldots$, with $\sum_{i=1}^{k} c_{ki} = 1$, such that, if $y_k = \sum_{i=1}^{k} c_{ki} x_i$, then $\|y_k - x\| \to 0$ as $k \to \infty$.*

For this important theorem we refer to S. Mazur [1], or M. Day [I, p. 45], or K. Yosida [I, p. 120].

As a particular case, let G be a fixed measurable subset of points $t = (t^1, \ldots, t^\nu) \in R^\nu$, $\nu \geq 1$. Let $X = L_1(G)$ denote the space of all L-integrable functions $h(t)$, $t \in G$, with norm $\|h\|_1 = \int_G |h(t)|\,dt$. We know that $X^* = L_\infty(G)$, that is, the dual space of X is the space of all real valued functions $y(t)$, $t \in G$, essentially bounded in G, and with norm $\|y\|_\infty = \text{ess sup}[|y(t)|, t \in G]$. The linear operation (y, h) is then $(y, h) = \int_G y(t)h(t)\,dt$. Then, a sequence of functions $h_k(t)$, $t \in G$, $k = 1, 2, \ldots$, of L-integrable functions is said to be convergent in $L_1(G)$, or strongly convergent in L_1, to an L_1-integrable function $h(t)$, $t \in G$, if

$$\lim_{k\to\infty} \int_G |h_k(t) - h(t)|\, dt = 0.$$

A sequence of functions $h_k(t)$, $t \in G$, $k = 1, 2, \dots$, of L-integrable functions is said to be weakly convergent in $L_1(G)$ to an L_1-integrable function $h(t)$, $t \in G$, provided

$$\lim_{k\to\infty} \int_G h_k(t)y(t)\, dt = \int_G h(t)y(t)\, dt$$

for all measurable and (essentially) bounded functions $y(t)$, $t \in G$. Such a function h is uniquely defined a.e. in G, and is said to be the weak limit of the sequence $[h_k]$ in G.

A family $\{h(t), t \in G\}$ of L-integrable functions in G is said to be sequentially strongly [weakly] relatively compact in L_1, provided every sequence $[h_k]$ of elements of $\{h\}$ contains a subsequence which is strongly [weakly] convergent in $L_1(G)$ (to an element $h \in L_1(G)$ which need not be in $\{h\}$). If we prescribe that any such strong limit [weak limit] h must be in $\{h\}$, then we say that the family is sequentially strongly compact [weakly compact].

Statement (10.1.i) for $X = L_1(G)$ now yields: If $h_k(t)$, $t \in G$, $k = 1, 2, \dots$, is a sequence of L-integrable functions on G, and $[h_k]$ converges weakly in L_1 to an L-integrable function $h(t)$, $t \in G$, then there is a system of real numbers $c_{ki} \geq 0$, $i = 1, \dots, k$, $k = 1, 2, \dots$, with $\sum_{i=1}^k c_{ki} = 1$ for all k, such that, if $H_k(t) = \sum_{i=1}^k c_{ki}h_i(t)$, $t \in G$, $k = 1, 2, \dots$, then $\int_G |H_k(t) - h(t)|\, dt \to 0$ as $k \to \infty$.

Definitions completely analogous hold for the spaces $X = L_1(G, R^n)$ of all vector valued functions $h(t) = (h^1, \dots, h^n)$, $t \in G$, with L_1-integrable components, and the dual space $X^* = L_\infty(G, R^n)$ of all vector valued functions $y(t) = (y^1, \dots, y^n)$, $t \in G$, with essentially bounded components. Then $\|h\|_1 = \int_G |h(t)|\, dt$, $\|y\|_\infty = \operatorname{ess\,sup}[|y(t)|, t \in G]$, $(y, h) = \int_G y(t) \cdot h(t)\, dt$, where $|\ |$ is the Euclidean norm and $y \cdot h = y^1h^1 + \cdots + y^nh^n = y^*h$ is the inner product in R^n.

If $h_k(t) = (h^1, \dots, h^n)$, $t \in G$, $k = 1, 2, \dots$, is a sequence of elements $h_k \in L_1(G, R^n)$ and $[h_k]$ is weakly convergent to an element $h(t) = (h^1, \dots, h^n)$, $t \in G$, $h \in L_1(G, R^n)$, then there is a system of numbers $c_{ki} \geq 0$, $i = 1, \dots, k$, $k = 1, 2, \dots$, with $\sum_{i=1}^k c_{ki} = 1$, such that, if $H_k(t) = \sum_{i=1}^k c_{ki}h_i(t)$, $t \in G$, $k = 1, 2, \dots$, then $\|H_k - h\|_1 \to 0$ as $k \to \infty$.

10.2 Absolute Integrability and Related Concepts

A function $x(t)$, $a \leq t \leq b$, is said to be AC or absolutely continuous, provided, given $\varepsilon > 0$, there is $\delta = \delta(x, \varepsilon) > 0$ such that

$$\sum_{i=1}^{N} |x(\beta_i) - x(\alpha_i)| \leq \varepsilon$$

for all finite systems of nonoverlapping intervals $[\alpha_i, \beta_i]$, $i = 1, \ldots, N$, in $[a, b]$ with $\sum_{i=1}^N (\beta_i - \alpha_i) \le \delta$.

A Lipschitz function $x(t)$, $a \le t \le b$, is certainly AC. That is, if there is a constant $L \ge 0$ such that $|x(t) - x(t')| \le L|t - t'|$ for all $t, t' \in [a, b]$, then x is AC, namely, given $\varepsilon > 0$, the requirement for absolute continuity above is satisfied for $\delta = \varepsilon/L$.

A family $\{x(t), a \le t \le b\}$ of AC functions is said to be equiabsolutely continuous if, given $\varepsilon > 0$, there is $\delta = \delta(\varepsilon) > 0$ such that

$$\sum_{i=1}^{N} |x(\beta_i) - x(\alpha_i)| \le \varepsilon$$

for all functions of the family, and for all finite systems of nonoverlapping intervals $[\alpha_i, \beta_i]$, $i = 1, \ldots, N$, in $[a, b]$ with $\sum_{i=1}^N (\beta_i - \alpha_i) \le \delta$. For the sake of brevity, we shall often say the functions x of the family are equiabsolutely continuous.

Any equiabsolutely continuous family x is also equicontinuous, as we see by taking $N = 1$.

If the functions x of the family are equi-Lipschitzian, that is, there is an $L \ge 0$ such that $|x(t) - x(t')| \le L|t - t'|$ for all elements x of the family and all $t, t' \in [a, b]$, then the family is certainly equiabsolutely continuous (and hence equicontinuous).

We know that any L-integrable function $h(t)$, $a \le t \le b$, is absolutely integrable, that is, given $\varepsilon > 0$, there is a $\delta = \delta(h, \varepsilon) > 0$ such that $\int_E |h(t)|\,dt \le \varepsilon$ for all measurable subsets E of $[a, b]$ with meas $E \le \delta$.

A family $\{h(t), a \le t \le b\}$ of L-integrable functions is said to be equiabsolutely integrable provided that, given $\varepsilon > 0$, there is $\delta = \delta(\varepsilon) > 0$ such that $\int_E |h(t)|\,dt \le \varepsilon$ holds for every element $h(t)$, $a \le t \le b$, of the family, and all measurable subsets E of $[a, b]$ with meas $E \le \delta$.

10.2.i. *If $\{x(t), a \le t \le b\}$ is a family of AC functions, then the family $\{x\}$ is equiabsolutely continuous if and only if the family of derivatives $\{x'\}$ is equiabsolutely integrable.*

Proof. Let us assume $\{x'\}$ to be equiabsolutely integrable, and let us prove that $\{x\}$ is equiabsolutely continuous. Indeed, given $\varepsilon > 0$, let $\delta = \delta(\varepsilon)$ be the number given in the definition of equiabsolute integrability of the derivatives, and let $[\alpha_i, \beta_i]$, $i = 1, \ldots, N$, be any system of nonoverlapping intervals in $[a, b]$ with $\sum_i (\beta_i - \alpha_i) \le \delta$. If $E = \bigcup_i [\alpha_i, \beta_i]$, then meas $E \le \delta$ and

$$\sum_i |x(\beta_i) - x(\alpha_i)| = \sum_i \left| \int_{\alpha_i}^{\beta_i} x'(t)\,dt \right| \le \int_E |x'(t)|\,dt \le \varepsilon,$$

and this holds for every element $x \in \{x\}$.

Conversely, assume that $\{x\}$ is equiabsolutely continuous, and let us prove that $\{x'\}$ is equiabsolutely integrable. Given $\varepsilon > 0$, let $\delta = \delta(\varepsilon/6)$ be the number given in the definition of equiabsolute continuity of $\{x\}$. Let $x(t)$, $a \le t \le b$, be any element of $\{x\}$, and let E be any measurable subset of $[a, b]$ of measure $\le \delta/2$. Let E^+, E^- be the subset of all $t \in E$ where $x'(t)$ is defined and $x'(t) \ge 0$, or $x'(t) \le 0$ respectively. Then both E^+ and E^- have measures $\le \delta/2$. Since $x'(t)$, $a \le t \le b$, is L-integrable, and hence, by itself, absolutely integrable, there is $\sigma = \sigma(\varepsilon; x) > 0$ such that $\int_F |x'(t)|\,dt \le \varepsilon/6$ for every measurable subset F with meas $F \le \sigma$. It is not restrictive to take $\sigma \le \delta/2$. Now E^+, with meas $E^+ \le \delta/2$, is certainly covered by some open set G with meas $G <$ meas $E^+ + \sigma$. Let (α_i, β_i), $i = 1, 2, \ldots$, denote the disjoint open intervals which are the components of

G, and note that

$$\sum_{i=1}^{\infty} (\beta_i - \alpha_i) = \text{meas } G \leq \text{meas } E^+ + \sigma \leq \frac{\delta}{2} + \frac{\delta}{2} = \delta.$$

Then, the same holds for the finite system (α_i, β_i), $i = 1, \ldots, N$, whose union we denote by G_N, N arbitrary. Thus, meas $G_N \leq \delta$, meas$(G_N - E^+) \leq \sigma$, and for N large enough also meas$(G - G_N) \leq \sigma$. Then

$$\begin{aligned}
\int_{E^+} |x'(t)|\, dt &= \left(\int_{E^+ \cap G_N} + \int_{E^+ \cap (G - G_N)}\right) x'\, dt \\
&= \left(\int_{E^+ \cap G_N} + \int_{G_N - E^+}\right) x'\, dt - \int_{G_N - E^+} x'\, dt + \int_{E^+ \cap (G - G_N)} x'\, dt \\
&\leq \int_{G_N} x'\, dt + \int_{G_N - E^+} |x'|\, dt + \int_{G - G_N} |x'|\, dt \\
&\leq \sum_{i=1}^{N} (x(\beta_i) - x(\alpha_i)) + \frac{\varepsilon}{6} + \frac{\varepsilon}{6} \leq \frac{\varepsilon}{6} + \frac{\varepsilon}{6} + \frac{\varepsilon}{6} = \frac{\varepsilon}{2}.
\end{aligned}$$

Analogously, we can prove that $\int_{E^-} |x'(t)|\, dt \leq \varepsilon/2$, and thus $\int_E |x'(t)|\, dt \leq \varepsilon$, and this holds for every measurable subset E of $[a, b]$ with meas $E < \delta/2$. This proves (10.2.i). □

10.2.ii. *If $\{x(t)\}$ is a class of equiabsolutely continuous functions $x(t)$, $a \leq t \leq b$, $t_0 \leq a < b \leq T$, t_0, T finite and fixed, then their total variations $V[x; a, b]$ are equibounded.*

Proof. Take $\varepsilon = 1$, and let $\delta > 0$ be such that for any element $x(t)$, $a \leq t \leq b$, of the class and for any finite system of nonoverlapping intervals $[\alpha_i, \beta_i] \subset [a, b]$ with $\sum_i (\beta_i - \alpha_i) \leq \delta$ we also have $\sum_i |x(\beta_i) - x(\alpha_i)| \leq 1$. Now divide $[t_0, T]$ into N equal parts each of length $\leq \delta$ by means of points $t_0, t_1, \ldots, t_N = T$. Finally, for any subdivision of $[a, b]$ into arbitrary intervals $[a_{j-1}, a_j]$ by means of points of subdivision $a_0 = a, a_1, \ldots, a_m = b$, we have $V = \sup S$ with $S = \sum_j |x(a_{j-1}) - x(a_j)|$. If we add the points t_i to the points of subdivision a_j, then we can associate the new intervals in at most N groups of consecutive intervals, each group covering an interval of length $\leq \delta$, and hence having a contribution ≤ 1 in S. Thus $S \leq N$, and $V[x; a, b] \leq N$. □

Together with a family $\{x\}$ of continuous functions $x(t)$, $a \leq t \leq b$, we may consider the larger family $\{y\}$ we obtain by adding to $\{x\}$ all its limit elements in the ρ-metric. Thus, $y(t)$, $a \leq t \leq b$, belongs to $\{y\}$ if either $y \in \{x\}$ or (more inclusively) there is a sequence $x_k(t)$, $a_k \leq t \leq b_k$, $k = 1, 2, \ldots$, of elements $x_k \in \{x\}$ with $\rho(x_k, y) \to 0$ as $k \to \infty$. In any case y is continuous in $[a, b]$.

10.2.iii. *If $\{x(t), a \leq t \leq b\}$ is an equiabsolutely continuous family, then the larger class $\{y\}$ of all the limit elements of $\{x\}$ in the ρ-metric is equiabsolutely continuous.*

Proof. Given $\varepsilon > 0$, let $\delta = \delta(\varepsilon)$ be the number given in the definition of equiabsolute continuity of the family $\{x\}$. Let $y(t)$, $a \leq t \leq b$, any element of $\{y\}$ which is not already an element of $\{x\}$. Then there is a sequence $x_k(t)$, $a_k \leq t \leq b_k$, $k = 1, 2, \ldots$, of elements of $\{x\}$ with $\rho(x_k, y) \to 0$ as $k \to \infty$. Hence, $a_k \to a$, $b_k \to b$ as $k \to \infty$. Let $[\alpha_s, \beta_s]$, $s = 1, \ldots, N$, be any finite system of nonoverlapping intervals contained in the open interval (a, b) with $\sum_s (\beta_s - \alpha_s) \leq \delta(\varepsilon)$. Then, for k sufficiently large, the intervals $[\alpha_s, \beta_s]$ are contained in the intervals $[a_k, b_k]$, and then $\sum_s |x_k(\beta_s) - x_k(\alpha_s)| \leq \varepsilon$ for all such k, and by taking the limit as $k \to \infty$, also $\sum_s |y(\beta_s) - y(\alpha_s)| \leq \varepsilon$, and this relation holds for system of intervals $[\alpha_s, \beta_s]$ contained in (a, b). Since y is continuous in $[a, b]$, the same relation holds for any such system contained in $[a, b]$. □

10.3 An Equivalence Theorem

We state here the following composite statement, in which, only to simplify notation, we assume that G is a bounded measurable subset of R^ν, $\nu \geq 1$.

10.3.i (AN EQUIVALENCE THEOREM). *Let $\{f(t), t \in G\}$ be a family of real valued L-integrable functions on a fixed bounded measurable subset G of R^ν. The following statements are equivalent:*

(a) *The family $\{f\}$ is sequentially weakly relatively compact in $L_1(G)$.*
(b) *The family $\{f\}$ is equiabsolutely integrable in G.*
(c) *There is a constant M and a real valued function $\phi(\zeta)$, $0 \leq \zeta < +\infty$, bounded below, such that $\phi(\zeta)/\zeta \to +\infty$ as $\zeta \to +\infty$, and $\int_G \phi(|f(t)|)\,dt \leq M$ for all $h \in \{h\}$ (growth condition (ϕ)).*
(d) *There is a real valued function $\psi(\zeta)$, $0 \leq \zeta < +\infty$, bounded below, such that $\psi(\zeta)/\zeta \to +\infty$ as $\zeta \to +\infty$, and the family $\{\psi(|f(t)|), t \in G\}$ is equiabsolutely integrable.*

In (c), (d) it is not restrictive to assume ϕ, ψ nonnegative, monotone nondecreasing, continuous, and convex in $[0, +\infty)$. Functions ϕ, or ψ, as above are often called Nagumo functions. The equivalence of (a) and (b) was proved by Dunford and Pettis (see, e.g., Edwards [I, p. 274]). The implication (b) ⇒ (c) was proved by de la Vallee-Poussin (see, e.g., Natanson [I, p. 164]). The implication (c) ⇒ (b) was proved by Tonelli [I] for some particular ϕ, and then by Nagumo in the general case (see, e.g. McShane [I, p. 176]). Then, (d) ⇒ (c) as a consequence of the implication (b) ⇒ (c). Finally, (b) ⇒ (d) can be proved as (b) ⇒ (c) by a suitable construction of the function ψ so that $\psi(\psi(\xi))$ has the property of ϕ in (c) (see, e.g., Candeloro and Pucci[1], and also Cesari and Pucci[1] for an elementary proof of (10.3.i)) for $\nu = 1$. The property expressed under (c) is often called the growth property (ϕ).

The growth property (ϕ) and analogous ones, will be discussed in Section 10.4 below and will be shown in Section 10.5 to guarantee properties of upper semicontinuity of certain relevant sets $\tilde{Q}^*(x)$, that is, of certain set valued maps $x \to \tilde{Q}^*(x)$. The equivalence of (a), (b), (c), (d), and the upper semicontinuity of the related sets $\tilde{Q}^*(x)$ will be then used in the proofs of the closure theorems in Section 10.6 and of the lower closure theorems in Section 10.7 all based on weak convergence. The same equivalence will also be used in the discussion of the sufficient conditions (Λ) in Section 10.7.

10.4 A Few Remarks on Growth Conditions

The implication (c) ⇒ (b) in Theorem (10.3.i) states that the growth condition, say (ϕ), in (c) implies equiabsolute integrability. When the functions for which the growth condition holds are the derivatives of AC functions x on an interval, then the functions x are equiabsolutely continuous. Moreover, there are analogous, though independent, growth conditions which have the same implication (say, below, (g1) = (ϕ), (g2), and (g3)). These conditions are rather common in the literature (see Bibliographical Notes) and are presented briefly below.

10.4.i. *If* $\{x(t), a \le t \le b\}$ *is any class of* AC *functions* $x(t) = (x^1, \ldots, x^n)$, *with* $-\infty < a_0 \le a < b \le b_0 < +\infty$, *and* (g1) (*or* ($\phi$)) *there is a scalar function* $\phi(\zeta)$, $0 \le \zeta < +\infty$, *bounded below, such that* $\phi(\zeta)/\zeta \to +\infty$ *as* $\zeta \to +\infty$, *and* $\int_a^b \phi(|x'(t)|)\,dt \le M$ *for some constant* M *and any element* x *of the class, then the class* $\{x\}$ *is equiabsolutely continuous and the class* $\{x'\}$ *is equiabsolutely integrable.*

This is a corollary of (10.3.i), implication (c) ⇒ (b), and (10.2.i).

10.4.ii. *If* $\{\eta(t), x(t), a \le t \le b\}$ *is any class of pairs of functions,* η *scalar, L-integrable,* $x(t) = (x^1, \ldots, x^n)$ AC, *with* $-\infty < a_0 \le a < b \le b_0 < +\infty$, a_0, b_0 *fixed, and with* $\int_a^b \eta(t)\,dt \le M$ *for some constant* $M \ge 0$ *and any element* (η, x) *of the class; and if* (g2) *given* $\varepsilon > 0$ *there is a locally L-integrable function* $\psi_\varepsilon(t) \ge 0$, *which may depend on* ε, *such that* $|x'(t)| \le \psi_\varepsilon(t) + \varepsilon\eta(t)$, $a \le t \le b$, *for every element* (η, x) *of the class, then the class* $\{x\}$ *is equiabsolutely continuous and the class* $\{x'\}$ *is equiabsolutely integrable.*

Proof. First, for $\varepsilon = 1$ we have $|x'(t)| \le \psi_1(t) + \eta(t)$, and hence $\eta(t) \ge -\psi_1(t)$ for all t and all pairs (η, x) of the family. Since $\psi_1(t)$ is L-integrable in $[a_0, b_0]$, we may take $M_0 = \int_{a_0}^{b_0} \psi_1(t)\,dt$. Now, given $\varepsilon > 0$, let $\sigma = \min[1, \varepsilon 2^{-1}(M + M_0 + 1)^{-1}]$. The function $\psi_\sigma(t)$ is L-integrable in $[a_0, b_0]$; hence, there is some $\delta > 0$ such that $\int_E \psi_\sigma\,dt \le \varepsilon/2$ for every measurable set E with meas $E < \delta$. Now let $\eta(t)$, $x(t)$, $a \le t \le b$, be any pair of the family, and take any measurable subset E of $[a, b]$ with meas $E \le \delta$. Then

$$\begin{aligned}\int_E |x'(t)|\,dt &\le \int_E [\psi_\sigma(t) + \sigma\eta(t)]\,dt \le \int_E [\psi_\sigma + \sigma(\eta + \psi_1)]\,dt \\ &\le \sigma \int_a^b \eta\,dt + \sigma \int_a^b \psi_1\,dt + \int_E \psi_\sigma\,dt \\ &\le \sigma(M_0 + M) + \int_E \psi_\sigma\,dt \le \frac{\varepsilon}{2} + \frac{\varepsilon}{2} = \varepsilon. \qquad \square\end{aligned}$$

10.4.iii. *If* $\{\eta(t), x(t), a \le t \le b\}$ *is any class of pairs of functions,* $\eta(t)$ *scalar, L-integrable,* $x(t) = (x^1, \ldots, x^n)$ AC *with* $-\infty < a_0 \le a < b \le b_0 < +\infty$, a_0,

b_0 fixed, with $\int_a^b \eta(t)\,dt \le M$ for some constant $M \ge 0$ and any element of the class; and if (g3) for every vector $p = (p_1, \ldots, p_n) \in R^n$ there is a locally L-integrable function $\phi_p(t) \ge 0$ which may depend on p, such that $\eta(t) \ge (p, x'(t)) - \phi_p(t)$, $a \le t \le b$, and for all elements (η, x) of the class, then the class $\{x\}$ is equiabsolutely continuous and the class of the derivatives $\{x'\}$ is equiabsolutely integrable.

Proof. We have denoted by (a, b) the inner product in R^n. Let $\phi(t)$, $\psi(t)$, $a_0 \le t \le b_0$, be the nonnegative L-integrable functions of assumption (g3) corresponding to the two unit vectors $p = u_1 = (1, 0, \ldots, 0)$ and $p = v_1 = (-1, 0, \ldots, 0)$. Then

$$x'^1(t) \le \eta(t) + \phi(t), \quad -x'^1(t) \le \eta(t) + \psi(t), \qquad a \le t \le b;$$

hence, $|x'^1(t)| \le \eta(t) + \phi(t) + \psi(t)$, $a \le t \le b$. Thus,

$$\eta(t) + \phi(t) + \psi(t) \ge 0, \qquad a \le t \le b. \tag{10.4.1}$$

Let $M_1 = \int_a^b (\phi(t) + \psi(t))\,dt$. Now, given $\varepsilon > 0$, let N be an integer such that $N^{-1}nM \le \varepsilon/3$, $N^{-1}nM_1 \le \varepsilon/3$. If u_i, v_i denote the unit vectors $u_i = (\delta_{ij}, j = 1, \ldots, n)$, $v_i = (-\delta_{ij}, j = 1, \ldots, n)$, then again by assumption (g3), for $p = Nu_i$ and $p = Nv_i$, there are functions $\Phi_i(t) \ge 0$, $\Psi_i(t) \ge 0$, L-integrable in $[a_0, b_0]$ such that

$$Nx'^i(t) \le \eta(t) + \Phi_i(t), \quad -Nx'^i(t) \le \eta(t) + \Psi_i(t), \qquad a \le t \le b, \quad i = 1, \ldots, n,$$

and hence

$$N|x'^i(t)| \le \eta(t) + \Phi_i(t) + \Psi_i(t), \qquad a \le t \le b,$$

for any pair (η, x) of the class. Then $\Phi(t) = \sum_{i=1}^n \Phi_i(t)$, $\Psi(t) = \sum_{i=1}^n \Psi_i(t)$ are L-integrable in $[a_0, b_0]$, and we also have

$$N|x'(t)| \le n\eta(t) + \Phi(t) + \Psi(t), \qquad a \le t \le b. \tag{10.4.2}$$

If E denotes any measurable subset of $[a, b]$, we have, from (10.4.1) and (10.4.2),

$$\begin{aligned}
\int_E |x'(t)|\,dt &\le N^{-1}n \int_E \eta(t)\,dt + N^{-1} \int_E (\Phi(t) + \Psi(t))\,dt \\
&\le N^{-1}n \int_E [\eta(t) + \phi(t) + \psi(t)]\,dt + N^{-1} \int_E (\Phi(t) + \Psi(t))\,dt \\
&\le N^{-1}n \int_a^b \eta(t)\,dt + N^{-1}n \int_a^b (\phi(t) + \psi(t))\,dt \\
&\quad + N^{-1} \int_E (\Phi(t) + \Psi(t))\,dt.
\end{aligned}$$

Since $\Phi + \Psi$ is L-integrable, there is $\delta > 0$ such that meas $E \le \delta$ implies $\int_E (\Phi + \Psi)\,dt \le \varepsilon/3$, and then

$$\int_E |x'(t)|\,dt \le \frac{\varepsilon}{3} + \frac{\varepsilon}{3} + \frac{\varepsilon}{3} = \varepsilon.$$

We have proved the equiabsolute integrability of the class $\{x'\}$. The equiabsolute continuity of the class $\{x\}$ follows from (10.2.i). □

Remark 1. We shall prove below that the growth conditions (g2) and (g3) are equivalent.

Remark 2. Under the conditions of any of the statements (10.4.i–iii), if in addition we know that the class $\{x\}$ is also equibounded, then the same class $\{x\}$ is sequentially relatively compact in the topology of the weak convergence of the derivatives (Section 2.14), that is, from any sequence of elements $[x_k]$ of the class $\{x\}$ there is a subsequence which converges uniformly (or in the ρ-metric), and whose derivatives are weakly convergent in L_1. The compactness in C is the Ascoli theorem, the weak compactness of the derivatives is the Dunford-Pettis theorem.

The conditions (g1–3) above have been expressed directly in terms of functions $\eta(t), x(t), x'(t), a \le t \le b$. Actually, we may think of these functions as solutions of orientor fields, say $(\eta(t), x'(t)) \in \tilde{Q}(t, x(t)), t \in [a,b]$ (a.e.), for classes $\{(\eta(t), x'(t)\}$ with $(t, x(t)) \in A, \int_a^b \eta(t)\,dt \le M$, A compact, and then the same properties (g1–3) can be derived from analogous geometric properties of the orientor field, that is, of the given sets $\tilde{Q}(t,x)$. We express these properties as local properties, and, for the sake of simplicity, we call them local properties (g1–3) of the orientor field.

We say that the local growth property (g1) is satisfied at $(\bar{t}, \bar{x}) \in A$, provided there are a neighborhood $N_\delta(\bar{t}, \bar{x})$ of $(\bar{t}, \bar{x})$ and a scalar function $\phi(\zeta)$, $0 \le \zeta < +\infty$, bounded below, such that $\phi(\zeta)/\zeta \to +\infty$ as $\zeta \to +\infty$, and $(t,x) \in N_\delta(\bar{t}, \bar{x}) \cap A, (y, z) \in \tilde{Q}(t,x)$ implies $y \ge \phi(|z|)$.

We say that the local growth property (g2) is satisfied at $(\bar{t}, \bar{x}) \in A$, provided there is a neighborhood $N_\delta(\bar{t}, \bar{x})$, and for every $\varepsilon > 0$ there is an L-integrable function $\psi_\varepsilon(t) \ge 0$, $\bar{t} - \delta \le t \le \bar{t} + \delta$, such that $(t,x) \in N_\delta(\bar{t}, \bar{x}) \cap A, (y,z) \in \tilde{Q}(t,x)$ implies $|z| \le \psi_\varepsilon(t) + \varepsilon y$.

We say that the local growth property (g3) is satisfied at $(\bar{t}, \bar{x}) \in A$, provided there is a neighborhood $N_\delta(\bar{t}, \bar{x})$, and for every vector $p \in R^n$ there is an L-integrable function $\phi_p(t) \ge 0, \bar{t} - \delta \le t \le \bar{t} + \delta$, such that $(t,x) \in N_\delta(\bar{t}, \bar{x}) \cap A, (y,z) \in \tilde{Q}(t,x)$ implies $y \ge (p,z) - \phi_p(t)$.

We may say that any of these condition is satisfied in A if it is satisfied at every point $(\bar{t}, \bar{x}) \in A$. For A compact, then some finite system of corresponding neighborhoods covers A, and for the classes $\{\eta(t), x'(t), a \le t \le b\}$ mentioned above, corresponding functions ϕ, or ψ_ε, or ϕ_p can be found for the whole interval $[a,b]$. We leave this derivation as an exercise for the reader. In the following statements (10.4.iv–vii) we refer to the local growth properties. The same holds for the properties in the large.

10.4.iv. (g1) *implies* (g2).

Let $\phi(\zeta)$ be the function as in (g1) such that $y \ge \phi(|z|)$ for all $(y,z) \in \tilde{Q}(t,x), (t,x) \in N_\delta(\bar{t}, \bar{x}) \cap A$. Let L be a real constant such that $\phi(\zeta) \ge L$ for all ζ, and thus $y \ge L$. Now, given $\varepsilon > 0$, let $M_\varepsilon > 0$ real be such $\phi(\zeta)/\zeta > \varepsilon^{-1}$ for all $\zeta \ge M_\varepsilon$, and take $\psi_\varepsilon(t) = M_\varepsilon + \varepsilon|L|$ for $\bar{t} - \delta \le t \le \bar{t} + \delta$. Then, for $(t,x) \in N_\delta(\bar{t}, \bar{x}), (y,z) \in \tilde{Q}(t,x)$ we have either $|z| \le M_\varepsilon$ and then $|z| \le M_\varepsilon + \varepsilon(y - L) \le \psi_\varepsilon + \varepsilon y$, or $|z| \ge M_\varepsilon$ and then $\phi(|z|)/|z| \ge \varepsilon^{-1}$, so $|z| \le \varepsilon\phi(|z|) \le \psi_\varepsilon(t) + \varepsilon y$.

10.4.v. (g2) *with* ψ_ε *constant implies* (g1).

For $\varepsilon=1$ we have $|z|\leq\psi_1+y$; hence $y\geq-\psi_1$ for all $(y,z)\in\tilde{Q}(t,x)$, $(t,x)\in N_\delta(\bar{t},\bar{x})\cap A$. If for all such t,x,y,z we have $|z|\leq M$ for some constant M, then we can take $\phi(\zeta)=-\psi_1$ for all $0\leq\zeta\leq M$, and take ϕ arbitrary for $\zeta>M$, say $\phi(\zeta)=-\psi_1+(\zeta-M)^2$. If $|z|$ can be as large as we want, then for every $s=1,2,\ldots$, there is a constant ψ_s such that $|z|\leq\psi_s+s^{-1}y$ for all $(y,z)\in\tilde{Q}(t,x)$, $(t,x)\in N_\delta(\bar{t},\bar{x})\cap A$, and it is not restrictive to take $\psi_s\geq 0$, $\psi_s\leq\psi_{s+1}$, $\psi_s\to+\infty$ as $s\to+\infty$. For $|z|\geq 1$ we have then $y/|z|\geq s-(s\psi_s)/|z|$, and hence $y/|z|\geq s/2$ for $|z|\geq 2\psi_s$. Now we take $\phi(\zeta)=-\psi_1$ for $\zeta\leq 2\psi_1$, and $\phi(\zeta)=(s/2)\zeta$ for $2\psi_s\leq\zeta<2\psi_{s+1}$. Then ϕ is bounded below, $\phi(\zeta)/\zeta\to+\infty$ as $\zeta\to+\infty$, and $y/|z|\geq s/2$ for $2\psi_s\leq|z|\leq 2\psi_{s+1}$ implies $y\geq\phi(|z|)$.

10.4.vi. (g3) *implies* (g2).

Indeed, suppose u_i, v_i denote the unit vectors $u_i=(\delta_{ij}, j=1,\ldots,n)$, $v_i=(-\delta_{ij}, j=1,\ldots,n)$, and for $\varepsilon>0$, let $p=n\varepsilon^{-1}u_i$ and $p=n\varepsilon^{-1}v_i$. Then by (g3), there are functions $\phi_i(t)\geq 0$, $\psi_i(t)\geq 0$, L-integrable in $[\bar{t}-\delta_i,\bar{t}+\delta_i]$, such that for $(t,x)\in N_{\delta_i}(\bar{t},\bar{x})\cap A$, $(y,z)\in\tilde{Q}(t,x)$, $z=(z^1,\ldots,z^n)$, we have

$$y\geq n\varepsilon^{-1}z^i-\phi_i(t),\quad y\geq -n\varepsilon^{-1}z^i-\psi_i(t),\qquad \bar{t}-\delta_i\leq t\leq\bar{t}+\delta_i,\quad i=1,\ldots,n.$$

For $\delta=\min\delta_i$, we have then

$$n\varepsilon^{-1}|z^i|\leq y+\phi_i(t)+\psi_i(t),\qquad \bar{t}-\delta\leq t\leq\bar{t}+\delta,\quad i=1,\ldots,n,$$

and for $\Phi(t)=\sum_{i=1}^n\phi_i(t)$, $\Psi(t)=\sum_{i=1}^n\psi_i(t)$, we also have

$$n\varepsilon^{-1}|z|\leq n\varepsilon^{-1}\sum_1^n|z^i|\leq ny+\Phi(t)+\Psi(t),\qquad \bar{t}-\delta\leq t\leq\bar{t}+\delta,$$

or

$$|z|\leq\varepsilon y+\varepsilon n^{-1}(\Phi(t)+\Psi(t)),\qquad \bar{t}-\delta\leq t\leq\bar{t}+\delta,$$

and this is (g2) for $\psi_\varepsilon=\varepsilon n^{-1}(\Phi+\Psi)$.

10.4.vii. (g2) *implies* (g3).

Indeed, given $p=(p_1,\ldots,p_n)\in R^n$, take $N=|p_1|+\cdots+|p_n|$, $\varepsilon=N^{-1}$. Then by (g2) we have $|z|\leq\psi_\varepsilon(t)+\varepsilon y$, hence

$$y\geq\varepsilon^{-1}|z|-\varepsilon^{-1}\psi_\varepsilon(t)=N|z|-\varepsilon^{-1}\psi_\varepsilon(t)$$

and finally for $\phi_p=\varepsilon^{-1}\psi_\varepsilon(t)$, also $y\geq(p,z)-\phi_p(t)$, and this is (g3).

10.5 The Growth Property (ϕ) Implies Property (Q)

The following theorem will be used in Sections 10.6 and 10.7 with the state variable x there having the same role as the variable x here. In some applications x may have the role of t or of (t,x).

10.5.i (THEOREM; CESARI [6, 7]). *Let A be any set of points $x\in R^h$, and for every $x\in A$, let $\tilde{Q}(x)$ denote a set of points $(y,z)=(y,z^1,\ldots,z^n)\in R^{1+n}$*

such that (a) $(y, z) \in \tilde{Q}(x)$, $y \le y'$, *implies* $(y', z) \in \tilde{Q}(x)$. *Let* $\phi(\zeta)$, $0 \le \zeta < +\infty$, *be a real valued function, bounded below and such that* $\phi(\zeta)/\zeta \to +\infty$ *as* $\zeta \to +\infty$. *For some* $\bar{x} \in A$ *let* $N_{\delta_0}(\bar{x})$ *be a neighborhood of* $\bar{x}$ *in* A, *and assume that* (b) $(y, z) \in \tilde{Q}(x)$, $x \in N_{\delta_0}(\bar{x})$, *implies* $y \ge \phi(|z|)$. *If the sets* $\tilde{Q}(x)$ *have property* (K) *at* $\bar{x}$, *and the set* $\tilde{Q}(\bar{x})$ *is convex, then the sets* $\tilde{Q}(x)$ *have property* (Q) *at* $\bar{x}$.

Alternatively, for every $x \in A$ *let* $Q_0(x)$ *be a subset of the z-space* R^n, *let* M_0 *denote the set* $M_0 = [(x, z) | x \in A, z \in Q_0(x)]$, *let* $T_0(x, z)$ *be a real valued lower semicontinuous function on* M_0, *and let* $\tilde{Q}(x)$ *denote the set* $\tilde{Q}(x) = [(y, z) | y \ge T_0(x, z), z \in Q_0(x)]$. *For some* $\bar{x} \in A$ *and neighborhood* $N_{\delta_0}(\bar{x})$ *of* $\bar{x}$ *in* A, *assume that* $x \in N_{\delta_0}(\bar{x})$, $z \in Q_0(x)$ *implies* $T_0(x, z) \ge \phi(|z|)$. *Then, if the sets* $Q_0(x)$ *have property* (K) *at* $\bar{x}$ *and the set* $\tilde{Q}(\bar{x})$ *is convex, then the sets* $\tilde{Q}(x)$ *have property* (Q) *at* $\bar{x}$.

This statement is a particular case of the following one.

10.5.ii (THEOREM; CESARI [6, 7]). *Let* A *be any set of points* $x \in R^h$, *and for every* $x \in A$, *let* $\tilde{Q}(x)$ *denote a set of points* $(\tilde{y}, z) = (y^0, y^1, \ldots, y^\mu, z^1, \ldots, z^n) \in R^{1+\mu+n}$ *such that* (a) $(\tilde{y}, z) \in \tilde{Q}(x)$, $\tilde{y} = (y^0, \ldots, y^\mu)$, $\tilde{y}' = (y^{0\prime}, \ldots, y^{\mu\prime})$, $y^i \le y^{i\prime}$, $i = 0, 1, \ldots, \mu$, *implies* $(\tilde{y}', z) \in \tilde{Q}(x)$. Let $\phi(\zeta)$, $0 \le \zeta < +\infty$, *be a real valued function, bounded below and such that* $\phi(\zeta)/\zeta \to +\infty$ *as* $\zeta \to +\infty$. *For some* $\bar{x} \in A$ *let* $N_{\delta_0}(\bar{x})$ *be a neighborhood of* $\bar{x}$ *in* A *and* L *a real constant, and assume that* (b) $(\tilde{y}, z) \in \tilde{Q}(x)$, $x \in N_{\delta_0}(\bar{x})$ *implies* $y^0 \ge \phi(|z|)$, $y^i \ge L$, $i = 1, \ldots, \mu$. *Then, if the sets* $\tilde{Q}(x)$ *have property* (K) *at* $\bar{x}$ *and the set* $\tilde{Q}(\bar{x})$ *is convex, then the sets* $\tilde{Q}(x)$ *have property* (Q) *at* $\bar{x}$.

Alternatively, for every $x \in A$ *let* $Q(x)$ *be a subset of the yz-space* $R^{\mu+n}$, *let* M *denote the set* $M = [(x, y, z) | x \in A, (y, z) \in Q(x)]$, *let* $T_0(x, y, z)$ *be a real valued lower semicontinuous function on* M, *and let* $\tilde{Q}(x)$ *denote the set* $\tilde{Q}(x) = [(y^0, y, z) | y^0 \ge T_0(x, y, z), (y, z) \in Q(x)] \subset R^{1+\mu+n}$, $y = (y^1, \ldots, y^\mu)$, $z = (z^1, \ldots, z^n)$. *For some* $\bar{x} \in A$ *and neighborhood* $N_{\delta_0}(\bar{x})$ *of* $\bar{x}$ *in* A *assume that* $x \in N_{\delta_0}(\bar{x})$, $(y, z) \in Q(x)$ *implies* $T_0(x, y, z) \ge \phi(|z|)$, $y^i \ge L$, $i = 1, \ldots, \mu$, *where* ϕ *is a function as above and* L *a constant. Then, if the sets* $Q(x)$ *have property* (K) *at* $\bar{x}$ *and the set* $\tilde{Q}(\bar{x})$ *is convex, then the sets* $\tilde{Q}(x)$ *have property* (Q) *at* $\bar{x}$.

Analogously, for every $x \in A$ *let* $Q_0(x)$ *be a subset of the z-space* R^n, *let* M_0 *denote the set* $M_0 = [(x, z) | x \in A, z \in Q_0(x)]$, *let* $T_i(x, z)$, $i = 0, 1, \ldots, \mu$, *be real valued lower semicontinuous functions on* M_0, *and let* $\tilde{Q}(x)$ *denote the set* $\hat{Q}(x) = [(y^0, y, z) | y^i \ge T_i(x, z), \ i = 0, 1, \ldots, \mu, \ z \in Q_0(x)]$, *where* $y = (y^1, \ldots, y^\mu)$. *For some* $\bar{x} \in A$ *and neighborhood* $N_{\delta_0}(\bar{x})$ *let us assume that* $x \in N_{\delta_0}(\bar{x})$, $z \in Q_0(x)$ *implies* $T_0(x, z) \ge \phi(|z|)$, $T_i(x, z) \ge L$, $i = 1, \ldots, \mu$, *for some function* ϕ *as above and some constant* L. *Then, if the sets* $Q_0(x)$ *have property* (K) *at* $\bar{x}$ *and the set* $\tilde{Q}(\bar{x})$ *is convex, then the sets* $\tilde{Q}(x)$ *have property* (Q) *at* $\bar{x}$.

Remark 1. In the statements (10.5.i, ii) we have implicitly assumed that the sets Q, $\tilde{Q}$ are not empty, and the functions T are finite everywhere on the sets M or M_0 as mentioned. Actually, the statements hold even if some of such sets are the empty set and the functions T are extended functions whose values are finite or $+\infty$, but now we assume that the same functions T are defined and lower semicontinuous in the whole space. Thus, whenever, say $T(\bar{x}) = +\infty$, it is required that $T(x) \to +\infty$ as $x \to \bar{x}$. Properties (K) and (Q) for maps $x \to Q(x)$ when we do not exclude empty sets are defined as usual, as mentioned in Section 8.5, Remark 1.

Proof of (10.5.ii). Let L_0 be a bound below for $\phi(\zeta)$. As usual, we denote by $\tilde{Q}(\bar{x}; \delta)$ the set $\bigcup \tilde{Q}(x)$ where $\bigcup$ is taken for all x in the δ-neighborhood $N_\delta(\bar{x})$ of $\bar{x}$ in A. We have to prove that if $(\bar{\tilde{y}}, \bar{z}) \in \bigcap_{\delta > 0}$ cl co $\tilde{Q}(\bar{x}; \delta)$, then we also have $(\bar{\tilde{y}}, \bar{z}) \in \tilde{Q}(\bar{x})$. First we note that for any $(\tilde{y}, z) \in \tilde{Q}(x)$, $\tilde{y} = (y^0, y^1, \ldots, y^\mu)$, we have $y^0 \geq \phi(|z|)$, $y^i \geq L$, $i = 1, \ldots, \mu$. By adding $|L| + 1$ to all y^i coordinates, $i = 1, \ldots, \mu$, and adding $|L_0| + 1$ to all y^0 coordinates and $\phi(\zeta)$, we make them all positive. We see that it is not restrictive to assume $\phi(\zeta) > 0$ and all $y^i > 0$, $i = 0, 1, \ldots, \mu$, for $(\tilde{y}, z) \in \tilde{Q}(x)$, $x \in N_{\delta_0}(\bar{x})$.

Now we have $(\bar{\tilde{y}}, \bar{z}) \in \bigcap_\delta$ cl co $\tilde{Q}(\bar{x}; \delta)$ as stated, and hence $(\bar{\tilde{y}}, \bar{z}) \in$ cl co $\tilde{Q}(\bar{x}; \delta)$ for all δ, $0 < \delta \leq \delta_0$.

Thus, there is a sequence of points $(\tilde{y}_k, z_k) \in$ co $\tilde{Q}(\bar{x}; \delta_k)$ with $0 < \delta_k \leq \delta_0$, $\delta_k \to 0$, $\tilde{y}_k \to \bar{\tilde{y}}$, $z_k \to \bar{z}$. By Carathéodory's theorem (8.4.iii) there are v sequences of points $(\tilde{y}_k^\gamma, z_k^\gamma) \in \tilde{Q}(x_k^\gamma)$, $k = 1, 2, \ldots, \gamma = 1, 2, \ldots, v$, and numbers λ_k^γ with $x_k^\gamma \in A$, $x_k^\gamma \to \bar{x}$ as $k \to \infty$, $\gamma = 1, \ldots, v$, and

$$(10.5.1) \qquad \begin{aligned} &0 \leq \lambda_k^\gamma \leq 1, \qquad 1 = \sum_\gamma \lambda_k^\gamma, \qquad y_k^i = \sum_\gamma \lambda_k^\gamma y_k^{i\gamma}, \quad i = 0, 1, \ldots, \mu, \\ &z_k = \sum_\gamma \lambda_k^\gamma z_k^\gamma, \qquad k = 1, 2, \ldots, \end{aligned}$$

where $\sum_\gamma$ ranges over all $\gamma = 1, \ldots, v$, and we can take $v = n + \mu + 2$. Here $\bar{\tilde{y}} = (\bar{y}^0, \bar{y}^1, \ldots, \bar{y}^\mu)$, $\tilde{y}_k = (y_k^0, y_k^1, \ldots, y_k^\mu)$, $\tilde{y}_k^\gamma = (y_k^{0\gamma}, y_k^{1\gamma}, \ldots, y_k^{\mu\gamma})$, $\gamma = 1, \ldots, v$, $k = 1, 2, \ldots$, and moreover $x_k^\gamma = (x_k^{1\gamma}, \ldots, x_k^{h\gamma})$, $z_k^\gamma = (z_k^{1\gamma}, \ldots, z_k^{n\gamma})$.

Since $0 \leq \lambda_k^\gamma \leq 1$ for all γ and k, there is a subsequence, say still $[k]$, such that $\lambda_k^\gamma \to \lambda^\gamma$ as $k \to \infty$, $\gamma = 1, \ldots, v$, and then $0 \leq \lambda^\gamma \leq 1$. From the second relation (10.5.1) we derive that $\sum_\gamma \lambda^\gamma = 1$, and thus at least one λ^γ is positive. By a suitable reindexing, we may well assume that for some α, $1 \leq \alpha \leq v$, we have $0 < \lambda^\gamma \leq 1$ for $\gamma = 1, \ldots, \alpha$, while $\lambda^\gamma = 0$ for $\gamma = \alpha + 1, \ldots, v$. We can even assume that $0 < 2^{-1}\lambda^\gamma \leq \lambda_k^\gamma \leq 1$ for $\gamma = 1, \ldots, \alpha$ and all k. Now in the third relation (10.5.1) all y_k^i, $y_k^{i\gamma}$, λ_k^γ are nonnegative numbers, and $y_k^i \to \bar{y}^i$ as $k \to \infty$, $i = 0, 1, \ldots, \mu$. Thus, there is a constant M such that $0 \leq \bar{y}^i$, $y_k^i \leq M$ for all i and k, and from the third relation (10.5.1) we derive, for $\gamma = 1, \ldots, \alpha$, that $2^{-1}\lambda^\gamma y_k^{i\gamma} \leq \lambda_k^\gamma y_k^{i\gamma} \leq y_k^i \leq M$, or $0 \leq y_k^{i\gamma} \leq 2(\lambda^\gamma)^{-1}M$, a finite number, and this holds for $\gamma = 1, \ldots, \alpha$, $i = 0, 1, \ldots, \mu$. Thus, there is a subsequence, say still $[k]$, such that $y_k^{i\gamma} \to y^{i\gamma}$ as $k \to \infty$, $\gamma = 1, \ldots, \alpha$, $i = 0, 1, \ldots, \mu$. For $\gamma = \alpha + 1, \ldots, v$, again from the third relation (10.5.1) we derive $0 \leq \lambda_k^\gamma y_k^{i\gamma} \leq M$, and by a further extraction we may assume that $\lambda_k^\gamma y_k^{i\gamma} \to A^{i\gamma} \geq 0$ as $k \to \infty$, and this holds for $\gamma = \alpha + 1, \ldots, v$ and $i = 0, 1, \ldots, \mu$.

If $\sum', \sum''$ denote sums ranging over all $\gamma = 1, \ldots, \alpha$ and $\gamma = \alpha + 1, \ldots, v$ respectively, then the third relation (10.5.1) yields as $k \to \infty$

$$\bar{y}^i = \sum{}' \lambda^\gamma y^{i\gamma} + \sum{}'' A^{i\gamma} \geq \sum{}' \lambda^\gamma y^{i\gamma}, \qquad i = 0, 1, \ldots, \mu.$$

Now, for $\gamma = 1, \ldots, \alpha$, $i = 0$, and all k, we also have

$$(10.5.2) \qquad 2(\lambda^\gamma)^{-1}M \geq y_k^{0\gamma} \geq \phi(|z_k^\gamma|), \qquad \gamma = 1, \ldots, \alpha, \quad k = 1, 2, \ldots.$$

By the growth property of ϕ we conclude that the α sequences $[z_k^\gamma, k = 1, 2, \ldots]$, $\gamma = 1, \ldots, \alpha$, are bounded in R^n. By a further extraction we can well assume that $z_k^\gamma \to z^\gamma$ as $k \to \infty$ where z^γ, $\gamma = 1, \ldots, \alpha$, are α points of R^n. Thus, for $\tilde{y}_k^\gamma = (y_k^{0\gamma}, \ldots, y_k^{\mu\gamma})$, $\tilde{y}^\gamma = (y^{0\gamma}, \ldots, y^{\mu\gamma})$, $\gamma = 1, \ldots, \alpha$, $k = 1, 2, \ldots$, we have $\tilde{y}_k^\gamma \to \tilde{y}^\gamma$ as $k \to \infty$, together with $z_k^\gamma \to z^\gamma$, $x_k^\gamma \to \bar{x}$. Now $(\tilde{y}_k^\gamma, z_k^\gamma) \in \tilde{Q}(x_k^\gamma)$ for all $\gamma = 1, \ldots, \alpha$, $k = 1, 2, \ldots$. Thus, given $\delta > 0$, we also have $(\tilde{y}_k^\gamma, z_k^\gamma) \in \tilde{Q}(\bar{x}; \delta)$ for all k sufficiently large. Hence, as $k \to \infty$, we have $(\tilde{y}^\gamma, z^\gamma) \in \text{cl } \tilde{Q}(\bar{x}; \delta)$, $\gamma = 1, \ldots, \alpha$, and this relation holds for all $\delta > 0$. Thus, $(\tilde{y}^\gamma, z^\gamma) \in \bigcap_\delta \text{cl } \tilde{Q}(\bar{x}; \delta)$, and by the property (K) of the sets $\tilde{Q}(x)$ at $\bar{x}$, we also have $(\tilde{y}^\gamma, z^\gamma) \in \tilde{Q}(\bar{x})$, $\gamma = 1, \ldots, \alpha$.

For $\gamma = \alpha + 1, \ldots, \nu$, the sequence $[y_k^{0\gamma}, k = 1, 2, \ldots]$ is either bounded or unbounded. Thus, by a suitable reindexing and extraction, we may well assume that, for some β, $1 \le \alpha \le \beta \le \nu$, the $\beta - \alpha$ sequences $[y_k^{0\gamma}, k = 1, 2, \ldots]$, $\gamma = \alpha + 1, \ldots, \beta$, converge to some $y^{0\gamma} \in R$ as $k \to \infty$, and that the $\nu - \beta$ sequences $[y_k^{0\gamma}, k = 1, 2, \ldots]$, $\gamma = \beta + 1, \ldots, \nu$, diverge, or $y_k^{0\gamma} \to +\infty$ as $k \to \infty$. Then, for $\gamma = \alpha + 1, \ldots, \beta$, the relations $y_k^{0\gamma} \ge \phi(|z_k^\gamma|)$ shows that the sequences $[z_k^\gamma, k = 1, 2, \ldots]$ also are bounded in R^n, and by a further extraction, they can be assumed to converge to some $z^\gamma \in R^n$, $\gamma = \alpha + 1, \ldots, \beta$. For $\gamma = \beta + 1, \ldots, \nu$, the same relations $y_k^{0\gamma} \ge \phi(|z_k^\gamma|)$, where now $y_k^{0\gamma} \to +\infty$, show that there are certain $\varepsilon_k^\gamma \in R^n$ with $z_k^\gamma = \varepsilon_k^\gamma y_k^{0\gamma}$ and $\varepsilon_k^\gamma \to 0$ as $k \to \infty$. If we now denote by $\sum''^*$ and $\sum'''$ summations ranging over all $\gamma = \alpha + 1, \ldots, \beta$ and $\gamma = \beta + 1, \ldots, \nu$ respectively, we can write the fourth relation (10.5.1) in the form

$$z_k = \sum' \lambda_k^\gamma z_k^\gamma + \sum''^* \lambda_k^\gamma z_k^\gamma + \sum''' \varepsilon_k^\gamma(\lambda_k^\gamma y_k^{0\gamma}),$$

where $\lambda_k^\gamma \to \lambda^\gamma > 0$, $z_k^\gamma \to z^\gamma$ in the first sum, where $\lambda_k^\gamma \to 0$, $z_k^\gamma \to z^\gamma$ in the second sum, and where $\varepsilon_k^\gamma \to 0$, $\lambda_k^\gamma y_k^{0\gamma} \to A^{0\gamma}$ in the third sum. Thus, as $k \to \infty$, we derive $\bar{z} = \sum' \lambda^\gamma z^\gamma$. We have now

$$0 < \lambda^\gamma \le 1, \ (\tilde{y}^\gamma, z^\gamma) \subset \tilde{Q}(\bar{x}), \quad \gamma = 1, \ldots, \alpha, \qquad 1 = \sum' \lambda^\gamma,$$
$$\bar{y}^i \ge \sum' \lambda^\gamma y^{i\gamma}, \quad \bar{z} = \sum' \lambda^\gamma z^\gamma, \qquad i = 0, 1, \ldots, \mu.$$

Since $\tilde{Q}(\bar{x})$ is convex, we also have

$$(\textstyle\sum' \lambda^\gamma y^\gamma, \sum' \lambda^\gamma z^\gamma) \in \tilde{Q}(\bar{x}),$$

and by the property (a) of the sets $\tilde{Q}(\bar{x})$, also $(\bar{y}, \bar{z}) \in \tilde{Q}(\bar{x})$, with $\bar{y} = (\bar{y}^0, \bar{y}^1, \ldots, \bar{y}^\mu)$. We have proved that $(\bar{y}, \bar{z}) \in \bigcap_\delta \text{cl co } \tilde{Q}(\bar{x}, \delta)$ implies $(\bar{y}, \bar{z}) \in \tilde{Q}(\bar{x})$. We have proved (10.5.ii) under the hypotheses that the sets $\tilde{Q}(x)$ have property (K) at $\bar{x}$. Here we have assumed that the set $\bigcap_{\delta > 0} \text{cl co } \tilde{Q}(\bar{x}; \delta)$ is not empty. If this set is empty, then certainly $\tilde{Q}(\bar{x})$ is empty, and property (Q) holds at $\bar{x}$.

Let us now consider the alternate cases of (10.5.ii). Assume that the sets $Q(x)$ have property (K) at $\bar{x}$. The argument is the same up to the statement $(\tilde{y}_k^\gamma, z_k^\gamma) \in \tilde{Q}(x_k)$ for all $\gamma = 1, \ldots, \alpha$, $k = 1, 2, \ldots$, with $\tilde{y}_k^\gamma = (y_k^{0\gamma}, y_k^{1\gamma}, \ldots, y_k^{\mu\gamma}) = (y_k^{0\gamma}, y_k^\gamma)$. From here we derive that $(y_k^\gamma, z_k^\gamma) \in Q(x_k) \subset R^{\mu+n}$, $\gamma = 1, \ldots, \alpha$, $k = 1, 2, \ldots$, with $x_k \to \bar{x}$, $y_k^\gamma \to y^\gamma$, $z_k^\gamma \to z^\gamma$ as $k \to \infty$. Hence, given $\delta > 0$, we also have $(y_k^\gamma, z_k^\gamma) \in Q(\bar{x}; \delta)$ for all k sufficiently large; and $(y^\gamma, z^\gamma) \in \text{cl } Q(\bar{x}; \delta)$. By property (K) we derive $(y^\gamma, z^\gamma) \in Q(\bar{x})$. On the other hand $y_k^{0\gamma} \ge T_0(x_k^\gamma, y_k^\gamma, z_k^\gamma)$, where T_0 is lower semicontinuous. Hence $y^{0\gamma} \ge T_0(\bar{x}, y^\gamma, z^\gamma)$, and we conclude that $(\tilde{y}^\gamma, z^\gamma) = (y^{0\gamma}, y^\gamma, z^\gamma) \in \tilde{Q}(\bar{x})$, $\gamma = 1, \ldots, \alpha$. The argument now proceeds as before.

Finally, we consider the case where the sets $Q_0(x) \subset R^n$ have property (K) at $\bar{x}$. Again, the argument proceeds as before, up to the statement $(\tilde{y}_k^\gamma, z_k^\gamma) \in \tilde{Q}(x_k^\gamma)$, $\gamma = 1, \ldots, \alpha$, $k = 1, 2, \ldots$. From here we derive that $z_k^\gamma \in Q_0(x_k^\gamma) \subset R^n$ for the same γ and k. Hence, by property (K) for these sets, we derive as before that $z^\gamma \in Q_0(\bar{x})$, $\gamma = 1, \ldots, \alpha$. On the other hand, $y_k^{i\gamma} \geq T_i(x_k^\gamma, z_k^\gamma)$, $i = 0, 1, \ldots, \mu$, and by the lower semicontinuity of these T_i, we derive $y^{i\gamma} \geq T_i(\bar{x}, z^\gamma)$, $i = 0, 1, \ldots, \mu$. Thus, $(\tilde{y}^\gamma, z^\gamma) \in \tilde{Q}(\bar{x})$, $\gamma = 1, \ldots, \alpha$, and the proof proceeds as before. Theorem (10.5.ii) is thereby proved. □

For $\mu = 0$ the main statement of (10.5.ii) reduces to the main statement of (10.5.i); for $\mu = 0$ the second and third alternatives of (10.5.ii) coalesce into the alternative case of (10.5.i).

It remains to prove the Remark 1. Now we allow the sets Q, $\tilde{Q}$ to be the empty set, but we assume the functions T to be extended functions defined and lower semicontinuous in the whole space.

In the proof above of (10.5.ii), first part, if for some $\bar{x}$, $\tilde{Q}(\bar{x}) = \emptyset = \bigcap_\delta \operatorname{cl} Q(x;\delta)$, then the argument is the same up to the point where we state that $(y^\gamma, z^\gamma) \in \operatorname{cl} Q(\bar{x};\delta)$, $\gamma = 1, \ldots, \alpha$. This is a contradiction since the second member is the empty set. Thus, no point $(\bar{\tilde{y}}, \bar{z})$ can belong to $\bigcap_\delta \operatorname{cl} \operatorname{co} \tilde{Q}(\bar{x};\delta)$. In the second part we assume now that $T_0(x, y, z)$ is an extended function defined and lower semicontinuous in $R^{h+\mu+n}$. If for some $\bar{x}$, $\tilde{Q}(\bar{x}) = \emptyset$, then either $Q(\bar{x}) = \emptyset$, or $Q(\bar{x}) \neq \emptyset$. If $Q(\bar{x}) = \emptyset = \bigcap_\delta \operatorname{cl} Q(\bar{x};\delta)$ the proof is the same up to the point where we state that $(y^\gamma, z^\gamma) \in Q(\bar{x})$, $\gamma = 1, \ldots, \alpha$, a contradiction, since $Q(\bar{x})$ is the empty set. If $Q(\bar{x}) \neq \emptyset$, $Q(\bar{x}) = \bigcap_\delta \operatorname{cl} Q(\bar{x};\delta)$, then $T_0(\bar{x}, y, z) = +\infty$ for all $(y, z) \in Q(\bar{x})$, and again the proof is the same up to the statement $y_k^{0\gamma} \geq T_0(x_k^\gamma, y_k^\gamma, z_k^\gamma)$, and now the second member approaches $+\infty$ as $k \to \infty$, while $y_k^{0\gamma} \to y^{0\gamma}$, a finite number, a contradiction.

In the proof of the third part, we assume now that all $T_i(x, z)$, $i = 0, 1, \ldots, \mu$, are extended functions defined and lower semicontinuous in R^{h+n}. If for some $\bar{x}$, $Q_0(\bar{x}) = \emptyset$, then either $Q_0(\bar{x}) = \emptyset$, or $Q_0(\bar{x}) \neq \emptyset$. If $Q_0(\bar{x}) = \emptyset = \bigcap_\delta \operatorname{cl} Q_0(\bar{x};\delta)$, then the proof is the same up to the statement $z^\gamma \in Q_0(\bar{x})$, $\gamma = 1, \ldots, \alpha$, a contradiction. If $Q_0(\bar{x}) \neq \emptyset$, $Q_0(\bar{x}) = \bigcap_\delta \operatorname{cl} Q_0(\bar{x};\delta)$, then for every $z \in Q_0(\bar{x})$ we must have $T_i(x, z) = +\infty$ for at least one i, and then in the statement $y_k^{i\gamma} \geq T_i(x_k, z_k)$, $i = 0, 1, \ldots, \mu$, $\gamma = 1, \ldots, \alpha$, the second member approaches $+\infty$ for at least one i, while $y_k^{i\gamma} \to y^{i\gamma}$ all finite numbers, again a contradiction. We have proved Remark 1.

Remark 2. Concerning the hypotheses of statements (10.5.i) and (10.5.ii), we note that the sets $\tilde{Q}(x)$ may have properties (Q) and (K) without the sets $Q_0(x)$ having either property. Indeed, take $A = [x \mid 0 \leq x \leq 1]$, $Q_0(0) = [z \mid z = 0]$, $Q_0(x) = [z \mid 0 < z \leq x]$, $0 < x \leq 1$, $\tilde{Q}(0) = [(z^0, z) \mid z = 0,\ 0 \leq z^0 < +\infty]$, $\tilde{Q}(x) = [(z^0, z) \mid z^0 \geq (xz)^{-1},\ 0 < z \leq x]$, $0 < x \leq 1$. These sets are all convex; the sets $\tilde{Q}(x)$ have property (Q) at every $\bar{x}$, $0 \leq \bar{x} \leq 1$; and the sets $Q_0(x)$, $0 < x \leq 1$, have neither property (K) nor (Q), since they are not closed.

Remark 3. The growth condition in (10.5.i) and (10.5.ii) can be simply expressed by saying: There is a neighborhood V_0 of x such that, given $\varepsilon > 0$, there is also a constant N such that $x \in V_0$, $|z| \geq N$, $(y^0, z) \in \tilde{Q}(x)$ [or $(y^0, y, z) \in \tilde{Q}(x)$] implies $|z| \leq \varepsilon y^0$.

Theorems (10.5.i) and (10.5.ii) are properties of orientor fields, that is, concern problems of optimal control when the control parameters are eliminated. Let us see here some of their implications in terms of the control parameters.

Let $x \to U(x)$, $x \in A \subset R^\nu$, $U(x) \subset R^m$ be a given set valued map, and let M denote its graph, or $M = [(x, u) \mid x \in A,\ u \in U(x)] \subset R^{\nu+m}$. Let $f_0(x, u)$, $g(x, u) = (g_1, \ldots, g_\mu)$, $f(x, u) = (f_1, \ldots, f_n)$ be functions defined on M. Let $\bar{x}$ be a point of A, and assume that there is a fixed δ_0-neighborhood $N_{\delta_0}(\bar{x})$ of $\bar{x}$ in A such that f and 1 are of slower growth than f_0 as $|u| \to \infty$ uniformly in $N_{\delta_0}(\bar{x})$. By this we mean that:

(G) Given $\varepsilon > 0$, there is $N = N(\varepsilon) \geq 0$ such that $x \in N_{\delta_0}(\bar{x})$, $u \in U(x)$, $|u| \geq N$ implies $|f(x, u)| \leq \varepsilon f_0(x, u)$, $1 \leq \varepsilon f_0(x, u)$.

Let y^0, y, z denote the variables $y^0 \in R$, $y = (y^1, \ldots, y^\mu) \in R^\mu$, $z = (z^1, \ldots, z^n) \in R^n$, and for $x \in A$ let $\tilde{Q}(x)$ denote the set of all (y^0, y, z) with $y^0 \geq f_0(x, u)$, $y^i \geq g_i(x, u)$, $i = 1, \ldots, \mu$, $z^i = f_i(x, u)$, $i = 1, \ldots, n$, $u \in U(x)$.

10.5.iii. *If M is closed, if the functions f_0, g_i, $i = 1, \ldots, \mu$, are nonnegative and lower semicontinuous, if the functions f_i, $i = 1, \ldots, n$, are continuous on M, and if f and 1 are of slower growth than f_0 as $|u| \to \infty$ uniformly on $N_{\delta_0}(\bar{x})$, then the sets $\tilde{Q}(x)$ have property* (K) *at $\bar{x}$, and if the set $\tilde{Q}(\bar{x})$ is convex, then the same sets $\tilde{Q}(x)$ also have property* (Q) *at $\bar{x}$.*

Proof. First let us prove that the sets $\tilde{Q}(x)$ have property (K) at $\bar{x}$. Let $(\bar{y}^0, \bar{y}, \bar{z}) \in \bigcap_{\delta > 0} \operatorname{cl} \bigcup [\tilde{Q}(x), x \in N_\delta(\bar{x})]$. Then there are points $x_k \in N_\delta(\bar{x})$ and points $(y_k^0, y_k, z_k) \in \tilde{Q}(x_k)$, $k = 1, 2, \ldots$, with $x_k \to \bar{x}$, $y_k^0 \to \bar{y}^0$, $y_k \to \bar{y}$, $z_k \to \bar{z}$ as $k \to \infty$. Hence, there are also points $u_k \in U(x_k)$ with $y_k^0 \geq f_0(x_k, u_k)$, $y_k^i \geq g_i(x_k, u_k)$, $i = 1, \ldots, \mu$, $z_k^i = f_i(x_k, u_k)$, $i = 1, \ldots, n$. Note that the sequence $[u_k]$ must be bounded, since in the opposite case there would be a subsequence, say still $[k]$, with $|u_k| \to \infty$, and hence $f_0(x_k, u_k) \to +\infty$, so $y_k^0 \to +\infty$, a contradiction. Since $[u_k]$ is bounded, there is a subsequence, say still $[k]$, with $u_k \to \bar{u} \in R^m$. Thus, $\bar{u} \in \bigcap_{\delta > 0} \operatorname{cl} \bigcup [U(x), x \in N_\delta(x)]$. The set M is closed by hypothesis; hence the sets $U(x)$ have property (K) at $\bar{x}$ by (8.5.iii) and subsequent Remark 2. The $(\bar{x}, \bar{u}) \in M$. By the lower semicontinuity of f_0 and g_i, and by the continuity of f, we have, as $k \to \infty$, $\bar{y}^0 \geq f_0(\bar{x}, \bar{u})$, $\bar{y}^i \geq g_i(\bar{x}, \bar{u})$, $i = 1, \ldots, \mu$, $\bar{z}^i = f_i(\bar{x}, \bar{u})$, $i = 1, \ldots, n$. Hence $(\bar{y}^0, \bar{y}, \bar{z}) \in \tilde{Q}(\bar{x})$, and the sets $\tilde{Q}(x)$ have property (K) at $\bar{x}$. To prove property (Q) we have first to prove the growth condition of Remark 3 above. Indeed, for the neighborhood $N_{\delta_0}(\bar{x})$ of $\bar{x}$, which we may suppose to be bounded, and $\varepsilon > 0$, there is $N = N(\varepsilon) > 0$ such that $|u| \geq N$, $u \in U(x)$, $x \in N_{\delta_0}(\bar{x})$ implies $1 \leq \varepsilon f_0$, $|f| \leq \varepsilon f_0$. For $|u| \leq N$, $x \in N_{\delta_0}(\bar{x})$, $u \in U(x)$, f is bounded, say $|f| \leq N_0$. Thus, $x \in N_{\delta_0}(x)$, $|z| = |f| > N_0$, $u \in U(x)$ implies $|u| \geq N$ and $|z| = |f| \leq \varepsilon f_0 \leq \varepsilon y^0$. By Remark 3 above, and thus by (10.5.i), the sets $\tilde{Q}(x)$ have property (Q) at $\bar{x}$. This proves (10.5.iii). □

Condition (G) is often used. Theorem (10.5.iii) holds also under analogous conditions, of equal practical interest, namely, either

(G′) There are constants $c, d \geq 0$ and a function $\phi(\zeta)$, $0 \leq \zeta < \infty$, bounded below, with $\phi(\zeta)/\zeta \to +\infty$ as $\zeta \to +\infty$, such that $|f(x, u)| \leq c|u| + d$ and $f_0(x, u) \geq \phi(|u|)$ for all $x \in N_\delta(\bar{x})$, $u \in U(x)$; or

(G″) (a) There is a function $\phi(\zeta)$, $0 \leq \zeta < \infty$, bounded below, with $\phi(\zeta)/\zeta \to +\infty$ as $\zeta \to +\infty$, such that $f_0(x, u) \geq \phi(|f(x, u)|)$ for all $x \in N_\delta(\bar{x})$, $u \in U(x)$; and (b) either $f_0(x, u) \to +\infty$ as $|u| \to +\infty$ uniformly in $N_\delta(\bar{x})$, or $|f(x, u)| \to +\infty$ as $|u| \to +\infty$ uniformly in $N_\delta(\bar{x})$.

Remark 4. In Chapters 11–16, whenever we shall be concerned with the generalized solutions of Section 1.14, we shall need properties of the new functions

$$f_i^* = f_i(x, p, v) = \sum_{j=1}^{h} p_j f_i(x, u^{(j)}), \quad i = 0, 1, \ldots, n, \qquad f^* = (f_1^*, \ldots, f_n^*),$$

$$g_i^* = g_i(x, p, v) = \sum_{j=1}^{h} p_j g_i(x, u^{(j)}), \quad i = 1, \ldots, r, \qquad g^* = (g_1^*, \ldots, g_r^*),$$

where h is any fixed integer ($h \geq n+2$), $p = (p_1, \ldots, p_h) \in \Gamma$, where Γ is the simplex $[p_j \geq 0,\ j = 1, \ldots, h,\ p_1 + \cdots + p_h = 1]$, and where $v = (u^{(1)}, \ldots, u^{(h)})$, $u^{(j)} \in U(x)$, $u^{(j)} \in R^m, j = 1, \ldots, h$, that is, $v \in (U(x))^h$. Thus, (p, v) is the new control variable and $V = \Gamma \times U^h$ is the control space. The corresponding sets $\tilde{Q}$ are here the sets

$$\tilde{R}(x) = [z^0 \geq f_0^*,\quad \mathfrak{z}^i \geq g_i^*, i = 1, \ldots, r,\quad z = f^*,\quad (p, v) \in \Gamma \times U^h].$$

The growth condition on f_0, f of (10.5.iii) is not inherited by the functions f_0^*, f^*, as the following example shows: take $f_0 = u^2$, $f = u$, $n = 1$, $f_0^* = p_1(u^{(1)})^2 + p_2(u^{(2)})^2$, $f^* = p_1 u^{(1)} + p_2 u^{(2)}$, $h = 2$, $p = (p_1, p_2)$. Then for $p_1 = 0$, $p_2 = 1$, $u^{(1)} = k$, $u^{(2)} = 0$, we have $f_0^* = 0$, $f^* = 0$, while $(0, 1, k, 0) \to \infty$ as $k \to \infty$. However, we shall prove that the growth condition of (10.5.iii) on the original functions f_0, f still guarantees that the sets $\tilde{R}(x)$ have properties (K) and (Q) at x_0.

First, let us prove that the sets $\tilde{R}(x)$ have property (K) at x_0. To this purpose let $(\bar{z}^0, \bar{\mathfrak{z}}, \bar{z})$ be a point of $\bigcap_\delta \operatorname{cl} \tilde{R}(x_0, \varepsilon)$. Then there is a sequence $(z_k^0, \mathfrak{z}_k, z_k)$, $k = 1, 2, \ldots$, with $z_k^0 \to \bar{z}^0$, $\mathfrak{z}_k \to \bar{\mathfrak{z}}$, $z_k \to \bar{z}$ as $k \to \infty$, $z_k^0 \geq \sum_j p_{jk} f_0(x_k, u_k^{(j)})$, $\mathfrak{z}_k^i \geq \sum_j p_{jk} g_i(x_k, u_k^{(j)})$, $z_k = \sum_j p_{jk} f(x_k, u_k^{(j)})$, $x_k \in A$, $x_k \to x_0$ as $k \to \infty$, where $\bar{\mathfrak{z}} = (\mathfrak{z}^1, \ldots, \mathfrak{z}^r)$, $\mathfrak{z}_k = (\mathfrak{z}_k^1, \ldots, \mathfrak{z}_k^r)$, $z = (z^1, \ldots, z^n)$, $z_k = (z_k^1, \ldots, z_k^n)$. Then $z_k^0 \geq 0$, $z^0 \geq 0$, $\mathfrak{z}_k^i \geq 0$, $\mathfrak{z}^i \geq 0$, and $[z_k^0]$ is a bounded sequence. Here we can extract a subsequence, say still $[k]$, and divide the indices j into two classes, according as the sequence $[u_k^{(j)}]$ is or is not bounded, and then we can even assume that $u_k^{(j)} \to \bar{u}^{(j)} \in R^m$ as $k \to \infty$, or alternatively $u_k^{(j)} \to \infty$ as $k \to \infty$. We can extract the subsequence in such a way that we also have $p_{jk} \to p_j$ as $k \to \infty$, $0 \leq p_j \leq 1, j = 1, \ldots, h$, and thus $\sum_j p_j = 1$. For any j of the second category we certainly have $f_0(x_k, u_k^{(j)}) \to +\infty$; hence $p_{jk} \to p_j = 0$, and $p_{jk} f_0(x_k, u_k^{(j)})$ bounded; hence, by another extraction, $p_{jk} f_0 \to c_j$ as $k \to \infty$. If $\sum'$, $\sum''$ denote sums extended over the two categories of indices j, we have $\sum' p_j = 1$, and for the j of the second category $f(x_k, u_k^{(j)}) = \varepsilon_{jk} f_0(x_k, u_k^{(j)})$, $\varepsilon_{jk} \in R^n$, with $\varepsilon_{jk} \to 0$ as $k \to \infty$. Thus,

$$\begin{aligned}
\bar{z} &= \lim z_k = \lim({\textstyle\sum'} + {\textstyle\sum''}) p_{jk} f(x_k, u_k^{(j)}) \\
&= \lim {\textstyle\sum'} p_{jk} f(x_k, u_k^{(j)}) + \lim {\textstyle\sum''} \varepsilon_{jk}(p_{jk} f_0(x_k, u_k^{(j)})) \\
&= {\textstyle\sum'} p_j f(x_0, u^{(j)}),
\end{aligned}$$

and analogously, since $f_0 \geq 0$, $g_i \geq 0$ are lower semicontinuous, we also have

$$\bar{z}^0 \geq {\textstyle\sum'} p_j f_0(x_0, u^{(j)}), \qquad \bar{\mathfrak{z}}^i \geq {\textstyle\sum'} p_j f_0(x_0, u^{(j)}), \quad i = 1, \ldots, r.$$

This shows that $(\bar{z}^0, \bar{\mathfrak{z}}, \bar{z}) \in \tilde{R}(x_0)$, that is, property (K) is proved.

Let us prove now that the sets $\tilde{R}(x) = \operatorname{co} \tilde{Q}(x)$, which are necessarily convex, have property (Q) at x_0. To this purpose we note that any set $\operatorname{co} \tilde{R}(x, \varepsilon)$ is the union of all points $(z^0, \mathfrak{z}, z) \in R^{1+r+n}$ of the form $(z^0, \mathfrak{z}, z) = \sum_\gamma \lambda_\gamma (z_\gamma^0, \mathfrak{z}_\gamma, z_\gamma)$ with $\lambda_\gamma \geq 0$, $\gamma = 1, \ldots, v$, $\sum_\gamma \lambda_\gamma = 1$, and $(z_\gamma^0, \mathfrak{z}_\gamma, z_\gamma) \in R(x_\gamma)$, $x_\gamma \in N_\varepsilon(x)$, $x \in A$. Thus, $z_\gamma^0 \geq f_0^*$, $\mathfrak{z}_\gamma^i \geq g_i^*$, $z_\gamma = f^*$, where f_0^*, g^*, f^* are computed at some $(x_\gamma, p_\gamma, v_\gamma)$, $p_\gamma \in \Gamma$, $v_\gamma \in (U(x_\gamma))^v$, and we can take any fixed integer $v \geq 2 + n + r$. For $p_\gamma = (p_{\gamma i}, \ldots, p_{\gamma h})$, $v_\gamma = (u_{\gamma 1}, \ldots, u_{\gamma h})$, $\sum_j p_{\gamma j} = 1$, $u_{\gamma j} \in U(x_\gamma), j = 1, \ldots, h$, we have

$$\begin{aligned}
z_\gamma^0 &\geq f_0^*(x_\gamma, p_\gamma, v_\gamma) = \sum_j p_{\gamma j} f_0(x_\gamma, u_{\gamma j}), \\
\mathfrak{z}_\gamma^i &\geq g_i^*(x_\gamma, p_\gamma, v_\gamma) = \sum_j p_{\gamma j} g_i(x_\gamma, u_{\gamma j}), \qquad i = 1, \ldots, r, \\
z_\gamma &= f^*(x_\gamma, p_\gamma, v_\gamma) = \sum_j p_{\gamma j} f(x_\gamma, u_{\gamma j}),
\end{aligned}$$

and we can take any fixed integer $h \geq 2 + n + r$, say $h = \nu$. Then

$$z^0 = \sum_\gamma \lambda_\gamma z_\gamma^0 = \sum_\gamma \sum_j \lambda_\gamma p_{\gamma j} f_0(x_\gamma, u_{\gamma j}),$$

$$\mathfrak{z}^i \geq \sum_\gamma \lambda_\gamma \mathfrak{z}_\gamma^i = \sum_\gamma \sum_j \lambda_\gamma p_{\gamma j} g^i(x_\gamma, u_{\gamma j}), \qquad i = 1, \dots, r,$$

$$z = \sum_\gamma \lambda_\gamma z_\gamma = \sum_\gamma \sum_j \lambda_\gamma p_{\gamma j} f(x_\gamma, u_{\gamma j}),$$

where $\sum_\gamma \sum_j \lambda_\gamma p_{\gamma j} = 1$. In other words co $\tilde{R}(x_0, \varepsilon)$ can be written in terms of the original functions f_0, g, f instead of the functions f_0^*, g^*, f^*, provided we take into consideration suitable convex combinations of ν^2 original points $(z_{\gamma j}^0, \mathfrak{z}_{\gamma j}, z_{\gamma j})$. By repeating the same argument as above, we prove that the sets $\tilde{R}(x)$ have property (Q) at x_0. The details of the proof, which is similar to the one for (10.5.i), are left as an exercise for the reader.

10.6 Closure Theorems for Orientor Fields Based on Weak Convergence

In the closure theorems of the present section, it is convenient to treat $t = (t^1, \dots, t^\nu)$ as ν-dimensional, $\nu \geq 1$, varying in a fixed measurable subset G of R^ν. We shall only require the convergence in measure in G of certain measurable functions $x(t) = (x^1, \dots, x^n)$ that we shall call trajectories, and the weak convergence in $L_1(G)$ of certain functions $\xi(t) = (\xi^1, \dots, \xi^r)$ that will take the place of the derivatives. This added generality does not complicate either the statements or the proofs of the theorems. These theorems, which we shall use here for more existence theorems for one dimensional problems, will be used elsewhere for multidimensional problems. Thus, for every $t \in G \subset R^\nu$ a subset $A(t)$ of the x-space R^r is assigned, and we take $A = [(t, x) | t \in G, x \in A(t)] \subset R^{\nu+n}$. For every $(t, x) \in A$ a subset $Q(t, x)$ of the z-space R^n is assigned and we take $M_0 = [(t, x, z) | (t, x) \in A, \ z \in Q(t, x)] \subset R^{\nu+n+r}$. We shall need in this section the Banach-Sachs-Mazur theorem (10.1.i).

10.6.i. *Let G be measurable with finite measure, and assume that, for almost all $\bar{t} \in G$, the set $A(\bar{t}) \subset R^n$ is closed, and the sets $Q(\bar{t}, x) \subset R^r$ are closed and convex, and have property (K) with respect to x in $A(\bar{t})$. Let $\xi(t)$, $x(t)$, $\xi_k(t)$, $x_k(t)$, $t \in G$, $k = 1, 2, \dots$, be measurable functions, $\xi, \xi_k \in (L_1(G))^r$, and*

$$x_k(t) \in A(t), \quad \xi_k(t) \in Q(t, x_k(t)), \qquad t \in G \text{ (a.e.)}, \quad k = 1, 2, \dots, \tag{10.6.1}$$

where $\xi_k \to \xi$ weakly in $(L_1(G))^r$ and $x_k(t) \to x(t)$ in measure in G as $k \to \infty$. Then

$$x(t) \in A(t), \quad \xi(t) \in Q(t, x(t)), \qquad t \in G \text{ (a.e.)}. \tag{10.6.2}$$

Remark 1. Statement (10.6.i) still holds even if we allow some of the sets $A(t)$ and $Q(t, x)$ to be empty. Precisely, we shall require that for almost all $\bar{t} \in G$ the sets $A(\bar{t})$ are not empty and closed, and we still require that for almost all $\bar{t} \in G$ the map $x \to Q(\bar{t}, x)$ has property (K) (with respect to x) in the closed set $A(\bar{t})$, thus involving possible sets $Q(\bar{t}, x)$ which may be empty. Still we require $x_k(t) \in A(t)$, $\xi_k(t) \in Q(t, x_k(t))$ a.e. in G, and thus the sets $Q(t, x_k(t))$ must be nonempty for a.a. $t \in G$. The conclusion is still $x(t) \in A(t)$, $\xi(t) \in Q(t, x(t))$ a.e. in G, and thus the sets $Q(t, x(t))$ must be nonempty for a.a.

$t \in G$. As we mentioned in Section 8.5, Remark 1, the definitions of properties (K) and (Q) for set valued maps hold unchanged even if some of the sets are empty.

Proof of (10.6.i). Let T_0 be the possible set of measure zero of all t such that $A(t)$ is not closed. By a suitable extraction we may well assume that $x_k(t) \to x(t)$ pointwise, a.e. in G. Let T_0' be the subset of measure zero of all $t \in G$ where $x_k(t)$ does not converge, or it does not converge to $x(t)$, or $x(t)$ is not finite. Then, $x(t) \in A(t)$ for all $t \in G - (T_0 \cup T_0')$. The sequence $\xi_k(t)$, $t \in G$, $k = 1, 2, \ldots$, converges weakly in $(L_1(G))^r$ to $\xi(t)$. By the Dunford–Pettis theorem (10.3.i) (equivalences of (a), (b), and (d)), there is a function $h(\zeta) \geq 0$, $0 \leq \zeta < +\infty$, with $h(\zeta)/\zeta \to +\infty$ as $\zeta \to +\infty$, such that the sequence of scalar functions $\rho_k(t) = h(|\xi_k(t)|) \geq 0$, $t \in G$, $k = 1, 2, \ldots$, is weakly convergent in $L_1(G)$, say to some function $\rho(t) \geq 0$, $t \in G$, $\rho \in L_1(G)$. As stated in Section 10.3, it is not restrictive to assume that h is monotone nondecreasing, continuous, and convex. Now, for any $s = 1, 2, 3, \ldots$, the sequence ρ_{s+k}, ξ_{s+k}, $k = 1, 2, \ldots$, converges weakly to ρ, ξ in $(L_1(G))^{r+1}$. By the Banach–Saks–Mazur theorem (10.1.i), there is a set of real numbers $c_{Nk}^{(s)} \geq 0$, $k = 1, 2, \ldots, N$, $N = 1, 2, \ldots$, with $\sum_{k=1}^{N} c_{Nk}^{(s)} = 1$, such that if

$$\rho_N^{(s)}(t) = \sum_{k=1}^{N} c_{Nk}^{(s)} \rho_{s+k}(t), \qquad \xi_N^{(s)}(t) = \sum_{k=1}^{N} c_{Nk}^{(s)} \xi_{s+k}(t), \qquad t \in G, \quad N = 1, 2, \ldots,$$

then $\xi_N^{(s)}(t) \to \xi(t)$ strongly in $(L_1(G))^r$ and $\rho_N^{(s)}(t) \to \rho(t)$ strongly in $L_1(G)$, and this is true for every $s = 1, 2, \ldots$. Then, for every s, there is also a subset T_s of measure zero of points $t \in G$, and a sequence of integers $N_\lambda^{(s)}$, $\lambda = 1, 2, \ldots$, with $N_\lambda^{(s)} \to \infty$, such that for $t \in G - T_s$, $\xi(t)$ and $\eta(t)$ are finite, and $\xi_{N\lambda(s)}^{(s)} \to \xi(t)$ and $\rho_{N\lambda(s)}^{(s)} \to \rho(t)$ as $\lambda \to \infty$. Let T denote the subset of measure zero in G which is the union of all sets T_0, T_0', T_s, $s = 1, 2, \ldots$.

Now let t_0 be any point $t_0 \in G - T$, and take $x_0 = x(t_0)$. Then $(t_0, x_k(t_0)) \to (t_0, x_0) \in A$, and given $\varepsilon > 0$ there is some s_0 such that $|x_s(t_0) - x_0| \leq \varepsilon$ for all $s \geq s_0$. For $s \geq s_0$ we have now

$$\xi_{s+k}(t_0) \in Q(t_0, x_{s+k}(t_0)), \quad |x_{s+k}(t_0) - x_0| \leq \varepsilon, \qquad k = 1, 2, \ldots.$$

We consider now the sets $Q^*(t, x) \subset R^{n+1}$ defined by

$$Q^*(t, x) = [(z^0, z) \,|\, z^0 \geq h(|z|),\ z \in Q(t, x)], \qquad (t, x) \in A.$$

Since $Q(t_0, x_0)$ is closed and convex, and h is monotone nondecreasing, continuous, and convex, then the set $Q^*(t_0, x_0)$ is also closed and convex. By (10.5.i), second part, the sets $Q^*(t_0, x_0)$ have property (Q) with respect to x at x_0. Now $\rho_k(t_0) = h(|\xi_k(t_0)|)$; hence

$$(\rho_{s+k}(t_0), \xi_{s+k}(t_0)) \in Q^*(t_0, x_{s+k}(t_0))$$
$$(\rho_N^{(s)}(t_0), \xi_N^{(s)}(t_0)) \in \operatorname{co} Q^*(t_0; x_0, \varepsilon),$$

where $Q^*(t_0; x_0, \varepsilon)$ denotes the union of all $Q^*(t_0, x)$ with $x \in A(t_0)$, $|x - x_0| \leq \varepsilon$, and the last relation holds for all N and $s \geq s_0$. Finally, for $N = N_\lambda$ and $\lambda \to \infty$ we have

$$(\rho(t_0), \xi(t_0)) \in \operatorname{cl} \operatorname{co} Q^*(t_0; x_0, \varepsilon)$$

where $\varepsilon > 0$ is arbitrary. Hence, by property (Q) of the sets $Q^*(t_0, x)$ with respect to x at x_0, we also have

$$(\rho(t_0), \xi(t_0)) \in \bigcap_{\varepsilon > 0} \operatorname{cl} \operatorname{co} Q^*(t_0; x_0, \varepsilon) = Q^*(t_0, x_0).$$

Hence,

$$\xi(t_0) \in Q(t_0, x_0).$$

We have proved that, for almost all $t \in G$, we have

$$x(t) \in A(t), \quad \xi(t) \in Q(t, x(t)), \qquad t \in G \text{ (a.e.)}. \qquad \square$$

The following statement, easier to prove than (10.6.i), will also be used.

10.6.ii. *With the same notation as for* (10.6.i), *let G be measurable and of finite measure, and let us assume that, for almost all $\bar{t}$, the sets $Q(\bar{t}, x)$, $x \in A(\bar{t})$, are closed and convex. Let $\xi(t)$, $x(t)$, $\xi_k(t)$, $\bar{\xi}_k(t)$, $t \in G$, $k = 1, 2, \ldots$, be measurable functions, $\xi, \xi_k, \bar{\xi}_k \in (L_1(G))^r$, $x(t) \in A(t)$, $\bar{\xi}_k(t) \in Q(t, x(t))$, $t \in G$ (a.e.), $k = 1, 2, \ldots$, with $\xi_k \to \xi$ weakly in $(L_1(G))^r$, $\delta_k(t) = \xi_k(t) - \bar{\xi}_k(t) \to 0$ weakly in $(L_1(G))^r$. Then $\xi(t) \in Q(t, x(t))$, $t \in G$ (a.e.).*

Proof. Since $\xi_k \to \xi$, $\delta_k \to 0$ weakly in $(L_1(G))^r$, then $\bar{\xi}_k \to \xi$ weakly in $(L_1(G))^r$, and $(\xi_k, \delta_k) \to (\xi, 0)$ in $(L_1(G))^{2r}$. By (10.1.i) there is a set of real numbers $c_{Nk} \geq 0$, $k = 1, \ldots, N$, $N = 1, 2, \ldots, \sum_{k=1}^{N} c_{Nk} = 1$, such that, if

$$\xi_N^*(t) = \sum_{k=1}^{N} c_{Nk}\xi_k(t), \quad \delta_N^*(t) = \sum_{k=1}^{N} c_{Nk}\delta_k(t), \qquad t \in G, \quad N = 1, 2, \ldots,$$

then $\xi_N^*(t) \to \xi(t)$, $\delta_N^*(t) \to 0$ strongly in $(L_1(G))^r$. Thus, there is a subsequence $[N_\lambda]$ such that $\xi_{N_\lambda}^*(t) \to \xi(t)$, $\delta_{N_\lambda}^*(t) \to 0$ pointwise in G. Let T_0 denote the set of measure zero of all t for which this does not occur, or $\xi(t)$ is not finite, or $Q(t, x(t))$ is not closed or is not convex. If $\bar{\xi}_N^*(t) = \sum_{k=1}^{N} c_{Nk}\bar{\xi}_k(t)$, $t \in G$, then, for $N = N_\lambda$, $t \in G - T_0$, we have $\bar{\xi}_N^*(t) = \xi_N^*(t) - \delta_N^*(t)$, $\bar{\xi}_N^*(t) \to \xi(t)$ as $N \to \infty$. Since $\bar{\xi}_k(t) \in Q(t, x(t))$ for all k, and $Q(t, x(t))$ is a closed and convex set, we also have $\bar{\xi}_N^*(t) \in Q(t, x(t))$ for $N = N_\lambda$ and as $\lambda \to \infty$, also $\xi(t) \in Q(t, x(t))$ for all $t \in G - T_0$, that is, a.e. in G. This proves (10.6.ii). $\square$

10.7 Lower Closure Theorems for Orientor Fields Based on Weak Convergence

We shall use essentially the same notation as in (10.6.i). Thus, points in R^ν, R^n, R^{1+r}, and R^{2+r} spaces will be denoted by $t = (t^1, \ldots, t^\nu)$, $x = (x^1, \ldots, x^n)$, $(z^0, z) = (z^0, z^1, \ldots, z^r)$ or $(\eta, \xi) = (\eta, \xi^1, \ldots, \xi^r)$, and (v, z^0, z) or (ρ, η, ξ). Let G be any measurable subset of the t-space R^ν of finite measure, for every $t \in G$ let $A(t)$ be a given nonempty subset of the x-space R^n, and let $A = [(t, x) | t \in G, x \in A(t)]$. For every $(t, x) \in A$ let $\tilde{Q}(t, x)$ be a given subset of the z^0z-space R^{r+1}. We denote by $x(t) = (x^1, \ldots, x^n)$, $(\eta(t), \xi(t)) = (\eta, \xi^1, \ldots, \xi^r)$, $t \in G$, given functions from G to R^n, R^{r+1} respectively.

10.7.i (A Lower Closure Theorem for Orientor Fields). *Let G be measurable and of finite measure, and assume that for almost all $\bar{t} \in G$, the set $A(\bar{t})$ is closed and that the sets $\tilde{Q}(\bar{t}, x)$ are closed and convex, and have property (K) with respect to x in $A(\bar{t})$. Let $\xi(t)$, $x(t)$, $\eta_k(t)$, $\xi_k(t)$, $x_k(t)$, $\lambda(t)$, $\lambda_k(t)$, $t \in G$, $k = 1, 2, \ldots$, be measurable functions, $\xi, \xi_k \in (L_1(G))^r$, $\eta_k \in L_1(G)$, with $x_k \to x$ in measure on G, $\xi_k \to \xi$ weakly in $(L_1(G))^r$ as $k \to \infty$,*

$$(10.7.1) \quad x_k(t) \in A(t), \quad (\eta_k(t), \xi_k(t)) \in \tilde{Q}(t, x_k(t)), \qquad t \in G \text{ (a.e.)}, \quad k = 1, 2, \ldots,$$

(10.7.2) $$-\infty < i = \liminf_{k\to\infty} \int_G \eta_k(t)\,dt < +\infty,$$

(10.7.3) $$\eta_k(t) \geq \lambda_k(t), \quad \lambda, \lambda_k \in L_1(G), \qquad \lambda_k \to \lambda \text{ weakly in } L_1(G).$$

Then there is a function $\eta(t)$, $t \in G$, $\eta \in L_1(G)$, such that

(10.7.4) $$x(t) \in A(t), \ (\eta(t), \xi(t)) \in \tilde{Q}(t, x(t)), \quad t \in G, \qquad \int_G \eta(t)\,dt \leq i.$$

Proof of (10.7.i). We give first the general proof in which certain auxiliary sets $\tilde{Q}'^*(t,x)$ are constructed which have property (Q) with respect to x. Immediately afterwards we sketch the drastically simpler proof for the case in which the original sets $\tilde{Q}(t,x)$ already have property (Q) with respect to x.

Let T_0 be the set of measure zero of all $t \in G$ for which $A(t)$ is not closed. Let $j_k = \int_G \eta_k(t)\,dt$, $k = 1, 2, \ldots$. By taking a suitable subsequence we may well assume that $j_k \to i$ and $x_k(t) \to x(t)$ pointwise a.e. in G as $k \to \infty$. Here $-\infty < i < +\infty$, so that if δ_s denotes the maximum of $|j_k - i|$ for $k \geq s+1$, we have $\delta_s \to 0$ as $s \to \infty$.

Let T_0' be the subset of measure zero of all $t \in G$ for which $A(t)$ is not closed, or where $x_k(t)$ does not converge, or it does not converge to $x(t)$, or $x(t)$ is not finite. Then $x(t) \in A(t)$ for all $t \in G - (T_0 \cup T_0')$. The sequence $\lambda_k(t)$, $\xi_k(t)$, $t \in G$, $k = 1, 2, \ldots$, converges weakly to $\lambda(t)$, $\xi(t)$ in $(L_1(G))^{r+1}$. By the equivalence theorem (10.3.i) there is a function $h(\zeta) \geq 0$, $0 \leq \zeta < +\infty$, with $h(\zeta)/\zeta \to +\infty$ as $\zeta \to +\infty$, and such that the sequence of scalar functions $\rho_k(t) = h(|\xi_k(t)| \geq 0$, $t \in G$, $k = 1, 2, \ldots$, also is weakly convergent in $L_1(G)$, say to some function $\rho(t) \geq 0$, $t \in G$, $\rho \in L_1(G)$.

It is not restrictive to assume that h is monotone nondecreasing, continuous, and convex. Now, for any $s = 1, 2, \ldots$, the sequence ρ_{s+k}, λ_{s+k}, ξ_{s+k}, $k = 1, 2, \ldots$, converges weakly to ρ, λ, ξ in $(L_1(G))^{r+2}$. By the Banach–Saks–Mazur theorem (10.1.i), there is a set of real numbers $c_{Nk}^{(s)} \geq 0$, $k = 1, \ldots, N$, $N = 1, 2, \ldots$, with $\sum_{k=1}^{N} c_{Nk}^{(s)} = 1$, such that if

$$\rho_N^{(s)}(t) = \sum_{k=1}^{N} c_{Nk}^{(s)} \rho_{s+k}(t), \qquad \lambda_N^{(s)}(t) = \sum_{k=1}^{N} c_{Nk}^{(s)} \lambda_{s+k}(t),$$

$$\xi_N^{(s)}(t) = \sum_{k=1}^{N} c_{Nk}^{(s)} \xi_{s+k}(t), \qquad t \in G, \quad N = 1, 2, \ldots$$

then $(\rho_N^{(s)}, \lambda_N^{(s)}, \xi_N^{(s)}) \to (\rho, \lambda, \xi)$ strongly in $(L_1(G))^{r+2}$, and this is true for every $s = 1, 2, \ldots$. Then, for every s, there is also a subset T_s of measure zero of points $t \in G$, and a sequence of integers $N_l^{(s)}$, $l = 1, 2, \ldots$, with $N_l^{(s)} \to \infty$, such that for $t \in G - T_s$, $\rho(t)$, $\lambda(t)$, $\xi(t)$ are finite and (with simplified notation) $(\rho_{N_l}^{(s)}(t), \lambda_{N_l}^{(s)}(t), \xi_{N_l}^{(s)}(t)) \to (\rho(t), \lambda(t), \xi(t))$ as $l \to \infty$. Let T denote the subset of measure zero in G which is the union of all T_0, T_0', T_s, $s = 1, 2, \ldots$. Let us take

$$\eta_N^{(s)}(t) = \sum_{k=1}^{N} c_{Nk}^{(s)} \eta_{s+k}(t),$$

and note that

$$\rho_k(t) = h[|\xi_k(t)|], \eta_k(t) \geq \lambda_k(t), \ t \in G, \quad \int_G \eta_k(t)\,dt = j_k, \qquad k = 1, 2, \ldots,$$

so that, for all $s = 1, 2, \ldots, N = 1, 2, \ldots$, we also have

$$\text{(10.7.5)} \qquad \eta_N^{(s)}(t) \geq \lambda_N^{(s)}(t), \quad t \in G, \qquad i - \delta_s \leq \int_G \eta_N^{(s)}(t)\,dt \leq i + \delta_s.$$

For $N = N_l^{(s)}$ and $l \to \infty$, the relations (10.7.5) and by Fatou's lemma (8.7.i) imply

$$\eta^{(s)}(t) = \liminf_{l\to\infty} \eta_{N_l}^{(s)}(t) \geq \lambda(t), \qquad t \in G - T,$$

$$\int_G \eta^{(s)}(t)\,dt \leq \liminf_{l\to\infty} \int_G \eta_{N_l}^{(s)}(t)\,dt \leq i + \delta_s, \qquad s = 1, 2, \ldots.$$

Thus, $\eta^{(s)}(t)$ is finite a.e. in G and of class $L_1(G)$.

Let T'_s denote the set of measure zero of all points $t \in G$ where $\eta^{(s)}(t)$ is not finite. Finally, if

$$\eta(t) = \liminf_{s\to\infty} \eta^{(s)}(t), \qquad t \in G,$$

then again we have $\eta(t) \geq \lambda(t)$, $t \in G$, $\int_G \eta(t)\,dt \leq i$.

Note that, for $t \in G - T_s$, $\lambda(t)$ is finite. Thus, for l sufficiently large, say $l \geq l_0(t, s)$, we certainly have $\eta_{N_l}^{(s)}(t) \geq \lambda(t) - 1$. In other words, we may drop from the sequence $N_l^{(s)}$, $l = 1, 2, \ldots$, enough initial terms (finitely many, depending on t and s) in such a way that the relation $\eta_N^{(s)}(t) \geq \lambda(t) - 1$ holds for l. Also, $\eta(t)$ is finite a.e. in G and of class $L_1(G)$. Let T''_0 denote the set of measure zero of all points $t \in G$ where η is not finite.

Let T^* denote the set of measure zero in G which is the union of all sets $T_0, T'_0, T''_0, T_s, T'_s$, $s = 1, 2, \ldots$. Let t_0 be any point $t_0 \in G - T^*$, and take $x_0 = x(t_0)$. Then $(t_0, x_k(t_0)) \to (t_0, x_0) \in A$, and, given $\varepsilon > 0$, there is some s_0 such that $|x_s(t_0) - x_0| \leq \varepsilon$ for all $s \geq s_0$. For $s \geq s_0$ we have

$$(\eta_{s+k}(t_0), \xi_{s+k}(t_0)) \in \tilde{Q}(t_0, x_{s+k}(t_0)), \quad |x_{s+k}(t_0) - x_0| \leq \varepsilon, \qquad k = 1, 2, \ldots.$$

We consider now the sets $Q^*(t, x) \subset R^{n+2}$ defined by

$$\tilde{Q}^*(t, x) = [(v, y, z) \ v \geq h(|z|), (y, z) \in \tilde{Q}(t, x)], \qquad (t, x) \in A,$$

and we also need the sets

$$\tilde{Q}'(t, x) = [(y, z) | y \geq \lambda(t) - 1, (y, z) \in \tilde{Q}(t, x)] \subset R^{n+1},$$
$$\tilde{Q}'^*(t, x) = [(v, y, z) \ v \geq h(|z|), y \geq \lambda(t) - 1, (y, z) \in \tilde{Q}(t, x)] \subset R^{n+2}.$$

For each t fixed, the sets $\tilde{Q}'(t, x)$, $\tilde{Q}'^*(t, x)$, $x \in A(t)$, are the intersections of $\tilde{Q}(t, x)$, $\tilde{Q}^*(t, x)$ with the fixed closed sets $[(y, z) | y \geq \lambda(t) - 1, z \in R^n]$, $[(v, y, z) | y \geq \lambda(t) - 1, (v, z) \in R^{n+1}]$. Then certainly the sets $\tilde{Q}'(t_0, x)$, $x \in A(t_0)$, have property (K) with respect to x at x_0, since the sets $\tilde{Q}(t_0, x)$ already have this property. Since h is monotone nondecreasing, continuous, and convex, the sets $\tilde{Q}^*(t_0, x_0)$ and $\tilde{Q}'^*(t_0, x_0)$ are convex. Finally, for $t = t_0$, we can apply the second statement of (10.5.ii) to the sets $Q'^*(t_0, x)$ with $\mu = 1$, $T_0(x, y, z) = h(|z|)$. The present variables v, y, z replace the variables (y^0, y, z) of the second part of (10.5.ii); the present sets $\tilde{Q}'(t_0, x)$ (for t_0 fixed) replace the sets $Q(x)$; and the present sets $\tilde{Q}'^*(t_0, x)$ for t_0 fixed replace the sets $\tilde{Q}(x)$. Also, the present function $h(|z|)$ and constant $\lambda(t_0) - 1$ replace the function $\phi(|z|)$ and constant L of the second part of (10.5.ii). Here $T_0(x, y, z) = h(|z|)$ is constant with respect to x, y and continuous in z, and thus certainly lower semicontinuous in (x, y, z) as required in (10.5.ii). We conclude that the sets $\tilde{Q}'^*(t_0, x)$, $x \in A(t_0)$, have property (Q) with respect to x at $x = x_0$. By (10.7.1) and the definitions of ρ_k and $\tilde{Q}^*(t, x)$ we have now, for $s \geq s_0$,

$$(\rho_{s+k}(t_0), \eta_{s+k}(t_0), \xi_{s+k}(t_0)) \in \tilde{Q}^*(t_0, x_{s+k}(t_0)), \qquad |x_{s+k}(t_0) - x_0| \leq \varepsilon,$$

and hence

$$\left(\sum_{k=1}^{N} c_{Nk}^{(s)} \rho_{s+k}(t_0), \sum_{k=1}^{N} c_{Nk}^{(s)} \eta_{s+k}(t_0), \sum_{k=1}^{N} c_{Nk}^{(s)} \xi_{s+k}(t_0)\right) \in \text{co}\, \tilde{Q}^*(t_0; x_0, \varepsilon).$$

Finally, for $N = N_l^{(s)}$, $l \geq l_0(t_0, s)$, we have $\eta_N^{(s)}(t_0) = \sum_{k=1}^{N} c_{Nk}^{(s)} \eta_{s+k}(t_0) \geq \lambda(t_0) - 1$, and hence

$$\left(\sum_{k=1}^{N} c_{Nk}^{(s)} \rho_{s+k}(t_0), \sum_{k=1}^{N} c_{Nk}^{(s)} \eta_{s+k}(t_0), \sum_{k=1}^{N} c_{Nk}^{(s)} \xi_{s+k}(t_0)\right) \in \text{co}\, \tilde{Q}'^*(t_0; x_0, \varepsilon). \tag{10.7.6}$$

As $l \to \infty$, the points in the first member of this relation form a sequence possessing $(\rho(t_0), \eta^{(s)}(t_0), \xi(t_0))$ as an element of accumulation in R^{n+2} (all $\rho(t_0)$, $\eta^{(s)}(t_0)$, $\xi(t_0)$ finite). Thus

$$(\rho(t_0), \eta^{(s)}(t_0), \xi(t_0)) \in \text{cl co}\, \tilde{Q}'^*(t_0; x_0, \varepsilon), \qquad s \geq s_0. \tag{10.7.7}$$

Note that $\eta(t_0) = \liminf_{s \to \infty} \eta^{(s)}(t_0)$ is finite, so that $(\rho(t_0), \eta(t_0), \xi(t_0))$ is a point of accumulation of the sequence in the first member of (10.7.7), while the second member is a closed set. Thus,

$$(\rho(t_0), \eta(t_0), \xi(t_0)) \in \text{cl co}\, \tilde{Q}'^*(t_0; x_0, \varepsilon).$$

Since $\varepsilon > 0$ is arbitrary, by property (Q) we have

$$(\rho(t_0), \eta(t_0), \xi(t_0)) \in \bigcap_{\varepsilon > 0} \text{cl co}\, \tilde{Q}'^*(t_0; x_0, \varepsilon) = \tilde{Q}'^*(t_0, x_0).$$

By the definition of $\tilde{Q}'^*(t_0, x_0)$ we then have

$$(\eta(t_0), \xi(t_0)) \in \tilde{Q}(t_0, x_0).$$

We have proved that for almost any $t \in G$ we have $x(t) \in A(t)$, $(\eta(t), \xi(t)) \in \tilde{Q}(t, x(t))$, and that $\eta \in L_1(G)$ with $\int_G \eta \, dt \leq i$. Theorem (10.7.i) is thereby proved. □

Second proof of (10.7.i). We assume here that for almost all $\bar{t}$ the sets $\tilde{Q}(\bar{t}, x)$ have property (Q) with respect to x. We proceed as in the proof above omitting the recourse to the equivalence theorem. Then we apply the Banach–Saks–Mazur theorem to the sequence λ_{s+k}, ξ_{s+k}, $k = 1, 2, \ldots$, obtaining $\lambda_N^{(s)}(t)$,

$\xi_N^{(s)}(t)$ as before, and then the functions $\eta_N^{(s)}(t)$ and $\eta(t)$. Now relations (10.7.6) hold as before, and then

$$\left(\sum_{k=1}^{N} c_{Nk}^{(s)}\eta_{s+k}(t_0),\ \sum_{k=1}^{N} c_{Nk}^{(s)}\xi_{s+k}(t_0)\right) \in \operatorname{co} \tilde{Q}(t_0; x_0, \varepsilon).$$

From here we derive in succession as before

$$(\eta^{(s)}(t), \xi(t_0)) \in \operatorname{cl} \operatorname{co} \tilde{Q}(t_0; x_0, \varepsilon),$$
$$(\eta(t_0), \xi(t_0)) \in \operatorname{cl} \operatorname{co} \tilde{Q}(t_0; x_0, \varepsilon).$$

By property (Q) then

$$(\eta(t_0), \xi(t_0)) \in \bigcap_{\varepsilon>0} \operatorname{cl} \operatorname{co} \tilde{Q}(t_0; x_0, \varepsilon) = \tilde{Q}(t_0, x_0)$$

which proves the theorem. □

Remark 1. Statement (10.7.i) still holds even if we allow some of the sets $A(t)$ and $Q(t, x)$ to be empty. Precisely, we shall require that for almost all $\bar{t} \in G$ the sets $A(\bar{t})$ are not empty and closed, and we still require that for almost all $\bar{t} \in G$ the map $x \to \tilde{Q}(\bar{t}, x)$ has property (K) (with respect to x) in the closed set $A(\bar{t})$, thus involving possible sets $\tilde{Q}(\bar{t}, x)$ which may be empty. Still we require $x_k(t) \in A(t)$, $(\eta_k(t), \xi_k(t)) \in \tilde{Q}(t, x_k(t))$ a.e. in G, and thus the sets $\tilde{Q}(t, x_k(t))$ must be nonempty for a.a. $t \in G$. The conclusion is still $x(t) \in A(t)$, $(\eta(t), \xi(t)) \in \tilde{Q}(t, x(t))$ a.e. in G, and thus the sets $\tilde{Q}(t, x(t))$ must be nonempty for a.a. $t \in G$. (Cf. the analogous Remark 1 of Section 10.6).

In Section 10.8 we shall prove a partial inverse of (10.7.i) showing, in particular, that (10.7.3) is not only a sufficient condition for lower closure, but also essentially a necessary one.

In this formulation of (10.7.i), $\lambda_k \to \lambda$ weakly in $L_1(G)$; hence $\|\lambda_k\|_1$ is a bounded sequence, and the part $i > -\infty$ of the requirement (10.7.2) is actually a consequence of (10.7.3). The lower closure theorem (10.7.i) will be used in situations where it is known that $-\infty < i < +\infty$ and where functions λ, λ_k satisfying (10.7.3) are easily found. Here we list simple alternative conditions, (Λ_1)–(Λ_7), under each of which functions λ, λ_k satisfying (10.7.3) can be immediately derived. Here we denote by (a, b) the inner product in R^n.

Λ_1. There is a real valued function $\psi(t) \geq 0$, $t \in G$, $\psi \in L_1(G)$, such that for every $(t, x) \in A$, $(y, z) \in \tilde{Q}(t, x)$, we have $y \geq -\psi(t)$.

Indeed, then we have $\eta_k(t) \geq \lambda_k(t) = -\psi(t)$, $t \in G$, $k = 1, 2, \ldots$, $\lambda = -\psi$.

Λ_2. There is a real valued function $\psi(t)$, $t \in G$, $\psi \in L_1(G)$, and a constant $\gamma \geq 0$ such that for all $(t, x) \in A$ and $(y, z) \in \tilde{Q}(t, x)$ we have $y \geq -\psi(t) - \gamma|z|$.

Indeed, then we have $\eta_k(t) \geq \lambda_k(t) = -\psi(t) - \gamma|\xi_k(t)|$, $t \in G$, $k = 1, 2, \ldots$. Since $\xi_k \to \xi$ weakly in $(L_1(G))^r$ by hypothesis, by the Dunford–Pettis theorem (cf. Section 10.3) we know that the same functions ξ_k are equiabsolutely integrable in G. Hence, by the same Dunford–Pettis theorem, the sequence $|\xi_k(t)|$, $t \in G$, $k = 1, 2, \ldots$, is weakly compact in $L_1(G)$, and there is, therefore, a subsequence $[k_s]$ such that $\lambda_{k_s}(t) = -\psi(t) - \gamma|\xi_{k_s}(t)|$, $t \in G$, $s = 1, 2, \ldots$, is weakly convergent in $L_1(G)$ to some function $-\psi(t) - \gamma\sigma(t)$, $\sigma(t) \geq 0$, $t \in G$, and we can apply (10.7.i).

Λ_3. $\xi_k \in (L_q(G))^r$, $1 \leq q \leq +\infty$, $\xi_k \to \xi$ weakly in $(L_q(G))^r$, and there are a real valued function $\psi(t) \geq 0$, $t \in G$, $\psi \in L_1(G)$, and an r-vector function $\phi(t)$, $t \in G$, $\phi \in (L_s(G))^r$, $1/s + 1/q = 1$, such that for all $(t, x) \in A$ and $(y, z) \in \tilde{Q}(t, x)$ we have $y \geq -\psi(t) - (\phi(t), z)$.

Note that if $\xi_k \to \xi$ weakly in $(L_q(G))^r$, then $\lambda_k(t) = -\psi(t) - (\phi(t), \xi_k(t))$ converges weakly in $L_1(G)$ to $-\psi - (\phi, \xi)$.

Λ_4. There are constants $\alpha \geq \beta$ real and $\gamma > 0$ such that (a) for every $(t, x) \in A$ and for every $|z| \leq \gamma$ there are points $(y, z) \in \tilde{Q}(t, x)$, and for all such points $y \geq \beta$; (b) for every $(t, x) \in A$ there is some point $(y_0, 0) \in \tilde{Q}(t, x)$ with $y_0 \leq \alpha$.

In other words, for every $(t, x) \in A$ the projection $Q(t, x)$ on the z-space R^n of $\tilde{Q}(t, x)$ contains the whole ball $|z| \leq \gamma$, and for all $(y, z) \in \tilde{Q}(t, x)$ with $|z| \leq \gamma$ we have $y \geq \beta$. Moreover, for every $(t, x) \in A$ there is some point $(y_0, 0) \in \tilde{Q}(t, x)$ with $y_0 \leq \alpha$. Now, if (y, z) is any point of $\tilde{Q}(t, x)$ with $|z| > \gamma$, then $z_1 = \gamma z/|z|$ has distance γ from the origin and is interior to the segment Oz between O and z, with $z_1 = (1 - \sigma)O + \sigma z$, where $\sigma = \gamma/|z|$. Then, by the convexity of $\tilde{Q}(t, x)$ there is some y_1 such that $(y_1, z_1) \in \tilde{Q}(t, x)$ and $\beta \leq y_1 \leq (1 - \sigma)y_0 + \sigma y$, or

$$y \geq \sigma^{-1} y_1 - \sigma^{-1}(1 - \sigma) y_0 \geq \beta\gamma^{-1}|z| - \alpha\gamma^{-1}|z|(1 - \gamma|z|^{-1}) = \alpha + \gamma^{-1}(\beta - \alpha)|z|,$$

and we are in the situation discussed under (Λ_2).

Λ_5. x, $x_k \in (L_p(G))^n$, $\|x_k - x\|_p \to 0$ as $k \to \infty$ for some p, $1 \leq p < +\infty$, $\xi, \xi_k \in (L_1(G))^r$, $\xi_k \to \xi$ weakly in $(L_1(G))^r$, and there are a real valued function $\psi(t) \geq 0$, $t \in G$, $\psi \in L_1(G)$, and constants γ, $\gamma' \geq 0$ such that for all $(t, x) \in A$, $(y, z) \in \tilde{Q}(t, x)$ we have $y \geq -\psi(t) - \gamma'|x|^p - \gamma|z|$.

The argument is similar to the one under (Λ_3), since $\|x_k - x\|_p \to 0$ implies $\||x_k|^p - |x|^p\|_1 \to 0$, and then the sequence $\lambda_k(t) = -\psi(t) - \gamma'|x_k(t)|^p - \gamma|\xi_k(t)|$, $t \in G$, $k = 1, 2, \ldots$, certainly possesses a weakly convergent subsequence in $L_1(G)$. Instead of the requirement x, $x_k \in (L_p(G))^n$, $\|x_k - x\|_p \to 0$, we may require $x_k^i, x^i \in L_{p_i}(G)$, $\|x_k^i - x^i\|_{p_i} \to 0$ as $k \to \infty$ for different p_i, $1 \leq p_i < \infty$, $i = 1, \ldots, n$. This remark holds throughout the present and next chapters.

Λ_6. $x_k \in (L_\infty(G))^n$, $\|x_k\|_\infty \leq L_0$, $\xi_k \in (L_\infty(G))^r$, $\|\xi_k\|_\infty \leq L_1$ for given constants L_0, L_1, and there are a real valued function $\psi(t) \geq 0$, $t \in G$, $\psi \in L_1(G)$, and a real valued monotone nondecreasing function $\sigma(\xi), 0 \leq \xi < +\infty$, such that for all $(t, x) \in A, (y, z) \in \tilde{Q}(t, x)$ we have $y \geq -\psi(t) - \sigma(|x| + |z|)$.

The argument is similar to the one above.

Λ_7. Let $1 \leq p < +\infty$, $1 < q \leq +\infty$, $x_k \in L_p(G)$, $\xi_k \in L_q(G)$, $\|x_k\|_p \leq L_1$, $\|\xi_k\|_q \leq L_2$ for some constants L_1, L_2, and assume that there are a constant $\delta \geq 0$, a real valued function $\psi(t) \geq 0$, $t \in G$, $\psi \in L_1(G)$, and a Borel measurable function $p(t, x): G \times R^n \to R^r$ such that for all $(t, x) \in A$, $(y, z) \in \tilde{Q}(t, x)$, and $1/s + 1/q = 1$, we have

$$y \geq -\psi(t) - \delta|x|^p - (p(t, x), z) \quad \text{and} \quad |p(t, x)|^s \leq \delta|x|^p + \psi(t).$$

Here we assume that $x_k \to x$ strongly in $(L_p(G))^n$ and that $\xi_k \to \xi$ weakly in $(L_q(G))^r$.

Since $x_k \to x$ strongly in $(L_p(G))^n$, then the functions $|x_k(t)|^p$ are equiabsolutely integrable, and so are the functions $|p(t, x_k(t))|^s$. Since $\xi_k \to \xi$ weakly in $(L_q(G))^r$, then $\|\xi_k\|_q$ is a bounded sequence, and by the Hölder inequality, the sequence $(p(t, x_k(t)), \xi_k(t))$, $t \in G$, $k = 1, 2, \ldots$, is also equiabsolutely integrable, and so is the sequence $\lambda_k(t) = -\psi(t) - \delta|x_k(t)|^p - (p(t, x_k(t)), \xi_k(t))$, $t \in G$, $k = 1, 2, \ldots$. Thus, $[\lambda_k]$ contains a subsequence which is weakly convergent in $L_1(G)$.

Remark 2. Note that in view of (10.3.i), whenever $\xi_k \to \xi$ weakly in L_1, there is some real valued Nagumo function $\Psi(\zeta), 0 \leq \zeta < +\infty$, with $\Psi(\zeta)/\zeta \to +\infty$ as $\zeta \to +\infty$, such that $\Psi(|\xi_k|)$ is weakly convergent in L_1, and then we could require that $z^0 \geq -\psi(t) - \delta|x|^p - \Psi(|z|)$.

Remark 3. The reader should note that there is a statement similar to (Λ_5) which is not true. Indeed, let us assume in (Λ_5) that $\xi_k \in (L_q(G))^r$ for some $q > 1$, and that (*) $y \geq -\psi(t) - \delta'|x|^p - \delta|z|^q$. Under this assumption we cannot conclude lower closure. This can be seen by the following example. Take $G = (0, 1)$, $A(t) = R$, $n = r = 1$, $q > 1$, $1/s + 1/q = 1$, $f_0(t, x, z) = s^{-1}|t^{-1}x|^s + t^{-1}xz$, and $\tilde{Q}(t, x) = [(y, z) | y \geq f_0, z \in R]$. By the elementary relation $\alpha\beta \leq s^{-1}\alpha^s + q^{-1}\beta^q$, $\alpha, \beta \geq 0$, we derive that for $\alpha = |t^{-1}x|$, $\beta = |z|$, that $s^{-1}|t^{-1}x|^s + q^{-1}|z|^q \geq |t^{-1}x||z|$, and hence $f_0 = s^{-1}|t^{-1}x|^s + t^{-1}xz \geq s^{-1}|t^{-1}x|^s - t^{-1}|x||z| \geq -q^{-1}|z|^q$. This shows that the relation (*) holds for $\psi \equiv 0$, $\delta' = 0$, $\delta = q^{-1}$.

On the other hand, let us take $x_k(t) = tk^{1/s}$, $\xi_k(t) = -k^{1/q}$ for $0 < t \leq k^{-1}$, and $x_k(t) = 0$, $\xi_k(t) = 0$ for $k^{-1} < t < 1$. Then $x_k \to 0$ uniformly in $(0, 1)$, and $\xi_k \to 0$ weakly in $L_q(0, 1)$. Then $\eta_k(t) = f_0(t, x_k(t), \xi_k(t)) = -q^{-1}k$ for $0 < t \leq k^{-1}$, $\eta_k(t) = 0$ for $k^{-1} < t < 1$, and $\int_0^1 \eta_k \, dt = -q^{-1} < 0$, for all k, while $\eta_0(t) = f_0(t, 0, 0) = 0$, and $\int_0^1 \eta_0 \, dt = 0$. Note that here $\xi_k \to \xi_0$ weakly in L_q does not imply that $|\xi_k|^q \to |\xi_0|^q$ weakly in L_1. On the other hand, if $\xi_k \to \xi_0$ strongly in L_q, then of course $|\xi_k|^q \to |\xi_0|^q$ strongly in L_1. The inequality (*) would be sufficient with $q = 1$.

Remark 4. In Theorem (10.7.i) the requirement (10.7.3) cannot be disregarded, even if we replace (10.7.2) by the stronger requirement $\eta_k(t) \leq M_0$. This can be shown by the following simple example. Take $\nu = n = r = 1$, $0 \le t \le 1$, $0 \le x \le 1$; $\tilde{Q}(t,x) = [(z^0, z) | z^0 \ge 0,\ z = 0]$ if $0 \le t < 1$, $0 \le x \le 1$, $t + x < 1$; $\tilde{Q}(t,x) = [(z^0, z) | z^0 \ge -x^{-1},\ z = 0]$ if $0 \le t < 1$, $0 < x \le 1$, $t + x \ge 1$; $\tilde{Q}(1,x) = R \times \{0\}$. Then, all sets $Q(t,x)$ are closed half straight lines, or lines, and have property (K), and even property (Q) with respect to x everywhere. Let us take $\xi_k(t) = \xi(t) = 0$, $x_k(t) = k^{-1}$, $x(t) = 0$, $0 \le t \le 1$, $\eta_k(t) = 0$ for $0 \le t \le 1 - k^{-1}$, and for $t = 1$, $\eta_k(t) = -k$ for $1 - k^{-1} < t < 1$. Then $\int_0^1 \eta_k(t)\,dt = -1$, $k = 1, 2, \dots$, $i = -1$. For $x(t) = 0$, $0 \le t \le 1$, we must have $\eta(t) \ge 0$ for all $0 \le t < 1$. Hence, $\int_0^1 \eta(t)\,dt \ge 0$, and the last relation (10.7.4) cannot be satisfied.

Remark 5. Note that in the lower closure theorem (10.7.i) *no property* (Q) was required for the given sets $\tilde{Q}(t,x)$ in R^{n+1}. This is possible because we have assumed the weak convergence in $L_1(G)$ of the functions $\xi_k(t)$, $t \in G$, $k = 1, 2, \dots$, and this implies, by the equivalence theorem (10.3.i), implication (a) $\Rightarrow$ (c), that there exists some function $\phi(\zeta)$, $0 \le \zeta < +\infty$, bounded below, with $\phi(\zeta)/\zeta \to +\infty$ as $\zeta \to +\infty$, and $\int_G \phi(|\xi_k(t)|)\,dt \le M$ for all k. In turn, having assumed that the functions $\xi_k(t)$, $t \in G$, $k = 1, 2, \dots$, have their values in a finite dimensional space R^r, Carathéodory's theorem (8.4.iii) holds, and by our theorem (10.5.i), the auxiliary sets $\tilde{Q}'^*(t,x) \subset R^{n+2}$ have property (Q) with respect to x in $A(t)$ (for almost all t).

We have shown that the proof of (10.7.i) is very much simplified if we know that the original sets $\tilde{Q}(t,x)$ already have property (Q) with respect to x.

The following statement, easier to prove than (10.7.i), will also be used:

10.7.ii. *With the same notation as for* (10.7.i), *let G be measurable and of finite measure, and let us assume that, for almost all $\bar{t} \in G$, the sets $\tilde{Q}(\bar{t},x)$, $x \in A(\bar{t})$, are closed and convex. Let $\xi(t)$, $x(t)$, $\eta_k(t)$, $\bar{\eta}_k(t)$, $\xi_k(t)$, $\bar{\xi}_k(t)$, $\lambda(t)$, $\lambda_k(t)$, $t \in G$, $k = 1, 2, \dots$, be measurable functions, $\xi, \xi_k, \bar{\xi}_k \in (L_1(G))^r$, $\eta_k, \bar{\eta}_k \in L_1(G)$, with $\xi_k(t) \to \xi(t)$ weakly in $(L_1(G))^r$, $\delta_k(t) = \xi_k(t) - \bar{\xi}_k(t) \to 0$ weakly in $(L_1(G))^r$, $\delta_k^0(t) = \eta_k(t) - \bar{\eta}_k(t) \to 0$ weakly in $L_1(G)$ as $k \to \infty$, $x(t) \in A(t)$, $(\bar{\eta}_k(t), \bar{\xi}_k(t)) \in \tilde{Q}(t, x(t))$, $t \in G$ (a.e.), $k = 1, 2, \dots$,*

$$-\infty < i = \liminf_{k \to \infty} \int_G \eta_k(t)\,dt < +\infty,$$

$$\eta_k(t) \ge \lambda_k(t), \qquad \lambda, \lambda_k \in L_1(G), \qquad \lambda_k \to \lambda \quad \textit{weakly in } L_1(G).$$

Then there is a function $\eta(t)$, $t \in G$, $\eta \in L_1(G)$, such that

$$(\eta(t), \xi(t)) \in \tilde{Q}(t, x(t)), \quad t \in G \text{ (a.e.)}, \qquad \int_G \eta(t)\,dt \le i.$$

Proof. Here $\xi_k \to \xi$, $\delta_k \to 0$ weakly in $(L_1(G))^r$, $\lambda_k \to \lambda$ weakly in $L_1(G)$; thus $(\xi_k, \delta_k, \delta_k^0, \lambda_k) \to (\xi, 0, 0, \lambda)$ weakly in $(L_1(G))^{2r+2}$, and by (10.1.i) there is a set of real numbers $c_{Nk} \ge 0$, $k = 1, \dots, N$, $N = 1, 2, \dots$, $\sum_{k=1}^N c_{Nk} = 1$, such that, if

$$(\xi_N^*(t), \delta_N^*(t), \delta_N^{0*}(t), \lambda_N^*(t)) = \sum_{k=1}^N c_{Nk}(\xi_k, \delta_k, \delta_k^0, \lambda_k), \quad t \in G, \quad N = 1, 2, \dots,$$

then $(\xi_N^*, \delta_N^*, \delta_N^{0*}, \lambda_N^*) \to (\xi, 0, 0, \lambda)$ strongly in $(L_1(G))^{2r+2}$. Then, there is also a subsequence N_λ such that $(\xi_{N_\lambda}^*, \delta_{N_\lambda}^*, \delta_{N_\lambda}^{0*}, \lambda_{N_\lambda}^*) \to (\xi, 0, 0, \lambda)$ pointwise a.e. in G as $\lambda \to \infty$. Let T_0 be the set of measure zero where this does not occur, or $\xi(t)$, or $\lambda(t)$ are not finite, or $\tilde{Q}(t, x(t))$ is not convex or not closed. If $\bar{\xi}_N^*(t)$, $\bar{\eta}_N^*(t)$ denote the functions $\sum_{k=1}^N c_{Nk}\bar{\xi}_k$ and $\sum_{k=1}^N c_{Nk}\bar{\eta}_k$ respectively, then for $t \in G - T_0$ and $N = N_\lambda$ we have $\bar{\xi}_N^* = \xi_N^* - \delta_N^* \to \xi(t)$, $\bar{\eta}_N^*(t) = \eta_N^*(t) - \delta_N^{0*}(t) \ge \lambda_N^*(t) - \delta_N^{0*}(t)$ with $\lambda_N^* - \delta_N^{0*} \to \lambda$ strongly in $L_1(G)$. By the remark after Fatou's Lemma (8.7.i), we know that $\eta(t) = \liminf_{\lambda\to\infty} \bar{\eta}_{N_\lambda}(t)$ is L-integrable in G with $\eta(t) \geq \lambda(t)$, $\int_G \eta(t)\,dt \leq i$. On the other hand, $(\bar{\eta}_k(t), \bar{\xi}_k(t)) \in \tilde{Q}(t, x(t))$ for all t and k, where $\tilde{Q}(t, x(t))$ is a closed convex set. Thus $(\bar{\eta}_N^*(t), \bar{\xi}_N^*(t)) \in \tilde{Q}(t, x(t))$ for all $t \in G - T_0$ and all N. For $N = N_\lambda$ and as $\lambda \to \infty$ we have now $(\eta(t), \xi(t)) \in \tilde{Q}(t, x(t))$, and this relation holds a.e. in G. □

10.8 Lower Semicontinuity in the Topology of Weak Convergence

A. Lower Semicontinuity of Integrals under Weak Convergence

From the lower closure theorem (10.7.i) in terms of orientor fields we shall now immediately derive a lower semicontinuity theorem (10.8.i) for the integral

$$(10.8.1)\quad I = \int_G F_0(t, x(t), \xi(t))\,dt, \qquad \xi(t) \in Q(t, x(t)), \qquad t \in G \text{ (a.e.)},$$

directly in terms of the function F_0.

The statement (10.8.i) actually concerns lower semicontinuity properties of multiple integrals. This added generality does not complicate the statement of the theorem.

We shall use essentially the same notation as before; in particular, the independent variable t, which is a ν-vector $t = (t^1, \ldots, t^\nu)$, $\nu \geq 1$, ranges over a bounded domain G of the t-space R^ν. For every $t \in G$ let $A(t)$ be a nonempty subset of the x-space R^n, $x = (x^1, \ldots, x^n)$, and let A be the set $A = [(t, x) | t \in G,\ x \in A(t)] \subset R^{\nu+n}$, whose projection on the t-space is G. For every $(t, x) \in A$ let $Q(t, x)$ be a given subset of the z-space R^r, $z = (z^1, \ldots, z^r)$, and let M be the set $M = [(t, x, z) | (t, x) \in A,\ z \in Q(t, x)] \subset R^{\nu+n+r}$. Let $F_0(t, x, z)$ be a given real valued function defined on M, and for every $(t, x) \in A$ let $\tilde{Q}(t, x)$ denote the set $\tilde{Q}(t, x) = [(z^0, z) | z^0 \geq F_0(t, x, z),\ z \in Q(t, x)]$. We may extend F_0 in $R^{\nu+n+r}$ by taking $F_0(t, x, z) = +\infty$ for $(t, x, z) \in R^{\nu+n+r} - M$. Then F_0 is said to be an extended function. For most applications it is sufficient to assume.

C. A closed, M closed, $F_0(t, x, z)$ continuous on M.

However, for the proof that follows, the following rather general assumption suffices:

C*. For every $\varepsilon > 0$ there is a compact subset K of G such that (a) meas$(G - K) < \varepsilon$, (b) the extended function $F_0(t, x, z)$ restricted to $K \times R^{n+r}$ is B-measurable, and (c) for almost all $\bar{t} \in G$ the extended function $F_0(\bar{t}, x, z)$ of (x, z) has values finite or $+\infty$, and is lower semicontinuous in R^{n+r}.

Under hypothesis (C), and for any pair of measurable functions $x(t) = (x^1, \ldots, x^n)$, $\xi(t) = (\xi^1, \ldots, \xi^r)$, $t \in G$, with $x(t) \in A(t)$, $\xi(t) \in Q(t, x(t))$, $t \in G$

(a.e.), then the function $F_0(t, x(t), \xi(t))$ is finite a.e. in G and measurable in G (cf. Hahn and Rosenthal [I, p. 122]). Under hypothesis (C*) and measurable functions $x(t)$, $\xi(t)$, $t \in G$, as above, again $F_0(t, x(t), \xi(t))$ is measurable in K. Since $\text{meas}(G - K) < \varepsilon$ and ε is arbitrary, we conclude that $F_0(t, x(t), \xi(t))$ is measurable in G.

Remark 1. Under hypothesis (C*) we may simply assume that F_0 is a given extended function in $R^{\nu+n+r}$, whose values for almost all $t \in G$ are finite or $+\infty$, and that for almost all $\bar{t} \in G$ the set $A(\bar{t}) = [\bar{x} \in R^n | F_0(\bar{t}, \bar{x}, z) \not\equiv +\infty] \neq \emptyset$. For any $(\bar{t}, \bar{x})$ let $Q(\bar{t}, \bar{x}) = [z \in R^r | F_0(\bar{t}, \bar{x}, z) \neq +\infty]$. Then A is any set of points (t, x) whose projection on the t-space is G and whose sections for almost all $t \in G$ are the sets $A(t)$.

Remark 2. If $F_0(t, x, z)$ is a Carathéodory function on $G \times R^{n+r}$, that is, F_0 is measurable in t for every (x, z), and continuous in (x, z) for almost all t, then, by (8.3.v), for every $\eta > 0$ there is a compact set $K \subset G$ with $\text{meas}(G - K) < \eta$ such that F_0 is continuous in $K \times R^{n+r}$. This shows that Carathéodory functions F_0 certainly have property (C*). A condition slightly more restrictive than (C*) is often used for the same purpose, namely (C_r^*), the same as (C*) where instead of (b), (c), $F_0(t, x, z)$ is required to be lower semicontinuous in $K \times R^{n+r}$ as a function of (t, x, z). This more restrictive condition (C_r^*), as proved by Ekeland and Temam [I, p. 216] and by Rockafellar [4, p. 176], is an equivalent form for the "normality" conditions required by these two authors.

In Section 10.7, as a comment on the lower closure theorem (10.7.i), we noted that the abstract condition (10.7.3) is certainly satisfied under the practical and easily verifiable alternative conditions (Λ_i). Here, in terms of the functional (10.8.1), some of the assumptions (Λ_i) are replaced by the following straightforward alternative assumptions (L_i):

(L_1) There is a real valued function $\psi(t) \geq 0$, $t \in G$, $\psi \in L_1(G)$, such that $F_0(t, x, u) \geq -\psi(t)$ for $(t, x, u) \in M$ and almost all t.

(L_2) There is a real valued function $\psi(t) \geq 0$, $t \in G$, $\psi \in L_1(G)$ and a constant $C \geq 0$ such that $F_0(t, x, u) \geq -\psi(t) - C|u|$ for $(t, x, u) \in M$ and almost all t.

(L_3) There is a real valued function $\psi(t) \geq 0$, $t \in G$, $\psi \in L_1(G)$ and an r-vector function $\phi(t) = (\phi_1, \ldots, \phi_r)$, $t \in G$, $\phi_i \in L_\infty(G)$, such that $F_0(t, x, u) \geq -\psi(t) - (\phi(t), u)$ for $(t, x, u) \in M$ and almost all t.

(L_4) There are constants $\alpha \geq \beta$ real and $\gamma > 0$ such that (a) for every $(t, x) \in A$ the set $Q(t, x)$ contains the ball $|z| \leq \gamma$ in R^r; and (b) $F_0(t, x, u) \geq \beta$ for all $(t, x) \in A$, $|u| \leq \gamma$, and $F_0(t, x, 0) \leq \alpha$. Here we assume explicitly that the sets $Q(t, x)$ are convex and that $F_0(t, x, u)$ is convex in u.

Under any one of these hypotheses (L_i), and for all measurable functions $x(t)$, $t \in G$, and L-integrable $\xi(t)$, $t \in G$, as before with $x(t) \in A(t)$, $\xi(t) \in Q(t, x(t))$, $t \in G$ (a.e.), then $F_0(t, x(t), \xi(t))$ is not only measurable in G but also not below some L-integrable function in G. Indeed, under (L_1) we have $F_0 \geq -\psi$; under (L_2) we have $F_0 \geq -\psi(t) - C|\xi(t)|$; under ($L_3$) we have $F_0 \geq -\psi(t) - (\phi(t), \xi(t)) \geq -\psi(t) - |\phi(t)|\,|\xi(t)|$, so $F_0 \geq -\psi(t) - L|\xi(t)|$, where $L = \text{ess sup}|\phi(t)|$. Under ($L_4$), for every $z \in R^r$, $z \in Q(t, x)$, $|z| \geq \gamma$ (if any), we take $z_1 = \gamma z/|z|$, so that $|z_1| = \gamma$ and $F_0(t, x, z_1) \geq \beta$. Moreover, for $\sigma = \gamma/|z|$,

$0 < \sigma < 1$, and by the convexity of Q and F_0 we have

$$F_0(t,x,z_1) \leq (1-\sigma)F_0(t,x,0) + \sigma F_0(t,x,z),$$

or

$$\begin{aligned} F_0(t,x,z) &\geq \sigma^{-1}F_0(t,x,z_1) - \sigma^{-1}(1-\sigma)F_0(t,x,0) \\ &\geq \gamma^{-1}|z|\beta - \gamma^{-1}|z|(1-\gamma|z|^{-1})\alpha \\ &= \alpha - \gamma^{-1}(\alpha-\beta)|z|, \end{aligned}$$

and we are in the situation (L_2). Note that under any of the hypotheses (L_i) with x measurable and $\xi \in (L_1(G))^r$, the Lebesgue integral $\int_G F_0(t, x(t), \xi(t))\,dt$ exists, finite or $+\infty$.

For G bounded and closed, A closed, $x(t)$, $t \in G$, continuous, and $x_k(t)$, $t \in G$, $k = 1, 2, \ldots$, converging uniformly to x in G, then it is enough to verify the conditions above only for $(t,x) \in A \cap \Gamma_\delta$, $(t,x,u) \in M$, where Γ_δ is a closed bounded neighborhood of the graph Γ of x in $R^{\nu+n}$. In particular, if F_0 is continuous, then condition (b) of (L_4) is always satisfied, since F_0 is bounded in the compact set $(A \cap \Gamma_\delta) \times [|u| \leq \gamma]$, say $|F_0| \leq c$, and we can take $\alpha = c$ and $\beta = -c$. More particularly, if $Q(t,x) = R^n$ for every (t,x), $M = A \times R^n$, then R^n certainly contains the ball $|u| \leq \gamma$, condition (a) of (L_4) is also satisfied, and (L_4) itself is satisfied.

10.8.i (A Lower Semicontinuity Theorem). *Let condition* (C) *or* (C*) *be satisfied, and assume that for almost all* $t \in G$ *and all* $x \in A(t)$ *the extended function* $F_0(t,x,z)$ *be convex in* z *(in* R^r *and hence the set* $\tilde{Q}(t,x)$ *is convex). Assume that any one of the conditions* (L_i) *holds. Let* $\xi(t)$, $x(t)$, $\xi_k(t)$, $x_k(t)$, $t \in G$, $k = 1, 2, \ldots$, *be measurable functions*, ξ, $\xi_k \in (L_1(G))^r$, *such that* $x_k \to x$ *in measure in* G, $\xi_k \to \xi$ *weakly in* $(L_1(G))^r$ *as* $k \to \infty$, *and* $x_k(t) \in A(t)$, $\xi_k(t) \in Q(t, x_k(t))$, $t \in G$ (a.e.), $k = 1, 2, \ldots$. *Then*, $x(t) \in A(t)$, $\xi(t) \in Q(t, x(t))$, $t \in G$ (a.e.), *and*

$$\int_G F_0(t, x(t), \xi(t))\,dt \leq \liminf_{k\to\infty} \int_G F_0(t, x_k(t), \xi_k(t))\,dt. \tag{10.8.2}$$

Proof. The integrals above exist, finite or $+\infty$. Let i denote the second member of (10.8.2). If $i = +\infty$ there is nothing to prove. Assume $i < +\infty$. Let us prove that $i > -\infty$. Under condition (L_1) this is evident. Since $\xi_k \to \xi$ weakly in $(L_1(G))^r$, then $\|\xi_k\|_{L_1}$ is bounded, say $\|\xi_k\|_{L_1} \leq N$. Under condition (L_2) then $F_0 \geq -\psi - C\xi_k$, and $\int_G F_0\,dt \geq -\int_G \psi\,dt - CN$. Under conditions (L_3) we have $F_0 \geq -\psi - L|\xi_k|$, where $L = \text{ess sup}|\psi|$, and then $\int_G F_0\,dt \geq -\int \psi\,dt - LN$. Under condition (L_4) we have again $F_0 \geq \gamma^{-1}(\beta - \alpha)|\xi_k| + \alpha$, and again $\int_G F_0\,dt \geq \gamma^{-1}(\beta - \alpha)N + \alpha \text{ meas } G$.

By taking $\eta_k(t) = F_0(t, x_k(t), \xi_k(t))$, $t \in G$, we have now $(\eta_k(t), \xi_k(t)) \in \tilde{Q}(t, x_k(t))$ with $\tilde{Q}(t,x) = [(z^0, z) \mid z^0 \geq F_0(t,x,z),\ z \in Q(t,x)]$. In order to apply (10.7.i) with cl $A(t)$ replacing $A(t)$, we need only prove that, for almost all $\bar{t} \in G$, these sets $\tilde{Q}(\bar{t}, x)$ have property (K) with respect to x in the closed set cl $A(\bar{t})$. Indeed, under condition (C) and for all $\bar{t} \in G$, the set $M(\bar{t}) = [(x,z) \mid x \in A(\bar{t}),\ z \in Q(\bar{t}, x)]$, the section of M with the hyperspace $t = \bar{t}$, is closed, and then

$$\tilde{M}(\bar{t}) = [(x,y,z) \mid x \in A(\bar{t}),\ y \geq F_0(\bar{t},x,z),\ z \in Q(\bar{t},x)]$$

is closed, since $F_0(\bar{t}, x, z)$ is continuous on the closed set $M(\bar{t})$ (hence lower semicontinuous on R^{n+r}), and $\tilde{M}(\bar{t})$ is closed because of (8.5.v). The closed set $\tilde{M}(\bar{t})$ is the graph of the sets $\tilde{Q}(\bar{t}, x)$ as x describes the closed set $A(\bar{t})$, and then the map $x \to Q(\bar{t}, x)$ has property (K) (with respect to x) on $A(\bar{t})$ by virtue of (8.5.iii) and subsequent Remark 2. From the orientor field relation $x_k(t) \in A(t)$, $(\eta_k(t), \xi_k(t)) \in \tilde{Q}(t, x_k(t))$, $t \in G$ (a.e.), $k = 1, 2, \ldots$, by applying (10.7.i) we derive that there is an L-integrable function $\eta(t)$, $t \in G$, such that $x(t) \in A(t)$, $(\eta(t), \xi(t)) \in \tilde{Q}(t, x(t))$, $t \in G$ (a.e.), and $\int_G \eta(t)\,dt \leq i$.

Under condition (C*) and for almost all $\bar{t} \in G$, the set

$$A(\bar{t}) = [x \in R^n \,|\, F_0(\bar{t}, x, z) \not\equiv +\infty]$$

is not empty by hypothesis, and the sets $Q(\bar{t}, x) = [z \in R^r \,|\, F_0(\bar{t}, x, z) \neq +\infty]$ are not empty for $x \in A(\bar{t})$. Now cl $A(\bar{t})$ is not empty and closed, but the sets $Q(\bar{t}, x)$ for $x \in (\text{cl } A(\bar{t}) - A(\bar{t}))$ are empty. Again, for almost all $\bar{t} \in G$, the extended function $F_0(\bar{t}, x, z)$ is lower semicontinuous in R^{n+r}; hence by (8.5.v) the sets $\tilde{Q}(\bar{t}, x) = \text{epi } F_0(\bar{t}, x, z)$ are closed. Again by (8.5.iii) and subsequent Remark 2, the map $x \to \tilde{Q}(\bar{t}, x)$ has property (K) (with respect to x) on the closed set cl $A(\bar{t})$, (and this involves also the empty sets $\tilde{Q}(\bar{t}, x)$ for $x \in (\text{cl } A(t) - A(t))$). From the orientor field relations

$$x_k(t) \in A(t), \quad (\eta_k(t), \xi_k(t)) \in \tilde{Q}(t, x_k(t)), \qquad t \in G \text{ (a.e.)}, \quad k = 1, 2, \ldots,$$

by applying (10.7.i) with cl $A(t)$ replacing $A(t)$ we derive that there is an L-integrable function $\eta(t)$, $t \in G$, such that $x(t) \in \text{cl } A(t)$, $(\eta(t), \xi(t)) \in \tilde{Q}(t, x(t))$, $t \in G$ (a.e.), and $\int_G \eta(t)\,dt \leq i$. Now for $x \in (\text{cl } A(t) - A(t))$ the set $\tilde{Q}(t, x)$ is empty. Thus, for almost all $t \in G$ we must have $\tilde{Q}(t, x(t))$ nonempty, hence $x(t) \in A(t)$, $t \in G$, (a.e.).

In any case $F_0(t, x(t), \xi(t)) \leq \eta(t)$, $t \in G$ (a.e.), and we know that $F_0(t, x(t), \xi(t))$ is measurable and not less than some L-integrable function in G. That is, F_0 is between two L-integrable functions, and then L-integrable. Moreover

$$\int_G F_0(t, x(t), \xi(t))\,dt \leq \int_G \eta(t)\,dt \leq i.$$

Theorem (10.8.i) is thereby proved. □

Remark 3. Note that, under the conditions of (10.8.i), the function $F_0(t, x(t), \xi(t))$ is certainly measurable, and because of the conditions (L_i), the Lebesgue integral $I[x, \xi] = \int_G F_0\,dt$ on the left hand side of (10.8.2) is either finite or $+\infty$. Theorem (10.8.i) can be completed with the statement that, if $I[x, \xi] = +\infty$, then the relation (10.8.2) is still valid in the sense that on the right hand side necessarily we have $\lim I[x_k, \xi_k] = +\infty$ as $k \to +\infty$. Indeed, otherwise, there would be a subsequence, say still $[k]$, with $-\infty < i = \lim I[x_k, \xi_k] < +\infty$, and by (10.7.i) there would be an L-integrable function $\eta(t)$, $t \in G$, with $\int_G \eta(t)\,dt \leq i$, and $\eta(t) \geq F_0(t, x(t), \xi(t))$, a.e. in G, a contradiction.

As a particular case, the theorem (10.8.i) contains, for $v = 1$, the case of integrals of the form

$$\int_{t_1}^{t_2} F_0(t, x(t), \xi(t))\,dt.$$

We need consider the situation in which $x_k(t)$, $\xi_k(t)$, $t_{1k} \le t \le t_{2k}$, may be defined on different intervals, $x(t)$, $\xi(t)$, $t_1 \le t \le t_2$, and that $t_{1k} \to t_1$, $t_{2k} \to t_2$ as $k \to \infty$. We shall assume that $\xi_k \to \xi$ weakly in L_2, and by this we understand that all ξ, ξ_k are extended to some large $[t_0, T]$ (containing all $[t_{1k}, t_{2k}]$) by taking them equal to zero outside their original intervals of definition, and that $\xi_k \to \xi$ weakly in $(L_1[t_0, T])^r$. We shall also assume that $x_k \to x$ in measure, and by this we understand that we have performed an analogous extension, and that $x_k \to x$ in measure in $[t_0, T]$. Alternatively, we may take $x(t) = x(t_1)$ for $t \le t_1$, $x(t) = x(t_2)$ for $t \ge t_2$, and analogously for x_k. This extension is more natural when all x, x_k are continuous in their intervals of definition and the convergence is in the ρ-metric. With these conventions, which will be used from now on, the following theorem holds.

10.8.ii (A LOWER SEMICONTINUITY THEOREM FOR AC TRAJECTORIES IN THE TOPOLOGY OF THE WEAK CONVERGENCE OF THE DERIVATIVES). *Let condition* (C), *or* (C*) *be satisfied, and assume that for almost all t and all $x \in A(t)$ the extended function $F_0(t, x, z)$ be convex in z (in R^n and hence the set $Q(t, x)$ is convex). Assume that any of the conditions* (L_i) *holds. Let $x(t)$, $\xi(t)$, $t_1 \le t \le t_2$, $x_k(t)$, $\xi_k(t)$, $t_{1k} \le t \le t_{2k}$, $k = 1, 2, \dots$, be measurable functions with $t_{1k} \to t_1$, $t_{2k} \to t_2$ as $k \to \infty$, and $x_k(t) \in A(t)$, $\xi_k(t) \in Q(t, x_k(t))$, $t \in [t_{1k}, t_{2k}]$* (a.e.). *Assume that $x_k \to x$ in measure, and $\xi_k \to \xi$ weakly in L_1. Then $x(t) \in A(t)$, $\xi(t) \in Q(t, x(t))$, $t \in [t_1, t_2]$* (a.e.), *and*

$$\int_{t_1}^{t_2} F_0(t, x(t), \xi(t))\, dt \le \liminf_{k \to \infty} \int_{t_{1k}}^{t_{2k}} F_0(t, x_k(t), \xi_k(t))\, dt.$$

In particular, if the functions x_k are AC *in $[t_{1k}, t_{2k}]$ and converge in the ρ-metric to a continuous function $x(t)$, $t_1 \le t \le t_2$, if $\xi_k(t) = x_k'(t)$, $t \in [t_{1k}, t_{2k}]$* (a.e.) *and $\xi_k \to \xi$ weakly in L_1, then x is* AC *and $\xi(t) = x'(t)$, $t \in [t_1, t_2]$* (a.e.).

Proof. First we note that, in case of a fixed interval ($t_{1k} = t_1$, $t_{2k} = t_2$ for all k), then (10.8.ii) is an immediate corollary of (10.8.i) with $G = [t_1, t_2]$.

In general, with $t_{1k} \to t_1$, $t_{2k} \to t_2$, we note that for any δ, $0 < \delta < 2^{-1}(t_2 - t_1)$, the interval $[t_1 + \delta, t_2 - \delta]$ is contained in all intervals $[t_{1k}, t_{2k}]$ with k sufficiently large. Under any of the conditions (L_i) we have seen that $F_k = F_0(t, x_k(t), \xi_k(t)) \ge -\psi(t) - C|\xi_k(t)|$ for some constant C. Since ξ_k converges weakly in L_1, we know from (10.3.i) (implication (a) $\Rightarrow$ (b)) that the sequence $[\xi_k]$ is equiabsolutely integrable. Thus, given $\varepsilon > 0$, we can take $\delta > 0$ sufficiently small so that $I_{k1} = \int_{t_{1k}}^{t_1+\delta} F_k\, dt \ge -\varepsilon$, $I_{k2} = \int_{t_2-\delta}^{t_{2k}} F_k\, dt \ge -\varepsilon$ for all k sufficiently large. For $I_k' = \int_{t_1+\delta}^{t_2-\delta} F_k\, dt$, we have now, as $k \to \infty$,

$$\begin{aligned} -\varepsilon + \liminf I_k' - \varepsilon &\le \liminf I_{k1} + \liminf I_k' + \liminf I_{k2} \\ &\le \liminf(I_{k1} + I_k' + I_{k2}) = i, \end{aligned}$$

or $\liminf I_k' \le i + 2\varepsilon$. By the above we have now

$$\int_{t_1+\delta}^{t_2-\delta} F_0(t, x(t), \xi(t))\, dt \le \liminf \int_{t_1+\delta}^{t_2-\delta} F_0(t, x_k(t), \xi_k(t))\, dt \le i + 2\varepsilon,$$

where $\delta > 0$ can be taken as small as we want. Again, $F_0(t, x(t), \xi(t)) \geq -\psi(t) - C|\xi(t)|$, an L-integrable function, so the limit of the first integral as $\delta \to 0+$ exists, is finite, and equals the L-integral $I[x]$. Since $\varepsilon > 0$ is arbitrary, we have $I[x] \leq i$. The last part of (10.8.i) follows from the remark that $t_{1k} \to t_1$, $t_{2k} \to t_2$, and if t^* is any point of (t_1, t_2), then $t_{1k} < t^* < t_{2k}$ for all k sufficiently large. Hence, $x_k(t) = x_k(t^*) + \int_{t^*}^{t} \xi_k(\tau)\, d\tau$, and as $k \to \infty$, we obtain $x(t) = x(t^*) + \int_{t^*}^{t} \xi(\tau)\, d\tau$. Thus, x is AC, and $\xi(t) = x'(t)$ a.e. in $[t_1, t_2]$.

Remark 4. Theorems (10.8.i) and (10.8.ii) are clearly corollaries of (10.7.i). If we know that for almost all t the sets $\tilde{Q}(t, x)$ have property (Q) with respect to x, then the simpler version of (10.7.i) is needed as mentioned at the beginning of the proof of (10.7.i).

Remark 5. As an example we see that $I[x] = \int_{t_1}^{t_2} |x'|^p\, dt$, $p \geq 1$, is by (10.8.ii) a lower semicontinuous functional. Note that if x, x_k are AC, $x_k \to x$ in the ρ-metric and $x'_k \to x'$ weakly in L_1, then $I[x] \leq \liminf I[x_k]$, and this is true for any $p \geq 1$. Thus, if x' is not L_p-integrable, then $I[x] = +\infty$, and $\lim I[x_k] = +\infty$ (cf. Remark 3 above).

Remark 6. If A is closed in R^{n+1}, if $Q(t, x) = R^n$ for every (t, x), and thus $M = A \times R^n$, if $F_0(t, x, z)$ is continuous on M and convex in z for every $(t, x) \in A$, then condition (L_4) is certainly satisfied as we mentioned above as a comment on condition (L_4). From (10.8.i) we conclude that the integral (10.8.3) is lower semicontinuous in the topology of the weak convergence of the derivatives (mode (b) of Section 2.14). Thus we have also proved here the sufficiency part in the statement (2.18.i).

B. Continuity of Linear Integrals under Weak Convergence

Before considering the question of the continuity of linear integrals, we shall prove two simple closure theorems for linear differential systems, which we shall also use in Section 11.4.

Let us consider the linear relation

$$y(t) = A_0(t, x(t)) + B(t, x(t))\xi(t),$$
$$y(t) = (y^1, \ldots, y^h), \qquad x(t) = (x^1, \ldots, x^n), \qquad \xi(t) = (\xi^1, \ldots, \xi^r),$$

where as usual all vectors x, y, ξ are thought of as column vectors, where $A_0(t, x)$ is an $h \times 1$ matrix, $B(t, x)$ is an $h \times r$ matrix, and all entries are defined in a subset A of $[t_0, T] \times R^n$. We may assume as usual that

(CL) A is a closed set and all entries are continuous in A. However, for what follows, the following much weaker assumption suffices:

(CL*) For every $\varepsilon > 0$ there is a compact subset K of $[t_0, T]$ such that (a) $\text{meas}([t_0, T] - K) < \varepsilon$, (b) the set $A_K = [(t, x) \in A \mid t \in K]$ is closed, and (c) all entries are continuous in A_K.

For instance the entries of the matrices A and B could be Carathéodory functions in $[t_0, T] \times R^n$, measurable in t for all x, and continuous in x for almost all t.

10.8.iii (A Lemma). *Let $x_k(t)$, $\xi_k(t)$, $t_{1k} \leq t \leq t_{2k}$, $k = 1, 2, \ldots$, and $x(t)$, $\xi(t)$, $t_1 \leq t \leq t_2$, be given functions, x, x_k measurable, ξ, ξ_k L_p-integrable for some*

$p \geq 1$, such that $t_{1k} \to t_1$, $t_{2k} \to t_2$, $x_k \to x$ in measure, $\xi_k \to \xi$ weakly in L_p as $k \to \infty$, and

$$y_k(t) = A_0(t, x_k(t)) + B(t, x_k(t))\xi_k(t), \qquad t \in [t_{1k}, t_{2k}] \text{ (a.e.)}, \qquad k = 1, 2, \ldots.$$

Assume that for some functions $\phi(t) \geq 0$, $\phi \in L_1$, and $\psi(t) \geq 0$, $\psi \in L_q$, $1/p + 1/q = 1$, we have

$$|A_0(t,x)| \leq \phi(t), \qquad |B(t,x)| \leq \psi(t), \qquad (t,x) \in [t_0, T] \times R^n,$$

with $\psi \in L_\infty$ if $p = 1$, $q = \infty$. Then $y_k \to y$ weakly in L_1 with

$$y(t) = A(t, x(t)) + B(t, x(t))\xi(t), \qquad t \in [t_1, t_2] \text{ (a.e.)}. \tag{10.8.3}$$

As agreed upon in Section (10.8A) we take $\xi(t)$ and all entries of the matrices $A_0(t, x(t))$, $B(t, x(t))$ to be equal zero outside $[t_1, t_2]$. Analogously we take $\xi_k(t)$ and all entries of the matrices $A(t, x_k(t))$, $B(t, x_k(t))$ to be equal zero outside $[t_{1k}, t_{2k}]$.

Proof. Since $|A_0(t, x_k(t))| \leq \phi(t)$, $\phi \in L_1$, it is only an exercise to prove that $A_0(t, x_k(t)) \to A_0(t, x(t))$ pointwise as well as strongly in L_1. Analogously, since $|B(t, x_k(t))| \leq \psi(t)$, $\psi \in L_p$, it is an exercise to show that $B(t, x_k(t)) \to B(t, x(t))$ pointwise as well as strongly in L_p. Finally, $B(t, x_k(t))\xi_k(t) \to B(t, x(t))\xi(t)$ weakly in L_1. Thus, $y_k \to y$ weakly in L_1, and (10.8.3) holds. Also note that $\int_{t_{1k}}^{t_{2k}} y_k(t)\,dt \to \int_{t_1}^{t_2} y(t)\,dt$. □

Let us consider now a system of linear differential equations of the form

$$\begin{aligned} x'(t) &= A_0(t, x(t)) + B(t, x(t))\xi(t), \\ x(t) &= (x^1, \ldots, x^n), \quad \xi(t) = (\xi^1, \ldots, \xi^r), \end{aligned} \tag{10.8.4}$$

where $A_0(t, x)$ is an $n \times 1$ matrix and $B(t, x)$ is an $n \times r$ matrix whose entries are defined on a subset A of $[t_0, T] \times R^n$ under the same general assumptions (CL) or (CL*). We assume that for given functions $\xi_k(t)$, $t_{1k} \leq t \leq t_{2k}$, $\xi_k \in L_p$, system (10.8.4) has certain AC solutions $x_k(t)$, $t_{1k} \leq t \leq t_{2k}$, and we prove under mild assumptions that there is subsequence, say still k, such that the sequence x_k, ξ_k has limit elements x, ξ satisfying (10.8.4).

10.8.iv (A LEMMA). *Let $x_k(t)$, $\xi_k(t)$, $t_0 \leq t_{1k} < t_{2k} \leq T$, $k = 1, 2, \ldots$, be given functions, x_k AC, $\xi_k \in L_p$ for some $p > 1$, $(t, x_k(t)) \in A$ for $t \in [t_{1k}, t_{2k}]$ (a.e.), satisfying (10.8.4), such that $\|\xi_k\|_p \leq \mu$ and $|x_k(t_k^*)| \leq N$ for some $t_k^* \in [t_{1k}, t_{2k}]$ and constants N, μ. Assume that for some constants $c, C \geq 0$, $p, q > 1$, $1/p + 1/q = 1$, and scalar functions $\phi(t), \psi(t) \geq 0$, $\phi \in L_1[t_0, T]$, $\psi \in L_q[t_0, T]$, we have*

$$|A_0(t,x)| \leq \phi(t) + c|x|, \quad |B(t,x)| \leq \psi(t) + C|x|.$$

Then there are functions $x(t)$, $\xi(t)$, $t_1 \leq t \leq t_2$, $x \in$ AC, $\xi \in L_p[t_1, t_2]$, and a subsequence, say still $[k]$, such that $t_{1k} \to t_1$, $t_{2k} \to t_2$, $x_k \to x$ in the ρ-metric, $\xi_k \to \xi$ weakly in L_p, $x_k' \to x'$ weakly in L_1, and x, ξ satisfy (10.8.4).

The same statement holds for $p = 1$, $q = \infty$, provided $\psi \in L_\infty$, and the condition $\|\xi_k\|_p \le \mu$ is replaced by the assumption that the sequence $[\xi_k(t)]$ is equiabsolutely integrable.

Proof. First, let us take a subsequence, say still $[k]$, such that $t_{1k} \to t_1$, $t_{2k} \to t_2$, $t_k^* \to t^*$, $x_k(t_k^*) \to x^*$ as $k \to \infty$. Now, since the sequence $\|\xi_k\|_p$ is bounded, there is some function $\xi(t)$, $t_1 \le t \le t_2$, and a further subsequence, say still k, such that $\xi_k \to \xi$ weakly in L_p. Now we have

$$\begin{aligned} x_k'(t) &= A_0(t, x_k(t)) + B(t, x_k(t))\xi_k(t), \\ |x_k'(t)| &\le \phi(t) + c|x_k(t)| + (\psi(t) + C|x_k(t)|)|\xi_k(t)| \\ &= (\phi(t) + \psi(t)|\xi_k(t)|) + (c + C|\xi_k(t)|)|x_k(t)|, \end{aligned} \tag{10.8.5}$$

where

$$\begin{aligned} \|\phi + \psi|\xi_k|\|_1 &\le \|\phi\|_1 + \|\psi\|_q\|\xi_k\|_p \le \|\phi\|_1 + \|\psi\|_q\mu = M_1, \\ \|c + C|\xi_k(t)|\|_p &\le c(T - t_0)^{1/p} + C\|\xi_k\|_p \le c(T - t_0)^{1/p} + C\mu = M_2. \end{aligned}$$

Thus

$$\begin{aligned} x_k(t) &= x_k(t_k^*) + \int_{t_k^*}^t x_k'(\tau)\,d\tau, \qquad |x_k(t_k^*)| \le N, \\ |x_k(t)| &\le N + M_1 + \int_{t_k^*}^t (c + C|\xi_k(\tau)|)|x_k(\tau)|\,d\tau, \end{aligned} \tag{10.8.6}$$

and by Gronwall's lemma (18.1.i),

$$\begin{aligned} |x_k(t)| &\le (N + M_1)\exp\left(\int_{t_k^*}^t (c + C|\xi_k(\tau)|\right)d\tau \\ &\le (N + M_1)\exp(c(T - t_0) + C(T - t)^{1/q}\mu) = M_3. \end{aligned}$$

Consequently, $|A_0(t, x_k(t))| \le \phi(t) + cM_3$, $|B(t, x_k(t))| \le \psi(t) + CM_3$, and

$$|x_k'(t)| \le (\phi(t) + cM_3) + (\psi(t) + CM_3)|\xi_k(t)|,$$

where $\phi \in L_1$, $\psi \in L_q$, $\xi_k \in L_p$, hence $x_k' \in L_1$ with norm and

$$\|x_k'\|_1 \le \|\phi\|_1 + cM_3(T - t_0) + (\|\psi\|_q + CM_3(T - t_0)^{1/q})\mu = M_4.$$

Moreover, x_k' is equiabsolutely integrable. Indeed if E is any measurable subset of $[t_{1k}, t_{2k}]$ then

$$\int_E |x_k'(t)|\,dt \le \int_E \phi\,dt + cM_3 \text{ meas } E + \left(\int_E (\psi(t) + CM_3)^q\,dt\right)^{1/q}\mu,$$

and the second member approaches zero as meas $E \to 0$ uniformly with respect to k. By the Dunford–Pettis theorem (10.3.i), there is an L_1-integrable function $\sigma(t)$ and a subsequence, say still $[k]$, such that $x_k' \to \sigma$ weakly in L_1. Then the sequence $[x_k]$ is equibounded and equiabsolutely continuous (10.2.i), and by Ascoli's theorem (9.1.i) there are an AC function $x(t)$, $t_1 \le t \le t_2$, and a further subsequence, say still $[k]$, such that $x_k \to x$ in the

ρ-metric. From (10.8.6) we derive now that $x(t) = x^* + \int_{t^*}^t \sigma(\tau)\,d\tau$, and hence $x'(t) = \sigma(t)$ a.e. Now we see that in the first relation (10.8.5) we have $|A_0(t, x_k(t))| \le \phi(t) + cM_3$, $|B(t, x_k(t))| \le \psi(t) + CM_3$, with $\phi(t) + cM_3 \in L_1$, $\psi(t) + CM_3 \in L_q$, and by (10.8.iii) we derive that $x_k' \to x'$ weakly in L_1 and hence $x'(t) = A(t, x(t)) + B(t, x(t))\xi(t)$, $t \in [t_1, t_2]$ (a.e.). For $p = 1$, and $[\xi_k]$ equiabsolutely integrable, by Dunford–Pettis (10.3.i) there is a subsequence, say still $[k]$, such that ξ converges weakly in L_1. The proof is now analogous. □

Remark. By a theorem of Krasnoselskii ([I, p. 27] and [1]), the requirement $|A_0(t,x)| \le \phi(t) + c|x|$, $\phi \in L_1$, is the necessary and sufficient condition in order that $A_0(\cdot, x(\cdot))$ be L_1 whenever $x(\cdot)$ is L_1. Analogously, the requirement $|B(t,x)| \le \psi(t) + C|x|$, $\psi \in L_q$, $1/p + 1/q = 1$, is the necessary and sufficient condition in order that $B(\cdot, x(\cdot))$ is L_q whenever $x(\cdot)$ is L_q.

We consider now linear integrals of the form

$$I[x, \xi] = \int_{t_1}^{t_2} \left(A_0(t, x) + \sum_{i=1}^{n} A_i(t, x)\xi_i \right) dt,$$

where the scalar functions $A_i(t, x)$, $i = 0, 1, \ldots, n$, are defined in a subset A of $[t_0, T] \times R^n$, under the general assumptions (CL) or (CL*). For $\xi(t) = x'(t)$, these integrals reduce to the usual linear integrals

$$I[x] = \int_{t_1}^{t_2} \left(A_0(t, x) + \sum_{i=1}^{n} A_i(t, x)x'^i \right) dt.$$

10.8.v (A Continuity Theorem for Linear Integrals). *Let $x_k(t)$, $\xi_k(t)$, $t_{1k} \le t \le t_{2k}$, $k = 1, 2, \ldots$, and $x(t)$, $\xi(t)$, $t_1 \le t \le t_2$, be given functions, x, x_k continuous, ξ, ξ_k L_p integrable for some $p \ge 1$, such that $x_k \to x$ in the ρ-metric, $\xi_k \to \xi$ weakly in L_p. Under condition* (CL) *we have $I[x_k, \xi_k] \to I[x, \xi]$ as $k \to \infty$.* (b) *Under condition* (CL*) *the same is true even if we know only that x, x_k are measurable, and $x_k \to x$ in measure, provided the graphs of x and x_k are in a subset A_0 of A where $|A_0(t, x)| \le \phi(t)$, $|A_i(t, x)| \le \psi(t)$, $i = 1, \ldots, n$, for some functions $\phi \in L_1$, $\psi \in L_q$ in $[t_0, T]$ ($\psi(t)$ a constant if $p = 1$, $q = \infty$).*

Thus, in particular, if the functions x, x_k are AC, if $x_k \to x$ in the ρ-metric, and $x_k' \to x'$ weakly in L_p, $p \ge 1$, then $I[x_k] \to I[x]$ under the assumptions (CL). Under the assumptions (CL*) the same is true provided the graphs of x and all x_k lie in a subset A_0 of A where we know that $|A_0(t, x)| \le \phi(t)$, $|A_i(t, x)| \le \psi(t)$, $i = 1, \ldots, n$, $\phi \in L_1$, $\psi \in L_q$ (ψ a constant if $p = 1$, $q = \infty$).

Proof of (10.8.v). Under conditions (CL) and all x, x_k continuous with $x_k \to x$ in the ρ-metric, then for any compact neighborhood N_δ of the graph Γ of x, then all x_k have their graphs in N_δ for k sufficiently large, and A_0 and all A_i are bounded in $A \cap N_\delta$. Then lemma (10.8.iii) applies. Under conditions

(CL*), the specific assumptions of the statement above make it possible to apply (10.8.iii) straightforwardly.

Another proof of (10.8.v) is as follows. By (10.8.ii) we derive that $I[x, \xi]$ is lower semicontinuous. But the same holds for $-I[x, \xi]$, that is, $I[x, \xi]$ is both lower and upper semicontinuous, that is, $I[x, \xi]$ is continuous.

We shall use lemma (10.8.iv) in Section 11.4 to prove existence theorems for linear problems of optimal control.

Exercise

Formulate a limit theorem for the integrals $I[x]$ when the trajectories are solutions of a differential system (10.8.4) containing arbitrary functions $\xi \in L_p$, $p > 1$ (controls).

10.9 Necessary and Sufficient Conditions for Lower Closure

A. Partial Converse of Lower Closure Theorem (10.7.i)

In Theorem (10.7.i) the requirement (10.7.3) not only cannot be disregarded, as the example in Remark 4 of Section 10.7 shows, but it is essential, as we shall see by proving a partial converse of (10.7.i). To this end a few comments and definitions are needed.

First, it is not restrictive to assume in (10.7.i) that $x_k \to x$ pointwise almost everywhere in G, since the present requirement that $x_k \to x$ in measure implies the existence of a subsequence convergent pointwise almost everywhere.

Analogously, in (10.7.3) we could merely require that the sequence $[\lambda_k]$ be only relatively weakly compact in $L_1(G)$, since again we can extract a subsequence which is convergent weakly to some $\lambda \in L_1(G)$.

Concerning the sets $\tilde{Q}(t, x)$, clearly it is enough that we assume them to be convex for $x = x(t)$, $t \in G$, only. Given any real valued function $\eta(t)$, $t \in G$, we take as usual $\eta^+(t) = 2^{-1}(|\eta| + \eta)$, $\eta^-(t) = 2^{-1}(|\eta| - \eta)$, $t \in G$, $\eta^+, \eta^- \geq 0$, $\eta = \eta^+ - \eta^-$.

10.9.i. *In* (10.7.i) *the requirement $\eta_k \geq \lambda_k$, $\lambda_k \to \lambda$ weakly in $L_1(G)$ is equivalent to the requirement that the sequence $\eta_k^-(t)$, $t \in G$, $k = 1, 2, \ldots$, be relatively weakly compact in $L_1(G)$.*

Proof. If $[\eta_k^-]$ is relatively weakly compact in $L_1(G)$, then we can find a subsequence such that η_k^- converges weakly in $L_1(G)$ toward some function $\eta^-(t)$, $t \in G$, $\eta^-(t) \geq 0$, $\eta^- \in L_1(G)$, and then we simply take $\lambda_k = -\eta_k^-$. Conversely, suppose that $\eta_k \geq \lambda_k$, $\lambda_k \to \lambda$ weakly in $L_1(G)$. Then by (10.3.i) the sequence $[\lambda_k(t), t \in G, k = 1, 2, \ldots]$ is equiabsolutely integrable. Then the same occurs for the sequence $\lambda_k^-(t)$, and since $0 \leq \eta_k^-(t) \leq \lambda_k^-(t)$, we see that the sequence $[\eta_k^-(t), t \in G, k = 1, 2, \ldots]$ also is equiabsolutely integrable, and thus weakly relatively compact in $L_1(G)$. This proves (10.9.i). □

Given a pair of measurable functions $x_0(t)$, $\xi_0(t)$, $t \in G$, $\xi_0 \in L_1(G)$, a sequence $(x_k(t), \xi_k(t), \eta_k(t))$, $t \in G$, $k = 1, 2, \ldots$, of measurable functions on G is said to be *admissible relatively to* x_0, ξ_0 provided (a) $x_k(t) \in A(t)$, $(\eta_k(t), \xi_k(t)) \in \tilde{Q}(t, x_k(t))$, $t \in G$ (a.e.), $k = 1, 2, \ldots$; (b) $-\infty < i = \liminf \int_G \eta_k(t)\,dt < +\infty$; (c) $x_k \to x_0$ in measure in G, and $\xi_k \to \xi_0$ weakly in $L_1(G)$.

Again, given a pair of measurable functions $x_0(t)$, $\xi_0(t)$, $t \in G$, $\xi_0 \in L_1(G)$, we say that the *lower closure property* holds at x_0, ξ_0 provided, for any sequence x_k, ξ_k, η_k admissible relatively to x_0, ξ_0, there exists some $\eta_0 \in L_1(G)$ such that $(\eta_0(t), \xi_0(t)) \in \tilde{Q}(t, x_0(t))$ a.e. in G, and $\int_G \eta_0(t)\,dt \leq i = \liminf \int_G \eta_k(t)\,dt$.

Also, we shall need the real valued function $T(t, x, z)$ defined in Section 1.13 for given sets $\tilde{Q}(t, x)$ in R^{n+1}, $(t, x) \in A$; also, we assume that $\tilde{Q}(t, x) = \text{epi}_z T(t, x, z)$.

As before, let $x_0(t)$, $\xi_0(t)$, $t \in G$, $\xi_0 \in L_1(G)$, be a given pair of measurable functions, and let us suppose this time that $T(t, x_0(t), \xi_0(t)) \in L_1(G)$. We say that the *lower compactness property* holds at x_0, ξ_0 provided for every sequence x_k, ξ_k, η_k, which is admissible relatively to x_0, ξ_0, the sequence $\eta_k^-(t)$, $t \in G$, $k = 1, 2, \ldots$, is weakly relatively compact in $L_1(G)$. We can state now the following proposition:

10.9.ii (THEOREM). *If $x_0(t)$, $\xi_0(t)$, $t \in G$, $\xi_0 \in L_1(G)$, is a given pair of measurable functions with $T(t, x_0(t), \xi_0(t)) \in L_1(G)$, then the lower compactness property holds at x_0, ξ_0 if and only if the lower closure property holds at x_0, ξ_0.*

Proof. If the sequence $[\eta_k^-]$ is weakly sequentially compact in L_1, then there is a subsequence, say still $[k]$, such that $\eta_k^- \to \eta^-$ weakly in L_1, and then, for $\lambda_k = -\eta_k^-$, the relation (10.7.3) holds, and statement (10.7.i) proves the lower closure property. Conversely, assume that lower closure property holds, let $[x_k, \xi_k, \eta_k]$ be any admissible sequence relative to x_0, ξ_0, and let us prove that $[\eta_k^-]$ is weakly relatively compact. The argument is by contradiction. Suppose that $[\eta_k^-]$ is not weakly relatively compact. Then $[\eta_k^-]$ is not equi-absolutely integrable. Thus, there is a $\delta > 0$, and for each integer $s = 1, 2, \ldots$, there is another integer $k_s \geq s$ and a measurable subset E_s of G, such that meas $E_s \leq k_s^{-1}$, $\eta_{k_s}(t) \leq 0$ for $t \in E_s$, and $\int_{E_s} \eta_{k_s}(t)\,dt \leq -\delta < 0$ for all $s = 1, 2, \ldots$. Let us define the sequence $\bar{x}_s(t)$, $\bar{\xi}_s(t)$, $\bar{\eta}_s(t)$, $t \in G$, $s = 1, 2, \ldots$, by taking

$$(\bar{x}_s(t), \bar{\xi}_s(t), \bar{\eta}_s(t)) = \begin{cases} (x_0(t), \xi_0(t), T(t, x_0(t), \xi_0(t))) & \text{for } t \in G - E_s, \\ (x_{k_s}(t), \xi_{k_s}(t), \eta_{k_s}(t)) & \text{for } t \in E_s. \end{cases}$$

Then $\bar{x}_s \to x_0$ in measure, $\bar{\xi}_s \to \xi_0$ weakly, and

$$\int_G \bar{\eta}_s(t)\,dt = \int_{G-E_s} T(t, x_0(t), \xi_0(t))\,dt + \int_{E_s} \eta_{k_s}(t)\,dt \leq \int_{G-E_s} T\,dt - \delta,$$

and the last member approaches $\int_G T\,dt - \delta$ as $s \to \infty$. Also, if η_0 denotes the function corresponding to (x_k, ξ_k, η_k) guaranteed by the lower closure property, then we have

$$\int_G \bar{\eta}_s\,dt \geq \int_{G-E_s} T\,dt + \int_{E_s} \eta_0\,dt$$

and the second member approaches $\int_G T\,dt$ as $s \to \infty$. Thus, $\liminf \int_G \bar{\eta}_s\,dt$ as $s \to \infty$ is finite, say j. Now $\bar{x}_s$, $\bar{\xi}_s$, $\bar{\eta}_s$ is admissible relative to x_0, ξ_0, and by the lower closure property there is a corresponding function $\bar{\eta} \in L_1(G)$ such that $\int_G \bar{\eta}(t)\,dt \leq j$, and $(\bar{\eta}(t), \xi_0(t)) \in \tilde{Q}(t, x_0(t))$ a.e. in G. By the definition of T we also have $\int_G \bar{\eta}\,dt \geq \int_G T\,dt$, so that

$$\int_G T\,dt \leq \int_G \bar{\eta}\,dt \leq j \leq \liminf \int_G \bar{\eta}_s\,dt \leq \int_G T\,dt - \delta,$$

a contradiction, since $\delta > 0$. Theorem (10.9.ii) is thereby proved. □

In view of (10.9.i) and (10.9.ii) we see that requirement (10.7.3) in Theorem (10.7.i) becomes essentially a necessary and sufficient condition. Theorem (10.7.i) with the remarks (10.9.i) and (10.9.ii) contains a result proved by Ioffe [1].

B. Lower Semicontinuity at Every Trajectory

Let G be a measurable subset of R, and let $f(t,x)$ be a Carathéodory function on $G \times R^n$, that is, $f(t,x)$ is measurable in t for every x and is continuous in x for almost all t. For any measurable function $x(t)$, $t \in G$, let us consider the function $f(t,x(t))$, $t \in G$; in other words, let F denote the Nemitskii operator defined by $x(t) \to f(t,x(t))$, or $F[x](t) = f(t,x(t))$, $t \in G$.

10.9.iii (Nemitskii). *If f is a Carathéodory function, then the Nemitskii operator F maps measurable functions into measurable functions. Moreover, if $x_k \to x_0$ in measure in G, then $F[x_k] \to F[x_0]$ in measure in G.*

We are interested in the case where F maps functions $x \in (L_p(G))^n$ into functions $F[x] \in L_q(G)$ for given p, q, $1 \leq p, q < +\infty$. For the sake of brevity we shall write L_p for $(L_p(G))^n$ for any p and n.

10.9.iv (Nemitskii, Krasnoselskii, Vainberg). *If f is a Carathéodory function, and $F: L_p \to L_q$, then F is continuous and bounded in the topologies of L_p and L_q. Moreover, $F: L_p \to L_q$ if and only if there are a constant M and a function $\psi(t) \geq 0$, $t \in G$, $\psi \in L_q$, such that $|f(t,x)| \leq \psi(t) + M|x|^{p/q}$.*

For proofs of statements (10.9.iii) and (10.9.iv) we refer to Krasnoselskii [I], where also ample bibliography is given on these important questions.

Whenever $f(t,x(t))$, $t \in G$, has a Lebesgue integral on G, finite or $+\infty$, then also the following functional $H: L_p \to R$ is defined:

$$H[x] = \int_G f(t,x(t))\,dt, \qquad x \in L_p.$$

This functional H is said to be lower semicontinuous in L_p provided (a) $H[x] \neq -\infty$, and (b) $x_k \to x$ in L_p implies $H[x] \leq \liminf_{k\to\infty} H[x_k]$. An analogous definition holds for weak lower semicontinuity.

10.9.v (Poljak [1]). *If p is any fixed number, $1 \leq p < \infty$, and $f(t,x)$ is a Carathéodory function, then the following statements are equivalent:*

(A) *$H[x]$ is lower semicontinuous in L_p;*
(B) *$H[x]$ is defined and $\neq -\infty$ for all $x \in L_p$;*
(C) *$H[x]$ is defined in L_p and $H[x] \geq \alpha - \beta\|x\|_p^p$ for suitable constants α, β;*
(D) *$f(t,x) \geq -a(t) - b|x|^p$ for some constant $b \geq 0$ and function $a(t) \geq 0$, $a \in L_p$.*

Proof. Obviously, (D) $\Rightarrow$ (C) $\Rightarrow$ (B). Now assume that (B) holds, and let f_- denote $f_-(t,x) = \min[0, f(t,x)]$. Let F_- denote the Nemitskii operator relative to f_-. Then, for $x \in L_p$, we have $-\infty \leq f_-(t,x(t)) \leq 0$, $f_-(t,x(t)) \in L_1$, that is, F_- maps L_p into L_1, and by (10.9.iv), $|f_-(t,x)| \leq a(t) + b|x|^p$ for some constant $b \geq 0$ and function $a(t) \geq 0$, $a \in L_p$. Thus,

$$f(t,x) \geq f_-(t,x) \geq -a(t) - b|x|^p, \qquad (t,x) \in G \times R^n.$$

We have proved that (B)$\Rightarrow$(D). Now assume that (A) holds. By definition of lower semicontinuity in L_p, we see that (A)$\Rightarrow$(B). Finally, assume that (D) holds and that $x_k \to x_0$ in L_p. For $f^*(t,x) = f(t,x) + a(t) + b|x|^p$, we have $f^*(t,x) \geq 0$ for all $(t,x) \in G \times R^n$. From $x_k \to x_0$ in L_p, it follows that $x_k \to x_0$ in measure in G, and by Nemitskii's theorem (10.9.iii), also $f^*(t, x_k(t)) \to f^*(t, x_0(t))$ in measure in G. By Fatou's lemma (8.7.i)

$$\int_G f^*(t, x_0(t))\, dt \leq \liminf_{k\to\infty} \int_G f^*(t, x_k(t))\, dt.$$

Thus,

$$\begin{aligned} H[x_0] &= \int_G f(t, x_0(t))\, dt = \int_G [f^*(t, x_0(t)) - a(t) - b|x_0(t)|^p]\, dt \\ &\leq \liminf_{k\to\infty} \int_G f^*(t, x_k(t))\, dt - \int_G a(t)\, dt - b\|x_0(t)\|_p^p \\ &= \liminf_{k\to\infty} \int_G f(t, x_k(t))\, dt + b\left[\lim_{k\to\infty} \|x_k\|_p^p - \|x_0\|_p^p\right] \\ &= \liminf_{k\to\infty} \int_G f(t, x_k(t))\, dt = \liminf_{k\to\infty} H[x_k]. \end{aligned}$$

We have proved that (D)$\Rightarrow$(A). Statement (10.9.v) is thereby proved. □

Let us now return to functions $T(t,x,z)$, $(t,x,z) \in G \times R^n \times R^r$, G measurable with finite measure in R^ν, T a Carathéodory function (that is, T measurable in t for all (x,z), and continuous in (x,z) for almost all $t \in G$). We are interested in the functional

$$I[x,z] = \int_G T(t, x(t), z(t))\, dt$$

defined for all measurable functions $x(t)$, $z(t)$, $t \in G$, for which the measurable function $T(t, x(t), z(t))$ has a Lebesgue integral, finite or $+\infty$. For the sake of simplicity we shall say that $I[x,z]$ is (s,s) lower semicontinuous in $L_1 \times L_1$ if $I[x,z]$ is defined and $\neq -\infty$ for all $x, z \in L_1$, and $I[x,z] \leq \liminf_{k\to\infty} I[x_k, z_k]$ whenever $x_k \to x$, $z_k \to z$ in L_1. Statement (10.9.v) for $p = 1$ then becomes:

10.9.vi. *If $T(t,x,z)$ is a Carathéodory function, then the following statements are equivalent:*

(a) *$I[x,z]$ is (s,s) lower semicontinuous in $L_1 \times L_1$;*
(b) *$I[x,z]$ is defined and $\neq -\infty$ for all $x, z \in L_1$;*
(c) *$I[x,z] \geq \alpha - \beta\|x\|_1 - \gamma\|z\|_1$ for suitable constants α, β, γ;*
(d) *$T(t,x,z) \geq -a(t) - b|x| - c|z|$ in $G \times R^n \times R^r$ for suitable constants $b, c \geq 0$ and a function $a(t) \geq 0$, $a \in L_1$.*

We may consider now sequences (x_k, z_k) of functions $x_k, z_k \in L_p$, with $x_k \to x$ strongly in L_p and $z_k \to z$ weakly in L_q. We shall say that (x_k, z_k) is (s,w)-$L^{p,q}$ convergent to (x,z) in $L_p \times L_q$.

10.9.vii. *If $T(t,x,z)$ is a Carathéodory function, then of the four statements*

(a*) *$I[x,z]$ is (s,w) lower semicontinuous at every $(x,z) \in L_1 \times L_1$,*
(b) *$I[x,z]$ is defined and $\neq -\infty$ for all $(x,z) \in L_1 \times L_1$.*
(c) *$I[x,z] \geq \alpha - \beta\|x\|_1 - \gamma\|z\|_1$ for suitable constants α, β, γ,*

(d*) *$T(t,x,z) \geq -a(t) - b|x| - c|z|$ in $G \times R^n \times R^r$, and $T(t,x,z)$ is convex in z for almost all $t \in G$ and all $x \in R^n$,*

(a*) *and* (d*) *are equivalent, and if T is convex as stated in* (d*), *then* (a*), (b), (c), (d*) *are equivalent.*

Proof. If (a*) holds, that is, $I[x,z]$ is (s,w) lower semicontinuous in $L_1 \times L_1$, then certainly $I[x,z]$ is (s,s) lower semicontinuous in $L_1 \times L_1$, and hence (b), (c) holds as well as the first part of (d*) by (10.9.vi). It remains to prove that (a*) implies the convexity of $T(t,x,z)$ in z, and this is essentially Tonelli's theorem (2.19.i) concerning convexity as a necessary condition for lower semicontinuity on every trajectory. The proof in the present situation can be obtained by the same argument as in (2.19.i) via Scorza-Dragoni's statement (8.3.v). We leave the details of the proof as an exercise for the reader. Conversely, if (d*) holds, then certainly $I[x,z]$ is (s,s) lower semicontinuous in $L_1 \times L_1$ by (10.9.vi), and we have only to prove that $I[x,z]$ is (s,w) lower semicontinuous in $L_1 \times L_1$. First, since $T(t,x,z)$ is convex in z for almost all $t \in G$ and all $x \in R^n$, then $I[x,z]$ is convex in $z \in L_1$ for every $x \in L_1$. Now let us assume that $I[x,z]$ is not (s,w)-$L^{1,1}$ lower semicontinuous. Then there is an $\varepsilon > 0$ and a sequence (x_k, z_k), $k = 1, 2, \ldots,$ which is (s,w)-$L^{1,1}$ convergent to some $(x,z) \in L_1 \times L_1$ and such that $I[x_k, z_k] \leq I[x,z] - \varepsilon$ for infinitely many k. By selection and relabeling we may assume that $I[x_k, z_k] \leq \lambda - \varepsilon$ for $k = 1, 2, \ldots,$ where $\lambda = I[x,z]$. By (10.1.i) there is a system of real numbers $c_{Ns} \geq 0$, $s = 1, \ldots, N$, $N = 1, 2, \ldots,$ such that $\sum_{s=1}^N c_{Ns} = 1$ and $\sum_{s=1}^N c_{Ns} z_s \to z$ strongly in L_1 as $N \to \infty$. Let N_k, $k = 1, 2, \ldots,$ be any sequence of integers with $N_k \to \infty$ as $k \to \infty$. Then $(x_k, \sum_{s=1}^{Nk} c_{N_k s} z_s)$, $k = 1, 2, \ldots,$ is (s,s)-$L^{1,1}$ convergent to (x,z) as $k \to \infty$. Hence, by the just stated (s,s)-lower semicontinuity of $I[x,z]$ and the convexity of $I[x,z]$ with respect to z, we have

$$\lambda = I[x,z] \leq \liminf_{k \to \infty} I\left[x_k, \sum_{s=1}^{N_k} c_{N_k s} z_s\right]$$

$$\leq \liminf_{k \to \infty} \sum_{s=1}^{N_k} c_{N_k s} I[x_k, z_s] \leq \lambda - \varepsilon,$$

a contradiction. We have proved that (d*) implies (a*). Statement (10.9.vii) is thereby proved. □

Remark. Theorem (10.9.vii) shows that the conditions (Λ) of Section 10.7, or (L) of Section 10.8, are not only sufficient but also, in a sense, necessary for lower closure. For more work along the lines of the present section we refer to Poljak [1], already cited, and to Rothe [1], Olech [1], and Ioffe [1] (see bibliographical notes for further references). The statements in Section 10.9 concern only strong and weak convergence in L_1. Analogous statements hold for strong and weak convergence in L_p. We shall present the material above more extensively, and including the analogous statements in L_p and in Sobolev spaces, in [IV], in connection with optimization with partial differential equations, where this material is essential. Let it be mentioned here that for (s,w) convergence in $L_p \times L_q$, then the statement of (10.9.vii) becomes

$$T(t,x,z) \geq (p(t,x), z) - c|x|^p - a(t),$$

where $a(t) \geq 0$, $a \in L_1$, and the map $t \to |p(t, x(t)|^{q'}$, $1/q + 1/q' = 1$, is relatively weakly compact in L_1.

C. Exercises

1. Prove statement (10.9.v) with weak convergence in L_p in (A) and $f(t, x)$ convex in x for almost all $t \in G$.
2. Prove in detail that, in (10.9.vii), (a*) implies the convexity of $T(t, x, z)$ with respect to z as stated in (d*).
3. Prove Theorems (10.6.ii) and (10.7.ii) by assuming that the relevant sets have the weak form of property (Q) stated in Remark 1 of Section 8.5. (The conclusions of (10.6.ii) and (10.7.ii) will be used in Chapter 14 under hypotheses which indeed imply this weak form of property (Q) as well as the hypotheses in the abovementioned theorems.)

Bibliographical Notes

References on the Banach–Saks–Mazur theorem (10.1.i) and on the Dunford–Pettis–Nagumo theorem (actually, the composite equivalence theorem (10.3.i)) have been already given in the text. Growth condition (g1), or (ϕ), in Section 10.4 is the usual Tonelli–Nagumo condition. Growth condition (g2) is slightly less demanding and was proposed by L. Cesari [35] (see also previous remark in L. Cesari, J. R. LaPalm, and T. Nishiura [1]). Condition (g3) has been proposed by R. T. Rockafellar [2], and as we have proved (10.4.vi–vii), (g3) and (g2) are equivalent. Growth condition (g1) or (ϕ), which is so closely related to weak convergence in L_1 as stated by the equivalence theorem (10.3.i), implies property (Q) of certain related set valued functions. This is the essence of theorems (10.5.i), (10.5.ii), which were proved by Cesari in various forms in [6, 7]. The same holds for statement (10.5.iii), and the analogous implications hold for the convex sets which occur with generalized solutions as proved in Remark 4 of Section 10.5. The closure and lower closure theorems (10.6.i) and (10.7.i), both based on weak convergence in L_1 of the "derivatives", are now a consequence of the Banach–Saks–Mazur theorem, of the Dunford–Pettis theorem, and of the implication stated by theorems (10.5.i–ii). Restricted forms of these closure and lower closure theorems have been proved independently and about at the same time by L. Cesari [13], M. F. Bidaut [1], and L. D. Berkovitz [1].

For further results on lower closure theorems, see L. Cesari [12, 13, 16, 17], L. Cesari and M. B. Suryanarayana [8], and C. Olech and A. Lasota [1, 2].

The very general lower closure theorem (10.7.i) concerns abstract integrals $\int \eta(t)\,dt$ under mere orientor field constraints $(\eta(t), \xi(t)) \in \tilde{Q}(t, x(t))$ for given convex subsets $\tilde{Q}(t, x)$ in R^{n+1}. The rather abstract requirement (10.7.3) in the lower closure theorem (10.7.i) is certainly verified under any of the alternative conditions (Λ_i) of Section 10.7, which are easy to verify and have therefore practical significance. Of these conditions (Λ_1) and (Λ_2) were noted by all of the aforementioned authors, condition (Λ_3) by L. D. Berkovitz (this condition is contained in (Λ_2)), and condition (Λ_4) by Cesari in [16]. The conclusion which has been reached recently—and which is embodied in statements (10.6.i), (10.7.i)—is that, in connection with "weak convergence of the derivatives" (weak convergence in $H^{1,1}$) there is no need for explicitly requiring any property (Q) (of the sets Q in R^n, or $\tilde{Q}$ in R^{n+1}) since the weak convergence implies a growth property by (10.3.i), and this in turn, by (10.5.i–ii), implies the property (Q) for certain auxiliary sets (the sets Q^* in R^{n+1} or $\tilde{Q}'^*$ in R^{n+2}). The lower closure theorem (10.7.i) and its immediate corollary, the lower semicontinuity theorem (10.8.i), are of course the main tools in the proof of the existence theorems of Chapter 11 based on "weak convergence of the derivatives". The required conditions in the lower closure theorems (10.7.i) and lower semicontinuity theorems (10.8.i,ii) are proved to be necessary in Section 10.9.

On the other hand there are existence theorems (see Chapter 12) for which mere "uniform convergence of the trajectories" is used, and for these theorems the corresponding closure and lower closure theorems based on mere uniform convergence (8.6.i), (8.8.i,ii) are needed and it appears that in these theorems some form of property (Q) is needed. Thus, the presentation, based on geometrical considerations, of lower closure and lower semicontinuity theorems in these Chapters 8 and 10 has a certain degree of uniformity.

Closure and lower closure theorems (10.6.ii) and (10.7.ii) have been proved by L. Cesari and M. B. Suryanarayana [1], and they will be used in connection with property (D) for the existence theorems of Chapter 13 covering a number of situations of practical significance some of which had been indicated by E. H. Rothe and L. D. Berkovitz. Each one of the specific conditions in the existence theorems of Chapter 13 implies property (D) and this in turn implies the weak form of property (Q) mentioned in Remark 1 of Section 10.5, which could be used in an alternate proof of the same existence theorems.

The lower semicontinuity theorems (10.8.i,ii) are corollaries of (10.7.i). The integrand function $F_0(t,x,z)$ is assumed either continuous on M, (condition (C) of Section 10.8), or briefly "measurable in t and lower semicontinuous in (x,z)", specifically, satisfying condition (C*) of Section 10.8. This condition is rather general among those proposed so far for such lower semicontinuity theorems (namely it is equivalent to the various forms of "normality" conditions used by R. T. Rockafellar [2] and by I. Ekeland and R. Temam [I], as proved by these authors).

The conditions (Λ_i) of practical interest in Section 10.7 are replaced in Section 10.8 by corresponding conditions (L_i), also easy to verify. Of these, condition (L_4), which corresponds to (Λ_4), is especially relevant, since (L_4) is always satisfied in classical free problems of the calculus of variations in one independent variable (Remark 3 of Section 10.8A). Thus, from (10.8.ii) and (2.19.i) it follows that if $F_0(t,x,z)$ is continuous in $A \times R^n$, then $\int F_0\,dt$ is lower semicontinuous with respect to weak convergence of the derivatives (weakly in $H^{1,1}$) if and only if $F_0(t,x,z)$ is convex in z, $x = (x^1, \ldots, x^n)$, $z = (z^1, \ldots, z^n)$, $n \geq 1$ (L. Cesari [17], and also, in this book, (2.18.i) and Remark 6 of Section 10.8). In particular, linear integrals, i.e. $\int F_0\,dt$, with $n \geq 1$, $F_0 = P(t,x) + \sum_{i=1}^n Q_i(t,x)z^i$, $P, Q_1, \ldots, Q_n$ continuous on A, are continuous in the topology of the weak convergence of the derivatives (namely, weak convergence in $H^{1,1}$).

For $n = 1$ and $F_0(t,x,z)$ continuous in $A \times R^1$ together with F_{0z}, F_{0zz}, F_{0tz}, Tonelli proved that $\int F_0\,dt$ is lower semicontinuous with respect to uniform convergence of the trajectories if and only if $F_{0zz} \geq 0$ (Tonelli [I, p. 400]). Tonelli also gave an example [I, p. 392] of a linear integral $\int F_0\,dt$, with $n = 1$ and $F_0 = P(t,x) + Q(t,x)z$, P, Q continuous on A, which is not continuous in the uniform topology, and McShane [5, p. 211] gave an example of a linear integral $\int F_0\,dt$ with $n = 2$ and $F_0 = P(t,x) + Q(t,x)z_1 + R(t,x)z_2$, P, Q, R of class C^∞, which is not continuous in the uniform topology. Tonelli [13] showed later that $\partial Q/\partial x_2 = \partial R/\partial x_1$ is a necessary condition for continuity in the uniform topology for such integrals.

M. Vidyasagar [1] proved a lemma analogous to (10.8.ii) requiring also a uniform Lipschitz condition on $A_0(t,x)$ and $B(t,x)$ with respect to x and uniform in t.

Section 10.9 is a brief presentation of further results concerning necessary and sufficient conditions for lower closure on all trajectories. These results are essentially based on the precise characterizations of properties of boundedness and continuity of Nemitskii and other operators, for which we refer to M. A. Krasnoselskii [I], to M. A. Krasnoselskii, P. Zabreiko, E. I. Pustylnik, and P. W. Sobolevskii [I]; to M. M. Vainberg [I], and to further work of L. V. Sragin [1] and of B. T. Poljak [1]. The part concerning the necessity of the convexity condition is of course Tonelli's theorem on lower semicontinuity on all trajectories (2.19.i). For further work on necessary and sufficient conditions for lower closure and lower semicontinuity theorems we refer to C. Oleach [1], A. D. Ioffe [1], V. I. Kazimirov [1], and V. S. Morozov and V. I. Plotnikov [1].

For independent work on lower semicontinuity in L_p spaces and Sobolev spaces we mention here also G. Fichera [1], and E. H. Rothe [1].

CHAPTER 11

Existence Theorems: Weak Convergence and Growth Conditions

11.1 Existence Theorems for Orientor Fields and Extended Problems

A. Some General Existence Theorems for Lagrange and Bolza Type Problems

Let A be a subset of the tx-space R^{n+1}, and let $A(t)$ denote its sections, that is, $A(t) = [x \in R^n | (t, x) \in A]$. For every $(t, x) \in A$ let $Q(t, x)$ be a given subset of the z-space R^n, $x = (x^1, \ldots, x^n)$, $z = (z^1, \ldots, z^n)$. Let M_0 be the set $M_0 = [(t, x, z) | (t, x) \in A,\ z \in Q(t, x)] \subset R^{1+2n}$, and let $F_0(t, x, z)$ be a given real valued function defined on M_0. Let B be a given subset of the $t_1 x_1 t_2 x_2$-space R^{2n+2}, and let $g(t_1, x_1, t_2, x_2)$ be a real valued function defined on B. Let $\Omega = \{x\}$ denote a nonempty collection of AC functions $x(t) = (x^1, \ldots, x^n)$, $t_1 \leq t \leq t_2$, such that

$$(11.1.1)\qquad \begin{aligned} &x(t) \in A(t), \quad x'(t) \in Q(t, x(t)) \quad \text{for } t \in [t_1, t_2] \text{ (a.e.)}, \\ &e[x] = (t_1, x(t_1), t_2, x(t_2)) \in B, \quad F_0(\cdot, x(\cdot), x'(\cdot)) \in L_1[t_1, t_2]. \end{aligned}$$

We are concerned with the existence of the minimum in Ω of the functional

$$(11.1.2)\qquad I[x] = g(t_1, x(t_1), t_2, x(t_2)) + \int_{t_1}^{t_2} F_0(t, x(t), x'(t))\, dt.$$

Any AC function x satisfying (11.1.1) will be called an admissible trajectory, or briefly a trajectory. For every $(t, x) \in A$ we denote as usual by $\tilde{Q}(t, x)$ the set $\tilde{Q}(t, x) = [(z^0, z) | z^0 \geq F_0(t, x, z),\ z \in Q(t, x)] \subset R^{n+1}$.

Concerning A, M_0, F_0 we repeat here the alternate assumptions of Section 10.8. For most applications it is sufficient to assume

C. A closed, M_0 closed, $F_0(t, x, z)$ continuous on M_0.

In this case we may think of F_0 as extended to the whole space R^{1+2n} by taking $F_0 = +\infty$ in $R^{1+2n} - M_0$. In the classical calculus of variations we usually have $Q(t, x) = R^n$ for all $(t, x) \in A$ and $M_0 = A \times R^n$. However, for the proof that follows the following rather general assumption suffices:

C*. $F_0(t, x, z)$ is a given extended function in R^{1+2n} and we assume that for every finite interval $[t_0, T]$ and $\varepsilon > 0$ there is a compact set $K \subset [t_0, T]$ such that (a) meas $([t_0, T] - K) < \varepsilon$, (b) the extended function $F_0(t, x, z)$ restricted to $K \times R^{2n}$ is B-measurable, and (c) for almost all $\bar{t} \in [t_0, T]$ the extended function $F_0(\bar{t}, x, z)$ of (x, z) has values finite or $+\infty$, and is lower semicontinuous in R^{2n}.

Under hypothesis (C*) we may simply assume that F_0 is a given extended function in R^{1+2n}, whose values for almost all $\bar{t} \in G$ are finite or $+\infty$, and we denote by A_0 the set of all $\bar{t} \in R$ such that $A(\bar{t}) = [\bar{x} \in R^n | F_0(\bar{t}, \bar{x}, z) \not\equiv +\infty] \neq \emptyset$. For any $(\bar{t}, \bar{x})$ take $Q(\bar{t}, \bar{x}) = [z \in R^r | F_0(t, x, z) \neq +\infty]$, and $\tilde{Q}(\bar{t}, \bar{x}) = \text{epi } F_0(\bar{t}, \bar{x}, z)$. Now A is any set of points (t, x) whose sections, for almost all $t \in A_0$, are the sets $A(t)$.

In any case it is clear that A, B, F_0 must be so related that the class of all AC functions $x(t)$ satisfying (11.1.1) is not empty. Note that under condition (C) as well as under condition (C*) (parts (a) and (b) suffice), then $F_0(t, x(t), z(t))$ is measurable for any two measurable functions $x(t)$, $z(t)$. However $F_0(t, x(t), z(t))$ may have the value $+\infty$ in a set of positive measure. Only if $x(t) \in A(t)$, $z(t) \in Q(t, x(t))$ a.e., then $F_0(t, x(t), z(t))$ is finite a.e..

The existence theorems of the present chapter will be based on the mode of convergence for AC trajectories x_k, x which in Section 2.14 we called "the weak convergence of the derivatives", that is, $x_k \to x$ in the ρ-metric, or uniformly, and $x_k' \to x'$ weakly in L_1 (mode (b) of Section 2.14).

We may be interested in the absolute minimum of $I[x]$ in the class Ω of all AC trajectories x satisfying (11.1.1). Alternatively, and as in Sections 9.2, 9.3, we may want to minimize $I[x]$ in a smaller class Ω of such AC functions, and in this case we need to know that Ω has a suitable closedness property. We say that Ω is Γ_w-closed provided: if $x_k(t)$, $t_{1k} \leq t \leq t_{2k}$, $k = 1, 2, \ldots$, are AC functions satisfying (11.1.1), all in the class Ω, if $x(t)$, $t_1 \leq t \leq t_2$, is an AC function satisfying (11.1.1), and if $x_k \to x$ in the weak convergence of the derivatives (that is, $x_k \to x$ in the ρ-metric and $x_k' \to x'$ weakly in L_1), then x is in the class Ω. The same usual classes of trajectories we have mentioned in Sections 9.2, 9.3 in connection with Γ_u-closure all have also the present Γ_w-closedness property.

In the existence theorems below we shall need alternate global "growth hypotheses". They are the usual ones we have already mentioned, but we state them again in the form they are needed here.

(γ1) there is a scalar function $\phi(\zeta)$, $0 \le \zeta < +\infty$, bounded below, such that $\phi(\zeta)/\zeta \to +\infty$ as $\zeta \to +\infty$, and $F_0(t, x, z) \ge \phi(|z|)$ for all $(t, x, z) \in M_0$;

(γ2) for any $\varepsilon > 0$ there is a locally integrable scalar function $\psi_\varepsilon(t) \ge 0$ such that $|z| \le \psi_\varepsilon(t) + \varepsilon F_0(t, x, z)$ for all $(t, x, z) \in M_0$;

(γ3) for every n-vector $p \in R^n$ there is a locally integrable function $\phi_p(t) \ge 0$ such that $F_0(t, x, z) \ge (p, z) - \phi_p(t)$ for all $(t, x, z) \in M_0$.

These are the conditions we anticipated in Section 2.20A, and we encountered in Section 10.4 both in their global and local forms. Condition (γ1) is the Tonelli-Nagumo condition. Condition (γ2) is a slight generalization of (γ1), and was discussed in Section 10.4. Condition (γ3) is actually equivalent to (γ2) as we proved in Section 10.4. The examples in Section 2.20B and those in Section 11.3 illustrate these conditions.

11.1.i (An Existence Theorem Based on Weak Convergence of the Derivatives). *Let A be bounded, B closed, and let condition* (C) *or* (C*) *be satisfied. Assume that for almost all t and all $x \in A(t)$ the extended function $F_0(t, x, z)$ be convex in z (in R^n and hence the sets $Q(t, x)$ and $\tilde{Q}(t, x)$ are convex). Let g be a lower semicontinuous function on B. Assume that any one of the growth conditions* (γ1), (γ2), (γ3) *is satisfied. Let Ω be any nonempty Γ_w-closed class of* AC *functions $x(t) = (x^1, \dots, x^n)$, $t_1 \le t \le t_2$, satisfying* (11.1.1). *Then the functional $I[x]$ in* (11.1.2) *has an absolute minimum in Ω.*

It is enough we limit ourselves to the nonempty part Ω_M of Ω of all elements $x \in \Omega$ with $I[x] \le M$ for some M.

For A not bounded see Section 11.2.

Proof. Here A is bounded, and thus, for every element $x(t)$, $t_1 \le t \le t_2$, of Ω we have $-M \le t_1 \le t_2 \le M$, $|x(t)| \le M$ for all $t \in [t_1, t_2]$ and some fixed M. Under condition (γ1), ϕ is bounded below; hence $F_0(t, x(t), x'(t)) \ge -M$ for all $t \in [t_1, t_2]$ and some constant M. Under condition (γ2) with $\varepsilon = 1$ we have $F_0(t, x(t), x'(t)) \ge -\psi_1(t)$, where $\psi_1 \ge 0$ is L-integrable in $[-M, M]$. Under condition (γ3) with $p = 0$ we have again $F_0 \ge -\phi_0(t)$. Thus condition (L_1) of Section 10.8 holds, and $J_1[x] = \int_{t_1}^{t_2} F_0(t, x(t), x'(t))\,dt \ge -M_1$ for all elements x of Ω and some constant M_1. Since $e[x] = (t_1, x(t_1), t_2, x(t_2)$ lies in $B \cap (\mathrm{cl}\, A \times \mathrm{cl}\, A)$, a compact set, and g is lower semicontinuous, we have $J_2[x] = g(e[x]) \ge -M_2$ for some constant M_2. Hence $I[x] = J_1 + J_2 \ge -M_1 - M_2$ is bounded below in Ω. If $i = \inf[I[x], x \in \Omega]$, then $-\infty < i < +\infty$. Let $x_k(t)$, $t_{1k} \le t \le t_{2k}$, $k = 1, 2, \dots$, be a minimizing sequence of elements $x_k \in \Omega$, that is, $I[x_k] \to i$ as $k \to +\infty$, with

$$x_k(t) \in A(t), \qquad x_k'(t) \in Q(t, x_k(t)), \qquad t \in [t_{1k}, t_{2k}] \text{ (a.e.)}, \qquad k = 1, 2, \dots.$$

We can assume that $i \le I[x_k] = J_1[x_k] + J_2[x_k] \le i + 1$. By a suitable selection we can even assume that $J_1[x_k] \to i_1$, $J_2[x_k] \to i_2$ as $k \to \infty$, with $i_1 + i_2 = i$. where i_1, i_2 are suitable numbers (neither of which need be the infimum of J_1 or J_2).

Under condition $(\gamma 1)$ we have $\phi(|x_k'(t)|) \leq F_0(t, x_k(t), x_k'(t))$, $t_{1k} \leq t \leq t_{2k}$; under condition $(\gamma 2)$ for any $\varepsilon > 0$ we have $|x_k'(t)| \leq \psi_\varepsilon(t) + \varepsilon F_0(t, x_k(t), x_k'(t))$, $t_{1k} \leq t \leq t_{2k}$; under condition $(\gamma 3)$ for any $p \in R^n$ we have $F_0(t, x_k(t), x_k'(t)) \geq (p, x_k'(t)) - \phi_p(t)$, $t_{1k} \leq t \leq t_{2k}$, $k = 1, 2, \ldots$. In any case, by virtue of (10.4.i,ii,iii) respectively, the equibounded sequence $[x_k]$ is equicontinuous and equiabsolutely continuous, and the sequence of derivatives $[x_k']$ is equiabsolutely integrable. By Ascoli's theorem (9.1.ii) there is a subsequence, say still $[k]$, such that $[x_k]$ converges uniformly to some continuous function $x(t)$, $t_1 \leq t \leq t_2$, namely, in the ρ-metric of Section 2.14 with $t_{1k} \to t_1$, $t_{2k} \to t_2$. By (10.2.iii) x is AC in $[t_1, t_2]$. By (10.3.i), (b) $\Rightarrow$ (a), we can choose the subsequence in such a way that $[x_k']$ is weakly convergent in L_1 to some L-integrable function $\xi(t)$.

By (10.8.ii) we have $\xi(t) = x'(t)$, $x(t) \in A(t)$, $x'(t) \in Q(t, x(t))$, $t \in [t_1, t_2]$ (a.e.), and $J_1[x] = \int_{t_1}^{t_2} F_0(t, x(t), x'(t))\,dt \leq i_1$. Since $e[x_k] \to e[x]$ in R^{2+2n} as $k \to \infty$, by the lower semicontinuity of g we derive that $J_2[x] = g(e[x]) \leq \lim J_2[x_k] = i_2$. Thus,

$$I[x] = J_1[x] + J_2[x] \leq i_1 + i_2 = i.$$

On the other hand Ω is Γ_w-closed, hence $x \in \Omega$ and $I[x] \geq i$. By comparison we have $I[x] = i$ and (11.1.i) is thereby proved. □

Note that in the present situation the sets $\tilde{Q}(t, x) = \text{epi } F_0(t, x, z)$ are closed and convex, and certainly for almost all $\bar{t}$ the sets $\tilde{Q}(\bar{t}, x)$ have property (K) with respect to x by (8.5.v) and (8.5.iii). Note that, moreover, for almost all $\bar{t}$ the sets $\tilde{Q}(\bar{t}, x)$ have property (Q) with respect to x as a consequence of the growth hypotheses and of theorem (10.5.i). Indeed, this is evident under condition $(\gamma 1)$. Under condition $(\gamma 2)$ we note that for fixed $\bar{t}$, then in the relation $|z| \leq \psi_\varepsilon(\bar{t}) + \varepsilon F_0(\bar{t}, x, z)$, $\psi_\varepsilon(\bar{t})$ is actually a constant, and by (10.4.v) this condition is equivalent to $(\gamma 1)$ with respect to x only. Condition $(\gamma 3)$ is equivalent to $(\gamma 2)$. Thus, our sets $\tilde{Q}(\bar{t}, x)$ have property (Q) with respect to x (for almost every $\bar{t}$), and then the above recourse to (10.8.ii), and in the last analysis to (10.7.i), is to be understood in the sense that we really need only the simpler versions of (10.7.i) and (10.8.i,ii) mentioned at the beginning of the proof of (10.7.i) and in Remark 4 of Section 10.8.

Remark 1. Concerning Theorem (11.1.i), we have already mentioned examples of Γ_w-closed classes Ω. Here we can add that we also obtain a Γ_w-closed class by a restriction of the form

$$C[x] = \int_{t_1}^{t_2} H(t, x(t), x'(t))\,dt \leq M$$

for some constant M, and an integrand H satisfying the conditions of (10.8.i), so that $C[x]$ is lower semicontinuous in the same topology we have been using in the proof of Theorem (11.1.i). Often $C[x]$ is called a comparison functional. If H already satisfies one of the growth conditions $(\gamma 1)$, $(\gamma 2)$, $(\gamma 3)$ much less needs to be required on f_0, as we shall see below. This happens for instance in the case of the familiar restriction

$$\int_{t_1}^{t_2} |x'|^p\,dt \leq M \quad \text{for } p > 1.$$

Remark 2. If the convexity condition in (11.1.i) is not satisfied, let us prove the existence of an optimal generalized solution. By a generalized solution (cf. Section 1.14) we mean a solution of the new problem in which as usual relations (11.1.1–2) are replaced by

$$J[x,p,v] = g(t_1, x(t_1), t_2, x(t_2)) + \int_{t_1}^{t_2} \sum_{s=1}^{\gamma} p_s(t) F_0(t, x(t), z^{(s)}(t))\, dt,$$

$$x'(t) = \sum_s p_s(t) z^{(s)}(t), \qquad \sum_s p_s(t) = 1, \qquad p_s(t) \geq 0, \qquad t \in [t_1, t_2]\,(\text{a.e.}),$$

$$(t, x(t)) \in A, \qquad (t_1, x(t_1), t_2, x(t_2)) \in B,$$

$$x(t)\ \text{AC}, \qquad p_s(t),\ z^{(s)}(t)\ \text{measurable}, \qquad \sum_s p_s(\cdot) F_0(\cdot, x(\cdot), z^{(s)}(\cdot)) \in L.$$

Actually this is a problem of optimal control, and the natural problem of the type (11.1.1–2) to associate to it is the following one

$$H[x] = g(t_1, x(t_1), t_2, x(t_2)) + \int_{t_1}^{t_2} T(t, x(t), x'(t))\, dt, \qquad x\ \text{AC},$$

$$(t, x(t)) \in A, \qquad (t_1, x(t_1), t_2, x(t_2)) \in B, \qquad T(\cdot, x(\cdot), x'(\cdot)) \in L,$$

$$T(t,x,z) = \inf \sum_s p_s F_0(t, x, z^{(s)}), \qquad z = \sum_s p_s z^{(s)}, \qquad \sum_s p_s = 1, \qquad p_s \geq 0.$$

Now let us consider the convex sets

$$\tilde{Q}^*(t,x) = \left[(z^0, z) \,\middle|\, z^0 \geq \sum_s p_s F_0(t, x, z^{(s)}),\ z = \sum_s p_s z^{(s)},\ \sum_s p_s = 1,\ p_s \geq 0 \right].$$

Let us prove that, for almost all t, the sets $\tilde{Q}^*(t,x)$ have properties (K) and (Q) with respect to x. Indeed, if F_0 satisfies condition ($\gamma 1$) then this is a consequence of Remark 4 of Section 10.5. If F_0 satisfies condition ($\gamma 2$), then for any given $\bar{t}$, in the relation $z \leq \psi_\varepsilon(\bar{t}) + \varepsilon F_0(\bar{t}, x, z)$, $\psi_\varepsilon(\bar{t})$ is a constant, and as above by (10.4.v) this condition is equivalent to ($\gamma 1$) with respect to x only. Condition ($\gamma 3$) is equivalent to ($\gamma 2$). Thus, for almost all $\bar{t}$, the sets $\tilde{Q}^*(t,x)$ have properties (K) and (Q) with respect to x in $A(\bar{t})$. As a consequence, for almost all $\bar{t}$, the sets $\tilde{Q}^*(t,x)$ are closed, and this implies that inf can be replaced by min in the definition of T, and the two problems relative to J and H are equivalent. Finally, for almost all $\bar{t}$, and by (8.5.iii) and (8.5.v), $T(\bar{t}, x, z)$ is lower semicontinuous in x, z. Now (11.1.i) applies to the functional H, and the existence of an optimal generalized solution is proved.

Remark 3. It is easy to see that the same proof of (11.1.i) above also proves that, under the same hypotheses of (11.1.i), the class Ω_M of all AC functions $x(t)$ satisfying (11.1.1) with $I[x] \leq M$ for any fixed M is closed and compact in the topology of weak convergence of the derivatives. That is, if $x_k \in \Omega$, $I[x_k] \leq M$, $k = 1, 2, \ldots$, then there is a subsequence, say still $[k]$, such that $x_k \to x$ in the ρ-metric, and $x_k' \to x'$ weakly in L_1, where x is AC, $x \in \Omega$, and $I[x] \leq M$.

In the proof of (11.1.i) the growth condition ($\gamma 1$) or ($\gamma 2$) or ($\gamma 3$) has several roles: it guarantees that condition (L_1) of Section 10.8 holds, it guarantees the boundedness below of $I[x]$, and it guarantees the relative compactness of the class $\Omega = \{x\}$ (in the topology of the weak convergence of the derivatives. Thus even if we assume straightforwardly that Ω is any given class possessing the weak compactness property above, we must still see to it that $I[x]$ is bounded below. Indirectly, we have to guarantee the existence of functions λ, λ_k as in (10.7.3). Actually, we have already stated in Section 10.8 to this effect, the alternate conditions (L_i) in terms of F_0. We shall use the conditions (L_i) again in this Section. For other alternate conditions see the exercises below.

11.1.ii (AN EXISTENCE THEOREM BASED ON WEAK CONVERGENCE OF THE DERIVATIVES). *Let A be bounded, B closed, and let condition* (C) *or* (C*) *be satisfied. Assume that for almost all t and all $x \in A(t)$ the extended function $F_0(t, x, z)$ is convex in z. Assume that one of the alternate conditions* (L_i) *holds. Let $g(t_1, x_1, t_2, x_2)$ be a lower semicontinuous scalar function on B. Let Ω be any nonempty Γ_w-closed class $\{x\}$ of* AC *functions $x(t) = (x^1, \ldots, x^n)$, $t_1 \leq t \leq t_2$, satisfying* (11.1.1), *and assume that the class of derivatives $\{x'(t), t_1 \leq t \leq t_2\}$ is equiabsolutely integrable. Then the functional $I[x]$ in* (11.1.1) *has an absolute minimum in Ω.*

It is enough we verify the above requirements for the nonempty class Ω_M of all $x \in \Omega$ satisfying $I[x] \leq M$ for some M.

The requirement on the class $\{x'\}$ is certainly verified if for instance we know that for some $p > 1$ and $D > 0$ we have $\int_{t_1}^{t_2} |x'(t)|^p \, dt \leq D$ for all elements x of the class Ω or Ω_M. This is a consequence of (10.3.i), (c) $\Rightarrow$ (b) with $\phi(\zeta) = \zeta^p$. Analogously, we satisfy the requirements on the class x' by a restriction of the form

$$C[x] = \int_{t_1}^{t_2} H(t, x(t), x'(t)) \, dt \leq M$$

for some constant M and where H satisfies the assumptions of (11.1.i), in particular one of the growth conditions ($\gamma 1$), ($\gamma 2$), ($\gamma 3$). Again by (10.8.ii) $C[x]$ is lower semicontinuous, and the classs so obtained is Γ_w-closed.

Theorem (11.1.i) is contained in (11.1.ii). Indeed, under either condition ($\gamma 1$), or ($\gamma 2$), or ($\gamma 3$), there is some integrable function $\psi(t) \geq 0$ such that $F_0(t, x(t), x'(t)) \geq -\psi(t)$ for all $x \in \Omega$ as we have shown in the proof of (11.1.i) and condition (L_1) holds. On the other hand, under condition ($\gamma 1$) there is some function $\phi(\zeta)$ bounded below with $\phi(\zeta)/\zeta \to +\infty$ as $\zeta \to +\infty$ and $F_0(t, x, z) \geq \phi(|z|)$. Then for every $x \in \Omega$ we have

$$M \geq \int_{t_1}^{t_2} F_0(t, x(t), x'(t)) \, dt \geq \int_{t_1}^{t_2} \phi(|x'(t)|) \, dt,$$

and the equiabsolute integrability of the class $\{x'\}$ follows from (10.1.i). Under condition ($\gamma 2$) and $\eta(t) = F(t, x(t), x'(t))$ we have $\int_{t_1}^{t_2} \eta(t)\,dt = \int_{t_1}^{t_2} F_0\,dt \le M$, and the equiabsolute integrability of $\{x'\}$ follows from (10.1.ii). Under ($\gamma 3$) the analogous conclusion follows from (10.1.iii).

Proof of (11.1.ii). Let M be a number such that Ω_M is not empty. It is not restrictive to search for the minimum of $I[x]$ in Ω_M. By (10.3.i), (b) $\Rightarrow$ (a), the class $\{x'\}$ is relatively sequentially weakly compact in L_1, and by (10.2.i) the class $\{x\}$ is equiabsolutely continuous. Since A is bounded, the same class $\{x\}$ is also equibounded, that is, $-M_0 \le t_1 < t_2 \le M_0$, $|x(t)| \le M_0$. Then, Ω_M is relatively sequentially weakly compact in the "topology of the weak convergence of the derivatives (mode (b) of Section 2.14). Again, g is lower semicontinuous in the compact set $B \bigcap (\text{cl } A \times \text{cl } A)$, hence $J_2 = g(e[x])$ is bounded below in Ω_M. Here $I = J_1 + J_2$ as before, and we do not know yet whether $J_1 = \int_{t_1}^{t_2} F_0\,dt$ is bounded below in Ω_M. Let $i = \inf[I[x], x \in \Omega_M]$, $-\infty \le i < +\infty$. Let $x_k(t)$, $t_{1k} \le t \le t_{2k}$, $k = 1, 2, \ldots, x_k \in \Omega_M$, be a minimizing sequence, that is, $I[x_k] \to i$ as $k \to \infty$. Here $[x_k]$ is a subset of Ω_M; hence there is a subsequence, say still $[k]$, such that $x_k' \to \xi$ weakly in L_1, and $x_k \to x$ in the ρ-metric, and thus $t_{1k} \to t_1$, $t_{2k} \to t_2$. As we have seen in the proof of (11.1.i), x is AC in $[t_1, t_2]$ and $x'(t) = \xi(t)$ a.e. in $[t_1, t_2]$. Thus, the sequence of L_1-norms $\|x_k'\|$ is bounded. We have seen in the proof of (10.8.i) that in this case each of the assumptions L_1, L_2, L_3, L_4 guarantees that $J_{1k} = \int_{t_{1k}}^{t_{2k}} F_0(t, x_k(t), x_k'(t))\,dt$ is bounded below, say $J_1[x] \ge -M_1$. As in the proof of (11.1.i) we also have $J_2[x] = g(e[x]) \ge -M_2$. Thus, $i \ge -M_1 - M_2$ is finite. As in the proof of (11.1.i) by a suitable selection we may assume that $J_1[x_k] \to i_1$, $J_2[x_k] \to i_2$ as $k \to \infty$, both i_1, i_2 finite with $i_1 + i_2 = i$. The proof is now the same as for (11.1.i). □

Remark 4. If the convexity assumption in (11.1.ii) is not satisfied, then as for (11.1.i) we can guarantee the existence of generalized solutions if the sets $\tilde{R}(t, x) = \text{co } \tilde{Q}(t, x)$ have property K with respect to x, where $\tilde{Q}(t, x) = \text{epi } F_0(t, x, z)$.

Statement (11.1.ii) for classical integrals with integrands which are continuous in $A \times R^n$ has a simpler form:

11.1.iii (AN EXISTENCE THEOREM BASED ON WEAK CONVERGENCE OF THE DERIVATIVES). *Let A be compact, B closed, and let $F_0(t, x, z)$ be continuous on $A \times R^n$, and convex in z for every $(t, x) \in A$. Let $g(t_1, x_1, t_2, x_2)$ be lower semicontinuous on B. Let $\Omega = \{x\}$ be any nonempty Γ_w-closed class of* AC *functions $x(t) = (x^1, \ldots, x^n)$, $t_1 \le t \le t_2$, satisfying* (11.1.1), *and suppose that the class $\{x'\}$ is equiabsolutely integrable. Then the functional $I[x]$ in* (11.1.2) *has an absolute minimum in Ω.*

It is enough we verify the above requirements for the nonempty class Ω_M of all $x \in \Omega$ satisfying $I[x] \le M$ for some M. This is a corollary of (11.1.ii) and (L_4).

B. An Existence Theorem for Mayer Problems

For $F_0 = 0$ the problem (11.1.1–2) reduces to a Mayer problem. Now we are concerned with the existence of the minimum in the class Ω of the functional $I[x] = g(e[x])$ under the constraints

$$\begin{aligned} x(t) \in A(t), \qquad x'(t) \in Q(t, x(t)) \quad \text{for } t \in [t_1, t_2] \text{ (a.e.)}, \\ e[x] = (t_1, x(t_1), t_2, x(t_2)) \in B. \end{aligned} \tag{11.1.3}$$

Instead of (11.1.ii) we have the simpler statement:

11.1.iv (AN EXISTENCE THEOREM FOR MAYER PROBLEMS). *Let A be bounded, B closed, and g lower semicontinuous on B. Let us assume that for almost all $\bar{t}$ the sets $A(\bar{t})$ are closed and that the sets $Q(\bar{t}, x)$ are all closed and convex and have property* (K) *on the closed set $A(\bar{t})$. Let $\Omega = \{x\}$ be any nonempty Γ_w-closed class of* AC *functions $x(t) = (x^1, \ldots, x^n)$, $t_1 \le t \le t_2$, satisfying* (11.1.3), *and assume that the class of derivatives $\{x'\}$ is equiabsolutely integrable. Then the functional $I[x] = g(e[x])$ has an absolute minimum in Ω.*

It is enough we verify the above requirements for the nonempty class Ω_M of all $x \in \Omega$ satisfying $I[x] \le M$ for some M. This is a corollary of (11.1.ii) and (L_1).

C. The Linear Integrals

Let us consider now a linear problem

$$I[x] = g(t_1, x(t_1), t_2, x(t_2)) + \int_{t_1}^{t_2} \left[A_0(t, x) + \sum_{i=1}^{n} A_i(t, x) x'^i \right] dt, \tag{11.1.4}$$

$$(t, x(t)) \in A, \qquad (t_1, x(t_1), t_2, x(t_2)) \in B,$$

where A is a subset of $[t_0, T] \times R^n$, t_0, T finite, $Q(t, x) = R^n$, $M = A \times R^n$. Let $H(t, x, z) = A_0(t, x) + \sum_{i=1}^n A_i(t, x) z^i$.

11.1.v (AN EXISTENCE THEOREM FOR LINEAR INTEGRALS). *Let A be a subset of $[t_0, T] \times R^n$, and assume that all $A_i(t, x)$, $i = 0, 1, \ldots, n$, satisfy the condition* (CL) *or* (CL*) *of Section* 10.8B. *Assume that for every $N > 0$ there are a function $\phi(t) \ge 0$, $\phi \in L_1[t_0, T]$ and a constant $C > 0$ such that $|A_0(t, x)| \le \phi(t)$, $|A_i(t, x)| \le C$ for all $(t, x) \in A$ with $|x| \le N$. Let $\Omega = \{x\}$ be any nonempty Γ_w-closed class of* AC *functions $x(t) = (x^1, \ldots, x^n)$, $t_1 \le t \le t_2$, with graph in A, such that each trajectory $x \in \Omega$ has at least a point $(t^*, x(t^*))$ in a given compact set P, and such that the class of derivatives $\{x'\}$ is equiabsolutely integrable. Let B be closed, and g lower semicontinuous and bounded below on B. Then the functional* (11.1.4) *has both an absolute minimum and an absolute maximum in Ω.*

In particular the same conclusion holds if, for some $p > 1$ and any $N > 0$ there are functions ϕ, $\psi \geq 0$, $\phi \in L_1$, $\psi \in L_q$, $1/p + 1/q = 1$, such that $|A_0(t,x)| \leq \phi(t)$, $|A_i(t,x)| \leq \psi(t)$ for all $(t,x) \in A$ with $|x| \leq N$, and if the class $\{x'\}$ is relatively sequentially weakly compact in L_p.

Note that if for some $p > 1$ and D we have $\int_{t_1}^{t_2} |x'|^p\,dt \leq D$ for all $x \in \Omega$, then the class $\{x'\}$ is certainly relatively sequentially weakly compact in L_p and in L_1, and equiabsolutely integrable.

Proof. By (10.2.ii) the total variations $V[x; t_1, t_2]$ of the elements $x \in \Omega$ are equibounded, say $V[x, t_1, t_2] \leq N_1$. Since P is bounded, then $|x(t^*)| \leq M_1$ and $|x(t)| \leq M_1 + N_1$ for all $x \in \Omega$. Thus all trajectories $x \in \Omega$ lie in the compact set $S = [t_0, T] \times [|x| \leq M_1 + N_1]$. If ϕ and C are the corresponding elements, then $|A_0(t,x)| \leq \phi(t)$, $|A_i(t,x)| \leq C$, $i = 1, \ldots, n$, for all $(t,x) \in S$. Then $|H(t,x,z)| \leq \phi(t) + nC|z|$ for all $(t,x) \in S$ and $z \in R^n$, and condition (L_2) holds. The theorem follows now from (11.1.ii).

Here is an alternate proof of the main statement. Let $i = \inf[I[x] \,|\, x \in \Omega]$, and let $x_k(t)$, $t_0 \leq t_{1k} \leq t \leq t_{2k} \leq T$, $k = 1, 2, \ldots$, be a minimizing sequence, so that $I[x_k] \to i$ and $(t_k^*, x_k(t_k^*)) \in P$ for some $t_{1k} \leq t_k^* \leq t_{2k}$. First, there is a subsequence, say still $[k]$, such that $t_{1k} \to t_1, t_{2k} \to t_2, (t_k^*, x_k(t_k^*)) \to (t^*, x^*) \in P$. Also x_k' is equiabsolutely integrable, hence by (10.3.i) there are a function $\xi \in L_1$ and a further subsequence, say still $[k]$, such that $x_k' \to \xi$ weakly in L_1. Now $\{x_k\}$ is absolutely equicontinuous, and also equibounded since P is bounded and the functions x_k have equibounded total variations. Thus, $\|x_k'\|_1 \leq \mu$, and there are $\phi \in L_1$ and constant C such that $|A_0(t, x_k(t))| \leq \phi(t)$, $|A_i(t, x_k(t))| \leq C$. Hence $\|H(t, x_k(t), x_k'(t))\|_1 \leq \|\phi\|_1 + nC\mu$. Thus, i is finite. From $x_k(t) = x_k(t_k^*) + \int_{t_k^*}^t x_k'(\tau)\,d\tau$ we derive that $x(t) = x^* + \int_{t^*}^t \xi(\tau)\,d\tau$, hence $x'(t) = \xi(t)$ a.e.. Finally, by (10.8.iii) we derive that $H(t, x_k(t), x_k'(t)) \to H(t, x(t), x'(t))$ weakly in L_1, and $I[x_k] \to I[x] = i$.

For the case $p > 1$ the proof is the same but we can take the subsequence in such a way that $x_k' \to \xi$ weakly in L_p and $\xi \in L_p$. Now there is some $\psi \in L_q$ such that $|B(t, x_k(t))| \leq \psi(t)$, and again by (10.8.iii), $H(t, x_k(t), x_k'(t)) \to H(t, x(t), x'(t))$ weakly in L_1. □

D. Existence Theorems with Comparison Functionals and Isoperimetric Problems

Problems with constraints of the form $C[x] \leq D$ (i.e., with a comparison functional), and of the form $C[x] = D$ (isoperimetric problems) can be written in terms of optimal control as we have already shown in Sections 1.5 and 4.8, and as such we shall discuss them again in Section 11.4. However we shall prove here a few statements directly. Therefore let us consider here functionals and constraints of the form

$$
\begin{aligned}
&I[x] = g(t_1, x(t_1), t_2, x(t_2)) + \int_{t_1}^{t_2} F_0(t, x(t), x'(t))\,dt, \\
&C[x] = \int_{t_1}^{t_2} H(t, x(t), x'(t))\,dt, \qquad x(t) = (x^1, \ldots, x^n), \\
&x(t) \in A(t), \qquad x'(t) \in Q(t, x(t)) \qquad \text{for } t \in [t_1, t_2] \text{ (a.e.)}, \\
&e[x] = (t_1, x(t_1), t_2, x(t_2)) \in B, \qquad F_0(\cdot, x(\cdot), x'(\cdot)) \in L_1, \\
&H(\cdot, x(\cdot), x'(\cdot)) \in L_1
\end{aligned}
\tag{11.1.5}
$$

where F_0 and H satisfy either condition (C) or (C*).

11.1.vi (An Existence Theorem with a Comparison Functional). *Let A be bounded, B closed, and g lower semicontinuous on B. Assume that both functions $F_0(t, x, z)$, $H(t, x, z)$ satisfy condition* (C) *or* (C*), *that are both convex in z, that one of them satisfies any of the growth conditions* ($\gamma 1$), ($\gamma 2$), ($\gamma 3$), *and that the other satisfies any of the conditions* (L_i). *Let Ω be any nonempty Γ_w-closed class of* AC *functions $x(t) = (x^1, \ldots, x^n)$, $t_1 \leq t \leq t_2$, satisfying $x(t) \in A(t)$, $e[x] \in B$, and both $F_0(\cdot, x(\cdot), x'(\cdot))$, $H(\cdot, x(\cdot), x'(\cdot))$ are L-integrable. Let D be a constant such that the class Ω_D of all $x \in \Omega$ with $C[x] \leq D$ is not empty. Then $I[x]$ has an absolute minimum in Ω_D.*

Also, let N be a constant such that the class Ω'_N of all $x \in \Omega$ with $I[x] \leq N$ is not empty. Then $C[x]$ has an absolute minimum in Ω'_N.

Proof. Let $i = \operatorname{Inf}[I[x], x \in \Omega_D]$ and $j = \operatorname{Inf}[C[x], x \in \Omega'_N]$. For the first part of the statement we take a minimizing sequence $[x_k]$ for $I[x]$ in Ω_D, that is, with $I[x_k] \to i$, $C[x_k] \leq D$. For the second part we take a minimizing sequence x_k for $C[x]$ in Ω'_N, that is, with $I[x_k] \leq N$, $C[x_k] \to j$. In either case, because of the growth conditions, the sequence of derivatives $[x'_k]$ is equiabsolutely integrable. As in (11.1.i) there is an AC function $x(t)$, $t_1 \leq t \leq t_2$, and a subsequence, say still $[x_k]$, such that $x_k \to x$ in the ρ-metric, and $x'_k \to x'$ weakly in L_1. Now both sequences $I[x_k]$ and $C[x_k]$ are bounded below, one because of the growth conditions, and one because of the properties (L_i), and thus both i and j are finite. Now the argument for (11.1.i) applies. In the first case we conclude that $I[x] \leq i$, $C[x] \leq D$; in the second case we conclude that $I[x] \leq N$, $C[x] \leq j$. Since Ω is Γ_w-closed we conclude that $x \in \Omega$. In the first case we conclude that $I[x] \leq i$, and by (10.8.i) we derive that $C[x] \leq D$. In the second case we conclude that $C[x] \leq j$; and by (10.8.i) we derive that $I[x] \leq N$. Since Ω is Γ_w-closed, we conclude that $x \in \Omega$. Thus, in the first case, $x \in \Omega_D$ and $I[x] \geq i$, hence $I[x] = i$, while in the second case, $x \in \Omega'_N$ and $C[x] \geq j$, hence $C[x] = j$. Both parts of (11.1.vi) are thereby proved. □

For a moment let $I[x]$ and $C[x]$ be any two functionals in a class Ω of admissible trajectories, and assume that, for all D in a certain range, $I[x]$ has always an absolute minimum in the class K_D of all $x \in \Omega$ with $C[x] = D$.

11.1.vii. *If* $\mathrm{Min}[I[x] | x \in \Omega,\ C[x] = D] = i_D$ *and* i_D *is a strictly increasing function of* D, *then* $\mathrm{Max}[C[x] | x \in \Omega,\ I[x] = i_D] = D$, *and the maximizing elements* $x_0 \in \Omega$ *are the same with* $I[x_0] = i_D$, $C[x_0] = D$. *An analogous statement holds by exchanging minima and maxima and the sense of the inequalities.*

Proof. First we see that for any minimizing element $x_0 \in K_D$ for I we have $I[x_0] = i_D$, $C[x_0] = D$, and x_0 belongs to the class K' of all elements $x \in \Omega$ with $I[x] = i_D$. If this class would contain an element $\bar{x}$ with $C[\bar{x}] = \bar{D} > D$, then in the class $K_{\bar{D}}$ of all $x \in \Omega$ with $C[x] = \bar{D}$, the minimum of $I[x]$ is given by an element $\bar{x}_0 \in K_{\bar{D}}$ with $I[\bar{x}_0] = i_{\bar{D}} \leq i_D$, while $I[\bar{x}_0] = i_D$, $C[\bar{x}_0] = \bar{D}$ shows that the minimum of $I[x]$ in K_D must be $< i_D$, a contradiction. □

Now we prove an existence theorem for isoperimetric problems with F_0 a general extended function and H a linear function in x'.

11.1.viii (An Existence Theorem for Isoperimetric Problems). *Let* A *be bounded,* B *closed, and* g *lower semicontinuous on* B. *Assume that* $F_0(t, x, z)$ *satisfies condition* (C) *or* (C*), *that* F_0 *satisfies any of the growth conditions* (γ1), (γ2), (γ3), *and* F_0 *is convex in* z. *Let* $C[x] = \int_{t_1}^{t_2} H\,dt$ *with* $H(t, x, x') = A_0(t, x) + \sum_1^n A_i(t, x)x'^i$ *and all* A_0, A_i *continuous on the closed* A. [*Alternatively, let us assume that* H *is* B*-measurable and that for almost all* $\bar{t}$ *the functions* $A_0(\bar{t}, x)$, $A_i(\bar{t}, x)$ *are continuous on the set* $A(\bar{t})$, *and that* $|A_0(t, x)| \leq \phi(t)$, $|A_i(t, x)| \leq C_i$ *for all* (t, x) *and suitable constants* C_i *and* L*-integrable function* $\phi(t)$.] *Let* Ω *be any nonempty* Γ_w*-closed class of* AC *functions* $x(t) = (x^1, \ldots, x^n)$, $t_1 \leq t \leq t_2$, *satisfying the requirements* (11.1.5). *Let* D *be a constant such that the class* K_D *of all* $x \in \Omega$ *with* $C[x] = D$ *is not empty. Then* $I[x]$ *has an absolute minimum in* K_D.

If A is unbounded but contained in a slab $[t_0 \leq t \leq T] \times R^n$, then theorem (11.1.viii) still holds provided we know that every trajectory $x \in K_D$ has at least a point $(t^*, x(t^*))$ on a given compact set P.

Actually, in the latter situation, we may relax the requirements above as follows.

First, we may only assume that H is B-measurable, that for almost all t the functions $A_0(t, x)$, $A_i(t, x)$ are continuous with respect to x on the set $A(t)$, and moreover that for any constant $N > 0$ there are constants C_i and an L-integrable function $\phi(t)$ such that $|A_0(t, x)| \leq \phi(t)$, $|A_i(t, x)| \leq C_i$ for all $(t, x) \in A$ with $|x| \leq N$.

Alternatively, we may only assume that F_0 satisfies one of the conditions (L_i) (instead of (γ1)–(γ3)), but we need to know that the class $\{x'\}$ of the derivatives of the elements $x \in K_D$ is equiabsolutely integrable.

Finally, we can only assume that F_0 satisfies one of the conditions (L_i), that for every $N > 0$ there are functions $\phi(t)$, $\psi_i(t)$, $\phi \in L_1$, $\psi_i \in L_q$, $1/p + 1/q = 1$, $p > 1$, such that $|A_0(t, x)| \leq \phi(t)$, $|A_i(t, x)| \leq \psi_i(t)$ for all $(t, x) \in A$ with $|x| \leq N$, and that the class $\{x'\}$ is relatively sequentially weakly compact in L_p.

Proof of (11.1.viii). For $i = \inf[I[x], x \in K_D]$ and any minimizing sequence $x_k(t), t_{1k} \le t \le t_{2k}, k = 1, 2, \ldots$, for $I[x]$ in K_D, we have $I[x_k] \to i, C[x_k] = D$. Under the main assumption of the theorem, then by the process used for the proof of (11.1.i) we derive that i is finite and that there are a subsequence, say still $[k]$, and an AC function $x(t)$, $t_1 \le t \le t_2$, such that $x_k \to x$ in the ρ-metric, $x'_k \to x'$ weakly in L_1, and that $I[x] \le i$. Since $C[x]$ is a continuous functional as we proved at the end of Section 10.8, we conclude that $C[x] = D$. Hence, by the closedness property of Ω, we derive that $x \in \Omega$, and then $x \in K_D$, $I[x] \ge i$, and by comparison $I[x] = i$.

If A is unbounded but contained in a slab as stated, then the sequence x'_k is equiabsolutely integrable, hence the total variations $V[x_k]$ are equibounded. Since the trajectories x_k have a point on the compact set P, the same trajectories x_k are also equibounded, say $|x_k(t)| \le N$ for some N and all t and k. Again $C[x]$ is a continuous functional and $C[x] = D$. An analogous argument holds in the other cases. □

It is possible to invert the role of F_0 and H in (11.1.viii) under some mild assumptions on the class Ω, on H and on F_0.

We say that an element $x(t)$, $t_1 \le t \le t_2$, of a given class Ω has property (π) at a point $(\bar{t}, \bar{x})$, $\bar{x} = x(\bar{t})$, provided $t_1 < \bar{t} < t_2$, $(\bar{t}, \bar{x})$ is in the interior of A, and there is $\delta > 0$ with $N_\delta(\bar{t}, \bar{x}) \subset A$ such that if we replace any arc $\lambda_0 : x = x(t)$, $\alpha \le t \le \beta$, contained in $N_\delta(\bar{t}, \bar{x})$, by any other arc $\lambda : x = \lambda(t)$, $\alpha \le t \le \beta$, $\lambda(\alpha) = x(\alpha)$, $\lambda(\beta) = x(\beta)$, also contained in $N_\delta(\bar{t}, \bar{x})$, then the new trajectory $\bar{x}(t)$, $t_1 \le t \le t_2$, belongs to Ω.

For $n = 1$, $H = A_0(t, x) + A_1(t, x)x'$, we assume that A_0, A_1 are of class C^1, and that every element x of Ω has at least one point $(\bar{t}, \bar{x})$, $\bar{x} = x(\bar{t})$, possessing property (π) and at which $A_{0x} \ne A_{1t}$.

Finally, we need some assumptions of F_0. Indeed for $n = 1$ we need to know that for all $(t, x) \in N_\delta(\bar{t}, \bar{x})$ the convex set $Q(t, x)$ coincides with R and F_0 is continuous in $N_\delta(\bar{t}, \bar{x}) \times R$.

11.1.ix. *Under the conditions of* (11.1.viii), *for* $n = 1$ *and the above assumptions, if* N *is such that the class* K'_N *of all* $x \in \Omega$ *with* $I[x] = N$ *is not empty, then* $C[x]$ *has an absolute minimum in* K'_N.

Proof. As usual let j denote the infimum of $C[x]$ in K'_N and, as in the proof of (11.1.vi), let $[x_k]$ be a minimizing sequence for $C[x]$, that is, $x_k \in K'_N$, $C[x_k] \to j$, $I[x_k] = N$. By the same argument as in (11.1.i) there is an AC function $x(t) = (x^1, \ldots, x^n)$, $t_1 \le t \le t_2$, and a subsequence, say still $[x_k]$, such that $x_k \to x$ in the ρ-metric, $x'_k \to x'$ weakly in L_1, and $I[x] \le N$. By the continuity of $C[x]$ as in Section 10.8, we also have $C[x] = j$. If $I[x] = N$, then $x \in K'_N$ and the proof is complete. We shall prove that $I[x] = N$. To this effect, we assume that $I[x] < N$ and we construct another trajectory $\bar{x} \in \Omega$, with $I[x] = N$ and $C[x] < j$, a contradiction.

For $n = 1$, let us assume that say, $A_{0x} > A_{1t}$ at $(\bar{t}, \bar{x})$. We can take $\delta > 0$ so small that $A_{0X} - A_{1T} > 0$ in $N_\delta(\bar{t}, \bar{x})$. Let $\lambda_0 : x = x(t)$, $\alpha \le t \le \beta$, be an arc of x contained in $N_\delta(\bar{t}, \bar{x})$ and such that $x'(\alpha)$ and $x'(\beta)$ exist and are finite, and let $\lambda : x = \lambda(t)$, $\alpha \le t \le \beta$, be a polygonal line also contained in $N_\delta(\bar{t}, \bar{x})$ with $\lambda(\alpha) = x(\alpha)$, $\lambda(\beta) = x(\beta)$, and completely above λ_0. Actually, we may think of the slopes of λ to be $\pm k$ with k large.

Because of the growth hypotheses on F_0, $I[\lambda]$ can be made as large as we want by taking k large. On the other hand, by taking β as close to α as needed, we may give to $I[\lambda]$ any value $\geq I[\lambda_0]$. Thus we can always arrange that $I[\lambda] = N - I[x]$. If $\bar{x}$ is the trajectory x with the arc λ replacing λ_0, then $I[\bar{x}] = N$. On the other hand, by Green's theorem, if $\sum$ denotes the region bounded by λ_0 and $-\lambda$, then

$$\int_{\lambda_0} H\,dt - \int_{\lambda} H\,dt = \int_{\partial\Sigma} A_0\,dt + A_1\,dx = \iint_{\Sigma} (A_{0x} - A_{1t})\,dt dx > 0.$$

Hence

$$C[\bar{x}] = C[x] + \left(\int_{\lambda} H\,dt - \int_{\lambda_0} H\,dt\right) < j,$$

a contradiction. If $A_{0x} < A_{1t}$ the argument is the same with the arc λ completely below the arc λ_0. This proves (11.1.ix) for $n = 1$. □

For $n > 1$, $H = A_0(t,x) + \sum_1^n A_i(t,x)x'^i$, we assume that $A_0, \ldots, A_n$ are of class C^1 in A, so that, if $V(t,x)$ denotes the $(n+1)$-vector $V(t,x) = (A_0, \ldots, A_n)$ and $n' = n(n+1)/2$, then the usual n'-vector curl V, constructed with the first order partial derivatives of $A_0, \ldots, A_n$, is a continuous function of (t,x) in A. Then we need the same assumption (π) and the assumption that curl $V \neq 0$ at some point $(\bar{t}, \bar{x})$ of the trajectory, that is, some of its components are not zero. Then the same theorem (11.1.ix) still holds under mild and generic further assumptions which for the sake of brevity we do not state here. We refer for the precise statement and proof to P. Pucci [1]. For $n > 1$ the proof is rather technical and differs from the previous one in many respects.

11.2 Elimination of the Hypothesis that A Is Bounded in Theorems (11.1.i–iv)

The hypothesis that A is bounded in the theorems (11.1.i–iii) can be easily removed as we have done in Chapter 9. All we have to do is to guarantee that a minimizing sequence can be contained in some bounded subset A_0 of A. After that the statements and the proofs are the same. Often all this results at a glance from the data and the geometrical configuration of the particular problem under consideration. However, we shall list here some general conditions for this to occur.

A. If the set A is not bounded, but A is contained in a fixed slab $[t_0 \leq t \leq T, x \in R^n]$ of R^{n+1}, then Theorem (11.1.i) is still valid if we know, for instance, that

(h_1) g is bounded below on B, say $g \geq -M_1$, and

(C_1) Ω is a given nonempty Γ_w-closed class of AC trajectories x each of which has at least one point $(t^*, x(t^*))$ on a given compact subset P of A (t^* may depend on the trajectory).

For instance, the curves $C: x = x(t)$, $t_1 \leq t \leq t_2$, may have the first end point (or the second end point) either fixed or on some fixed compact set P of A.

It is enough to consider only those elements $x \in \Omega$ with $I[x] \leq M$ for some M; hence $\int_{t_1}^{t_2} F_0\,dt \leq M + M_1$. Then, under conditions (C_1) and ($\gamma 1$), we have $\phi(\zeta) \geq -\nu$, $\nu \geq 0$, a constant, and there is some $N \geq 0$ such that $\phi(\zeta) \geq \zeta$ for all $\zeta \geq N$. If $x \in \Omega$, if

E^* is the subset of $[t_1, t_2]$ where $|x'(t)| \geq N$, and if $E = [t_1, t_2] - E^*$, then

$$F_0(t, x(t), x'(t)) + v \geq 0$$

for all t,

$$F_0(t, x(t), x'(t)) \geq |x'(t)|$$

for $t \in E^*$, $|x'(t)| \leq N$ for $t \in E$, and

$$\begin{aligned} M + M_1 &\geq \int_{t_1}^{t_2} [F_0(t, x(t), x'(t)) + v]\, dt - \int_{t_1}^{t_2} v\, dt \\ &\geq \int_{E^*} [F_0(t, x(t), x'(t)) + v]\, dt - v(T - t_0) \\ &\geq \int_{E^*} |x'(t)|\, dt - v(T_0 - t_0) + \int_E [|x'(t)| - N]\, dt \\ &= \int_{t_1}^{t_2} |x'(t)|\, dt - (v + N)(T - t_0), \end{aligned}$$

or $\int_{t_1}^{t_2} |x'(t)|\, dt \leq M + (v + N)(T - t_0)$, a fixed number. Thus, the curves $C: x = x(t)$, $t_1 \leq t \leq t_2$, under consideration have total variation below a fixed number, and since they contain a point of the bounded set P, they are contained in some fixed cylinder $[(t, x) | t_0 \leq t \leq T, |x| \leq M_0]$.

Under condition (C_1) and $(\gamma 2)$, taking $\varepsilon = 1$, we have $|z| \leq \psi_1(t) + F_0$, or

$$M + M_1 \geq \int_{t_1}^{t_2} F_0(t, x(t), x'(t))\, dt \geq \int_{t_1}^{t_2} [|x'(t)| - \psi_1(t)]\, dt \geq \int_{t_1}^{t_2} |x'(t)|\, dt - \int_{t_0}^{T} \psi_1\, dt,$$

and again the curves under consideration have length below a fixed number.

The same conclusion holds under conditions (C_1) and $(\gamma 3)$.

If A is neither bounded nor contained in any slab as above, then theorem (11.1.i) is still valid if we know for instance that (h_1) and (C_1) hold, and in addition that

C_2. There are constants $\mu > 0$, $R_0 \geq 0$ such that $F_0(t, x, z) \geq \mu > 0$ for all (t, x, z) with $|t| \geq R_0$.

Indeed, any part of the curve $C: x = x(t)$ lying in the slab $[-R_0 \leq t \leq R_0, x \in R^n]$ has an integral bounded below, say $\geq -M_1$ for some constant M_1. We assume R_0 large enough so that P is completely contained in the slab. Again let $I[x] \leq M$. Now, if $R_1 = R_0 + \mu^{-1}(M_1 + M)$, then any of the curves C above must be contained in the slab $[-R_1 \leq t \leq R_1, x \in R^n]$, since otherwise, such a curve would contain at least an arc with $-R_1 \leq t \leq -R_0$, or $R_0 \leq t \leq R_1$, and then $I[x] \geq \mu(R_1 - R_0) - M_1 > M$.

B. Concerning Theorem (11.1.ii), if A is not bounded but contained in a slab $[t_0 \leq t \leq T] \times R^n$, t_0, T finite, we may again assume that (h_1) and (C_1) hold, and that the Γ_w-closed class $\Omega = \{x(t), t_1 \leq t \leq t_2\}$ of AC trajectories is such that the class $\{x'(t), t_1 \leq t \leq t_2\}$ is equiabsolutely integrable. Indeed, by (10.2.ii), the total variations $V[x; t_1, t_2]$ are equibounded, say $V[x] \leq N_1$. Since any trajectory x in Ω has at least one point $(t^*, x(t^*)) \in P$ in a given compact set P, then $|x(t^*)| \leq M_1$ and $|x(t)| \leq M_1 + N_1$, $t_1 \leq t \leq t_2$, for every element x of Ω.

It may be of interest to know that the same conclusion can be derived from assumptions (h_1), (C_1), and the following one:

C_3. There are constants $c > 0$, $R \geq 0$ and a locally integrable function $\psi(t) \geq 0$, $t \in R$, such that $F_0(t, x, z) \geq -\psi(t) + c|z|$ for all (t, x, z) with $|x| \geq R$.

Indeed, we may take R so large that P is completely contained in the cylinder $\tilde{A} = [(t,x)|t_0 \le t \le T, |x| \le R]$. Now let x be an element of Ω with $I[x] \le M$. The parts of C inside $\tilde{A}$ contribute to the value of $I[x]$ an amount certainly above some constant $-\mu$. Now take the cylinder $\tilde{A}_1 = [(t,x)|t_0 \le t \le T, |x| \le R + R_1]$. If E^* denotes the set of all $t \in [t_1, t_2]$ with $|x(t)| \ge R$, then

$$\begin{aligned}\int_{E^*} |x'(t)|\,dt &\le c^{-1} \int_{E^*} [F_0(t, x(t), x'(t)) + \psi(t)]\,dt \\ &\le c^{-1}\left[I[x] + \mu + \int_{t_0}^{T} \psi(t)\,dt\right] \\ &\le c^{-1}(M + \mu + M_2),\end{aligned}$$

a fixed constant. Thus, C is completely inside $\tilde{A}_1$ if R_1 is larger then the above constant.

If A is not bounded nor contained in any slab as above, then (11.1.ii) is still valid provided we know, for instance, that (h_1), (C_1), (C_2), (C_3) hold.

Concerning Theorem (11.1.iii), the same remarks above hold as for (11.1.ii).

C. Concerning Theorem (11.1.iv) for Mayer problems, then for A closed and contained in a slab as above, the following condition may be used:

h_2. $g(t_1, x_1, t_2, x_2) \to +\infty$ *as* $|x_1| + |x_2| \to \infty$ *uniformly with respect to* t_1, t_2.

For A not compact or contained in any slab as above, we may use the condition:

h_3. $g(t_1, x_1, t_2, x_2) \to +\infty$ as $|t_1| + |x_1| + |t_2| + |x_2| \to +\infty$.

11.3 Examples

Many simple examples concerning Lagrange problems of the calculus of variations have been anticipated in Section 2.20B and they all concerned Theorem (11.1.i) with $g = 0$ and F_0 continuous in $A \times R^n$. Here are a few more examples concerning Theorems (11.1.i–iv) without restrictions.

1. Take $n = 1$, $F_0 = 0$ for $|x'| \le 1$, $F_0 = x'^2$ for $|x'| > 1$, $g(x_2) = 4$ if $|x_2| < 2$, $g(x_2) = 0$ if $|x_2| \ge 2$, $t_1 = 0$, $x_1 = 0$, $t_2 = 1$, x_2 undetermined. Here $A = [0,1] \times R$, $B = (0,0,1) \times R$, g is lower semicontinuous, $F_0(t,x,x')$ satisfies growth condition (gl) and, for every $\bar{t}$, $F_0(\bar{t},x,x')$ is lower semicontinuous in (x,x'). Conditions (h_1) and (C_1) hold. By (11.1.i), $I[x]$ has an absolute minimum in the class Ω of all trajectories under the constraints.

2. Take $n = 1$, $F_0 = t^{1/2}(1-t)^{1/2}x'^2$, $g(x_2) = 0$ for $-\infty < x_2 \le 1$, $g(x_2) = 1$ for $1 < x_2 < +\infty$, $t_1 = 0$, $x_1 = 2$, $t_2 = 1$, x_2 undetermined. Here $A = [0,1] \times R$, $B = (0,2,1) \times R$, g is lower semicontinuous. Let us prove that F_0 satisfies growth condition (g2). To this effect, given $\varepsilon > 0$, take $\psi_\varepsilon(t) = \varepsilon^{-1}t^{-1/2}(1-t)^{-1/2}$, $0 < t < 1$, an L-integrable function in $(0,1)$. Then, for $|z| \le \varepsilon^{-1}t^{-1/2}(1-t)^{-1/2}$ we have $|z| \le \psi_\varepsilon(t) \le \psi_\varepsilon(t) + F_0$; for $|z| > \varepsilon^{-1}t^{-1/2}(1-t)^{-1/2}$ we have $\varepsilon t^{1/2}(1-t)^{1/2}|z| > 1$ and $|z| < \varepsilon t^{1/2}(1-t)^{1/2}z^2$, hence $|z| < \varepsilon F_0 \le \psi_\varepsilon(t) + \varepsilon F_0$. Also, for every $\bar{t}$, $F_0(\bar{t},x,z)$ is a continuous function of (x,z). Let Ω denote the class of all AC trajectories satisfying the above constrains and also satisfying the further constraint $|x(1/2)| \le 1$. The class Ω is Γ_w-closed. By (11.1.i), $I[x]$ has an absolute minimum in the class Ω.

3. Take $n = 2$, $F_0 = \sin t + (2 + \cos tx)x'^2 + (2 + \sin ty)y'^2$ if $0 \le t \le 1$, $F_0 = 1 + x'^2 + y'^2$ if $t > 1$, $t_1 = 0$, $x_1 = 0$, $y_1 = 0$, (t_2, x_2, y_2) on the locus $\Gamma = [(t-1)(x^2 + y^2) = 1, t > 1]$, $g(t_2, x_2, y_2) = (t_2 - 1)^{-1}$, $A = [0, +\infty) \times R^2$, $B = (0,0,0) \times \Gamma$. Conditions (g1), ($h_1$) and ($C_1$) hold. By (11.1.i), $I[x]$ has an absolute minimum in the class Ω of all trajectories under the constraints.

4. Take $n = 1$, $F_0 = |t - x|^{-1/2}x'^2$ if $|t - x| \le 1$, and $x' \ge -t^{-1}$, $F_0 = +\infty$ otherwise. In particular $F_0 = +\infty$ for $x = t$, $0 \le t \le 1$. Take $g = 0$, $t_1 = 0$, $x_1 = 0$, $t_2 = 1$, $x_2 = 1$. Here $A = [(t,x)|0 \le t \le 1, |x - t| \le 1]$, $B = (0,0,1,1)$. For every $\bar{t} \in [0,1]$ the extended function $F_0(\bar{t}, x, z)$ is lower semicontinuous in (x, z), and for every $(\bar{t}, \bar{x}) \in A$ the extended function $F_0(\bar{t}, \bar{x}, z)$ is convex in z. Here $|t - x| \le 1$, hence $|t - x|^{-1/2} \ge 1$, and $F_0 \ge x'^2$. Thus, (g1) holds with $\phi(\zeta) = \zeta^2$. Conditions (h_1) and (C_1) hold. The class Ω of all trajectories satisfying the data is not empty. Indeed, let us prove that $x(t) = t + \varepsilon t(1 - t)$, $0 \le t \le 1$, for any $0 < \varepsilon \le 1$ belongs to Ω. First $|x(t) - t| \le 1$ and thus $(t, x(t)) \in A$ for all $0 \le t \le 1$. Moreover, $x'(t) = 1 + \varepsilon - 2\varepsilon t > -t^{-1}$ for $0 < t \le 1$; hence $x'(t) \in Q(t, x(t))$ since $Q(t,x) = [z | z \ge -t^{-1}]$. Finally,

$$|t - x(t)|^{-1/2}x'^2(t) = \varepsilon^{-1/2}t^{-1/2}(1 - t)^{-1/2}(1 + \varepsilon - 2\varepsilon t)^2$$

is L-integrable in $(0, 1)$. By (11.1.i), $I[x]$ has an absolute minimum in Ω.

5. Minimum of $\int_{t_1}^{t_2} (x' - 2t + 1)^2\, dt$ under the constraints $t_1 = 0$, $x(t_1) = 1$, $(t_2, x(t_2))$ on the locus $\Gamma = [t^2 + x^2 = 4, t \ge 0]$, and $\int_{t_1}^{t_2} t|x'|\, dt \le 1/2$. Here $n = 1$, $A = [t \ge 0, t^2 + x^2 \le 4]$, $F_0 = (x' - 2t + 1)^2$, $H = t|x'|$, F_0 satisfies ($\gamma 1$), H satisfies (L_1). The minimum exists by (11.1.vi).

6. Minimum of $\int_{t_1}^{t_2} (t^2|x'| + x')\, dt$ under the constraints $t_1 = -1$, $x(t_1) = -1$, t_2 undetermined, $x(t_2) = 1$, and $\int_{t_1}^{t_2} (1 + x'^2)\, dt \le 5$. Here $H = 1 + x'^2 \ge 1$; hence $t_2 \le 4$, and we can take $A = [-1, 4] \times [-1, 1]$. Then $F_0 = t^2|x'| + x'$ satisfies (L_2), and $H = 1 + x'^2$ satisfies ($\gamma 1$). The minimum exists by (11.1.vi).

7. Minimum of $\int_{t_1}^{t_2} x'^2\, dt$ under the constraints $t_1 = -1$, $x(t_1) = 0$, $t_2 = 1$, $x(t_2) = 1$, and $\int_{t_1}^{t_2} (tx + (\operatorname{sgn} t + x)x')\, dt = 2$. Here $n = 1$, $F_0 = x'^2$, $H = tx + (\operatorname{sgn} t + x)x'$, $A_0 = tx$, $A_1 = \operatorname{sgn} t + x$, $|A_0| \le |x|$, $A_1 \le 1 + |x|$. The minimum exists by (11.1.viii).

8. Minimum of $\int_{t_1}^{t_2} x^2\, dt$ under the constraints t_1, $x(t_1)$, t_2, $x(t_2)$ fixed, $t_1 < t_2$, and $\int_{t_1}^{t_2} x'^2\, dt = C$ for a given $C > (x(t_2) - x(t_1))^2(t_2 - t_1)^{-1}$. Here $n = 1$, $F_0 = x^2$, $A_0 = x^2$, $A_1 = 0$, $A_{0x} - A_{1t} = 2x \ne 0$ for all $x \ne 0$, $H = x'^2$ satisfies ($\gamma 1$). The minimum exists by (11.1.ix). Here the same integral has also a maximum under the same constraints.

Exercises

Show that some of the Existence Theorems of Sections 11.1–2 apply:

1. Take $n = 1$, $g = 0$. For $t \le 0$ take $F_0(t, x, x') = (1 + t^2 + x^2)x'^2$ if $x' \le 0$, $F_0 = +\infty$ if $x' > 0$; for $t > 0$ take $F_0 = (1 + t^2 + x^2)x'^2$ if $x' \ge 0$, $F_0 = +\infty$ if $x' < 0$. Here $Q(t, x)$ are the sets $Q(t, x) = [-\infty < z \le 0]$ if $t \le 0$; $Q(t, x) = [0 \le z < +\infty]$ if $t > 0$. Let Ω denote the class of all admissible trajectories with t_1, x_1, t_2, x_2 fixed, $t_1 < 0 < t_2$, $x_1 \ne x_2$.
2. Take $n = 1$, $F_0 = (1 + t + x)x'$, $g = 0$, $t_1 < t_2$, x_1, x_2 fixed. Let Ω denote the class of all AC functions $x(t)$, $t_1 \le t \le t_2$, with $\int_{t_1}^{t_2} x'^2\, dt \le D$ where D is any number larger than $(x_2 - x_1)^2(t_2 - t_1)^{-1}$.
3. Take $n = 2$, $F_0 = t^{-1/2} + (1 + t + x)x' + (1 - t - y)y'$ a linear integrand, $g = 0$, $t_1 = 0$, $x_1 = 1$, $y_1 = 1$, $t_2 = 1$, (x_2, y_2) on the locus $\Gamma = [x^2 + y^2 = 4]$. Let Ω denote the class of all admissible trajectories with $\int_0^1 (x'^2 + y'^2)\, dt \le 4$.

4. Take $n = 2$, $\varphi(t) = \operatorname{sgn} t$, $F_0 = t^{-1/2}\sin(x+y) + \varphi(t)(2x+y)x' + \varphi(t)(x+2y)y'$, $g = 0$, $t_1 = -1$, $x_1 = 1$, $y_1 = 1$, $t_2 = 1$, (x_2, y_2) undetermined. Let Ω denote the class all admissible trajectories with $\int_0^1 (x'^2 + y'^2)\,dt \le 2$.

11.4 Existence Theorems for Problems of Optimal Control with Unbounded Strategies

A. Existence Theorems for Lagrange and Bolza Problems of Optimal Control

Essentially, most of the existence theorems for Lagrange and Bolza problems of optimal control presented here are corollaries of the existence theorems of Section 11.1 for extended free problems. However, new remarks are needed in connection with a different emphasis and different possible applications.

We are concerned here with the problem of the absolute minimum of the functional

$$I[x,u] = g(t_1, x(t_1), t_2, x(t_2)) + \int_{t_1}^{t_2} f_0(t, x(t), u(t))\,dt \tag{11.4.1}$$

with constraints, boundary conditions, and differential equations

$$\begin{gathered} x(t) \in A(t), \qquad (t_1, x(t_1), t_2, x(t_2)) \in B, \qquad u(t) \in U(t, x(t)), \\ dx/dt = f(t, x(t), u(t)), \qquad t \in [t_1, t_2] \text{ (a.e.)}, \\ f_0(\cdot, x(\cdot), u(\cdot)) \in L_1[t_1, t_2], \qquad x(t) \text{ AC}, \\ u(t) \text{ measurable in } [t_1, t_2], \end{gathered} \tag{11.4.2}$$

where $x = (x^1, \dots, x^n)$, $f = (f_1, \dots, f_n)$, $u = (u^1, \dots, u^m)$. Here A is a given subset of the tx-space R^{n+1}, B a given subset of the $t_1x_1t_2x_2$-space R^{2n+2}, and for every $(t,x) \in A$, $U(t,x)$ is a given subset of the u-space R^m. Let $g(t_1, x_1, t_2, x_2)$ be a given real valued function on B. Let A_0 denote the projection of A on the t-axis. For every $t \in A_0$ let $A(t)$ denote the corresponding section of A, or $A(t) = [x \mid (t,x) \in A] \subset R^n$. Let M denote the set of all (t,x,u) with $(t,x) \in A$, $u \in U(t,x)$. Let $f_0(t,x,u)$, $f(t,x,u) = (f_1, \dots, f_n)$ be given functions defined on M, and for every $(t,x) \in A$ let $\tilde{Q}(t,x)$ denote the set of all $(z^0, z) \in R^{n+1}$ with $z^0 \ge f_0(t,x,u)$, $z = f(t,x,u)$, $u \in U(t,x)$. Then the projection $Q(t,x)$ of $\tilde{Q}(t,x)$ on the z-space R^n is the set $Q(t,x) = [z \mid z = f(t,x,u), u \in U(t,x)]$, or $Q(t,x) = f(t,x,U(t,x))$. Below we shall assume A_0 to be an interval of the t-axis, finite or infinite.

As usual, we say that a pair $x(t)$, $u(t)$, $t_1 \le t \le t_2$, is admissible for the problem (11.4.1–2) if x is AC, u is measurable, and the requirements (11.4.2) are satisfied. A function $x(t)$, $t_1 \le t \le t_2$, is said to be admissible for the problem (11.4.1–2) if there is some u such that (x,u) is admissible. Given a class $\Omega = \{(x, u)\}$ of admissible pairs for (11.4.1–2) we may denote by $\Omega_x = \{x\}$ the class of corresponding trajectories, or $\Omega_x = \{x\} = \{x | (x, u) \in \Omega\}$. Given any class Ω of admissible pairs for (11.4.1) we shall denote by i the infimum of $I[x, u]$ in Ω.

We may be concerned with the problem of the minimum of $I[x, u]$ in the class Ω of all admissible pairs. Alternatively, and as in Sections 9.2, 9.3, 11.1, we may want to minimize $I[x, u]$ in a smaller class Ω of admissible pairs, and in this case we need to know that Ω has a mild closedness property. We say that Ω is Γ_{0w}-closed provided (a) Ω_x is Γ_w-closed (Section 11.1), and (b) $x \in \Omega_x$ and (x, u) admissible for (11.4.2), implies that $(x, u) \in \Omega$. In other words, the class Ω is Γ_{0w}-closed provided: if $x_k(t)$, $u_k(t)$, $t_{1k} \leq t \leq t_{2k}$, $k = 1, 2, \ldots$, are admissible pairs all in Ω, if $x(t)$, $u(t)$, $t_1 \leq t \leq t_2$, is an admissible pair, and if $x_k \to x$ in the weak convergence of the derivatives, (that is, $x_k \to x$ in the ρ-metric and $x_k' \to x'$ weakly in L_1), then (x, u) belongs to Ω.

Actually somewhat less than (b) is needed. Namely, in proving the existence of the minimum it would be enough to know that (a) holds, and that (b′) if $x \in \Omega_x$ and (x, u) is admissible, then either $(x, u) \in \Omega$, or there is some $\bar{u}$ such that $(x, \bar{u}) \in \Omega$ and $I[x, \bar{u}] \leq I[x, u]$.

Let M_0 denote the set of all $(t, x, z) \in R^{2n+1}$ with $(t, x) \in A$, $z \in Q(t, x)$. We shall need the function $T(t, x, z)$, $-\infty \leq T < \infty$, defined on M_0 by taking

$$\begin{aligned} T(t, x, z) &= \inf[z^0 | (z^0, z) \in \tilde{Q}(t, x)] \\ &= \inf[z^0 | z^0 \geq f_0(t, x, u),\ z = f(t, x, u),\ u \in U(t, x)]. \end{aligned}$$

We may extend T to all of R^{2n+1} by taking $T(t, x, z) = +\infty$ for $T(t, x, z) \in R^{2n+1} - M_0$. Then $Q(t, x)$, $\tilde{Q}(t, x)$ are the empty sets for $(t, x) \in R^{n+1} - A$.

In the discussion below we shall reduce the problem of the absolute minimum of the functional (11.4.1) under the constraints (11.4.2) to the problem of the absolute minimum of the functional

$$J[x] = g(t_1, x(t_1), t_2, x(t_2)) + \int_{t_1}^{t_2} T(t, x(t), x'(t))\, dt, \tag{11.4.3}$$

under the constraints

$$\begin{gathered} (t, x(t)) \in A, \quad (t_1, x(t_1), t_2, x(t_2)) \in B, \quad x'(t) \in Q(t, x(t)), \\ t \in [t_1, t_2] \text{ (a.e.)}, \quad T(\cdot, x(\cdot), x'(\cdot)) \in L_1[t_1, t_2], \end{gathered} \tag{11.4.4}$$

a problem we have studied in Section 11.1.

For most applications it would be enough to assume

C′. *A closed, M closed, and f_0, f continuous on M.*

However much less is needed. For instance, if A and M are products of intervals (possibly infinite) $A = I_t \times I_x$, $M = I_t \times I_x \times I_u$, $I_t \subset R$, $I_x \subset R^n$, $I_u = U \subset R^m$, then we could simply assume that f_0 and f are Carathéodory functions on M, namely, measurable in t for every (x, u), and such that for almost all $\bar{t}$, $f_0(\bar{t}, x, u)$, $f(\bar{t}, x, u)$ are continuous functions of (x, u). Then such functions would have the Lusin property as stated in (8.3.v), namely, given $\varepsilon > 0$, there is a closed subset K of I_t such that f_0 and f, restricted to $K \times I_x \times I_u$, are continuous and meas$(I_t - K) < \varepsilon$.

But it may well be that either A or M or both are not products of intervals, namely, the sections $A(t)$ of A may well depend on t, and the control sets $U(t, x)$ may well depend on t and x. In this situation we could require that the relevant set valued maps $t \to A(t)$ and $(t, x) \to U(t, x)$ are measurable (Section 8.3), and structure accordingly the whole argument. All this is unnecessary. All we need on A, M, f_0, and f, for the proof that follows, is the following Carathéodory type property:

C′*. For every $\varepsilon > 0$ there is a closed subset K of A_0 such that meas$(A_0 - K) < \varepsilon$, the sets $A_K = [(t, x) \in A \mid t \in K]$, $M_K = [(t, x, u) \in M \mid t \in K]$ are closed, and both $f_0(t, x, u)$, $f(t, x, u) = (f_1, \ldots, f_n)$ are continuous on M_K.

It is easy to see that in any case the function $T(t, x, z)$ is either B-measurable, or at least it satisfies requirements (a), (b) of condition (C*) of Section 11.1. Indeed, if M is closed, and f_0, f are continuous on M, then the B-measurability of T was proved in Exercise 4 of Section 8.2C. Under the alternate assumption (C′*) for every $\varepsilon = s^{-1}$, $s = 1, 2, \ldots$, there is a closed subset K_s of A_0 such that meas$(A_0 - K_s) < s^{-1}$ and f_0, f are continuous on M_K, and then T is B-measurable on $(K_s \times R^{2n})$.

As mentioned in Section 11.1, this is enough to guarantee that for any AC function $x(t)$, $t_1 \le t \le t_2$, then $T(t, x(t), x'(t))$ is measurable in $[t_1, t_2]$. If $\Omega = \{(x, u)\}$ is any nonempty class of admissible pairs for the problem (11.4.1–2), then $x \in \Omega_x$ is AC, u is measurable, and $f_0(t, x(t), u(t)) \ge T(t, x(t), x'(t))$, $t \in [t_1, t_2]$ (a.e.), where f_0 is L-integrable and T is measurable. Thus $T(t, x(t), x'(t))$ is L-integrable in the sense that its L-integral is either finite or $-\infty$. Thus $J[x]$ is defined in Ω_x and we denote by j the infimum of $J[x]$ in Ω_x. Certainly $-\infty \le j \le i < +\infty$.

Remark 1. It may well occur that for an AC trajectory $x \in \Omega_x$ the measurable function $T(t, x(t), x'(t))$ has actually L-integral $-\infty$. Take $f_0(t, x, u) = -t^{-1}u^2(1 + u^2)^{-1}$, $f(t, x, u) = \sin u$, $U(t, x) = [u \mid 0 \le u < +\infty]$ for $0 < t \le 1, x \in R$; $f_0 = 0, f = 0$, $U(0, x) = \{0\}$ for $t = 0$, $x \in R$. Then $A = [0, 1] \times R$, $M = M_1 \cup M_2$ is the union of $M_1 = (0, 1] \times R \times [|z| \le 1]$ and of $M_2 = \{0\} \times R \times \{0\}$. Also, $T(t, x, z) = -t^{-1}$ on M_1, $T(t, x, z) = 0$ on M_2. Let Ω consists of all pairs $x(t)$, $u(t)$, $0 \le t \le 1$, x AC, u measurable, $n = m = 1$, $x(0) = 0$, $u(t) \in U$. Then $x(t) = 0$, $u(t) = 0$, $0 \le t \le 1$, is an admissible pair. Yet, $T(t, x(t), x'(t)) = -t^{-1}$ is not L-integrable, that is, the AC trajectory $x(t) = 0$ does not satisfy the requirements (11.4.4).

Remark 2. It may well occur that for an AC function x satisfying (11.4.4) there is no u such that (x, u) is admissible for (4.4.1–2) and $f_0(t, x(t), u(t)) = T(t, x(t), x'(t))$. Indeed, take $f_0(t, x, u) = (1 + u^2)^{-1}$, $f(t, x, u) = \sin u$, $U(t, x) = [u \mid 0 \le u < +\infty]$. Then M_0 is the set $[(t, x, z) \mid (t, x) \in R^2, |z| \le 1]$, and $T = 0$ on M_0. Now for the AC function $x(t) = 0$, $0 \le t \le 1$, we have $T(t, x(t), x'(t)) = 0$ and for no measurable u we can have $f_0(t, x(t), u(t)) = 0$, $x'(t) = 0 = f(t, x(t), u(t))$ for almost all $t \in [0, 1]$.

In the existence theorems below we shall need alternate "growth hypotheses" of the scalar function f_0 with respect to the vector function $f = (f_1, \ldots, f_n)$. They are the usual growth conditions but expressed directly in terms of f_0 and f.

(g1′) There is a scalar function $\phi(\zeta)$, $0 \le \zeta < +\infty$, bounded below, such that $\phi(\zeta)/\zeta \to +\infty$ as $\zeta \to +\infty$, and $f_0(t, x, u) \ge \phi(|f(t, x, u)|)$ for all $(t, x, u) \in M$.

(g2′) For any $\varepsilon > 0$ there is a locally integrable scalar function $\psi_\varepsilon(t) \ge 0$ such that $|f(t, x, u)| \le \psi_\varepsilon(t) + \varepsilon f_0(t, x, u)$ for all $(t, x, u) \in M$.

(g3′) For every n-vector $p \in R^n$ there is a locally integrable function $\phi_p(t) \ge 0$ such that $f_0(t, x, u) \ge (p, f(t, x, u)) - \phi_p(t)$ for all $(t, x, u) \in M$.

11.4.i (An Existence Theorem Based on Weak Convergence of the Derivatives). *Let A be bounded, B closed, and let condition* (C′) *or* (C′*) *be satisfied. Assume that* (k) *for almost all $\bar{t} \in A_0$ the sets $\tilde{Q}(\bar{t}, x)$ are convex and have property* (K) *with respect to x only on the closed set $A(\bar{t})$. Let g be a lower semicontinuous scalar function on B. Assume that any one of the growth conditions* (g1′), (g2′), (g3′) *is satisfied. Let Ω be a nonempty Γ_{0w}-closed class of admissible pairs x, u. Then the functional* (11.4.1) *has an absolute minimum in Ω.*

For A not bounded see Section 11.5. It is enough we limit ourselves to the nonempty part Ω_M of Ω of all elements $(x, u) \in \Omega$ with $I[x, u] \le M$ for some M. We just recall from (8.5.vii) that, if for almost all $\bar{t}$, the functions $f_0(\bar{t}, x, u)$, $f(\bar{t}, x, u)$ are continuous in (x, u) and either $f_0(\bar{t}, x, u) \to +\infty$, or $|f(\bar{t}, x, u)| \to +\infty$ as $|u| \to +\infty$ locally uniformly with respect to x, then the sets $\tilde{Q}(\bar{t}, x)$ certainly have property (K) with respect to x.

Proof of (11.4.i). First note that conditions (g1′), (g2′), (g3′) for problem (11.4.1–2) imply conditions (γ1), (γ2), (γ3) respectively for problem (11.4.3–4).

Under hypothesis (g1′) the function f_0 is bounded below, say $f_0 \ge -M_1$; under hypothesis (g2′) we have $f_0 \ge -\psi_1(t)$; under hypothesis (g3′) we have $f_0 \ge -\phi_0(t)$; in any case $f_0(t, x, u) \ge -\psi(t)$ where ψ is a nonnegative locally integrable function. By virtue of (8.5.v) and of (8.5.iii) and subsequent Remark 2, and under condition (k) above, then for a.a. $\bar{t} \in G$ the sets $\tilde{Q}(\bar{t}, x)$ are closed, hence $\tilde{Q}(\bar{t}, x) = \mathrm{epi}_z T(\bar{t}, x, z)$, and the extended function $T(\bar{t}, x, z)$ is lower semicontinuous in (x, z) and convex in z, or equivalently for almost all $\bar{t}$, the sets $M(\bar{t}) = [(x, z^0, z) \mid x \in A(\bar{t}),\ z^0 \ge f_0(\bar{t}, x, u),\ z = f(\bar{t}, x, u),\ u \in U(\bar{t}, x)]$ are closed and convex. Since we have already proved that T is B-measurable on the sets $K \times R^{2n}$, we see that, under either condition (C′) or (C′*) on the functions f_0, f, the extended function $T(t, x, z)$ satisfies the condition (C*) of Section 11.1. Also, there is at most a set of measure zero of points $\bar{t}$ where the function $T(\bar{t}, x, z)$ may take the value $-\infty$ for some (x, z). Otherwise, the extended function T takes only finite values or $+\infty$.

For any admissible pair $(x, u) \in \Omega$, that is, $x \in \Omega_x$, we have now $f_0(t, x(t), u(t)) \geq T(t, x(t), x'(t)) \geq -\psi(t)$. Hence, the measurable function $T(t, x(t), x'(t))$ is L-integrable as lying between two functions having the same property. Moreover, $-\infty < j \leq i < +\infty$. Finally, the class Ω_x is Γ_w-closed. All conditions of Theorem (11.1.i) are satisfied. By (11.1.i) then there is an element $x \in \Omega_x$ such that

$$(T(t, x(t), x'(t)),\ x'(t)) \in \tilde{Q}(t, x(t)), \quad t \in [t_1, t_2] \text{ (a.e.)},$$

$$I[x] = g(t_1, x(t_1), t_2, x(t_2)) + \int_{t_1}^{t_2} T(t, x(t), x'(t))\, dt = j \leq i.$$

Let $R^+ = [0 \leq v < +\infty]$, and take $\tilde{U}(t, x) = U(t, x) \times R^+$, $\tilde{f}(t, x, u, v) = (\tilde{f}_0, f)$ with $\tilde{f}_0 = f_0(t, x, u) + v$. With these definitions and for almost all t we have

$$\begin{aligned}\tilde{Q}(t, x) &= [(z^0, z) \,|\, z^0 \geq f_0(t, x, u),\ z = f(t, x, u),\ u \in U(t, x)] \\ &= [(z^0, z) \,|\, z^0 = f_0(t, x, u) + v,\ z = f(t, x, u),\ u \in U(t, x),\ v \in R^+] \\ &= [(z^0, z) = \tilde{f}(t, x, u, v),\ (u, v) \in \tilde{U}(t, x)].\end{aligned}$$

For every $\varepsilon = s^{-1}$, $s = 1, 2, \ldots$, we take K_s closed such that meas$(A_0 - K_s) < s^{-1}$, A_K is closed, M_K is closed, and f_0, f are continuous on M_K. Then, $\tilde{f}$ is continuous on the closed set $M_K \times R^+$. By the implicit function theorem (8.2.iii) there are measurable functions $u_s(t) \in U(t, x(t))$, $v_s(t) \geq 0$, $t \in K_s$, such that

$$(T(t, x(t), x'(t)),\ x'(t)) = \tilde{f}(t, x(t), u_s(t), v_s(t)), \quad t \in K_s \text{(a.e.)},$$

that is,

$$T(t, x(t), x'(t)) = f_0(t, x(t), u_s(t)) + v_s(t),\ \ x(t) = f(t, x(t), u_s(t)).$$

Since $T \leq f_0$ we must have $v_s = 0$ and $T(t, x(t), x'(t)) = f_0(t, x(t), u_s(t))$, $t \in K_s$ (a.e.). If we take $u(t) = u_s(t)$ for $t \in K_s - (K_1 \cup \cdots \cup K_{s-1})$, $s = 1, 2, \ldots$, then $u(t)$ is defined a.e. in $[t_1, t_2]$, is measurable, $u(t) \in U(t, x(t))$, $T(t, x(t), x'(t)) = f_0(t, x(t), u(t))$, and $x'(t) = f(t, x(t), u(t))$, $t \in [t_1, t_2]$ (a.e.). In other words, (x, u) is an admissible pair for problem (11.4.1–2) and $I[x, u] = J[x]$.

Since Ω is Γ_{0w}-closed, by the part (b) of the definition of Γ_{0w}-closedness, we conclude that (x, u) belongs to Ω and hence $I[x, u] \geq i$. Thus $i \leq I[x, u] = J[x] = j \leq i$, where i is finite, equality holds throughout, $I[x, u] = i$ and (11.4.i) is proved. Under the alternate assumption (b′) then there is a measurable $\bar{u}$ such that $(x, \bar{u}) \in \Omega$, $I[x, \bar{u}] \leq I[x, u]$. Thus, $i \leq I[x, \bar{u}] \leq I[x, u] = J[x] = j \leq i$, and equality sign holds throughout. Theorem (11.4.i) is thereby proved. □

Remark 3. If the convexity condition in (11.4.i) is not satisfied, there still exists an optimal generalized solution. The argument is similar to the one in Remark 2 of Section 11.1A.

To state the next theorem, we shall denote, as usual, by Ω a class of admissible pairs x, u, and by Ω_x the corresponding class of trajectories. Analogously let Ω_M denote the subclass of all $(x, u) \in \Omega$ with $I[x, u] \le M$, and $\Omega_{x,M}$ the corresponding class of trajectories x.

Also we shall need certain alternative conditions similar to the conditions (L_i) of Section 10.8:

(L'_1) There is a locally integrable real valued function $\psi(t) \ge 0$, $t \in R$, such that $f_0(t, x, u) \ge -\psi(t)$ for all $(t, x, u) \in M$.

(L'_2) There are a locally integrable real valued function $\psi(t) \ge 0$, $t \in R$, and a constant $c > 0$ such that $f_0(t, x, u) \ge -\psi(t) - c|f(t, x, u)|$ for all $(t, x, u) \in M$.

(L'_3) There are a locally integrable real valued function $\psi(t) \ge 0$, $t \in R$, and an n-vector valued bounded measurable function $\varphi(t) = (\varphi_1, \ldots, \varphi_n)$, $t \in R$, such that $f_0(t, x, u) \ge -\psi(t) - (\varphi(t), f(t, x, u))$ for all $(t, x, u) \in M$.

(L'_4) There are constants $\alpha \ge \beta$ real and $\gamma > 0$ such that (a) for every $(t, x) \in A$ the set $Q(t, x)$ contains the ball $B_0 = [z \in R^n \,|\, |z| \le \gamma]$ and $f_0(t, x, u) \ge \beta$ for all $(t, x) \in A$ and all $u \in U(t, x)$ with $|f(t, x, u)| \le \gamma$; (b) $f_0(t, x, u) \le \alpha$ for all $u \in U(t, x)$ with $f(t, x, u) = 0$. These requirements on f_0, f are certainly satisfied if M is closed, $f_0(t, x, u)$, $f(t, x, u)$ are continuous on M, and $|f(t, x, u)| \to +\infty$ as $|u| \to +\infty$ uniformly for $(t, x) \in A$.

11.4.ii (An Existence Theorem Based on the Weak Convergence of the Derivatives). *Let A be bounded, B closed, and let condition* (C′) *or* (C′*) *be satisfied. Assume that* (k) *for almost all $\bar{t} \in A_0$ the sets $\tilde{Q}(\bar{t}, x)$ are convex and have property* (K) *with respect to x only on the closed set $A(\bar{t})$. Let g be a lower semicontinuous scalar function on B. Assume that any one of the conditions* (L'_i) *holds. Let $\Omega = \{(x, u)\}$ be a nonempty Γ_{0w}-closed class of admissible pairs, and assume that the class $\{x'\}$ of the derivatives of the trajectories x in Ω is equiabsolutely integrable. Then the functional $I[x, u]$ in* (11.4.1) *has an absolute minimum in Ω.*

It is enough that we limit ourselves to the nonempty part Ω_M of Ω of all elements $(x, u) \in \Omega$ with $I[x, u] \le M$ for some M. For A not bounded see Section 11.5. A sufficient condition for requirement (k) was mentioned after statement (11.4.i).

Proof. Each of the conditions (L'_i) for the problem (11.4.1–2) implies the corresponding condition (L_i) of Section 11.1 for the problem (11.1.3–4). Under condition (L'_1) we have the same type of lower bound for f_0 and T we had in the proof of (11.4.i), or $f_0(t, x(t), u(t)) \ge T(t, x(t), x'(t)) \ge -\psi(t)$, where $\psi \ge 0$ is a fixed locally integrable function, and $x'(t) = f(t, x(t), u(t))$ a.e. in $[t_1, t_2]$. The argument is therefore exactly the same as for (11.4.i).

Under conditions (L'_2), (L'_3), (L'_4) for (11.4.1–2), conditions (L_2), (L_3), (L_4) hold for (11.1.3–4), and we have seen in Sections 10.8 and 11.1, that (L_3), (L_4) are actually particular cases of (L_2), that is, in any case, there are a locally integrable function $\psi(t) \geq 0$ and a constant $c \geq 0$ such that

$$f_0(t, x, u) \geq -\psi(t) - c|f(t, x, u)|, \qquad \text{or } T(t, x, z) \geq -\psi(t) - c|z|.$$

Thus, there is at most a set of measure zero of points $\bar{t}$ where $T(\bar{t}, x, z)$ may take the value $-\infty$ for some (x, z). Otherwise, the extended function $T(t, x, z)$ takes only finite value or $+\infty$. As in the proof of (11.4.i), the problem (11.4.3–4) satisfies condition (C*) of Section 11.1 and all requirements of (11.1.ii). The proof now continues as for (11.4.i). □

Remark 4. If the convexity condition in (11.4.ii) is not satisfied, there still exists an optimal generalized solution, provided the sets $\tilde{R}(t, x) = \text{co } \tilde{Q}(t, x)$ have property (K) with respect to x. Cf. the analogous Remark 4 of Section 11.1A.

In applications often the following simple corollary of (11.4.ii) suffices:

11.4.iii (An Existence Theorem Based on Weak Convergence of the Derivatives). *Let A be compact, B closed, M closed, f_0 and f continuous on M, with either $f_0(t, x, u) \to +\infty$, or $|f(t, x, u)| \to +\infty$ as $|u| \to +\infty$, or both, locally uniformly for $(t, x) \in A$. Assume that for almost all $\bar{t}$ the sets $Q(\bar{t}, x)$, $x \in A(\bar{t})$, contain a fixed ball $B_0 = [z \in R^n \,|\, |z| \leq \gamma]$, and that the sets $\tilde{Q}(\bar{t}, x)$ are convex. Let g be lower semicontinuous on B. Let $\Omega = \{(x, u)\}$ be a nonempty Γ_{0w}-closed class of admissible pairs, and assume that the class $\{x'\}$ of the derivatives of the corresponding trajectories is equiabsolutely integrable. Then the functional $I[x, u]$ in (11.4.1) has an absolute minimum in Ω.*

It is enough that we limit ourselves to the nonempty part Ω_M of Ω of all elements $(x, u) \in \Omega$ with $I[x, u] \leq M$. From (8.5.vii) the sets $\tilde{Q}(t, x)$, $x \in A(\bar{t})$, have property (K) with respect to x on the closed set $A(\bar{t})$. Theorem (11.4.iii) is now a corollary of (11.4.ii) and (L_4').

B. Existence Theorems for Problems of Optimal Control with a Comparison Functional

We consider now the problem (11.4.1–2) when a comparison functional, say

$$C[x, u] = \int_{t_1}^{t_2} H(t, x(t), u(t))\, dt, \tag{11.4.5}$$

is assigned, and we consider classes Ω of admissible pairs $x(t)$, $u(t)$, $t_1 \leq t \leq t_2$, for which both $I[x, u]$ and $C[x, u]$ are finite and $C[x, u] \leq M$ for a given constant M. For any $(t, x) \in A$ we shall denote by $\tilde{Q}_H(t, x)$ the set $\tilde{Q}_H(t, x) = [(\mathfrak{z}, z^0, z) \,|\, \mathfrak{z} \geq H(t, x, u),\ z^0 \geq f_0(t, x, u),\ z = f(t, x, u)]$, $u \in U(t, x)$. Also, let $\tilde{f}$ denote the $(n+1)$-vector function $\tilde{f}(t, x, u) = (f_0, f) = (f_0, f_1, \ldots, f_n)$.

As usual, we shall consider the problem of the minimum of $I[x, u]$ in the class Ω of all admissible pairs x, u satisfying $C[x, u] \leq M$. Alternatively, we may consider the same problem in a smaller class Ω of such admissible pairs, and in this case we shall require a mild closure property on the class Ω. We say that Ω is Γ_{0w}-closed with respect to H, f_0, f provided the following variant of our definition at the beginning of this section holds: (a) Ω_x is Γ_w-closed (Section 11.1), and (b) $x \in \Omega_x$, (x, u) admissible, $C[x, u] \leq M$, implies that $(x, u) \in \Omega$. Actually, it would be enough to assume instead of (b) the following less demanding requirement: (b′) $x \in \Omega_x$, (x, u) admissible, $C[x, u] \leq M$, implies that either $(x, u) \in \Omega$, or that there is some other $\bar{u}$ such that $(x, \bar{u})$ is admissible, $C[x, \bar{u}] \leq M$, $I[x, \bar{u}] \leq I[x, u]$, and $(x, \bar{u}) \in \Omega$. Also, we shall denote by $Q_H^*(t, x)$ the set of all $(\mathfrak{z}, z^0, z)$ with $\mathfrak{z} \geq H(t, x, u)$, $z^0 = f_0(t, x, u)$, $z = f(t, x, u)$, $u \in U(t, x)$.

11.4.iv (An Existence Theorem with a Dominant Comparison Functional and Isoperimetric Problems). *Let f_0, H satisfy condition* (C′) *or* (C′*). *Let A be bounded, B closed, and g lower semicontinuous on B. Assume that H satisfies one of the growth conditions* (g1′), (g2′), (g3′) *with respect to $\tilde{f} = (f_0, f)$, and that f_0 satisfies one of the conditions* (L_i') *with respect to f. Assume that* (k) *for almost all $\bar{t} \in A_0$ the sets $\tilde{Q}_H(\bar{t}, x)$ are convex and have property* (K) *with respect to x on the closed set $A(\bar{t})$. Let Ω be a nonempty Γ_{0w}-closed class of admissible pairs.*

(a) *Let D be a constant such that the subclass Ω_D of all $(x, u) \in \Omega$ with $C[x, u] \leq D$ is not empty. Then $I[x, u]$ has an absolute minimum in Ω_D.*

(b) *Let N be a constant such that the subclass Ω_N' of all $(x, u) \in \Omega$ with $I[x, u] \leq N$ is not empty, then $C[x, u]$ has an absolute minimum in Ω_N'.*

(c) *Here we require for g to be continuous on B, we assume that* (k) *holds for the sets $Q_H^*(t, x)$, and we do not require that f_0 satisfies a condition* (L_i') *with respect to f. Let N be a constant such that the subclass K_N' of all (x, u) with $I[x, u] = N$ is not empty. Then $C[x, u]$ has an absolute minimum in K_N'.*

Proof. For a proof of (11.4.iv) we introduce the additional state variable x^0 with $dx^0/dt = f_0(t, x, u)$, $x^0(t_1) = 0$, so that $I[x, u] = g(e[x]) + x^0(t_2)$. Let $\tilde{x} = (x^0, x)$, $\tilde{f} = (f_0, f)$. We only sketch a proof of the various parts.

For part (a) let $i = \inf[I[x, u] \,|\, (x, u) \in \Omega_D]$. For a minimizing sequence x_k, u_k we have $e[x_k] \in B$, $g(e[x_k]) + x_k^0(t_{2k}) \to i$, and for $\eta_k(t) = H(t, x_k(t), u_k(t))$, $\tilde{\xi}_k(t) = d\tilde{x}_k/dt = \tilde{f}(t, x_k(t), u_k(t))$, $\tilde{\xi} = (\xi_k^0, \ldots, \xi_k^n)$, we have

$$(\eta_k(t), \tilde{\xi}_k(t)) \in \tilde{Q}_H(t, x_k(t)), \qquad t \in [t_{1k}, t_{2k}] \text{ (a.e.)},$$

$$\int_{t_{1k}}^{t_{2k}} \eta_k(t)\, dt \leq D, \qquad \int_{t_{1k}}^{t_{2k}} \xi_k^0(t)\, dt = x_k^0(t_{2k}), \qquad k = 1, 2, \ldots.$$

Because of the growth properties of H with respect to $\tilde{f} = (f_0, f)$ we derive that the functions $\tilde{\xi}_k(t)$ are equiabsolutely integrable, and because A is bounded there are an AC $(n+1)$-vector function $\tilde{x}(t)$, $t_1 \leq t \leq t_2$, an L-integrable $(n+1)$-vector function $\tilde{\xi}(t) = (\xi^0, \xi)$, and a subsequence, say still $[k]$, such that $x_k \to x$ in the ρ-metric, $\tilde{\xi}_k \to \tilde{\xi}$ weakly in L_1. Finally, we can apply the same process as for (10.7.i) where now the needed bounds below for $\eta_k(t)$ are provided by the growth properties of H as usual. Then there is an L-integrable scalar function $\eta(t)$ such that $(\eta(t), \tilde{\xi}(t)) \in \tilde{Q}_H(t, x(t))$,

$t \in [t_1, t_2]$ (a.e.), and $\int_{t_1}^{t_2} H\,dt \le D$. Hence, $e[x_k] \to e[x], x_k^0(t_{2k}) \to x^0(t_2)$, $\int_{t_1}^{t_2} \xi^0(t)\,dt = x^0(t_2)$, and $g(e[x]) + x^0(t_2) \le i$ because of the lower semicontinuity of g. By the implicit function theorem we derive now the existence of a measurable function $u(t)$ with $u(t) \in U(t, x(t))$, such that

$$\eta(t) \ge H(t, x(t), u(t)), \qquad \xi^0(t) \ge f_0(t, x(t), u(t)),$$
$$\xi(t) = f(t, x(t), u(t)), \qquad t \in [t_1, t_2] \text{ (a.e.)}.$$

Thus, $f(t, x(t), u(t))$ is L-integrable as being equal to $\xi(t)$, $H(t, x(t), u(t))$ is L-integrable as being between $\eta(t)$ above and the L-integrable functions below which are provided by the growth conditions, and $f_0(t, x(t), u(t))$ is L-integrable as being between $\xi^0(t)$ above and the L-integrable functions below which are provided by the properties (L_i). Finally, $C[x, u] = \int_{t_1}^{t_2} H\,dt \le D$, and

$$I[x, u] = g(e[x]) + \int_{t_1}^{t_2} f_0(t, x(t), u(t))\,dt \le g(e[x]) + x^0(t_2) \le i.$$

Now $(x, u) \in \Omega_D$, hence $I[x, u] \ge i$, and by comparison $I[x, u] = i$. Part (a) is proved.

For part (b), let $j = \inf[C[x, u] \,|\, (x, u) \in \Omega'_N]$, and let $[x_k, u_k]$ be a minimizing sequence, that is, $g(e[x_k]) + x^0(t_{2k}) \le N$, $C[x_k, u_k] \to j$. The proof is now the same as before, and for the limit element $\tilde{x}$ and $\tilde{\xi} = d\tilde{x}/dt$ we have $g(e[x]) + x^0(t_2) \le N$, and finally $I[x, u] \le N$, $C[x, u] \le j$. Thus $(x, u) \in \Omega'_N$, $C[x, u] \ge j$, and finally $C[x, u] = j$. Part (b) is proved.

For part (c), we have $g(e[x_k]) + x_k(t_{2k}) = N$, and now $x_k \to x$, $e[x_k] \to e[x]$ imply that $g(e[x]) + x^0(t_2) = N$ since g is continuous on B. Because of the definition of the sets $Q_H^*(t, x)$, by the implicit function theorem we derive now the existence of a measurable function $u(t)$ with $u(t) \in U(t, x(t))$ such that

$$\eta(t) \ge H(t, x(t), u(t)), \qquad \xi^0(t) = f_0(t, x(t), u(t)),$$
$$\xi(t) = f(t, x(t), u(t)), \qquad t \in [t_1, t_2] \text{ (a.e.)}.$$

Thus, both $f_0(t, x(t), u(t))$ and $f(t, x(t), u(t))$ are L-integrable because $\tilde{\xi}(t)$ has this property, and $H(t, x(t), u(t))$ is L-integrable because of the growth properties. Finally, $C[x, u] = \int_{t_1}^{t_2} H\,dt \le j$, and $I[x, u] = g(e[x]) + \int_{t_1}^{t_2} f_0\,dt = g(e[x]) + x^0(t_2) = N$. Part (c) is proved and so is Theorem (11.4.iv). □

For $f_0 = 0$, the problem (11.4.1–2) with comparison functional $C[x, u]$ required to satisfy $C[x, u] \le M$ reduces to a Mayer problem with a comparison functional $C[x, u]$, and we have the following simpler statement, where the sets $Q(t, x)$ now replace the sets $\tilde{Q}(t, x)$.

11.4.v (An Existence Theorem for Mayer Problems with a Comparison Functional). *Let A, H, f satisfy condition* (C′) *or* (C′*). *Let A be bounded, B closed, and g lower semicontinuous on B. Assume that H satisfies one of the growth conditions* (g1′), (g2′), (g3′) *with respect to f. Assume that for almost all $\bar{t} \in A_0$ the sets $\tilde{Q}_H(t, x)$ are convex and have property* (K) *with respect to*

x on the closed set $A(\bar{t})$. Let Ω be a nonempty class of admissible pairs x, u for which in addition $C[x,u] \le M$. Let Ω be Γ_{0w}-closed with respect to H, f. Then the functional $I[x,u] = g(e[x])$ has an absolute minimum in Ω.

C. Existence Theorems for Problems of Optimal Control with Differential System Linear in u

We consider now functionals of the same type (11.4.1) with systems of differential equations which are linear in u. In other words we consider control problems of the form

$$I[x,u] = g(t_1, x(t_1), t_2, x(t_2)) + \int_{t_1}^{t_2} f_0(t, x(t), u(t))\, dt,$$

$$(11.4.6) \qquad \begin{gathered} x'(t) = A_0(t, x(t)) + B(t, x(t))u(t), \\ x \text{ AC}, \qquad u \text{ measurable}, \qquad f_0(\cdot, x(\cdot), u(\cdot)) \in L_1, \\ (t, x(t)) \in A = [t_0, T] \times R^n, \qquad (t_1, x(t_1), t_2, x(t_2)) \in B, \\ x(t) = (x^1, \ldots, x^n), \qquad u(t) = (u^1, \ldots, u^m), \qquad u(t) \in U = R^m, \end{gathered}$$

where A_0 is an $n \times 1$ matrix and $B(t,x)$ is an $n \times m$ matrix.

For such problem, any pair of functions $x(t)$, $u(t)$, $t_1 \le t \le t_2$, x AC, u measurable, satisfying the above relations, is said to be admissible. If $\Omega = \{x, u\}$ is any class of such pairs, we denote as usual by Ω_x the class of all trajectories x, or $\Omega_x = [x \mid (x,u) \in \Omega]$, and also we denote by Ω_u the class of all control functions, or $\Omega_u = [u \mid (x,u) \in \Omega]$.

Obviously, all statements of Section 11.4 holds also in the present situation and we do not repeat them.

We shall consider first the case in which $f_0(t,x,u)$ is convex in u for all (t,x), and satisfies one of the conditions (L_i) of Section 10.8A, $i = 1, 2, 3, 4$.

Alternatively, we shall consider the case in which $f_0(t,x,u)$ satisfies one of the growth conditions ($\gamma 1$), ($\gamma 2$), ($\gamma 3$) with respect to u of Section 11.1.

In either case it is convenient to assume a topology for the control functions u, say L_p for some $p \ge 1$. Thus we shall consider classes $\Omega = \{x, u\}$ of admissible pairs x, u with x AC and $u \in L_p$.

In this situation we shall say that Ω is Γ'_{0w}-complete provided: If $(x_k, u_k) \in \Omega$, $k = 1, 2, \ldots$, is a sequence of pairs in Ω, and $x_k \to x$ in the ρ-metric, x AC, $x'_k \to x'$ weakly in L_1, $u_k \to u$ weakly in L_p, $u \in L_p$, and the pair (x,u) is admissible, then $(x,u) \in \Omega$. The class of all admissible pairs is of course Γ'_{0w}-complete.

11.4.vi (An Existence Theorem with a Topology on the Control Functions). *Let A be a subset of $[t_0, T] \times R^n$, $U = R^m$, B closed, and g lower semicontinuous and bounded below in B. Let $f_0(t,x,u)$, $(t,x,u) \in A \times R^m$, be a function satisfying either* (C′) *or* (C′*), *satisfying one of the conditions* (L_i), *and convex in u for every (t,x). Let $A_0(t,x)$, $B(t,x)$ be matrices of the types $n \times n$ and $n \times m$ respectively, and satisfying either* (CL) *or* (CL*) *of*

Section 10.8B. *Assume that for some functions* ϕ, $\psi \geq 0$, $\phi \in L_1$, $\psi \in L_q$, $p, q > 1, 1/p + 1/q = 1$ (or $p = 1$, $q = \infty$), *and constants* c, C *we have*

$$|A_0(t,x)| \leq \phi(t) + c|x|, \qquad |B(t,x)| \leq \psi(t) + C|x|, \qquad (t,x) \in A.$$

Assume that the class Ω *of admissible pairs* x, u *is nonempty and* Γ'_{0w}*-closed, that each trajectory* $x \in \Omega_x$ *has some point* $(t^*, x(t^*))$ *on a given compact set* P, *and that* Ω_u *is known to be relatively sequentially weakly compact in* L_p. *Then the functional* $I[x,u]$ *in* (11.4.6) *has an absolute minimum in* Ω.

Proof. Let $i = \inf[I[x,u], (x,u) \in \Omega]$, $-\infty \leq i < +\infty$. Note that g is bounded below by hypothesis, but at present we have no bound below for the integral in (11.4.6). Let x_k, u_k, $k = 1, 2, \ldots$, be a minimizing sequence, that is, $(x_k, u_k) \in \Omega$, $I[x_k, u_k] \to i$. Thus, the elements u_k belong to Ω_u, hence, there are a function $u(t)$, $t_1 \leq t \leq t_2$, $u \in L_p$, and a subsequence, say still $[k]$, such that $u_k \to u$ weakly in L_p. Then, $\|u_k\|_p$ is bounded, say $\|u_k\|_p \leq \mu$. By the conditions (L_i) we conclude now that the integral $\int_{t_{1k}}^{t_{2k}} f_0(t, x_k(t), u_k(t))\,dt$ is bounded below. Then $I[x_k, u_k]$ is also bounded below, and i is finite. We have

$$x_k'(t) = A_0(t, x_k(t)) + B(t, x_k(t))u_k(t), \quad t \in [t_{1k}, t_{2k}] \text{ (a.e.)}, \quad k = 1, 2, \ldots.$$

By (10.8.iv) there are functions $x(t)$, $u(t)$, $t_1 \leq t \leq t_2$, x AC, $u \in L_p$, and a subsequence, say still $[k]$, such that $x_k \to x$ in the ρ-metric, $u_k \to u$ weakly in L_p, $x_k' \to x'$ weakly in L_1, and

$$x'(t) = A_0(t, x(t)) + B(t, x(t))u(t), \qquad t \in [t_1, t_2] \text{ (a.e.)}.$$

By (10.8.i) we have now $I[x,u] \leq \liminf I[x_k, u_k]$, or $I[x] \leq i$. By the closure property of Ω we derive that $(x,u) \in \Omega$, hence $I[x,u] \geq i$, and finally $I[x,u] = i$.

□

11.4.vii. *The same as* (11.4.vi) *where now we assume that* $f_0(t,x,u)$ *satisfies one of the growth conditions* ($\gamma 1$), ($\gamma 2$), ($\gamma 3$) *with respect to* u, *and is convex in* u *for every* (t,x). *Assume that* $A_0(t,x)$ *and* $B(t,x)$ *are as in* (11.4.vi) *with* $p = 1$. *For the class* Ω *of admissible pairs* x, u, $u \in L_1$, *we assume only the* Ω *is nonempty and* Γ'_{0w}*-closed, and that each trajectory* $x \in \Omega_x$ *has some point* $(t^*, x(t^*))$ *on a given compact set* P. *Then the functional* $I[x,u]$ *in* (11.4.6) *has an absolute minimum in* Ω.

Proof. The proof is analogous to the one for (11.4.vi), but here we derive from Section 10.4 that the sequence $[u_k]$ is equiabsolutely integrable. Then there is a subsequence, say still $[k]$, such that $t_{1k} \to t_1$, $t_{2k} \to t_2$, $t_0 \leq t_1 \leq t_2 \leq T$, and by Dunford–Pettis there is an L_1-integrable function $u(t)$, $t_1 \leq t \leq t_2$, and a further subsequence, say still $[k]$, such that $u_k \to u$ weakly in L_1. Directly by the growth conditions we know that there is an L_1-integrable function ψ in $[t_0, T]$ such that $f_0(t,x,u) \geq -\psi(t)$; hence $I[x,u]$ is bounded below and i is finite. By (10.8.iv), case $p = 1$, we know that there are an AC function $x(t)$, $t_1 \leq t \leq t_2$, and a further subsequence, say still $[k]$, such that

$x_k \to x$ in the ρ-metric, $u_k \to u$ weakly in L_1, $x'_k \to x'$ weakly in L_1, and $x'(t) = A(t, x(t)) + B(t, x(t))u(t)$, $t \in [t_1, t_2]$ (a.e.). By (10.8.i) we have now $I[x, u] \leq \liminf I[x_k, u_k]$, or $I[x, u] \leq i$. By the closure property Γ'_{0w} for $p = 1$ we conclude that $(x, u) \in \Omega$, hence $I[x, u] \geq i$, and finally $I[x, u] = i$. □

As an important particular case we consider now quadratic integrals with linear differential equations:

$$I[x, u] = 2^{-1} \int_{t_1}^{t_2} [x^*(t)P(t)x(t) + u^*(t)R(t)u(t)]\, dt,$$

$$(11.4.7) \qquad x'(t) = F(t)x(t) + G(t)u(t), \qquad t \in [t_1, t_2], \quad u(t) \in U = R^m,$$

$$A = [t_0, T] \times R^n, \quad (t_1, x(t_1), t_2, x(t_2)) \in B,$$

where B is a closed set, and P, R, F, G are $n \times n, m \times m, n \times n, n \times m$ matrices respectively, $P^* = P$, $R^* = R$, all with say continuous entries in $[t_0, T]$. We assume that Ω is the class of all admissible pairs x, u with x AC and $u \in L_2$, and that Ω is nonempty.

If we assume that P is positive semidefinite for all t, and R is positive definite for all t, in the sense that there is some constant $\lambda > 0$ such that $u^*R(t)u \geq \lambda|u|^2$ for all $t \in [t_0, T]$ and all $u \in R^m$, then

$$f_0(t, x, u) = x^*Px + u^*Ru \geq \lambda|u|^2.$$

In other words, f_0 satisfies growth condition $(\gamma 1)$ with respect to u, f_0 is convex in u, and (11.4.vii) applies. The integral $I[x, u]$ in (11.4.7) has an absolute minimum in Ω.

If we only know that $R(t)$ is semidefinite positive but $\Omega = \{x, u\}$ is the class of all admissible pairs with $\int_{t_1}^{t_2} |u|^2\, dt$ for some $D > 0$, then the class Ω is relatively weakly compact in L_2, (11.4.vi) applies, and $I[x, u]$ has an absolute minimum in Ω. For instance, for $R(t) \equiv 0$, still $I[x, u]$ has an absolute minimum in Ω.

If U is any compact and convex set of R^m, none of the specific conditions above is needed, and $I[x, u]$ has an absolute minimum in Ω because of the theorems of Chapter 9.

Thus, the following problems with $n = 1$, $m = 1$, $x(1)$ undetermined,

$$I = \int_0^1 (x^2 + u^2)\, dt, \quad x' = u, \quad u \in R, \quad x(0) = x_0,$$

$$I = \int_0^1 x^2\, dt, \quad x' = u, \quad |u| \leq 1, \quad x(0) = x_0,$$

$$I = \int_0^1 x^2\, dt, \quad x' = u, \quad u \in R, \quad \int_0^1 u^2\, dt \leq 1, \quad x(0) = x_0,$$

have an absolute minimum. Instead, the problem

$$(11.4.8) \qquad I = \int_0^1 x^2\, dt, \quad x' = u, \quad u \in R, \quad x(0) = x_0,$$

has obviously no absolute minimum if $x_0 \neq 0$, while for $x_0 = 0$ it has the optimal solution $x(t) \equiv 0$.

Remark 5. We consider here in some detail problem (11.4.7) with t_1, t_2 fixed, $x(t_1) = x_0$, $R(t)$ only positive semidefinite (possibly identically zero) and no weak L_2-compactness on Ω_u. The following operational approach is of interest. Let $V = [t_1, t_2]$.

If $\Phi(t)$ denote any fundamental $n \times n$ matrix of the linear differential system $dy/dt = F(t)y$, we take $\phi(t, \tau) = \Phi(t)\Phi^{-1}(\tau)$. Then the differential equation $x' = F(t)x + G(t)u(t)$ with $x(t_1) = x_0$ has the unique solution

$$x(t) = \phi(t, t_1)x_0 + \int_{t_1}^{t} \phi(t, \tau)G(\tau)u(\tau)\,d\tau = Sx_0 + Tu, \tag{11.4.9}$$

where S and T are linear operators $S: R^n \to (L_2(V))^n$, $T:(L_2(V))^m \to (L_2(V))^n$. Then

$$T^*v = G^*(t) \int_{t}^{t_2} \phi^*(\tau, t)v(\tau)\,d\tau, \quad T^*:(L_2(V))^n \to (L_2(V))^m,$$

is the dual operator of T in the sense that, if $(x, y) = \int_{t_1}^{t_2} x^*y\,dt$ denotes the usual inner product in $(L_2(V))^n$ and $(L_2(V))^m$, then $(Tu, y) = (u, T^*y)$ for all $u \in (L_2(V))^m$ and $y \in (L_2(V))^n$. Indeed,

$$\begin{aligned}(Tu, y) &= \int_{t_1}^{t_2} \left(\int_{t_1}^{t} \phi(t, \tau)G(\tau)u(\tau)\,d\tau \right)^* y(t)\,dt \\ &= \int_{t_1}^{t_2} \left(\int_{\tau}^{t_2} u^*(\tau)G^*(\tau)\phi^*(t, \tau)y(t)\,dt \right) d\tau \\ &= \int_{t_1}^{t_2} u^*(\tau)G^*(\tau)\,d\tau \int_{\tau}^{t_2} \phi^*(t, \tau)y(t)\,dt = (u, T^*y).\end{aligned}$$

By substituting (11.4.9) in the expression of $I[x, u]$ we have the new functional

$$J[u] = 2^{-1}(u, Au) + (u, w) + j_0,$$
$$A = T^*PT + R, \qquad w = T^*PSx_0, \qquad j_0 = 2^{-1}(Sx_0, PSx_0).$$

We have obtained a functional depending on u alone, $u \in (L_2(V))^m$, and no other parameter or constraint. It is easy to see that $A:(L_2(V))^m \to (L_2(V))^m$ is selfadjoint.

11.4.viii. *The quadratic functional $J[u]$ has an absolute minimum in $(L_2(V))^m$ if and only if* (a) *A is positive semidefinite, and* (b) *w is in the range of A, that is, $w = Au$ for some u.*

Proof. Since A is linear, it is easy to see that for all $u, h \in (L_2(V))^m$ we have

$$J[u + h] - J[u] = (Au + w, h) + (Ah, h).$$

If J has an absolute minimum at u, then $J[u + h] - J[u] \geq 0$ for all h. Hence, by the usual argument we must have $(Au + w, h) = 0$ for all h, and then $Au + w = 0$, or $w = -Au = A(-u)$, and also $(Ah, h) \geq 0$ for all h. That is, w is in the range of A, and A is positive semidefinite. The sufficiency is obvious. □

If we know that the linear map A is bounded and coercive, that is, $\|Au\|_2 \leq C\|u\|_2$, $(Au, u) \geq c\|u\|_2^2$ for some constants $c, C > 0$ and all $u \in (L_2(V))^m$, then A is onto (and has an inverse A^{-1} which is also bounded and coercive) and both (a) and (b) are certainly satisfied.

It is interesting to apply (11.4.viii) to the problem (11.4.8). For this case $n = m = 1$, $P = 1, R = 0, F = 0, G = 1, Tu = \int_0^t u(\tau)\,d\tau, T^*x = \int_t^1 x(\tau)\,d\tau, w = T^*x_0, A = T^*T,$ and $(Au, u) = (T^*Tu, u) = (Tu, Tu) \geq 0$, that is, A is certainly positive semidefinite. Now

$w = Au$ means here

$$\int_t^1 \left(\int_0^s u(\tau)\,d\tau \right) ds = \int_t^1 x_0\,ds,$$

that is

$$\int_t^1 \left(\int_0^s u(\tau)\,d\tau - x_0 \right) ds = 0.$$

Then by differentiation, $\int_0^s u(\tau)\,d\tau = x_0$, $u(s) = 0$, $x_0 = 0$.

11.5 Elimination of the Hypothesis that A Is Bounded in Theorems (11.4.i–v)

The hypothesis that A is bounded in the theorems (11.4.i–v) can be easily removed as we have done in Section 11.2 and, before, in Chapter 9. As already stated, all we have to do is to guarantee that a minimizing sequence can be contained in some compact subset A_0 of A.

1. If the set A is not bounded, but A is contained in a fixed slab $[t_0 \leq t \leq T_0, x \in R^n]$ of R^{n+1}, then Theorem (11.4.i) is still valid if we know for instance that (h'_1) g is bounded below on B, say $g \geq -M_1$, and (C'_1) Ω is a given nonempty Γ_{0w}-closed class of AC trajectories x, each of which has at least one point $(t^*, x(t^*))$ on a given compact subset P of A, and where t^* may depend on the trajectory. The proof is the same as in Section 11.3.
2. If A is neither bounded nor contained in any slab as above, then Theorem (11.4.i) is still valid if we know for instance that (h'_1) and (C'_1) hold, and in addition that (C'_2) there are constants $\mu > 0$, $R_0 \geq 0$, such that $f_0(t, x, u) \geq \mu > 0$ for all $(t, x, u) \in M$ with $t \geq R_0$. The proof is the same as in Section 11.3.

Concerning Theorem (11.4.ii), if we assume that condition (g1′), or (g2′), or (g3′) hold, then the same considerations hold as for Theorem (11.4.i). If we do not want to invoke conditions (g′), then the following holds. If A is not compact but closed and contained in a slab $[t_0 \leq t \leq T, x \in R^n]$, then (11.4.ii) is still valid, provided we know that (h'_1), (C'_1) hold, and for instance (C'_3) there are constants $c > 0$, $R \geq 0$, and a locally integrable function $\psi(t) \geq 0$, $t \in R$, such that $f_0(t, x, u) \geq -\psi(t) + c|f(t, x, u)|$ for all $(t, x, u) \in M$ with $|x| \geq R$.

If A not compact, nor contained in any slab as above, then (11.4.ii) is still valid, provided we know, for instance, that (h'_1), (C'_1), (C'_2), (C'_3) hold.

For Theorem (11.4.iii) the same considerations hold as for (11.4.ii).

For Theorem (11.4.iv) it is enough that we transfer to H the assumptions we have made for f_0 in (11.4.i). For (11.4.v) the same considerations hold as for (11.1.iii), that is, we may require the additional assumption (h_2) of Section 11.3 for A contained in a slab, or (h_3) for A unbounded and contained in no slab.

Alternatively, instead of condition (C'_3) we may assume that

C′. There is a constant $C \geq 0$ such that $x^1 f_1 + \cdots + x^n f_n \leq C(|x|^2 + 1)$ for all $(t, x, u) \in M$.

A particular case of (C′) is of course that $|f(t, x, u)| \leq C(|x| + 1)$ for all $(t, x, u) \in M$.

Instead of (C′) we may consider the more general condition

(C″) There is a scalar function $V(t,x)$ of class C^1 in A and a positive constant C such that

$$\operatorname{grad}_x V(t,x)\cdot f(t,x,u) + \partial V/\partial t \le CV(t,x)$$

for all $(t,x,u)\in M$, and, moreover, for every a, b, α real, the set $[x \mid V(t,x)\le \alpha$ for some $(t,x)\in A,\ a\le t\le b]$ is compact.

The argument is the same as in Sections 9.4, 9.5. The reader may also extend to the present situation the remaining remarks in Sections 9.4, 9.5.

11.6 Examples

1. $I[x,u]=\int_0^{t_2}(1+tu+u^2)\,dt$, with system $x'=u$, $u\in U=R$, $t_1=0$, $n=m=1$, $x(0)=1$, $0\le t_2\le 1$, $(t_2,x(t_2))$ on the locus $\Gamma: t=(1+x^2)^{-1}$, $-\infty<x<+\infty$ (a Lagrange problem of the calculus of variations written as a problem of control). Take for A the slab $[0\le t\le 1, x\in R]$, a closed set; then $M=A\times R$ is also closed, and so is $B=\{0\}\times\{1\}\times\Gamma$. The sets $\tilde{Q}(t,x)$ are all closed and convex in R^2. Also $f_0\ge(|u|-1)^2$, and thus the growth condition (g1′) of Section 11.4 holds with $\phi(\zeta)=(\zeta-1)^2$. Here A is a closed slab, and condition (C_1') holds, since $(t_1,x_1)=(0,1)$ is fixed.

2. $I[x,u]=\int_0^1(1+t)u^2\,dt$ with system $x'=u$, $u\in U=R$, $n=1$, $m=1$, $t_1=0$, $t_2=1$, $x(0)=1$, $x(1)=0$. Here we can take $A=[0,1]\times R$ closed, $M=A\times R$ closed, $B=\{0\}\times\{1\}\times\{1\}\times\{0\}$ compact. The sets $Q(t,x)$ are all convex and closed. Here $f_0\ge u^2$, so growth condition (g1′) holds with $\phi(\zeta)=\zeta^2$, and condition (C_1') holds, since B is compact. The functional $I[x,u]$ has an absolute minimum under the constraints.

3. $I[x,u]=\int_0^1 t^\alpha u^2\,dt$, $0\le\alpha<1$, with system $x'=u$, $u\in U=R$, $t_1=0$, $t_2=1$, $x(0)=1$, $x(1)=0$ (a Lagrange problem of the calculus of variations written as a problem of control). Here we take $A=[0\le t\le 1, x\in R]$ closed, $n=1$, $m=1$, $M=A\times R$ also closed, $B=\{0\}\times\{1\}\times\{1\}\times\{0\}$ compact. Growth condition (g2′) of Section 11.4 holds. To prove this, note that, for any $\varepsilon>0$, the function $\psi_\varepsilon(t)=\varepsilon^{-1}t^{-\alpha}$ is L-integrable in $[0,1]$. Now, either $|u|\le\varepsilon^{-1}t^{-\alpha}$, and then $|f|=|u|\le\varepsilon^{-1}t^{-\alpha}\le\psi_\varepsilon(t)+\varepsilon f_0$; or $|u|>\varepsilon^{-1}t^{-\alpha}$, and then $1<\varepsilon t^\alpha|u|$ and $|f|=|u|\le\varepsilon t^\alpha u^2\le\psi_\varepsilon(t)+\varepsilon f_0$. In any case $|f|\le\psi_\varepsilon(t)+\varepsilon f_0$, and condition (g2′) holds. The sets $\tilde{Q}(t)$ are here $\tilde{Q}(t)=[(z^0,z)\mid z^0\ge t^\alpha z^2, -\infty<z<+\infty]$, $0\le t\le 1$, and they are all closed and convex. Because of t^α being positive everywhere in $(0,1]$, we see that for all $\bar{t}$, $0<\bar{t}\le 1$, the sets $Q(\bar{t},x)$ have growth property (ϕ) of Section 10.5 with respect to x only; hence, the same sets have property (K), as well as property (Q), with respect to x only (see (10.5.iii) with $g=0$, or (G′) of Section 10.5). Here, A is a closed slab, and condition (C_1') is satisfied, since the initial point $(0,1)$ is fixed. The functional has an absolute minimum under the constraints. (See also Exercise 5 in Section 2.20C).

4. $I[x,u]=\int_0^{t_2}[(1+x^2)u^2+1]\,dt$ with system $x'=(1+x)u$, $u\in U=R$, $m=n=1$, $t_1=0$, $t_2\ge 0$, $x(0)=0$, (t_2,x_2) on the locus $\Gamma: tx=1$, $t>0$. Here we can take $A=[0\le t<+\infty, x\in R]$ closed, $M=A\times R$ closed, $B=\{0\}\times\{0\}\times\Gamma$ also closed. A is not contained in any slab, but condition (C_2') is satisfied, since $f_0\ge 1$. Condition (C_1') is satisfied, since $(t_1,x_1)=(0,0)$ is fixed. Finally, $(1+x)^2\le 2+2x^2$, $f_0=(1+x^2)u^2+1\ge 2^{-1}|f|^2$, and the growth condition (g1′) of Section 11.4 holds with $\phi(\zeta)=2^{-1}\zeta^2$. The functional has an absolute minimum under the constraints.

5. $I[x,u]=\int_0^1(1+u^2)^\varepsilon\,dt$, $\varepsilon>\frac{1}{2}$, with system $x'=u$, $u\in U=R$, $t_1=0$, $0\le t_2\le 1$, $x(0)=0$, (t_2,x_2) on the segment $\Gamma=[0\le t\le 1, x=1]$, $A=[0\le t\le 1, 0\le x\le 1]$ (a

problem of the calculus of variations written as a problem of control). Here A is compact, $M = A \times R$ is closed, $B = \{0\} \times \{0\} \times \Gamma$ is compact, the fixed set $\tilde{Q}$ is closed and convex. The growth condition (g1′) is satisfied with $\phi(\zeta) = \zeta^{2\varepsilon}$. The functional has an absolute minimum under the constraints. (This is not the case for $\varepsilon = \frac{1}{2}$. See Example 2 in Section 11.7 below).

6. $I[x, y, u] = \int_0^1 (t^2x^2y^2 + x^2 + y^2)|u|\,dt$ with system $x' = (x + y)u$, $y' = u$, $u \in U = R$, $x(0) = 1$, $y(1) = 0$, in the class Ω of all admissible systems $x(t)$, $y(t)$, $u(t)$, $0 \le t \le 1$ with $\int_0^1 (x'^2 + y'^2)\,dt \le 4$. Here $n = 2$, $m = 1$, $A = [0, 1] \times R^2$ is closed, $M = A \times R$ is also closed, $x(1)$, $y(0)$ are undetermined, B closed. Also $\int_0^1 |y'|\,dt \le (\int_0^1 (x'^2 + y'^2)\,dt)^{1/2} \le 2$, so that the point $(x(0), y(0))$ is certainly in the compact set $P = [(x, y)|x = 1, |y| \le 2]$, and condition (C_1') holds. Here $|f| \to +\infty$ as $u \to +\infty$; hence, the sets $\tilde{Q}(t, x, y)$ certainly have property (K) by (8.5.vii). Note that for no t do the convex sets $\tilde{Q}(t, x, y) = [(t^2x^2y^2 + x^2 + y^2)|u|, z^1 = (x + y)u, z^2 = u, u \in R]$ satisfy property (Q) with respect to (x, y). However, the functional has an absolute minimum under the constraints by (11.4.ii) and conditions (C1′) and (K).

7. $I[x, u] = \int_0^1 f_0(t, x, u)\,dt$ with differential system $x' = f(t, x, u)$, $x(0) = x_1$, $x(1) = x_2$, $u \in U = R$, with $f_0(t, x, u) = a_0(t, x)|u| + b_0(t, x)$, $f(t, x, u) = a(t, x)u + b(t, x)$, $a_0(t, x) \ge 0$, $a(t, x) \ne 0$, $(t, x) \in A = [0, 1] \times R$, a, b, a_0, b_0 continuous functions on A, Ω the class of all admissible pairs $x(t)$, $u(t)$, $0 \le t \le 1$, with $\int_0^1 x'^2\,dt \le c$ for $c > (x_2 - x_1)^2$. Here $m = n = 1$, A is closed, $M = A \times R$ is closed, $B = \{0\} \times \{x_1\} \times \{1\} \times \{x_2\}$ is compact, and condition (C_1') is satisfied. Note that $f_0(t, x, u) \ge b_0(t, x)$ and b_0 is bounded below in each compact part of A. Thus, a condition (L_1') holds in each compact part of A. Also, $|f| \to +\infty$ as $|u| \to +\infty$ uniformly in each compact part of A, and thus the sets $\tilde{Q}(t, x)$ have property (K) in (t, x). Note that for no t do the same convex sets $\tilde{Q}(t, x)$ satisfy property (Q) with respect to x. However, the functional $I[x, u]$ has an absolute minimum under the constraints by (11.4.ii).

8. $I[x, y, u, v] = \int_0^1 f_0(t, x, y, u, v)\,dt$ with differential system $x' = f_1$, $y' = f_2$, $(u, v) \in U = R^2$, $x(0) = y(0) = 1$, x_2, y_2 undetermined, $t_2 = 1$, in the class Ω of all admissible systems $x(t)$, $y(t)$, $u(t)$, $v(t)$, $0 \le t \le 1$, with $\int_0^1 (x'^2 + y'^2)\,dt \le 4$. Here $f_0(t, x, y, u, v) = a_0(t, x, y)|u| + b_0(t, x, y)|v|$, $f(t, x, y, u, v) = (f_1, f_2)$, $f_i = a_i(t, x, y)u + b_i(t, x, y)v$, $i = 1, 2$, and there are positive constants c, c_0, d, D such that $a_0 \ge c_0$, $b_0 \ge c_0$, $D(u^2 + v^2)^{1/2} \ge |f| \ge d(u^2 + v^2)^{1/2}$. Then $|f| \to \infty$ as $(u^2 + v^2)^{1/2} \to \infty$, and by (8.5.vii) the sets $\tilde{Q}(t, x, y)$ have property (K). Also, for $(u^2 + v^2)^{1/2} \ge R$ at least one of $|u|$ and $|v|$ is $\ge R2^{-1/2}$ and we have $f_0 = a_0|u| + b_0|v| \ge c_0 2^{-1/2} R \ge 2^{-1/2} c_0 D^{-1} |f|$. Conditions (C_1'), (C_3') hold. Note that for no t do the sets $\tilde{Q}(t, x, y)$ satisfy property (Q) with respect to (x, y) (unless a_1, b_1, a_2, b_2 are constant with respect to (x, y)). However, the functional $I[x, y, u, v]$ has an absolute minimum under the constraints by (11.4.ii).

11.7 Counterexamples

1. $I[x, u] = \int_0^1 tu^2\,dt$ with system $x' = u$, $n = m = 1$, $x(0) = 1$, $x(1) = 0$, $u \in U = R$, $A = [0 \le t \le 1, x \in R]$. Here A is compact, $M = A \times R$ is closed, $B = \{0\} \times \{1\} \times \{1\} \times \{0\}$ is compact, the sets $\tilde{Q}(t, x) = [(z^0, u)|z^0 \ge tu^2, u \in R]$ are all closed and convex, and the growth condition (g1′) holds at every $0 < t \le 1$, but does not hold at $t = 0$. Here $I \ge 0$; hence $i \ge 0$. On the other hand, for the admissible pairs $x_k(t)$, $u_k(t)$, $0 \le t \le 1$, $k = 2, 3, \dots$, defined by $x_k(t) = 1$, $u_k(t) = 0$ for $0 \le t < k^{-1}$, $x_k(t) = -(\log t)/(\log k)$, $u_k(t) = -t^{-1}/\log k$ for $k^{-1} \le t \le 1$, we have $I[x_k, u_k] = (\log k)^{-1}$. Hence $I[x_k, u_k] \to 0$

as $k \to \infty$, and thus $i \le 0$. Finally, $i = 0$, and I cannot have the value zero, since that would imply $tu^2 = 0$ a.e. in $[0, 1]$, $u = 0$ a.e. in $[0, 1]$, x a constant. The functional $I[x, u]$ above has no absolute minimum under the constraints. (A control version of Section 1.5, no. 4).

2. $I[x, u] = \int_0^{t_2} (1 + u^2)^{1/2} dt$, with system $x' = u$, $u \in U = R$, $t_1 = 0$, $0 \le t_2 \le 1$, $x(0) = 0$, (t_2, x_2) on the segment $\Gamma = [0 \le t \le 1, x = 1]$, $A = [0 \le t \le 1, 0 \le x \le 1]$ (the problem of a path C in nonparametric form, $C: x = x(t)$, $0 \le t \le t_2$, x AC in $[0, t_2]$, of minimum length joining $(0, 0)$ to Γ). Obviously, the problem has no absolute minimum. The growth condition (g1′) is nowhere satisfied. (A control version of a remark in Section 1.5, no. 3).

3. $I[x, u] = \int_0^{2\pi} |x - 2t| dt$ with differential system $x' = |x - 2t|u$, $u \in U = [u | u \ge 1]$, $x(0) = 0$, $x(2\pi) = 4\pi$, in the class Ω of all admissible pairs $x(t)$, $u(t)$, $0 \le t \le 2\pi$, with $\int_0^{2\pi} x'^2 dt \le 10\pi$. Here $n = m = 1$, and we can take $A = [0, 2\pi] \times R$ closed, $M = A \times R$ closed, $B = \{0\} \times \{0\} \times \{2\pi\} \times \{4\pi\}$ compact. Also $f_0 \ge 0$; thus condition (L_1') is satisfied with $\psi(t) = 0$. Let $i \ge 0$ be the infimum of $I[x, u]$ in Ω. Note that for the elements $x_k(t) = 2t - k^{-1} \sin kt$, $u_k(t) = k(2 - \cos kt)|\sin kt|^{-1}$, $0 \le t \le 2\pi$, $k = 1, 2, \ldots,$ we have $x_k'(t) = 2 - \cos kt = f(t, x(t), u_k(t))$, $u_k(t) \ge 1$, measurable and finite almost everywhere in $[0, 2\pi]$. Also, $\int_0^{2\pi} x_k'^2 dt = 9\pi$, and thus $(x_k, u_k) \in \Omega$ for all k. Finally, $I[x_k, u_k] = 4k^{-1} \to 0$ as $k \to \infty$, and hence $i = 0$. The functional does not take the value zero, since this would require $x(t) = 2t$, $x'(t) = 2 = |x(t) - 2t|u(t)$, $t \in [0, 2\pi]$ (a.e.), and this is impossible. Condition (C_1') is satisfied, since both initial and terminal points are fixed. The functional has no absolute minimum. Note that here the sets Q and $\tilde{Q}$ are all closed and convex. Indeed, if $x = 2t$, then $Q(t, x) = \{0\}$, $\tilde{Q}(t, x) = [(z^0, z) | z^0 \ge 0, z = 0] = R^+$; if $|x - 2t| = a > 0$, then $Q(t, x) = [z = au, u \ge 1]$, $\tilde{Q}(t, x) = [(z^0, z) | z^0 \ge a, z = au, u \ge 1]$. The sets $\tilde{Q}(t, x)$ do not have properties (K) and (Q) at $x = 2t$.

Bibliographical Notes

The treatment of the existence theorem of this chapter, and connected lower semicontinuity and lower closure theorems of Sections 10.7–8, reflects a number of remarks which have been made in the last years.

In this connection and for the use of mere uniform convergence of the trajectories the upper semicontinuity properties (K) and (Q) of the relevant sets $Q(t, x)$, $\tilde{Q}(t, x)$ with respect to (t, x), (that is, the same properties for the set valued maps, or multifunctions $(t, x) \to \tilde{Q}(t, x)$ and similar ones), were most natural (L. Cesari [6], 1966) (Section 8.8), and lower closure theorems take the place of lower semicontinuity theorems.

There was then the remark that, in connection with weak convergence of the derivatives, or weak convergence in $H^{1,1}$, the same properties with respect to x only suffice in the proof of lower closure theorems, that is, the same properties for the set valued maps $x \to \tilde{Q}(\bar{t}, x)$ and analogous ones for almost every $\bar{t}$ (L. Cesari [13], M. F. Bidaut [1], L. D. Berkovitz [1], independently around 1975) (Section (10.7)).

Then there was the remark that the property (K) requirement really suffices, since weak convergence implies a growth property which in turn, by a remark of Cesari [7] (Section 10.5), implies property (Q) for certain auxiliary sets, or maps $x \to \tilde{Q}^*(\bar{t}, x)$, with essentially no further change in the argument of either lower closure or lower semicontinuity theorems (A. D. Ioffe [1], 1977, I. Ekeland and R. Temam [1], L. Cesari and M. B. Suryanarayana [7, 8, 9]). It appears that property (Q) in some weak form, or equivalent properties, are needed in the proof of the underlying lower closure or lower semicontinuity theorems, whether we name explicitly these properties or not.

Then there was the remark that the terminology is somewhat simplified by a consistent use of the "Lagrangian", or extended function $T(t, x, x')$ (Section 1.12) for which the value $+\infty$ is allowed, and which is related to the sets $\tilde{Q}$ by the simple relation $\text{epi}_z\, T(t, x, z) = \text{cl}\, \tilde{Q}(t, x)$ (R. T. Rockafellar [6] in connection with his approach in terms of Convex Analysis and duality) (We shall cover these ideas in Chapter 17). However, in terms of the "Lagrangian" T the proof of the lower semicontinuity does not change much, and the same auxiliary sets $\tilde{Q}^*$ are needed with property (Q) in some form.

Actually, property (K) with respect to x of the sets $\tilde{Q}(t, x)$ is equivalent to the lower semicontinuity with respect to (x, z) of the extended function $T(t, x, z)$, and to the closure of the set epi $T(t, \cdot, \cdot)$. Analogously, property (Q) of the sets $\tilde{Q}$ can be equivalently expressed in terms of "seminormality" properties (in the sense of Tonelli and McShane of the function T (L. Cesari [8, 10, 11]) (Section 17.5), as well as in terms of Convex Analysis (G. S. Goodman [1]) (Section 17.6).

In this Chapter we have taken full advantage of all these steps. Indeed we proved the lower closure theorem (10.7.i) with full use of the auxiliary sets $\tilde{Q}^*$, we immediately derived the corresponding lower semicontinuity theorem (10.8.i), which is actually equivalent to (10.7.i), and we proved first the existence theorems (11.1.i) and (11.1.ii) in terms of the Lagrangian T. Then in Section 11.4 we derived the existence theorems for optimal control from the previous ones, that is, by actual use of their corresponding Lagrangian T, but the hypotheses are in terms of original problems of optimal control, or criteria are given.

In general, in applications, it appears that in problems where the Lagrangian is given, the terminology in terms of Lagrangian is more suitable (Section 11.3); in problems of optimal control where certain functions f_0, f are given containing the control parameters (Section 11.4), the terminology in terms of these functions is more suitable, or the one in terms of sets immediately defined from them, better than in terms of the Lagrangian function $T(t, x, x')$ which seldom can be written explicitly. We tried to show in this book the essential equivalence of the different terminologies.

Theorem (11.1.i) can be thought of as a present day form and far reaching extension of the fundamental 1914 theorem of Tonelli [4, 6, and I, vol. 2, p. 282]. We proved it first by relying on the lower semicontinuity theorem (10.8.i) and therefore on the equivalent lower closure theorem (10.7.i). We noted that under the growth conditions of (11.1.i) the sets $\tilde{Q}$ already have property (Q) with respect to x, hence for this theorem much simpler version of (10.7.i) and (10.8.i) suffice (Cf. second proof of (10.7.i)).

We have completed Section 11.1 with existence theorems for problems with a comparison functional, for problems with an integrand linear in the derivatives, for isoperimetric problems in which one of the integrals is linear, for problems with optimal generalized solutions. Theorem (11.1.vi) can be traced in E. J. McShane [18]; Remark (11.1.vii) and the Theorem (11.1.ix) for $n = 1$ are in L. Tonelli [I].

The problems of optimal control with unbounded controls are covered in Section 11.4, and their proof is given in terms of the related function $T(t, x, z)$. Most of those theorems had been proved already in terms of orientor fields anyhow, and the proofs are essentially the same. Theorems (11.4.i) and (11.4.ii) correspond essentially to the theorems (11.1.i), (11.1.ii) respectively. The existence theorem (11.4.iv) for problems with a comparison functional and for isoperimetric problems appears to be somehow more elaborated and comprehensive than in other presentations. The existence theorem (11.4.vi) for linear integrals based on the topology of weak convergence in L_p, $p > 1$, includes a number of previous statements, in particular the result of M. Vidyasagar [1], who assumed a uniform Lipschitz condition with respect to x.

The ideas underlying the present Chapter have been shown to be relevant in other situations. M. B. Suryanarayana [3, 4, 5] has studied problems of optimization with canonic hyperbolic equations and with linear total partial differential equations. T. S. Angell [1, 2, 4] has studied problems of optimal control with functional differential equations, with hereditary equations, and with nonlinear Volterra equations. H. S. Hou [1–4] has studied problems monitored by abstract nonlinear equations, including parabolic partial differential equations. Both T. S. Angell and H. S. Hou use properties (K) and (Q) in proving the existence of solutions (controllability) as well as in proving the existence of optimal solutions. R. F. Baum [1, 3] has studied stochastic control problems, and problems monitored by partial differential equations in R^n with lower dimensional controls. For extensions to problems in Banach spaces cf. L. Cesari [18]. For an existence theory for Pareto problems, that is, problems with functionals having their values in R^n or in Banach spaces, cf. L. Cesari and M. B. Suryanarayana [4, 5, 6, 7]. For further work on Pareto problems see also C. Olech [2, 5], N. O. Dacunha and E. Polak [1], P. L. Yu [1], P. L. Yu and G. Leitmann [1], L. A. Zadeh [1].

For Remark 5 of Section 11.4 we refer to W. F. Powers, B. D. Cheng and E. R. Edge [1] where a further analysis is made for the characterization of the singular solutions and for the elaboration of rapidly convergent methods for the numerical determination of the solutions.

Along the same lines discussed in this Chapter we mention here the extensive work of E. J. McShane [5–7, 10, 18], C. Olech [8, 9], C. Olech and A. Lasota [1, 2], A. Lasota and F. H. Szaframiec [1], E. O. Roxin [1], M. Q. Jacobs [1].

Many more ideas in existence theorems for one dimensional problems will be discussed in the next Chapters 12, 13, 14, 15, 16 in connection with different topologies and different viewpoints.

S. Cinquini [1–6] discussed problems of the calculus of variations for curves and surfaces depending on derivatives of higher order.

Only mention can be made here of the extensive work of C. B. Morrey [I] on multiple integrals, existence and regularity of the solutions of elliptic partial differential equations, and continuous surfaces of finite area.

We have already mentioned that the concept of generalized solutions was introduced by L. C. Young [1] in 1936 in terms of functional analysis. We refer to L. C. Young [I, 1–9], W. H. Fleming [1–4], W. H. Fleming and L. C. Young [1–2], E. J. McShane [12, 13, 14, 18] for work on generalized solutions in one and more variables. In particular W. H. Fleming (loc. cit), in the same frame of reference, developed a theory for solutions of stochastic partial differential equations.

Finally, only mention can be made here of the fundamental work of J. L. Lions [I] for multidimensional problems in the frame of differential inequalities, covering quadratic functionals for problems monitored by elliptic, parabolic, and hyperbolic linear partial differential equations.

For parametric problems of the calculus of variations on surfaces S in R^3, under sole continuity assumptions and finiteness of the Lebesgue area, L. Cesari [20, 21] discussed the Weierstrass condition as a necessary and also as a sufficient condition for lower semicontinuity of the parametric integrals $I[S, f_0]$ of the calculus of variations with respect to the topology of the Fréchet distance (uniform topology). On the basis of these results, and surface area theory, L. Cesari [22] proved the existence of a parametric surface S_0 for which $I[S, f_0]$ has a minimum value among all surfaces S of finite area and spanning a given simple continuous curve in R^3 (if any such surface exists). This is the Plateau problem for general parametric integrals of the calculus of variations. For such surfaces (merely continuous and of finite Lebesgue area, with no differentiability assumptions), the concept of the integral $I[S, f_0]$ was discussed by L. Cesari [19], in the spirit of surface area theory, as a Weierstrass integral. Later the same integral was discussed by L. Cesari [23, 24] in an abstract form, in connection with any quasiadditive vector valued set function (instead of a mere signed area function), showing that the property of quasiadditivity is preserved by the nonlinear parametric integrand f_0 and that the Weierstrass integral can be defined both as a Burkill type integral, and as a Lebesgue–Stieltjes integral with respect to the area measure defined by the surface and with the classical Jacobians replaced by Radon–Nikodym derivatives of the relevant set functions. This work has been continued by many authors (G. W. Warner, J. C. Breckenridge, A. W. J. Stoddart, L. H. Turner, T. Nishiura, A. Averna, C. Bardaro, M. Boni, P. Brandi, D. Candeloro, C. Gori, P. Pucci, M. Ragni, A. Salvadori, C. Vinti), and will be presented in III. Abstract lower semicontinuity theorems and a great many other properties of such integrals have been proved. In this connection extensive work has been done by E. Silverman on the lower semicontinuity of integrals on k-dimensional manifolds in R^n (cf. III).

CHAPTER 12

Existence Theorems: The Case of an Exceptional Set of No Growth

12.1 The Case of No Growth at the Points of a Slender Set. Lower Closure Theorems

Any of the growth conditions ($\gamma 1$), ($\gamma 2$), ($\gamma 3$) in (11.1.i) can be remarkably reduced. Indeed, we may assume that on the points (t, x) of a "slender" subset S of A no growth condition holds. We shall see that this will lead to a notable enlargement of the class of problems for which we can prove the existence of an optimal solution.

Given a fixed set S of the tx-space, $x = (x^1, \dots, x^n)$, and any set α of the t-axis, we shall denote by $\beta_i = S^i(\alpha)$ the set of all real numbers ξ such that for some $(t, x) \in S$ we have $t \in \alpha$, $x^i = \xi$, $i = 1, \dots, n$. We shall say that $S^i(\alpha)$ is the image of α on the x^i-axis relative to the set S. Note that when S is the graph of a curve $x = g(t)$, $t_1 \le t \le t_2$, in the tx-space, or $x^i = g_i(t)$, $t_1 \le t \le t_2$, $i = 1, \dots, n$, then $S^i(\alpha) = g_i(\alpha)$ is exactly the image of α on the x^i-axis by means of the component g_i of g.

A subset S of the tx-space R^{1+n}, $x = (x^1, \dots, x^n)$, is said to be *slender* if the following property holds: (S) For every set α of measure zero on the t-axis, the sets $S^i(\alpha)$ also have measure zero on the x^i-axis, $i = 1, \dots, n$. In other words, S is slender if meas $\alpha = 0$ implies meas $S^i(\alpha) = 0$, $i = 1, \dots, n$.

Any finite set S is slender. Any set S contained in a countable family of straight lines parallel to the t-axis is slender. Now consider any set S contained on countably many curves C in R^{1+n} of the type $C: x = x(t)$, $t \in I$, where $x(t)$, $t \in I$, is any AC n-vector function on an interval I of the t-axis. Any such set S also is slender.

Also, if F is the product in R^n of sets of measure zero F_i on the x^i-axis, $i = 1, \dots, n$, and S is contained in the family of straight lines parallel to the

t-axis $x = c$, $c \in F$ (that is, $x^i = c_i$, $c_i \in F_i$, $i = 1, \ldots, n$, $c = (c_1, \ldots, c_n)$ with $c \in F = F_1 \times \cdots \times F_n$, then F is slender.

Here are examples of functions $F_0(t, x, x')$ which satisfy the local growth condition (γ1) (Section 11.1) at all points $(\bar{t}, \bar{x}) \in A = R^{1+n}$ but the points $(\bar{t}, \bar{x})$ of an exceptional set S. For each example the set S is stated.

(a) $n = 1$, $F_0 = (t^2 + x^2)x'^2$, $S = [(0, 0)]$ slender;
(b) $n = 1$, $F_0 = |x^2 - t^2|x'^2$, $S = [(t, x)|x = \pm t, t \in R]$ slender;
(c) $n = 1$, $F_0 = x^2 \sin^2(x^{-1})x'^2$ if $x \neq 0$, $F_0 = 0$ if $x = 0$, $S = [(t, x)|x = 0$, $x = (k\pi)^{-1}$, k integer, $t \in R]$ slender;
(d) $n = 1$, $F_0 = |x|x'^2$, $S = [(t, 0), t \in R]$ slender;
(e) $n = 2$, $F_0 = |t^2 - x^2 - y^2|(x'^2 + y'^2)$, $S = [(t, x, y)|t = \pm(x^2 + y^2)^{1/2}$, $(x, y) \in R^2]$ not slender;
(f) $n = 2$, $F_0 = [|x - t| + |y - t^2|](x'^2 + y'^2)$, $S = [(t, x, y)|x = t, y = t^2, t \in R]$ slender.

Of course, for $F_0 = (1 + x'^2)^{1/2}$, $F_0 = x(1 + x'^2)^{1/2}$, $F_0 = (x - a)^{-1/2}(1 + x'^2)^{1/2}$ all points (t, x) are exceptional. These are slow growth integrands, for which we refer to Chapter 14, Section 14.4, Examples 1,3, and Counterexamples 2,3.

12.1.i (A LOWER CLOSURE THEOREM WITH A SLENDER SET OF EXCEPTIONAL POINTS). *Let A be compact in the tx-space R^{1+n}, $x = (x^1, \ldots, x^n)$, for every $(t, x) \in A$, let $\tilde{Q}(t, x)$ be a set of points $(z^0, z) \in R^{1+n}$, $z = (z^1, \ldots, z^n)$, and let $Q(t, x)$ denote the projection of $\tilde{Q}(t, x)$ on the z-space. Let us assume that* (a) *if $(z^0, z) \in \tilde{Q}(t, x)$ and $z^{0'} \geq z^0$, then $(z^{0'}, z) \in \tilde{Q}(t, x)$. Let S be a closed slender subset of A, and let us assume that* (b) *for every point $(\bar{t}, \bar{x}) \in S$ there is a neighborhood $N_\delta(\bar{t}, \bar{x})$ and real numbers $v > 0$, r and $b = (b_1, \ldots, b_n)$, such that $(t, x) \in N_\delta(\bar{t}, \bar{x}) \cap A$, $(z^0, z) \in \tilde{Q}(t, x)$ implies $z^0 \geq r + b \cdot z + v|z|$;* (c) *for every point $(\bar{t}, \bar{x}) \in A - S$ there is a neighborhood $N_\delta(\bar{t}, \bar{x})$ and, for every $\varepsilon > 0$, an L-integrable function $\psi_\varepsilon(t)$, $\bar{t} - \delta \leq t \leq \bar{t} + \delta$, such that $(t, x) \in N_\delta(\bar{t}, \bar{x}) \cap A$, $(z^0, z) \in \tilde{Q}(t, x)$ implies $|z| \leq \psi_\varepsilon(t) + \varepsilon z^0$. Let $\eta_k(t)$, $x_k(t)$, $t_{1k} \leq t \leq t_{2k}$, $k = 1, 2, \ldots$, be a sequence of functions, $\eta_k(t)$ scalar L-integrable, $x_k(t) = (x^1, \ldots, x^n)$* AC *in $[t_{1k}, t_{2k}]$, such that*

$$(t, x_k(t)) \in A, \quad (\eta_k(t), x_k'(t)) \in \tilde{Q}(t, x_k(t)), \quad t_{1k} \leq t \leq t_{2k}, \text{(a.e.)}, k = 1, 2, \ldots,$$

$$\liminf_{k \to \infty} \int_{t_{1k}}^{t_{2k}} \eta_k(t)\, dt < +\infty.$$

Then, the trajectories x_k are equicontinuous and have equibounded lengths. Also, there is a subsequence $[k_s]$ such that x_{k_s} converges in the ρ-metric toward a continuous and AC *function $x(t)$, $t_1 \leq t \leq t_2$.*

In addition, if $i = \liminf \int_{t_{1k}}^{t_{2k}} \eta_k(t)\, dt$, and we know that for almost all $\bar{t} \in [t_1, t_2]$ the sets $\tilde{Q}(t, x)$ have property (Q) *with respect to (t, x) at $(\bar{t}, x(\bar{t}))$, then there is a real valued function $\eta(t)$, $t_1 \leq t \leq t_2$, such that $(t, x(t)) \in A$, $(\eta(t), x'(t)) \in \tilde{Q}(t, x(t))$, $t_1 \leq t \leq t_2$* (a.e.), *and $-\infty < \int_{t_1}^{t_2} \eta(t)\, dt \leq i < +\infty$.*

In the proof of (12.1.i) below we shall denote by Ω^* the class of all pairs $\eta(t)$, $x(t)$, $t_1 \le t \le t_2$, $\eta(t)$ scalar L-integrable, $x(t) = (x^1, \dots, x^n)$ AC in $[t_1, t_2]$ with $(t, x(t)) \in A$, $(\eta(t), x'(t)) \in \tilde{Q}(t, x(t))$, $t \in [t_1, t_2]$ (a.e.).

For the proof of this theorem we need a simple preparatory lemma. To this effect, let $g(t)$, $a \le t \le b$, denote any real valued continuous function, let m and M be the minimum and maximum of g in $[a, b]$, and let V be the total variation of g in $[a, b]$, $0 \le V \le +\infty$. For every y real, let $N(y)$ denote the number of distinct points $t \in [a, b]$ where $g(t) = y$, so that $0 \le N(y) \le +\infty$ for all y, $N(y) = 0$ for $y < m$ and for $y > M$, $N(y) \ge 1$ for $m \le y \le M$.

12.1.ii (Banach). *The function $N(y)$ is measurable, and for its Lebesgue integral (finite or $+\infty$) we have the identity*

$$+\infty \ge \int_{-\infty}^{+\infty} N(y)\,dt = \int_m^M N(y)\,dy = V \ge M - m \ge 0.$$

For proofs of this statement we refer to Banach [1], Saks [I, p. 280], and Cesari [25]. We shall not need this statement in such generality. What we need is the following related statement which we shall prove directly below.

12.1.iii. *If the real valued continuous function $g(t)$, $a \le t \le b$, is AC, $g(a) = c$, $g(b) = d$, $c < d$, if G is any open subset of $[c, d]$, and if $E = g^{-1}(G) = [t \in [a, b] \,|\, g(t) \in G]$ and $E' = [a, b] - E$, then $\int_{E'} |g'(t)|\,dt \ge d - c - \operatorname{meas} G$.*

Proof. First let us assume that G is the union of finitely many disjoint intervals (c_i, d_i), $i = 1, \dots, N$. It is not restrictive, in this case, to assume that

$$c = d_0 \le c_1 < d_1 \le c_2 < \cdots \le c_N < d_N \le c_{N+1} = d.$$

Let us consider the $N + 1$ intervals $[d_i, c_{i+1}]$, $i = 0, 1, \dots, N$, disregarding those, if any, which are reduced to single points. Also, note that $E = [t \in [a, b] \,|\, g(t) \in G]$ is a set, open in $[a, b]$, whose components, are disjoint intervals (a_j, b_j), $j = 1, 2, \dots$, in $[a, b]$. Now, for each $i = 0, 1, \dots, N$, and interval $[d_i, c_{i+1}]$, there is at least one interval $[p_i, q_i] \subset [a, b]$ such that $g(t)$ spans exactly $[d_i, c_{i+1}]$ as t spans $[p_i, q_i]$ (not necessarily monotonically). There are at most finitely many of such intervals, and we choose one of them we denote $[p_i, q_i]$. The intervals (p_i, q_i) are disjoint, and if $E^* = \bigcup_{i=0}^N (p_i, q_i)$, then $E^* \cap E = \emptyset$, or $[a, b] - E \supset E^*$. Thus,

$$\int_{[a,b]-E} |g'(t)|\,dt \ge \int_{E^*} |g'(t)|\,dt \ge \sum_{i=0}^{N} \left| \int_{p_i}^{q_i} g'(t)\,dt \right|$$

$$= \sum_{i=0}^{N} |g(q_i) - g(p_i)| = d - c - \operatorname{meas} G.$$

Now let G be an arbitrary open subset of $[c, d]$, and let g_i, $i = 1, 2, \dots,$ denote the disjoint open intervals which are the components of G. Let $E = [t \in [a, b] \,|\, g(t) \in G]$, and let $[e_j, j = 1, 2, \dots]$ denote the components of E. For every m let us consider the open sets $G_m = g_1 \cup g_2 \cup \cdots \cup g_m$ and

$E_m = [t \in [a,b] | g(t) \in G_m]$, noting that each E_m is made up of a collection of the components e_j. By the previous estimates we have

$$\int_{[a,b]-E_m} |g'(t)| \, dt \geq d - c - \text{meas } G_m.$$

Now $G_m \subset G$, $E_m \subset E$, $[a,b] - E_m \supset [a,b] - E$, and as $m \to +\infty$, we have $G_m \uparrow G$, $[a,b] - E_m \downarrow [a,b] - E$, and hence

$$\int_{[a,b]-E} |g'(t)| \, dt \geq d - c - \text{meas } G. \qquad \square$$

Proof of (12.1.i).

(a) We shall denote by $N_h(\bar{t}, \bar{x})$ the open ball of center $(\bar{t}, \bar{x})$ in R^{n+1} and radius $h > 0$. Since $S \subset A$, S closed, A compact, then S is compact. Let $\sigma = 2\sqrt{n+1} + 1$. For every $(t,x) \in S$ let $\delta > 0$, $v > 0$, r, $b = (b_1, \ldots, b_n)$ be the numbers such that $(t,x) \in N_{\sigma\delta}(\bar{t}, \bar{x}) \cap A$, $(z^0, z) \in \tilde{Q}(t,x)$ implies $z^0 \geq r + b \cdot z + v|z|$. Actually, we shall consider the smaller neighborhoods $N_\delta(\bar{t}, \bar{x})$, which still form an open cover of S, and finitely many of them, therefore, cover S, say $N_{\delta_i}(t_i, x_i)$, $i = 1, \ldots, M$. Let $\delta = \min \delta_i$. We divide the tx-space R^{1+n} into cubes of side length δ (and therefore of diameter $\sqrt{n+1}\delta$), by means of the hyperplanes $t = h\delta$, $x^i = l_i\delta$, $i = 1, \ldots, n$, $h, l_i = 0, \pm 1, \pm 2, \ldots$. Finitely many of these cubes have points in common with A, and they cover A; let us denote them by Q, $l = (l_1, \ldots, l_n)$. Any one of these cubes has side length δ and diameter $\sqrt{n+1}\delta$. Some of these cubes, say Q_{hl}, may have points in common with balls $N_{\delta_i}(t_i, x_i)$, but then $\delta \leq \delta_i$, $\delta_i + \sqrt{n+1}\delta < \sigma\delta_i$, and then Q_{hl} is completely contained in the larger ball $N_{\sigma\delta_i}(t_i, x_i)$. We associate to such Q_{hl} the expression $\xi_{hl} = r_i + b_i \cdot z + v_i|z|$ relative to $N_{\sigma\delta_i}(t_i, x_i)$, and we denote it by $\xi_{hl} = r_{hl} + b_{hl} \cdot z + v_{hl}|z|$.

(b) We shall now refine the partition $\{Q_{hl}\}$. Note that the cubes Q_{hl}, which all together cover A, when projected on the t-axis, are contained in a minimal interval $[p\delta, q\delta]$ with $p < q$ integers. For every $t_0 \in [p\delta, q\delta]$, $\{t_0\}$ has measure zero on the t-axis; hence $S^i(\{t_0\})$ has measure zero, $i = 1, \ldots, n$, and can be covered by an open set F_i of measure $\leq \mu$ for any $0 < \mu < \delta/2$. If $H(t_0)$ denotes the hyperspace $t = t_0$, and $\bar{F} = F_1 \times \cdots \times F_n \subset R^n$ is the Cartesian product of these open subsets, then $F_0 = [(t_0, x) | x \in \bar{F}]$, is an open subset of $H(t_0)$, and $(H(t_0) - F_0) \cap A$ is compact and free of points of S. In other words, $(H(t_0) - F_0) \cap A \subset A - S$. Since S is closed, and $(H(t_0) - F_0) \cap A$ compact and free of points of S, the minimum distance of the two sets S and $(H(t_0) - F_0) \cap A$ is positive. Thus, there is a $\rho_0 > 0$ such that $(t_0 - \rho_0, t_0 + \rho_0) \times (H(t_0) - F_0) \cap A$ also is free of points of S. In other words, the set of points of S in the slab $(t_0 - \rho_0, t_0 + \rho_0) \times R^n$ is contained in $(t_0 - \rho_0, t_0 + \rho_0) \cap \bar{F}$. For each $(t_0, \bar{x}) \in (H(t_0) - F_0) \cap A$, and given $N > 0$, there are ρ, $0 < \rho \leq \rho_0$, and an L-integrable function $\psi_0(t) \geq 0$, $t_0 - \rho \leq t \leq t_0 + \rho$, such that $(t,x) \in N_{3\rho}(t_0, \bar{x}) \cap A$, $(z^0, z) \in \tilde{Q}(t, \bar{x})$ implies $|z| \leq \psi_0(t) + N^{-1}z^0$. The compact set $(H(t_0) - F_0) \cap A$ can be covered, therefore, by finitely many of these balls, $N_{\rho_s}(t_0, x_s)$, $s = 1, \ldots, M_2$, and we take $\bar{\rho}=\min[\rho_s, s=1,\ldots,M_2]$ and also $\psi_N(t)=\max[\psi_{0s}(t), s=1,\ldots,M]$, where $\psi_{0s}(t)$ is the function ψ_0 relative to $N_{\rho_s}(t_0,x_s)$.

In this manner we have associated an open interval of the form $(t-\rho, t+\rho)$ with each $t\in[p\delta, q\delta]$, thereby obtaining an open covering of $[p\delta, q\delta]$. Thus, finitely many of them cover $[p\delta, q\delta]$. By a suitable contraction, these finitely many intervals can be used to define a partition $P:p\delta = t_0 < t_1 < \cdots < t_R = q\delta$ of $[p\delta, q\delta]$, and it may be assumed without loss of generality that the points $s\delta$ for $p \le s \le q$, s integer, are all used in the partition. Now we can refine the partition of A into parts Q_{hl} by means of the hyperplanes $t = t_j$, $j = 1,\ldots, R$. The new parts are intervals, which we still denote by Q_{jl}. We shall call them cubes, for the sake of simplicity. Their sides parallel to the x^i-axis have all length δ; their sides parallel to the t-axis have lengths $t_j - t_{j-1} \le \delta$.

Summarizing, the following type of partition has been obtained. Given $\mu > 0$, N integer, there are expressions ξ_{jl} as above and a partition of the tx-space into cubes Q_{jl} as above, whose edges in the x^i-direction have length $\delta > 0$ independent of μ and N, such that (a) $z^0 \ge -N\psi_N(t) + N|z|$ for all $(t,x)\in A$, $(z^0,z)\in \tilde{Q}(t,x)$, and (t,x) in any of the cubes Q_{jl} of the slab $(t_{j-1},t_j)\times R^n$ minus the set $H_j = (t_{j-1},t_j)\times F_0$; and (b) $z^0 \ge \xi_{jl}$ for all $(t,x)\in A$, $(z^0,z)\in\tilde{Q}(t,x)$, and $(t,x)\in H_j = (t_{j-1},t_j)\times F_0$. In this second case then the relation $z^0 \ge \xi_{jl}$ holds also in any one of the $3^n - 1$ cubes not in H_j of the same section $(t,x)\in A$, $t_{j-1}\le t\le t_j$. The projection of H_j on each of the x^i-axes has measure $\le \mu$. Note that the constants δ, r_{jl}, b_{jl}, v_{jl} are independent of μ and N. Let $r = \max|r_{jl}|$, $b_0 = \max|b_{jl}|$, $v = \min v_{jl}$, and take $0 < \mu \le \delta/2$ and $N > 2b_0(1 + 4\sqrt{n+1})$.

(c) Let $\eta(t)$, $x(t)$, $a \le t \le b$, be any element of Ω^*. Let $C_j: x = x(t)$, $t_{j-1} \le t \le t_j$, denote the part of $C: x = x(t)$, $a \le t \le b$ (if any) defined in $[t_{j-1}, t_j]$. Divide C_j into more subarcs $C_{j1},\ldots, C_{jT_j}$ as follows: the first end of C_{j1} is $x(t_{j-1})$ [or $x(a)$ if $t_{j-1} < a < t_j$]; the second end point is either the first point where C_j leaves the $3^n - 1$ cubes in the section $[(t,x)\in A, t_{j-1}\le t < t_j]$ adjacent to the cube containing $x(t_{j-1})$, or $x(t_j)$ if C_j does not leave these 3^n cubes (or $x(b)$ if $t_{j-1} < b < t_j$). Continuing in this manner, C_j is broken up into arcs C_{js}, $s = 1,\ldots, T_j$. This process must terminate after a finite number of steps, since each arc C_{js}, except the first and the last one, has length $\ge \delta$.

Let Λ_{js} be the set of all t in the domain of C_{js} where $x(t)\in H_j$; let Λ'_{js} be the complement of Λ_{js} in this domain. Let $\lambda_{js} = \int_{\Lambda_{js}} |x'(t)|\,dt$, $\lambda'_{js} = \int_{\Lambda'_{js}} |x'(t)|\,dt$. We have

$$
\begin{aligned}
I = \int_a^b \eta(t)\,dt &= \sum\left(\int_{\Lambda_{js}} + \int_{\Lambda'_{js}}\right)\eta(t)\,dt \\
&\ge \sum \int_{\Lambda_{js}} [r_{js} + b_{js}\cdot x'(t) + v_{js}|x'(t)|]\,dt \\
&\quad + \sum \int_{\Lambda'_{js}} [-N\psi_N(t) + N|x'(t)|]\,dt \\
&\ge -r(b-a) - \sum b_0 \left|\int_{\Lambda_{js}} x'(t)\,dt\right| + v\lambda + N\lambda' - \int_a^b N\psi_N(t)\,dt,
\end{aligned}
$$

where $\lambda = \sum \lambda_{js}$, $\lambda' = \sum \lambda'_{js}$. On the other hand

$$\left| \left(\int_{A_{js}} + \int_{A'_{js}} \right) x'(t)\,dt \right| \leq 2\delta\sqrt{n+1},$$

$$\left| \int_{A_{js}} x'(t)\,dt \right| - \int_{A'_{js}} |x'(t)|\,dt \leq 2\delta\sqrt{n+1},$$

$$\left| \int_{A_{js}} x'(t)\,dt \right| \leq 2\delta\sqrt{n+1} + \lambda'_{js}$$

for all j and s. Moreover $\lambda'_{js} \geq \delta - \mu \geq \delta - \delta/2 = \delta/2$ for $s = 1, \ldots, T_j - 1$, $\lambda'_{jT_j} \geq 0$. Let D denote the diameter of A. Then

$$\begin{aligned} -b_0 \sum \left| \int_{A_{js}} x'(t)\,dt \right| &\geq \sum_{j=1}^{R} \sum_{s=1}^{T_j - 1} [-b_0(\lambda'_{js} + 2\delta\sqrt{n+1})] \\ &\quad - b_0 \sum_{j=1}^{R} (\lambda'_{jT_j} + 2\delta\sqrt{n+1}) \\ &\geq -b_0 \sum_{j=1}^{R} \sum_{s=1}^{T_j - 1} [\lambda'_{js} + 4\sqrt{n+1}\lambda'_{js}] \\ &\quad - 2b_0 R\delta\sqrt{n+1} - b_0 \sum_{j=1}^{R} \lambda'_{jT_j} \\ &\geq -b_0(1 + 4\sqrt{n+1})\lambda' - 2b_0 R\delta\sqrt{n+1}. \end{aligned}$$

Finally,

$$\begin{aligned} I &\geq -rD - 2b_0 R\delta\sqrt{n+1} + v\lambda \\ &\quad + [N - b_0(1 + 4\sqrt{n+1})]\lambda' - \int_{p\delta}^{q\delta} N\psi_N(t)\,dt \\ &\geq -rD - 2b_0 R\delta\sqrt{n+1} + v\lambda + 2^{-1}N\lambda' - \int_{p\delta}^{q\delta} N\psi_N(t)\,dt, \end{aligned}$$

and if $v_0 = \min[v, 2^{-1}N]$ and $I = \int_a^b \eta(t)\,dt \leq M_0$, we also have

$$M_0 \geq I \geq -rD - 2b_0 R\delta\sqrt{n+1} - \int_{p\delta}^{q\delta} N\psi_N(t)\,dt + v_0 \int_a^b |x'(t)|\,dt.$$

Thus, given any constant M_0, for any pair η, x in Ω_0 with $\int_a^b \eta(t)\,dt \leq M_0$, the trajectory x has uniformly bounded total variation $V[x]$, and thus uniformly bounded length. Let L be a bound for the lengths of the trajectories x of the collection Ω_0^* of the pairs η, x in Ω^* with $\int_a^b \eta(t)\,dt \leq M_0$.

Note that $I = \int_a^b \eta(t)\,dt$ is also bounded below in Ω_0^*. This can be derived from the last inequality, which indeed yields

$$I \geq -rD - 2b_0 R\delta\sqrt{n+1} - \int_{p\delta}^{q\delta} N\psi_N(t)\,dt.$$

Actually, a stronger statement can be proved: For every element $\eta(t)$, $x(t)$, $a \leq t \leq b$, of Ω_0^* and measurable subset E of $[a, b]$ we have $\int_E \eta(t)\,dt \geq -rD -$

$b_0L - \int_{p\delta}^{q\delta} N\psi_N(t)\,dt$. Indeed, if $E_1 = [t \in E | (t, x(t)) \in H_j$ for some $j]$ and $E_2 = E - E_1$, then $\eta(t) \geq -N\psi_N(t)$ for $t \in E_2$, and $\eta(t) \geq -r - b_0|x'(t)|$ for $t \in E_1$, and

$$\int_E \eta(t)\,dt \geq \int_{E_2} (-N\psi_N(t))\,dt + \int_{E_1} (-r - b_0|x'(t)|)\,dt$$
$$\geq -rD - b_0L - \int_{p\delta}^{q\delta} N\psi_N(t)\,dt.$$

Let Z denote this last number. Thus, in particular $I = \int_a^b \eta(t)\,dt \geq Z$ for every pair η, x in Ω_0^*.

(d) Let us consider again the family Ω_0^* of all pairs $\eta(t)$, $x(t)$, $a \leq t \leq b$, contained in Ω^* with $\int_a^b \eta(t)\,dt \leq M_0$, and let us prove that the trajectories $\{x(t)\}$ are equicontinuous. If they are not, then there is an $\varepsilon > 0$ such that for every positive integer k there is some pair $\eta_k(t)$, $x_k(t)$, $a_k \leq t \leq b_k$, in the class Ω_0^* and two points $t_{k1}, t_{k2} \in [a_k, b_k]$ such that $0 < t_{k2} - t_{k1} < k^{-1}$, $|x_k(t_{k2}) - x_k(t_{k1})| > \varepsilon$, and $I_k = \int_{a_k}^{b_k} \eta_k(t)\,dt \leq M_0$. Let us suppose, without loss of generality, that $t_{k1} \to t_0$, $t_{k2} \to t_0$, $x_k(t_{k1}) \to x_1$, $x_k(t_{k2}) \to x_2$ as $k \to \infty$. Then $|x_2 - x_1| \geq \varepsilon$. The sets $S^i(\{t_0\})$ have measure zero. Hence, they can be covered by open sets F_i of measure $\leq \mu$, for any $0 < \mu < \varepsilon/4n$. Let F denote the set $F = [(t_0, x) | x^i \in F_i, i = 1, \ldots, n]$. Then F is open in the hyperplane $H(t_0): t = t_0$. Let $N \geq (4n/\varepsilon)[M_0 + 1 + |2Z - 2r\delta - b_0L - 1|]$, where r, δ, and b_0 are the constants defined above. The set $(H(t_0) - F) \cap A$ is compact, and for every point $(t_0, \bar{x}) \in (H(t_0) - F) \cap A$ there is some $\bar{\rho} > 0$ and L-integrable function $\psi_N(t)$ such that $(t, x) \in N_{\bar{\rho}}(t_0, \bar{x}) \cap A$, $(z^0, z) \in \tilde{Q}(t, x)$ implies $|z| \leq \psi_N(t) + N^{-1}z^0$. A finite number of these neighborhoods cover $(H(t_0) - F) \cap A$. Let ρ be the minimum $\bar{\rho}$ for such finite covering. It is not restrictive to assume $\rho \leq \delta$. Divide the curve $C: x = x_k(t)$, $a_k \leq t \leq b_k$, into three parts C_{k1}, C_{k2}, C_{k3} according as $a_k \leq t \leq t_{k1}$, $t_{k1} \leq t \leq t_{k2}$, $t_{k2} \leq t \leq b_k$. Divide the interval $[t_{k1}, t_{k2}]$ into two subsets, say $E_2 = [t | x(t) \in F]$, $E_1 = [t_{k1}, t_{k2}] - E_2$. Then, for some k_0 and all $k \geq k_0$, we have $|t_{k1} - t_0| \leq \rho$, $|t_{k2} - t_0| \leq \rho$, and $|x_k(t_{k1}) - x_k(t_{k2})| \geq \varepsilon/2$. For $k \geq k_0$ we also have

$$I_k = \int_{a_k}^{b_k} \eta_k(t)\,dt = I_{k1} + I_{k2} + I_{k3} \geq 2Z + I_{k2}$$
$$\geq 2Z + \left(\int_{E_1} + \int_{E_2}\right)\eta_k(t)\,dt$$
$$\geq 2Z + \int_{E_1} (-N\psi_N(t) + N|x_k'(t)|)\,dt + \int_{E_2} [-r - b_0|x_k'(t)|]\,dt$$
$$\geq 2Z - 2r\delta - b_0L + N\int_{E_1} |x_k'(t)|\,dt - \int_{t_{k1}}^{t_{k2}} N\psi_N(t)\,dt.$$

Here E_2 is an open subset of $[t_{k1}, t_{k2}]$ and we know that $|x(t_{k1}) - x(t_{k2})| \geq \varepsilon/2$. Thus, for at least one component, say x^1, we also have $|x^1(t_{k1}) - x^1(t_{k2})| \geq \varepsilon/2n$, with $\varepsilon/2n > \mu \geq$ meas F_1. Thus, by (12.1.iii),

$$\int_{E_1} |x'(t)|\,dt \geq \int_{E_1} |x'^1(t)|\,dt \geq \frac{\varepsilon}{2n} - \text{meas } F_1 > \frac{\varepsilon}{2n} - \mu.$$

Thus,

$$I_k = \int_{a_k}^{b_k} \eta_k(t)\,dt \geq 2Z - 2r\delta - b_0 L + N\left(\frac{\varepsilon}{2n} - \mu\right) - \int_{t_{k1}}^{t_{k2}} N\psi_N(t)\,dt.$$

Since $t_{k2} - t_{k1} \to 0$, we can take k_0 so that the last integral is ≤ 1, and because of the choice of μ and N, we have $\varepsilon/2n - \mu \geq \varepsilon/4n$, and

$$I_k = \int_{a_k}^{b_k} \eta_k(t)\,dt \geq 2Z - 2r\delta - b_0 L - 1$$
$$+ [M_0 + 1 + |2Z - 2r\delta - b_0 L - 1|] \geq M_0 + 1,$$

a contradiction. Thus, for the pairs η, x in Ω^* with $\int_a^b \eta\,dt \leq M_0$, the trajectories x are equicontinuous.

(e) Let $\eta_k(t)$, $x_k(t)$, $a_k \leq t \leq b_k$, $k = 1, 2, \ldots$, be a sequence of pairs from Ω^* with $\int_{a_k}^{b_k} \eta_k(t)\,dt \leq M_0$. Then the sequence x_k is equicontinuous and the total variations $V[x_k]$ are bounded. By a suitable extraction there is a subsequence which is convergent in the ρ-metric to a continuous function $x(t)$, $a \leq t \leq b$. Let us prove that x is AC. Suppose that x is not AC. Let s, $0 \leq s \leq l = l(C)$, denote the usual arc length parameter for the curve $C: x = x(t)$, $a \leq t \leq b$, thought of as a path curve in the tx-space R^{n+1}. Note that given any measurable set $E \subset [a, b]$ the usual Lebesgue measure $|E|$ of E is the infimum of the number $\sum_i (\beta_i - \alpha_i)$ for any countable covering (α_i, β_i), $i = 1, 2, \ldots$, of E. Analogously, we can define another measure (length measure) $l(E)$ by taking the infimum of the numbers $\sum_i (s(\beta_i) - s(\alpha_i))$ for all the same open coverings of E. Obviously, $|E| \leq l(E)$. If x is not AC, then there is some set E of Lebesgue measure zero on $[a, b]$ which has positive length measure, or $|E| = 0$, $l(E) = \lambda > 0$. Now the n sets $S^i(E)$, $i = 1, \ldots, n$, have all zero Lebesgue measure. If $P = [(t, x) | t \in E, x = x(t)]$, then $P \cap S$ has projection of zero Lebesgue measure on each coordinate axis. As a consequence, there is some subset E' of E with $|E'| = 0$, $l(E') > \lambda/2$, and $(t, x(t)) \notin S$ for $t \in E'$, or $P' \cap S = \emptyset$ with $P' = \{(t, x) | t \in E', x = x(t)\}$. Let 2ρ be the distance of the two sets P' and S. Let $N = (2/\lambda)(|M_0| + 1 + |Z - 1|)$. Then there is an L-integrable function $\psi_N(t) \geq 0$ such that for (t, x) at a distance $\leq \rho$ from P' and $(z^0, z) \in \tilde{Q}(t, x)$ we have $|z| \leq \psi_N(t) + N^{-1} z^0$. Since E' is compact, $|E'| = 0$, it may be covered by a finite set of open intervals (α_j, β_j), $j = 1, \ldots, R$, such that if $F = \bigcup_{j=1}^R (\alpha_j, \beta_j)$ we have $\int_F N\psi_N(t)\,dt < 1$, and x maps F into the ρ-neighborhood of P'. Let k_0 be such that $\int_F |x_k'(t)|\,dt > \lambda/2$ for all $k \geq k_0$. Finally

$$I = \int_{a_k}^{b_k} \eta_k(t)\,dt \geq Z + \int_F \eta_k(t)\,dt$$
$$\geq Z + \int_F (-N\psi_N(t)\,dt + N|x_k'(t)|)\,dt$$
$$\geq Z - 1 + N\lambda/2$$
$$\geq Z - 1 + (|M_0| + 1 + |Z - 1|) \geq |M_0| + 1 \geq M_0 + 1,$$

a contradiction. We have proved that x is AC.

We have proved the first part of (12.1.i). The second part is a corollary of (8.8.i) and Remark 2 of Section 8.8. □

12.2 Existence Theorems for Extended Free Problems with an Exceptional Slender Set

We are interested here in existence theorems for the minimum of Bolza and Lagrange problems

$$I[x] = g(t_1, x(t_1), t_2, x(t_2)) + \int_{t_1}^{t_2} F_0(t, x(t), x'(t))\, dt \tag{12.2.1}$$

under the usual constraints

$$(t, x(t)) \in A, \qquad (t_1, x(t_1), t_2, x(t_2)) \in B, \qquad x'(t) \in Q(t, x(t)),$$

where $Q(t, x)$ are given subsets of R^n, $B \subset R^{2+2n}$, $A \subset R^{n+1}$. Then, for every $(t, x) \in A$ we denote by $\tilde{Q}(t, x)$ the set $\tilde{Q}(t, x) = [(z^0, z) | z^0 \geq F_0(t, x, z),\ z \in Q(t, x)]$. The existence theorems of this section can be thought of as variants of (11.1.i) where the global growth conditions ($\gamma 1$) or ($\gamma 2$) or ($\gamma 3$) are replaced by the local growth condition (g1) of Section 10.4, which is assumed to hold at all points $(t, x) \in A$ but those of a slender subset S of A. At the points $(\bar{t}, \bar{x}) \in S$ we shall need a much milder condition. On the other hand, the present theorems will be based on uniform convergence of trajectories (mode (a) of Section 2.14), and in this situation it appears convenient to assume F_0 continuous in its arguments.

The local condition which will be assumed at the points of S is only a geometrical transcription of condition (b) in (12.1.i):

β. We say that the local condition (β) is satisfied at the point $(\bar{t}, \bar{x})$ of A provided that there are a neighborhood $N_\delta(\bar{t}, \bar{x})$, a vector $b = (b^1, \ldots, b^n) \in R^n$, and numbers r real and $v > 0$ such that $(t, x) \in N_\delta(\bar{t}, \bar{x}) \cap A$, $z \in Q(t, x)$ implies $F_0(t, x, z) \geq r + b \cdot z + v|z|$.

Here $b \cdot z = (b, z)$ denotes the inner product in R^n. Summarizing, let A be a subset of the tx-space R^{1+n}, and for every $(t, x) \in A$ let $Q(t, x)$ be a given subset of the z-space R^n, $x = (x^1, \ldots, x^n)$, $z = (z^1, \ldots, z^n)$. Let M_0 be the set $M_0 = [(t, x, z) | (t, x) \in A,\ z \in Q(t, x)] \subset R^{1+2n}$, let $F_0(t, x, z)$ be a given real-valued function defined on M_0, and let us extend F_0 to all of R^{1+2n} by taking $F_0 = +\infty$ on $R^{1+2n} - M_0$. Let B be a given subset of the $t_1 x_1 t_2 x_2$-space R^{2n+2}, and let $g(t_1, x_1, t_2, x_2)$ be a real valued function on B. Let Ω be the class of all AC functions $x(t) = (x^1, \ldots, x^n)$, $t_1 \leq t \leq t_2$, satisfying

$$\begin{aligned} &(t, x(t)) \in A, \quad x'(t) \in Q(t, x(t)), \qquad t \in [t_1, t_2] \text{ (a.e.)}, \\ &(t_1, x(t_1), t_2, x(t_2)) \in B, \qquad F_0(\cdot, x(\cdot), x'(\cdot)) \in L_1[t_1, t_2]. \end{aligned} \tag{12.2.2}$$

12.2.i (AN EXISTENCE THEOREM FOR EXTENDED FREE PROBLEMS WITH A SLENDER SET OF EXCEPTIONAL POINTS). *Let A be compact, B closed, M_0 closed, g lower semicontinuous on B, $F_0(t, x, z)$ continuous on M_0, and assume that the sets $\tilde{Q}(t, x)$ satisfy property* (*Q*) *with respect to* (t, x) *at every point* $(\bar{t}, \bar{x}) \in A$. *Let S be a closed slender subset of A, and assume that* (a) *for every point* $(\bar{t}, \bar{x}) \in A - S$ *the local growth condition* (g1) *holds*; (b) *for every point* $(\bar{t}, \bar{x},) \in S$ *condition* (β) *holds. Then the functional $I[x]$ in* (12.2.1) *has an absolute minimum in the class Ω of all* AC *functions* $x(t) = (x^1, \ldots, x^n)$, $t_1 \leq t \leq t_2$, *satisfying* (12.2.2).

Remark 1. In Theorem (12.2.i) the required condition (Q) of the sets $\tilde{Q}(t, x)$ at the points $(t, x) \in A$ is, under mild assumptions, an immediate consequence of the other hypotheses. Indeed, if we assume that the extended function $T(t, x, z)$ is lower semicontinuous in (t, x, z), and convex in z for every $(t, x) \in A$, then the sets $\tilde{Q}(t, x)$ are convex and have property (K) with respect to (t, x) in A. They have property (Q) at every point $(t, x) \in A - S$ as a consequence of property (g1) and (10.5.i). Concerning the points $(t, x) \in S$, let us assume that (X) for every $(\bar{t}, \bar{x}) \in S$, $\bar{z} \in Q(t, x)$, and $\varepsilon > 0$ there are constants $r, b = (b_1, \ldots, b_n)$ real, $\delta > 0$ such that $T(\bar{t}, \bar{x}, \bar{z}) < r + b \cdot \bar{z} + \varepsilon$, and $(t, x) \in N_\delta(\bar{t}, \bar{x})$ implies $T(t, x, z) \geq r + b \cdot z$ for all z. Also we assume that (α) for every $(\bar{t}, \bar{x}) \subset S$, then $(\bar{z}^0, \bar{z}) \in \bigcap_\delta \operatorname{cl} \operatorname{co} \tilde{Q}(\bar{t}, \bar{x}, \delta)$ implies that $\bar{z} \in Q(\bar{t}, \bar{x})$. Here, as usual, $\tilde{Q}(\bar{t}, \bar{x}, \delta) = [\bigcup Q(t, x), (t, x) \in N_\delta(\bar{t}, \bar{x})]$. As we prove in (17.5.i), property (Q) at $(\bar{t}, \bar{x})$ is equivalent to properties (α) and (X) together.

Remark 2. In Theorem (12.2.i) the sets $\tilde{Q}(t, x)$ are defined by $\tilde{Q}(t, x) = \operatorname{epi}_z F_0(t, x, z)$. Note that having assumed that A is compact, the assumption "M_0 closed and F_0 continuous on M_0" corresponds to condition **C** of Section 11.1. Again for A compact, statement (12.2.i) holds even under the alternative assumption "F_0 satisfies condition **C***" of Section 11.1.

Proof of (12.2.i). At the points $(\bar{t}, \bar{x}) \in A - S$ condition (c) of (12.1.i) holds, since this is condition (g2), and we know from (10.4.iv) that (g1) implies (g2). The sets $\tilde{Q}(t, x)$ satisfy all conditions required in (12.1.i). The set A is compact, and thus there is some M such that $(t, x) \in A$ implies $|t| \leq M$, $|x| \leq M$. Thus, for every AC trajectory $x(t)$, $t_1 \leq t \leq t_2$, satisfying (12.2.2) we have $-M \leq t_1 < t_2 \leq M$, $|x(t)| \leq M$. Since Ω is assumed to be nonempty, then for $i = \inf I[x]$, we certainly have $-\infty \leq i < +\infty$. Note that $(t_1, x(t_1), t_2, x(t_2)) \in B \cap (A \times A)$ for every $x \in \Omega$, and the lower semicontinuous function g has a minimum $-M_1$ in the compact set $B \cap (A \times A)$. Let $x_k(t)$, $t_{1k} \leq t \leq t_{2k}$, $k = 1, 2, \ldots$, be any sequence of elements x_k of Ω such that $I[x_k] = g(e[x_k]) + \int_{t_{1k}}^{t_{2k}} F_0(t, x_k(t), x_k'(t))\, dt \to i$ as $k \to \infty$. If I_{k1} and I_{k2} are the two terms whose sum is $I[x_k]$, we have $I_{k1} + I_{k2} \to i$ and $I_{k1} \geq -M_1$. There is a subsequence, say still $[k]$ for the sake of simplicity, such that $I_{k1} \to i_1 \geq -M_1$, $I_{k2} \to i_2$, i_1 finite, $-\infty \leq i_2 < +\infty$, $i_1 + i_2 = i$.

As usual we take $\eta_k(t) = F_0(t, x_k(t), x_k'(t))$, $t_{1k} \leq t \leq t_{2k}$, so that $(\eta_k(t), x_k'(t)) \in \tilde{Q}(t, x_k(t))$, $t \in [t_{1k}, t_{2k}]$ (a.e.). From (12.1.i) we derive that i_2 too is finite, and that there are an L-integrable function $\eta(t)$, an AC function $x(t)$, $t_1 \leq t \leq t_2$, and a subsequence which we still denote as $[k]$, such that $x_k \to x$ in the

ρ-metric (in particular, $t_{1k} \to t_1$, $t_{2k} \to t_2$, and $e[x_k] \to e[x]$), and

$$(12.2.3)\qquad \begin{aligned} &(\eta(t), x'(t)) \in \tilde{Q}(t, x(t)), \qquad \int_{t_1}^{t_2} \eta(t)\,dt \le i_2 < +\infty, \\ &F_0(t, x(t), x'(t)) \le \eta(t), \qquad g(e[x]) \le \liminf_{k\to\infty} g(e[x_k]) = i_1. \end{aligned}$$

Let us prove that $F_0(t, x(t), x'(t))$ is L-integrable in $[t_1, t_2]$. Indeed the graph $\Gamma = [(t, x(t)), t_1 \le t \le t_2]$ of x is a compact set, and for each of its points $(\bar{t}, x(\bar{t}))$ there is a neighborhood $N_\delta(\bar{t}, x(\bar{t}))$, say a ball σ of center $(\bar{t}, x(\bar{t}))$ and some radius 2ρ, for which either (g1) or (β) holds. The concentric balls σ' of radius ρ form a cover of Γ. Hence finitely many of such balls σ' (say $\sigma'_1, \sigma'_2, \ldots, \sigma'_N$, of radii $\rho_1, \ldots, \rho_N$), cover Γ. Let $\rho_0 = \min[\rho_i, i = 1, \ldots, N]$. Now it is easy to divide Γ into at most N arcs $\Gamma_i: x = x(t)$, $t'_i \le t \le t''_i$, each of length $\ge \rho_0$, each contained in only one of the balls $\sigma_1, \ldots, \sigma_N$ of radii $2\rho_1, \ldots, 2\rho_N$ respectively. Now in the balls σ_i corresponding to property (g1), the corresponding function ϕ is bounded below, and so is $F_0(t, x(t), x'(t))$ for $t'_i \le t \le t''_i$. In the balls σ_i corresponding to property (β), then $F_0(t, x(t), x'(t)) \ge -r + b \cdot x'(t) + v|x'(t)|$ for $t'_i \le t \le t''_i$, certainly an L-integrable function. Thus, F_0 has a Lebesgue integral $\int_{t_1}^{t_2} F_0\,dt$ in $[t_1, t_2]$ which either is finite, or $+\infty$. The latter case, however, is excluded by relations (12.2.3), or $F_0 \le \eta$ with $\eta \in L_1[t_1, t_2]$. This proves that $F_0(t, x(t), x'(t))$ is L-integrable in $[t_1, t_2]$; hence $x \in \Omega$, and

$$i \le I[x] = g(e[x]) + \int_{t_1}^{t_2} F_0\,dt \le g(e[x]) + \int_{t_1}^{t_2} \eta(t)\,dt \le i_1 + i_2 = i.$$

Since $i = i_1 + i_2$ is finite, equality must hold throughout, or $I[x] = i$. Theorem (12.2.i) is thereby proved. □

12.3 Existence Theorems for Problems of Optimal Control with an Exceptional Slender Set

We are interested here in existence theorems for the minimum in Bolza and Lagrange problems of optimal control

$$(12.3.1)\qquad I[x, u] = g(t_1, x(t_1), t_2, x(t_2)) + \int_{t_1}^{t_2} f_0(t, x(t), u(t))\,dt,$$

$$(12.3.2)\qquad \begin{aligned} &dx/dt = f(t, x(t), u(t)), \qquad t \in [t_1, t_2] \text{ (a.e.)}, \\ &u(t) \in U(t, x(t)), \qquad (t_1, x(t_1), t_2, x(t_2)) \in B. \end{aligned}$$

The existence theorems of this section can be thought of as variants of (11.4.i) where the global growth conditions (g1′), or (g2′), or (g3′) are replaced by a local condition (g1′) which is assumed to hold at all points $(\bar{t}, \bar{x}) \in A$ except those of a slender subset S of A. At the points $(\bar{t}, \bar{x})$ of S we shall need a much milder condition (β). Again, as in Section 12.2, the theorems are based on uniform convergence, and we assume that f_0 and f are continuous functions of their arguments.

Summarizing, let A be a subset of the tx-space R^{1+n}, for every $(t,x) \in A$ let $U(t,x)$ be a given subset of the u-space R^m, and let $M = [(t,x,u) | (t,x) \in A, u \in U(t,x)]$. Let $f_0(t,x,u)$, $f(t,x,u) = (f_1, \ldots, f_n)$ be given functions on M, and for every $(t,x) \in A$ let $Q(t,x) = [z | z = f(t,x,u), u \in U(t,x)] \subset R^n$ and $\tilde{Q}(t,x) = [(z^0, z) | z^0 \geq f_0(t,x,u),\ z = f(t,x,u),\ u \in U(t,x)] \subset R^{1+n}$. Let B be a given subset of the $t_1x_1t_2x_2$-space R^{2n+2}, and let $g(t_1,x_1,t_2,x_2)$ be a real valued function on B. Here are the conditions (g1′) and (β') we need:

(g1′) We say that the local condition (g1′) is satisfied at the point $(\bar{t},\bar{x})$ of A provided there are a neighborhood $N_\delta(\bar{t},\bar{x})$, a function $\phi(\zeta)$, $0 \leq \zeta < +\infty$, bounded below, such that $\phi(\zeta)/\zeta \to +\infty$ as $\zeta \to +\infty$, and such that for all $(t,x) \in N_\delta(\bar{t},\bar{x}) \cap A$, $u \in U(t,x)$ we have $f_0(t,x,u) \geq \phi(|f(t,x,u)|)$.

(β') We say that the local condition (β) is satisfied at the point $(\bar{t},\bar{x})$ of A provided there are a neighborhood $N_\delta(\bar{t},\bar{x})$, a vector $b = (b^1, \ldots, b^n) \in R^n$, and numbers r real and $v > 0$ such that for all $(t,x) \in N_\delta(\bar{t},\bar{x}) \cap A$, $u \in U(t,x)$ we have $f_0(t,x,u) \geq -r + b \cdot f(t,x,u) + v|f(t,x,u)|$.

12.3.i (AN EXISTENCE THEOREM FOR PROBLEMS OF OPTIMAL CONTROL WITH A SLENDER SET OF EXCEPTIONAL POINTS). *Let A be compact, B closed, M closed, g lower semicontinuous on B, $f_0(t,x,u)$, $f(t,x,u)$ continuous on M, and assume that the sets $\tilde{Q}(t,x)$ are convex and have property* (Q) *with respect to (t,x) in A. Let S be a closed slender subset of A, and assume that* (a) *for every point $(\bar{t},\bar{x}) \in A - S$ the local growth condition* (g1′) *holds*; (b) *for every point $(\bar{t},\bar{x}) \in S$ condition* (β') *holds. Then the functional $I[x,u]$ in* (12.3.1) *has an absolute minimum in the class Ω of all admissible pairs $x(t) = (x^1, \ldots, x^n)$, $u(t) = (u^1, \ldots, u^m)$, $t_1 \leq t \leq t_2$, satisfying* (12.3.2).

The proof is left as an exercise for the reader.

12.4 Examples

A.

The integrands below satisfy the conditions of (12.2.i). For each we take for A a compact subset of R^{1+n}, $Q = R^n$, and we indicate the slender set S of the tx-space R^{1+n}.

1. $F_0 = |x|x'^2 + (1 + x'^2)^{1/2}$, $n = 1$, $S = [x = 0, t \in R]$;
2. $F_0 = x^2(1-x)^2|x'|^q + (1 + x'^2)^{1/2}$, $n = 1$, $q > 1$, $S = [(t,x) | t \in R,\ x = 0$ and $x = 1]$.
3. $F_0 = (x^2 + y^2)(x'^2 + y'^2) + (1 + x'^2 + y'^2)^{1/2}$, $n = 2$, $S = [x = 0, y = 0, t \in R]$.
4. $F_0 = (x - t)^2 x'^2 + (1 + x'^2)^{1/2}$, $n = 1$, $S = [x = t, t \in R]$.
5. $F_0 = (x \sin(x^{-1}))^2 x'^2 + (1 + x'^2)^{1/2}$ for $x \neq 0$, $F_0 = (1 + x'^2)^{1/2}$ for $x = 0$, $n = 1$, $S = [(t,x) | t \in R, x = 0$ and $x = (k\pi)^{-1}, k = \pm 1, \pm 2, \ldots]$, countably many straight lines parallel to the t-axis.

6. $F_0 = [(x^2 - t^2)^2 + y^2](x'^2 + y'^2) + (1 + x'^2 + y'^2)^{1/2}$, $n = 2$, $S = [(t, x, y)|x = \pm t$, $y = 0$, $t \in R]$, two straight lines.

7. $F_0 = |x|x'^2 + (1 + x'^2)^{1/2} - 2x'$, $n = 1$, $S = [x = 0, t \in R]$.

8. $F_0 = (x^2 + y^2)(x'^2 + y'^2) + (1 + x'^2 + y'^2)^{1/2} - 3x' - 5y' - 1$, $n = 2$, $S = [x = 0$, $y = 0$, $t \in R]$, a straight line parallel to the t-axis.

B.

The functions f_0, f below satisfy the conditions of (12.3.i). For each we take for A a compact subset of R^{1+n}, $Q = R^n$, and we indicate the slender set S of the tx-space R^{1+n}.

1. $f_0 = (x^2 + y^2)(u^2 + v^2) + (1 + u^2 + v^2)^{1/2} + 2^{-1}(1 + x^2 + y^2)^{-1}(u + v)$, $f_1 = u + v$, $f_2 = u - v$, $n = 2$, $m = 2$, $U = R^2$, $S = [(t, 0, 0), t \in R]$.

2. $f_0 = [(x - t)^2 + (y - t^2)^2](u^2 + v^2) + |u| + |v|$, $f_1 = u + v$, $f_2 = u - v$, $n = 2$, $m = 2$, $U = R^2$, $S = [(t, x, y)|x = t, y = t^2, t \in R]$.

3. $f_0 = (|x||x - t| + |y + t|)(|u| + |v|)^2 + |u - v|$, $f_1 = u + v$, $f_2 = u - v$, $n = 2$, $m = 2$, $U = R^2$, $S = [(t, x, y)|(x = t, y = -t)$ and $(x = 0, y = -t), t \in R]$.

4. $f_0 = |x|u^4 + (1 + u^4)^{1/2}$, $f = u^2$, $n = 1$, $m = 1$, $u \in R$, $S = [x = 0, t \in R]$.

5. $f_0 = |x|u^4 + (1 + u^4)^{1/2} - 3u^2$, $f = u^2$, $n = 1$, $u \in R$, $S = [x = 0, t \in R]$, $m = 1$.

6. $f_0 = x^3u^2 - x^2u$, $f = x^2u$, $n = 1$, $m = 1$, $x \geq 0$, $u \in R$, $S = [x = 0, t \in R]$.

7. $f_0 = (1 - x^2)^2u^2 + |1 - x^2||u|$, $f = (1 - x^2)u$, $n = 1$, $m = 1$, $u \in R$, $S = [x = -1$, $x = 1, t \in R]$.

12.5 Counterexamples

1. $n = 2$, $F_0 = xy' - yx'$. All curves joining $(0, 0, 0)$ and $(2\pi, 0, 0)$, $A = R^3$. All points of A are exceptional. For the sequence $x_k(t) = k^{-1} \sin k^4t$, $y_k(t) = k^{-1} \cos k^4t - k^{-1}$, $0 \leq t \leq 2\pi$, we have $I[x_k, y_k] = -2\pi k^2 \to -\infty$ as $k \to \infty$. The problem has no absolute minimum.

2. $n = 1$, $F_0 = (x^2 + x'^2)^{1/2}$. All curves joining $(0, 0)$, $(1, 1)$. (We have seen in Section 1.5, Example 5 that this problem has no minimum.) All points of $A = R^2$ are exceptional.

3. $n = 1$, $F_0 = (x'y - y'x) + (x^2 + y^2)^3(x'^2 + y'^2)$. All curves joining $(0, 0, 0)$ to $(2\pi, 0, 0)$. The singular set is $S = [(t, 0, 0), t \in R]$, a slender set. Condition (β) is not satisfied on the set S. For the sequence $x_k(t) = k^{-1} \cos k^4t - k^{-1}$, $y_k(t) = k^{-1} \sin k^4t$, $0 \leq t \leq 2\pi$, $k = 1, 2, \ldots$, we have $I[x_k, y_k] \to -\infty$ as $k \to \infty$. This problem has no absolute minimum.

4. $n = 1$, $m = 1$, $f_0 = (x + u)^2t$, $f = x + u$, $U = R$, $S = [(0, x)|x \in R]$ is not slender. For $0 \leq t \leq 1$, $x(0) = 1$, $x(1) = 0$, $T(t, x, z) = \min[f_0(t, x, u)|z = x + u, u \in R] = tz^2$. The problem is equivalent to $I = \int_0^1 tx'^2\,dt$, $x(0) = 1$, $x(1) = 0$, which we know has no minimum (Section 1.5, Example 4).

Bibliographical Notes

L. Tonelli [II; [10] and E. J. McShane [10] have proved existence theorems for classical Lagrange problems of the calculus of variations in which the growth assumption was allowed to fail in sets made up of a finite number of points, or contained in one or finitely many straight lines, or contained in one or finitely many smooth curves. The concept of a "slender" set of points, including all these cases, can be found in L. H. Turner [1] for Lagrange problems, and in L. Cesari, J. R. LaPalm, and D. A. Sanchez [1] for problems of optimal control. The present exposition is modeled on the latter paper. The lower closure theorem (12.1.i) is difficult to prove, because it also proves that in the present situation a minimizing sequence of trajectories is equicontinuous with integrals bounded below, possesses an AC limit, and the lower closure property holds. The existence theorem (12.2.i) is based therefore on mere uniform convergence, and this requires property (Q) with respect to (t, x). This property however can be guaranteed under mild additional assumptions in the present situation. The existence theorem (12.2.i) includes most of the results obtained by L. Tonelli and E. J. McShane for specific forms of the exceptional slender set. The existence theorem (12.3.i) is the expected extension of (12.2.i) to problems of optimal control.

CHAPTER 13
Existence Theorems: The Use of Lipschitz and Tempered Growth Conditions

13.1 An Existence Theorem under Condition (D)

In this section we are interested in problems of optimal control

$$I[x,u] = \int_{t_1}^{t_2} f_0(t,x,u)\,dt, \qquad x(t) = (x^1,\ldots,x^n), \quad u(t) = (u^1,\ldots,u^m),$$

under the constraints $dx/dt = f(t,x,u)$, $u \in U(t) \subset R^m$, and where f_0 and f satisfy any one of a great many alternative conditions (conditions of the F, G, and H type below), all easy to verify and of some practical interest. Of course, the result will apply also to extended free problem $\int_{t_1}^{t_2} f_0(t,x,x')\,dt$, or in the notation above $I = \int_{t_1}^{t_2} f_0(t,x,u)\,dt$ with $dx/dt = u$, $u \in U(t) \subset R^n$ $m = n$, (or $u \in U = R^n$ as in classical free problems). On the other hand, all the alternative conditions (of the F, G, or H type) are used only to guarantee that a single analytical condition holds (condition (D) below), and this condition leads to rather straightforward proofs of the existence theorems, e.g., (13.1.iii) below (cf. Cesari and Suryanarayana [1], 1975).

Initial work covering some of the alternative conditions was done by Rothe [1] in 1966.

The same conditions of the F, G, or H-type have been investigated also more recently, and claims have been made that they lead to existence theorems free of property (Q). Actually, the same conditions imply property (D), as already stated, and this in turn implies a weak form of property (Q) (see (13.1.i) below) which could well be used to prove the existence theorems by a variant of the usual argument of Chapter 11. This shows that the claim that the new theorems are free of property (Q) is not true.

The analytical relevance of condition (D) will be illustrated by Theorems (13.1.i) and (13.1.ii) below, and by the consequent existence theorems. In proving these we will have occasion to use the lower closure theorem (10.7.ii).

Points in R^ν, R^n, R^m, R^r spaces will be denoted by $t = (t^1, \ldots, t^\nu)$, $x = (x^1, \ldots, x^n)$, $u = (u^1, \ldots, u^m)$, and $\xi = (\xi^1, \ldots, \xi^r)$ or $z = (z^1, \ldots z^r)$. Also, A is a given subset of the tx-space $R^{\nu+n}$, A_0 is the projection of A on R^ν, and for $t \in A_0$ we let $A(t) = [x \in R^n \,|\, (t,x) \in A]$. For every $t \in A_0$ let $U(t)$ be a given subset of R^m, and let $M = [(t,x,u) \,|\, (t,x) \in A,\ u \in U(t)] \subset R^{\nu+n+m}$. Let $f(t,x,u) = (f_1, \ldots, f_r)$ be a given function defined on M. Let G be a given subset of A_0. We shall assume that G is measurable with finite measure.

We shall need the Carathéodory type condition (C′) similar to the conditions we have already encountered in Sections 8.3 and 10.8.

Condition (C′). Given G of finite measure, and A, M, f as stated, we say that condition (C′) holds provided, given $\varepsilon > 0$, there is a compact subset K of G such that $\text{meas}(G - K) < \varepsilon$, the sets $A_K = [(t,x) \in A \,|\, t \in K]$ and $M_K = [(t,x,u) \in M \,|\, t \in K]$ are closed, and the function $f(t,x,u)$, restricted to M_K, is continuous on M_K.

Whenever f has property (C′) and $x(t)$, $u(t)$, $t \in G$, are measurable with $x(t) \in A(t)$, $u(t) \in U(t)$, then $\xi(t) = f(t, x(t), u(t))$, $t \in G$, also is measurable.

With $U(t)$ depending on t only, as stated, and A, M, f satisfying condition (C′), let $x(t)$, $x_k(t)$, $u_k(t)$, $t \in G$, $k = 1, 2, \ldots$, be a given sequence of measurable functions for which we assume that $x(t)$, $x_k(t) \in A(t)$, $u_k(t) \in U(t)$, $t \in G$ (a.e.), $k = 1, 2, \ldots$, and $x_k(t) \to x(t)$ in measure on G as $k \to \infty$.

Condition (D). We say that f satisfies condition (D) with respect to the sequence x_k, u_k, provided the differences

$$\delta_k(t) = f(t, x_k(t), u_k(t)) - f(t, x(t), u_k(t)), \qquad t \in G, \quad k = 1, 2, \ldots,$$

belong to $(L_1(G))^r$ and approach zero strongly in $(L_1(G))^r$, that is, $\|\delta_k\|_1 \to 0$ as $k \to \infty$.

13.1.i (Condition (D) Implies a Weak Form of Property (Q)). *Let A, M, f satisfy condition (C′), let G be a measurable subset of A_0 of finite measure, let $U(t)$ depend on t only, and let $x(t)$, $x_k(t)$, $u_k(t)$, $t \in G$, $k = 1, 2, \ldots$, be measurable functions with $x(t)$, $x_k(t) \in A(t)$, $u_k(t) \in U(t)$, $t \in G$ (a.e.), such that $x_k(t) \to x(t)$ in measure in G as $k \to \infty$. Let $\xi_k(t) = f(t, x_k(t), u_k(t))$, $\bar{\xi}_k(t) = f(t, x(t), u_k(t))$, $\delta_k(t) = \xi_k(t) - \bar{\xi}_k(t)$, $t \in G$, $k = 1, 2, \ldots$. Let us assume that the sets $Q(t, x(t))$ are closed and convex. If the differences $\delta_k(t)$ are of class $L_1(G))^r$ and $\|\delta_k\|_1 \to 0$, then there is a subsequence $[k_s]$ such that the weak property (Q) of Remark 1 of Section 8.5 holds a.e. in G, that is,*

$$Q(t, x(t)) \supset \bigcap_{h=1}^{\infty} \text{cl co} \left\{ \bigcup_{s=h}^{\infty} \xi_{k_s}(t) \right\}, \qquad t \in G \text{ (a.e.)}.$$

Proof. There is a subsequence $[k_s]$ such that $\delta_{k_s}(t) \to 0$ pointwise a.e. in G, and now we have only to apply (8.5.viii). □

The following statement is also of interest:

13.1.ii (A NECESSARY AND SUFFICIENT CONDITION FOR PROPERTY (D)). *Let A, M, f satisfy condition (C′), let G be a measurable set of finite measure, let $U(t)$ depend on t only, and let $x(t), x_k(t), u_k(t), t \in G, k = 1, 2, \ldots$, be measurable functions with $x(t)$, $x_k(t) \in A(t)$, $u_k(t) \in U(t)$, $t \in G$, such that $u_k \in (L_1(G))^m$, $\|u_k\|_1 \le M_0$ for some constant M_0, $x_k(t) \to x(t)$ in measure in G as $k \to \infty$, and the differences $\delta_k(t)$ are of class $(L_1(G))^r$. Then $\delta_k \to 0$ strongly in $(L_1(G))^r$ if and only if the same functions $\delta_k(t)$, $t \in G$, $k = 1, 2, \ldots$, are equiabsolutely integrable.*

Proof. *Necessity*: If $\delta_k \to 0$ strongly in $(L_1(G))^r$, then $\int_G |\delta_k(t)|\,dt \to 0$ as $k \to \infty$, and obviously the functions δ_k are equiabsolutely integrable.

Sufficiency: Let us assume that the functions δ_k are equiabsolutely integrable in G. Then, given $\varepsilon > 0$, there is $\sigma > 0$ such that $\int_H |\delta_k(t)|\,dt \le \varepsilon$ for all $k = 1, 2, \ldots$, and any measurable subset H of G with meas $H \le \sigma$. By property (C′), there is a compact subset K of G with meas$(G - K) < \sigma/4$ such that the set $M_K = [(t, x, u) \in M \mid t \in K]$ is closed and $f(t, x, u)$ is continuous on M_K. Since x is measurable (and finite a.e. in G), there is an integer $\lambda > 0$ such that, if $\Lambda = [t \in G \,|\, |x(t)| \le \lambda]$, then meas$(G - \Lambda) \le \sigma/8$. Now the set M' of all $(t, x, u) \in M$ with $t \in K$, $|x| \le \lambda + 1$, $|u| \le 4M_0/\sigma$ is certainly compact, as it is a bounded subset of the closed set M_K, and f is uniformly continuous on M'. Hence, there is η, $0 < \eta \le 1$, such that $(t, x, u), (t, y, u) \in M'$, $|x - y| \le \eta$ implies $|f(t, x, u) - f(t, y, u)| \le \varepsilon(\text{meas } G)^{-1}$. If $N = 4M_0/\sigma$, then for any k, and for the set P_K of all $t \in G$ with $|u_k(t)| \ge N$, we have N meas $P_k \le M_0$, since $\|u_k\|_1 \le M_0$. Hence, meas $P_k < M_0/N = \sigma/4$. Finally, since $x_k \to x$ in measure, there is some integer k_0 such that for $k \ge k_0$ and for the set $\Lambda_k = [t \in G \,|\, |x_k(t) - x(t)| \le \eta]$, we have meas$(G - \Lambda_k) \le \sigma/4$. Then, for the set $\Lambda'_k = \Lambda_k \cap (G - P_k) \cap K \cap \Lambda$ we also have meas$(G - \Lambda'_k) \le \sigma$. We have now $\int_{G - \Lambda'_k} |\delta_k(t)|\,dt \le \varepsilon$. On the other hand, for $t \in \Lambda'_k$ we also have $t \in K$, $|x_k(t) - x(t)| \le \eta \le 1$, $|x(t)| \le \lambda$, $|x_k(t)| \le \lambda + 1$, $|u_k(t)| \le N = 4M_0/\sigma$, $(t, x(t), u_k(t))$, $(t, x_k(t), u_k(t)) \in M'$, and $|f(t, x(t), u_k(t)) - f(t, x_k(t), u_k(t))| \le \varepsilon(\text{meas } G)^{-1}$, and this holds for all $t \in \Lambda'_k$, $k \ge k_0$, with meas$(G - \Lambda'_k) \le \sigma$. We have proved that $\delta_k \to 0$ in measure. On the other hand, $\int_{\Lambda'_k} |\delta_k(t)|\,dt \le \varepsilon$, and finally, $\int_G |\delta_k(t)|\,dt \le 2\varepsilon$ for all $k \ge k_0$. Thus, $\|\delta_k\|_1 \to 0$ as $k \to \infty$. This proves (13.1.ii). □

Remark. The sufficiency part in (13.1.ii) can also be stated as follows. Let A, M, f satisfy property (C′). Let $S = \{x(t), u(t), t \in G\}$ denote a class of measurable functions x, u in G with the properties: (i) $x(t) \in A(t)$, $u(t) \in U(t)$, $t \in G$; (ii) for every $\sigma > 0$ there are constants $M_1(\sigma)$, $M_2(\sigma) > 0$ such that the sets $[t \in G \,|\, |x(t)| \le M_1(\sigma)]$ and $[t \in G \,|\, |u(t)| \le M_2(\sigma)]$ have measures

$\geq$meas $G - \sigma$; (iii) for any two pairs (x, u), (y, u) in S the differences $\delta(t) = f(t, x(t), u(t)) - f(t, y(t), u(t))$, $t \in G$, belong to $(L_1(G))^r$ and are equiabsolutely integrable in G. Then $\|\delta\|_1 \to 0$ uniformly in S as $x - y \to 0$ in measure. The proof is the same as for statement (13.1.ii). If condition (iii) is not satisfied, then $\delta(t) \to 0$ in measure uniformly in S as $x - y \to 0$ in measure. Note that condition (ii) is certainly satisfied if $x \in (L_1(G))^n$, $\|x\|_1 \leq L_0$, $u \in (L_1(G))^m$, $\|u\|_1 \leq L_1$, for given constants L_0, L_1. The equiabsolute integrability of the differences $\delta(t)$ is guaranteed by the inequalities on δ in the conditions (F), (G), (H) below.

13.1.iii (AN EXISTENCE THEOREM BASED ON WEAK CONVERGENCE OF THE DERIVATIVES AND PROPERTY (D)). *Let B be closed and g lower semicontinuous on g. Let A be compact, let $U(t)$ be independent of x, and let $M = [(t, x, u) | (t, x) \in A,\ u \in U(t)]$. Let f_0, $f = (f_1, \ldots, f_n)$ be defined on M, and assume that A, M, f_0, f satisfy condition* (C). *Let us assume that for almost all $t \in A_0$, the sets $\tilde{Q}(t, x)$, $x \in A(t)$, are closed and convex. Let us also assume that f_0, f satisfy one of the conditions* (L_i') *of Section* 11.4. *Let Ω be a nonempty Γ_{0w}-closed class of admissible pairs x, u, and assume that, for every minimizing sequence $x_k(t)$, $u_k(t)$, $t_{1k} \leq t \leq t_{2k}$, $k = 1, 2, \ldots$, of elements of Ω there is also a subsequence $[k_s]$ such that, for some* AC *trajectory $x(t)$, $t_1 \leq t \leq t_2$, we have*

(a) *$x_{k_s} \to x$ in the ρ-metric;*
(b) *$x'_{k_s} \to x'$ weakly in L_1;*
(c) *$f(t, x_{k_s}(t), u_{k_s}(t)) - f(t, x(t), u_{k_s}(t)) \to 0$ weakly in L_1^r;*
(d) *$f_0(t, x_{k_s}(t), u_{k_s}(t)) - f_0(t, x(t), u_{k_s}(t)) \to 0$ weakly in L_1.*

Then the functional $I[x, u]$ has an absolute minimum in Ω.

Relations (a), (b) certainly hold if we assume that any nonempty class, say Ω_{0M}, of all pairs $(x, u) \in \Omega$ with $I[x, u] \leq M$ is nonempty and relatively compact with respect to "weak convergence of the derivatives" (mode (b) of Section 2.14). Relations (c), (d) certainly hold if we assume that f_0, f satisfy any of the conditions of the F, G, or H type below (not necessarily the same, but all implying property (D), and so also (c) and (d)).

For A not compact but closed, see Section 11.5.

For Mayer problems, that is, $f_0 = 0$, the same statement holds, with (d) omitted, and the sets $\tilde{Q}$ replaced by the sets $Q(t, x) = [z = f(t, x, u),\ u \in U(t)]$.

Proof. To simplify the proof, we assume $g = 0$, that is, we consider a Lagrange problem. For any g, that is, for a Bolza problem, the proof below is the same with modifications as in the proof of (11.1.i).

First, let us define as usual the sets $Q(t, x) = f(t, x, U(t)) \subset R^r$, the sets $\tilde{Q}(t, x) \subset R^{r+1}$, and the extended scalar function $T(t, x, z)$.

Let $i = \inf I[x,u]$, $-\infty \le i < +\infty$, and let $x_k(t)$, $u_k(t)$, $t_{1k} \le t \le t_{2k}$, $k = 1, 2, \ldots$, be a minimizing sequence, so that $I[x_k, u_k] \to i$ as $k \to \infty$. By (a), (b) there is a subsequence, say still $[k]$, and an AC function $x(t)$, $t_1 \le t \le t_2$, such that, for $x_k'(t) = f(t, x_k(t), u_k(t))$, we have $x_k \to x$ in the ρ-metric, $x_k' = \xi_k \to x' = \xi$ weakly in L_1 as $k \to \infty$. Let

$$\eta_k(t) = f_0(t, x_k(t), u_k(t)), \qquad \bar{\eta}_k(t) = f_0(t, x(t), u_k(t)),$$

$$\bar{\xi}_k(t) = f(t, x(t), u_k(t)), \qquad I[x_k, u_k] = \int_{t_{1k}}^{t_{2k}} \eta_k(t)\, dt,$$

so that, by (c) and (d),

$$\eta_k(t) - \bar{\eta}_k(t) \to 0 \text{ weakly in } L_1,\ \xi_k(t) - \bar{\xi}_k(t) \to 0 \text{ weakly in } (L_1)^r \quad \text{as } k \to \infty.$$

Moreover

$$(\eta_k(t), \xi_k(t)) \in \tilde{Q}(t, x_k(t)), \qquad (\bar{\eta}_k(t), \bar{\xi}_k(t)) \in \tilde{Q}(t, x(t)).$$

From relations (L_i) there is a constant c such that $\eta_k(t) \ge -\psi(t) - c|\xi_k(t)|$, so that $I[x_k, u_k]$ is bounded below; hence i is finite, $-\infty < i < +\infty$. Moreover, the functions $\lambda_k(t)$, $\lambda(t)$ are immediately constructed satisfying the requirements of (10.7.ii). Indeed, we take $\lambda_k(t) = -\psi(t) - c|\xi_k(t)|$, and note that $\xi_k \to \xi$ weakly in L_1; hence by the Dunford–Pettis theorem (Section 10.3) the functions ξ_k, and so the functions $|\xi_k|$ and λ_k, are equiabsolutely integrable. Again, by Dunford–Pettis, there is a subsequence, say still $[k]$, such that $|\xi_k| \to \xi$ weakly in L_1 for some $\xi \in L_1$ and $\lambda_k \to \lambda = -\psi - c\xi$ weakly in L_1. By (10.7.ii) there is a function $\eta(t)$, $t \in [t_1, t_2]$, $\eta \in L_1$, such that

$$(\eta(t), x'(t)) \in \tilde{Q}(t, x(t)), \quad t \in [t_1, t_2] \text{ (a.e.)}, \qquad \int_{t_1}^{t_2} \eta(t)\, dt \le i.$$

By hypothesis, for almost all $t \in [t_1, t_2]$, all sets $\tilde{Q}(t, x)$, $x \in A(t)$, are closed; hence, $\eta(t) \ge T(t, x(t), x'(t))$, $(T(t, x(t), x'(t)), x'(t)) \in \tilde{Q}(t, x(t))$, $t \in [t_1, t_2]$ (a.e.). By the usual argument, $T(t, x(t), u(t))$ is measurable in $[t_1, t_2]$. By Section 8.2C, Exercise 7, there is a measurable function $u(t)$, $t_1 \le t \le t_2$, such that $u(t) \in U(t)$, $x'(t) = f(t, x(t), u(t))$, $T(t, x(t), x'(t)) = f_0(t, x(t), u(t))$, $t \in [t_1, t_2]$ (a.e.). Note that $\eta(t) \ge T(t, x(t), x'(t)) \ge -\psi(t) - c|x'(t)|$, that is, $T(t, x(t), x'(t))$ is between L-integrable functions and is measurable, and hence is L-integrable in $[t_1, t_2]$. Thus

$$i \ge \int_{t_1}^{t_2} \eta(t)\, dt \ge \int_{t_1}^{t_2} T(t, x(t), x'(t))\, dt = \int_{t_1}^{t_2} f_0(t, x(t), u(t))\, dt = I[x, u].$$

Here x, u is now an admissible pair, and since Ω is Γ_{0w}-closed, then $(x, u) \in \Omega$, and $I[x, u] \ge i$. By comparison we have $I[x, u] = i$ and (13.1.iii) is proved. □

For Mayer problems, that is, $f = 0$, the proof is the same with the sets Q replacing the sets $\tilde{Q}$, and obvious simplifications.

13.2 Conditions of the F, G, and H Types Each Implying Property (D) and Weak Property (Q)

We shall now discuss alternative hypotheses, each of which guarantees that $\delta_k \in (L_1(G))^r$ and $\delta_k \to 0$ strongly in $(L_1(G))^r$ as $k \to \infty$. We shall always assume below that f satisfies the Carathéodory type continuity condition (C′).

Note that, under the conditions (F) below, we assume that $x_k \in (L_p(G))^n$ but the u_k are only measurable; under the conditions (G), we assume that $x_k \in (L_p(G))^n$ and $u_k \in (L_q(G))^m$; under the conditions (H), we assume that $u_k \in (L_q(G))^m$ but the x_k are only measurable. We state the conditions for Mayer problems and thus only f is involved. For Lagrange and Bolza problems both f_0 and f are required to satisfy the conditions.

A. Lipschitz Type Conditions (F).

F_p (Geometric Viewpoint). (i) For $1 \le p < \infty$, $x, x_k \in (L_p(G))^n$, $\|x_k - x\|_p \to 0$, and (ii) $|f(t, x_k(t), u_k(t)) - f(t, x(t), u_k(t))| \le F_k(t)h(|x_k(t) - x(t)|)$, $t \in G$, $k = 1, 2, \ldots$, where $h(\zeta)$, $0 \le \zeta < +\infty$, is a given monotone nondecreasing function with $h(0+) = 0$, $h(\zeta) \le c|\zeta|^\gamma$, $c \ge 0$, $0 < \gamma \le p$, for all $\zeta \ge \zeta_0 \ge 0$ (c, γ, ζ_0 given constants), and $F_k(t) \ge 0$, $t \in G$, $F_k \in L_{p'}(G)$, given functions with $p' = p/(p - \gamma)$ ($p' = \infty$ if $\gamma = p$), and $\|F_k\|_{p'} \le C$, a given constant.

Let us prove that conditions (C′) and (F_p) imply $\|\delta_k\|_1 \to 0$ as $k \to \infty$. Given $\varepsilon > 0$, let $\sigma > 0$ be so chosen that $C\sigma^{\gamma/p}h(\zeta_0) \le \varepsilon$. It is to be noted that, since $\gamma > 0$, we have $p' > 1$. If $p' < \infty$ let p'' be defined by $1/p' + 1/p'' = 1$, so that $p'' = p/\gamma$ and then, by the Hölder inequality, we have for any measurable subset A of G, $\int_A |F_k(t)| \le |A|^{\gamma/p}C$, where $|A|$ is the measure of A. In particular, $\|F_k\|_1 \le C'$, where $C' = |G|^{\gamma/p}C$. If $p' = \infty$, then $C' = |G| \cdot C$. Let $\eta > 0$ be now so chosen that $h(\eta)C' \le \varepsilon$. Now, if $\|x_k - x\|_p \to 0$, then $\|x_k - x\|_1 \to 0$ and also $x_k - x \to 0$ in measure. Thus, there is some k_0 such that, for all $k \ge k_0$, we have $\|x_k - x\|_p < (\varepsilon/Cc)^{1/\gamma}$ and the set $G_{1k} = [t \in G \,|\, |x_k(t) - x(t)| \le \eta]$ has measure $> |G| - \sigma$. We consider also the sets $G_{2k} = [t \in G \,|\, \eta < |x_k(t) - x(t)| < \zeta_0]$ and $G_{3k} = [t \in G \,|\, |x_k(t) - x(t)| \ge \zeta_0]$. Then $|G_{2k}| < \sigma$, $|G_{3k}| < \sigma$; and, by the Hölder inequality and the definitions above, we have for $k \ge k_0$

$$\begin{aligned}\int_G |\delta_k(t)|\,dt &= \left(\int_{G_{1k}} + \int_{G_{2k}} + \int_{G_{3k}}\right)|\delta_k(t)|\,dt \\ &\le h(\eta)\int_G F_k(t)\,dt + |G_{2k}|^{\gamma/p}Ch(\zeta_0) + c\int_G F_k|x_k - x|^\gamma\,dt \\ &\le C'h(\eta) + \sigma^{\gamma/p}Ch(\zeta_0) + cC(\|x_k - x\|_p)^\gamma \le 3\varepsilon.\end{aligned}$$

We have proved that if $\|x_k - x\|_p \to 0$, then $\|\delta_k\|_1 \to 0$ under the hypotheses (C) and (F_p).

It is to be noted that the above proof uses the fact that the functions F_k, $k = 1, 2, \ldots$, are equiabsolutely integrable under the stated conditions, in particular $\gamma > 0$. If $\gamma = 0$, that is, $p' = 1$, then we need to assume that F_k are equiabsolutely integrable, and it is not enough to assume $\|F_k\|_1 \le C$. In this case, δ_k automatically are equiabsolutely integrable (because of the condition on f) and, by statement (13.1.i), $\|\delta_k\|_1 \to 0$ as $k \to \infty$.

The most relevant particular case is of course $\gamma = p = 1$, $F_k(t) = F(t) = 1$, $h(\zeta) = c\zeta$, $0 \le \zeta < +\infty$, and then (i) and (ii) reduce to $x, x_k \in (L_1(G))^n$, $\|x_k - x\|_1 \to 0$, $|f(t, x_k(t), u_k(t)) - f(t, x(t), u_k(t))| \le c|x_k(t) - x(t)|$, $t \in G$, $k = 1, 2, \ldots$.

F_p FUNCTIONAL VIEWPOINT, $1 \le p < \infty$). Let $\{u(t)\}$ be a family of control functions, and assume that for every control function $u = u(t)$ of the class there is a function $F_u(t)$ with the following properties. First, let $p, c, \gamma, \zeta_0, p', h$ be as in (F_p) above, and let $F_u(t) \ge 0$, $t \in G$, $F_u \in L_{p'}(G)$, $\|F_u\|_{p'} \le M$, and

$$|f(t, x_1, u(t)) - f(t, x_2, u(t))| \le F_u(t)h(|x_1 - x_2|).$$

Let $x(t), x_k(t), u_k(t), \xi_k(t), \bar{\xi}_k(t), t \in G, k = 1, 2, \ldots$, be measurable functions as in (13.1.iii) with $u_k \in \{u(t)\}$ for all k; assume that $\|x_k - x\|_p \to 0$ as $k \to \infty$. Then weak property (Q) holds as in (13.1.i).

In particular, F may be simply a function of (t, u), say, $F = F(t, u)$ or $F_u(t) = F(t, u(t))$, and in this case all we have to require is that $\int_G |F(t, u(t))|^p \, dt \le M$ for all $u \in \{u(t)\}$.

We shall list here a few more cases of property (F_p), limiting ourselves for the sake of simplicity to the geometric viewpoint.

Note that property (F_p) does not guarantee that the single functions $f(t, x(t), u_k(t))$, $f(t, x_k(t), u_k(t))$, $t \in G$, $k = 1, 2, \ldots$, are of class $(L_1(G))^r$ [only the differences $\delta_k \in (L_1(G))^r$]. We may add to (F_p) the requirement below.

F'_p. There is a function $\psi(t) \ge 0$, $t \in G$, $\psi \in L_1(G)$, and a constant $c \ge 0$ such that, for all $(t, x) \in A_G$ and $u \in U(t)$, we have

$$|f(t, x, u)| \le \psi(t) + c|x|^p.$$

We shall denote by (F_p^*) the union of condition (F_p) above (in any form), and of (F'_p). Under condition (F_p^*), then certainly the single functions $f(t, x(t), x_k(t))$, $f(t, x_k(t), u_k(t))$ are of class $(L_1(G))^r$.

F_∞. $x, x_k \in L_\infty(G)$, $\|x_k - x\|_\infty \to 0$, and

$$|f(t, x_k(t), u_k(t)) - f(t, x(t), u_k(t))| \le F_k(t)h(|x_k(t) - x(t)|), \qquad t \in G,$$

$k = 1, 2, \ldots$, where $h(\zeta) \ge 0, 0 \le \zeta < +\infty$, is a given monotone nondecreasing function with $h(0+) = 0$, and $F_k(t) \ge 0$, $t \in G$, $F_k \in L_1(G)$, are given functions with $\|F_k\|_1 \le C$, a given constant.

Let us prove that conditions (C) and (F_∞) imply $\|\delta_k\|_1 \to 0$. Let $\varepsilon > 0$ be given, let us choose $\eta > 0$ so that $Ch(\eta) < \varepsilon$, and let us choose k_0 so that $k \ge k_0$ implies $\|x_k - x\|_\infty < \eta$; hence, $|x_k(t) - x(t)| \le \|x_k - x\|_\infty \le \eta$ a.e. in G. For $k \ge k_0$ we have been

$$\int_G |\delta_k| \, dt \le \int_G F_k(t)h(\|x_k - x\|_\infty) \, dt \le Ch(\eta) < \varepsilon.$$

As above, we may also denote by (F_∞^*) the union of (F_p) and of the following further requirement, analogous to (F'_p).

F'_∞. There are a function $\psi(t) \ge 0$, $t \in G$, $\psi \in L_1(G)$, and a monotone nondecreasing function $\sigma(\zeta) \ge 0$, $0 \le \zeta < +\infty$, such that for all $(t, x, u) \in M$, we have

$$|f(t, x, u)| \le \psi(t)\sigma(|x|).$$

Note that for f linear in x, that is, of the form $f(t, x, u) = B(t, u)x + C(t, u)$, $B = [b_{ij}(t, u)]$, $C = [c_i(t, u)]$ matrices of the types $r \times n$, $r \times 1$, with $|b_{ij}(t, u)| \le \Phi(t)$, for

$\Phi(t) \geq 0$, $t \in G$, $\Phi \in L_{p'}(G)$, as in (F_p) with $\gamma = 1$, $1 \leq p < \infty$, then condition (F_p) is certainly satisfied. For $p = \infty$, $\Phi \in L_1(G)$, then condition (F_∞) is satisfied.

B. Growth Type Conditions (G)

$\mathbf{G}_{pq}$. (i) *There exists a continuous function* $\phi(u)$, $\phi\colon R^m \to R^m$, *with* $|\phi(u)| \to \infty$ *as* $|u| \to \infty$, *such that for* $1 \leq p,q < \infty$, $x, x_k \in (L_p(G))^n$, u_k *measurable*, $\phi(u_k) \in (L_q(G))^m$, $\|x\|_p$, $\|x_k\|_p \leq L_0$, $\|\phi(u_k)\|_q \leq L$, (L_0, L *given constants*), $x_k(t) \to x(t)$ *in measure in G as* $k \to \infty$, *and* (ii) *there are constants* c, c', α, β *with* $c, c' \geq 0$, $0 < \alpha \leq p$, $0 < \beta \leq q$, *and a function* $\psi(t) \geq 0$, $t \in G$, $\psi \in L_1(G)$, *such that for all* (t, x, u), $(t, y, u) \in M$, *we have*

$$\begin{aligned} &|f(t, x_k(t), u_k(t)) - f(t, x(t), u_k(t))| \\ &\quad \leq \psi(t) + c(|x_k(t)|^{p-\alpha} + |x(t)|^{p-\alpha}) + c'|\phi(u_k(t))|^{q-\beta}. \end{aligned}$$

Let us prove that conditions (C) and (G_{pq}) imply $\|\delta_k\|_1 \to 0$. Indeed, let p'', q'' be defined by $1/p'' + (p - \alpha)/p = 1$, $1/q'' + (q - \beta)/q = 1$, so that $p'' = p/\alpha$, $q'' = q/\beta$. For any measurable subset H of G we have, by the Hölder inequality,

$$\begin{aligned} \int_H |\delta_k|\,dt &\leq \int_H \psi(t)\,dt + c\int_H (|x_k(t)|^{p-\alpha} + |x(t)|^{p-\alpha})\,dt + c'\int_H |\phi(u_k(t))|^{q-\beta}\,dt \\ &\leq \int_H \psi(t)\,dt + c|H|^{\alpha/p}(\|x_k\|_p^{p-\alpha} + \|x\|_p^{p-\alpha}) + c'|H|^{\beta/q}\|\phi(u_k(t))\|_q^{q-\beta} \\ &\leq \int_H \psi(t)\,dt + 2cL_0^{p-\alpha}|H|^{\alpha/p} + c'L^{q-\beta}|H|^{\beta/q}. \end{aligned}$$

The last member obviously approaches zero as $|H| \to 0$, and the equiabsolute integrability of the functions $\delta_k(t)$, $t \in G$, is proved. Then $\|\delta_k\|_1 \to 0$ by statement (13.1.i) and the Remark of Section 13.1.

Note that the inequality (G_{pq}) is certainly satisfied if we know that $p > 1$ and that $|f(t, x, u) - f(t, y, u)| \leq \psi(t) + c|x - y| + c'|u|^{q-\beta}$.

Remark. The following statement is of interest. Let $S = \{x(t), u(t), t \in G\}$ denote any class of functions with $x(t) \in A(t)$, $u(t) \in U(t)$, $t \in G$ (a.e.), $x \in (L_p(G))^m$, $u \in (L_q(G))^m$, $\|x\|_p \leq L_0$, $\|u\|_q \leq L$ (L_0, L given constants), and let f satisfy conditions (C) and (G_{pq}) above. For any two pairs (x, u), (y, u) in S let us consider the differences $\delta(t) = f(t, x(t), u(t)) - f(t, y(t), u(t))$, $t \in G$. Then $\delta \in (L_1(G))^r$, and the same $\delta(t)$, $t \in G$, are equiabsolutely integrable, and $\|\delta\|_1 \to 0$ uniformly in S as $x(t) - y(t) \to 0$ in measure. The proof is the same as above, where the Remark of Section 13.1 is used.

$\mathbf{G}_{\infty q}$. (i) There exists a continuous function $\phi(u)$, $\phi\colon R^m \to R^m$, with $|\phi(u)| \to \infty$ as $|u| \to \infty$, such that for $1 \leq q < \infty$, x, $x_k \in (L_\infty(G))^n$, $\phi(u_k) \in (L_q(G))^m$, $\|x\|_\infty$, $\|x_k\|_\infty \leq L_0$, $\|\phi(u_k)\|_q \leq L$ (L, L_0 given constants), $x_k(t) \to x(t)$ in measure in G as $k \to \infty$, and (ii) there are constants c', β, $0 < \beta \leq q$, a function $\psi(t) \geq 0$, $t \in G$, $\psi \in L_1(G)$, and a monotone nondecreasing function $\sigma(\zeta) \geq 0, 0 \leq \zeta < +\infty$, such that for all (t, x, u), $(t, y, u) \in M$ we have

$$|f(t, x, u) - f(t, y, u)| \leq \psi(t)\sigma(|x| + |y|) + c'|\phi(u)|^{q-\beta}.$$

Let us prove that conditions (C) and $(G_{\infty q})$ imply $\|\delta_k\|_1 \to 0$. Indeed, note that the requirements $\|x_k\|_\infty$, $\|x\|_\infty \leq L_0$ imply $|x_k(t)|$, $|x(t)| \leq L_0$, $t \in G$ (a.e.). From this and

$\|\phi(u_k)\|_q \le L$, we derive as above, by the Hölder inequality, that

$$\int_H |\delta_k(t)|\,dt \le \sigma(2L_0)\int_H \psi(t)\,dt + c'L^{q-\beta}|H|^{\beta/q}.$$

Here ψ is a fixed L-integrable function on G, hence absolutely integrable, and the last relation proves, therefore, that the functions $\delta_k(t)$, $t \in G$, $k = 1, 2, \ldots$, are equiabsolutely integrable. Then $\|\delta_k\|_1 \to 0$ by (13.1.i).

Here too inequality $(G_{\infty q})$ is certainly satisfied if we know that

$$|f(t,x,u) - f(t,y,u)| \le \psi(t)\sigma(|x-y|) + c'|\phi(u)|^{q-\beta}.$$

Note that, in (G_{pq}), $(G_{\infty q})$, we do not assume $\sigma(0+) = 0$. These conditions (G_{pq}), $(G_{\infty q})$ are only growth conditions, the continuity property of f (property (C)) having the main role in the proof that $\delta_k \to 0$ strongly in $(L_1(G))^r$.

We leave to the reader to state conditions $(G_{p\infty})$ and $(G_{\infty\infty})$ analogous to the ones above.

Note that none of the properties (G_{pq}) above, $1 \le p, q \le \infty$, can guarantee that the single functions $f(t, x(t), u_k(t)), f(t, x_k(t), u_k(t))$, $t \in G$, $k = 1, 2, \ldots$, are of class $(L_1(G))^r$ (but the differences δ_k are in this class). When needed, we may denote by (G_{pq}^*), $1 \le p$, $q \le \infty$, the union of (G_{pq}) and of the further requirement that, for all $(t, x, u) \in M$, we have

(G'_{pq}) $\quad |f(t,x,u)| \le \psi(t) + c|x|^p + c'|\phi(u)|^q \quad$ if $1 \le p, q < \infty$,

$(G'_{\infty q})$ $\quad |f(t,x,u)| \le \psi(t)\sigma(|x|) + c'|\phi(u)|^q \quad$ if $p = \infty$, $1 \le q < \infty$,

and analogous requirements $(G'_{p\infty})$, $(G'_{\infty\infty})$.

Note that, if relation (G_{pq}) holds as stated, and relation (G'_{pq}) holds for all u and a fixed $x(t)$, $t \in G$, $x \in (L_p(G))^n$, then (G'_{pq}) holds for all $(t, x, u) \in M$ as stated. Indeed,

$$\begin{aligned}|f(t,x,u)| &\le |f(t,x(t),u)| + |f(t,x,u) - f(t,x(t),u)| \\ &\le [2\psi(t) + c|x(t)|^p + c|x(t)|^{p-\alpha} + c + c'] \\ &\quad + c'|\phi(u)|^q + c(|x|^{p-\alpha} - 1) + c'(|\phi(u)|^{q-\beta} - 1) \\ &\le \psi_0(t) + c|x|^p + 2c'|\phi(u)|^q.\end{aligned}$$

An analogous remark holds for the other conditions.

C. Growth Type Conditions (H)

H_q. (i) For $1 \le q < \infty$, x, x_k measurable, we have $x_k(t) \to x(t)$ in measure in G as $k \to \infty$, $u_k \in (L_q(G))^m$, $\|u_k\|_q \le L$, a constant; and (ii) there are other constants c', β, $0 < \beta \le q$, and a function $\psi(t) \ge 0$, $t \in G$, $\psi \in L_1(G)$, such that for all (t, x, u), $(t, y, u) \in M$, we have

$$|f(t,x,u) - f(t,y,u)| \le \psi(t) + c'|u|^{q-\beta}.$$

The proof that conditions (C) and (H_q) imply $\|\delta_k\|_1 \to 0$ is the same as for $(G_{\infty q})$.

H_∞. (i) x, x_k measurable, $x_k(t) \to x(t)$ in measure in G as $k \to \infty$, $u_k \in (L_\infty(G))^m$, $\|u_k\|_\infty \le L$, and (ii) there are a function $\psi(t) \le 0$, $t \in G$, $\psi \in L_1(G)$, and a monotone nondecreasing function $\sigma(\zeta) \ge 0$, $0 \le \zeta < +\infty$, such that for all (t, x, u), $(t, y, u) \in M$, we have $|f(t,x,u) - f(t,y,u)| \le \psi(t)\sigma(|u|)$.

When needed, we shall denote by (H_q^*) the union of (H_q) and of the further requirement that for all $(t, x, u) \in M$, we have

$$|f(t, x, u)| \le \psi(t) + c'|u|^q.$$

D. The Case of f Linear in u

For $f(t, x, u)$ linear in u, that is, of the form

$$f(t, x, u) = B(t, x)u + C(t, x),$$

$B = [b_{ij}(t, x)]$, $C = [c_i(t, x)]$ matrices of the types $r \times m$, $r \times 1$, suitable conditions can be stated in terms of the matrices B and C. For instance:

GL$_{pq}$. (i) For $1 \le p < \infty$, $1 < q < \infty$, x, $x_k \in (L_p(G))^n$, $u_k \in (L_q(G))^m$, $\|x\|_p$, $\|x_k\|_p \le L_0$, $\|u_k\|_q \le L$, we have $x_k(t) \to x(t)$ in measure, and (ii) there are constants c, α, $0 < \alpha \le p$, and a function $\psi(t) \ge 0$, $t \in G$, $\psi \in L_1(G)$, such that for all (t, x), $(t, y) \in A_G$, we have

$$|b_{ij}(t, x)| \le c, \qquad |c_i(t, x) - c_i(t, y)| \le \psi(t) + c(|x|^{p-\alpha} + |y|^{p-\alpha}).$$

We need only to verify that f satisfies condition (G_{pq}). We have

$$\begin{aligned} |f(t, x_k(t), u_k(t)) - f(t, x(t), u_k(t))| &\le \sum_i \sum_j |b_{ij}(t, x_k(t)) - b_{ij}(t, x(t))||u_k^j(t)| \\ &\quad + \sum_i |c_i(t, x_k(t)) - c_i(t, x(t))| \\ &\le 2rmc|u_k(t)| + 2r\psi(t) + rc(|x_k(t)|^{p-\alpha} + |x(t)|^{p-\alpha}), \end{aligned}$$

and (G_{pq}) is satisfied with $\phi(u) = u$, $\beta = q - 1$, and α as given.

Note that, under condition (GL_{pq}), the functions $B(t, x_k(t))u_k(t)$, $t \in G$, $k = 1, 2, \ldots$, are of class $(L_q(G))^r$ and have L_q-norms $\le rmcL$. Thus, they always possess a subsequence which is weakly convergent in $(L_q(G))^r$, and then also in $(L_1(G))^r$. If, in addition to (GL_{pq}), we know that (GL'_{pq}) $|c_i(t, x)| \le \psi(t) + c|x|^p$, then certainly the functions $\xi_k(t) = f(t, x_k(t), u_k(t))$, $k = 1, 2, \ldots$, belong to $(L_1(G))^r$. If we also know that the functions $c_i(t, x_k(t))$, $t \in G$, $k = 1, 2, \ldots$, are equiabsolutely integrable, then there is certainly a subsequence $[k_s]$ such that ξ_{k_s} converges weakly in $(L_1(G))^r$. Analogous remarks hold also under the conditions which we list below, but we shall omit them for the sake of brevity.

GL$_{p1}$. For $1 \le p < \infty$, $q = 1$, the same as (GL_{pq}) with $q = 1$, where now $u_k \in (L_1(G))^m$, and the sequence u_k is known to be equiabsolutely integrable in G.

The latter requirement is certainly satisfied if the sequence u_k is known to converge weakly in $(L_1(G))^m$.

To prove that conditions (C) and (GL_{p1}) imply that $\|\delta_k\|_1 \to 0$, note that

$$\begin{aligned} |\delta_k(t)| &= |f(t, x_k(t), u_k(t)) - f(t, x(t), u_k(t))| \\ &\le 2rmc|u_k(t)| + 2r\psi(t) + rc(|x_k(t)|^{p-\alpha} + |x(t)|^{p-\alpha}), \end{aligned}$$

and the sequence $\delta_k(t)$, $t \in G$, $k = 1, 2, \ldots$, is then equiabsolutely integrable in G. Finally, $\|\delta_k\|_1 \to 0$ by (13.1.ii)

($GL_{\infty q}$) (i) For $1 < q < \infty$, the same as in (GL_{pq}) except that now x, $x_k \in L_\infty(G)$, $u_k \in (L_q(G))^m$, $x_k(t) \to x(t)$ in measure, $\|u_k\|_q \leq L$, and (ii) there are a constant c, a function $\psi(t)$, $t \in G$, $\psi \in L_1(G)$, and a monotone nondecreasing function $\sigma(\zeta) \geq 0$, $0 \leq \zeta < +\infty$, such that for all $(t, x) \in A_G$, we have $|b_{ij}(t, x)| \leq c$, $|c_i(t, x)| \leq \psi(t)\,\sigma(|x|)$.

($GL_{\infty 1}$) The same as ($GL_{\infty q}$) with $q = 1$, and the sequence u_k is known to be equiabsolutely integrable.

($GL_{\infty\infty}$) The same as (GL_{pq}) except that now x, $x_k \in (L_\infty(G))^n$, $u_k \in (L_\infty(G))^m$, $x_k(t) \to x(t)$ in measure, $\|x\|_\infty$, $\|x_k\|_\infty \leq L_0$, $\|u_k\|_\infty \leq L$, $\xi_k \to \xi$ weakly in $(L_1(G))^r$.

(HL_q) For $1 \leq q < \infty$, x, x_k measurable, we have $x_k(t) \to x(t)$ in measure in G, $u_k \in (L_q(G))^m$, $\|u_k\|_q \leq L$, a constant, and (ii) there are constants c and a function $\psi(t)$, $t \in G$, $\psi \in L_1(G)$, such that for all (t, x), $(t, y) \in A_G$ we have $|b_{ij}(t, x)| \leq c$, $|c_i(t, x) - c_i(t, y)| \leq \psi(t)$. For $q = 1$, we must also require explicitly that the functions $u_k(t)$, $t \in G$, $k = 1, 2, \ldots$, be equiabsolutely integrable in G.

(HL_∞) The same as (HL_q) except that now $u_k \in (L_\infty(G))^m$, $\|u_k\|_\infty \leq L$, and still x, x_k are measurable and $x_k(t) \to x(t)$ in measure in G as $k \to \infty$.

13.3 Examples

1. Let us consider the Mayer problem with $n = 2$, $m = 1$, state variables $x = (x^1, x^2)$, control variable u, $t_1 = 0$, $t_2 = 1$, $I = g = x^1(1) + x^2(1)$, $dx^1/dt = f_1 = u$, $dx^2/dt = f_2 = x^1 u$, $x^1(0) = 0$, $0 \leq x^2(0) \leq 1$, $u \in U = R$, constraint $\int_0^1 u^2\,dt \leq 1$. Then

$$|x^1(t)| = \left|\int_0^t u\,dt\right| \leq \left(\int_0^t u^2\,dt\right)^{1/2} \leq 1,$$

$$|x^2(t)| = \left|x^2(0) + \int_0^t x^1 u\,dt\right| \leq 1 + \int_0^1 |u|\,dt \leq 2.$$

Thus, it is not restrictive to take for A the compact set $A = [0 \leq t \leq 1, |x^1| \leq 1, |x^2| \leq 2]$. Here $f = (f_1, f_2)$, and

$$|f(t, x, u) - f(t, y, u)| = |(0, (x^1 - y^1)u)| \leq |x - y|\,|u|.$$

Condition (F_∞) holds with $F(t, u) = |u|$, $h(\zeta) = \zeta$, $0 \leq \zeta < +\infty$, and

$$\int_0^1 F(t, u)\,dt = \int_0^1 |u|\,dt \leq \left(\int_0^1 u^2\,dt\right)^{1/2} \leq 1,$$

$$\int_0^1 (dx^1/dt)^2\,dt = \int_0^1 u^2\,dt \leq 1, \qquad \int_0^1 (dx^2/dt)^2\,dt \leq \int_0^1 (x^1(t)u(t))^2\,dt \leq 1.$$

Hence, $(dx^1/dt, dx^2/dt)$ belong to the unit ball in $(L_2(0, 1))^2$ which is weakly compact. If $[x_k]$ is a minimizing sequence, then there is a subsequence, say still $[k]$, such that $x_k' \to \xi$ weakly in $(L_2(0, 1))^2$, with x_k' equiabsolutely integrable, x_k equiabsolutely continuous, and $x_k \to x$ uniformly. Then $x' = \xi$ a.e. in $[0, 1]$. The sets $Q(x) = [z = (z^1, z^2) \mid z^1 = u, z^2 = x^1 u]$ are straight lines, and hence closed and convex. From (13.1.iii) the problem has an absolute minimum.

2. We seek the minimum of $\int_0^1 xx'^2\,dt$ with $x(0) = 0$, $x(1) = 1$, constraints $x(t) \geq 0$, $\int_0^1 x'^2\,dt \leq C$ for some $C > 1$. Equivalently, we have the Lagrange problem of the minimum of $I[x, u] = \int_0^1 xu^2\,dt$ with $x(0) = 0$, $x(1) = 1$, $u \in U = R$, $\int_0^1 u^2\,dt \leq C$, $n = 1$, $m = 1$, $dx/dt = f = u$, $f_0 = xu^2$, $\tilde{f} = (f_0, f)$. Hence, $|x(t)| = |\int_0^t u\,dt| \leq C^{1/2}$ and we can take for A the compact set $A = [0 \leq t \leq 1, 0 \leq x \leq C^{1/2}]$. Here

$$|\tilde{f}(t, x, u) - \tilde{f}(t, y, u)| = |(xu^2 - yu^2, 0)| \leq |x - y|u^2,$$

and for $F(t, u) = u^2$, we have $\int_0^1 F(t, u)\,dt \leq C$. Property ($F_\infty$) holds with $h(\zeta) = \zeta$. By (13.1.iii) the problem has an optimal solution.

3. We seek the minimum of $\int_0^1 (t + x + x'^2)\,dt$, $x(0) = 0$, $x(1) = 1$, with $\int_0^1 x'^2\,dt \leq C$, a constant $C > 1$. Equivalently, we have the Lagrange minimum problem with $f_0 = t + x + u^2$, $f = u$, $t_1 = 0$, $t_2 = 1$, $x(0) = 0$, $x(1) = 1$, $\int_0^1 u^2\,dt \leq C$, $u \in U = R$. Here $\tilde{f} = (f_0, f)$ satisfies condition (G_{22}) with $\alpha = \beta = 1$, since $|\tilde{f}(t, x, u) - \tilde{f}(t, y, u)| \leq |x| + |y|$. If x_k, u_k is a minimizing sequence, then there is a subsequence, say still $[k]$, such that $x_k' \to \xi$ weakly in L_1, $x_k \to x$ uniformly, x AC with $x' = \xi$, and the sets $\tilde{Q}(t, x) = [(z^0, z) | z^0 \geq 1 + x + u^2, z = u, u \in R]$ are convex and closed.

The same holds for

$$f_0(t, x, u) = t + \sin tx + (1 + \cos t)x + (\sin t)u^2 + (\sin x)u,$$

with

$$|f_0(t, x, u) - f_0(t, y, u)| \leq 2 + 2|x| + 2|y| + 2|u|,$$

and (G_{pq}) holds with $\psi = 2$, $c = 2$, $p = q = 2$, $c' = 2$, $\alpha = \beta = 1$, $\phi(u) = u$. The sets $\tilde{Q}(t, x) = [(z^0, z) | z^0 \geq f_0(t, x, u), z = u, u \in R]$ are obviously closed and convex. From (13.1.iii) the problem has an absolute minimum.

4. We seek the minimum of $\int_0^1 f_0(t, x, u)\,dt$ with $x(0) = 1$, $x(1) = 0$, $\int_0^1 u^2\,dt \leq C$, $f = u$, $C > 1$, and $f_0(t, x, u) = (\sin t)(e^{tx} + \sin tx) + (1 + \sin tx)u + u^2$. Here

$$|(f_0(t, x, u) - f_0(t, y, u)| \leq (e^{|x|} + 1) + (e^{|y|} + 1) + 2|u|,$$

and condition $G_{\infty q}$ is satisfied with $\sigma(\zeta) = e^{\zeta} + 1$, $\phi(u) = u$, $\psi(t) = 0$, $q = 2$, $\beta = 1$. As before, if $[x_k, u_k]$ is a minimizing sequence, then $\int_0^1 x_k'^2\,dt = \int_0^1 u_k^2\,dt \leq C$, $L = C^{1/2}$, and there is a subsequence, say still $[k]$, such that $x_k' \to \xi$ weakly in L_1, $x_k \to x$ uniformly, with x AC and $x' = \xi$. From (13.1.iii) the problem has an optimal solution.

5. We seek the minimum of $\int_0^1 f_0(t, x, u, v)\,dt$ with $x(0) = 0$, $x(1) = 1$, $n = 1$, $m = 2$, scalar state variable x, control variables u, v, $A = [(t, x) | 0 \leq t \leq 1, x \in R]$, $U = R^2$, $f = u$, constraint $\int_0^1 (u^2 + v^2)\,dt \leq C$, a constant $C > 1$, and $f_0 = -|x|\,|v|^{1/2} + 4^{-1}x^2|v|$. Here A is closed, $M = [(t, x, u, v) | 0 \leq t \leq 1, (x, u, v) \in R^3]$ is closed, and $Q(t, x) = R$. Note that $\phi(\zeta) = -a\zeta + 4^{-1}a^2\zeta^2$, $0 \leq \zeta < +\infty$, has minimum -1 at $\zeta = 2/a$ if $a \neq 0$, and $\phi(\zeta) \equiv 0$ if $a = 0$. Thus, for the sets $\tilde{Q}(t, x) = [(z^0, z) | z^0 \geq f_0, z = u \in R]$ we have $\tilde{Q}(t, x) = [z^0 \geq -1, z \in R]$ if $x \neq 0$; $\tilde{Q}(t, 0) = [z^0 \geq 0, z \in R]$ if $x = 0$. The sets $\tilde{Q}(t, x)$ are all convex and closed, but they do *not* have property (K) with respect to x at $x = 0$. Note that the graph $\tilde{M}$ of the sets $\tilde{Q}(t, x)$ is the set of all (t, x, z^0, z) with $(t, x, z) \in [0, 1] \times R^2$, and $z^0 \geq -1$ if $x \neq 0$, $z^0 \geq 0$ if $x = 0$, and $\tilde{M}$ is *not* closed. Note that the function $T(t, x, z)$ is defined in $[0, 1] \times R^2$, with $T = -1$ for $x \neq 0$, $T = 0$ for $x = 0$, and T is *not* lower semicontinuous in $[0, 1] \times R^2$. However, T is bounded below, and so is $I[x, u, v]$. If $[x_k, u_k, v_k]$ is a minimizing sequence, then

$$|x_k(t)| \leq \int_0^1 |u_k(t)|\,dt \leq \left(\int_0^1 u_k^2\,dt\right)^{1/2} \leq C^{1/2}, \qquad \int_0^1 x_k'^2\,dt = \int_0^1 u_k^2\,dt \leq C.$$

We can restrict A to the compact set $A_0 = [0 \le t \le 1, |x| \le C^{1/2}]$, and there is a subsequence, say $[k]$, such that $x_k \to x$ uniformly in $[0, 1]$, $x'_k \to \xi$ weakly in L_1, x AC and $x' = \xi$. Also,

$$\begin{aligned} |f_0(t, x, u, v) - f_0(t, y, u, v)| &\le (|x| + |y|)|v|^{1/2} + 4^{-1}(x^2 + y^2)|v| \\ &\le 2C^{1/2} \cdot 2^{-1}(1 + |v|) + 2^{-1}C|v| \\ &= C^{1/2} + (C^{1/2} + 2^{-1}C)|v|, \end{aligned}$$

and $G_{\infty q}$ holds with $q = 2$, $\beta = 1$, $\psi(t) = C^{1/2}$, $\sigma(\zeta) = 1$, $c' = C^{1/2} + 2^{-1}C$, $\phi(v) = |v|$. By (13.1.iii) this problem has an absolute minimum.

Remark. Examples 1–4 could have been handled by the theorems of Chapter 11 as well, but not Example 5, whose sets $\tilde{Q}(t, x)$ have not property (K) with respect to x.

Bibliographical Notes

The existence theorems in this Chapter 13 have been proved and are presented here directly in terms of problems of optimal control (cf. Cesari and Suryanarayana [1]. They are based on the remark that differences such as $\delta_k(t) = f(t, x_k(t), u_k(t)) - f(t, x(t), u_k(t))$, where $x_k(t) \to x(t)$ in measure and under suitable limitations on the behavior of the control functions $u_k(t)$, converge strongly to zero in L_1 if and only if they are equiabsolutely integrable. Moreover, $\delta_k \to 0$ strongly in L_1 implies a suitable weak form of property (Q), which by itself could be used to prove lower closure theorems and consequent existence theorems. There are a great number of natural hypotheses which guarantee that $\delta_k \to 0$ strongly in L_1, easy to verify and actually satisfied in many applications. They all imply therefore the existence of an optimal solution. One of them, namely (F_p^*) in Section 13.2A, was noticed about at the same time by L. D. Berkovitz, who used it to claim that it did not involve property (Q), while instead it implies a weak form of that property, as do all the others listed here. It is relevant that these conditions are easily expressed also in terms of problems with space variable and controls in Banach spaces, as observed by L. Cesari and M. B. Suryanarayana [4, 5, 6, 7]. We shall make more systematic use of them in other expositions covering problems with partial differential equations (distributed parameters) and problems in Banach spaces (cf. Cesari's forthcoming book IV). Lipschitz type conditions and G-type conditions have been proposed by E. H. Rothe [1] in connection with lower semicontinuity theorems for free problems. Analogous conditions have been proposed by Serrin in connection with different modes of convergence. (See C. B. Morrey [I] for more bibliographical information.)

CHAPTER 14

Existence Theorems: Problems of Slow Growth

We discuss here existence theorems for the usual integrals $I[x] = \int_{t_1}^{t_2} F_0(t, x(t), x'(t))\, dt$ as in Section 11.1, but where $F_0(t, x, x')$ does not satisfy any of the growth conditions we have considered in Chapters 11, 12, 13. Well known problems are of this kind (cf. Section 3.12). There are a number of methods to cope with these problems; we mention here one based on their reduction to equivalent "parametric problems" (Sections 14.1–2). In Section 14.3 we state a number of existence theorems for the usual integrals $I[x]$, and in Section 14.4 we present many examples of problems for which the existence theorems in Section 14.3 hold.

14.1 Parametric Curves and Integrals

A. Parametric Curves

The concept of a parametric curve $\mathfrak{C}$ in R^n occurs when we agree to consider a suitable equivalence concept between n-vector continuous maps $x = x(\tau)$, $a \le \tau \le b$, and $y = y(\sigma)$, $c \le \sigma \le d$, $x = (x^1, \ldots, x^n)$, $y = (y^1, \ldots, y^n)$. A parametric curve $\mathfrak{C}$ is then a class of equivalent maps. The concept of equivalence will leave unchanged the sense in which the curve is traveled, and thus we shall speak of oriented parametric curves.

The concept of Lebesgue equivalence is a natural one and must be mentioned: Two continuous maps x and y as above are said to be Lebesgue equivalent if there is a strictly increasing continuous map $\sigma = h(\tau)$, $a \le \tau \le b$, $h(a) = c$, $h(b) = d$ (or homeomorphism) such that $y(h(\tau)) = x(\tau)$ for all $a \le \tau \le b$. For technical reasons only a slightly more general concept is needed, namely the concept of Fréchet equivalence. Two continuous maps x and y as above are said to be Fréchet equivalent, or F-equivalent, if for every $\varepsilon > 0$ there is some homeomorphism $h: \sigma = h(\tau)$, $a \le \tau \le b$, $h(a) = c$, $h(b) = d$, such that $|y(h(\tau)) - x(\tau)| \le \varepsilon$ for all $a \le \tau \le b$. If we represent this relation by writing $x \sim y$, it is easily seen that (a) $x \sim x$; (b) $x \sim y$ implies $y \sim x$; (c) $x \sim y$, $y \sim z$ implies $x \sim z$. Then a class of F-equivalent maps is called a *parametric* curve $\mathfrak{C}$, or a Fréchet curve (briefly, an F-curve), and every element of the class is said to be a representation of $\mathfrak{C}$, in symbols, $\mathfrak{C}: x = x(\tau)$, $a \leq \tau \leq b$. The main concepts associated with a Fréchet curve $\mathfrak{C}$ must be F-invariant, that is, must be shared by each of its representations.

For any given F-curve $\mathfrak{C}: x = x(\tau)$, $a \leq \tau \leq b$, the set $[\mathfrak{C}] = [x] = [x \in R^n | x = x(\tau), a \leq \tau \leq b]$ is said to be the set covered by $\mathfrak{C}$ in R^n. Then $x \sim y$ implies $[x] = [y]$, and this is certainly a compact subset of R^n.

The Jordan length $L[\mathfrak{C}]$ of a Fréchet curve $\mathfrak{C}$ is then defined as a total variation,

$$L[\mathfrak{C}] = V[x] = \sup \sum_{i=1}^{N} |x(\tau_i) - x(\tau_{i-1})|,$$

where sup is taken with respect to all subdivisions $a = \tau_0 < \tau_1 < \cdots < \tau_N = b$ of $[a, b]$. If $x \sim y$, one can see easily that $V[x] = V[y]$, and thus $L[\mathfrak{C}] = V[x]$ is independent of the representation of $\mathfrak{C}$. An F-curve $\mathfrak{C}: x = x(\tau)$, $a \leq \tau \leq b$, is said to be rectifiable if $L[\mathfrak{C}] < +\infty$. A rectifiable curve $\mathfrak{C}: x = x(\tau)$, $a \leq \tau \leq b$, covers a set $[\mathfrak{C}]$, or $[x]$, of measure zero in R^n. The following further properties of rectifiable F-curves $\mathfrak{C}$ are relevant:

14.1.i. (a) $\mathfrak{C}: x = x(\tau) = (x^1, \ldots x^n)$, $a \leq \tau \leq b$, *is rectifiable if and only if the n continuous functions* $x^i(t)$, $i = 1, \ldots n$, *are of bounded variation* (BV) (*Jordan*, 1889).

(b) *If* $\mathfrak{C}: x = x(t) = (x^1, \ldots, x^n)$, $a \leq t \leq b$, *is rectifiable, then* $x(t)$ *is* BV *in* $[a, b]$, *the n derivatives* $x'(t) = (x'^1, \ldots, x'^n)$ *exist* a.e. *in* $[a, b]$ *and are L-integrable in* $[a, b]$, *and* $L[\mathfrak{C}] = L[x] \geq \int_a^b |x'(t)|\, dt$ (*Lebesgue integral of the Euclidean norm of* x') (*Tonelli*, 1912).

(c) *The equality holds if and only if x is* AC, *that is, the n functions* x^i *are* AC *in* $[a, b]$ (*Tonelli*, 1912).

(d) *If* $\mathfrak{C}$ *is rectifiable, it always possesses* AC *representations. In particular, the arc length parameter s*, $0 \leq s \leq L$, *yields a unique* AC *representation* $x = X(s)$, $0 \leq s \leq L$, *with* $|X'(s)| = 1$ a.e. *in* $[0, L]$.

Given two F-curves $\mathfrak{C}_1: x = x_1(\tau)$, $a \leq \tau \leq b$, and $\mathfrak{C}_2: x = x_2(\sigma)$, $c \leq \sigma \leq d$, it is useful to have a concept of distance $d(\mathfrak{C}_1, \mathfrak{C}_2)$ between $\mathfrak{C}_1$ and $\mathfrak{C}_2$. The F-distance is defined by $d(\mathfrak{C}_1, \mathfrak{C}_2) = d(x_1, x_2) = \inf[|x_2(h(\tau)) - x_1(\tau)|, a \leq \tau \leq b]$, where inf is taken over all homeomorphisms $\sigma = h(\tau)$, $a \leq \tau \leq b$, $h(a) = c$, $h(b) = d$. It is immediately seen that if $x_1 \sim y_1$, $x_2 \sim y_2$, then $d(x_1, x_2) = d(y_1, y_2)$ (that is, d depends only on the two F-curves $\mathfrak{C}_1$, $\mathfrak{C}_2$), and that $d(x_1, x_2) = 0$ if and only if $x_1 \sim x_2$. It can also be proved that (a) $d(\mathfrak{C}_1, \mathfrak{C}_2) \geq 0$, and $=$ sign holds if and only if $\mathfrak{C}_1 = \mathfrak{C}_2$; i.e., $x_1 \sim x_2$. (b) $d(\mathfrak{C}_1, \mathfrak{C}_2) = d(\mathfrak{C}_2, \mathfrak{C}_1)$; (c) $d(\mathfrak{C}_1, \mathfrak{C}_3) \leq d(\mathfrak{C}_1, \mathfrak{C}_2) + d(\mathfrak{C}_2, \mathfrak{C}_3)$. Thus, the F-curves in R^n form a metric space with distance function d. A sequence of F-curves $\mathfrak{C}_k$, $k = 1, 2, \ldots$, in R^n is said to be convergent to an F-curve $\mathfrak{C}$ provided $d(\mathfrak{C}_k, \mathfrak{C}) \to 0$ as $k \to \infty$. It is relevant to note here that $\mathfrak{C}_k \to \mathfrak{C}$ if and only if there are representations, say $\mathfrak{C}_k: x = x_k(\tau)$, $0 \leq \tau \leq 1$, $\mathfrak{C}: x = x(\tau)$, $0 \leq \tau \leq 1$, with $x_k \to x$ uniformly in $[0, 1]$.

The Jordan length is lower semicontinuous with respect to this type of convergence, that is:

14.1.ii. *If* $\mathfrak{C}$, $\mathfrak{C}_k$, $k = 1, 2, \ldots$, *are F-curves in* R^n *and* $d(\mathfrak{C}_k, \mathfrak{C}) \to 0$ *as* $k \to \infty$, *then* $L[\mathfrak{C}] \leq \liminf L[\mathfrak{C}_k]$.

The following compactness theorem for F-curves is most useful:

14.1.iii. (Hilbert). *If K is any closed bounded subset of R^n, and $\{\mathfrak{C}\}$ any family of F-curves lying in K and with equibounded Jordan lengths (e.g., $[\mathfrak{C}] \subset K$, $L[\mathfrak{C}] \leq M$ for some M and all $\mathfrak{C} \in \{\mathfrak{C}\}$), then the family $\{\mathfrak{C}\}$ is relatively sequentially compact with respect to the F-distance d. In other words, any sequence $\mathfrak{C}_k$, $k = 1, 2, \ldots$, of elements of $\{\mathfrak{C}\}$ possesses a subsequence $[\mathfrak{C}_{k_s}]$ which is convergent in the d-metric toward some F-curve $\mathfrak{C}$, and $L[\mathfrak{C}] \leq \liminf_s L[\mathfrak{C}_{k_s}] \leq M$.*

Proof. By representing the curves by means of their arc length parameter we have a family $\{X(s), 0 \leq s \leq L\}$, $L = L[\mathfrak{C}]$, which is equibounded and equicontinuous, namely Lipschitzian of constant one. The statement is now a corollary of Ascoli's Theorem in the form (9.1.ii). □

Remark 1. (Existence of a geodesic between two points on a manifold). If M is a continuous manifold in R^n we say that a path curve $\mathfrak{C}$ is on M if all its points are on M and $\mathfrak{C}$ has a continuous representation in terms of the local representation parameters of M. If there are rectifiable curves joining two points P, Q on a closed continuous manifold M, then there is also a curve joining P, Q on M and of minimum length. For a sketch of proof, let Ω denote the class of all rectifiable curves joining P and Q on M and let i be the infimum of the Jordan lengths of the curves in Ω. Let $\mathfrak{C}_k$, $k = 1, 2, \ldots$, be a minimizing sequence, that is, $L[\mathfrak{C}_k] \to i$, and $\mathfrak{C}_k \in \Omega$. By (14.1.iii) there is a path curve $\mathfrak{C}_0$ in R^n, and a subsequence, say still $[k]$, such that $d(\mathfrak{C}_k, \mathfrak{C}_0) \to 0$ as $k \to \infty$, and by (14.1.ii) $L[\mathfrak{C}_0] \leq i$. Now the points of $\mathfrak{C}_0$ are on M since M is closed, and because of the closure of Ω with respect to the distance function d, we have $\mathfrak{C}_0 \in \Omega$. Then $L[\mathfrak{C}_0] \geq i$. and by comparison also $L[\mathfrak{C}_0] = i$. (See III for details).

Remark 2. The concept of parametric *closed* curve $\mathfrak{C}_0$ can be introduced as above by identifying logically the end points a and b of the intervals $[a, b]$ taken into consideration (which implies $x(a) = x(b)$). Then $[a, b]$ becomes equivalent to a circumference, and the class of orientation preserving homeomorphisms h is now much larger.) The same concepts can be introduced as before (closed F-curves, distance of two closed F-curves), and the same theorems hold.

B. Parametric Integrals

The concept of integral over a rectifiable F-curve $\mathfrak{C}: x = x(t) = (x^1, \ldots, x^n)$, $a \leq t \leq b$, in any of its AC representations x, can be defined as usual as

$$I[\mathfrak{C}] = I[x] = \int_a^b f_0(x, x')\, dt, \tag{14.1.1}$$

and this integral is independent of the chosen AC representation if and only if (a) f_0 does not depend on t, and (b) f_0 is positive homogeneous of degree one in x', that is, $f_0(x, kx') = kf_0(x, x')$ for all $k \geq 0$. Thus, let A_1 be a closed subset of R^n, and let $f_0(x, x')$ be a continuous function on $A_1 \times R^n$ satisfying (a) and (b); then the formula above defines $I[\mathfrak{C}]$ for any F-curve $\mathfrak{C}$ with $[\mathfrak{C}] \subset A_1$ and for any of the AC representations of $\mathfrak{C}$. Then, $I[\mathfrak{C}]$ is called a parametric integral and $f_0(x, x')$ a parametric integrand.

For instance, for $n = 2$, and by writing x, y, x', y' instead of x^1, x^2, x'^1, x'^2, the following functions f_0 all are parametric integrands:

$$(14.1.2)\qquad \begin{aligned} f_0 &= (x'^2 + y'^2)^{1/2},\\ f_0 &= (2x'^2 + 3y'^2)^{1/2} - (x'^2 + y'^2)^{1/2},\\ f_0 &= (1 + x^2 + y^2)[2(x'^2 + y'^2)^{1/2} - |x'|],\\ f_0 &= (1 + x^2 + y^2)(x'^2 + y'^2)^{1/2} + xy' + x'y,\\ f_0 &= 2(x^2 + y^2)^{-1/2}(x'^2 + y'^2)^{1/2} + (x^2 + y^2)^{-1}(xy' - x'y),\\ f_0 &= 2(1 + x^2 + y^2)^{3/2}(x'^2 + y'^2)^{1/2} + x^3x' + y^3y'. \end{aligned}$$

Note that, because of the positive homogeneity of f_0 in x', the usual sets $\tilde{Q}(x) = [(z^0, z), z^0 \geq f_0(x, x'), z \in R^n]$ are here cones with vertex the origin in R^n.

One might think that lower semicontinuity theorems and existence theorems for the minimum of parametric integrals could all be particular cases of those we have proved in Chapters 8–13 for the nonparametric case. This is true only for the simplest theorems. Because of a different emphasis and a great many technicalities, new proofs are needed for new results which have no parallel in the nonparametric case. We refer to [III], which includes also a new approach to the unexpectedly rich class of "parametric problems of optimal control". Here we list only a few existence theorems for the minimum of $I[\mathfrak{C}]$.

C. Existence Theorems for Parametric Integrals

Here A_1 denotes a subset of the x-space R^n, B_0 a subset of the x_1x_2-space R^{2n}, and V_0 the unit sphere in the z-space, or $V_0 = [z \in R^n \,|\, |z| = 1]$. We consider the problem of the minimum of $I[\mathfrak{C}]$ in the class Ω of all rectifiable F-curves $\mathfrak{C}: x = x(t), a \leq t \leq b, [\mathfrak{C}] \subset A_1$, and satisfying given boundary conditions which, for the parametric case, are of the form $(x(a), x(b)) \in B_0$. Alternatively, we may want to minimize $I[\mathfrak{C}]$ in any d-closed class Ω of rectifiable F-curves $\mathfrak{C}$. The following existence theorem is rather typical of the parametric case.

14.1.iv. *If A_1 is compact, B_0 is closed, $f_0(x, x')$ is continuous on $A_1 \times R^n$ and convex in x', and there is a real valued monotone nondecreasing function $\phi(\zeta)$, $-\infty < \zeta < +\infty$, such that*

$$(14.1.3)\qquad L[\mathfrak{C}] \leq \phi(I[\mathfrak{C}])$$

for all F-curves $\mathfrak{C}$ lying in A_1, then $I[\mathfrak{C}]$ has an absolute minimum in the class Ω of all rectifiable F-curves $\mathfrak{C}$ in A_1.

Here is a short proof of (14.1.iv), which is only a repetition of proof II for the Filippov theorem (9.3.i). First, it is not restrictive to assume that $I[\mathfrak{C}] \leq M_1$ for all $\mathfrak{C}$ in Ω and some constant M_1. Then, by (14.1.i), also $L[\mathfrak{C}] \leq \phi(I[\mathfrak{C}]) \leq \phi(M_1) = M_2$. Let W_0 denote the closed unit ball in R^n, or $W_0 = [z \in R^n \,|\, |z| \leq 1]$. Then $f_0(x, x')$ is continuous in the compact set $A_1 \times W_0$, hence bounded there, say $|f_0| \leq M$. For any $\mathfrak{C}$ in Ω and the arc length representation of $\mathfrak{C}: x = X(s)$, $0 \leq s \leq L$, $L = L[\mathfrak{C}] \leq M_2$, we have $|X'(s)| = 1$ a.e., and hence $|I[\mathfrak{C}]| \leq LM \leq M_2M$. Thus, $i = \inf_\Omega I[\mathfrak{C}]$ is finite. Let $\mathfrak{C}_k: x = X_k(s)$, $0 \leq s \leq L_k$, $L_k = L[\mathfrak{C}_k] \leq M_2$, $k = 1, 2, \ldots$, be any minimizing sequence, so that $I[\mathfrak{C}_k] \to i$ as $k \to \infty$. Also, $X_k(s) \in A_1$, a compact set, and $|X_k'(s)| \leq 1$, so $X_k'(s) \in W_0$, a fixed compact convex set. The functions X_k are equibounded and Lipschitzian of constant one, and hence equicontinuous. By Ascoli's theorem (9.1.ii) there is a subsequence, say still $[k]$, such that $X_k \to x_0$ in the ρ-metric, or uniformly, and $x_0(s)$, $0 \le s \le L_0 \le M_2$, is Lipschitz of constant one and hence AC, and $|x_0'(s)| \le 1$, so $x_0'(s) \in W_0$ a.e. in $[0, L]$. By the same argument as in proof II of the Filippov's theorem (9.3.i), we have the lower semicontinuity property $I[x_0] \le \liminf_k I[x_k]$. In other words, for the F-curve defined by $\mathfrak{C}: x = x_0(s)$, $0 \le s \le L_0$, we have $I[\mathfrak{C}_0] = I[x_0] \le i$. Since $L[\mathfrak{C}_0] \le \liminf_k L[\mathfrak{C}_k] \le M_1$, $\mathfrak{C}_0$ belongs to Ω; hence $I[\mathfrak{C}_0] \ge i$, and $I[\mathfrak{C}_0] = i$ and (14.1.iv) is proved.

Many existence theorems for the parametric case are actually criteria for (14.1.3) to hold. Here are some of these criteria ((a) to (g), which are due to Tonelli and McShane, and we shall prove them in III):

(a) $f_0(x, x') > 0$ for all $x \in A_1$ and $x' \in V_0$.
(b) There is a fixed vector $b = (b^1, \ldots, b^n)$ such that $f_0(x, x') + b \cdot x' > 0$ for all $(x, x') \in A_1 \times V_0$.
(c) There is a continuous n-vector gradient function $G(x) = (G_1, \ldots, G_n)$, $x \in A_1$, such that $f_0(x, x') + G(x) \cdot x' > 0$ for all $(x, x') \in A_1 \times V_0$. (Here G is called a gradient function if there is a function $F(x)$ of class C^1 in A_1 such that $G_i = \partial F/\partial x^i$, $i = 1, \ldots, n$, in A_1).

For instance, all the examples in (14.1.2) satisfy (a). Analogously

$$f_0 = (x'^2 + y'^2)^{1/2} + 5x'$$

satisfies (b) with $b = (-5, 0)$, and the following examples satisfy (c) with the functions F and G which are indicated:

$$f_0 = (x'^2 + y'^2)^{1/2} - (xy' + x'y), \qquad G = (y, x), \quad F = xy,$$
$$f_0 = (x'^2 + y'^2)^{1/2} - (x^2 + y^2)^{-1}(xy' - x'y), \qquad G = (x^2 + y^2)^{-1}(-y, x),$$
$$f_0 = 2(x'^2 + y'^2)^{1/2} - |x'| - (x^3x' + y^3y'), \qquad G = (x^3, y^3), \quad F = 4^{-1}(x^4 + y^4).$$

In the last but one example, A_1 could be any simply connected region in R^2 not containing the origin, and then $F = \arctan y/x$).

A point $x_0 \in R^n$ is said to be a zero for $f_0(x, x')$ if $f_0(x_0, x') = 0$ for some $x' \in R^n$, $x' \ne 0$. For instance both

$$f_0 = (x^2 + y^2)^{1/2}(x'^2 + y'^2)^{1/2}, \qquad f_0 = (x'^2 + y'^2)^{1/2} - x'(1 + x^2 + y^2)^{-1}$$

have a zero at $(0, 0)$; the first one vanishes there for all (x', y'), the second one for all $(x', 0)$ with $x' \ge 0$. The functions

$$f_0 = x^2y^2(x'^2 + y'^2)^{1/2}, \qquad f_0 = (x'^2 + y'^2)^{1/2} - y'(1 + x^2y^2)^{-1}$$

have a zero at every point (x, y) of the x- and of the y-axes; the first one vanishes there for all (x', y'), the second one for all $(0, y')$ with $y' \ge 0$. The function $f_0 = (x'^2 + y'^2)^{1/2} - x'$ has a zero at every point of the xy-plane, since f_0 vanishes for every $(x', 0)$ with $x' \ge 0$.

We are now in a position to add to the criteria (a)–(c) above for (14.1.3) to hold, the following one:

(d) $f_0(x, x') \ge 0$ in $A_1 \times R^n$, and there is a fixed unit vector $v = (v^1, \ldots, v^n)$, $|v| = 1$, and a number $\delta > 0$ such that $f_0(x, x') > 0$ for $v \cdot x' \ge \delta$, $(x, x') \in A_1 \times V_0$.

This condition simply states that f_0 may vanish at any point $x \in A$, but only in directions x' forming an angle $0 \le \sigma \le \pi/2 - \delta'$ with the fixed direction v and

$\cos(\pi/2 - \delta') = \delta$. The following functions satisfy (d), namely, they vanish at most in the one direction which is indicated:

$f_0 = (1 + x^2 + y^2)[(x'^2 + y'^2)^{1/2} - x']$, $f_0 = 0$ at all (x, y) in the sole direction $(1,0)$;
$f_0 = (1 + |x| + |y|)[(x'^2 + y'^2)^{1/2} - (x')^{1/3}(x'^2 + y'^2)^{1/3}]$, $f_0 = 0$ at all (x, y) in the sole direction $(1,0)$;
$f_0 = (x'^2 + y'^2)^{1/2} - (1 + x^2y^2)^{-1}x'$, $f_0 = 0$ at all points $(x,0)$ and $(0, y)$ in the sole direction $(1,0)$.

Two further criteria are as follows:

(e) $f_0(x, x') \geq 0$ in $A_1 \times R^n$; f_0 possesses a finite number of zeros in A_0, say x_i, $i = 1, \ldots, N$; and for each i there is a vector $v_i = (v_1^1, \ldots, v_n^n)$, $|v_i| = 1$, and number $\delta_i > 0$ such that $f_0(x_i, x') > 0$ for $v_i \cdot x' \geq \delta_i$, $|x'| = 1$.
(f) $f_0(x, x') \geq 0$ in $A_1 \times R^n$, f_0 possesses a set Z of zeros, which are all contained on the finite union of simple continuous curves in A_0, which may have points in common, but form no closed curve in R^n. Moreover, for every $x \in Z$ there is also a vector v, $|v| = 1$, and a number $\delta > 0$ (both of which may depend on x) such that $f_0(x, x') > 0$ for $v \cdot x' > \delta$, $|x'| = 1$.

For instance

$$f_0 = (x'^2 + y'^2)^{1/2} - 2x(1 + x^2 + y^2)^{-1}x'$$

satisfies (e) with only two zeros: $(1,0)$ in the direction $(1,0)$, and $(-1,0)$ in the direction $(-1,0)$. For instance

$$f_0 = |x'| + |y'| - 2x(1 + x^2)^{-1}x' - 2y(1 + y^2)^{-1}y'$$

satisfies (e) with only four zeros $(1,0)$, $(-1,0)$, $(0,1)$, $(0,-1)$, and f_0 vanishes there in the corresponding directions $(1,0)$, $(-1,0)$, $(0,1)$, $(0,-1)$.

For instance, $f_0 = (x'^2 + y'^2)^{1/2} - (1 + x^2y^2)^{-1}x'$, already mentioned above, satisfies (f). The restriction on the zeros stated in (e) and (f) can be removed provided f_0 behaves suitably around such zeros. Here is one criterion:

(g) $f_0(x, x') \geq 0$ in $A_1 \times R^n$, f_0 possesses a finite number of zeros in A_1, all interior to A_1, say x_i, $i = 1, \ldots, N$, where $f_0(x_i, x') = 0$ for all $x' \in R^n$. However, each point x_i, $i = 1, \ldots, N$, has the following property: in a neighborhood N of x_i, we have $c|x - x_i|^\gamma \leq f_0(x, x') \leq C|x - x_i|^\gamma$ for all $x \in N$, $x' \in R^n$, $|x'| = 1$, and constants $0 < c < C < \infty$, $\gamma > 0$. For instance, all functions below satisfy (g) with A_1 as indicated:

$f_0 = (x^2 + y^2)(x'^2 + y'^2)^{1/2}$, $\quad A_1 = [(x, y)| -1 \leq x, y \leq 1]$,
$f_0 = (a^2x^2 + b^2y^2)(x'^2 + y'^2)^{1/2}$, $\quad a, b > 0$, $\quad A_1 = [(x, y)| -1 \leq x, y \leq 1]$,
$f_0 = (x^2 + y^2)^{1/3}(x'^2 + y'^2)^{1/2}$, $\quad A_1 = [(x, y)| -1 \leq x, y \leq 1]$,
$f_0 = [(x^2 - 1)^2 + y^2](x'^2 + y'^2)^{1/2}$, $\quad A_1 = [(x, y)| -2 \leq x, y \leq 2]$,
$f_0 = [(x^2 - 1)^2 + (y^2 - 1)^2]^{1/3}(x'^2 + y'^2)^{1/2}$, $\quad A_1 = [(x, y)| -2 \leq x, y \leq 2]$
$f_0 = (x^2 + y^2)[(1 + x^2 + y^2)(2x'^2 + 3y'^2)^{1/2} - (x'^2 + y'^2)^{1/2}]$,
$\quad A_1 = [(x, y)| -1 \leq x, y \leq 1]$
$f_0 = ((x^2 - 1)^2 + (y^2 - 1)^2 + (z^2 - 1)^2)^{1/4}(x'^2 + y'^2 + z'^2)^{1/2}$,
$\quad n = 3$, $\quad A_1 = [(x, y, z)| x^2 + y^2 + z^2 \leq 4]$.

For proofs of criteria (a)–(g) and details see, e.g., [III].

14.2 Transformation of Nonparametric into Parametric Integrals

A. The Parametric Integral $\mathfrak{J}[\mathfrak{C}]$ Associated to $I[x]$

We present here existence theorems for free nonparametric problems concerning the usual integral

$$I[x] = \int_{t_1}^{t_2} F_0(t, x(t), x'(t))\, dt \tag{14.2.1}$$

as in Sections 11.1–3, but where $F_0(t, x, x')$ does not satisfy any of the growth properties (g) of Sections 11.1–3.

First let us note that, if in (14.2.1) we think of t as an increasing AC function of a new variable, $t = t(\tau)$, $\tau_1 \le \tau \le \tau_2$, with $t_1 = t(\tau_1)$, $t_2 = t(\tau_2)$, then for $X(\tau) = x(t(\tau))$, $p(\tau) = t'(\tau)$, we have

$$\begin{aligned} I[x] &= \int_{\tau_1}^{\tau_2} F_0(t(\tau), x(t(\tau)), x'(t(\tau)))t'(\tau)\, d\tau \\ &= \int_{\tau_1}^{\tau_2} F_0(t(\tau), X(\tau), X'(\tau)/p(\tau))p(\tau)\, d\tau, \end{aligned}$$

with obvious conventions whenever $p(\tau) = 0$. We are going to recognize the last integral as a parametric integral in R^{n+1}.

Given a set A in the tx-space R^{n+1}, $x = (x^1, \dots, x^n)$, and a scalar function $F_0(t, x, u)$, $(t, x, u) \in A \times R^n$, $u = (u^1, \dots, u^n)$, we introduce an auxiliary variable $p > 0$ and the new integrand function

$$G_0(t, x, p, u) = pF_0(t, x, u/p) \tag{14.2.2}$$

with $G_0(t, x, 1, u) = F_0(t, x, u)$. It is convenient to think of (t, x) as a new "space" variable, or $(n + 1)$-vector $\mathfrak{z} = (z^0, z^1, \dots, z^n)$, and of (p, u) as a new "direction" variable, or $(n + 1)$-vector $w = (p, u) = (p, u^1, \dots, u^n)$, or $\mathfrak{z}' = (z'^0, z') = (z'^0, z'^1, \dots, z'^n)$, so that $G_0(t, x, p, u)$ becomes $G_0(\mathfrak{z}, \mathfrak{z}')$ with $\mathfrak{z} \in A \subset R^{n+1}$, $\mathfrak{z}' \in R^{n+1}$, $z'^0 > 0$. Note that G_0 is positive homogeneous of degree one in $w = (p, u)$, or $\mathfrak{z}'$, that is, $G_0(\mathfrak{z}, k\mathfrak{z}') = kG_0(\mathfrak{z}, \mathfrak{z}')$ for all $k > 0$. When possible, the scalar function $F_0(t, x, p, u)$ will be extended by continuity into a function G_0 defined for all $(t, x) \in A$, $(p, u) \in R^{n+1}$, $p \ge 0$.

We shall assume $F_0(t, x, u)$ to be continuous in $A \times R^n$ and then $G_0(t, x, p, u)$ is certainly continuous in $A \times (0, +\infty) \times R^n$. The function G_0 may or may not be extendable into a continuous function in $A \times [0, +\infty) \times R^n$. If G_0 admits of such a continuous extension, then

$$\mathfrak{J}[\mathfrak{C}] = \int_{\tau_1}^{\tau_2} G_0(\mathfrak{z}(\tau), \mathfrak{z}'(\tau))\, d\tau \tag{14.2.3}$$

is a parametric integral such as we have considered in Section 14.1 and which is defined for rectifiable F-curves $\mathfrak{C}$ in R^{n+1} lying in A as the value of (14.2.3) for any AC representation of $\mathfrak{C}$, namely $\mathfrak{z}(\tau)$, $\tau_1 \le \tau \le \tau_2$, $\mathfrak{z}$ AC, or $t(\tau)$, $x(\tau)$,

$\tau_1 \le \tau \le \tau_2$, t, x AC, with $t'(\tau) \ge 0$ (that is, t monotone nondecreasing). In this situation, if $x(t)$, $t_1 \le t \le t_2$, is any AC n-vector function with $(t, x(t)) \in A$ for all $t \in [t_1, t_2]$, and $f_0(\cdot, x(\cdot), x'(\cdot)) \in L_1[t_1, t_2]$, then $t = t$, $x = x(t)$, $t_1 \le t \le t_2$, is a parametric rectifiable curve—actually a particular representation of a rectifiable F-curve $\mathfrak{C}$. If $t = t(\tau)$, $x = x(\tau)$, $\tau_1 \le \tau \le \tau_2$, is any AC representation of the same curve, then we derive

$$\begin{aligned}\mathfrak{J}[\mathfrak{C}] &= \int_{\tau_1}^{\tau_2} G_0(t(\tau), x(\tau), t'(\tau), x'(\tau))\, d\tau \\ &= \int_{t_1}^{t_2} G_0(t, x(t), 1, x'(t))\, dt \\ &= \int_{t_1}^{t_2} F_0(t, x(t), x'(t))\, dt = I[x].\end{aligned}$$

Actually it may well occur that G_0 cannot be extended by continuity in $A \times [0, +\infty) \times E^n$, and yet (14.2.3) preserves its character of a parametric integral. The parametric integral $\mathfrak{J}[\mathfrak{C}]$ is said to be *associated* to the nonparametric integral $I[x]$. Statements concerning the equality $\mathfrak{J}[\mathfrak{C}] = I[x]$ will be proved in [III]. Also, we shall need statements guaranteeing that a parametric curve $t = t(\tau)$, $x = X(\tau)$, $\tau_1 \le \tau \le \tau_2$, possesses a representation $t = t$, $x(t)$, $t_1 \le t \le t_2$, or simply a nonparametric representation $x(t)$, $t_1 \le t \le t_2$.

B. Examples

We give here a few usual integrands $F_0(t, x, u)$ and the corresponding "parametric integrands" $G_0(t, x, p, u)$:

1. $n = 1$, $F_0 = (1 + u^2)^{1/2}$, $G_0 = (p^2 + u^2)^{1/2}$.
2. $n = 2$, $F_0 = (1 + u^2 + v^2)^{1/2}$, $G_0 = (p^2 + u^2 + v^2)^{1/2}$.
3. $n = 2$, $F_0 = (1 + u^2 + v^2)^{1/4}$, $G_0 = p^{1/2}(p^2 + u^2 + v^2)^{1/4}$.
4. $n = 1$, $F_0 = u^2$, $G_0 = p^{-1}u^2$.

We shall often need below the partial derivative G_{0p} of G_0 with respect to p (under differentiability assumptions on F_0), or

$$\begin{aligned}G_{0p}(t, x, p, u) &= \partial G_0/\partial p = (\partial/\partial p)[pF_0(t, x, u/p)] \\ &= F_0(t, x, u/p) - p^{-1} \sum_{i=1}^{n} u^i F_{0x'^i}(t, x, u/p).\end{aligned}$$

For instance, in Examples 1–4 above we have, respectively,

1: $G_{0p} = p(p^2 + u^2)^{-1/2}$;
2: $G_{0p} = p(p^2 + u^2 + v^2)^{-1/2}$;
3: $G_{0p} = 2^{-1}p^{-1/2}(p^2 + u^2 + v^2)^{-3/4}(2p^2 + u^2 + v^2)$;
4: $G_{0p} = -p^{-2}u^2$.

14.3 Existence Theorems for (Nonparametric) Problems of Slow Growth

We begin with some remarks concerning the parametric integrand $G_0(t, x, p, u)$, or $G_0(\tilde{z}, \tilde{z}')$, we have just associated to the usual (nonparametric) integrand $F_0(t, x, x')$.

It can be proved that, if $F_0(t, x, x')$ is convex in x' in R^n, then for all $(t, x) \in A$, $G_0(t, x, p, u)$ is convex in (p, u) in the open half space $(0, +\infty) \times R^n$. Also, it can be proved that if A is closed, $F_0(t, x, x')$ is continuous in $A \times R^n$ and convex in x' for every $(t, x) \in A$, then $G_0(t, x, p, u)$ is lower semicontinuous in $A \times [0, +\infty) \times R^n$ (and continuous in $A \times (0, +\infty) \times R^n$). Moreover, if $G_0(t, x, 0, u)$ happens to be finite everywhere and continuous in (t, x, u), then $G_0(t, x, p, u)$ is (finite everywhere and) continuous in $A \times [0, +\infty) \times R^n$.

Since $G_0(\tilde{z}, k\tilde{z}') = kG_0(\tilde{z}, \tilde{z}')$ for all $k > 0$ and $\tilde{z} \in A$, $\tilde{z}' \in (0, +\infty) \times R^n$, we may define $G_0(\tilde{z}, 0)$ to be zero, so that the homogeneity property alone holds for all $k \geq 0$. Furthermore, note that for F_0 convex in x', if we keep (t, x, u) fixed and we allow $p > 0$ to approach zero, then $G_0(t, x, p, u)$, a convex function of p, must approach a finite limit or $+\infty$. This limit, finite or $+\infty$, will be taken as a definition of $G_0(t, x, 0, u)$, though this function, so defined, may not be continuous at $p = 0$. For $u = 0$, then $G_0(z, x, p, 0) = pF_0(t, x, 0)$ approaches zero as $p \to 0+$, and thus the value zero for G_0 at $\tilde{x}' = 0$ coincides with the value we have already agreed upon on the basis of homogeneity above.

Concerning the possibility of inverting an AC function $t(s)$, $a \leq s \leq b$, into an AC function $s(t)$, $c \leq t \leq d$, we mention here the statement: If $t(s)$ is an AC monotone nondecreasing function, $a \leq s \leq b$, $c = t(a)$, $d = t(b)$, then the inverse function $s(t)$, $c \leq t \leq d$, of $t(s)$ exists and is continuous and AC in $[c, d]$ if and only if $t'(s) > 0$ a.e. in $[a, b]$.

As usual we consider the problem of the minimum of the integral

$$I[x] = \int_{t_1}^{t_2} F_0(t, x(t), x'(t))\, dt \tag{14.3.1}$$

in the class Ω of all AC functions $x(t) = (x^1, \ldots, x^n)$, $t_1 \leq t \leq t_2$, with $(t, x(t)) \in A$, $(t_1, x(t_1), t_2, x(t_2)) \in B$, and $F(\cdot, x(\cdot), x'(\cdot))$ L-integrable. We say that these are the admissible trajectories. The following existence theorems, and related implications, are proved in [III].

14.3.i (Existence Theorem for Usual Integrals). *Let $A = [t_0, T] \times A_1$, where A_1 is a compact subset of the x-space R^n; let $B = B_1 \times B_2$, where B_1, B_2 are closed subsets of the tx-space R^{n+1} such that for every $(t_1, x_1, t_2, x_2) \in B_1 \times B_2$ we have $t_1 < t_2$; let $F_0(t, x, x')$ be of class C^1 in $A \times R^n$ and convex in x' for every $(t, x) \in A$. Let us assume that the associated parametric integrand $G_0(t, x, p, u)$ is continuous in $A \times [0, +\infty) \times R^n$, and that* (a) *$G_{0p}(t, x, 0, u) = -\infty$ for $(t, x, u) \in A \times R^n$, $|u| = 1$, and* (b) *there are constants $M_1, M_2, \delta > 0$ such that for all $(t, x) \in A$, $\tilde{z}' = (p, u) \in R^{n+1}$ with $p \geq 0$,*

$|\bar{z}| = 1$, and t^* with $|t^* - t| < \delta$ we have $|G_{0t}(t^*, x, p, u)| \leq M_1 G_0(t, x, p, u) + M_2$. Let us further assume that (λ) there is a monotone nondecreasing function $\phi(\zeta)$, $-\infty < \zeta < +\infty$, such that $L[\mathfrak{C}] \leq \phi(\mathfrak{J}[\mathfrak{C}])$ for all rectifiable parametric curves $\mathfrak{C}: t = t(s)$, $X = X(s)$, $0 \leq s \leq L$, with graph in A, $t(0)$, $X(0)$, $t(L)$, $X(L) \in B$, $t(s)$ monotone nondecreasing, and s the arc length parameter. Then, $I[x]$ has an absolute minimum in the class of all admissible trajectories.

For $n = 1$ condition (λ) is certainly satisfied if, say, for every $(\bar{t}, \bar{x}) \in A$ there are a vector $b = (b^0, b^1) \in R^2$ and constants $v > 0$, $\delta > 0$ such that $G_0(t, x, p, u) + b^0 p + b^1 u \geq v$ for all $(p, u) \in R^2$, $p \geq 0$, and all $(t, x) \in A$ at a distance $\leq \delta$ from $(\bar{t}, \bar{x})$.

For $n \geq 1$, condition (λ) is certainly satisfied if, say, G_0 is continuous in $A \times [0, \infty) \times R^n$ and G_0 satisfies any of the conditions (a)–(g) of Section 14.1C.

For $n > 1$ condition (λ) is certainly satisfied if (λ1) for every $(\bar{t}, \bar{x}) \in A$ there are a vector $b = (b^0, b^1, \ldots, b^n) \in R^{n+1}$ and constants $v > 0$, $\delta > 0$ such that $G_0(t, x, p, u) + b^0 p + \sum_1^n b^i u^i \geq v$ for all $(p, u) \in R^{n+1}$, $p \geq 0$, and all $(t, x) \in A$ at a distance $\leq \delta$ from $(\bar{t}, \bar{x})$; (λ2) G_0 is bounded below; (λ3) there is a constant $\mu > 0$ such that all rectifiable parametric curves $\mathfrak{C}_0$ lying in any hyperplane $t =$ constant, or $\mathfrak{C}_0: t = c$, $x = x(s)$, $0 \leq s \leq L_0$, with graph in A *and* $\mathfrak{J}[\mathfrak{C}_0] = 0$ have length $\leq \mu$.

Conditions (λ2), (λ3) are certainly satisfied if $G_0(t, x, 0, u) > 0$ for all $u \in R^n$, $u \neq 0$, and all $(t, x) \in A$ except at most for a countable set E_c on any hyperspace $t = c$.

14.3.ii (Existence Theorem for Usual Integrals). *Let A, $B = B_1 \times B_2$ as in (14.3.i), let $F_0(x, x')$ be independent of t, continuous and bounded below on $A_1 \times R^n$, and convex in x' for every $x \in A_1$. Let us assume that the associated parametric integrand $G_0(x, p, u)$ is continuous in $A \times [0, +\infty] \times R^n$ with continuous partial derivative G_{0p} in the same set $A \times [0, +\infty] \times R^n$ with $(p, u) \neq (0, 0)$. Let us further assume that* (a′) *for every $x \in A$, $u \in R^n$, $u \neq 0$ we have $G_{0p}(x, p, u) = 0$ if and only if $p = 0$; and if $n > 1$,* (b′) *the same as (λ) in* (14.3.i). *Then, $I[x]$ has an absolute minimum in the class Ω of all admissible trajectories.*

The following statement concerns integrals of the form

$$I^*[x] = \int_{t_1}^{t_2} (\psi(x))^{-1} f_0(x(t), x'(t))\, dt \tag{14.3.2}$$

under the same general assumptions as for (14.3.i or ii).

14.3.iii. *Let $\psi(x) \geq 0$, $x \in R^n$, be a continuous function. Let Z denote the closed subset of R^n where $\psi(x) = 0$, and let us assume that $G_0(x, p, u) > 0$ for all x in some neighborhood U of Z, all $p > 0$ and $u \in R^n$. Then, the integral $I^*[x]$ has an absolute minimum in the class Ω of all admissible trajectories.*

Remark. The following statement concerning slow growth integrals is revealing of what actually may occur in situations not covered by theorems (14.3.i–iii). We consider here the nonparametric integral with $n = 1$

$$I[x] = \int_{t_1}^{t_2} \Phi(t)(1 + x'^2)^{1/2}\, dt, \tag{14.3.3}$$

where Φ is positive and continuously differentiable in $[t_1, t_2]$. We take $A = [t_1, t_2] \times R$, and we consider the case of both end points fixed: $1 = (t_1, x_1)$, $2 = (t_2, x_2)$, $t_1 < t_2$. Note that $\Phi(t)(1 + x'^2)^{1/2}$ is L-integrable in $[t_1, t_2]$ for all AC real valued functions x.

14.3.iv. *The integral* (14.3.3) *has an absolute minimum in the class* Ω *of all* AC *functions* x *with* $x(t_1) = x_1$, $x(t_2) = x_2$, t_1, t_2, x_1, x_2 *fixed, if and only if*

$$|x_2 - x_1| \le \int_{t_1}^{t_2} m[\Phi^2(t) - m^2]^{-1/2}\, dt \tag{14.3.4}$$

where $m = \min[\Phi(t),\ t_1 \le t \le t_2]$.

For instance, if $t_1 = 1$, $t_2 = 2$, $\Phi(t) = t$, $m = 1$, then $I[x] = \int_1^2 t(1 + x'^2)^{1/2}$ has an absolute minimum in Ω if and only if $|x_2 - x_1| \le \int_1^2 (t^2 - 1)^{-1/2}\, dt = \log(2 + 3^{1/2}) = 1.317$. On the other hand, for any $t_1 < t_2$, $\Phi(t) = 1$, $m = 1$, then $I[x] = \int_{t_1}^{t_2} (1 + x'^2)^{1/2}\, dt$ has an absolute minimum for all x_1, x_2 (namely, the segment $s = 12$, as we know), and indeed the integral in (14.3.4) is $+\infty$ in this situation. For the proof of (14.3.iv) we refer to P. Kaiser [1, 2].

14.4 Examples

1. $F_0 = (1 + x'^2)^{1/2}$ satisfies the conditions of (14.3.ii). Note that $F_0 \ge 0$, $G_0(p, u) = (p^2 + u^2)^{1/2}$ is continuous for $p \ge 0$, $u \in R$; $G_{0p} = p(p^2 + u^2)^{-1/2}$ is continuous for all $p \ge 0$, $u \in R$, $(p, u) \ne (0, 0)$; and $G_{0p} = 0$ with $u \ne 0$ if and only if $p = 0$. Thus, $I[x]$ has an absolute minimum in the class of all AC curves joining say, any two points $1 = (t_1, x_1)$, $2 = (t_2, x_2)$, $t_1 < t_2$; or 1 to any curve $\Gamma: x = g(t)$, $t' \le t \le t''$ with $t_1 < t'$; or any two sets B_1 compact, B_2 closed, with the property that $(t_1, x_1) \in B_1$, $(t_2, x_2) \in B_2$ implies $t_1 < t_2$. Here $I[x]$ is the length integral.

2. $F_0 = (1 + x'^2)^{1/2} - (1 + x'^2)^{1/4}$, satisfies the conditions of (14.3.i). Note that $F_0(x') \ge 0$, $G_0(p, u) = (p^2 + u^2)^{1/2} - p^{1/2}(p^2 + u^2)^{1/4} \ge 0\ (p \ge 0, u \in R)$, $G_{0p}(0, u) = -\infty$. Because of $F_{0t} = 0$, condition (b) of (14.3.i) is satisfied, and by Section 14.1C(d) condition (λ) of (14.3.i) is satisfied. Indeed, $G_0(t, x, \cos\theta, \sin\theta) = 1 - (\cos\theta)^{1/2}$, $-\pi/2 \le \theta \le \pi/2$, and $G_0 = 0$ only for $\theta = 0$. $I[x]$ has an absolute minimum in classes of AC curves as described in Example 1.

3. $F_0 = (x - a)^{-1/2}(1 + x'^2)^{1/2}$, $x \ge a$, satisfies the conditions of (14.3.iii) with $\psi(x) = (x - a)^{1/2} \ge 0$ continuous and $f_0 = (1 + x'^2)^{1/2}$ satisfying those of (14.3.ii). Here F_0 corresponds to the problem of minimum time of descent (brachistochrone, Section 3.12). Thus, the problems of minimum time of descent from $1 = (t_1, x_1)$ to $2 = (t_2, x_2)$, $t_2 > t_1, x_2 > x_1$, or from $1 = (t_1, x_1)$ to a curve $B_2 = [x = x(t), t' \le t \le t'']$, $t_1 < t'$, $x(t') > x_1$, have always an optimal solution.

Counterexamples

1. This example shows that the minimum may not exist if $B = B_1 \times B_2$, B_1, B_2 closed, and $(t_1, x_1, t_2, x_2) \in B$ does not imply $t_1 < t_2$. Indeed, for $F_0 = (1 + x'^2)^{1/2}$, $n = 1$, the problem of an AC curve $x = x(t)$, $t_1 \le t \le t_2$, of minimum length joining $(0,0)$ to $B_2 = [(t,x), t \ge 0, x = 1]$ has no optimal solution.

2. In Section 3.13 we have discussed at length the integral $F_0 = x(1 + x'^2)^{1/2}$, $x \ge 0$, $n = 1$, (problem of minimum area of a surface of revolution), and we have seen that, given two fixed points $1 = (t_1, x_1)$, $2 = (t_2, x_2)$, $t_1 < t_2$, $x_1 > 0$, $x_2 > 0$, the existence or nonexistence of an absolute minimum of $I[x]$ for AC curves $x = x(t)$, $t_1 \le t \le t_2$, joining 1 and 2, depends on whether 2 is above or below a suitable curve $x = \Gamma(t)$, $t_1 \le t < +\infty$ ($\Gamma(t_1) = 0$, $0 < \Gamma(t) < \Gamma(t')$ for $t_1 < t < t'$, $\Gamma(+\infty) = +\infty$). The corresponding parametric integral $G_0 = x(p^2 + u^2)^{1/2}$ is continuous in $[0, +\infty) \times R^2$ and has $x = 0$ as the only zero, but G_0 does not satisfy any of the local conditions stated in (14.3.i–iii). For instance, $G_{0p} = xp(p^2 + u^2)^{-1/2}$ and $G_{0p} = 0$ for all p if $x = 0$.

3. $F_0 = (x^2 + x'^2)^{1/2}$, $n = 1$. Here $G_0(x, p, u) = (p^2 x^2 + u^2)^{1/2}$ is continuous in $R \times [0, +\infty) \times R$. For $x \ne 0$ we have $G_{0p}(x, p, u) = 0$ if and only if $p = 0$. For $x = 0$, however, we have $G_{0p}(0, p, u) = 0$ identically. In Section 1.6, Example 5 we mentioned that the problem of minimum with F_0 as integrand and fixed end points $(0,0)$, $(1,1)$ has no optimal solution.

4. Here is an example of a problem with $n = 2$ and no optimal solution, where condition ($\lambda 3$) of (14.3.i) is not satisfied. Take $A = [(t, x, y) | 0 \le t \le 1, x^2 + y^2 \le 1]$, $F_0(x, y, x', y') = 2(yx' - xy') + (1 + x^2 + y^2)(1 + x'^2 + y'^2)^{1/2}$, and consider the problem of the minimum of $I[x, y] = \int F_0\, dt$ in the class of all AC trajectories $x(t)$, $y(t)$, $0 \le t \le 1$, joining $(0, 1, 0)$ to $(1, 1, 0)$. Here

$$G_0(x, y, p, u, v) = 2(yu - xv) + (1 + x^2 + y^2)(p^2 + u^2 + v^2)^{1/2}.$$

By the elementary inequality $1 + x^2 + y^2 \ge 2(x^2 + y^2)^{1/2}$ in A, and the Schwarz inequality (in R^2) we derive

$$F_0 \ge 2[(yx' - xy') + (x^2 + y^2)^{1/2}(1 + x'^2 + y'^2)^{1/2}] > 0,$$
$$G_0 \ge 2[(yu - xv) + (x^2 + y^2)^{1/2}(p^2 + u^2 + v^2)^{1/2}] \ge 0,$$

and equality $G_0 = 0$ holds if and only if $x^2 + y^2 = 1$, $p = 0$, $xu + yv = 0$. Obviously, condition (b) of (14.3.i) holds. No minimizing curve exists. First, for the infimum i we obviously have $i \ge 0$. On the other hand, if we consider the sequence

$$C_k: x = x_k(t) = \cos 2k\pi t, \quad y = y_k(t) = \sin 2k\pi t, \qquad 0 \le t \le 1, \quad k = 1, 2, \dots,$$

joining $(0, 1, 0)$ to $(1, 1, 0)$, we have

$$I[C_k] = I[x_k, y_k] = 2(-2k\pi + (1 + 4k^2\pi^2)^{1/2})$$

which tends to zero as $k \to \infty$. Thus, $i = 0$.

For no trajectory x, y we can have $I = 0$, since $F_0 > 0$, and for no curve $\mathfrak{C}$ we can have $\mathfrak{J} = 0$, since this would imply $t' = 0$ a.e., or $t =$ constant, while we need to join points with $t_1 = 0$, $t_2 = 1$. Condition ($\lambda 3$) of (14.3.i) is not satisfied, since for curves $\mathfrak{C}: t = c$, $x = \cos 2k\pi\tau$, $y = \sin 2k\pi\tau$, $0 \le \tau \le 1$, we have $\mathfrak{J}[\mathfrak{C}] = 0$, and $L[\mathfrak{C}] = 2k\pi$ as large as we want.

Bibliographical Notes

In Sections 14.3–4 we have briefly mentioned existence theorems for problems for which no growth conditions hold of any kind, such as the length integral or the integral in the classical brachistochrone problem. We have followed the approach proposed by E. J. McShane ([5], 1934), which consists in reducing the given usual (nonparametric) problem to a parametric one, showing that the parametric problem has an optimal solution, and proving, under hypotheses, that such parametric solution has an AC representation as a nonparametric curve in R^{n+1}, and actually is an optimal solution of the original nonparametric problem (cf. III for details and proofs).

An alternate approach has been proposed by L. Tonelli [10], and another one by E. J. McShane [7].

In III we shall present further work by C. Vinti concerning the underplaying between usual integrals and their associated parametric ones (cf. Bibliography at the end of Chapter 11). We shall also present in III recent work on parametric problems of optimal control (R. M. Goor [1, 2]). This work, with connection of our problems with questions of algebra and topology, shows also that the class of parametric optimal control problems is much larger than one would expect.

The concept of parametric, or Fréchet, curves, the lower semicontinuity theorems for parametric problems with respect to Fréchet distance (uniform topology), and the existence theorems will be discussed in more details elsewhere (Cesari [III]). The existence theorems for parametric problems can be traced back to L. Tonelli [II] for $n = 2$ and under some smoothness assumptions on the data, and to E. J. McShane [10] for any n.

CHAPTER 15

Existence Theorems: The Use of Mere Pointwise Convergence on the Trajectories

The existence theorems of this chapter are based on pointwise convergence of the trajectories and on the use of Helly's selection theorem. To present these existence theorems we first state Helly's theorem in Section 15.1, and then we state and prove some closure theorems based on mere pointwise convergence in Section 15.2 which are similar to those we proved in Section 8.6 for uniform convergence. Corresponding existence theorems for Lagrange problems of the Calculus of Variations and for problems of optimal control are stated and proved in Sections 15.3 and 15.4. Examples are given in Section 15.5.

15.1 The Helly Theorem

15.1.i. *Given any family* $\{x\}$ *of equibounded functions* $x(t) = (x^1, \dots, x^n)$, $a \le t \le b$, *with equibounded total variations* $V[x]$, *then any sequence* $[x_k]$ *of elements* $x_k \in \{x\}$ *contains a subsequence* $[x_{k_s}]$ *which converges pointwise everywhere in* $[a, b]$ *toward a* BV *function* $x_0(t)$, $a \le t \le b$, *with* $V[x_0] \le \liminf_{s\to\infty} V[x_{k_s}]$.

For this theorem we refer to Liusternik and Sobolev [I, p. 119]. The proof of this theorem is similar to the proof of Ascoli's theorem (9.1.i).

In (15.1.i) we have assumed, for the sake of simplicity, that the functions $x(t)$, $a \le t \le b$, are defined on the same interval $[a, b]$. If the intervals of definition $[a, b]$ are different, but they are all contained in a fixed finite interval $[a_0, b_0]$, say $-\infty < a_0 \le a \le b \le b_0 < +\infty$, then the same functions x can be all extended to the interval $[a_0, b_0]$ by taking $x(t) = x(a)$ for $t \le a$, and $x(t) = x(b)$ for $t \ge b$. Then, given the sequence $x_k(t)$, $a_k \le t \le b_k$, it is enough to apply (15.1.i) to the same functions extended to $[a_0, b_0]$, obtain the sequence $[x_{k_s}]$ for $[a_0, b_0]$, and then we may even further select the subsequence so that a_{k_s} and b_{k_s} also converge, say $a_{k_s} \to a$, $b_{k_s} \to b$ as $s \to \infty$, $a_0 \le a \le b \le b_0$.

15.2 Closure Theorems with Components Converging Only Pointwise

We shall need a variant of the closure theorem (8.6.i). We think of the space R^n as the product space $R^s \times R^{n-s}$; hence $x = (y, z)$ with $y \in R^s$, $z \in R^{n-s}$. Analogously, let A_0 be a given subset of the ty-space R^{s+1}, and let $A = A_0 \times R^{n-s}$, so that A is as usual a subset of the tx-space R^{n+1}. Finally, we assume that the orientor field in A has the form

$$dx/dt \in Q(t, y); \tag{15.2.1}$$

in other words, the set Q depends on t and y only, and not on z. A solution of this orientor field is then an AC n-vector function $x(t) = (y(t), z(t))$, $t_1 \le t \le t_2$, with $(t, x(t)) \in A$, that is, $(t, y(t)) \in A_0$ for every $t \in [t_1, t_2]$, and

$$dx/dt \in Q(t, y(t)), \qquad t \in [t_1, t_2] \text{ (a.e.)},$$

that is,

$$(y'(t), z'(t)) \in Q(t, y(t)). \tag{15.2.2}$$

15.2.i (A Closure Theorem). *Let A_0 be a closed subset of the ty-space R^{s+1}, and $A = A_0 \times R^{n-s}$; for every $(t, y) \in A_0$ let $Q(t, y)$ denote a closed subset of R^n, and assume that the sets $Q(t, y)$ are convex and closed and have property* (Q) *with respect to (t, y) at every point $(t, y) \in A_0$ with the exception perhaps of a set of points whose t coordinate lies on a set H of measure zero on the t-axis. Let $x_k(t)$, $t_{1k} \le t \le t_{2k}$, $k = 1, 2, \ldots$, be a sequence of solutions of the orientor field* (15.2.2) *$x_k(t) = (y_k(t), z_k(t))$, for which we assume that the s-vector $y_k(t)$ converges in the ρ-metric toward an* AC *vector function $y(t)$, $t_1 \le t \le t_2$, and that the $(n - s)$-vector $z_k(t)$ converges pointwise for all t, $t_1 < t < t_2$, toward a vector $z(t)$ which admits of a decomposition $z(t) = Z(t) + S(t)$, where $Z(t)$ is an* AC *vector function in $[t_1, t_2]$, and $S'(t) = 0$* a.e. *in $[t_1, t_2]$, that is, $S(t)$ is a singular function. Then the* AC *n-vector $X(t) = [y(t), Z(t)]$, $t_1 \le t \le t_2$, is a solution of the orientor field* (15.2.2).

In other words, we know that each $x_k(t) = (y_k(t), z_k(t))$, $t_{1k} \le t \le t_{2k}$, $k = 1, 2, \ldots$, is AC, that $(t, y_k(t)) \in A_0$ for every $t \in [t_{1k}, t_{2k}]$, and that $(y_k'(t), z_k'(t)) \in Q(t, y_k(t))$ a.e. in $[t_{1k}, t_{2k}]$; we know that $\rho(y_k, y) \to 0$, and hence $t_{1k} \to t_1$, $t_{2k} \to t_2$ as $k \to \infty$, that $z_k(t) \to z(t) = Z(t) + S(t)$ pointwise for every $t \in (t_1, t_2)$, that $S(t)$ is singular, and $(y(t), Z(t))$ is AC; and we want to prove that $(t, y(t)) \in A_0$ for every $t \in [t_1, t_2]$ and that $(y'(t), Z'(t)) \in Q(t, y(t))$ a.e. in $[t_1, t_2]$. For $s = n$ this closure theorem (15.2.i) reduces to Theorem (8.6.i).

Proof. The proof that $(t, y(t)) \in A_0$ for every $t \in [t_1, t_2]$ is the same as for the closure theorem (8.6.i).

Let us prove the remaining part of (15.2.i), where we shall need to know only that $z_k(t) \to z(t)$ for almost all $t \in (t_1, t_2)$.

For almost all $t \in [t_1, t_2]$ the derivative $X'(t) = [y'(t), Z'(t)]$ exists and is finite, $S'(t)$ exists and $S'(t) = 0$, and $z_k(t) \to z(t)$. Let t_0 be such a point with $t_1 < t_0 < t_2$. Then there is a $\sigma > 0$ with $t_1 < t_0 - \sigma < t_0 + \sigma < t_2$, and for some k_0 and all $k \ge k_0$, also $t_{1k} < t_0 - \sigma < t_0 + \sigma < t_{2k}$. Let $x_0 = X(t_0) = (y_0, Z_0)$, or $y_0 = y(t_0)$, $Z_0 = Z(t_0)$. Let $z_0 = z(t_0)$, $S_0 = S(t_0)$. We have $S'(t_0) = 0$, hence $z'(t_0)$ exists and $z'(t_0) = Z'(t_0)$. Also, we know that $z_k(t_0) \to z(t_0)$.

We have $y_k(t) \to y(t)$ uniformly in $[t_0 - \sigma, t_0 + \sigma]$, and all functions $y(t)$, $y_k(t)$ are continuous in the same interval. Thus, they are equicontinuous in $[t_0 - \sigma, t_0 + \sigma]$. Given $\varepsilon > 0$, there is $\delta > 0$ such that $t, t' \in [t_0 - \sigma, t_0 + \sigma]$, $|t - t'| \le \delta$, $k \ge k_0$ implies

$$|y(t) - y(t')| \le \varepsilon/2, \qquad |y_k(t) - y_k(t')| \le \varepsilon/2.$$

We can assume $0 < \delta < \sigma$, $\delta \le \varepsilon$. For any h, $0 < h \le \delta$, let us consider the averages

$$m_h = h^{-1} \int_0^h X'(t_0 + s)\,ds = h^{-1}[X(t_0 + h) - X(t_0)], \tag{15.2.3}$$

$$m_{hk} = h^{-1} \int_0^h x_k'(t_0 + s)\,ds = h^{-1}[x_k(t_0 + h) - x_k(t_0)], \tag{15.2.4}$$

where $X = (y, Z)$, $x_k = (y_k, z_k)$.

Given $\eta > 0$ arbitrary, we can fix h, $0 < h \le \delta < \sigma$, so small that

$$|m_h - X'(t_0)| \le \eta, \tag{15.2.5}$$

$$|S(t_0 + h) - S(t_0)| < \eta h/4. \tag{15.2.6}$$

This is possible because $h^{-1} \int_0^h X'(t_0 + s)\,ds \to X'(t_0)$ and $[S(t_0 + h) - S(t_0)]h^{-1} \to 0$ as $h \to 0+$. Also, we can choose h in such a way that $z_k(t_0 + h) \to z(t_0 + h)$ as $k \to +\infty$. This is possible because $z_k(t) \to z(t)$ for almost all $t_1 < t < t_2$.

Having so fixed h, let us take $k_1 \ge k_0$ so large that

$$|y_k(t_0) - y(t_0)|,\ |y_k(t_0 + h) - y(t_0 + h)| \le \min[\eta h/4, \varepsilon/2],$$
$$|z_k(t_0) - z(t_0)|,\ |z_k(t_0 + h) - z(t_0 + h)| \le \eta h/8.$$

This is possible because $y_k(t) \to y(t)$, $z_k(t) \to z(t)$ both at $t = t_0$ and $t = t_0 + h$. Then, we have

$$\begin{aligned} &|h^{-1}[y_k(t_0 + h) - y_k(t_0)] - h^{-1}[y(t_0 + h) - y(t_0)]| \\ &\quad \le |h^{-1}[y_k(t_0 + h) - y(t_0 + h)]| + |h^{-1}[y_k(t_0) - y(t_0)]| \\ &\quad \le h^{-1}(\eta h/4) + h^{-1}(\eta h/4) = \eta/2. \end{aligned}$$

Analogously, since $z = Z + S$, we have

$$\begin{aligned} &|h^{-1}[z_k(t_0 + h) - z_k(t_0)] - h^{-1}[Z(t_0 + h) - Z(t_0)]| \\ &\quad = |h^{-1}[z_k(t_0 + h) - z_k(t_0)] - h^{-1}[z(t_0 + h) - z(t_0)] + h^{-1}[S(t_0 + h) - S(t_0)]| \\ &\quad \le |h^{-1}[z_k(t_0 + h) - z(t_0 + h)]| + |h^{-1}[z_k(t_0) - z(t_0)]| + |h^{-1}[S(t_0 + h) - S(t_0)]| \\ &\quad \le h^{-1}(\eta h/8) + h^{-1}(\eta h/8) + h^{-1}(\eta h/4) = \eta/2. \end{aligned}$$

Finally, we have

$$\begin{aligned} |m_{hk} - m_h| &= |h^{-1}[x_k(t_0 + h) - x_k(t_0)] - h^{-1}[X(t_0 + h) - X(t_0)]| \\ &\le |h^{-1}[y_k(t_0 + h) - y_k(t_0)] - h^{-1}[y(t_0 + h) - y(t_0)]| \\ &\quad + |h^{-1}[z_k(t_0 + h) - z_k(t_0)] - h^{-1}[Z(t_0 + h) - Z(t_0)]| \\ &\le \eta/2 + \eta/2 = \eta. \end{aligned} \tag{15.2.7}$$

We conclude that for the chosen value of h, $0 < h \le \delta < \sigma$, and every $k \ge k_1$ we have

$$|m_h - X'(t_0)| \le \eta, \qquad |m_{hk} - m_h| \le \eta, \qquad |y_k(t_0) - y(t_0)| \le \varepsilon/2. \tag{15.2.8}$$

For $0 \le \tau \le h$ we have now

$$\begin{gathered} |y_k(t_0 + \tau) - y(t_0)| \le |y_k(t_0 + \tau) - y_k(t_0)| + |y_k(t_0) - y(t_0)| \le \varepsilon/2 + \varepsilon/2 = \varepsilon, \\ |(t_0 + \tau) - t_0| \le h \le \delta \le \varepsilon, \end{gathered} \tag{15.2.9}$$

$$x_k'(t_0 + \tau) = (y_k'(t_0 + \tau), z_k'(t_0 + \tau)) \in Q(t_0 + \tau, y_k(t_0 + \tau)), \qquad \tau \in [0, h] \text{ (a.e.)}.$$

Hence, for almost all τ, $0 \le \tau \le h$,

$$x_k'(t_0 + \tau) = (y_k'(t_0 + \tau), z_k'(t_0 + \tau)) \in Q(t_0, y_0, 2\varepsilon),$$

and consequently

$$x_k'(t_0 + \tau) = (y_k'(t_0 + \tau), z_k'(t_0 + \tau)) \in \mathrm{cl\,co}\, Q(t_0, y_0, 2\varepsilon), \qquad \tau \in [0, h] \text{ (a.e.)}.$$

The average m_{hk} as defined by (15.2.4) is then also a point of the same closed and convex set (cf. Section 8.4B), or

$$m_{hk} \in \mathrm{cl\,co}\, Q(t_0, y_0, 2\varepsilon)$$

for the chosen h and every $k \geq k_1$. By relations (15.2.5) and (15.2.8) we deduce

$$|X'(t_0) - m_{hk}| \leq |X'(t_0) - m_h| + |m_h - m_{hk}| \leq 2\eta,$$

and hence

$$X'(t_0) \in [\mathrm{cl\,co}\, Q(t_0, y_0, 2\varepsilon)]_{2\eta}.$$

Here $\eta > 0$ is an arbitrary number, and the set in brackets is closed; hence

$$X'(t_0) \in \mathrm{cl\,co}\, Q(t_0, y_0, 2\varepsilon)$$

for every $\varepsilon > 0$. By property (Q) we have

$$X'(t_0) \in \bigcap_{\varepsilon} \mathrm{cl\,co}\, Q(t_0, y_0, 2\varepsilon) = Q(t_0, y_0),$$

where $y_0 = y(t_0)$ and $X'(t_0) = (y'(t_0), Z'(t_0))$. We have proved that for almost all $t \in [t_1, t_2]$ we have

$$dX/dt \in Q(t, y(t)).$$

Theorem (15.2.i) is thereby proved. □

The following example illustrates the closure theorem (15.2.i). Let $n = 2$, $s = 1$, $n - s = 1$, $A = R^3$, $Q = Q(t, y) = [z = (z^1, z^2) | z^2 \geq 0, \; -1 \leq z^1 \leq 1]$. If $\phi(t)$, $0 \leq t \leq 1$, denotes a singular continuous monotone function with $\phi(0) = 0$, $\phi(1) = 1$, $\phi'(t) = 0$ a.e. in $[0, 1]$, let us define $\phi(t)$ in $(-\infty, +\infty)$ by taking $\phi = 0$ for $t \leq 0$ and $\phi = 1$ for $t \geq 1$. Let $z_k(t) = k \int_t^{t+k^{-1}} \phi(t)\, d\tau$, $0 \leq t \leq 1$, $k = 1, 2, \ldots$. Here the scalar functions $z_k(t)$ are absolutely continuous, monotone nondecreasing, with $z_k'(t) \geq 0$ and $z_k(t) \to z(t) = \phi(t)$ uniformly in $[0, 1]$ as $k \to \infty$. Let us take $Z(t) = 0$, $y(t) = 0$, $y_k(t) = 0$, $0 \leq t \leq 1$, $k = 1, 2, \ldots$, and then $z(t) = Z(t) + \phi(t)$, $Z(t)$ absolutely continuous, $\phi(t)$ singular. Here (y_k, z_k) converges uniformly toward (y, z) in $[0, 1]$. All pairs (y_k, z_k) are solutions of the orientor field $(y', z') \in Q$; (y, Z) is a solution of the same orientor field, but (y, z) is not since z is not AC.

15.3 Existence Theorems for Extended Problems Based on Pointwise Convergence

A. The Statements

To present the next existence theorems let us introduce the following specific notation. Let α, n, $0 \leq \alpha \leq n$, be given integers, and for every $x = (x^1, \ldots, x^n)$, let y, z denote $y = (x^1, \ldots, x^\alpha)$, $z = (x^{\alpha+1}, \ldots, x^n)$, so that $x = (y, z)$. Let A_0 be a closed subset of the

ty-space $R^{\alpha+1}$, let I be a subset of the z-space $R^{n-\alpha}$ which is the product of $n-\alpha$ closed intervals, or $I=[a_{\alpha+1},b_{\alpha+1}]\times\cdots\times[a_n,b_n]$, and let $A=A_0\times I\subset R^{n+1}$.

Let B_0 be a given subset of the $t_1x_1^1\cdots x_1^n t_2x_2^1\cdots x_2^\alpha$-space $R^{2+n+\alpha}$, and let B be the set $B=B_0\times R^{n-\alpha}\subset R^{2+2n}$. Let $g(t_1,x_1,t_2,x_2)$ be a real valued function defined on B which is monotone nondecreasing with respect to each of the variables $x_2^{\alpha+1},\ldots,x_2^n$.

For every $(t,y)\in A_0$ let $Q(t,y)$ denote a given subset of the $\mathfrak{z}$-space R^α, $\mathfrak{z}=(\mathfrak{z}^1,\ldots,\mathfrak{z}^\alpha)$, and let M_0 denote the set of all $(t,y,\mathfrak{z})\in R^{2\alpha+1}$ with $(t,y)\in A_0$, $\mathfrak{z}\in Q(t,y)$. Let $F_0(t,y,\mathfrak{z})$, $T_{\alpha+1}(t,y,\mathfrak{z}),\ldots,T_n(t,y,\mathfrak{z})$ be $n-\alpha+1$ real valued functions on M_0.

We consider here the problem of the minimum of the functional

$$I[x]=g(t_1,x(t_1),t_2,x(t_2))+\int_{t_1}^{t_2}F_0(t,y(t),y'(t))\,dt, \tag{15.3.1}$$

with the constraints

$$\begin{aligned}&(t,x(t))\in A=A_0\times I,\qquad e[x]=(t_1,x(t_1),t_2,x(t_2))\in B=B_0\times R^{n-\alpha},\\ &x(t)=(y(t),z(t)),\quad dz^i/dt=T_i(t,y(t),y'(t)),\quad i=\alpha+1,\ldots,n,\quad t\in[t_1,t_2]\ \text{(a.e.)},\\ &y(t),z(t)\ \text{AC in }[t_1,t_2],\qquad F_0(\cdot,y(\cdot),y'(\cdot))\ L\text{-integrable in }[t_1,t_2].\end{aligned} \tag{15.3.2}$$

For every $(t,y)\in A_0$ let $\tilde{Q}(t,y)$ denote the set of all $(z^0,\mathfrak{z},\bar{\mathfrak{z}})\in R^{n+1}$, or $(z^0,\mathfrak{z}^1,\ldots,\mathfrak{z}^\alpha,\mathfrak{z}^{\alpha+1},\ldots,\mathfrak{z}^n)$ with $\mathfrak{z}=(\mathfrak{z}^1,\ldots,\mathfrak{z}^\alpha)\in Q(t,y)$, $z^0\geq F_0(t,y,\mathfrak{z})$, $\mathfrak{z}^{\alpha+1}\geq T_{\alpha+1}(t,y,\mathfrak{z}),\ldots,$ $\mathfrak{z}^n\geq T_n(t,y,\mathfrak{z})$. It is convenient to consider also the extended functions $F_0,T_{\alpha+1},\ldots,T_n$ defined in $R^{2\alpha+1}$ by taking them equal to $+\infty$ in $R^{2\alpha+1}-M_0$.

15.3.i (An Existence Theorem Based on Pointwise Convergence of Some Components). *Let $1\leq\alpha\leq n$, let $A_0\subset R^\alpha$ be compact and $A=A_0\times I$, let $B_0\subset R^{n+2+a}$ be closed and $B=B_0\times R^{n-\alpha}$, and let g be lower semicontinuous on B and monotone nondecreasing in each of the variables $x_2^{\alpha+1},\ldots,x_2^n$. Let us assume that the extended functions $F_0(t,y,\mathfrak{z})$, $T_i(t,y,\mathfrak{z})$, $i=\alpha+1,\ldots,n$, are bounded below and lower semicontinuous in $R^{2\alpha+1}$, and also convex in $\mathfrak{z}$ for every (t,y). Let us further assume that the function F_0 satisfies the growth condition $(\gamma 1)$ of Section 11.1. Let Ω be the family of all AC trajectories $x(t)=(y(t),z(t))$, $t_1\leq t\leq t_2$, satisfying (15.3.2). Then the functional (15.3.1) has an absolute minimum in Ω.*

The same conclusion holds if F_0 satisfies one of the conditions $(\gamma 2)$ or $(\gamma 3)$ of Section 11.1, provided we know that the sets $\tilde{Q}(t,y)$ have property (Q) with respect to (t,y) at every point $(t,y)\in A_0$ with the exception perhaps of a set of points whose t-coordinate lies on a set of measure zero on the t-axis.

Proof. Here A_0 is bounded, and thus for every element $x(t)=(y(t),z(t))$, $t_1\leq t\leq t_2$, in Ω we have $-M\leq t_1\leq t_2\leq M$, $|y(t)|\leq M$ for some M and all $t\in[t_1,t_2]$. Here F_0 is bounded below, so that $F_0(t,y(t),y'(t))>-M$ and $J_1=\int_{t_1}^{t_2}F_0\,dt>-M_1$ for some constant M_1. Here g is lower semicontinuous in the compact set $B\cap(A\times A)$, hence bounded below, and $J_2=g(e[x])\geq -M_2$ for some constant M_2, and $I=J_1+J_2$ is bounded below in Ω. Let $i=\inf[I[x],x\in\Omega]$, $-\infty<i<+\infty$. Let $x_k(t)=(y_k(t),z_k(t))$, $t_{1k}\leq t\leq t_{2k}$, $k=1,2,\ldots$, be a minimizing sequence of elements $x_k\in\Omega$; thus $I[x_k]\to i$ as $k\to\infty$. We may well assume that

$$i\leq g(t_{1k},x_k(t_{1k}),t_{2k},x(t_{2k}))+\int_{t_{1k}}^{t_{2k}}F_0(t,y_k(t),y_k'(t))\,dt\leq i+k^{-1}\leq i+1,$$

$k=1,2,\ldots,$ and, by a suitable extraction, that $J_1[x_k]\to i_1$, $J_2[x_k]\to i_2$ for suitable numbers i_1,i_2 with $i_1+i_2=i$. The same conclusion holds under conditions $(\gamma 2)$ or $(\gamma 3)$.

By condition (γ1), (γ2), or (γ3) and statements (10.4.i), (10.4.ii), (10.4.iii), we derive as usual that the equibounded sequence $[y_k]$ is equicontinuous and equiabsolutely continuous, and that the sequence $[y_k']$ is equiabsolutely integrable. Thus, there is a subsequence, say still $[k]$, such that $[y_k]$ converges uniformly, namely in the ρ-metric (Section 2.14), to an absolutely continuous function $y(t)$, $t_1 \le t \le t_2$, and that $[y_k']$ converges weakly in L_1 toward an integrable function $\xi(t)$, with $y'(t) = \xi(t)$ a.e. in $[t_1, t_2]$.

To simplify the argument, let us assume for a moment that all functions F_0 and T_i are nonnegative. Let $z_k^0(t) = \int_{t_{1k}}^t F_0(\tau, y_k(\tau), y_k'(\tau))\,d\tau$, $t_{1k} \le t \le t_{2k}$, $k = 1, 2, \ldots$. Then we see that $(z_k^0(t), y_k(t), z_k(t))$ are AC solutions of the orientor field

$$(z_k'^0(t), y_k'(t), z_k'(t)) \in \tilde{Q}(t, y_k(t)), \qquad t \in [t_{1k}, t_{2k}] \text{ (a.e.)}, \quad k = 1, 2, \ldots,$$

or equivalently,

$$\begin{aligned} &z_k'^0(t) \ge F_0(t, y_k(t), y_k'(t)), \quad z_k'^i(t)) \ge T_i(t, y_k(t), y_k'(t)), \qquad i = \alpha + 1, \ldots, n, \\ &(t, y_k(t)) \in A_0, \quad y_k'(t) \in Q(t, y_k(t)), \quad t \in [t_{1k}, t_{2k}] \text{ (a.e.)}, \qquad k = 1, 2, \ldots. \end{aligned}$$

Here the functions F_0 and T_i are nonnegative, and the functions $z_k^0(t)$, $z_k^i(t)$, $i = \alpha + 1, \ldots, n$, are monotone nondecreasing in $[t_{1k}, t_{2k}]$ with values $0 \le z^0(t) \le i_1 + 1$, $a_i \le z_k^i(t) \le b_i$, $i = \alpha + 1, \ldots, n$. By Helly's theorem (15.1.i) we can extract a further subsequence, say still k, in such a way that each of the $n - \alpha + 1$ sequences $[z_k^0(t), k = 1, 2, \ldots]$, $[z_k^i(t), k = 1, 2, \ldots]$, $i = \alpha + 1, \ldots, n$, converges pointwise everywhere to some monotone nondecreasing function $z^0(t)$, $z^i(t)$, $t_1 \le t \le t_2$. Thus $0 \le z^0(t) \le i_2$, $a_i \le z^i(t) \le b_i$, $t \in [t_1, t_2]$, $i = \alpha + 1, \ldots, n$, with $z^0(t_1) = 0$. Moreover, $z^0(t) = Z^0(t) + S^0(t)$, $z^i(t) = Z^i(t) + S^i(t)$, $i = \alpha + 1, \ldots, n$, Z^0, S^0, Z^i, S^i monotone nondecreasing, $S^0(t) \ge 0$, $S^i(t) \ge 0$, $S^0(t_1) = 0$, $S^i(t_1) = 0$, $Z^0(t_1) = 0$, $Z^i(t_1) = z^i(t_1)$, all Z^0, Z^i AC in $[t_1, t_2]$, all S^0, S^i singular functions, that is, $dS^0/dt = 0$, $dS^i/dt = 0$, $t \in [t_1, t_2]$ (a.e.). Also, $0 \le Z^0(t_2) \le z^0(t_2) \le i_1 + 1$, $a_i \le Z^i(t_2) \le z^i(t_2) \le b_i$. Let $Z(t) = (Z^{\alpha+1}, \ldots, Z^n)$, $S(t) = (S^{\alpha+1}, \ldots, S^n)$, $t \in [t_1, t_2]$. Then

$$\begin{aligned} -M_2 &\le g(t_1, y(t_1), Z(t_1), t_2, y(t_2), Z(t_2)) \\ &\le g(t_1, y(t_1), Z(t_1), t_2, y(t_2), Z(t_2) + S(t_2)) \\ &= \liminf_{k \to \infty} g(t_{1k}, x_k(t_{1k}), t_{2k}, x_k(t_{2k})) = i_2. \end{aligned}$$

Here the extended functions $F_0(t, y, \mathfrak{z})$, $T_i(t, y, \mathfrak{z})$ are lower semicontinuous in $(t, y, \mathfrak{z})$. Hence, by (8.5.v) and Remark 3 of Section 8.5 with (t, y) replacing x, the sets $\tilde{Q}(t, y)$ have property (K) with respect to (t, y) in A_0. Under condition (γ1) the same sets $\tilde{Q}(t, y)$ have also property (Q) in A by virtue of (10.5.i). Hence, by the closure theorem (15.2.i), we derive that

$$\begin{aligned} &(t, y(t)) \in A_0, \qquad (t, x(t)) = (t, y(t), Z(t)) \in A_0 \times I, \\ &(Z'^0(t), x'(t)) = (Z'^0(t), y'(t), Z'(t)) \in \tilde{Q}(t, y(t)), \qquad y'(t) \in Q(t, y(t)), \\ &Z'^0(t) \ge F_0(t, y(t), y'(t)), \quad Z'^i(t) \ge T_i(t, y(t), y'(t)), \qquad t \in [t_1, t_2] \text{ (a.e.)}. \end{aligned}$$

Under conditions (γ2) or (γ3) we have explicitly assumed that property (Q) holds in A and the same conclusion follows. Thus,

$$\begin{aligned} J_1[y] &= \int_{t_1}^{t_2} F_0(t, y(t), y'(t))\,dt \le \int_{t_1}^{t_2} Z'^0(t)\,dt = Z^0(t_2) \\ &\le Z^0(t_2) + S^0(t_2) = z^0(t_2) = \lim_{k \to \infty} z_k^0(t_2) = \lim_{k \to \infty} J_1[y_k] = i_1. \end{aligned}$$

Now we consider the AC functions

$$X^i(t) = x^i(t_1) + \int_{t_1}^{t} T_i(\tau, y(\tau), y'(\tau))\, d\tau, \qquad i = \alpha + 1, \ldots, n,$$

with $X^i(t_1) = x^i(t_1) = z^i(t_1)$, $X^i(t) \leq Z^i(t)$, $t_1 \leq t \leq t_2$, in particular $X^i(t_2) \leq Z^i(t_2)$. Then, for $X(t) = (X^{\alpha+1}, \ldots, X^n)$, we have

$$g(t_1, y(t_1), X(t_1), t_2, y(t_2), X(t_2)) \leq g(t_1, y(t_1), Z(t_1), t_2, y(t_2), Z(t_2)) \leq i_2,$$
$$dX^i/dt = T_i(t, y(t), y'(t)), \qquad t \in [t_1, t_2] \text{ (a.e.)}, \quad i = \alpha + 1, \ldots, n,$$

or $J_2(y, X) \leq J_2(y, Z) \leq i_2$. Thus, for $x(t) = (y, X)$ we have $I[x] = J_1 + J_2 \leq i_1 + i_2 = i$. Since $x \in \Omega$, we also have $I[x] \geq i$, and hence $I[x] = i$ with $x = (y, Z) \in \Omega$. This proves (15.3.i) for F_0, $T_i \geq 0$. For F_0, T_i of arbitrary signs, but bounded below (say $F_0 \geq -M$, $T_i \geq -M_i$), the argument is the same provided we write, say, $F_0 = (F_0 + M) - M$ with $F_0 + M \geq 0$, and hence $z^0(t) = \bar{z}^0(t) - M(t - t_1)$ with $\bar{z}^0$ monotone nondecreasing as before. Analogous argument holds for each T_i. Theorem (15.3.i) is thereby proved. □

Remark 1. Theorem (15.3.i) holds with obvious changes even if the lower semicontinuous functions F_0, T_i are not bounded below, but there are locally integrable functions $\psi(t) \geq 0$, $\psi_i(t) \geq 0$ such that $F_0 \geq -\psi(t)$, $T_i \geq -\psi_i(t)$ for all $(t, y, \mathfrak{z}) \in M_0$.

Theorem (15.3.i) holds even if, not F_0, but at least one of the functions T_i, $i = \alpha + 1, \ldots, n$, satisfies the growth property $(\gamma 1)$ with respect to $\mathfrak{z}$.

Theorem (15.3.i) holds even if neither F_0 nor any of the functions T_i satisfies property $(\gamma 1)$, but we know that the sets $\tilde{Q}(t, y)$ satisfy property (Q) with respect to (t, y) in A, and we know that Ω is the family of all AC trajectories $x(t) = (y(t), z(t))$ with z as before, and y belonging to a Γ_w-closed class Ω_0 of AC trajectories y in R^α which is sequentially relatively compact in the topology of the weak convergence of the derivatives. For instance, it may well occur that the class of all y satisfying a relation $I[y] \leq M$ has this property.

Remark 2. For $\alpha = n$, then $x = y$, $A = A_0$, $B = B_0$, the vector z and relative constraints are nonexistent, $g(t_1, x_1, t_2, x_2)$ is lower semicontinuous on B with no other requirement, and (15.3.i) reduces essentially to statement (11.1.i), for which therefore we have just given a new proof.

In statement (15.3.i) we have assumed $1 \leq \alpha \leq n$. For $\alpha = 0$, then we understand that $x = z$, $F_0 = 0$, the growth conditions do not apply, and we have a Mayer problem. The statement is still valid if we know that the sets $\tilde{Q}(t)$ have property (Q) with respect to t.

For $1 \leq \alpha \leq n$ and $F_0 = 0$, again we have a Mayer problem. The statement is still valid if there is a comparison functional $J = \int_{t_1}^{t_2} H\, dt$ where H has the same properties we have required for F_0 in (15.3.i) and we restrict Ω to only those AC elements (y, z) with $J \leq M$ for a given M.

Remark 3. Theorem (15.3.i) can be interpreted as an existence theorem for the AC α-vector function $y(t)$, $t_1 \leq t \leq t_2$, with $n - \alpha$ comparison functionals

$$\int_{t_1}^{t_2} T_i(t, y(t), y'(t))\, dt \leq b_i, \qquad i = \alpha + 1, \ldots, n,$$

for given functions $T_i(t, y, y') \geq 0$. For $a_i = 0$ then the $n - \alpha$ requirements $a_i \leq z_i(t) \leq b_i$ are trivially satisfied.

Here none of the functions F_0 and T_i need be dominant.

B. Examples

1. We seek the minimum of $I = \int_0^1 tx'^2\,dt$ under the constraints $J = \int_0^1 x'^2\,dt \le 2$ and $x(0) = 1$, $x(1) = 0$.

2. We seek the minimum of $I = \int_0^1 x'^2\,dt$ under the constraints $J = \int_0^1 tx'^4\,dt \le 2$, and $x(0) = 1$, $x(1) = 0$.

In both cases take $n = 2$, $\alpha = 1$.

3. We seek the minimum of $I = \int_0^1 (-t + |x|x'^2)\,dt$ under the constraints $J_1 = \int_0^1 x'^2\,dt \le 3$, $J_2 = \int_0^1 |x|\,dt \le 4$, and $x(0) = 1$, $x(1) = 0$.

Take $n = 3$, $\alpha = 1$.

15.4 Existence Theorems for Problems of Optimal Control Based on Pointwise Convergence

As in Section 15.3, let α, n, $1 \le \alpha \le n$, be given integers, and for every $x = (x^1, \dots, x^n)$, let $y = (x^1, \dots, x^\alpha)$, $z = (x^{\alpha+1}, \dots, x^n)$, so that $x = (y, z)$. Let A_0 be a compact subset of the ty-space $R^{\alpha+1}$, and let $I = [a_{\alpha+1}, b_{\alpha+1}] \times \cdots \times [a_n, b_n]$ be a subset of the z-space $R^{n-\alpha}$, so that $A = A_0 \times I \subset R^{n+1}$ is also compact. For every $(t, y) \in A_0$ let $U(t, y)$ be a given subset of the u-space R^m, and let M_0 denote the set of all (t, y, u) with $(t, y) \in A_0$, $u \in U(t, y)$. Let $f_0(t, y, u)$, $f(t, y, u) = (f_1, \dots, f_n)$ be given functions defined on M_0. Let B_0 be a given subset of the $t_1x_1^1 \cdots x_1^n t_2 x_2^1 \cdots x_2^\alpha$-space $R^{2+n+\alpha}$, and let B be the set $B = B_0 \times R^{n-\alpha} \subset R^{2n+2}$. Let $g(t_1, x_1, t_2, x_2)$ be a real valued function defined on B, which is monotone nondecreasing with respect to each variable $x_2^{\alpha+1}, \dots, x_2^n$.

We consider here the problem of minimum of the functional

$$I[x,u] = g(t_1, x(t_1), t_2, x(t_2)) + \int_{t_1}^{t_2} f_0(t, y(t), u(t))\,dt \tag{15.4.1}$$

with the constraints

$$\begin{gathered}(t, x(t)) \in A = A_0 \times I, \qquad e([x]) = (t_1, x(t_1), t_2, x(t_2)) \in B = B_0 \times R^{n-\alpha}, \\ x(t) = (y(t), z(t)), \qquad dx/dt = f(t, y(t), u(t)), \\ f_0(\cdot, y(\cdot), u(\cdot)) \in L[t_1, t_2], \qquad u(t) \in U(t, y(t)), \quad t \in [t_1, t_2] \text{ (a.e.)}, \\ y(t),\ z(t) \text{ AC}, \qquad u(t) \text{ measurable in } [t_1, t_2].\end{gathered} \tag{15.4.2}$$

For every $(t, y) \in A_0$ we denote by $\tilde{Q}(t, y)$ the set of all $(\mathfrak{z}^0, \mathfrak{z}^1, \dots, \mathfrak{z}^n) \in R^{n+1}$ with $\mathfrak{z}^0 \ge f_0(t, y, u)$, $\mathfrak{z}^i = f_i(t, y, u)$, $i = 1, \dots, \alpha$, $\mathfrak{z}^i \ge f_i(t, y, u)$, $i = \alpha + 1, \dots, n$, $u \in U(t, y)$, and by $Q(t, y)$ its projection on the $\mathfrak{z}$-space R^α, $\mathfrak{z} = (\mathfrak{z}^1, \dots, \mathfrak{z}^\alpha)$, or $Q(t, y) = [(\mathfrak{z}^1, \dots, \mathfrak{z}^\alpha) | \mathfrak{z}^i = f_i(t, y, u), i = 1, \dots, \alpha, u \in U(t, y)]$. We denote by Ω_0 the class of all admissible pairs $x(t)$, $u(t)$, $t_1 \le t \le t_2$, for the problem above, namely, x is AC, u is measurable, and requirements (15.4.2) are satisfied.

15.4.i (AN EXISTENCE THEOREM BASED ON POINTWISE CONVERGENCE OF SOME COMPONENTS). *Let $1 \le \alpha \le n$, let $A_0 \subset R^{\alpha+1}$ be compact and $A = A_0 \times I$, let $B_0 \subset R^{n+2+\alpha}$ be closed and $B = B_0 \times R^{n-\alpha}$, and let g be lower semicontinuous on B and monotone* *nondecreasing in each of the variables* $x_2^{\alpha+1}, \ldots, x_2^n$. *Let* M_0 *be closed,* $f_0, f_{\alpha+1}, \ldots, f_n$ *nonnegative and lower semicontinuous on* M_0, *and* $f_1, \ldots, f_\alpha$ *continuous on* M. *Let us further assume that the sets* $\tilde{Q}(t, y)$ *have property* (K) *with respect to* (t, y), *and the functions* $f_0, f = (f_1, \ldots, f_\alpha)$ *satisfy condition* (gl′) *of Section* 11.4. *Then the functional* $I[x, u]$ *in* (15.4.1) *has an absolute minimum in* Ω_0.

The same conclusion holds under the growth hypothesis (g2′) *or* (g3′) *of Section* 11.4, *provided we know that the sets* $\tilde{Q}(t, y)$ *have property* (Q) *with respect to* (t, y) *at every point* $(t, y) \in A_0$ *with the exception perhaps of a set of points whose t-coordinate lies on a set of measure zero on the t-axis.*

This statement is a corollary of (15.3.i).

Here comments can be made which are analogous to Remarks 1–3 of Section 15.3. Thus, it is enough that any one of the functions $f_0, f_{\alpha+1}, \ldots, f_n$ satisfies growth condition (gl′) with respect to $(f_1, \ldots, f_\alpha)$. Alternatively, no condition (gl′) is needed provided we know that the sets $\tilde{Q}(t, y)$ have property (Q) with respect to (t, y) in A, and we know that the AC functions y belong to a class Ω_0 which is Γ_w-closed and sequentially relatively compact with respect to the topology of the weak convergence of the derivatives. For instance, it may well occur that the class of all y satisfying a relation $I \le M$ has this property.

Finally, Theorem (15.4.i) can be interpreted as the problem of optimal control concerning the AC α-vector function $y(t)$, $t_1 \le t \le t_2$, corresponding control $u(t)$, and functional

$$I[y, u] = g(e[y]) + \int_{t_1}^{t_2} f_0(t, y(t), u(t))\, dt,$$

with constraints

$$u(t) \in U(t, y(t)), \quad e[y] \in B, \quad dy^i/dt = f_i(t, y(t), u(t)), \qquad t_1 \le t \le t_2, \quad i = 1, \ldots, \alpha,$$

and comparison functionals

$$\int_{t_1}^{t_2} f_i(t, y(t), u(t))\, dt \le b_i, \qquad i = \alpha + 1, \ldots, n.$$

Here again none of the functions $f_0, f_{\alpha+1}, \ldots, f_n$ need be dominant with respect to the others as we had to assume in Chapter 11.

As a last remark we mention that the condition that A must be compact in the existence theorems of this section can be removed as we have done in Chapters 9 and 11.

15.5 Exercises

The following problems have an absolute minimum.

1. $I = \int_0^1 (1 + x^2 + t^2)u^2\, dt$, $dx/dt = x + u$, $dy/dt = u^4$, $u \in U = R$, $n = 2$, $x(0) = y(0) = 0$, $x(1) = 1$, $y(1) \le 1$, $\alpha = 1$, $m = 1$, $n = 2$.

2. $I = \int_{-1}^1 |t|^{1/2}(u^2 + v^2)\, dt + |x(1)| + |y(t)|$, $dx/dt = x + u - v - 1$, $dy/dt = x^2 + u^4 + v^4$, $x^2(-1) + y^2(-1) = 1$, $x(1)$, $y(1)$ undetermined, $n = 2$, $m = 2$, $\alpha = 1$.

3. $I = \int_0^1 tu^2\, dt$ with $dy/dt = t + u$, $y(0) = 0$, $y(1) = 1$, and $J = \int_0^1 u^4\, dt \le 2$, $u \in U = R$, $n = 2$, $\alpha = 1$, $m = 1$.

4. $I = \int_0^1 u^2\, dt$ with $dy/dt = t + u$, $y(0) = 0$, $y(1) = 1$, and $J = \int_0^1 tu^2\, dt \le 3$, $u \in U = R$, $n = 2$, $\alpha = 1$, $m = 1$.

Bibliographical Notes

In this chapter we have presented existence theorems in which some component of the space variable in a minimizing sequence are proved to be of uniform bounded variation and therefore a subsequence exists which converges pointwise everywhere by Helly's theorem. The possible singular parts in the limit, under hypotheses, can be disregarded, yielding the existence of an AC optimal solution as usual. For theorems of this sort we refer to E. J. McShane [18] and L. Cesari [6, 7]. Wide extensions of the present viewpoint have been obtained in problems where actually discontinuous solutions are sought, or, in different notations, where control functions are sought in a space of measures. This extended approach will be discussed elsewhere.

CHAPTER 16
Existence Theorems: Problems with No Convexity Assumptions

The existence theorems of this Chapter concern control systems as well as problems of the calculus of variations which are linear in the state variables, but not necessarily linear in the controls and no convexity assumptions are required. Theorems of this type were first noted by L. W. Neustadt, and they are based on set theoretical considerations due to A. A. Lyapunov. We first prove in Section 16.1 some theorems of the Lyapunov type, and we use them in Section 16.2 to prove Neustadt type existence theorems for the bounded case. In this situation we prove in Section 16.3 that there always are bang-bang solutions. In Sections 16.5–6 we handle the unbounded case, and in Section 16.7 problems of the calculus of variations.

16.1 Lyapunov Type Theorems

Below, A will denote any measurable subset of some Euclidean space R^p with finite Lebesgue measure in R^p. In the applications in this book, however, A will be always an interval $[a, b]$ of R.

16.1.i. *Let $f(t) = (f_1, \ldots, f_n)$ be any function defined on A whose components are real valued L-integrable functions on A. Let $w(t)$, $t \in A$, be any real valued function, $0 \leq w(t) \leq 1$. Then there is a measurable subset E of A such that*

$$\int_A f(t)w(t)\,dt = \int_E f(t)\,dt. \tag{16.1.1}$$

In particular, for every α, $0 \leq \alpha \leq 1$, there is a measurable subset $E(\alpha)$ of A with $\alpha \int_A f(t)\,dt = \int_{E(\alpha)} f(t)\,dt$.

Proof. The proof is by induction on n. We give here the general induction step. The proof of the initial step $n = 1$ is the same and is left to the reader. Let us consider the subset X of $L_\infty(A)$ made up of all real valued functions $\rho(t)$, $t \in A$, $0 \le \rho(t) \le 1$. Then

$$T\rho = \int_A f(t)\rho(t)\,dt$$

defines a map $T: X \to R^n$ from X to R^n. We take in X the weak topology of $L_\infty(A)$, and in R^n the usual topology. Then, if $a = Tw$, $T^{-1}a$ is a nonempty subset of X. This set $T^{-1}a$ is convex. Indeed, if $\rho_1, \rho_2 \in T^{-1}a$, or $a = \int_A f\rho_1\,dt$, $a = \int_A f\rho_2\,dt$, $0 \le \rho_1(t)$, $\rho_2(t) \le 1$, then for $\rho(t) = \alpha\rho_1 + (1-\alpha)\rho_2$ we have $0 \le \rho(t) \le 1$, and $\int_A f\rho\,dt = a$. The set $T^{-1}a$ is weakly compact in $L_\infty(A)$. Indeed, let $[\rho_k]$ be a sequence of elements of $T^{-1}a$, then $0 \le \rho_k(t) \le 1$ for all $t \in A$ and k; hence, there is a subsequence, say still $[k]$, such that $\rho_k \to \rho$ weakly in $L_\infty(G)$ for some element ρ of $L_\infty(G)$. Necessarily $0 \le \rho(t) \le 1$, and $\int_A f\rho_k\,dt = a$ implies, by the weak convergence, that $\int_A f\rho\,dt = a$. Thus, $T^{-1}a$ is a compact convex subset of $L_\infty(G)$, and by (8.4.vii), $T^{-1}a$ possesses at least one extreme point $\theta(t)$, $t \in A$, $0 \le \theta(t) \le 1$. Let us prove that θ has values 0 and 1 almost everywhere in A. Suppose this is not the case; then there is some $\varepsilon > 0$ and a measurable subset E of A of positive measure such that $\varepsilon \le \theta(t) \le 1 - \varepsilon$ in E. Let E_1, E_2 be any decomposition of E into two subsets E_1, E_2 both of positive measure, $E_1 \cap E_2 = \emptyset$, $E_1 \cup E_2 = E$. By the induction hypothesis there are measurable subsets $F_1 \subset E_1$, $F_2 \subset E_2$ such that

$$2^{-1}\int_{E_1} f_i\,dt = \int_{F_1} f_i\,dt, \quad 2^{-1}\int_{E_2} f_i\,dt = \int_{F_2} f_i\,dt, \qquad i = 1, \dots, n-1.$$

Let $h_1(t)$, $t \in A$, be the function defined by taking $h_1 = 1$ in F_1, $h_1 = -1$ in $E_1 - F_1$, $h_1 = 0$ otherwise, and let $h_2(t)$, $t \in A$, be defined analogously. Then

$$\int_{E_1} f_i h_1\,dt = \int_{F_1} f_i\,dt - \int_{E_1 - F_1} f_i\,dt = 2\int_{F_1} f_i\,dt - \int_{E_1} f_i\,dt = 0,$$

Similarly,

$$\int_{E_2} f_i h_2\,dt = 0, \qquad i = 1, \dots, n-1.$$

Also, since $E_1 \cap E_2 = \emptyset$ and $h_j = 0$ in $A - E_j$, we have

$$\int_{E_2} f_i h_1\,dt = 0, \quad \int_{E_1} f_i h_2\,dt = 0, \qquad i = 1, \dots, n-1.$$

Now there are numbers α, β, not both zero, $|\alpha| \le \varepsilon$, $|\beta| \le \varepsilon$, such that, if we take $h(t) = \alpha h_1(t) + \beta h_2(t)$, $t \in A$, we have

$$\int_E f_n h\,dt = \alpha\int_{E_1} f_n h_1\,dt + \beta\int_{E_2} f_n h_2\,dt = 0.$$

Moreover, $h(t) = 0$ in $A - E$, $|h(t)| \le \varepsilon$ in E, and we still have

$$\int_E f_i h\,dt = 0, \qquad i = 1, \dots, n-1.$$

Thus,

$$\int_A f(\theta \pm h)\,dt = \int_A f\theta\,dt \pm \int_A fh\,dt = a,$$

and both functions $\theta \pm h$ have values between 0 and 1, that is, belong to $T^{-1}(a)$. Since θ is the middle point of the segment between $\theta + h$ and $\theta - h$, θ is not an extreme point of $T^{-1}a$, a contradiction. We have proved that θ has the only values 0 and 1, a.e. in A; hence, θ is the characteristic function of a set F for which (16.1.1) holds, and (16.1.i) is proved. □

16.1.ii. *Given any vector function $f(t) = (f_1, \dots, f_n)$, $t \in A$, whose components are L-integrable in A, two fixed subsets E and F of A, and any α, $0 \le \alpha \le 1$, then there is a* *measurable subset* $C(\alpha)$ *of* $E \cup F$ *with* $C(0) = E$, $C(1) = F$, *such that*

$$\int_{C(\alpha)} f\,dt = (1-\alpha)\int_E f\,dt + \alpha\int_F f\,dt.$$

Proof. Let us apply (16.1.i) to the two disjoints sets $E - F$ and $F - E$, and let $C' \subset E - F$, $C'' \subset F - E$ denote the sets such that

$$\int_{C'} f\,dt = \alpha\int_{E-F} f\,dt, \qquad \int_{C''} f\,dt = \alpha\int_{F-E} f\,dt.$$

Then, for $C = C(\alpha) = (E - F - C') \cup C'' \cup (E \cap F)$, we have

$$\begin{aligned}\int_C f\,dt &= \int_{E-F} f\,dt - \alpha\int_{E-F} f\,dt + \alpha\int_{E\cap F} f\,dt + (1-\alpha)\int_{E\cap F} f\,dt + \alpha\int_{F-E} f\,dt \\ &= (1-\alpha)\left(\int_{E-F} + \int_{E\cap F}\right) f\,dt + \alpha\left(\int_{F-E} + \int_{E\cap F}\right) f\,dt \\ &= (1-\alpha)\int_E f\,dt + \alpha\int_F f\,dt.\end{aligned}$$

□

16.1.iii Theorem (Lyapunov [1]). *Given any vector function* $f(t) = (f_1, \ldots, f_n)$, $t \in A$, *whose components are L-integrable on A, then*

$$\mu(E) = \int_E f(t)\,dt$$

describes a convex subset H of R^n *as E describes all measurable subsets of A.*

Proof. If μ_1, μ_2 are any two points of H, then there are measurable subsets E_1, E_2 of A such that $\mu_1 = \int_{E_1} f\,dt$, $\mu_2 = \int_{E_2} f\,dt$, and by (16.1.ii), for every α, $0 \le \alpha \le 1$, there is some measurable set $C = C(\alpha) \subset E_1 \cup E_2$ such that

$$\mu = \int_C f\,dt = (1-\alpha)\int_{E_1} f\,dt + \alpha\int_{E_2} f\,dt = (1-\alpha)\mu_1 + \alpha\mu_2.$$

□

This proves that E is convex. Note that the set H is certainly bounded, since $|\mu| \le \int_A |f|\,dt$. We shall prove in (16.1.v) that H is closed, and thus H is a compact convex subset of R^n.

16.1.iv. *Let* $f(t) = (f_1, \ldots, f_n)$, $g(t) = (g_1, \ldots, g_n)$, $t \in A$, *be any two vector functions, whose components are L-integrable on A. Then, for every measurable subset E of A take* $F = A - E$ *and define a function* $h_E(t)$, $t \in A$, *by taking* $h_E(t) = f(t)$ *for* $t \in E$, $h_E(t) = g(t)$ *for* $t \in F$. *Then*

$$\mu(E) = \int_A h_E\,dt = \int_E f\,dt + \int_F g\,dt$$

describes a convex subset H of R^n *as E describes all measurable subsets of A.*

Proof. Here we have

$$\mu(E) = \int_A h_E\,dt = \int_E f\,dt + \int_F g\,dt = \int_E (f-g)\,dt + \int_A g\,dt.$$

By (16.1.iii), $\int_E (f-g)\,dt$ describes a convex subset H' of R^n as E describes all measurable subsets of A. Thus, our set H is the translation of H' in R^n by the displacement $\int_A g\,dt$.

□

Let $f^{(j)}(t) = (f_1^{(j)}, \ldots, f_n^{(j)})$, $t \in A$, $j = 1, \ldots, h$, be given vector functions whose components are all L-integrable on A. Let us consider arbitrary decompositions $E_1, \ldots, E_h$

of A into disjoint measurable subsets, $E_u \cap E_v = \emptyset$, $u, v = 1, \ldots, h$, $u \neq v$, $\bigcup_u E_u = A$. Then

$$\mu = \mu(E_1, \ldots, E_h) = \int_{E_1} f^{(1)}\,dt + \cdots + \int_{E_h} f^{(h)}\,dt$$

describes a set H of R^n when $E_1, \ldots, E_h$ describe all possible decompositions of A into measurable subsets E_j of A, $j = 1, \ldots, h$.

Analogously, let us consider arbitrary measurable weight functions $p_j(t)$, $t \in A$, with $0 \leq p_j(t) \leq 1$, $j = 1, \ldots, h$, and $p_1(t) + \cdots + p_h(t) = 1$. Then

$$\nu = \nu(p_1, \ldots, p_h) = \int_A p_1 f^{(1)}\,dt + \cdots + \int_A p_h f^{(h)}\,dt$$

describes a set K of R_n when $p_1, \ldots, p_h$ describe all possible systems of measurable functions $p_j(t)$, $t \in A$, $0 \leq p_j(t) \leq 1$, $j = 1, \ldots, h$, $\sum_j p_j(t) = 1$.

16.1.v (THEOREM). *The subsets H and K of R^n are convex and compact, and $H = K$.*

Proof (a) H is convex. For $h = 2$ this was proved in (16.1.iv). Here is a proof for any $h \geq 2$. If $\mu_1, \mu_2 \in H$, then there are corresponding decompositions $E_1, \ldots, E_h$ and $F_1, \ldots, F_h$ of A such that

$$\mu_1 = \sum_j \int_{E_j} f^{(j)}\,dt, \qquad \mu_2 = \sum_j \int_{F_j} f^{(j)}\,dt,$$

with $E_j \cap E_k = \emptyset$, $F_j \cap F_k = \emptyset$, $j \neq k$, $\bigcup_j E_j = A$, $\bigcup_k F_k = A$. Let us consider the decomposition of A into the h^2 disjoint sets $A_{jk} = E_j \cap F_k$, $j, k = 1, \ldots h$. Given α, $0 \leq \alpha \leq 1$, let us apply (16.1.iv) to each set A_{jk}, $j \neq k$, and the two vector functions $f^{(j)}$ and $f^{(k)}$. Then there is a decomposition A'_{jk}, A''_{jk} of A_{jk} into measurable parts such that $A'_{jk} \cap A''_{jk} = \emptyset$, $A'_{jk} \cup A''_{jk} = A_{jk}$, and such that

$$\int_{A'_{jk}} f^{(j)}\,dt + \int_{A''_{jk}} f^{(k)}\,dt = (1 - \alpha)\int_{A_{jk}} f^{(j)}\,dt + \alpha \int_{A_{jk}} f^{(k)}\,dt, \qquad j \neq k.$$

Let us take now

$$G_j = A_{jj} \cup \Big(\bigcup_{k \neq j} A'_{jk}\Big) \cup \Big(\bigcup_{k \neq j} A''_{kj}\Big), \qquad j = 1, \ldots, h.$$

Then the h sets G_j, $j = 1, \ldots, h$, form a decomposition of A into disjoint measurable sets. We have now

$$\begin{aligned}
\sum_j \int_{G_j} f^{(j)}\,dt &= \sum_j \int_{A_{jj}} f^{(j)}\,dt + \sum_{j,k,j\neq k} \int_{A'_{jk}} f^{(j)}\,dt + \sum_{j,k,j\neq k} \int_{A''_{kj}} f^{(j)}\,dt \\
&= \sum_j \int_{A_{jj}} f^{(j)}\,dt + \sum_{j,k,j\neq k} \int_{A'_{jk}} f^{(j)}\,dt + \sum_{j,k,j\neq k} \int_{A''_{jk}} f^{(k)}\,dt \\
&= \sum_j \int_{A_{jj}} f^{(j)}\,dt + (1 - \alpha) \sum_{j,k,j\neq k} \int_{A_{jk}} f^{(j)}\,dt + \alpha \sum_{j,k,j\neq k} \int_{A_{jk}} f^{(k)}\,dt \\
&= (1 - \alpha) \sum_j \Big[\int_{A_{jj}} f^{(j)}\,dt + \sum_{k \neq j} \int_{A_{jk}} f^{(j)}\,dt\Big] \\
&\quad + \alpha \sum_j \Big[\int_{A_{jj}} f^{(j)}\,dt + \sum_{k \neq j} \int_{A_{kj}} f^{(j)}\,dt\Big] \\
&= (1 - \alpha) \sum_j \int_{E_j} f^{(j)}\,dt + \alpha \sum_j \int_{F_j} f^{(j)}\,dt = (1 - \alpha)\mu_1 + \alpha\mu_2.
\end{aligned}$$

(b) K is convex. Indeed, if $v_1, v_2 \in K$, $0 \leq \alpha \leq 1$, and $v = (1 - \alpha)v_1 + \alpha v_2$, where

$$v_1 = v(p_1, \ldots, p_h) = \sum_{j=1}^{h} \int_A p_j f^{(j)} \, dt,$$

$$v_2 = v(q_1, \ldots, q_h) = \sum_{j=1}^{h} \int_A q_j f^{(j)} \, dt,$$

then

$$v = (1 - \alpha)v_1 + \alpha v_2 = \sum_{j=1}^{h} \int_A ((1 - \alpha)p_j + \alpha q_j) f^{(j)} \, dt \in K.$$

(c) K is compact. First, K is bounded, since

$$|v| \leq \sum_{j=1}^{n} \int_A |f^{(j)}| \, dt = M$$

for all $v \in K$ and a fixed M. We have only to prove that K is closed in R^n. Let v_k, $k = 1, 2, \ldots$, be points of K with $v_k \to v \in R^n$ as $k \to \infty$, say

$$v_k = \sum_{j=1}^{h} \int_A p_{kj}(t) f^{(j)}(t) \, dt, \qquad k = 1, 2, \ldots,$$

$0 \leq p_{kj}(t) \leq 1$, $t \in A$, $j = 1, \ldots, h$, $\sum_j p_{kj}(t) = 1$. By weak compactness of the unit ball in $L_\infty(A)$, there are measurable functions $p_j(t)$, $t \in A$, $j = 1, \ldots, h$, $0 \leq p_j(t) \leq 1$, and a subsequence, say still $[k]$, such that $p_{kj} \to p_j$ weakly in $L_\infty(\alpha)$ as $k \to \infty$, $j = 1, \ldots, h$, and then $\sum_j p_j(t) = 1$, a.e. in A, and

$$v = \sum_{j=1}^{h} \int_A p_j(t) f^{(j)}(t) \, dt$$

because of the weak convergence in $L_\infty(A)$. Thus, $v \in K$ and K is closed.

(d) $H \subset K$. Indeed, if $\mu \in H$, then $\mu = \sum_j \int_{E_j} f^{(j)} \, dt$ for suitable subsets E_j of A, $E_j \cap E_s = \emptyset$, $j \neq s$, $\bigcup E_j = A$. We now define the functions $p_j(t)$, $t \in A$, by taking $p_j(t) = 1$, for $t \in E_j$, $p_j(t) = 0$ otherwise, $j = 1, \ldots, h$. Then $0 \leq p_j(t) \leq 1$, $\sum_j p_j(t) = 1$, $p = (p_1, \ldots, p_h)$ is a system of weights for the set K, and

$$v = \sum_j \int_{E_j} f^{(j)} \, dt = \sum_j \int_A p_j f^{(j)} \, dt \in K.$$

(e) extr $K \subset H$. Let $v = v(p_1, \ldots, p_h)$ be any extreme point of K. Let us prove that the functions p_j have the values zero and one almost everywhere in A. Indeed, in the contrary case, there would be an index i, a number $\varepsilon > 0$, and a measurable subset E of A with meas $E > 0$ and $\varepsilon \leq p_i(t) \leq 1 - \varepsilon$ in E. But then there has to be another index j having the analogous property on some subset of E and some other $\varepsilon > 0$. It is not restrictive to assume $i = 1, j = 2$, and thus there is some $\varepsilon > 0$ and a measurable subset, say E, of A with meas $E > 0$, $\varepsilon \leq p_1(t), p_2(t) \leq 1 - \varepsilon$ in E. Now we take $q(t) = (q_1, \ldots, q_h)$, $r(t) = (r_1, \ldots, r_h)$, $t \in A$, with $q_i = p_i = r_i$, $i = 3, \ldots, h$, in $A - E$, and $q_1 = p_1 - \varepsilon$, $q_2 = p_2 + \varepsilon$, $r_1 = p_1 + \varepsilon$, $r_2 = p_2 - \varepsilon$ in E, $\varepsilon > 0$. Then $2^{-1}(q + r) = p$ everywhere in A, $0 \leq q_j \leq 1$, $0 \leq r_j \leq 1$, $\sum_j q_j = 1$, $\sum_j r_j = 1$. Then

$$v_1 = \sum_j \int_A q_j f^{(j)} \, dt, \qquad v_2 = \sum_j \int_A r_j f^{(j)} \, dt$$

are points of K with $v = 2^{-1}(v_1 + v_2)$, a contradiction, since v is an extreme point of A.

(f) $H = K$. Indeed, $H \subset K$, extr $K \subset H$, $H = \text{co } H \subset \text{co } K \subset \text{co(extr } K) \subset \text{co } H$; hence, equality must hold in this relation, or $H = \text{co } H = \text{co } K = \text{co(extr } K) = K$. Statement (16.1.v) is thereby proved. □

16.2 The Neustadt Theorem for Mayer Problems with Bounded Controls

We are concerned here with the problem of the absolute minimum of the functional

$$(16.2.1) \qquad I[x, u] = g(t_1, x(t_1), t_2, x(t_2)),$$

with differential system, boundary conditions, and constraints

$$(16.2.2) \quad dx/dt = f(t, x, u) = D(t)x(t) + C(t, u(t)), \qquad t_0 \le t_1 \le t \le t_2 \le T,$$

and

$$(16.2.3) \quad \begin{array}{ll} (t_1, x(t_1), t_2, x(t_2)) \in B, \quad u(t) \in U(t), & t_1 \le t \le t_2 \text{(a.e.)}, \\ x = (x^1, \ldots, x^n), \quad u = (u^1, \ldots, u^m), & B \subset R^{2n+2}, \quad U(t) \subset R^m. \end{array}$$

Let M_0 denote the set of all (t, u) with $t_0 \le t \le T$, $u \in U(t)$, and assume M_0 to be compact and $C(t, u) = (C_1, \ldots, C_n)$ a given continuous function on M_0. Here $D(t) = [d_{ij}(t)]$ is a given $n \times n$ matrix with entries $d_{ij}(t)$ continuous on $[t_0, T]$. We state and prove here an existence theorem with no convexity assumptions for systems with bounded controls.

16.2.i (Neustadt's Existence Theorem for Mayer Problems). *Let $U(t)$ depend only on t, let M_0 be compact and B closed, let $D(t)$ be continuous on $[t_0, T]$, let g be continuous on B, and let $C(t, u)$ continuous on M_0. Let P be some compact subset of $A = [t_0, T] \times R^n$, and let Ω be the class of all admissible pairs $x(t), u(t), t_1 \le t \le t_2$, such that every trajectory x has at least one point on the compact set P, or $(t^*, x(t^*)) \in P$, where t^* may depend on the trajectory. Then the functional* (16.2.1) *has an absolute minimum in Ω.*

Proof. Let $i = \inf I[x, u]$. First, instead of problem (16.2.1–3) we consider the problem for generalized solutions

$$(16.2.4) \quad \begin{gathered} I[y, p, v] = g(t_1, y(t_1), t_2, y(t_2)), \\ dy/dt = D(t)y(t) + \sum_{j=1}^{\nu} p_j(t)C(t, u^{(j)}(t)), \qquad y(t) = (y^1, \ldots, y^n), \\ (t_1, y(t_1), t_2, y(t_2)) \in B, \qquad u^{(j)}(t) \in U(t), \qquad p(t) \in \Gamma, \\ t \in [t_1, t_2] \text{ (a.e.)}, \end{gathered}$$

where $v = (u^{(1)}, \ldots, u^{(v)})$, $p = (p_1, \ldots, p_v)$, $0 \le p_j(t) \le 1$, $t \in [t_1, t_2]$, $j = 1, \ldots, v$, $\sum_j p_j(t) = 1$, where, by Caratheodory's theorem (8.4.iii) we may well take $v = n + 1$, and where Γ is the usual fundamental simplex $p_j \ge 0, j = 1, \ldots, v$, $\sum_j p_j = 1$. Here the sets Q are replaced by the subsets $Q^*(t, y)$ in R^n defined by

$$Q^*(t, y) = \left[z \in R^n \,\middle|\, z = D(t)y + \sum_{j=1}^{v} p_j C(t, u^{(j)}),\ p \in \Gamma,\ u^{(j)} \in U(t) \right].$$

If Ω^* denotes the class of all generalized systems y, p, v such that each trajectory y has at least one point on P, or $(t^*, y(t^*)) \in P$, then $\Omega^* \supset \Omega$, and if $j = \inf I[y, p, v]$ in Ω^*, then $j \le i$.

Here A is the set $A = [(t, y) \mid t_0 \le t \le T, y \in R^n]$, and for $t \in [t_0, T]$, $y \in R^n$, the sets $Q^*(t, y)$ are convex and compact. Actually, they are the same set $Q_0^*(t) = [z = \sum_j p_j C(t, u^{(j)})]$ but for the displacement $z_0 = A(t)y$ which depends continuously on t and y. The set $A = [t_0, T] \times R^n$ is closed, not compact, and actually a slab in R^{1+n}. Here M_0 is replaced by the set M_0^* of all (t, p, v) with $t \in [t_0, T]$, $p \in \Gamma$, $v \in (U(t))^v$; hence M_0^* is compact, and $\sum_j p_j C(t, u^{(j)})$ is bounded on M_0^*, say $|\sum_j p_j C(t, u^{(j)})| \le L$. If b is a common bound for the entries $d_{ij}(t)$ of D then the second members f_i of equations (16.2.4) satisfy

$$\left| \sum_i y_i f_i \right| = \left| y^* D(t) y + \left(\sum_j p_j C(t, u^{(j)}(t)) \right)^* y \right| \le n^2 |y|^2 b + nL|y| \le N(|y|^2 + 1)$$

for a suitable constant N. By Filippov's existence theorem (9.2.i) and conditions (a), (b), (c) of Section 9.4 we conclude that problem (16.2.4) has an optimal solution $y(t)$, $p(t)$, $v(t)$. If $\Phi(t^*, t)$ denotes the fundamental $n \times n$ matrix of the homogeneous system $y' = Dy$ with $\Phi(t^*, t^*) = I$, the identity matrix, then y is given by

$$y(t) = \Phi(t^*, t) \left[y(t^*) + \int_{t^*}^{t} \Phi^{-1}(t^*, \alpha) \sum_{j=1}^{v} p_j(\alpha) C(\alpha, u^{(j)}(\alpha))\, d\alpha \right]$$

for $t^* \le t \le t_2$, and an analogous relation holds in $[t_1, t^*]$. In particular, for $t = t_2$.

$$\text{(16.2.5)} \quad y(t_2) = \Phi(t^*, t_2) \left[y(t^*) + \int_{t^*}^{t_2} \Phi^{-1}(t^*, \alpha) \sum_{j=1}^{v} p_j(\alpha) C(\alpha, u^{(j)}(\alpha))\, d\alpha \right]$$

and analogously for $t = t_1$.

Now, by force of (16.1.v) there is a finite decomposition of $[t^*, t_2]$ into measurable subsets $F_1, \ldots, F_v$, such that the integral in (16.2.5) is equal to the corresponding sum of integrals on $F_1, \ldots, F_v$ of the functions $\Phi^{-1}(t^*, \alpha) C(\alpha, u^{(j)}(\alpha))$, $j = 1, \ldots, v$, respectively, or

$$y(t_2) = \Phi(t^*, t_2) \left[y(t^*) + \sum_{j=1}^{v} \int_{F_j} \Phi^{-1}(t^*, \alpha) C(\alpha, u^{(j)}(\alpha))\, d\alpha \right]$$

and there is an analogous decomposition $E_1, \ldots, E_v$ of $[t_1, t^*]$.

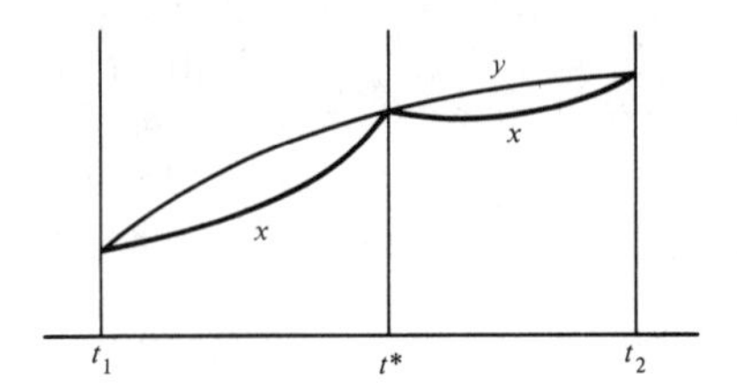

If we now take $\bar{u}(t) = u^{(j)}(t)$ for $t \in F_j$ in $[t^*, t_2]$, and $\bar{u}(t) = u^{(j)}(t)$ for $t \in E_j$ in $[t_1, t^*]$, then we also have

$$y(t_2) = \Phi(t^*, t_2)\left[y(t^*) + \int_{t^*}^{t_2} \Phi^{-1}(t^*, \alpha)C(\alpha, \bar{u}(\alpha))\, d\alpha \right],$$

and an analogous relation holds in $[t_1, t^*]$. In other words, the solution $\bar{x}(t)$, $t_1 \le t \le t_2$, of the system $\bar{x}'(t) = D(t)\bar{x} + C(t, \bar{u}(t))$ with $\bar{x}(t^*) = y(t^*)$ has values at $t = t_2$ and $t = t_1$ given by $\bar{x}(t_1) = y(t_1)$, $\bar{x}(t_2) = y(t_2)$. Then the functional $I[x, u]$ for $x = \bar{x}$, $u = \bar{u}$ has the value $j = I[y, p, v]$, the minimum of $I[y, p, v]$ in Ω^*. Then $i \le I[\bar{x}, \bar{u}] = j = I[y, p, v] \le i$, and hence $i = I[\bar{x}, \bar{u}] = I[y, p, v] = j$. We have proved (16.2.i). □

Remark. In Neustadt's theorem (16.2.i) as stated, the set of the tx-space we usually denote by A is $A = [t_0, T] \times R^n$, and thus is a slab of the tx-space. The usual conditions (a)–(c) of Section 9.4 for the present situation are already incorporated in the theorem, in particular, (b) is there explicitly, and (c), and even (c_0), are certainly satisfied, as mentioned in the proof.

In case $[t_0, T]$ in (16.2.i) is replaced by $(-\infty, +\infty)$, then Theorem (16.2.i) still holds, provided we now require: (d*) For every N, the set M_{0N} of all (t, u) with $|t| \le N$, $u \in U(t)$, is compact; and (d) as stated in Section 9.4. We may replace condition (d) by the weaker requirement (d′).

16.3 The Bang-Bang Theorem

We are concerned here with the problem of the absolute minimum for the functional, differential system, and constraints

$$I[x, u] = g(t_1, x(t_1), t_2, x(t_2)), \tag{16.3.1}$$

$$dx/dt = D(t)x(t) + C(t)u(t), \qquad t_0 \le t_1 \le t \le t_2 \le T, \tag{16.3.2}$$

$$(t_1, x(t_1), t_2, x(t_2)) \in B, \qquad u(t) \in U, \tag{16.3.3}$$

with D, B, g as in Section 16.2, where now U is a fixed subset of R^m, and $C(t) = [c_{ij}(t)]$ is an $n \times m$ matrix with entries continuous on $[t_0, T]$. The existence theorem (16.2.i) holds without change. However, a slight reinforcement in the conditions will guarantee that there always are optimal solutions with control $u(t)$ whose values are in the extreme set of the convex set U. These solutions are usually called "bang-bang".

16.3.i (THEOREM: EXISTENCE OF BANG-BANG SOLUTIONS). *Let U be a fixed compact set in R^m, let B be closed, let $D(t)$, $C(t)$ be continuous on $[t_0, T]$, and let g be continuous on B. Let P be some compact subset of $A = [t_0, T] \times R^n$, let Ω denote the class of all admissible pairs $x(t)$, $u(t)$, $t_1 \le t \le t_2$, such that every trajectory x in Ω has at least one point on P, or $(t^*, x(t^*)) \in P$. Let us assume that Ω is nonempty. Then the functional* (16.3.1) *with constraints* (16.3.2), (16.3.3) *has an absolute minimum in Ω. Moreover, there always is an optimal solution $x(t)$, $u(t)$, $t_1 \le t \le t_2$, with $u(t) \in \text{extr co } U$ (a bang-bang optimal solution).*

As we showed in Section 4.2, Remark 7, by an example, there may well be optimal solutions which are not bang-bang. However, if the optimal solution is unique, it must be bang-bang (under the conditions of (16.3.i)).

Proof. We know from (8.4.viii) that extr co $U \subset U$. We prove the theorem under the restriction that extr co U is closed (cf. Remark below). By (16.2.i) we know that problem (16.3.1–3) has some optimal solution $x(t)$, $u(t)$, $t_1 \le t \le t_2$; hence $u(t) \in U$, $t \in [t_1, t_2]$, $x'(t) = D(t)x(t) + C(t)u(t)$. Now by (8.4.iii), (8.4.vii) and (8.4.viii), every $u(t) \in U$ certainly is a convex combination

$$u(t) = \sum_{j=1}^{\mu} q_j(t)u^{(j)}(t), \qquad u^{(j)}(t) \in \text{extr co } U \subset U,$$

with $0 \le q_j(t) \le 1, j = 1, \ldots, \mu, \sum_j q_j(t) = 1$, and we can take $\mu = m + 1$, but we have no guarantee that the functions q_j, $u^{(j)}$ are measurable. However, if we take the spaces $I = [t_1, t_2]$, $M = I \times \Gamma \times (\text{extr co } U)^\mu$, $N = I \times R^m$, and we denote by $f: M \to N$ the continuous map $(t, q, v) \to (t, \sum q_j u^{(j)})$ and by $\rho: I \to N$ the measurable map $t \to (t, u(t))$, then by (8.2.iii) there is a measurable map $\psi: t \to (t, q(t), v(t))$ with $q(t) = (q_1, \ldots, q_\mu) \in \Gamma$, $v(t) = (u^{(1)}, \ldots, u^{(\mu)})$, $u^{(j)}(t) \in U$, $t \in I, j = 1, \ldots, \mu$. Now the functions q_j are measurable, we have

$$x'(t) = D(t)x(t) + C(t) \sum_{j=1}^{\mu} q_j(t)u^{(j)}(t),$$

$$x(t_2) = \Phi(t^*, t_2)\left[x(t^*) + \int_{t^*}^{t_2} \Phi^{-1}(t^*, \alpha)C(\alpha) \sum_{j=1}^{\mu} q_j(\alpha)u^{(j)}(\alpha)\, d\alpha \right],$$

and an analogous relation holds for the interval $[t_1, t^*]$. By (16.1.v) there is a decomposition of $[t^*, t_2]$ into disjoint measurable sets $F_1, \ldots, F_\mu$ such that the integral above is equal to a sum of the integrals over F_j of the functions

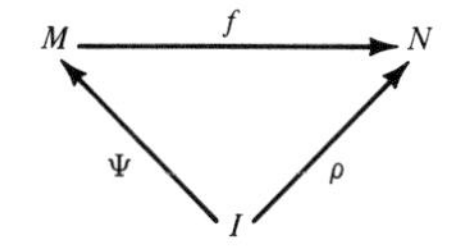

$\Phi^{-1}(t^*, \alpha)C(\alpha)u^{(j)}(\alpha)$, $j = 1, \ldots, \mu$, or

$$x(t_2) = \Phi(t^*, t_2)\left[x(t^*) + \sum_{j=1}^{\mu} \int_{F_j} \Phi^{-1}(t^*, \alpha)C(\alpha)u^{(j)}(\alpha)\, d\alpha\right].$$

There is an analogous decomposition $E_1, \ldots, E_\mu$ of $[t_1, t^*]$. If now we take $\bar{u}(t) = u^{(j)}(t)$ for $t \in F_j$ and for $t \in E_j$, we also have

$$x(t_2) = \Phi(t^*, t_2)\left[x(t^*) + \int_{t^*}^{t_2} \Phi^{-1}(t^*, \alpha)C(\alpha)\bar{u}(\alpha)\, d\alpha\right],$$

and an analogous relation holds in $[t_1, t^*]$.

In other words, the solution $\bar{x}(t)$, $t_1 \leq t \leq t_2$, of the differential system $x' = D(t)x + C(t)\bar{u}(t)$ with initial value $\bar{x}(t^*) = x(t^*)$, has values at $t = t_2$ and $t = t_1$, given by $\bar{x}(t_1) = x(t_1)$, $\bar{x}(t_2) = x(t_2)$. Then $\bar{x}(t)$, $\bar{u}(t)$, $t_1 \leq t \leq t_2$, is also optimal for the given problem (16.3.1–3), and $\bar{u}(t)$ has its values in extr co U. Theorem (16.3.i) is thereby proved. □

Examples. Most of the problems listed in Section 6.1 are linear problems of the type considered here, with U compact and convex, and extr U compact. The optimal solutions determined there are bang-bang. The absolute minimum of the functional, and the bang-bang optimal solutions determined there, are not modified if we replace the compact convex set U by the compact nonconvex set extr U.

Exercise. It is left as an exercise for the reader to prove directly the statement (16.3.i) with extr U replaced by bd U. Here bd U is certainly closed, and there is no need of the requirement concerning extr U.

Remark. Theorems (16.2.i), (16.3.i) will be extended in Section 16.5 to situations where U may be unbounded, and M_0 only closed. Statement (16.3.i) holds even without the requirement extr co U closed. This extension calls for the use of selection theorems for measurable set valued functions as (8.3.iv) (cf. second proof of (8.3.vi)).

16.4 The Neustadt Theorem for Lagrange and Bolza Problems with Bounded Controls

We are concerned here with the problem of the absolute minimum of the functional

(16.4.1) $$I[x, u] = g(t_1, x(t_1), t_2, x(t_2)) + \int_{t_1}^{t_2} [d(t)x(t) + c(t, u(t))]\, dt,$$

with differential system, boundary conditions, and constraints

(16.4.2) $$dx/dt = D(t)x(t) + C(t, u(t)), \qquad t_0 \leq t_1 \leq t \leq t_2 \leq T,$$

and

$$(16.4.3)\qquad \begin{aligned} &(t_1, x(t_1), t_2, x(t_2)) \in B, \qquad u(t) \in U(t), \quad t_1 \leq t \leq t_2 \text{ (a.e.)}, \\ &x = (x^1, \ldots, x^n), \quad u = (u^1, \ldots, u^m), \quad B \subset R^{2n+2}, \quad U(t) \subset R^m. \end{aligned}$$

Let M_0 denote the set of all (t, u) with $t_0 \leq t \leq T$, $u \in U(t)$, let us assume that M_0 is compact and that $C(t, u) = (C_1, \ldots, C_n)$ and $c(t, u)$ (scalar) are continuous on M_0. Here $D(t) = [d_{ij}(t)]$ is an $n \times n$ matrix, $d(t) = \text{row}(d_1, \ldots, d_n)$ is a $1 \times n$ matrix, and we assume that all entries $d_{ij}(t)$, $d_i(t)$ are continuous on $[t_0, T]$.

16.4.i. *Let M_0 be compact and B closed, let $D(t)$, $d(t)$ be continuous on $[t_0, T]$, let g be continuous on B, and let $C(t, u)$, $c(t, u)$ be continuous on M_0. Let P be some compact subset of $A = [t_0, T] \times R^n$, and let Ω be the class of all admissible pairs $x(t)$, $u(t)$, $t_1 \leq t \leq t_2$, such that every trajectory x has at least one point on the compact set P, or $(t^*, x(t^*)) \in P$. Then the functional* (16.4.1) *with constraints* (16.4.2), (16.4.3) *has an absolute minimum in Ω.*

This statement can be proved by reducing the problem to a Mayer problem as we have done for (9.3.i) and then applying Neustadt's theorem (16.2.i). The details of the proof are left as an exercise for the reader. If $[t_0, T]$ is replaced by $(-\infty, +\infty)$, then (16.4.i) still holds provided additional conditions are satisfied, as, for instance, those in Section 9.5.

Let us consider now the problem of the absolute minimum for the functional, differential equations, and constraints

$$(16.4.4)\qquad I[x, u] = g(t_1, x(t_1), t_2, x(t_2)) + \int_{t_1}^{t_2} [d(t)x(t) + c(t)u(t)]\, dt,$$

$$(16.4.5)\qquad dx/dt = D(t)x(t) + C(t)u(t), \qquad t_0 \leq t_1 \leq t \leq t_2 \leq T,$$

$$(16.4.6)\qquad (t_1, x(t_1), t_2, x(t_2)) \in B, \qquad u(t) \in U, \quad t \in [t_1, t_2] \text{ (a.e.)},$$

where $D(t) = [d_{ij}]$, $C(t) = [c_{ij}]$, $d(t) = (d_1, \ldots, d_n)$, $c(t)$ are $n \times n$, $n \times m$, $1 \times n$, $1 \times m$ matrices respectively with entries defined on $[t_0, T]$; $g(t_1, x_1, t_2, x_2)$ is a given function on B; and U is a fixed subset of R^m.

16.4.ii. *Let U be a fixed compact set in R^m, let B be a closed subset of R^{2+2n}, let $D(t)$, $C(t)$, $d(t)$, $c(t)$ be matrices with continuous entries on $[t_0, T]$, and let g be continuous on B. Let P be some compact subset of $A = [t_0, T] \times R^n$; let Ω denote the class of all admissible pairs $x(t)$, $u(t)$, $t_1 \leq t \leq t_2$, such that every trajectory x in Ω has at least one point on P, or $(t^*, x(t^*)) \in P$. Assume that Ω is nonempty. Then the functional* (16.4.4) *with constraints* (16.4.5), (16.4.6) *has an absolute minimum in Ω. Moreover, there always is an optimal solution $x(t)$, $u(t)$, $t_1 \leq t \leq t_2$, with $u(t) \in \text{extr co } U \subset U$ (a bang-bang optimal solution).*

This is a corollary of (16.3.i) and (16.4.i). The details of the proof are left to the reader. If $[t_0, T]$ is replaced by $(-\infty, +\infty)$, Theorem (16.4.ii) still holds under such additional conditions as those listed in Section 9.5.

16.5 The Case of Unbounded Controls

A. The Lagrange and Bolza Problems

We consider again problem (16.4.1–3):

$$(16.5.1) \quad I[x,u] = g(t_1, x(t_1), t_2, x(t_2)) + \int_{t_1}^{t_2} [d(t)x(t) + c(t,u(t))]\, dt,$$

$$(16.5.2) \quad dx/dt = D(t)x(t) + C(t,u(t)), \qquad t_0 \le t_1 \le t \le t_2 \le T,$$

with

$$(16.5.3) \quad \begin{gathered} (t_1, x(t_1), t_2, x(t_2)) \in B, \quad u(t) \in U(t),\ t_1 \le t \le t_2, \quad (t^*, x(t^*)) \in P, \\ x = (x^1, \dots, x^n), \quad u = (u^1, \dots, u^m), \qquad B \subset R^{2n+2}, \quad U(t) \subset R^m, \\ C(t,u) = (C_1, \dots, C_n), \qquad D(t) = [d_{ij}(t),\, i,j = 1, \dots, n]. \end{gathered}$$

Let $A = [t_0, T] \times R^n \subset R^{n+1}$, $M_0 = [(t,u) \mid t \in [t_0, T],\, u \in U(t)] \subset R^{m+1}$, and $f_0(t,x,u) = d(t)x + c(t,u)$, $f(t,x,u) = D(t)x + C(t,u)$.

16.5.i (An Existence Theorem with No Convexity Conditions and Unbounded Controls). *Let t_0, T be finite, let B be a closed subset of R^{2n+2}, let g be bounded below and lower semicontinuous on B, and let P be a compact subset of the tx-space R^{1+n}. Let us assume* (a) *that $d(t)$, $D(t)$, $c(t,u)$, $C(t,u)$ are given $1 \times n$, $n \times n$, 1×1, $n \times 1$ matrices with continuous entries in $[t_0, T]$ and M_0 respectively. Let $c(t,u)$ be nonnegative. Let us further assume* (b) *that given $\varepsilon > 0$ there is $N = N(\varepsilon) \ge 0$ such that $|C(t,u)| \le \varepsilon c(t,u)$ and $1 \le \varepsilon c(t,u)$ for all $t \in [t_0, T]$, $u \in U(t)$, $|u| \ge N$. Let Ω be the class, which we suppose nonempty, of all admissible pairs $x(t)$, $u(t)$, $t_1 \le t \le t_2$, whose trajectories x contain at least one point on P, namely $(t^*, x(t^*)) \in P$. Then the functional* (16.5.1) *under the constraints* (16.5.2–3) *has an absolute minimum in Ω.*

This theorem holds also under a set of alternate and much weaker hypotheses replacing (a) and (b). First let us assume that $d(t)$, $D(t)$ have entries which are measurable and bounded in $[t_0, T]$, and that $c(t,u)$, $C(t,u)$ satisfy condition (C′) of Section 11.4, that is, given $\varepsilon > 0$, there is a compact subset K of $[t_0, T]$ with meas $\{[t_0, T] - K\} < \varepsilon$, such that the set $M_K = [t \in K, u \in U(t)]$ is closed and $c(t,u)$, $C(t,u)$ are continuous in M_K. Secondly, we need to know that for almost all $\bar{t}$ the sets $\tilde{Q}(\bar{t}, x) = [(z^0, z) \mid z^0 \ge d(\bar{t})x + c(\bar{t}, u),\, z = D(\bar{t})x + C(\bar{t}, u),\, u \in U(\bar{t})]$, $x \in R^n$, have property (K) with respect to x in R^n. This certainly occurs if, in addition to what we have already assumed, we know that either $c(t,u) \to +\infty$, or $|C(t,u)| \to +\infty$ as $u \to +\infty$, $u \in U(\bar{t})$. Finally, we need to know that $c(t,u)$, $C(t,u)$ satisfy one of the growth conditions (g1′), (g2′), (g3′) of Section 11.4, i.e., in the present case,

(g1′): there is a scalar function $\phi(\zeta)$, $0 \le \zeta < +\infty$, bounded below, such that $\phi(\zeta)/\zeta \to +\infty$ as $\zeta \to +\infty$, and $\phi(|C(t,u)|) \le c(t,u)$ for all $(t,u) \in M$;

(g2′): Given $\varepsilon > 0$ there is a locally integrable scalar function $\psi_\varepsilon(t)$, which may depend on ε, such that $|C(t,u)| \leq \psi_\varepsilon(t) + \varepsilon c(t,u)$ for all $(t,u) \in M$;
(g3′): For every n-vector $p \in R^n$ there is a locally integrable scalar function $\phi_p(t) \geq 0$ such that $c(t,u) \geq (p, C(t,u)) - \phi_p(t)$ for all $(t,u) \in M$.

Proof. We prove the theorem under the alternate hypotheses only.

(a) Let $I_1 = g$, $I_2 = \int_{t_1}^{t_2} [dx + c]\, dt$, where I_1 is bounded below, say $I_1 \geq -M_1$. In search of the minimum of $I[x,u]$ we may limit ourselves to those admissible pairs $x(t)$, $u(t)$, $t_1 \leq t \leq t_2$, for which $I[x,u] \leq M$ for some constant M. Then

$$I_2[x,u] = \int_{t_1}^{t_2} [d(t)x(t) + c(t,u)]\, dt = I[x,u] - g \leq M_1 + M,$$

$$x(t) = \Phi(t^*,t)\left[x(t^*) + \int_{t^*}^{t} \Phi^{-1}(t^*,\alpha)C(\alpha,u(\alpha))\, d\alpha\right], \tag{16.5.4}$$

$$|x(t)| \leq \|\Phi(t^*,t)\| \left[|x(t^*)| + \int_{t^*}^{t} \|\Phi^{-1}(t^*,\alpha)\| \, |C(\alpha,u(\alpha)|\, d\alpha\right].$$

For suitable constants L_0, L, L', D we certainly have $t_2 - t_1, |t - t^*| \leq D$, $\|\Phi\| \leq L$, $\|\Phi^{-1}\| \leq L'$, $|x(t^*)| \leq L_0$, and then

$$|x(t)| \leq L\left[L_0 + L' \int_{t^*}^{t} |C(\alpha,u(\alpha))|\, d\alpha\right]. \tag{16.5.5}$$

Under hypothesis (g1′), ϕ is bounded below, say $\phi \geq -M_2$; hence $c(t,u) \geq -M_2$. Given $\varepsilon > 0$ there is $N = N(\varepsilon)$ such that $\phi(\zeta)/\zeta \geq 1/\varepsilon$ for $\zeta \geq N(\varepsilon)$, and then $|\phi(\zeta)| \leq \bar{M} = \bar{M}(\varepsilon)$ for $0 \leq \zeta \leq N(\varepsilon)$ and some other constant $\bar{M}(\varepsilon)$. If E', E'' denote the subsets of $[t_1,t_2]$ where $|C(t,u(t))| < N(\varepsilon)$ and $|C(t,u(t))| \geq N(\varepsilon)$ respectively, then $\zeta \leq \varepsilon\phi(\zeta)$ for $\zeta \geq N(\varepsilon)$ and

$$\begin{aligned} &|C(t,u(t))| \leq \varepsilon\phi(|C(t,u(t))|) \leq \varepsilon c(t,u(t)) \quad \text{for } t \in E'', \\ &|C(t,u(t))| \leq \bar{N} = N(\varepsilon) \quad \text{for } t \in E'. \end{aligned} \tag{16.5.6}$$

Thus, from (16.5.5) and (16.5.6), we also have

$$\begin{aligned} |x(t)| &\leq LL_0 + LL'\left(\int_{E'} + \int_{E''}\right)|C(t,u(t))|\, dt \\ &\leq LL_0 + LDL'\bar{N} + LL'\varepsilon \int_{t_1}^{t_2} |c(t,u(t))|\, dt \end{aligned} \tag{16.5.7}$$

and moreover, for $|d(t)| \leq l$,

$$\begin{aligned} \int_{t_1}^{t_2} |c(t,u(t))|\, dt &\leq \int_{t_1}^{t_2} [c(t,u(t)) + 2M_2]\, dt \\ &= I_2[x,u] - \int_{t_1}^{t_2} d(t)x(t)\, dt + 2M_2(t_2 - t_1) \\ &\leq M_1 + M + 2M_2 D + l \int_{t_1}^{t_2} |x(t)|\, dt \\ &\leq M_1 + M + 2M_2 D + l(LL_0 + LDL'\bar{N})D + lLDL'\varepsilon \int_{t_1}^{t_2} |c(t,u(t))|\, dt. \end{aligned}$$

By choosing $\varepsilon = \varepsilon_0$ so that $lLDL'\varepsilon_0 \leq \frac{1}{2}$, we have, for $\bar{N} = N(\varepsilon_0)$,

$$\int_{t_1}^{t_2} |c(t,u(t))|\, dt \leq 2[M_1 + M + 2M_2 D + l(LL_0 + LDL'\bar{N})D] = M_3,$$

where now M_3 is a fixed constant which depends only on M. Finally, from (16.5.7),

$$|x(t)| \leq LL_0 + LDL'\bar{N} + LL'M_3\varepsilon_0 = M_4,$$

another constant which depends only on M.

(b) Again under hypothesis (gl′), let us prove that the trajectories $x(t)$, $t_1 \leq t \leq t_2$, for which $I[x,u] \leq M$ are equiabsolutely continuous and their derivatives x' are equiabsolutely integrable. Indeed, let $\varepsilon > 0$ again be any positive number, and let us consider the constants $N(\varepsilon)$, $\bar{M}(\varepsilon)$ above, and the decomposition E', E'' of $[t_1, t_2]$ defined in (a). If E denotes any measurable subset of $[t_1, t_2]'$ we have, for $\|D(t)\| \leq L_1$,

$$\begin{aligned}\int_E |x'(t)|\,dt &= \int_E |D(t)x(t) + C(t, u(t))|\,dt \\ &\leq \int_E [\|D(t)\|\,|x(t)| + |C(t, u(t))|]\,dt \\ &\leq L_1 M_4 \text{ meas } E + \left(\int_{E\cap E'} + \int_{E\cap E''}\right)|C(t, u(t))|\,dt \\ &\leq [L_1 M_4 + N(\varepsilon)] \text{ meas } E + \varepsilon \int_{t_1}^{t_2} c(t, u(t))\,dt \\ &\leq [L_1 M_4 + N(\varepsilon)] \text{ meas } E + \varepsilon M_3.\end{aligned}$$

Given $\eta > 0$ we can determine $\varepsilon = \varepsilon_1$ so that $\varepsilon M_3 \leq \eta/2$, and then take > 0 so that $[L_1 M_3 + N(\varepsilon_1)]\sigma \leq \eta/2$. For meas $E \leq \sigma$ we then have $\int_E |x'(t)|\,dt \leq \eta/2 + \eta/2 = \eta$, and our claim under (b) is proved.

(c) Under hypothesis (g2′), given $\varepsilon > 0$ we have $|C(t, u)| \leq \psi_\varepsilon(t) + \varepsilon c(t, u)$ and hence $c(t, u) \geq -\psi_1(t)$, and from (16.5.5), and by defining $\bar{M} = \int_{t_0}^T \psi_1(t)\,dt$, $M(\varepsilon) = \int_{t_0}^T \psi_\varepsilon(t)\,dt$

$$|x(t)| \leq LL_0 + LL'M(\varepsilon) + LL'\varepsilon \textstyle\int_{t_1}^{t_2} |c(t, u(t))|\,dt,$$

$$\begin{aligned}\int_{t_1}^{t_2} |c(t, u(t))|\,dt &\leq \int_{t_1}^{t_2} [c(t, u(t)) + 2\psi_1(t)]\,dt \\ &\leq I_2[x, u] - \int_{t_1}^{t_2} d(t)x(t)\,dt + 2\bar{M} \\ &\leq M_0 + M_1 + 2\bar{M} + l \int_{t_1}^{t_2} |x(t)|\,dt \\ &\leq M_0 + M_1 + 2\bar{M} + lL(L_0 + L'M(\varepsilon))D \\ &\quad + lLDL'\varepsilon \int_{t_1}^{t_2} |c(t, u(t))|\,dt.\end{aligned}$$

As before, for $\varepsilon = \varepsilon_0$, $lLDL'\varepsilon_0 \leq \frac{1}{2}$, and we have

$$\int_{t_1}^{t_2} |c(t, u(t))|\,dt \leq M_3, \qquad |x(t)| \leq M_4,$$

where the constants M_3, M_4 depend only on M. Furthermore, again for any arbitrary $\varepsilon > 0$ we have

$$\begin{aligned}\int_E |x'(t)|\,dt &\leq \int_E [\|D(t)\|\,|x(t)| + |C(t, u(t))|]\,dt \\ &\leq L_1 M_4 \text{ meas } E + \int_E \psi_\varepsilon(t)\,dt + \varepsilon \int_{t_1}^{t_2} |c(t, u(t))|\,dt \\ &\leq L_1 M_4 \text{ meas } E + \int_E \psi_\varepsilon(t)\,dt + \varepsilon M_3.\end{aligned}$$

Given any $\eta > 0$ we first take $\varepsilon = \varepsilon_1$ so that $\varepsilon M_3 \le \eta/3$, then we take $\delta > 0$ so small that $L_1 M_4 \delta < \eta/3$ and such that, for meas $E \le \delta$, we also have $\int_E \psi_{\varepsilon_1}(t)\,dt \le \eta/3$. Then $\int_E |x'(t)|\,dt \le \eta$ for meas $E \le \delta$.

The same conclusions hold also under hypothesis (g3′) since we have seen that (g3′) is equivalent to (g2′) (Section 10.4).

(d) Before proceeding, we remark as in Section 10.3 that it is not restrictive to assume in Theorem (16.5.i) that $\phi(\zeta)$, $0 \le \zeta < +\infty$, is bounded below, monotone nondecreasing, and convex in $[0, +\infty)$. We shall now consider the problem of the generalized solutions $y(t)$, $p(t)$, $v(t)$, $t_1 \le t \le t_2$, corresponding to problem (16.5.1–3). That is, we consider the auxiliary problem

$$(16.5.8)\quad J[y,p,v] = g(t_1, y(t_1), t_2, y(t_2)) + \int_{t_1}^{t_2} \left[d(t)y(t) + \sum_{j=1}^{h} p_j(t)c(t, u^{(j)}(t)) \right] dt,$$

$$(16.5.9)\qquad dy/dt = D(t)y(t) + \sum_{j=1}^{h} p_j(t)C(t, u^{(j)}(t)), \qquad (t^*, y(t^*)) \in P,$$

$$(16.5.10)\quad (t_1, y(t_1), t_2, y(t_2)) \in B, \qquad u^{(j)}(t) \in U(t), \quad t \in [t_1, t_2],\ j = 1, \dots, h,$$

for some fixed h, say $h = n + 2$, where $y = (y^1, \dots, y^n)$, $p = (p_1, \dots, p_h)$, $p_j \ge 0$, $p_1 + \cdots + p_h = 1$, $v = (u^{(1)}, \dots, u^{(h)})$. Here p, v are new control variables, say $p \in \Gamma$, $v \in V(t) = (U(t))^h$. Let us note that, for

$$C(t,p,v) = \sum_{j=1}^{h} p_j C(t, u^{(j)}), \qquad c(t,p,v) = \sum_{j=1}^{h} p_j c(t, u^{(j)}),$$

we have, under hypothesis (g1′) and because of the monotonicity and convexity of ϕ,

$$\begin{aligned}\phi(|C(t,p,v)|) &= \phi\left(\left|\sum_j p_j C(t,u^{(j)})\right|\right) \le \phi\left(\sum_j p_j |C(t,u^{(j)})|\right)\\ &\le \sum_j p_j \phi(|C(t,u^{(j)})|) \le \sum_j p_j c(t,u^{(j)}) = c(t,p,v).\end{aligned}$$

Analogously, under hypothesis (g2′)

$$\begin{aligned}|C(t,p,v)| &= \left|\sum_j p_j C(t,u^{(j)})\right| \le \sum_j p_j |C(t,u^{(j)})|\\ &\le \sum_j p_j[\psi_\varepsilon(t) + \varepsilon c(t,u^{(j)})] = \psi_\varepsilon(t) + \varepsilon \sum_j p_j c(t,u^{(j)})\\ &= \psi_\varepsilon(t) + \varepsilon c(t,p,v);\end{aligned}$$

and under hypothesis (g3′) for $q \in R^n$,

$$\begin{aligned}(q, C(t,p,v)) &= \left(q, \sum_j p_j C(t,u^{(j)})\right) = \sum_j p_j (q, C(t,u^{(j)}))\\ &\le \sum_j p_j[c(t,u^{(j)}) + \phi_q(t)]\\ &= \phi_q(t) + \sum_j p_j c(t,u^{(j)})) = \phi_q(t) + c(t,p,v).\end{aligned}$$

In other words, system (16.5.8–10) has the same properties we have observed for system (16.5.1–3). Thus, $J(t,p,v) \le M$ implies $|y(t)| \le M_4$ and $\int_{t_1}^{t_2} |c(t,p(t),v(t))\,dt \le M_3$

for suitable constants which depend only on M. Moreover, the trajectories y are equibounded and equiabsolutely continuous, and their derivatives y' are equiabsolutely integrable. In other words, the class of trajectories y corresponding to $J \leq M$ is relatively compact (in the topology of the weak convergence of the derivatives).

(e) Let $f_0(t, y, p, v) = d(t)y + c(t, p, v)$, $f(t, y, p, v) = D(t)y + C(t, p, v)$. Under hypothesis (g1′) we take $\phi_1(\zeta) = \phi(\zeta - L_1 M_4) - lM_4$ for $\zeta > L_1 M_4$, and $\phi_1(\zeta) = \phi(0) - lM_3$ for $0 \leq \zeta \leq L_1 M_4$. Note that ϕ_1 is still convex. Now for $|y| \leq M_4$, $|d(t)| \leq l$, $\|D(t)\| \leq L_1$, we have

$$\begin{aligned}\phi_1(|f|) &= \phi_1(|D(t)y + C(t, p, v)|) \leq \phi_1(L_1 M_4 + |C(t, p, v)|) \\ &= \phi(|C(t, p, v)|) - lM_4 \leq c(t, p, v) - lM_4 = f_0 - d(t)y - lM_4 \leq f_0.\end{aligned}$$

In other words, there is a function $\phi_1(\zeta)$, bounded below, monotone nondecreasing, convex, with $\phi_1(\zeta)/\zeta \to +\infty$ as $\zeta \to +\infty$, such that $\phi_1(|f|) < f_0$ for all (t, p, v) with $|y| \leq M_4$.

Analogously, under hypothesis (g2′) we have, for $\bar{\psi}_\varepsilon(t) = (L_1 + l)M_4 + \psi_\varepsilon(t)$,

$$\begin{aligned}|f| &= |D(t)y + C(t, p, v)| \\ &\leq L_1 M_4 + \psi_\varepsilon(t) + \varepsilon c(t, p, v) = L_1 M_4 + \psi_\varepsilon(t) + \varepsilon[f_0 - d(t)y] \\ &\leq L_1 M_4 + \psi_\varepsilon(t) + lM_4 + \varepsilon f_0 = \bar{\psi}_\varepsilon(t) + \varepsilon f_0;\end{aligned}$$

and analogously under hypothesis (g3′) for $q \in R^n$, and $\bar{\phi}_q(t) = (L_1|q| + l)M_4 + \phi_q(t)$,

$$\begin{aligned}(q, f) &= (q, D(t)y + C(t, p, v)) = (q, D(t)y) + (q, C(t, p, v)) \\ &\leq L_1 M_4|q| + \phi_q(t) + c(t, p, v) = L_1 M_4|q| + \phi_q(t) + f_0 - d(t)y \\ &\leq L_1 M_4|q| + \phi_q(t) + lM_4 + f_0 = \bar{\phi}_q(t) + f_0.\end{aligned}$$

We see that f_0, f satisfy the growth condition (γ1), or (γ2), or (γ3) of Section 11.1. We are now in a position to apply the existence theorem (11.1.i) with $A = [(t, y) | t_0 \leq t \leq T,\ |y| \leq M_4]$, $M = [(t, y, p, v) | (t, y) \in A,\ p \in \Gamma,\ |v| \in V]$, and with the constraint $(t^*, y(t^*)) \in P$, and of course $e[y] \in B$. By (11.1.i) problem (16.5.8–10) has an absolute minimum $y(t)$, $p(t)$, $v(t)$, $t_1 \leq t \leq t_2$.

(f) Note that if i denotes the infimum of I, and j the infimum of J under the constraints, then $-\infty < j \leq i < +\infty$, $J([y, p, v] = j$. By repeating now the same argument at the end of the proof of (16.2.i) based on Lyapunov type theorems, we obtain an admissible pair x, u (usual solution), such that $I[x, u] = J[y, p, v]$. Hence, $i \leq I[x, u] = J[y, p, v] = j \leq i$, or $j = i$, and (16.5.i) is proved. □

Remark. Note that if there is a locally integrable function $\phi(t)$ such that $|C(t, u)| \leq \phi(t)$ for all $(t, u) \in M_0$, then the equiboundedness of the trajectories and the equiabsolute integrability of their derivatives follow directly from our argument in (a) and (b) without any growth hypothesis. If in addition we know that $c(t, u) \geq -\phi(t)$, then the existence of the minimum follows from the usual arguments.

B. The Mayer Problem with a Comparison Functional

We consider now the problem

(16.5.11) $$I[x, u] = g(t_1, x(t_1), t_2, x(t_2)),$$

(16.5.12) $$dx/dt = D(t)x(t) + C(t, u(t)), \qquad t_0 \leq t_1 \leq t \leq t_2 \leq T.$$

(16.5.13) $(t_1, x(t_1), t_2, x(t_2)) \in B, \qquad u(t) \in U(t), \quad t_1 \le t \le t_2, \qquad (t^*, x(t^*)) \in P,$

(16.5.14) $$J[x, u] = \int_{t_1}^{t_2} [d(t)x(t) + c(t, u(t))]\, dt \le M,$$

where we use the same notation as in Subsection A.

16.5.ii (An Existence Theorem for Mayer Problems with No Convexity Assumptions and a Comparison Functional). *Under the same assumptions as in* (16.5.i), *let* Ω *denote the class of all admissible pairs* $x(t)$, $u(t)$, $t_1 \le t \le t_2$, *whose trajectory contains at least one point on the compact set* P, *or* $(t^*, x(t^*)) \in P$, *and for which* $J[x, u] \le M$ *for some constant* M. *If* Ω *is not empty, then the functional* (16.5.11) *with constraints* (16.5.12–14) *has an absolute minimum in* Ω.

The proof is a variant of the one for (16.5.i).

C. Lagrange and Bolza Problems with a Comparison Functional

We consider here again problem (16.5.1–3) with a comparison functional

(16.5.15) $$J[x, u] = \int_{t_1}^{t_2} [b(t)x(t) + h(t, u(t))]\, dt,$$

where we make the same assumptions on $b(t)$ as for $d(t)$ in (16.5.i), and the same assumptions on $h(t, u)$ as for $c(t, u)$ in (16.5.i).

16.5.iii (An Existence Theorem for Lagrange and Bolza Problems with No Convexity Assumptions and a Comparison Functional). *Let us make the same general assumptions as in the text of* (16.5.i) (*or alternates*), *and on* $b(t)$ *the same as on* $d(t)$, *and on* $h(t, u)$ *as on* $c(t, u)$. *Let us assume that* (1) C *and* h *satisfy the same hypotheses as* C, c *in* (16.5.i), *and* (2) $c(t, u) \ge -\psi(t)$ *for all* t, u *where* $\psi(t) \ge 0$ *is a given locally integrable function. Let* Ω *be the class, which we assume to be nonempty, of all admissible pairs* $x(t)$, $u(t)$, $t_1 \le t \le t_2$, *whose trajectories have at least one point on* P, *for which also* J *is finite and* $J[x, u] \le M$ *for a given constant* M. *Then the functional* I *has an absolute minimum in* Ω.

The proof is a variant of the one for (16.5.i).

D. Pointwise Convergence on Some Components

In the next statement we shall need the notation $x = (y, z)$, $y = (x^1, \dots, x^\alpha)$, $z = (x^{\alpha+1}, \dots, x^n)$. Then P will be a compact subset of $A_0 = [t_0, T] \times R^\alpha$, and as usual $A = [t_0, T] \times R^n$, $M_0 = [(t, u) \mid t \in [t_0, T], u \in U(t)]$.

Also, let $d(t) = (d_1(t), \dots, d_\alpha(t), 0, \dots, 0)$, that is, $d_{\alpha+1} = \cdots = d_n = 0$. Analogously, let $D(t) = [d_{ij}(t)]$ with $d_{ij}(t) = 0$ for all $j = \alpha + 1, \dots, n$, $i = 1, \dots, n$. Finally, let $b(t) = 0$, that is, the comparison functional is $J = \int_{t_1}^{t_2} h(t, u(t))\, dt$.

16.5.iv (An Existence Theorem with No Convexity Conditions and Pointwise Convergence of Some Components). *Under the general assumptions as in* (16.5.iii), *let*

$1 \leq \alpha \leq n$, $d_j(t) = 0$, $d_{ij}(t) = 0$, $b(t) = 0$, $j = \alpha + 1, \ldots, n$, $i = 1, \ldots, n$. *Let us assume that*

($g\alpha 1$) *Given $\varepsilon > 0$ there is $N(\varepsilon) \geq 0$ such that $|C_i(t, u)| \leq \varepsilon h(t, u)$, $1 \leq \varepsilon h(t, u)$, $i = 1, \ldots, \alpha$, for all $t \in [t_0, T]$, $u \in U(t)$, $|u| \geq N$.*

($g\alpha 2$) *There are a constant $K \geq 0$ and an L-integrable function $\psi(t)$, $t_0 \leq t \leq T$, such that $C_i(t, u) \leq \psi(t) + Kh(t, u))$, $i = \alpha + 1, \ldots, n$, for all $t \in [t_0, T]$, $u \in U(t)$. Also we assume $c(t, u) \geq 0$.*

($g\alpha 3$) *$B = B_0 \times R^{n-\alpha}$, where B_0 is a closed subset of the $t_1 y_1 z_1 t_2 y_2$-space $R^{1+n+\alpha}$ such that the corresponding set $\{z_1\}_B$ is bounded, and $g(t_1, y_1, z_1, t_2, y_2, z_2)$ is a real valued continuous function which is monotone nondecreasing with respect to each component of z_2, or $x_2^{\alpha+1}, \ldots, x_2^n$. (Here $\{z_2\}_B = R^{n-\alpha}$.)*

Let Ω be the class of all admissible pairs $x(t)$, $u(t)$, $t_1 \leq t \leq t_2$, with $x(t)$ AC, $x(t) = (y(t), z(t))$, $t_0 \leq t_1 \leq t \leq t_2 \leq T$, whose trajectories x contain at least one point $(t^*, y(t^*)) \in P$, and for which $J[x, u] \leq M$ for some constant M. If Ω is not empty, then the functional (16.5.1) (*with constraints* (16.5.2), (16.5.3) *and $J \leq M$) has an absolute minimum in Ω.*

Proof. The proof is a variant of the one for (16.5.i). Here condition ($g\alpha 1$) guarantees that the trajectories $y(t)$ are equiabsolutely continuous, equibounded, with derivatives $y'(t)$ equiabsolutely integrable. Condition ($g\alpha 2$) guarantees that the trajectories $z(t)$ are of equibounded total variation. By taking $dx^0/dt = d(t)x(t) + c(t, u(t))$, $x^0(t_1) = 0$, the functions $x^0(t)$ are also of equibounded total variation. Now we consider the sets

$$Q(t) = [(\mathfrak{z}^0, \mathfrak{z}) \mid \mathfrak{z}^0 \geq c(t, u),\ \mathfrak{z}^i = C_i(t, u),\ i = 1, \ldots, \alpha,\ \mathfrak{z}^i \geq C_i(t, u),\ i = \alpha + 1, \ldots, n],$$

where $\mathfrak{z} = (\mathfrak{z}^1, \ldots, \mathfrak{z}^n)$, and we consider also their convex hulls

$$R(t) = \Big[(\mathfrak{z}^0, \mathfrak{z}) \mid \mathfrak{z}^0 \geq \sum_s p_s c(t, u^{(s)}),\ \mathfrak{z}^i = \sum_s p_s C_i(t, u^{(s)}),\ i = 1, \ldots, \alpha,$$

$$\mathfrak{z}^i \geq \sum_s p_s C_i(t, u^{(s)}),\ i = \alpha + 1, \ldots, n,\ p \in \Gamma,\ v \in V(t) \Big],$$

where $\sum_s$ ranges over $s = 1, \ldots, h$, where h is a fixed number (say, $h = n + 2$), where $p = (p_1, \ldots, p_h) \in \Gamma$, $(u^{(1)}, \ldots, u^{(h)}) = v \in V(t) = (U(t))^h$. By the last remark in Section 8.5 we derive that these convex sets have property (Q) with respect to t in $[t_0, T]$. Finally, if

$$f_0^*(t, y, p, v) = \sum_{j=1}^{\alpha} d_j(t) y^j + \sum_{s=1}^{h} p_s c(t, u^{(s)}),$$

$$f_i(t, y, p, v) = \sum_{j=1}^{\alpha} d_{ij}(t) y^j + \sum_{s=1}^{h} p_s C_i(t, u^{(s)}), \qquad i = 1, \ldots, n,$$

we see that the usual sets

$$R^*(t, y) = [(\mathfrak{z}^0, \mathfrak{z}) \mid \mathfrak{z}^0 \geq f_0^*(t, y, p, v),\ \mathfrak{z}^i = f_i(t, y, p, v),\ i = 1, \ldots, \alpha,$$
$$\mathfrak{z}^i \geq f_i(t, y, p, v),\ i = \alpha + 1, \ldots, n,\ p \in \Gamma,\ v \in V(t)]$$

are a simple parallel displacement of the sets $R(t)$ with continuous coefficients $d_i(t)$, $d_{ij}(t)$, that is, with displacements which depend continuously on y and t. Thus, the same sets $R^*(t, y)$ have property (Q) with respect to (t, y) in A_0. Statement (16.5.iv) now follows from the arguments for (16.5.i) and those of Chapter 15. □

16.6 Examples for the Unbounded Case

1. Let us consider the Lagrange problem with $n = 2, m = 1$, system $x' = (1 - t)x + y$, $y' = u$, $U(t) = R$, $t_1 = 0$, $t_2 = 1$, $x(0) = x_1$, $y(0) = y_1$, $x(1) = x_2$, $y(1) = y_2$ fixed, and functional $I[x, y, u] = \int_0^1 |1 - u^2|\,dt$, that is $d(t) = 0$, $c(t, u) = |1 - u^2|$. The set $M_0 = [0, 1] \times R$ is closed, $B = \{(0, x_1, y_1, 1, x_2, y_2)\}$ is a singleton, here $d(t) = 0$, $c(t, u) = |1 - u^2| \geq 0$, the 2×2 matrix $D(t) = [1 - t, 1, 0, 0]$ has continuous entries, and $C(t, u) = (C_1, C_2)$, $C_1 = 0$, $C_2 = u$. The growth condition (b) holds. Note that the sets $\tilde{Q}(t, x, y) = [(z^0, z^1, z^2) | z^0 \geq |1 - u^2|, z^1 = (1 - t)x + y, z^2 = u, u \in R]$ are closed but certainly not convex; P is the singleton $\{(x_1, y_1)\}$. It is easy to see that Ω is not empty. Theorem (16.5.i) applies, and the functional I has an absolute minimum under the constraints.

2. Let us consider the Mayer problem with $n = 2$, $m = 1$, system

$$x' = (1 - t)x + t^2 y + t^{-1/2} \cos u,$$
$$y' = -t^2 x + (1 + t)y + t^{-1/2} \sin u,$$

$U(t) = R$, $t_1 = 0$, $t_2 \geq 0$, $x(0) = a$, $y(0) = b$, and functional $I[x, y, u] = (x_2^2 + y_2^2)^{1/2} + t_2$, where as usual $x_2 = x(t_2)$, $y_2 = y(t_2)$. Since $I \to +\infty$ as $t_2 \to +\infty$ uniformly, we can limit ourselves to admissible pairs (or usual solutions) $x(t)$, $y(t)$, $u(t)$, $0 \leq t \leq t_2$, with $0 \leq t_2 \leq T$ and some fixed $T > 0$. Here $M_0 = [0, T] \times R$, $B = (0, a, b) \times [0, T] \times R^2$ are closed sets. The 2×2 matrix $D(t)$ has continuous entries. Here $C(t, u) = (C_1, C_2)$, $C_1 = t^{-1/2} \cos u$, $C_2 = t^{-1/2} \sin u$, $|C_1|, |C_2| \leq \varphi(t) = t^{-1/2}$, an L-integrable function in $[0, T]$. Here the Remark of Section 16.5A applies, $P = \{(a, b)\}$ is a singleton, and I has an absolute minimum.

3. Let us consider the Mayer problem with $n = 2$, $m = 1$, system

$$x' = (1 - t)x + t^2 y + t(1 - t)u,$$
$$y' = -t^2 x + (1 + t)y + t(1 - t)|u|,$$

with $U(t) = R$, $t_1 = 0$, $t_2 = 1$, $x(0) = a$, $y(0) = b$, $\int_0^1 t^2 u^2\,dt \leq M$, a, b, M constants, $I[x, y, u] = x_2 + y_2 = x(1) + y(1)$. Here the 2×2 matrix $D(t)$ has continuous entries, $C(t, u) = (C_1, C_2)$, $C_1 = t(1 - t)u$, $C_2 = t(1 - t)|u|$, and the sets $Q(t) = [(z^1, z^2)\ z^2 = |z^1|, -\infty < z^1 < +\infty]$, $t \in (0, 1)$, are not convex. We have now $d(t) = 0$, $c(t, u) = t^2 u^2$, and

$$\begin{aligned} |C_1(t, u)|, |C_2(t, u)| &\leq t|u| = 2(2^{-1}\varepsilon^{-1/2})(\varepsilon^{1/2} t|u|) \\ &\leq 2^{-2}\varepsilon^{-1} + \varepsilon t^2 u^2 = \psi_\varepsilon(t) + \varepsilon c(t, u), \end{aligned}$$

where $\psi_\varepsilon(t) = 2^{-2}\varepsilon^{-1}$. Assumption (g2') of (16.5.i) is satisfied, $P = \{(a, b)\}$, statement (16.5.ii) applies, and I has an absolute minimum.

4. Let us consider the Lagrange problem with $n = 2$, $m = 1$, the same system as in Example 3, with $t_1 = 0$, $t_2 = 1$, $x(0) = y(0) = 0$, $U(t) = R$, and functional $I[x, y, u] = \int_0^1 [tx + (1 - t)y + (t^2 + 1)u^2]\,dt$. Here the sets $M_0 = [0, 1] \times R$, $B = (0, 0, 0, 1) \times R^2$ are closed, $P = \{(0, 0)\}$, (16.5.i) applies, and I has an absolute minimum.

5. Let us consider the Lagrange problem with $n = 2$, $m = 1$, system $x' = tx + y$, $y' = x - ty + u$, $t_1 = 0$, $t_2 = 1$, $x(0) = a$, $y(0) = b$, $U(t) = R$, with functional $I[x, y, u] = \int_0^1 (1 + t)(1 - u^2)^2\,dt$. Here $d(t) = 0$, $c(t, u) = (1 + t)(1 - u^2)^2$, the 2×2 matrix $D(t)$ has continuous entries, $C(t, u) = (C_1, C_2)$, $C_1 = 0$, $C_2 = u$, $P = \{(a, b)\}$, $M_0 = [0, 1] \times R$, $B = (0, a, b, 1) \times R^2$, (16.5.i) applies, and I has an absolute minimum.

6. Let us consider the Mayer problem with $n = 3$, $m = 1$, $t_1 = 0$, $t_2 = 1$, system

$$x' = (1 - t)x + t^2 y + t(1 - t)u,$$
$$y' = -t^2 x + (1 + t)y + t(1 - t)|u|,$$
$$z' = tx + (1 - t)y + (t^2 + 1)u^2,$$

with $U(t) = R$, $x(0) = y(0) = z(0) = 0$, $z(1) \leq 0$, functional $I[x, y, z, u] = -x^2(1) - y^2(1) - z^2(1)$, and $J[x, y, z, u] = \int_0^1 (t^2 + 1)u^2\, dt \leq 1$. We have here $C(t, u) = (C_1, C_2, C_3)$, $C_1 = t(1 - t)u$, $C_2 = t(1 - t)|u|$, $C_3(t, u) = H(t, u) = (t^2 + 1)u^2$, $\psi = 0$, $K = 1$, $M_0 = [0, 1] \times R$, $U(t) = R$, $B = (0, 0, 0, 0, 1) \times R^2 \times (-\infty, 0]$. Theorem (16.5.iv) applies with $\alpha = 2$, and I has an absolute minimum.

7. Let us consider the Bolza problem as in Example 6 with $I[x, y, z, u] = -x^2(1) - y^2(1) + \int_0^1 t^2(1 - t^2)|u|\, dt$. Theorem (16.4.iv) applies with $\alpha = 2$, and I has an absolute minimum.

8. The same as Example 7 with $I[x, y, z, u] = -x^2(1) - y^2(1) + \int_0^1 t^3(1 + 2t)(1 - |u|)^2\, dt$. Theorem (16.4.iv) applies with $\alpha = 2$, and I has an absolute minimum.

16.7 Problems of the Calculus of Variations without Convexity Assumptions

We consider here the problem of the minimum of the functional

$$I[x] = \int_{t_1}^{t_2} [d(t)x(t) + c(t, x'(t))]\, dt, \qquad t_0 \leq t_1 \leq t \leq t_2 \leq T.$$
$$e[x] = (t_1, x(t_1), t_2, x(t_2)) \in B, \qquad x = (x^1, \ldots, x^n),$$

or equivalently

$$I[x, u] = \int_{t_1}^{t_2} [d(t)x(t) + c(t, u(t))]\, dt, \qquad t_0 \leq t_1 \leq t_2 \leq T,$$
$$dx/dt = u(t), \qquad (t_1, x(t_1), t_2, x(t_2)) \in B, \quad u(t) \in R^n,$$
$$x = (x^1, \ldots, x^n), \qquad u = (u^1, \ldots, u^n),$$

where t_0, T are finite, $d(t)$ is a $1 \times n$ matrix with continuous entries, $c(t, u)$ is a scalar continuous function on $M_0 = [t_0, T] \times R^n$, and $f_0(t, x, u) = d(t)x + c(t, u)$.

16.7.i (An Existence Theorem for Problems with No Convexity Assumptions). *Let t_0, T be finite, let B be closed, let $d(t)$ be continuous on $[t_0, T]$, and let $c(t, u)$ be continuous on M_0. Let us assume that there is a function $\phi(\zeta)$, $0 \leq \zeta < +\infty$, bounded below, such that $\phi(\zeta)/\zeta \to +\infty$ as $\zeta \to +\infty$, and $c(t, u) \geq \phi(|u|)$ for all $(t, u) \in M_0$. Let P be a compact subset of $A = [t_0, T] \times R^n$. Let Ω be the class of all AC functions $x(t)$ such that $e[x] \in B$, $c(\cdot, u(\cdot))$ is L-integrable, and $(t^*, x(t^*)) \in P$ for some t^* which may depend on x. Then $I[x]$ has an absolute minimum in Ω.*

This is a corollary of (16.5.i).

Examples

1. $f_0 = x + (1 - u^2)^2, n = 1.$
2. $f_0 = tx + (1 + t^2)y + (1 - u^2 - v^2)^2, n = 2.$
3. $f_0 = tx + t^{1/2}x + y + (2 + \cos t)(u - u^2)^2 + v^2, n = 2.$
4. $f_0 = x + y + (2 + t^2)(1 - u^2 - v^2)^2, n = 2.$

For instance, for $n = 1$, (16.7.i) applies to the integral $I = \int_0^1 (1 - x'^2(t))^2\, dt$ with boundary conditions $x(0) = x(1) = 0$. Then $x_0(t) = 2^{-1} - |t - 2^{-1}|$, $0 \leq t \leq 1$, is one of the infinitely many optimal solutions, and $I[x_0] = 0$. Actually, any AC function $x(t)$, $0 \leq t \leq 1$, with $x(0) = x(1) = 0$ and $x'(t) = \pm 1$ a.e., is optimal. For the corresponding generalized problem, we may take

$$J[y, p, v] = \int_0^1 [p_1(t)(1 - u_1^2(t))^2 + p_2(t)(1 - u_2^2(t))^2]\, dt$$

with $dy/dt = p_1(t)u_1 + p_2(t)u_2$, $y(0) = y(1) = 0$, where $p = (p_1, p_2)$, $v = (u_1, u_2)$, $0 \leq p_i \leq 1$, $i = 1, 2$, $p_1(t) + p_2(t) = 1$, $v \in R^2$. Then $y(t) = 0$, $p_1(t) = p_2(t) = 2^{-1}$, $u_1(t) = 1$, $u_2(t) = -1$, $0 \leq t \leq 1$, is one of the many generalized optimal solutions. This shows that there are optimal generalized trajectories which are not usual trajectories.

Bibliographical Notes

The first proof of the statement that "the range of any nonatomic vector valued measure is convex" is in A. A. Lyapunov [1]. It was L. W. Neustadt [1] who first realized that Lyapunov's result could be used to prove existence theorems for problems of optimal control which are linear in the space variables, with compact control space and under no convexity assumptions. Other proofs and remarks followed: D. Blackwell [1], H. Halkin [1, 3], H. Hermes [1], J. Kurzweil [1], H. Hermes and J. P. LaSalle [1], J. F. C. Kingman and A. P. Robertson [1], M. de Wilde [1], P. R. Halmos [1]. The last paper contained an error, which was corrected in P. R. Halmos [2].

The presentation in Section 16.1 of Lyapunov's results follows L. Cesari [14], particularly because of the proof of the conclusive theorem (16.1.v) which can be immediately applied to problems of control. The existence theorems in Sections 16.2–3 for problems of optimal control which are linear in the state variable and with compact control space are obtained by first proving the existence of optimal generalized solutions and then showing the existence of "equivalent" usual solutions by using Theorem (16.1.v). The same argument also proves the bang-bang theorem. The existence theorems in Section 16.4, for problems of optimal control which are linear in the state variables and with possibly unbounded control functions, are from L. Cesari [15]. The same general process of proof is used. These theorems apply as well (Section 16.7) to problems of the calculus of variations which are linear in the space variables, under no convexity assumptions, and with AC but possibly non-Lipschitzian solutions.

T. S. Angell [3, 5] has proved existence theorems for problems of optimal control monitored by Volterra equations without convexity assumptions. M. B. Suryanarayana [4, 5] has proved existence theorems for systems monitored by linear hyperbolic equations and by total partial differential equations.

CHAPTER 17

Duality and Upper Semicontinuity of Set Valued Functions

17.1 Convex Functions on a Set

We shall consider here a real valued function $F(u)$ defined and everywhere finite on a set U of R^n. If U is convex (Section 8.4), then F is said to be convex in U if $u_1, u_2 \in U, 0 \leq \alpha \leq 1$, implies $F(\alpha u_1 + (1-\alpha)u_2) \leq \alpha F(u_1) + (1-\alpha)F(u_2)$. The function F is said to be extended (Section 8.5) if we take $F = +\infty$ in $R^n - U$. With obvious conventions the convexity of F in R^n is equivalent to the statement that U is convex and F is convex in U. As mentioned in Section 8.5, the set $\tilde{Q} = [(z^0, u) | +\infty > z^0 > F(u), u \in U]$ is said to be the epigraph of F, or epi F.

17.1.i. *The extended function F is convex in R^n if and only if* epi F *is convex.*

17.1.ii. *The extended function F is lower semicontinuous in R^n if and only if* epi F *is closed* (cf. (8.5.v)).

17.1.iii. *If U is a convex set in R^n and $F(u)$, $u \in U$, a given real valued function, then $F(u)$ is convex if and only if $u_j \in U$, $\lambda_j \geq 0$, $j = 1, \ldots, v$, v finite, $\lambda_1 + \cdots + \lambda_v = 1$, $u_0 = \sum_{j=1}^{n} \lambda_j u_j$ implies $F(u_0) \leq \sum_{j=1}^{n} \lambda_j F(u_j)$.*

This is a corollary of (8.4.i). Note that F is said to be concave in U if U is convex and $u_1, u_2 \in U$, $0 \leq \alpha \leq 1$, implies $F(\alpha u_1 + (1-\alpha)u_2) \geq \alpha F(u_1) + (1-\alpha)F(u_2)$, that is, $-F$ is convex. From (17.1.iii) we derive that a function $F(u)$, $u \in U$, on a convex set U, is "affine", that is, of the form $F(u) = r + \sum_i b_i u^i$, if and only if it is both convex and concave in U.

Note that if $p_i \geq 0$, $i = 1, \ldots, N$, $N \geq 2$, are arbitrary numbers with $p_1 + \cdots + p_N > 0$, then the relation above for convex functions can be written in the equivalent form

$$F\left(\frac{p_1u_1 + \cdots + p_Nu_N}{p_1 + \cdots + p_N}\right) \le \frac{p_1F(u_1) + \cdots + p_NF(u_N)}{p_1 + \cdots + p_N},$$

which is sometimes referred to as Jensen's inequality.

A linear (affine) scalar function $z(u) = r + b_1u^1 + \cdots + b_nu^n$, $u \in R^n$, is said to be a (nonvertical) supporting hyperplane (or plane) of $F(u)$, $u \in U$, at a point $\bar{u} \in U$, provided $F(\bar{u}) = z(\bar{u})$ and $F(u) \ge z(u)$ for all $u \in U$.

As usual in this book, n-vectors $b = (b_1, \ldots, b_n), u = (u^1, \ldots, u^n)$ are thought of as column vectors, and the inner product therefore is written in any of the forms $\sum_j b_ju^j = b^*u = b \cdot u$.

17.1.iv. *If U is a convex subset of R^n, and $F(u)$, $u \in U$, a given real valued convex function, then $F(u)$ has a supporting plane at every* interior *point $\bar{u}$ of U.*

Proof. We know already that the set $\tilde{Q} = [(z, u) | z \ge F(u), u \in U] \subset R^{n+1}$ is convex, and by (8.4.iv) there exists some supporting hyperplane to the convex set $\tilde{Q}$ at the point $(\bar{z}, \bar{u})$, $\bar{z} = F(\bar{u})$, say $p^0z + p \cdot u - c = 0$ with p^0, $p = (p^1, \ldots, p^n)$ real, $p^0\bar{z} + p \cdot \bar{u} - c = 0$, and $p^0z + p \cdot u - c \ge 0$ for all $u \in U$ and $z \ge F(u)$. Let us prove that $p^0 \ne 0$. Indeed, if $p^0 = 0$, then we have $p \cdot \bar{u} - c = 0$, $p \cdot u - c \ge 0$ for all $u \in U$. If $u_1 \ne \bar{u}$ is any point of R^n and ε real, then for $u(\varepsilon) = \varepsilon u_1 + (1 - \varepsilon)\bar{u}$ we have $u(\varepsilon) \to \bar{u}$ as $\varepsilon \to 0$, and $p \cdot u(\varepsilon) - c = \varepsilon p \cdot (u_1 - \bar{u})$. Since $\bar{u} \in \text{int } U$, then both $u(\varepsilon)$, $u(-\varepsilon)$ belong to U for $|\varepsilon|$ sufficiently small, and yet one of the two numbers $p \cdot u(\pm\varepsilon) - c$ is negative, a contradiction. We have proved that $p^0 \ne 0$. Actually, we must have $p^0 > 0$, since $p^0z + p \cdot u - c \ge 0$ for all $z \ge F(u)$. Finally, if we take $z(u) = (-p \cdot u + c)/p^0$, then $z(\bar{u}) = F(\bar{u})$ and $F(u) \ge z(u)$ for all $u \in U$. □

Given a set U, we denote as usual by int U the subset of its interior points.

If U has no interior points, that is, int $U = \emptyset$, statement (17.1.iv) has the following implication. First, let us denote by R the linear variety of R^n of minimum dimension r containing U. Then, $U \subset R \subset R^n$, $0 \le r \le n$. If U is reduced to a single point, then $R = U$ and $r = 0$. Otherwise, $1 \le r \le n$, and we denote by Rint U the certainly nonempty set of points of U which are interior to U with respect to R. Thus, int $U \subset R$int $U \subset U \subset R \subset R^n$. Statement (17.1.iv) has the following corollary.

17.1.v. *Under the same hypotheses as in* (17.1.iv), *$F(u)$ has a supporting plane at every point $\bar{u} \in R$int U.*

17.1.vi. *Under the same hypotheses as in* (17.1.iv), *$F(u)$ is continuous at every point $\bar{u} \in R$int U. In particular, if $U = R^n$, then F is continuous in R^n.*

Proof. We may well assume that U is not a single point; that is, $1 \le r \le n$, and Rint $U \ne \emptyset$. Let $\bar{u}$ be any point $\bar{u} \in R$int U, and let $z = c + p \cdot u$ be some supporting plane at $\bar{u}$, so that $F(\bar{u}) = c + p \cdot \bar{u}$. Assume, if possible, that for some $\sigma > 0$ and some sequence of points $u_k \in R\text{int } U$ with $u_k \to \bar{u}$ as $k \to \infty$, we have $F(u_k) - F(\bar{u}) \leq -\sigma$ for all k. Then, $F(u_k) \geq c + p \cdot u_k$, and hence $-\sigma \geq F(u_k) - F(\bar{u}) \geq p \cdot (u_k - \bar{u})$. As $k \to \infty$, we have $-\sigma \geq 0$, a contradiction. Assume now, if possible, that for some $\sigma > 0$ and sequence of points $u_k \in R\text{int } U$ with $u_k \to \bar{u}$ as $k \to \infty$, we have $F(u_k) - F(\bar{u}) \geq \sigma$ for all k. Then we can choose r points $v_j \in U$, $j = 1, \ldots, r$, independent in R^n, such that $|v_j - \bar{u}| = \delta > 0$, $j = 1, \ldots, r$, and $\bar{u} = \sum_j r^{-1} v_j$. Since $u_k \to \bar{u}$, we have $u_k = \sum_j \lambda_{jk} v_j$ with $\lambda_{jk} \to r^{-1}$ as $k \to \infty$, $j = 1, \ldots, r$. If $\lambda_k = \min[\lambda_{jk}, j = 1, \ldots, r]$ then $0 \leq \lambda_k \leq r^{-1}$, $u_k = \sum_j (\lambda_{jk} - \lambda_k) v_j + (\lambda_k r)\bar{u}$, and hence

$$F(u_k) \leq \sum_j (\lambda_{jk} - \lambda_k) F(v_k) + (\lambda_k r) F(\bar{u}),$$

where $\lambda_{jk} - \lambda_k \to 0$, $\lambda_k r \to 1$. For all k sufficiently large, we have then $F(u_k) \leq F(\bar{u}) + \sigma/2$, a contradiction. This proves that F is continuous at every point of $R\text{int } U$. □

Statements (17.1.iv) and (17.1.v) cannot be made stronger so as to include points of $U - R\text{int } U$. Indeed, the function $F(u)$, $-1 \leq u \leq 1$, defined by $F(u) = 0$ for $-1 < u < 1$ and $F(-1) = F(1) = 1$ is convex but not continuous at the end points $u = 0$ and $u = 1$. The function $F(u) = -(1 - u^2)^{1/2}$, $-1 \leq u \leq 1$, is convex and continuous on $[-1, 1]$ but has no "supporting plane" (of the form $z = p \cdot u + c$) at the end points, $u = \pm 1$.

17.1.vii. *Under the same hypotheses as in* (17.1.iv), *$F(u)$ is bounded below on every bounded part K of U.*

Proof. Indeed, if K contains more than one point, then K contains some point $\bar{u} \in R\text{int } U$, and if $z(u) = p \cdot u + c$ is a supporting plane at $\bar{u}$, then $F(u) \geq p \cdot u + c$ for all $u \in K \subset U$, and $p \cdot u + c$ has a finite lower bound on K. □

17.1.viii. *Under the same hypotheses as in* (17.1.iv), *$F(u)$ is upper semicontinuous at every $\bar{u} \in U - R\text{int } U$ along any segment s issuing from $\bar{u}$ and contained in U.*

Proof. Let s be the segment $s = \bar{u}u_0$, $s \subset U$. Assume, if possible, that there is a sequence of points $u_k \in s \subset U$, $u_k \to \bar{u}$ as $k \to \infty$, with $F(u_k) \geq F(\bar{u}) + \sigma$ for all k for some $\sigma > 0$. Then all points interior to the segment s are certainly points of $R\text{int } U$, say $u = (1 - \alpha)\bar{u} + \alpha u_0$, $0 < \alpha < 1$, and since $F(u) \leq (1 - \alpha)F(\bar{u}) + \alpha F(u_0)$, we see that F is bounded above on s. Since $h_k = u_k - \bar{u} \to 0$ as $k \to \infty$, there is a sequence of numbers $\beta_k > 1$ with $\beta_k \to \infty$, $\beta_k h_k \to 0$ as $k \to \infty$. Hence, the points $u'_k = \bar{u} + \beta_k(u_k - \bar{u})$, $k = 1, 2, \ldots$, are on the half straight line from $\bar{u}$ containing s, and $u'_k \to \bar{u}$ as $k \to \infty$. Thus, $u'_k \in s$, $u'_k \in R\text{int } U$ for all k sufficiently large, and the following relations hold:

$$u_k = \beta_k^{-1} u'_k + \beta_k^{-1}(\beta_k - 1)\bar{u},$$

$$F(u_k) \leq \beta_k^{-1} F(u'_k) + \beta_k^{-1}(\beta_k - 1)F(\bar{u}),$$

$$F(u'_k) \geq \beta_k F(u_k) - (\beta_k - 1)F(\bar{u}) \geq F(\bar{u}) + \beta_k \sigma.$$

Hence $F(u'_k) \to +\infty$ as $k \to \infty$, a contradiction since F is bounded above on s. We have proved that F is upper semicontinuous at $\bar{u}$ along s. □

17.1.ix. *If U is a convex subset of R^n, if $F(u)$, $u \in U$, is a given real valued convex function on U, and if its epigraph $\tilde{Q} = [(z^0, u) | z^0 \geq F(u), u \in U] = \text{epi } F \subset R^{n+1}$, is closed and convex, then the function $F(u)$ is lower semicontinuous at every point $\bar{u} \in U - R\text{int } U$, and therefore continuous on every segment s issuing from $\bar{u}$ and contained in U.*

Proof. Assume, if possible, that there is a number $\sigma > 0$ and points $\bar{u}$, u_k, $k = 1, 2, \dots$, with $\bar{u} \in U - R\text{int } U$, $u_k \in U$, $F(u_k) < F(\bar{u}) - \sigma$ for all k. Take $\bar{z}^0 = F(\bar{u})$, and note that all points $(\bar{z}^0 - \sigma, u_k)$ are in $\tilde{Q} = \text{epi } F$. Then, as $k \to \infty$, we see that $(\bar{z}^0 - \sigma, \bar{u})$ is in the closed set $\tilde{Q}$, a contradiction, since $(z, \bar{u}) \in \tilde{Q}$ if and only if $z \geq \bar{z}^0 = F(\bar{u})$. The last part of the statement is now a consequence of (17.1.viii) □

17.1.x. *If U is a convex subset of R^n, if $F(u)$, $u \in U$, is a given real valued function on U, and if* epi F *is closed and convex, then F is convex and lower semicontinuous.*

Proof. Because of (17.1.i), F is convex if and only if epi F is convex. If epi F is convex, then F is continuous at every $\bar{u} \in R\text{int } U$ by (17.1.v), and if epi F is also closed, then F is lower semicontinuous at every $\bar{u} \in U - R\text{int } U$ by (17.1.ix), and thus F is lower semicontinuous in all of U. □

A function $F(u)$, $u \in U$, convex on a convex set U, may not be continuous at the points of $U - R\text{int } U$, even if the set $\tilde{Q}$ is closed, as the following example shows. Take $U = [(u, v) | 0 \leq u \leq 1,\ v \geq 0,\ (u - 1)^2 + v^2 \leq 1]$; $F(u, v) = v$ if $0 \leq u \leq 1$, $0 \leq v \leq u$; $F(u, v) = (2u)^{-1}(u^2 + v^2)$ if $0 < u < 1$, $u \leq v \leq (1 - (1 - u)^2)^{1/2}$. Obviously, U is convex, F is convex in (u, v), but F is not continuous at $(0, 0)$, since $F(0, 0) = 0$,

$$F(u, (1 - (1 - u)^2)^{1/2}) = 1 \text{ for all } 0 < u < 1.$$

Given a convex set $U \subset R^n$ and a scalar function $F(u)$, $u \in U$, we say that F is convex at the point $\bar{u} \in U$ provided $F(\bar{u}) \leq \sum_{j=1}^{\nu} \lambda_j F(u_j)$ for any convex combination $\bar{u} = \sum_{j=1}^{\nu} \lambda_j u_j$ of points $u_j \in U$, $j = 1, \dots, \nu$ ($\lambda_j \geq 0$, $\lambda_1 + \cdots + \lambda_\nu = 1$, $\nu \geq 2$ any integer).

17.1.xi. *If U is a convex subset of R^n, and $F(u)$, $u \in U$, a given real valued function, then $F(u)$ is convex at an interior point $\bar{u}$ of U if and only if $F(u)$ has a supporting plane at $\bar{u}$.*

Proof. Suppose F is convex at the point $\bar{u} \in \text{int } U$. Then, the smallest convex set co $\tilde{Q}$ containing $\tilde{Q} = [(z, u) | z \geq F(u),\ u \in U] = \text{epi } F \subset R^{n+1}$ is the set of all points $(z, u) = \sum_{j=1}^{\nu} \lambda_j (z_j, u_j)$ with $(z_j, u_j) \in \tilde{Q}$, $\lambda_j \geq 0$, $\lambda_1 + \cdots + \lambda_\nu = 1$, ν finite. Now, $(z, \bar{u}) \notin \text{co } \tilde{Q}$ if $z < F(\bar{u})$, since for every convex combination $(z, u) = \sum_{j=1}^{\nu} \lambda_j (z_j, u_j)$ with $u = \bar{u}$, $u = \sum_{j=1}^{\nu} \lambda_j u_j$, we have $z = \sum_j \lambda_j z_j \geq \sum_j \lambda_j F(u_j) \geq F(\bar{u})$, so $z \geq F(\bar{u})$. Hence, $(F(\bar{u}), \bar{u})$ is a boundary point of co $\tilde{Q}$. Then there is a hyperplane $V = [(z, u) | p_0 z + p \cdot u - c = 0] \subset R^{n+1}$ such that $p_0 F(\bar{u}) + p \cdot \bar{u} - c = 0$ and $p_0 z + p \cdot u - c \geq 0$ for all $(z, u) \in \text{co } \tilde{Q}$.

For every convex combination $\bar{u} = \sum_j \lambda_j u_j$ and numbers $z_j \geq F(u_j)$, we have $(z_j, u_j) \in \text{co } \tilde{Q}$, and $p_0 z_j + p \cdot u_j - c \geq 0$. Therefore, $p_0[\sum_j \lambda_j z_j] + p \cdot \bar{u} - c \geq 0$, $p_0 F(\bar{u}) + p \cdot \bar{u} - c = 0$, and $p_0[\sum \lambda_j z_j - F(\bar{u})] \geq 0$. Since the expression in brackets is nonnegative because of the convexity of F at u, we conclude that $p_0 \geq 0$. But $p_0 = 0$ implies $p \cdot u - c \geq 0$ for all $u \in U$ with $p \cdot \bar{u} - c = 0$; hence $p \cdot (u - \bar{u}) \geq 0$ for all $u \in U$, which is impossible. Thus, $p_0 > 0$, and the hyperplane V can be written in the form $z = b \cdot u + r$, with $b = -p/p_0$, $r = c/p_0$, and $z \geq b \cdot u + r$ for all $(z, u) \in \text{co } \tilde{Q}$, $F(\bar{u}) = b \cdot \bar{u} + r$. Thus, $z(u) = b \cdot u + r$ is a supporting plane for $F(u)$ at $u = \bar{u}$. Conversely, if $F(u)$ has a supporting plane $z(u) = b \cdot u + r$ at $\bar{u} \in U$, then for every convex combination $\bar{u} = \sum_j \lambda_j u_j$ of points $u_j \in U$ we have $\sum_j \lambda_j F(u_j) \geq \sum_j \lambda_j z(u_j) = \sum_j [b \cdot u_j + r]\lambda_j = b \cdot \bar{u} + r = F(\bar{u})$, and F is convex at $\bar{u}$. Statement (17.1.xi) is thereby proved. □

Remark. The sufficiency part of (17.1.xi) can be stated in a stronger form as follows: If $F(\bar{u}) \leq \sum_{j=1}^{\nu} \lambda_j F(u_j)$ for any convex combination $\bar{u} = \sum_{j=1}^{\nu} \lambda_j u_j$ of points $u_j \in U$, $j = 1, \dots, \nu$, $\lambda_j \geq 0$, $\lambda_1 + \cdots + \lambda_\nu = 1$, and all possible ν, $2 \leq \nu \leq n + 2$, then $F(u)$ has a supporting plane at $\bar{u}$. Indeed, in the proof above, and by force of Carathéodory's theorem (8.4.iii) in R^{n+1}, we can restrict ourselves to the convex combinations with $v \le n + 2$.

17.2 The Function $T(x; z)$

For the sake of simplicity we denote by x the vector variable which in other sections we denote by (t, x). Let A be a given subset of the x-space R^v, for every $x \in A$ let $Q(x)$ be a given subset of the z-space R^n, and let M denote the set $M = [(x, z) | x \in A,\ z \in Q(x)] \subset R^{v+n}$. For every $x \in A$ let $\tilde{Q}(x)$ denote a subset of R^{n+1} whose projection on the z-space is $Q(x)$, and assume that (a) for every $(z^0, z) \in \tilde{Q}(x)$ and $z'^0 \ge z^0$ we also have $(z'^0, z) \in \tilde{Q}(x)$. Let $\tilde{M}$ denote the set $\tilde{M} = [(x, z^0, z) | (z^0, z) \in \tilde{Q}(x),\ x \in A] \subset R^{v+n+1}$.

For every $x \in A$ and $z \in R^n$, let $T(x, z)$ denote

$$(17.2.1) \quad T(x, z) = \inf[z^0 | (z^0, z) \in \tilde{Q}(x)], \qquad -\infty \le T(x, z) \le +\infty$$

Then $T(x, z) = +\infty$ for every $x \in A$, $z \in R^n - Q(x)$; and $-\infty \le T(x, z) < +\infty$ for every $x \in A$, $z \in Q(x)$. Thus, $T(x, z)$ is defined in M, and $T(x, z) < +\infty$ everywhere in M. Note that the graph of $Q(x)$ is M and the graph of $\tilde{Q}(x)$ is $\tilde{M}$. Obviously $\tilde{M} \subset \operatorname{epi} T$. If all sets $\tilde{Q}(x)$ are closed and bounded below, then for every $x \in A$, $T(x, z)$ is finite, $T(x, z) \in \tilde{Q}(x)$, min holds instead of inf in (17.2.1), and $\tilde{M} = \operatorname{epi} T$. We shall also consider the extended function T defined in the whole of R^{v+n} by taking $T(x, z) = +\infty$ everywhere in $R^{v+n} - M$.

Note that the convexity of $\tilde{Q}(x)$ implies the convexity of $Q(x)$, but $Q(x)$ may not be closed even if $\tilde{Q}$ is closed. For instance, $\tilde{Q} = [(z^0, z) | z^0 \ge (\tan z)^2,\ -\pi/2 < z < \pi/2] \subset R^2$ is convex and closed, while $Q = [z | -\pi/2 < z < \pi/2] \subset R^1$ is convex but not closed.

As before, we denote by $R = R(x)$ a linear variety in R^n of minimum dimension r containing $Q(x)$; thus, $Q(x) \subset R \subset R^n$, $0 \le r \le n$. As usual we denote by int $Q(x)$ the set of all $z \in R^n$ which are interior to $Q(x)$ with respect to R^n, and by Rint $Q(x)$ the set of all points z which are interior to $Q(x)$ with respect to R; thus

$$\operatorname{int} Q(x) \subset R\operatorname{int} Q(x) \subset Q(x) \subset R \subset R^n.$$

The results of Sections 8.5B and 8.5C apply here, in particular the final statement of Remark 4 of Section 8.5B: If we take for $\tilde{Q}(x)$ to be the empty sets for $x \in R^v - A$, then the set epi T is closed if and only if $T(x, u)$ is lower semicontinuous in R^{v+n}, and if and only if the sets $\tilde{Q}(x)$, $x \in R^v$ have property (K) in R^v.

Having in view the properties of the sets A and M, the following more detailed statement holds.

17.2.i. (a) *If the sets $\tilde{Q}(x)$ have property* (K) *in A, then the real valued function $T(x, z)$, $(x, z) \in M$, is lower semicontinuous in M. The converse is also true if M is closed.* (b) *If the extended function T is lower semicontinuous in $R^v \times R^n$, and the sets $\tilde{Q}(x)$ are closed, then the sets $\tilde{Q}(x)$ have property* (K) *in A. The* *converse is also true if A is closed.* (c) *Finally, if A and M are closed, then the extended function T is lower semicontinuous in $R^{\nu+n}$ if and only if the set* epi *T is closed, and if and only if the sets $\tilde{Q}(x)$ have property* (K) *in A.*

Proof. The first part could be derived from the second one. However, we prove the two parts independently. Let us prove the first part.

Let us assume that the sets $\tilde{Q}(x)$ have property (K) in A, and let us prove that T is lower semicontinuous everywhere in M. Let $(\bar{x}, \bar{z})$ be a point of M. If $T(\bar{x}, \bar{z}) = -\infty$ there is nothing to prove. Let $\bar{z}^0 = T(\bar{x}, \bar{z})$ be finite, and let us assume, if possible, that T is not lower semicontinuous at $(\bar{x}, \bar{z})$. Then, there is a $\sigma > 0$ and a sequence of points $(x_k, z_k) \in M$ with $(x_k, z_k) \to (\bar{x}, \bar{z})$ as $k \to \infty$, and $T(x_k, z_k) < \bar{z}^0 - \sigma$ for all k. By property (a) we derive that $(\bar{z}^0 - \sigma, z_k) \in \tilde{Q}(x_k)$ for all k, and, given $\varepsilon > 0$, also $(\bar{z}^0 - \sigma, z_k) \in \tilde{Q}(\bar{x}, 2\varepsilon)$ for all k sufficiently large, where $\tilde{Q}(\bar{x}, 2\varepsilon)$ is the union of all $\tilde{Q}(x)$, $x \in A$, with $|x - \bar{x}| \leq 2\varepsilon$ (cf. Section 8.5). Then $(\bar{z}^0 - \sigma, \bar{z}) \in \text{cl } \tilde{Q}(\bar{x}, 2\varepsilon)$. By property (K) we derive $(\bar{z}^0 - \sigma, \bar{z}) \in \tilde{Q}(\bar{x})$, a contradiction, since $\bar{z}^0 = T(\bar{x}, \bar{z})$. An analogous argument holds if $T(\bar{x}, \bar{z}) = +\infty$. We have proved that T is lower semicontinuous in M.

Conversely, assume that T is lower semicontinuous in M. Let x_0 be a point of A, and let us prove that the sets $\tilde{Q}(x)$ have property (K) at x_0. Let (z_0^0, z_0) be a point of $\bigcap_\delta \text{cl } \tilde{Q}(x_0, \delta)$. Then there is a sequence of points (z_k^0, z_k, x_k) with $z_k^0 \to z_0^0$, $z_k \to z_0$, $x_k \to x_0$, $x_k \in A$, $(z_k^0, z_k) \in \tilde{Q}(x_k)$, $(x_k, z_k) \in M$, $T(x_k, z_k) \leq z_k^0$, and by the lower semicontinuity of T at $(x_0, z_0) \in \text{cl } M = M$, we have $T(x_0, z_0) \leq z_0^0$, or $(z_0^0, z_0) \in \tilde{Q}(x_0)$. We have proved that the sets $\tilde{Q}(x)$ have property (K) in A. Thus we have proved (a).

Let us prove the second part of (17.2.i). Here the graph of $\tilde{Q}$ is epi T, and by Remark 2 after (8.5.iii) we know that $x \to \tilde{Q}(x)$ certainly has property (K) in A if the graph of $\tilde{Q}(x)$ (that is, epi T) is closed, and that the converse is also true if A is closed. On the other hand, from (17.1.ii), we know that epi T is closed if and only if the extended function T is lower semicontinuous in $R^{\nu+n}$. This proves (b). The last part of (17.2.i) is only a corollary of (a), (b), and (17.1.ii). With $\tilde{Q}(x)$ the empty set and $Q(x) = R^n$ for $x \in R^\nu - A$, then parts (a), (b), (c) hold for $A = R^\nu$ as necessary and sufficient conditions with no restrictions. □

17.2.ii. *If $\tilde{Q}(\bar{x})$ is convex, then either $T(\bar{x}, z) = -\infty$ for all $z \in \text{Rint } Q(\bar{x})$; or $T(\bar{x}, z)$ is finite everywhere in $Q(\bar{x})$ and a convex function of z in $Q(\bar{x})$, $T(\bar{x}, z)$ is bounded below on every bounded subset of $Q(\bar{x})$, and $T(\bar{x}, z)$ is continuous on the convex set* Rint *$Q(\bar{x})$, open with respect to R. Finally, if $Q(\bar{x})$ is convex and closed and $T(\bar{x}, \bar{z}) > -\infty$ for all $z \in Q(\bar{x})$, then $T(\bar{x}, z)$ is lower semicontinuous at every point $z \in Q(\bar{x}) - \text{Rint } Q(\bar{x})$, hence everywhere in $\tilde{Q}(x)$.*

Proof. If $Q(\bar{x})$ is a single point, then $r = 0$, $\text{Rint } Q(\bar{x}) = \emptyset$, and nothing has to be proved. Assume that $Q(\bar{x})$ is not a single point. Then, $1 \leq r \leq n$, and $\text{Rint } Q(\bar{x}) \neq \emptyset$. Let $\bar{z}$ be any point $\bar{z} \in \text{Rint } Q(\bar{x})$. Assume that, at some point $z_1 \in Q(\bar{x})$, $z_1 \neq \bar{z}$, we have $T(\bar{x}, z_1) = -\infty$, and let us prove that $T(\bar{x}, \bar{z}) = -\infty$. For any integer k, there are points $(z_k^0, z_1) \in \tilde{Q}(\bar{x})$ with $z_k^0 < -k$, $k = 1, 2, \ldots$. Take $\lambda = z_1 - \bar{z}$, and choose $\delta > 0$ so small that $z_2 = \bar{z} - \lambda\delta \in \text{Rint } Q(\bar{x})$. Take any point $(z_2^0, z_2) \in \tilde{Q}(\bar{x})$, and note that all points

$$(\alpha z_2^0 + (1 - \alpha)z_k^0, \alpha z_2 + (1 - \alpha)z_1), \qquad 0 \leq \alpha \leq 1,$$

belong to $\tilde{Q}(\bar{x})$. In particular, for $\alpha = (1 + \delta)^{-1}$, we have

$$\begin{aligned} \alpha z_2 + (1 - \alpha)z_1 &= \alpha(\bar{z} - \lambda\delta) + (1 - \alpha)z_1 \\ &= \bar{z} - (1 - \alpha)(\bar{z} - z_1) - \alpha\lambda\delta = \bar{z} + \lambda(1 - \alpha - \alpha\delta) = \bar{z}, \end{aligned}$$

$$T(\bar{x}, \bar{z}) \leq \alpha z_2^0 + (1 - \alpha)z_k^0 \leq (1 + \delta)^{-1}z_2^0 - (1 - (1 + \delta)^{-1})k,$$

where the last term approaches $-\infty$ as $k \to \infty$; hence $T(\bar{x}, \bar{z}) = -\infty$. Since $\bar{x}$ is any point of Rint $Q(\bar{x})$, we have proved the first part of (17.2.ii).

The remaining parts of (17.2.ii) are now a consequence of the definitions and statements (17.1.vi, vii, ix). □

In the next few lines we show by examples that the cases considered in (17.2.ii) can actually occur, and in particular they can occur in the situation which interests control theory, where $f_0(t, x, u)$, $f(t, x, u) = (f_1, \ldots, f_n)$ are continuous functions of (t, x) in A and of a control variable u. Precisely, let A be a given subset of the tx-space R^{1+n}, for every $(t, x) \in A$ let $U(t, x)$ be a given subset of the u-space R^m, let M denote the set of all $(t, x, u) \in R^{2+n+m}$ with $(t, x) \in A$, $u \in U(t, x)$, let f_0 and f be defined on M, and take

$$\tilde{Q}(t, x) = [(z^0 z) \,|\, z^0 \geq f_0(t, x, u),\ z = f(t, x, u),\ u \in U(t, x)] \subset R^{n+1},$$
$$Q(t, x) = [z \,|\, z = f(t, x, u),\ u \in U(t, x)] \subset R^n.$$

Let A be closed, M closed, f_0 and f continuous on M. The first of the two cases mentioned in (17.2.ii) may actually occur even in situations where the sets $\tilde{Q}$ have property (Q) at $\bar{x}$. Indeed, take $m = n = 1$, $f_0 = u$, $f = 0$, $U = R$. Then $Q = [z \,|\, z = 0]$, $\tilde{Q} = [(z^0, z) \,|\, z^0 \in R,\ z = 0]$, and $T = -\infty$. As another example, take $n = 1$, $m = 2$, u, v control variables, $f_0 = u$, $f = \sin v$, $U = [(u, v) \in R^2]$. Then $Q = [z \,|\, -1 \leq z \leq 1]$, $\tilde{Q} = [(z^0, z) \,|\, z^0 \in R,\ -1 \leq z \leq 1]$, and $T(z) = -\infty$ for all $-1 \leq z \leq 1$. In both cases, Q and $\tilde{Q}$ are fixed, closed, convex sets, and certainly have property (Q). As a third example, take $n = 1$, $m = 2$, u, v control variables, $f_0 = (1 - \sin^2 v)u$, $f = \sin v$, $U = [(u, v) \in R^2]$. Then $Q = [z \,|\, -1 \leq z \leq 1]$ and $\tilde{Q} = [(z^0, z) \,|\, z^0 \in R$ if $-1 < z < 1$; $z^0 \geq 0$ if $z = \pm 1]$. Finally, $T(z) = -\infty$ for $-1 < z < 1$, $T(z) = 0$ for $z = \pm 1$.

The following example shows that $T(\bar{x}, z)$ may not be lower semicontinuous on $Q(\bar{x})$ if the set $\tilde{Q}(\bar{x})$ is not closed. As usual, we shall denote by $[g(P)]_h$ the function of P which has the value $g(P)$ if $g(P) < h$, and the value h if $g(P) \geq h$. Now, take $n = 1$, $m = 2$, u, v control variables, $f_0 = [(1 - \sin^2 v)u]_{-1}$, $f = \sin v$, $U = [(u, v) \in R^2]$. Then, $Q = [z \,|\, -1 \leq z \leq 1]$, $\tilde{Q} = [(z^0, z) \,|\, z^0 \geq -1$ if $-1 < z < 1$; $z^0 \geq 0$ if $z = \pm 1]$. Finally, $T(z) = -1$ for $-1 < z < 1$, $T(z) = 0$ for $z = \pm 1$, and the set $\tilde{Q}$ is not closed.

The following example shows that, even if the sets $\tilde{Q}(x)$ is closed and convex, the function $T(\bar{x}, z)$ may not be continuous at the points $z \in Q(\bar{x}) - R$int $Q(\bar{x})$. Let Q be the convex set $[(\xi, \eta) \,|\, 0 \leq \xi \leq 1,\ \eta \geq 0,\ (\xi - 1)^2 + \eta^2 \leq 1]$, and let $T(\xi, \eta)$ be defined by taking $T = \eta$ for $0 \leq \xi \leq 1$, $0 \leq \eta \leq \xi$, $T = (2\xi)^{-1}(\xi^2 + \eta^2)$ for $0 < \xi < 1$, $\xi \leq \eta \leq (1 - (1 - \xi)^2)^{1/2}$. As we have seen in Section 17.1, $T(\xi, \eta)$ is convex and bounded in Q, and continuous in Q except at the point $(\xi = 0, \eta = 0)$. Now let us define f_0, f, U. First, let U be the union of the two disjoint sets

$$U_1 = [(u, v, w) \,|\, 0 \leq u \leq 1,\ -1 \leq v \leq u - 1,\ w \geq 0],$$
$$U_2 = [(u, v, w) \,|\, 0 \leq u \leq 1,\ u \leq v \leq (1 - (1 - u)^2)^{1/2},\ w \geq 0].$$

Let $\sigma(w) = (w + 1)^{-1}$, $w \geq 0$. Finally, let us define the functions $f_0(u, v, w)$, $f_1(u, v, w)$, $f_2(u, v, w)$, continuous on $U = U_1 \cup U_2$, by taking $f_1 = u$, $f_2 = v + 1$, $f_0 = v + 1$ on U_1, and

$$f_1 = u, \qquad f_2 = v, \qquad f_0 = \frac{\sigma(w) + (1 - \sigma(w))(u^2 + v^2)}{\sigma(w) + 2(1 - \sigma(w))u}$$

on U_2. Then, if $\tilde{Q}$ denotes the corresponding set

$$\tilde{Q} = [(z^0, \xi, \eta) \,|\, z^0 \geq f_0,\ \xi = f_1,\ \eta = f_2,\ (u, v, w) \in U = U_1 \cup U_2]$$

and

$$T(\xi,\eta) = \inf[z^0 | (z^0,\xi,\eta) \in \tilde{Q}],$$

then T is exactly the convex function defined above on Q, and Q is convex and closed, but T is discontinuous.

The following example shows that, at a point $\bar{z} \in Q(\bar{x}) - R\text{int}\, Q(\bar{x})$ the supporting plane of $\tilde{Q}(\bar{x})$ may be vertical even if $\tilde{Q}(\bar{x})$ is convex and closed, $Q(\bar{x})$ is convex and compact, and $T(\bar{x}, z)$ continuous on $Q(\bar{x})$. Indeed, take

$$Q = [(u,v) | u^2 + v^2 \leq 1], \qquad T = -(1 - u^2 - v^2)^{1/2},$$
$$U = Q, \quad f_1 = u, \quad f_2 = v, \quad f_0 = T,$$
$$\tilde{Q} = [(z^0, u, v) | z^0 \geq T, (u,v) \in U].$$

17.3 Seminormality

Let $A \subset R^\nu$, $Q(x) \subset R^n$, $\tilde{Q}(x) \subset R^{n+1}$ be the sets introduced in the previous sections, and $T(x,z)$, $x \in A$, $z \in R^n$, the corresponding (extended) real valued function defined by (17.2.1).

For every $\bar{x} \in A$ and $\delta > 0$ let $\tilde{Q}(\bar{x};\delta)$ denote the set

$$\tilde{Q}(\bar{x};\delta) = \bigcup_{x \in N_\delta(\bar{x})} \tilde{Q}(x),$$

where $N_\delta(\bar{x})$ is the set of all $x \in A$ with $|x - \bar{x}| \leq \delta$. We say that condition (α) is satisfied at a point $(\bar{x}, \bar{z})$, $\bar{x} \in A$, $\bar{z} \in R^n$, provided

$$(\alpha) \qquad \text{if} \quad (z^0, \bar{z}) \in \bigcap \text{cl co}\, \tilde{Q}(\bar{x};\delta), \quad \text{then} \quad \bar{z} \in Q(\bar{x}).$$

Thus, condition (α) is a necessary condition for property (Q).

Note that whenever $Q(x) = R^n$ for every $x \in A$, this condition (α) is trivially satisfied. This case, $Q(x) = R^n$ for all $x \in A$, is the usual case for classical problems of the calculus of variations, with x replaced by (t,x), T replaced by $f_0(t,x,x')$, and f_0 defined in $A \times R^n$.

We shall now introduce the following condition (X) at a point $(\bar{x}, \bar{z})$, $\bar{x} \in A$, $\bar{z} \in Q(\bar{x})$:

X. For every $\varepsilon > 0$ there are numbers $\delta > 0$ and r real, and a real vector $p = (p_1, \ldots, p_n)$, such that

(X_1) $T(x,z) \geq r + p \cdot z$ for all $z \in Q(x)$ and all $x \in N_\delta(\bar{x}) \cap A$;
(X_2) $T(\bar{x}, \bar{z}) < r + p \cdot \bar{z} + \varepsilon$.

For short, we shall say that T is *seminormal* at $(\bar{x}, \bar{z})$ if properties (α) and (X) hold at $(\bar{x}, \bar{z})$. We say that T is seminormal at $\bar{x} \in A$ if properties (α) and (X) hold at the points $(\bar{x}, \bar{z})$ for all $\bar{z} \in Q(\bar{x})$. Seminormality in A then means that properties (α), (X) hold at all $(\bar{x}, \bar{z})$, $\bar{x} \in A$, $\bar{z} \in Q(\bar{x})$.

Finally, we say that property (X′) holds at $(\bar{x}, \bar{z})$ provided

X′. For every $\varepsilon > 0$ there are numbers $\delta > 0$, $v > 0$, r real, and a real vector $p = (p_1, \ldots, p_n)$ such that

(X'_1) $T(x,z) \geq r + p \cdot z + v|z - \bar{z}|$ for all $z \in Q(x)$, and all $x \in N_\delta(\bar{x})$ in A;
(X'_2) $T(\bar{x}, \bar{z}) < r + p \cdot \bar{z} + \varepsilon$.

Again, we shall say that T is *normal* at $(\bar{x}, \bar{z})$ if properties (α) and (X') hold at $(\bar{x}, \bar{z})$. We say that T is normal at $\bar{x} \in A$ if properties (α) and (X') hold at the points $(\bar{x}, \bar{z})$ for all $\bar{z} \in Q(\bar{x})$. We say that T is normal in A if properties (α) and (X') hold at all points $(\bar{x}, \bar{z})$, $\bar{x} \in A$, $\bar{z} \in Q(\bar{x})$.

17.4 Criteria for Property (Q)

As above, A is a closed subset of the x-space R^ν, and for each $x \in A$ a subset $Q(x)$ of R^n is given. Let M denote the set $M = [(x, z) | x \in A, z \in Q(x)] \subset R^{\nu+n}$. For every $x \in A$ let $\tilde{Q}(x)$ denote a subset of R^{n+1} whose projection on the z-space R^n is $Q(x)$, and such that, if $(z^0, z) \in \tilde{Q}(x)$, $z'^0 > z^0$, then $(z'^0, z) \in \tilde{Q}(x)$. For every $x \in A$ let $T(x, z) = \inf[z^0 | (z^0, z) \in \tilde{Q}(x)]$, $-\infty \leq T(x, z) \leq +\infty$, $z \in R^n$. Then $T(x, z) < +\infty$ for $x \in A$, $z \in Q(x)$; $T(x, z) = +\infty$ for $x \in A$, $z \in R^n - Q(x)$.

Criterion 1. *Let A be closed, $T(x, z)$ lower semicontinuous on M, and $\tilde{Q}(x) = [(z^0, z) | z^0 \geq T(x, z)]$. If there is a real valued function $\phi(\zeta)$, $0 \leq \zeta < +\infty$, bounded below, such that $\phi(\zeta)/\zeta \to +\infty$ as $\zeta \to +\infty$, $(z^0, z) \in \tilde{Q}(x)$ implies $z^0 \geq \phi(|z|)$, and the set $\tilde{Q}(\bar{x})$ is convex, then the sets $\tilde{Q}(x)$ have property* (Q) *at $\bar{x}$.*

This is a corollary of (10.5.i).

Criterion 2. *Let A be closed, and for any $x \in A$ let $Q(x)$, $\tilde{Q}(x)$ be given sets in R^n and R^{n+1} such that $Q(x)$ is the projection of $\tilde{Q}(x)$ on the z-space. Let $\tilde{Q}(x) = [(z^0, z) | z^0 \geq T(x, z)$, $z \in Q(x)]$, $x \in A$. If T satisfies properties* (α) *and* (X) *at a point $\bar{x} \in A$, then the sets $\tilde{Q}(x)$ have property* (Q) *at $\bar{x}$.*

Proof. We assume that, for a given $\bar{x} \in A$, T satisfies conditions (α) and (X) at every $(\bar{x}, \bar{z})$, $\bar{z} \in Q(\bar{x})$, and we prove that the sets $\tilde{Q}(x)$ satisfy condition (Q) at $\bar{x}$ (and hence $\tilde{Q}(\bar{x})$ is closed and convex). We have only to prove that, if $\bar{\tilde{z}} = (\bar{z}^0, \bar{z}) \in \bigcap_\delta \operatorname{cl} \operatorname{co} \tilde{Q}(\bar{x}; \delta)$, then $\bar{\tilde{z}} = (\bar{z}^0, \bar{z}) \in \tilde{Q}(\bar{x})$. From condition (α) we know already that $\bar{z} \in Q(\bar{x})$.

For $\bar{\tilde{z}} = (\bar{z}^0, \bar{z}) \in \bigcap_\delta \operatorname{cl} \operatorname{co} \tilde{Q}(\bar{x}; \delta)$ and any $\delta > 0$ we certainly have $\bar{\tilde{z}} \in \operatorname{cl} \operatorname{co} Q(\bar{x}; \delta)$ and thus there are points $\tilde{z} = (z^0, z) \in \operatorname{co} \tilde{Q}(\bar{x}, \delta)$ at a distance as small as we want from $\bar{\tilde{z}} = (\bar{z}^0, \bar{z})$. Thus, there is a sequence of numbers $\delta_k > 0$ and of points $\tilde{z}_k = (z_k^0, z_k) \in \operatorname{co} \tilde{Q}(\bar{x}; \delta_k)$ such that $\delta_k \to 0$, $\tilde{z}_k \to \bar{\tilde{z}}$ as $k \to \infty$. In other words, for every integer k, there is a system of points $x_k^\gamma \in N_{\delta_k}(\bar{x})$, $\tilde{z}_k^\gamma = (z_k^{0\gamma}, z_k^\gamma) \in \tilde{Q}(x_k^\gamma)$, and numbers $\lambda_k^\gamma \geq 0$, $\gamma = 1, \ldots \mu$, such that

$$1 = \sum_\gamma \lambda_k^\gamma, \quad \tilde{z}_k = \sum_\gamma \lambda_k \tilde{z}_k^\gamma, \quad z_k^0 = \sum_\gamma \lambda_k^\gamma z_k^{0\gamma}, \quad z_k = \sum_\gamma \lambda_k^\gamma z_k^\gamma, \tag{17.4.1}$$
$$z_k^{0\gamma} \geq T(x_k^\gamma, z_k^\gamma), \qquad z_k^\gamma \in Q(x_k^\gamma),$$

where $\sum_\gamma$ ranges over $\gamma = 1, \ldots, \mu$, and $x_k^\gamma \to \bar{x}$, $\tilde{z}_k \to \bar{\tilde{z}}$, $z_k^0 \to \bar{z}^0$, $z_k \to \bar{z}$ as $k \to \infty$, $\gamma = 1, \ldots, \mu$. By Carathéodory's theorem we may take $\mu = n + 2$. Given $\varepsilon > 0$, by conditions (X_1) and (X_2) there is a neighborhood $N_\delta(x)$ of $\bar{x}$ in A, and numbers r, $b = (b_1, \ldots, b_n)$, such that

$$\bar{T}(x, z) = T(x, z) - r - b \cdot z \geq 0 \quad \text{for all } x \in N_\delta(\bar{x}) \text{ and } z \in Q(x); \tag{17.4.2}$$

$$\bar{T}(\bar{x}, \bar{z}) = T(\bar{x}, \bar{z}) - r - b \cdot \bar{z} \leq \varepsilon. \tag{17.4.3}$$

For k sufficiently large, so that $|x_k^\gamma - \bar{x}| \le \delta$, $\gamma = 1, \dots, \mu$, we have now from (17.4.1), (17.4.2)

$$z_k^0 = \sum_\gamma \lambda_k^\gamma z_k^{0\gamma} \ge \sum_\gamma \lambda_k^\gamma T(x_k^\gamma, z_k^\gamma) \ge \sum_\gamma \lambda_k^\gamma [r + b \cdot z_k^\gamma]$$

$$= r + b \cdot \sum_\gamma \lambda_k^\gamma z_k^\gamma = r + b \cdot z_k.$$

As $k \to \infty$, we obtain $\bar{z}^0 \ge r + b \cdot \bar{z}$; hence, by (17.4.3),

$$\bar{z}^0 \ge r + b \cdot \bar{z} \ge T(\bar{x}, \bar{z}) - \varepsilon.$$

Here $\varepsilon > 0$ is arbitrary; hence $\bar{z}^0 \ge T(\bar{x}, \bar{z})$. This shows that $\bar{\tilde{z}} = (\bar{z}^0, \bar{z}) \in \tilde{Q}(\bar{x})$. We have proved that the sets $\tilde{Q}(x)$ satisfy property (Q) at $\bar{x}$. □

The following criteria are better expressed in terms of control theory. Here A is a subset of the x-space R^ν, for every $x \in A$ a subset $U(x)$ is given in the u-space R^m, $u = (u^1, \dots, u^m)$, and M denotes the set $[(x,u) | x \in A,\ u \in U(x)] \subset R^{\nu+m}$. Let $f_0(x,u)$, $f(x,u) = (f_1, \dots, f_n)$ be given functions on M.

Criterion 3. *Let A be closed, M closed, $f_0(x,u)$, $f(x,u) = (f_1, \dots, f_n)$ continuous on M, and assume that 1 and f are of slower growth than f_0 as $|u| \to +\infty$ uniformly in a closed neighborhood $N_{\delta_0}(\bar{x})$ of $\bar{x}$ in A. If the set $\tilde{Q}(\bar{x})$ is convex, then the sets $\tilde{Q}(x)$ have property (Q) at $\bar{x}$.*

The proof is analogous to the one for (10.5.i) and is left as an exercise for the reader. Here we say that 1 and f are of slower growth than f_0 as $|u| \to +\infty$ uniformly in $N_{\delta_0}(\bar{x})$ provided, given $\varepsilon > 0$, there is N such that for all $|u| \ge N$ and $x \in N_{\delta_0}(\bar{x})$ we have $1 \le \varepsilon f_0(x,u)$, $|f(x,u)| < \varepsilon f_0(x,u)$.

In the following Criterion 4 we shall assume $U = R^m$, $M = A \times R^m$, and, as in Section 17.3, we shall say that the real valued function $f_0(x,u)$, $u \in R^m$, $x \in A$, is is seminormal in u at a point $\bar{x} \in A$ provided for every $\bar{u} \in R^m$ the following condition (X) holds: Given $\varepsilon > 0$ there are $\delta > 0$, and r, $b = (b_1, \dots, b_m)$ real (which may all depend on $\bar{u}$ and ε), such that $f_0(\bar{x}, \bar{u}) < r + b \cdot \bar{u} + \varepsilon$, and $f_0(x,u) \ge r + b \cdot u$ for all $u \in R^m$ and $x \in N(\bar{x}) \cap A$. Note that if $f_0(x,u)$ is seminormal in u at a point $\bar{x} \in A$, and r_0, $b_0 = (b_{01}, \dots, b_{0m})$ are real numbers, then also $f_0(x,u) - r_0 - b_0 \cdot u$ is seminormal in u at the point $\bar{x}$.

Criterion 4. *Let A be closed, $U = R^m$, $M = A \times R^m$, $f_0(x,u)$ continuous on M, and $f = B(x)u + C(x)$, where the entries of the matrices B and C are continuous on A. If f_0 is seminormal in u at a point $\bar{x} \in A$, and there are numbers r_0, $b_0 = (b_{01}, \dots, b_{0m})$ real and $\delta_0 > 0$, $\sigma > 0$, such that $f_0(x,u) \ge r_0 + b_0 \cdot u + \sigma|u|$ for all $x \in N_{\delta_0}(\bar{x})$ and $u \in R^m$, then the sets $\tilde{Q}(x) = [(z^0, z) | z^0 \ge f_0(x,u),\ z = f(x,u),\ u \in R^m] \subset R^{n+1}$ have property (Q) at $\bar{x}$.*

Proof. We know that $f_0(x,u) - r_0 - b_0 \cdot u \ge \sigma|u|$ for all $x \in N_{\delta_0}(\bar{x})$ and $u \in R^m$. By replacing f_0 with $f_0 - r_0 - b_0 \cdot u$ if necessary, we see that it is not restrictive to assume $f_0 \ge \sigma|u|$ for all $x \in N_{\delta_0}(\bar{x})$ and $u \in R^m$. We have to prove that $\tilde{z} = (z^0, z) \in \bigcap_\delta \operatorname{cl} \operatorname{co} \tilde{Q}(\bar{x}; \delta)$ implies $\tilde{z} \in \tilde{Q}(\bar{x})$. Let $\bar{\tilde{z}}$ be a given point $\bar{\tilde{z}} = (\bar{z}^0, \bar{z}) \in \bigcap_\delta \operatorname{cl} \operatorname{co} \tilde{Q}(\bar{x}; \delta)$, and let us prove that $\bar{\tilde{z}} \in Q(\bar{x})$. For every $\delta > 0$ we have $\bar{\tilde{z}} \in \operatorname{cl} \operatorname{co} \tilde{Q}(\bar{x}; \delta)$, and thus, for every $\delta > 0$, there are points $\tilde{z} = (z^0, z) \in \operatorname{co} \tilde{Q}(\bar{x}; \delta)$ at a distance as small as we want from $\bar{\tilde{z}} = (\bar{z}^0, \bar{z})$. Thus, there is a sequence of numbers $\delta_k > 0$ and points $\tilde{z}_k = (z_k^0, z_k) \in \operatorname{co} \tilde{Q}(\bar{x}; \delta_k)$ such that $\delta_k \to 0$, $\tilde{z}_k \to \bar{\tilde{z}}$ as $k \to \infty$. In other words, for every integer k, there are a system of points $x_k^\gamma \in N_{\delta_k}(\bar{x})$, $\gamma = 1, \ldots, v$, say $v = n + 2$, corresponding points $\tilde{z}_k^\gamma = (z_k^{0\gamma}, z_k^\gamma) \in \tilde{Q}(x_k^\gamma)$, points $u_k^\gamma \in R^m$, and numbers λ_k^γ, $0 \le \lambda_k^\gamma \le 1$, $\gamma = 1, \ldots, v$, such that

$$1 = \sum_\gamma \lambda_k^\gamma, \quad \tilde{z}_k = \sum_\gamma \lambda_k^\gamma \tilde{z}_k^\gamma, \quad z_k^0 = \sum_\gamma \lambda_k^\gamma z_k^0, \quad z_k = \sum_\gamma \lambda_k^\gamma z_k^\gamma,$$

(17.4.4)

$$z_k^{0\gamma} \ge f_0(x_k^\gamma, u_k^\gamma), \qquad z_k^\gamma = f(x_k^\gamma, u_k^\gamma) = B(x_k^\gamma)u_k^\gamma + C(x_k^\gamma),$$

where $\gamma = 1, \ldots, v$, $k = 1, 2, \ldots$, where $\sum_\gamma$ ranges over $\gamma = 1, \ldots, v$, $x_k^\gamma \in N_{\delta_k}(\bar{x})$, and where $x_k^\gamma \to \bar{x}$, $\tilde{z}_k \to \tilde{\bar{z}}$, $z_k^0 \to \bar{z}^0$, $z_k \to \bar{z}$ as $k \to \infty$, $\gamma = 1, \ldots, v$.

By hypothesis $f_0(x, u) \ge \sigma|u|$ for all $x \in N_{\delta_0}(\bar{x})$. If k is sufficiently large so that $\delta_k \le \delta_0$, and hence $|x_k^\gamma - \bar{x}| \le \delta_k < \delta_0$, then because $\zeta = \sigma|u|$ is a convex function in u, we have

$$z_k^0 = \sum_\gamma \lambda_k^\gamma z_k^{0\gamma} \ge \sum_\gamma \lambda_k^\gamma f_0(x_k^\gamma, u_k^\gamma) \ge \sum_\gamma \lambda_k^\gamma \sigma|u_k^\gamma| \ge \sigma\left|\sum_\gamma \lambda_k^\gamma u_k^\gamma\right|. \tag{17.4.5}$$

Thus, $|\sum_\gamma \lambda_k^\gamma u_k^\gamma| \le \sigma^{-1} z_k^0$, where $z_k^0 \to \bar{z}^0$ as $k \to \infty$. This proves that $\sum_\gamma \lambda_k^\gamma u_k^\gamma$, $k = 1, 2, \ldots$, is a bounded sequence of points of R^m. By a suitable extraction, there is a subsequence, say still $[k]$, such that $u_k = \sum_\gamma \lambda_k^\gamma u_k^\gamma \to \bar{u} \in R^m$ as $k \to \infty$.

From the third relation (17.4.4) where $z_k^0 \to \bar{z}^0$, $z_k^{0\gamma} \ge 0$, $0 \le \lambda_k^\gamma \le 1$, we derive that each of the v sequences $[\lambda_k^\gamma z_k^{0\gamma}, k = 1, 2, \ldots]$, $\gamma = 1, \ldots, v$, is bounded. From the fifth relation (17.4.4) we then derive that

$$\lambda_k^\gamma z_k^{0\gamma} \ge \lambda_k^\gamma f_0(x_k^\gamma, u_k^\gamma) \ge \lambda_k^\gamma \sigma|u_k|, \tag{17.4.6}$$

and hence $\lambda_k^\gamma|u_k^\gamma| \le \sigma^{-1}\lambda_k^\gamma z_k^{0\gamma}$. Thus, each of the v sequences $[\lambda_k^\gamma u_k^\gamma, k = 1, 2, \ldots]$, $\gamma = 1, \ldots, v$, is bounded.

If we denote by Δ_k^γ the expression

$$\begin{aligned}\Delta_k^\gamma &= \lambda_k^\gamma[B(x_k^\gamma)u_k^\gamma + C(x_k^\gamma)] - \lambda_k^\gamma[B(\bar{x})u_k^\gamma + C(\bar{x})] \\ &= [B(x_k^\gamma) - B(\bar{x})]\lambda_k^\gamma u_k^\gamma + \lambda_k^\gamma[C(x_k^\gamma) - C(\bar{x})],\end{aligned}$$

and because of the continuity of B and C, since $x_k \to \bar{x}$, $0 \le \lambda_k \le 1$, we conclude that $\Delta_k^\gamma \to 0$ as $k \to \infty$, $\gamma = 1, \ldots, v$.

Given $\varepsilon > 0$, by the seminormality of $f_0(x, u)$ in u at $\bar{x}$, and for the point $\bar{z} \in R^m$ determined above, there are numbers $\delta' > 0$ and r, $b = (b_1, \ldots, b^m)$ real such that

$$\begin{aligned}&f_0(x, u) \ge r + b \cdot u \quad \text{for all } x \in N_{\delta'}(\bar{x}),\ u \in R^m,\\ &f_0(\bar{x}, \bar{u}) \le r + b \cdot \bar{u} + \varepsilon.\end{aligned}$$

Now we have, for k sufficiently large,

$$\begin{aligned}z_k^0 &= \sum_\gamma \lambda_k^\gamma z_k^{0\gamma} \ge \sum_\gamma \lambda_k^\gamma f_0(x_k^\gamma, u_k^\gamma) \ge \sum_\gamma \lambda_k^\gamma[r + b \cdot u_k^\gamma] = r + b \cdot u_k \\ &= r + b \cdot \bar{u} + b \cdot (u_k - \bar{u}) \ge f_0(\bar{x}, \bar{u}) + b \cdot (u_k - \bar{u}) - \varepsilon,\\ z_k &= \sum_\gamma \lambda_k^\gamma z_k^\gamma = \sum_\gamma \lambda_k^\gamma[B(x_k)u_k^\gamma + C(x_k)] \\ &= \sum_\gamma \lambda_k^\gamma[B(\bar{x})u_k^\gamma + C(\bar{x})] + \sum_\gamma \Delta_k^\gamma = B(\bar{x})u_k + C(\bar{x}) + \sum_\gamma \Delta_k^\gamma.\end{aligned}$$

At the limit as $k \to \infty$, we obtain

$$\bar{z}^0 \ge f_0(\bar{x}, \bar{u}) - \varepsilon, \qquad \bar{z} = B(\bar{x})\bar{u} + C(\bar{x}),$$

and because $\varepsilon > 0$ is arbitrary, also $\bar{z}^0 \ge f_0(\bar{x}, \bar{u})$, $\bar{z} = f(x, u)$; hence $\tilde{\bar{z}} = (\bar{z}^0, \bar{z}) \in Q(\bar{x})$. Criterion 4 is thereby proved. □

Criterion 5. *Let A be closed, $U = R^m$, $M = A \times R^m$, $f_0(x,u)$, $f(x,u)$ continuous on M. Let $\bar{x} \in A$, and $N_\delta(\bar{x})$ be a neighborhood of $\bar{x}$ in A such that:* (1) *for every $\varepsilon > 0$ there is a constant $\mu_\varepsilon > 0$ such that $|f(x,u)| \le \mu_\varepsilon + \varepsilon f_0(x,u)$ for all $x \in N_\delta(\bar{x})$, and* (2) *there is an increasing function $\Lambda(\zeta)$, $0 \le \zeta < +\infty$, with $\Lambda(\zeta) \to +\infty$ as $\zeta \to \infty$, such that $f_0(x,u) \ge \Lambda(|u|)$ for all $x \in N_\delta(\bar{x})$ and $u \in R^m$. If $\tilde{Q}(\bar{x})$ is convex, then the sets $\tilde{Q}(x)$ have property* (Q) *at $\bar{x}$.*

Proof. The proof proceeds as for Criterion 4 up to relation (17.4.5), which becomes here

$$z_k^0 = \sum_\gamma \lambda_k^\gamma z_k^{0\gamma} \ge \sum_\gamma \lambda_k^\gamma f_0(x_k^\gamma, u_k^\gamma) \ge \sum_\gamma \lambda_k^\gamma \Lambda(|u_k^\gamma|). \tag{17.4.7}$$

Let us divide the v sequences $[u_k^\gamma, k = 1, 2, \ldots]$ or indices $\gamma = 1, \ldots, v$ into two categories. The first category is the one for which $[u_k^\gamma]$ is bounded; then by an extraction we can assume that $u_k^\gamma \to u^\gamma \in R^m$ as $k \to \infty$. Let us put in the second category the remaining ones, for which $[u_k^\gamma]$ is unbounded, and then by a further extraction we can assume that $|u_k^\gamma| \to \infty$ as $k \to \infty$. Let sums $\sum_\gamma'$, $\sum_\gamma''$ range over the two categories of indices γ. We may well assume by a further extraction that the sequences $[\lambda_k^\gamma]$ have a limit λ^γ as $k \to \infty$, $0 \le \lambda^\gamma \le 1$, $\gamma = 1, \ldots, v$.

For the terms of the second category in the last member of (17.4.7) we have $\Lambda(|u_k^\gamma|) \to +\infty$, while the first member in (17.4.7) is bounded, and the remaining terms remain bounded. Hence, $\lambda_k^\gamma \to \lambda^\gamma = 0$ as $k \to \infty$ for γ of the second category. Furthermore $\lambda_k^\gamma \Lambda(|u_k^\gamma|)$ is nonnegative and bounded; hence $\lambda_k^\gamma f_0(x_k^\gamma, u_k^\gamma) \ge 0$ for k sufficiently large, and (17.4.7) yields $z_k^0 \ge \sum_\gamma' \lambda_k^\gamma f_0(x_k^\gamma, u_k^\gamma)$. At the limit as $k \to \infty$ we have

$$\bar{z}^0 \ge \sum{}' \lambda^\gamma f_0(\bar{x}, u^\gamma). \tag{17.4.8}$$

We also have for all k

$$z_k = \sum_\gamma \lambda_k^\gamma f(x_k^\gamma, u_k^\gamma) = \left(\sum_\gamma{}' + \sum_\gamma{}''\right)(\lambda_k^\gamma f(x_k^\gamma, u_k^\gamma)), \tag{17.4.9}$$

and for the indices of the second category and k sufficiently large we have

$$\lambda_k^\gamma |f(x_k^\gamma, u_k^\gamma)| \le \lambda_k^\gamma \mu_\varepsilon + \varepsilon(\lambda_k^\gamma f_0(x_k^\gamma, u_k^\gamma)), \tag{17.4.10}$$

where the terms in parentheses are nonnegative and bounded, say $\le M$ for all k, and $\lambda_k^\gamma \mu_\varepsilon \to 0$ as $k \to \infty$. Thus, for k sufficiently large we have $\lambda_k^\gamma |f(x_k^\gamma, u_k^\gamma)| \le 2M\varepsilon$. Here $\varepsilon > 0$ is arbitrary; thus, $\lambda_k^\gamma |f(x_k^\gamma, u_k^\gamma)| \to 0$ as $k \to \infty$ for every index γ of the second category. Relation (17.4.9) yields now, as $k \to \infty$,

$$\bar{z} = \sum{}' \lambda^\gamma f(\bar{x}, u^\gamma). \tag{17.4.11}$$

From (17.4.10) and (17.4.11) we conclude that $(\bar{z}^0, \bar{z}) \in \operatorname{co} \tilde{Q}(\bar{x})$ and hence $(\bar{z}^0, \bar{z}) \in \tilde{Q}(\bar{x})$, since $\tilde{Q}(\bar{x})$ is convex. Property (Q) at $\bar{x}$ is thereby proved. □

Criterion 6. *Let A be closed, $U = R^m$, $M = A \times R^m$, $f_0(x,u)$ continuous in M, and $f(x,u) = Bu + C(x)$, where B is a constant matrix with* rank $B = m$, *and the entries of the matrix C are continuous in A. Let f_0 be seminormal in u at a point $\bar{x} \in A$. Then the sets $\tilde{Q}(x)$ have property* (Q) *at $\bar{x}$.*

Criterion 7. *Let A be closed, $U = R^m$, $M = A \times R^m$, $f_0(x,u)$ continuous in M, and $f(x,u) = B(x)u + C(x)$, where the entries of the matrices B and C are continuous in A. Let f_0 be* *seminormal in u at a point $\bar{x} \in A$. Let us further assume that there are numbers r_0, $b_0 = (b_{01}, \ldots, b_{0n})$ and $\delta_0 > 0$, $\sigma > 0$ such that $f_0(x, u) \geq r_0 + b_0 \cdot f(x, u) + \sigma|u|$ for all $x \in N_{\delta_0}(\bar{x})$ and all $u \in R^m$. Then the sets $\tilde{Q}(x)$ have property* (Q) *at $\bar{x}$.*

Criterion 8. *Let A be closed, $U = R^m$, $M = A \times R^m$, $f_0(x, u)$ continuous in M, and $f(x, u) = B(x)u + C(x)$, where the entries of the matrices B and C are continuous in A. Let f_0 be seminormal in u at a point $\bar{x} \in A$. Let us further assume that $f_0(x, u) \to +\infty$ as $|u| \to +\infty$ uniformly in some compact neighborhood $N_{\delta_0}(\bar{x})$ of $\bar{x}$ in A. Then the sets $\tilde{Q}(x)$ have property* (Q) *at $\bar{x}$.*

Criterion 9. *Let A be closed, $U = R^m$, $M = A \times R^m$, $f_0(x, u)$ continuous in M, and $f(x, u) = B(x)u + C(x)$, where the entries of the matrices B and C are continuous in A. Let f_0 be normal in u at a point $\bar{x} \in A$. Let us further assume that* rank $B(\bar{x}) = m$. *Then the sets $\tilde{Q}(x)$ have property* (Q) *at $\bar{x}$.*

Criteria 1, 2, 3, 4 were proved by Cesari [8], and Criterion 5 by Rupp [1]. Criteria 6, 7, 8, 9 were proved by Kaiser [3], and we refer to this paper for their proofs and for critical examples.

17.5 A Characterization of Property (Q) for the Sets $\tilde{Q}(t, x)$ in Terms of Seminormality

As usual, let A be a given closed subset of the x-space R^n, for every $x \in A$ let $Q(x)$ be a given subset of the z-space R^n, and let $\tilde{Q}(x)$ be a subset of the z^0z-space R^{n+1} whose projection on the z-space is $Q(x)$. For every $(x, z) \in A \times R^n$ let $T(x, z) = \inf[z^0 | (z^0, z) \in \tilde{Q}(x)]$. We have now the following characterization of property (Q):

17.5.i (Cesari [14]). *If $T(\bar{x}, z) > -\infty$ in $Q(\bar{x})$, and $\tilde{Q}(\bar{x}) = [(z^0, z) | z^0 \geq T(\bar{x}, z), z \in Q(\bar{x})]$, then the sets $\tilde{Q}(x)$ have property* (Q) *at $\bar{x}$ if and only if properties* (α) *and* (X) *hold at the point $\bar{x}$.*

Proof. First we note that if the set $\tilde{Q}(\bar{x})$ is closed, then $\tilde{Q}(\bar{x}) = [(z^0, z) | z^0 \geq T(\bar{x}, z), z \in Q(\bar{x})]$; in other words, $T(\bar{x}, z)$ is a minimum and not a mere inf as in its definition.

For fixed $\bar{x} \in A$ and $\delta > 0$ let us consider the sets

$$\tilde{Q}(\bar{x}; \delta) = \bigcup_{x \in N_\delta(\bar{x})} \tilde{Q}(x) \subset R^{n+1},$$

$$\tilde{Q}^*(\bar{x}; \delta) = \text{co } \tilde{Q}(\bar{x}; \delta) \subset \text{co}\left[\bigcup_{x \in N_\delta(\bar{x})} \tilde{Q}(x)\right] \subset R^{n+1}, \tag{17.5.1}$$

and projection on the z-space R^n,

$$Q^*(\bar{x}; \delta) = \text{co } Q(\bar{x}; \delta) = \text{co}\left[\bigcup_{x \in N_\delta(\bar{x})} Q(x)\right] \subset R^n.$$

Both sets $\tilde{Q}^*(\bar{x};\delta)$ and $Q^*(\bar{x};\delta)$ are convex, and

$$\tilde{Q}(\bar{x}) \subset \tilde{Q}^*(\bar{x};\delta),$$
$$Q(\bar{x}) \subset Q^*(x;\delta).$$

As before, we consider the function T^* analogous to T, or

$$T^*(\bar{x},\delta;z) = \inf[z^0 | (z^0, z) \in \tilde{Q}^*(\bar{x},\delta)], \tag{17.5.2}$$

so that $T^*(\bar{x},\delta,z) = +\infty$ whenever $z \in R^n - Q^*(\bar{x};\delta)$, and $-\infty \le T^*(\bar{x},\delta;z) < +\infty$ for $z \in Q^*(\bar{x};\delta)$. Moreover, we have $T^*(\bar{x},\delta;z) \le T(\bar{x};z)$.

We have already proved in Section 17.4 (Criterion 2) that conditions (α) and (X) are enough to guarantee property (Q). Let us assume that for a given $\bar{x} \in A$, $T(\bar{x},z) > -\infty$ for all $z \in Q(\bar{x})$, and that the sets $\tilde{Q}(x)$ have property (Q) at $\bar{x}$. We have to prove that T satisfies conditions (α) and (X) at all $z \in Q(\bar{x})$. We have already noticed that condition (α) is a necessary condition for property (Q). Also, we know that the set $\tilde{Q}(x)$ is closed and convex. Thus, $\tilde{Q}(\bar{x}) = [(z^0, z) | z^0 \ge T(\bar{x},z), z \in Q(\bar{x})]$.

Since $T(\bar{x},z) > -\infty$ for all $z \in Q(\bar{x})$ by hypothesis, we know from (17.2.ii) that $T(\bar{x},z)$ is a lower semicontinuous convex function of z in the convex set $Q(\bar{x})$. We have already noticed that $-\infty \le T^*(\bar{x},z;\delta) \le T(\bar{x},z) < +\infty$ for all $z \in Q(\bar{x})$ and $\delta > 0$.

Now take any point $\bar{z} \in Q(\bar{x})$, and let $\bar{z}^0 = T(\bar{x},\bar{z})$. Then, as noticed, the point $(\bar{z}^0,\bar{z})$ belongs to $\tilde{Q}(\bar{x})$. Given $\varepsilon > 0$, the point $\bar{P} = (\bar{z}^0 - \varepsilon, \bar{z})$ is not on the closed set $\tilde{Q}(\bar{x})$, and hence has a minimum distance η from this set, with $0 < \eta \le \varepsilon$. Since $T(\bar{x},z)$ is lower semicontinuous at $\bar{z}$, there is some η', $0 < \eta' \le \eta/2$, such that $T(\bar{x},z) > T(\bar{x},\bar{z}) - \eta/3$ for all $z \in Q(\bar{x})$ with $|z - \bar{z}| \le \eta'$.

Let σ be the closed ball in R^{n+1} with center $\bar{P} = (\bar{z}^0 - \varepsilon, \bar{z})$ and radius $\eta'/3$. Let σ_0 denote the projection of σ on the z-space; thus, σ_0 is the closed ball in R^n with center $\bar{z}$ and radius $\eta'/3$. We shall denote also by σ_1 the closed ball in R^n with center $\bar{z}$ and radius $2\eta'/3$. Now let us consider the convex sets $\tilde{Q}^*(\bar{x};\delta) = \mathrm{co}\,\tilde{Q}(\bar{x};\delta)$ defined in (17.5.1) and their relative function $T^*(\bar{x},\delta;z)$ defined in (17.5.2).

Let us prove that there is some $\delta_0 > 0$ such that

$$0 \le T(\bar{x},z) - T^*(\bar{x},\delta;z) \le \eta/3 \tag{17.5.3}$$

for all $0 < \delta \le \delta_0$ and $z \in \sigma_1 \cap Q^*(\bar{x};\delta)$. Indeed, in the contrary case there would be numbers $\delta_k > 0$ and points $z_k \in \sigma_1 \subset R^n$, $k = 1, 2, \ldots$, with $\delta_k \to 0$ as $k \to \infty$ and $T^*(\bar{x},\delta_k,z_k) < T(\bar{x},\bar{z}) - \eta/3$, and hence points $(z_k^0, z_k) \in \mathrm{co}\,\tilde{Q}^*(\bar{x};\delta_k)$ with $z_k^0 \le T(\bar{x},\bar{z}) - \eta/3 = \bar{z}^0 - \eta/3$. Hence, for every $\delta > 0$ we have $(z_k^0, z_k) \in \mathrm{co}\,\tilde{Q}(\bar{x},\delta)$ for all k sufficiently large, and then also $(\bar{z}^0 - \eta/3, z_k) \in \mathrm{co}\,\tilde{Q}(\bar{x},\delta)$. If $\bar{z}'$ is any point of accumulation of $[z_k]$, we have $\bar{z}' \in \sigma_1$, $(\bar{z}^0 - \eta/3, \bar{z}') \in \mathrm{cl\,co}\,\tilde{Q}(\bar{x};\delta)$, and by property (Q) also $(\bar{z}^0 - \eta/3, \bar{z}') \in \tilde{Q}(\bar{x}) = \bigcap_\delta \mathrm{cl\,co}\,\tilde{Q}(\bar{x};\delta)$. This implies $T(\bar{x},\bar{z}') \le \bar{z}^0 - \eta/3$ with $\bar{z}' \in \sigma_1$, $|\bar{z}' - \bar{z}| \le 2\eta'/3 < \eta'$, a contradiction. We have proved that, for some $\delta_0 > 0$, relation (17.5.3) holds for all $0 < \delta \le \delta_0$ and $z \in \sigma_1 \cap Q^*(\bar{x};\delta)$.

Let us prove that any two points

$$P = (z^0, z) \in \sigma \quad \text{and} \quad P' = (z'^0, z') \in \tilde{Q}^*(\bar{x}; \delta_0)$$

have a distance $\{P, P'\} \geq \eta'/3$. Indeed, either P' is outside the cylinder $[z^0 \in R, z \in \sigma_1]$, and then

$$\{P', P\} \geq |z' - z| \geq |z' - \bar{z}| - |z - \bar{z}| \geq 2\eta'/3 - \eta'/3 = \eta'/3,$$

or P' is inside the cylinder above, and then by (17.5.3), for $0 < \delta < \delta_0$,

$$\begin{aligned} z'^0 &\geq T^*(\bar{x}, \delta; z') > T(\bar{x}, \bar{z}) - 2\eta/3 = \bar{z}^0 - 2\eta/3, \\ \{P', P\} &\geq z'^0 - z^0 = [\bar{z}^0 - (\bar{z}^0 - \varepsilon)] + [z'^0 - \bar{z}^0] + [\bar{z}^0 - \varepsilon - z^0] \\ &\geq \varepsilon - 2\eta/3 - \eta'/3 \geq \eta/3 - \eta'/3 \geq \eta'/3. \end{aligned}$$

Thus, the convex sets σ and $\tilde{Q}^*(\bar{x}, \delta)$ have a distance $\geq \eta'/3$, and the same occurs for the closed convex sets σ and cl $\tilde{Q}^*(\bar{x}, \delta)$, σ compact. We conclude that there is some hyperplane π in R^{n+1} separating the two convex sets σ and cl $\tilde{Q}^*(\bar{x}, \delta)$.

This hyperplane π must intersect the vertical segment $[\bar{z}^0 - \varepsilon + \eta/3 \leq z^0 \leq \bar{z}^0, z = \bar{z}]$ at some point $(z^0 = r, z = \bar{z})$, and π cannot be parallel to the z^0-axis, otherwise all points of the straight line $z = \bar{z}$ would be on π; in particular the center P of the ball σ is on π, and not all of σ can be separated from $\tilde{Q}^*(\bar{x}; \delta)$. Thus, π is of the form

$$\pi : z = r + b \cdot (z - \bar{z}) = (r - b \cdot \bar{z}) + b \cdot z;$$

$\tilde{Q}(\bar{x})$ as well as cl $\tilde{Q}^*(\bar{x}; \delta)$ is above π, and thus $(z^0, z) \in$ cl $\tilde{Q}^*(x; \delta)$ implies $z^0 \geq (r - b \cdot \bar{z}) + b \cdot z$. In other words, for $0 < \delta \leq \delta_0$, $x \in N_\delta(\bar{x})$, $x \in A$, we have $T(x, z) \geq (r - b \cdot \bar{z}) + b \cdot z$. On the other hand, $T(\bar{x}, \bar{z}) = \bar{z}^0 = (\bar{z}^0 - \varepsilon) + \varepsilon < r + \varepsilon = (r - b \cdot \bar{z}) + b \cdot \bar{z} + \varepsilon$. We have proved that property (X) holds at the point $\bar{x} \in A$. Statement (17.5.i) is thereby proved. □

17.6 Duality and Another Characterization of Property (Q) in Terms of Duality

A. The Dual Operation

We consider here extended real valued functions Tu, $u \in R^n$, that is, we allow T to have the values $+\infty$ and $-\infty$. In applying the usual definition of convexity, $T((1 - \alpha)u_1 + \alpha u_2) \leq (1 - \alpha)Tu_1 + \alpha Tu_2$ for all $0 \leq \alpha \leq 1$, $u_1, u_2 \in R^n$, we may encounter some difficulties, since forms such as $0(\pm\infty)$ and $+\infty - \infty$ may occur. For such functions it would be more convenient to say that T is convex provided its epigraph is a convex subset of R^{n+1}, where epigraph is defined as usual by epi $T = [(y, u) | Tu \leq y < +\infty, y \neq -\infty, u \in R^n]$. However, we shall soon limit ourselves to functions T which never take the value $-\infty$, which may take the value $+\infty$, though Tu is not identically equal to $+\infty$ (that is, $Tu \neq -\infty$, $Tu \not\equiv +\infty$, $u \in R^n$). For these functions the usual definition of convexity applies with the natural conventions ($r + \infty = +\infty$, $r(+\infty) = +\infty$ for all $r \geq 0$, and $+\infty + \infty = +\infty$).

For instance, for $n = 1$, $T_1 u = +\infty$ for $u < 0$, $T_1 u = -\infty$ for $u \geq 0$, then epi $T_1 = [(y, u) | 0 \leq u < +\infty, -\infty < y < +\infty]$ is convex and closed; for $n = 1$, $T_2 u = +\infty$ for $u \leq 0$, $T_2 u = -\infty$ for $u > 0$, and then epi $T_2 = [(y, u) | 0 < u < +\infty, -\infty < y < +\infty]$ is convex and open; for $n = 1$, $T_3 u = +\infty$ for $u < 0$, $T_3 u = 0$ for $u \geq 0$, and then epi $T_3 = [(y, u) | 0 \leq u < \infty, 0 \leq y < +\infty]$ is convex and closed.

The statements we have proved in Section 17.1 concerning convex functions in a set apply here, with obvious changes. Statements (17.1.i,ii) will be most relevant here, namely,

17.6.i. *If Tu, $u \in R^n$, is an extended real valued function, then* epi T *is convex if and only if T is convex, and* epi T *is closed if and only if T is lower semicontinuous in R^n.*

In the following we will need the operation of closure of a function Tu, $u \in R^n$. We denote by cl T the function defined by the relation epi(cl T) = cl(epi T). Thus, for $n = 1$, $Tu = +\infty$ for $u \leq 0$, $Tu = 0$ for $u > 0$, we have (cl T)(u) $= +\infty$ for $u < 0$ and (cl T)(u) $= 0$ for $u \geq 0$.

Given T as before, we consider all pairs r, p, $r \in R$, $p \in R^n$, such that $-r + p \cdot u \leq Tu$ for all u. In other words, we consider all half spaces $S_0^+ = [z^0 \geq -r + p \cdot u, u \in R^n]$ with epi $T \subset S_0^+$.

17.6.ii. *If Tu, $u \in R^n$, is an extended real valued function, $Tu \not\equiv +\infty$, $Tu \neq -\infty$ for all u, and T is convex and lower semicontinuous in R^n, then for every $\bar{u} \in R^n$,*

$$T\bar{u} = \sup[-r + p \cdot \bar{u} | -r + p \cdot u \leq Tu \textit{ for all } u \in R^n]$$

or equivalently epi $T = \bigcap S_0^+$.

Proof. Let $\{S^+\}$ be the family of all half spaces $S^+ = [(z^0, u) | p_0 z^0 + p \cdot u + c \geq 0]$ containing cl(epi T), and let $\{S_0^+\}$ be the family of all half spaces considered above. Thus $\{S^+\} \supset \{S_0^+\}$, and hence, by (8.4.vi), cl(epi T) $= \bigcap S^+ \subset \bigcap S_0^+$. On the other hand, the sole half spaces in $\{S^+\} - \{S_0^+\}$ are those of the form $[(z^0, u) | pu + c \geq 0]$, that is, $p_0 = 0$, or $R \times S_{00}^+$ where $S_{00}^+ = [u | pu + c \geq 0] \subset R^n$ are the half spaces in R^n whose intersection is the convex set cl U, and these do not affect epi T. Thus, since epi T is closed by (17.6.i), we have epi T = cl(epi T) $= \bigcap S^+ = \bigcap S_0^+$. This proves (17.6.ii).

□

If Tu, $u \in R^n$, is any extended real valued function in R^n, then for every $y \in R^n$, we consider all r, if any, such that $-r + y \cdot u \leq Tu$ for all $u \in R^n$, and we take $T^*y = \inf r$. In other words, we take

$$T^*y = \sup[y \cdot u - Tu \mid u \in R^n], \qquad y \in R^n. \tag{17.6.1}$$

Indeed, if $r = T^*y$, then $r \geq y \cdot u - Tu$ for all u, that is, $Tu \geq -r + y \cdot u$ for all u, and $r = T^*y$ is the inf of all numbers r for which this holds. Note that if $-r + y \cdot u \leq Tu$ for all u holds for no $r \in R$, that is, the class of such r is empty, then $T^*y = +\infty$. We say that the extended function T^* is the *dual* of T, and that the passage from T to T^* is the *dual* operation.

The following examples may clarify: (a) Let $n = 1$, $Tu = 0$ if $-1 \leq u \leq 1$, $Tu = +\infty$ if $|u| > 1$, and then $T^*y = |y|$ for all $-\infty < y < +\infty$. (b) Let $n = 1$, $Tu = u$, $-\infty < u < +\infty$, and then $T^*y = 0$ if $y = 1$, $T^*y = +\infty$ if $y \neq 1$.

17.6.iii. *If Tu, $u \in R^n$, is an extended real valued function, $Tu \not\equiv +\infty$, $Tu \neq -\infty$ for all u, and Tu is convex and lower semicontinuous in R^n, then*

$$\text{epi } T = \bigcap_{y \in R^n} [(z^0, u) \mid z^0 \geq y \cdot u - T^*y]. \tag{17.6.2}$$

Moreover T^*y, $y \in R^n$, is also an extended real valued function, $T^*y \not\equiv +\infty$, $T^*y \neq -\infty$ for all y, and T^*y is convex and lower semicontinuous in R^n.

Proof. The first part is a corollary of (17.6.ii). Let us prove that $T^*y \not\equiv +\infty$. Indeed, $Tu \neq -\infty$ for all u; hence epi $T = \bigcap S_0^+$ for a nonempty class $\{S_0^+\}$. If $S_0^+ = [(z^0, u) \mid -\bar{r} + \bar{y} \cdot u, u \in R^n]$ is one of these half spaces, then $T^*\bar{y} \leq \bar{r}$, or $T^*\bar{y} \neq +\infty$, and thus $T^*y \not\equiv +\infty$. Let us prove that $T^*y \neq -\infty$ for all y. Indeed, if $T^*y = -\infty$, it means that epi T is above any half space $-r + y \cdot u$, or $Tu \equiv +\infty$, a contradiction. Let us prove that T^*y is lower semicontinuous in R^n. Let $\bar{r} = T^*\bar{y}$, and assume $\bar{r}$ finite. This means that for any $\varepsilon > 0$ we have $Tu \geq -\bar{r} - \varepsilon + \bar{y} \cdot u$ for all $u \in R^n$, while it is not true that $Tu \geq -\bar{r} + \varepsilon + \bar{y} \cdot u$ for all u. Thus, there is some $\bar{u}$ such that

$$-\bar{r} - \varepsilon + \bar{y} \cdot \bar{u} \leq T\bar{u} \leq -\bar{r} + \varepsilon + \bar{y} \cdot \bar{u}. \tag{17.6.3}$$

Now take $\delta > 0$ such that $\delta|\bar{u}| < \varepsilon$, and any $y \in R^n$, $|y - \bar{y}| \leq \delta$. Then, for $r = y \cdot \bar{u} - T\bar{u}$ we certainly have $+r - y \cdot \bar{u} = -T\bar{u}$, and by addition with (17.6.3), also $-\bar{r} + r - \varepsilon + (\bar{y} - y) \cdot \bar{u} \leq 0 \leq -\bar{r} + r + \varepsilon + (\bar{y} - y) \cdot \bar{u}$. Here $|(\bar{y} - y) \cdot \bar{u}| \leq |\bar{y} - y|\,|\bar{u}| \leq \delta|\bar{u}| < \varepsilon$, and therefore $|\bar{r} - r| \leq 2\varepsilon$. Since $-r + y \cdot \bar{u} = T\bar{u}$, we must have

$$T^*y \geq r = \bar{r} + (r - \bar{r}) \geq T^*\bar{y} - 2\varepsilon,$$

and this holds for all $y \in R^n$, $|y - \bar{y}| \leq \delta$. We have proved that T is lower semicontinuous at any $\bar{y}$ with $T\bar{y}$ finite. An analogous argument holds at any $\bar{y}$ with $T\bar{y} = +\infty$.

Let us prove that T^* is convex in R^n. Let y_1, y_2 be two points of R^n, let $0 \leq \alpha \leq 1$, and $y = (1 - \alpha)y_1 + \alpha y_2$. Let $r_1 = T^*y_1$, $r_2 = T^*y_2$, and assume both r_1, r_2 finite. Then given $\varepsilon > 0$ we have

$$Tu \geq -r_1 - \varepsilon + y_1 \cdot u, \qquad Tu \geq -r_2 - \varepsilon + y_2 \cdot u$$

for all u, and hence

$$Tu \geq -(1-\alpha)r_1 - \alpha r_2 - \varepsilon + [(1-\alpha)y_1 + \alpha y_2] \cdot u,$$

again for all u. This implies that

$$T^*y = T^*[(1-\alpha)y_1 + \alpha y_2] \leq (1-\alpha)r_1 + \alpha r_2 = (1-\alpha)T^*y_1 + \alpha T^*y_2.$$

This proves the convexity of T^* between any two points y_1, y_2 where T^* is finite. If one or both of r_1, r_2 are $+\infty$, the argument is analogous. This proves (17.6.iii). □

By (17.6.ii) applied to T^*, we have now

$$T^*\bar{y} = \sup[-r + p \cdot \bar{y} \,|\, -r + p \cdot y \leq T^*y \text{ for all } y \in R^n].$$

Moreover, we can repeat the process and define

$$T^{**}z = \sup[z \cdot y - T^*y \,|\, y \in R^n], \qquad z \in R^n,$$

and again $T^{**}z \not\equiv +\infty$, $T^{**}z \neq -\infty$ for all z, and $T^{**}$is convex and lower semicontinuous in R^n. Moreover

17.6.iv. $T^{**} = T$.

Proof. For every $u \in R^n$, we know that $Tu = \sup[y \cdot u - T^*y \,|\, y \in R^n]$; thus $Tu \geq y \cdot u - T^*y$ for all y, and given $\varepsilon > 0$ there is some $\bar{y}$ such that $Tu < \bar{y} \cdot u - T^*\bar{y} + \varepsilon$. Now $T^{**}z = \sup[z \cdot y - T^*y \,|\, y \in R^n]$. Hence, $T^{**}z \geq z \cdot y - T^*y$ for all $y \in R^n$, in particular $T^{**}u \geq u \cdot y - T^*y$ for all y, and $T^{**}u \geq u \cdot \bar{y} - T^*\bar{y}$. Finally,

$$Tu < \bar{y} \cdot u - T^*\bar{y} + \varepsilon \leq \bar{y} \cdot u + (-u \cdot \bar{y} + T^{**}u) + \varepsilon,$$

or $Tu < T^{**}u + \varepsilon$, where ε is arbitrary. Thus $Tu \leq T^{**}u$. Analogously, we have $T^{**}u \leqq u \cdot \bar{\bar{y}} - T^*\bar{\bar{y}} + \varepsilon$ for some $\bar{\bar{y}}$; hence $T^{**}u \leqq u \cdot y + \varepsilon + (-y \cdot u + Tu)$, or $T^{**}u \leqq Tu + \varepsilon$ and finally $T^{**}u \leqq Tu$. By comparison, we have $Tu = T^{**}u$ for all $u \in R^n$, where $r = Tu$, $\rho = T^{**}u$ are finite. An analogous proof holds in the cases where one of these numbers is $+\infty$. □

If we denote by Γ_0 the family of all extended real valued functions Tu, $u \in R^n$, with $Tu \not\equiv +\infty$, $Tu \neq -\infty$ for all $u \in R^n$, T convex and lower semicontinuous in R^n, then we see that the operation $T \to T^*$ maps Γ_0 into Γ_0, and because of (17.6.iv) we conclude that this operation is onto and 1–1. It is to be noted that the dual operation $T \to T^*$ defined by (17.6.1) is the operation by means of which we pass from the Lagrangian f to the Hamiltonian M in the calculus of variations and optimal control theory (see Section 17.7). This operation can be traced back to Legendre.

Remark. We shall encounter extended integrand functions $T(t, x, z)$, which are finite everywhere on a measurable set M of the txz-space R^{1+2n}, which are $+\infty$ in $R^{1+2n} - M$, which are measurable in t for all (x, z), and such that, for almost all $\bar{t}$, $T(\bar{t}, x, z)$ is a lower semicontinuous function in (x, z) and convex in z. Let A_0 denote the projection of M on the t-axis, which we assume for the sake of simplicity to be an interval, finite or infinite. Then, for almost all $\bar{t} \in A_0$, $T(\bar{t}, x, z)$ is not identically $+\infty$, and never $= -\infty$. Let $T^*(t, x, y)$ be the dual of $T(t, x, z)$ with respect to z, or $T^*(t, x, y) = \sup_z[y \cdot z - T(t, x, z)]$. It is easy to see, in the frame of (17.6.iii), that $T^*(t, x, y)$ is measurable in t for all (x, y), and that, for almost all $\bar{t} \in A_0$, $T^*(\bar{t}, x, y)$ is not identically $+\infty$, never $= -\infty$, and $T^*(t, x, y)$ is lower semicontinuous in (x, y) and convex in y.

B. The Operations $\bigwedge$ and $\bigvee$

Given a family $T_i u$, $u \in R^n$, of functions $T_i \in \Gamma_0$ depending on an index i ranging on an arbitrary index set I, we define the following basic "lattice" operations $\bigvee_i$ and $\bigwedge_i$:

$$\bigvee_i T_i(u) = \sup_i T_i(u),$$

$$\bigwedge_i T_i(u) = \sup[-r + p \cdot u \mid -r + p \cdot u \leqq T_i(u) \text{ for all } u \in R^n \text{ and } i \in I].$$

On the other hand, if Q_i, $i \in I$, denotes a family of closed convex sets in R^{n+1}, then we define the analogous operations $\mathring{\bigvee}_i$ and $\mathring{\bigwedge}_i$:

$$\mathring{\bigvee_i} Q_i = \text{cl co} \bigcup_i Q_i,$$

$$\mathring{\bigwedge_i} Q_i = \bigcap_i Q_i.$$

With this notation, and for functions $T_i \in \Gamma_0$, we have

$$\mathring{\bigwedge_i} (\text{epi } T_i) = \text{epi}\left(\bigvee_i T_i\right), \qquad \mathring{\bigvee_i} (\text{epi } T_i) = \text{epi}\left(\bigwedge_i T_i\right). \tag{17.6.4}$$

We may well have $(\bigvee_i T_i)(u) \equiv +\infty$, as it happens for $I = \{1, 2\}$, $T_1 u = +\infty$ for $u < 1$, $T_1 u = 0$ for $u \geqq 1$, $T_2 u = +\infty$ for $u > -1$, $T_2 u = 0$ for $u \leqq -1$. Analogously, we may well have $(\bigwedge_i T_i)u = -\infty$ for some u. Indeed, for $T_i u = -i$, $u \in R^n$, $i = 1, 2, \ldots$, we have $(\bigwedge_i T_i)(u) = -\infty$ for all $u \in R^n$. However

17.6.v. *If all $T_i \in \Gamma_0$, $i \in I$, and $(\bigvee_i T_i)u \not\equiv +\infty$, then $\bigvee_i T_i \in \Gamma_0$. If $(\bigwedge_i T_i)u \neq -\infty$ for all u, then $\bigwedge_i T_i \in \Gamma_0$.*

Proof. If $(\bigvee_i T_i)u \not\equiv +\infty$, then the intersection of the convex closed sets epi T_i, $i \in I$, is nonempty and thus necessarily closed and convex. Since $T_i u \neq -\infty$ for all i and u, then $(\bigvee_i T_i)u$ has the same property, and $\bigvee_i T_i \in \Gamma_0$. Analogously, if $(\bigwedge_i T_i)u \neq -\infty$ for all u, then consider the intersection of all half spaces $S_0^+ = [z^0 \geqq -r + p \cdot u, u \in R^n]$ with the property that $T_i u \geqq -r + p \cdot u$ for all u and i. This intersection is not empty and necessarily closed and convex. As before, $\bigwedge_i T_i \in \Gamma_0$. □

17.6.vi. *For $T_i \in \Gamma_0$ we have*

$$\left(\bigwedge_i T_i\right)^* = \bigvee_i T_i^*, \quad \left(\bigvee_i T_i\right)^* = \bigwedge_i T_i^*. \tag{17.6.5}$$

Proof. Here we have

$$\left(\bigwedge_i T_i\right)u = \sup[-r + p \cdot u \mid -r + p \cdot u \leqq T_i u \text{ for all } u \text{ and } i],$$

$$\left(\bigwedge_i T_i\right)^*(p) = \inf[r \mid -r + pu \leqq T_i u \text{ for all } u \text{ and } i].$$

On the other hand

$$T_i^* p = \inf[r \mid -r + p \cdot u \leqq T_i u \text{ for all } u],$$

$$\begin{aligned}\left(\bigvee_i T_i^*\right)(p) &= \sup_i \inf[r \mid -r + p \cdot u \leqq T_i u \text{ for all } u] \\ &= \inf[r \mid -r + p \cdot u \leqq T_i u \text{ for all } u \text{ and } i].\end{aligned}$$

This proves the first part of (17.6.vi). The proof of the second part is analogous. □

C. The Case of an Ordered Index Set I

We consider now the case in which the index set I is ordered by an ordered relation $\prec$ possessing a "least" element ω. Concerning the relation we assume the following: (1) $i \prec j$, $j \prec k$ implies $i \prec k$; (2) given $i, j \in I$, $i, j \neq \omega$, there is $k \in I$, $k \neq \omega$, such that $k \prec i$, $k \prec j$.

Then, instead of the operation "sup" we have used above, we may use the operation "lim sup" in terms of the operations $\bigvee$ and $\bigwedge$, or $\mathring{\bigvee}$ and $\mathring{\bigwedge}$. Thus, instead of, say, $\mathring{\bigvee}_i Q_i = \text{cl co} \bigcup_i Q_i$ we shall take

$$\mathring{\bigwedge_{\lambda \neq \omega}} \mathring{\bigvee_{i \prec \lambda}} Q_i = \bigcap_{\lambda \neq \omega} \text{cl co} \bigcup_{i \prec \lambda} Q_i$$

and instead of $\mathring{\bigwedge}_i T_i$ we shall take

$$\mathring{\bigvee_{\lambda \neq \omega}} \mathring{\bigwedge_{i \prec \lambda}} T_i = \sup_{\lambda \neq \omega} \sup[-r + p \cdot u \mid -r + pu \leqq T_i(u), u \in R^n, i \prec \lambda]$$

In this connection we note that

$$\tilde{T}_\lambda u = \left(\bigvee_{i \prec \lambda} T_i\right)(u) = \sup\left[\bigwedge_i T_i(u), i \prec \lambda\right], \qquad u \in R^n,$$

is a convex lower semicontinuous function of u which depends monotonically on λ, that is, $\lambda' \prec \lambda$ implies $\tilde{T}_{\lambda'}(u) \leqq \tilde{T}_\lambda(u)$, $u \in R^n$. Moreover, by the use of the operation cl we have defined in Subsection A above, we also have:

17.6.vii. *If the index set I is ordered by an order relation as stated, then*

$$\text{cl} \inf_\lambda \tilde{T}_\lambda u = \bigwedge_\lambda \tilde{T}_\lambda u. \tag{17.6.6}$$

Proof. First, let us prove that $\tau(u) = \inf_\lambda \tilde{T}_\lambda u$ is a convex function of u in R^n. Indeed, if $u_1, u_2 \in R^n$, and $z_i^0 = \inf_\lambda \tilde{T}_\lambda u_i$, $i = 1, 2$, then given $\varepsilon > 0$ there are $\lambda_i \in I$ such that $z_i^0 \leqq \tilde{T}_{\lambda_i} u_i < z_i^0 + \varepsilon$, and since $\tilde{T}_\lambda u$ is monotone in λ, then we also have $z_i^0 \leqq \tilde{T}_\lambda u_i < z_i^0 + \varepsilon$ for all $\lambda \prec \lambda_i$, $i = 1, 2$. Finally, by property (2) of the order relation $\prec$, there is some $\lambda' \prec \lambda_i$, $i = 1, 2$, $\lambda' \neq \omega$, such that

$$z_i^0 \leqq \tilde{T}_{\lambda'} u_i < z_i^0 + \varepsilon, \qquad i = 1, 2.$$

For any α, $0 \leqq \alpha \leqq 1$, then

$$\begin{aligned}\left(\inf_\lambda \tilde{T}_\lambda\right)((1-\alpha)u_1 + \alpha u_2) &\leqq \tilde{T}_{\lambda'}((1-\alpha)u_1 + \alpha u_2)\\ &\leqq (1-\alpha)\tilde{T}_{\lambda'} u_1 + \alpha \tilde{T}_{\lambda'} u_2\\ &\leqq (1-\alpha) z_1^0 + \alpha z_2^0 + \varepsilon,\end{aligned}$$

or $\tau((1-\alpha)u_1 + \alpha u_2) \leqq (1-\alpha)\tau(u_1) + \alpha\tau(u_2) + \varepsilon$, and since ε is arbitrary we have $\tau\,(1-\alpha)u_1 + \alpha u_2) \leqq (1-\alpha)\tau(u_1) + \alpha\tau(u_2)$.

Now note that, for all λ, we have epi $\tilde{T}_\lambda \subset \operatorname{epi}(\inf_\lambda \tilde{T}_\lambda)$. Hence

$$\bigcup_\lambda \operatorname{epi} \tilde{T}_\lambda \subset \operatorname{epi}\left(\inf_\lambda \tilde{T}_\lambda\right),$$

and since the last set is convex, we also have

$$\operatorname{co} \bigcup_\lambda \operatorname{epi} \tilde{T}_\lambda \subset \operatorname{epi}\left(\inf_\lambda \tilde{T}_\lambda\right).$$

By applying the operation cl defined in Subsection A, we have

$$\operatorname{cl} \operatorname{co} \bigcup_\lambda \operatorname{epi} \tilde{T}_\lambda \subset \operatorname{cl} \operatorname{epi}\left(\inf_\lambda \tilde{T}_\lambda\right).$$

This is the same as saying that

$$\bigwedge \tilde{T}_\lambda \leqq \operatorname{cl} \inf_\lambda \tilde{T}_\lambda. \tag{17.6.7}$$

To prove (17.6.6) we have only to prove that

$$\operatorname{epi} \inf_\lambda \tilde{T}_\lambda \subset \operatorname{cl} \operatorname{co} \bigcup_\lambda \operatorname{epi} \tilde{T}_\lambda, \tag{17.6.8}$$

since, by taking the closure of the set at the left, we have the relation opposite to (17.6.7), and thus equality must hold in (17.6.7). Suppose, if possible, that (17.6.8) is not true. Then we could strictly separate some point in the set on the left from the closed convex set on the right by means of a nonvertical closed hyperplane. Hence, for some constant c and some point u we would have

$$\inf_\lambda \tilde{T}_\lambda u < c < T_\lambda u$$

for all λ. Taking the infimum over λ on the right then leads to a contradiction. This proves (17.6.vii). □

D. Characterization of Property (Q) in Terms of Duality

Let us consider the situation where we have a family of sets $\tilde{Q}(x) = \text{epi } T(x,u) = [(z^0, u) | z \geqq T(x,u), u \in R^n]$, indexed by $x \in A$. If $x_0 \in A$ is a given point and $x \in A$ are preordered by their distance $d(x, x_0)$ from x_0—that is, we say that $x \prec x'$ if $x, x' \in A$, $d(x, x_0) < d(x', x_0)$—then x_0 is the "least" element, and properties (1) and (2) hold. With this understanding, the "lim sup", in the sense mentioned in Subsection C, for the sets $\tilde{Q}(x)$ at x_0 is given by

$$\mathring{\bigwedge_{\delta>0}} \mathring{\bigvee}[\tilde{Q}(x) | d(x, x_0) \leqq \delta], \tag{17.6.9}$$

or equivalently

$$\bigcap_{\delta<0} \text{cl co} \bigcup [\tilde{Q}(x) | d(x, x_0) \leqq \delta].$$

Thus, property (Q) at x_0 reduces to

$$\hat{Q}(x_0) = \mathring{\bigwedge_{\delta>0}} \mathring{\bigvee}[\tilde{Q}(x) | d(x, x_0) \leqq \delta],$$

or equivalently

$$\text{epi } T(x_0, u) = \mathring{\bigwedge_{\delta>0}} \mathring{\bigvee}[\text{epi } T(x,u) | d(x, x_0) \leqq \delta],$$

or

$$T(x_0, u) = \bigvee_{\delta>0} \bigwedge [T(x,u) | d(x, x_0) \leqq \delta], \tag{17.6.10}$$

that is, property (Q) is the "upper semicontinuity" of the same sets with respect to the operations $\mathring{\bigvee}$ and $\mathring{\bigwedge}$ (or the analogous property for the functions $T(x,u)$, $u \in R^n$, in terms of the operators $\bigwedge$ and $\bigvee$).

17.6.viii (Theorem; Goodman [1]). *If $T(x,u) > -\infty$ for all x and u, then the sets $\tilde{Q}(x)$ have property* (Q) *at x if and only if the dual function $T^*(x,y)$ satisfies the relation*

$$T^*(x_0, y) = \bigwedge_{\delta} \bigvee [T^*(x,y) | d(x, x_0) \leqq \delta]. \tag{17.6.11}$$

Indeed, by taking the dual of both members of (17.6.10) and using relations (17.6.5) we have

$$\begin{aligned} T^*(x_0, u) &= \left(\bigvee_{\delta>0} \bigwedge [T(x,u), d(x, x_0) \leqq \delta] \right)^* \\ &= \bigwedge_{\delta>0} (\bigwedge [T(x,u), d(x, x_0) \leqq \delta])^* \\ &= \bigwedge_{\delta>0} \bigvee [T^*(x,y), d(x, x_0) \leqq \delta]. \end{aligned}$$

17.6.ix (Theorem; Goodman [1]). *If $T(x,u) > -\infty$ for all x and u, then the sets $\tilde{Q}(x)$ have property (Q) at x if and only if*

$$T^*(x_0, y) = \text{cl}\left[\limsup_{x \to x_0} T^*(x,y) \right]. \tag{17.6.12}$$

Indeed, by applying (17.6.6) to (17.6.11), we have

$$\begin{aligned} T^*(x_0, y) &= \bigwedge_{\delta} \{\bigvee[T^*(x, y) | d(x, x_0) \leqq \delta]\} \\ &= \text{cl} \inf_{\delta} \{\bigvee[T^*(x, y) | d(x, x_0) \leqq \delta]\} \\ &= \text{cl}\left[\limsup_{x \to x_0} T^*(x, y)\right]. \end{aligned}$$

It is to be noted that if the term in brackets in (17.6.12) is finite for all y, then it is necessarily closed (that is, the epigraph is closed), and in that case relation (17.6.12) just means upper semicontinuity of $T^*(x, y)$ at x_0.

17.6.x (THEOREM; GOODMAN [1]). *If $T(x, u) > -\infty$ for all x and u, then the sets $\tilde{Q}(x)$ have property* (Q) *at x_0 if and only if*

$$T(x_0, y) = \left[\limsup_{x \to x_0} T^*(x_0, y)\right]^*.$$

Indeed, by taking the dual of both members of (17.6.12), we have

$$\begin{aligned} T(x_0, y) = T^{**}(x_0, y) &= \left(\text{cl}\left[\limsup_{x \to x_0} T^*(x, y)\right]\right)^* \\ &= \left[\limsup_{x \to x_0} T^*(x, y)\right]^* \end{aligned}$$

with the stated conventions.

17.7 Characterization of Optimal Solutions in Terms of Duality

Let E be a linear space, let E^* be its dual, so that $\langle u, u^* \rangle$ is a bilinear symmetric operator, or $E \times E^* \to R$. Let Tu be any real (extended) function defined on E, not identically equal to $+\infty$, never equal to $-\infty$, and convex in u. As in Section 17.6, we define the dual T^* of T by taking, for every $u^* \in E^*$,

$$T^*u^* = \sup_u (\langle u, u^* \rangle - Tu). \tag{17.7.1}$$

Then T^*u^* is also a real extended function in E^*. For every $u \in E$ let $\partial Tu = \{u^*\}$ denote the collection of all elements $u^* \in E^*$ such that $Tz \geq Tu + \langle z - u, u^* \rangle$ for all $u \in E$. In other words ∂Tu is the set of all u^* such that $h(z) = Tu + \langle z - u, u^* \rangle$ is a supporting hyperplane to T at $u \in E$.

17.7.i. *Given $u^* \in E^*$, then T^*u^*, the supremum defined by* (17.7.1), *is actually attained if and only if there is some $u \in E$ such that*

$$T^*u^* = \langle u, u^* \rangle - Tu,$$

or equivalently, $u^* \in \partial Tu$.

Proof. Note that

$$Tu + T^*u^* \geq \langle u, u^* \rangle \quad \text{for all } u \in E^*, u \in E,$$

and that

$$Tu + T^*u^* = \langle u, u^* \rangle$$

whenever T^*u^* is the maximum of $\langle u, u^* \rangle - Tu$ for all $u \in E$ (and also, whenever Tu is the maximum of $\langle u, u^* \rangle - T^*u^*$ for all $u^* \in E^*$). Note, in particular, that for $u^* = 0$, $T^*(0) = \sup_u[-Tu] = -\inf Tu$, and that $T^*(0)$ is attained, that is, T attains its minimum at some $u \in E$ if and only if $T^*(0) = -Tu$ and $0 \in \partial Tu$. □

Let us consider now the Lagrange problem of optimal control concerning the minimum of the integral $I[x, u]$ below with the constraints as indicated:

$$I[x, u] = \int_{t_1}^{t_2} f_0(t, x(t), u(t))\, dt,$$

(17.7.2)
$$dx/dt = f(t, x(t), u(t)),$$
$$(t, x(t)) \in A, \quad u(t) \in U(t, x(t)), \qquad x(t_1) = x_1, \quad x(t_2) = x_2,$$
$$x = (x^1, \ldots, x^n), \qquad u = (u^1, \ldots, u^m),$$

for which we have defined

(17.7.3)
$$F_0(t, x, z) = \inf[z^0 \,|\, (z^0, z) \in \tilde{Q}(t, x)]$$
$$= \inf[z^0 \,|\, z^0 \geq f_0(t, x, u),\ z = f(t, x, u),\ u \in U(t, x)],$$

and then (17.7.2) is transformed into the problem

(17.7.4)
$$I[x] = \int_{t_1}^{t_2} F_0(t, x(t), x'(t))\, dt, \qquad x'(t) \in Q(t, x(t)),$$
$$x(t_1) = x_1, \quad x(t_2) = x_2, \qquad x = (x^1, \ldots, x^n),$$

where $Q(t, x) = f(t, x, U(t, x))$, the projection of $\tilde{Q}(t, x)$ on the z-space R^n, though we do not yet claim equivalence between problems (17.7.2) and (17.7.3). (In Chapters 9, 11 we have proved this equivalence under a number of very general assumptions.) To simplify our discussion here, we assume that A is a closed subset of the tx-space R^{n+1}, and that for almost all $\bar{t}$ of some interval A_0 (finite or infinite), with $[t_1, t_2] \subset A_0$, we have $F_0(\bar{t}, x, z) > -\infty$ for all x, z. This certainly occurs if, for instance, $F_0(t, x, z) \geq -\psi(t) - c|z|$ for some constant c and a locally integrable function $\psi(t) \geq 0$, a hypothesis we have often assumed in our existence theorems. We shall also assume that for every AC function $x(t)$, $t_1 \leq t \leq t_2$, t_1, t_2 finite, $[t_1, t_2] \subset A_0$, the integral $\int_{t_1}^{t_2} F_0(t, x(t), x'(t))\, dt$ exists, finite or $+\infty$. This certainly occurs under the hypothesis $F_0 \geq -\psi(t) - c|z|$ above, and also it occurs necessarily if F_0 is convex in z and continuous in (t, x, z), as we have seen in (2.17.i).

Note that in (17.7.4), and when searching for the minimum of the functional, the requirements $(t, x(t)) \in A$, $x'(t) \in Q(t, x(t))$ need not be actually stated explicitly, since whenever the same relations are not satisfied a.e. in $[t_1, t_2]$, then $F_0 = +\infty$ in a set of positive measure, and $I[x, u] = +\infty$.

Note that the requirement that for any $(\bar{t}, \bar{x}) \in A$ the set $\tilde{Q}(\bar{t}, \bar{x})$ be closed is equivalent to the hypothesis that min holds in (17.7.3) instead of inf. Analogously, the assumption that, for a given $\bar{t}$, the sets $\tilde{Q}(t, x)$ have property (K) with respect to x on $A(\bar{t})$ is equivalent to the assumption that the (extended) function $F_0(\bar{t}, x, z)$ is lower semicontinuous in (x, z) in R^{2n} (cf. Section 17.2).

The Hamiltonian H and corresponding function M in Chapter 5 for problem (17.7.2) are (for $\lambda_0 = 1$)

$$H(t, x, u, \lambda) = f_0(t, x, u) + \lambda \cdot f(t, x, u), \qquad \lambda = (\lambda_1, \ldots, \lambda_n),$$
$$M(t, x, z) = \inf[f_0(t, x, u) + \lambda \cdot f(t, x, u), u \in U(t, x)],$$

and these relations, in terms of problem (17.7.4), become

$$\begin{aligned} H(t, x, z, \lambda) &= F_0(t, x, z) + \lambda \cdot z, \qquad z = (z^1, \ldots, z^n), \\ M(t, x, \lambda) &= \inf[F_0(t, x, z) + \lambda \cdot z, z \in R^n]. \end{aligned} \tag{17.7.5}$$

Again, we have written $z \in R^n$ instead of $z \in Q(t, x)$, since the extended function F_0 is $+\infty$ for $z \notin Q(t, x)$, and so is M.

Now let W_0 denote the class of all integrands $F_0(t, x, z)$, F_0 an extended function, measurable in t for all (x, z), such that for almost all $\bar{t}$, $F_0(\bar{t}, x, z)$ is lower semicontinuous in (x, z) and convex in z. Then, by (17.6.iii), $F_0^*(t, x, p)$ is also an integrand of the same class W_0, and since $F_0^{**} = F_0$, we have

$$F_0^*(t, x, p) = \sup_{z \in R^n} [p \cdot z - F_0(t, x, z)], \tag{17.7.6}$$

$$F_0(t, x, z) = \sup_{p \in R^n} [p \cdot z - F_0^*(t, x, p)]. \tag{17.7.7}$$

Since $F_0(t, x, z) + \lambda \cdot z = H(t, x, z, \lambda)$, we have also

$$\begin{aligned} F_0^*(t, x, -p) &= \sup[-p \cdot z - F_0(t, x, z), z \in R^n] \\ &= -\inf[F_0(t, x, z) + p \cdot z, z \in R^n] \\ &= -\inf[H(t, x, z, p), z \in R^n] = -M(t, x, p), \end{aligned}$$

or

$$F_0^*(t, x, -p) = -M(t, x, p). \tag{17.7.8}$$

Thus, we have established the relation between the notation in Chapter 5 and that of the present chapter. In Chapter 5, as usual in the theory of optimal control, H is called the Hamiltonian; some authors would prefer to call $F_0^*(t, x, p)$ or $M(t, x, \lambda)$ the Hamiltonian. Note that whenever the sup in (17.7.6) is attained for almost all $\bar{t}$ by measurable functions $p(t)$, $x'(t)$, $t_1 \leq t \leq t_2$, then we have the relation

$$F_0^*(t, x(t), p(t)) = p(t) \cdot x'(t) - F_0(t, x(t), x'(t)) \quad \text{(a.e.)},$$

or briefly, using the notation xy for $x \cdot y$, also $F_0 = px' - F_0^*$, or $F_0^* + F_0 = px'$.

As before, we consider now the analogous class W of all integrands $F_0(t, x, z)$ with F_0 an extended function which is measurable in t for all (x, z), and such that for almost all $\bar{t}$ of some interval $I_0 \subset R$, $F_0(\bar{t}, x, z)$ is lower semicontinuous and convex in (x, z) in R^{2n}. Then we can take duals in terms of the variable (x, z), that is,

$$G_0(t, y, w) = \sup_{(x,z)} [w \cdot x + y \cdot z - F_0(t, x, z)]. \tag{17.7.9}$$

Then G_0 is in the same class W, and by duality, also

$$F_0(t, x, z) = \sup_{(y,w)} [w \cdot x + y \cdot z - G_0(t, y, w)]. \tag{17.7.10}$$

By only changing notation, relations (17.7.9–10) become

$$G_0(t, p, p') = \sup_{(x,x')} [p' \cdot x + p \cdot x' - F_0(t, x, x')].$$

Hence

$$G_0(t, p, p') + F_0(t, x, x') \geq p' \cdot x + p \cdot x'$$

for all p, p', x, x', while equality holds if and only if max holds instead of sup in (17.7.9) or in (17.7.10).

Let $[t_1, t_2]$ be fixed numbers, and for any two points $a, b \in R^n$ let $\omega = \Omega(a, b)$ be the class of all AC functions $x(t) = (x^1, \ldots, x^n)$, $t_1 \leq t \leq t_2$, with $x(t_1) = a$, $x(t_2) = b$. Also, let $J[x]$ denote the integral $J[x] = \int_{t_1}^{t_2} G_0(t, x(t), x'(t))\, dt$. For any two AC functions $p(t), x(t)$ in $[t_1, t_2]$, say $x \in \Omega(a, b)$, $p \in \Omega(c, d)$, we have then

$$G_0(t, p(t), p'(t)) + F_0(t, x(t), x'(t)) \geq x(t) \cdot p'(t) + x'(t) \cdot p(t),$$

and by integration

$$J[p] + I[x] \geq b \cdot d - a \cdot c.$$

Let $\omega(a, b)$, $\omega_1(c, d)$ denote

$$\omega(a, b) = \inf_{\Omega(a,b)} I[x], \qquad \omega_1(c, d) = \inf_{\Omega(b,d)} J[p],$$

which we assume to be finite or $+\infty$. Then

$$\omega_1(c, d) + \omega(a, b) \geq b \cdot d - a \cdot c$$

or

$$\omega_1(c, d) \geq \sup_{(a,b)} [a \cdot (-c) + b \cdot d - \omega(a, b)].$$

This is the dual operation on $\omega(a, b)$. Indeed, the following theorem holds.

17.7.ii. *The extended real functions in R^{2n}, $\omega(a, b)$, and $\omega_1(c, d)$ are lower semicontinuous and convex in R^{2n}, never equal to $-\infty$, and not identically $+\infty$, and moreover*

$$\omega_1(c, d) = \sup_{(a,b)} [a \cdot (-c) + b \cdot d - \omega(a, b)] = \omega^*(-c, d),$$

$$\omega(a, b) = \sup_{(c,d)} [(-a) \cdot c + b \cdot d - \omega_1(c, d)] = \omega^*(-a, b).$$

The following statement is now a corollary of the above considerations.

17.7.iii (Theorem). *For any given system of points t_1, t_2, a, b, c, d, then $I[x]$ attains its minimum if and only if $J[p]$ attains its minimum, and if and only if for some AC functions $x(t)$, $p(t)$, $t_1 \leq t < t_2$, we have*

$$J[p] = (b \cdot d - a \cdot c) - I[x],$$

$$G_0(t, p(t), p'(t)) = p'(t)x(t) + p(t)x'(t) - F_0(t, x(t), x'(t)),$$

$$(p'(t), p(t)) \in \partial F_0(t, x(t), x'(t)), \qquad t \in [t_1, t_2] \text{ (a.e.)}.$$

Note that if F_0 is differentiable, then $\partial F_0(t, x(t), x'(t))$ is the linear operator defined by

$$F_{0x}(t, x(t), x'(t)) \cdot h(t) + F_{0x'}(t, x(t), x'(t)) \cdot h'(t),$$

where $F_{0x} = (\partial F_0/\partial x^i, i = 1, \ldots, n)$, $F_{0x'} = (\partial F_0/\partial x'^i, i = 1, \ldots, n)$, that is,

$$(\partial I[x])h = \int_{t_1}^{t_2} [F_{0x} \cdot h + F_{0x'} \cdot h']\, dt,$$

or

$$p'(t) = F_{0x}(t, x(t), x'(t)), \quad p(t) = F_{0x'}(t, x(t), x'(t)), \qquad t \in [t_1, t_2] \text{ (a.e.)},$$

and hence $(d/dt)F_{0x'} = F_{0x}$, that is,

$$\frac{d}{dt} F_{0x'^i} = F_{0x^i}, \qquad i = 1, \ldots, n, \quad t \in [t_1, t_2] \text{ (a.e.)}.$$

On the other hand, if f_0, f satisfy the conditions of one of the implicit function theorems, say (8.2.iii), or (8.3.vi) and subsequent remarks, then there is a measurable function $u(t)$ such that, for $\lambda = -p(t)$, we have

$$\begin{aligned}
F_0(t, x(t), x'(t)) &= f_0(t, x(t), u(t)), \qquad u(t) \in U(t, x(t)),\\
x'^i(t) &= f_i(t, x(t), u(t)), \qquad t \in [t_1, t_2] \text{ (a.e.)}, \quad i = 1, \ldots, n,\\
\lambda_i'(t) &= -p_i'(t) = -F_{0x^i}(t, x(t), x'(t))\\
&= -[(\partial/\partial x^i)F_0(t, x, x'(t))]_{x = x(t)}\\
&= -\left[(\partial/\partial x^i)\left\{f_0(t, x, u(t)) + \sum_j \lambda_j(t) f_j(t, x, u(t))\right\}\right]_{x = x(t)}\\
&= -\left[f_{0x^i}(t, x(t), u(t)) + \sum_j \lambda_j(t) f_{jx^i}(t, x(t), u(t))\right]\\
&= -H_{x^i}(t, x(t), u(t), \lambda(t)).
\end{aligned}$$

Thus, we have also the canonical equations

$$\frac{dx^i}{dt} = \frac{\partial H}{\partial \lambda_i}, \quad \frac{d\lambda_i}{dt} = -\frac{\partial H}{\partial x^i}, \qquad i = 1, \ldots, n, \quad t \in [t_1, t_2] \text{ (a.e.)}.$$

17.8 Property (Q) as an Extension of Maximal Monotonicity

Let X be a real Hilbert space with inner product (x, y) and norm $\|x\| = (x, x)^{1/2}$. A set valued function $Q(x)$, $x \in A \subset X$, $Q(x) \subset X$, is said to be monotone in A provided $x_1, x_2 \in A$, $z_1 \in Q(x_1)$, $z_2 \in Q(x_2)$ implies $(z_1 - z_2, x_1 - x_2) \geq 0$. Here A is said to be the domain of the set valued function $x \to Q(x)$. Thus, for $X = R$, $Q(x) = f(x)$ a single valued, real valued function, we have the usual definition of monotone nondecreasing functions.

A set valued function $x \to Q(x)$, $x \in A \subset X$, $Q(x) \subset X$, is said to be maximal monotone provided it is monotone and for any other monotone function $x \to R(x)$, $x \in B \subset X$, $R(x) \subset X$, with $A \subset B$, $Q(x) \subset R(x)$, we have $A = B$, $Q(x) = R(x)$ for all $x \in A = B$. In other words, neither the domain nor any of the sets $Q(x)$ can be enlarged preserving monotonicity.

Here are a few examples of monotone maps with $X = R$: (1) The map $x \to f(x)$ with $0 \leq x \leq 1$, $f(x) = x$ is monotone but not maximal monotone. (2) The map $x \to f(x)$

with $-\infty < x < +\infty$, $f(x) = x$ is maximal monotone. (3) The map $x \to g(x)$ with $-\infty < x < +\infty$ with $g(x) = x$ for $0 \leq x \leq 1$, $g(x) = 1$ for $x \geq 1$, $g(x) = 0$ for $x \leq 0$ is maximal monotone. (4) The map $x \to Q(x)$, $0 \leq x \leq 1$, $Q(x) = \{0\}$ for $0 < x < 1$, $Q(0) = [-\infty < z \leq 0]$, $Q(1) = [0 \leq z < +\infty]$, is maximal monotone. (5) The map $x \to f(x)$, $-\infty < x < +\infty$, with $f(x) = 1 + x$ for $x \geq 0$, $f(x) = -1 + x$ for $x < 0$ is monotone but not maximal monotone. (6) The map $x \to Q(x)$, with $Q(x) = \{1 + x\}$ for $x > 0$, $Q(x) = \{-1 + x\}$ for $x < 0$, $Q(0) = [-1 \leq z \leq 1]$ is maximal monotone. (7) The map $x \to f(x)$, $-1 < x < +1$, $f(x) = x(1 - x^2)^{-1}$ is maximal monotone.

We just mention here the following statement which summarizes a number of important properties of maximal monotone functions.

17.8.i. *For a maximal monotone function $x \to Q(x)$, $x \in A \subset X$, $Q(x) \subset X$, we have* (a) *A is convex*; (b) *For $x \in A$ the set $Q(x)$ is convex and closed*; (c) *For $x_0 \in \operatorname{int} A$, there are $\delta_0 > 0$ and $M_0 \geq 0$, which may depend on x_0 such that $x \in A$, $\|x - x_0\| \leq \delta_0$, $z \in Q(x)$ implies $\|z\| \leq M_0$*; (d) *for $x_1, x_2 \in A$, $x_1 \neq x_2$, then* $\operatorname{int} Q(x_1) \cap \operatorname{int} Q(x_2) = \emptyset$.

For proofs of the various parts of this statement we refer to Brezis [1] and Fitzpatrick, Hess, and Kato [1]. The following theorem, which we state and prove only for $X = R^n$ for the sake of simplicity, shows that property (Q) is essentially an extension of maximal monotonicity.

17.8.ii (Suryanarayana [8, 10]). *Let $x \to Q(x)$, $x \in A \subset R^n$, $Q(x) \subset R^n$, be any maximal monotone set valued function. Then, for every $x_0 \in R\operatorname{int} A$, the map $x \to Q(x)$ has property* (Q) *at x_0 in A. In particular, for $A = R^n$, the map has property* (Q) *in R^n.*

Proof. Let δ_0 and M_0 be the constants relative to property (c) of (17.8.i). Let $z_0 \in \bigcap_\delta \operatorname{cl}\operatorname{co} Q(x_0; \delta)$, where $Q(x_0; \delta) = \bigcup[Q(x) \mid x \in A, |x - x_0| \leq \delta]$. Then $z_0 \in \operatorname{cl}\operatorname{co} Q(x_0; \delta)$ for all $0 < \delta \leq \delta_0$, and thus there are sequences δ_k, z_k, $k = 1, 2, \ldots$, with $\delta_k > 0$, $\delta_k \to 0$, $z_k \to z_0$ as $k \to \infty$, $z_k \in \operatorname{co} Q(x_0; \delta_k)$. By Carathéodory's theorem, for each z_k there are convex combinations of points z_k^γ with $z_k^\gamma \in Q(x_k^\gamma)$, $x_k^\gamma \in A$, $|x_k^\gamma - x_0| \leq \delta_k \leq \delta_0$, and

$$z_k = \sum_\gamma \lambda_k^\gamma z_k^\gamma, \quad \sum_\gamma \lambda_k^\gamma = 1, \quad \lambda_k^\gamma \geq 0, \quad |z_k^\gamma| \leq M_0, \quad \gamma = 1, \ldots, \nu,$$

where ν is a fixed integer that we can take $\nu = n + 1$.

Here $0 \leq \lambda_r^\gamma \leq 1$, so each sequence $[\lambda_k^\gamma, k = 1, 2, \ldots]$ is bounded; hence there is a subsequence, say still $[k]$, such that $\lambda_k^\gamma \to \lambda^\gamma$ as $k \to \infty$, $\gamma = 1, \ldots, \nu$, and thus $0 \leq \lambda^\gamma \leq 1$, $\sum_\gamma \lambda^\gamma = 1$. Each sequence $[z_k^\gamma, k = 1, 2, \ldots]$ also is bounded in R^n; hence we may take the subsequence in such a way that $z_k^\gamma \to z^\gamma \in R^n$ as $k \to \infty$, $\gamma = 1, \ldots, \nu$, and finally, $z_0 = \sum_\gamma \lambda^\gamma z^\gamma$, $\lambda^\gamma \geq 0$, $\sum_\gamma \lambda^\gamma = 1$. Now for any $\gamma = 1, \ldots, \nu$ and $k = 1, 2, \ldots$, we have $z_k^\gamma \in Q(x_k^\gamma)$, and for any $\bar{x} \in A$ and $\bar{z} \in Q(\bar{x})$ we also have $(z_k^\gamma - \bar{z}, x_k^\gamma - \bar{x}) \geq 0$. By the continuity of the inner product, we have, as $k \to \infty$, $(z^\gamma - \bar{z}, x_0 - \bar{x}) \geq 0$ for all $\bar{x} \in A$ and $\bar{z} \in Q(\bar{x})$. Since $x \to Q(x)$ is maximal monotone, we must have $z^\gamma \in Q(x_0)$, $\gamma = 1, \ldots, \nu$. Finally, $Q(x_0)$ is convex, and hence $z_0 = \sum_\gamma \lambda^\gamma z^\gamma$ belongs to $Q(x_0)$, or $x_0 \in A$, $z_0 \in Q(x_0)$. This proves property (Q) at x_0 in A. □

Remark. The proof of (17.8.ii) we have given here is similar to the proofs in Section 10.5. A different proof of this theorem has been given by M. B. Suryanarayana [8, 10] for a maximal monotone set valued map $x \to Q(x)$, $x \in X$, $Q(x) \subset X$, in any real Hilbert space X. Moreover, Suryanarayana has shown that property (Q) is also the extension of a large class of monotonicity type properties of set valued functions.

Bibliographical Notes

The material in Section 17.4 and in particular Criteria 1–4 for property (Q) are taken from L. Cesari [8, 9]. Criterion 5 is due to R. D. Rupp [1]. The remaining criteria are taken for from P. J. Kaiser [3].

The concepts of normality and seminormality of scalar functions $T(x, z)$, continuous in (x, z) and convex in z, were introduced by L. Tonelli in his early work in 1914 [I], and used systematically by him and later by E. J. McShane [6–10]. We mention here that for A compact, $Q(x) = R^n$ for all $x \in A$, and a scalar function $T(x, z)$, $x \in A$, $z \in Q(x) = R^n$, continuous in $A \times R^n$ and convex in z, the following holds: For $\bar{x} \in A$, then $T(x, z)$ is normal in z at the point $\bar{x}$ if and only if the graph G of $z^0 = T(\bar{x}, z)$, $z \in R^n$, contains no straight line (see L. Tonelli [I] for T of class C^1, and L. H. Turner [1] for T merely continuous). A proof of this statement is also in L. Cesari [8, p. 126] (see also L. Cesari [9, 10, 11]).

The necessary and sufficient condition (17.5.i) for property (Q) in terms of seminormality is due to L. Cesari [8, p. 134].

The necessary and sufficient conditions for property (Q) in terms of duality in Section 17.6 are due to G. S. Goodman [1]. Because of these characterizations, results of C. Olech [8, 9], R. T. Rockafellar [6], and A. D. Ioffe [1] in terms of duality can be equivalently expressed in terms of property (Q) by the use of Goodman's results.

We have presented the results we needed on convexity in Sections 2.16–17, 8.4–5, and 17.1–3, and on duality in Section 17.6. For convex theory in general and duality in particular we must refer here to T. Bonnesen and W. Fenchel [I], I. Ekeland and R. Temam [I], W. Fenchel [1], C. Olech [3], C. Olech and V. Klee [1], R. T. Rockafellar [I; 1–10], and F. A. Valentine [I].

The concept of duality can be traced back to Legendre, and we have shown in Section 17.7 that duality is the same operation with which we pass from the Lagrangian to the Hamiltonian in the calculus of variations and in optimal control theory—an operation which had its origin and motivation in theoretical mechanics (see e.g. P. Appell [I]). Concerning Theorem (17.7.iii) we refer to R. T. Rockafellar [6].

In Section 17.8 we show that property (Q) is a generalization of the concept of maximal monotonicity of G. Minty and H. Brezis. The result is due to M. B. Suryanarayana [8], who proved it for maps in a real Hilbert space. The simple proof in Section 17.8 for finite dimensional spaces is similar to the one for (10.5.i). For maximal monotone maps see, e.g., H. Brezis [I]. Actually, M. B. Suryanarayana [10] has considered a large class of concepts of maximal monotone multifunctions, including the one above and defined by means of analytic properties or in terms of lattice theory. All these properties imply property (Q). For lattice theory we refer to G. Birkhoff [I].

CHAPTER 18
Approximation of Usual and of Generalized Solutions

In this chapter we cover first the question of approximating absolute continuous solutions by means of C^1 solutions for Lipschitzian problems (Section 18.1). In particular we state and proved Angell's theorem, which includes most previous results, in terms of the same property (D) we have used in Chapter 13 for a far different purpose. In Section 18.2 we then present a simple example due to Manià which shows that the just-mentioned approximation, in general, may not be possible (Lavrentiev's phenomenon). In Section 18.3 we present a proof that generalized solutions for Lipschitzian problems can be approximated by means of usual absolutely continuous solutions, and in Section 18.4 we show by means of an example that such an approximation, in general, may not be possible.

18.1 The Gronwall Lemma

The following statement is often used in the theory of ordinary differential equations, and will be used in Section 18.2 below.

18.1.i (GRONWALL'S LEMMA). *If* $u(t) \geq 0$, $v(t) \geq 0$, $0 \leq t < +\infty$, *are given functions,* $u(t)$ *continuous,* $v(t)$ *L-integrable in every finite interval, and if for some nonnegative constant* C *we have*

$$u(t) \leq C + \int_0^t u(\alpha)v(\alpha)\,d\alpha, \qquad t \geq 0, \tag{18.1.1}$$

then we have also

$$u(t) \leq C \exp \int_0^t v(\alpha)\,d\alpha, \qquad t \geq 0. \tag{18.1.2}$$

Proof. If $C > 0$, by algebraic manipulation of (18.1.1) we have

$$uv\left(C + \int_0^t uv\,d\alpha\right)^{-1} \leq v,$$

and by integration

$$\log\left(C + \int_0^t uv\,d\alpha\right) - \log C \leq \int_0^t v(\alpha)\,d\alpha,$$

or

$$u \leq C + \int_0^t uv\,d\alpha \leq C\exp\int_0^t v(\alpha)\,d\alpha.$$

If $C = 0$, then (18.1.1) holds for every constant $C_1 > 0$ and then we have $0 \leq u(t) \leq C_1 \exp \int_0^t v(\alpha)\,d\alpha$, $t \geq 0$. This relation, as $C_1 \to 0$, implies $u(t) \equiv 0$. Thereby (18.1.i) is proved for all $C \geq 0$. □

18.2 Approximation of AC Solutions by Means of C^1 Solutions

We shall use here the same notation as for Mayer problems, with A a subset of the tx-space, but $U(t)$ depending on t only, and M and $f(t, x, u)$ as usual. We disregard for a moment the set B, that is, we disregard the specific boundary conditions which may be associated to a given Mayer problem.

Any Lagrange problem with functional $I = \int_{t_1}^{t_2} f_0\,dt$, can be reduced to a Mayer problem with the usual addition of the variable x^0, differential equation $dx^0/dt = f_0$, and condition $x^0(t_1) = 0$, and then we have a Mayer problem for the state variables $\tilde{x} = (x^0, x) = (x^0, x^1, \ldots, x^n)$, and differential system $d\tilde{x}/dt = \tilde{f}$, $\tilde{f} = (f_0, f) = (f_0, f_1, \ldots, f_n)$ as before.

A. The Bounded Case

18.2.i. *Let A be closed, let $U(t)$ be a closed set independent of x, and let $M = [(t, x, u) | (t, x) \in A,\ u \in U(t)]$ be closed. Let $f(t, x, u)$ be continuous on M and locally Lipschitzian with respect to x in M. Let $x_0(t)$, $u_0(t)$, $t_1 \leq t \leq t_2$, be a usual solution whose control u_0 is bounded, $|u_0(t)| \leq L$ in $[t_1, t_2]$, and whose trajectory x_0 lies in the interior of A. Then, given $\varepsilon > 0$ there is $\delta > 0$ such that for any measurable function $u(t)$, $t_1 \leq t \leq t_2$, with $u(t) \in U(t)$, $|u(t)| \leq L$ in $[t_1, t_2]$, and $\int_{t_1}^{t_2} |u(t) - u_0(t)|\,dt < \delta$, the corresponding trajectory $x(t)$, $t_1 \leq t \leq t_2$, with $x(t_1) = x_0(t_1)$, $dx/dt = f(t, x(t), u(t))$, exists in all of $[t_1, t_2]$ and lies in A, and $|x(t) - x_0(t)| < \varepsilon$ for all $t_1 \leq t \leq t_2$.*

Remark 1. Under the conditions of (18.2.i), if u is continuous on $[t_1, t_2]$, then x is of class C^1 in $[t_1, t_2]$. In other words, whenever we can approximate the control u_0 by means of continuous controls in the L_1-norm, then the trajectory x_0 can be approximated by means of trajectories of class C^1 in the uniform topology. This is certainly the case if U is say a fixed interval in R^m. If $g(t_1, x_1, t_2, x_2)$ is a continuous function as we usually assume in Mayer problems, the value of the functional $I[x_0, u_0] = g(e[x_0])$ can be approximated together with the trajectory.

Proof of (18.2.i). Let $\Gamma = [(t, x_0(t)),\ t_1 \leq t \leq t_2]$ be the graph of x_0, and consider a compact neighborhood V of Γ in A, $\Gamma \subset V \subset A$. Let ε_0 be the distance of the points of Γ from the boundary of V. Let M_0 be the set of all $(t, x, u) \in M$ with $(t, x) \in V$, $|u| \leq L_0$. Then M_0 is compact. For each point $(\bar{t}, \bar{x}, \bar{u}) \in M_0$ there is a neighborhood $\bar{W}$ of $(\bar{t}, \bar{x}, \bar{u})$ in M_0 and a constant $\bar{L}$ such that $|f(t, x, u) - f(t, y, u)| \leq \bar{L}|x - y|$ for all (t, x, u), $(t, y, v) \in \bar{W}$, since f is locally Lipschitzian on M. Since M_0 is compact, finitely many neighborhoods $W_1, \ldots, W_n$ as above cover M_0. If L is the maximum of the corresponding constants $L_1, \ldots, L_n$, then by a standard argument we can show that for a new constant L we have

$$|f(t, x, u) - f(t, y, u)| \leq L|x - y|$$

for all (t, x, u), $(t, y, u) \in M_0$. Also, f is continuous on M_0, and hence $|f(t, x, u| \leq K$ for all $(t, x, u) \in M_0$ and some constant $K \geq 0$.

Let $\varepsilon' = \min[\varepsilon, \varepsilon_0]$, and let $\sigma > 0$ be a number such that $\sigma(2K + t_2 - t_1)\exp(L(t_2 - t_1)) < \varepsilon'$. Because of the uniform continuity of f on the compact set M_0, there is some $\eta > 0$ such that $|f(t, x, u) - f(t, x, v)| < \sigma$ for all (t, x, u), $(t, x, v) \in M_0$ with $|u - v| < \eta$. Finally, let $\delta = \sigma\eta$. If $|u(t)| \leq L_0$ and $\int_{t_1}^{t_2} |u(t) - u_0(t)| \leq \delta = \sigma\eta$, let E'' be the set of all $t \in [t_1, t_2]$ where $|u(t) - u_0(t)| > \eta$. Then

$$\eta(\text{meas } E'') \leq \int_{E''} |u - u_0|\, dt \leq \int_{t_1}^{t_2} |u - u_0|\, dt \leq \delta = \sigma\eta,$$

and hence meas $E'' \leq \sigma$. If $E' = [t_1, t_2] - E''$, then $|u(t) - u_0(t)| \leq \eta$ for all $t \in E'$. For every $t \in [t_1, t_2]$ let $E'_t = E' \cap [t_1, t]$, $E''_t = E'' \cap [t_1, t]$. We have

$$x'_0(t) = f(t, x_0(t), u_0(t)), \qquad t \in [t_1, t_2] \text{ (a.e.)},$$
$$x'(t) = f(t, x(t), u(t)), \qquad x(t_1) = x_0(t_1),$$

and $x(t)$ certainly exists in some right neighborhood of t_1, since $f(t, x, u(t))$ is uniformly Lipschitzian with respect to x in a neighborhood of $(t_1, x(t_1))$. Thus, for all $t \geq t_1$ of at least a right neighborhood of t_1 we have

$$x(t) - x_0(t) = \int_{t_1}^{t} [f(\tau, x(\tau), u(t)) - f(\tau, x_0(\tau), u_0(t))]\, d\tau,$$

$$\begin{aligned}
|x(t) - x_0(t)| \leq{}& \int_{t_1}^{t} |f(\tau, x(\tau), u(\tau)) - f(\tau, x_0(\tau), u(\tau))|\, d\tau \\
&+ \int_{E'_t} |f(\tau, x_0(\tau), u(\tau)) - f(\tau, x_0(\tau), u_0(\tau))|\, d\tau \\
&+ \int_{E''_t} |f(\tau, x_0(\tau), u_0(\tau))|\, d\tau + \int_{E''_t} |f(\tau, x_0(\tau), u(\tau))|\, d\tau \\
\leq{}& L \int_{t_1}^{t} |x(\tau) - x_0(\tau)|\, d\tau + \sigma(\text{meas } E'_t) + 2K(\text{meas } E''_t).
\end{aligned}$$

Since meas $E_t'' \leq$ meas $E'' \leq \sigma$, meas $E_t' \leq t_2 - t_1$, we have

$$|x(t) - x_0(t)| \leq (2K + t_2 - t_1)\sigma + L\int_{t_1}^{t} |x(\tau) - x_0(\tau)| d\tau,$$

and by Gronwall's lemma (18.1.1)

$$|x(t) - x_0(t)| \leq (2K + t_2 - t_1)\sigma \exp(L(t_2 - t_1)) < \varepsilon' = \min[\varepsilon, \varepsilon_0].$$

This shows that the trajectory x remains in the neighborhood V of Γ and hence in the interior of A. Thus, x exists in all of $[t_1, t_2]$, and moreover $|x(t) - x_0(t)| \leq \varepsilon$ for all $t_1 \leq t \leq t_2$. This proves (18.2.i). □

Remark 2. Theorem (18.2.i) applies immediately to the case where the system is linear in x, that is,

$$f(t, x, u) = A(t)x + B(t)u + C(t), \tag{18.2.1}$$

where A, B, C are matrices with entries continuous on some fixed interval $[t_0, T]$ as in Chapters 6 and 16. More generally Theorem (18.2.i) applies to the case where

$$f(t, x, u) = A(t)x + C(t, u), \tag{18.2.2}$$

where A, C are matrices with continuous entries on $[t_0, T]$ and $[t_0, T] \times U$ respectively, and U is a compact set as in Chapters 6 and 16.

B. The Unbounded Case

Unbounded control functions $u_0(t)$, $u(t)$, $t_1 \leq t \leq t_2$, are allowed in statement (18.2.i) only under additional hypotheses. We shall see counterexamples in Section 18.4 below. But we have to expect difficulties in the unbounded case, as the following considerations show.

First of all, the simple requirements $u(t)$ measurable (even $u(t)$ L-integrable), and $u(t) \in U(t)$, $t_1 \leq t \leq t_2$, do not assure that the differential equation $dx/dt = f(t, x, u(t))$ has any solution at all, in particular any solution starting at some given point $(t_1, x_1) \in$ int A. For instance, for $f(t, x, u) = u^2$, $u(t) = t^{-1/2}$, $0 < t \leq 1$, we have $f(t, x, u(t)) = t^{-1}$, which is not L-integrable in any neighborhood of $t = 0$, and thus there are no AC solutions through points $(0, x_1)$ with $u(t) = t^{-1/2}$.

Secondly, $u(t)$ and $u_0(t)$ may be measurable and L-integrable, and very close in the L_1-norm, and yet $f(t, x, u(t))$ and $f(t, x, u_0(t))$ may be quite different. For instance, for $f(t, x, u) = u^2$, $u_0(t) = 0$, $u(t) = \varepsilon t^{-1/2}$, we have $\int_0^1 |u(t) - u_0(t)|\, dt = 2\varepsilon$, and yet $f(t, x, u_0(t)) = 0$, $f(t, x, u(t)) = t^{-1}$ is not L-integrable in $[0, 1]$.

Thus, a theorem analogous to (18.2.i) for the unbounded case must contain provisions in order that the solutions x, x_0 (a) exist, and (b) are close.

Let A, $U(t)$, M, f be as in (a), and let $x_0(t)$, $u_0(t)$, $t_1 \leq t \leq t_2$, be a given usual solution. Thus, x_0 is AC, u_0 is measurable, $u_0(t) \in U(t)$, $x_0'(t) = f(t, x_0(t), u_0(t))$, $t \in [t_1, t_2]$ (a.e.), and we assume as in (a) that the graph Γ of x_0 is in the interior of A. Let us consider a class $\mathscr{U}$ of measurable functions $u(t)$, $t_1 \leq t \leq t_2$, with $u(t) \in U(t)$, $t \in [t_1, t_2]$ (a.e.), and with the following properties:

(1) $u_0(t)$, $t_1 \leq t \leq t_2$, is an element of $\mathscr{U}$.
(2) There is a number $\sigma_0 > 0$, an L-integrable function $S(t) \geq 0$, $t_1 \leq t \leq t_2$, and a positive continuous function $\phi(\zeta)$, $0 \leq \zeta < +\infty$, which is not L-integrable in $[0, +\infty)$,

such that for every element $u(t)$, $t_1 \leq t \leq t_2$, of $\mathscr{U}$, we have

$$|f(t,x,u(t))| \leq \phi^{-1}(|x|)S(t), \qquad t_1 \leq t \leq t_2, \quad (t,x) \in A, \tag{18.2.3}$$

$$|f(t,x,u(t)) - f(t,x_0(t),u(t))| \leq |x - x_0(t)|S(t). \tag{18.2.4}$$

Here $f(t,x,u(t))$ is continuous in x for every $t \in [t_1,t_2]$, and measurable in t for every x. By a known existence theorem of ordinary differential equation theory (see, e.g., E. J. McShane [I, p. 342]), then for every point $(t_1,x_1) \in$ int A and element $u \in \mathscr{U}$ there is an AC solution $x(t)$ of the equation $dx/dt = f(t,x,u(t))$ with $x(t_1) = x_1$, and x exists in a right neighborhood of t_1.

18.2.ii. *The same as* (18.2.i), *with* $u_0(t)$, $u(t)$, $t_1 \leq t \leq t_2$, *possibly unbounded, but in a class* $\mathscr{U}$ *with properties* (1) *and* (2) *as above.*

Proof. Let Γ, V, ε_0, and $\varepsilon' = \min[\varepsilon,\varepsilon_0]$ be as in the proof of (18.2.i). Let $K = \int_{t_1}^{t_2} S(t)\,dt$; let $d = \max[|x_0(t)|,\ t_1 \leq t \leq t_2]$, and $c = \min[\phi(\zeta),\ 0 \leq \zeta \leq d]$, so that $c > 0$, and $|f(t,x_0(t),u(t))| \leq c^{-1}S(t)$, $t_1 \leq t \leq t_2$. Let $\sigma > 0$ be a number such that $\sigma(4c^{-1} + t_2 - t_1)\exp K < \varepsilon' = \min[\varepsilon,\varepsilon_0]$. Let $\chi > 0$ be a number such that $\int_H S(t)\,dt < \sigma$ for every measurable subset H of $[t_1,t_2]$ with meas $H \leq \chi$. Let $L > 0$ be a constant so large that if H_0 is the set of all $t \in [t_1,t_2]$ with $|u_0(t)| \geq L$, then meas $H_0 < \chi$. Now, let M_0 be the compact set of all $(t,x,u) \in M$ with $(t,x) \in V$, $|u| \leq L$ as in the proof of (18.2.i). Let $\eta > 0$ be a number such that $|f(t,x,u) - f(t,x,v)| < \sigma$ for all $(t,x,v) \in M_0$ with $|u - v| < \eta$. Let $\delta = \chi\eta$, and let $u(t)$, $t_1 \leq t \leq t_2$, be an element of $\mathscr{U}$ with $|u(t) - u_0(t)|$ L-integrable in $[t_1,t_2]$ and $\int_{t_1}^{t_2} |u(t) - u_0(t)|dt \leq \delta = \chi\eta$. If $E'' = [t | t_1 \leq t \leq t_2,\ |u(t) - u_0(t)| > \eta]$ and $E' = [t_1,t_2] - E'$, then, as in the proof of (18.2.i), we have meas $E'' \leq \chi$ and $|u(t) - u_0(t)| \leq \eta$ for all $t \in E'$. For every $t \in [t_1,t_2]$ we take now $E_t'' = (E'' \cup H_0) \cap [t_1,t]$, $E_t' = [t_1,t] - E_t''$.

Finally, as in the proof of (18.2.i), we have

$$\begin{aligned}
|x(t) - x_0(t)| \leq& \int_{t_1}^{t} |f(\tau,x(\tau),u(\tau)) - f(\tau,x_0(\tau),u(\tau))|\,d\tau \\
&+ \int_{E'_t} |f(\tau,x_0(\tau),u(\tau)) - f(\tau,x_0(\tau),u_0(\tau))|\,d\tau \\
&+ \int_{E_t''} |f(\tau,x_0(\tau),u(\tau))|\,d\tau + \int_{E_t''} |f(\tau,x_0(\tau),u_0(\tau))|\,d\tau \\
\leq& \int_{t_1}^{t} S(\tau)|x(\tau) - x_0(\tau)|\,d\tau \\
&+ \sigma(\text{meas } E_t') + 2c^{-1}\int_{E_t''} S(\tau)\,d\tau,
\end{aligned}$$

where meas $E_t' \leq t_2 - t_1$, meas $E_t'' \leq$ meas E'' + meas $H_0 \leq 2\chi$, and hence

$$|x(t) - x_0(t)| \leq \sigma(4c^{-1} + t_2 - t_1) + \int_{t_1}^{t} S(\tau)|x(\tau) - x_0(\tau)|\,d\tau.$$

By Gronwall's lemma then

$$|x(t) - x_0(t)| \leq \sigma(4c^{-1} + t_2 - t_1)\exp K < \varepsilon' = \min[\varepsilon,\varepsilon_0].$$

The argument proceeds now as for the proof of (18.2.i). □

Remark 3. Note that in statement (18.2.i) we have approximated the given trajectory $x_0(t)$, $t_1 \leq t \leq t_2$, by trajectories $x(t)$ of class C^1 and with $x(t_1) = x_0(t_1)$. In other words we have matched exactly the initial data. The question whether we can match both end data at the same time is not easy, and the following example shows that it may well be impossible, even in the simple bounded case. Let us consider the differential problem

$$dx/dt = y, \quad dy/dt = u, \qquad 0 \le t \le t_2,$$
$$x(0) = 1, \quad y(0) = 0, \quad x(t_2) = 0, \quad y(t_2) = 0, \quad u \in U = [-1 \le u \le 1].$$

This is exactly the problem we have studied in Section 6.1. Let $x_0(t)$, $y_0(t)$, $u_0(t)$, $0 \le t \le t_{20}$, be the particular unique solution we have characterized in Section 6.1, for which t_{20} is the minimum time for the system to transfer $(a, b) = (1, 0)$ to $(0, 0)$. Since $(1, 0)$ is not on the switching locus $\Gamma = [x = -y^2 \operatorname{sgn} y]$ we have discussed in Section 6.1, we know that $u_0(t)$ takes the values -1 and $+1$ in two intervals $[0, t)$ and $(t, t_{20}]$, $0 \le t \le t_{20}$, and the trajectory x_0, y_0 has a corner point at $t = t$. For any other solution $x(t)$, $y(t)$, $u(t)$, $0 \le t \le t_{20}$, with $x(0) = 1$, $y(0) = 0$ and class C^1, we must have $(x(t_{20}), y(t_{20})) \ne (0, 0)$. We have shown however that the distance between these two points can be made as small as possible, under the hypotheses in Theorems (18.2.i,ii).

Without some assumption, it may well occur that the infimum i_0 of the functional in the class of all C^1 trajectories is larger that the infimum i of the same functional in the class of all AC trajectories, and that the trajectories themselves cannot be uniformly approximated, as the Lavrentiev phenomenon shows (Section 18.5).

18.3 The Brouwer Fixed Point Theorem

The following statement has been the point of departure of a great deal of research.

18.3.i (BROUWER). *Any continuous map $\phi: C \to C$, of the closed unit cube C in R^n into itself possesses at least one fixed point x, that is, $x \in C$ with $x = \phi(x)$.*

For a proof we refer to Kelley [I]. For C we can take in (18.3.i) any subset of R^n topologically equivalent to the unit cube.

An equivalent form of (18.3.i) is as follows, where instead it is a matter of convenience to refer to the unit cube C and to its n pairs of opposite faces, say F_i', F_i, $i = 1, \dots, n$.

18.3.ii. *If $f(x) = (f_1, \dots, f_n)$, $x \in C$, is a continuous n-vector function on the unit cube in R^n, or $f: C \to R^n$, and f_i has constant and opposite signs on F_i and on F_i' (that is, e.g., $f_i \ge 0$ on F_i, $f_i \le 0$ on F_i'), then there is at least one point $x \in C$ where $f(x) = 0$ (that is, $f_i(x) = 0$, $i = 1, \dots, n$).*

Obviously, the theorem holds even up to a permutation $(s_1 s_2 \cdots s_n)$ on the faces (that is, e.g., if $f_i \ge 0$ on F_{s_i}, $f_i \le 0$ on F_{s_i}', $(s_1 s_2 \cdots s_n)$ being a permutation of $(12 \cdots n)$). For a proof of the equivalence of statements (18.3.i) and (18.3.ii) see C. Miranda [1].

18.4 Further Results Concerning the Approximation of AC Trajectories by Means of C^1 Trajectories

We shall mainly consider here free problems

$$J[x] = \int_{t_1}^{t_2} F_0(t, x(t), x'(t))\,dt, \qquad (t, x(t)) \in A,$$

where A is a given subset of the tx space R^{1+n} and $F_0(t, x, z)$ is a given real valued function defined on the set $M = A \times R^n \subset R^{1+2n}$. Let $x_0(t) = (x^1, \dots, x^n)$, $t_1 \le t \le t_2$, be a given trajectory for which

a. $x_0(t)$ is AC, $(t, x_0(t)) \in A$ for $t \in [t_1, t_2]$, $F_0(\cdot, x_0(\cdot), x_0'(\cdot)) \in L_1[t_1, t_2]$.

As is intrinsic to the problem of approximation, we need that some neighborhood of the graph of x_0 is contained in A. Thus, for some $d > 0$, let $A_d = [(t, x) | t \in [t_1, t_2], |x - x_0(t)| \le d]$, $M_d = A_d \times R^n$, and assume that

b. $A_d \subset A$, $M_d \subset M$, and F is continuous on M_d.

An assumption slightly weaker than (b) is the following one:

c. $A_d \subset A$, $M_d \subset M$, and for every $\lambda > 0$ there is a closed subset K of $I = [t_1, t_2]$ such that meas$(I - K) < \lambda$, and F_0 is continuous on the set $M_K = [(t, x, z) \in M_d | t \in K]$. Moreover, F_0 maps bounded subsets of M into bounded subsets of R.

Finally, we shall assume that F_0 has the property (D) defined in Section 13.1. We shall use here the following form of condition (D). We shall say that F_0 has the property (D) with respect to a sequence of AC functions $[x_k]$ provided $\|\delta_k\|_1 \to 0$ as $k \to \infty$, where $\delta_k(t) = F_0(t, x_k(t), x_k'(t)) - F_0(t, x(t), x_k'(t))$, $t_1 \le t \le t_2$, $k = 1, 2, \dots$. We know from Section 13.1 that if $\|x_k'\|_1 \le M_0$, and $x_k(t) \to x(t)$ in measure in $[t_1, t_2]$, then $\|\delta_k\|_1 \to 0$ if and only if the functions $\delta_k(t)$, $t_1 \le t \le t_2$, $k = 1, 2, \dots$, are equiabsolutely integrable.

18.4.i. THEOREM: T. S. ANGELL [6]. *Under hypotheses* (a) *and* (b) [*or* (a) *and* (c)] *and* (D), *given* $\varepsilon > 0$, *there is a trajectory* $y(t)$, $t \in I$, *continuous with piecewise continuous derivative* y', *such that* $y(t_1) = x_0(t_1)$, $y(t_2) = x_0(t_2)$, $|y(t) - x_0(t)| \le \varepsilon$ *for all* $t \in I$, *and* $|J[y] - J[x_0]| < \varepsilon$.

Proof. (a) Since x_0 is AC, we can take $R > 0$ so that, if $E = [t \in I | |x_0'^i(t)| \le R, i = 1, \dots, n]$, then meas $E > 0$. Let $m > R$ denote any integer and take $E_m = [t \in I | |x_0'^i(t)| \le m, i = 1, \dots, n]$. Thus, $E \subset E_m \subset I$. For any constant n-vector $\alpha = (\alpha_1, \dots, \alpha_n)$, let $x_m'(t; \alpha)$, $t \in I$, be the function defined by taking $x_m'(t, \alpha) = 0$ if $t \in I - E_m = \hat{E}_m$; $x_m'(t; \alpha) = x_0'(t) + \alpha$ if $t \in E$; $x_m'(t; \alpha) = x_0'(t)$ if $t \in E_m - E$. Also, let us take

$$x_m(t; \alpha) = x_0(t_1) + \int_{t_1}^{t} x_m'(s; \alpha)\, ds, \qquad t \in I = [t_1, t_2].$$

Since x_0 satisfies the analogous relation in I, by difference and by components we have, for $I_t = [0 \le s \le t]$,

$$\begin{aligned} x_m^i(t; \alpha) - x_0^i(t) &= \int_{t_1}^{t} [x_m'^i(s; \alpha) - x_0'^i(s)]\, ds = \int_{E \cap I_t} \alpha_i\, ds - \int_{\hat{E}_m \cap I_t} x_0'^i(s)\, ds \\ &= \alpha_i \operatorname{meas}(E \cap I_t) - \int_{\hat{E}_m \cap I_t} x_0'^i(s)\, ds \\ &= |t_2 - t_1| \alpha_i \theta_1^i(t) + \left[\int_{\hat{E}_m} |x_0'^i(s)|\, ds \right] \theta_2^i(t), \end{aligned}$$

where, for each $i = 1, \ldots, n$, $0 \le \theta_1^i(t) \le 1$, $-1 \le \theta_2^i(t) \le 1$. In particular,

$$x_m^i(t_2;\alpha) - x_0^i(t_2) = \alpha_i \text{ meas } E + \theta_2^i(t_2) \int_{\hat{E}_m} |x_0^i(s)|\, ds,$$

and for $\alpha_i > 0$ and m sufficiently large,

$$\text{(18.4.1)} \qquad \int_{\hat{E}_m} |x_0'^i(s)|\, ds < \alpha_i \text{ meas } E, \qquad i = 1, \ldots, n.$$

Let $\alpha_{10}, \ldots, \alpha_{n0} > 0$ be arbitrary numbers, and let $\alpha_0 = (\alpha_{10}, \ldots, \alpha_{n0})$. Let $\bar{m}$ be sufficiently large so that (18.4.1) holds for $\alpha_i = \alpha_{i0}$ and $m = \bar{m}$. We may also assume $\bar{m} > |\alpha_0|^{-1}$. For any $i = 1, \ldots, n$, let $\alpha = (\alpha_1, \ldots, \alpha_{i0}, \ldots, \alpha_n)$, $\hat{\alpha} = (\alpha_1, \ldots, -\alpha_{i0}, \ldots, \alpha_n)$, with $|\alpha_j| \le \alpha_{j0}$, $j = 1, \ldots, n$, $j \ne i$. Then, for any $m \ge \bar{m}$ we have

$$x_m^i(t_2;\alpha) - x_0^i(t_2) > 0, \qquad x_m^i(t_2;\hat{\alpha}) - x_0^i(t_2) < 0,$$

for any α, $\hat{\alpha}$ as above. Having fixed m as stated, we see that these relations hold for $i = 1, \ldots, n$. By (18.3.ii), there is a point $\bar{\alpha} = (\bar{\alpha}_1, \ldots, \bar{\alpha}_n)$, $-\alpha_{i0} < \bar{\alpha}_i \le \alpha_{i0}$, $i = 1, \ldots, n$, such that

$$x_m^i(t_2;\bar{\alpha}) - x_0^i(t_2) = 0, \qquad i = 1, \ldots, n.$$

This shows that the functions $x_m(t;\bar{\alpha})$, $t_1 \le t \le t_2$, satisfy the same end conditions as x_0.

Note that, as $|\alpha_0| \to 0$, then $m \to \infty$, and meas $\hat{E}_m \to 0$. Moreover, meas $\hat{E}_m \le$ meas $\hat{E}_{\bar{m}}$ whenever $m \ge \bar{m}$. Finally, since $|x_m^i(t;\alpha) - x_0^i(t)| \le \alpha^i|t_2 - t_1| + \int_{\hat{E}_m} |x_0'^i(t)|\, dt$, we see that $|x_m^i(t;\alpha) - x_0^i(t)| \to 0$ as $|\alpha| \to 0$, and so we can find $\delta > 0$ for which, if $|\alpha| \le \delta$, then $|x_m^i(t,\alpha) - x_0^i(t)| \le \varepsilon$ for all $t \in I$, and for some δ_0 we also have $|x_m(t;\alpha) - x_0(t)| \le d$ for $|\alpha| \le \delta_0$, that is, the graph of $x_m(t)$ lies in A_d.

(b) Concerning the functional J we note that $x_m'(t;\alpha)$ is bounded, and since F_0 in any case maps bounded sets into bounded sets, certainly the measurable function $F_0(t, x_m(t;\bar{\alpha}), x_m'(t;\bar{\alpha}))\, dt$ is bounded and hence integrable in $[t_1, t_2]$. We have now

$$\begin{aligned}
|\Delta J| &= \left| \int_I F_0(t, x_m(t;\bar{\alpha}), x_m'(t;\bar{\alpha}))\, dt - \int_I F_0(t, x_0(t), x_0'(t))\, dt \right| \\
&\le \int_I |F_0(t, x_m(t;\bar{\alpha}), x_m'(t;\bar{\alpha})) - F_0(t, x_0(t), x_m'(t;\bar{\alpha}))|\, dt \\
&\quad + \int_E |F_0(t, x_0(t), x_m'(t;\bar{\alpha})) - F_0(t, x_0(t), x_0'(t))|\, dt \\
&\quad + \int_{\hat{E}_m - E} |F_0(t, x_0(t), x_m'(t;\bar{\alpha})) - F_0(t, x_0(t), x_0'(t))|\, dt \\
&\quad + \int_{\hat{E}_m} |F_0(t, x_0(t), x_m'(t;\bar{\alpha})) - F_0(t, x_0(t), x_0'(t))|\, dt \\
&= \Delta_1 + \Delta_2 + \Delta_3 + \Delta_4.
\end{aligned}$$

On $\hat{E}_m - E$ we have $x_m'(t;\alpha) = x_0'(t)$, and hence $\Delta_3 = 0$.

Let $[\alpha_k]$ be a sequence of vectors $\alpha_k = (\alpha_{k1}, \ldots, \alpha_{kn})$, $\alpha_{ki} > 0$, $i = 1, \ldots, n$, with $|\alpha_k| \to 0$ as $k \to \infty$. We may assume that $|\alpha_k| \le \delta$, $|\alpha_k| \le \delta_0$ for all k, and we denote by $m_k > 0$ the corresponding integer, $m_k \to \infty$ as $k \to \infty$. As we have seen, to each k there correspond an $\bar{\alpha}_k$ and hence a function $x_{m_k} = x_{m_k}(t;\bar{\alpha}_k)$, $t \in I$, with the same end points as x_0. The functions x_{m_k} converge uniformly to x_0 in I as $k \to \infty$. Moreover, $|x_{m_k}'(t)| \le |x_0'(t)| + |\bar{\alpha}_k|$ for all $t \in I$. Here x_0' is L-integrable, and $\bar{\alpha}_k \to 0$. Thus the functions x_{m_k}' are equiabsolutely integrable. By property (D) and Theorem (13.1.ii) we conclude that $\Delta_1 \to 0$ as $k \to \infty$. There is therefore an index k_1 such that $\Delta_1 \le \varepsilon/4$ for $k \ge k_1$.

For any m we have $x_m'(t) = 0$ on $\hat{E}_m$ and thus

$$\Delta_4 \leq \int_{\hat{E}_m} |F_0(t, x_0(t), 0)|\, dt + \int_{\hat{E}_m} |F_0(t, x_0(t), x_0'(t))|\, dt.$$

Here $[(t, x_0(t), 0) | t \in I]$ is a bounded subset in the domain of F_0. Since F_0 maps bounded sets into bounded sets, the function $F_0(t, x_0(t), 0)$ is bounded on I. The function $F_0(t, x_0(t), x_0'(t))$ is L-integrable by hypothesis. Since meas $\hat{E}_{m_k} \to 0$ as $k \to \infty$, there is an integer k_4 such that $\Delta_4 \leq \varepsilon/4$ for $k \geq k_4$.

Concerning Δ_2, note that the functions $x_{m_k}'(t, \alpha_k)$ are equibounded on E; hence, the functions $h_{m_k}(t) = |F_0(t, x_0(t), x_{m_k}'(t; \alpha_k)) - F(t, x_0(t), x_0'(t))|$ are equiabsolutely integrable on E. Hence, there is $\delta' > 0$ such that $\int_H h_{m_k}(t)\, dt \leq \varepsilon/4$ for all k whenever meas $H \leq \delta'$. By condition (C), there is a compact subset K of I such that meas$(I - K) \leq \delta'$, and F_0 is continuous on M_k and hence uniformly continuous on any bounded subset of M_K. Since $x_{m_k} \to x_0$ and $x_{m_k}' \to x_0'$ uniformly on E, there is an index k_2 such that $\int_{K \cap E} |h_{m_k}(t)|\, dt \leq \varepsilon/4$ for $k \geq k_2$, and then

$$\Delta_2 = \left(\int_{K \cap E} + \int_{(I-K) \cap E}\right) |F_0(t, x_0(t), x_{m_k}'(t; \alpha_k)) - F_0(t, x_0(t), x_0'(t))|\, dt \leq \varepsilon/2.$$

Combining these results, we have $|\Delta J| \leq \varepsilon/4 + \varepsilon/2 + 0 + \varepsilon/4 = \varepsilon$ for $k \geq \max[k_1, k_2, k_4]$.

We have proved that $|x_{m_k}(t; \alpha_k) - x_0(t)| \leq \varepsilon$, $t_1 \leq t \leq t_2$, for some k, say $k = \bar{k}$, with $\bar{x}(t) = x_{m_k}(t; \alpha_k)$ certainly AC in $[t_1, t_2]$, with bounded derivative $\bar{x}'(t)$, and thus $x(t)$ is Lipschitzian of some constant C. Moreover $|J[\bar{x}] - J[x]| \leq \varepsilon$.

(c) We shall now in turn approximate $\bar{x}(t)$ with a function $y_N(t)$ which is continuous with sectionally continuous derivative, namely, the polygonal line obtained by dividing I into N equal parts $I_s = [\tau_{s-1}, \tau_s]$, $s = 1, \ldots, N$, y_N varying linearly in each part between the values of $\bar{x}(t)$ at the end points of I_s. Then each y_N is still Lipschitzian of constant C as $\bar{x}$, $|y_N'(t)| \leq C$, and $y_N \to \bar{x}$ as $N \to \infty$ uniformly in I. Now $\bar{x}'$ is L-integrable; therefore, by Lusin's theorem, given $\eta > 0$ (say $0 < \eta < \varepsilon/2$), there is a compact subset K' of I with meas$(I - K') < \eta$ such that $\bar{x}'$ is continuous restricted to K'. Thus, there is $\delta > 0$ (say $\delta \leq \varepsilon$) such that $|\bar{x}(t') - \bar{x}(t'')| \leq \eta$ whenever $t', t'' \in K'$, $|t' - t''| \leq \delta$. Let $L = t_2 - t_1$, take N sufficiently large so that each I_s has length $\leq \delta$, $0 < \sigma < 1$, $\sigma \leq \varepsilon L^{-1}$, and $\sigma < \varepsilon(4C)^{-1}$. Let $\{I_s\}'$ be the collection of those intervals I_s with meas$(I_s \cap K') \geq (1 - \sigma)$ meas I_s, and let $\{I_s\}''$ be the remaining intervals. If $\sum', \bigcup'$ $[\sum'', \bigcup'']$ denote sums and unions ranging over all $I_s \in \{I_s\}'$ $[I_s \in \{I_s\}'']$, then

$$\begin{aligned}
\text{meas}(K' \cap \textstyle\bigcup'' I_s) &= \text{meas} \textstyle\bigcup''(I_s \cap K') = \sum'' \text{meas}(I_s \cap K') \leq \sigma \sum'' \text{meas } I_s \leq L\sigma, \\
\text{meas}(\textstyle\bigcup'(I_s \cap K')) &\geq \text{meas}(K' \cap (\textstyle\bigcup' I_s)) = \text{meas}[K' - (K' \cap (\textstyle\bigcup'' I_s))] \\
&\geq L - \eta - L\sigma \geq L - 2\varepsilon.
\end{aligned}$$

For $t \in I_s \cap K'$, $I_s \in \{I_s\}'$ we have

$$\begin{aligned}
|y_N'(t) - \bar{x}'(t)| &\leq (\tau_s - \tau_{s-1})^{-1} \int_{I_s} |\bar{x}'(\tau) - \bar{x}'(t)|\, d\tau \\
&= (\tau_s - \tau_{s-1})^{-1} \left(\int_{I_s \cap K'} + \int_{I_s - K'}\right) |\bar{x}'(\tau) - \bar{x}'(t)|\, d\tau \\
&\leq (\tau_s - \tau_{s-1})^{-1} [\eta\, \text{meas}(I_s \cap K') + 2C\, \text{meas}(I_s - K')] \\
&\leq \eta + 2C\sigma \leq \varepsilon/2 + \varepsilon/2 = \varepsilon.
\end{aligned}$$

Thus, $|y_N'(t) - \bar{x}'(t)| \leq \varepsilon$ for all $t \in E_N = \bigcup' I_s \cap K'$, $|y_N'(t) - \bar{x}'(t)| \leq 2C$ for all $t \in I - E_N$ with meas$(I - E_N) < 2\varepsilon$.

By property (C) there is a compact subset K of I with meas$(I - K) < \varepsilon$ and F_0 is continuous on M_K. Thus,

$$\begin{aligned}|\Delta' J| &= |J[y_N] - J[\bar{x}]| \\ &\leq \left(\int_{I-E_N} + \int_{I-K} + \int_{E_N \cap K}\right)|F_0(t, y_N(t), y_N'(t)) - F_0(t, \bar{x}(t), \bar{x}'(t))|\, dt,\end{aligned}$$

where the integrand is bounded, say $\leq 2M$, independently of N, in $I - E_N$ and in $I - K$, and is $\leq \lambda$, for any given λ, in $E_N \cap K$, for N sufficiently large. Then $|\Delta' J| \leq 2M\varepsilon + 2M\varepsilon + L\lambda$, and this number can be made as small as we want. □

Remark. Theorem (18.4.i) unifies a number of separated sufficient conditions some as (α) and (β) below proposed by Tonelli and Manià. Indeed, the approximation property stated in (18.4.i) certainly holds if either of the following simple assumptions is valid in place of property (D):

(α) F_0 is continuous with $|F_0(t, x, z)| \leq \lambda + \Lambda|z|$ for all $(t, x, z) \in M_d$ and for some constants λ, Λ; or

(β) F_0 is continuous with continuous partial derivatives F_{0x} satisfying $|F_{0x}(t, x, z)| \leq \lambda + \Lambda|z|$ for all $(t, x, z) \in M_d$ and constants λ, Λ; or

(γ) F_0 is continuous with continuous partial derivatives F_{0x} satisfying $|F_{0x}(t, x, z)| \leq \lambda + \Lambda\mu(t, z)$ for all $(t, x, z) \in M_d$ and some constants λ, Λ, and where $\mu(t, z) = \min[F_0(t, x, z), |x - x_0(t)| \leq d]$.

The same holds if F_0 is a sum of functions $F_0 = F_1 + F_2 + F_3$ which satisfy conditions (α), (β), (γ) respectively. Indeed, we shall simply prove that each of these assumptions implies that F_0 has property (D) with respect to the sequence, say $[x_k]$, we have constructed in the proof of (18.3.i).

Indeed, under condition (α) and for any measurable subset H of $[t_1, t_2]$ we have

$$\begin{aligned}\int_H |\delta_k(t)|\, dt &= \int_H |F_0(t, x_k(t), x_k'(t)) - F_0(t, x_0(t), x_k'(t))|\, dt \\ &\leq \int_H |F_0(t, x_k(t), x_k'(t))|\, dt + \int_H |F_0(t, x_0(t), x_k'(t))|\, dt \\ &\leq 2\int_H [\lambda + \Lambda|x_k'(t)|]\, dt,\end{aligned}$$

where $|x_k'(t)| \leq |x_0'(t)| + \alpha_0$. Thus $\int_H |\delta_k(t)|\, dt \leq (2\lambda + \Lambda|\alpha_0|)$ meas $H + 2\int_H |x_0'(t)|\, dt$, and this last expression $\to 0$ as meas $H \to 0$. This proves the equiabsolute integrability of the functions $\delta_k(t)$.

Under condition (β) we have analogously

$$\int_H |\delta_k(t)|\, dt \leq \int_H [\lambda + \Lambda|x_k'(t)|]|x_k(t) - x_0(t)|\, dt$$

Here, given $\eta > 0$, we have $|x_k(t) - x_0(t)| \leq \eta$ for all k sufficiently large, and also $|x_k'(t)| \leq |x_0'(t)| + |\alpha_0|$. Thus $\int_H |\delta_k(t)|\, dt \leq \eta[(\lambda + \Lambda|\alpha_0|)$ meas $H + \Lambda \int_H |x_0'(t)|\, dt]$, and again the last expression approaches zero as meas $H \to 0$.

Under condition (γ), we have analogously

$$\int_H |\delta_k(t)|\, dt \leq \int_H [\lambda + \Lambda\mu(t, x_k'(t))]|x_k(t) - x_0(t)|\, dt.$$

As above, given $\eta > 0$ and for k sufficiently large we have

$$\int_H |\delta_k(t)|\,dt \le \eta\left[\lambda \text{ meas } H + \Lambda \int_H \mu(t, x_k'(t))\,dt\right]$$

$$\le \eta\left[\lambda \text{ meas } H + \Lambda \int_H \left|F_0(t, x_0(t), x_k'(t))\right| dt\right]$$

$$\le \eta\lambda \text{ meas } H + \eta\Lambda \int_H \left|F_0(t, x_0(t), x_0'(t))\right| dt$$

$$+\eta\Lambda\left(\int_{E\cap H} + \int_{(E_k - E)\cap H} + \int_{E_k\cap H}\right)\left|F_0(t, x_0(t), x_k'(t)) - F_0(t, x_0(t), x_0'(t))\right| dt$$

$$= \delta_1 + \delta_2 + \delta_3 + \delta_4 + \delta_5.$$

Obviously $\delta_1 \to 0$ as meas $H \to 0$, and $\delta_2 \to 0$ also, since the integrand is a fixed L-integrable function. Here $\delta_4 = 0$, and $\delta_5 \to 0$ as meas $H \to 0$, since the integrands are the sum of equibounded functions and of a fixed L-integrable one. Finally, concerning δ_3, we just mentioned that on E we have $|x_k'(t)| \le |x_0'(t)| + |\alpha_0| \le R + |\alpha_0|$. Thus, the integrands in δ_2 are equibounded functions on E, and therefore $\delta_3 \to 0$ as meas $H \to 0$. We have proved that F_0 has property (D) with respect to the sequence $[x_k]$ in the proof of (18.3.i).

Examples

1. The function $F_0 = |x|^{1/2}z + z^2$ is the sum of two function F_1 satisfying (α) and F_2 satisfying (β). Here is a direct proof that F_0 has property (D) with respect to the sequence in the proof of (18.3.i). Indeed,

$$|\delta_k(t)| = \left||x_k(t)|^{1/2} - |x_0(t)|^{1/2}\right||x_k'(t)| \le |x_k(t) - x_0(t)|^{1/2}|x_k'(t)|.$$

Here $x_k \to x$ uniformly; hence given $\eta > 0$ we have $|x_k(t) - x_0(t)| \le \eta$ for all k sufficiently large, while $|x_k'(t)| \le |x_0'(t)| + |\alpha_0|$. Thus, the functions $\delta_k(t)$ are equiabsolutely integrable.

2. $F_0 = (|x|^{1/2} + 1)\exp z^2$ does not satisfy any of the conditions above. However, it is easy to see that it has property (D) with respect to the sequence $[x_k]$ in the proof of (18.3.i). Indeed,

$$\begin{aligned}
|\delta_k(t)| &= \left||x_k(t)|^{1/2} - |x_0(t)|^{1/2}\right| \exp(x_k'(t))^2 \\
&\le |x_k(t) - x_0(t)|^{1/2} \exp(x_k'(t))^2 \\
&\le |x_k(t) - x_0(t)|^{1/2}(1 + |x_0(t)|^{1/2}) \exp(x_k'(t))^2 \\
&\le |x_k(t) - x_0(t)|^{1/2}|F_0(x_0(t), x_k'(t))|.
\end{aligned}$$

Here $x_k \to x_0$ uniformly; hence, given $\eta > 0$, for k sufficiently large, and for any measurable subset H of $[t_1, t_2]$, we have

$$\int_H |\delta_k(t)|\,dt \le \eta \int_H |F_0(x_0(t), x_k'(t))|\,dt,$$

and we can proceed in the proof that the functions $\delta_k(t)$ are equiabsolutely integrable as we have done under hypothesis (γ).

18.5 The Infimum for AC Solutions Can Be Lower than the One for C^1 Solutions

A. The Approximation May Not Preserve Both End Points Conditions

We consider now an actual Lagrange or Mayer problem with a given set of boundary conditions, say $(t_1, x(t_1), t_2, x(t_2)) \in B$, and we ask whether it is possible to approximate uniformly a given AC trajectory $x_0(t)$, $t_1 \leq t \leq t_2$, by means of trajectories $x(t)$, $t_1 \leq t \leq t_2$, of class C^1 satisfying exactly the same system of boundary conditions. In general, the answer is negative, as the following example shows. Consider the nonlinear differential system and boundary conditions with $n = 2$, $m = 1$, $x' = u$, $y' = (x^2 - t^2)^2$, $-1 \leq t \leq 1$, $u(t) \in U = R$, $y(-1) = 0$, $y(1) = 0$, and take $x_0(t) = |t|$, $y(t) = 0$, $u(t) = \operatorname{sgn} t$, $-1 \leq t \leq 1$, with x_0 absolutely continuous but not of class C^1. For any trajectory $x(t)$, $y(t)$, $-1 \leq t \leq 1$, of class C^1, the expression $(x^2 - t^2)^2$ must be >0 in a set of positive measure, and then y cannot satisfy both conditions $y(-1) = y(1) = 0$. Note that $u_0(t) = \pm 1$ is bounded in $[-1, 1]$ and that $f_1 = u$, $f_2 = (x^2 - t^2)^2$ are polynomials in t, x, u.

B. Lavrentiev's Phenomenon

Let us consider a Lagrange problem of the Calculus of Variation

$$I[x] = \int_{t_1}^{t_2} f_0(t, x(t), x'(t))\,dt, \qquad x = (x^1, \ldots, x^n),$$

with boundary conditions $x(t_1) = x_1$, $x(t_2) = x_2$, $A = R^{n+1}$, $f_0(t, x, u)$ continuous in $A \times R^n$. Let Ω be the class of all AC functions $x(t)$, $t_1 \leq t \leq t_2$, satisfying boundary conditions and such that $f_0(t, x(t), x'(t)) \in L_1[t_1, t_2]$. Let Ω_0 be the class of all elements x of Ω which are of class C^1. Then $\Omega \supset \Omega_0$. If i and i_0 denote the infimum of $I[x]$ in Ω and Ω_0 respectively, we have $i \leq i_0$. M. Lavrentiev [1] showed that we may well have $i < i_0$. The example we give below is easier than the one exhibited by Lavrentiev. The example is due to B. Maniá [7], with f_0 a polynomial in t, x, x'.

Let $n = 1$, let $m \geq 3$ be a fixed integer, let

$$I[x] = \int_0^1 (x^3 - t)^2 x'^{2m}\,dt,$$

$$x(0) = 0, \qquad x(1) = 1,$$

and take $x_0(t) = t^{1/3}$, $0 \leq t \leq 1$. Then $I[x_0] = 0$. Let us prove that for any other element $x(t)$, $0 \leq t \leq 1$, of Ω with x AC and x' bounded, we have

$I[x] \geq \eta > 0$, where η is a constant which depends only on m. In particular for $x \in \Omega$ of class C^1 we certainly have $I[x] \geq \eta$.

Let C_0 denote the curve $C_0\!:x = t^{1/3}, 0 \leq t \leq 1$, in the tx-plane, let Γ_1, Γ_2 be the curves $\Gamma_1\!:x = 2^{-1}t^{1/3}, 0 \leq t \leq 1$, and $\Gamma_2\!:x = 4^{-1}t^{1/3}, 0 \leq t \leq 1$, and for any ξ, $0 < \xi \leq 2^{-1}$, let R_ξ denote the region of the tx-plane bounded by Γ_1, Γ_2, and the straight lines $t = \xi$ and $t = 2\xi$. It is easy to see that in this region R_ξ the expression $(x^3 - t)^2$ has an absolute minimum given by $\varepsilon(\xi) = \frac{49}{64}\xi^2$. Let $\gamma\!:x = x(t)$, $t_1 \leq t \leq t_2$, be an absolute continuous arc lying in R_ξ with $(t_1, x(t_1)) = (\xi, 4^{-1}\xi^{1/3})$. Then $I_\gamma = \int_{t_1}^{t_2} (x^3 - t)x'^{2m}\,dt \geq \frac{49}{64}\xi^2 \int_{t_1}^{t_2} x'^{2m}\,dt$. We know that the last functional, in the class of all AC functions with given end values x_1, x_2 at t_1, t_2, takes on its minimum value for x linear in $[t_1, t_2]$ between x_1 and x_2 (see exercises at the end of Section 2.11D. Hence

$$I_\gamma \geq \tfrac{49}{64}\xi^2 (x_2 - x_1)^{2m}(t_2 - t_1)^{-2m+1}.$$

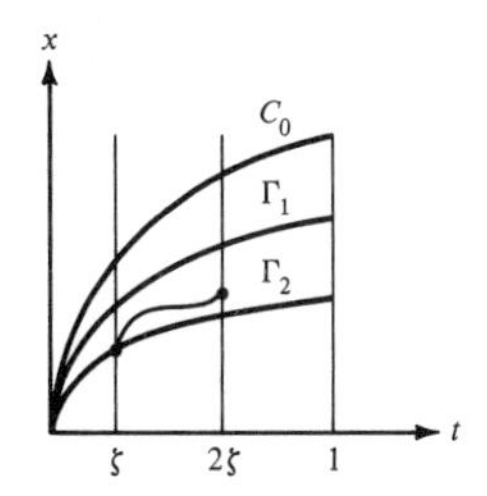

If $(t_1, x_1) = (\xi, 4^{-1}\xi^{1/3})$ as stated, and $(t_2, x_2) = (2\xi, x_2)$, $4^{-1}(2\xi)^{1/3} \leq x_2 \leq 2^{-1}(2\xi)^{1/3}$, then

$$\begin{aligned} I_\gamma &\geq \tfrac{49}{64}\xi^2 \cdot \xi^{-2m+1}(4^{-1}(2\xi)^{1/3} - 4^{-1}\xi^{1/3})^{2m} \\ &= 4^{-2m}(\tfrac{49}{64})(2^{1/3} - 1)^{2m} \cdot \xi^{3^{-1}(9-4m)}. \end{aligned}$$

The last expression is positive and approaches $+\infty$ as $\xi \to 0+$; hence, there is a constant η_1 such that $I_\gamma \geq \eta_1$ in the situation taken into consideration.

If $(t_1, x_1) = (\xi, 4^{-1}\xi^{1/3})$ as stated, and $(t_2, x_2) = (t_2, 2^{-1}t_2^{1/3})$, $\xi \leq t_2 \leq 2\xi$, then

$$\begin{aligned} I_\gamma &\geq \tfrac{49}{64}\xi^2 \cdot (t_2 - \xi)^{-2m+1}(2^{-1}t_2^{1/3} - 4^{-1}\xi^{1/3})^{2m} \\ &= \tfrac{49}{64}\xi^2 \Lambda(t_2; \xi). \end{aligned}$$

For any fixed ξ, $0 < \xi \leq 2^{-1}$, we see that $\Lambda(t_2; \xi) \to +\infty$ as $t_2 \to \xi + 0$, and thus the minimum of Λ must be either at $t_2 = 2\xi$, or at some $t_2 = \alpha\xi$, $1 < \alpha < 2$. In the first case we have again $I_\gamma \geq \eta_1$. In the second case we have

$$\{(d/dt_2)[(t_2 - \xi)^{-2m+1}(2^{-1}t_2^{1/3} - 4^{-1}\xi^{1/3})^{2m}]\}_{t_2 = \alpha\xi} = 0,$$
$$\{m(t_2 - \xi) - 3(2m - 1)t_2^{2/3}(2^{-1}t_2^{1/3} - 4^{-1}\xi^{1/3})\}_{t_2 = \alpha\xi} = 0,$$

and α must satisfy the equation, independent of ξ,

$$m(\alpha - 1) - 3(2m - 1)\alpha^{2/3}(2^{-1}\alpha^{1/3} - 4^{-1}) = 0.$$

In this situation, then

$$I_\gamma \geq \tfrac{49}{64}(\alpha - 1)^{-2m+1}(2^{-1}\alpha^{1/3} - 4^{-1})^{2m}\xi^{3^{-1}(9-4m)},$$

where again the last expression approaches $+\infty$ as $\xi \to 0+$. Thus, $I_\gamma \geq \eta_2$, where $\eta_2 > 0$ is a positive constant depending on m only.

Finally, if $\gamma: x = x(t)$, $2^{-1} \leq t_1 < t_2 \leq 1$, is any AC arc lying between Γ_1 and Γ_2, with end points on Γ_2 and Γ_1 respectively, then

$$4^{-1}t^{1/3} \leq x(t) \leq 2^{-1}t^{1/3}, \quad t_1 \leq t \leq t_2, \qquad x(t_1) = 4^{-1}t_1^{1/3}, \quad x(t_2) = 2^{-1}t_2^{1/3},$$
$$I_\gamma \geq \tfrac{49}{64}2^{-2}(t_2 - t_1)^{-2m+1}(2^{-1}t_2^{1/3} - 4^{-1}t_1^{1/3})^{2m},$$

and again $I_\gamma \geq \eta_3$ where η_3 is a positive constant. Let $\eta = \min[\eta_1, \eta_2, \eta_3] > 0$.

Now, if $C': x = x(t)$, $0 \leq t \leq 1$, is any AC arc with $x(0) = 0$, $x(1) = 1$, and $x'(t)$ bounded, then C' is below Γ_1 in some right neighborhood of the origin. Therefore C' has a maximal arc $\gamma: x = x(t)$, $t_1 \leq t \leq t_2$, lying between Γ_1 and Γ_2 and end points on Γ_2 and Γ_1 respectively. If $0 < t_1 \leq 2^{-1}$, then $I[x] \geq \min[\eta_1, \eta_2]$; if $2^{-1} \leq t_1 < 1$, then $I[x] \geq \eta_3$. In any case

$$I[x] = \int_0^1 (x^3(t) - t)^2 x'^{2m}(t)\, dt \geq \eta > 0$$

for all absolutely continuous x with $x(0) = 0$, $x(1) = 1$, with bounded x'—in particular, for all x of class C^1. This proves our contention.

C. The Lavrentiev Phenomenon in Optimal Control

The Lavrentiev phenomenon as described in Subsection B for a Lagrange problem of the Calculus of Variations must be reckoned with also in general Lagrange and Mayer problems of optimal control with unbounded controls. Examples of the same kind can be used to show that in such Mayer or Lagrange problems, a given AC trajectory cannot be approximated uniformly by means of trajectories of class C^1 together with the cost functional.

For instance, let us take $m = 1$, $n = 2$,

$$I[x, y, u] = y(1),$$
$$dx/dt = u, \quad dy/dt = (x^3 - t)^2 u^{2m}, \qquad 0 \leq t \leq 1,$$
$$x(0) = 0, \qquad x(1) = 1, \qquad y(0) = 0,$$
$$(t, x, y) \in A = R^3, \qquad u \in U = R,$$

and let us consider the admissible system $x_0(t) = t^{1/3}$, $y_0(t) = 0$, $u_0(t) = 3^{-1}t^{-2/3}$, $0 \leq t \leq 1$. We see that for any other admissible system $x(t)$, $y(t)$, $u(t)$, $0 \leq t \leq 1$, with $x(0) = 0$, $x(1) = 1$, $y(0) = 0$, with x, y of class C^1 and u continuous, no matter how much x is close to x_0 in the uniform topology, then y will remain far from y_0, and actually $y(1) \geq \eta > 0$, where η is the constant determined in Subsection B, while $y_0(1) = 0$.

18.6 Approximation of Generalized Solutions by Means of Usual Solutions

As we pointed out in Section 1.14A it is relevant that generalized solutions can be thought of as limits of usual solutions, and that in the same time, the value of the functional computed on any generalized solution can be thought of as the limit of the corresponding values taken by the functional on usual solutions. Since the infimum j of the functional on the class of all generalized solutions is certainly less than or equal to the infimum i of the functional on the class of usual solutions, we then would be able to conclude that actually $j = i$, as pointed out in Section 1.14A, and this is what occurs in all situations considered in Chapter 16.

We consider here the Mayer problem with functional

$$I[x,u] = g(t_1, x(t_1), t_2, x(t_2)),$$

and with constraints, differential equations, and boundary conditions

$$(t_1, x(t)) \in A, \quad u(t) \in U(t), \qquad dx/dt = f(t, x(t), u(t)),$$
$$(t_1, x(t_1), t_2, x(t_2)) \in B.$$

Here $x(t) = (x^1, \dots, x^n)$, $u(t) = (u^1, \dots, u^m)$, $U(t) \subset R^m$, $A \subset R^{1+n}$, $B \subset R^{2+2n}$.

The corresponding generalized solution problem concerns functional, constraints, differential equations, and boundary conditions

$$I[y,p,v] = g(t_1, y(t_1), t_2, y(t_2)),$$

$$(t, y(t)) \in A, \qquad u^{(j)}(t) \in U(t), \qquad p_j(t) \geq 0, \quad j = 1, \dots, \nu,$$

$$\frac{dy}{dt} = \sum_{j=1}^{\nu} p_j(t) f(t, y(t), u^{(j)}(t)), \quad \sum_{j=1}^{\nu} p_j(t) = 1, \qquad (t_1, y(t_1), t_2, y(t_2)) \in B.$$

Here $y = (y^1, \dots, y^n)$, $v(t) = (u^{(1)}, \dots, u^{(\nu)})$, $p(t) = (p_1, \dots, p_\nu)$, $u^j(t) = (u_{j1}, \dots, u_{jm})$, and it is not restrictive to take $\nu = n + 1$.

We consider a generalized solution, or system $y(t)$, $p(t)$, $v(t)$, $t_1 \leq t \leq t_2$, and we attempt to approximate uniformly the trajectory $y(t)$, $t_1 \leq t \leq t_2$, by the trajectories $x_k(t)$ of a sequence of usual solutions $x_k(t)$, $u_k(t)$, $t_1 \leq t \leq t_2$, $k = 1, 2, \dots$, satisfying the same constraints and (as well as possible) the boundary conditions. Here we shall require that the usual trajectories x_k have the same initial values as the generalized trajectory, or $x_k(t_1) = y(t_1) = x_1$, $k = 1, 2, \dots$. (We could as well require instead $x_k(t_2) = y(t_2) = y_2$.) We shall prove by an example at the end of this Section that in general both end values cannot be matched, under the sole hypothesis of our theorem (18.6.i). We shall assume that the graph G of the generalized trajectory is interior to A, so that certainly the solutions x_k lie in the interior of A for k sufficiently large.

18.6.i. *Let A be closed, let $U(t)$ be closed and independent of x, let $M = [(t,x,u) \,|\, (t,x) \in A, u \in U(t)]$ be closed, and let $f(t,x,u)$ be continuous on M*

and locally Lipschitzian with respect to x. Let $y(t)$, $p(t)$, $v(t)$, $t_1 \leq t \leq t_2$, be a given generalized solution whose trajectory y lies in the interior of A and whose controls $u^{(j)}(t)$, $t_1 \leq t \leq t_2$, $j = 1, \ldots, v$, are bounded. Then there is a sequence $x_k(t)$, $u_k(t)$, $t_1 \leq t \leq t_2$, $k = 1, 2, \ldots$, of usual solutions such that $x_k(t) \to y(t)$ uniformly in $[t_1, t_2]$ as $k \to \infty$, and $x_k(t_1) = y(t_1)$.

Actually, for every t, $u_k(t)$ has one of the values $u^{(j)}(t)$, $j = 1, \ldots, v$, and then certainly $u_k(t) \in U(t)$, $t_1 \leq t \leq t_2$. However, $x_k(t_2)$ may well be different from $y(t_2)$. We only need below that for some $\delta > 0$ the set $\Gamma_\delta = [(t, x) | t_1 \leq t \leq t_2, |x - y(t)| \leq \delta]$ is contained in A. Concerning the Lipschitz hypothesis, we understand that if $|u^{(j)}(t)| \leq S$ for $t \in [t_1, t_2]$, $j = 1, \ldots, v$, then there is a constant $L \geq 0$ such that $|f(t, x, u) - f(t, y, u)| \leq L|x - y|$ for all (t, x), $(t, y) \in A$ and $u = u^{(j)}(t)$. Actually, all we need is that

$$|f(t, x, u^{(j)}(t)) - f(t, y(t), u^{(j)}(t)| \leq L|x - y(t)|$$

for $(t, x) \in \Gamma_\delta$, $j = 1, \ldots, v$.

Proof. Given $\varepsilon > 0$, let $\varepsilon_0 = \min[\varepsilon, \delta]$. Let us divide $[t_1, t_2]$ into N equal parts each of length $h = (t_2 - t_1)/N$ by means of the points $\tau_i = t_1 + ih$, $i = 0, 1, \ldots, N$. In each interval $[\tau_{i-1}, \tau_i]$ we now apply Theorem (16.1.v). Then $[\tau_{i-1}, \tau_i]$ can be decomposed into v disjoint measurable subsets E_{ij}, $j = 1, \ldots, v$, such that

$$\int_{\tau_{i-1}}^{\tau_i} \sum_j p_j(t) f(t, y(t), u^{(j)}(t))\, dt = \sum_j \int_{E_{ij}} f(t, y(t), u^{(j)}(t))\, dt.$$

Thus, if we take $\bar{u}(t) = u^{(j)}(t)$ for $t \in E_{ij}$, $j = 1, \ldots, v$, then $\bar{u}(t)$ is measurable in $[\tau_{i-1}, \tau]$, $\bar{u}(t) \in U(t)$, and

$$\int_{\tau_{i-1}}^{\tau_i} \sum_j p_j(t) f(t, y(t), u^{(j)}(t))\, dt = \int_{\tau_{i-1}}^{\tau_i} f(t, y(t), \bar{u}(t))\, dt.$$

This holds for all $i = 1, \ldots, N$, and thus $\bar{u}(t)$ is defined and measurable in $[t_1, t_2]$, $\bar{u}(t) \in U(t)$, and if we take

$$\phi(t) = \int_{t_1}^{t} \left[\sum_j p_j(\tau) f(\tau, y(\tau), u^{(j)}(\tau)) - f(\tau, y(\tau), \bar{u}(\tau)) \right] d\tau, \qquad t_1 \leq t \leq t_2,$$

then $\phi(\tau_i) = 0$, $i = 0, 1, \ldots, N$. If K is a bound for $|f(t, y(t), u^{(j)}(t)|$ in $[t_1, t_2]$, $j = 1, \ldots, v$, then the integrand in the definition of ϕ is in absolute value $\leq 2K$. Since each point $t \in [t_1, t_2]$ has a distance $h/2$ from at least one point τ_i, we have

$$|\phi(t)| = |\phi(t) - \phi(\tau_i)| \leq 2K(h/2) = Kh, \qquad t_1 \leq t \leq t_2.$$

Now, for the solution $\bar{x}(t)$ of the differential equation $dx/dt = f(t, x, \bar{u}(t))$ with initial data $x(t_1) = y(t_1)$, we have

$$\bar{x}(t) = y(t_1) + \int_{t_1}^{t} f(\tau, \bar{x}(\tau), \bar{u}(\tau))\, d\tau,$$

while

$$y(t) = y(t_1) + \int_{t_1}^{t} \sum_{j=1}^{v} p_j(\tau) f(\tau, y(\tau), u^{(j)}(\tau))\, d\tau.$$

Then

$$|y(t) - \bar{x}(t)| = \left| \int_{t_1}^{t} \left[\sum_{j-1}^{\nu} p_j(\tau) f(\tau, y(\tau), u^{(j)}(\tau)) - f(\tau, y(\tau), \bar{u}(\tau)) \right] d\tau \right.$$

$$\left. + \int_{t_1}^{t} [f(\tau, y(\tau), \bar{u}(\tau)) - f(\tau, \bar{x}(\tau), \bar{u}(\tau))] \, d\tau \right|$$

$$\leq |\phi(t)| + \int_{t_1}^{t} |f(\tau, y(\tau), \bar{u}(\tau)) - f(\tau, \bar{x}(\tau), \bar{u}(\tau))| \, d\tau$$

$$\leq Kh + \int_{t_1}^{t} L|y(\tau) - \bar{x}(\tau)| \, d\tau.$$

By Gronwall's lemma we derive

$$|y(t) - \bar{x}(t)| \leq Kh \exp L(t_2 - t_1),$$

and this relation is valid at $t = t_1$ and in any right neighborhood of t_1 where $(t, \bar{x}(t)) \in \Gamma_\delta$. If we choose $h = (t_2 - t_1)/N$ so that $Kh \exp L(t_2 - t_1) \leq \varepsilon_0$, then the relation above holds in all of $[t_1, t_2]$, and $|y(t) - \bar{x}(t)| \leq \varepsilon_0 \leq \varepsilon$, $t_1 \leq t \leq t_2$. By taking $\varepsilon = k^{-1}$, $k = 1, 2, \ldots$, we obtain the sequence stated in (18.6.i). Note that $x_k(t) \to y(t)$ uniformly, hence $g(t_1, x_k(t_1), t_2, x_k(t_2)) \to g(t, y(t_1), t_2, y(t_2))$ as $k \to \infty$. □

Remark. Under the conditions of (18.6.i) alone, there may be no usual solution satisfying *both* end conditions at t_1 and t_2. This is shown by the differential problem

$$x' = u, \quad y' = x^2, \qquad u(t) \in U = \{+1\} \cup \{-1\},$$

with $n = 2$, $m = 1$, and boundary data $y(0) = y(1) = 0$. The generalized solution

$$x(t) = 0, \quad y(t) = 0, \qquad u^{(1)}(t) = 1, \quad u^{(2)}(t) = -1,$$
$$p_1(t) = 2^{-1}, \quad p_2(t) = 2^{-1}, \qquad 0 \leq t \leq 1,$$

satisfies both end data. For any usual solution $x(t)$, $y(t)$, $u(t)$, $0 \leq t \leq 1$, with $x(t_1) = 0$, we have $x'(t) = u(t) = \pm 1$ a.e. in $[t_1, t_2]$. Hence x is not identically zero in $[0, 1]$, and $y(1) > 0$.

18.7 The Infimum for Generalized Solutions Can Be Lower than the One for Usual Solutions

Let σ, c be positive constants, $c \leq \frac{1}{8}, .\sigma < 1$. Let U denote the set made up of only the three points $u = 0, \frac{1}{2}, 1$. Let V denote the interval $V = [-c, c]$. Let $F(u) = 1 - 2|u - \frac{1}{2}|$, so that $F(0) = F(1) = 0$, $F(\frac{1}{2}) = 1$. Let us consider the problem of minimizing the integral

$$I[x, y, z, u, v] = \int_0^1 F(u) \, dt$$

under the constraints

$$dx/dt = (\sigma + t)^{-1}(u - x), \qquad dy/dt = v, \qquad dz/dt = (x - y)^2,$$
$$x(0) = y(0) = \tfrac{1}{2}, \quad z(0) = z(1) = 0, \qquad u \in U, \quad v \in V, \qquad 0 \leq t \leq 1.$$

Here $n = 3$, $m = 2$, x, y, z are the state variables, u, v the control variables, $u \in U$, $v \in V$, $A = [0,1] \times R^3$.

Let us prove that the only usual solution of this problem is $u(t) = \frac{1}{2}$, $x(t) = y(t) = \frac{1}{2}$, $z(t) = 0$, $v(t) = 0$, and then $I = 1$. Indeed, $z'(t) \geq 0$, $z(0) = z(1) = 0$. Hence, $z'(t) = 0$, $x(t) = y(t)$, $z(t) = 0$, and $x'(t) = y'(t) = v$, $|x'(t)| \leq c \leq \frac{1}{8}$, and $x(t)$ is Lipschitzian of constant $\frac{1}{8}$. Since $x(0) = \frac{1}{2}$, we have $|x(t) - \frac{1}{2}| \leq \frac{1}{8}$, or $\frac{1}{2} - \frac{1}{8} \leq x(t) \leq \frac{1}{2} + \frac{1}{8}$. On the other hand, $|u(t) - x(t)| = |(\sigma + t)x'(t)| \leq 2c \leq \frac{1}{4}$. Thus, $u(t)$ can take only the value $u = \frac{1}{2}$, and the equation for x reduces to $(\sigma + t)x' + x = \frac{1}{2}$, or $(d/dt)((\sigma + t)x) = \frac{1}{2}$; hence $(\sigma + t)x(t) = t/2 + C$. For $t = 0$, we have $C = \sigma/2$, and $x(t) = \frac{1}{2}$, $0 \leq t \leq 1$.

Let us consider the generalized problem corresponding to the one above, with

$$\mu = 2, \qquad p = (p_1, p_2), \quad p_1 \geq 0, \quad p_2 \geq 0, \quad p_1 + p_2 = 1,$$
$$\tilde{u} = (u^{(1)}, u^{(2)}), \quad \tilde{v} = (v^{(1)}, v^{(2)}),$$

$$J[x, y, z, p, \tilde{u}, \tilde{v}] = \int_0^1 [p_1 F(u^{(1)}) + p_2 F(u^{(2)})]\, dt,$$

$$dx/dt = (\sigma + t)^{-1}[p_1 u^{(1)} + p_2 u^{(2)} - x],$$
$$dy/dt = p_1 v^{(1)} + p_2 v^{(2)}, \qquad dz/dt = (x - y)^2,$$
$$x(0) = y(0) = \tfrac{1}{2}, \qquad z(0) = z(1) = 0,$$
$$u^{(1)} \in U, \quad u^{(2)} \in U, \qquad v^{(1)} \in V, \quad v^{(2)} \in V, \qquad 0 \leq t \leq 1.$$

Again we must have $x(t) = y(t)$, $z(t) = 0$, $|p_1 v^{(1)} + p_2 v^{(2)}| \leq c$, and $|dx/dt| = |dy/dt| \leq c$. We now take $p_1 = \frac{1}{2}$, $p_2 = \frac{1}{2}$, $u^{(1)} = 0$, $u^{(2)} = 1$, $v^{(1)} = 0$, $v^{(2)} = 0$, $p_1 u^{(1)} + p_2 u^{(2)} = \frac{1}{2}$, and $x(t) = y(t) = \frac{1}{2}$, $0 \leq t \leq 1$. Now

$$J = \int_0^1 [\tfrac{1}{2}F(0) + \tfrac{1}{2}F(1)]\, dt = 0.$$

We see that the infimum j for generalized solutions can be far below the infimum i for usual solutions; namely, we have here $j = 0 < i = 1$.

Bibliographical Notes

The question of approximating AC solutions and the value of the corresponding integral by means of C^1 solutions became relevant since AC solutions started being used in the calculus of variations (cf. Chapter 1). In 1927 M. Lavrentiev [1] found the first example of a problem of the calculus of variations for which the infimum for AC solutions is lower than the infimum for C^1 solutions. In Section 18.5 we have presented a much simpler example which is due to B. Manià [6, 7]. In Section 18.4 we presented a theorem due to T. S. Angell [6] showing that various known criteria, under which AC solutions and corresponding integrals can be approximated by means of C^1 solutions, can be unified in terms of the property (D) we have already used in Chapter 13 for existence theorems for optimal solutions.

For the proof in Section 18.6 concerning the approximation of generalized solutions and corresponding integrals by means of AC usual solutions, see R. V. Gamkrelidze [1] and L. Cesari [6]. The example in Section 18.7, showing that the infimum for generalized solutions can be lower than for AC usual solutions, is new.

Questions of approximations with multifunctions were discussed by M. Q. Jacobs [4].

Bibliography

Ahmed, N. U. and K. L. Teo, [1] Comments on selector theorems in Banach spaces, *J. Optimization Theory Appl.* **19**, 1976, 117–118.

Akhiezer, N. I., [I] *The Calculus of Variations*, Blaisdell 1962, viii + 247.

Angell, T. S., [1] Existence theorems for optimal control problems involving functional differential equations, *J. Optimization Theory Appl.* **7**, 1971, 148–169. [2] Existence theorems for hereditary Lagrange and Mayer problems of optimal control, *SIAM J. Control Optimization* **14**, 1976, 1–18. [3] Control for linear Volterra systems without convexity. *Dynamical Systems, An International Symposium* (L. Cesari, J. K. Hale, J. P. LaSalle, eds.), Academic Press 1976, vol. 1, 311–315. [4] On the optimal control of systems governed by nonlinear Volterra equations, *J. Optimization Theory Appl.* **19**, 1976, 29–45. [5] Existence of optimal control without convexity and a bang-bang theorem for linear Volterra equations, *J. Optimization Theory Appl.* **19**, 1976, 63–79. [6] A note on approximation of optimal solutions of free problems of the calculus of variations, *Rend. Circ. Mat. Palermo.* (2) 28, 1979, 258–272. [7] Mayer problems with singular components for functional differential equations, *Annali Mat. Pura Appl.* **127**, 1981, 13–24. [8] On controllability for nonlinear hereditary systems: a fixed point approach, *J. Nonlinear Analysis* (*4*) 1980, 529–545. [9] The controllability problem for nonlinear Volterra systems, *J. Optimization Theory Appl.*, to appear.

Appell, P., [I] *Traité de Mécanique Rationelle*, 6th ed., Gauthier Villars 1953, 3 vols.

Athans, M. and P. L. Falb, [I] *Optimal Control*, McGraw-Hill 1966, xiv + 879.

Aubin, J. P., [1] A Pareto minimum principle. *Differential Games and Related Topics* (H. W. Kuhn and S. P. Szego, eds.), North-Holland 1971, 147–175.

Baiada, E., [1] Sulle funzioni continue separatamente rispetto alle variabili e gli integrali curvilinei, *Rend. Sem. Mat. Univ. Padova* **17**, 1948, 201–218. [2] Il problema isoperimetrico del calcolo delle variazioni, *Ann. Scuola Norm. Sup. Pisa.* (*2*) **15**, 1946, 97–112 (1950). [3] Sopra una classe particolare di problemi di calcolo delle variazioni, *Ann. Scuola Norm. Sup. Pisa* (*3*) **6**, 1952, 173–186. [4] La variazione totale, la lunghezza di una curva, e l'integrale del calcolo delle variazioni, *Rend. Cl. Sc. Fis. Mat. Nat. Accademia Naz. Lincei* (*8*) **22**, 1957, 584–588.

Balakrishnan, A. V., [1] (ed.) *Control Theory and the Calculus of Variations*, Academic Press 1969, xiii + 422. [II] *Introduction to Optimization Theory in Hilbert Spaces*. Lecture Notes in Oper. Res. and Math. Systems, No. 42, Springer-Verlag 1971, ii + 153.

Banach, S., [I] *Théorie des Opérations Linéaires* Hafner, Warsaw 1932, vii + 254. [1] Sur les lignes rectifiables et les surfaces dont l'aire est finie, *Fund. Math.* **6**, 1924, 170–188.

Banks, H. T., [1] Necessary conditions for control problems with variable time lags, *SIAM J. Control* **6**, 1968, 9–47. [2] A maximum principle for optimal control problems with functional differential systems, *Bull. Am. Math. Soc.* **75**, 1969, 158–161. [3] Representations for solutions of linear functional differential equations, *J. Diff. Equations* **5**, 1969, 309–409.

[4] Variational problems involving functional differential equations, *SIAM J. Control* **7**, 1969, 1–17.

Baum, R. F., [1] An existence theorem for optimal control systems with state variable in *C* and stochastic control problems, *J. Optimization Theory Appl.* **5**, 1970, 335–346. [2] Optimal control systems with stochastic boundary conditions, *J. Operations Research* **20**, 1972, 875–887. [3] Existence theorems for multidimensional control systems with lower dimensional controls, *SIAM J. Control* **10**, 1972, 623–638. [4] Existence theorems for Lagrange control problems with unbounded time domain, *J. Optimization Theory Appl.* **19**, 1976, 89–116. [5] Necessary conditions for distributed parameter systems with controls of fewer variables than state variables, *J. Optimization Theory Appl.* **30**, 1980, 663–681.

Bell, D. J. and D. H. Jacobson, [I] *Singular Optimal Control Problems*, Academic Press 1975, xi + 190.

Bellman, R., [I] *Stability Theory of Differential Equations*, McGraw-Hill 1954, xiii + 166. [II] *Dynamic Programming*, Princeton University Press 1957, xxv + 342. [III] *Adaptive Control Processes, a Guided Tour*, Princeton University Press 1961, xvi + 255.

Berge, C., [I] *Espaces Topologiques, Fonctions Multivoques*, Dunod 1965, xii + 283.

Berger, M. S., [I] *Nonlinearity and Functional Analysis*, Academic Press 1977, xix + 417.

Berkovitz, L. D., [I] *Optimal Control Theory*, Springer-Verlag, Applied Math. Sciences No. 12, 1974, ix + 304. [1] Existence theorems in problems of optimal control, *Studia Math.* **44**, 1972, 275–285. [2] Existence theory for optimal control problems. *Optimal Control and Differential Equations* (A. B. Schwarzkopf, W. G. Kelley, S. B. Eliason, eds.), Academic Press 1978, 107–130. [3] Lower semicontinuity of integral functionals, *Trans. Amer. Math. Soc.* **192**, 1974, 51–57.

Bidaut, M. F., [1] Quelques résultats d'existence pour des problèmes de contrôle optimal, *C. R. Acad. Sci. Paris* **274**, 1972, 62–65.

Birkhoff, G., [I] *Lattice Theory*, Amer. Math. Soc. Coll. Publ. No. 25, 2nd ed., 1949, xiv + 283.

Blackwell, D. [1] The range of certain vector integrals, *Proc. Amer. Math. Soc.* **2**, 1951, 390–395.

Bliss, G. A., [I] *Calculus of Variations*, Math. Assoc. Amer. 1925, xiii + 189. [II] *Lectures on the Calculus of Variations*, Univ. of Chicago Press 1946, ix + 296.

Boltyanskii, V. G., [1] Sufficient conditions for optimality and the justification of the dynamic programming method, *SIAM J. Control* **4**, 1966, 326–361.

Bolza, O., [I] *Vorlesungen über Variationsrechnung*, Teubner 1909 (Koehler and Amelang 1949, ix + 705. [II] *Lectures on the Calculus of Variations*, Stechert 1946, xi + 271.

Bonnesen, T. and W. Fenchel, [1] *Theorie der konvexen Koerper*, Springer-Verlag, Ergebnisse Math. u. Grenzgeb. No. 3, 1934.

Brezis, H., [I] *Opérateurs Maximaux Monotones*, North-Holland, Math. Studies No. 5, 1973, iv + 183.

Browder, F. E., [1] Remarks on the direct method of the calculus of variations, *Arch. Rational Mech. Anal.* **20**, 1965, 251–258.

Candeloro, D., and P. Pucci, [1] Un criterio di compattezza debole alla Dunford-Pettis, *Rend. Accad. Naz. Lincei (8)* **64**, 1978, 124–129.

Carathéodory, C., [I] *Calculus of Variations and Partial Differential Equations of the First Order*, Teubner, Berlin, 2 vols., 1935; Engl. transl.: Holden Day, 1965, xvi + 174; 1967, 175–398.

Castaing, C. and M. Valadier, [I] *Convex Analysis and Measurable Multifunctions*, Springer-Verlag, Lecture Notes in Math. No. 580, 1977, vii + 278.

Cesari, L., [I] *Surface Area*, Princeton University Press 1956, x + 595. [II] *Asymptotic Behavior and Stability Problems in Ordinary Differential Equations*, Springer-Verlag, Ergebn. der Math. u. ih. Grenzgebiete No. 16, 1959, vii + 271 (2nd ed. 1963; 3rd ed. 1971; Russian ed. MIR 1964). [III] *Parametric Optimization and Related Topics*, Springer-Verlag, to appear. [IV] *Optimization with Partial Differential Equations*, to appear.

[1] Problemi di Lagrange con vincoli unilaterali, Convegno Lagrangiano, *Atti Accad. Sci. Torino*, **98**, 1964, 88–119. [2] Semicontinuità e convessità nel calcolo delle variazioni *Ann. Scuola Norm. Sup. Pisa* (3) 18, 1964, 389–423. [3] Un teorema di esistenza in problemi di control ottimi, *Ann. Scuola Norm. Sup. Pisa (3)* **19**, 1965, 35–78. [4] An existence theorem in problems of optimal control, *SIAM J. Control* **3**, 1965, 7–22. [5] Existence theorems for optimal solutions in Pontryagin and Lagrange problems, *SIAM J. Control* **3**, 1965, 475–498. [6] Existence theorems for weak and usual optimal solutions in Lagrange problems with

unilateral constraints, I and II, *Trans. Amer. Math. Soc.* **124**, 1966, 369–412, 413–429. [7] Existence theorems for optimal controls of the Mayer type, *SIAM J. Control* **6**, 1968, 517–552. [8] Seminormality and upper semicontinuity in optimal control, *J. Optimization Theory Appl.* **6**, 1970, 114–137. [9] Closure, lower closure, and semicontinuity theorems in optimal control, *SIAM J. Control* **9**, 1971, 287–315. [10] Lagrange problems of optimal control and convex sets not containing any straight line, *Atti Sem. Mat. Fis. Univ. Modena* **23**, 1974, 118–139. [11] Convexity and seminormality in the calculus of variations, *Accad. Naz. Sci. Let. Arti Modena Atti Memorie* (*6*) **14**, 1972, 119–133. [12] Closure theorems for orientor fields, *Bull. Amer. Math. Soc.* **79**, 1973, 684–689. [13] Closure theorems for orientor fields and weak convergence, *Arch. Rational Mech. Anal.* **55**, 1974, 332–356. [14] Convexity of the range of certain integrals, *SIAM J. Control* **13**, 1975, 666–676. [15] An existence theorem without convexity conditions, *SIAM J. Control* **12**, 1974, 319–331. [16] Lower semicontinuity and lower closure theorems without seminormality conditions, *Ann. Mat. Pura Appl.* **98**, 1974, 381–397. [17] A necessary and sufficient condition for lower semicontinuity, *Bull. Amer. Math. Soc.* **80**, 1974, 467–472. [18] Geometric and analytic views in existence theorems for optimal control in Banach spaces, I, II, and III, *J. Optimization Theory Appl.* **14**, 1974, 505–520; **15**, 1975, 467–497; **19**, 1976, 185–214. [19] La nozione di integrale sopra una superficie in forma parametrica, *Ann. Scuola Norm. Sup. Pisa* (*2*) **13**, 1944, 78–117. [20] Condizioni sufficienti per la *semicontinuità degli integrali sopra una superficie*, *Ann. Scuola Norm. Sup. Pisa* (*2*) **14**, 1945, 47–75. [21] Condizioni necessarie per la semicontinuita degli integrali sopra una superficie in forma parametrica, *Ann. Mat. Pura Appl.* (*4*) **29**, 1949, 199–223. [22] An existence theorem of calculus of variations for integrals on parametric surfaces, *Amer. J. Math.* **14**, 1952, 265–295. [23] Quasi additive set functions and the concept of integral over a variety, *Trans. Amer. Math. Soc.* **102**, 1962, 94–113. [24] Extension problem for quasi additive set functions and Radon–Nikodym derivatives, *Trans. Amer. Math. Soc.* **102**, 1962, 114–146. [25] Variation, multiplicity, and semicontinuity, *Amer. Math. Monthly* **65**, 1958, 317–332. [26] Rectifiable curves and the Weierstrass integral, *Amer. Math. Monthly* **65**, 1958, 485–500. [27] Recent results in surface area theory, *Amer. Math. Monthly* **66**, 1959, 173–192. [28] Sulle funzioni a variazione limitata, *Ann. Scuola Norm Sup. Pisa* (*2*) **5**, 1936, 299–313. [29] Existence theorems for multidimensional problems of optimal control. *Mayaguez 1965 Conference*, *Differential Equations and Dynamic Systems*, Academic Press 1967, 115–132. [30] Existence theorems for multidimensional Lagrange problems, *J. Optimization Theory Appl.* **1**, 1967, 87–112. [31] Multidimensional Lagrange and Pontryagin problems. *Los Angeles 1967 Conference*, *Mathematical Theory of Control*, Academic Press 1967, 272–284. [32] Sobolev spaces and multidimensional Lagrange problems of optimization, *Ann. Scuola Norm. Sup. Pisa* **22**, 1963, 193–227. [33] Multidimensional Lagrange problems of optimization in a fixed domain and an application to a problem of magnetohydrodynamics, *Arch. Rational Mech. Anal.* **29**, 1968, 81–104. [34] Optimization with partial differential equations in Dieudonné–Rashevsky form and conjugate problems, *Arch. Rational Mech. Anal.* **33**, 1969, 339–357. [35] Existence theorems for Lagrange problems in Sobolev spaces, *Proceedings of Symposia*, *Amer. Math. Soc.* **18/1**, 1970, 39–49. [36] Existence theorems for abstract multidimensional control problems, *J. Optim. Theory Appl.* **6**, 1970, 210–236. [37] Existence theorems for problems of optimization with distributed and boundary controls. *Proceedings International Congress of Mathematicians Nice* 1970, vol. 3, 157–161. Symposium Amer. Math. Soc., Berkeley 1971.

Cesari, L. and R. F. Baum, [1] A recent proof of Pontryagin's necessary condition, *SIAM J. Control* **10**, 1972, 56–75.

Cesari, L. and D. E. Cowles, [1] Existence theorems for optimization problems with distributed and boundary controls, *Arch. Rational Mech. Anal.* **46**, 1972, 321–355.

Cesari, L. and P. J. Kaiser, [1] Closed operators and existence theorems in multidimensional problems of the calculus of variations, *Bull. Amer. Math. Soc.* **80**, 1974, 473–478.

Cesari, L., J. R. LaPalm, and T. Nishiura [1] Remarks on some existence theorems for optimal control, *J. Optimization Theory Appl.* **3**, 1969, 296–305.

Cesari, L., J. R. LaPalm, and D. A. Sanchez, [1] An existence theorem for Lagrange problems with unbounded controls and a slender set of exceptional points, *SIAM J. Control* **9**, 1971, 590–605, and **11**, 1973, 677.

Cesari, L., and P. Pucci, [1] An elementary proof of an equivalence theorem, *Amer. Math. Monthly*, to appear.

Cesari, L. and M. B. Suryanarayana, [1] Closure theorems without seminormality conditions, *J. Optimization Theory Appl.* **15**, 1975, 441–465. [2] Convexity and property (Q) in optimal control theory, *SIAM J. Control* **12**, 1974, 705–720. [3] Nemitsky's operators and lower closure theorems, *J. Optimization Theory Appl.* **19**, 1976, 165–184. [4] Existence theorems for Pareto optimization in Banach spaces, *Bull. Amer. Math. Soc.* **82**, 1975, 306–308. [5] Existence theorems for Pareto problems of optimization. *Calculus of Variations and Control Theory* (D. Russel, ed.), Academic Press 1976, 139–154. [6] Existence theorems for Pareto optimization. Multivalued and Banach space valued functionals, *Trans. Amer. Math. Soc.* **244**, 1978, 37–65. [7] An existence theorem for Pareto problems, *Nonlinear Anal.* **2**, 1978, 225–233. [8] Uppersemicontinuity properties of set valued functions, *Nonlinear Anal.* **4**, 1980, 639–656. [9] On recent existence theorems in the theory of optimization, *J. Optimization Theory Appl.*, **31**, 1980, 397–416.

Chong, F. V., [1] On a smooth selection theorem and its applications to multivalued integral equations. *Mat. Sb.* 105 (147), 1978, 622–637.

Cinquini, S., [1] Sopra l'esistenza della soluzione nei problemi di calcolo delle variazioni di ordine *n*, *Ann. Scuola Norm. Sup. Pisa* (*2*) **5**, 1936, 169–190. [2] Nuovi teoremi di esistenza dell' estremo in campi illimitati per i problemi di calcolo delle variazioni di ordine *n*, *Ann. Scuola Norm. Sup. Pisa* (*2*) **6**, 1937, 191–209. [3] L'estremo assoluto degli integrali doppi dipendenti dalle derivate di ordine superiore, *Ann. Scuola Norm. Sup. Pisa* **10**, 1941, 215–248. [4] Sopra i problemi variazionali in forma parametrica dipendenti dalle derivate di ordine superiore. *Ann. Scuola Norm. Sup. Pisa* (*2*) **13**, 1944, 19–49. [5] Sopra gli integrali doppi del calcolo variazioni dipendenti dalle derivate del secondo ordine, *Rend. Istituto Lombardo* **84**, 1951, 327–336. [6] Sopra l'esistenza dell'estremo per una classe di integrali curvilinei in forma parametrica, *Ann. Mat. Pura Appl.* (*4*) **49**, 1960, 25–71. [7] Sopra la continuità di una classe di integrali del calcolo delle variazioni, *Riv. Mat. Univ. Parma* (*3*) **3**, 1974, 139–161.

Cole, J. K., [1] A selector theorem in Banach spaces, *J. Optimization Theory Appl.* **7**, 1971, 170–172.

Conti, R., [I] *Problemi di Controllo e di Controllo Ottimale*, Unione Editr. Torinese 1974, v+239.

Dacunha, N. O. and E. Polak, [1] Constrained minimization under vector valued criteria in linear topological spaces. *Mathematical Theory of Control* (A. V. Balakrishnan and L. W. Neustadt, eds.), Academic Press 1967, 96–108.

Day, M., [I] *Normed Linear Spaces*, 3rd ed., Springer-Verlag, Ergebn. Math. u. ih. Grenzgeb. No. 21, 1973.

Dienstel, J., [I] *Geometry of Banach Spaces. Selected Topics.* Springer-Verlag, Lecture Notes in Math. No. 485, 1975, xi+282.

Dunford, N. and J. T. Schwartz, [I] *Linear Operators*, 3 vols., Interscience, xiv+2592. 1958, 1963, 1971,

Edwards, R. E., [I] *Functional Analysis*, Holt, Rinehart and Winston 1965, xiii+781.

Egorov, A. I., [I] Optimal control in a Banach space, *Math. Systems Theory* **1**, 1962, 347–352.

Ekeland, I. and R. Temam, [I] *Analyse Convexe et Problèmes Variationnels*, Dunod, 1974, ix+340.

Elsgolc, L. E., [I] *Calculus of Variations*, Pergamon Press, Addison-Wesley, 1962, 178.

Faedo, S., [1] Condizioni necessarie per la semicontinuità di un nuovo tipo di funzionali, *Annali Mat. Pura Appl.* (*4*) **23**, 1944, 69–121. [2] Un nuovo tipo di funzionali continui, *Rend. Mat. e appl.* (*5*) **4**, 1943, 223–249. [3] Sulle condizioni di Legendre e di Weierstrass per gli integrali di Fubini-Tonelli, *Annali Scuola Norm. Sup. Pisa* (*2*) **15**, 1946, 127–135 (1950).

Fenchel, W., [1] On conjugate convex functions, *Canad. J. Math.* **1**, 1949, 73–77.

Fichera, G., [1] Semicontinuity of multiple integrals in ordinary form. *Arch. Rat. Mech. Anal.* **17**, 1964, 339–352. [2] Sull'ubicazione e l'unicità delle estremanti del polinomiale quadratico nella sfera di Hilbert, *Publ. Ist. Naz. Appl. Calcolo*, no. 160, 1944. [3] Sui funzionali continui con la metrica di Fréchet, *Rend. Accad. Naz. Lincei* (*8*) **2**, 1947, 174–177.

Filippov, A. F., [1] On certain questions in the theory of optimal control, *SIAM J. Control* **1**, 1962, 76–84. [2] Differential equations with multivalued discontinuous right hand side, *Dokl. Akad. Nauk SSSR* **151**, 1963, 65–68.

Fitzpatrick, P. M., P. Hess, and T. Kato, [1] Local boundedness of monotone-type operators, *Proc. Japan Acad.* **48**, 1972, 275–277.

Fleming, W. H., [1] Some Markovian optimization problems. *J. Math. and Mech.*, **12**, 1963, 131–140. [2] Duality and a priori estimates in Markovian optimization problems, *J. Math. Anal. Appl.* **16**, 1966, 254–279. [3] Stochastic Lagrange multipliers, *Proc. Conf. Math. Control Theory*, Los Angeles, Jan. 1967. [4] The Cauchy problem for a nonlinear first order partial differential equation, *J. Diff. Equations* **5**, 1969, 515–530. [5] Optimal continuous parameter stochastic control, *SIAM Review* **11**, 1969, 470–509. [6] Stochastically perturbed dynamical systems, *Rocky Mountain Math. J.*, **4**, 1974, 407–433.

Fleming, W. H. and R. W. Rishel, [I] *Deterministic and Stochastic Optimal Control*, Springer-Verlag 1975, vii+222.

Fleming, W. H., and L. C. Young, [1] Representations of general surfaces as mixtures, *Rend. Circ. Mat. Palermo* (2) **5**, 1956, 117–144. [2] Generalized surfaces with prescribed elementary boundary, *ibid.* 320–340.

Forsyth, A. R., [I] *Calculus of Variations*, (Constable 1926), Dover 1960.

Fox, C., [I] *An Introduction to the Calculus of Variations*, Oxford Univ. Press 1950, viii+271.

Gamkrelidze, R. V., [1] On sliding optimal states, *Dokl. Akad. Nauk SSSR* **143**, 1962, 1243–1245 = *Soviet Math. Dokl.* **3**, 1962, 559–561. [2] On some extremal problems in the theory of differential equations with applications to the theory of optimal control, *SIAM J. Control* **3**, 1965, 106–128.

Gelfand, I. M. and S. V. Fomin, [I] *Calculus of Variations*, Prentice-Hall 1963, xii+232.

Gilbert, E. G., [1] An iterative procedure for computing the minimum of a quadratic form on a convex set, *SIAM J. Control* **4**, 1966, 61–80. [2] Vehicle cruise: improved fuel economy by periodic control, *Automatica* **12**, 1976, 159–166. [3] Optimal periodic control: a general theory of necessary conditions, *SIAM J. Control* **15**, 1977, 717–745.

Gilbert, E. G. and R. O. Barr, [1] Some efficient algorithms for a class of abstract optimization problems arising in optimal control, *IEEE Trans. Autom. Control* **AC-14**, 1969, 640–652.

Gilbert, E. G. and D. S. Bernstein, [1] Second order necessary conditions in optimal control accessory-problem results without normality conditions, *J. Optimization Theory Appl.*, to appear.

Goh, B. S., [1] The second variation for the singular Bolza problem, *SIAM J. Control* **4**, 1966, 309–325. [2] Necessary conditions for singular extremals involving multiple control variables, *SIAM J. Control* **4**, 1966, 716–731.

Goldstein, S., [I] Classical Mechanics, Addison-Wesley 1959.

Goldstine, H. H., [I] *A History of the Calculus of Variations from the 17th through the 19th Century*, Springer-Verlag 1980, xi+410.

Goodman, G. S., [1] The duality of convex functions and Cesari's property (Q), *J. Optimization Theory Appl.* **19**, 1976, 17–23 (also, Proc. Conference Zakopane 1974). [2] On a theorem of Scorza Dragoni and its application to optimal control. *Mathematical Theory of Control* (A. V. Balakrishnan and L. W. Neustadt, eds.), Academic Press 1967, 222–233.

Goor, R. M., [1] Gradient approximation of vector fields, *J. Approximation Theory* **12**, 1974, 385–395. [2] Existence theorems for parametric problems in the calculus of variations and approximation, *Trans. Amer. Math. Soc.* **223**, 1976, 347–365.

Graffi, D., [I] *Elementi di Meccanica Razionale*, Casa Editrice R. Patron, Bologna, 1964, viii + 505.

Graves, L. M., [I] *The Theory of Functions of Real Variables*, McGraw-Hill 1946, x+300. [1] On the existence of the absolute minimum in space problems of the calculus of variations, *Ann. Math.* **28**, 1927, 153–170. [2] On the problem of Lagrange, *Amer. J. Math.* **53**, 1931, 547–554. [3] On the existence theorem of the calculus of variations. *Monatsh. Math. Phys.* **39**, 1932, 101–104. [4] The existence of an extremum in problems of Mayer, *Trans. Amer. Math. Soc.* **39**, 1936, 456–471. [5] The Weierstrass condition for multiple integral variation problems, *Duke Math. J.* **5**, 1939, 656–660.

Gumowski, I. and C. Mira, [I] *Optimization in Control Theory and Practice*, Cambridge Univ. Press 1968, viii+237.

Hadley, G. and M. C. Kemp, [I] *Variational Methods in Economics*, North-Holland 1971, ix+376.

Hahn, H. and A. Rosenthal, [I] *Set Functions*, The University of New Mexico Press, Albuquerque 1948, ix+324.

Halkin, H., [1] Lyapunov's theorem on the range of a vector measure and Pontryagin's maximum principle, *Arch. Rational Mech. Anal.* **10**, 1962, 296–304. [2] Method of convex ascent. *Computing methods in Optimization Problems* (Proc. Conf. Univ. Calif., Los Angeles, 1964),

Academic Press 1965, 211–239. [3] Optimal control for systems described by differential equations. *Advances in Control Systems*, Academic Press 1964, 173–196. [4] Some further generalizations of a theorem of Lyapunov, *Arch. Rational Mech. Anal.* **17**, 1964, 272–277. [5] Topological aspects of optimal control of dynamical polysystems, *Contributions to Differential Equations* **3**, 1964, 377–385. [6] On the necessary condition for optimal control of nonlinear systems, *J. Analyse Math.* **12**, 1964, 1–82. [7] A generalization of LaSalle's bang-bang principle, *SIAM J. Control* **2**, 1964, 199–202. [8] Conditional integrability over measure spaces, *Bull. Amer. Math. Soc.* **71**, 1965, 680–681. [9] On a generalization of a theorem of Lyapunov, *J. Math. Anal. Appl.* **10**, 1965, 325–329. [10] Convexity and control theory. *Functional Analysis and Optimization*, Academic Press 1966, 85–97. [11] Necessary and sufficient condition for a convex set to be closed, *Amer. Math. Monthly* **73**, 1966, 628–630. [12] On the uniform limit of multiple balayage of vector integrals, *Amer. Math. Monthly* **73**, 1966, 733–735. [13] An abstract framework for the theory of process of optimization, *Bull. Amer. Math. Soc.* **72**, 1966, 677–678. [14] A property of nonseparated convex sets, *Proc. Amer. Math. Soc.* **17**, 1966, 1389–1395. [15] A maximum principle of the Pontryagin type for systems described by nonlinear difference equations, *SIAM J. Control* **4**, 1966, 90–111. [16] Mathematical foundation for systems optimization. *Topics in Optimization*, Academic Press 1967, 197–262. [17] Nonlinear nonconvex programming in an infinite dimensional space. *Mathematical Theory of Control* (Proc. Conf., Los Angeles), Academic Press 1967, 10–25. [18] Finitely convex sets of nonlinear differential equations, *Math. Systems Theory* **1**, 1967, 51–53. [19] Implicit functions and optimization problems without continuous differentiability of the data, *SIAM J. Control* **12**, 1974, 229–236. [20] Brouwer fixed point theorem versus contraction mapping theorem in optimal control theory. *International Conference on Differential Equations*, Academic Press 1975, 337–344. [21] Extremal properties of biconvex contingent equations. *Proc. NRL-MRC Conf.*, Academic Press 1972, 109–119. [22] Necessary conditions for optimal control problems with infinite horizon, *Econometrica* **42**, 1974, 267–272. [23] Mathematical programming without differentiability. *Calculus of Variations and Control Theory* (D. Russell, ed.), Academic Press 1976, 279–287.

Halkin, H. and L. W. Neustadt, [1] General necessary conditions for optimization problems, *Proc. Nat. Acad. Sci. U.S.A.* **56**, 1966, 1066–1071. [2] Control as programming in general normed linear spaces. *Lecture Notes in Operations Research and Mathematical Economics*, vols. 11 and 12, Springer-Verlag 1969.

Halmos, P. R., [1] On the set of values of a finite measure, *Bull. Amer. Math. Soc.* **53**, 1947, 138–141. [2] The range of a vector measure, *Bull. Amer. Math. Soc.* **54**, 1948, 416–421.

Hausdorff, F., [I] *Mengenlehre*, Dover 1944, 307 pp. (1st ed., *Grundzüge der Mengenlehre* 1914, 2nd ed. 1927, 3rd ed. 1935.)

Haynes, G. W., [1] On the optimality of a totally singular vector control: an extension of the Green's theorem approach to higher dimensions, *SIAM J. Control 4*, 662–677.

Hermes, H., [1] A note on the range of a vector measure; application to the theory of optimal control, *J. Math. Anal. Appl.* **8**, 1964, 78–83.

Hermes, H. and J. P. LaSalle, [I] *Functional Analysis and Time Optimal Control*, Academic Press 1969, viii + 136.

Hestenes, M. R., [I] *Calculus of Variations and Optimal Control Theory*, Wiley 1966, xii + 405. [II] *Optimization Theory. The Finite Dimensional Case*, Wiley 1975, xiii + 447. [1] A general problem in the calculus of variations with applications to paths of least time. ASTIA Document No. AD 112382, 1950.

Hestenes, M. R. and E. J. McShane, [1] A theorem on quadratic forms and its applications in the calculus of variations, *Trans. Amer. Math. Soc.* **47**, 1940, 501–512.

Himmelberg, C. J., [1] Fixed points of compact multifunctions, *J. Math. Anal. Appl.* **38**, 1972, 205–207.

Himmelberg, C. J., M. Q. Jacobs, and F. S. Van Vleck, [1] Measurable multifunctions, selector theorems, and Filippov's implicit function lemma, *J. Math. Anal. Appl.* **25**, 1969, 276–284.

Hocking, J. G. and G. S. Young, [I] *Topology*, Addison-Wesley 1961, ix + 374.

Hou, S. H., [1] Implicit function theorems in topological spaces, *Applicable Analysis* **13**, 1982, 209–218. [2] On property (Q) and other semicontinuity properties of multifunctions, *Pacific Journal*, to appear. [3] Controllability and feedback systems, *Nonlinear Analysis*, to appear. [4] Existence theorems of optimal control in Banach spaces, *Nonlinear Analysis*, to appear.

Intriligator, M. D., [I] *Mathematical Optimization and Economic Theory*, Prentice-Hall 1971, xix + 508.

Ioffe, A. D., [1] On lower semicontinuity of integral functionals, I and II, *SIAM J. Control Optimization* **15**, 1977, 521–538, 991–1000. [2] Survey of measurable selection theorems: Russian literature supplement, *SIAM J. Control Optimization 16*, 1978, 728–732. [3] An existence theorem for problems of the calculus of variations, *Dokl. Nauk SSSR* 205, 1972, 277–280 = *Soviet Math. Dokl.* **13**, 1972, 919–923.

Ioffe, A. D. and V. M.Tihomirov, [I] *Theory of Extremal Problems*, North-Holland 1979, xii + 460.

Jacobs, M. Q., [1] Some existence theorems for linear optimal control problems, *SIAM J. Control* **5**, 1967, 418–437. [2] Remarks on some recent extensions of Filippov's implicit function lemma, *SIAM J. Control* **5**, 1967, 622–627. [3] Attainable sets in systems with unbounded controls, *J. Differential Equations* **4**, 1968, 408–423. [4] On the approximation of integrals of multivalued functions, *SIAM J. Control* **7**, 1969, 158–177.

Jacobson, D. H., [1] A new necessary condition of optimality for singular control problems, Harvard Univ., Div. Engin. and Appl. Phys. Report No. 576, 1968. [2] Sufficient conditions for nonnegativity of the second variation in singular and nonsingular control problems, *Ibid.*, No. 596, 1969. [3] On conditions of optimality for singular control problems, *IEEE Trans. Autom. Control* **AC-15**, 1970, 109–110.

Kaiser, P. J., [1] Length and variation with respect to a measure, *Atti Sem. Mat. Fis. Univ. Modena* **24**, 1975, 221–235. [2] A problem of slow growth in the calculus of variations, *Atti. Sem. Mat. Fis. Univ. Modena* **24**, 1975, 236–246. [3] Seminormality properties of convex sets, *Rend. Circ. Mat. Palermo* (2) **28**, 1979, 161–182.

Kaiser, P. J. and M. B. Suryanarayana, [1] Orientor field equations in Banach spaces, *J. Optimization Theory Appl* **19**, 1976, 141–164.

Kazemi-Dehkordi, M. [1] Necessary conditions for optimality of systems governed by ordinary and partial differential equations, A Ph.D. thesis at the University of Michigan (Mathematics) 1982.

Kazimirov, V. I., [1] On the semicontinuity of integrals of the calculus of variations, *Uspehi Mat. Nauk* **11**, 1956, 125–129.

Kelley, J. L., [I] *General Topology*, Van Nostrand 1955, xiv + 298.

Kelley, H. J., [1] A second variation test for singular extremals, *AIAA J.* **2**, 1964, 1380–1382. [2] A transformation approach to singular subarcs in optimal trajectory and control problems, *SIAM J. Control* **2**, 1964, 234–240.

Kelley, H. J., R. E. Kopp, and H. G. Moyer [I] Singular extremals, *Topics in Optimization* (G. Leitmann ed.), Academic Press 1967, 67–101.

Kimball, W. S., [I] *Calculus of Variations*, Butterworth Publ., London 1952, viii + 543.

Kingman, J. F. C. and A. P. Robertson, [I] On a theorem of Lyapunov, *J. London Math. Soc.* **43**, 1968, 347–351.

Kopp, R. E. and H. G. Moyer, [1] Necessary conditions for singular extremals, *AIEE J.* **3**, 1965, 1439–1444.

Krasnoselskii, M. A., [I] *Topological Methods in the Theory of Nonlinear Integral Equations*, Pergamon Press 1964, xi + 395. [1] The continuity of a certain operator, *Dokl. Akad. Nauk USSR* **77**, 1951, 185–188.

Krasnoselskii, M. A., and Ya. B. Rutitzkii, [I] *Convex Functions and Orlicz Spaces*, Nordhoff 1961.

Krasnoselskii, M. A., P. P. Zabreiko, E. I. Pustylnik, and P. W. Sobolevskii, [I] *Integral Operators in Spaces of Summable Functions*, Nauka, Moscow 1966.

Kuhn, H. W. and G. P. Szegö (eds.), [I] *Mathematical Systems, Theory and Economics*, Springer-Verlag, Lecture Notes in Operational Research and Mathematical Economics No. 11, 1969.

Kuratowski, K., [1] Les fonctions semi-continues dans l'espace des ensembles fermés, *Fund.* Math. **18**, 1932, 148–166.

Kuratowski, K., and C. Ryll-Nardzewski, [1] A general theorem on selectors, *Bull. Acad. Polon. Sci.* **13**, 1965, 397–403.

Kurzweil, J., [1] On the linear theory of optimal control systems, *Časopis Pěst. Mat.* **89**, 1964, 99–101 (Russian).

Lanczos, C., [I] *The Variational Principles of Mechanics*, Univ. of Toronto Press 1949, xxv + 307.

LaPalm, J. R., [1] Remarks on certain growth conditions in problems of optimal control, *J. Optimization Theory Appl.* **4**, 1969, 321–329. [2] A recent significant growth condition in optimal control theory, *J. Optimization Theory Appl.* **4**, 1969, 378–385.

Lasota, A., and F. H. Szaframiec, [1] An axiomatic approach to the problem of the closedness of the set of trajectories in control theory, *Bull. Acad. Polon. Sci.*, **17**, 1969, 733–738.

Lavrentiev, M., [1] Sur quelques problèmes du calcul des variations, *Ann. Mat. Pura Appl.* 4, 1926, 7–28.

Lebesgue, H., [1] Integrale, longeur, aire, *Annali Mat. Pura Appl.* **7**, 1902, 231–359. [2] Sur le problème de Dirichlet, *Rend. Circ. Mat. Palermo* **24**, 1907, 371–402.

Lee, E. B. and L. Markus, [I] *Foundations of Optimal Control Theory*, Wiley 1967, x + 576.

Leitmann, G., [I] *An Introduction to Optimal Control*, McGraw-Hill, 1966, xi + 163. [II] (ed.) *Topics in Optimization*, Academic Press 1967, xv + 469.

Leitmann, G. and P. L. Yu, [1] Nondominated decisions and cone convexity in dynamic multicriteria decision problems, *J. Optimization Theory Appl.* **14**, 1974, 573–584.

Lions, J. L., [I] Contrôle Optimal de Systèmes gouvernés par des équations aux dérivées partielles, Dunod 1968, xii + 426.

Liusternik, L. A. and V. J. Sobolev, [I] *Elements of Functional Analysis*, Ungar 1961, ix ∓ 227.

Lyapunov, A. A., [1] Sur les fonctions-vecteurs complètement additives, *Izv. Akad. Nauk SSSR, Ser. Mat.* **8**, 1940, 465–478.

Magenes, E., [1] Sui teoremi di Tonelli per la semicontinuità nei problemi di Mayer e di Lagrange, *Ann. Scuola Norm. Sup. Pisa (2)* **15**, 1940, 113–125. [2] Intorno agli integrali di Fubini-Tonelli, I and II, *Annali Scuola Norm. Sup. Pisa (3)* **2**, 1948, 1–38; *(3)* **3**, 1948, 95–128. [3] Sulle equazioni di Eulero relative ai problemi di calcolo delle variazioni degli integrali di Fubini-Tonelli, *Sem. Mat. Univ. Padova* **19**, 1950, 62–102. [4] Sulle estremanti dei polinomiali nella sfera di Hilbert, *Ibid.*, **20**, 1951, 24–47. [5] Sul minimo relativo nei problemi di calcolo delle variazioni d'ordine n, *Ibid.* 21, 1952, 1–24.

Manià, B., [1] Esistenza dell'estremo assoluto in un classico problema di Mayer, *Ann. Scuola Norm. Sup. Pisa (2)* **2**, 1933, 343–354. [2] Sulla curva di massima velocita finale, *Ann. Scuola Norm. Sup. Pisa (2)* **3**, 1934, 317–334. [3] Sui problemi di Lagrange e di Mayer, *Rend. Circ. Mat. Palermo* **58**, 1934, 285–310. [4] Sopra una classe di problemi di Mayer considerati come limiti di ordinari problemi di minimo, *Rend. Sem. Mat. Univ. Padova* 1934, 99–121. [5] Sopra un problema di navigazione di Zermelo, *Math. Ann.* **113**, 1936, 584–599. [6] Sull'approssimazione delle curve e degli integrali, *Boll. Un. Mat. Ital.* **13**, 1934, 7–28. [7] Sopra un esempio di Lavrentieff, *Boll. Un. Mat. Ital.* **13**, 1934, 147–153.

Mazur, S., [1] Ueber konvexe Mengen in linearen normierten Räumen, *Studia Math.* **4**, 1933, 70–84.

McClamroch, N. H., [1] Duality and bounds for the matrix Riccati equation, *J. Math. Anal. Appl.* **25**, 1969, 622–627. [2] A sufficient condition for discrete time optimal processes, *Intern. J. Control (1)* **12**, 1970, 157–161. [3] A general adjoint relation for functional differential and Volterra integral equations with applications to control, *J. Optimization Theory Appl.* **7**, 1971, 346–356. [4] Some input-output properties of nonlinear multivariable feedback systems, *IEEE Trans. Aut. Control* **AC-21**, 1976, 567–572.

McShane, E. J., [I] *Integration*, Princeton Univ. Press 1947, viii + 394. [1] On the necessary condition of Weierstrass in the multiple integral problem of the calculus of variations, I and II, *Ann. Math.* **32**, 1931, 578–590, 723–733. [2] On the semicontinuity of integrals in the calculus of variations. *Ann. Math.* **33**, 1932, 460–484. [3] Parametrization of saddle surfaces with applications to the problem of Plateau, *Trans. Amer. Math. Soc.* **35**, 1933, 716–733. [4] Extension of range of functions, *Bull. Amer. Math. Soc.* **40**, 1934, 837–842. [5] Existence theorems for ordinary problems of the calculus of variations. *Ann. Scuola Norm. Sup. Pisa* [2] **3**, 1934, 181–211, 287–315. [6] Semicontinuity of integrals in the calculus of variations, *Duke Math. J.* **2**, 1936, 597–616. [7] Some existence theorems for problems in the calculus of variations, *Duke Math. J.* **4**, 1938, 132–156. [8] A navigation problem in the calculus of variations, *Amer. Math. J.* **59**, 1937, 327–334. [9] Recent developments in the calculus of variations. *Amer. Math. Soc. Semicentennial Publ. 2*, 1938, 69–97. [10] Some existence theorems in the calculus of variations, I to V, *Trans. Amer. Math. Soc.* **44**, 1938, 428–438, 439–453; **45**, 1939, 151–171, 173–196, 197–216 [11] On multipliers for Lagrange problems, *Amer. J. Math.* **61**, 1939, 809–819. [12] Curve-space topologies associated with variational problems, *Ann. Scuola Norm. Sup. Pisa (2)* **9**, 1940, 45–60. [13] Generalized curves, *Duke Math. J.* **6**, 1940, 513–536. [14] Necessary conditions in the generalized-curve problem of the calculus of variations, *Duke Math. J.* **7**, 1940, 1–27. [15] Existence theorems for Bolza problems in the calculus of variations, *Duke Math. J.* **7**, 1940, 28–61. [16] A metric in the space of generalized curves, *Ann. Math.* **52**, 1950, 328–349. [17] A generalization of convexity and martingales in linear spaces, *Proc. Nat. Acad. Sci. U.S.A.* **54**, 1965, 37–40. [18] Relaxed controls and variational problems, *SIAM J. Control* **5**, 1967, 438–485. [19] On the necessary condition of Weierstrass in the multiple integral problem of the calculus of variations, *Rend. Circ. Mat. Palermo* (2) **16**, 1967, 321–345. [20] Optimal controls, relaxed and ordinary. *Mathematical Theory of Control* (A. V. Balakrishnan and L. W. Neustadt, eds.), Academic Press 1967, 1–9.

McShane, E. J. and R. B. Warfield, [1] On Filippov's implicit function lemma, *Proc. Amer. Math. Soc.* **18**, 1967, 41–47.

Miele, A., [I] *Flight Mechanics*, 2 vols., Addison-Wesley 1962. [1] Extremization of linear integrals by Green's theorem, *Optimization Techniques* (G. Leitmann ed.), Academic Press 1962, 69–98. [2] The calculus of variations in applied aerodynamics and flight mechanics, ibid. 99–170. [3] Problemi di minimo tempo nel volo non stazionario degli aeroplani, *Atti Accad. Sci. Torino* **85**, 1950–51, 41–52.

Miranda, C., [1] Un'osservazione su un teorema di Brouwer, *Boll. Unione Mat. Ital.* (2) **3**, 1940, 5–7.

Morozov, V. S. and V. I. Plotnikov, [1] Necessary and sufficient conditions for the continuity and semicontinuity of functionals of the calculus of variations, *Mat. Sb.* **57** (99), 1962, 265–280.

Morrey, C. B., [I] *Multiple Integrals in the Calculus of Variations*, Springer-Verlag 1966, ix + 506.

Nagumo, N., [1] Über die gleichmässige Summierbarkeit und ihre Anwendung auf ein Variationsproblem, *Japan. J. Math.* **6**, 1929, 173–182.

Natanson, I. P., [I] *Theory of functions of a real variable I*, Ungar 1961, pp. 277.

Nelson, W. C., and E. E. Loft, [I] *Space Mechanics*, Prentice-Hall 1962, x + 245.

Neustadt, L. W., [I] *Optimization, A Theory of Necessary Conditions*, Princeton Univ. Press 1976, xiii + 424. [1] The existence of optimal controls in the absence of convexity conditions, *J. Math. Anal. Appl.* **7**, 1963, 110–117. [2] A survey of certain aspects of control systems. *Mathematics of the Decision Sciences II* (G. B. Dantzig and A. F. Veinott, eds.) Amer. Math. Soc. 1968, 3–16. [3] An abstract variational theory with applications to a broad class of optimization problems. I. General theory, II. Applications, *SIAM J. Control* **4**, 1966, 505–527; **5**, 1967, 90–137. [4] Optimal control problems as mathematical programming in an unorthodox space. *Control Theory and the Calculus of Variations* (A. V. Balakrishnan, ed.), Academic Press 1969, 175–207. [5] A general theory of extremals, *J. Comput. System Sci.* **3**, 1969, 57–92. [6] Optimal control problems with operator equation restrictions. *Symposium on Optimization, Nice* (A. V. Balakrishnan et al., eds.), Springer-Verlag 1970, 292–306. [7] On the solutions of certain integral-like operator equations, existence, uniqueness, and dependence theorems, *Arch. Rational Mech. Anal.* **38**, 1970, 131–160.

Newman, P., [I] *Readings in Mathematical Economics*, Johns Hopkins Press 1968, 2 vols., xii + 532, xi + 358.

Olech, C., [1] A contribution to the time optimal control problem. *Dritte Konferenz über Nichtlineare Schwingungen*, Abh. Deutsch. Akad. Wiss. Berlin 1965, 438–446. [2] A note concerning set valued measurable functions, *Bull. Acad. Polon. Sci.* **13**, 1965, 317–321. [3] A note concerning extremal points of a convex set, *Bull. Acad. Polon. Sci.* **13**, 1965, 347–351. [4] Extremal solutions of a control system, *J. Differential Equations* **2**, 1966, 74–101. [5] Existence theorems for optimal problems with vector valued cost functions, *Trans. Amer. Math. Soc.* **136**, 1969, 157–180. [6] Existence theorems for optimal control problems involving multiple integrals, *J. Differential Equations* **6**, 1969, 512–526. [7] Lexicographical order, range of integrals, and bang-bang principle. *Mathematical Theory of Control* (A. V. Balakrishnan and L. W. Neustadt, eds.), Academic Press 1967, 35–45. [8] Existence theory in optimal control problems, the underlying ideas. *Int. Conference on Differential Equations* (H. A. Antosiewicz, ed.), Academic Press 1975, 612–629. [9] Weak lower semicontinuity of integral functionals, *J. Optimization Theory Appl.* **19**, 1976, 3–16.

Olech. C. and V. Klee, [1] Characterization of a class of convex sets, *Math. Scand.* **20**, 1967, 290–296.

Olech, C. and A. Lasota, [1] On the closedness of the set of trajectories of a control system, *Bull. Acad. Polon. Sci.* **14**, 1966, 615–621. [2] On Cesari's semicontinuity condition for set valued mappings, *Bull. Acad. Polon. Sci.* **16**, 1968, 711–716.

Pallu de la Barriere, R., [I] *Optimal Control Theory*, Saunders 1967, xii + 412.

Plis, A., [1] Trajectories and quasitrajectories for an orientor field, *Bull. Acad. Polon. Sci.* **11**, 1963, 369–370. [2] Remarks on measurable set valued functions, *Bull. Acad. Polon. Sci.* **9**, 1961, 857–859. [3] Measurable orientor fields, *Bull. Acad. Polon. Sci.* **13**, 1965, 565–569.

Poljak, B. T., [1] Semicontinuity of integral functionals and existence theorems on extremal problems, *Mat. Sb.* **78**, 1969, 65–84 = *Math. USSR Sbornik* **7**, 1969, 59–77. [2] Existence theorems and convergence of minimizing sequences in extremum problems with restrictions, *Dokl. Akad. Nauk SSSR* **166**, 72–75.

Pontryagin, L. S., [1] Optimal regulation processes, *Uspehi Math. Nauk SSR* **14**, 1959, 3–20 = *Amer. Math. Soc. Translations* (*2*) **18**, 1961, 321–339 = *Automation Express* **2**, 1959, 15–17, 26–30.

Pontryagin, L. S., V. G. Boltyanskii, R. V. Gamkrelidze, and E. F. Mishchenko, [I] *The Mathematical Theory of Optimal Processes*, Interscience 1962, viii + 360.

Powers, W. F. and J. P. McDanell, [1] Necessary conditions for joining optimal singular and nonsingular subarcs, *SIAM J. Control* **9**, 1971, 161–173. [2] New Jacobi-type necessary and sufficient conditions for singular optimization problems, *AIAA J.* **8**, 1970, 1416–1420. [3] Switching conditions and synthesis technique for singular Saturn guidance problem, *J. Spacecraft and Rockets* **8**, 1971, 1027–1032.

Powers, W. F. and E. R. Edge, [1] Function space quasi Newton algorithm for optimal control problems with bounded control and singular arcs, *J. Optimization Theory Appl.* **20**, 1976, 455–479.

Powers, W. F., B. D. Cheng, and E. R. Edge, [1] Singular optimal control computation, *J. Guidance and Control* **1**, 1978, 83–89.

Pucci, P. [1] On an existence theorem for isoperimetric problems. *Applicable Analysis*, to appear.

Reid, T. W., [1] A matrix differential equation of Riccati type, *Amer. J. Math.* **68**, 1946, 237–246; **70**, 1948, p. 250. [2] Oscillation criteria for linear differential systems with complex coefficients, *Pacific J. Math.* **6**, 1956, 733–751. [3] An elementary sufficiency proof of an absolute minimum for a nonparametric variation problem, *J. Optimization Theory Appl.* **18**, 1976, 335–349.

Rham, G. de, [I] *Variétés Différentiables*, Actualités Scient. et Ind., Herman, Paris 1955, vii + 196.

Riesz, F. and B. Sz.-Nagy, [I] Functional Analysis, Ungar 1955, xii + 468.

Robertson, A. P., [1] On measurable selections, *Proc. Roy. Soc. Edinburgh* **72**, 1972–73, 1–7.

Rockafellar, R. T., [I] *Convex Analysis*, Princeton Univ. Press 1970. [1] Measurable dependence of convex sets and functions on parameters, *J. Math. Anal. Appl.* **28**, 1969, 4–25. [2] Integrals which are convex functionals, I and II, *Pacific J. Math.* **24**, 1968, 525–539; **39**, 1971, 439–469. [3] Convex integrals functions and duality. *Contributions to Nonlinear Functional Analysis*, Academic Press 1971, 215–236. [4] Integral functionals, normal integrands and measurable selection. *Nonlinear Operators and the Calculus of Variations*, Bruxelles 1975; Springer-Verlag, Lecture Notes in Math. No 543, 1976, 157–207. [5] Weak compactness of level sets of integral functionals. *Troisième Colloque d'Analyse Fonctionelle* (H. G. Garnir, ed.), CBRM, Liège, 1971, 85–98. [6] Existence theorems for general control problems of Bolza and Lagrange, *Advances in Math.* **15**, 1975, 312–333. [7] Level sets and continuity of conjugate convex functions, *Trans. Amer. Math. Soc.* **123**, 1966, 46–63. [8] Existence and duality theorems for convex problems of Bolza, *Trans. Amer. Math. Soc.* **159**, 1971, 1–40. [9] Convex functions and duality, *Optimization Problems and Dynamics*, 1969, 117–141. [10] Convex functions, monotone operators and variational inequalities. *Theory and Applications of Monotone Operators* (Proc. NATO Advanced Study Inst., Venice, 1968), Oderisi, Gubbio 1969, 35–65. [11] Convex functions and duality in optimization problems and dynamics. *Mathematical Systems Theory and Economics*, I (H. W. Kuhn and G. P. Szegö, eds.), (Int. Summer School, Varenna, 1967), Springer-Verlag 1969, 117–141. [12] Minimax theorems and conjugate saddle functions, *Math. Scand.* **14**, 1964, 151–173. [13] Local boundedness of nonlinear monotone operators, *Michigan Math. J.* **16**, 1969, 397–407.

Rothe, E. H., [1] An existence theorem in the calculus of variations, *Arch. Rational Mech. Anal.* **21**, 1966, 151–162.

Roxin, E. O., [1] The existence of optimal controls, *Michigan Math. J.* **9**, 1962, 109–119. [2] Pontryagin's maximum principle. *International Symposium on Nonlinear Differential Equations and Nonlinear Mechanics* (J. P. LaSalle and S. Lefschetz, eds.), Academic Press 1963, 303–324.

Rozonoer, L. I., [1] The maximum principle of L. S. Pontryagin in optimal system theory, *Automatic Remote Control* **20**, 1959, 1288–1302, 1405–1421.

Rupp, R. D., [1] Hypotheses implying Cesari's property (Q). *J. Optimization Theory Appl.* **19**, 1976, 119–124.

Saks, S., [I] *Theory of the Integral*, Hafner, Warsaw 1937, vi + 347.

Sard, A., [1] The measure of the critical values of differentiable maps, *Bull. Amer. Math. Soc.* **48**, 1942, 863–897.

Schmeidler, D., [1] Fatou's lemma in several dimensions, *Proc. Amer. Math. Soc.* **24**, 1970, 300–306.

Schuur, J. D. and S. N. Chow, [1] An existence theorem for ordinary differential equations in Banach spaces, *Bull. Amer. Math. Soc.* **77**, 1971, 1018–1020. [2] Fundamental theory of contingent differential equations in Banach spaces, *Trans. Amer. Math. Soc.* **179**, 1973, 133–144.

Schwartz, J. T., [I] *Nonlinear Functional Analysis*, Gordon and Breach 1969, vii + 236.

Scorza-Dragoni, G., [1] Un teorema sulle funzioni continue rispetto ad una e misurabili rispetto ad un altra variable, *Rend. Sem. Mat. Univ. Padova* 17, 1948, 102–106.

Sragin, L. V., [1] Weak continuity of the Nemitskii operator, *Moskov. Oblast. Ped. Inst. Učen. Zap.* **57**, 1957, 73–79.

Strauss, A., [I] *An Introduction to Optimal Control Theory*, Lecture Notes in Operational Research and Mathematical Economics No. 3, Springer-Verlag, 1968, v + 153.

Suryanarayana, M. B., [1] On multidimensional integral equations of the Volterra type, *Pacific J. Math.* **41**, 1972, 809–828. [2] Necessary conditions for optimization problems with hyperbolic partial differential equations, SIAM J. Control 11, 1973, 130–147. [3] Existence theorems for optimization problems with hyperbolic partial differential equations, *J. Optimization Theory Appl.* **15**, 1975, 361–392. [4] Existence theorems for optimization problems concerning linear hyperbolic partial differential equations without convexity, *J. Optimization Theory Appl.* **19**, 1976, 47–62. [5] Linear control problems with total differential equations, *Trans. Amer. Math. Soc.* **200**, 1974, 233–249. [6] A Sobolev space and a Darboux problem, *Pacific J. Math.* **69**, 1977, 535–550. [7] Remarks on lower semicontinuity and lower closure, *J. Optimization Theory Appl.* **19**, 1976, 125–140. [8] Monotonicity and upper semicontinuity, *Bull. Amer. Math. Soc.* **82**, 1976, 936–938. [9] Remarks on existence theorems for Pareto optimality. *Dynamical Systems* (A. R. Bednarek and L. Cesari, eds.), Academic Press 1977, 335–347. [10] Monotonicity and upper semicontinuity of multifunctions, to appear.

Tonelli, L., [I] Fondamenti di Calcolo delle Variazioni, Zanichelli, 2 vols., (a), (b), 1921–23, vii + 466, viii + 660. [II] *Opere Scelte*, Cremonese 1962, 4 vols. (Volumes 3 and 4 collect all Tonelli's papers in the calculus of variations, but not the book Fondamenti di Calcolo delle Variazioni, vol. 1 in real analysis, vol. 4 in applications.) [1] Sulla rettificazione delle curve, *Atti Accad. Sci. Torino* **43**, 1908, 783–800 = *Opere Scelte I*, 52–68. [2] Sulla lunghezza di una curva, *Atti Accad. Sci. Torino* **47**, 1912, 1067–1075 = *Opere Scelte I*, 227–235. [3] Sui massimi e minimi assoluti del calcolo delle variazioni, *Rend. Circ. Mat. Palermo* **32**, 1911, 297–333 = *Opere Scelte 2*, 10–66. [4] Sur une methode directe du calcul des variations, *C. R. Acad. Sci. Paris* **158**, 1914, 1776–1778, 1983–1985 = *Opere Scelte* **3**, 177–182. [5] Sul caso regolare nel calcolo delle variazioni, *Rend. Circ. Mat. Palermo* **35**, 1913, 49–73 = *Opere Scelte* **2**, 101–133. [6] Sur une methode directe du calcul des variations, *Rend. Circ. Mat. Palermo* **39**, 1915, 233–264 = *Opere Scelte 2*, 289–333. [7] La semicontinuità nel calcolo delle variazioni, *Rend. Circ. Mat. Palermo* **44**, 1920, 167–249 = *Opere Scelte 2*, 340–442. [8] Sur la semi-continuité des integrales doubles du calcul des variations, *Acta Math.* **53**, 1929, 325–346 = *Opere Scelte 3*, 27–49. [9] L'estremo assoluto degli integrali doppi, *Ann. Scuola Norm. Sup. Pisa (2)* **2**, 1933, 89–130 = *Opere Scelte 3*, 108–159. [10] Sugli integrali del calcolo delle variazioni in forma ordinaria, *Ann. Scuola Norm. Sup. Pisa (2)* **3**, 1934, 401–450 = *Opere Scelte 3*, 192–354. [11] Sulle proprieta delle estrementi, *Ann. Scuola Norm. Sup. Pisa* **3**, 1934, 213–237 = *Opere Scelte* **3**, 160–191. [12] Sulla semicontinuità nei problemi di Mayer e di Lagrange, *Rend. Accad. Naz. Lincei (6)* **24**, 1936, 399–404 = *Opere Scelte 3*, 342–348. [13] Su gli integrali continui del calcolo delle variazioni. *Scritti matem. offerti a L. Berzolari*, Ist. Mat. Univ. Pavia 1936, 283–289 = *Opere Scelte* 3, 349–356. [14] Su l'esistenza delle estremanti assolute per gli integrali doppi, *Ann. Scuola Norm. Sup. Pisa* **8**, 1939, 161–165 = *Opere Scelte 3*, 387–393. [15] L'analisi funzionale nel calcolo delle variazioni, *Ann. Scuola Norm. Sup. Pisa* **9**, 1940, 289–302 = *Opere Scelte 3*, 418–435. [16] Sulle funzioni d'intervallo, *Ann. Scuola Norm. Sup. Pisa* **8**, 1939, 309–321 = *Opere Scelte 1*, 554–569. [17] Sur une question du calcul des variations, *Rec. Math. Moscou* **1**, 1926, 87–98 = *Opere Scelte 3*, 1–15.

Turner, L. H., [1] The direct method in the calculus of variations. Thesis, Purdue University, Lafayette, Indiana 1957. [2] An invariant property of Cesari's surface integral, *Proc. Amer. Math. Soc.* **9**, 1958, 920–925. [3] Sufficient conditions for semicontinuous surface integrals, *Michigan Math. J.* **10**, 1963, 193–206.

Vainberg, M. M., [I] *Variational Methods for the Study of Nonlinear Operators*, GITTL, Moscow 1956; English transl., Holden Day 1964.

Valentine, F. A., [I] *Convex Sets*, McGraw-Hill 1964, ix+238.

Vidyasagar, M., [1] On the existence of optimal controls, *Journ. Optim. Theory Appl.* **12**, 1975, 273–278.

Vinti, C., [1] L'integrale di Weierstrass, *Rend. Ist. Lombardo Sci. Lett.* (*A*) **92**, 1958, 423–434. [2] L'integrale di Weierstrass e l'integrale del Calcolo delle Variazioni in forma ordinaria, *Atti Accad. Sci. Lett. Palermo* (*4*) **19**, 1958–59, 51–82. [3] Proprietà di alcuni integrali del Calcolo delle Variazioni, *Annali Scuola Norm. Sup. Pisa* (*3*) **20**, 1966, 173–195. [4] L'integrale di Fubini Tonelli nel senso di Weierstrass, *Ibid.* (*3*) **22**, 1968, 229–263 and 355–376.

Vitali, G., [1] Sui gruppi di punti e sulle funzioni di variabili reali, *Atti Accad. Sci. Torino* **43**, 1908, 75–92.

Volterra, V., [I] *Lecons sur les Fonctions de Lignes*, Gauthier Villars, Paris 1913. [II] *La Lutte pour la Vie*, Gauthier Villars, Paris 1927. [III] *Opere Matematiche*, Accad. Naz. Lincci, Roma 1954. [1] Sopra le funzioni che dipendono da altre funzioni, *Rend. Accad. Naz. Lincei* (*4*) **3**, 1887, 97–105, 141–146, 153–158. [2] Sopra le funzioni dipendenti da linee, *Rend. Accad. Naz. Lincei* (*4*) **3**, 1887, 225–230, 274–281.

Volterra, V. and J. Peres, [I] *Théorie Générale des Fonctionnelles*, Gauthier Villars, Paris 1936, xii+359.

Wagner, D. H., [1] Survey of measurable selection theorems, *SIAM J. Control Optimization* **15**, 1977, 859–903.

Warga, J., [I] *Optimal Control of Differential and Functional Equations*, Academic Press 1972, xv+531. [1] Relaxed variational problems, *J. Math. Anal. Appl.* **4**, 1962, 111–128. [2] Necessary conditions for minimum in relaxed variational problems, *J. Math. Anal. Appl.* **4**, 1962, 129–145. [3] Functions of relaxed controls, *SIAM J. Control* **5**, 1967, 628–641. [4] Minimizing variational curves restricted to a preassigned set, *Trans. Amer. Math. Soc.* **112**, 1964, 432–455. [5] On a class of minimax problems in the calculus of variations, *Michigan Math. J.* **12**, 1965, 289–311. [6] Unilateral variational problems with several inequalities, *Michigan Math. J.* **12**, 1965, 449–460. [7] Variational problems with unbounded controls, *SIAM J. Control* **3**, 1966, 424–438. [8] Control problems with functional restrictions, *SIAM J. Control* **8**, 1970, 360–371. [9] Conflicting and minimax controls, *J. Math. Anal. Appl.* **33**, 1971, 655–673.

Wazewski, T., [1] Systèmes de commande et équations au contingent. *Bull. Acad. Polon. Sci.* **9**, 1961, 151–155. [2] Sur une condition d'existence des fonctions implicites mesurables, *Bull. Acad. Polon. Sci.* **9**, 1961, 861–864. [3] Sur une généralisation de la notion des solutions d'une équation au contingent, *Bull. Acad. Polon. Sci.* **10**, 1962, 11–15. [4] Sur les systèmes de commande nonlinéaires dont le contredomaine de commande n'est pas forcement convex, *Bull. Acad. Polon. Sci.* **10**, 1962, 17–21. [5] On an optimal control problem. *Differential Equations and Their Applications* (Proc. Conference, Prague, 1962), Academic Press 1963, 229–242.

Weinstock, R., [I] *Calculus of Variations*, McGraw-Hill 1952, x+326.

Whitney, H., [1] Analytic extensions of differential functions defined in closed sets, *Trans. Amer. Math. Soc.* **36**, 1934, 63–89. [2] Derivatives, difference quotients, and Tayler's formula, I and II, *Bull. Amer. Math. Soc.* **40**, 1934, 89–94; *Ann. Math.* **35**, 1934, 476–481. [3] Functions differentiable on the boundary of regions, *Ann. Math.* **35**, 1934, 482–485. [4] Differentiable functions defined in closed sets, I and II, *Ann. Math.* **36**, 1934, 369–387; **40**, 1936, 309–317

Wilde, M. de, [1] Sur un théorème de Lyapunov, *Bull. Soc. Roy. Sci. Liège* **38**, 1969, 96–100. [2] A note on the bang-bang principle, *J. London Math. Soc.* (*2*) **1**, 1969, 753–759.

Yosida, K., [I] *Functional Analysis*, 4th ed., Springer-Verlag 1974, xi+496.

Young, L. C., [I] *Lectures on the Calculus of Variations and Optimal Control Theory*, Saunders 1969, xi+331. [1] Generalized curves and the calculus of variations, *C. R. Soc. Sci. Varsovie* (*3*) **30**, 1937, 211–234. [2] On approximation by polygons in the calculus of variations, *Proc. Royal Soc.* (*A*) **141**, 1933, 325–341. [3] Necessary conditions in the calculus of variations, *Acta Math.* **69**, 1938, 239–258. [4] Generalized surfaces in the calculus of variations, I and II. *Ann. Math.* **43**, 1942, 84–103, 530–544. [5] Some applications of the Dirichlet integral to the theory of surfaces, *Trans. Amer. Math. Soc.* **64**, 1948, 317–335. [6] Surfaces parametriques generalisées, *Bull. Soc. Math. France* **79**, 1951, 59–85. [7] Contours on generalized and extremal varieties, *J. Math. Mech.* **11**, 1962, 615–646. [8] Generalized varieties as limits, *J. Math. Mech.* **13**, 1964, 673–692. [9] A theory of boundary values, *Proc. London Math. Soc.* (*3*) **14A**, 1965, 300–314.

Yu, P. L., [1] Cone convexity, cone extreme points, and nondominated solutions in decision problems with multiobjectives, *J. Optimization Theory Appl.* **14**, 1974, 319–377.

Zachnissan, L. E., [1] Deparametrization of the Pontryagin maximum principle. *Mathematical Theory of Control* (A. V. Balakrishnan and L. W. Neustadt, eds.), Academic Press 1967, 234–245.

Zadeh, L. A., [1] Optimality and nonscalar valued performance criteria, *IEEE Trans. Auto. Control* **AC-8**, 1963, 59–60.

Zaremba, S. K., [1] Sur les équations au paratingent, *Bull. Sci. Math.* (*2*) **60**, 1936, 139–160.

Author Index

Subject Index

Statements

Conditions

Conditions Concerning Unbounded Domains